T0190237

OT 37
Operator Theory: Advances and Applications
Vol. 37

Springer Basel AG

OT 37
Operator Theory: Advances and Applications
Vol. 37

Springer Basel AG

Konrad Schmüdgen

Unbounded Operator Algebras and Representation Theory

1990

Springer Basel AG

Author's address:

Prof. Konrad Schmüdgen
Sektion Mathematik
Karl-Marx Universität
Karl-Marx Platz
Leipzig 7010 - DDR

Library of Congress Cataloging in Publication Data

Schmüdgen, Konrad:
Unbounded operator algebras and representation theory / Konrad Schmüdgen.
p. cm. - - (Operator theory, advances and applications ; vol. 37)
Bibliography: p.
Includes index.

1. Operator algebras. 2. Representations of algebras. I. Title.
II. Series: Operator theory, advances and applications ; v. 37.
QA326.S35 1990
512'.55 - - dc20 89-32477 CIP

CIP-Titelaufnahme der Deutschen Bibliothek

Schmüdgen, Konrad:
Unbounded operator algebras and representation theory /
Konrad Schmüdgen. — Basel ; Boston ; Berlin : Birkhäuser, 1990
(Operator theory ; Vol. 37)

NE: GT

ISBN 978-3-0348-7471-7 ISBN 978-3-0348-7469-4 (eBook)
DOI 10.1007/978-3-0348-7469-4

Originally published by Akademie Verlag, Berlin in 1990.
Softcover reprint of the hardcover 1st edition 1990

To Katja and Alexander

Les théories ont leurs commencements: des allusions vagues, des essais inachevés, des problèmes particuliers; et même lorsque ces commencements importent peu dans l'état actuel de la Science, on aurait tort de les passer sous silence.

F. Riesz,
Les systèmes d'équations linéaires a une infinité d'inconnues,
Paris, 1913, p. 1.

Scientific subjects do not progress necessarily on the lines of direct usefulness. Very many applications of the theories of pure mathematics have come many years, sometimes centuries, after the actual discoveries themselves. The weapons were at hand, but the men were not able to use them.

A. R. Forsyth,
Perry's Teaching of Mathematics,
London, 1902, p. 35.

Preface

$*$-algebras of unbounded operators in Hilbert space, or more generally algebraic systems of unbounded operators, occur in a natural way in unitary representation theory of Lie groups and in the Wightman formulation of quantum field theory. In representation theory they appear as the images of the associated representations of the Lie algebras or of the enveloping algebras on the Garding domain and in quantum field theory they occur as the vector space of field operators or the $*$-algebra generated by them. Some of the basic tools for the general theory were first introduced and used in these fields. For instance, the notion of the weak (bounded) commutant which plays a fundamental role in the general theory had already appeared in quantum field theory early in the sixties. Nevertheless, a systematic study of unbounded operator algebras began only at the beginning of the seventies. It was initiated by (in alphabetic order) Borchers, Lassner, Powers, Uhlmann and Vasiliev. From the very beginning, and still today, representation theory of Lie groups and Lie algebras and quantum field theory have been primary sources of motivation and also of examples. However, the general theory of unbounded operator algebras has also had points of contact with several other disciplines. In particular, the theory of locally convex spaces, the theory of von Neumann algebras, distribution theory, single operator theory, the moment problem and its non-commutative generalizations and noncommutative probability theory, all have interacted with our subject.

This book is an attempt to provide a treatmant of $*$-algebras of unbounded operators in Hilbert space (the so-called O*-algebras) and of (unbounded) $*$-representations of general $*$-algebras. Roughly speaking, an *O*-algebra* is a $*$-algebra $\mathcal{A}$ of linear operators defined on a common dense linear subspace $\mathcal{D}$ of a Hilbert space and leaving $\mathcal{D}$ invariant. The multiplication in $\mathcal{A}$ is the composition of operators, which makes sense because of the invariance of the domain $\mathcal{D}$, and the involution $a \to a^+$ in $\mathcal{A}$ is defined by letting a^+ be the restriction to $\mathcal{D}$ of the usual Hilbert space adjoint a^*. We always assume that an O*-algebra on $\mathcal{D}$ contains the identity map of $\mathcal{D}$. A *$*$-representation* of a general $*$-algebra with unit is a $*$-homomorphism of the $*$-algebra onto some O*-algebra. Moreover, we also consider some more general families of closable linear operators (O-families, O-vector spaces, O-algebras, O*-families and O*-vector spaces) which are always defined on a common dense domain $\mathcal{D}$.

Our objective is threefold. First, the book gives a thorough treatment of certain of the basic concepts involved in the theory of O*-algebras and $*$-representations. These mainly concern notions like the graph topology, closed and self-adjoint $*$-representations, closed and self-adjoint O*-algebras, weak and strong (bounded) commutants, strongly positive and completely strongly positive $*$-representations, to name the most important, which have proved to be useful and fundamental in the theory. We also develop concepts like directed O-families, commutatively dominated O*-algebras, weak and strong unbounded commutants, form commutants, induced extensions and strongly n-positive $*$-representations with the anticipation that these will be useful in future research. Secondly, we aim to prove some of the more involved results of the existing theory. As a sample, results in Sections 2.4, 4.3, 5.3, 5.4, 6.2, 7.3, 9.2, 9.4, 10.2, 10.4, 10.5, 11.2, 12.3 and 12.4 could be mentioned in this respect. Thirdly, the book presents many examples and counter-examples that help to delimit the general theory. These sometimes require more involved constructions and arguments than many of the positive results in the theory. For instance, we construct a self-adjoint $*$-representation of the polynomial algebra in two variables, the bounded commutant of which is a given properly infinite von Neumann algebra in separable Hilbert space.

The scope of this book is, of course, dictated by the stage of the existing theory. Thus, for instance, the topological theory of O*-algebras occupies a relatively large space in this monograph, simply because it is much more developed than other parts of the theory. The choice of the material contained in this book also depends on the author's personal view of the existing theory and on his particular research interests. Some topics such as GB*-algebras, Hilbert algebras, tensor algebras and applications in physics are not included. Often the original proofs of the results have been improved, errors have been corrected or the result has been generalized. Frequently the terminology and the notation have been changed, we hope for the better. Also several new concepts are introduced.

Apart from the preliminary chapter, the book consists of two parts which are independent to a large extent (see also the introduction to Part II). In Part I O*-algebras and topologies on the domains and the algebras are studied, while Part II is concerned with $*$-representations of general $*$-algebras. Those topics in the theory of $*$-representations that primarily involve the study of topologies or the structure of O*-algebras are treated in part I. Such topics are the continuity of $*$-representations, the realization of the generalized Calkin algebra and the abstract characterization of the $*$-algebras $\mathcal{L}^+(\mathcal{D}_i : i \in I)$. Chapter 10 gives a rather thorough treatment of integrable representations of Lie algebras resp. enveloping algebras. This chapter stands almost entirely by itself; it requires only a few general definitions and facts from earlier sections.

Almost no bibliographical comments are given in the body of the text; they are gathered in a section entitled "Notes" at the end of each chapter. There, the sources of the main results, basic concepts and some examples are cited (of course, as far as the author is aware), but no attempt has been made to be encyclopaedic. Some of these sections contain a list of references dealing with problems similar to those in the text.

The first two digits in the number of a theorem, proposition, lemma, definition or example refer to the section and the third digit to the position of the item. Remarks and formulas are numbered and quoted consecutively within the sections. When a reference to a formula in another section is made, the number of the section is added; for instance, 3.2/(1) means formula (1) in Section 3.2. The end of a proof is marked by □ and of an example by ○. The reader should also note that we often fix assumptions or notations at the beginning of a chapter, section or subsection which keep in force throughout the whole chapter, section or subsection. Further, the proofs of facts stated in the examples are frequently merely sketched and sometimes they are omitted altogether.

I am grateful to Dr. Jürgen Friedrich and Dr. Klaus-Detlef Kürsten for their critical reading of large parts of the manuscript and for many valuable suggestions. I am also very indebted to Professor Paul S. Muhly for his help in writing this book. Last but not least, I wish to thank R. Helle, Dr. R. Höppner and G. Reiher of the Akademie-Verlag for their patience and help in preparing this book.

Leipzig, Fall 1987

K. Schmüdgen

Contents

1. Preliminaries

In this chapter we summarize some basic definitions, notation and results that will be required in this monograph. Some, but not all, of them are standard or well known. General terminology which is used essentially in one chapter, section or subsection will be introduced therein.

First we collect some general notation. Throughout, $\mathbb{C}$ denotes the complex numbers, $\mathbb{T}$ the complex numbers of modulus one, $\mathbb{R}$ the real numbers, $\mathbb{Z}$ the integers, $\mathbb{N}$ the positive integers and $\mathbb{N}_0$ the non-negative integers. For $t = (t_1, \ldots, t_d) \in \mathbb{R}^d$ and $n = (n_1, \ldots, n_d) \in \mathbb{N}_0^d$, t^n is the usual multi-index notation, i.e., $t^n := t_1^{n_1} \ldots t_d^{n_d}$, where $t_k^0 := 1$ for $k = 1, \ldots, d$. The abbreviations l.h. and c.l.h. mean the linear hull and the closed linear hull, respectively. Sequences and nets are written as $(x_n : n \in \mathbb{N})$ resp. $(x_i : i \in I)$ or simply as (x_n) resp. (x_i). In general, sets are denoted by braces such as $\{x_n : n \in \mathbb{N}\}$. For an open or closed subset M of $\mathbb{R}^d$, $L^p(M)$ is the L^p-space with respect to the Lebesgue measure on M. If M is a C^∞-manifold (with or without boundary) and $n \in \mathbb{N} \cup \{\infty\}$, then $C^n(M)$ is the set of all complex functions of class C^n on M. We denote by $C_0^\infty(M)$ the set of all functions in $C^\infty(M)$ whose support is a compact subset of M. The continuous complex functions on a topological space M are denoted by $C(M)$. For a and b in $\mathbb{R}$, we shall write $C^\infty[a, b]$ for $C^\infty([a, b])$, $C[a, b]$ for $C([a, b])$, $C_0^\infty(a, b)$ for $C_0^\infty((a, b))$ and $L^p(a, b)$ for $L^p((a, b))$. As usual, δ_{nm} is the Kronecker symbol. The closed unit ball of a normed space E is denoted by $\mathcal{U}_E$.

1.1. Locally Convex Spaces

As general references for the theory of locally convex spaces we shall use the textbooks Schäfer [1], Köthe [1], [2] and Jarchow [1].

All considered vector spaces are either over the real field $\mathbb{R}$ or over the complex field $\mathbb{C}$. When we speak about a vector space or a locally convex space without specifying the field, we always mean spaces over $\mathbb{C}$. Let U and M be subsets of a vector space E over $\mathbb{K}$. Then U *absorbs* M if there is an $\alpha > 0$ such that $M \subseteq \lambda U$ for all $\lambda \in \mathbb{K}$, $|\lambda| \geqq \alpha$, and U is *absorbing* if it absorbs every singleton $\{\varphi\}$, $\varphi \in E$. The absolutely convex hull of U is denoted by aco U.

If τ is a topology on a set E, then we write $E[\tau]$ for the corresponding topological space. The induced topology on a subset F of E is denoted by $\tau \upharpoonright F$ or simply again by τ if no confusion can arise. If τ_1 and τ_2 are topologies on E, then $\tau_1 \subseteq \tau_2$ means that τ_1 is coarser (weaker) than τ_2.

A *locally convex space* is a (not necessarily Hausdorff) topological vector space over $\mathbb{K} = \mathbb{R}$ or over $\mathbb{K} = \mathbb{C}$ which has a 0-neighbourhood base $\boldsymbol{U}$ satisfying the following conditions:

(i) For $U_1, U_2 \in \boldsymbol{U}$, there is a $U \in \boldsymbol{U}$ such that $U \subseteqq U_1 \cap U_2$.

(ii) If $U \in \boldsymbol{U}$, then $\lambda U \in \boldsymbol{U}$ for all $\lambda \in \mathbb{K}$, $\lambda \neq 0$.

(iii) Each $U \in \boldsymbol{U}$ is absolutely convex and absorbing.

If $\boldsymbol{U}$ is a non-empty family of subsets of a real or complex vector space which fulfills (i), (ii) and (iii), then there is a unique topology τ on E such that $E[\tau]$ is a locally convex space and $\boldsymbol{U}$ is a 0-neighbourhood base for τ. By a *locally convex topology* on a vector space E we mean a topology τ on E for which $E[\tau]$ is a locally convex space. Let Γ be a non-empty family of seminorms on a vector space E. The collection $\boldsymbol{U}$ of all sets

$$\{\varphi \in E : p_n(\varphi) \leqq \varepsilon \text{ for } n = 1, \ldots, k\}, \text{ where } p_1, \ldots, p_k \in \Gamma,\ k \in \mathbb{N} \text{ and } \varepsilon > 0,$$

satisfies (i)—(iii); so $\boldsymbol{U}$ is a 0-neighbourhood base for a unique locally convex topology τ on E. We then say that τ is *generated* (or *defined* or *determined*) by Γ. The family Γ is *directed* if, given $p_1, p_2 \in \Gamma$, there is a $p \in \Gamma$ such that $p_1 \leqq p$ and $p_2 \leqq p$.

In what follows we suppose that E is a locally convex Hausdorff space.

Let $E^{\prime}$ denote the dual of E. The *weak topology* $\sigma \equiv \sigma(E, E^{\prime})$ is the locally convex topology on E defined by the seminorms $\varphi \to |\varphi^{\prime}(\varphi)|$, $\varphi^{\prime} \in E^{\prime}$. A sequence $(\varphi_n : n \in \mathbb{N})$ in E *converges weakly* to $\varphi \in E$ if it converges in the locally convex space $E[\sigma]$ to φ, i.e. if $\lim_n \varphi^{\prime}(\varphi_n) = \varphi^{\prime}(\varphi)$ for all $\varphi^{\prime} \in E^{\prime}$. The *weak*-topology* $\sigma^{\prime} \equiv \sigma(E^{\prime}, E)$ on $E^{\prime}$ is generated by the seminorms $\varphi^{\prime} \to |\varphi^{\prime}(\varphi)|$, $\varphi \in E$. The *strong topology* on $E^{\prime}$ is denoted by β; it is generated by the family of seminorms

$$r_M(\varphi^{\prime}) := \sup_{\varphi \in M} |\varphi^{\prime}(\varphi)|, \quad \varphi^{\prime} \in E^{\prime},$$

where M ranges over the bounded subsets of E. The vector space E becomes a linear subspace of $(E^{\prime}[\beta])^{\prime}$ by identifying $\varphi \in E$ with the linear functional $\varphi^{\prime} \to \varphi^{\prime}(\varphi)$ on $E^{\prime}$. E is *semireflexive* if $E = (E^{\prime}[\beta])^{\prime}$ under this identification, and E is *reflexive* if E is semireflexive and if the topology of E coincides with the strong topology on $(E^{\prime}[\beta])^{\prime}$.

A *Frechet space* is a complete metrizable locally convex space. The locally convex space E is said to be a *quasi-Frechet space* (or briefly, a *QF-space*) if for every bounded set M in E there is a subspace G of E which is a Frechet space in the induced topology of E and which contains M. It is obvious that each Frechet space is a QF-space.

The space E is *barrelled* if every barrel in E (i.e., every closed absolutely convex absorbing subset of E) is a 0-neighbourhood in E. E is a *semi-Montel space* if each bounded subset of E is relatively compact. A *Montel space* is a barrelled semi-Montel space. The space E is *bornological* if every absolutely convex set in E that absorbs each bounded set in E is a 0-neighbourhood of E. The *bornological topology associated with the topology of E* is the coarsest bornological topology on E which is finer than the topology of E.

A *fundamental system* of bounded sets in E is a family S of bounded sets such that every bounded subset of E is contained in some set of S. The space E is a *DF-space* if it admits a countable fundamental system of bounded sets and if it has the following property: If the intersection of a sequence of closed absolutely convex 0-neighbourhoods in E absorbs all bounded sets, then it is itself a 0-neighbourhood in E.

A *precompact set* in E is a set which is relatively compact in the completion of E. (Often these sets are called *totally bounded*.)

Lemma 1.1.1. *Let Γ be a directed family of seminorms which generates the topology of E, and let M be a subset of E. Suppose that, given $p \in \Gamma$ and $\varepsilon > 0$, there exists a bounded set $M_{p,\varepsilon}$ contained in a finite dimensional subspace of E such that for each $\varphi \in M$ there is a $\psi \in M_{p,\varepsilon}$ satisfying $p(\varphi - \psi) \leqq \varepsilon$. Then M is a precompact set in E.*

Proof. Without loss of generality we can assume that E is already complete and $M_{p,\varepsilon}$ is closed. We have to show that the closure $\overline{M}$ of M is compact. For let W be an ultrafilter on M. Fix $p \in \Gamma$ and $\varepsilon > 0$. Set $V := \{\varphi \in E : p(\varphi) \leqq \varepsilon\}$. The set $M_{p,\varepsilon}$ is compact, so there exists a finite set N in E such that $M_{p,\varepsilon} \subseteqq N + V$. By assumption, $M \subseteqq M_{p,\varepsilon} + V$. The set $M_{p,\varepsilon} + V$ is closed in E (because $M_{p,\varepsilon}$ is compact). Hence $\overline{M} \subseteqq M_{p,\varepsilon} + V \subseteqq (N + V) + V = N + 2V = \bigcup_{\psi \in N} (\psi + 2V)$. Because W is an ultrafilter, this implies that $(\psi + 2V) \in W$ for some $\psi \in N$. Since $(\psi + 2V) - (\psi + 2V) = 4V$ and the sets $4V$ form a 0-neighbourhood base on E, this shows that W is a Cauchy filter on E. Hence W is convergent and M is compact. □

The locally convex space E admits the *approximation property* if the identity map of E can be approximated, uniformly on every precompact subset of E, by continuous linear mappings of finite rank.

Suppose that the topology of E is generated by a directed family, say Γ, of norms on E. Then E is called a *Schwartz space* if for every $p \in \Gamma$ there is a $q \in \Gamma$ such that the set $\{\varphi \in E : q(\varphi) \leqq 1\}$ is precompact in the normed linear space (E, p).

Let E and F be locally convex Hausdorff spaces. We define the two main topologies on the algebraic tensor product $E \otimes F$ of E and F. For seminorms p and q on E and F, respectively, let $p \otimes_\pi q$ denote the seminorm on $E \otimes F$ which is defined by

$$p \otimes_\pi q(z) = \inf \left\{ \sum_{n=1}^{k} p(\varphi_n)\, q(\psi_n) \right\}, \quad z \in E \otimes F,$$

where the infimum is taken over all representations $z = \sum_{n=1}^{k} \varphi_n \otimes \psi_n$ in $E \otimes F$. Suppose Γ_E and Γ_F are directed families of seminorms which generate the topologies of E and F, respectively. The *projective tensor topology* on $E \otimes F$ is defined by the family of seminorms $\{p \otimes_\pi q : p \in \Gamma_E \text{ and } q \in \Gamma_F\}$. Equipped with it, the space $E \otimes F$ is called the *projective tensor product* and denoted by $E \otimes_\pi F$. Let $E \mathbin{\widehat{\otimes}_\pi} F$ be the completion of $E \otimes_\pi F$. We denote by $\mathfrak{E}(E)$ and $\mathfrak{E}(F)$ the equicontinuous subsets of E' and F', respectively. For $M \in \mathfrak{E}(E)$ and $N \in \mathfrak{E}(F)$, let

$$\varepsilon_{M,N}(z) := \sup_{\varphi' \in M} \sup_{\psi' \in N} \left| \sum_{n=1}^{k} \varphi'(\varphi_n)\, \psi'(\psi_n) \right|, \quad z = \sum_{n=1}^{k} \varphi_n \otimes \psi_n \in E.$$

The *injective tensor topology* is generated by the family of seminorms $\{\varepsilon_{M,N} : M \in \mathfrak{E}(E)$ and $N \in \mathfrak{E}(F)\}$. The injective tensor product $E \otimes_\varepsilon F$ is the vector space $E \otimes F$ endowed with this topology. The completion of $E \otimes_\varepsilon F$ is denoted by $E \mathbin{\widehat{\otimes}_\varepsilon} F$.

The following result is occasionally called the *Mittag-Leffler theorem.*

Lemma 1.1.2. *Let $(E_n : n \in \mathbb{N}_0)$ be a sequence of Banach spaces. Suppose that for each $n \in \mathbb{N}_0$, E_{n+1} is a dense linear subspace of E_n and the embedding map of E_{n+1} into E_n is continuous. Then $E_\infty := \bigcap_{n \in \mathbb{N}_0} E_n$ is dense in each space E_k, $k \in \mathbb{N}_0$.*

Proof. There is no loss of generality to assume that $k = 0$. Suppose $\varphi \in E_0$ and $\varepsilon > 0$ Let $\|\cdot\|_n$, $n \in \mathbb{N}_0$, denote the norm of E_n. Since the embedding of E_{n+1} into E_n is con tinuous, there exists a constant $\alpha_n > 0$ such that $\|\cdot\|_n \leqq \alpha_n \|\cdot\|_{n+1}$ on E_{n+1} for $n \in \mathbb{N}_0$. Upon replacing $\|\cdot\|_n$ by $\alpha_1 \alpha_2 \ldots \alpha_{n-1} \|\cdot\|_n$ for $n \in \mathbb{N}$, we can assume without loss of generality that $\|\cdot\|_n \leqq \|\cdot\|_{n+1}$ for $n \in \mathbb{N}_0$. Set $\varphi_0 := \varphi$. Since E_{n+1} is dense in E_n, we can construct inductively a sequence $(\varphi_n : n \in \mathbb{N}_0)$ of elements $\varphi_n \in E_n$ such that $\|\varphi_{n+1} - \varphi_n\|_{n+1} \leqq \varepsilon 2^{-n-1}$ for $n \in \mathbb{N}_0$. Then we have

$$\|\varphi_{m+n+r} - \varphi_{m+n}\|_m \leqq \sum_{l=1}^{r} \|\varphi_{m+n+l} - \varphi_{m+n+l-1}\|_m \leqq$$

$$\sum_{l=1}^{r} \|\varphi_{m+n+l} - \varphi_{m+n+l-1}\|_{m+n+l} \leqq \sum_{l=1}^{r} \varepsilon 2^{-m-n-l} < \varepsilon 2^{-n} \tag{1}$$

for $m, n \in \mathbb{N}_0$ and $r \in \mathbb{N}$. From this we conclude that the sequence $(\varphi_{m+n} : n \in \mathbb{N}_0)$ is a Cauchy sequence in the Banach space E_m, $m \in \mathbb{N}_0$. Let ψ denote the limit of the sequence $(\varphi_{0+n} : n \in \mathbb{N}_0)$ in E_0. Then, of course, ψ is also the limit of $(\varphi_{m+n} : n \in \mathbb{N}_0)$ in E_m for all $m \in \mathbb{N}_0$. Hence $\psi \in E_\infty$. Setting $m = n = 0$ and letting $r \to +\infty$ in (1), we obtain $\|\psi - \varphi\|_0 \leqq \varepsilon$ which shows that E_∞ is dense in E_0. □

1.2. Spaces of Linear Mappings and Spaces of Sesquilinear Forms

First let E and F be vector spaces. We denote by E^- the complex conjugate vector space of E. That is, E^- is equal to E as a set, the addition in E^- is the same as in E, but the multiplication by scalars is replaced in E^- by the mapping $(\lambda, \varphi) \to \bar{\lambda}\varphi$, $\lambda \in \mathbb{C}$ and $\varphi \in E$. Let $L(E, F)$ be the vector space of all linear mappings of E into F, and let $B(E, F)$ denote the vector space of all sesquilinear forms on $E \times F$. We set $L(E) := L(E, E)$ and $B(E) := B(E, E)$. A *sesquilinear form* on $E \times F$ is a mapping of $E \times F$ into $\mathbb{C}$ which is linear in the first and conjugate linear in the second variable. For $\mathfrak{c} \in B(E, F)$, define $\mathfrak{c}^+(\psi, \varphi) := \overline{\mathfrak{c}(\varphi, \psi)}$, $\varphi \in E$ and $\psi \in F$; then $\mathfrak{c}^+ \in B(F, E)$. If $\mathfrak{c} \in B(E, E)$ and $\varphi, \psi \in E$, then we have the so-called *polarization identity*

$$4\mathfrak{c}(\varphi, \psi) = \mathfrak{c}(\varphi + \psi, \varphi + \psi) - \mathfrak{c}(\varphi - \psi, \varphi - \psi) + i\mathfrak{c}(\varphi + i\psi, \varphi + i\psi) - i\mathfrak{c}(\varphi - i\psi, \varphi - i\psi). \tag{1}$$

It is proved by computing the right-hand side of (1).

From now on we assume in this section that E and F are locally convex spaces. Since the vector spaces E and E^- have the same convex sets, they have the same locally convex topologies. We also denote by E^- the vector space E^- equipped with the topology of E. We shall write E^+ for the conjugate vector space $(E')^-$ of the dual E' of E. Let $\mathfrak{L}(E, F)$ denote the vector space of continuous linear mappings of E into F. Set $\mathfrak{L}(E) := \mathfrak{L}(E, E)$.

A sesquilinear form $\mathfrak{c}$ on $E \times F$ is said to be *separately continuous* if $\overline{\mathfrak{c}(\varphi, \cdot)} \in F'$ for each $\varphi \in E$ and $\mathfrak{c}(\cdot, \psi) \in E'$ for each $\psi \in F$; $\mathfrak{c}$ is called *continuous* if it is a continuous mapping of $E \times F$ into $\mathbb{C}$, when $E \times F$ carries the product topology. We denote the vector spaces of all separately continuous sesquilinear forms and of all continuous sesquilinear forms by $\mathfrak{B}(E, F)$ and $\mathscr{B}(E, F)$, respectively. From the theory of locally convex spaces (see Schäfer [1], III, 5.1) it is known that $\mathfrak{B}(E, F) = \mathscr{B}(E, F)$ if E and F are Frechet spaces or if E and F are barrelled (DF)-spaces.

For $x \in L(E, F^+)$ and $y \in L(F^+, E)$, we define

$$\mathfrak{c}_x(\varphi, \psi) = \overline{(x\varphi)(\psi)}, \quad \varphi \in E \text{ and } \psi \in F \tag{2}$$

and

$$\mathfrak{d}_y(\varphi^{\mathsf{I}}, \psi^{\mathsf{I}}) = \varphi^{\mathsf{I}}(y\psi^{\mathsf{I}}), \quad \varphi^{\mathsf{I}} \in E^{\mathsf{I}} \text{ and } \psi^{\mathsf{I}} \in F^{\mathsf{I}}. \tag{3}$$

Then, obviously, $\mathfrak{c}_x \in B(E, F)$ and $\mathfrak{d}_y \in B(E^{\mathsf{I}}, F^{\mathsf{I}})$.

Lemma 1.2.1. *For $x \in L(E, F^+)$ and $y \in L(F^+, E)$, we have:*

(i) $x \in \mathfrak{L}(E, F^+[\sigma^{\mathsf{I}}])$ if and only if $\mathfrak{c}_x \in \mathfrak{B}(E, F)$.

(ii) $y \in \mathfrak{L}(F^+[\sigma^{\mathsf{I}}], E[\sigma])$ if and only if $\mathfrak{d}_y \in \mathfrak{B}(E^{\mathsf{I}}[\sigma^{\mathsf{I}}], F^{\mathsf{I}}[\sigma^{\mathsf{I}}])$.

The mappings $x \to \mathfrak{c}_x$ and $y \to \mathfrak{d}_y$ are linear bijections of $\mathfrak{L}(E, F^+[\sigma^{\mathsf{I}}])$ on $\mathfrak{B}(E, F)$ and of $\mathfrak{L}(F^+[\sigma^{\mathsf{I}}], E[\sigma])$ on $\mathfrak{B}(E^{\mathsf{I}}[\sigma^{\mathsf{I}}], F^{\mathsf{I}}[\sigma^{\mathsf{I}}])$, respectively.

Proof. We prove (i). Suppose $x \in \mathfrak{L}(E, F^+[\sigma^{\mathsf{I}}])$ and let $\varphi \in E$. Since $x\varphi \in F^+$, $\overline{\mathfrak{c}_x(\varphi, \cdot)} \equiv (x\varphi)(\cdot) \in F^{\mathsf{I}}$. (Recall that F^+ is equal to F^{I} as a set.) From $x \in \mathfrak{L}(E, F^+[\sigma^{\mathsf{I}}])$ it follows that $\mathfrak{c}_x(\cdot, \psi) \equiv \overline{(x\cdot)(\psi)} \in E^{\mathsf{I}}$ for each $\psi \in F$. Thus $\mathfrak{c}_x \in \mathfrak{B}(E, F)$. Conversely, assume that $\mathfrak{c}_x \in \mathfrak{B}(E, F)$. Then $(x\varphi)(\cdot) \equiv \overline{\mathfrak{c}_x(\varphi, \cdot)} \in F^{\mathsf{I}}$ for each $\varphi \in E$, so that $x(E) \subseteq F^+$. Further, $\overline{(x\cdot)(\psi)} = \mathfrak{c}_x(\cdot, \psi) \in E^{\mathsf{I}}$ for all $\psi \in F$ which means that x maps E continuously into $F^+[\sigma^{\mathsf{I}}]$, i.e., $x \in \mathfrak{L}[E, F^+[\sigma^{\mathsf{I}}])$.

It is clear that the mapping $x \to \mathfrak{c}_x$ is linear and injective. To prove that it is surjective, let $\mathfrak{c} \in \mathfrak{B}(E, F)$. If $\varphi \in E$, then $\overline{\mathfrak{c}(\varphi, \cdot)} \in F^{\mathsf{I}}$. That is, there is a unique $\psi^{\mathsf{I}}_\varphi \in F^{\mathsf{I}}$ such that $\psi^{\mathsf{I}}_\varphi(\psi) = \overline{\mathfrak{c}(\varphi, \psi)}$ for all $\psi \in F$. Define $x\varphi := \psi^{\mathsf{I}}_\varphi$. We then obtain a linear mapping x of E into F^+ which satisfies $\mathfrak{c}_x = \mathfrak{c}$ by construction. From (i), $x \in \mathfrak{L}(E, F^+[\sigma^{\mathsf{I}}])$.

The assertions concerning y and $\mathfrak{d}_y$ follow in a similar way. □

Lemma 1.2.2. *If $x \in L(E, F^+)$ and $\mathfrak{c}_x \in \mathscr{B}(E, F)$, then $x \in \mathfrak{L}(E, F^+[\beta])$.*

Proof. Since $\mathfrak{c}_x \in \mathscr{B}(E, F)$, there are continuous seminorms p and q on E and F, respectively, such that $|\mathfrak{c}_x(\varphi, \psi)| \leqq p(\varphi)\, q(\psi)$, $\varphi \in E$ and $\psi \in F$. Let M be a bounded subset of F. Then $\lambda := \sup \{q(\psi) : \psi \in M\} < \infty$ and $\sup_{\psi \in M} |(x\varphi)(\psi)| = \sup_{\psi \in M} |\mathfrak{c}_x(\varphi, \psi)| \leqq \lambda p(\varphi)$ for $\varphi \in E$. This shows that $x \in \mathfrak{L}(E, F^+[\beta])$. □

Suppose that $x \in \mathfrak{L}(E, F^+[\sigma^{\mathsf{I}}])$ and $y \in \mathfrak{L}(F^+[\sigma^{\mathsf{I}}], E[\sigma])$. By Lemma 1.2.1 we have $\mathfrak{c}_x \in \mathfrak{B}(E, F)$ and $\mathfrak{d}_y \in \mathfrak{B}(E^{\mathsf{I}}[\sigma^{\mathsf{I}}], F^{\mathsf{I}}[\sigma^{\mathsf{I}}])$; so $(\mathfrak{c}_x)^+ \in \mathfrak{B}(F, E)$ and $(\mathfrak{d}_y)^+ \in \mathfrak{B}(F^{\mathsf{I}}[\sigma^{\mathsf{I}}], E^{\mathsf{I}}[\sigma^{\mathsf{I}}])$. Applying the reversed directions in Lemma 1.2.1, there are elements $x^+ \in \mathfrak{L}(F, E^+[\sigma^{\mathsf{I}}])$ and $y^+ \in \mathfrak{L}(E^+[\sigma^{\mathsf{I}}], F[\sigma])$ such that $(\mathfrak{c}_x)^+ = \mathfrak{c}_{x^+}$ and $(\mathfrak{d}_y)^+ = \mathfrak{d}_{y^+}$.

Now we define some locally convex topologies on certain spaces of linear mappings. They are needed in Sections 3.1 and 3.3.

I. The Equicontinuous Topology τ_e on $\mathfrak{L}(F^+[\sigma^{\mathsf{I}}], E[\sigma])$

If M is an equicontinuous subset of E^{I} and N is an equicontinuous subset of F^{I}, we define $p_{M,N}(y) := \sup_{\varphi^{\mathsf{I}} \in M} \sup_{\psi^{\mathsf{I}} \in N} |\varphi^{\mathsf{I}}(y\psi^{\mathsf{I}})|$, $y \in \mathfrak{L}(F^+[\sigma^{\mathsf{I}}], E[\sigma])$. It can be shown (see e.g. Schäfer [1], III, 5.5) that $p_{M,N}(\cdot)$ is finite on $\mathfrak{L}(F^+[\sigma^{\mathsf{I}}], E[\sigma])$, so that $p_{M,N}$ is a seminorm. Let τ_e denote the locally convex topology on $\mathfrak{L}(F^+[\sigma^{\mathsf{I}}], E[\sigma])$ generated by the family of all such seminorms $p_{M,N}$. The topology τ_e is called the *equicontinuous topology* on the space $\mathfrak{L}(F^+[\sigma^{\mathsf{I}}], E[\sigma])$.

II. The Bounded Topology τ_b on $\mathfrak{L}(E, F^+[\beta])$

Let S and T be non-empty families of bounded subsets of E and F, respectively. For $M \in S$ and $N \in T$, let

$$p_{M,N}(x) := \sup_{\varphi \in M} \sup_{\psi \in N} |(x\varphi)(\psi)|, \quad x \in \mathfrak{L}(E, F^+[\beta]).$$

Since $p_{M,N}(x) = \sup_{\varphi \in M} r_N(x\varphi)$ and $x \in \mathfrak{L}(E, F^+[\beta])$, $p_{M,N}(\cdot)$ is finite and hence a seminorm on $\mathfrak{L}(E, F^+[\beta])$. The family of seminorms $\{p_{M,N}: M \in S \text{ and } N \in T\}$ gives rise to a locally convex topology on $\mathfrak{L}(E, F^+[\beta])$ which we denote by $\tau_{S,T}$. If S and T are the families of all bounded subsets of E and F, respectively, then the corresponding topology $\tau_{S,T}$ is called the *bounded topology* on $\mathfrak{L}(E, F^+[\beta])$ and denoted by τ_b.

III. The Inductive Topology τ_{in}

Assume that L is a linear subspace of $L(E, F^+)$ such that $\mathfrak{c}_x \in \mathscr{B}(E, F)$ for all $x \in L$.

Let Γ_E and Γ_F be directed families of seminorms which generate the locally convex topologies of E and F, respectively. Suppose $p \in \Gamma_E$ and $q \in \Gamma_F$. Let $L_{p,q}$ denote the set of all $x \in L$ for which there exists a non-negative number λ such that

$$|\mathfrak{c}_x(\varphi, \psi)| \equiv |(x\varphi)(\psi)| \leqq \lambda p(\varphi)\, q(\psi) \quad \text{for all} \quad \varphi \in E \text{ and } \psi \in F. \tag{4}$$

For $x \in L_{p,q}$, let $\mathfrak{l}_{p,q}(x)$ be the infimum over all $\lambda \geqq 0$ satisfying (4). It is easily seen that $L_{p,q}$ is a linear subspace of L and $\mathfrak{l}_{p,q}$ is a norm on $L_{p,q}$. Since $\mathfrak{c}_x \in \mathscr{B}(E, F)$ for all $x \in L$ by assumption and the families Γ_E and Γ_F are directed, we have $L = \bigcup_{p \in \Gamma_E} \bigcup_{q \in \Gamma_F} L_{p,q}$. Further, if $p, p_1 \in \Gamma_E$ and $q, q_1 \in \Gamma_F$ satisfy $p \leqq p_1$ on E and $q \leqq q_1$ on F, then $(L_{p,q}, \mathfrak{l}_{p,q})$ is a linear subspace of $(L_{p_1,q_1}, \mathfrak{l}_{p_1,q_1})$, and the corresponding embedding map is continuous. Therefore, the topology of the inductive limit of the family of normed spaces $\{(L_{p,q}, \mathfrak{l}_{p,q}): p \in \Gamma_E \text{ and } q \in \Gamma_F\}$ is well-defined on L (cf. SCHÄFER [1], II, 6.3). This topology is denoted by τ_{in} and called the *inductive topology* on L. It is not difficult to check that this topology does not depend on the families Γ_E and Γ_F.

If $\varphi \in E$ and $\psi \in F$, then the set $\{x \in L: |\mathfrak{c}_x(\varphi, \psi)| \leqq 1\}$ contains a 0-neighbourhood in the normed linear space $(L_{p,q}, \mathfrak{l}_{p,q})$ for any $p \in \Gamma_E$ and $q \in \Gamma_F$; hence it is a 0-neighbourhood in $L[\tau_{in}]$. This implies that $L[\tau_{in}]$ is a Hausdorff space. Being the inductive limit of normed spaces and Hausdorff, $L[\tau_{in}]$ is a bornological space.

Since $L \subseteq \mathfrak{L}(E, F^+[\beta])$ by Lemma 1.2.2, the topology τ_b is also defined on L. We show that $\tau_b \subseteq \tau_{in}$ on L. For let M and N be bounded sets in E and F, respectively. Then $\lambda_{p,q} := \sup_{\varphi \in M} p(\varphi) \sup_{\psi \in N} q(\psi) < \infty$ and $p_{M,N}(x) \leqq \lambda_{q,p}\mathfrak{l}_{p,q}(x)$ for $x \in L$, $p \in \Gamma_E$ and $q \in \Gamma_F$. Thus $p_{M,N}$ is continuous on each normed space $(L_{p,q}, \mathfrak{l}_{p,q})$ and hence on $L[\tau_{in}]$ which proves that $\tau_b \subseteq \tau_{in}$.

Since in particular $L \subseteq \mathfrak{L}(E, F^+[\sigma'])$, $L^+ := \{x^+ : x \in L\}$ also satisfies the above assumptions, so that τ_{in} is defined on L^+. From $(L_{p,q})^+ = (L^+)_{q,p}$ and $\mathfrak{l}_{p,q}(x) = \mathfrak{l}_{q,p}(x^+)$ for $x \in L_{p,q}$, $p \in \Gamma_E$ and $q \in \Gamma_F$ it follows that $x \to x^+$ is a continuous mapping of $L[\tau_{in}]$ onto $L^+[\tau_{in}]$. We denote by $\mathscr{U}_{p,q}$ the unit ball of $(L_{p,q}, \mathfrak{l}_{p,q})$.

Remark 1. The three topologies defined above are closely related to various standard topologies from the theory of locally convex spaces. If we identify y and $\mathfrak{d}_y$, then τ_e is the topology of bi-equicontinuous convergence on $\mathfrak{B}(E'[\sigma'], F'[\sigma'])$; see e.g. SCHÄFER [1], III, 5.5. The topology τ_b is precisely the topology of uniform convergence on bounded sets on the space $\mathfrak{L}(E, F^+[\beta])$: see e.g. SCHÄFER [1], III, § 3. Under the isomorphism $x \to \mathfrak{c}_x$, τ_b goes into the topology of bi-bounded convergence on $\mathscr{B}(E, F)$; see e.g. SCHÄFER [1], IV, 9.7. If we identify an element $x \in L(E, F^+)$ with the linear functional on $E \otimes F^-$ defined by $\sum_n \varphi_n \otimes \psi_n \to \sum_n \mathfrak{c}_x(\varphi_n, \psi_n)$, then the topology τ_{in} on the maximal space $L_{max} := \{x \in L(E, F^+): \mathfrak{c}_x \in \mathscr{B}(E, F)\}$ coincides with Beresanskii's topology η on the dual $(E \otimes_\pi F^-)'$; see e.g. BERESANSKII [1] or JARCHOW [1], 10.3.

1.3. Ordered $*$-Vector Spaces

For ordered vector spaces we refer to Chapter V of SCHÄFER [1] and also to PERESSINI [1].

$*$-Vector Spaces

An *involution* on a (complex) vector space L is a mapping $x \to x^+$ of L into L satisfying $(\alpha x + \beta y)^+ = \bar{\alpha} x^+ + \bar{\beta} y^+$ and $(x^+)^+ = x$ for all $x, y \in L$ and $\alpha, \beta \in \mathbb{C}$. A *$*$-vector space* is a (complex) vector space equipped with an involution. The involution of $*$-vector spaces (and so, in particular, of $*$-algebras) is always denoted by $x \to x^+$. If x is a Hilbert space operator, then x^+ should not be confused with the adjoint operator of x which we shall denote by x^*.

Suppose L is a $*$-vector space. A *$*$-vector subspace* of L is a linear subspace of L which is invariant under the involution. An element x of L is called *hermitian* if $x = x^+$. The real vector space $L_h := \{x \in L: x = x^+\}$ is called the *hermitian part* of L. Each element $x \in L$ can be expressed uniquely of the form $x = x_1 + ix_2$ with $x_1, x_2 \in L_h$. (Indeed, $x_1 := \frac{1}{2}(x^+ + x)$ and $x_2 := \frac{1}{2} i(x^+ - x)$ have the desired properties. If x_1' and x_2' are elements of L_h such that $x = x_1' + ix_2'$, then $x^+ = x_1' - ix_2'$ and hence $x_1 = x_1'$ and $x_2 = x_2'$.) Thus we have $L = L_h + iL_h$. The vector space of all (complex) linear functionals on the vector space L is denoted by L^*. For $f \in L^*$, define $f^+(x) := \overline{f(x^+)}$, $x \in L$. Then the map $f \to f^+$ is an involution on the vector space L^*. Hence L^* is also a $*$-vector space, and the terminology of the preceding paragraph applies to L^* as well. That is, a linear functional f on L is said to be *hermitian* if $f = f^+$, i.e., if $f(x) = \overline{f(x^+)}$ for all $x \in L$ or equivalently if f is real-valued on L_h. Further, L_h^* is the real vector space of hermitian linear functionals on L.

The following simple lemma is temporarily used in the text.

Lemma 1.3.1. *If g is a (real) linear functional on the real vector space L_h, then there is a unique (complex) linear functional f on L such that $g = f \restriction L_h$.*

Proof. Let $x \in L$. We write x as $x = x_1 + ix_2$ with unique elements $x_1, x_3 \in L_h$ and define $f(x) := g(x_1) + ig(x_2)$. Suppose $\lambda = \lambda_1 + i\lambda_2 \in \mathbb{C}$ with $\lambda_1, \lambda_2 \in \mathbb{R}$. By the real linearity of g and the definition of f, we have $f(\lambda x) = f(\lambda_1 x_1 - \lambda_2 x_2 + i(\lambda_1 x_2 + \lambda_2 x_1))$ $= g(\lambda_1 x_1 - \lambda_2 x_2) + ig(\lambda_1 x_2 + \lambda_2 x_1) = \lambda_1 g(x_1) - \lambda_2 g(x_2) + i\lambda_1 g(x_2) + i\lambda_2 g(x_1) =$ $(\lambda_1 + i\lambda_2)(g(x_1) + ig(x_2)) = \lambda f(x)$. Clearly, $g = f \restriction L_h$. The uniqueness of f is obvious. □

Ordered Vector Spaces

A *wedge* in a (real or complex) vector space E is a non-void subset K of E such that $K + K \subseteq K$ and $\lambda K \subseteq K$ for all $\lambda > 0$. A wedge K is called a *cone* if $K \cap (-K) = \{0\}$.

An *ordered vector space* is a real vector space E equipped with a reflexive transitive relation "$\geq$" satisfying the following two conditions:

(i) $x \geq y$ implies $x + z \geq y + z$ for all $z \in E$,

(ii) $x \geq y$ implies $\lambda x \geq \lambda y$ for all $\lambda > 0$.

We shall denote the ordered vector space by $(E, \geq)$. The set $K := \{x \in E : x \geq 0\}$ is then a wedge in E which is called the *positive wedge* of the ordered vector space $(E, \geq)$. Conversely, if K is a wedge in a real vector space E, then the definition "$x \geq y$ if and only if $x - y \in K$" yields a relation "$\geq$" such that $(E, \geq)$ is an ordered vector space with positive wedge K.

Suppose $(E, \geq)$ is an ordered vector space with positive wedge K. By definition $y \leq x$ means that $x \geq y$. The sets $[x, y] := \{z \in E : x \leq z \leq y\}$ are called the *order intervals* of $(E, \geq)$. A subset U of E is said to be *K-saturated* if $U = \bigcup_{x,y \in U} [x, y]$ or equivalently if $U = (U + K) \cap (U - K)$. The wedge K is said to be *normal* for a locally convex topology τ on E if τ admits a 0-neighbourhood base of K-saturated sets. The finest locally convex on E for which every order interval of $(E, \geq)$ is bounded is called the *order topology* of $(E, \geq)$.

Ordered $*$-Vector Spaces

Let L be a $*$-vector space and let K be a wedge in L_h. A linear functional f on L is called *K-positive* if $f(x) \geqq 0$ for all $x \in K$. We denote the set of all K-positive linear functionals by K^*. Obviously, K^* is a wedge in the vector space L^*. We say that a $*$-vector subspace L_1 of L is *cofinal* in L *with respect to* K if for every $x \in L_h$ there is a $y \in (L_1)_h$ such that $y \in K$ and $y - x \in K$.

An *ordered $*$-vector space* L is a $*$-vector space L together with a wedge K in the real vector space L_h. By the canonical one-to-one correspondence between orderings and wedges in real vector spaces mentioned in the preceding subsection, one can also say that an ordered $*$-vector space is a $*$-vector space for which the hermitian part is an ordered vector space. Suppose L is an ordered $*$-vector space with positive wedge K. A $*$-vector subspace L_1 of L is called *cofinal* in L if L_1 is cofinal in L with respect to K. A subset C of K is called *order-dominating* for L if for each $x \in L_h$ there exist $y \in C$ and $\lambda > 0$ such that $\lambda y - x \in K$.

Let K be a convex set in a real or complex vector space. A point x in K is called an *extreme point* of K if $x = \lambda x_1 + (1 - \lambda) x_2$ with $x_1, x_2 \in K$ and $0 < \lambda < 1$ always implies that $x_1 = x_2$. The set of extreme points of K is denoted by ex K. If K is a wedge, then the following concept is of more interest. An *extremal point* of a wedge K is a point x in K such that $y \in K$ and $x - y \in K$ imply that $y = \lambda x$ for some $\lambda \in [0, 1]$.

Lemma 1.3.2. *Suppose that L is an ordered $*$-vector space with positive wedge K and L_0 is a cofinal $*$-vector subspace of L. Set $K_0 := K \cap L_0$. If f_0 is a K_0-positive linear functional on L_0, then there exists a K-positive linear functional f on L which extends f_0. If f_0 is an extremal point of K_0^*, then f can be chosen to be an extremal point of K^*.*

Proof. It is sufficient to prove the assertion in case where L_0 has codimension 1 in L. A standard application of Zorn's lemma then gives the result in the general case. If f_0 is an extremal point of K_0^*, then we apply Zorn's lemma to the set of all extremal extensions of f.

Since L_0 has codimension 1 in L, there is an element $x \in L_h \setminus L_0$ such that L is spanned by x and L_0. Let "$\geq$" denote the ordering of L. Because L_0 is cofinal in L, there are $y_1, y_2 \in (L_0)_h$ such that $y_1 \leq x \leq y_2$. Hence $\delta := \inf \{f_0(v) : v \in (L_0)_h$ and $v \geq x\}$ is well-defined. By $f_0 \in K_0^*$, we have $\delta \geqq f_0(y_1)$ and so $\delta \in \mathbb{R}$. Each $z \in L$ is uniquely expressable as $z = \alpha x + y$ with $\alpha \in \mathbb{C}$ and $y \in L_0$. Therefore, $f(z) := \alpha\delta + f_0(y)$ defines unambiguously a linear functional on L which extends f_0.

We show that f is K-positive. Suppose $z = \alpha x + y \in K$. Since $K \subseteq L_h$ and $x \in L_h$, we have $0 = z - z^+ = (\alpha - \bar{\alpha})\, x + y - y^+$. Since $x \notin L_0$, this implies that α is real. If $\alpha = 0$, then $y \in K_0$ and so $f(z) = f_0(y) \geqq 0$. Now suppose $\alpha > 0$. Then $x \geq -\alpha^{-1}y$. If $v \in (L_0)_h$ and $v \geq x$, then $v \geq -\alpha^{-1}y$ and hence $f_0(v) \geqq f_0(-\alpha^{-1}y)$. This yields $\delta = f(x) \geqq f_0(-\alpha^{-1}y)$ and $f(z) \geqq 0$. If $\alpha < 0$, the proof is similar. Thus $f \in K^*$.

Now suppose that f_0 is an extremal point of K_0^*. We prove that f is an extremal point of K^*. Let $g \in K^*$ be such that $g(z) \leqq f(z)$ for all $z \in K$. Since $f_0 \equiv f \upharpoonright L_0$ is an extremal point of K_0^*, there exists a $\lambda \in [0, 1]$ such that $g(y) = \lambda f(y)$ for all $y \in L_0$. The proof is complete if we have shown that $g(x) = \lambda f(x)$. If $v \in (L_0)_h$ and $v \geq x$, then $\lambda f(v) = \lambda f_0(v) = g(v) \geqq g(x)$ and $0 \leqq f(v - x) - g(v - x) = (1 - \lambda)\, f_0(v) - \delta + g(x)$, hence $\lambda\delta \geqq g(x)$ and $0 \leqq (1 - \lambda)\, \delta - \delta + g(x)$. Therefore, $g(x) = \lambda\delta = \lambda f(x)$. □

Remark 1. The preceding proof showed that the assertions of the lemma remain valid if the above definition of cofinality is replaced by the weaker requirement that for given $x \in L_h$ there is a $y \in (L_0)_h$ such that $y - x \in K$.

1.4. $*$-Algebras and Topological $*$-Algebras

An *algebra* is a vector space A in which a mapping $(a, b) \to ab$ of $\mathsf{A} \times \mathsf{A}$ into A is defined that satisfies the following axioms:

(i) $a(bc) = (ab)\, c$,

(ii) $(a + b)\, c = ac + bc$ and $a(b + c) = ab + ac$,

(iii) $\alpha(ab) = (\alpha a)\, b = a(\alpha b)$

for all $a, b, c \in \mathsf{A}$ and $\alpha \in \mathbb{C}$. The element ab is called the *product* of a and b. Suppose A is an algebra. An element $1 \in \mathsf{A}$ is called a *unit element* of A if it satisfies $1a = a1 = a$ for all $a \in \mathsf{A}$. The unit elements of abstract algebras are always denoted by the symbol 1. If 1 is a unit element of A, then we set $a^0 := 1$ for each $a \in \mathsf{A}$ and we frequently write α instead of $\alpha \cdot 1$ for $\alpha \in \mathbb{C}$. A *character* on A is a linear functional f on A such that $f \not\equiv 0$ and $f(ab) = f(a)\, f(b)$ for all $a, b \in \mathsf{A}$.

A $*$-*algebra* is an algebra A with an involution $a \to a^+$ on A that also satisfies $(ab)^+ = b^+a^+$ for $a, b \in \mathsf{A}$. Since a $*$-algebra is in particular a $*$-vector space, the terminology from Section 1.3 also applies to $*$-algebras. A $*$-algebra A is said to be *symmetric if A* has a unit and for every $a \in \mathsf{A}_h$ and $\alpha \in \mathbb{C} \setminus \mathbb{R}$ the element $a - \alpha$ is invertible in A.

Suppose A is a $*$-algebra. An ideal J of A is called a $*$-*ideal* if $x^+ \in \mathsf{J}$ when $x \in \mathsf{J}$. Let $\mathcal{P}(\mathsf{A})$ denote the set of all finite sums $\sum_{n=1}^{k} a_n^+ a_n$ with $a_1, \ldots, a_k \in \mathsf{A}$ and $k \in \mathbb{N}$. A linear functional f on A is called *positive* if $f(a^+a) \geqq 0$ for all $a \in$ A. It is obvious that $\mathcal{P}(\mathsf{A})$ is a wedge in A_h and that the positive linear functionals on A are precisely the $\mathcal{P}(\mathsf{A})$-positive linear functionals or equivalently the functionals in $\mathcal{P}(\mathsf{A})^*$. If A has a unit, then a *state* of A is a positive linear functional f on A which satisfies $f(1) = 1$. The set of all states of A is denoted by $\mathcal{Z}(\mathsf{A})$. An *m-admissible wedge* in a $*$-algebra A is a wedge $\mathcal{K}$ in the hermitian part A_h such that $\mathcal{P}(\mathsf{A}) \subseteqq \mathcal{K}$ and $a^+xa \in \mathcal{K}$ for all $a \in$ A and $x \in \mathcal{K}$. It is easy to verify that $\mathcal{P}(\mathsf{A})$ is the smallest m-admissible wedge in A. If the $*$-algebra A has a unit, then a wedge $\mathcal{K}$ in A_h is m-admissible if and only if $1 \in \mathcal{K}$ and $a^+xa \in \mathcal{K}$ for all $a \in$ A and $x \in \mathcal{K}$. (Indeed, the necessity is clear, since $1 = 1^+1 \in \mathcal{P}(\mathsf{A})$. For the sufficiency, we note that $a^+a = a^+1a \in \mathcal{K}$ for every $a \in$ A and hence $\mathcal{P}(\mathsf{A}) \subseteqq \mathcal{K}$.)

The inequality occuring in the following lemma is called the *Cauchy-Schwarz inequality*. It will be often used in the sequel.

Lemma 1.4.1. *Suppose that f is a positive linear functional on a $*$-algebra* A. *Then* $|f(b^+a)^2| \leqq f(a^+a)\, f(b^+b)$ *for all* $a, b \in$ A. *If* A *has a unit, then the functional f is hermitian and so* $\mathcal{P}(\mathsf{A})^* \subseteqq \mathsf{A}_h^*$.

Proof. For arbitrary $\alpha, \beta \in \mathbb{C}$, we have $f\big((\alpha a + \beta b)^+ (\alpha a + \beta b)\big) =$

$$\bar{\alpha}\alpha f(a^+a) + \bar{\alpha}\beta f(a^+b) + \alpha\bar{\beta} f(b^+a) + \bar{\beta}\beta f(b^+b) \geqq 0. \tag{1}$$

From this we see that $\bar{\alpha}\beta f(a^+b) + \alpha\bar{\beta} f(b^+a)$ is real for $\alpha, \beta \in \mathbb{C}$, so

$$\overline{f(a^+b)} = f(b^+a). \tag{2}$$

The expression in (1) is a positive semi-definite quadratic form, hence its principal minors are non-negative. Combined with (2), this gives $f(a^+a)\, f(b^+b) - |f(b^+a)|^2 \geqq 0$. If A admits a unit, then (2) in case $b = 1$ shows that f is hermitian. □

A *topological algebra* is an algebra A equipped with a locally convex topology τ such that the multiplication in A is separately continuous, i.e., for each $a \in$ A the mappings $x \to xa$ and $x \to ax$ are continuous in $\mathsf{A}[\tau]$. If even the map $(a, b) \to ab$ of $\mathsf{A}[\tau] \times \mathsf{A}[\tau]$ into $\mathsf{A}[\tau]$ is continuous, then we shall say that the multiplication is *jointly continuous* in $\mathsf{A}[\tau]$. If $\mathsf{A}[\tau]$ is a topological algebra for which $\mathsf{A}[\tau]$ is a Frechet space or a barrelled DF-space, then the multiplication is automatically jointly continuous in $\mathsf{A}[\tau]$. This follows at once from the general continuity theorems for bilinear mappings mentioned in Section 1.2. By a *topological isomorphism* of two topological algebras we mean an algebraic isomorphism which is also a homeomorphism.

A *topological $*$-algebra* is a $*$-algebra with a locally convex topology τ such that $\mathsf{A}[\tau]$ is a topological algebra and the involution of A is continuous in $\mathsf{A}[\tau]$. In fact, it suffices to assume the continuity of the involution and of all left (or right) multiplications; the continuity of the right (or left) multiplications follows then from the identity $ab = (b^+a^+)^+$, $a, b \in$ A.

1.5. The Topologies τ_F, τ_n, τ_0 and τ^F, τ^n, τ^0

In this section we develop some locally convex topologies on ordered $*$-vector spaces resp. on $*$-algebras which are related to order properties. These topologies are used in Section 3.3.

The Topologies τ_F, τ_n, τ_0 on an Ordered $*$-Vector Space

In this subsection L denotes an ordered $*$-vector space. Let K be the corresponding wedge in L_h and "$\geq$" the associated order relation.

Lemma 1.5.1. *Suppose U is an absolutely convex subset of the real vector space L_h, and let* aco U *denote its absolutely convex hull in the complex vector space L. Then* (aco U) $\cap L_h = U$. *If U is absorbing in L_h, then* aco U *is absorbing in L.*

Proof. Let $x \in (\text{aco } U) \cap L_h$. Then there are $\lambda_1, \ldots, \lambda_n \in \mathbb{C}$ and $x_1, \ldots, x_k \in U$ such that $x = \sum_{n=1}^{k} \lambda_n x_n$ and $\sum_{n=1}^{k} |\lambda_n| \leqq 1$. From $x = x^+$ and $U \subseteq L_h$, $x = \sum_{n=1}^{k} (\text{Re } \lambda_n) x_n$. Since $\sum_{n=1}^{k} |\text{Re } \lambda_n| \leqq 1$ and U is absolutely convex in L_h, this yields $x \in U$. Since trivially $U \subseteq (\text{aco } U) \cap L_h$, $(\text{aco } U) \cap L_h = U$.

Now suppose that U is absorbing in L_h. Let $x \in L$. We write x as $x = x_1 + ix_2$ with $x_1, x_2 \in L_h$. For $k = 1, 2$, there is a number $\alpha_k > 0$ such that $x_k \in \lambda_k U$ for all $\lambda_k \in \mathbb{R}$, $|\lambda_k| \geqq \alpha_k$. Set $\alpha := 2(\alpha_1 + \alpha_2)$. If $\lambda \in \mathbb{C}$ and $|\lambda| \geqq \alpha$, then $2x_k \in \alpha U$ for $k = 1, 2$ and so $x = x_1 + ix_2 \in \text{aco } (\alpha U) = \alpha \text{ aco } U \subseteq \lambda \text{ aco } U$. Hence aco U is absorbing in L. □

Let $\boldsymbol{U}_{h,n}$ be the collection of all absolutely convex absorbing subsets of L_h which are K-saturated, and let $\boldsymbol{U}_{h,0}$ denote the family of all absolutely convex subsets of L_h that absorb all order intervals of $(L_h, \geq)$. Each $U \in \boldsymbol{U}_{h,0}$ is also absorbing in L_h, since $x \in [x, x]$ for $x \in L_h$. Obviously, $\boldsymbol{U}_{h,n}$ and $\boldsymbol{U}_{h,0}$ satisfy the conditions (i)—(iii) in Section 1.1; hence there are locally convex topologies $\tau_{h,n}$ and $\tau_{h,0}$ on the real vector space L_h such that $\boldsymbol{U}_{h,n}$ and $\boldsymbol{U}_{h,0}$ are 0-neighbourhood bases for $\tau_{h,n}$ and $\tau_{h,0}$, respectively. From these definitions it is clear that $\tau_{h,n}$ is the finest locally convex topology on L_h for which K is normal and that $\tau_{h,0}$ is the order topology of $(L_h, \geq)$.

We denote by $\boldsymbol{U}_n$ resp. $\boldsymbol{U}_0$ the family of all absolutely convex sets U in L for which $U \cap L_h$ belongs to $\boldsymbol{U}_{h,n}$ resp. $\boldsymbol{U}_{h,0}$. Let U be a set from $\boldsymbol{U}_{h,n}$ or $\boldsymbol{U}_{h,0}$. Since $U \cap L_h$ is absorbing in L_h, U is absorbing in L by Lemma 1.5.1. Therefore, the families $\boldsymbol{U}_n$ and $\boldsymbol{U}_0$ also satisfy the conditions (i)—(iii) in 1.1. Hence there exist locally convex topologies τ_n and τ_0 on the complex vector space L such that $\boldsymbol{U}_n$ is a 0-neighbourhood base for τ_n and $\boldsymbol{U}_0$ is a 0-neighbourhood base for τ_0. We call τ_0 the *order topology* of the ordered $*$-vector space L. Some basic properties of these topologies are collected in

Proposition 1.5.2. (i) $\tau_n \subseteq \tau_0$.

(ii) $\tau_n \upharpoonright L_h = \tau_{h,n}$ *and* $\tau_0 \upharpoonright L_h = \tau_{h,0}$.

(iii) *The involution of L is continuous in $L[\tau_n]$ and in $L[\tau_0]$.*

(iv) *τ_n is the finest locally convex topology τ on L for which K is normal in $L_h[\tau]$.*

(v) *τ_0 is the finest locally convex topology τ on L such that each order interval of $(L_h, \geq)$ is bounded in $L[\tau]$.*

(vi) *If the topology τ_0 is Hausdorff, then $L[\tau_0]$ is a bornological locally convex space.*

Proof. (i) Suppose $U \in \boldsymbol{U}_{h,n}$. Let $x, y \in L$. Since U is absorbing on L_h, there is a $\lambda > 0$ such that $x \in \lambda U$ and $y \in \lambda U$. Hence $[x, y] \subseteq \lambda U$, since U is K-saturated. Therefore, $U \in \boldsymbol{U}_{h,0}$ and so $\tau_n \subseteq \tau_0$.

(ii) Let $U \in \boldsymbol{U}_{h,n}$. Since (aco U) $\cap L_h = U$ by Lemma 1.5.1, aco $U \in \boldsymbol{U}_n$ and so $\tau_{h,n} \subseteq \tau_n \upharpoonright L_h$. From the definition it is obvious that $\tau_n \upharpoonright L_h \subseteq \tau_{h,n}$; hence $\tau_{h,n} = \tau_n \upharpoonright L_h$. The proof for τ_0 is the same.

(iii) follows from the fact that the sets in $\boldsymbol{U}_n$ and $\boldsymbol{U}_0$ are invariant under the involution of L.

(iv) Since $\tau_n \upharpoonright L_h = \tau_{h,n}$ by (ii), K is normal in $L_h[\tau_n]$. Let τ be a locally convex topology on L for which K is normal in $L_h[\tau]$, and let V be an absolutely convex 0-neighbourhood for τ. Since K is normal in $L_h[\tau]$, there is a set $U \in \boldsymbol{U}_{h,n}$ such that $V \cap L_h \supseteq U$. Since aco $U \in \boldsymbol{U}_n$ by Lemma 1.5.1 and $V \supseteq \text{aco}\,(V \cap L_h) \supseteq \text{aco}\,U$, V is a 0-neighbourhood for τ_n.

(v) follows directly from the definition.

(vi) Let U be an absolutely convex subset of L which absorbs each τ_0-bounded subset of L. By definition all order intervals of $(L_h, \geq)$ are τ_0-bounded, so $U \cap L_h$ absorbs all order intervals. Hence $U \cap L_h \in \boldsymbol{U}_{h,0}$ and $U \in \boldsymbol{U}_0$. If τ_0 is Hausdorff, then the preceding shows that $L[\tau_0]$ is bornological. □

Next we give another description of the topology τ_n and we define the topologies $\tau_{\boldsymbol{F}}$. Let $\boldsymbol{F}_{\max}$ denote the collection of all weakly bounded subsets of K^*, i.e., $\boldsymbol{F}_{\max}$ is the family of all sets M of linear functionals on L which are non-negative on K satisfying $\sup\{|f(x)| : f \in M\} < \infty$ for all $x \in L$. For $M \in \boldsymbol{F}_{\max}$, we define a seminorm on L by

$$r_M(x) := \sup_{f \in M} |f(x)|, \quad x \in L.$$

Let $\boldsymbol{F}$ be a non-empty subset of $\boldsymbol{F}_{\max}$, and let $\tau_{\boldsymbol{F}}$ denote the locally convex topology on L which is generated by the family of seminorms $\{r_M : M \in \boldsymbol{F}\}$.

Lemma 1.5.3. *K is normal in $L_h[\tau_{\boldsymbol{F}}]$.*

Proof. Without loss of generality we can assume that $M_1 \cap M_2 \in \boldsymbol{F}$ and $\lambda M_1 \in \boldsymbol{F}$ for $M_1, M_2 \in \boldsymbol{F}$ and $\lambda > 0$. Then the sets $W_M := \{x \in L_h : r_M(x) \leq 1\}$, $M \in \boldsymbol{F}$, form a 0-neighbourhood base for the topology $\tau_{\boldsymbol{F}} \upharpoonright L_h$. It suffices to check that each set W_M is K-saturated. We suppose $x, y \in W_M$ and $z \in [x, y]$. Then $z - x \in K$ and $y - z \in K$. Since $M \subseteq K^*$, $f(z - x) \geq 0$ and $f(y - z) \geq 0$ which leads to $\text{Re}\, f(x) \leq \text{Re}\, f(z) \leq \text{Re}\, f(y)$ and $\text{Im}\, f(x) = \text{Im}\, f(z) = \text{Im}\, f(y)$ for $f \in M$. Therefore, $|f(z)| \leq \max\left(|f(x)|, |f(y)|\right) \leq 1$, i.e., $z \in W_M$ and W_M is K-saturated. □

Proposition 1.5.4. *Suppose that τ is a locally convex topology on L such that K is normal in $L_h[\tau]$ and the involution of L is continuous in $L[\tau]$.*

(i) *Then there exists a subset $\boldsymbol{F}$ of $\boldsymbol{F}_{\max}$ such that $\tau = \tau_{\boldsymbol{F}}$ on L.*

(ii) *If f is a continuous linear functional on $L[\tau]$, then there are K-positive continuous linear functionals f_1, f_2, f_3, f_4 on $L[\tau]$ such that $f = (f_1 - f_2) + \mathrm{i}(f_3 - f_4)$.*

Proof. (i) From Schäfer's duality theorem (see e.g. Schäfer [1], V, 3.3) it follows that there exists a family $\boldsymbol{F}_{\mathbb{R}}$ of equicontinuous sets of real linear functionals on the real locally convex space $L_h[\tau]$ with non-negative values on K such that the family of seminorms $\left\{r_N(x) = \sup_{g \in N} |g(x)| : N \in \boldsymbol{F}_{\mathbb{R}}\right\}$ on L_h generates the topology $\tau \upharpoonright L_h$. Let $N \in \boldsymbol{F}_{\mathbb{R}}$. By Lemma 1.3.1, each $g \in N$ extends to a linear functional f_g on the complex vector space L. The set $M(N) := \{f_g : g \in N\}$ is weakly bounded, since we have $\sup\{|f_g(x_1 + \mathrm{i}x_2)| : g \in N\} = \sup\{|g(x_1) + ig(x_2)| : g \in N\} \leq r_N(x_1) + r_N(x_2) < \infty$ for $x_1, x_2 \in L_h$. Setting $\boldsymbol{F} := \{M(N) : N \in \boldsymbol{F}_{\mathbb{R}}\}$, we have $\boldsymbol{F} \subseteq \boldsymbol{F}_{\max}$ and $\tau \upharpoonright L_h = \tau_{\boldsymbol{F}} \upharpoonright L_h$. Since the functionals $g \equiv f_g \upharpoonright L_h$ are real on L_h, we have $r_{M(N)}(x) = r_{M(N)}(x^+)$ for $x \in L$

and $M \in \boldsymbol{F}$. Hence the involution is continuous in $L[\tau_F]$. By assumption the involution is also continuous in $L[\tau]$; so we obtain $\tau = \tau_F$ on L.

(ii) By the continuity of the involution in $L[\tau]$, it suffices to assume that f is hermitian. Since K is normal in $L_h[\tau]$, $g := f \upharpoonright L_h$ can be written as $g = g_1 - g_2$, where g_1 and g_2 are (real) continuous linear functionals on $L_h[\tau]$ with non-negative values on K (SCHÄFER [1], V, 3.3, Corollary 3). Then the extension f_k of g_k to L (by Lemma 1.3.1) is continuous on $L[\tau]$ and K-positive for $k = 1, 2$. By the uniqueness of this extension, $f = f_1 - f_2$. □

Corollary 1.5.5. *$\tau_n = \tau_{F_{\max}}$ on L.*

Proof. Since τ_n satisfies the assumptions of Proposition 1.5.4, $\tau_n \subseteq \tau_{F_{\max}}$. By Lemma 1.5.3 and the characterization of τ_n given in Proposition 1.5.2, (iv), $\tau_{F_{\max}} \subseteq \tau_n$. □

The Topologies τ_F, τ_n, τ_0 and τ^F, τ^n, τ^0 on a $*$-Algebra

In this subsection we assume that A is a $*$-algebra with unit and K is a fixed wedge in A_h which contains $\mathcal{P}(A)$. Let "$\geq$" denote the order relation on A_h associated with the wedge K. We retain the notation from the preceding subsection.

Since $\mathcal{P}(A) \subseteq K$ and A has a unit element, the functionals of K^* are hermitian by Lemma 1.4.1. Hence $r_M(x) = r_M(x^+)$ for all $x \in A$ and $M \in \boldsymbol{F}_{\max}$. Therefore, if $\boldsymbol{F}$ is a non-empty subset of $\boldsymbol{F}_{\max}$, the involution of A is continuous in $A[\tau_F]$.

Lemma 1.5.6. *Suppose that τ is a locally convex topology on A such that for every $a \in A$ the mapping $x \to a^+xa$ is continuous in $A[\tau]$. Then $A[\tau]$ is a topological algebra.*

Proof. Let $a \in A$. The continuity of the mappings $x \to ax$ and $x \to xa$ in $A[\tau]$ follows from the identities

$$ax = \frac{1}{4}\{(a+1)\,x(a+1)^+ - (a-1)\,x(a-1)^+ + i(a+i\cdot 1)\,x(a+i\cdot 1)^+ - i\,(a-i\cdot 1)\,x(a-i\cdot 1)^+\}$$

and

$$xa = \frac{1}{4}\{(a+1)^+\,x(a+1) - (a-1)^+\,x(a-1) + i\,(a+i\cdot 1)^+\,x(a+i\cdot 1) - i\,(a-i\cdot 1)^+\,x(a-i\cdot 1)\}$$

which hold for arbitrary a and x in A. □

A subset $\boldsymbol{F}$ of $\boldsymbol{F}_{\max}$ is said to be A-*invariant* if the set $M_a := \{f_a(\cdot) := f(a^+\cdot a) : f \in M\}$ belongs to $\boldsymbol{F}$ for each $M \in \boldsymbol{F}$ and $a \in A$.

Lemma 1.5.7. *If $\boldsymbol{F}$ is an A-invariant non-empty subset of $\boldsymbol{F}_{\max}$, then $A[\tau_F]$ is a topological $*$-algebra.*

Proof. From the definition of M_a it is clear that $r_M(a^+xa) = r_{M_a}(x)$ for all $a, x \in A$ and $M \in \boldsymbol{F}$. Since F is assumed to be A-invariant, this shows that the mapping $x \to a^+xa$ is continuous in $A[\tau_F]$ for each $a \in A$. By Lemma 1.5.6, $A[\tau_F]$ is a topological algebra. The continuity of the involution in $A[\tau_F]$ has been already mentioned above. □

Proposition 1.5.8. *Suppose that the wedge K in A_h is m-admissible. Then $\mathsf{A}[\tau_n]$ and $\mathsf{A}[\tau_0]$ are topological $*$-algebras.*

Proof. We first prove the assertion for τ_n. By Corollary 1.5.5, $\tau_n = \tau_{\boldsymbol{F}_{\max}}$ on A. Therefore, by Lemma 1.5.7, it suffices to show that $\boldsymbol{F}_{\max}$ is A-invariant. Take $M \in \boldsymbol{F}_{\max}$ and $a \in \mathsf{A}$. Since K is m-admissible by assumption, $f_a(\cdot) = f(a^+ \cdot a)$ is also in K^* for each $f \in M$. Thus $M_a \subseteq K^*$. From $f_a(x) = f(a^+xa)$ for $x \in \mathsf{A}$ it is clear that M_a is weakly bounded. Hence $M_a \in \boldsymbol{F}_{\max}$ and $\boldsymbol{F}_{\max}$ is A-invariant.

Now we show that $\mathsf{A}[\tau_0]$ is a topological $*$-algebra. From Proposition 1.5.2, (iii), the involution of A is continuous in $\mathsf{A}[\tau_0]$. Suppose that $U \in \boldsymbol{U}_0$ and $a \in \mathsf{A}$. Put $V := \{x \in \mathsf{A} : a^+xa \in U\}$. Obviously, V is absolutely convex in A. We prove that $V \cap L_h$ absorbs all order intervals of $(\mathsf{A}_h, \succeq)$. Let $x, y \in \mathsf{A}_h$ and let $z \in [x, y]$. Then $z - x \in K$. Since K is m-admissible, $a^+(z - x)\, a \in K$ and hence $a^+za \succeq a^+xa$. Similarly, $a^+za \preceq a^+ya$. This shows that $a^+[x, y]\, a \subseteq [a^+xa, a^+ya]$. Since $U \cap L_h$ absorbs the order intervals, $[a^+xa, a^+ya] \subseteq \lambda(U \cap L_h)$ for some $\lambda > 0$. Hence $a^+[x, y]\, a \subseteq \lambda(U \cap L_h)$, so that $[x, y] \subseteq \lambda(V \cap L_h)$ according to the definition of V. Therefore, $V \in \boldsymbol{U}_0$. Further, the preceding shows that the mapping $x \to a^+xa$ is continuous in $\mathsf{A}[\tau_0]$. By Lemma 1.5.6, $\mathsf{A}[\tau_0]$ is a topological algebra. □

Now we turn to the topologies $\tau^{\boldsymbol{F}}$, τ^n and τ^0. Suppose $M \in \boldsymbol{F}_{\max}$. From $\mathcal{P}(\mathsf{A}) \subseteq K$, we have $f(x^+x) \geq 0$ for $x \in \mathsf{A}$ and $f \in M$. We define

$$r^M(x) := r_M(x^+x)^{1/2} \equiv \sup_{f \in M} f(x^+x)^{1/2}, \quad x \in \mathsf{A}. \tag{1}$$

Since $\mathcal{P}(\mathsf{A}) \subseteq K$, the functionals f in M satisfy the Cauchy-Schwarz inequality. This implies that

$$r_M(x^+y) \leq r^M(x)\, r^M(y) \quad \text{for} \quad x, y \in \mathsf{A}. \tag{2}$$

We show that r^M is a seminorm on A. It clearly suffices to verify the triangle inequality. Let $x, y \in \mathsf{A}$. Using (1) and (2), we have

$$\begin{aligned} r^M(x + y)^2 &= r_M\big((x + y)^+ (x + y)\big) \leq r_M(x^+x) + r_M(x^+y) + r_M(y^+x) + r_M(y^+y) \\ &\leq r^M(x)^2 + 2r^M(x)\, r^M(y) + r^M(y)^2 = \big(r^M(x) + r^M(y)\big)^2; \end{aligned}$$

so r^M is a seminorm on A.

If $\boldsymbol{F}$ is a non-empty subset of $\boldsymbol{F}_{\max}$, let $\tau^{\boldsymbol{F}}$ denote the locally convex topology on A which is defined by the family of seminorms $\{r^M : M \in \boldsymbol{F}\}$. We write τ^n for $\tau^{\boldsymbol{F}_{\max}}$.

Proposition 1.5.9. *Suppose that $\boldsymbol{F}$ is a non-empty subset of $\boldsymbol{F}_{\max}$.*

(i) $\tau_{\boldsymbol{F}} \subseteq \tau^{\boldsymbol{F}}$, $\tau_n \subseteq \tau^n$ *and* $\tau^{\boldsymbol{F}} \subseteq \tau^n$.

(ii) *The multiplication of A is jointly continuous on $\mathsf{A}[\tau_{\boldsymbol{F}}]$ if and only if $\tau_{\boldsymbol{F}} = \tau^{\boldsymbol{F}}$.*

Proof. (i) Suppose $M \in \boldsymbol{F}$. By (2), $r_M(x) = r_M(1^+x) \leq r^M(1)\, r^M(x)$ for $x \in \mathsf{A}$. This shows that $\tau_{\boldsymbol{F}} \subseteq \tau^{\boldsymbol{F}}$. In case $\boldsymbol{F} = \boldsymbol{F}_{\max}$ we get $\tau_n \subseteq \tau^n$. $\tau^{\boldsymbol{F}} \subseteq \tau^n$ is trivial.

(ii) Without loss of generality we assume that the family of seminorms $\{r_M : M \in \boldsymbol{F}\}$ is directed. First suppose that the multiplication is jointly continuous in $\mathsf{A}[\tau_{\boldsymbol{F}}]$. Suppose $M \in \boldsymbol{F}$. Since the family $\{r_N\}$ is directed, there are an $N \in \boldsymbol{F}$ and a $\lambda > 0$ such that $r_M(xy) \leq \lambda r_N(x)\, r_N(y)$ for all $x, y \in \mathsf{A}$. Letting $x = y^+$ and using (1), we obtain $r^M(y)^2 = r_M(y^+y) \leq \lambda r_N(y^+)\, r_N(y) = \lambda r_N(y)^2$. Therefore, $\tau^{\boldsymbol{F}} \subseteq \tau_{\boldsymbol{F}}$. Since $\tau_{\boldsymbol{F}} \subseteq \tau^{\boldsymbol{F}}$ by (i), we have $\tau_{\boldsymbol{F}} = \tau^{\boldsymbol{F}}$.

Conversely, suppose that $\tau_F = \tau^F$. Let $M \in F$. From $\tau_F = \tau^F$ and from the continuity of the involution in $\mathsf{A}[\tau_F]$ it follows that r^M and $r_+^M(x) := r^M(x^+)$, $x \in \mathsf{A}$, are continuous seminorms on $\mathsf{A}[\tau_F]$. By (2), we have $r_M(xy) \leqq r_+^M(x)\, r^M(y)$ for all $x, y \in \mathsf{A}$. This shows that the multiplication is jointly continuous in $\mathsf{A}[\tau_F]$. □

In case $F = F_{\max}$ Proposition 1.5.9, (ii), and Corollary 1.5.5 give

Corollary 1.5.10. *The multiplication of* A *is jointly continuous in* $\mathsf{A}[\tau_n]$ *if and only if* $\tau_n = \tau^n$.

Now we define the topology τ^0. Let U^0 denote the collection of all absolutely convex subsets of A which absorb each set $R_a := \{x \in \mathsf{A} : a^+a - x^+x \in K\}$, $a \in \mathsf{A}$. The sets in U^0 are absorbing, since $a \in R_a$ for $a \in \mathsf{A}$. Obviously, U^0 satisfies the conditions (i)—(iii) in 1.1; so U^0 is a 0-neighbourhood base for a locally convex topology on A which we denote by τ^0. By definition, τ^0 is the finest locally convex topology on A for which each set R_a, $a \in \mathsf{A}$, is bounded. If the topology τ^0 is Hausdorff, then the locally convex space $\mathsf{A}[\tau^0]$ is bornological. This follows exactly in the same way as assertion (vi) of Proposition 1.5.2 if we replace the order intervals by the set R_a, $a \in \mathsf{A}$.

Proposition 1.5.11. $\tau^n \subseteqq \tau^0$ *and* $\tau_0 \subseteqq \tau^0$ *on* A.

Proof. By definition, a 0-neighbourhood base for the topology τ^n is given by the absolutely convex sets $W^M := \{x \in \mathsf{A} : r^M(x) \leqq 1\}$, $M \in F_{\max}$. Fix $M \in F_{\max}$. Let $a \in \mathsf{A}$. Since $\mathcal{P}(\mathsf{A}) \subseteqq K$ by assumption, we have $r^M(x) \leqq r^M(a)$ for all $x \in R_a$. This implies that W^M absorbs R_a; so $W^M \in U^0$. This proves that $\tau^n \subseteqq \tau^0$.

In order to show that $\tau_0 \subseteqq \tau^0$, we first prove that

$$4R_a \subseteqq [-a^+a - 4\cdot 1, a^+a + 4\cdot 1] + \mathrm{i}[-a^+a - 4\cdot 1, a^+a + 4\cdot 1] \quad \text{for each } a \in \mathsf{A}. \qquad (3)$$

Let $x \in R_a$. We write x as $x = x_1 + \mathrm{i}x_2$ with $x_1, x_2 \in \mathsf{A}_h$. For arbitrary $y \in \mathsf{A}$ we have the identity

$$4y = (y+1)^+(y+1) - (y-1)^+(y-1) + \mathrm{i}(y + \mathrm{i}\cdot 1)^+(y + \mathrm{i}\cdot 1) - \mathrm{i}(y - \mathrm{i}\cdot 1)^+(y - \mathrm{i}\cdot 1). \qquad (4)$$

Setting $y = x - 1$ in (4) and comparing the real parts on both sides, we get $4(x_1 - 1) = x^+x - (x - 2\cdot 1)^+(x - 2\cdot 1)$. Since $\mathcal{P}(\mathsf{A}) \subseteqq K$ and $x \in R_a$, this yields $4x_1 \leq x^+x + 4\cdot 1 \leq a^+a + 4\cdot 1$. Similarly we obtain $4x_1 \geq -a^+a - 4\cdot 1$ if we put $y = x + 1$ into (4). Thus $4x_1 \in [-a^+a - 4\cdot 1, a^+a + 4\cdot 1]$. From $x \in R_a$, $-\mathrm{i}x \in R_a$. Replacing x by $-\mathrm{i}x$ in the preceding, it follows that $4x_2 \in [-a^+a - 4\cdot 1, a^+a + 4\cdot 1]$. This gives (3).

Now let $U \in U_0$. Since U absorbs all order intervals, it follows from (3) that U absorbs the sets R_a, $a \in \mathsf{A}$. Thus $U \in U^0$. This shows that $\tau_0 \subseteqq \tau^0$. □

1.6. Operators on Hilbert Space

The theory of Hilbert space operators is developed in many textbooks such as Birman/Solomjak [1], Kato [1], Reed/Simon [1], [2], Riesz/Sz.-Nagy [1] and Weidmann [1]. For von Neumann algebras we refer to Dixmier [1], Kadison/Ringrose [1], [2], Stratila/Zsido [1] and Takesaki [1].

In this book all Hilbert spaces are complex. In general, they are denoted by $\mathcal{H}$, $\mathcal{H}_1$, $\mathcal{H}_2$ or $\mathcal{K}$. If not stated otherwise, scalar product and norm of the underlying Hilbert space are denoted by $\langle \cdot, \cdot \rangle$ and $\|\cdot\|$, respectively. We assume the scalar product to be linear in the first variable and conjugate-linear in the second.

Throughout the following we assume that $\mathcal{H}$ is a Hilbert space. The vector space of all bounded linear operators of $\mathcal{H}$ into another Hilbert space $\mathcal{K}$ is denoted by $\mathbf{B}(\mathcal{H}, \mathcal{K})$, and $\mathbf{B}(\mathcal{H}, \mathcal{H})$ is abbreviated by $\mathbf{B}(\mathcal{H})$. For linear subspaces $\mathcal{D}_1$ and $\mathcal{D}_2$ of $\mathcal{H}$, $\mathbf{F}(\mathcal{D}_2, \mathcal{D}_1)$ is the set of all finite rank operators x in $\mathbf{B}(\mathcal{H})$ satisfying $x\mathcal{H} \subseteq \mathcal{D}_1$ and $x^*\mathcal{H} \subseteq \mathcal{D}_2$. We write $\mathbf{F}(\mathcal{D}_1)$ for $\mathbf{F}(\mathcal{D}_1, \mathcal{D}_1)$. In particular, $\mathbf{F}(\mathcal{H})$ is the set of finite rank operators in $\mathbf{B}(\mathcal{H})$. If ψ and φ are vectors in $\mathcal{H}$, then $\psi \otimes \varphi$ is the operator $\langle \cdot, \psi \rangle \varphi$ on $\mathcal{H}$, and $\psi \perp \varphi$ means that $\langle \psi, \varphi \rangle = 0$. If $\mathcal{M}$ is a subset of $\mathcal{H}$, then $\mathcal{M}^\perp := \{\psi \in \mathcal{H} : \psi \perp \varphi \text{ for all } \varphi \in \mathcal{M}\}$ is the orthogonal complement of $\mathcal{M}$. A *projection* on $\mathcal{H}$ is a self-adjoint idempotent in $\mathbf{B}(\mathcal{H})$. If $\mathcal{K}$ is a closed linear subspace of $\mathcal{H}$ and $x \in \mathbf{B}(\mathcal{H})$, then $P_{\mathcal{K}}$ denotes the projection on $\mathcal{H}$ with range $\mathcal{K}$ and $\mathrm{pr}_{\mathcal{K}}\, x$ denotes the restriction $P_{\mathcal{K}} x \restriction \mathcal{K}$ of $P_{\mathcal{K}} x$ to $\mathcal{K}$. We frequently omit the subscript $\mathcal{K}$ and write $\mathrm{pr}\, x$ when no confusion is possible. For $x \in \mathbf{B}(\mathcal{H})$, $\mathrm{Re}\, x := \frac{1}{2}(x^* + x)$ and $\mathrm{Im}\, x := \frac{1}{2}\mathrm{i}(x^* - x)$. The identity map of $\mathcal{H}$ is denoted by I or by $I_{\mathcal{H}}$. If $\lambda \in \mathbf{C}$, we often write simply λ instead of $\lambda \cdot I$. Further, we set $\mathbf{C} \cdot I := \{\lambda \cdot I : \lambda \in \mathbf{C}\}$.

By an *operator* in $\mathcal{H}$ we mean a linear mapping a of a linear subspace of $\mathcal{H}$, called the *domain* of a and denoted by $\mathcal{D}(a)$, into $\mathcal{H}$. Suppose a is an operator on $\mathcal{H}$. If b is another operator on $\mathcal{H}$, then $a \subseteq b$ means that b is an extension of a, i.e., $\mathcal{D}(a) \subseteq \mathcal{D}(b)$ and $a\varphi = b\varphi$ for $\varphi \in \mathcal{D}(a)$. We write $\|\cdot\|_a$ for the seminorm $\|a \cdot\|$ on $\mathcal{D}(a)$, $\ker a$ for the null space of a, $\sigma(a)$ for the spectrum of a and $a \restriction \mathcal{D}$ for the restriction of a to $\mathcal{D}$. We set $\mathcal{D}^\infty(a) := \bigcap_{n \in \mathbf{N}} \mathcal{D}(a^n)$. The expression a^0 is always interpreted to be the identity map. The *graph* of a is the linear subspace $\mathrm{gr}\, a := \{(\varphi, a\varphi) : \varphi \in \mathcal{D}(a)\}$ of the Hilbert direct sum $\mathcal{H} \oplus \mathcal{H}$ equipped with scalar product and norm of $\mathcal{H} \oplus \mathcal{H}$. The operator a is called *closed* when $\mathrm{gr}\, a$ is closed in $\mathcal{H} \oplus \mathcal{H}$. Note that we do not assume closed operators to be densely defined. If a admits a closable extension, then a is said to be *closable*. In this case there exists a minimal closed extension of a which is called the *closure* of a and denoted by $\bar{a}$. The *adjoint* a^* of a densely defined operator a is defined on the domain $\mathcal{D}(a^*)$ of all vectors $\varphi \in \mathcal{H}$ for which there exists a vector $\psi \in \mathcal{H}$ such that $\langle a\eta, \varphi \rangle = \langle \eta, \psi \rangle$ for all $\eta \in \mathcal{D}(a)$; for such vectors ψ, $a^*\varphi := \psi$. A *core* for a closable operator a is a linear subspace $\mathcal{D}$ of $\mathcal{D}(a)$ such that $a \subseteq \overline{a \restriction \mathcal{D}}$. Equivalent conditions for the latter are that $\mathcal{D}$ is dense in $\mathcal{D}(a)$ relative to the norm $\|\cdot\|_a + \|\cdot\|$ or that the graph of $a \restriction \mathcal{D}$ is dense in the graph of a.

A densely defined operator a is called *symmetric* if $a \subseteq a^*$ (or equivalently, if $\langle a\varphi, \psi \rangle = \langle \varphi, a\psi \rangle$ for all $\varphi, \psi \in \mathcal{D}(a)$) and *skew-symmetric* if $a \subseteq -a^*$. A symmetric operator a is said to be *positive* if $\langle a\varphi, \varphi \rangle \geqq 0$ for all $\varphi \in \mathcal{D}(a)$. We then write $a \geqq 0$. A *self-adjoint* operator is a densely defined operator a such that $a = a^*$. The positive square root of a positive self-adjoint operator a is denoted by $a^{1/2}$. An operator is called *essentially self-adjoint* if it is closable and its closure is self-adjoint. By a *formally normal* operator we mean a densely defined operator a such that $\mathcal{D}(a) \subseteq \mathcal{D}(a^*)$ and $\|a\varphi\| = \|a^*\varphi\|$ for all $\varphi \in \mathcal{D}(a)$. A *normal* operator is a formally normal operator a such that $\mathcal{D}(a) = \mathcal{D}(a^*)$. A densely defined closed operator a is normal if and only if $aa^* = a^*a$ (see e.g. WEIDMANN [1], 5.6).

Let a be a densely defined closed operator on $\mathcal{H}$. We set $|a| := (a^*a)^{1/2}$. There exists a unique partial isometry u on $\mathcal{H}$ such that $a = u\,|a|$ and $\ker u = \ker |a|$. The formula

$a = u\,|a|$ is called the *polar decomposition* of a. The following properties of this decomposition (cf. KATO [1], VI, § 2.7.) are used later. We have $|a| = u^*a = a^*u$, $|a^*| = u\,|a|\,u^*$, $\mathcal{D}(|a|) = \mathcal{D}(a)$ and $\|a\varphi\| = \|\,|a|\,\varphi\|$ for $\varphi \in \mathcal{D}(a) = \mathcal{D}(|a|)$.

Let a be a symmetric operator on $\mathcal{H}$. The closed linear subspaces $\mathcal{H}_+ := \ker(a^* - \mathrm{i}) \equiv ((a + \mathrm{i})\,\mathcal{D}(a))^\perp$ and $\mathcal{H}_- := \ker(a^* - \mathrm{i}) \equiv ((a - \mathrm{i})\,\mathcal{D}(a))^\perp$ are called the *deficiency spaces* of a. The dimensions d_+ and d_- of these spaces or the couple (d_+, d_-) are said to be the *deficiency indices* of a. If a is closed, then the *Cayley transform* of a is the isometric linear mapping u of $(a + \mathrm{i})\,\mathcal{D}(a) \equiv \mathcal{H} \ominus \mathcal{H}_+$ onto $(a - \mathrm{i})\,\mathcal{D}(a) \equiv \mathcal{H} \ominus \mathcal{H}_-$ which is defined by $u(a + \mathrm{i})\,\varphi := (a - \mathrm{i})\,\varphi$, $\varphi \in \mathcal{D}(a)$.

We state some well-known facts (cf. WEIDMANN [1], 5.3) which are frequently used in the sequel. Suppose a is a symmetric operator on $\mathcal{H}$ and α_1 and α_2 are complex numbers with $\operatorname{Im}\alpha_1 > 0$ and $\operatorname{Im}\alpha_2 < 0$. Then a is essentially self-adjoint if and only if $(a - \alpha_1)\,\mathcal{D}(a)$ and $(a - \alpha_2)\,\mathcal{D}(a)$ are both dense in $\mathcal{H}$. Other equivalent conditions are that a has deficiency indices $(0, 0)$ or that $\bar{a} = a^*$. If a is closed, then a is self-adjoint if and only if $(a - \alpha_1)\,\mathcal{D}(a) = (a - \alpha_2)\,\mathcal{D}(a) = \mathcal{H}$. A linear subspace $\mathcal{D}$ of $\mathcal{D}(a)$ is a core for a if and only if $(a - \alpha)\,\mathcal{D}$ is dense in $(a - \alpha)\,\mathcal{D}(a)$ in the norm of $\mathcal{H}$ for some (and then for all) $\alpha \in \mathbb{C} \setminus \mathbb{R}$. If a is self-adjoint, then $\mathcal{D}$ is a core for a if and only if $(a - \alpha)\,\mathcal{D}$ is dense in $\mathcal{H}$ for some (all) $\alpha \in \mathbb{C} \setminus \sigma(a)$.

Proposition 1.6.1. *Let a be a closed symmetric operator on a Hilbert space $\mathcal{H}$. Suppose that at least one of the deficiency indices is finite. Then $\mathcal{D}^\infty(a)$ is a core for each power a^k, $k \in \mathbb{N}_0$, of a. In particular, $\mathcal{D}^\infty(a)$ is dense in $\mathcal{H}$.*

Proof. There is no loss of generality to assume that $\mathcal{H}_+ \equiv \ker(a^* - \mathrm{i})$ is finite dimensional. (Otherwise we replace a by $-a$.) Let u be the Cayley transform of a. We extend u to the whole $\mathcal{H}$ by defining it to be the zero operator on $\mathcal{H}_+$. By a slight abuse of notation, we denote this operator again by u. For $n \in \mathbb{N}_0$, let q_{n+1} be the projection of $\mathcal{H}$ onto the finite dimensional linear subspace $\mathcal{G}_{n+1} := \mathcal{H}_+ + u^*H_+ + \cdots + (u^*)^n\,\mathcal{H}_+$, and let $\|\cdot\|_n$ denote the norm $\|(a + \mathrm{i})^n \cdot\|$ on $\mathcal{D}(a^n)$.

Our first objective is to show that $\mathcal{D}(a^n) = (I - u)^n\,(I - q_n)\,\mathcal{H}$ for $n \in \mathbb{N}$. We prove this by induction on n. For $n = 1$ the assertion follows at once from the definition of the Cayley transform. Assume that this is true for some $n \in \mathbb{N}$. Let $\varphi \in \mathcal{D}(a^{n+1})$. Then $\varphi \in \mathcal{D}(a^n)$, so that $\varphi = (I - u)^n\,\zeta$ for some $\zeta \in (I - q_n)\,\mathcal{H}$. Further, $(2\mathrm{i})^n\,\zeta = (a + \mathrm{i})^n\,\varphi \in \mathcal{D}(a)$ and so $\zeta = (I - u)\,\eta$ with $\eta \in (I - q_1)\,\mathcal{H}$. Since $\eta \perp q_1\mathcal{H} = \mathcal{H}_+$ and $\zeta = (I - u)\,\eta \perp \mathcal{H}_+, u^*\mathcal{H}_+, \ldots, (u^*)^{n-1}\,\mathcal{H}_+$, it follows that $\eta \perp \mathcal{H}_+, u^*\mathcal{H}_+, \ldots, (u^*)^n\,\mathcal{H}_+$, i.e., $\eta \in (I - q_{n+1})\,\mathcal{H}$ and $\varphi = (I - u)^n\,\zeta = (I - u)^{n+1}\,\eta \in (I - u)^{n+1}\,(I - q_{n+1})\,\mathcal{H}$. Conversely, it is easy to check that the latter set is contained in $\mathcal{D}(a^{n+1})$; so $\mathcal{D}(a^{n+1}) = (I - u)^{n+1}\,(I - q_{n+1})\,\mathcal{H}$ and the induction proof is complete.

We want to apply Lemma 1.1.2 in case $E_n := (\mathcal{D}(a^n), \|\cdot\|_n)$, $n \in \mathbb{N}_0$. From $\mathcal{D}(a^n) = (I - u)^n\,(I - q_n)\,\mathcal{H}$ and $\|\varphi\|_n = 2^n\,\|(I - q_n)\,\psi\|$ for $\varphi = (I - u)^n\,(I - q_n)\,\psi \in \mathcal{D}(a^n)$ we conclude that the normed space E_n is complete for $n \in \mathbb{N}$. Further, E_0 is the Hilbert space $\mathcal{H}$ itself and so complete. We have $\|\cdot\|_n \leqq \|\cdot\|_{n+1}$ on E_{n+1}; hence the embedding of E_{n+1} into E_n is continuous. We check that E_{n+1} is dense in E_n for each $n \in \mathbb{N}_0$. First note that E_n is a Hilbert space relative to the scalar product $\langle\cdot,\cdot\rangle_n := \langle(a + \mathrm{i})^n\cdot, (a + \mathrm{i})^n\cdot\rangle$ on E_n. Thus it is sufficient to show that the orthogonal complement of E_{n+1} in $(E_n, \langle\cdot,\cdot\rangle_n)$ consists only of the zero vector. We suppose that $\varphi \in \mathcal{D}(a^n)$ satisfies $\langle\varphi, \eta\rangle_n = 0$ for all $\eta \in \mathcal{D}(a^{n+1})$. Writing φ as $\varphi = (I - u)^n\,\psi$ with $\psi \in (I - q_n)\,\mathcal{H}$, this gives $0 = \langle\varphi, (I - u)^{n+1}\,(I - q_{n+1})\,\zeta\rangle_n = 4^n\langle\psi, (I - u)\,(I - q_{n+1})\,\zeta\rangle = 4^n\langle(I - u^*)\,\psi, (I - q_{n+1})\zeta\rangle$

for all $\zeta \in \mathcal{H}$, so that $(I - u^*)\,\psi \in q_{n+1}\mathcal{H} \equiv \mathcal{S}_{n+1}$. From the definition of the spaces $\mathcal{S}_k$, $k \in \mathbb{N}$, it is clear that the vector $(I - u^*)\,\psi$ of $q_{n+1}\mathcal{H}$ is of the form $(I - u^*)\,\xi + \xi_+$ with $\xi \in q_n\mathcal{H}$ and $\xi_+ \in \mathcal{H}_+$. Then $\langle \psi - \xi, (I - u)(I - q_1)\,\zeta\rangle = \langle (I - u^*)(\psi - \xi), (I - q_1)\,\zeta\rangle = \langle \xi_+, (I - q_1)\,\zeta\rangle = 0$ for all $\zeta \in \mathcal{H}$, so that $(\psi - \xi) \perp (I - u)(I - q_1)\,\mathcal{H} = \mathcal{D}(a)$. By the definition of a symmetric operator (see above), $\mathcal{D}(a)$ is dense in $\mathcal{H}$. Therefore, $\psi - \xi = 0$. Since $\psi \perp \xi$ by $\psi \in (I - q_n)\,\mathcal{H}$, $\psi = 0$ and so $\varphi = 0$. Thus we have shown that the sequence $(E_n : n \in \mathbb{N}_0)$ of normed spaces satisfies the assumptions of Lemma 1.1.2. By Lemma 1.1.2, $\mathcal{D}^\infty(a) = \bigcap_{n \in \mathbb{N}_0} E_n$ is dense in each normed space $E_k = (\mathcal{D}(a^k), \|\cdot\|_k)$. Hence $\mathcal{D}^\infty(a)$ is a core for each operator a^k, $k \in \mathbb{N}_0$. In case $k = 0$ this means that $\mathcal{D}^\infty(a)$ is dense in $\mathcal{H}$. □

Remark 1. Actually the preceding proof yields the following stronger statement. If a is a closed symmetric operator on $\mathcal{H}$ such that the space $\mathcal{S}_n := \mathcal{H}_+ \dotplus u^*\mathcal{H}_+ + \cdots + (u^*)^{n-1}\mathcal{H}_+$ is closed in $\mathcal{H}$ for all $n \in \mathbb{N}$, then $\mathcal{D}^\infty(a)$ is a core for any a^k, $k \in \mathbb{N}_0$.

Let a be an arbitrary operator on $\mathcal{H}$ and let x be in $\mathbb{B}(\mathcal{H})$. We say that x *commutes with* a if $xa \subseteq ax$, i.e., if $x\varphi \in \mathcal{D}(a)$ and $xa\varphi = ax\varphi$ for all $\varphi \in \mathcal{D}(a)$. Suppose that a is self-adjoint. Then $xa \subseteq ax$ if and only if x commutes with all spectral projections of a. Further, if $xa \subseteq ax$, then x also commutes with all measurable functions (with respect to the spectral measure) of a.

We say that two normal operators a and b on $\mathcal{H}$ *strongly commute* provided that the spectral projections of a and b mutually commute. (Recall that each normal operator has a unique spectral resolution, cf. Rudin [1], 13.33.)

Lemma 1.6.2. *Let a and b be normal operators on $\mathcal{H}$. Suppose $\sigma(a) \neq \mathbb{C}$, and let $\alpha \in \mathbb{C} \setminus \sigma(a)$. Then the operators a and b strongly commute if and only if $(a - \alpha)^{-1}\, b \subseteq b(a - \alpha)^{-1}$.*

Proof. Let $e(\cdot)$ and $f(\cdot)$ denote the spectral projections of the normal operators $(a - \alpha)^{-1}$ and b, respectively. From the properties of the spectral resolution (see Rudin [1], Theorem 13.33) it is well-known that $(a - \alpha)^{-1}\, b \subseteq b(a - \alpha)^{-1}$ if and only if $(a - \alpha)^{-1} f(\delta) = f(\delta)\,(a - \alpha)^{-1}$ for all $\delta \in \mathbb{C}$. By the same result applied to the normal operator $(a - \alpha)^{-1}$, the latter is equivalent to $e(\gamma)\, f(\delta) = f(\delta)\, e(\gamma)$ for all $\gamma, \delta \in \mathbb{C}$. Since $e\big((\lambda - \alpha)^{-1}\big)$ is obviously the spectral projection of a at λ for $\lambda \in \mathbb{C}$ and $\lambda \neq \alpha$, the last statement means that a and b strongly commute. □

Let $\mathcal{N}$ be a von Neumann algebra on $\mathcal{H}$ and let a be a closed operator on $\mathcal{H}$. We say that a is *affiliated with* $\mathcal{N}$ when $xa \subseteq ax$ for all x in the commutant $\mathcal{N}'$ of $\mathcal{N}$. We denote by $\mathbb{A}(\mathcal{N})$ the set of all densely defined closed operators on $\mathcal{H}$ which are affiliated with $\mathcal{N}$. If $a = u\,|a|$ is the polar decomposition of a, then $a \in \mathbb{A}(\mathcal{N})$ if and only if $u \in \mathcal{N}$ and $|a| \in \mathbb{A}(\mathcal{N})$ (Dixmier [1], p. 16).

If a is self-adjoint, then $a \in \mathbb{A}(\mathcal{N})$ if and only if all spectral projections of a are in $\mathcal{N}$ or equivalently if $(a - \alpha)^{-1} \in \mathcal{N}$ for some (and then for all) $\alpha \in \mathbb{C} \setminus \sigma(a)$.

Lemma 1.6.3. *Suppose that $\mathcal{N}$ is an abelian von Neumann algebra.*

(i) *Each operator $a \in \mathbb{A}(\mathcal{N})$ is normal and each symmetric operator $a \in \mathbb{A}(\mathcal{N})$ is self-adjoint.*

(ii) *For arbitrary operators $a, a_1, \ldots, a_n \in \mathbb{A}(\mathcal{N})$ and $n \in \mathbb{N}$, $\mathcal{D}(a) \cap \mathcal{D}(a_1) \cap \cdots \cap \mathcal{D}(a_n)$ is a core for a.*

(iii) $\mathbb{A}(\mathcal{N})$ *forms a commutative* $*$*-algebra with unit I under the operations* $a \mathbin{\hat{+}} b := \overline{a + b}$ *for addition,* $a \mathbin{\hat{\cdot}} b := \overline{ab}$ *for multiplication and the usual scalar multiplication.*

Proof. KADISON/RINGROSE [1], Theorem 5.6.15. □

1.7. Lie Groups, Lie Algebras and Enveloping Algebras

This section is mainly a preliminary section for Chapter 10. Proofs of the facts stated here and further details can be found (for instance) in VARADARAJAN [1].

Suppose that G is a real (finite dimensional) Lie group. Let e be the identity element of G, G_0 the connected component of e in G and μ a left Haar measure on G.

We denote by $\mathfrak{g}$ the Lie algebra of G. That is, $\mathfrak{g}$ is the tangent space to G at e endowed with the Lie bracket $[\cdot,\cdot]$ defined by formula (2) below. Let $x \to \exp x$ denote the exponential map of $\mathfrak{g}$ into G. For $x \in \mathfrak{g}$, let $\tilde{x}$ be the right-invariant vector field on G defined by

$$(\tilde{x}f)(g) = \frac{\mathrm{d}}{\mathrm{d}t} f(\exp(-tx)\,g)\Big|_{t=0}, \quad f \in C^\infty(G). \tag{1}$$

The Lie bracket in $\mathfrak{g}$ is defined such that

$$\widetilde{[x, y]} = \tilde{x}\tilde{y} - \tilde{y}\tilde{x}, \quad x, y \in \mathfrak{g}, \tag{2}$$

where the multiplication on the right hand side is the composition of operators. If $x, y \in \mathfrak{g}$, set $\operatorname{ad} x(y) := [x, y]$. For $g \in G$, $\operatorname{Ad} g(\cdot)$ is defined as the differential of the inner automorphism $h \to ghg^{-1}$ of G. We have

$$\exp \operatorname{Ad} g(x) = g \exp x\, g^{-1}, \quad x \in \mathfrak{g}, g \in G, \tag{3}$$

and

$$\operatorname{Ad} \exp x(y) = \sum_{n=0}^{\infty} \frac{(\operatorname{ad} x)^n (y)}{n!}, \quad x, y \in \mathfrak{g}, \tag{4}$$

where the series in (4) converges in any locally convex topology on the finite dimensional real vector space $\mathfrak{g}$.

Let $\mathcal{E}(\mathfrak{g})$ denote the universal enveloping algebra of the complexification $\mathfrak{g}_{\mathbb{C}}$ of the Lie algebra $\mathfrak{g}$. We simply refer to $\mathcal{E}(\mathfrak{g})$ as the *enveloping algebra* of $\mathfrak{g}$. The algebra $\mathcal{E}(\mathfrak{g})$ is defined as the quotient algebra of the tensor algebra over $\mathfrak{g}_{\mathbb{C}}$ by the two-sided ideal generated by the elements $x \otimes y - y \otimes x - [x, y]$, where $x, y \in \mathfrak{g}$. As usual, we consider $\mathfrak{g}$ as a linear subspace of $\mathcal{E}(\mathfrak{g})$ by identifying $\mathfrak{g}$ with its image under the quotient map.

Let $\{x_1, \ldots, x_d\}$ be a basis for $\mathfrak{g}$. For a multi-index $n = (n_1, \ldots, n_d) \in \mathbb{N}_0^d$, we set $|n| := n_1 + \cdots + n_d$ and $x^n := x_1^{n_1} \ldots x_d^{n_d}$, where x_k^0 is the unit element 1 of the algebra $\mathcal{E}(\mathfrak{g})$. The *Poincaré-Birkhoff-Witt theorem* asserts that the elements x^n, $n \in \mathbb{N}_0^d$, form a basis for the vector space $\mathcal{E}(\mathfrak{g})$. For $m \in \mathbb{N}_0$, let $\mathcal{E}_m(\mathfrak{g})$ denote the linear span of the elements x^n, where $n \in \mathbb{N}_0^d$, $|n| \leq m$. The element $\Delta := x_1^2 + \cdots + x_d^2$ of $\mathcal{E}(\mathfrak{g})$ is called the *Nelson Laplacian* relative to the basis $\{x_1, \ldots, x_d\}$.

Let A be an (associative complex) algebra. By a homomorphism of the Lie algebra $\mathfrak{g}$ into A we mean a map θ of $\mathfrak{g}$ into A such that $\theta(\alpha x + \beta y) = \alpha\theta(x) + \beta\theta(y)$ and $\theta([x, y]) = \theta(x)\,\theta(y) - \theta(y)\,\theta(x)$ for $x, y \in \mathfrak{g}$ and $\alpha, \beta \in \mathbb{R}$. The enveloping algebra $\mathcal{E}(\mathfrak{g})$ has the following important *universal property*: If A is any algebra with unit and θ is any homomorphism of $\mathfrak{g}$ into A, then there exists a unique identity preserving homomorphism of the algebra $\mathcal{E}(\mathfrak{g})$ into the algebra A which extends θ. For notational simplicity this homomorphism will also be denoted by θ. A similar remark applies to antihomorphisms of $\mathfrak{g}$ into A.

The following facts are based on this universal property. Let $\mathfrak{D}(G)$ denote the algebra of all right-invariant differential operators on G defined on $C^\infty(G)$. By (2), the map $x \to \tilde{x}$ is a homomorphism of $\mathfrak{g}$ into $\mathfrak{D}(G)$. It extends to an isomorphism $x \to \tilde{x}$ of the algebras $\mathcal{E}(\mathfrak{g})$ and $\mathfrak{D}(G)$. For $g \in G$, $\operatorname{Ad} g(\cdot)$ is an automorphism of the Lie algebra $\mathfrak{g}$, so it has a unique extension to an automorphism of $\mathcal{E}(\mathfrak{g})$. The map $x \to x^+ := -x$ is an antiisomorphism of $\mathfrak{g}$. Its unique extension to an antiisomorphism of $\mathcal{E}(\mathfrak{g})$ is an involution for the algebra $\mathcal{E}(\mathfrak{g})$. We equip $\mathcal{E}(\mathfrak{g})$ with this involution, so $\mathcal{E}(\mathfrak{g})$ becomes a $*$-algebra with unit.

A *unitary representation* U of G on a Hilbert space $\mathcal{H}(U)$ is a homomorphism $g \to U(g)$ of G into the group of unitaries of $\mathcal{H}(U)$ such that $U(e) = I$ and such that the map $g \to U(g)\,\varphi$ of G into $\mathcal{H}(U)$ is continuous for each vector $\varphi \in \mathcal{H}(U)$.

Notes

1.1. The notion of a QF-space was introduced by Kürsten [2].

1.4. The concept of an m-admissible wedge is due to Powers [2].

1.5. The assertion concerning the topology τ_0 in Proposition 1.5.8 was obtained independently by Kunze [1] and for operator algebras by Jurzak [2].

1.6. Proposition 1.6.1 is due to Schmüdgen [14].

The first part of this monograph is devoted to a study of algebras of unbounded operators in Hilbert space (O*-algebras) with the emphasis on related topologies on the domain as well as the algebra itself.

In Chapter 2 basic notions on O-families and O*-algebras are introduced and the graph topology on the domain is investigated. In Chapter 3 and 4 we study topologies on O*-algebras or more generally on spaces of sesquilinear forms associated with them. Chapter 5 deals with linear functionals which are defined by trace class operators in the ordinal. In Chapter 6 we consider two special types of *-algebras, the generalized [illegible] algebras and the maximal O*-algebra $\mathcal{L}^+(\mathcal{D})$ on a domain $\mathcal{D}$. Chapter 7 is concerned with commutants of O*-algebras, a subject which is also important for the study of *-representations in Part II.

Part I.
O*-Algebras and Topologies

The first part of this monograph is devoted to a study of $*$-algebras of unbounded operators in Hilbert space (O*-algebras) with the emphasis on related topologies on the domain as well on the algebra itself.

In Chapter 2 basic notions on O-families and O*-algebras are introduced and the graph topology on the domain is investigated. In Chapter 3 and 4 we study topologies on O*-algebras or more generally on spaces of sesquilinear forms associated with them. Chapter 5 deals with linear functionals which are defined by trace class operators in the predual. In Chapter 6 we consider two special types of $*$-algebras, the generalized Calkin algebra and the maximal O*-algebra $\mathcal{L}^+(\mathcal{D})$ on a domain $\mathcal{D}$. Chapter 7 is concerned with commutants of O*-algebras, a subject which is also important for the study of $*$-representations in Part II.

2. O-Families and Their Graph Topologies

In this chapter, some basic concepts of O-families are developed, and the graph topologies of O-families are studied in detail. An O-family is a set of closable linear operators defined on a common (dense) domain in a Hilbert space which contains the identity map. By means of the graph seminorms, each O-family $\mathcal{A}$ gives rise to a locally convex topology on its domain, the graph topology of $\mathcal{A}$. The corresponding locally convex space is denoted by $\mathcal{D}_{\mathcal{A}}$. Most of the material in this chapter is directly related to the graph topology.

Section 2.1 introduces basic notions like O-families, O-vector spaces, O-algebras, O*-families, O*-vector spaces, O*-algebras and $\mathcal{L}^+(\mathcal{D})$. Section 2.2 is concerned with directed O-families, closed O-families and commutatively dominated O-families. In Section 2.3 we take up a more detailed study of the locally convex space $\mathcal{D}_{\mathcal{A}}$. In case where $\mathcal{D}_{\mathcal{A}}$ is a quasi-Frechet space, the structure of the bounded sets in $\mathcal{D}_{\mathcal{A}}$ can be described in a rather explicit way. This is done in Section 2.4. In Section 2.6 we deal with the order relation defined by the positive cone of an O*-algebra. In Section 2.5 we discuss a number of examples and counter-examples of O-families and especially of O*-algebras.

2.1. O-Families, O*-Families and O*-Algebras

Throughout this section $\mathcal{D}$ is a dense linear subspace of a Hilbert space $\mathcal{H}$. We call such a space $\mathcal{D}$ a *domain* in $\mathcal{H}$ or simply a domain. The identity map of $\mathcal{D}$ is denoted by $I_{\mathcal{D}}$ or by I if no confusion can arise.

Definition 2.1.1. An *O-family* on $\mathcal{D}$ is a set of closable linear operators with domain $\mathcal{D}$ which contains the identity map $I_{\mathcal{D}}$. We call $\mathcal{D}$ the *domain* of the O-family.

If $\mathcal{A}$ is an O-family, we write $\mathcal{D}(\mathcal{A})$ for the domain of $\mathcal{A}$. Thus, by definition, $\mathcal{D}(a) = \mathcal{D}(\mathcal{A})$ when a is in $\mathcal{A}$. It is obvious that the set of all closable linear operators with domain $\mathcal{D}$ is the largest O-family on $\mathcal{D}$. This set is denoted by $\mathcal{C}(\mathcal{D}, \mathcal{H})$.

Definition 2.1.2. An *O-vector space* is an O-family $\mathcal{A}$ such that the operator $\alpha a + \beta b$ is in $\mathcal{A}$ for arbitrary operators a, b in $\mathcal{A}$ and complex numbers α, β.

Recall that ab denotes the composition of operators a and b. That is, if a and b are operators on $\mathcal{D}$ and $b\mathcal{D} \subseteq \mathcal{D}$, then ab is the operator with domain $\mathcal{D}$ defined by $ab\varphi = a(b\varphi)$, $\varphi \in \mathcal{D}$.

Definition 2.1.3. An *O-algebra* is an O-vector space $\mathcal{A}$ such that $b\mathcal{D}(\mathcal{A}) \subseteq \mathcal{D}(\mathcal{A})$ and $ab \in \mathcal{A}$ for all a, b in $\mathcal{A}$.

With the addition and scalar multiplication of operators, each O-vector space is a (complex) vector space. An O-algebra is an algebra with the product defined by the composition of operators. Note that the identity map I (which is contained in any O-family by the above definition) is the unit element of this algebra.

Definition 2.1.4. An *O*-family* on $\mathcal{D}$ is a set $\mathcal{A}$ of linear operators with domain $\mathcal{D}$ such that $I_{\mathcal{D}} \in \mathcal{A}$, $\mathcal{D} \subseteq \mathcal{D}(a^*)$ and $a^+ := a^* \upharpoonright \mathcal{D}$ belongs to $\mathcal{A}$ whenever a is in $\mathcal{A}$.

Let $\mathcal{A}$ be an O*-family on $\mathcal{D}$. Then $\mathcal{A}$ is an O-family on the domain $\mathcal{D}(\mathcal{A}) = \mathcal{D}$. (Indeed, since $\mathcal{D} \subseteq \mathcal{D}(a^*)$ and $\mathcal{D}$ is dense in $\mathcal{H}$, each operator $a \in \mathcal{A}$ is closable.) Further, if $a \in \mathcal{A}$, then

$$\langle a\varphi, \psi\rangle = \langle \varphi, a^+\psi\rangle \quad \text{for all} \quad \varphi, \psi \in \mathcal{D} \tag{1}$$

and hence $a = (a^+)^+$. From the latter we see in particular that $a \to a^+$ is a bijective mapping of $\mathcal{A}$.

Definition 2.1.5. An *O*-vector space* is an O-vector space which is also an O*-family.

If $\mathcal{A}$ is an O*-vector space, then it is clear from the preceding remarks that the map $a \to a^+$ is an involution on the vector space $\mathcal{A}$. With the involution $a \to a^+$, each O*-vector space is a $*$-vector space. The set $\mathcal{C}^+(\mathcal{D}, \mathcal{H}) := \{a \in \mathcal{C}(\mathcal{D}, \mathcal{H}) : \mathcal{D} \subseteq \mathcal{D}(a^*)\}$ is obviously the largest O*-family on the domain $\mathcal{D}$. It is even an O*-vector space. (That $a + b \in \mathcal{C}^+(\mathcal{D}, \mathcal{H})$ when $a, b \in \mathcal{C}^+(\mathcal{D}, \mathcal{H})$ follows from $\mathcal{D}\big((a+b)^*\big) \supseteq \mathcal{D}(a^*) \cap \mathcal{D}(b^*) \supseteq \mathcal{D}$.)

Definition 2.1.6. An *O*-algebra* is an O-algebra that is also an O*-family.

A slight reformulation of the preceding three definitions is given in

Lemma 2.1.7. *An O-family [resp. O-vector space, O-algebra] $\mathcal{A}$ is an O*-family [resp. O*-vector space, O*-algebra] if and only if for each $a \in \mathcal{A}$ there exists a $b \in \mathcal{A}$ (depending, of course, on a) such that*

$$\langle a\varphi, \psi\rangle = \langle \varphi, b\psi\rangle \quad \textit{for all} \quad \varphi, \psi \in \mathcal{D}(\mathcal{A}). \tag{2}$$

Moreover, if (2) *is fulfilled, then $a = b^+$ and $b = a^+$.*

Proof. The only if part is clear, since if $b := a^+$, (1) gives (2). We verify the if part. From (2) we conclude that $\mathcal{D}(\mathcal{A}) = \mathcal{D}(b) \subseteq \mathcal{D}(a^*)$ and $b \subseteq a^*$, that is, $a^+ = a^* \upharpoonright \mathcal{D}(\mathcal{A}) = b$. Since $b \in \mathcal{A}$ by assumption, $a^+ \in \mathcal{A}$, and the if part is proved. Since $a^+ = b$, $a = (a^+)^+ = b^+$. □

Let $\mathcal{L}^+(\mathcal{D})$ denote the set of all linear operators a in the Hilbert space $\mathcal{H}$ with domain $\mathcal{D}$ for which $a\mathcal{D} \subseteq \mathcal{D}$, $\mathcal{D} \subseteq \mathcal{D}(a^*)$ and $a^*\mathcal{D} \subseteq \mathcal{D}$.

Proposition 2.1.8. *$\mathcal{L}^+(\mathcal{D})$ is the largest O*-algebra on the domain $\mathcal{D}$.*

Proof. We first check that $\mathcal{L}^+(\mathcal{D})$ is an O*-family. Let $a \in \mathcal{L}^+(\mathcal{D})$. We have to show that $a^+ = a^* \upharpoonright \mathcal{D}$ belongs to $\mathcal{L}^+(\mathcal{D})$ as well. But this is true, because $a^+\mathcal{D} = a^*\mathcal{D} \subseteq \mathcal{D}$, $(a^+)^* = (a^* \upharpoonright \mathcal{D})^* \supseteq a^{**} \supseteq a$ and hence $(a^+)^*\mathcal{D} = a\mathcal{D} \subseteq \mathcal{D}$. We next prove that $\mathcal{L}^+(\mathcal{D})$ is an O-algebra. Suppose $a, b \in \mathcal{L}^+(\mathcal{D})$. It is clear that $\lambda a \in \mathcal{L}^+(\mathcal{D})$ if $\lambda \in \mathbb{C}$. From $\mathcal{D}\big((a+b)^*\big) \supseteq \mathcal{D}(a^*) \cap \mathcal{D}(b^*) \supseteq \mathcal{D}$ and $(a+b)^*\,\mathcal{D} = (a^* + b^*)\,\mathcal{D} \subseteq \mathcal{D}$ we see that $a + b \in \mathcal{L}^+(\mathcal{D})$. We show that $ab \in \mathcal{L}^+(\mathcal{D})$. Let $\varphi \in \mathcal{D}$ and $\psi \in \mathcal{D}$. By (1), $\langle ab\varphi, \psi\rangle = \langle b\varphi, a^+\psi\rangle$. Since $a^+\mathcal{D} \subseteq \mathcal{D}$ as just shown, (1) applies once more and yields $\langle ab\varphi, \psi\rangle = \langle \varphi, b^+a^+\psi\rangle$. Therefore, $b^+a^+ \subseteq (ab)^*$ which gives $\mathcal{D} \subseteq \mathcal{D}\big((ab)^*\big)$ and $(ab)^*\,\mathcal{D} = b^+a^+\mathcal{D}$

$\subseteq \mathcal{D}$. Thus $ab \in \mathcal{L}^+(\mathcal{D})$. By the preceding, we have shown that $\mathcal{L}^+(\mathcal{D})$ is an O*-algebra. Moreover, $b^+a^+ = (ab)^* \upharpoonright \mathcal{D} = (ab)^+$.

In order to prove that $\mathcal{L}^+(\mathcal{D})$ is the largest O*-algebra with domain $\mathcal{D}$, let $\mathcal{A}$ be any O*-algebra on $\mathcal{D}(\mathcal{A}) = \mathcal{D}$. Let $a \in \mathcal{A}$. Since $\mathcal{A}$ is an O-algebra, $a\mathcal{D} \subseteq \mathcal{D}$ by Definition 2.1.3. Since $\mathcal{A}$ is an O*-family, we have $a^+ \in \mathcal{A}$ by Definition 2.1.5. Hence $a^*\mathcal{D} = a^+\mathcal{D} \subseteq \mathcal{D}$. This proves $\mathcal{A} \subseteq \mathcal{L}^+(\mathcal{D})$. □

Corollary 2.1.9. *Let $\mathcal{A}$ be an O*-algebra. With the addition, scalar multiplication and product of linear operators on $\mathcal{D}(\mathcal{A})$ and with the involution $a \to a^+$, $\mathcal{A}$ is a *-algebra with unit $I = I_{\mathcal{D}(\mathcal{A})}$, and $\mathcal{A}$ is a *-subalgebra of $\mathcal{L}^+(\mathcal{D}(\mathcal{A}))$.*

Proof. We already noted above that $\mathcal{A}$ is an algebra and a *-vector space. In the proof of Proposition 2.1.8 it was shown that $(ab)^+ = b^+a^+$ for $a, b \in \mathcal{A}$. Therefore, $a \to a^+$ is an algebra involution on $\mathcal{A}$, so that $\mathcal{A}$ is a *-algebra. The last statement is obvious. □

By Corollary 2.1.9, the O*-algebras with domain $\mathcal{D}$ are precisely the *-subalgebras of $\mathcal{L}^+(\mathcal{D})$ that contain $I_\mathcal{D}$. This characterization could be also taken as the definition of an O*-algebra.

Remark 1. Let us add a few words concerning our terminology. By an O-family in a Hilbert space $\mathcal{H}$ we mean an O-family whose domain is a dense linear subspace of $\mathcal{H}$. In general, the letter $\mathcal{D}$ is used to denote dense linear subspaces of a Hilbert space; for instance, we shall speak about O-families on a domain $\mathcal{D}$. But the symbol $\mathcal{D}$ can be also considered as an assignment which associates with every O-family $\mathcal{A}$ the domain $\mathcal{D}(\mathcal{A})$ of $\mathcal{A}$ (or, in the notation of Definition 2.2.1, the locally convex space $\mathcal{D}_\mathcal{A}$).

Remark 2. If $\mathcal{A}$ is an O-family in the Hilbert space $\mathcal{H}$ with $\mathcal{D}(\mathcal{A}) = \mathcal{H}$, then it follows immediately from the closed graph theorem that each operator in $\mathcal{A}$ is bounded. This implies that $\mathcal{C}(\mathcal{H}, \mathcal{H}) = \mathcal{C}^+(\mathcal{H}, \mathcal{H}) = \mathcal{L}^+(\mathcal{H}) = \mathbb{B}(\mathcal{H})$.

Next we prove a few general results about O-families and O*-algebras.

Proposition 2.1.10. *If there exists an operator $a \in \mathcal{L}^+(\mathcal{D})$ which is closed on $\mathcal{D}$, then $\mathcal{D} = \mathcal{H}$ and hence $\mathcal{L}^+(\mathcal{D}) = \mathbb{B}(\mathcal{H})$.*

Proof. Let $\mathcal{H}_1$ be the domain $\mathcal{D}$ equipped with the scalar product $\langle \varphi, \psi \rangle_1 := \langle \varphi, \psi \rangle + \langle a\varphi, a\psi \rangle$, $\varphi, \psi \in \mathcal{D}$. Since we assumed that the operator a is closed, $\mathcal{H}_1$ is a Hilbert space. From the definition of $\mathcal{H}_1$, it is clear that $\langle a\cdot, \eta \rangle$ is a continuous linear functional on the Hilbert space $\mathcal{H}_1$ for each $\eta \in \mathcal{H}$. By the Riesz theorem, there exists a vector $\zeta_\eta \in \mathcal{H}_1$ such that $\langle a\varphi, \eta \rangle = \langle \varphi, \zeta_\eta \rangle_1$, $\varphi \in \mathcal{H}_1$. Using the fact that $a \in \mathcal{L}^+(\mathcal{D})$, we obtain $\langle a\varphi, \eta \rangle = \langle \varphi, \zeta_\eta \rangle + \langle a\varphi, a\zeta_\eta \rangle = \langle \varphi, (I + a^+a)\, \zeta_\eta \rangle$ for all $\varphi \in \mathcal{D}$. This implies $\eta \in \mathcal{D}(a^*)$. Thus $\mathcal{D}(a^*) = \mathcal{H}$. From the closed graph theorem, a^* and hence a are bounded operators. Because a is also closed, the latter gives $\mathcal{D} = \mathcal{H}$. By Remark 2, $\mathcal{L}^+(\mathcal{H}) = \mathbb{B}(\mathcal{H})$. □

A result in a similar spirit is

Proposition 2.1.11. *Let a be a symmetric operator in $\mathcal{L}^+(\mathcal{D})$. Suppose that there exists a norm $\|\cdot\|_1$ on $\mathcal{D}$ which is stronger than the norm of $\mathcal{H}$* (i.e., $\|\cdot\| \leqq \|\cdot\|_1$ on $\mathcal{D}$) *such that a is bounded relative to this norm (i.e., there is an $\alpha > 0$ such that $\|a\varphi\|_1 \leqq \alpha \|\varphi\|_1$ for $\varphi \in \mathcal{D}$). Then a is a bounded operator on $\mathcal{D}$ with respect to the norm of $\mathcal{H}$.*

Proof. Let $\varphi \in \mathcal{D}$. Since $a\mathcal{D} \subseteq \mathcal{D}$, it follows from the assumptions that $\|a^n\varphi\| \leqq \|a^n\varphi\|_1 \leqq \alpha^n \|\varphi\|_1$ for all $n \in \mathbb{N}$. This shows that each $\varphi \in \mathcal{D}$ is an analytic vector for the symmetric operator a. From Nelson's lemma (see Proposition 10.3.4), a is essentially self-adjoint. Let $\bar{a} = \int \lambda \, de(\lambda)$ be the spectral decomposition of the self-adjoint operator $\bar{a}$. Fix $\varphi \in \mathcal{D}$ and $k \in \mathbb{N}$, $k > \alpha$. From the spectral theorem, $k^n \|(I - e(-k, k)) \varphi\| \leqq \|\bar{a}^n (I - e(-k, k)) \varphi\| \leqq \|a^n \varphi\| \leqq \|a^n \varphi\|_1 \leqq \alpha^n \|\varphi\|_1$ for $n \in \mathbb{N}$. Since $k > \alpha$, the latter can only be true for all $n \in \mathbb{N}$ if $(I - e(-k, k)) \varphi = 0$. Since $\mathcal{D}$ is dense in $\mathcal{H}$, this yields $I = e(-k, k)$. Hence $\bar{a}$ is a bounded operator on $\mathcal{H}$. □

Recall that a *division algebra* is an algebra with unit in which each non-zero element is invertible. From elementary algebra we know that it suffices to assume that each non-zero element has a left inverse. (Indeed, let b be a left inverse of a. Since b has also a left inverse, say c, we have $ab = 1ab = (cb)ab = c(ba)b = cb = 1$, so a is invertible.)

Proposition 2.1.12. *Suppose $\mathcal{A}$ is a $*$-subalgebra of $\mathcal{L}^+(\mathcal{D})$ which is a division algebra. Then $\mathcal{A}$ consists only of scalar multiples of the unit, i.e., $\mathcal{A} = \{\lambda \cdot 1 : \lambda \in \mathbb{C}\}$.*

Proof. Upon replacing $\mathcal{D}$ by $\mathcal{D}_1 := 1(\mathcal{D})$, we can assume without loss of generality that $1 = I_{\mathcal{D}} \equiv I$, that is, $\mathcal{A}$ is an O*-algebra on $\mathcal{D}$. Suppose $a = a^+ \in \mathcal{A}$. For $\lambda \in \mathbb{C}$, let a_λ denote the inverse of $a - \lambda I$ in $\mathcal{A}$ (of course, provided that $a - \lambda I \neq 0$). Since $(a - \lambda I) \mathcal{D} \supseteqq (a - \lambda I) a_\lambda \mathcal{D} = \mathcal{D}$ for any $\lambda \in \mathbb{C} \setminus \mathbb{R}$ and since $\mathcal{D}$ is dense in $\mathcal{H}$, the operator $\bar{a}$ is self-adjoint. Let $e(t)$, $t \in \mathbb{R}$, denote the spectral projections of this operator. To prove the assertion, it is sufficient to show that $a = \lambda I$ for some $\lambda \in \mathbb{R}$. Assume the contrary, i.e., $a - \lambda I \neq 0$ for all $\lambda \in \mathbb{R}$. Let $\lambda \in \mathbb{R}$ and $\varphi \in \mathcal{D}$. If $\psi \in \ker(\bar{a} - \lambda I)$, then $\langle \eta, \psi \rangle = \langle (a - \lambda I) a_\lambda \eta, \psi \rangle = \langle a_\lambda \eta, (\bar{a} - \lambda I) \psi \rangle = 0$ for all $\eta \in \mathcal{D}$; hence $\psi = 0$ and the operator $\bar{a} - \lambda I$ is invertible. From $(a - \lambda I) a_\lambda = (\bar{a} - \lambda I)(\bar{a} - \lambda I)^{-1} \upharpoonright \mathcal{D}$ and from $\ker(\bar{a} - \lambda I) = \{0\}$ we conclude that $a_\lambda = (\bar{a} - \lambda I)^{-1} \upharpoonright \mathcal{D}$. Hence $\varphi \in \mathcal{D}((\bar{a} - \lambda I)^{-1})$. From the spectral theorem, we have

$$\|(\bar{a} - \lambda I)^{-1} \varphi\|^2 = \int_{-\infty}^{+\infty} (t - \lambda)^{-2} \, d\|e(t) \varphi\|^2 \geqq \int_{\lambda}^{\lambda+\varepsilon} \varepsilon^{-2} \, d\|e(t) \varphi\|^2$$

$$= \varepsilon^{-2}(\|e(\lambda + \varepsilon) \varphi\|^2 - \|e(\lambda) \varphi\|^2)$$

and similarly

$$\|(\bar{a} - \lambda I)^{-1} \varphi\|^2 \geqq \varepsilon^{-2}(\|e(\lambda)\varphi\|^2 - \|e(\lambda - \varepsilon) \varphi\|^2) \text{ for all } \varepsilon > 0.$$

From these inequalities it follows at once that the function $t \to \|e(t) \varphi\|^2$ is differentiable on $\mathbb{R}$ and its derivative vanishes identically on $\mathbb{R}$, so that the function is constant. Since $e(-\infty) = 0$, $e(t) \varphi = 0$ and so $e(t) = 0$ for all $t \in \mathbb{R}$. Hence $a = 0$ which is a contradiction. Thus $a = \lambda I$ for some $\lambda \in \mathbb{R}$. □

Though $*$-representations are the main subject of Part II of this monograph, at least the definition is already needed in Part I.

Definition 2.1.13. Suppose A is an (abstract) $*$-algebra with unit. A *$*$-representation* of A on $\mathcal{D}$ is a $*$-homomorphism π of A into $\mathcal{L}^+(\mathcal{D})$ such that $\pi(1) = I$. We then call $\mathcal{D}$ the *domain* of π and write $\mathcal{D}(\pi)$ for $\mathcal{D}$. A $*$-representation π of A is called *faithful* or a realization of A if $\pi(a) = 0$ for $a \in$ A implies $a = 0$.

Equivalently, a $*$-representation of a $*$-algebra A with unit on $\mathcal{D}$ is a $*$-homomorphism π of A *onto* an O*-algebra on $\mathcal{D}$. In order to see that this is equivalent to Definition 2.1.13, it suffices to check that the latter implies that $\pi(1) = I$. Indeed, since π maps A *onto* an O*-algebra, there exists $a \in$ A such that $\pi(a) = I$. Then $I = \pi(a) = \pi(a1) = \pi(a)\,\pi(1) = I\pi(1) = \pi(1)$.

2.2. The Graph Topology

Suppose $\mathcal{A}$ is an O-family in a Hilbert space $\mathcal{H}$.

Definition 2.1.1. The *graph topology* of $\mathcal{A}$ is the locally convex topology $t_{\mathcal{A}}$ on the domain $\mathcal{D}(\mathcal{A})$ defined by the family of seminorms $\{\|\cdot\|_a := \|a\cdot\|: a \in \mathcal{A}\}$. The locally convex space $\mathcal{D}(\mathcal{A})\,[t_{\mathcal{A}}]$ is denoted by $\mathcal{D}_{\mathcal{A}}$. In the cases $\mathcal{A} = \mathcal{C}(\mathcal{D}, \mathcal{H})$ and $\mathcal{A} = \mathcal{L}^+(\mathcal{D})$ we write t_c and t_+, respectively, in place of $t_{\mathcal{A}}$.

Remark 1. Since $I \in \mathcal{A}$, the graph topology $t_{\mathcal{A}}$ is always finer than the topology on $\mathcal{D}(\mathcal{A})$ determined by the norm of the Hilbert space $\mathcal{H}$. It is clear that the graph topology is generated by the Hilbert space norm on $\mathcal{D}(\mathcal{A})$ if and only if each operator in $\mathcal{A}$ is bounded.

Remark 2. The graph topology $t_{\mathcal{A}}$ is the weakest locally convex topology on $\mathcal{D}(\mathcal{A})$ relative to which each operator in $\mathcal{A}$ is a continuous mapping of $\mathcal{D}(\mathcal{A})$ into the Hilbert space $\mathcal{H}$. Another slight reformulation is the following. The graph topology $t_{\mathcal{A}}$ is the weakest locally convex topology on $\mathcal{D}(\mathcal{A})$ which makes the embedding of $\mathcal{D}(\mathcal{A})$ into the normed space $\left(\mathcal{D}(\mathcal{A}), \|\cdot\|_a + \|\cdot\|\right)$ continuous for each $a \in \mathcal{A}$. The latter means that $t_{\mathcal{A}}$ is a *projective topology* in the sense of the theory of locally convex spaces (see e.g. SCHÄFER [1], II, § 5).

Lemma 2.2.2. *If $\mathcal{A}$ is an O-algebra, then $\mathcal{A} \subseteq \mathfrak{L}(\mathcal{D}_{\mathcal{A}})$, i.e., each $a \in \mathcal{A}$ is a continuous mapping of the locally convex space $\mathcal{D}_{\mathcal{A}}$ into itself.*

Proof. We have $\|a\varphi\|_b = \|ba\varphi\| = \|\varphi\|_{ba}$ if $\varphi \in \mathcal{D}(\mathcal{A})$ and $a, b \in \mathcal{A}$. Since $\mathcal{A}$ is an O-algebra, $ab \in \mathcal{A}$; so the preceding proves that $a \in \mathfrak{L}(\mathcal{D}_{\mathcal{A}})$. □

If $\mathcal{A}$ is an O*-algebra, we denote by $\mathcal{L}^+(\mathcal{D}_{\mathcal{A}})$ the set of all operators in $\mathcal{L}^+(\mathcal{D})$ for which $x \in \mathfrak{L}(\mathcal{D}_{\mathcal{A}})$ and $x^+ \in \mathfrak{L}(\mathcal{D}_{\mathcal{A}})$. From the next proposition we see in particular that $\mathcal{L}^+(\mathcal{D}_{\mathcal{A}}) = \mathcal{L}^+(\mathcal{D}_{\mathcal{B}})$ for $\mathcal{B} := \mathcal{L}^+(\mathcal{D}_{\mathcal{A}})$.

Proposition 2.2.3. *For any O*-algebra $\mathcal{A}$, $\mathcal{L}^+(\mathcal{D}_{\mathcal{A}})$ is an O*-algebra on the domain $\mathcal{D}(\mathcal{A})$. It is the largest O*-algebra on $\mathcal{D}(\mathcal{A})$ whose graph topology coincides with the graph topology of $\mathcal{A}$. In particular, $\mathcal{A} \subseteq \mathcal{L}^+(\mathcal{D}_{\mathcal{A}})$.*

Proof. It follows immediately from the above definition that $\mathcal{L}^+(\mathcal{D}_{\mathcal{A}})$ is an O-algebra and that $\mathcal{L}^+(\mathcal{D}_{\mathcal{A}})$ is invariant under the involution $a \to a^+$; that is, $\mathcal{L}^+(\mathcal{D}_{\mathcal{A}})$ is an O*-algebra. From Lemma 2.2.2, $\mathcal{A} \subseteq \mathcal{L}^+(\mathcal{D}_{\mathcal{A}})$. If $\mathcal{B}$ is an O*-algebra on $\mathcal{D}(\mathcal{B}) = \mathcal{D}(\mathcal{A})$ with $t_{\mathcal{B}} = t_{\mathcal{A}}$, then $\mathcal{B} \subseteq \mathcal{L}^+(\mathcal{D}_{\mathcal{B}}) = \mathcal{L}^+(\mathcal{D}_{\mathcal{A}})$. It remains to check that the graph topologies of $\mathcal{L}^+(\mathcal{D}_{\mathcal{A}})$ and $\mathcal{A}$ are equal. The graph topology of $\mathcal{L}^+(\mathcal{D}_{\mathcal{A}})$ is finer than $t_{\mathcal{A}}$, since $\mathcal{A} \subseteq \mathcal{L}^+(\mathcal{D}_{\mathcal{A}})$. It is coarser than $t_{\mathcal{A}}$, because each operator $x \in \mathcal{L}^+(\mathcal{D}_{\mathcal{A}})$ maps $\mathcal{D}_{\mathcal{A}}$ continuously into $\mathcal{D}_{\mathcal{A}}$ and hence into $\mathcal{H}$. □

Definition 2.2.4. An O-family $\mathcal{A}$ is called *directed* if the family of seminorms $\{\|\cdot\|_a: a \in \mathcal{A}\}$ on $\mathcal{D}(\mathcal{A})$ is directed, that is, given two operators $a, b \in \mathcal{A}$, there is an operator $c \in \mathcal{A}$ such that $\|\cdot\|_a \leqq \|\cdot\|_c$ and $\|\cdot\|_b \leqq \|\cdot\|_c$ on $\mathcal{D}(\mathcal{A})$.

Remark 3. One advantage of this notion is the following fact. If $\mathcal{A}$ is a directed O-family, then a linear mapping, say T, of $\mathcal{D}_{\mathcal{A}}$ into a locally convex space E is continuous if and only if for each continuous seminorm p on E there are an operator $a \in \mathcal{A}$ and a constant λ such that $p(T\varphi) \leqq \lambda \|a\varphi\|$ for all $\varphi \in \mathcal{D}(\mathcal{A})$.

Example 2.2.5. Let $\mathcal{D} := C_0^\infty(\mathbb{R})$, considered as a domain in the Hilbert space $\mathcal{H} = L^2(\mathbb{R})$. Let $\mathcal{A}$ be a subset of $L^2_{\mathrm{loc}}(\mathbb{R})$ containing the function that is identically 1. We let the functions in $\mathcal{A}$ act as multiplication operators with domain $\mathcal{D}$ in $\mathcal{H}$; so $\mathcal{A}$ becomes an

O-family on $\mathcal{D}$. The O-family $\mathcal{A}$ is directed if and only if for arbitrary $f, g \in \mathcal{A}$ there is an $h \in \mathcal{A}$ such that $|f(t)| \leqq |h(t)|$ and $|g(t)| \leqq |h(t)|$ a.e. on $\mathbb{R}$ relative to the Lebesgue measure on $\mathbb{R}$. For instance, $L^2_{\mathrm{loc}}(\mathbb{R})$, is a directed O*-vector space with domain $\mathcal{D}$. ○

Proposition 2.2.6. *Each O*-algebra $\mathcal{A}$ is a directed O-family. More precisely, we have $\|\cdot\|_{a_k} \leqq \|\cdot\|_{I+a_1^+a_1+\cdots+a_n^+a_n}$ on $\mathcal{D}(\mathcal{A})$ for all $a_1, \ldots, a_n \in \mathcal{A}$.*

Proof. If $a_1, \ldots, a_n \in \mathcal{A}$ and $\varphi \in \mathcal{D}(\mathcal{A})$, then

$$\begin{aligned}\|\varphi\|^2_{I+a_1^+a_1+\cdots+a_n^+a_n} &= \|(I + a_1^+a_1 + \cdots + a_n^+a_n)\, \varphi\|^2 \\ &= \|(a_1^+a_1 + \cdots + a_n^+a_n)\, \varphi\|^2 + \|\varphi\|^2 \\ &\quad + 2 \operatorname{Re} \langle (a_1^+a_1 + \cdots + a_n^+a_n)\, \varphi, \varphi\rangle \\ &\geqq \|a_1\varphi\|^2 + \cdots + \|a_n\varphi\|^2 = \|\varphi\|^2_{a_1} + \cdots + \|\varphi\|^2_{a_n}\end{aligned}$$

which gives the assertion. □

Lemma 2.2.7. *Let $\mathcal{A}$ be an O*-algebra such that the locally convex space $\mathcal{D}_{\mathcal{A}}$ is metrizable. Let $(\delta_n : n \in \mathbb{N})$ be a given sequence of positive numbers. Then there is a sequence $(a_n : n \in \mathbb{N})$ of symmetric operators in $\mathcal{A}$ such that $a_1 = \delta_1 I$, $\delta_n^2\|a_n\varphi\| \leqq \delta_n\|a_n^2\varphi\| \leqq \|a_{n+1}\varphi\|$ for all $\varphi \in \mathcal{D}(\mathcal{A})$ and $n \in \mathbb{N}$ and such that the graph topology $\mathrm{t}_{\mathcal{A}}$ on $\mathcal{D}(\mathcal{A})$ is generated by the family of seminorms $\{\|\cdot\|_{a_n} : n \in \mathbb{N}\}$.*

Proof. Since $\mathcal{D}_{\mathcal{A}}$ is metrizable, there is a sequence $(b_n : n \in \mathbb{N})$ of operators in $\mathcal{A}$ such that the graph topology $\mathrm{t}_{\mathcal{A}}$ is determined by the family of seminorms $\{\|\cdot\|_{b_n} : n \in \mathbb{N}\}$. The sequence (a_n) will be defined inductively. Set $a_1 := \delta_1 I$. If the operators $a_1, \ldots, a_n \in \mathcal{A}$ are chosen, then we define $a_{n+1} := I + \delta_{n+1}^2 I + \delta_n^2 a_n^4 + b_n^+ b_n$. From the inequality in Proposition 2.2.6 we conclude that the sequence (a_n) has the desired properties. □

One of the fundamental concepts about O-families is that of a closed O-family. We next define and study this notion.

Definition 2.2.8. An O-family $\mathcal{A}$ is said to be *closed* if the locally convex space $\mathcal{D}_{\mathcal{A}}$ is complete. A domain $\mathcal{D}$ is called *closed* if the O*-algebra $\mathcal{L}^+(\mathcal{D})$ on $\mathcal{D}$ is closed.

Lemma 2.2.9. *Suppose $\mathcal{A}$ is an O-family such that $\mathcal{D}(\mathcal{A}) = \bigcap_{a \in \mathcal{A}} \mathcal{D}(\bar{a})$. Then the locally convex space $\mathcal{D}_{\mathcal{A}}$ is complete and $\mathcal{A}$ is closed.*

Proof. Let $(\varphi_i : i \in I)$ be a Cauchy net in the locally convex space $\mathcal{D}_{\mathcal{A}}$. Then, for each $a \in \mathcal{A}$, $(a\varphi_i)$ is a Cauchy net in the Hilbert space $\mathcal{H}$, so that there exists a vector $\varphi_a \in \mathcal{H}$ such that $\varphi_a = \lim a\varphi_i$ in $\mathcal{H}$. Put $\varphi := \varphi_I$. Let $a \in \mathcal{A}$. Since the operator a is closable, $\varphi = \lim \varphi_i$ and $\varphi_a = \lim a\varphi_i$ in $\mathcal{H}$ imply that $\varphi \in \mathcal{D}(\bar{a})$ and $\varphi_a = \bar{a}\varphi$. Thus $\varphi \in \bigcap_{a \in \mathcal{A}} \mathcal{D}(\bar{a}) = \mathcal{D}(\mathcal{A})$. From $\lim \|a(\varphi_i - \varphi)\| = 0$ for each $a \in \mathcal{A}$ it follows that $\varphi = \lim \varphi_i$ in the locally convex space $\mathcal{D}_{\mathcal{A}}$. □

Suppose $\mathcal{A}$ is an O-family. Define $\overline{\mathcal{D}}(\mathcal{A}) := \bigcap_{a \in \mathcal{A}} \mathcal{D}(\bar{a})$ and $\bar{\mathcal{A}} := \{\bar{a} \upharpoonright \overline{\mathcal{D}}(\mathcal{A}) : a \in \mathcal{A}\}$. Then $\bar{\mathcal{A}}$ is also an O-family with domain $\overline{\mathcal{D}}(\mathcal{A})$ which obviously satisfies the assumptions of Lemma 2.2.9. Therefore, by Lemma 2.2.9, $\bar{\mathcal{A}}$ is a closed O-family.

Let $\hat{\mathcal{D}}(\mathcal{A})$ denote the closure of $\mathcal{D}(\mathcal{A})$ in the locally convex space $\mathcal{D}_{\bar{\mathcal{A}}}$ and let $\hat{\mathcal{A}} := \{\hat{a} := \bar{a} \upharpoonright \hat{\mathcal{D}}(\mathcal{A}) : a \in \mathcal{A}\}$. Since $\mathcal{D}_{\bar{\mathcal{A}}}$ is complete by Lemma 2.2.9, $\mathcal{D}_{\hat{\mathcal{A}}}$ is complete as well and hence $\hat{\mathcal{A}}$ is a closed O-family. From the definition of $\hat{\mathcal{A}}$ it is clear that $\mathcal{D}_{\hat{\mathcal{A}}}$ is the completion of the locally convex space $\mathcal{D}_{\mathcal{A}}$. Moreover, for each $a \in \mathcal{A}$ the operator $\hat{a} \in \mathfrak{L}(\mathcal{D}_{\hat{\mathcal{A}}}, \mathcal{H})$ is the continuous extension to $\hat{\mathcal{D}}(\mathcal{A})$ of the operator $a \in \mathfrak{L}(\mathcal{D}_{\mathcal{A}}, \mathcal{H})$.

Definition 2.2.10. The O-family $\hat{\mathcal{A}}$ on the domain $\hat{\mathcal{D}}(\mathcal{A})$ is called the *closure* of the O-family $\mathcal{A}$.

Proposition 2.2.11. *Suppose that $\mathcal{A}$ is an O-vector space [resp. O*-vector space, O-algebra, O*-algebra]. Then $\hat{\mathcal{A}}$ is also an O-vector space [resp. O*-vector space, O-algebra, O*-algebra]. The map $a \to \hat{a}$ is a bijective linear mapping [resp. bijective involution preserving linear mapping, an isomorphism, a $*$-isomorphism] of $\mathcal{A}$ onto $\hat{\mathcal{A}}$.*

Proof. Since $\hat{a} \in \mathfrak{L}(\mathcal{D}_{\hat{\mathcal{A}}}, \mathcal{H})$ is the continuous extension of $a \in \mathfrak{L}(\mathcal{D}_{\mathcal{A}}, \mathcal{H})$ for $a \in \mathcal{A}$ as noted above, $\hat{\mathcal{A}}$ is an O-vector space and the map $a \to \hat{a}$ preserves the linear structure. It is obvious that this map is bijective and that it preserves the involution when $\mathcal{A}$ is an O*-family. Thus it suffices to prove the assertion when $\mathcal{A}$ is an O-algebra. Suppose $a, b \in \mathcal{A}$. Let $\varphi \in \hat{\mathcal{D}}(\mathcal{A})$. Then there exists a net $(\varphi_i : i \in I)$ of vectors in $\mathcal{D}(\mathcal{A})$ such that $\varphi = \lim \varphi_i$ in $\mathcal{D}_{\bar{\mathcal{A}}}$. In particular, this gives $\lim \|\varphi - \varphi_i\|_{\bar{b}} = \lim \|\bar{b}\varphi - b\varphi_i\| = 0$ and $\lim \|\varphi - \varphi_i\|_{\overline{ab}} = \lim \|\overline{ab}\varphi - ab\varphi_i\| = 0$. Since the operator a is closable, it follows that $\bar{b}\varphi \in \mathcal{D}(\bar{a})$ and $\overline{ab}\varphi = \bar{a}\bar{b}\varphi$. Since $a \in \mathcal{A}$ was arbitrary, $\bar{b}\varphi \in \bigcap_{a \in \mathcal{A}} \mathcal{D}(\bar{a}) = \overline{\mathcal{D}}(\mathcal{A})$. For $a \in \mathcal{A}$, we have $\lim \|\bar{b}\varphi - b\varphi_i\|_{\bar{a}} = \lim \|\bar{a}(\bar{b}\varphi - b\varphi_i)\| = \lim \|\overline{ab}\varphi - ab\varphi_i\| = 0$. From this we conclude that the vector $\bar{b}\varphi$ belongs to the closure of the set $\{b\varphi_i : i \in I\}$ in the locally convex space $\mathcal{D}_{\bar{\mathcal{A}}}$. Since $\mathcal{A}$ is an O-algebra, $b\mathcal{D}(\mathcal{A}) \subseteq \mathcal{D}(\mathcal{A})$, so that $b\varphi_i \in \mathcal{D}(\mathcal{A})$ for each $i \in I$. Therefore, $\bar{b}\varphi \in \hat{\mathcal{D}}(\mathcal{A})$. That is, $\bar{b}\hat{\mathcal{D}}(\mathcal{A}) \subseteq \hat{\mathcal{D}}(\mathcal{A})$ for all $b \in \mathcal{A}$. Further, from the preceding proof, $\hat{a}\hat{b}\varphi = \bar{a}\bar{b}\varphi = \overline{ab}\varphi = \widehat{ab}\varphi$ for $\varphi \in \hat{\mathcal{D}}(\mathcal{A})$ and $a, b \in \mathcal{A}$. Since $\hat{\mathcal{A}}$ is an O-vector space and the map $a \to \hat{a}$ is linear as noted above, we have shown that $\hat{\mathcal{A}}$ is an O-algebra and that the map $a \to \hat{a}$ provides an isomorphism of the algebras $\mathcal{A}$ and $\hat{\mathcal{A}}$. □

For general O-families $\mathcal{A}$ it may happen that $\hat{\mathcal{D}}(\mathcal{A}) \neq \overline{\mathcal{D}}(\mathcal{A})$, that is, $\mathcal{D}(\mathcal{A})$ is not dense in $\mathcal{D}_{\bar{\mathcal{A}}}$; see Example 2.5.10. The next proposition shows that there is no difference between $\hat{\mathcal{A}}$ and $\bar{\mathcal{A}}$ (or equivalently, between $\hat{\mathcal{D}}(\mathcal{A})$ and $\overline{\mathcal{D}}(\mathcal{A})$) if the O-family $\mathcal{A}$ is directed, in particular, if $\mathcal{A}$ is an O*-algebra.

Proposition 2.2.12. *Suppose that $\mathcal{A}$ is a directed O-family. Let $\mathcal{A}_0$ be a subset of $\mathcal{A}$ such that the family of seminorms $\{\|\cdot\|_a : a \in \mathcal{A}_0\}$ is directed and generates the graph topology of $\mathcal{A}$. Then $\hat{\mathcal{D}}(\mathcal{A}) = \overline{\mathcal{D}}(\mathcal{A}) = \bigcap_{a \in \mathcal{A}_0} \mathcal{D}(\bar{a})$ and $\hat{\mathcal{A}} = \bar{\mathcal{A}}$. The O-family $\mathcal{A}$ is closed if and only if $\mathcal{D}(\mathcal{A}) = \bigcap_{a \in \mathcal{A}_0} \mathcal{D}(\bar{a})$.*

Proof. The final assertion follows immediately from the first one. Since always $\hat{\mathcal{D}}(\mathcal{A}) \subseteq \overline{\mathcal{D}}(\mathcal{A}) \subseteq \bigcap_{a \in \mathcal{A}_0} \mathcal{D}(\bar{a})$ by definition, the proof of the proposition will be complete if we have shown that $\bigcap_{a \in \mathcal{A}_0} \mathcal{D}(\bar{a}) \subseteq \hat{\mathcal{D}}(\mathcal{A})$. We suppose $\varphi \in \bigcap_{a \in \mathcal{A}_0} \mathcal{D}(\bar{a})$. Let $a \in \mathcal{A}$ and let $\varepsilon > 0$. Since $\varphi \in \mathcal{D}(\bar{a})$, there is a vector $\varphi_{a,\varepsilon} \in \mathcal{D}(\mathcal{A})$ such that $\|\varphi - \varphi_{a,\varepsilon}\|_{\bar{a}} = \|\bar{a}(\varphi - \varphi_{a,\varepsilon})\| < \varepsilon$. Since $\{\|\cdot\|_{\bar{a}} ; a \in \mathcal{A}_0\}$ is a directed (!) family of seminorms which generates the graph topology $t_{\bar{\mathcal{A}}}$, the preceding shows that φ belongs to the closure of $\mathcal{D}(\mathcal{A})$ in $\mathcal{D}_{\bar{\mathcal{A}}}$. By the definition of $\hat{\mathcal{D}}(\mathcal{A})$, this means that $\varphi \in \hat{\mathcal{D}}(\mathcal{A})$. Thus $\bigcap_{a \in \mathcal{A}_0} \mathcal{D}(\bar{a}) \subseteq \hat{\mathcal{D}}(\mathcal{A})$. □

Proposition 2.2.13. *For every O-family $\mathcal{A}$ there exists a directed O*-vector space $\mathcal{A}_1$ on the domain $\mathcal{D}(\mathcal{A}_1) \equiv \mathcal{D}(\mathcal{A})$ such that the graph topologies of $\mathcal{A}$ and of $\mathcal{A}_1$ on $\mathcal{D}(\mathcal{A})$ coincide, i.e., $\mathcal{D}_{\mathcal{A}} \equiv \mathcal{D}_{\mathcal{A}_1}$.*

Proof. Suppose $\mathcal{B} = \{b_1, \ldots, b_n\}$ is a finite subset of $\mathcal{A}$ containing I. Then $\mathcal{B}$ is an O-family on the domain $\mathcal{D}(\mathcal{B}) = \mathcal{D}(\mathcal{A})$. Let $\hat{\mathcal{B}} = \{\hat{b}_1, \ldots, \hat{b}_n\}$ be its closure. Define a positive sesquilinear form $h_{\mathcal{B}}$ on $\hat{\mathcal{D}}(\mathcal{B})$ by $h_{\mathcal{B}}(\varphi, \psi) = \langle \hat{b}_1\varphi, \hat{b}_1\psi \rangle + \cdots + \langle \hat{b}_n\varphi, {}_n\psi \rangle$,

$\varphi, \psi \in \hat{\mathcal{D}}(\mathcal{B})$. Clearly, the norm $h_{\mathcal{B}}(\varphi, \varphi)^{1/2} = (\|\hat{b}_1\varphi\|^2 + \cdots + \|\hat{b}_n\varphi\|^2)^{1/2}$, $\varphi \in \hat{\mathcal{D}}(\mathcal{B})$, generates the graph topology of $\mathcal{B}$. Therefore, because $\mathcal{D}_{\hat{\mathcal{B}}}$ is complete, the form $h_{\mathcal{B}}$ is closed. From the form representation theorem (KATO [1], VI, § 2, Theorem 2.23), there is a self-adjoint operator $A_{\mathcal{B}}$ with domain $\mathcal{D}(A_{\mathcal{B}}) = \hat{\mathcal{D}}(\mathcal{B})$ such that $h_{\mathcal{B}}(\cdot, \cdot) \equiv \langle A_{\mathcal{B}}\cdot, A_{\mathcal{B}}\cdot\rangle$. The linear span $\mathcal{A}_1$ of all operators $a_{\mathcal{B}} := A_{\mathcal{B}} \upharpoonright \mathcal{D}(\mathcal{A})$, where $\mathcal{B}$ is a finite subset of $\mathcal{A}$ with $I \in \mathcal{B}$, is an O*-vector space on $\mathcal{D}(\mathcal{A})$. By construction, we have $\|a_{\mathcal{B}}\cdot\|^2 \equiv \|b_1\cdot\|^2 + \cdots + \|b_n\cdot\|^2$ on $\mathcal{D}(\mathcal{A})$. From this we see that $t_{\mathcal{A}} = t_{\mathcal{A}_1}$. From this formula it follows also that $\|a_{\mathcal{B}_1}\cdot\| \leqq \|a_{\mathcal{B}_2}\cdot\|$ when $\mathcal{B}_1 \subseteq \mathcal{B}_2$, so that the family of all such seminorms $\|\cdot\|_{a_{\mathcal{B}}}$ is directed. This implies that the O*-vector space $\mathcal{A}_1$ is directed. □

Remark 4. If $\mathcal{A}$ and $\mathcal{A}_1$ are O-families on the same domain $\mathcal{D}(\mathcal{A}) \equiv \mathcal{D}(\mathcal{A}_1)$ such that $t_{\mathcal{A}} = t_{\mathcal{A}_1}$, then $\hat{\mathcal{D}}(\mathcal{A}) = \hat{\mathcal{D}}(\mathcal{A}_1)$ and $t_{\hat{\mathcal{A}}} = t_{\hat{\mathcal{A}}_1}$. We prove this assertion. Having shown that $\hat{\mathcal{D}}(\mathcal{A}) = \hat{\mathcal{D}}(\mathcal{A}_1)$, the equality $t_{\hat{\mathcal{A}}} = t_{\hat{\mathcal{A}}_1}$ follows by continuity from $t_{\mathcal{A}} = t_{\mathcal{A}_1}$. Thus it suffices to prove that $\hat{\mathcal{D}}(\mathcal{A}) = \hat{\mathcal{D}}(\mathcal{A}_1)$. We let $\varphi \in \hat{\mathcal{D}}(\mathcal{A})$. Then φ is the limit of a net $(\varphi_i : i \in I)$ of vectors $\varphi_i \in \mathcal{D}(\mathcal{A})$ in $\mathcal{D}_{\hat{\mathcal{A}}}$. Since $t_{\mathcal{A}} = t_{\mathcal{A}_1}$, $(\varphi_i : i \in I)$ is a Cauchy net in $\mathcal{D}_{\mathcal{A}_1}$. Hence there exists a vector $\psi \in \hat{\mathcal{D}}(\mathcal{A}_1)$ such that $\psi = \lim \varphi_i$ in $\mathcal{D}_{\hat{\mathcal{A}}_1}$. Since the topologies $t_{\mathcal{A}}$ and $t_{\hat{\mathcal{A}}_1}$ are stronger than the norm topology of $\mathcal{H}$, φ and ψ are also the limits of (φ_i) in $\mathcal{H}$. Thus $\varphi = \psi \in \hat{\mathcal{D}}(\mathcal{A}_1)$; so $\hat{\mathcal{D}}(\mathcal{A}) \subseteq \hat{\mathcal{D}}(\mathcal{A}_1)$. The reversed inclusion follows by symmetry.

By an *O-subfamily*, resp. *O*-subalgebra*, of an O-family, resp. O*-algebra, $\mathcal{A}$ we mean an O-family, resp. O*-algebra, on $\mathcal{D}(\mathcal{A})$ which is contained in $\mathcal{A}$.

We now introduce an important class of O*-algebras.

Definition 2.2.14. We say that an O*-algebra $\mathcal{A}$ in the Hilbert space $\mathcal{H}$ is *commutatively dominated* if there exist a directed O-subfamily $\mathcal{A}_0$ of $\mathcal{A}$ and a commutative von Neumann algebra $\mathcal{N}$ in $\mathcal{H}$ such that the graph topologies $t_{\mathcal{A}}$ and $t_{\mathcal{A}_0}$ coincide and such that the operator $\bar{a}$ is affiliated with $\mathcal{N}$ for each a in $\mathcal{A}_0$.

The following lemma shows that there is no loss of generality to assume in Definition 2.2.14 that $\mathcal{A}_0$ is an O*-subalgebra of $\mathcal{A}$.

Lemma 2.2.15. *Let $\mathcal{A}$ be a commutatively dominated O*-algebra, and let $\mathcal{A}_0$ and $\mathcal{N}$ be as in Definition 2.2.14. Then the closure of each operator in the O*-algebra generated by $\mathcal{A}_0$ is affiliated with $\mathcal{N}$.*

Proof. Let b be an operator in the O*-algebra which is generated by $\mathcal{A}_0$. Then there are numbers $n, k \in \mathbb{N}$, $k \leqq n$, a polynomial $p \in \mathbb{C}[x_1, \ldots, x_n]$ and operators $a_1, \ldots, a_n \in \mathcal{A}_0$ such that $b = p(a_1, \ldots, a_k, a_{k+1}^+, \ldots, a_n^+)$. Note that $a_1, \ldots, a_k, a_{k+1}^+, \ldots, a_n^+$ commute in $\mathcal{L}^+(\mathcal{D}(\mathcal{A}))$, since $\bar{a}_1, \ldots, \bar{a}_n$ belong to the commutative $*$-algebra $\mathbb{A}(\mathcal{N})$; see Lemma 1.6.3. Let $\breve{b}$ be the polynomial $p(\bar{a}_1, \ldots, \bar{a}_k, a_{k+1}^*, \ldots, a_n^*)$ formed in the commutative $*$-algebra $\mathbb{A}(\mathcal{N})$. Obviously, $\bar{b} \subseteq \breve{b}$. Since $\mathcal{A}_0$ is a directed O-family and $t_{\mathcal{A}} = t_{\mathcal{A}_0}$, there exist an operator $a \in \mathcal{A}_0$ and a constant λ such that $\|b\varphi\| \leqq \lambda\|a\varphi\|$, $\varphi \in \mathcal{D}(\mathcal{A})$. From this and the fact that $\bar{b} \subseteq \breve{b}$ we conclude that $\mathcal{D}(\bar{a}) \subseteq \mathcal{D}(\bar{b})$ and that $\mathcal{D}$ is a core for $\breve{b} \upharpoonright \mathcal{D}(\bar{a})$ $(= \bar{b} \upharpoonright \mathcal{D}(\bar{a}))$. Since $\bar{a} \in \mathbb{A}(\mathcal{N})$ by definition and $\breve{b} \in \mathbb{A}(\mathcal{N})$ by construction, Lemma 1.6.3, (iii), says that $\mathcal{D}(\bar{a}) \cap \mathcal{D}(\breve{b}) \equiv \mathcal{D}(\bar{a})$ is a core for $\breve{b}$. These two facts imply that $\mathcal{D}$ is a core for $\breve{b}$. Hence $\bar{b} = \breve{b}$, and $\bar{b}$ is affiliated with $\mathcal{N}$. □

Example 2.2.16. Suppose A is a (bounded or unbounded) self-adjoint operator in a Hilbert space $\mathcal{H}$. Let $A = \int \lambda \, dE(\lambda)$ be the spectral resolution of A. Suppose $(h_n : n \in \mathbb{N})$ is a sequence of measurable and a.e. finite real functions on the real line satisfying

$$h_1(t) \geqq 1 \text{ and } h_n(t)^2 \leqq h_{n+1}(t) \text{ a.e. on } \mathbb{R} \text{ for all } n \in \mathbb{N}. \tag{1}$$

Here and throughout the further investigations based on this example (in Sections 2.4, 3.4, 4.3 and 6.2) we assume that measure theoretic notions always refer to the spectral measure of A. For instance, a function is called *measurable* if it is $\langle E(\cdot)\,\varphi, \varphi\rangle$-measurable for all $\varphi \in \mathcal{H}$, and a.e. means $\langle E(\cdot)\,\varphi, \varphi\rangle$-almost everywhere for all $\varphi \in \mathcal{H}$. By the functional calculus of self-adjoint operators (see RIESZ/SZ.-NAGY [1], IX, 128.), $h_n(A)$ is a self-adjoint operator in $\mathcal{H}$ for each $n \in \mathbb{N}$. Define

$$\mathcal{D}(\mathcal{A}) = \bigcap_{n\in\mathbb{N}} \mathcal{D}\big(h_n(A)\big). \tag{2}$$

From the properties of the functional calculus of self-adjoint operators it is easily seen that $\mathcal{D}\big(h_{n+1}(A)\big)$ is a dense linear subspace of the normed space $\big(\mathcal{D}(h_n(A)), \|\cdot\|_{h_n(A)}\big)$, $n \in \mathbb{N}$. (This follows also from Lemma 1.6.3.) Therefore, by Lemma 1.1.2, $\mathcal{D}(\mathcal{A})$ is dense in $\big(\mathcal{D}(h_n(A)), \|\cdot\|_{h_n(A)}\big)$ and hence a core for $h_n(A)$, $n \in \mathbb{N}$. In particular, $\mathcal{D}(\mathcal{A})$ is dense in $\mathcal{H}$.

From (1) it follows that $h_n(A)\,\mathcal{D}(\mathcal{A}) \subseteq \mathcal{D}(\mathcal{A})$ for $n \in \mathbb{N}$; so $h_n(A) \restriction \mathcal{D}(A)$ is in $\mathcal{L}^+\big(\mathcal{D}(\mathcal{A})\big)$. Suppose $\mathcal{A}$ is an O*-algebra on the domain $\mathcal{D}(\mathcal{A})$ defined by (2) such that

$$a_n := h_n(A) \restriction \mathcal{D}(\mathcal{A}) \quad \text{is in } \mathcal{A} \text{ for all } n \in \mathbb{N}. \tag{3}$$

Then $\mathcal{A}$ is a commutatively dominated O-algebra and $\mathcal{D}_{\mathcal{A}}$ is a Frechet space.*

Proof. From (1) we conclude that $\mathcal{A}_0 := \{I, a_n : n \in \mathbb{N}\}$ is a directed O-subfamily of $\mathcal{A}$. Since, as noted above, $\mathcal{D}(\mathcal{A})$ is a core for each operator $h_n(A)$, we have $\overline{a_n} = h_n(A)$, $n \in \mathbb{N}$. Hence $\overline{a_n}$ is affiliated with the commutative von Neumann algebra $\mathcal{N} := \{E(\lambda) : \lambda \in \mathbb{R}\}''$ and $\mathcal{D}(\mathcal{A}) = \bigcap_{n\in\mathbb{N}} \mathcal{D}(\overline{a_n})$. From the latter, $\mathcal{A}_0$ is closed, so that $\mathcal{D}_{\mathcal{A}_0}$ is a Frechet space. Therefore, $t_{\mathcal{A}} = t_{\mathcal{A}_0} = t_+$ on $\mathcal{D}(\mathcal{A})$. □ ○

Proposition 2.2.17. *If $\mathcal{A}$ is a commutatively dominated O*-algebra in the Hilbert space $\mathcal{H}$ such that $\mathcal{D}_{\mathcal{A}}$ is a Frechet space, then $\mathcal{A}$ is of the form described in Example 2.2.16. That is, there are a self-adjoint operator A in $\mathcal{H}$ and measurable a.e. finite real functions h_n, $n \in \mathbb{N}$, on $\mathbb{R}$ such that (1), (2) and (3) are valid.*

Proof. By Lemma 2.2.15 there is an O*-subalgebra $\mathcal{A}_0$ of $\mathcal{A}$ having the properties stated in Definition 2.2.14. Since $t_{\mathcal{A}} = t_{\mathcal{A}_0}$ is metrizable, it follows from Lemma 2.2.7 that there exists a sequence $(a_n : n \in \mathbb{N})$ of symmetric operators in $\mathcal{A}_0$ such that $a_1 = I$, $\|a\varphi\| \leqq \|a_n^2\varphi\| \leqq \|a_{n+1}\varphi\|$ for $\varphi \in \mathcal{D}(\mathcal{A})$ and $n \in \mathbb{N}$ and such that $t_{\mathcal{A}}$ is generated by the directed family of seminorms $\{\|\cdot\|_{a_n} : n \in \mathbb{N}\}$. By assumption, the closed symmetric operators $\overline{a_n}$, $n \in \mathbb{N}$, are affiliated with the commutative von Neumann algebra $\mathcal{N}$. Hence (by Lemma 1.6.3) these operators are self-adjoint and their spectral projections mutually commute. Therefore, there are a self-adjoint operator A in $\mathcal{H}$ and measurable a.e. finite real functions h_n, $n \in \mathbb{N}$, such that $\overline{a_n} = h_n(A)$ (RIESZ/SZ.-NAGY [1], IX, 130.). Since $\mathcal{A}$ is closed, we conclude from Proposition 2.2.12 that $\mathcal{D}(\mathcal{A}) = \bigcap_{n\in\mathbb{N}} \mathcal{D}(\overline{a_n}) = \bigcap_{n\in\mathbb{N}} \mathcal{D}\big(h_n(A)\big)$, so (2) is proved. (3) is obvious from the construction. We verify (1). Put $f_n(t) := h_n(t)^2$ on $\mathbb{R}$ for $n \in \mathbb{N}$. From $\|a_n^2\varphi\| \leqq \|a_{n+1}\varphi\|$, we have $\|f_n(A)\,\varphi\| \leqq \|h_{n+1}(A)\,\varphi\|$ for $\varphi \in \mathcal{D}(\mathcal{A})$. Because $\overline{a_{n+1}} = h_{n+1}(A)$, $\mathcal{D}(\mathcal{A})$ is a core for $h_{n+1}(A)$, so that the latter extends to all vectors $\varphi \in \mathcal{D}\big(h_{n+1}(A)\big) \subseteq \mathcal{D}\big(f_n(A)\big)$. But then the properties of the functional calculus (as discussed in RIESZ/SZ.-NAGY [1], IX) yield $f_n(t) \equiv h_n(t)^2 \leqq h_{n+1}(t)$ a.e. on $\mathbb{R}$. Since $a_1 = I$, we can take $h_1(t) \equiv 1$, and (1) is shown. □

2.3. The Locally Convex Space $\mathcal{D}_{\mathcal{A}}$

If $\mathcal{A}$ is an O-family, $\mathcal{A}(I)$ will denote the set of all operators a in $\mathcal{A}$ which satisfy $\|\cdot\| \leqq \|\cdot\|_a$ on $\mathcal{D}(\mathcal{A})$.

First we show that $\mathcal{D}_{\hat{\mathcal{A}}}$ is the projective limit of a family of Hilbert spaces. We refer to JARCHOW [1], 2.6, or to SCHÄFER [1], II, § 5, for the facts about projective limits used in the following discussion.

Suppose that $\mathcal{A}$ is a *directed* O-family. We equip the set $\mathcal{A}(I)$ with the following relation: $a \lesssim b$ if and only if $\|\cdot\|_a \leqq \|\cdot\|_b$. Since $I \in \mathcal{A}$ and $\mathcal{A}$ is a directed O-family, $\mathcal{A}(I)$ is a non-empty directed set. For $a \in \mathcal{A}(I)$, the domain $\mathcal{D}(\bar{a})$ endowed with the scalar product $\langle \cdot, \cdot \rangle_{\bar{a}} := \langle \bar{a} \cdot, \bar{a} \cdot \rangle$ is a Hilbert space. This space will be denoted by $\mathcal{H}_a$. Suppose $a, b \in \mathcal{A}(I)$ and $a \lesssim b$. Then $\mathcal{H}_b \subseteq \mathcal{H}_a$ and $\|\cdot\|_{\bar{a}} \leqq \|\cdot\|_{\bar{b}}$ on $\mathcal{H}_b$; hence the embedding map $g_{a,b}$ of $\mathcal{H}_b$ into $\mathcal{H}_a$ is a continuous linear map. It is obvious that $g_{a,a}$, $a \in \mathcal{A}(I)$, is the identity map and $g_{a,c} = g_{a,b} g_{b,c}$ if $a, b, c \in \mathcal{A}(I)$, $a \lesssim b$, $b \lesssim c$. Therefore, the family of Hilbert spaces $\{\mathcal{H}_a : a \in \mathcal{A}(I)\}$ and the family of linear mappings $\{g_{a,b} : a, b \in \mathcal{A}(I)$ and $a \lesssim b\}$ form a projective system. Let $\lim\limits_{a \in \mathcal{A}(I)} \operatorname{proj} \mathcal{H}_a$ denote the projective limit of this system. As a linear space, $\lim\limits_{a \in \mathcal{A}(I)} \operatorname{proj} \mathcal{H}_a$ consists of all elements (φ_a) of the product $\prod\limits_{a \in \mathcal{A}(I)} \mathcal{H}_a$ which satisfy $g_{a,b}\varphi_b = \varphi_a$ whenever $a, b \in \mathcal{A}(I)$ and $a \lesssim b$. From the definitions of the mappings $g_{a,b}$ it is clear that $(\varphi_a) \to \varphi_I$ is an isomorphism of the vector spaces $\lim\limits_{a \in \mathcal{A}(I)} \operatorname{proj} \mathcal{H}_a$ and $\bigcap\limits_{a \in \mathcal{A}(I)} \mathcal{H}_a = \bigcap\limits_{a \in \mathcal{A}} \mathcal{D}(\bar{a})$. Since $\mathcal{A}$ is directed, Proposition 2.2.12 shows that the latter space is $\hat{\mathcal{D}}(\mathcal{A})$. For notational simplicity, we identify the vector spaces $\lim\limits_{a \in \mathcal{A}(I)} \operatorname{proj} \mathcal{H}_a$ and $\hat{\mathcal{D}}(\mathcal{A})$ via this isomorphism. The topology of the projective limit $\lim\limits_{a \in \mathcal{A}(I)} \operatorname{proj} \mathcal{H}_a \ (\equiv \hat{\mathcal{D}}(\mathcal{A}))$ is defined as the weakest locally convex topology for which all embedding maps of $\hat{\mathcal{D}}(\mathcal{A})$ into $\mathcal{H}_a$, $a \in \mathcal{A}(I)$, are continuous. But this is, of course, the graph topology of $\hat{\mathcal{A}}$ (see Remark 2 in 2.2). Thus $\mathcal{D}_{\hat{\mathcal{A}}} = \lim\limits_{a \in \mathcal{A}(I)} \operatorname{proj} \mathcal{H}_a$ as locally convex spaces. This proves the first statement in

Proposition 2.3.1. *If $\mathcal{A}$ is a directed O-family, then $\mathcal{D}_{\hat{\mathcal{A}}} = \lim\limits_{a \in \mathcal{A}(I)} \operatorname{proj} \mathcal{H}_a$. If $\mathcal{A}$ is an arbitrary O-family, then the locally convex space $\mathcal{D}_{\hat{\mathcal{A}}}$ is the projective limit of a family of Hilbert spaces.*

Proof. We prove the second assertion. By Proposition 2.2.13, there is a directed O-family on $\mathcal{D}(\mathcal{A}_1) = \mathcal{D}(\mathcal{A})$ such that $t_{\mathcal{A}} = t_{\mathcal{A}_1}$. By Remark 4 in 2.2, $\mathcal{D}_{\hat{\mathcal{A}}} = \mathcal{D}_{\hat{\mathcal{A}}_1}$ and the first assertion applies. □

Corollary 2.3.2. (*i*) *For each O-family $\mathcal{A}$, the locally convex space $\mathcal{D}_{\mathcal{A}}$ has the approximation property.*

(ii) *Suppose $\mathcal{A}$ is a closed O-family. Then the locally convex space $\mathcal{D}_{\mathcal{A}}$ is semireflexive. The space $\mathcal{D}_{\mathcal{A}}$ is reflexive if and only if it is barrelled. If $\mathcal{D}_{\mathcal{A}}$ is a Frechet space, then $\mathcal{D}_{\mathcal{A}}$ is reflexive.*

Proof. (i): From Proposition 2.3.1, $\mathcal{D}_{\hat{\mathcal{A}}}$ is the projective limit of a family of Hilbert spaces. Therefore, its subspace $\mathcal{D}_{\mathcal{A}}$ has the approximation property (SCHÄFER [1], III, 9.2).

(ii): Because $\mathcal{A}$ is closed, $\mathcal{D}_\mathcal{A}$ itself is the projective limit of a family of Hilbert spaces. Using this fact, all assertions follow directly from standard results about locally convex spaces (SCHÄFER [1], IV, 5.8 and 5.5, II, 7.1). □

Proposition 2.3.3. *Let $\mathcal{A}$ be an O-family in a separable Hilbert space $\mathcal{H}$. If the graph topology of $\mathcal{A}$ is metrizable, then the locally convex space $\mathcal{D}_\mathcal{A}$ is separable.*

Proof. Because of Proposition 2.2.13, we can assume that the O-family $\mathcal{A}$ is directed. Then, since $\mathcal{D}_\mathcal{A}$ is metrizable, there exists a sequence $(a_n : n \in \mathbb{N})$ of operators in $\mathcal{A}$ such that $\|\cdot\| \leqq \|\cdot\|_{a_n} \leqq \|\cdot\|_{a_{n+1}}$ for $n \in \mathbb{N}$ and such that the graph topology of $\mathcal{A}$ is determined by the family of seminorms $\{\|\cdot\|_{a_n} : n \in \mathbb{N}\}$. Fix $n \in \mathbb{N}$. Then $|\overline{a_n}|$ is a self-adjoint operator in the separable Hilbert space $\mathcal{H}$. From the spectral theorem it follows easily that the Hilbert space $\big(\mathcal{D}(|\overline{a_n}|), \|\cdot\|_{|\overline{a_n}|}\big)$ is separable. Since $\mathcal{D}(\overline{a_n}) = \mathcal{D}(|\overline{a_n}|)$ and $\|\overline{a_n}\varphi\| = \||\overline{a_n}|\, \varphi\|$ for $\varphi \in \mathcal{D}(\overline{a_n})$, $\big(\mathcal{D}(\overline{a_n}), \|\cdot\|_{\overline{a_n}}\big)$ and so its dense linear subspace $\big(\mathcal{D}(\mathcal{A}), \|\cdot\|_{a_n}\big)$ are separable. The union of countable dense subsets of the spaces $\big(\mathcal{D}(\mathcal{A}), \|\cdot\|_{a_n}\big)$ is, of course, a countable dense subset of $\mathcal{D}_\mathcal{A}$. □

Remark 1. The O*-algebra $\mathcal{A}$ in Example 2.5.8 is closed and $\mathcal{D}_\mathcal{A}$ is not reflexive. There even exists a domain $\mathcal{D}$ in a separable Hilbert space for which the locally convex space $\mathcal{D}[t_+]$ is complete (i.e., $\mathcal{L}^+(\mathcal{D})$ is closed), but neither reflexive nor separable; cf. Example 2.5.7. The latter shows (for instance) that Proposition 2.3.3 is no longer valid if the assumption that $t_\mathcal{A}$ is metrizable is omitted.

Now we investigate the continuous linear functionals on the locally convex space $\mathcal{D}_\mathcal{A}$ and the dual $\mathcal{D}'_\mathcal{A}$ of $\mathcal{D}_\mathcal{A}$. More correctly, we shall prefer to work with the conjugate vector space $\mathcal{D}^+_\mathcal{A}$ of the dual $\mathcal{D}'_\mathcal{A}$ rather than the dual itself. This is due to the fact that, in contrast to the space $\mathcal{D}'_\mathcal{A}$, the canonical embedding of the Hilbert space $\mathcal{H}$ into $\mathcal{D}^+_\mathcal{A}$ is linear, and we can identify $\mathcal{H}$ with a linear subspace of $\mathcal{D}^+_\mathcal{A}$.

Before turning to the space $\mathcal{D}^+_\mathcal{A}$, we develop some general facts and notation needed later. Let a be a closable linear operator with domain $\mathcal{D}$ in a Hilbert space $\mathcal{H}$. Suppose that $\|\cdot\| \leqq \|\cdot\|_a$ on $\mathcal{D}$. Then $\mathcal{H}_a \equiv \big(\mathcal{D}(\bar{a}), \|\cdot\|_{\bar{a}}\big)$ is a Hilbert space with scalar product $\langle\cdot,\cdot\rangle_{\bar{a}} \equiv \langle \bar{a}\,\cdot, \bar{a}\,\cdot\rangle$. Let $\mathcal{H}^a$ be the conjugate space of the dual of the normed space $(\mathcal{D}, \|\cdot\|_a)$. We denote by $\|\varphi'\|^a$ the norm of a functional $\varphi' \in \mathcal{H}^a$. Then we have by definition $|\varphi'(\varphi)| \leqq \|\varphi'\|^a\, \|\varphi\|_a$ for all $\varphi' \in \mathcal{H}^a$ and all $\varphi \in \mathcal{D}$. Let $\mathcal{V}_a$ and $\mathcal{V}^0_a$ be the unit balls of the normed spaces $(\mathcal{D}, \|\cdot\|_a)$ and $(\mathcal{H}^a, \|\cdot\|^a)$, respectively. Since $\mathcal{D}$ is dense in $\mathcal{H}_a$, $\mathcal{H}^a$ is canonically isomorphic to the conjugate space of the dual of the Hilbert space $\mathcal{H}_a$. Therefore, by the Riesz representation theorem of continuous linear functionals on a Hilbert space, the mapping $\xi \to \langle\cdot, \xi\rangle_{\bar{a}}$ is an isometric isomorphism of the normed spaces $\mathcal{H}_a$ and $\mathcal{H}^a$. From this we see in particular that $(\mathcal{H}^a, \|\cdot\|^a)$ is a Hilbert space. Since $\|\cdot\| \leqq \|\cdot\|_a$ and $\mathcal{D}$ is dense in $\mathcal{H}$, $\psi \to \langle\cdot, \psi\rangle$ is an injective linear mapping of $\mathcal{H}$ into $\mathcal{H}^a$. For notational simplicity, we identify $\mathcal{H}$ with its image under this mapping. Retaining the above notation, we have

Lemma 2.3.4. (i) $\mathcal{V}^0_a = \{\langle\cdot, \xi\rangle_{\bar{a}} : \xi \in \mathcal{U}_{\mathcal{H}_a}\} = \{\langle a\,\cdot, \zeta\rangle : \zeta \in \mathcal{U}_\mathcal{H}\}$.
(ii) $\mathcal{D}^a = \{\langle\cdot, \xi\rangle_{\bar{a}} : \xi \in \mathcal{H}_a\} = \{\langle a\,\cdot, \zeta\rangle : \zeta \in \mathcal{H}\}$.
(iii) *$\mathcal{D}$ is dense in $(\mathcal{H}^a, \|\cdot\|^a)$.*

Proof. (i): From the isometric isomorphism of $\mathcal{H}_a$ and $\mathcal{H}^a$ mentioned above, we obtain the first equality. We check the second equality. It is obvious that $\langle a\,\cdot, \zeta\rangle \in \mathcal{V}^0_a$ if $\zeta \in \mathcal{U}_\mathcal{H}$. Conversely, let $\varphi'(\cdot) \equiv \langle\cdot, \xi\rangle_{\bar{a}} \in \mathcal{V}^0_a$ with $\xi \in \mathcal{U}_{\mathcal{H}_a}$. Putting $\zeta = a\xi$, we have $\varphi'(\cdot) \equiv \langle a\,\cdot, \zeta\rangle$ and $\zeta \in \mathcal{U}_\mathcal{H}$.

(ii) follows immediately from (i).

(iii): Suppose $\varphi^{\mathsf{l}} \in \mathcal{H}^a$. By (ii), there is a $\zeta \in \mathcal{H}$ such that $\varphi^{\mathsf{l}}(\cdot) \equiv \langle a\cdot, \zeta\rangle$. Let $\varepsilon > 0$ be given. Since the operator a is closable, the domain $\mathcal{D}(a^*)$ is dense in $\mathcal{H}$. Hence we can find a vector $\eta \in \mathcal{D}(a^*)$ such that $\|\zeta - \eta\| < \varepsilon$. Because $\mathcal{D}$ is dense in $\mathcal{H}$, there exists $\psi \in \mathcal{D}$ such that $\|a^*\eta - \psi\| < \varepsilon$. For $\varphi \in \mathcal{D}$, we have

$$\begin{aligned}|\varphi^{\mathsf{l}}(\varphi) - \langle\varphi, \psi\rangle| &= |\langle a\varphi, \zeta\rangle - \langle\varphi, \psi\rangle| \\ &= |\langle a\varphi, \zeta\rangle - \langle a\varphi, \eta\rangle + \langle\varphi, a^*\eta\rangle - \langle\varphi, \psi\rangle| \\ &\leqq \|\varphi\|_a \|\zeta - \eta\| + \|\varphi\| \|a^*\eta - \psi\| \leqq 2\varepsilon\|\varphi\|_a.\end{aligned}$$

Here we also used that $\|\cdot\| \leqq \|\cdot\|_a$. Hence $\|\varphi^{\mathsf{l}} - \psi\|^a \leqq 2\varepsilon$. This proves that $\mathcal{D}$ is dense in $\mathcal{H}^a$. □

Now let $\mathcal{A}$ be an O-family in a Hilbert space $\mathcal{H}$. From the above definitions it is clear that for each $a \in \mathcal{A}(I)$ $\mathcal{H}^a$ is a linear subspace of the vector space $\mathcal{D}_{\mathcal{A}}^+$ and $\mathcal{V}_a^0$ is the polar of $\mathcal{V}_a$ in the dual $\mathcal{D}_{\mathcal{A}}^{\mathsf{l}}$. As explained before Lemma 2.3.4, we always consider the Hilbert space $\mathcal{H}$ as a linear subspace of $\mathcal{D}^a$ for any $a \in \mathcal{A}(I)$ and hence of the vector space $\mathcal{D}_{\mathcal{A}}^+$ by identifying the vector $\psi \in \mathcal{H}$ with the functional $\langle\cdot, \psi\rangle$ on $\mathcal{D}(\mathcal{A})$. In other words, if a functional $\varphi^{\mathsf{l}} \in \mathcal{D}_{\mathcal{A}}^+$ belongs to $\mathcal{H}$, then for each $\varphi \in \mathcal{D}(\mathcal{A})$ the value $\varphi^{\mathsf{l}}(\varphi)$ is simply the scalar product $\langle\varphi, \varphi^{\mathsf{l}}\rangle$ and $\overline{\varphi^{\mathsf{l}}(\varphi)}$ equals $\langle\varphi^{\mathsf{l}}, \varphi\rangle$. This suggests the following notational convention which extends these equalities by definition to general functionals in $\mathcal{D}_{\mathcal{A}}^+$. We define

$$\langle\varphi, \varphi^{\mathsf{l}}\rangle := \varphi^{\mathsf{l}}(\varphi) \quad\text{and}\quad \langle\varphi^{\mathsf{l}}, \varphi\rangle := \overline{\varphi^{\mathsf{l}}(\varphi)} \quad\text{for}\quad \varphi \in \mathcal{D}(\mathcal{A}) \quad\text{and}\quad \varphi^{\mathsf{l}} \in \mathcal{D}_{\mathcal{A}}^+. \tag{1}$$

This notation, which strongly resembles the scalar product notation, will be frequently used throughout the next four chapters. Its advantages will be seen later (see Remark 5 in 3.2). Some basic properties of the space $\mathcal{D}_{\mathcal{A}}^+$ are collected in

Proposition 2.3.5. *Suppose $\mathcal{A}$ is a directed O-family in the Hilbert space $\mathcal{H}$.*

(i) *The vector space $\mathcal{D}_{\mathcal{A}}^+$ is the union of the directed family $\{\mathcal{H}^a : a \in \mathcal{A}(I)\}$ of vector subspaces.*

(ii) $\mathcal{D}_{\mathcal{A}}^+ = \{\langle\cdot, \xi\rangle_{\bar{a}} : a \in \mathcal{A}(I)$ *and* $\xi \in \mathcal{H}_a\}$
$= \{\langle a\cdot, \zeta\rangle : a \in \mathcal{A}(I)$ *and* $\zeta \in \mathcal{H}\}$.

(iii) *$\mathcal{D}(\mathcal{A})$ is dense in $\mathcal{D}_{\mathcal{A}}^+[\beta]$. More precisely, for each $\varphi^{\mathsf{l}} \in \mathcal{D}_{\mathcal{A}}^+$ there is a sequence of vectors in $\mathcal{D}(\mathcal{A})$ which converges to φ^{l} in $\mathcal{D}_{\mathcal{A}}^+[\beta]$.*

Proof. (i): As already noted, $\mathcal{H}^a$ is a linear subspace of $\mathcal{D}_{\mathcal{A}}^+$ for $a \in \mathcal{A}(I)$. Obviously, $\mathcal{H}^a \subseteqq \mathcal{H}^b$ if $a, b \in \mathcal{A}(I)$ and $a \lesssim b$. Hence the family $\{\mathcal{H}^a : a \in \mathcal{A}(I)\}$ is directed. Since $\mathcal{A}$ is directed, $\{\|\cdot\|_a : a \in \mathcal{A}(I)\}$ is a directed family of seminorms generating $\mathsf{t}_{\mathcal{A}}$; hence each $\varphi^{\mathsf{l}} \in \mathcal{D}_{\mathcal{A}}^+$ is contained in $\mathcal{H}^a$ for some $a \in \mathcal{A}(I)$.

(ii): The set $\mathcal{D}_{\mathcal{A}}^+$ is the union of all sets $\mathcal{H}^a$, $a \in \mathcal{A}(I)$, by (i). Thus both equalities follow directly from Lemma 2.3.4, (ii).

(iii): Suppose $\varphi^{\mathsf{l}} \in \mathcal{D}_{\mathcal{A}}^+$. By (i), $\varphi^{\mathsf{l}} \in \mathcal{H}^a$ for some $a \in \mathcal{A}(I)$. From Lemma 2.3.4, (iii), there is a sequence $(\psi_n : n \in \mathbb{N})$ of vectors in $\mathcal{D}(\mathcal{A})$ such that $\varphi^{\mathsf{l}} = \lim \psi_n$ in $\mathcal{H}^a$. Let $\mathcal{M}$ be a bounded subset of $\mathcal{D}_{\mathcal{A}}$. Then $\lambda := \sup\{\|\varphi\|_a : \varphi \in \mathcal{M}\} < \infty$. From

$$r_{\mathcal{M}}(\varphi^{\mathsf{l}} - \psi_n) = \sup_{\varphi \in \mathcal{M}} |\langle\varphi, \varphi^{\mathsf{l}} - \psi_n\rangle| \leqq \sup_{\varphi \in \mathcal{M}} \|\varphi\|_a \|\varphi^{\mathsf{l}} - \psi_n\|^a \leqq \lambda\|\varphi^{\mathsf{l}} - \psi_n\|^a$$

we see that $\varphi^{\mathsf{l}} = \lim_n \psi_n$ in $\mathcal{D}_{\mathcal{A}}^+[\beta]$. □

Remark 2. Let $\mathcal{A}$ be an arbitrary O-family in $\mathcal{H}$. As just discussed, $\mathcal{H}$ is a linear subspace of $\mathcal{D}_\mathcal{A}^+$. Thus we have the following chain of locally convex spaces

$$\mathcal{D}_\mathcal{A} \subsetneqq \mathcal{H} \subseteqq \mathcal{D}_\mathcal{A}^+[\beta],$$

where the two embedding maps are continuous and each space is a dense linear subspace of its successor. (Indeed, the continuity of the embeddings is an immediate consequence of the fact that the graph topology $t_\mathcal{A}$ is finer than the topology determined by the norm of $\mathcal{H}$. To prove the density of $\mathcal{H}$ in $\mathcal{D}_\mathcal{A}^+[\beta]$, there is no loss of generality by Proposition 2.2.13 to assume that $\mathcal{A}$ is a directed O-family. But then it follows from Proposition 2.3.5, (iii).). Therefore, the triplet $\{\mathcal{D}_\mathcal{A}, \mathcal{H}, \mathcal{D}'_\mathcal{A}[\beta]\}$ is what is commonly called a *Gelfand triplet* or a *rigged Hilbert space*.

Next we use some properties of the dual space of $\mathcal{D}_\mathcal{A}$ in order to give another characterization of the domain $\hat{\mathcal{D}}(\mathcal{A})$ for O-families $\mathcal{A}$ with metrizable graph topologies. It will be derived from the following proposition.

Proposition 2.3.6. *Let $\mathcal{A}$ be an O-family. Suppose $(\varphi_n : m \in \mathbb{N})$ is a bounded sequence in the locally convex space $\mathcal{D}_\mathcal{A}$ and φ is a vector in $\mathcal{H}$. If $\lim_n \langle \varphi_n, \psi\rangle = \langle \varphi, \psi\rangle$ for all $\psi \in \mathcal{D}(\mathcal{A})$, then $\varphi \in \hat{\mathcal{D}}(\mathcal{A})$ and φ is the limit of the sequence $(\varphi_n : n \in \mathbb{N})$ in the weak topology of the locally convex space $\mathcal{D}_{\hat{\mathcal{A}}}$.*

Proof. By Proposition 2.2.13 and Remark 4 in 2.2 we can assume without loss of generality that the O-family $\mathcal{A}$ is directed. Suppose $a \in \mathcal{A}(I)$. Since $\{\varphi_n : n \in \mathbb{N}\}$ is bounded in $\mathcal{D}_\mathcal{A}$, $\lambda_a := \sup\{\|a\varphi_n\| : n \in \mathbb{N}\} < \infty$. Let $\varphi' \in \mathcal{H}^a$ and let $\varepsilon > 0$. Since $\mathcal{D}(\mathcal{A})$ is dense in $\mathcal{H}^a$ by Lemma 2.3.4, (iii), there is a $\psi \in \mathcal{D}(\mathcal{A})$ such that $\|\varphi' - \psi\|^a \leqq \varepsilon$. Then we have $|\varphi'(\varphi_n - \varphi_m) - \langle \varphi_n - \varphi_m, \psi\rangle| \leqq \|\varphi_n - \varphi_m\|_a \|\varphi' - \psi\|^a \leqq 2\lambda_a\varepsilon$ for $n, m \in \mathbb{N}$. Since the sequence $(\langle \varphi_n, \psi\rangle : n \in \mathbb{N})$ converges, we conclude from the preceding that $(\varphi_n : n \in \mathbb{N})$ is a weak Cauchy sequence in the Hilbert space $\mathcal{H}_a$. Let $\varphi_a \in \mathcal{H}_a$ be its limit in the weak topology of $\mathcal{H}_a$. Since $\lim_n \langle \psi, \varphi_n\rangle = \langle \psi, \varphi\rangle = \langle \psi, \varphi_a\rangle$ for $\psi \in \mathcal{D}(\mathcal{A})$, we obtain $\varphi = \varphi_a$, so that $\varphi \in \mathcal{D}(\bar{a})$. Thus $\varphi \in \bigcap_{a \in \mathcal{A}(I)} \mathcal{D}(\bar{a}) = \hat{\mathcal{D}}(\mathcal{A})$ by Proposition 2.2.12, since $\mathcal{A}$ is directed. By the definition of $\varphi_a \equiv \varphi$, we have $\varphi'(\varphi) = \lim_n \varphi'(\varphi_n)$ for all $\varphi' \in \mathcal{H}^a$ and $a \in \mathcal{A}(I)$. Since $\mathcal{D}'_{\hat{\mathcal{A}}}$ is the union of all $\mathcal{H}^{\bar{a}} \equiv \mathcal{H}^a$, $a \in \mathcal{A}(I)$, by Proposition 2.3.5, (i), this means that $\varphi = \lim_n \varphi_n$ in the topology $\sigma(\mathcal{D}_{\hat{\mathcal{A}}}, \mathcal{D}'_{\hat{\mathcal{A}}})$. □

Corollary 2.3.7. *Suppose that $\mathcal{A}$ is an O-family on $\mathcal{H}$ with metrizable graph topology. For each vector φ in $\mathcal{H}$, the following three statements are equivalent:*

(i) $\varphi \in \hat{\mathcal{D}}(\mathcal{A})$.

(ii) *There is a bounded sequence $(\varphi_n : n \in \mathbb{N})$ in $\mathcal{D}_\mathcal{A}$ which converges weakly in the Hilbert space $\mathcal{H}$ to φ.*

(iii) *There is a bounded sequence $(\varphi_n : n \in \mathbb{N})$ in $\mathcal{D}_\mathcal{A}$ such that $\lim_n \langle \varphi_n, \psi\rangle = \langle \varphi, \psi\rangle$ for all $\psi \in \mathcal{D}(\mathcal{A})$.*

Proof. We verify (i) → (ii). We let $\varphi \in \hat{\mathcal{D}}(\mathcal{A})$. Since $t_\mathcal{A}$ is metrizable and hence is $t_{\hat{\mathcal{A}}}$, there exists a sequence (!) of vectors in $\mathcal{D}(\mathcal{A})$ which converges to φ in $\mathcal{D}_{\hat{\mathcal{A}}}$. Clearly, this sequence has the properties stated in (ii). (ii) → (iii) is trivial, and (iii) → (i) has been shown in Proposition 2.3.6. □

Remark 3. A by-product of the preceding results is the following fact. Let $\mathcal{A}$ be an O-family, $(\varphi_n : n \in \mathbb{N})$ a sequence of vectors in $\mathcal{D}(\mathcal{A})$ and φ a vector in $\mathcal{D}(\mathcal{A})$. Then we have $\varphi = \lim_n \varphi_n$ in the weak topology of $\mathcal{D}_{\mathcal{A}}$ if and only if the sequence (φ_n) is bounded in $\mathcal{D}_{\mathcal{A}}$ and $\lim_n \langle \varphi_n, \psi \rangle = \langle \varphi, \psi \rangle$ for all $\psi \in \mathcal{D}(\mathcal{A})$. The sufficiency of the latter condition follows from Proposition 2.3.6. We verify its necessity. Let $a \in \mathcal{A}$. Since (φ_n) converges weakly to φ in $\mathcal{D}_{\mathcal{A}}$, we have $\lim_n \langle a\varphi_n, \eta \rangle = \langle a\varphi, \eta \rangle$ for all $\eta \in \mathcal{H}$. This implies that $\{a\varphi_n : n \in \mathbb{N}\}$ is bounded in $\mathcal{H}$; so $\{\varphi_n : n \in \mathbb{N}\}$ is bounded in $\mathcal{D}_{\mathcal{A}}$. The second condition is obvious.

Now we compare the graph topologies and the corresponding bounded sets of different O-families acting on the same domain.

Lemma 2.3.8. *Let $\mathcal{A}$ be an O-family in $\mathcal{H}$, and let b be a closable linear operator on the domain $\mathcal{D}(\mathcal{A})$. Then the set $\mathcal{V}_b := \{\varphi \in \mathcal{D}(\mathcal{A}) : \|\varphi\|_b \leqq 1\}$ is a barrel in the locally convex space $\mathcal{D}_{\mathcal{A}}$.*

Proof. We abbreviate $\mathcal{U} := \{\psi \in \mathcal{D}(b^*) : \|\psi\| \leqq 1\}$. Since b is closable, $\mathcal{D}(b^*)$ is dense in $\mathcal{H}$. Therefore,

$$\mathcal{V}_b = \bigcap_{\psi \in \mathcal{U}} \{\varphi \in \mathcal{D}(\mathcal{A}) : |\langle b\varphi, \psi \rangle| \leqq 1\} = \bigcap_{\psi \in \mathcal{U}} \{\varphi \in \mathcal{D}(\mathcal{A}) : |\langle \varphi, b^*\psi \rangle| \leqq 1\}.$$

It is clear that for each $\psi \in \mathcal{U}$ the set $\{\varphi \in \mathcal{D}(\mathcal{A}) : |\langle \varphi, b^*\psi \rangle| \leqq 1\}$ is closed in $\mathcal{D}_{\mathcal{A}}$. Hence $\mathcal{V}_b$ is closed in $\mathcal{D}_{\mathcal{A}}$. Since $\mathcal{V}_b$ is obviously absolutely convex and absorbing, this proves that $\mathcal{V}_b$ is a barrel in $\mathcal{D}_{\mathcal{A}}$. □

Proposition 2.3.9. *Suppose that $\mathcal{A}$ is an O-family which satisfies at least one of the following three conditions:*

(i) *$\mathcal{D}_{\mathcal{A}}$ is barrelled.*

(ii) *$\mathcal{A}$ is closed and $\mathcal{D}_{\mathcal{A}}$ is bornological.*

(iii) *$\mathcal{A}$ is closed and $\mathcal{D}_{\mathcal{A}}$ is reflexive.*

Then $t_{\mathcal{A}} = t_c$ on $\mathcal{D}(\mathcal{A})$. If $\mathcal{B}$ is another O-family on the domain $\mathcal{D}(\mathcal{A})$, then $t_{\mathcal{B}} \subseteqq t_{\mathcal{A}}$. If, in addition, $\mathcal{A}$ is an O-algebra, then $t_{\mathcal{A}} = t_+$ on $\mathcal{D}(\mathcal{A})$.*

Proof. Since each complete bornological space is barrelled (Schäfer [1], II, 8.4), (ii) implies (i). From Corollary 2.3.2, (iii) implies (i). Thus it suffices to prove the assertions in case where (i) is fulfilled. Let $\mathcal{B}$ be an O-family on $\mathcal{D}(\mathcal{B}) = \mathcal{D}(\mathcal{A})$. In order to prove that $t_{\mathcal{B}} \subseteqq t_{\mathcal{A}}$, we can assume without loss of generality by Proposition 2.2.13 that $\mathcal{B}$ is a directed O-vector space. Suppose $b \in \mathcal{B}$. From Lemma 2.3.8, $\mathcal{V}_b$ is a barrel in $\mathcal{D}_{\mathcal{A}}$. Because $\mathcal{D}_{\mathcal{A}}$ is barrelled by assumption (i), $\mathcal{V}_b$ is a 0-neighbourhood in $\mathcal{D}_{\mathcal{A}}$. Since $\mathcal{B}$ is a directed O-vector space, the collection of sets $\mathcal{V}_b$, where $b \in \mathcal{B}$, forms a 0-neighbourhood base in $\mathcal{D}_{\mathcal{B}}$. Thus we have shown that $t_{\mathcal{B}} \subseteqq t_{\mathcal{A}}$ on $\mathcal{D}(\mathcal{A})$.

Next we prove that $t_{\mathcal{A}} = t_c$. Applying the preceding in case $\mathcal{B} := \mathcal{C}(\mathcal{D}(\mathcal{A}), \mathcal{H})$ we obtain $t_c \subseteqq t_{\mathcal{A}}$. Since $\mathcal{C}(\mathcal{D}(\mathcal{A}), \mathcal{H})$ is the largest O-family on $\mathcal{D}(\mathcal{A})$, we trivially have that $t_{\mathcal{A}} \subseteqq t_c$. Thus $t_{\mathcal{A}} = t_c$. If $\mathcal{A}$ is an O*-algebra, then a similar reasoning proves that $t_{\mathcal{A}} = t_+$. □

Remark 4. If $\mathcal{A}$ is an O-family [resp. O*-algebra] such that $\mathcal{D}_{\mathcal{A}}$ is a Frechet space then all three conditions in Proposition 2.3.9 are fulfilled and hence $t_{\mathcal{A}} = t_c$ [resp. $t_{\mathcal{A}} = t_+$] on $\mathcal{D}(\mathcal{A})$. This also follows at once from the closed graph theorem.

If $\mathcal{A}$ and $\mathcal{B}$ are closed O-families on the same domain, then the graph topologies $t_{\mathcal{A}}$ and $t_{\mathcal{B}}$ may be different; see Example 2.5.8. Nevertheless, the locally convex spaces $\mathcal{D}_{\mathcal{A}}$ and $\mathcal{D}_{\mathcal{B}}$ have the same bounded sets as we show now. This fact is important for the topologization of O-vector spaces (see Proposition 3.3.1).

Proposition 2.3.10. *Let $\mathcal{A}$ and $\mathcal{B}$ be O-families on the domain $\mathcal{D}(\mathcal{A}) = \mathcal{D}(\mathcal{B})$. If the O-family $\mathcal{A}$ is closed, then each bounded set in $\mathcal{D}_{\mathcal{A}}$ is bounded in $\mathcal{D}_{\mathcal{B}}$ as well.*

Proof. By Proposition 2.2.13, there is no loss of generality to assume that $\mathcal{B}$ is a directed O-vector space. Let $\mathcal{M}$ be a bounded subset of $\mathcal{D}_{\mathcal{A}}$. Our aim is to prove that $\mathcal{M}$ is bounded in $\mathcal{D}_{\mathcal{B}}$. Therefore, we can assume without loss of generality that $\mathcal{M}$ is absolutely convex and closed in $\mathcal{D}_{\mathcal{A}}$. (Otherwise we replace $\mathcal{M}$ by the closure of its absolutely convex hull in $\mathcal{D}_{\mathcal{A}}$; this set is also bounded in $\mathcal{D}_{\mathcal{A}}$.) Since $\mathcal{A}$ is closed, $\mathcal{D}_{\mathcal{A}}$ is complete. Hence $\mathcal{M}$ is complete in the induced topology of $\mathcal{D}_{\mathcal{A}}$. By the Banach-Mackey theorem (see e.g. SCHÄFER [1], II, 8.5), each barrel in a locally convex space absorbs the absolutely convex complete bounded subsets of the space. For each $b \in \mathcal{B}$, $\mathcal{U}_b$ is a barrel in the locally convex space $\mathcal{D}_{\mathcal{A}}$ by Lemma 2.3.8. Therefore, $\mathcal{U}_b$ absorbs $\mathcal{M}$. Since the family of sets $\{\mathcal{U}_b : b \in \mathcal{B}\}$ is a 0-neighbourhood base in $\mathcal{D}_{\mathcal{B}}$ (recall that $\mathcal{B}$ is assumed to be a directed O-vector space), this means that the set $\mathcal{M}$ is bounded in $\mathcal{D}_{\mathcal{B}}$. □

An immediate consequence of Proposition 2.3.10 is

Corollary 2.3.11. *If $\mathcal{A}$ and $\mathcal{B}$ are closed O-families on the same domain $\mathcal{D}(\mathcal{A}) = \mathcal{D}(\mathcal{B})$, then the locally convex spaces $\mathcal{D}_{\mathcal{A}}$ and $\mathcal{D}_{\mathcal{B}}$ have the same families of bounded sets.*

Proposition 2.3.12. *Suppose $\mathcal{A}$ is an O-family. If the locally convex space $\mathcal{D}_{\mathcal{A}}$ is a QF-space, then it is semireflexive.*

Proof. First of all, note that a locally convex Hausdorff space F is semireflexive if and only if each bounded subset of F is relatively $\sigma(F, F')$-compact (SCHÄFER [1], IV, 5.5). Suppose that $\mathcal{M}$ is a bounded subset of $\mathcal{D}_{\mathcal{A}}$. Because $\mathcal{D}_{\mathcal{A}}$ is a QF-space, there is a Frechet linear subspace $\mathcal{E}$ of $\mathcal{D}_{\mathcal{A}}$ which contains $\mathcal{M}$. Obviously, the graph topology $t_{\mathcal{A}}$ is also generated by the directed family of Hilbert norms $(\|a_1 \cdot\|^2 + \cdots + \|a_n \cdot\|^2)^{1/2}$, where $a_1, \ldots, a_n \in \mathcal{A}$, $a_1 = I$ and $n \in \mathbb{N}$. This implies that $\mathcal{E}$ is the projective limit of Hilbert spaces and hence semireflexive (SCHÄFER [1], IV, 5.8). Therefore, $\mathcal{M}$ is relatively $\sigma(\mathcal{E}, \mathcal{E}')$-compact. Since the topologies $\sigma(\mathcal{E}, \mathcal{E}')$ and $\sigma(\mathcal{D}_{\mathcal{A}}, \mathcal{D}_{\mathcal{A}}')$ coincide on $\mathcal{E}$ (by the Hahn-Banach theorem), $\mathcal{M}$ is relatively $\sigma(\mathcal{D}_{\mathcal{A}}, \mathcal{D}_{\mathcal{A}}')$-compact. This proves that $\mathcal{D}_{\mathcal{A}}$ is semireflexive. □

Before stating the next proposition, we prove an auxiliary lemma.

Lemma 2.3.13. *Suppose b is a closable linear operator in the Hilbert space $\mathcal{H}$ satisfying $\|b\varphi\| \geqq \|\varphi\|$, $\varphi \in \mathcal{D}(b)$. Then the embedding map of the Hilbert space $\mathcal{H}_b \equiv (\mathcal{D}(\bar{b}), \|\cdot\|_{\bar{b}})$ into $\mathcal{H}$ is compact if and only if there exists a compact operator c on $\mathcal{H}$ such that* $\ker c = \{0\}$ *and $b \subseteqq c^{-1}$.*

Proof. The if part is obvious. We verify the only if part. Suppose that the embedding of H_b into $\mathcal{H}$ is compact. We can assume that $\bar{b}$, and so $|\bar{b}|$, is an unbounded operator, since otherwise $\mathcal{H}$ is finite dimensional and then the assertion is trivial. By means of the spectral theorem, we write the unbounded self-adjoint operator $|\bar{b}|$ as a direct sum of unbounded self-adjoint operators b_n, $n \in \mathbb{N}$, in mutually orthogonal closed subspaces $\mathcal{H}_n$, $n \in \mathbb{N}$, of $\mathcal{H}$. Then $\mathcal{D}(b_n) \neq \mathcal{H}_n$, so that we can take a vector $\varphi_n \in \mathcal{H}_n$ such that

$\varphi_n \notin \mathcal{D}(b_n)$ for $n \in \mathbb{N}$. Let $\mathcal{G}_1 := \text{c.l.h.}\ \{\varphi_n : n \in \mathbb{N}\}$. It is easy to check that $\mathcal{G}_1 \cap \mathcal{D}(\bar{b}) \equiv \mathcal{G}_1 \cap \mathcal{D}(|\bar{b}|) = \{0\}$. On the other hand, the Hilbert space $\mathcal{H}$ is separable, since its dense subset $\mathcal{D}(\bar{b})$ is the range of the compact embedding map of $\mathcal{H}_b$ into $\mathcal{H}$. Hence $\mathcal{G}_0 := \left(\bar{b}\mathcal{D}(\bar{b})\right)^\perp$ is separable. Thus there exists a compact operator c_0 of the Hilbert space $\mathcal{G}_0$ into $\mathcal{G}_1$ with trivial kernel. Define $c(\bar{b}\varphi + \psi) := \varphi + c_0\psi$ for $\varphi \in \mathcal{D}(\bar{b})$ and $\psi \in \mathcal{G}_0$. Note that $\bar{b}\mathcal{D}(\bar{b})$ is a closed subspace of $\mathcal{H}$, since $\|b\varphi\| \geqq \|\varphi\|$, $\varphi \in \mathcal{D}(b)$. Using the compactness of the embedding map, we conclude easily that c is a compact operator on $\mathcal{H}$. From $\mathcal{G}_1 \cap \mathcal{D}(\bar{b}) = \{0\}$ and $\ker c_0 = \{0\}$ we obtain that $\ker c = \{0\}$. By construction, $b \subseteqq c^{-1}$. $\square$

Proposition 2.3.14. *Suppose that $\mathcal{A}$ is a directed O-family. Consider the following assertions:*

(i) *$\mathcal{D}_\mathcal{A}$ is a Schwartz space.*

(ii) *For each $a \in \mathcal{A}(I)$ there exists $b \in \mathcal{A}(I)$ satisfying $a \lesssim b$ such that the embedding map of the Hilbert space $\mathcal{H}_b$ into the Hilbert space $\mathcal{H}_a$ is compact.*

(iii) *There exist an operator $b \in \mathcal{A}$ and a compact operator c on $\mathcal{H}$ such that* $\ker c = \{0\}$ *and $b \subseteqq c^{-1}$.*

Then (i) $\leftrightarrow$ (ii) $\rightarrow$ (iii). *If in addition $\mathcal{A}$ is an O-algebra (in particular if $\mathcal{A}$ is an O*-algebra), then all three statements are equivalent.*

Proof. Since $\mathcal{A}$ is a directed O-family, $\{\|\cdot\|_a : a \in \mathcal{A}(I)\}$ is a directed family of norms which generates the topology $t_\mathcal{A}$. We apply the definition of a Schwartz space (cf. p. 15) to this family. For $a, b \in \mathcal{A}(I)$, it is clear that $\mathcal{V}_b := \{\varphi \in \mathcal{D}(\mathcal{A}) : \|\varphi\|_b \leqq 1\}$ is precompact in $\left(\mathcal{D}(\mathcal{A}), \|\cdot\|_a\right)$ if and only if the embedding of $\mathcal{H}_b$ into $\mathcal{H}_a$ is compact. By the definition of a Schwartz space this gives the equivalence of (i) and (ii).

(ii) $\rightarrow$ (iii): Set $a = I$ in (ii) and apply Lemma 2.3.13 to the corresponding operator b.

Suppose now that $\mathcal{A}$ is an O-algebra and also that $\mathcal{A}$ is a directed O-family. It remains to prove (iii) $\rightarrow$ (i). Let $a \in \mathcal{A}(I)$, and let b and c be as in (iii). There is no loss of generality to assume that $\|c\| \leqq 1$. Then $b \in \mathcal{A}(I)$ and so $ba \in \mathcal{A}(I)$. Since $ba\mathcal{V}_{ba} \subseteqq \mathcal{U}_\mathcal{H}$ and c is a compact operator, $c(ba\mathcal{V}_{ba}) = a\mathcal{V}_{ba}$ is relatively compact in $\mathcal{H}$. Since $a \in \mathcal{A}(I)$, $\mathcal{V}_{ba}$ is relatively compact in $\mathcal{H}_a$ and hence precompact in $\left(\mathcal{D}(\mathcal{A}), \|\cdot\|_a\right)$. This shows that $\mathcal{D}_\mathcal{A}$ is a Schwartz space. $\square$

Proposition 2.3.15. *If the graph topology $t_\mathcal{A}$ of an O*-algebra $\mathcal{A}$ is normable, then every operator in $\mathcal{A}$ is bounded.*

Proof. Suppose $t_\mathcal{A}$ is generated by a norm $\|\cdot\|_1$ on $\mathcal{D}(\mathcal{A})$. Then, by Lemma 2.2.2, each operator in $\mathcal{A}$ is continuous in the normed space $\left(\mathcal{D}(\mathcal{A}), \|\cdot\|_1\right)$. Therefore, by Proposition 2.1.11, every symmetric operator and hence every operator in $\mathcal{A}$ is bounded. $\square$

2.4. Bounded Sets in Quasi-Frechet Domains

For a linear subspace $\mathcal{D}$ of a Hilbert space $\mathcal{H}$, let $\mathbb{B}(\mathcal{D})_+$ denote the set of all positive self-adjoint operators in $\mathbb{B}(\mathcal{H})$ which map $\mathcal{H}$ into $\mathcal{D}$. (In Section 3.1 this set is studied in detail.) Recall that a QF-space is a locally convex space in which every bounded set is contained in some Frechet subspace.

The following theorem is the central result in this section. It will be seen later (see e.g. Sections 3.4 and 5.4) that it is a powerful tool in studying topological questions.

Theorem 2.4.1. *Let $\mathcal{A}$ be an O-family in the Hilbert space $\mathcal{H}$. Suppose that $\mathcal{D}_{\mathcal{A}}$ is a QF-space. Then for each bounded subset $\mathcal{M}$ of $\mathcal{D}_{\mathcal{A}}$ there exists an operator $c \in \mathbb{B}(\mathcal{D}(\mathcal{A}))_+$ such that $\mathcal{M} \subseteq c\mathcal{U}_{\mathcal{H}}$. If $\mathcal{D}_{\mathcal{A}}$ is a Frechet space and $\mathcal{H}$ is separable, then c can be chosen such that in addition* $\ker c = \{0\}$.

The crucial step in the proof of the theorem is contained in the following lemma. It will be used in Section 5.4 as well.

Lemma 2.4.2. *Suppose $\mathcal{A}$ is an O-family in the Hilbert space $\mathcal{H}$ and $\mathcal{E}$ is a Frechet subspace of the locally convex space $\mathcal{D}_{\mathcal{A}}$. Let $(a_n : n \in \mathbb{N})$ be a sequence of operators in $\mathcal{A}$ with $a_1 = I$ such that the induced topology on $\mathcal{E}$ of the graph topology $t_{\mathcal{A}}$ is generated by the family of seminorms $\{\|\cdot\|_{a_n} : n \in \mathbb{N}\}$. Let $\delta = (\delta_n : n \in \mathbb{N})$ be a sequence of positive numbers. Define $\mathcal{D}_\delta := \left\{\varphi \in \mathcal{E} : \mathfrak{h}_\delta(\varphi) := \sum_{n=1}^{\infty} \delta_n \|a_n\varphi\|^2 < \infty\right\}$. Let $\mathcal{H}_\delta$ be the closure of $\mathcal{D}_\delta$ in $\mathcal{H}$. Then there exists an operator c on $\mathcal{H}$ such that the following is satisfied.*

(i) $c \in \mathbb{B}(\mathcal{D}_\delta)_+$, $c\mathcal{H} = \mathcal{D}_\delta$ *and* $\ker c = (\mathcal{H}_\delta)^\perp$.

(ii) $\sum_{n=1}^{\infty} \delta_n \|a_n c\psi\|^2 = \|\psi\|^2$ *for all* $\psi \in \mathcal{H}_\delta$.

(iii) *Suppose in addition that $a_n \in \mathcal{L}^+(\mathcal{D}(\mathcal{A}))$ for $n \in \mathbb{N}$. If $\varphi \in \mathcal{D}_\delta$ and the series $\sum_{n=1}^{\infty} \delta_n a_n^+ a_n \varphi$ converges in $\mathcal{H}$, then $c^2 \sum_{n=1}^{\infty} \delta_n a_n^+ a_n \varphi = \varphi$.*

Proof. Since $\|a_n(\varphi + \psi)\|^2 \leqq 2(\|a_n\varphi\|^2 + \|a_n\psi\|^2)$ for $\varphi, \psi \in \mathcal{E}$ and $n \in \mathbb{N}$, $\mathcal{D}_\delta$ is a vector space. Define $\mathfrak{h}(\varphi, \psi) := \sum_{n=1}^{\infty} \delta_n \langle a_n\varphi, a_n\psi\rangle$ for $\varphi, \psi \in \mathcal{D}_\delta$. From the inequality $|\langle a_n\varphi, a_n\psi\rangle| \leqq \|a_n\varphi\|^2 + \|a_n\psi\|^2$ we see that $\mathfrak{h}(\varphi, \psi)$ is finite for all $\varphi, \psi \in \mathcal{D}_\delta$. Therefore, $\mathfrak{h}$ is a positive sesquilinear form with domain $\mathcal{D}_\delta$ in the Hilbert space $\mathcal{H}_\delta$. We prove that this form is closed. Since $\mathfrak{h}(\varphi, \varphi) \geqq \delta_1 \|a_1\varphi\|^2 \equiv \delta_1 \|\varphi\|^2$ for $\varphi \in \mathcal{D}_\delta$, we have to show that the domain $\mathcal{D}_\delta$ is complete in the norm $\|\cdot\|_{\mathfrak{h}} := \mathfrak{h}(\cdot, \cdot)^{1/2}$. Let $(\varphi_n : n \in \mathbb{N})$ be a Cauchy sequence in the normed space $(\mathcal{D}_\delta, \|\cdot\|_{\mathfrak{h}})$. From $\|\varphi_k - \varphi_l\|_{a_n} \leqq \delta_n^{-1/2} \|\varphi_k - \varphi_l\|_{\mathfrak{h}}$ for $k, l, n \in \mathbb{N}$ and from the assumptions concerning (a_n) and $\mathcal{E}$ it follows that $(\varphi_n : n \in \mathbb{N})$ is a Cauchy sequence in the Frechet space $\mathcal{E}$. (Recall that $\mathcal{E}$ carries the induced topology of $\mathcal{D}_{\mathcal{A}}$.) Hence there is a vector $\varphi \in \mathcal{E}$ such that $\varphi = \lim \varphi_n$ in $\mathcal{E}$. We check that $\varphi \in \mathcal{D}_\delta$ and $\varphi = \lim \varphi_n$ in $(\mathcal{D}_\delta, \|\cdot\|_{\mathfrak{h}})$. Let $\varepsilon > 0$ be given. Since (φ_n) is a Cauchy sequence in $(\mathcal{D}_\delta, \|\cdot\|_{\mathfrak{h}})$, there is a $n(\varepsilon) \in \mathbb{N}$ such that $\|\varphi_k - \varphi_l\|_{\mathfrak{h}}^2 = \mathfrak{h}(\varphi_k - \varphi_l, \varphi_k - \varphi_l) = \sum_{n=1}^{\infty} \delta_n \|a_n(\varphi_k - \varphi_l)\|^2 < \varepsilon$ if $k \geqq n(\varepsilon)$ and $l \geqq n(\varepsilon)$. Letting $l \to \infty$, we get $\sum_{n=1}^{\infty} \delta_n \|a_n(\varphi_k - \varphi)\|^2 \leqq \varepsilon$ if $k \geqq n(\varepsilon)$. This gives $\varphi_k - \varphi \in \mathcal{D}_\delta$ for $k \geqq n(\varepsilon)$ and hence $\varphi \in \mathcal{D}_\delta$. Because $\|\varphi_k - \varphi\|_{\mathfrak{h}}^2 \equiv \sum_{n=1}^{\infty} \delta_n \|a_n(\varphi_k - \varphi)\|^2 \leqq \varepsilon$ for $k \geqq n(\varepsilon)$, the preceding shows that $\varphi = \lim \varphi_n$ in $(\mathcal{D}_\delta, \|\cdot\|_{\mathfrak{h}})$. Thus $\mathfrak{h}$ is closed. From the representation theorem of closed positive sesquilinear forms (in the formulation given in Kato [1], VI, § 2, Theorem 2.23), there is a positive self-adjoint operator T on the Hilbert space $\mathcal{H}_\delta$ such that $\mathcal{D}(T^{1/2}) = \mathcal{D}(\mathfrak{h}) \equiv \mathcal{D}_\delta$ and $\mathfrak{h}(\varphi, \psi) = \langle T^{1/2}\varphi, T^{1/2}\psi\rangle$ for all $\varphi, \psi \in \mathcal{D}_\delta$. Since $\mathfrak{h}(\varphi, \varphi) \geqq \delta_1 \|\varphi\|^2$ for $\varphi \in \mathcal{D}_\delta$, $T^{1/2}$ has a bounded inverse on the Hilbert space $\mathcal{H}_\delta$. We define $c := (T^{1/2})^{-1} \oplus 0$ relative to the decomposition $\mathcal{H} = \mathcal{H}_\delta \oplus (\mathcal{H}_\delta)^\perp$. Then c is a bounded operator on $\mathcal{H}$ which obviously satisfies (i).
To prove (ii), fix $\psi \in \mathcal{H}_\delta$. Letting $\varphi := c\psi$, we have by definition $\varphi \in \mathcal{D}(T^{1/2})$, $\psi = T^{1/2}\varphi$

and

$$\sum_{n=1}^{\infty} \delta_n \|a_n c\psi\|^2 = \mathfrak{h}(c\psi, c\psi) = \mathfrak{h}(\varphi, \varphi) = \langle T^{1/2}\varphi, T^{1/2}\varphi \rangle = \|\psi\|^2.$$

Finally, we verify (iii). Suppose that the assumptions in (iii) are fulfilled. Suppose that $\zeta = \sum_{n=1}^{\infty} \delta_n a_n^+ a_n \varphi$ in $\mathcal{H}$. Then, for all $\psi \in \mathcal{D}_\delta$, $\mathfrak{h}(\varphi, \psi) = \sum_{n=1}^{\infty} \delta_n \langle a_n^+ a_n \varphi, \psi \rangle = \langle \zeta, \psi \rangle$. But $\mathfrak{h}(\varphi, \psi) = \langle T^{1/2}\varphi, T^{1/2}\psi \rangle$ for all $\psi \in \mathcal{D}_\delta \equiv \mathcal{D}(T^{1/2})$. Combining both formulas we conclude that $T^{1/2}\varphi \in \mathcal{D}\big((T^{1/2})^*\big) \equiv \mathcal{D}(T^{1/2})$ and hence $\mathfrak{h}(\varphi, \psi) = \langle T\varphi, \psi \rangle$ for $\psi \in \mathcal{D}_\delta$. Combined with $\mathfrak{h}(\varphi, \psi) = \langle \zeta, \psi \rangle$, this gives $T\varphi = P_{\mathcal{H}_\delta}\zeta$, since $T\varphi \in \mathcal{H}_\delta$. By definition, $cP_{\mathcal{H}_\delta}\zeta = c\zeta$ and so $c^2\zeta = c(cT^{1/2})\, T^{1/2}\varphi = cT^{1/2}\varphi = \varphi$ which proves (iii). □

Proof of Theorem 2.4.1. Let $\mathcal{M}$ be a bounded subset of $\mathcal{D}_{\mathcal{A}}$. Since $\mathcal{D}_{\mathcal{A}}$ is a QF-space, it follows immediately from the definition of a QF-space that there exist a sequence $(a_n : n \in \mathbb{N})$ and a space $\mathcal{E}$ satisfying the assumptions of Lemma 2.4.2 such that $\mathcal{M}$ is contained in $\mathcal{E}$. Because $\mathcal{M}$ is bounded in $\mathcal{D}_{\mathcal{A}}$, there are positive numbers δ_n such that $\delta_n\Big(\sup_{\varphi \in \mathcal{M}} \|a_n\varphi\|^2\Big) \leqq 2^{-n}$ for $n \in \mathbb{N}$. Put $\delta := (\delta_n : n \in \mathbb{N})$. Then we have

$$\mathfrak{h}_\delta(\varphi) \equiv \sum_{n=1}^{\infty} \delta_n \|a_n\varphi\|^2 \leqq 1 \quad \text{for all } \varphi \in \mathcal{M}. \tag{1}$$

If $\mathcal{D}_{\mathcal{A}}$ is a Frechet space and $\mathcal{H}$ is separable, then we can set $\mathcal{E} := \mathcal{D}_{\mathcal{A}}$ and there exists a countable subset $\{\psi_k : k \in \mathbb{N}\}$ of $\mathcal{D}(\mathcal{A})$ which is dense in $\mathcal{H}$. In this case we choose δ_n such that in addition $\delta_n \|a_n \psi_k\|^2 \leqq 2^{-n}$ for all $k, n \in \mathbb{N}$, $k < n$. Then

$$\mathfrak{h}_\delta(\psi_k) \leqq \sum_{n=1}^{k} \delta_n \|a_n \psi_k\|^2 + \sum_{n=k+1}^{\infty} 2^{-n} < \infty \qquad \text{for } k \in \mathbb{N}. \tag{2}$$

Now let c be the operator of Lemma 2.4.2. Since $\mathcal{D}_\delta \subseteqq \mathcal{D}(\mathcal{A})$, $c \in \mathbf{B}\big(\mathcal{D}(\mathcal{A})\big)_+$. We show that $\mathcal{M} \subseteqq c\mathcal{U}_{\mathcal{H}}$. Suppose $\varphi \in \mathcal{M}$. Since $\mathfrak{h}_\delta(\varphi) \leqq 1$ by (1), $\varphi \in \mathcal{D}_\delta$, so that $\varphi \in c\mathcal{H}$ by Lemma 2.4.2, (i). That is, $\varphi = c\psi$ with $\psi \in \mathcal{H}_\delta$. By (1) and Lemma 2.4.2, (ii), $\mathfrak{h}_\delta(\varphi) = \sum_{n=1}^{\infty} \delta_n \|a_n c\psi\|^2 = \|\psi\|^2 \leqq 1$, that is, $\psi \in \mathcal{U}_{\mathcal{H}}$. If $\mathcal{D}_{\mathcal{A}}$ is a Frechet space and $\mathcal{H}$ is separable, then (2) shows that the dense set $\{\psi_k : k \in \mathbb{N}\}$ in $\mathcal{H}$ is contained in $\mathcal{D}_\delta$; so $\ker c = (\mathcal{H}_\delta)^\perp = \{0\}$ by Lemma 2.4.2, (i). □

Theorem 2.4.3. *Let $\mathcal{A}$ be an O-family in the Hilbert space $\mathcal{H}$ such that $\mathcal{D}_{\mathcal{A}}$ is a Frechet space. Suppose that there exists a sequence $(a_n : n \in \mathbb{N})$ in $\mathcal{A}$ with $a_1 = I$ such that $\{\|\cdot\|_{a_n} : n \in \mathbb{N}\}$ is a directed family of seminorms which generates the graph topology $\mathfrak{t}_{\mathcal{A}}$ on $\mathcal{D}(\mathcal{A})$. Suppose that $\mathcal{N}$ is a von Neumann algebra on $\mathcal{H}$ such that each operator $\overline{a_n}$, $n \in \mathbb{N}$, is affiliated with $\mathcal{N}$. Then the operator c in Theorem 2.4.1 can be chosen in the von Neumann algebra $\mathcal{N}$. That is, for each bounded subset $\mathcal{M}$ of $\mathcal{D}_{\mathcal{A}}$ there is an operator c in $\mathcal{N} \cap \mathbf{B}\big(\mathcal{D}(\mathcal{A})\big)_+$ such that $\mathcal{N} \subseteqq c\mathcal{U}_{\mathcal{H}}$.*

Proof. Since $\mathcal{D}_{\mathcal{A}}$ is a Frechet space, we can take $\mathcal{E} = \mathcal{D}_{\mathcal{A}}$ in the proof of Theorem 2.4.1 and so in Lemma 2.4.2. From the preceding proof of Theorem 2.4.1, it suffices to show that under the above assumption the operator c constructed in the proof of Lemma 2.4.2 belongs to $\mathcal{N}$.

Suppose U is a unitary operator in the commutant $\mathcal{N}'$. Since $\overline{a_n}$ is affiliated with $\mathcal{N}$,

$U\overline{a_n} \subseteq \overline{a_n}U$ for each $n \in \mathbb{N}$. In particular, this gives $U\mathcal{D}(\mathcal{A}) \subseteq \bigcap_{n\in\mathbb{N}} \mathcal{D}(\overline{a_n})$. Because of the above assumptions about the sequence (a_n), Proposition 2.2.12 applies with $\mathcal{A}_0 := \{a_n : n \in \mathbb{N}\}$ and yields $\hat{\mathcal{D}}(\mathcal{A}) = \bigcap_{n\in\mathbb{N}} \mathcal{D}(\overline{a_n})$. Since $\mathcal{D}_\mathcal{A}$ is a Frechet space, $\mathcal{D}(\mathcal{A}) = \hat{\mathcal{D}}(\mathcal{A})$. Hence $U\mathcal{D}(\mathcal{A}) \subseteq \mathcal{D}(\mathcal{A})$. Let $\varphi \in \mathcal{D}_\delta$. (Throughout this proof, we freely use the notation of the proof of Lemma 2.4.2.) The unitary U commutes with $\overline{a_n}$, $n \in \mathbb{N}$, and leaves $\mathcal{D}(\mathcal{A})$ invariant; thus $U\varphi \in \mathcal{D}(\mathcal{A})$ and $\|a_n U\varphi\| = \|Ua_n\varphi\| = \|a_n\varphi\|$, $n \in \mathbb{N}$. This implies $U\varphi \in \mathcal{D}_\delta$, so that $U\mathcal{D}_\delta \subseteq \mathcal{D}_\delta$. Similarly, if $\varphi, \psi \in \mathcal{D}_\delta$, then $\langle a_n U\varphi, a_n U\psi\rangle = \langle a_n\varphi, a_n\psi\rangle$ for $n \in \mathbb{N}$ and hence $\mathfrak{h}(U\varphi, U\psi) = \mathfrak{h}(\varphi, \psi)$. Since $\mathcal{N}'$ is a von Neumann algebra, we can replace U by U^* and obtain $U^*\mathcal{D}_\delta \subseteq \mathcal{D}_\delta$. Fix $\psi \in \mathcal{D}(T)$. Let $\varphi \in \mathcal{D}_\delta$. Then we have $U^*\varphi \in \mathcal{D}_\delta \equiv \mathcal{D}(T^{1/2})$ and

$$\begin{aligned}\langle \varphi, UT\psi\rangle &= \langle T^{1/2}U^*\varphi, T^{1/2}\psi\rangle = \mathfrak{h}(U^*\varphi, \psi) = \mathfrak{h}(UU^*\varphi, U\psi) = \mathfrak{h}(\varphi, U\psi)\\ &= \langle T^{1/2}\varphi, T^{1/2}U\psi\rangle.\end{aligned}$$

Since $\varphi \in \mathcal{D}(T^{1/2}) \equiv \mathcal{D}_\delta$ was arbitrary, this implies that $T^{1/2}U\psi \in \mathcal{D}\big((T^{1/2})^*\big) \equiv D(T^{1/2})$ and $UT\psi = T^{1/2}T^{1/2}U\psi = TU\psi$. That is, we have shown that $UD(T) \subseteq \mathcal{D}(T)$ and $UT\psi = TU\psi$ for $\psi \in \mathcal{D}(T)$. By definition, $c^2 = T^{-1} \oplus 0$ with respect to the orthogonal decomposition $\mathcal{H} = \mathcal{H}_\delta \oplus (\mathcal{H}_\delta)^\perp$. Therefore, it follows from the latter that $Uc^2 = c^2U$. Since c is a positive self-adjoint operator on $\mathcal{H}$, this yields $Uc = cU$. Hence $c \in (\mathcal{N}')' = \mathcal{N}$. □

Remark 1. It is easily seen (and stated in Corollary 3.1.3) that for any O-family $\mathcal{A}$ and operator $c \in \mathbb{B}\big(\mathcal{D}(\mathcal{A})\big)$ the set $c\mathcal{U}_\mathcal{H}$ is bounded in $\mathcal{D}_\mathcal{A}$. From this and Theorem 2.4.1 we conclude that $\big\{c\mathcal{U}_\mathcal{H} : c \in \mathbb{B}\big(\mathcal{D}(\mathcal{A})\big)_+\big\}$ *is a fundamental system of bounded sets in* $\mathcal{D}_\mathcal{A}$ *provided that* $\mathcal{D}_\mathcal{A}$ *is a QF-space.* The same holds for the family of sets $\big\{c\mathcal{U}_\mathcal{H} : c \in \mathcal{N} \cap \mathbb{B}\big(\mathcal{D}(\mathcal{A})\big)_+\big\}$ under the assumptions of Theorem 2.4.3.

Example 2.4.4. Suppose that $\mathcal{A}$ is a commutatively dominated O*-algebra and $\mathcal{D}_\mathcal{A}$ is a Frechet space. In that case the assertion of Theorem 2.4.1 and the preceding remark take a more explicit form which we will describe now. By Proposition 2.2.17, we can assume that $\mathcal{A}$ is as set out in Example 2.2.16. We shall retain the notation from this example. Let $\mathfrak{F}_\infty$ denote the collection of all measurable non-negative functions h on $\mathbb{R}$ for which the functions $h(t)\,h_n(t)$, $n \in \mathbb{N}$, are essentially bounded on $\mathbb{R}$. *Then the family of sets* $\{h(A)\,\mathcal{U}_\mathcal{H} : h \in \mathfrak{F}_\infty\}$ *is a fundamental system of bounded sets in* $\mathcal{D}_\mathcal{A}$.

Proof. If $h \in \mathfrak{F}_\infty$, then $h(A)\,\mathcal{U}_\mathcal{H}$ is a bounded subset of $\mathcal{D}_\mathcal{A}$, since hh_n is essentially bounded for all $n \in \mathbb{N}$, $\mathcal{D} = \bigcap_{n\in\mathbb{N}} \mathcal{D}\big(h_n(A)\big)$ and the topology $\mathfrak{t}_\mathcal{A}$ is generated by the seminorms $\{\|\cdot\|_{a_n} : n \in \mathbb{N}\}$. In order to prove that each bounded set $\mathcal{M}$ in $\mathcal{D}_\mathcal{A}$ is contained in $h(A)\,\mathcal{U}_\mathcal{H}$ for some $h \in \mathfrak{F}_\infty$, we proceed as in the proof of Theorem 2.4.1. We only explain the necessary modifications in that proof. Define $g(t) := \left(\sum_{n=1}^\infty \delta_n h_n(t)^2\right)^{1/2}$ and $h(t) := g(t)^{-1}$ for $t \in \mathbb{R}$, where $(+\infty)^{1/2} := +\infty$ and $(+\infty)^{-1} := 0$. Then, obviously, $h \in \mathfrak{F}_\infty$. If $\varphi \in \mathcal{M}$, then, by (1),

$$\mathfrak{h}_\delta(\varphi) = \sum_{n=1}^\infty \delta_n \|h_n(A)\,\varphi\|^2 = \sum_{n=1}^\infty \delta_n \int h_n(\lambda)^2\,\mathrm{d}\,\|E(\lambda)\,\varphi\|^2 = \int g(\lambda)^2\,\mathrm{d}\,\|E(\lambda)\,\varphi\|^2 \leqq 1;$$

so $\varphi \in \mathcal{D}\big(g(A)\big)$ and $\|g(A)\,\varphi\| \leqq 1$. Putting $\psi := g(A)\,\varphi$, we have $\varphi = h(A)\,\psi$ and $\psi \in \mathcal{U}_\mathcal{H}$.

This proves that $\mathcal{M} \subseteq h(A)\,\mathcal{U}_{\mathcal{H}}$. (Note that in general the function g is not a.e. finite, and $\mathcal{D}(g(A))$ is not dense in $\mathcal{H}$.) □

If the underlying Hilbert space $\mathcal{H}$ is separable, then the last assertion can be also derived from Theorem 2.4.3 applied to the von Neumann algebra $\mathcal{N} := \{E(\lambda) : \lambda \in \mathbb{R}\}''$. We sketch this proof. By Theorem 2.4.3, $\mathcal{M} \subseteq c\,\mathcal{U}_{\mathcal{H}}$ for some $c \in \mathcal{N} \cap \mathbb{B}(\mathcal{D}(\mathcal{A}))_+$. Since $\mathcal{H}$ is separable, a result due to J. v. Neumann (see RIESZ/SZ.-NAGY [1], IX, 129.) says that each operator in $\mathcal{N}$ is a (measurable) function of A. Thus $c = h(A)$ for some measurable function h on $\mathbb{R}$. Since $c \geqq 0$, we can take h to be non-negative. Since $c\mathcal{H} \subseteq \mathcal{D}(h_n(A))$, it follows from Lemma 2.4.2, (ii), or from the closed graph theorem that $h_n(A)\,c \equiv h_n(A)\,h(A)$ is a bounded operator for each $n \in \mathbb{N}$. From this we conclude that $h_n(A)\,h(A) = (h_n h)\,(A)$ and hence $h \in \mathfrak{F}_\infty$. ○

2.5. Examples and Counter-Examples

First we shall discuss a few typical examples of O*-algebras.

Example 2.5.1. *The O*-algebra* $\mathbb{C}[x_1, \ldots, x_n]$

Suppose $n \in \mathbb{N}$. Let $\mathbb{C}[\mathsf{x}_1, \ldots, \mathsf{x}_n]$ denote the abstract commutative polynomial algebra, that is, $\mathbb{C}[\mathsf{x}_1, \ldots, \mathsf{x}_n]$ is the free commutative $*$-algebra with unit element of n hermitian generators $\mathsf{x}_1, \ldots, \mathsf{x}_n$. As a vector space $\mathbb{C}[\mathsf{x}_1, \ldots, \mathsf{x}_n]$ has a canonical algebraic basis $\{\mathsf{x}^k := \mathsf{x}_1^{k_1} \ldots \mathsf{x}_n^{k_n};\, k = (k_1, \ldots, k_n) \in \mathbb{N}_0^n\}$, where $\mathsf{x}_l^0 := 1$ for $l = 1, \ldots, n$. The multiplication in $\mathbb{C}[\mathsf{x}_1, \ldots, \mathsf{x}_n]$ is the usual multiplication of polynomials, and the involution is uniquely determined by the requirement $\mathsf{x}_l^+ = \mathsf{x}_l$, $l = 1, \ldots, n$.

Now let $\mathcal{D}$ be a dense linear subspace of a Hilbert space $\mathcal{H}$. Suppose that $x_1, \ldots, x_n$ are operators in $\mathcal{L}^+(\mathcal{D})$ satisfying

$$x_l^+ = x_l \quad \text{and} \quad x_l x_m = x_m x_l \quad \text{for} \quad l, m = 1, \ldots, n. \tag{1}$$

We denote by $\mathbb{C}[x_1, \ldots, x_n]$ the O*-algebra on $\mathcal{D}$ which is generated by the set $\{x_1, \ldots, x_n\}$. Moreover, the definition $\pi(\mathsf{x}_l) := x_l$, $l = 1, \ldots, n$, uniquely determines a $*$-representation π of the $*$-algebra $\mathbb{C}[\mathsf{x}_1, \ldots, \mathsf{x}_n]$ on $\mathcal{D}$ such that $\mathbb{C}[x_1, \ldots, x_n] = \pi(\mathbb{C}[\mathsf{x}_1, \ldots, \mathsf{x}_n])$. It is clear that any $*$-representation of $\mathbb{C}[\mathsf{x}_1, \ldots, \mathsf{x}_n]$ arises in that way.

We illustrate the preceding by taking multiplication operators for $x_1, \ldots, x_n$. Let μ be a positive regular Borel measure on $\mathbb{R}^n$, and let $\mathcal{D}$ be the domain $\{\varphi \in L^2(\mathbb{R}^n; \mu)$: $t^k\varphi(t) \in L^2(\mathbb{R}^n; \mu)$ for all $k \in \mathbb{N}_0^n\}$ in the Hilbert space $\mathcal{H} := L^2(\mathbb{R}^n; \mu)$. Define $(x_l\varphi)\,(t) := t_l\varphi(t)$ for $t = (t_1, \ldots, t_n) \in \mathbb{R}^n$, $\varphi \in \mathcal{D}$ and $l = 1, \ldots, n$. Then the operators $x_1, \ldots, x_n$ are in $\mathcal{L}^+(\mathcal{D})$, and they satisfy (1). In this case, $\mathbb{C}[x_1, \ldots, x_n]$ is a closed O*-algebra. ○

Example 2.5.2. *The O*-algebra* $\mathbb{A}(p_1, q_1, \ldots, p_n, q_n)$

Suppose $n \in \mathbb{N}$. We let $\mathbb{A}(\mathsf{p}_1, \mathsf{q}_1, \ldots, \mathsf{p}_n, \mathsf{q}_n)$ denote the abstract $*$-algebra with unit which is generated by $2n$ hermitian generators $\mathsf{p}_1, \mathsf{q}_1, \ldots, \mathsf{p}_n, \mathsf{q}_n$ satisfying the commutation relations

$$\mathsf{p}_k\mathsf{q}_l - \mathsf{q}_l\mathsf{p}_k = -i\delta_{kl}, \quad \mathsf{p}_k\mathsf{p}_l = \mathsf{p}_l\mathsf{p}_k \ \text{and} \ \mathsf{q}_k\mathsf{q}_l = \mathsf{q}_l\mathsf{q}_k \ \text{for} \ k, l = 1, \ldots, n. \tag{2}$$

We call the $*$-algebra $\mathbb{A}(\mathsf{p}_1, \mathsf{q}_1, \ldots, \mathsf{p}_n, \mathsf{q}_n)$ the *Weyl algebra*. The set

$$\{\mathsf{p}_1^{k_1}\mathsf{q}_1^{l_1} \ldots \mathsf{p}_n^{k_n}\mathsf{q}_n^{l_n};\, (k_1, l_1, \ldots, k_n, l_n) \in \mathbb{N}_0^{2n}\} \tag{3}$$

is a basis of the vector space $\mathbb{A}(\mathsf{p}_1, \mathsf{q}_1, \ldots, \mathsf{p}_n, \mathsf{q}_n)$.

Let $\mathcal{D}$ be the Schwartz space $\mathcal{S}(\mathbb{R}^n)$ of rapidly decreasing C^∞-functions on $\mathbb{R}^n$. We consider $\mathcal{D} = \mathcal{S}(\mathbb{R}^n)$ as a domain in the Hilbert space $\mathcal{H} = L^2(\mathbb{R}^n)$. For $l = 1, \ldots, n$, $\varphi \in \mathcal{D}$ and $t = (t_1, \ldots, t_n) \in \mathbb{R}^n$, define $(p_l\varphi)(t) = -\mathrm{i}\frac{\partial\varphi}{\partial t_l}(t)$ and $(q_l\varphi)(t) = t_l\varphi(t)$.

Obviously, $p_1, q_1, \ldots, p_n, q_n$ are operators in $\mathcal{L}^+(\mathcal{D})$. Let $\mathbb{A}(p_1, q_1, \ldots, p_n, q_n)$ be the O*-algebra on $\mathcal{D}$ generated by these operators. It is easily seen that $\mathbb{A}(p_1, q_1, \ldots, p_n, q_n)$ is the O*-algebra of all differential operators with polynomial coefficients on $\mathcal{D} = \mathcal{S}(\mathbb{R}^n)$. Similarly as in case of the polynomial algebra, the O*-algebra $\mathbb{A}(p_1, q_1, \ldots, p_n, q_n)$ can be considered as the image of the abstract Weyl algebra by a $*$-representation. Indeed, since $p_1, q_1, \ldots, p_n, q_n$ are symmetric operators in $\mathcal{L}^+(\mathcal{D})$ which satisfy (2) (of course with p_m, q_m in place of $\mathsf{p}_m, \mathsf{q}_m$, $m = 1, \ldots, n$), the definition $\pi(\mathsf{p}_l) := p_l$ and $\pi(\mathsf{q}_l) := q_l$, $l = 1, \ldots, n$, extends uniquely to a $*$-representation π of the Weyl algebra. Then we have $\mathbb{A}(p_1, q_1, \ldots, p_n, q_n) = \pi(\mathbb{A}(\mathsf{p}_1, \mathsf{q}_1, \ldots, \mathsf{p}_n, \mathsf{q}_n))$. Moreover, π is faithful.

The operators $\overline{p_l}$ and $\overline{q_l}$ are self-adjoint. These operators are of great importance in quantum physics. In the non-relativistic quantum mechanics, $\overline{q_1}, \ldots, \overline{q_n}$ are the *position operators* and $\overline{p_1}, \ldots, \overline{p_n}$ are the *momentum operators* of a free particle. The operators $p_1, q_1, \ldots, p_n, q_n$ form the so-called Schrödinger representation of abstract canonical commutation relations (2). We shall call π the *Schrödinger representation* of the Weyl algebra $\mathbb{A}(\mathsf{p}_1, \mathsf{q}_1, \ldots, \mathsf{p}_n, \mathsf{q}_n)$. Recall that the "usual locally convex topology" of the space $\mathcal{S}(\mathbb{R}^n)$ is generated by the directed family of seminorms

$$\left\{q_m(\varphi) := \sup_{|k|\leq m} \sup_{t\in\mathbb{R}^n} (1 + |t|^2)^m \left|\left(\frac{\partial}{\partial t}\right)^k \varphi(t)\right| : m \in \mathbb{N}\right\},$$

where $|t| := (t_1^2 + \cdots + t_n^2)^{1/2}$, $\left(\frac{\partial}{\partial t}\right)^k := \left(\frac{\partial}{\partial t_1}\right)^{k_1} \cdots \left(\frac{\partial}{\partial t_n}\right)^{k_n}$ and $|k| := k_1 + \cdots + k_n$ for $t = (t_1, \ldots, t_n) \in \mathbb{R}^n$ and $k = (k_1, \ldots, k_n) \in \mathbb{N}_0^n$. Equipped with this topology, $\mathcal{S}(\mathbb{R}^n)$ is a Frechet space. The graph topology of the O*-algebra $\mathbb{A}(p_1, q_1, \ldots, p_n, q_n)$ coincides with this topology. It is already generated by the family of seminorms $\{\|\cdot\|_{a^m} : m \in \mathbb{N}\}$, where a is the operator $I + p_1^2 + q_1^2 + \cdots + p_n^2 + q_n^2$. (These facts follow at once from Reed/Simon [1], Appendix to V.3.) Further, the operator $A := \bar{a}$ is self-adjoint, and we have $\mathcal{D} = \mathcal{S}(\mathbb{R}^n) = \bigcap_{m\in\mathbb{N}} \mathcal{D}(A^m)$. From this we see that $\mathcal{D}$ is of the form described in Example 2.2.16, and that each O*-algebra on $\mathcal{D}$ that contains $\mathbb{A}(p_1, q_1, \ldots, p_n, q_n)$ is a commutatively dominated O*-algebra. ○

Example 2.5.3. *Differential Operators on $C_0^\infty(\mathbb{R})$*

Let $\mathcal{D} := C_0^\infty(\mathbb{R})$, considered as a dense linear subspace of the Hilbert space $\mathcal{H} := L^2(\mathbb{R})$. Suppose F is a linear subspace of $C^\infty(\mathbb{R})$ which contains the constant functions and which has the property that $f^{(k)} \in F$ for all $k \in \mathbb{N}$ when $f \in F$. By $\mathcal{A}_F$ we denote the set of all differential operators

$$a = \sum_{n=0}^{k} f_n(t) \left(\frac{\mathrm{d}}{\mathrm{d}t}\right)^n$$

acting on $\mathcal{D}$, where $k \in \mathbb{N}_0$ and $f_0, \ldots, f_k \in F$. For $m \in \mathbb{N}$, let η_m be a fixed function in $C^\infty(\mathbb{R})$ such that $\eta_m(t) = 0$ if $|t| < m - 1$ and $\eta_m(t) = 1$ if $|t| \geqq m$. Let $\mathcal{B}$ be the set of all differential operators

$$a = \sum_{m=1}^{\infty} \eta_m(t) \sum_{n=0}^{k_m} f_{mn}(t) \left(\frac{\mathrm{d}}{\mathrm{d}t}\right)^n$$

on $\mathcal{D}$, where all f_{mn} are in $C^\infty(\mathbb{R})$ and $(k_m : m \in \mathbb{N})$ is an arbitrary sequence of non-negative integers. Then $\mathcal{A}_F$ and $\mathcal{B}$ are O*-algebras and $\mathcal{A}_F \subseteq \mathcal{B}$. For instance, if F is the polynomial, then $\mathcal{A}_F$ is the restriction to $\mathcal{D} \equiv C_0^\infty(\mathbb{R})$ of the O*-algebra $\mathbb{A}(p_1,q_1)$ considered in the preceding example.

It can be shown (with some work) that the graph topology $t_\mathcal{B}$ coincides with the topology of the inductive limit on $C_0^\infty(\mathbb{R})$ of the family $\{C_0^\infty(-k, k): k \in \mathbb{N}\}$ of Frechet spaces, where the topology of $C_0^\infty(-k, k)$ is generated by the seminorms $q_{k,n}(\varphi) := \sup\{|\varphi^{(n)}(t)| : t \in (-k, k)\}$, $n \in \mathbb{N}_0$. Thus, $\mathcal{D}_\mathcal{B}$ is barrelled (as the inductive limit of Frechet spaces) and so $t_\mathcal{B} = t_+$ by Proposition 2.3.9.

Suppose that $F = C^\infty(\mathbb{R})$. Then $\mathcal{A}_F$ is a closed O*-algebra and $t_\mathcal{A} \neq t_\mathcal{B}$ on $\mathcal{D}$. That is, $\mathcal{A}_F$ and $\mathcal{B}$ are closed O*-algebras on the same domain with different graph topologies. ○

Example 2.5.4. *Sequence Spaces*

Let $\mathcal{A}$ be a subset of $\mathbb{C}^\mathbb{N}$, the vector space of all complex sequences. Suppose that $\mathcal{A}$ contains the sequence $1 := (1, 1, \ldots)$. Define $\mathcal{D}(\mathcal{A}) := \{(\varphi_n) \in \mathbb{C}^\mathbb{N} : (a_n\varphi_n) \in l^2(\mathbb{N})$ for all $(a_n) \in \mathcal{A}\}$. Then $\mathcal{D}(\mathcal{A})$ is a dense linear subspace of the Hilbert space $l^2(\mathbb{N})$. Note that $\mathcal{D}(\mathcal{A})$ is a "gestufter Raum" of order 2 in the sense of KÖTHE [1], § 30,8.

Let $\mathcal{A}^s$ be the set of all $(b_n) \in \mathbb{C}_\mathbb{N}$ for which there exist finitely many sequences, say $(a_{1n}), \ldots, (a_{ln})$, in $\mathcal{A}$ such that $|b_n| \leqq |a_{1n}| + \cdots + |a_{ln}|$ for all $n \in \mathbb{N}$. Then $\mathcal{A}^s$ is the smallest solid linear subspace of $\mathbb{C}^\mathbb{N}$ which contains $\mathcal{A}$. Moreover, $\mathcal{D}(\mathcal{A}) = \mathcal{D}(\mathcal{A}^s)$. (We say that a subset $\mathcal{B}$ of $\mathbb{C}^\mathbb{N}$ is solid if $(b_n) \in \mathcal{B}$, $(c_n) \in \mathbb{C}^\mathbb{N}$ and $|c_n| \leqq |b_n|$ for all $n \in \mathbb{N}$ imply that $(c_n) \in \mathcal{B}$.) Each $a = (a_n) \in \mathcal{A}^s$ defines a diagonal operator on the domain $\mathcal{D}(\mathcal{A})$ by $a(\varphi_n) := (a_n\varphi_n)$, $(\varphi_n) \in \mathcal{D}(\mathcal{A})$. We also denote by $\mathcal{A}$ and $\mathcal{A}^s$ the corresponding sets of diagonal operators on the domain $\mathcal{D}(\mathcal{A}) = \mathcal{D}(\mathcal{A}^s)$. Then $\mathcal{A}$ and $\mathcal{A}^s$ are closed O-families. For any such set $\mathcal{A}$, $\mathcal{A}^s$ is a directed O*-vector space. It is obvious that $\mathcal{A}^s$ is an O*-algebra if, given two sequences (a_n) and (b_n) in $\mathcal{A}$, there is a sequence $(c_n) \in \mathcal{A}$ such that $|a_nb_n| \leqq |c_n|$ for all $n \in \mathbb{N}$. If $\mathcal{A}^s$ is an O*-algebra, then $\mathcal{A}^s$ is commutatively dominated.

We mention two important special cases.

First let $\mathcal{A} := \{(k^n : n \in \mathbb{N}) : k \in \mathbb{N}\}$. Then $\mathcal{D}(\mathcal{A})$ is the space

$$s := \left\{(\varphi_n) \in \mathbb{C}^\mathbb{N} : q_k\big((\varphi_n)\big) := \sum_{n=1}^\infty k^n\,|\varphi_n| < \infty \text{ for all } k \in \mathbb{N}\right\}$$

of so-called rapidly decreasing sequences. The graph topology $t_\mathcal{A} \equiv t_{\mathcal{A}^s}$ coincides with the "usual topology" of s, i.e., with the locally convex topology on s generated by the family of seminorms $\{q_k : k \in \mathbb{N}\}$. Clearly, $\mathcal{A}^s$ is an O*-algebra. Moreover, the vector space $\mathcal{A}^s$ coincides with the sequence space s', the dual of s. Let $\mathcal{A} := \{(e^{kn} : n \in \mathbb{N}) : k \in \mathbb{N}\}$. Then $\mathcal{A}^s$ is also an O*-algebra, and $\mathcal{D}(\mathcal{A})$ is nothing else than the space of all sequences which occur in the power series expansion of holomorphic functions on the complex plane. ○

Example 2.5.5. *The Arens Algebra* $L^\omega(0, 1)$

Let $L^\omega(0, 1) := \bigcap_{p>1} L^p(0, 1)$. In this example let $\|\cdot\|_p$ denote the norm of $L^p(0, 1)$. By the Hölder inequality we have $\|fg\|_p \leqq \|f\|_{2p}\,\|g\|_{2p}$ for $f, g \in L^\omega(0, 1)$ and $p > 1$. From this we conclude that $fg \in L^\omega(0, 1)$ whenever $f, g \in L^\omega(0, 1)$. Thus $L^\omega(0, 1)$ is a $*$-algebra with the pointwise algebraic operations and with the involution defined by $(f^+)(t) := \overline{f(t)}$,

$t \in (0, 1)$. We equip $L^\omega(0, 1)$ with the locally convex topology defined by the seminorms $\|\cdot\|_p$, $p > 1$. Then, again by the Hölder inequality, the multiplication is continuous in $L^\omega(0, 1)$, so $L^\omega(0, 1)$ is a commutative Frechet topological $*$-algebra with unit. It is usually called the *Arens algebra*. We prove that $L^\omega(0, 1)$ has no characters. Assume the contrary, that is, there exists a character, say ϑ, on $L^\omega(0, 1)$. Its restriction to $C[0, 1]$ is a character on $C[0, 1]$. Hence there exists $t_0 \in [0, 1]$ such that $\vartheta(f) = f(t_0)$ for all $f \in C[0, 1]$. Set $g(t) := \log 2\,|t - t_0|$ on $(0, 1)$, $h(t) := (\log 2\,|t - t_0|)^{-1}$ if $t \in [0, 1]$, $t \neq t_0$, and $h(t_0) := 0$. Then $g \in L^\omega(0, 1)$, $h \in C[0, 1]$ and $gh = 1$ in $L^\omega(0, 1)$, so that $1 = \vartheta(1) = \vartheta(gh) = \vartheta(g)\,\vartheta(h) = \vartheta(g)\,h(t_0) = 0$ which is the desired contradiction.

Let $\mathcal{D}$ be the domain $\{\varphi \in L^2(0, 1) : f\varphi \in L^2(0, 1)$ for all $f \in L^\omega(0, 1)\}$ in the Hilbert space $\mathcal{H} := L^2(0, 1)$. Then $\pi(f)\,\varphi := f\varphi$, $f \in L^\omega(0, 1)$ and $\varphi \in \mathcal{D}$, defines a faithful $*$-representation of the $*$-algebra $L^\omega(0, 1)$ on $\mathcal{D}$. ○

Example 2.5.6. *$*$-Algebras of Continuous Functions*

Suppose X is a locally compact Hausdorff space and μ is a regular Borel measure on X. By the regularity of the measure μ, the linear space $\mathcal{D} := \{\varphi \in L^2(X;\mu) : f\varphi \in L^2(X;\mu)$ for $f \in C(X)\}$ is dense in the Hilbert space $\mathcal{H} := L^2(X;\mu)$. Define $\pi(f)\,\varphi := f\varphi$ for $f \in C(X)$ and $\varphi \in \mathcal{D}$. Then π is a faithful $*$-representation of the $*$-algebra $C(X)$ (with the usual algebraic operations) on the domain $\mathcal{D}$ in the Hilbert space $\mathcal{H}$. ○

Without carrying out the details we mention some other methods which can be used for the construction of O*-algebras. They occur in unitary representation theory of Lie groups as the images $\mathrm{d}U(\mathcal{E}(\mathfrak{g}))$ of the enveloping algebras $\mathcal{E}(\mathfrak{g})$ under the infinitesimal representations $\mathrm{d}U$; see Section 10.1 for details. The O*-algebras generated by the field operators in quantum field theory give other important examples. Rather general sources are obtained if we use (linear) differential operators with C^∞-coefficients on open subsets of $\mathbb{R}^n$ or more generally on C^∞-manifolds, or unbounded operators which are affiliated with von Neumann algebras.

The following examples are mainly intended as counter-examples. We begin with a somewhat more involved example which is stated without proof.

Example 2.5.7. *A Non-Reflexive Non-Separable Domain in a Separable Hilbert Space*

There exists a dense linear subspace $\mathcal{D}$ of a separable Hilbert space such that:

(i) $\mathcal{D}[t_+]$ is complete and semi-reflexive.

(ii) Each bounded set in $\mathcal{D}[t_+]$ is contained in a finite dimensional linear subspace of $\mathcal{D}$. In particular, $\mathcal{D}[t_+]$ is a QF-space.

(iii) $\mathcal{D}[t_+]$ is not separable.

(iv) $\mathcal{D}[t_+]$ is not reflexive.

The construction of $\mathcal{D}$ and the proofs of these facts can be found in Kürsten [1]. ○

Example 2.5.8. *A Closed O*-Algebra $\mathcal{A}$ such that $t_\mathcal{A} \neq t_+$*

Suppose that $\mathcal{B}$ is an O*-algebra in the Hilbert space $\mathcal{H}$ for which $\mathcal{D}_\mathcal{B}$ is a non-normable Frechet space. For instance, we may take the O*-algebra in Example 2.5.2.

Let $\tilde{\mathcal{H}} := \sum_{n \in \mathbb{N}} \oplus\, \mathcal{H}_n$, where $\mathcal{H}_n := \mathcal{H}$ for $n \in \mathbb{N}$. Let $\mathcal{D}(\mathcal{A})$ be the set of all vectors $(\varphi_n) \in \tilde{\mathcal{H}}$ such that $\varphi_n \in \mathcal{D}(\mathcal{B})$ for all $n \in \mathbb{N}$ and such that the set $\{n \in \mathbb{N} : \varphi_n \neq 0\}$ is finite. If $(b_n : n \in \mathbb{N})$ is a sequence of operators in $\mathcal{B}$, then we denote by (b_n) the operator on $\mathcal{D}(\mathcal{A})$ defined by $(b_n)\,(\varphi_n) := (b_n\varphi_n)$, $(\varphi_n) \in \mathcal{D}(\mathcal{A})$. Let $\mathcal{A}$ be the set of all operators $(\lambda_n b)$, where $(\lambda_n : n \in \mathbb{N})$ is an arbitrary complex sequence and $b \in \mathcal{B}$. Clearly, $\mathcal{A}$ is an O*-algebra with domain $\mathcal{D}(\mathcal{A})$. We show that $\mathcal{A}$ is closed. Let (φ_n) be a vector in $\hat{\mathcal{D}}(\mathcal{A}) = \bigcap_{a \in \mathcal{A}} \mathcal{D}(\bar{a})$. Let $k \in \mathbb{N}$. Taking $a = (\delta_{kn} b)$ with $b \in \mathcal{B}$, we get $\varphi_k \in \bigcap_{b \in \mathcal{B}} \mathcal{D}(\bar{b}) = \mathcal{D}(\mathcal{B})$, since $\mathcal{B}$ is closed. We still have to show that the set $\mathbb{N}' := \{n \in \mathbb{N} : \varphi_n \neq 0\}$ is finite. If not, then $(\varphi_n) \notin \mathcal{D}\big(\overline{(\lambda_n I)}\big)$, where $\lambda_n := \|\varphi_n\|^{-1}$ if $n \in \mathbb{N}'$ and $\lambda_n := 0$ otherwise. This shows that $(\varphi_n) \in \mathcal{D}(\mathcal{A})$ and $\mathcal{A}$ is closed.

Next we prove that $t_\mathcal{A} \neq t_+$ on $\mathcal{D}(\mathcal{A})$. Let $(b_n : n \in \mathbb{N})$ be a sequence in $\mathcal{B}$ such that the topology of the Frechet space $\mathcal{D}_\mathcal{B}$ is generated by the family of seminorms $\{\|\cdot\|_{b_n} : n \in \mathbb{N}\}$. If we had $t_\mathcal{A} = t_+$, then there would exist an operator $b \in \mathcal{B}$ and a complex sequence $(\lambda_n : n \in \mathbb{N})$ such that $\|(\varphi_n)\|_{(b_n)} \leqq \|(\varphi_n)\|_{(\lambda_n b)}$ for all $(\varphi_n) \in \mathcal{D}(\mathcal{A})$. This obviously implies that the graph topology $t_\mathcal{B}$ is generated by the single norm $\|\cdot\|_b$ which contradicts the above assumption. Thus we have $t_\mathcal{A} \neq t_+$ on $\mathcal{D}(\mathcal{A})$. ○

Example 2.5.9. *A Frechet-Montel Space $\mathcal{D}_\mathcal{A}$ which is not a Schwartz Space*

For $k \in \mathbb{N}$, let $\mathfrak{x}_k$ denote the $\mathbb{N} \times \mathbb{N}$-matrix $[x_{nm}^{(k)}]_{n,m \in \mathbb{N}}$ defined by $x_{nm}^{(k)} := m^k$ if $n = 1, \dots, k$ and $x_{nm}^{(k)} := n^k$ if $n \in \mathbb{N}$, $n \geqq k + 1$. We denote the corresponding diagonal operator in the Hilbert space $l^2(\mathbb{N}^2)$ by x_k, that is, $x_k(\varphi_{nm}) := (x_{nm}^{(k)} \varphi_{nm})$. Let $\mathcal{D}$ be the intersection of the domains of all finite products of the operators x_k, $k \in \mathbb{N}$. Letting $a_k := x_k \upharpoonright \mathcal{D}$, we clearly have $a_k \in \mathcal{L}^+(\mathcal{D})$ for $k \in \mathbb{N}$. Let $\mathcal{A}$ be the O*-algebra on $\mathcal{D}(\mathcal{A}) := \mathcal{D}$ which is generated by the set $\{a_k : k \in \mathbb{N}\}$. It is obvious that $\mathcal{D}_\mathcal{A}$ is a Frechet space.

Let a be an operator of the form $a_{k_1}^{n_1} \dots a_{k_l}^{n_l}$, where $l, n_1, \dots, n_l, k_1, \dots, k_l \in \mathbb{N}$. Since $\bar{a} = x_{k_1}^{n_1} \dots x_{k_l}^{n_l}$, it follows immediately from the special form of the operators x_k, $k \in \mathbb{N}$, that the embedding map of $\mathcal{H}_a \equiv \big(\mathcal{D}(\bar{a}), \|\cdot\|_{\bar{a}}\big)$ into $\mathcal{H}$ is not compact. Since the graph topology $t_\mathcal{A}$ is generated by the directed family of seminorms $\|\cdot\|_a$, where a is of the above form, this implies that condition (ii) in Proposition 2.3.14 is not fulfilled; hence $\mathcal{D}_\mathcal{A}$ is not a Schwartz space.

But the Frechet space $\mathcal{D}_\mathcal{A}$ is a Montel space. This follows directly from the criterion as stated in Köthe [1], § 30, 9. Another possibility to check this goes as follows. Let $\varepsilon > 0$, and let $\mathcal{M}$ be a bounded set in $\mathcal{D}_\mathcal{A}$. From the special form of the operators x_k it is not difficult to show that for any $a \in \mathcal{A}$ there is a bounded subset $\mathcal{M}_{a,\varepsilon}$ of $\mathcal{D}_\mathcal{A}$ contained in a finite dimensional subspace of $\mathcal{D}_\mathcal{A}$ such that for each $\varphi \in \mathcal{M}$ there is a $\psi \in \mathcal{M}_{a,\varepsilon}$ satisfying $\|\varphi - \psi\|_a \leqq \varepsilon$. By Lemma 1.1.1, $\mathcal{M}$ is precompact and hence relatively compact, since $\mathcal{D}_\mathcal{A}$ is complete. Since $\mathcal{D}_\mathcal{A}$ is a Frechet space, this proves that $\mathcal{D}_\mathcal{A}$ is a Montel space. ○

Example 2.5.10. *An O-Family $\mathcal{A}$ with $\hat{\mathcal{D}}(\mathcal{A}) \neq \overline{\mathcal{D}}(\mathcal{A})$*

Let A and B be the multiplication operators on the Hilbert space $\mathcal{H} := L^2(1, +\infty)$ by the functions $f(t) = t$ and $g(t) = t(t-1)^{-1}$, respectively. Then A, B and $A + B$ are positive self-adjoint operators with bounded inverses. Let Q denote the rank one projection $\zeta \otimes \zeta$ on $\mathcal{H}$, where $\zeta(t) := t^{-1}$. Define $\mathcal{D}(\mathcal{A}) := A^{-1}B^{-1}(I - Q)\,\mathcal{H}$, $a := A \upharpoonright \mathcal{D}(\mathcal{A})$ and $b := B \upharpoonright \mathcal{D}(\mathcal{A})$. Let $\mathcal{A}$ be the O-family $\{I, a, b\}$ on $\mathcal{D}(\mathcal{A})$.

Since $\zeta \notin B^{-1}\mathcal{H}$, we have $\ker\,(I - Q)\,B^{-1} = \ker\big(B^{-1}(I - Q)\big)^* = \{0\}$, so that $A\mathcal{D}(\mathcal{A}) = B^{-1}(I - Q)\,\mathcal{H}$ is dense in $\mathcal{H}$. Because $A^{-1} \in \mathbf{B}(\mathcal{H})$, this implies that $\mathcal{D}(\mathcal{A})$ is a core for A. Thus $\bar{a} = A$. A similar reasoning, based on $\zeta \notin A^{-1}\mathcal{H}$, shows that $\bar{b} = B$. Therefore, $\overline{\mathcal{D}}(\mathcal{A}) = \mathcal{D}(\bar{a}) \cap \mathcal{D}(\bar{b}) = \mathcal{D}(A) \cap \mathcal{D}(B) = \mathcal{D}(A + B)$, where the last equality follows from the special form of the functions f and g.

On the other hand, from the inequality $\|(a+b)\cdot\| \leqq \|\cdot\|_a + \|\cdot\|_b$ it follows that $\hat{\mathcal{D}}(A) \subseteqq \mathcal{D}(\overline{a+b})$. Since $(A+B)\,\mathcal{D}(A) = (A+B)A^{-1}B^{-1}(I-Q)\mathcal{H} = (B^{-1}+A^{-1})(I-Q)\mathcal{H} = (I-Q)\,\mathcal{H}$ is not dense in $\mathcal{H}$ and $(A+B)^{-1} \in \mathbb{B}(\mathcal{H})$, $\mathcal{D}(\mathcal{A})$ is not a core for the self-adjoint operator $A+B$. Hence $\overline{a+b} \subsetneqq A+B$; so $\hat{\mathcal{D}}(\mathcal{A}) \subseteqq \mathcal{D}(\overline{a+b}) \subsetneqq \mathcal{D}(A+B)$. Combined with the preceding, this proves that $\hat{\mathcal{D}}(\mathcal{A}) \subsetneqq \overline{\mathcal{D}}(\mathcal{A})$.

2.6. The Positive Cone of an O*-Algebra

In this section, $\mathcal{A}$ denotes an O*-algebra.

Definition 2.6.1. The set $\mathcal{A}_+ := \{a \in \mathcal{A}_{\rm h} : \langle a\varphi, \varphi\rangle \geqq 0 \text{ for } \varphi \in \mathcal{D}(\mathcal{A})\}$ is the *positive cone* of the O*-algebra $\mathcal{A}$. A linear functional f on $\mathcal{A}$ is called *strongly positive* if $f(a) \geqq 0$ for all $a \in \mathcal{A}_+$. A $*$-representation π of the $*$-algebra $\mathcal{A}$ is said to be *strongly positive* if $\pi(\mathcal{A}_+) \subseteqq \pi(\mathcal{A})_+$.

The terminology "cone" for the set $\mathcal{A}_+$ is justified by the first statement in

Lemma 2.6.2. (i) *$\mathcal{A}_+$ is an m-admissible cone in the real vector space $\mathcal{A}_{\rm h}$. In particular, $\mathcal{P}(\mathcal{A}) \subseteqq \mathcal{A}_+$.*

(ii) $\mathcal{A}_{\rm h} = \mathcal{A}_+ - \mathcal{A}_+$.

Proof. (i): Straightforward. (ii) follows from the identity $4a = (a+I)^2 - (a-I)^2$, $a \in \mathcal{A}_{\rm h}$. □

We define an order relation "$\geqq$" on the hermitian part $\mathcal{A}_{\rm h}$ of $\mathcal{A}$ by $a \geqq b$ if and only if $a - b \in \mathcal{A}_+$, $a, b \in \mathcal{A}_{\rm h}$.

Since $\mathcal{A}_+$ is a cone, the relation "$\geqq$" is reflexive, antisymmetric and transitive. Thus $(\mathcal{A}_{\rm h}, \geqq)$ is an ordered vector space with positive cone $\mathcal{A}_+$, and the O*-algebra $\mathcal{A}$ becomes an ordered $*$-vector space.

Remark 1. A very useful property of the cone $\mathcal{A}_+$ is that $\mathcal{B}_+ = \mathcal{B} \cap \mathcal{A}_+$ for any O*-subalgebra $\mathcal{B}$ of $\mathcal{A}$. A similar assertion for the cone $\mathcal{P}(\mathcal{A})$ is not true in general as simple examples show.

Remark 2. An important fact in C^*-algebra theory is the equality $\mathcal{P}(\mathcal{A}) = \mathcal{A}_+$ which holds for every C^*-algebra $\mathcal{A}$. The following example shows that for O*-algebras (or even for incomplete $*$-subalgebras of $\mathbb{B}(\mathcal{H})$) this is no longer true in general.

Example 2.6.3. Let $\mathcal{A}$ be the $*$-subalgebra $\mathbb{C}[x]$ of $\mathbb{B}(\mathcal{H})$, where $\mathcal{H} := L^2(0,1)$ and $(x\varphi)(t) := t\varphi(t)$, $t \in (0,1)$, for $\varphi \in \mathcal{H}$. Obviously, $f(p) := p(2)$, $p(x) \in \mathcal{A} = \mathbb{C}[x]$, defines a positive linear functional f on the $*$-algebra $\mathcal{A}$. Since $1 - x \in \mathcal{A}_+$ and $1 - x \notin \mathcal{P}(\mathcal{A})$, $\mathcal{P}(\mathcal{A}) \neq \mathcal{A}_+$. From $f(1-x) = -1$ we conclude that the linear functional f is not strongly positive on the O*-algebra $\mathcal{A}$. ○

The following two easy lemmas indicate the close link between order-domination in $\mathcal{A}$ and generation of the graph topology $t_{\mathcal{A}}$.

Lemma 2.6.4. *For any subset $\mathcal{B}$ of $\mathcal{A}$, the following three conditions are equivalent:*

(i) *The family of seminorms $\{\|\cdot\|_b : b \in \mathcal{B}\}$ generates the graph topology $t_{\mathcal{A}}$.*

(ii) *The set $\mathcal{B}_S := \{b_1^+b_1 + \cdots + b_k^+b_k : b_1, \ldots, b_k \in \mathcal{B} \text{ and } k \in \mathbb{N}\}$ is order-dominating for $\mathcal{A}$.*

(iii) *The vector space B_C spanned by the operators b^+b, where $b \in \mathcal{B}$, is cofinal in $\mathcal{A}$.*

Proof. (i) → (ii): Suppose $a \in \mathcal{A}_h$. Since $t_\mathcal{A}$ is generated by the seminorms $\|\cdot\|_b$, $b \in \mathcal{B}$, there are operators $b_1, \ldots, b_k \in \mathcal{B}$ and a $\lambda > 0$ such that

$$\|a\varphi\| + \|\varphi\| \leqq \lambda(\|b_1\varphi\| + \cdots + \|b_k\varphi\|) \quad \text{for} \quad \varphi \in \mathcal{D}(\mathcal{A}).$$

Hence

$$\langle a\varphi, \varphi\rangle \leqq (\|a\varphi\| + \|\varphi\|)^2 \leqq \lambda^2 k(\|b_1\varphi\|^2 + \cdots + \|b_k\varphi\|^2)$$

for all $\varphi \in \mathcal{D}(\mathcal{A})$ which gives $a \leqq \lambda^2 k(b_1^+ b_1 + \cdots + b_k^+ b_k)$.
(ii) → (iii) is trivial.
(iii) → (i): Suppose $a \in \mathcal{A}$. That $\mathcal{B}_C$ is cofinal in $\mathcal{A}$, implies that there is a $b \in (\mathcal{B}_C)_h$ such that $a^+a \leqq b$. The operator b is of the form $b = \lambda_1 b_1^+ b_1 + \cdots + \lambda_k b_k^+ b_k$ with $\lambda_1, \ldots, \lambda_k \in \mathbb{R}$ and $b_1, \ldots, b_k \in \mathcal{B}$. Taking $\lambda > 0$ such that $\lambda_n \leqq \lambda$ for all $n = 1, \ldots, k$, we have

$$\|a\varphi\|^2 = \langle a^+a\varphi, \varphi\rangle \leqq \lambda\langle (b_1^+ b_1 + \cdots + b_k^+ b_k)\, \varphi, \varphi\rangle \leqq \lambda(\|b_1\varphi\| + \cdots + \|b_k\varphi\|)^2$$

for $\varphi \in \mathcal{D}(\mathcal{A})$.

From this (i) follows. □

Lemma 2.6.5. *If $\mathcal{B}$ is a subset of $\mathcal{A}_h$ such that the (complex) linear span of $\mathcal{B}$ is cofinal in $\mathcal{A}$, then the graph topology $t_\mathcal{A}$ is already generated by the family of seminorms $\{\|\cdot\|_b : b \in \mathcal{B}\}$.*

Proof. Suppose $a \in \mathcal{A}$. The assumption implies that we can find real numbers $\lambda_1, \ldots, \lambda_{n+k}$ and operators $b_1, \ldots, b_{n+k} \in \mathcal{B}$, $n, k \in \mathbb{N}$, such that $I \leqq \lambda_1 b_1 + \cdots + \lambda_n b_n$ and $a^+a \leqq \lambda_{n+1} b_{n+1} + \cdots + \lambda_{n+k} b_{n+k}$. Set $\lambda := \max\{|\lambda_l| : l = 1, \ldots, n + k\}$. Then

$$\langle I\varphi, \varphi\rangle \leqq \langle (\lambda_1 b_1 + \cdots + \lambda_n b_n)\, \varphi, \varphi\rangle \leqq \lambda(|\langle b_1\varphi, \varphi\rangle| + \cdots + |\langle b_n\varphi, \varphi\rangle|)$$
$$\leqq \lambda(\|b_1\varphi\| + \cdots + \|b_n\varphi\|)\, \|\varphi\|,$$

so that $\|\varphi\| \leqq \lambda(\|b_1\varphi\| + \cdots + \|b_n\varphi\|)$ for $\varphi \in \mathcal{D}(\mathcal{A})$. From the latter and $a^+a \leqq \lambda_{n+1} b_{n+1} + \cdots + \lambda_{n+k} b_{n+k}$ it follows that

$$\|a\varphi\|^2 = \langle a^+a\varphi, \varphi\rangle \leqq \lambda(\|b_{n+1}\varphi\| + \cdots + \|b_{n+k}\varphi\|)\, \|\varphi\| \leqq \lambda^2(\|b_1\varphi\| + \cdots + \|b_{n+k}\varphi\|)^2$$

for $\varphi \in \mathcal{D}(\mathcal{A})$ which yields the assertion. □

The next two corollaries follow directly from these lemmas.

Corollary 2.6.6. *If $\mathcal{B}$ is an O*-subalgebra of $\mathcal{A}$, then the graph topologies $t_\mathcal{B}$ and $t_\mathcal{A}$ coincide if and only if $\mathcal{B}$ is cofinal in $\mathcal{A}$.*

Corollary 2.6.7. *The following three conditions are equivalent:*

(i) *The locally convex space $\mathcal{D}_\mathcal{A}$ is metrizable.*
(ii) *There exists a countable subset of $\mathcal{A}_+$ which is order-dominating for $\mathcal{A}$.*
(iii) *There exists a countable subset of $\mathcal{A}_h$ such that its (complex) linear span is cofinal in $\mathcal{A}$.*

Corollary 2.6.8. *Suppose that π is a strongly positive *-representation of $\mathcal{A}$. If $\mathcal{B}$ is a subset of $\mathcal{A}$ such that the seminorms $\|\cdot\|_b$, $b \in \mathcal{B}$, generate the graph topology $t_\mathcal{A}$, then the family of seminorms $\{\|\cdot\|_{\pi(b)} : b \in \mathcal{B}\}$ determines the graph topology $t_{\pi(\mathcal{A})}$.*

Proof. By Lemma 2.6.4, (i) → (ii), $\mathcal{B}_S$ is order-dominating for $\mathcal{A}$. Since π is strongly positive, $\pi(\mathcal{B}_S)$ is order-dominating for $\pi(\mathcal{A})$. Because π is a *-representation, Lemma 2.6.4, (ii) → (i), gives the assertion. □

Corollary 2.6.9. *Suppose $\mathcal{B}$ is an O*-subalgebra of the O*-algebra $\mathcal{A}$ such that $t_{\mathcal{B}} = t_{\mathcal{A}}$. Then every strongly positive linear functional on $\mathcal{B}$ can be extended to a strongly positive linear functional on $\mathcal{A}$.*

Proof. Combine Corollary 2.6.6 with Lemma 1.3.2. □

Corollary 2.6.10. *If $\mathcal{B}$ is an O*-algebra on $\mathcal{D} = \mathcal{D}(\mathcal{B})$ such that the locally convex space $\mathcal{D}_{\mathcal{B}}$ is barrelled, then each strongly positive linear functional on $\mathcal{B}$ has an extension to a strongly positive linear functional on $\mathcal{L}^+(\mathcal{D})$.*

Proof. By Proposition 2.3.9 we have $t_{\mathcal{B}} = t_+$ on $\mathcal{D}$; so Corollary 2.6.9 applies with $\mathcal{A} := \mathcal{L}^+(\mathcal{D})$. □

In the next example we consider the cones $\mathcal{P}(\mathcal{A})$ and $\mathcal{A}_+$ in case of the polynomial algebra $\mathbb{C}[x_1, \dots, x_n]$ and we indicate the relation to the n-dimensional classical moment problem.

Example 2.6.11. Suppose $n \in \mathbb{N}$. Let $\mathcal{A}$ denote the O*-algebra $\mathbb{C}[x_1, \dots, x_n]$ on the domain $\mathcal{D} := \{\varphi \in L^2(\mathbb{R}^n) : t^k\varphi(t) \in L^2(\mathbb{R}^n)$ for $k \in \mathbb{N}_0^n\}$ in the Hilbert space $L^2(\mathbb{R}^n)$, where x_l, $l = 1, \dots, n$, is the multiplication operator with domain $\mathcal{D}$ defined by $(x_l\varphi)(t) := t_l\varphi(t)$ for $\varphi \in \mathcal{D}$ and $t = (t_1, \dots, t_n) \in \mathbb{R}^n$; see also Example 2.5.1. In this case,

$$\mathcal{A}_+ = \{p \in \mathbb{C}[x_1, \dots, x_n] : p(t_1, \dots, t_n) \geqq 0 \text{ for all } (t_1, \dots, t_n) \in \mathbb{R}^n\}. \tag{1}$$

We denote by $M_+(\mathbb{R}^n)$ the set of all positive regular Borel measures μ on $\mathbb{R}^n$ which have moments of all order. By definition, the latter means that the function $\varphi_k(t) := t^k$, $t \in \mathbb{R}^n$, is in $L^1(\mathbb{R}^n; \mu)$ for all $k \in \mathbb{N}_0^n$. Let $M(\mathbb{R}^n)$ be the set of all complex Borel measures μ on $\mathbb{R}^n$ which are of the form $\mu = (\mu_1 - \mu_2) + \mathrm{i}(\mu_3 - \mu_4)$ with $\mu_1, \mu_2, \mu_3, \mu_4 \in M_+(\mathbb{R}^n)$. A standard result from the theory of the moment problem (see SHOHAT/TAMARKIN [1], ch. I, Theorem 1.1) reformulates in the present context as follows.

Statement 1: *A linear functional f on $\mathcal{A}$ is strongly positive if and only if there is a measure $\mu \in M_+(\mathbb{R}^n)$ such that $f(p) = \int_{\mathbb{R}^n} p(t)\, \mathrm{d}\mu(t)$ for all $p \in \mathbb{C}[x_1, \dots, x_n]$.*
In other words, the strongly positive linear functionals on the O*-algebra $\mathcal{A}$ are precisely the solutions of the Hamburger moment problem on $\mathbb{R}^n$. We next discuss the relation between $\mathcal{P}(\mathcal{A})$ and $\mathcal{A}_+$. From the fundamental theorem of algebra we easily conclude that each polynomial $p \in \mathbb{C}[x]$ which is non-negative on $\mathbb{R}$ can be written as $p = q^+q$ with $q \in \mathbb{C}[x]$. (It suffices to note that real roots of p have an even multiplicity and complex roots of p appear in conjugate pairs.) Therefore, if $n = 1$, then $\mathcal{P}(\mathcal{A}) = \mathcal{A}_+$, and positive linear functionals are always strongly positive. This is no longer true if $n \geqq 2$. From now on assume that $n \in \mathbb{N}$, $n \geqq 2$.

Statement 2: *The polynomial $p_0(x_1, \dots, x_n) := x_1^2x_2^2(x_1^2 + x_2^2 - 1) + 1$ is in $\mathcal{A}_+$, but not in $\mathcal{P}(\mathcal{A})$.*

Proof. Let $t = (t_1, \dots, t_n) \in \mathbb{R}^n$. If $t_1^2 + t_2^2 \geqq 1$, then obviously $p_0(t) \geqq 0$. If $t_1^2 + t_2^2 \leqq 1$, then $t_1^2t_2^2 \leqq 1$ and hence $p_0(t) = 1 - t_1^2t_2^2(1 - t_1^2 - t_2^2) \geqq 0$. Thus $p_0 \in \mathcal{A}_+$.

In order to prove that $p_0 \notin \mathcal{P}(\mathcal{A})$, we assume the contrary, that is, $p_0 = \sum_{l=1}^{k} q_l^+ q_l$ for some $q_1, \dots, q_k \in \mathbb{C}[x_1, \dots, x_n]$. Since $p_0(0, t_2, t_3, \dots, t_n) = p(t_1, 0, t_3, \dots, t_n) = 1$ for all $t_1, t_2, \dots, t_n \in \mathbb{R}$, it follows that each q_l is of the form $\lambda_l + x_1x_2p_l$, where $\lambda_l \in \mathbb{C}$ and p_l is a linear polynomial from $\mathbb{C}[x_1, \dots, x_n]$. Comparing coefficients in $p_0 = \sum_{l=1}^{k} q_l^+ q_l$, we obtain the equality $\sum_{l=1}^{k} t_1^2 t_2^2 |p_l(t)|^2 = t_1^2 t_2^2 (t_1^2 + t_2^2 - 1)$ for $t = (t_1, t_2, \dots, t_n) \in \mathbb{R}^n$ which is impossible. This proves that p_0 is not in $\mathcal{P}(\mathcal{A})$. $\square$

The assertion that $p_0 \notin \mathcal{P}(\mathcal{A})$ follows from Statement 3 as well. We define a bijection $m(\cdot,\cdot)$ of $\mathbb{N}_0^2$ onto $\mathbb{N}$ by setting $m(0,0) = 1$, $m(1,2) = 2$, $m(2,1) = 3$, $m(1,1) = 4$, $m(1,0) = 5$, $m(0,1) = 6$, $m(2,0) = 7$, $m(0,2) = 8$, $m(3,0) = 9$, $m(0,3) = 10$ and $m(k,l) = (k+l)(k+l+1)/2 + l + 1$ for $(k,l) \in \mathbb{N}_0^2$, $k + l \geqq 4$. Let f_0 be the linear functional on $\mathcal{A}$ defined by $f_0(p(x_1, \dots, x_n)) = f_0(p(x_1, x_2, 0, \dots, 0))$ for $p \in \mathbb{C}[x_1, \dots, x_n]$, $f_0(x_1^k x_2^l) = 0$ if k or l is an odd number and $f_0(x_1^k x_2^l) = g_{m(k/2, l/2)}$ otherwise, where $g_r = 1$ if $r = 1, 2, 3$, $g_4 = 4$ and $g_r = r!^{(r+1)!}$ if $r \in \mathbb{N}$, $r \geqq 5$.

Statement 3: *f_0 is a positive linear functional on $\mathcal{A}$ which is not strongly positive.*

Proof. Since $p_0 \in \mathcal{A}_+$ (by Statement 2) and $f_0(p_0) = -1$ we see that f_0 is not strongly positive. We prove that f_0 is a positive linear functional on $\mathcal{A}$. By the definition of f_0 there is no loss of generality to assume that $n = 2$. Put $\lambda_{m(k,l),m(r,s)} := f_0(x_1^{k+r} x_2^{l+s})$ for $(k,l), (r,s) \in \mathbb{N}_0^2$. Suppose $p \in \mathbb{C}[x_1, x_2]$. Writing p as a finite sum $\sum_{(k,l)} \alpha_{k,l} x_1^k x_2^l$, we have

$$f_0(p^+p) = \sum_{(k,l),(r,s)} \alpha_{k,l} \overline{\alpha_{r,s}} \lambda_{m(k,l),m(r,s)}.$$

Therefore, it is sufficient to prove that the matrix $\Lambda_j := [\lambda_{k,l}]_{k,l=1,\dots,j}$ is positive definite for each $j \in \mathbb{N}$. We show by induction that $\det \Lambda_j \geqq 1$ for all $j \in \mathbb{N}$. A direct calculation proves that $\det \Lambda_j \geqq 1$ for $r = 1, 2, 3, 4$. Now suppose $j \in \mathbb{N}$, $j \geqq 5$. Assume that $\det \Lambda_{j-1} \geqq 1$. A simple computation shows that $\max(m(k,l), m(r,s)) > m((k+r)/2, (l+s)/2)$ for $(k,l), (r,s) \in \mathbb{N}_0^2$ provided that the right hand side is defined and $(k,l) \neq (r,s)$. This implies that $|\lambda_{k,l}| \leqq g_{j-1}$ if $k \leqq j$, $l \leqq j$, $(k,l) \neq (j,j)$ and $j, k, l \in \mathbb{N}$. Developing the determinant $\det \Lambda_j$ by the j-th row and using these inequalities and the induction hypothesis $\det \Lambda_{j-1} \geqq 1$, we obtain

$$\det \Lambda_j \geqq (\det \Lambda_{j-1})\, g_j - (j-1)\,(j-1)!\, g_{j-1}^j \geqq g_j - j!\, g_{j-1}^j + 1$$
$$\geqq j!^{(j+1)!} - j!(j-1)!^{j!} + 1 \geqq 1. \quad \square \bigcirc$$

Example 2.6.12. Let $\mathcal{A}$ be the O*-algebra $\mathbf{A}(p_1, q_1)$ from Example 2.5.2. Set $N := (p_1^2 + q_1^2 - I)/2$. Then $\overline{N}$ is a self-adjoint operator with spectrum $\mathbb{N}_0$ (cf. REED/SIMON [1], Appendix to V.3). Hence $(N-1)(N-2) \in \mathcal{A}_+$.

On the other hand, if $p \in \mathbb{C}[x]$, then the operator $p(N)$ belongs to $\mathcal{P}(\mathcal{A})$ if and only if there are polynomials $q_0, q_1, \dots, q_{n-1} \in \mathbb{C}[x]$, $n \in \mathbb{N}$, such that $p(N) = q_0(N)^+ q_0(N) + N q_1(N)^+ q_1(N) + \dots + N(N-1) \cdots (N-n+2)\, q_{n-1}(N)^+ q_{n-1}(N)$. (A proof of this statement can be found in FRIEDRICH/SCHMÜDGEN [1].) From this it follows easily that $(N-k_1) \cdots (N-k_r)$ is not in $\mathcal{P}(\mathcal{A})$ when $k_1, \dots, k_r$ are pairwise different positive integers. In particular, $(N-1)(N-2) \notin \mathcal{P}(\mathcal{A})$. Thus $\mathcal{P}(\mathcal{A}) \neq \mathcal{A}_+$. $\bigcirc$

Remark 3. Let $\mathcal{A}$ be either the O*-algebra of Example 2.6.11, with $n \geqq 2$, or the O*-algebra of Example 2.6.12. Since $\mathcal{P}(\mathcal{A}) \neq \mathcal{A}_+$, the existence of a positive linear functional on $\mathcal{A}$ which is not strongly positive follows easily from Corollary 11.6.4 by using a separation theorem for convex sets (see the proof of Corollary 11.6.2). The functional f_0 in Statement 3 above is an explicit example of this kind.

Notes

A pioneering paper for the systematic study of unbounded operator algebras and their topologization is LASSNER [1] which appeared already in 1969 as a preprint. From the beginning the general theory of these algebras was developed parallel to and strongly interacting with the theory of (unbounded) $*$-representations; so one should also compare the historical comments in the notes after Chapter 8.

2.1. O*-algebras and the maximal O*-algebra $\mathcal{L}^+(\mathcal{D})$ were introduced and investigated by LASSNER [1] who used the name "Op$*$-algebras". Propositions 2.1.10 and 2.1.11 can be found in LASSNER [1]. Proposition 2.1.12 seems to be new.

2.2. The graph topology was introduced independently by LASSNER [1] and POWERS [1]. Also the closure of an O*-algebra resp. a $*$-representation was defined by these authors, and Proposition 2.2.12 (in these cases) was established. Commutatively dominated O*-algebras have been first studied (without mentioning this name) by SCHMÜDGEN [9].

2.3. Some of the basic properties of the locally convex spaces $\mathcal{D}_{\mathcal{A}}$ follow immediately from standard theory on locally convex spaces combined with the fact that $\mathcal{D}_{\hat{\mathcal{A}}}$ is the projective limit of a family of Hilbert spaces. The latter fact was observed in SCHMÜDGEN [4] and in FRIEDRICH/LASSNER [1].

Proposition 2.3.6 (in a somewhat weaker version) is in SCHMÜDGEN [20]. Proposition 2.3.10 is from SCHMÜDGEN [4] and Proposition 2.3.12 from KÜRSTEN [2]. The main part of Proposition 2.3.14 is implicit in SCHMÜDGEN [5].

It is still an open problem whether or not there exists an O*-algebra $\mathcal{A}$ such that $\mathcal{D}_{\mathcal{A}}$ is a nuclear Frechet space without basis; cf. MITJAGIN [1], p. 228. Note that there exist nuclear Frechet spaces without basis, see MITJAGIN/ZOBIN [1].

2.4. In the case where $\mathcal{A}$ is an O*-algebra Theorem 2.4.1 was proved by KÜRSTEN [2], [5]. Theorem 2.4.3 is due to the author.

2.5. The Examples 2.5.1—2.5.6 are more or less standard. The Arens algebra $L^\omega(0, 1)$ was introduced by ARENS [1]. A generalization of this algebra has been defined and studied by INOUE [1], [2].

Example 2.5.7 is due to KÜRSTEN [1]. Examples with the properties of Example 2.5.8 appeared in FRIEDRICH/LASSNER [1] and in SCHMÜDGEN [4]. Example 2.5.9 is an adaption of an example in KÖTHE [1], § 30.

2.6. The assertions of Statements 2 and 3 in Example 2.6.11 have a long history. Hilbert proved in 1888 that there exists a nonnegative polynomial in two variables which is not a sum of squares; cf. HILBERT [1] or GELFAND/WILENKIN [1], II, § 7.2. The simple example in Statement 2 is taken from BERG/CHRISTENSEN/JENSEN [1]. That there exist positive linear functionals on $\mathbb{C}[x_1, x_2]$ which are not strongly positive was shown independently by SCHMÜDGEN [6] and BERG/CHRISTENSEN/JENSEN [1]. The example in Statement 3 is from FRIEDRICH [1].

Additional References:

2.1. ASCOLI/EPIFANIO/REVISTO [1].
2.3. LASSNER/TIMMERMANN [2].
2.4. JUNEK [2].
2.5. BROOKS [1].

3. Spaces of Linear Mappings Associated with O-Families and Their Topologization

This chapter is concerned with some fundamental spaces of linear mappings which are associated with O-families (in a sense defined below) and with some methods of their topologization. Though the case we are mainly interested in is when the O-families are O*-algebras, we give most of the basic definitions and facts in the more general context of O-families. Suppose that $\mathcal{A}$ and $\mathcal{B}$ are O-families in the Hilbert space $\mathcal{H}$.

The most important object in this and in the following chapters is the vector space $\mathcal{L}(\mathcal{D}_{\mathcal{A}}, \mathcal{D}_{\mathcal{B}}^{+})$ of all linear mappings x of the domain $\mathcal{D}(\mathcal{A})$ into $\mathcal{D}_{\mathcal{B}}^{+}$, the conjugate vector space of the dual of the locally convex space $\mathcal{D}_{\mathcal{B}}$, for which the associated sesquilinear form $\langle x\cdot, \cdot\rangle$ is continuous on $\mathcal{D}_{\mathcal{A}} \times \mathcal{D}_{\mathcal{B}}$. In Section 3.2 we begin the study of this space. Two algebras, denoted by $\mathcal{L}(\mathcal{D}_{\mathcal{B}}^{+}, \mathcal{D}_{\mathcal{A}})$ and $\mathbb{B}(\mathcal{D}(\mathcal{B}), \mathcal{D}(\mathcal{A}))$, of linear mappings of $\mathcal{D}_{\mathcal{B}}^{+}$ resp. $\mathcal{H}$ into the domain $\mathcal{D}(\mathcal{A})$ are considered in Section 3.1. They are very useful tools for many topological problems concerning the space $\mathcal{L}(\mathcal{D}_{\mathcal{A}}, \mathcal{D}_{\mathcal{B}}^{+})$.

Sections 3.3 and 3.5 are devoted to the topologization of linear subspaces of $\mathcal{L}(\mathcal{D}_{\mathcal{A}}, \mathcal{D}_{\mathcal{B}}^{+})$ and more generally of O*-algebras. Section 3.3 deals with various possible methods of defining locally convex topologies which can all be considered as generalizations of the operator norm topology in C^*-algebra theory. There are two basic general topological concepts (the bounded topology and the inductive topology) and some locally convex topologies related to the order structure. In Section 3.5 we briefly consider the weak and strong-operator topologies and the ultraweak and ultrastrong topologies.

Several kinds of density results are contained in Section 3.4. They are used later for different purposes. For instance, it is shown that the algebra $\mathbb{B}(\mathcal{D}(\mathcal{B}), \mathcal{D}(\mathcal{A}))$ is dense in $\mathcal{L}(\mathcal{D}_{\mathcal{A}}, \mathcal{D}_{\mathcal{B}}^{+})$ relative to the bounded topology provided that $\mathcal{A}$ and $\mathcal{B}$ are O*-algebras for which $\mathcal{D}_{\mathcal{A}}$ and $\mathcal{D}_{\mathcal{B}}$ are QF-spaces. The continuity of $*$-representations and of positive linear functionals on topological $*$-algebras are investigated in Section 3.6.

If not specified further by additional assumptions we assume throughout this chapter that $\mathcal{A}$ and $\mathcal{B}$ are *O-families on the same Hilbert space* $\mathcal{H}$.

3.1. The Algebras $\mathbb{B}(\mathcal{D}_2, \mathcal{D}_1)$ and $\mathcal{L}(\mathcal{D}_{\mathcal{B}}^{+}, \mathcal{D}_{\mathcal{A}})$

In the first subsection we consider bounded linear operators which, together with their adjoints, map the whole Hilbert space into given dense domains. The second subsection deals with linear mappings of $\mathcal{D}_{\mathcal{B}}^{+}$ into the domain $\mathcal{D}(\mathcal{A})$.

The Algebra $\mathbb{B}(\mathcal{D}_2, \mathcal{D}_1)$

Definition 3.1.1. If $\mathcal{D}_1$, $\mathcal{D}_2$ and $\mathcal{D}$ are dense linear subspaces of a Hilbert space $\mathcal{H}$, we define

$$\mathbb{B}(\mathcal{D}_2, \mathcal{D}_1) := \{c \in \mathbb{B}(\mathcal{H}) : c\mathcal{H} \subseteq \mathcal{D}_1 \text{ and } c^*\mathcal{H} \subseteq \mathcal{D}_2\},$$

$$\mathbb{B}(\mathcal{D}) := \mathbb{B}(\mathcal{D}, \mathcal{D}) \quad \text{and} \quad \mathbb{B}(\mathcal{D})_+ := \{c \in \mathbb{B}(\mathcal{D}) : c \geqq 0\}.$$

From this definition it is obvious that $\mathbb{B}(\mathcal{D}_2, \mathcal{D}_1)$ is a subalgebra of $\mathbb{B}(\mathcal{H})$ and that $\mathbb{B}(\mathcal{D}_2, \mathcal{D}_1)^* = \mathbb{B}(\mathcal{D}_1, \mathcal{D}_2)$. Thus $\mathbb{B}(\mathcal{D})$ is a $*$-subalgebra of $\mathbb{B}(\mathcal{H})$. Moreover, $\mathbb{B}(\mathcal{D}_2, \mathcal{D}_1) = \mathbb{B}(\mathcal{D}_2, \mathcal{H}) \cap \mathbb{B}(\mathcal{H}, \mathcal{D}_1)$.

Lemma 3.1.2. *Let a be a closable linear operator on $\mathcal{H}$. Suppose that $c \in \mathbb{B}(\mathcal{H})$ and $c\mathcal{H} \subseteq \mathcal{D}(a)$. Then $ac \in \mathbb{B}(\mathcal{H})$, c^*a^* is bounded on $\mathcal{D}(a^*)$ and $\overline{c^*a^*} \in \mathbb{B}(\mathcal{H})$. If $a \in \mathcal{L}^+(\mathcal{D})$ for $\mathcal{D} := \mathcal{D}(a)$, then c^*a^+ is bounded on $\mathcal{D}$ and $\overline{c^*a^+} \in \mathbb{B}(\mathcal{H})$.*

Proof. We have that $(ac)^* \supseteq c^*a^*$. Since a is closable, $\mathcal{D}(a^*)$ is dense in $\mathcal{H}$. Hence the adjoint of the operator ac is densely defined. Therefore, ac is a closed linear operator defined on the whole Hilbert space $\mathcal{H}$. By the closed graph theorem, $ac \in \mathbb{B}(\mathcal{H})$. Thus $(ac)^* \in \mathbb{B}(\mathcal{H})$ which implies that c^*a^* is bounded on $\mathcal{D}(a^*)$. The other assertions are clear. □

Corollary 3.1.3. *Let $\mathcal{A}$ be an O-family on $\mathcal{D}$ and let $c \in \mathbb{B}(\mathcal{H})$. Suppose that $c\mathcal{H} \subseteq \mathcal{D}$. Then $c \in \mathfrak{L}(\mathcal{H}, \mathcal{D}_{\mathcal{A}})$ and $c\mathcal{U}_{\mathcal{H}}$ is a bounded subset of the locally convex space $\mathcal{D}_{\mathcal{A}}$.*

Proof. Let $a \in \mathcal{A}$. Since a is closable by Definition 2.1.1, Lemma 3.1.2 yields $ac \in \mathbb{B}(\mathcal{H})$. Since $\|ac\varphi\| \leqq \|ac\| \, \|\varphi\|$ for $\varphi \in \mathcal{D}$, $c \in \mathfrak{L}(\mathcal{H}, \mathcal{D}_{\mathcal{A}})$. Since $\sup \{\|\varphi\|_a : \varphi \in c\mathcal{U}_{\mathcal{H}}\} \leqq \|ac\|$, $c\mathcal{U}_{\mathcal{H}}$ is bounded in $\mathcal{D}_{\mathcal{A}}$. □

Proposition 3.1.4. *Let $\mathcal{A}$ be an O*-algebra on the domain $\mathcal{D}(\mathcal{A})$ of the Hilbert space $\mathcal{H}$ such that the locally convex space $\mathcal{D}_{\mathcal{A}}$ is sequentially complete. Let c and d be operators of $\mathbb{B}(\mathcal{H})$. Suppose that there exist positive numbers α and δ such that $\|d^*\varphi\| \leqq \alpha \, \| \, |c^*|^\delta \, \varphi\|$ for all φ in $\mathcal{H}$. Suppose that $c\mathcal{H} \subseteq \mathcal{D}(\mathcal{A})$. Then $d\mathcal{H} \subseteq \mathcal{D}(\mathcal{A})$. In particular, $|c^*|^\varepsilon \, \mathcal{H} \subseteq \mathcal{D}(\mathcal{A})$ for all $\varepsilon > 0$.*

Proof. Fix $x \in \mathcal{A}$. Our first step is to show that $|c^*|^\varepsilon \, \mathcal{H} \subseteq \mathcal{D}(\bar{x})$ for each positive number ε. If we write ε as $\varepsilon = \varepsilon' + n$ with $n \in \mathbb{N}_0$ and $0 < \varepsilon' \leqq 1$, we have $|c^*|^\varepsilon \, \mathcal{H} \subseteq |c^*|^{\varepsilon'} \, \mathcal{H}$. Therefore, it suffices to prove $|c^*|^\varepsilon \, \mathcal{H} \subseteq \mathcal{D}(\bar{x})$ for all $\varepsilon \in \mathbb{R}$, $0 < \varepsilon \leqq 1$. Fix such an ε and take a $k \in \mathbb{N}$ such that $(2\varepsilon)^{-1} \leqq k$. Let y denote the positive self-adjoint operator $\bar{x}^*\bar{x}$. Put $a := (y + I)^{1/2\varepsilon}$. From $x^+x \subseteq y$ we get $\mathcal{D}(\mathcal{A}) = \mathcal{D}((x^+x)^k) \subseteq \mathcal{D}(y^k) = \mathcal{D}((y + I)^k) \subseteq \mathcal{D}(a)$. Thus $c\mathcal{H} \subseteq \mathcal{D}(a)$. Since $a = a^*$, Lemma 3.1.2 shows that c^*a is bounded on $\mathcal{D}(a)$, i.e., $\|c^*a\varphi\| \leqq \|c^*a\| \, \|\varphi\|$ for $\varphi \in \mathcal{D}(a)$. Since a has a bounded inverse and hence $a\mathcal{D}(a) = \mathcal{H}$, the latter gives $\|c^*\psi\| \leqq \|c^*a\| \, \|a^{-1}\psi\|$ for all $\psi \in \mathcal{H}$. Hence $|c^*|^2 \leqq \|c^*a\|^2 \, a^{-2}$. Because $0 < \varepsilon \leqq 1$, the Kato-Heinz inequality (see e.g. Reed/Simon [1], VIII, Exercise 51) applies and yields $|c^*|^{2\varepsilon} \leqq \|c^*a\|^{2\varepsilon} \, a^{-2\varepsilon}$, i.e., $\| \, |c^*|^\varepsilon \, \varphi\| \leqq \|c^*a\|^\varepsilon \, \|a^{-\varepsilon}\varphi\|$ for $\varphi \in \mathcal{H}$. Therefore, if $\zeta \in \mathcal{H}$, then $|\langle |c^*|^\varepsilon \, \zeta, a^\varepsilon\psi\rangle| \leqq \|\zeta\| \, \| \, |c^*|^\varepsilon \, a^\varepsilon\psi\| \leqq \|\zeta\| \, \|c^*a\|^\varepsilon \, \|\psi\|$ for all $\psi \in \mathcal{D}(a^\varepsilon)$. Hence $|c^*|^\varepsilon \, \zeta \in \mathcal{D}((a^\varepsilon)^*) = \mathcal{D}(a^\varepsilon) = \mathcal{D}((y + I)^{1/2}) = \mathcal{D}(y^{1/2}) = \mathcal{D}(|\bar{x}|) = \mathcal{D}(\bar{x})$. Thus $|c^*|^\varepsilon \, \mathcal{H} \subseteq \mathcal{D}(\bar{x})$.

Next we prove that $|c^*|^\varepsilon \mathcal{H} \subseteq \mathcal{D}(\mathcal{A})$ for $\varepsilon > 0$. Let e_n, $n \in \mathbb{N}$, denote the spectral projection of the positive self-adjoint operator $|c^*|$ associated to the interval $[0, 1/n]$. Suppose $\varphi \in \mathcal{H}$. Define $\varphi_n := |c^*|^\varepsilon (I - e_n) \varphi$, $n \in \mathbb{N}$. Since $|c^*|^\varepsilon (I - e_n) \mathcal{H} \subseteq |c^*|^2 \mathcal{H} = cc^*\mathcal{H} \subseteq \mathcal{D}(\mathcal{A})$, we have $\varphi_n \in \mathcal{D}(\mathcal{A})$ for $n \in \mathbb{N}$. Let $x \in \mathcal{A}$. By Lemma 3.1.2, the operator $\bar{x} |c^*|^\varepsilon$ is bounded, since $|c^*|^\varepsilon \mathcal{H} \subseteq \mathcal{D}(\bar{x})$ as shown above. Therefore,

$$\|\varphi_n - \varphi_m\|_x = \|x(\varphi_n - \varphi_m)\| = \|x\, |c^*|^\varepsilon (e_n - e_m) \varphi\| \leqq \|\bar{x}\, |c^*|^\varepsilon\| \, \|(e_n - e_m) \varphi\| \to 0$$
$$\text{if} \quad n \to \infty \quad \text{and} \quad m \to \infty.$$

This shows that the sequence $(\varphi_n : n \in \mathbb{N})$ is a Cauchy sequence in $\mathcal{D}_\mathcal{A}$. Since $\mathcal{D}_\mathcal{A}$ is assumed to be sequentially complete, this sequence has a limit, say φ_0, in $\mathcal{D}_\mathcal{A}$. From $\lim_n \varphi_n = |c^*|^\varepsilon \varphi$ in $\mathcal{H}$ we obtain $|c^*|^\varepsilon \varphi = \varphi_0 \in \mathcal{D}(\mathcal{A})$. This proves that $|c^*|^\varepsilon \mathcal{H} \subseteq \mathcal{D}(\mathcal{A})$.

Finally, we show that $d\mathcal{H} \subseteq \mathcal{D}(\mathcal{A})$. From the assumptions concerning c and d it follows that there exists an operator $b \in \mathbf{B}(\mathcal{H})$ such that $d^* = b\, |c^*|^\delta$. Hence $d\mathcal{H} = |c^*|^\delta b^*\mathcal{H} \subseteq |c^*|^\delta \mathcal{H} \subseteq \mathcal{D}(\mathcal{A})$. □

Corollary 3.1.5. *Let $\mathcal{D}_1$, $\mathcal{D}_2$ and $\mathcal{D}$ be dense linear subspaces of the Hilbert space $\mathcal{H}$. Suppose that there are O*-algebras $\mathcal{A}_1$, $\mathcal{A}_2$ and $\mathcal{A}$ on $\mathcal{D}_1 = \mathcal{D}(\mathcal{A}_1)$, $\mathcal{D}_2 = \mathcal{D}(\mathcal{A}_2)$ and $\mathcal{D} = \mathcal{D}(\mathcal{A})$, respectively, such that the locally convex spaces $\mathcal{D}_{\mathcal{A}_1}$, $\mathcal{D}_{\mathcal{A}_2}$ and $\mathcal{D}_\mathcal{A}$ are sequentially complete.*

(i) *If $c \in \mathbf{B}(\mathcal{D}_2, \mathcal{D}_1)$, then $|c|^\varepsilon \in \mathbf{B}(\mathcal{D}_2)_+$ and $|c^*|^\varepsilon \in \mathbf{B}(\mathcal{D}_1)_+$ for each $\varepsilon > 0$.*

(ii) *Let $c, d \in \mathbf{B}(\mathcal{H})$. Suppose that $\|d\varphi\| \leqq \|c\varphi\|$ and $\|d^*\varphi\| \leqq \|c^*\varphi\|$ for $\varphi \in \mathcal{H}$. If $c \in \mathbf{B}(\mathcal{D}_2, \mathcal{D}_1)$, then $d \in \mathbf{B}(\mathcal{D}_2, \mathcal{D}_1)$.*

(iii) *Suppose $c, d \in \mathbf{B}(\mathcal{H})$ and $0 \leqq d \leqq c$. If $c \in \mathbf{B}(\mathcal{D})_+$, then $d \in \mathbf{B}(\mathcal{D})_+$.*

(iv) *Suppose $c = c^* \in \mathbf{B}(\mathcal{D})$. Let f be a bounded function on $\mathbb{R}$ which is measurable with respect to the spectral measure of c. Suppose that there are positive numbers α and δ such that $|f(\lambda)| \leqq \alpha\, |\lambda|^\delta$ for $\lambda \in \mathbb{R}$. Then $f(c) \in \mathbf{B}(\mathcal{D})$.*

Proof. (i) is already contained in Proposition 3.1.4. Using that $\|b\varphi\| = \|\, |b|\, \varphi\|$ for $b \in \mathbf{B}(\mathcal{H})$ and $\varphi \in \mathcal{H}$, (ii) follows from Proposition 3.1.4 applied in case $\alpha = \delta = 1$. We verify (iii). The assumptions of Proposition 3.1.4 are fulfilled with $d^{1/2}$ in place of d and $\delta = 1/2$, $\alpha = 1$. Therefore, $d^{1/2} \in \mathbf{B}(\mathcal{D})$, so that $d \in \mathbf{B}(\mathcal{D})$. (iv) follows by letting $d = f(c)$. □

Corollary 3.1.6. *Suppose $\mathcal{D}_1$ and $\mathcal{D}_2$ satisfy the assumptions of Corollary 3.1.5. For each operator $c \in \mathbf{B}(\mathcal{D}_2, \mathcal{D}_1)$, there exist operators $c_1 \in \mathbf{B}(\mathcal{D}_2, \mathcal{D}_1)$, $c_2 \in \mathbf{B}(\mathcal{D}_2)_+$, $c_3 \in \mathbf{B}(\mathcal{D}_1)_+$ and $c_4 \in \mathbf{B}(\mathcal{D}_2, \mathcal{D}_1)$ such that $c = c_1c_2 = c_3c_4$.*

Proof. Let $c = u\, |c|$ be the polar decomposition of $c \in \mathbf{B}(\mathcal{D}_2, \mathcal{D}_1)$. Set $c_1 := u\, |c|^{1/2}$ and $c_2 := |c|^{1/2}$. By Corollary 3.1.5, (i), $c_2 \in \mathbf{B}(\mathcal{D}_2)_+$. This gives $c_1^*\mathcal{H} = |c|^{1/2} u^*\mathcal{H} = c_2u^*\mathcal{H} \subseteq \mathcal{D}_2$. We prove that $c_1\mathcal{H} \subseteq \mathcal{D}_1$. We have $c_1c_1^* = u\, |c|\, u^* = c^*$, where the last equality follows from the properties of the polar decomposition (see p. 29). Thus $\|\, |c_1^*|\, \varphi\|^2 = \langle c_1c_1^*\varphi, \varphi\rangle = \|\, |c^*|^{1/2} \varphi\|^2$ for $\varphi \in \mathcal{H}$. Therefore, Proposition 3.1.4 applies with $d = c_1$, $\delta = 1/2$, $\alpha = 1$ and yields $c_1\mathcal{H} \subseteq \mathcal{D}_1$. This proves $c_1 \in \mathbf{B}(\mathcal{D}_2, \mathcal{D}_1)$. The assertions concerning c_3 and c_4 follow if we apply the preceding with c^* in place of c. (Of course, one can also define $c_3 = |c^*|^{1/2}$ and $c_4 = |c^*|^{1/2} u$ and verify the above properties directly.) □

Corollary 3.1.7. *Let $\mathcal{D}_1$, $\mathcal{D}_2$ and $\mathcal{D}$ be as in Corollary 3.1.5.*

(i) *If $c \in \mathbf{B}(\mathcal{D}_2, \mathcal{D}_1)$, then $\overline{a_1 c a_2} \in \mathbf{B}(\mathcal{D}_2, \mathcal{D}_1)$ for all $a_1 \in \mathcal{L}^+(\mathcal{D}_1)$ and $a_2 \in \mathcal{L}^+(\mathcal{D}_2)$.*

(ii) *$\mathbf{B}(\mathcal{D}) \upharpoonright \mathcal{D}$ is a two-sided $*$-ideal of the $*$-algebra $\mathcal{L}^+(\mathcal{D})$.*

Proof. (i): Let $c \in \mathbf{B}(\mathcal{D}_2, \mathcal{D}_1)$, $a_1 \in \mathcal{L}^+(\mathcal{D}_1)$ and $a_2 \in \mathcal{L}^+(\mathcal{D}_2)$. By Corollary 3.1.6, there are operators $c_1 \in \mathbf{B}(\mathcal{D}_2, \mathcal{D}_1)$ and $c_2 \in \mathbf{B}(\mathcal{D}_2)_+$ such that $c = c_1 c_2$. From Lemma 3.1.2, $a_1 c_1 \in \mathbf{B}(\mathcal{H})$ and $\overline{c_2 a_2} \in \mathbf{B}(\mathcal{H})$. From $a_1 c a_2 = a_1 c_1 c_2 a_2$ on $\mathcal{D}_2$ we get $\overline{a_1 c a_2} = a_1 c_1 \cdot \overline{c_2 a_2}$ $\in \mathbf{B}(\mathcal{H})$, $\overline{a_1 c a_2}\mathcal{H} \subseteq a_1 c_1 \mathcal{H} \subseteq a_1 \mathcal{D}_1 \subseteq \mathcal{D}_1$, $(\overline{a_1 c a_2})^* = (c_2 a_2)^* (a_1 c_1)^* = a_2^+ c_2 (a_1 c_1)^*$ and $(\overline{a_1 c a_2})^* \mathcal{H} \subseteq a_2^+ c_2 \mathcal{H} \subseteq \mathcal{D}_2$. This proves that $\overline{a_1 c a_2} \in \mathbf{B}(\mathcal{D}_2, \mathcal{D}_1)$.

(ii) follows immediately from (i) by letting $\mathcal{D} = \mathcal{D}_1 = \mathcal{D}_2$. □

Remark 1. If $\mathcal{D}_1$ and $\mathcal{D}_2$ are arbitrary dense linear subspaces of the Hilbert space $\mathcal{H}$, then the operator $a_1 c a_2$ is bounded on its domain $\mathcal{D}_2$ for each $c \in \mathbf{B}(\mathcal{D}_2, \mathcal{D}_1)$, $a_1 \in \mathcal{L}^+(\mathcal{D}_1)$ and $a_2 \in \mathcal{L}^+(\mathcal{D}_2)$. This follows at once from Corollary 3.1.7 applied to the linear subspaces $\tilde{\mathcal{D}}_1 := \mathcal{D}(\hat{\mathcal{A}}_1)$ and $\tilde{\mathcal{D}}_2 := \mathcal{D}(\hat{\mathcal{A}}_2)$, where $\mathcal{A}_1 := \mathcal{L}^+(\mathcal{D}_1)$ and $\mathcal{A}_2 := \mathcal{L}^+(\mathcal{D}_2)$.

The Algebra $\mathcal{L}(\mathcal{D}_{\mathcal{B}}^+, \mathcal{D}_{\mathcal{A}})$

We shall use the symbols σ and $\sigma^|$ to denote the weak-topology $\sigma(\mathcal{D}_{\mathcal{A}}, \mathcal{D}_{\mathcal{A}}^|)$ and the weak*-topology $\sigma(\mathcal{D}_{\mathcal{B}}^|, \mathcal{D}_{\mathcal{B}})$, respectively, for arbitrary O-families $\mathcal{A}$ and $\mathcal{B}$.

Definition 3.1.8. Let $\mathcal{L}(\mathcal{D}_{\mathcal{B}}^+, \mathcal{D}_{\mathcal{A}})$ be the vectorspace $\mathfrak{L}(\mathcal{D}_{\mathcal{B}}^+[\sigma^|], \mathcal{D}(\mathcal{A})[\sigma])$ of all continuous linear mappings of $\mathcal{D}_{\mathcal{B}}^+[\sigma^|]$ into $\mathcal{D}(\mathcal{A})[\sigma]$. We write $\mathcal{L}(\mathcal{D}_2^+, \mathcal{D}_1)$ for $\mathcal{L}(\mathcal{D}_{\mathcal{B}}^+, \mathcal{D}_{\mathcal{A}})$ if $\mathcal{A} = \mathcal{L}^+(\mathcal{D}_1)$, $\mathcal{D}_1 = \mathcal{D}(\mathcal{A})$ and $\mathcal{B} = \mathcal{L}^+(\mathcal{D}_2)$, $\mathcal{D}_2 = \mathcal{D}(\mathcal{B})$.

We recall some facts which have been stated in Section 1.2 for general locally convex spaces in the special case $\mathcal{L}(\mathcal{D}_{\mathcal{B}}^+, \mathcal{D}_{\mathcal{A}})$. Let y be a linear mapping of $\mathcal{D}_{\mathcal{B}}^+$ into $\mathcal{D}(\mathcal{A})$. By 1.2/(3) and 2.3/(1), the associated sesquilinear form $\mathfrak{d}_y$ on $\mathcal{D}_{\mathcal{A}}^| \times \mathcal{D}_{\mathcal{B}}^|$ is defined by

$$\mathfrak{d}_y(\varphi^|, \psi^|) = \langle y\psi^|, \varphi^| \rangle, \ \varphi^| \in \mathcal{D}_{\mathcal{A}}^| \quad \text{and} \quad \psi^| \in \mathcal{D}_{\mathcal{B}}^|. \tag{1}$$

By Lemma 1.2.1, (ii), $y \in \mathcal{L}(\mathcal{D}_{\mathcal{B}}^+, \mathcal{D}_{\mathcal{A}})$ if and only if $\mathfrak{d}_y \in \mathfrak{B}(\mathcal{D}_{\mathcal{A}}^|[\sigma^|], \mathcal{D}_{\mathcal{B}}^|[\sigma^|])$, i.e., if $\langle y\psi^|, \cdot \rangle$ $\in (\mathcal{D}_{\mathcal{A}}^|[\sigma^|])^| \equiv \mathcal{D}(\mathcal{A})$ for each $\psi^| \in \mathcal{D}_{\mathcal{B}}^|$ and $\overline{\langle y \cdot, \varphi^| \rangle} \in (\mathcal{D}_{\mathcal{B}}^|[\sigma^|])^| \equiv \mathcal{D}(\mathcal{B})$ for each $\varphi^| \in \mathcal{D}_{\mathcal{A}}^|$. The mapping $y \to \mathfrak{d}_y$ is a vector space isomorphism of $\mathcal{L}(\mathcal{D}_{\mathcal{B}}^+, \mathcal{D}_{\mathcal{A}})$ and $\mathfrak{B}(\mathcal{D}_{\mathcal{A}}^|[\sigma^|], \mathcal{D}_{\mathcal{B}}^|[\sigma^|])$. For each $y \in \mathcal{L}(\mathcal{D}_{\mathcal{B}}^+, \mathcal{D}_{\mathcal{A}})$, there is a unique element $y^+ \in \mathcal{L}(\mathcal{D}_{\mathcal{A}}^+, \mathcal{D}_{\mathcal{B}})$ such that $(\mathfrak{d}_y)^+$ $= \mathfrak{d}_{y^+}$. By (1) and 2.3/(1), y^+ is characterized by the relation

$$\langle y\psi^|, \varphi^| \rangle = \langle \psi^|, y^+\varphi^| \rangle, \ \varphi^| \in \mathcal{D}_{\mathcal{A}}^| \quad \text{and} \quad \psi^| \in \mathcal{D}_{\mathcal{B}}^|. \tag{2}$$

The map $y \to y^+$ is a conjugate-linear bijection of $\mathcal{L}(\mathcal{D}_{\mathcal{B}}^+, \mathcal{D}_{\mathcal{A}})$ on $\mathcal{L}(\mathcal{D}_{\mathcal{A}}^+, \mathcal{D}_{\mathcal{B}})$.

Lemma 3.1.9. *If $y \in \mathcal{L}(\mathcal{D}_{\mathcal{B}}^+, \mathcal{D}_{\mathcal{A}})$, then $y \upharpoonright \mathcal{H} \in \mathbf{B}(\mathcal{D}(\mathcal{B}), \mathcal{D}(\mathcal{A}))$ and $(y \upharpoonright \mathcal{H})^* = y^+ \upharpoonright \mathcal{H}$.*

Proof. Setting $c := y \upharpoonright \mathcal{H}$ and $d := y^+ \upharpoonright \mathcal{H}$, (2) yields $\langle c\varphi, \psi \rangle = \langle \varphi, d\psi \rangle$ for $\varphi, \psi \in \mathcal{H}$. Here on both sides $\langle \cdot, \cdot \rangle$ means the scalar product of $\mathcal{H}$, since $y\mathcal{H} \subseteq \mathcal{D}(\mathcal{A})$ and $y^+\mathcal{H}$ $\subseteq \mathcal{D}(\mathcal{B})$. Consequently, $(y \upharpoonright \mathcal{H})^* \equiv c^* = d \equiv y^+ \upharpoonright \mathcal{H}$. Since c^* is defined on the whole $\mathcal{H}$, $c \in \mathbf{B}(\mathcal{H})$ by the closed graph theorem. Since $c\mathcal{H} = y\mathcal{H} \subseteq \mathcal{D}(\mathcal{A})$ and $c^*\mathcal{H} = y^+\mathcal{H}$ $\subseteq \mathcal{D}(\mathcal{B})$, $c \in \mathbf{B}(\mathcal{D}(\mathcal{B}), \mathcal{D}(\mathcal{A}))$. □

Suppose y and z are elements of $\mathcal{L}(\mathcal{D}_{\mathcal{B}}^+, \mathcal{D}_{\mathcal{A}})$. The composition yz of y and z is a linear mapping of $\mathcal{D}_{\mathcal{B}}^+$ into $\mathcal{D}(\mathcal{A})$. We show that $yz \in \mathcal{L}(\mathcal{D}_{\mathcal{B}}^+, \mathcal{D}_{\mathcal{A}})$. Indeed, since $\mathcal{D}(\mathcal{B}) \subseteq \mathcal{D}_{\mathcal{A}}^{|}$, we have $\sigma^{|} \upharpoonright \mathcal{D}(\mathcal{A}) \subseteq \sigma$. From $y, z \in \mathfrak{L}(\mathcal{D}_{\mathcal{B}}^+[\sigma^{|}], \mathcal{D}(\mathcal{A})[\sigma])$ it therefore follows that $y \upharpoonright \mathcal{D}(\mathcal{A}) \in \mathfrak{L}(\mathcal{D}(\mathcal{A})[\sigma], \mathcal{D}(\mathcal{A})[\sigma])$ and so $yz \in \mathfrak{L}(\mathcal{D}_{\mathcal{B}}^+[\sigma^{|}], \mathcal{D}(\mathcal{A})[\sigma]) = \mathcal{L}(\mathcal{D}_{\mathcal{B}}^+, \mathcal{D}_{\mathcal{A}})$. With the product just defined, $\mathcal{L}(\mathcal{D}_{\mathcal{B}}^+, \mathcal{D}_{\mathcal{A}})$ is an algebra. Moreover, we have $(yz)^+ = z^+y^+$ for $y, z \in \mathcal{L}(\mathcal{D}_{\mathcal{B}}^+, \mathcal{D}_{\mathcal{A}})$, where z^+y^+ is the product of z^+ and y^+ in $\mathcal{L}(\mathcal{D}_{\mathcal{A}}^+, \mathcal{D}_{\mathcal{B}})$. In particular, we conclude that $\mathcal{L}(\mathcal{D}_{\mathcal{A}}^+, \mathcal{D}_{\mathcal{A}})$ with the involution $y \to y^+$ is a $*$-algebra.

Proposition 3.1.10. *Suppose that $\mathcal{A}$ and $\mathcal{B}$ are O*-algebras in $\mathcal{H}$ such that the locally convex spaces $\mathcal{D}_{\mathcal{A}}$ and $\mathcal{D}_{\mathcal{B}}$ are sequentially complete. Then the mapping $y \to y \upharpoonright \mathcal{H}$ is an isomorphism of the algebras $\mathcal{L}(\mathcal{D}_{\mathcal{B}}^+, \mathcal{D}_{\mathcal{A}})$ and $\mathbf{B}(\mathcal{D}(\mathcal{B}), \mathcal{D}(\mathcal{A}))$. Moreover, $y^+ \upharpoonright \mathcal{H} = (y \upharpoonright \mathcal{H})^*$ for all $y \in \mathcal{L}(\mathcal{D}_{\mathcal{B}}^+, \mathcal{D}_{\mathcal{A}})$.*

Proof. From Lemma 3.1.9 we know already that $y \upharpoonright \mathcal{H} \in \mathbf{B}(\mathcal{D}(\mathcal{B}), \mathcal{D}(\mathcal{A}))$ and $y^+ \upharpoonright \mathcal{H} = (y \upharpoonright \mathcal{H})^*$ for $y \in \mathcal{L}(\mathcal{D}_{\mathcal{B}}^+, \mathcal{D}_{\mathcal{A}})$. Thus it is clear that the map $y \to y \upharpoonright \mathcal{H}$ is an algebra homomorphism of $\mathcal{L}(\mathcal{D}_{\mathcal{B}}^+, \mathcal{D}_{\mathcal{A}})$ into $\mathbf{B}(\mathcal{D}(\mathcal{B}), \mathcal{D}(\mathcal{A}))$. Since $\mathcal{H}$ is dense in $\mathcal{D}_{\mathcal{B}}^+[\sigma^{|}]$ by Proposition 2.3.5, this map is injective. It remains to show that it is also surjective. Let $c \in \mathbf{B}(\mathcal{D}(\mathcal{B}), \mathcal{D}(\mathcal{A}))$. By Corollary 3.1.6, there are operators $c_1 \in \mathbf{B}(\mathcal{D}(\mathcal{B}), \mathcal{D}(\mathcal{A}))$ and $c_2 \in \mathbf{B}(\mathcal{D}(\mathcal{B}))_+$ such that $c = c_1c_2$. Since $c_1 \in \mathfrak{L}(\mathcal{H}, \mathcal{D}_{\mathcal{A}})$ by Corollary 3.1.3, for any $\varphi^{|} \in \mathcal{D}_{\mathcal{A}}^{|}$ the map $\varphi \to \varphi^{|}(c_1\varphi)$ is a continuous linear functional on $\mathcal{H}$, so that $c_1 \in \mathfrak{L}(\mathcal{H}[\sigma], \mathcal{D}(\mathcal{A})[\sigma])$. Similarly, $c_2 \in \mathfrak{L}(\mathcal{H}[\sigma], \mathcal{D}(\mathcal{B})[\sigma])$. For the Hilbert space $\mathcal{H}$ it is obvious that $\mathcal{H}[\sigma] \equiv \mathcal{H}^+[\sigma^{|}]$, so $c_2 \in \mathfrak{L}(\mathcal{H}^+[\sigma^{|}], \mathcal{D}(\mathcal{B})[\sigma]) = \mathcal{L}(\mathcal{H}^+, \mathcal{D}_{\mathcal{B}})$. Therefore, $c_2^+ \in \mathcal{L}(\mathcal{D}_{\mathcal{B}}^+, \mathcal{H}) = \mathfrak{L}(\mathcal{D}_{\mathcal{B}}^+[\sigma^{|}], \mathcal{H}[\sigma])$ and hence $y := c_1c_2^+ \in \mathfrak{L}(\mathcal{D}_{\mathcal{B}}^+[\sigma^{|}], \mathcal{D}(\mathcal{A})[\sigma]) = \mathcal{L}(\mathcal{D}_{\mathcal{B}}^+, \mathcal{D}_{\mathcal{A}})$. By Lemma 3.1.9, $c_2^+ \upharpoonright \mathcal{H} = c_2^* = c_2$. Hence $y \upharpoonright \mathcal{H} = c_1c_2^+ \upharpoonright \mathcal{H} = c_1c_2 = c$ which proves that the map is surjective. □

Corollary 3.1.11. *Suppose $\mathcal{A}$ is an O*-algebra on $\mathcal{D}(\mathcal{A})$ such that $\mathcal{D}_{\mathcal{A}}$ is sequentially complete. Then the mapping $y \to y \upharpoonright \mathcal{H}$ is a $*$-isomorphism of the $*$-algebras $\mathcal{L}(\mathcal{D}_{\mathcal{A}}^+, \mathcal{D}_{\mathcal{A}})$ and $\mathbf{B}(\mathcal{D}(\mathcal{A}))$.*

Proof. Set $\mathcal{A} = \mathcal{B}$ in Proposition 3.1.10. □

Corollary 3.1.12. *Let $\mathcal{A}$ and $\mathcal{B}$ be as in Proposition 3.1.10. Then for each $y \in \mathcal{L}(\mathcal{D}_{\mathcal{B}}^+, \mathcal{D}_{\mathcal{A}})$ there are $y_1, y_4 \in \mathcal{L}(\mathcal{D}_{\mathcal{B}}^+, \mathcal{D}_{\mathcal{A}})$, $y_2 = y_2^+ \in \mathcal{L}(\mathcal{D}_{\mathcal{B}}^+, \mathcal{D}_{\mathcal{B}})$ and $y_3 = y_3^+ \in \mathcal{L}(\mathcal{D}_{\mathcal{A}}^+, \mathcal{D}_{\mathcal{A}})$ such that $y = y_1y_2 = y_3y_4$.*

Proof. Combine Proposition 3.1.10 and Corollary 3.1.6. □

Corollary 3.1.13. *Let $\mathcal{A}$ and $\mathcal{B}$ be as in Proposition 3.1.10. Then $\mathcal{L}(\mathcal{D}_{\mathcal{B}}^+, \mathcal{D}_{\mathcal{A}}) \subseteq \mathfrak{L}(\mathcal{D}_{\mathcal{B}}^+[\beta], \mathcal{D}_{\mathcal{A}})$ and $\mathfrak{B}(\mathcal{D}_{\mathcal{A}}^{|}[\sigma^{|}], \mathcal{D}_{\mathcal{B}}^{|}[\sigma^{|}]) \subseteq \mathcal{B}(\mathcal{D}_{\mathcal{A}}^{|}[\beta], \mathcal{D}_{\mathcal{B}}^{|}[\beta])$. For each $y \in \mathcal{L}(\mathcal{D}_{\mathcal{B}}^+, \mathcal{D}_{\mathcal{A}})$, $\mathfrak{d}_y$ is in $\mathcal{B}(\mathcal{D}_{\mathcal{A}}^{|}[\beta], \mathcal{D}_{\mathcal{B}}^{|}[\beta])$.*

Proof. Let $y \in \mathcal{L}(\mathcal{D}_{\mathcal{B}}^+, \mathcal{D}_{\mathcal{A}})$. By Corollary 3.1.12, there are $y_1 \in \mathcal{L}(\mathcal{D}_{\mathcal{B}}^+, \mathcal{D}_{\mathcal{A}})$ and $y_2 = y_2^+ \in \mathcal{L}(\mathcal{D}_{\mathcal{B}}^+, \mathcal{D}_{\mathcal{B}})$ such that $y = y_1y_2$. Put $c_1 := y_1 \upharpoonright \mathcal{H}$ and $c_2 := y_2 \upharpoonright \mathcal{H}$. By Lemma 3.1.2, ac_1 is bounded for $a \in \mathcal{A}$. If $\varphi^{|} \in \mathcal{D}_{\mathcal{A}}^{|}$ and $\psi^{|} \in \mathcal{D}_{\mathcal{B}}^{|}$, we have

$$\|y\psi^{|}\|_a = \|ac_1y_2\psi^{|}\| \leqq \|ac_1\| \sup_{\psi \in \mathcal{U}_{\mathcal{H}}} |\langle y_2\psi^{|}, \psi\rangle| = \|ac_1\|\, r_{c_2\mathcal{U}_{\mathcal{H}}}(\psi^{|})$$

and

$$|\mathfrak{d}_y(\varphi^{|}, \psi^{|})| = |\langle y\psi^{|}, \varphi^{|}\rangle| = |\langle y_2\psi^{|}, y_1^+\varphi^{|}\rangle| \leqq \|y_2\psi^{|}\|\, \|y_1^+\varphi^{|}\|$$

$$= \sup_{\psi \in \mathcal{U}_{\mathcal{H}}} |\langle y_2\psi^{|}, \psi\rangle| \cdot \sup_{\varphi \in \mathcal{U}_{\mathcal{H}}} |\langle y_1^+\varphi^{|}, \varphi\rangle| = r_{c_2\mathcal{U}_{\mathcal{H}}}(\psi^{|})\, r_{c_1\mathcal{U}_{\mathcal{H}}}(\varphi^{|}).$$

Since $c_1 \in \mathbb{B}(\mathcal{D}(\mathcal{B}), \mathcal{D}(\mathcal{A}))$ and $c_2 \in \mathbb{B}(\mathcal{D}(\mathcal{B}))$, $c_1\mathcal{U}_{\mathcal{H}}$ and $c_2\mathcal{U}_{\mathcal{H}}$ are bounded sets in $\mathcal{D}_{\mathcal{A}}$ and $\mathcal{D}_{\mathcal{B}}$, respectively, by Corollary 3.1.3. Therefore, the preceding inequalities show that $y \in \mathfrak{L}(\mathcal{D}^+_{\mathcal{B}}[\beta], \mathcal{D}_{\mathcal{A}})$ and $\mathfrak{d}_y \in \mathcal{B}(\mathcal{D}'_{\mathcal{A}}[\beta], \mathcal{D}'_{\mathcal{B}}[\beta])$. Since each $\mathfrak{d} \in \mathfrak{B}(\mathcal{D}'_{\mathcal{A}}[\sigma'], \mathcal{D}'_{\mathcal{B}}[\sigma'])$ is of the form $\mathfrak{d} = \mathfrak{d}_y$ with $y \in \mathcal{L}(\mathcal{D}^+_{\mathcal{B}}, \mathcal{D}_{\mathcal{A}})$ by Lemma 1.2.1, the proof is complete. □

Corollary 3.1.14. *Suppose $\mathcal{A}$ and $\mathcal{B}$ are O*-algebras in $\mathcal{H}$ such that the spaces $\mathcal{D}_{\mathcal{A}}$ and $\mathcal{D}_{\mathcal{B}}$ are semireflexive. Then $\mathcal{L}(\mathcal{D}^+_{\mathcal{B}}, \mathcal{D}_{\mathcal{A}}) = \mathfrak{L}(\mathcal{D}^+_{\mathcal{B}}[\beta], \mathcal{D}_{\mathcal{A}}) = \mathfrak{L}(\mathcal{D}^+_{\mathcal{B}}[\beta], \mathcal{D}(\mathcal{A})[\sigma])$ and $\mathfrak{B}(\mathcal{D}'_{\mathcal{A}}[\sigma'], \mathcal{D}'_{\mathcal{B}}[\sigma']) = \mathfrak{B}(\mathcal{D}'_{\mathcal{A}}[\beta], \mathcal{D}'_{\mathcal{B}}[\beta]) = \mathcal{B}(\mathcal{D}'_{\mathcal{A}}[\beta], \mathcal{D}'_{\mathcal{B}}[\beta])$.*

Proof. Since each semireflexive locally convex space is sequentially complete (SCHÄFER [1], IV, 5.5), the assumptions of Corollary 3.1.13 are fulfilled, so that $\mathcal{L}(\mathcal{D}^+_{\mathcal{B}}, \mathcal{D}_{\mathcal{A}}) \subseteq \mathfrak{L}(\mathcal{D}^+_{\mathcal{B}}[\beta], \mathcal{D}_{\mathcal{A}})$ and $\mathfrak{B}(\mathcal{D}'_{\mathcal{A}}[\sigma'], \mathcal{D}'_{\mathcal{B}}[\sigma']) \subseteq \mathcal{B}(\mathcal{D}'_{\mathcal{A}}[\beta], \mathcal{D}'_{\mathcal{B}}[\beta])$. Combining the latter with the obvious relations $\mathfrak{L}(\mathcal{D}^+_{\mathcal{B}}[\beta], \mathcal{D}_{\mathcal{A}}) \subseteq \mathfrak{L}(\mathcal{D}^+_{\mathcal{B}}[\beta], \mathcal{D}(\mathcal{A})[\sigma])$ and $\mathcal{B}(\mathcal{D}'_{\mathcal{A}}[\beta], \mathcal{D}'_{\mathcal{B}}[\beta]) \subseteq \mathfrak{B}(\mathcal{D}'_{\mathcal{A}}[\beta], \mathcal{D}'_{\mathcal{B}}[\beta])$ we see that it is sufficient to show that $\mathfrak{L}(\mathcal{D}^+_{\mathcal{B}}[\beta], \mathcal{D}(\mathcal{A})[\sigma]) \subseteq \mathcal{L}(\mathcal{D}^+_{\mathcal{B}}, \mathcal{D}_{\mathcal{A}})$ and $\mathfrak{B}(\mathcal{D}'_{\mathcal{A}}[\beta], \mathcal{D}'_{\mathcal{B}}[\beta]) \subseteq \mathfrak{B}(\mathcal{D}'_{\mathcal{A}}[\sigma'], \mathcal{D}'_{\mathcal{B}}[\sigma'])$. Both inclusions follow immediately from the semireflexivity of $\mathcal{D}_{\mathcal{A}}$ and $\mathcal{D}_{\mathcal{B}}$. We verify the first one. Suppose $y \in \mathfrak{L}(\mathcal{D}^+_{\mathcal{B}}[\beta], \mathcal{D}(\mathcal{A})[\sigma])$. From the continuity of y, we have $\overline{\langle y\cdot, \varphi'\rangle} \in (\mathcal{D}'_{\mathcal{B}}[\beta])'$ for $\varphi' \in \mathcal{D}'_{\mathcal{A}}$. Since $\mathcal{D}_{\mathcal{B}}$ is semireflexive, $(\mathcal{D}'_{\mathcal{B}}[\beta])' = \mathcal{D}(\mathcal{B})$, so $\overline{\langle y\cdot, \varphi'\rangle} \in \mathcal{D}(\mathcal{B})$. From $y(\mathcal{D}^+_{\mathcal{B}}) \subseteq \mathcal{D}(\mathcal{A})$, $\langle y\psi', \cdot\rangle \in \mathcal{D}(\mathcal{A})$ for $\psi' \in \mathcal{D}'_{\mathcal{B}}$. This proves that $y \in \mathcal{L}(\mathcal{D}^+_{\mathcal{B}}, \mathcal{D}_{\mathcal{A}})$. □

Unter the assumptions of Proposition 3.1.10, for each operator $c \in \mathbb{B}(\mathcal{D}(\mathcal{B}), \mathcal{D}(\mathcal{A}))$ there is a unique $y \in \mathcal{L}(\mathcal{D}^+_{\mathcal{B}}, \mathcal{D}_{\mathcal{A}})$ such that $c = y \upharpoonright \mathcal{H}$. We shall denote this element y by $\hat{c}$.

Since $\mathcal{L}(\mathcal{D}^+_{\mathcal{B}}, \mathcal{D}_{\mathcal{A}}) = \mathfrak{L}(\mathcal{D}^+_{\mathcal{B}}[\sigma'], \mathcal{D}(\mathcal{A})[\sigma])$, the equicontinuous topology τ_e (cf. p. 17) is defined on $\mathcal{L}(\mathcal{D}^+_{\mathcal{B}}, \mathcal{D}_{\mathcal{A}})$.

Suppose $\mathcal{A}$ and $\mathcal{B}$ are O*-algebras. Let $a \in \mathcal{A}$ and $b \in \mathcal{B}$. By Remark 1 and Lemma 3.1.9, the operator ayb is bounded on $\mathcal{D}(\mathcal{B})$ for each y in $\mathcal{L}(\mathcal{D}^+_{\mathcal{B}}, \mathcal{D}_{\mathcal{A}})$. Hence $\|\cdot\|_{a,b} := \|a\cdot b\|$ is a seminorm on $\mathcal{L}(\mathcal{D}^+_{\mathcal{B}}, \mathcal{D}_{\mathcal{A}})$.

Proposition 3.1.15. *Let $\mathcal{A}$ and $\mathcal{B}$ be O*-algebras in the Hilbert space $\mathcal{H}$.*

(i) *The equicontinuous topology τ_e on $\mathcal{L}(\mathcal{D}^+_{\mathcal{B}}, \mathcal{D}_{\mathcal{A}})$ is generated by the directed family of seminorms $\{\|\cdot\|_{a,b}: a = a^+ \in \mathcal{A}(I)$ and $b = b^+ \in \mathcal{B}(I)\}$*

(ii) *$\mathcal{L}(\mathcal{D}^+_{\mathcal{B}}, \mathcal{D}_{\mathcal{A}})[\tau_e]$ is a topological algebra with jointly continuous multiplication. The mapping $y \to y^+$ is a homeomorphism of $\mathcal{L}(\mathcal{D}^+_{\mathcal{B}}, \mathcal{D}_{\mathcal{A}})[\tau_e]$ on $\mathcal{L}(\mathcal{D}^+_{\mathcal{A}}, \mathcal{D}_{\mathcal{B}})[\tau_e]$. $\mathcal{L}(\mathcal{D}^+_{\mathcal{A}}, \mathcal{D}_{\mathcal{A}})[\tau_e]$ is a topological *-algebra.*

Proof. (i): Suppose $a \in \mathcal{A}(I)$ and $b \in \mathcal{B}(I)$. Let $\mathcal{M} := \mathcal{V}^0_a$ be the polar of the 0-neighbourhood $\mathcal{V}_a \equiv \{\varphi \in \mathcal{D}(\mathcal{A}): \|\varphi\|_a \leq 1\}$ in $\mathcal{D}_{\mathcal{A}}$. By Lemma 2.3.4, (i), $\mathcal{V}^0_a = \{\langle a\cdot, \zeta\rangle: \zeta \in \mathcal{U}_{\mathcal{H}}\}$. Similarly, the polar $\mathcal{N} := \mathcal{V}^0_b$ is the set $\{\langle b\cdot, \eta\rangle: \eta \in \mathcal{U}_{\mathcal{H}}\}$. Using this description and (2), we have for $y \in \mathcal{L}(\mathcal{D}^+_{\mathcal{B}}, \mathcal{D}_{\mathcal{A}})$

$$\begin{aligned} p_{\mathcal{M},\mathcal{N}}(y) &= \sup_{\varphi' \in \mathcal{M}} \sup_{\psi' \in \mathcal{N}} |\langle y\psi', \varphi'\rangle| = \sup_{\psi' \in \mathcal{N}} \sup_{\zeta \in \mathcal{U}_{\mathcal{H}}} |\langle ay\psi', \zeta\rangle| \\ &= \sup_{\psi' \in \mathcal{N}} \sup_{\zeta \in \mathcal{U}_{\mathcal{D}(\mathcal{A})}} |\langle \psi', y^+a^+\zeta\rangle| = \sup_{\eta \in \mathcal{U}_{\mathcal{H}}} \sup_{\zeta \in \mathcal{U}_{\mathcal{D}(\mathcal{A})}} |\langle \eta, by^+a^+\zeta\rangle| \\ &= \sup_{\eta \in \mathcal{U}_{\mathcal{D}(\mathcal{B})}} \sup_{\zeta \in \mathcal{U}_{\mathcal{D}(\mathcal{A})}} |\langle ayb^+\eta, \zeta\rangle| = \|ayb^+\| = \|y\|_{a,b^+}, \end{aligned} \tag{3}$$

where $\mathcal{U}_{\mathcal{D}(\mathcal{A})}$ and $\mathcal{U}_{\mathcal{D}(\mathcal{B})}$ are the unit balls of the normed spaces $(\mathcal{D}(\mathcal{A}), \|\cdot\|)$ and $(\mathcal{D}(\mathcal{B}), \|\cdot\|)$, respectively. Since $\{\mathcal{V}^0_a: a = a^+ \in \mathcal{A}(I)\}$ and $\{\mathcal{V}^0_b: b = b^+ \in \mathcal{B}(I)\}$ are fundamental systems for the equicontinuous subsets of $\mathcal{D}'_{\mathcal{A}}$ and $\mathcal{D}'_{\mathcal{B}}$, respectively, it follows from (3) that the family of seminorms $\{\|\cdot\|_{a,b}: a = a^+ \in \mathcal{A}(I) \text{ and } b = b^+ \in \mathcal{B}(I)\}$ is directed and generates the topology τ_e.

(ii): For $a \in \mathcal{A}$, $b \in \mathcal{B}$ and $y_1, y_2 \in \mathcal{L}(\mathcal{D}^+_{\mathcal{B}}, \mathcal{D}_{\mathcal{A}})$, we have

$$\|y_1y_2\|_{a,b} = \|ay_1y_2b\| \leqq \|ay_1 \upharpoonright \mathcal{H}\| \, \|y_2b\| = \|y_1\|_{a,I} \, \|y_2\|_{I,b}.$$

Combined with (i), this proves that the multiplication is jointly continuous in $\mathcal{L}(\mathcal{D}^+_{\mathcal{B}}, \mathcal{D}_{\mathcal{A}})\,[\tau_e]$. By (2), $p_{\mathcal{M},\mathcal{N}}(y) = p_{\mathcal{N},\mathcal{M}}(y^+)$ for $y \in \mathcal{L}(\mathcal{D}^+_{\mathcal{B}}, \mathcal{D}_{\mathcal{A}})$ and for equicontinuous subsets $\mathcal{M}$ and $\mathcal{N}$ of $\mathcal{D}'_{\mathcal{A}}$ and $\mathcal{D}'_{\mathcal{B}}$, respectively. Therefore, $y \to y^+$ is a homeomorphism of $\mathcal{L}(\mathcal{D}^+_{\mathcal{B}}, \mathcal{D}_{\mathcal{A}})\,[\tau_e]$ onto $\mathcal{L}(\mathcal{D}^+_{\mathcal{A}}, \mathcal{D}_{\mathcal{B}})\,[\tau_e]$. □

Remark 2. For any O-family $\mathcal{A}$ in $\mathcal{H}$, we have $\mathcal{L}(\mathcal{H}^+, \mathcal{D}_{\mathcal{A}}) = \mathbf{B}(\mathcal{H}, \mathcal{D}(\mathcal{A}))$.

Remark 3. The main assumption for most of the results in this section is that the locally convex spaces $\mathcal{D}_{\mathcal{A}}$ and $\mathcal{D}_{\mathcal{B}}$ are sequentially complete. There are at least two important classes of O-families $\mathcal{A}$ for which the space $\mathcal{D}_{\mathcal{A}}$ is sequentially complete. This is the case if the O-family $\mathcal{A}$ is closed or if $\mathcal{D}_{\mathcal{A}}$ is a QF-space.

3.2. The Vector Space $\mathcal{L}(\mathcal{D}_{\mathcal{A}}, \mathcal{D}^+_{\mathcal{B}})$

Recall from Section 1.2 that with each linear mapping x of $\mathcal{D}(\mathcal{A})$ into $\mathcal{D}^+_{\mathcal{B}}$ we associated a sesquilinear form $\mathfrak{c}_x$ on $\mathcal{D}(\mathcal{A}) \times \mathcal{D}(\mathcal{B})$. Combining the formulas 1.2/(2) and 2.3/(1), we obtain

$$\mathfrak{c}_x(\varphi, \psi) = \langle x\varphi, \psi \rangle \quad \text{for} \quad \varphi \in \mathcal{D}(\mathcal{A}) \quad \text{and} \quad \psi \in \mathcal{D}(\mathcal{B}). \tag{1}$$

Definition 3.2.1. $\mathcal{L}(\mathcal{D}_{\mathcal{A}}, \mathcal{D}^+_{\mathcal{B}}) := \{x \in L(\mathcal{D}(\mathcal{A}), \mathcal{D}^+_{\mathcal{B}}): \mathfrak{c}_x \in \mathcal{B}(\mathcal{D}_{\mathcal{A}}, \mathcal{D}_{\mathcal{B}})\}$. In the case where $\mathcal{A} = \mathcal{L}^+(\mathcal{D}_1)$, $\mathcal{D}_1 = \mathcal{D}(\mathcal{A})$ and $\mathcal{B} = \mathcal{L}^+(\mathcal{D}_2)$, $\mathcal{D}_2 = \mathcal{D}(\mathcal{B})$ we write $\mathcal{L}(\mathcal{D}_1, \mathcal{D}^+_2)$ in place of $\mathcal{L}(\mathcal{D}_{\mathcal{A}}, \mathcal{D}^+_{\mathcal{B}})$.

By Lemma 1.2.1, the mapping $x \to \mathfrak{c}_x$ provides an isomorphism of the vector spaces $\mathcal{L}(\mathcal{D}_{\mathcal{A}}, \mathcal{D}^+_{\mathcal{B}})$ and $\mathcal{B}(\mathcal{D}_{\mathcal{A}}, \mathcal{D}_{\mathcal{B}})$. Let x be a linear mapping of $\mathcal{D}(\mathcal{A})$ into $\mathcal{D}^+_{\mathcal{B}}$. By the above definition x is in $\mathcal{L}(\mathcal{D}_{\mathcal{A}}, \mathcal{D}^+_{\mathcal{B}})$ if and only if there are continuous seminorms p and q on $\mathcal{D}_{\mathcal{A}}$ and $\mathcal{D}_{\mathcal{B}}$, respectively, such that $|\langle x\varphi, \psi\rangle| = |\mathfrak{c}_x(\varphi, \psi)| \leqq p(\varphi)\, q(\psi)$ for all $\varphi \in \mathcal{D}(\mathcal{A})$ and $\psi \in \mathcal{D}(\mathcal{B})$. If $\mathcal{A}$ and $\mathcal{B}$ are directed O-vector spaces, then $x \in \mathcal{L}(\mathcal{D}_{\mathcal{A}}, \mathcal{D}^+_{\mathcal{B}})$ if and only if there are operators $a \in \mathcal{A}$ and $b \in \mathcal{B}$ such that $|\langle x\varphi, \psi\rangle| \leqq \|a\varphi\|\,\|b\psi\|$ for $\varphi \in \mathcal{D}(\mathcal{A})$ and $\psi \in \mathcal{D}(\mathcal{B})$.

Remark 1. There is one case where a confusion between the spaces $\mathcal{L}(\mathcal{D}^+_{\mathcal{B}}, \mathcal{D}_{\mathcal{A}})$ and $\mathcal{L}(\mathcal{D}_{\mathcal{A}_1}, \mathcal{D}^+_{\mathcal{B}_1})$ would be possible, namely, when $\mathcal{D}^+_{\mathcal{B}} = \mathcal{D}(\mathcal{A}_1)$ and $\mathcal{D}(\mathcal{A}) = \mathcal{D}^+_{\mathcal{B}_1}$. But in this case $\mathcal{D}(\mathcal{A}) = \mathcal{D}(\mathcal{A}_1) = \mathcal{H}$ and both $\mathcal{L}(\mathcal{D}^+_{\mathcal{B}}, \mathcal{D}_{\mathcal{A}})$ and $\mathcal{L}(\mathcal{D}_{\mathcal{A}_1}, \mathcal{D}^+_{\mathcal{B}_1})$ are equal to $\mathbf{B}(\mathcal{H})$, so that no ambiguity can arise.

Remark 2. If $\mathcal{B}$ consists of bounded operators only, then $\mathcal{D}^+_{\mathcal{B}} = \mathcal{H}$. In this case $\mathcal{L}(\mathcal{D}_{\mathcal{A}}, \mathcal{D}^+_{\mathcal{B}}) = \mathcal{L}(\mathcal{D}_{\mathcal{A}}, \mathcal{H})$ is simply the space $\mathfrak{L}(\mathcal{D}_{\mathcal{A}}, \mathcal{H})$ of all continuous linear mappings of $\mathcal{D}_{\mathcal{A}}$ into $\mathcal{H}$. If in addition $\mathcal{D}(\mathcal{A}) = \mathcal{H}$, the operators of $\mathcal{A}$ are also bounded (see Remark 2 in 2.1) and hence $\mathcal{L}(\mathcal{D}_{\mathcal{A}}, \mathcal{D}^+_{\mathcal{B}}) = \mathbf{B}(\mathcal{H})$.

Remark 3. From the definition it is clear that $\mathfrak{L}(\mathcal{D}_{\mathcal{A}}, \mathcal{H})$ is a linear subspace of $\mathcal{L}(\mathcal{D}_{\mathcal{A}}, \mathcal{D}^+_{\mathcal{B}})$ for any O-family in $\mathcal{H}$. In particular, $\mathcal{A} \subseteqq \mathcal{L}(\mathcal{D}_{\mathcal{A}}, \mathcal{D}^+_{\mathcal{B}})$ for every O-family $\mathcal{B}$ in $\mathcal{H}$.

Remark 4. Let us adopt two notational conventions which will be often used in the sequel. First note that $x \upharpoonright \mathcal{D}(\mathcal{A})$ is in $\mathcal{L}(\mathcal{D}_{\mathcal{A}}, \mathcal{D}^+_{\mathcal{B}})$ for each $x \in \mathbf{B}(\mathcal{H})$. By abuse of notation, we simply write $x \in \mathcal{L}(\mathcal{D}_{\mathcal{A}}, \mathcal{D}^+_{\mathcal{B}})$ if $x \in \mathbf{B}(\mathcal{H})$ and we consider $\mathbf{B}(\mathcal{H})$ as a linear subspace of $\mathcal{L}(\mathcal{D}_{\mathcal{A}}, \mathcal{D}^+_{\mathcal{B}})$ (although, strictly speaking, we mean $x \upharpoonright \mathcal{D}(\mathcal{A}) \in \mathcal{L}(\mathcal{D}_\mathcal{A}, \mathcal{D}_\mathcal{B}^+)$ and $\mathbf{B}(\mathcal{H}) \upharpoonright \mathcal{D}(\mathcal{A})$). Similar notation is used for $\mathcal{L}^+(\mathcal{D}_\mathcal{A})$ if $\mathcal{A}$ is an O*-algebra. That is, if $x \in \mathbf{B}(\mathcal{H})$ and $x \upharpoonright \mathcal{D}(\mathcal{A})$ is in $\mathcal{L}^+(\mathcal{D}_\mathcal{A})$, then we shall write simply $x \in \mathcal{L}^+(\mathcal{D}_\mathcal{A})$. In this way, $\mathbf{B}(\mathcal{D}(\mathcal{A}))$ becomes a *-subalgebra of $\mathcal{L}^+(\mathcal{D}_\mathcal{A})$.

Before studying the structure of the vector space $\mathcal{L}(\mathcal{D}_\mathcal{A}, \mathcal{D}_\mathcal{B}^+)$, we briefly discuss the relation between $\mathcal{L}(\mathcal{D}_\mathcal{A}, \mathcal{D}_\mathcal{B}^+)$ and $\mathfrak{L}(\mathcal{D}_\mathcal{A}, \mathcal{D}_\mathcal{B}^+[\beta])$. By Lemma 1.2.2, we always have $\mathcal{L}(\mathcal{D}_\mathcal{A}, \mathcal{D}_\mathcal{B}^+) \subseteq \mathfrak{L}(\mathcal{D}_\mathcal{A}, \mathcal{D}_\mathcal{B}^+[\beta]) \subseteq \mathfrak{L}(\mathcal{D}_\mathcal{A}, \mathcal{D}_\mathcal{B}^+[\sigma'])$. From Example 2.5.8 we know that there is a closed O*-algebra $\mathcal{A}$ with $\mathfrak{t}_\mathcal{A} \neq \mathfrak{t}_+$. Combined with assertion (ii) of the next lemma, this shows that $\mathcal{L}(\mathcal{D}_\mathcal{A}, \mathcal{D}_\mathcal{B}^+) \neq \mathfrak{L}(\mathcal{D}_\mathcal{A}, \mathcal{D}_\mathcal{B}^+[\beta])$ in general.

Lemma 3.2.2. (i) *If $\mathfrak{B}(\mathcal{D}_\mathcal{A}, \mathcal{D}_\mathcal{B}) = \mathcal{B}(\mathcal{D}_\mathcal{A}, \mathcal{D}_\mathcal{B})$ (in particular, if $\mathcal{D}_\mathcal{A}$ and $\mathcal{D}_\mathcal{B}$ are Frechet spaces), then $\mathcal{L}(\mathcal{D}_\mathcal{A}, \mathcal{D}_\mathcal{B}^+) = \mathfrak{L}(\mathcal{D}_\mathcal{A}, \mathcal{D}_\mathcal{B}^+[\sigma']) = \mathfrak{L}(\mathcal{D}_\mathcal{A}, \mathcal{D}_\mathcal{B}^+[\beta])$.*

(ii) *Suppose $\mathcal{A}$ is an O*-algebra. If $\mathcal{A}$ is closed, then $\mathcal{L}^+(\mathcal{D}(\mathcal{A})) \subseteq \mathfrak{L}(\mathcal{D}_\mathcal{A}, \mathcal{D}_\mathcal{A}^+[\beta])$. If $\mathfrak{t}_\mathcal{A} \neq \mathfrak{t}_+$ on $\mathcal{D}(\mathcal{A})$, then $\mathcal{L}^+(\mathcal{D}(\mathcal{A})) \nsubseteq \mathcal{L}(\mathcal{D}_\mathcal{A}, \mathcal{D}_\mathcal{A}^+)$.*

Proof. (i): As already noted on p. 16, we have $\mathfrak{B}(\mathcal{D}_\mathcal{A}, \mathcal{D}_\mathcal{B}) = \mathcal{B}(\mathcal{D}_\mathcal{A}, \mathcal{D}_\mathcal{B})$ if $\mathcal{D}_\mathcal{A}$ and $\mathcal{D}_\mathcal{B}$ are Frechet spaces. Suppose now that $\mathfrak{B}(\mathcal{D}_\mathcal{A}, \mathcal{D}_\mathcal{B}) = \mathcal{B}(\mathcal{D}_\mathcal{A}, \mathcal{D}_\mathcal{B})$. Then $\mathcal{L}(\mathcal{D}_\mathcal{A}, \mathcal{D}_\mathcal{B}^+) = \{x \in L(\mathcal{D}(\mathcal{A}), \mathcal{D}_\mathcal{B}^+): \mathfrak{c}_x \in \mathfrak{B}(\mathcal{D}_\mathcal{A}, \mathcal{D}_\mathcal{B})\} = \mathfrak{L}(\mathcal{D}_\mathcal{A}, \mathcal{D}_\mathcal{B}^+[\sigma'])$, where the last equality comes from Lemma 1.2.1, (i). Combined with the inclusions $\mathcal{L}(\mathcal{D}_\mathcal{A}, \mathcal{D}_\mathcal{B}^+) \subseteq \mathfrak{L}(\mathcal{D}_\mathcal{A}, \mathcal{D}_\mathcal{B}^+[\beta]) \subseteq \mathfrak{L}(\mathcal{D}_\mathcal{A}, \mathcal{D}_\mathcal{B}^+[\sigma'])$, the assertion follows.

(ii): First suppose $\mathcal{A}$ is closed. Let $x \in \mathcal{L}^+(\mathcal{D}(\mathcal{A}))$, and let $\mathcal{M}$ be a bounded subset of $\mathcal{D}_\mathcal{A}$. By Proposition 2.3.10, $\mathcal{M}$ is bounded in $\mathcal{D}(\mathcal{A})[\mathfrak{t}_+]$, so that $\lambda := \sup\{\|x^+\psi\|: \psi \in \mathcal{M}\} < \infty$. From the inequality

$$r_\mathcal{M}(x\varphi) \equiv \sup_{\psi \in M} |\langle x\varphi, \psi\rangle| \leqq \lambda \|\varphi\| \quad \text{for} \quad \varphi \in \mathcal{D}(\mathcal{A})$$

we see that $x \in \mathfrak{L}(\mathcal{D}_\mathcal{A}, \mathcal{D}_\mathcal{A}^+[\beta])$. Suppose now that $\mathfrak{t}_\mathcal{A} \neq \mathfrak{t}_+$ on $\mathcal{D}(\mathcal{A})$. In order to prove that $\mathcal{L}^+(\mathcal{D}(\mathcal{A})) \nsubseteq \mathcal{L}(\mathcal{D}_\mathcal{A}, \mathcal{D}_\mathcal{A}^+)$, we assume the contrary, that is, $\mathcal{L}^+(\mathcal{D}(\mathcal{A})) \subseteq \mathcal{L}(\mathcal{D}_\mathcal{A}, \mathcal{D}_\mathcal{A}^+)$. Let $x \in \mathcal{L}^+(\mathcal{D}(\mathcal{A}))$. Then $x^+x \in \mathcal{L}(\mathcal{D}_\mathcal{A}, \mathcal{D}_\mathcal{A}^+)$, so that there exists an operator $a \in \mathcal{A}$ such that $|\langle x^+x\varphi, \psi\rangle| \leqq \|a\varphi\| \|a\psi\|$ for all $\varphi, \psi \in \mathcal{D}(\mathcal{A})$. In case $\varphi = \psi$ this gives $\|x\varphi\| \leqq \|a\varphi\|$, $\varphi \in \mathcal{D}(\mathcal{A})$. Therefore, $\mathfrak{t}_\mathcal{A} = \mathfrak{t}_+$ which is the desired contradiction. □

As explained in Section 1.2 for each $x \in \mathfrak{L}(\mathcal{D}_\mathcal{A}, \mathcal{D}_\mathcal{B}^+[\sigma'])$ there is a unique mapping $x^+ \in \mathfrak{L}(\mathcal{D}_\mathcal{B}, \mathcal{D}_\mathcal{A}^+[\sigma'])$ such that $(\mathfrak{c}_x)^+ = \mathfrak{c}_{x^+}$. By (1) and 2.3/(1), we have

$$\langle x\varphi, \psi\rangle = \langle \varphi, x^+\psi\rangle \quad \text{for} \quad \varphi \in \mathcal{D}(\mathcal{A}) \quad \text{and} \quad \psi \in \mathcal{D}(\mathcal{B}). \tag{2}$$

Now let $x \in \mathcal{L}(\mathcal{D}_\mathcal{A}, \mathcal{D}_\mathcal{B}^+)$. Since $\mathcal{L}(\mathcal{D}_\mathcal{A}, \mathcal{D}_\mathcal{B}^+) \subseteq \mathfrak{L}(\mathcal{D}_\mathcal{A}, \mathcal{D}_\mathcal{B}^+[\sigma'])$, x^+ is well-defined by the preceding formula. Since $\mathfrak{c}_x \in \mathcal{B}(\mathcal{D}_\mathcal{A}, \mathcal{D}_\mathcal{B})$ obviously implies $\mathfrak{c}_{x^+} \equiv (\mathfrak{c}_x)^+ \in \mathcal{B}(\mathcal{D}_\mathcal{B}, \mathcal{D}_\mathcal{A})$, we have $x^+ \in \mathcal{L}(\mathcal{D}_\mathcal{B}, \mathcal{D}_\mathcal{A}^+)$. Thus $x \to x^+$ is a conjugate-linear mapping of $\mathcal{L}(\mathcal{D}_\mathcal{A}, \mathcal{D}_\mathcal{B}^+)$ onto $\mathcal{L}(\mathcal{D}_\mathcal{B}, \mathcal{D}_\mathcal{A}^+)$. Moreover, $(x^+)^+ = x$ for all $x \in \mathcal{L}(\mathcal{D}_\mathcal{A}, \mathcal{D}_\mathcal{B}^+)$.

Of course, the two special cases $\mathcal{B} = \mathbf{B}(\mathcal{H})$ and $\mathcal{A} = \mathcal{B}$ of the spaces $\mathcal{L}(\mathcal{D}_\mathcal{A}, \mathcal{D}_\mathcal{B}^+)$ are of particular interest. The first one was mentioned in Remark 2. We now briefly specialize to the second case which is even more important. That is, we consider the space $\mathcal{L}(\mathcal{D}_\mathcal{A}, \mathcal{D}_\mathcal{A}^+)$. In this case, the map $x \to x^+$ is an involution of the vector space $\mathcal{L}(\mathcal{D}_\mathcal{A}, \mathcal{D}_\mathcal{A}^+)$. With this involution, *$\mathcal{L}(\mathcal{D}_\mathcal{A}, \mathcal{D}_\mathcal{A}^+)$ is a *-vector space.* When $x \in \mathcal{L}(\mathcal{D}_\mathcal{A}, \mathcal{D}_\mathcal{A}^+)$, we have the *polarization identity*

$$\begin{aligned} 4\langle x\varphi, \psi\rangle = {} & \langle x(\varphi + \psi), \varphi + \psi\rangle - \langle x(\varphi - \psi), \varphi - \psi\rangle \\ & + \mathrm{i}\langle x(\varphi + \mathrm{i}\psi), \varphi + \mathrm{i}\psi\rangle - \mathrm{i}\langle x(\varphi - \mathrm{i}\psi), \varphi - \mathrm{i}\psi\rangle \end{aligned} \tag{3}$$

for $\varphi, \psi \in \mathcal{D}(\mathcal{A})$. It is merely formula 1.2/(1) applied in case $c = c_x$. If $x \in \mathcal{L}(\mathcal{D}_\mathcal{A}, \mathcal{D}_\mathcal{A}^+)_\mathrm{h}$, it follows from 2.3/(1) and (2) that $\langle x\varphi, \varphi\rangle$ is real for all $\varphi \in \mathcal{D}(\mathcal{A})$. Therefore, if $x, y \in \mathcal{L}(\mathcal{D}_\mathcal{A}, \mathcal{D}_\mathcal{A}^+)_\mathrm{h}$, we can define

$$x \geqq y \quad \text{if and only if} \quad \langle x\varphi, \varphi\rangle \geqq \langle y\varphi, \varphi\rangle \quad \text{for all} \quad \varphi \in \mathcal{D}(\mathcal{A}). \tag{4}$$

Suppose that $\mathcal{L}$ is a $*$-vector subspace of $\mathcal{L}(\mathcal{D}_\mathcal{A}, \mathcal{D}_\mathcal{A}^+)$. Then $\mathcal{L}_+ := \{x \in \mathcal{L}_\mathrm{h} : x \geqq 0\}$ is a cone in the real vector space $\mathcal{L}_\mathrm{h}$. (The property $\mathcal{L}_+ \cap (-\mathcal{L}_+) = \{0\}$ is an immediate consequence of (3).) The order relation on $\mathcal{L}_\mathrm{h}$ associated with the cone $\mathcal{L}_+$ is nothing but the relation "$\geqq$" defined by (4), i.e., $x \geqq y$ is equivalent to $y - x \in \mathcal{L}_+$ for $x, y \in \mathcal{L}_\mathrm{h}$. Thus $(\mathcal{L}_\mathrm{h}, \geqq)$ is an ordered vector space, and $\mathcal{L}$ *is an ordered $*$-vector space*. Following the terminology of Section 2.6, we call linear functionals on $\mathcal{L}$ with non-negative values on $\mathcal{L}_+$ *strongly positive*.

Remark 5. The advantage of the notational convention 2.3/(1) is that basic formulas for elements of $\mathcal{L}(\mathcal{D}_\mathcal{A}, \mathcal{D}_\mathcal{B}^+)$ (for instance, (2), (3) and (4)) are quite similar to the corresponding formulas for operators in O*-vector spaces. That is, we can consider these formulas or parts of it in the Hilbert space language (with $\langle\cdot,\cdot\rangle$ denoting the scalar product) if, roughly speaking, all ingredients make sense in the Hilbert space. We illustrate this remark by two simple examples. If x is an operator of $\mathfrak{L}(\mathcal{D}_\mathcal{A}, \mathcal{H})$ such that $\mathcal{D}(\mathcal{B}) \subseteq \mathcal{D}(x^*)$, then the mapping $x^+ \in \mathcal{L}(\mathcal{D}_\mathcal{B}, \mathcal{D}_\mathcal{A}^+)$ in (2) is the restriction to $\mathcal{D}(\mathcal{B})$ of the Hilbert space adjoint x^* of x. Let a be an unbounded operator in $\mathcal{A}$, let $\xi \in \mathcal{H}$ with $\xi \notin \mathcal{D}(a^*)$ and let $\eta \in \mathcal{D}(\mathcal{A})$, $\eta \neq 0$. Define $x\varphi = \langle a\varphi, \xi\rangle\, \eta$ for $\varphi \in \mathcal{D}(\mathcal{A})$. Then x is a Hilbert space operator contained in $\mathcal{L}(\mathcal{D}_\mathcal{A}, \mathcal{D}_\mathcal{B}^+)$ for which $x^+ \big(\in \mathcal{L}(\mathcal{D}_\mathcal{B}, \mathcal{D}_\mathcal{A}^+)\big)$ is not a Hilbert space operator. That is, $x^+\big(\mathcal{D}(\mathcal{B})\big) \nsubseteq \mathcal{H}$, and the expression $\langle \varphi, x^+\psi\rangle$ on the right-hand side of (2) does not mean the scalar product of $\mathcal{H}$. Moreover, $\mathcal{D}(x^*) = \{\eta\}^\perp$, so that x is an operator in $\mathfrak{L}(\mathcal{D}_\mathcal{A}, \mathcal{H})$ which is not closable.

Let us return to the general space $\mathcal{L}(\mathcal{D}_\mathcal{A}, \mathcal{D}_\mathcal{B}^+)$. We denote by $\mathcal{F}(\mathcal{D}_\mathcal{A}, \mathcal{D}_\mathcal{B}^+)$ the set of finite rank mappings in $\mathfrak{L}(\mathcal{D}_\mathcal{A}, \mathcal{D}_\mathcal{B}^+[\beta])$, i.e., the set of those $x \in \mathfrak{L}(\mathcal{D}_\mathcal{A}, \mathcal{D}_\mathcal{B}^+[\beta])$ for which the vector space $x\big(\mathcal{D}(\mathcal{A})\big)$ is finite dimensional. For $z = \sum_{n=1}^{k} \varphi_n^\dagger \otimes \psi_n^\dagger$ in the algebraic tensor product $\mathcal{D}_\mathcal{A}^\dagger \otimes \mathcal{D}_\mathcal{B}^+$, we define $\chi(z)\,\varphi = \sum_{n=1}^{k} \langle \varphi, \varphi_n^\dagger\rangle\, \psi_n^\dagger$, $\varphi \in \mathcal{D}(\mathcal{A})$. By standard arguments from the theory of locally convex spaces (see e.g. JARCHOW [1], p. 330) it follows that $\chi(\cdot)$ is an isomorphism of the vector spaces $\mathcal{D}_\mathcal{A}^\dagger \otimes \mathcal{D}_\mathcal{B}^+$ and $\mathcal{F}(\mathcal{D}_\mathcal{A}, \mathcal{D}_\mathcal{B}^+)$. Since, of course, $\chi(z) \in \mathcal{L}(\mathcal{D}_\mathcal{A}, \mathcal{D}_\mathcal{B}^+)$ for $z \in \mathcal{D}_\mathcal{A}^\dagger \otimes \mathcal{D}_\mathcal{B}^+$, we conclude that $\mathcal{F}(\mathcal{D}_\mathcal{A}, \mathcal{D}_\mathcal{B}^+)$ is also the set of finite rank mappings in $\mathcal{L}(\mathcal{D}_\mathcal{A}, \mathcal{D}_\mathcal{B}^+)$. For simplicity of notation, we identify $z \in \mathcal{D}_\mathcal{A}^\dagger \otimes \mathcal{D}_\mathcal{B}^+$ with $\chi(z) \in \mathcal{F}(\mathcal{D}_\mathcal{A}, \mathcal{D}_\mathcal{B}^+)$; that is, we let $\varphi^\dagger \otimes \psi^\dagger$ denote the mapping $\langle\cdot, \varphi^\dagger\rangle\, \psi^\dagger$ of $\mathcal{L}(\mathcal{D}_\mathcal{A}, \mathcal{D}_\mathcal{B}^+)$ for $\varphi^\dagger \in \mathcal{D}_\mathcal{A}^\dagger$ and $\psi^\dagger \in \mathcal{D}_\mathcal{B}^\dagger$. Then $\mathcal{F}(\mathcal{D}_\mathcal{A}, \mathcal{D}_\mathcal{B}^+)$ is the linear span of $\varphi^\dagger \otimes \psi^\dagger$, where $\varphi^\dagger \in \mathcal{D}_\mathcal{A}^\dagger$ and $\psi^\dagger \in \mathcal{D}_\mathcal{B}^\dagger$. Following the corresponding notation for $\mathcal{L}(\mathcal{D}_\mathcal{A}, \mathcal{D}_\mathcal{B}^+)$, we write $\mathcal{F}(\mathcal{D}_1, \mathcal{D}_2^+)$ for $\mathcal{F}(\mathcal{D}_\mathcal{A}, \mathcal{D}_\mathcal{B}^+)$ if $\mathcal{A} = \mathcal{L}^+(\mathcal{D}_1)$, $\mathcal{D}_1 = \mathcal{D}(\mathcal{A})$ and $\mathcal{B} = \mathcal{L}^+(\mathcal{D}_2)$, $\mathcal{D}_2 = \mathcal{D}(\mathcal{B})$.

Remark 6. From the definitions it is clear that the spaces $\mathcal{L}(\mathcal{D}_\mathcal{B}^+, \mathcal{D}_\mathcal{A})$, $\mathcal{L}(\mathcal{D}_\mathcal{A}, \mathcal{D}_\mathcal{B}^+)$ and $\mathcal{F}(\mathcal{D}_\mathcal{A}, \mathcal{D}_\mathcal{B}^+)$ introduced so far in this and the previous section and the space $\mathcal{V}(\mathcal{D}_\mathcal{A}, \mathcal{D}_\mathcal{B}^+)$ which will be defined in Section 6.1 depend only on the graph topologies $\mathrm{t}_\mathcal{A}$ and $\mathrm{t}_\mathcal{B}$ rather than on the O-families $\mathcal{A}$ and $\mathcal{B}$. Therefore, by Proposition 2.2.13, when dealing with one of these spaces, we can assume without loss of generality that $\mathcal{A}$ and $\mathcal{B}$ are directed O*-vector spaces.

We introduce some more notation which will be frequently used. Let $\mathcal{L}$ be a linear subspace of $\mathcal{L}(\mathcal{D}_\mathcal{A}, \mathcal{D}_\mathcal{B}^+)$. Suppose $a \in \mathcal{A}$ and $b \in \mathcal{B}$. We write $(\mathcal{L}_{a,b}, \mathfrak{l}_{a,b})$ for the normed space $(\mathcal{L}_{\|\cdot\|_a, \|\cdot\|_b}, \mathfrak{l}_{\|\cdot\|_a, \|\cdot\|_b})$ defined in Section 1.2. That is, $\mathcal{L}_{a,b}$ is the set of all x in $\mathcal{L}$ for which there exists a $\lambda \geqq 0$ such that $|\langle x\varphi, \psi\rangle| \leqq \lambda \, \|\varphi\|_a \, \|\psi\|_b$ for all $\varphi \in \mathcal{D}(\mathcal{A})$ and $\psi \in \mathcal{D}(\mathcal{B})$, and $\mathfrak{l}_{a,b}(x)$ is the smallest number $\lambda \geqq 0$ which has this property. We then have

$$|\langle x\varphi, \psi\rangle| \leqq \mathfrak{l}_{a,b}(x) \, \|a\varphi\| \, \|b\psi\| \quad \text{for} \quad x \in \mathcal{L}_{a,b}, \varphi \in \mathcal{D}(\mathcal{A}) \quad \text{and} \quad \psi \in \mathcal{D}(\mathcal{B}). \tag{5}$$

Further, let $\mathcal{U}_{a,b} := \mathcal{U}_{\|\cdot\|_a, \|\cdot\|_b}$, i.e.,

$$\mathcal{U}_{a,b} = \{x \in \mathcal{L}(\mathcal{D}_\mathcal{A}, \mathcal{D}_\mathcal{B}^+) \colon |\langle x\varphi, \psi\rangle| \leqq \|a\varphi\| \, \|b\psi\| \quad \text{for} \quad \varphi \in \mathcal{D}(\mathcal{A}) \quad \text{and} \quad \psi \in \mathcal{D}(\mathcal{B})\}.$$

In other words, $\mathcal{U}_{a,b}$ is the unit ball in the normed space $(\mathcal{L}(\mathcal{D}_\mathcal{A}, \mathcal{D}_\mathcal{B}^+)_{a,b}, \mathfrak{l}_{a,b})$. Recall that $\mathcal{D}_a \equiv (\mathcal{D}(\mathcal{A}), \|\cdot\|_a)$ and $\mathcal{H}_a \equiv (\mathcal{D}(\bar{a}), \|\cdot\|_{\bar{a}})$.

Proposition 3.2.3. *Suppose $a \in \mathcal{A}(I)$ and $b \in \mathcal{B}(I)$. For each $x \in \mathcal{U}_{a,b}$ there exists an operator $y \in \mathbf{B}(\mathcal{H})$ with $\|y\| \leqq 1$ such that $\langle x\varphi, \psi\rangle = \langle ya\varphi, b\psi\rangle$ for $\varphi \in \mathcal{D}(\mathcal{A})$ and $\psi \in \mathcal{D}(\mathcal{B})$.*

Proof. Since $x \in \mathcal{U}_{a,b}$, $\mathfrak{c}_x(\cdot,\cdot) \equiv \langle x\cdot, \cdot\rangle$ is a continuous sesquilinear form on $\mathcal{D}_a \times \mathcal{D}_b$. Let $\bar{\mathfrak{c}}_x$ denote its continuous extension to $\mathcal{H}_a \times \mathcal{H}_b$. There exists a bounded operator z of the Hilbert space $\mathcal{H}_a$ into $\mathcal{H}_b$ such that $\bar{\mathfrak{c}}_x(\varphi, \psi) = \langle z\varphi, \psi\rangle_{\bar{b}}$ for $\varphi \in \mathcal{H}_a$ and $\psi \in \mathcal{H}_b$. Since $|\langle z\varphi, \psi\rangle_{\bar{b}}| \leqq \|\varphi\|_{\bar{a}} \, \|\psi\|_{\bar{b}}$ because of $x \in \mathcal{U}_{a,b}$ for $\varphi \in \mathcal{H}_a$ and $\psi \in \mathcal{H}_b$, we have $\|z\varphi\|_{\bar{b}} \leqq \|\varphi\|_{\bar{a}}$, $\varphi \in \mathcal{H}_a$. Therefore, the equation $y(\bar{a}\varphi) := \bar{b}z\varphi$, $\varphi \in \mathcal{H}_a$, defines an operator of the closed subspace $\bar{a}\mathcal{D}(\bar{a})$ into $\mathcal{H}$ satisfying $\|y\psi\| \leqq \|\psi\|$ for $\psi \in \bar{a}\mathcal{D}(\bar{a})$. Set $y\psi = 0$ if $\psi \in (\bar{a}\mathcal{D}(\bar{a}))^\perp$. Then we have $\|y\| \leqq 1$ and $\langle ya\varphi, b\psi\rangle = \langle \bar{b}z\varphi, \bar{b}\psi\rangle = \bar{\mathfrak{c}}_x(\varphi, \psi) = \langle x\varphi, \psi\rangle$ for $\varphi \in \mathcal{D}(\mathcal{A})$ and $\psi \in \mathcal{D}(\mathcal{B})$. □

Let $a \in \mathcal{A}(I)$ and $b \in \mathcal{B}(I)$. If $y \in \mathbf{B}(\mathcal{H})$, then $\langle ya\cdot, b\cdot\rangle$ is a continuous sesquilinear form on $\mathcal{D}_\mathcal{A} \times \mathcal{D}_\mathcal{B}$; hence there exists an element x_y of $\mathcal{L}(\mathcal{D}_\mathcal{A}, \mathcal{D}_\mathcal{B}^+)$ such that $\langle x_y\varphi, \psi\rangle = \langle ya\varphi, b\psi\rangle$ for $\varphi \in \mathcal{D}(\mathcal{A})$ and $\psi \in \mathcal{D}(\mathcal{B})$. Let Q_a and Q_b denote the projections on $\mathcal{H}$ whose ranges are the closures of $a\mathcal{D}(\mathcal{A})$ and $b\mathcal{D}(\mathcal{B})$ in $\mathcal{H}$, respectively. It is clear that $x_y \in \mathcal{L}(\mathcal{D}_\mathcal{A}, \mathcal{D}_\mathcal{B}^+)$ and $\mathfrak{l}_{a,b}(x_y) = \|Q_b y Q_a\|$. We define a mapping $R_{a,b}$ of $Q_b\mathbf{B}(\mathcal{H})\,Q_a$ into $\mathcal{L}(\mathcal{D}_\mathcal{A}, \mathcal{D}_\mathcal{B}^+)_{a,b}$ by $R_{a,b}(y) := x_y$, $y \in Q_b\mathbf{B}(\mathcal{H})\,Q_a$. We show that $R_{a,b}$ is surjective. Let $x \in \mathcal{L}(\mathcal{D}_\mathcal{A}, \mathcal{D}_\mathcal{B}^+)_{a,b}$, $x \neq 0$. Then $\mathfrak{l}_{a,b}(x)^{-1}\,x \in \mathcal{U}_{a,b}$, so that $\langle \mathfrak{l}_{a,b}(x)^{-1}\,x\cdot, \cdot\rangle = \langle y_1 a\cdot, b\cdot\rangle$ for some $y_1 \in \mathbf{B}(\mathcal{H})$ by Proposition 3.2.3. Letting $y := \mathfrak{l}_{a,b}(x)\,Q_b y_1 Q_a$, we obviously have $R_{a,b}(y) = x$. We summarize this discussion in

Corollary 3.2.4. *If $a \in \mathcal{A}(I)$ and $b \in \mathcal{B}(I)$, then the mapping $R_{a,b}$ defined above is an isometric isomorphism of the normed space $(Q_b\mathbf{B}(\mathcal{H})\,Q_a, \|\cdot\|)$ onto $(\mathcal{L}(\mathcal{D}_\mathcal{A}, \mathcal{D}_\mathcal{B}^+)_{a,b}, \mathfrak{l}_{a,b})$.*

Remark 7. Proposition 3.2.3 and the map $R_{a,b}$ are useful tools which allow us to transform problems of $\mathcal{L}(\mathcal{D}_\mathcal{A}, \mathcal{D}_\mathcal{B}^+)$ into problems in $\mathbf{B}(\mathcal{H})$; see the Theorems 4.4.2 and 4.4.5 for some typical applications. In case $\mathcal{A} = \mathcal{B}$ and $a = b$ the mapping $R_a := R_{a,a}$ also preserves the involution (that is, $R_a(y^*) = R_a(y)^+$ for $y \in Q_a\mathbf{B}(\mathcal{H})\,Q_a$) and the order relation (that is, $R_a(y) \geqq 0$ for $y \in (Q_a\mathbf{B}(\mathcal{H})\,Q_a)_h$ if and only if $y \geqq 0$ on $\mathcal{H}$). An application of this remark is given in the next corollary.

Corollary 3.2.5. *Each element $x \in \mathcal{L}(\mathcal{D}_\mathcal{A}, \mathcal{D}_\mathcal{A}^+)_\mathrm{h}$ can be written as $x = x_1 - x_2$ with $x_1, x_2 \in \mathcal{L}(\mathcal{D}_\mathcal{A}, \mathcal{D}_\mathcal{A}^+)_+$.*

Proof. By Remark 6, we can assume without loss of generality that $\mathcal{A}$ is a directed O-vector space. Then $x \in \mathcal{U}_{a,b}$ for some $a \in \mathcal{A}(I)$ and $b \in \mathcal{B}(I)$. By Corollary 3.2.4 and Remark 7, $x = R_{a,a}(y)$ for some $y \in (Q_a\mathbf{B}(\mathcal{H})\,Q_a)_\mathrm{h}$. Writing y as $y = y_1 - y_2$ with $y_1, y_2 \in \mathbf{B}(\mathcal{H})_+$ and letting $x_k := R_{a,b}(Q_a y_k Q_a)$ for $k = 1, 2$, the proof is complete. □

The next lemma gives another perspective on the normed space $(\mathcal{L}(\mathcal{D}_\mathcal{A}, \mathcal{D}_\mathcal{B}^+)_{a,b}, \mathfrak{l}_{a,b})$.

Lemma 3.2.6. *If $a \in \mathcal{A}(I)$ and $b \in \mathcal{B}(I)$, then $\mathcal{L}(\mathcal{D}_\mathcal{A}, \mathcal{D}_\mathcal{B}^+)_{a,b} = \mathfrak{L}(\mathcal{D}_a, \mathcal{H}^b)$, and $\mathfrak{l}_{a,b}(x)$ is the operator norm of $x \in \mathfrak{L}(\mathcal{D}_a, \mathcal{H}^b)$, i.e., $\mathfrak{l}_{a,b}(x) = \sup\{\|x\varphi\|^b \colon \varphi \in \mathcal{D}(\mathcal{A})$ and $\|\varphi\|_a = 1\}$.*

Proof. Suppose $x \in \mathcal{L}(\mathcal{D}_{\mathcal{A}}, \mathcal{D}_{\mathcal{B}}^{+})$. From (5) we see that $x\varphi \in \mathcal{H}^b$ and $\|x\varphi\|^b \leqq \mathfrak{l}_{a,b}(x) \|\varphi\|_a$ for $\varphi \in \mathcal{D}(\mathcal{A})$, i.e., $x \in \mathfrak{L}(\mathcal{D}_a, \mathcal{H}^b)$ and $\|x\|_{a,b} \leqq \mathfrak{l}_{a,b}(x)$. Here $\|x\|_{a,b}$ denotes the operator norm of $x \in \mathfrak{L}(\mathcal{D}_a, \mathcal{H}^b)$. Conversely, if $x \in \mathfrak{L}(\mathcal{D}_a, \mathcal{H}^b)$, then $|\langle x\varphi, \psi\rangle| \leqq \|x\varphi\|^b \|\psi\|_b \leqq \|x\|_{a,b} \|\varphi\|_a \|\psi\|_b$ for $\varphi \in \mathcal{D}(\mathcal{A})$ and $\psi \in \mathcal{D}(\mathcal{B})$. Hence $x \in \mathcal{L}(\mathcal{D}_{\mathcal{A}}, \mathcal{D}_{\mathcal{B}}^{+})_{a,b}$ and $\mathfrak{l}_{a,b}(x) \leqq \|x\|_{a,b}$. Thus $\mathfrak{l}_{a,b}(x) = \|\varphi x\|_{a,b}$. □

In the remainder of this section, we assume that $\mathcal{A}$ and $\mathcal{B}$ are O*-algebras in the Hilbert space $\mathcal{H}$. Our next aim is to define a "multiplication" on $\mathcal{L}(\mathcal{D}_{\mathcal{A}}, \mathcal{D}_{\mathcal{B}}^{+})$ by operators of $\mathcal{L}^{+}(\mathcal{D}_{\mathcal{A}})$ from the right and by operators of $\mathcal{L}^{+}(\mathcal{D}_{\mathcal{B}})$ from the left.

Suppose that $x \in \mathcal{L}(\mathcal{D}_{\mathcal{A}}, \mathcal{D}_{\mathcal{B}}^{+})$, $u \in \mathcal{L}^{+}(\mathcal{D}_{\mathcal{A}})$ and $v \in \mathcal{L}^{+}(\mathcal{D}_{\mathcal{B}})$. Then there are operators $a, a_1 \in \mathcal{A}(I)$ and $b, b_1 \in \mathcal{B}(I)$ such that $x \in \mathcal{L}(\mathcal{D}_{\mathcal{A}}, \mathcal{D}_{\mathcal{B}}^{+})_{a,b}$, $u \in \mathfrak{L}(\mathcal{D}_{a_1}, \mathcal{D}_a)$ and $v^{+} \in \mathfrak{L}(\mathcal{D}_{b_1}, \mathcal{D}_b)$.

By Lemma 3.2.6, $x \in \mathfrak{L}(\mathcal{D}_a, \mathcal{H}^b)$. Therefore, xu, the composition of x and u, is in $\mathfrak{L}(\mathcal{D}_{a_1}, \mathcal{H}^b)$. Applying Lemma 3.2.6 once more, we get $xu \in \mathcal{L}(\mathcal{D}_{\mathcal{A}}, \mathcal{D}_{\mathcal{B}}^{+})_{a_1,b}$.

By (2), $x^{+} \in \mathcal{L}(\mathcal{D}_{\mathcal{B}}, \mathcal{D}_{\mathcal{A}}^{+})_{b,a}$. Therefore, by the preceding, $x^{+}v^{+} \in \mathcal{L}(\mathcal{D}_{\mathcal{B}}, \mathcal{D}_{\mathcal{A}}^{+})_{b_1,a}$. Hence $(x^{+}v^{+})^{+} \in \mathcal{L}(\mathcal{D}_{\mathcal{A}}, \mathcal{D}_{\mathcal{B}}^{+})_{a,b_1}$. We define $v \circ x := (x^{+}v^{+})^{+}$. Applying (2) twice, we get

$$\langle (v \circ x)\varphi, \psi\rangle = \langle x\varphi, v^{+}\psi\rangle \quad \text{for} \quad \varphi \in \mathcal{D}(\mathcal{A}) \quad \text{and} \quad \psi \in \mathcal{D}(\mathcal{B}). \tag{6}$$

This formula characterizes the mapping $v \circ x$ of $\mathcal{L}(\mathcal{D}_{\mathcal{A}}, \mathcal{D}_{\mathcal{B}}^{+})$; it could be taken also as a definition of $v \circ x$. Moreover, (6) shows that $v \circ x$ does not depend on the operators a, b, b_1 as chosen above. Further, since xu and $v \circ x$ are both in $\mathcal{L}(\mathcal{D}_{\mathcal{A}}, \mathcal{D}_{\mathcal{B}}^{+})$, $v \circ (xu)$ and $(v \circ x)\, u$ are again well-defined elements of $\mathcal{L}(\mathcal{D}_{\mathcal{A}}, \mathcal{D}_{\mathcal{B}}^{+})$. As stated in Lemma 3.2.7, (i), below, $v \circ (xu) = (v \circ x)\, u$. We call the elements xu, $v \circ x$ and $v \circ xu := v \circ (xu)$ of $\mathcal{L}(\mathcal{D}_{\mathcal{A}}, \mathcal{D}_{\mathcal{B}}^{+})$ *partial products*, and we refer to the corresponding operations as *partial multiplication* in $\mathcal{L}(\mathcal{D}_{\mathcal{A}}, \mathcal{D}_{\mathcal{B}}^{+})$.

Remark 8. If $x \in \mathcal{L}(\mathcal{D}_{\mathcal{A}}, \mathcal{D}_{\mathcal{B}}^{+})$ maps $\mathcal{D}(\mathcal{A})$ into $\mathcal{D}(\mathcal{B})$ ($\subseteq \mathcal{D}_{\mathcal{B}}^{+}$), then we see from (6) that $v \circ x$ is simply the composition vx of v and x. In particular, if u, v and x are in $\mathcal{L}^{+}(\mathcal{D}_{\mathcal{A}})$, then the partial product $v \circ xu$ in $\mathcal{L}(\mathcal{D}_{\mathcal{A}}, \mathcal{D}_{\mathcal{A}}^{+})$ is nothing but the usual product vxu in $\mathcal{L}^{+}(\mathcal{D}_{\mathcal{A}})$.

Remark 9. We show by two examples how earlier considerations can be reformulated in terms of the partial products. Since $\mathbf{B}(\mathcal{H}) \subseteq \mathcal{L}(\mathcal{D}_{\mathcal{A}}, \mathcal{D}_{\mathcal{B}}^{+})$ (see Remark 4) $b^{+} \circ ya$ is well-defined in $\mathcal{L}(\mathcal{D}_{\mathcal{A}}, \mathcal{D}_{\mathcal{B}}^{+})$ for $y \in \mathbf{B}(\mathcal{H})$, $a \in \mathcal{A}(I)$ and $b \in \mathcal{B}(I)$. Then Proposition 3.2.3 states that $\mathcal{U}_{a,b} = b^{+} \circ \mathcal{U}_{\mathbf{B}(\mathcal{H})} a$, and the mapping $R_{a,b}$ defined above takes the form $R_{a,b}(y) = b^{+} \circ ya$, $y \in Q_b \mathbf{B}(\mathcal{H})\, Q_a$. Moreover, $\mathcal{L}(\mathcal{D}_{\mathcal{A}}, \mathcal{D}_{\mathcal{B}}^{+})_{a,b} = b^{+} \circ \mathbf{B}(\mathcal{H})\, a$. In order to explain the second example, we recall from Section 3.1 that for $c \in \mathbf{B}\big(\mathcal{D}(\mathcal{B})\big)$, $\hat{c}$ denotes the unique extension of c to an element of $\mathcal{L}(\mathcal{D}_{\mathcal{B}}^{+}, \mathcal{D}_{\mathcal{B}})$ if the assumptions of Proposition 3.1.10 are valid. Thus the composition $\hat{c}x$ of $\hat{c}$ and $x \in \mathcal{L}(\mathcal{D}_{\mathcal{A}}, \mathcal{D}_{\mathcal{B}}^{+})$ is well-defined. Since $c \in \mathcal{L}^{+}(\mathcal{D}_{\mathcal{B}})$ (see again Remark 4), it follows from $\widehat{(c^{*})} = (\hat{c})^{+}$ by Proposition 3.1.10 and from (6) that $\hat{c}x = c \circ x$. However, we shall prefer the notation $\hat{c}x$ in this case; see, for instance, Section 3.4.

Lemma 3.2.7. *If* $x \in \mathcal{L}(\mathcal{D}_{\mathcal{A}}, \mathcal{D}_{\mathcal{B}}^{+})$, $u \in \mathcal{L}^{+}(\mathcal{D}_{\mathcal{A}})$ *and* $v, v_1, v_2 \in \mathcal{L}^{+}(\mathcal{D}_{\mathcal{B}})$, *then*

(i) $v \circ (xu) = (v \circ x)\, u$,

(ii) $v_1 \circ (v_2 \circ x) = (v_1 v_2) \circ x$ and $I \circ x = x$,

(iii) $(v \circ xu)^{+} = u^{+} \circ x^{+}v^{+}$.

Proof. The assertion follows by straightforward computations based on (2) and (6). □

The partial multiplication in $\mathcal{L}(\mathcal{D}_{\mathcal{A}}, \mathcal{D}_{\mathcal{B}}^{+})$ fits into the general context of A-modules. We recall the necessary definitions.

Definition 3.2.8. Suppose that A is an algebra. A linear space X is said to be a *left* A-*module* if a bilinear mapping $(a, x) \to a \cdot x$ of $A \times X$ into A is specified which satisfies

$$\text{(l)} \qquad a_1 \cdot (a_2 \cdot x) = (a_1 a_2) \cdot x \quad \text{for} \quad a_1, a_2 \in A \quad \text{and} \quad x \in X.$$

X is called a *right* A-*module* if a bilinear mapping $(a, x) \to x \cdot a$ of $A \times X$ into X is specified such that

$$\text{(r)} \qquad (x \cdot a_1) \cdot a_2 = x \cdot (a_1 a_2) \quad \text{for } a_1, a_2 \in A \quad \text{and} \quad x \in X.$$

X is called a A-*bimodule* if it is both a left A-module and a right A-module and the module operations satisfy the following axiom:

$$\text{(b)} \qquad a_1 \cdot (x \cdot a_2) = (a_1 \cdot x) \cdot a_2 \quad \text{for } a_1, a_2 \in A \quad \text{and} \quad x \in X.$$

Then the linear space $\mathcal{L}(\mathcal{D}_\mathcal{A}, \mathcal{D}_\mathcal{B}^+)$ becomes a right $\mathcal{L}^+(\mathcal{D}_\mathcal{A})$-module and a left $\mathcal{L}^+(\mathcal{D}_\mathcal{B})$-module with the module operations defined by $x \cdot a := xa$ and $b \cdot x := b \circ x$, respectively. (Indeed, (l) and (b) follow from Lemma 3.2.7, (ii) and (i); (r) is obvious.) In particular, $\mathcal{L}(\mathcal{D}_\mathcal{A}, \mathcal{D}_\mathcal{A}^+)$ is a $\mathcal{L}^+(\mathcal{D}_\mathcal{A})$-bimodule.

Remark 10. Formula (i) in Lemma 3.2.7 can be considered as an associative law for the partial product. One might ask about the following more general version of the associative law. Suppose $a \in \mathcal{L}^+(\mathcal{D}_\mathcal{A})$ and $x, y \in \mathcal{L}(\mathcal{D}_\mathcal{A}, \mathcal{D}_\mathcal{A}^+)$. If the operators xa and $a \circ y$ of $\mathcal{L}(\mathcal{D}_\mathcal{A}, \mathcal{D}_\mathcal{A}^+)$ are even in $\mathcal{L}^+(\mathcal{D}_\mathcal{A})$, then the partial products $(xa) \circ y$ and $x(a \circ y)$ make sense and are elements of $\mathcal{L}(\mathcal{D}_\mathcal{A}, \mathcal{D}_\mathcal{A}^+)$. Is $(xa) \circ y = x(a \circ y)$? The following example shows that the answer is negative in general.

Example 3.2.9. Suppose that there exists a symmetric operator b in the O*-algebra $\mathcal{A}$ which is not essentially self-adjoint. Upon replacing b by $-b$ if necessary, we can assume that $\ker(b^* + \mathrm{i}) \neq \{0\}$. Take a non-zero vector $\xi \in \ker(b^* + \mathrm{i})$. Put $a := b + \mathrm{i}$ and $y := \xi \otimes \xi$. Define a bounded operator x on $\mathcal{H}$ by $x(\bar{b} + \mathrm{i})\,\varphi = \varphi$ for $\varphi \in \mathcal{D}(\bar{b})$ and $x\psi = 0$ for $\psi \in ((\bar{b} + \mathrm{i})\,\mathcal{D}(\bar{b}))^\perp$. Then $xa = I \in \mathcal{L}^+(\mathcal{D}_\mathcal{A})$ and hence $(xa) \circ y = y \neq 0$. But $\langle (a \circ y)\,\varphi, \psi \rangle = \langle y\varphi, a^+\psi \rangle = \langle \varphi, \xi \rangle \langle \xi, (b - \mathrm{i})\,\psi \rangle = 0$ for $\varphi, \psi \in \mathcal{D}(\mathcal{A})$ which gives $a \circ y = 0 \in \mathcal{L}^+(\mathcal{D}_\mathcal{A})$ and so $x(a \circ y) = 0$. ○

3.3. Topologies Generalizing the Operator Norm Topology

In the present section we develop various processes for topologizing linear subspaces of $\mathcal{L}(\mathcal{D}_\mathcal{A}, \mathcal{D}_\mathcal{B}^+)$. They all have in common the feature that the corresponding topologies are generated by the operator norm whenever the O-families $\mathcal{A}$ and $\mathcal{B}$ consist of bounded operators only. The topological concepts discussed in this section are closely related to standard procedures of topologizing spaces of linear mappings and spaces of sesquilinear forms in the theory of locally convex spaces or to standard notions in the theory of ordered vector spaces. The first two subsections are concerned with two fundamental topological concepts, the bounded topology τ_b and the inductive topology τ_{in}, for general spaces $\mathcal{L}(\mathcal{D}_\mathcal{A}, \mathcal{D}_\mathcal{B}^+)$. In Section 1.2 we defined these topologies in case of general locally convex spaces. In the third and the fourth subsection we specialize to $*$-invariant linear subspaces of $\mathcal{L}(\mathcal{D}_\mathcal{A}, \mathcal{D}_\mathcal{A}^+)$ and to O*-algebras, and we investigate the topologies $\tau_\mathcal{D}$, $\tau_\mathcal{N}$, $\tau_\mathcal{O}$ and $\tau^\mathcal{D}$, $\tau^\mathcal{N}$, $\tau^\mathcal{O}$, τ^*, respectively. Except for τ^*, these topologies are special cases of the topologies τ_F, τ_n, τ_0, τ^F, τ^n, τ^0 which we have studied in Section 1.5 in a more general setting.

The Bounded Topology τ_b

Since $\mathcal{L}(\mathcal{D}_{\mathcal{A}}, \mathcal{D}_{\mathcal{B}}^+) \subseteq \mathfrak{L}(\mathcal{D}_{\mathcal{A}}, \mathcal{D}_{\mathcal{B}}^+[\beta])$ by Lemma 1.2.2, the topologies $\tau_{S,T}$ and τ_b from Section 1.2 are defined on $\mathcal{L}(\mathcal{D}_{\mathcal{A}}, \mathcal{D}_{\mathcal{B}}^+)$. We recall their definitions in the present setting. Let S and T be nonempty families of bounded subsets of $\mathcal{D}_{\mathcal{A}}$ and $\mathcal{D}_{\mathcal{B}}$, respectively. Then $\tau_{S,T}$ is the locally convex topology defined by the seminorms

$$p_{\mathcal{M},\mathcal{N}}(x) = \sup_{\varphi \in \mathcal{M}} \sup_{\psi \in \mathcal{N}} |\langle x\varphi, \psi\rangle|, \quad x \in \mathcal{L}(\mathcal{D}_{\mathcal{A}}, \mathcal{D}_{\mathcal{B}}^+),$$

where $\mathcal{M} \in S$ and $\mathcal{N} \in T$. The bounded topology τ_b is the topology $\tau_{S,T}$ if S and T contain all bounded subsets of $\mathcal{D}_{\mathcal{A}}$ and $\mathcal{D}_{\mathcal{B}}$, respectively.

Remark 1. In addition to the topology τ_b there are other topologies $\tau_{S,T}$ which are important. If S and T are the families of all finite subsets, then $\tau_{S,T}$ is the weak operator topology; see also Section 3.5. The topology $\tau_{S,T}$, where S and T are the families of all precompact subsets, is used in Section 5.3.

The bounded topology τ_b always refers to a fixed space $\mathcal{L}(\mathcal{D}_{\mathcal{A}}, \mathcal{D}_{\mathcal{B}}^+)$. Now let $\mathcal{A}_1$ and $\mathcal{B}_1$ be two other O-families such that $\mathcal{D}(\mathcal{A}_1) = \mathcal{D}(\mathcal{A})$ and $\mathcal{D}(\mathcal{B}_1) = \mathcal{D}(\mathcal{B})$. It is natural to ask when the bounded topologies of $\mathcal{L}(\mathcal{D}_{\mathcal{A}}, \mathcal{D}_{\mathcal{B}}^+)$ and of $\mathcal{L}(\mathcal{D}_{\mathcal{A}_1}, \mathcal{D}_{\mathcal{B}_1}^+)$ coincide on the intersection $\mathcal{L}(\mathcal{D}_{\mathcal{A}}, \mathcal{D}_{\mathcal{B}}^+) \cap \mathcal{L}(\mathcal{D}_{\mathcal{A}_1}, \mathcal{D}_{\mathcal{B}_1}^+)$. As shown by Example 3.3.2 below, this is not true in general. A rather general sufficient condition is given in

Proposition 3.3.1. *If $\mathcal{A}$, $\mathcal{A}_1$, $\mathcal{B}$ and $\mathcal{B}_1$ are closed O-families on domains $\mathcal{D}(\mathcal{A}_1) = \mathcal{D}(\mathcal{A})$ and $\mathcal{D}(\mathcal{B}_1) = \mathcal{D}(\mathcal{B})$, respectively, in the Hilbert space $\mathcal{H}$, then the bounded topologies of $\mathcal{L}(\mathcal{D}_{\mathcal{A}}, \mathcal{D}_{\mathcal{B}}^+)$ and of $\mathcal{L}(\mathcal{D}_{\mathcal{A}_1}, \mathcal{D}_{\mathcal{B}_1}^+)$ induce the same topology on $\mathcal{L}(\mathcal{D}_{\mathcal{A}}, \mathcal{D}_{\mathcal{B}}^+) \cap \mathcal{L}(\mathcal{D}_{\mathcal{A}_1}, \mathcal{D}_{\mathcal{B}_1}^+)$.*

Proof. By the definition of the bounded topology it clearly suffices to show that the spaces $\mathcal{D}_{\mathcal{A}}$ and $\mathcal{D}_{\mathcal{A}_1}$ and the spaces $\mathcal{D}_{\mathcal{B}}$ and $\mathcal{D}_{\mathcal{B}_1}$ have the same families of bounded sets. But this follows from Corollary 2.3.11. □

Example 3.3.2. Let $\mathcal{A}$ be the closed O*-algebra $C(\mathbb{R})$ on $\mathcal{D}(\mathcal{A}) := \{\varphi \in L^2(R) : \psi\cdot\varphi \in L^2(\mathbb{R})$ for all $\psi \in C(\mathbb{R})\}$ in the Hilbert space $L^2(\mathbb{R})$, where the functions of $C(\mathbb{R})$ act as multiplication operators on $\mathcal{D}(\mathcal{A})$. Let $\mathcal{A}_1 := \mathbb{C} \cdot I$ and $\mathcal{D}(\mathcal{A}_1) := \mathcal{D}(\mathcal{A})$. Fix a function $\zeta \in C(\mathbb{R})$, $\zeta \neq 0$, and define $\zeta_n(t) := \zeta(t - n)$, $t \in \mathbb{R}$ and $n \in \mathbb{N}$. Then the bounded topology of $\mathcal{L}(\mathcal{D}_{\mathcal{A}}, \mathcal{D}_{\mathcal{A}}^+)$ on $\mathcal{A}$ is generated by the seminorms $p_k(\psi) := \sup\{|\psi(t)| : t \in (-k, k)\}$, $k \in \mathbb{N}$. Hence $0 = \lim_n \zeta_n$ in $\mathcal{L}(\mathcal{D}_{\mathcal{A}}, \mathcal{D}_{\mathcal{A}}^+)[\tau_b]$. But the sequence $(\zeta_n : n \in \mathbb{N})$ does not converge in $\mathcal{L}(\mathcal{D}_{\mathcal{A}_1}, \mathcal{D}_{\mathcal{A}_1}^+)[\tau_b]$, since the topology τ_b of $\mathcal{L}(\mathcal{D}_{\mathcal{A}_1}, \mathcal{D}_{\mathcal{A}_1}^+)$ is determined by the operator norm. ○

Lemma 3.3.3. *If $\mathcal{D}_{\mathcal{A}}$ and $\mathcal{D}_{\mathcal{B}}$ are Frechet spaces, then the locally convex space $\mathcal{L}(\mathcal{D}_{\mathcal{A}}, \mathcal{D}_{\mathcal{B}}^+)[\tau_b]$ is complete.*

Proof. Since $\mathcal{L}(\mathcal{D}_{\mathcal{A}}, \mathcal{D}_{\mathcal{B}}^+) = \mathfrak{L}(\mathcal{D}_{\mathcal{A}}, \mathcal{D}_{\mathcal{B}}^+[\beta])$ under the above assumptions by Lemma 3.2.2, (i), the assertion follows at once from general results in the theory of locally convex spaces (KÖTHE [2], § 39, 6.). □

Now we turn to the continuity of the algebraic operations.

Proposition 3.3.4. (i) *The involution $x \to x^+$ is a continuous mapping of $\mathcal{L}(\mathcal{D}_{\mathcal{A}}, \mathcal{D}_{\mathcal{B}}^+)[\tau_b]$ onto $\mathcal{L}(\mathcal{D}_{\mathcal{B}}, \mathcal{D}_{\mathcal{A}}^+)[\tau_b]$.*

(ii) *Suppose $\mathcal{A}$ and $\mathcal{B}$ are O*-algebras. If $a \in \mathcal{L}^+(\mathcal{D}_\mathcal{A})$ and $b \in \mathcal{L}^+(\mathcal{D}_\mathcal{B})$, then $x \to b \circ xa$ is a continuous mapping of $\mathcal{L}(\mathcal{D}_\mathcal{A}, \mathcal{D}_\mathcal{B}^+)\,[\tau_b]$ into itself.*

Proof. (i) follows immediately from the equation $p_{\mathcal{M},\mathcal{N}}(x) = p_{\mathcal{N},\mathcal{M}}(x^+)$, $x \in \mathcal{L}(\mathcal{D}_\mathcal{A}, \mathcal{D}_\mathcal{B}^+)$.

(ii): Let $\mathcal{M}$ and $\mathcal{N}$ be bounded sets in $\mathcal{D}_\mathcal{A}$ and $\mathcal{D}_\mathcal{B}$, respectively. Since $a \in \mathfrak{L}(\mathcal{D}_\mathcal{A})$ and $b^+ \in \mathfrak{L}(\mathcal{D}_\mathcal{B})$, the sets $a\mathcal{M}$ and $b^+\mathcal{N}$ are also bounded. We have $p_{\mathcal{M},\mathcal{N}}(b \circ xa) = p_{a\mathcal{M},b^+\mathcal{N}}(x)$ for $x \in \mathcal{L}(\mathcal{D}_\mathcal{A}, \mathcal{D}_\mathcal{B}^+)$, from which the assertion follows. $\square$

Proposition 3.3.5. *Suppose $y \in \mathcal{L}(\mathcal{D}_\mathcal{A}^+, \mathcal{D}_\mathcal{A})$ and $z \in \mathcal{L}(\mathcal{D}_\mathcal{B}^+, \mathcal{D}_\mathcal{B})$. Then $zxy \in \mathcal{L}(\mathcal{D}_\mathcal{A}^+, \mathcal{D}_\mathcal{B})$ and $zxy \restriction \mathcal{H} \in \mathbf{B}(\mathcal{D}(\mathcal{A}), \mathcal{D}(\mathcal{B}))$ for all $x \in \mathcal{L}(\mathcal{D}_\mathcal{A}, \mathcal{D}_\mathcal{B}^+)$, and $\|z \cdot y \restriction \mathcal{H}\|$ is a continuous seminorm on $\mathcal{L}(\mathcal{D}_\mathcal{A}, \mathcal{D}_\mathcal{B}^+)\,[\tau_b]$. If $\mathcal{A}$ and $\mathcal{B}$ are O*-algebras, then $x \to zxy$ is a continuous mapping of $\mathcal{L}(\mathcal{D}_\mathcal{A}, \mathcal{D}_\mathcal{B}^+)\,[\tau_b]$ into $\mathcal{L}(\mathcal{D}_\mathcal{A}^+, \mathcal{D}_\mathcal{B})\,[\tau_e]$.*

Proof. Let $x \in \mathcal{L}(\mathcal{D}_\mathcal{A}, \mathcal{D}_\mathcal{B}^+)$. We have $\langle zxy\varphi', \psi'\rangle = \langle \varphi', y^+x^+z^+\psi'\rangle$ for $\varphi' \in \mathcal{D}'_\mathcal{A}$ and $\psi' \in \mathcal{D}'_\mathcal{B}$. Since $zxy(\mathcal{D}_\mathcal{A}^+) \subseteq \mathcal{D}(\mathcal{B})$ and $y^+x^+z^+(\mathcal{D}_\mathcal{B}^+) \subseteq \mathcal{D}(\mathcal{A})$, this gives $\mathfrak{d}_{zxy} \in \mathfrak{B}(\mathcal{D}'_\mathcal{B}[\sigma'], \mathcal{D}'_\mathcal{A}[\sigma'])$. Thus we have $zxy \in \mathcal{L}(\mathcal{D}_\mathcal{A}^+, \mathcal{D}_\mathcal{B})$. By Lemma 3.1.9, $zxy \restriction \mathcal{H}$ is in $\mathbf{B}(\mathcal{D}(\mathcal{A}), \mathcal{D}(\mathcal{B}))$. Put $c := y \restriction \mathcal{H}$ and $d := z^+ \restriction \mathcal{H}$. From Lemma 3.1.9 and Corollary 3.1.3, $c\mathcal{U}_\mathcal{H}$ and $d\mathcal{U}_\mathcal{H}$ are bounded subsets of $\mathcal{D}_\mathcal{A}$ and $\mathcal{D}_\mathcal{B}$, respectively. Hence $p_{c\mathcal{U}_\mathcal{H},d\mathcal{U}_\mathcal{H}}(\cdot) \equiv \|z \cdot y \restriction \mathcal{H}\|$ is a continuous seminorm on $\mathcal{L}(\mathcal{D}_\mathcal{A}, \mathcal{D}_\mathcal{B}^+)\,[\tau_b]$.

Suppose now that $\mathcal{A}$ and $\mathcal{B}$ are O*-algebras. We prove the continuity of the mapping $x \to zxy$. Let $\mathcal{M}$ and $\mathcal{N}$ be equicontinuous subsets of $\mathcal{D}'_\mathcal{A}$ and $\mathcal{D}'_\mathcal{B}$, respectively. Since $\langle zxy\varphi', \psi'\rangle = \langle xy\varphi', z^+\psi'\rangle$ for $\varphi' \in \mathcal{M}$ and $\psi' \in \mathcal{N}$, the proof is complete if we have shown that $y\mathcal{M}$ and $z^+\mathcal{N}$ are bounded subsets of $\mathcal{D}_\mathcal{A}$ and $\mathcal{D}_\mathcal{B}$, respectively. Since $\mathcal{N}$ is equicontinuous, there exists an operator $a_1 \in \mathcal{A}(I)$ such that $|\langle \varphi', \varphi\rangle| \leqq \|a_1\varphi\|$ for all $\varphi' \in \mathcal{M}$ and $\varphi \in \mathcal{D}(\mathcal{A})$. If $a \in \mathcal{A}$, then

$$\sup_{\varphi' \in \mathcal{M}} \|ay\varphi'\| = \sup_{\varphi' \in \mathcal{M}} \sup_{\zeta \in \mathcal{U}_{\mathcal{D}(\mathcal{A})}} |\langle \varphi', y^+a^+\zeta\rangle| \leqq \sup_{\zeta \in \mathcal{U}_{\mathcal{D}(\mathcal{A})}} \|a_1y^+a^+\zeta\|.$$

The latter is finite, since $y^+ \restriction H \in \mathbf{B}(\mathcal{D}(\mathcal{A}))$ and the operator $a_1y^+a \equiv a_1(y^+ \restriction \mathcal{H})\,a$ is bounded by Remark 1 in 3.1. This shows that $y\mathcal{M}$ is bounded in $\mathcal{D}_\mathcal{A}$. The proof for $z^+\mathcal{N}$ is similar. $\square$

Corollary 3.3.6. *Suppose $\mathcal{A}$ and $\mathcal{B}$ are O*-algebras such that $\mathcal{D}_\mathcal{A}$ and $\mathcal{D}_\mathcal{B}$ are sequentially complete spaces. If $c \in \mathbf{B}(\mathcal{D}(\mathcal{A}))$ and $d \in \mathbf{B}(\mathcal{D}(\mathcal{B}))$, then $\hat{d}x c \in \mathbf{B}(\mathcal{D}(\mathcal{A}), \mathcal{D}(\mathcal{B}))$ for all $x \in \mathcal{L}(\mathcal{D}_\mathcal{A}, \mathcal{D}_\mathcal{B}^+)$ and $\|\hat{d} \cdot c\|$ is a continuous seminorm on $\mathcal{L}(\mathcal{D}_\mathcal{A}, \mathcal{D}_\mathcal{B}^+)\,[\tau_b]$.*

Proof. Set $y := c$ and $z := d$ in Proposition 3.3.5. $\square$

We briefly discuss the two special cases $\mathcal{L}(\mathcal{D}_\mathcal{A}, \mathcal{D}_\mathcal{A}^+)$ and $\mathcal{L}(\mathcal{D}_\mathcal{A}, \mathcal{H})$ of the spaces $\mathcal{L}(\mathcal{D}_\mathcal{A}, \mathcal{D}_\mathcal{B}^+)$ separately, since they are of particular interest.

Special Case 1: $\mathcal{L}(\mathcal{D}_\mathcal{A}, \mathcal{D}_\mathcal{A}^+)$

We denote the bounded topology τ_b of $\mathcal{L}(\mathcal{D}_\mathcal{A}, \mathcal{D}_\mathcal{A}^+)$ by $\tau_\mathcal{D}$. Suppose $\mathcal{M}$ is a bounded subset of $\mathcal{D}_\mathcal{A}$. We define seminorms $p_\mathcal{M}$ and $p'_\mathcal{M}$ on $\mathcal{L}(\mathcal{D}_\mathcal{A}, \mathcal{D}_\mathcal{A}^+)$ by

$$p_\mathcal{M}(x) := p_{\mathcal{M},\mathcal{M}}(x) \equiv \sup_{\varphi,\psi \in \mathcal{M}} |\langle x\varphi, \psi\rangle| \quad \text{and} \quad p'_\mathcal{M}(x) := \sup_{\varphi \in \mathcal{M}} |\langle x\varphi, \varphi\rangle|.$$

Obviously, $p'_{\operatorname{aco}\mathcal{M}} \leqq p_{\operatorname{aco}\mathcal{M}} = p_\mathcal{M}$ on $\mathcal{L}(\mathcal{D}_\mathcal{A}, \mathcal{D}_\mathcal{A}^+)$. From the polarization formula 3.2/(3) we conclude that $p_\mathcal{M} \leqq 4p'_{\operatorname{aco}\mathcal{M}}$. That is, we have

$$p'_{\operatorname{aco}\mathcal{M}}(x) \leqq p_\mathcal{M}(x) \leqq 4p'_{\operatorname{aco}\mathcal{M}}(x) \quad \text{for all } x \in \mathcal{L}(\mathcal{D}_\mathcal{A}, \mathcal{D}_\mathcal{A}^+). \tag{1}$$

Let $\mathbf{S}$ be a non-empty family of bounded sets of $\mathcal{D}_{\mathcal{A}}$. We write $\tau_{\mathbf{S}}$ for the topology $\tau_{\mathbf{S},\mathbf{S}}$. Suppose that the family $\mathbf{S}$ is directed, i.e., given $\mathcal{M}_1, \mathcal{M}_2 \in \mathbf{S}$, there is an $\mathcal{M}_3 \in \mathbf{S}$ such that $\mathcal{M}_1 \cup \mathcal{M}_2 \subseteq \mathcal{M}_3$. Then the topology $\tau_{\mathbf{S}}$ is generated by the (directed) family of seminorms $\{p_{\mathcal{M}} : \mathcal{M} \in \mathbf{S}\}$. From (1) we see that $\tau_{\mathbf{S}}$ is also generated by $\{p'_{\mathrm{aco}\,\mathcal{M}} : \mathcal{M} \in \mathbf{S}\}$. In particular this shows that the topology $\tau_{\mathcal{D}}$ is generated by each of the families $\{p_{\mathcal{M}}\}$ and $\{p'_{\mathcal{M}}\}$, where $\mathcal{M}$ ranges over the bounded sets in $\mathcal{D}_{\mathcal{A}}$.

Proposition 3.3.7. *For any non-empty family* $\mathbf{S}$ *of bounded sets in* $\mathcal{D}_{\mathcal{A}}$ *the positive cone* $\mathcal{L}(\mathcal{D}_{\mathcal{A}}, \mathcal{D}_{\mathcal{A}}^+)_+$ *is normal in* $\mathcal{L}(\mathcal{D}_{\mathcal{A}}, \mathcal{D}_{\mathcal{A}}^+)_{\mathrm{h}}\,[\tau_{\mathbf{S}}]$. *In particular,* $\mathcal{L}(\mathcal{D}_{\mathcal{A}}, \mathcal{D}_{\mathcal{A}}^+)_+$ *is normal in* $\mathcal{L}(\mathcal{D}_{\mathcal{A}}, \mathcal{D}_{\mathcal{A}}^+)_{\mathrm{h}}\,[\tau_{\mathcal{D}}]$.

Proof. There is no loss of generality to assume that $\mathbf{S}$ is directed. (Otherwise we replace $\mathbf{S}$ by the family $\tilde{\mathbf{S}}$ of all finite unions of sets from $\mathbf{S}$. Then $\tilde{\mathbf{S}}$ is directed and $\tau_{\mathbf{S}} = \tau_{\tilde{\mathbf{S}}}$.) Then, by the preceding, $\tau_{\mathbf{S}}$ is generated by the directed family of seminorms $\{p'_{\mathrm{aco}\,\mathcal{M}} : \mathcal{M} \in \mathbf{S}\}$. Therefore, the sets $\{x \in \mathcal{L}(\mathcal{D}_{\mathcal{A}}, \mathcal{D}_{\mathcal{A}}^+)_{\mathrm{h}} : p'_{\mathrm{aco}\,\mathcal{M}}(x) \leqq \varepsilon\}$, where $\mathcal{M} \in \mathbf{S}$ and $\varepsilon > 0$, form a 0-neighbourhood base for the topology $\tau_{\mathbf{S}}$, and these sets are obviously absolutely convex and $\mathcal{L}(\mathcal{D}_{\mathcal{A}}, \mathcal{D}_{\mathcal{A}}^+)_+$-saturated. □

Special Case 2: $\mathcal{L}(\mathcal{D}_{\mathcal{A}}, \mathcal{H})$

We shall denote the topology τ_{b} of $\mathcal{L}(\mathcal{D}_{\mathcal{A}}, \mathcal{H})$ by $\tau^{\mathcal{D}}$. (Recall that $\mathcal{L}(\mathcal{D}_{\mathcal{A}}, \mathcal{H}) = \mathcal{L}(\mathcal{D}_{\mathcal{A}}, \mathcal{D}_{\mathcal{B}}^+)$ for any O-family $\mathcal{B}$ of bounded operators.) For any bounded subset $\mathcal{M}$ of $\mathcal{D}_{\mathcal{A}}$ we define a seminorm $p^{\mathcal{M}}$ on $\mathcal{L}(\mathcal{D}_{\mathcal{A}}, \mathcal{H})$ by $p^{\mathcal{M}}(x) := p_{\mathcal{M},U_{\mathcal{H}}}(x) = \sup_{\varphi \in \mathcal{M}} \|x\varphi\|$.

Suppose $\mathbf{S}$ is a non-empty family of bounded subsets of $\mathcal{D}_{\mathcal{A}}$. We write $\tau^{\mathbf{S}}$ for the topology $\tau_{\mathbf{S},\mathbf{T}}$, where $\mathbf{T}$ is the singleton $\{\mathcal{U}_{\mathcal{H}}\}$. Then the topology $\tau^{\mathbf{S}}$ is determined by the family of seminorms $\{p^{\mathcal{M}} : \mathcal{M} \in S\}$. In particular, the topology $\tau^{\mathcal{D}}$ on $\mathcal{L}(\mathcal{D}_{\mathcal{A}}, \mathcal{H})$ is generated by the family of all seminorms $p^{\mathcal{M}}$, where $\mathcal{M}$ is a bounded subset of $\mathcal{D}_{\mathcal{A}}$.

The Inductive Topology τ_{in}

Suppose that $\mathcal{L}$ is a linear subspace of $\mathcal{L}(\mathcal{D}_{\mathcal{A}}, \mathcal{D}_{\mathcal{B}}^+)$. According to Definition 3.2.1, we have $\mathfrak{c}_x \in \mathcal{B}(\mathcal{D}_{\mathcal{A}}, \mathcal{D}_{\mathcal{B}})$ for all $x \in \mathcal{L}$. Therefore, the inductive topology τ_{in} (see Section 1.2, III) is defined on $\mathcal{L}$. Recall from Section 1.2 that $\tau_{\mathrm{b}} \subseteqq \tau_{\mathrm{in}}$ on $\mathcal{L}$.

As noted in Remark 6 in 3.2, there is no loss of generality to assume that $\mathcal{A}$ and $\mathcal{B}$ are directed O-families. Then $\{\|\cdot\|_a : a \in \mathcal{A}\}$ and $\{\|\cdot\|_b : b \in \mathcal{B}\}$ are directed families of seminorms, so that $\mathcal{L}[\tau_{\mathrm{in}}]$ is the inductive limit of the family of normed spaces $\{(\mathcal{L}_{a,b}, \mathfrak{l}_{a,b}) : a \in \mathcal{A}$ and $b \in \mathcal{B}\}$.

Proposition 3.3.8. (i) *The mapping* $x \to x^+$ *of* $\mathcal{L}[\tau_{\mathrm{in}}]$ *into* $\mathcal{L}^+[\tau_{\mathrm{in}}]$ *is continuous.*

(ii) *Suppose that* $\mathcal{A}$ *and* $\mathcal{B}$ *are O*-algebras,* $a \in \mathcal{L}^+(\mathcal{D}_{\mathcal{A}})$ *and* $b \in \mathcal{L}^+(\mathcal{D}_{\mathcal{B}})$. *Then* $x \to b \circ xa$ *maps* $\mathcal{L}(\mathcal{D}_{\mathcal{A}}, \mathcal{D}_{\mathcal{B}}^+)\,[\tau_{\mathrm{in}}]$ *continuously into itself.*

Proof. (i) was already shown in Section 1.2. (ii): Suppose $a_1 \in \mathcal{A}$ and $b_1 \in \mathcal{B}$. If $x \in \mathcal{L}_{a_1,b_1}$, then, by 3.2/(5) and 3.2/(6), $|\langle b \circ xa\varphi, \psi\rangle| = |\langle xa\varphi, b^+\psi\rangle| \leqq \mathfrak{l}_{a_1,b_1}(x)\, \|a_1 a\varphi\|\, \|b_1 b^+\psi\|$ for $\varphi \in \mathcal{D}(\mathcal{A})$ and $\psi \in \mathcal{D}(\mathcal{B})$; so $b \circ xa \in \mathcal{L}_{a_1a,b_1b^+}$ and $\mathfrak{l}_{a_1a,b_1b^+}(b \circ xa) \leqq \mathfrak{l}_{a_1,b_1}(x)$. This shows that $x \to b \circ xa$ maps the normed space $\mathcal{L}_{a_1,b_1}$ continuously into the normed space $\mathcal{L}_{a_1a,b_1b^+}$. By general properties of the inductive limit (see e.g. SCHÄFER [1], II, § 6) the assertion follows. □

We now consider the two special cases $\mathcal{L}(\mathcal{D}_{\mathcal{A}}, \mathcal{D}_{\mathcal{A}}^+)$ and $\mathcal{L}(\mathcal{D}_{\mathcal{A}}, \mathcal{H})$.

Special Case 1: $\mathcal{L}(\mathcal{D}_{\mathcal{A}}, \mathcal{D}_{\mathcal{A}}^{+})$

Suppose $a \in \mathcal{A}$. We abbreviate $\mathcal{L}_a := \mathcal{L}_{a,a}$, $\mathcal{U}_a := \mathcal{U}_{aa}$, $\mathfrak{l}_a := \mathfrak{l}_{a,a}$, and $\mathfrak{l}'_a(x) := \inf\{\lambda \geqq 0 : |\langle x\varphi, \varphi\rangle| \leqq \lambda \|x\varphi\|^2$ for all $\varphi \in \mathcal{D}(\mathcal{A})\}$ if $x \in \mathcal{L}_a$. For notational simplicity we set $\mathfrak{l}_a(x) = \mathfrak{l}'_a(x) := +\infty$ if $x \notin \mathcal{L}_a$ and $x \in \mathcal{L}$. The following lemma shows that $\mathfrak{l}_a$ and $\mathfrak{l}'_a$ are equivalent norms on $\mathcal{L}_a$.

Lemma 3.3.9. *For arbitrary* $a \in \mathcal{A}$ *and* $x \in \mathcal{L}$, $\mathfrak{l}_a(x) \leqq 4\mathfrak{l}'_a(x) \leqq 4\mathfrak{l}_a(x)$.

Proof. It is trivial that $\mathfrak{l}'_a \leqq \mathfrak{l}_a$ on $\mathcal{L}$ and $\mathfrak{l}_a(x) \leqq 4\mathfrak{l}'_a(x)$ if $x \notin \mathcal{L}_a$. Suppose $x \in \mathcal{L}_a$. Then $|\langle x\varphi, \varphi\rangle| \leqq \mathfrak{l}'_a(x) \|a\varphi\|^2$ for all $\varphi \in \mathcal{D}(\mathcal{A})$. Therefore, by 3.2/(3), $|\langle x\varphi, \psi\rangle| \leqq 4\mathfrak{l}'_a(x) \|a\varphi\| \|a\psi\|$ for all $\varphi, \psi \in \mathcal{D}(\mathcal{A})$. This gives $\mathfrak{l}_a(x) \leqq 4\mathfrak{l}'_a(x)$. □

From Lemma 3.3.9 we conclude easily that $\mathcal{L}[\tau_{\text{in}}]$ is the inductive limit of the family of normed spaces $\{(\mathcal{L}_a, \mathfrak{l}'_a) : a \in \mathcal{A}\}$ if $\mathcal{A}$ is a directed O-family.

Remark 2. The main advantage of the norms $\mathfrak{l}'_a$ is that they are better related to the order structure, as the following simple observation shows. If $\mathcal{A}$ is an O*-algebra, $a \in \mathcal{A}$, and $x = x^+ \in \mathcal{L}$, then $\mathfrak{l}'_a(x) \leqq 1$ is equivalent to $x \leqq a^+a$ and $-x \leqq a^+a$.

As indicated in Remark 2, there is a link between the topology τ_{in} and the order structure. We now make this more precise by showing that $\tau_{\text{in}} = \tau_{\mathcal{O}}$ on $\mathcal{L}$ under certain assumptions. For $a, b \in \mathcal{L}(\mathcal{D}_{\mathcal{A}}, \mathcal{D}_{\mathcal{A}}^+)_{\text{h}}$, let $[a, b]_{\mathcal{L}}$ denote the set $\{z \in \mathcal{L}_{\text{h}} : a \leqq z \leqq b\}$. Recall that, as usual, $[x, y]$ is the order interval $\{z \in \mathcal{L}_{\text{h}} : x \leqq z \leqq y\}$ if $x, y \in \mathcal{L}_{\text{h}}$.

Proposition 3.3.11. *Suppose $\mathcal{A}$ is an O*-algebra and $\mathcal{L}$ is a *-vector subspace of $\mathcal{L}(\mathcal{D}_{\mathcal{A}}, \mathcal{D}_{\mathcal{A}}^+)$. Then the topology $\tau_{\mathcal{O}}$ is finer than τ_{in} on $\mathcal{L}$. If for each operator $a \in \mathcal{A}_+$ there exist elements $x, y \in \mathcal{L}_{\text{h}}$ such that $[-a, a]_{\mathcal{L}} \subseteqq [x, y]$, then the topologies τ_{in} and $\tau_{\mathcal{O}}$ of $\mathcal{L}$ coincide. In particular, we have $\tau_{\text{in}} = \tau_{\mathcal{O}}$ on $\mathcal{L}$ if $\mathcal{L}$ is cofinal in the ordered *-vector space $\mathcal{L}(\mathcal{D}_{\mathcal{A}}, \mathcal{D}_{\mathcal{A}}^+)$ or if $\mathcal{A} \subseteqq \mathcal{L}$.*

Proof. We first show that $\tau_{\text{in}} \subseteqq \tau_{\mathcal{O}}$ on $\mathcal{L}$. Let $\mathcal{U}$ be an absolutely convex 0-neighbourhood for τ_{in} and let $x, y \in \mathcal{L}_{\text{h}}$. There is an operator $a \in \mathcal{A}$ such that $x \in \mathcal{U}_{a,a}$ and $y \in \mathcal{U}_{a,a}$. If $z \in [x, y]$, then

$$|\langle z\varphi, \varphi\rangle| \leqq |\langle x\varphi, \varphi\rangle| + |\langle y\varphi, \varphi\rangle| \leqq 2 \|a\varphi\|^2 \quad \text{for} \quad \varphi \in \mathcal{D}(\mathcal{A}),$$

i.e., $z \in \mathcal{L}_a$ and $\mathfrak{l}_a'(z) \leqq 2$. Since $\mathcal{U}$ absorbs the set $\{z \in \mathcal{L}_a : \mathfrak{l}_a'(z) \leqq 2\}$, it absorbs the order interval $[x, y]$. Hence $\mathcal{U} \in \boldsymbol{U}^0$, and $\tau_{\text{in}} \subseteqq \tau_{\mathcal{O}}$.

Now assume that the above condition concerning the order intervals is satisfied. Let $\mathcal{U} \in \boldsymbol{U}^0$ and let $a \in \mathcal{A}$. By assumption, there are $x, y \in \mathcal{L}_{\text{h}}$ such that $[-a^+a, a^+a]_{\mathcal{L}} \subseteqq [x, y]$. Since $\mathcal{U}$ absorbs order intervals, $2\delta[x, y] \subseteqq \mathcal{U}$ for some $\delta > 0$. Suppose $z \in \mathcal{L}_a$ and $\mathfrak{l}_a'(z) \leqq \delta$. Since $\mathcal{L}$ is a *-vector space, we can write z as $z = z_1 + iz_2$ with $z_1, z_2 \in \mathcal{L}_{\text{h}}$. From $\mathfrak{l}_a'(z_k) \leqq \mathfrak{l}_a'(z) \leqq \delta$ we obtain $z_k \leqq \delta a^+a$ and $-z_k \leqq \delta a^+a$, so that $2z_k \in 2\delta[-a^+a, a^+a]_{\mathcal{L}} \subseteqq 2\delta[x, y] \subseteqq \mathcal{U}$ for $k = 1, 2$. Hence $z \in \mathcal{U}$. This proves that $\mathcal{U} \cap \mathcal{L}_a$ contains a 0-neighbourhood of the normed space $(\mathcal{L}_a, \mathfrak{l}_a')$. Since $\mathcal{L}[\tau_{\text{in}}]$ is the inductive limit of the normed spaces $(\mathcal{L}_a, \mathfrak{l}_a')$, $a \in \mathcal{A}$, $\mathcal{U}$ is a 0-neighbourhood for τ_{in}. Thus $\tau_{\mathcal{O}} \subseteqq \tau_{\text{in}}$ on $\mathcal{L}$. Together with the preceding, we have shown that $\tau_{\text{in}} = \tau_{\mathcal{O}}$ on $\mathcal{L}$.

It is clear that the above assumption concerning the order intervals is fulfilled, if $\mathcal{L}$ is cofinal in $\mathcal{L}(\mathcal{D}_{\mathcal{A}}, \mathcal{D}_{\mathcal{A}}^+)$, and thus, in particular, if $\mathcal{L}$ contains $\mathcal{A}$. □

We illustrate the previous proposition by a simple example.

Example 3.3.12. Let $\mathcal{A}$ be the O*-algebra $\mathbb{C}[x]$ on $\mathcal{D}(\mathcal{A}) := \{\varphi \in L^2(\mathbb{R}) : t^n\varphi(t) \in L^2(\mathbb{R})$ for all $n \in \mathbb{N}\}$ in the Hilbert space $L^2(\mathbb{R})$, where the polynomials act as multiplication operators on $\mathcal{D}(\mathcal{A})$. We consider $\mathcal{L} := L^\infty(\mathbb{R})$ as a subspace of $\mathcal{L}(\mathcal{D}_{\mathcal{A}}, \mathcal{D}_{\mathcal{A}}^+)$ by identifying elements of $L^\infty(\mathbb{R})$ and the corresponding multiplication operators on the domain $\mathcal{D}(\mathcal{A})$. Then $\tau_{\mathcal{O}} \neq \tau_{\text{in}}$ on $\mathcal{L}$, since $\tau_{\mathcal{O}}$ is obviously generated by the norm $\|\cdot\|_\infty$ of $L^\infty(\mathbb{R})$, but the norm is not continuous on $\mathcal{L}[\tau_{\text{in}}]$. (Otherwise $\|\cdot\|_\infty$ would be continuous on $(\mathcal{L}_a, \mathfrak{l}_a)$, where a is the polynomial $1 + t^2$; that is, there would be a $\lambda > 0$ such that $\|f\|_\infty \leqq \lambda \|f(t)(1 + t^2)^{-1}\|_\infty$ for all $f \in L^\infty(\mathbb{R})$. This is impossible.) Let $\mathcal{L}_1$ be the set of all $f \in \mathcal{L} \equiv L^\infty(\mathbb{R})$ which are supported in $[0, 1]$. Clearly, $\mathcal{L}_1$ satisfies the condition in Proposition 3.3.11, so that $\tau_{\mathcal{O}} = \tau_{\text{in}}$ on $\mathcal{L}_1$. Note that $\mathcal{L}$ and $\mathcal{L}_1$ are both not cofinal in $\mathcal{L}(\mathcal{D}_{\mathcal{A}}, \mathcal{D}_{\mathcal{A}}^+)$. ○

Remark 3. The topology $\tau_{\mathcal{D}}$ on $\mathcal{L}$ is one of the topologies $\tau_{\boldsymbol{F}}$ defined in Section 1.5. Since $\mathcal{L}_+$ is normal in $\mathcal{L}_{\text{h}}[\tau_{\mathcal{D}}]$ and the involution is continuous in $\mathcal{L}[\tau_{\mathcal{D}}]$, this follows at once from Proposition 1.5.4. We verify this directly. Suppose $\mathcal{M}$ is a bounded subset of $\mathcal{D}_{\mathcal{A}}$. Let $\tilde{\mathcal{M}} := \{\omega_\varphi : \varphi \in \mathcal{M}\}$, where ω_φ is the linear functional on $\mathcal{L}$ defined by $\omega_\varphi(\cdot) = \langle \cdot\varphi, \varphi\rangle$. We check that $\tilde{\mathcal{M}} \in \boldsymbol{F}_{\max}$. (We use the notation of 1.5.) Suppose $x \in \mathcal{L}$. Then there is a continuous seminorm p on $\mathcal{D}_{\mathcal{A}}$ such that $|\langle x\varphi, \varphi\rangle| \leqq p(\varphi)^2$, $\varphi \in \mathcal{D}(\mathcal{A})$. Then $\sup\{|\omega_\varphi(x)| : \varphi \in \mathcal{M}\} \leqq \sup\{p(\varphi)^2 : \varphi \in \mathcal{M}\} < \infty$, since $\mathcal{M}$

is bounded in $\mathcal{D}_{\mathcal{A}}$. Therefore, $\tilde{\mathcal{M}}$ is weakly bounded on $\mathcal{L}$. (This also follows from $\mathcal{L} \subseteq \mathfrak{L}(\mathcal{D}_{\mathcal{A}}, \mathcal{D}^+_{\mathcal{A}}[\beta])$; cf. Lemma 1.2.2.) Since obviously $\tilde{\mathcal{M}} \subseteq \mathcal{L}^*_+$, we have $\tilde{\mathcal{M}} \in \boldsymbol{F}_{\max}$. Further, $p'_{\mathcal{M}}(x) = \sup_{\varphi \in \mathcal{M}} |\omega_\varphi(x)| = r_{\tilde{\mathcal{M}}}(x)$ for $x \in \mathcal{L}$. Let $\boldsymbol{F}_b$ denote the family of all $\tilde{\mathcal{M}}$, where $\mathcal{M}$ is a bounded subset of $\mathcal{D}_{\mathcal{A}}$. Since the topologies $\tau_{\mathcal{D}}$ and $\tau_{\boldsymbol{F}_b}$ are generated by the seminorms $p'_{\mathcal{M}}$ and $r_{\tilde{\mathcal{M}}}$, respectively, we see that $\tau_{\mathcal{D}} = \tau_{\boldsymbol{F}_b}$ on $\mathcal{L}$.

The Topologies $\tau^{\mathcal{D}}$, $\tau^{\mathcal{N}}$ $\tau^{\mathcal{O}}$, τ^*

Suppose that $\mathcal{A}$ is an O*-algebra. We denote by $\tau^{\mathcal{N}}$ and $\tau^{\mathcal{O}}$ the topologies τ^n and τ^0, respectively, on $\mathcal{A}$ as defined in Section 1.5 in case $\mathsf{A} := \mathcal{A}$ and $K := \mathcal{A}_+$.

Remark 4. In Remark 3 (applied with $\mathcal{L} = \mathcal{A}$) we have seen that $\tau_{\mathcal{D}} = \tau_{\boldsymbol{F}_b}$ on $\mathcal{A}$. (We retain the notation of Remark 3 and Section 1.5.) If $\mathcal{M}$ is a bounded subset of $\mathcal{D}_{\mathcal{A}}$, then

$$p^{\mathcal{M}}(x) = \sup_{\varphi \in \mathcal{M}} \|x\varphi\| = \sup_{\varphi \in \mathcal{M}} \omega_\varphi(x^+x)^{1/2} = r^{\tilde{\mathcal{M}}}(x), \quad x \in \mathcal{A}.$$

This shows that the topology $\tau^{\mathcal{D}}$ on $\mathcal{A}$ coincides with the topology $\tau^{\boldsymbol{F}_b}$.

Using the fact that $\tau_{\mathcal{D}} = \tau_{\boldsymbol{F}_b}$ and $\tau^{\mathcal{D}} = \tau^{\boldsymbol{F}_b}$ on $\mathcal{A}$, we restate some facts from Section 1.5 in the present setting. From (2) and from Propositions 1.5.9 and 1.5.11 we obtain the following diagram which describes the relations between the various topologies on $\mathcal{A}$:

$$\begin{array}{ccccc} \tau_{\mathcal{D}} & \subseteq & \tau_{\mathcal{N}} & \subseteq & \tau_{\mathcal{O}} \\ \cap\!\!| & & \cap\!\!| & & \cap\!\!| \\ \tau^{\mathcal{D}} & \subseteq & \tau^{\mathcal{N}} & \subseteq & \tau^{\mathcal{O}} \end{array} \tag{3}$$

Proposition 1.5.9 and Corollary 1.5.10 yield

Proposition 3.3.13. *Let $\mathcal{A}$ be an O*-algebra.*

(i) *The multiplication is jointly continuous in $\mathcal{A}[\tau_{\mathcal{D}}]$ if and only if $\tau_{\mathcal{D}} = \tau^{\mathcal{D}}$.*

(ii) *The multiplication is jointly continuous in $\mathcal{A}[\tau_{\mathcal{N}}]$ if and only if $\tau_{\mathcal{N}} = \tau^{\mathcal{N}}$.*

In case the graph topology $t_{\mathcal{A}}$ is metrizable, a similar assertion for the topologies $\tau_{\mathcal{O}}$ and $\tau^{\mathcal{O}}$ will be proven in Section 4.2.

Proposition 3.3.14. *For each O*-algebra $\mathcal{A}$ the topology $\tau^{\mathcal{O}}$ coincides with the inductive topology τ_{in} on $\mathcal{A}$ when $\mathcal{A}$ is considered as a subspace of $\mathcal{L}(\mathcal{D}_{\mathcal{A}}, \mathcal{H})$.*

Proof. By definition the collection $\boldsymbol{U}^0$ of all absolutely convex subsets of $\mathcal{A}$ that absorb each set $R_a = \{x \in \mathcal{A} : x^+x \leqq a^+a\}$, $a \in \mathcal{A}$, is a 0-neighbourhood base for $\tau^{\mathcal{O}}$. The set R_a is nothing but the unit ball of the normed space $(\mathcal{L}^a, \mathfrak{l}^a)$ in case $\mathcal{L} := \mathcal{A}$. Therefore, $\boldsymbol{U}^0$ is also a 0-neighbourhood base for the topology τ_{in} (which is the topology of the inductive limit of the normed spaces $(\mathcal{L}^a, \mathfrak{l}^a)$, $a \in \mathcal{A}$, with $\mathcal{L} = \mathcal{A}$) on $\mathcal{A}$. $\square$

Next we introduce one more topology. Again we suppose that $\mathcal{A}$ is an O*-algebra. For a in $\mathcal{A}$ and a bounded set $\mathcal{M}$ in $\mathcal{D}_{\mathcal{A}}$, we define seminorms $p^{a,\mathcal{M}}$ and $p_+^{a,\mathcal{M}}$ on $\mathcal{L}^+(\mathcal{D}_{\mathcal{A}})$ by

$$p^{a,\mathcal{M}}(x) = \sup_{\varphi \in \mathcal{M}} \|ax\varphi\| \quad \text{and} \quad p_+^{a,\mathcal{M}}(x) = \sup_{\varphi \in \mathcal{M}} \|ax^+\varphi\|. \tag{4}$$

Note that these quantities are finite and hence are seminorms on $\mathcal{L}^+(\mathcal{D}_{\mathcal{A}})$, since $\mathcal{L}^+(\mathcal{D}_{\mathcal{A}}) \subseteq \mathfrak{L}(\mathcal{D}_{\mathcal{A}})$. The topology on $\mathcal{L}^+(\mathcal{D}_{\mathcal{A}})$ that is induced by the topology of uniform convergence of bounded sets on $\mathfrak{L}(\mathcal{D}_\mathcal{A})$ is determined by the family of all seminorms $p^{a,\mathcal{M}}$. Endowed with this topology, $\mathcal{L}^+(\mathcal{D}_\mathcal{A})$ becomes a topological algebra, but the involution of $\mathcal{L}^+(\mathcal{D}_\mathcal{A})$ is not continuous in general. Let τ^* denote the locally convex topology on $\mathcal{L}^+(\mathcal{D}_\mathcal{A})$ which is defined by the family of seminorms $p^{a,\mathcal{M}}$ and $p_+^{a,\mathcal{M}}$, where $a \in \mathcal{A}$ and $\mathcal{M}$ is a bounded subset of $\mathcal{D}_\mathcal{A}$. From this definition it is clear that τ^* is the coarsest locally convex topology on $\mathcal{L}^+(\mathcal{D}_\mathcal{A})$ that is finer than the topology of uniform convergence on bounded sets of $\mathfrak{L}(\mathcal{D}_\mathcal{A})$ and that makes the involution of $\mathcal{L}^+(\mathcal{D}_\mathcal{A})$ continuous. Since $p^\mathcal{M} \equiv p^{I,\mathcal{M}}$ for each bounded set $\mathcal{M}$, we have $\tau^\mathcal{D} \subseteqq \tau^*$ on $\mathcal{L}^+(\mathcal{D}_\mathcal{A})$.

Now assume that the O*-algebra $\mathcal{A}$ is closed. Set $\mathcal{D} := \mathcal{D}(\mathcal{A})$. Then the definition of τ^* can be extended to the whole space $\mathcal{L}^+(\mathcal{D})$. Indeed, let $\mathcal{M}$ be a bounded set in $\mathcal{D}_\mathcal{A}$. By Proposition 2.3.10, $\mathcal{M}$ is bounded in $\mathcal{D}[t_+]$, so that $\sup \{\|ax\varphi\| : \varphi \in \mathcal{M}\} < \infty$ for $a \in \mathcal{A}$ and for all $x \in \mathcal{L}^+(\mathcal{D})$. Therefore, the above formulas in (4) define seminorms $p^{a,\mathcal{M}}$ and $p_+^{a,\mathcal{M}}$ on $\mathcal{L}^+(\mathcal{D})$. We denote by $\tau^*(\mathcal{A})$ the locally convex topology on $\mathcal{L}^+(\mathcal{D})$ generated by the family of seminorms $p^{a,\mathcal{M}}$ and $p_+^{a,\mathcal{M}}$, where, as above, $a \in \mathcal{A}$ and $\mathcal{M}$ is a bounded subset of $\mathcal{D}_\mathcal{A}$.

Proposition 3.3.15. *Suppose $\mathcal{A}$ is an O*-algebra on $\mathcal{D} := \mathcal{D}(\mathcal{A})$.*

(i) *$\mathcal{L}^+(\mathcal{D}_\mathcal{A})\,[\tau^*]$ and $\mathcal{A}[\tau^*]$ are topological *-algebras.*

(ii) *The topology τ^* on $\mathcal{A}$ is the coarsest among all locally convex topologies τ on $\mathcal{A}$ for which $\tau^\mathcal{D} \subseteqq \tau$ on $\mathcal{A}$ and $\mathcal{A}[\tau]$ is a topological *-algebra.*

(iii) *If $\tau_\mathcal{D} = \tau^\mathcal{D}$ on $\mathcal{A}$, then $\tau_\mathcal{D} = \tau^*$ on $\mathcal{A}$.*

(iv) *If the O*-algebra $\mathcal{A}$ is closed, then the locally convex space $\mathcal{L}^+(\mathcal{D})\,[\tau^*(\mathcal{A})]$ is complete.*

Proof. (i): The continuity of the involution in $\mathcal{L}^+(\mathcal{D}_\mathcal{A})\,[\tau^*]$ is obvious. Let $a, b \in \mathcal{A}$ and let $\mathcal{M}$ be a bounded subset of $\mathcal{D}_\mathcal{A}$. Since $b^+ \in \mathfrak{L}(\mathcal{D}_\mathcal{A}, \mathcal{D}_\mathcal{A})$, $b^+\mathcal{M}$ is also bounded in $\mathcal{D}_\mathcal{A}$. Thus the continuity of the left multiplication $x \to bx$ in $\mathcal{L}^+(\mathcal{D}_\mathcal{A})\,[\tau^*]$ follows from the identities $p^{a,\mathcal{M}}(bx) = p^{ab,\mathcal{M}}(x)$ and $p_+^{a,\mathcal{M}}(bx) = p_+^{a,b^+\mathcal{M}}(x)$, $x \in \mathcal{L}^+(\mathcal{D}_\mathcal{A})$. Therefore, $\mathcal{L}^+(\mathcal{D}_\mathcal{A})\,[\tau^*]$ and so $\mathcal{A}[\tau^*]$ are topological *-algebras.

(ii): As noted above or shown by (i), $\tau^\mathcal{D} \subseteqq \tau^*$ and $\mathcal{A}[\tau^*]$ is a topological *-algebra. Now let τ be a locally convex topology on $\mathcal{A}$ such that $\tau^\mathcal{D} \subseteqq \tau^*$ and $\mathcal{A}[\tau^*]$ is a topological *-algebra. Let $a \in \mathcal{A}$, and let $\mathcal{M}$ be a bounded subset of $\mathcal{D}_\mathcal{A}$. From $\tau^\mathcal{D} \subseteqq \tau^*$, $p^\mathcal{M}$ is a continuous seminorm on $\mathcal{A}[\tau]$. Since left multiplications and involution are continuous in $\mathcal{A}[\tau]$, $p^{a,\mathcal{M}}(\cdot) = p^\mathcal{M}(a\,\cdot)$ and $p_+^{a,\mathcal{M}}(\cdot)$ are continuous seminorms on $\mathcal{A}[\tau]$. This proves $\tau^* \subseteqq \tau$.

(iii): By Proposition 3.3.10, $\mathcal{A}[\tau_\mathcal{D}]$ is a topological *-algebra; so if $\tau_\mathcal{D} = \tau^\mathcal{D}$ on $\mathcal{A}$, then (ii) yields $\tau_\mathcal{D} = \tau^\mathcal{D} = \tau^*$ on $\mathcal{A}$.

(iv): Suppose $(x_i : i \in I)$ is a Cauchy net in $\mathcal{L}^+(\mathcal{D})\,[\tau^*(\mathcal{A})]$. Then, for each $\varphi \in \mathcal{D}$, $(x_i\varphi : i \in I)$ and $(x_i^+\varphi : i \in I)$ are Cauchy nets in $\mathcal{D}_\mathcal{A}$. Since $\mathcal{A}$ is closed, $\mathcal{D}_\mathcal{A}$ is complete, and there are vectors $\zeta_\varphi \in \mathcal{D}$ and $\zeta_\varphi^+ \in \mathcal{D}$ such that $\zeta_\varphi = \lim_i x_i\varphi$ and $\zeta_\varphi^+ = \lim_i x_i^+\varphi$ in $\mathcal{D}_\mathcal{A}$. From $\langle x_i\varphi, \psi\rangle = \langle \varphi, x_i^+\psi\rangle$ for $i \in I$ we conclude that $\langle \zeta_\varphi, \psi\rangle = \langle \varphi, \zeta_\psi^+\rangle$ for all $\varphi, \psi \in \mathcal{D}$. Therefore, the equation $x\varphi := \zeta_\varphi$, $\varphi \in \mathcal{D}$, defines an operator x in $\mathcal{L}^+(\mathcal{D})$. We have $x^+\varphi = \zeta_\varphi^+$ for $\varphi \in \mathcal{D}$. It is straightforward to verify that $x = \lim_i x_i$ in $\mathcal{L}^+(\mathcal{D})\,[\tau^*(\mathcal{A})]$. □

For O*-algebras on QF-domains there is an interesting and very useful description of the topologies τ_b, $\tau_\mathcal{D}$, $\tau^\mathcal{D}$, τ^* and of the strong topology on $\mathcal{D}'_\mathcal{A}$.

Theorem 3.3.16. *Suppose that $\mathcal{A}$ and $\mathcal{B}$ are O*-algebras in the Hilbert space $\mathcal{H}$ such that $\mathcal{D}_{\mathcal{A}}$ and $\mathcal{D}_{\mathcal{B}}$ are QF-spaces. Then the following families of seminorms are directed and generating for the corresponding topologies.*

(i) $\{q_{c,d}(x) := \|\hat{d}xc\| : c \in \mathbf{B}(\mathcal{D}(\mathcal{A}))_+$ *and* $d \in \mathbf{B}(\mathcal{D}(\mathcal{B}))_+\}$ *for the bounded topology* τ_b *on* $\mathcal{L}(\mathcal{D}_{\mathcal{A}}, \mathcal{D}_{\mathcal{B}}^+)$.

(ii) $\{q_c(x) := \|\hat{c}xc\| : c \in \mathbf{B}(\mathcal{D}(\mathcal{A}))_+\}$ *for the topology* $\tau_{\mathcal{D}}$ *on* $\mathcal{L}(\mathcal{D}_{\mathcal{A}}, \mathcal{D}_{\mathcal{A}}^+)$.

(iii) $\{q^c(x) := \|xc\| : c \in \mathbf{B}(\mathcal{D}(\mathcal{A}))_+\}$ *for the topology* $\tau^{\mathcal{D}}$ *on* $\mathcal{L}(\mathcal{D}_{\mathcal{A}}, \mathcal{H})$.

(iv) $\{q^{a,c}(x) := \|axc\| + \|cxa\| : a \in \mathcal{A}_h$ *and* $c \in \mathbf{B}(\mathcal{D}(\mathcal{A}))_+\}$ *for the topology* τ^* *on* $\mathcal{L}^+(\mathcal{D}_{\mathcal{A}})$.

(v) $\{s_c(\varphi') := \|\hat{c}\varphi'\| : c \in \mathbf{B}(\mathcal{D}(\mathcal{A}))_+\}$ *for the strong topology* β *on* $\mathcal{D}_{\mathcal{A}}'$.

Recall that $\hat{c}$ *and* $\hat{d}$ *are the extensions of* $c \in \mathbf{B}(\mathcal{D}(\mathcal{A}))_+$ *and* $d \in \mathbf{B}(\mathcal{D}(\mathcal{B}))_+$ *to elements of* $\mathcal{L}(\mathcal{D}_{\mathcal{A}}^+, \mathcal{D}_{\mathcal{A}})$ *and* $\mathcal{L}(\mathcal{D}_{\mathcal{B}}^+, \mathcal{D}_{\mathcal{B}})$, *respectively; see Section* 3.1.

Proof. By Theorem 2.4.1 and Corollary 3.1.3, the families $\{c\mathcal{U}_{\mathcal{H}} : c \in \mathbf{B}(\mathcal{D}(\mathcal{A}))_+\}$ and $\{d\mathcal{U}_{\mathcal{H}} : d \in \mathbf{B}(\mathcal{D}(\mathcal{B}))_+\}$ are fundamental systems of bounded sets in $\mathcal{D}_{\mathcal{A}}$ and $\mathcal{D}_{\mathcal{B}}$, respectively. All assertions are derived from this fact. Set $\mathcal{M}_c := c\mathcal{U}_{\mathcal{H}}$ and $\mathcal{N}_d := d\mathcal{U}_{\mathcal{H}}$ for $c \in \mathbf{B}(\mathcal{D}(\mathcal{A}))_+$ and $d \in \mathbf{B}(\mathcal{D}(\mathcal{B}))_+$. Using the fact that $\hat{d} = (\hat{d})^+$ by Proposition 3.1.10, we obtain

$$p_{\mathcal{M}_c,\mathcal{N}_d}(x) = \sup_{\zeta,\eta \in \mathcal{U}_{\mathcal{H}}} |\langle xc\zeta, d\eta\rangle| = \sup_{\zeta,\eta\in\mathcal{U}_{\mathcal{H}}} |\langle \hat{d}xc\zeta, \eta\rangle| = \|\hat{d}xc\| = q_{c,d}(x)$$

for $x \in \mathcal{L}(\mathcal{D}_{\mathcal{A}}, \mathcal{D}_{\mathcal{B}}^+)$. Since the family of seminorms $\{p_{\mathcal{M}_c,\mathcal{N}_d} : c \in \mathbf{B}(\mathcal{D}(\mathcal{A}))_+$ and $d \in \mathbf{B}(\mathcal{D}(\mathcal{B}))_+\}$ is directed and generates the topology τ_b on $\mathcal{L}(\mathcal{D}_{\mathcal{A}}, \mathcal{D}_{\mathcal{B}}^+)$, this proves the assertion of (i). (ii) and (iii) follow similarly, since $p_{\mathcal{M}_c} \equiv p_{\mathcal{M}_c,\mathcal{M}_c} = q_{c,c} \equiv q_c$ on $\mathcal{L}(\mathcal{D}_{\mathcal{A}}, \mathcal{D}_{\mathcal{A}}^+)$ and $p^{\mathcal{M}_c}(x) = \|xc\| = q^c(x)$ for $x \in \mathcal{L}(\mathcal{D}_{\mathcal{A}}, \mathcal{H})$. We prove (iv). If $a \in \mathcal{A}$ and $c \in \mathbf{B}(\mathcal{D}(\mathcal{A}))_+$, then $p^{a,\mathcal{M}_c}(x) = \|axc\|$ and $p_+^{a,\mathcal{M}_c}(x) = p^{a,\mathcal{M}_c}(x^+) = \|ax^+c\| = \|(ax^+c)^*\| = \|cxa^+\|$ for $x \in \mathcal{L}^+(\mathcal{D}_{\mathcal{A}})$. This gives (iv), because the topology τ^* is generated by the directed family of seminorms $p^{a,\mathcal{M}} + p_+^{a,\mathcal{M}}$, where $a \in \mathcal{A}_h$ and $\mathcal{M}$ ranges over a fundamental system of bounded sets in $\mathcal{D}_{\mathcal{A}}$. Finally, we verify (v). By $\hat{c} = (\hat{c})^+$, we get

$$r_{\mathcal{M}_c}(\varphi') = \sup_{\zeta\in\mathcal{U}_{\mathcal{H}}} |\langle \varphi', c\zeta\rangle| = \sup_{\zeta\in\mathcal{U}_{\mathcal{H}}} |\langle \hat{c}\varphi', \zeta\rangle| = \|\hat{c}\varphi'\| = s_c(\varphi')$$

for $\varphi' \in \mathcal{D}_{\mathcal{A}}'$ and $c \in \mathbf{B}(\mathcal{D}(\mathcal{A}))_+$. This yields (v). □

Remark 5. The preceding proof actually shows the following slightly stronger statement which can be useful in concrete cases. Suppose $\mathbf{B}_{\mathcal{A}}$, resp. $\mathbf{B}_{\mathcal{B}}$, is a subset of $\mathbf{B}(\mathcal{D}(\mathcal{A}))_+$, resp. $\mathbf{B}(\mathcal{D}(\mathcal{B}))_+$, such that the family $\{c\mathcal{U}_{\mathcal{H}} : c \in \mathbf{B}_{\mathcal{A}}\}$, resp. $\{d\mathcal{U}_{\mathcal{H}} : d \in \mathbf{B}_{\mathcal{B}}\}$, is a fundamental system of bounded sets in $\mathcal{D}_{\mathcal{A}}$, resp. $\mathcal{D}_{\mathcal{B}}$. Let $\mathcal{A}_0$ be a subset of $\mathcal{A}_h$ such that the family of seminorms $\{\|\cdot\|_a : a \in \mathcal{A}_0\}$ is directed and generates the graph topology $t_{\mathcal{A}}$. Then the assertions of Theorem 3.3.16 remain valid if we replace in (i)–(v) the sets $\mathbf{B}(\mathcal{D}(\mathcal{A}))_+$, $\mathbf{B}(\mathcal{D}(\mathcal{B}))_+$ and $\mathcal{A}_h$ by $\mathbf{B}_{\mathcal{A}}$, $\mathbf{B}_{\mathcal{B}}$ and $\mathcal{A}_0$, respectively. We shall use this remark in the next example.

Example 3.3.17. We continue the investigations of Examples 2.2.16 and 2.4.4, and we retain the notation introduced in these examples. Assume that $\mathcal{A}$ is an O*-algebra as set out in Example 2.2.16. Recall that by Proposition 2.2.17 each commutatively dominated O*-algebra $\mathcal{A}$ for which $\mathcal{D}_{\mathcal{A}}$ is a Frechet space is of this form. We have shown in Example 2.4.4 that $\{h(A)\,\mathcal{U}_{\mathcal{H}} : h \in \mathfrak{F}_\infty\}$ is a fundamental system of bounded sets in $\mathcal{D}_{\mathcal{A}}$. Therefore, it follows from Theorem 3.3.16 and from the preceding remark that the topologies $\tau_{\mathcal{D}}$, $\tau^{\mathcal{D}}$, τ^* and β on $\mathcal{L}(\mathcal{D}_{\mathcal{A}}, \mathcal{D}_{\mathcal{A}}^+)$, $\mathcal{L}(\mathcal{D}_{\mathcal{A}}, \mathcal{H})$, $\mathcal{L}^+(\mathcal{D}_{\mathcal{A}})$ and $\mathcal{D}_{\mathcal{A}}'$, respectively, are generated by the following directed families of seminorms:

$$\tau_{\mathcal{D}}: \{q_{h(A)}(x) = \|\widehat{h(A)}\, x h(A)\| : h \in \mathfrak{F}_\infty\},$$

$$\tau^{\mathcal{D}}: \{q^{h(A)}(x) = \|x h(A)\| : h \in \mathfrak{F}_\infty\},$$

$$\tau^*: \{q^{h_n(A), h(A)}(x) = \|h_n(A)\, x h(A)\| + \|h(A)\, x h_n(A)\| : h \in \mathfrak{F}_\infty \text{ and } n \in \mathbb{N}\},$$

$$\beta: \{s_{h(A)}(\varphi') = \|\widehat{h(A)}\, \varphi'\| : h \in \mathfrak{F}_\infty\}. \bigcirc$$

We state a by-product of Theorem 3.3.16, (v), as

Corollary 3.3.18. *If $\mathcal{A}$ is an O*-algebra and $\mathcal{D}_\mathcal{A}$ is a QF-space, then the locally convex space $\mathcal{D}'_\mathcal{A}\,[\beta]$ has the approximation property.*

Proof. It suffices to show that each seminorm s_c, $c \in \mathbb{B}(\mathcal{D}(\mathcal{A}))_+$, on $\mathcal{D}'_\mathcal{A}$ is a Hilbertian seminorm (KÖTHE [2], § 43, 1., (4)). If $c \in \mathbb{B}(\mathcal{D}(\mathcal{A}))_+$, then $c^{1/2} \in \mathbb{B}(\mathcal{D}(\mathcal{A}))_+$ by Corollary 3.1.5 and s_c is the seminorm associated with the semi-scalar product $\langle \widehat{c^{1/2}}\cdot, \widehat{c^{1/2}}\cdot \rangle$ on $\mathcal{D}'_\mathcal{A}$. □

We gather a few general remarks concerning the topologies defined in this section.

Remark 6. The topology τ_{in} of a linear subspace $\mathcal{L}$ of $\mathcal{L}(\mathcal{D}_\mathcal{A}, \mathcal{D}^+_\mathcal{B})$ does not coincide in general with the topology which is induced by the topology τ_{in} of $\mathcal{L}(\mathcal{D}_\mathcal{A}, \mathcal{D}^+_\mathcal{B})$, see also Remark 4 in 4.5. Note that the latter topology is always coarser.

Remark 7. The topologies τ_b, τ_{in}, $\tau_\mathcal{D}$, $\tau^\mathcal{D}$, τ^* depend, in general, on the underlying space $\mathcal{L}(\mathcal{D}_\mathcal{A}, \mathcal{D}^+_\mathcal{B})$, $\mathcal{L}(\mathcal{D}_\mathcal{A}, \mathcal{D}^+_\mathcal{A})$, $\mathcal{L}(\mathcal{D}_\mathcal{A}, \mathcal{H})$ resp. *-algebra $\mathcal{L}^+(\mathcal{D}_\mathcal{A})$ where these topologies are defined. For instance, if an O*-algebra $\mathcal{A}_1$ is contained in different spaces $\mathcal{L}(\mathcal{D}_\mathcal{A}, \mathcal{D}^+_\mathcal{A})$, then the corresponding topologies $\tau_\mathcal{D}$ do not coincide on $\mathcal{A}_1$ in general, see Example 3.3.2 and Proposition 3.3.1. Therefore, if confusion can arise, we write $\tau_\mathcal{D}(\mathcal{A})$, $\tau^\mathcal{D}(\mathcal{A})$ and $\tau^*(\mathcal{A})$ for the topologies $\tau_\mathcal{D}$, $\tau^\mathcal{D}$ and τ^* on $\mathcal{L}(\mathcal{D}_\mathcal{A}, \mathcal{D}^+_\mathcal{A})$, $\mathcal{L}(\mathcal{D}_\mathcal{A}, \mathcal{H})$ and $\mathcal{L}^+(\mathcal{D}_\mathcal{A})$, respectively. (For $\tau^*(\mathcal{A})$ this is in accordance with the above notation.) Note that if $\mathcal{A}$ is an O*-algebra and $\mathcal{A}_1 := \mathcal{L}^+(\mathcal{D}_\mathcal{A})$, then $\mathrm{t}_\mathcal{A} = \mathrm{t}_{\mathcal{A}_1}$ on $\mathcal{D}(\mathcal{A})$ and hence $\tau_\mathcal{D}(\mathcal{A}) = \tau_\mathcal{D}(\mathcal{A}_1)$ on $\mathcal{L}(\mathcal{D}_\mathcal{A}, \mathcal{D}^+_\mathcal{A}) = \mathcal{L}(\mathcal{D}_{\mathcal{A}_1}, \mathcal{D}^+_{\mathcal{A}_1})$, $\tau^\mathcal{D}(\mathcal{A}) = \tau^\mathcal{D}(\mathcal{A}_1)$ on $\mathcal{L}(\mathcal{D}_\mathcal{A}, \mathcal{H}) = \mathcal{L}(\mathcal{D}_{\mathcal{A}_1}, \mathcal{H})$ and $\tau^*(\mathcal{A}) = \tau^*(\mathcal{A}_1)$ on $\mathcal{L}^+(\mathcal{D}_\mathcal{A}) = \mathcal{L}^+(\mathcal{D}_{\mathcal{A}_1})$. Further, we shall adopt the following notational convention. Whenever we speak about the topology $\tau_\mathcal{D}$ (resp. $\tau^\mathcal{D}$, τ^*) on an O*-algebra $\mathcal{A}_1$ without specifying the space $\mathcal{L}(\mathcal{D}_\mathcal{A}, \mathcal{D}^+_\mathcal{A})$ (resp. $\mathcal{L}(\mathcal{D}_\mathcal{A}, \mathcal{H})$, $\mathcal{L}^+(\mathcal{D}_\mathcal{A})$), we always mean the topology $\tau_\mathcal{D}$ (resp. $\tau^\mathcal{D}$, τ^*) relative to $\mathcal{L}(\mathcal{D}_{\mathcal{A}_1}, \mathcal{D}^+_{\mathcal{A}_1})$ (resp. $\mathcal{L}(\mathcal{D}_{\mathcal{A}_1}, \mathcal{H})$, $\mathcal{L}^+(\mathcal{D}_{\mathcal{A}_1})$).

Remark 8. In contrast to the topologies mentioned in the preceding remark, the topologies $\tau_\mathcal{N}$, $\tau_\mathcal{O}$, $\tau^\mathcal{N}$, $\tau^\mathcal{O}$ on a *-vector subspace $\mathcal{L}$ of $\mathcal{L}(\mathcal{D}_\mathcal{A}, \mathcal{D}^+_\mathcal{A})$, resp. an O*-algebra $\mathcal{A}_1$, are intrinsic topologies in the sense that they depend only on $\mathcal{L}$, resp. $\mathcal{A}_1$, and the corresponding positive cones $\mathcal{L}_+$ and $(\mathcal{A}_1)_+$.

Recall that for each O*-algebra $\mathcal{A}$ the topologies $\tau_\mathcal{D}$, $\tau_\mathcal{N}$, $\tau_\mathcal{O}$, $\tau^\mathcal{D}$, $\tau^\mathcal{N}$, $\tau^\mathcal{O}$ and τ^* are well-defined on $\mathcal{A}$. Some basic relations are described by (3). It is natural to ask when some of these topologies coincide. Results about the equality of the topologies $\tau_\mathcal{D}$, $\tau_\mathcal{N}$ and $\tau_\mathcal{O}$ are established in the next chapter. Here we only show that except for bounded O*-algebras $\mathcal{A}$ the topologies $\tau_\mathcal{D}$, $\tau^\mathcal{D}$ and τ^* on $\mathcal{L}^+(\mathcal{D}_\mathcal{A})$ are different.

Proposition 3.3.19. *Suppose $\mathcal{A}$ is an O*-algebra which contains at least one unbounded operator.*

(i) *In none of the topologies $\tau_\mathcal{D}$, $\tau^\mathcal{D}$ and τ^* of the O*-algebra $\mathcal{L}^+(\mathcal{D}_\mathcal{A})$ is the multiplication of $\mathcal{L}^+(\mathcal{D}_\mathcal{A})$ jointly continuous.*

(ii) *On $\mathcal{L}^+(\mathcal{D}_\mathcal{A})$, $\tau_\mathcal{D} \neq \tau^\mathcal{D}$ and $\tau^\mathcal{D} \neq \tau^*$.*

(iii) *$\mathcal{L}^+(\mathcal{D}_\mathcal{A})\,[\tau_\mathcal{D}]$ and $\mathcal{L}^+(\mathcal{D}_\mathcal{A})\,[\tau^\mathcal{D}]$ are not complete.*

Proof. (i): We first prove the assertion concerning τ^*. Assume to the contrary that the multiplication is jointly continuous in $\mathcal{L}^+(\mathcal{D}_{\mathcal{A}})[\tau^*]$. Fix a unit vector ψ in $\mathcal{D}(\mathcal{A})$ and put $\mathcal{N} := \{\psi\}$. Then there exist an operator $a \in \mathcal{A}$ and a bounded subset $\mathcal{M}$ of $\mathcal{D}_{\mathcal{A}}$ such that, in particular,

$$p^{I,\mathcal{N}}(x(\psi \otimes \varphi)) = \|x\varphi\| \leqq p^{a,\mathcal{M}}(x)\, p^{a,\mathcal{M}}(\psi \otimes \varphi) = p^{a,\mathcal{M}}(x)\, r_{\mathcal{M}}(\psi)\, \|a\varphi\|$$

for all $x \in \mathcal{L}^+(\mathcal{D}_{\mathcal{A}})$ and $\varphi \in \mathcal{D}(\mathcal{A})$. This shows that the graph topology $\mathrm{t}_{\mathcal{A}}$ on $\mathcal{D}(\mathcal{A})$ is generated by the single norm $\|\cdot\|_a$. Therefore, by Proposition 2.3.15, all operators of $\mathcal{A}$ must be bounded which contradicts our assumption and completes the proof for τ^*. Setting $a = I$ in the preceding, the proof for the topology $\tau^{\mathcal{D}}$ is the same. If the multiplication were jointly continuous in $\mathcal{L}^+(\mathcal{D}_{\mathcal{A}})[\tau_{\mathcal{D}}]$, then $\tau_{\mathcal{D}} = \tau^{\mathcal{D}}$ on $\mathcal{L}^+(\mathcal{D}_{\mathcal{A}})$ by Proposition 3.3.13, and the multiplication would be jointly continuous in $\mathcal{L}^+(\mathcal{D}_{\mathcal{A}})[\tau^{\mathcal{D}}]$. As we have just seen, this is not true.

(ii): Since the multiplication is not jointly continuous in $\mathcal{L}^+(\mathcal{D}_{\mathcal{A}})[\tau_{\mathcal{D}}]$ by (i), Proposition 3.3.13 ensures that $\tau_{\mathcal{D}} \neq \tau^{\mathcal{D}}$ on $\mathcal{L}^+(\mathcal{D}_{\mathcal{A}})$. The proof of the second assertion in (ii) will be indirect. Suppose to the contrary that $\tau^{\mathcal{D}} = \tau^*$ on $\mathcal{L}^+(\mathcal{D}_{\mathcal{A}})$. Suppose $a \in \mathcal{A}$, and let ψ and $\mathcal{N}$ be as in (i). From $\tau^{\mathcal{D}} = \tau^*$ it follows that there exists a bounded subset $\mathcal{M}$ of $\mathcal{D}_{\mathcal{A}}$ such that for all $\varphi \in \mathcal{D}(\mathcal{A})$,

$$p^{a,\mathcal{N}}(\psi \otimes \varphi) = \|a\varphi\| \leqq p^{\mathcal{M}}(\psi \otimes \varphi) = r_{\mathcal{M}}(\psi)\, \|\varphi\|.$$

Hence a is bounded on $\mathcal{D}(\mathcal{A})$. Since $a \in \mathcal{A}$ is arbitrary, this contradicts the assumption.

(iii): Since there is an unbounded operator in $\mathcal{A}$, $\mathcal{D}(\mathcal{A}) \neq \mathcal{H}$, and there exists a sequence $(\varphi_n : n \in \mathbb{N})$ of vectors in $\mathcal{D}(\mathcal{A})$ which converges in $\mathcal{H}$ to a vector $\varphi \notin \mathcal{D}(\mathcal{A})$. Then $(\varphi_n \otimes \varphi_n : n \in \mathbb{N})$ is a sequence which is Cauchy in both $\mathcal{L}^+(\mathcal{D}_{\mathcal{A}})[\tau_{\mathcal{D}}]$ and $\mathcal{L}^+(\mathcal{D}_{\mathcal{A}})[\tau^{\mathcal{D}}]$ but has no limit in either space. □

From Proposition 3.3.19,(iii), if the maximal O*-algebra $\mathcal{L}^+(\mathcal{D})$ on a domain $\mathcal{D}$ contains unbounded operators, then $\mathcal{L}^+(\mathcal{D})[\tau_{\mathcal{D}}]$ is not complete. This shows that the completion of an O*-algebra $\mathcal{A}[\tau_{\mathcal{D}}]$ is not necessarily an O*-algebra. A sufficient condition is given in the next proposition.

Proposition 3.3.20. *Suppose that $\mathcal{A}$ is a closed O*-algebra for which the multiplication is jointly continuous in $\mathcal{A}[\tau_{\mathcal{D}}]$. Then there exists an O*-algebra $\tilde{\mathcal{A}}$ on $\mathcal{D}(\tilde{\mathcal{A}}) = \mathcal{D}(\mathcal{A})$ such that $\tilde{\mathcal{A}}[\tau_{\mathcal{D}}(\tilde{\mathcal{A}})]$ is the completion of $\mathcal{A}[\tau_{\mathcal{D}}(\mathcal{A})]$.*

The proof requires a simple lemma.

Lemma 3.3.21. *Let A be a subalgebra of an algebra B. Suppose that τ is a locally convex topology on B such that the multiplication of A is jointly continuous in $\mathsf{A}[\tau]$. Then the closure $\tilde{\mathsf{A}}$ of A in $\mathsf{B}[\tau]$ is an algebra.*

Proof. Let p be a continuous seminorm on $\mathsf{B}[\tau]$. The assumption implies that there exists a continuous seminorm q on $\mathsf{B}[\tau]$ such that $p(xy) \leqq q(x)\, q(y)$ for all $x, y \in \mathsf{A}$. By continuity, this inequality extends to arbitrary elements x and y of $\tilde{\mathsf{A}}$. If $a, b \in \tilde{\mathsf{A}}$ and $\varepsilon > 0$, then there exist elements $a_0 \in \mathsf{A}$ and $b_0 \in \mathsf{B}$ such that $q(a - a_0)\, q(b) < \varepsilon$ and $q(a_0)\, q(b - b_0) < \varepsilon$. Then we obtain $p(ab - a_0 b_0) \leqq p\big((a - a_0)\, b\big) + p\big(a_0(b - b_0)\big) \leqq q(a - a_0)\, q(b) + q(a_0)\, q(b - b_0) < 2\varepsilon$. This proves that $ab \in \tilde{\mathsf{A}}$. □

Proof of Proposition 3.3.20.
Set $\mathcal{D} := \mathcal{D}(\mathcal{A})$. Since the multiplication is jointly continuous in $\mathcal{A}[\tau_{\mathcal{D}}]$, we have $\tau_{\mathcal{D}}(\mathcal{A}) = \tau^*(\mathcal{A})$ on $\mathcal{A}$ by combining Propositions 3.3.13 and 3.3.19, (iii). Therefore, $\mathsf{A} := \mathcal{A}$ and $\mathsf{B}[\tau] := \mathcal{L}^+(\mathcal{D})\,[\tau^*(\mathcal{A})]$ satisfy the assumption of Lemma 3.3.21; hence the closure $\tilde{\mathcal{A}}$ of $\mathcal{A}$ in $\mathcal{L}^+(\mathcal{D})\,[\tau^*(\mathcal{A})]$ is a subalgebra of $\mathcal{L}^+(\mathcal{D})$. Because the involution is continuous in $\mathcal{L}^+(\mathcal{D})\,[\tau^*(\mathcal{A})]$, $\tilde{\mathcal{A}}$ is $*$-invariant and hence an O*-algebra on $\mathcal{D}(\mathcal{A}) \equiv \mathcal{D}$. From Proposition 3.3.19, (iv), $\mathcal{L}^+(\mathcal{D})\,[\tau^*(\mathcal{A})]$ is complete and hence is $\tilde{\mathcal{A}}[\tau^*(\mathcal{A})]$. Therefore $\tilde{\mathcal{A}}[\tau^*(\mathcal{A})]$ is the completion of $\mathcal{A}[\tau^*(\mathcal{A})] \equiv \mathcal{A}[\tau_{\mathcal{D}}(\mathcal{A})]$. Thus our proof will be complete once we have shown that $\tau_{\mathcal{D}}(\tilde{\mathcal{A}}) = \tau^*(\mathcal{A})$ on $\tilde{\mathcal{A}}$.

We first note that $\tau_{\mathcal{D}}(\tilde{\mathcal{A}}) \subseteq \tau^*(\mathcal{A})$ on $\tilde{\mathcal{A}}$. Indeed, since each bounded subset of $\mathcal{D}_{\tilde{\mathcal{A}}}$ is trivially bounded in $\mathcal{D}_{\mathcal{A}}$, we have $\tau^{\mathcal{D}}(\tilde{\mathcal{A}}) \subseteq \tau^*(\mathcal{A})$ and hence $\tau_{\mathcal{D}}(\tilde{\mathcal{A}}) \subseteq \tau^*(\mathcal{A})$ on $\tilde{\mathcal{A}}$. Since $\mathcal{A}$ and $\tilde{\mathcal{A}}$ are closed O*-algebras on $\mathcal{D}$, $\tau_{\mathcal{D}}(\mathcal{A}) = \tau_{\mathcal{D}}(\tilde{\mathcal{A}})$ on $\mathcal{A}$ by Proposition 3.3.1. As stated above, $\tau^*(\mathcal{A}) = \tau_{\mathcal{D}}(\mathcal{A})$ on $\mathcal{A}$; so $\tau^*(\mathcal{A}) = \tau_{\mathcal{D}}(\tilde{\mathcal{A}})$ on $\mathcal{A}$. Since $\tau_{\mathcal{D}}(\tilde{\mathcal{A}}) \subseteq \tau^*(\mathcal{A})$ on $\tilde{\mathcal{A}}$ as just shown, $\mathcal{A}$ is dense in $\tilde{\mathcal{A}}$ relative to both topologies $\tau^*(\mathcal{A})$ and $\tau_{\mathcal{D}}(\tilde{\mathcal{A}})$. Therefore, the equality $\tau^*(\mathcal{A}) = \tau_{\mathcal{D}}(\tilde{\mathcal{A}})$ extends to $\tilde{\mathcal{A}}$. □

3.4 Some Density Results

For a linear subspace $\mathcal{D}$ of the Hilbert space $\mathcal{H}$, let $\mathbb{P}(\mathcal{D})$ denote the set of all projections on $\mathcal{H}$ whose range is contained in $\mathcal{D}$.

Theorem 3.4.1. *Suppose that $\mathcal{A}$ and $\mathcal{B}$ are O*-algebras in the same Hilbert space $\mathcal{H}$ such that $\mathcal{D}_{\mathcal{A}}$ and $\mathcal{D}_{\mathcal{B}}$ are QF-spaces.*

(i) *Suppose $\mathcal{R}$ is a bounded subset of $\mathcal{L}(\mathcal{D}_{\mathcal{A}}, \mathcal{D}_{\mathcal{B}}^+)\,[\tau_b]$. Then for any continuous seminorm p on $\mathcal{L}(\mathcal{D}_{\mathcal{A}}, \mathcal{D}_{\mathcal{B}}^+)\,[\tau_b]$ there exist projections $e \in \mathbb{P}(\mathcal{D}(\mathcal{A}))$ and $f \in \mathbb{P}(\mathcal{D}(\mathcal{B}))$ such that $p(x - \hat{f}xe) \leqq 1$ for all x in $\mathcal{R}$. In case $\mathcal{A} = \mathcal{B}$ we can take $e = f$. Moreover, $\hat{f}\mathcal{R}e$ is a bounded subset of $\mathbb{B}(\mathcal{D}(\mathcal{A}), \mathcal{D}(\mathcal{B}))\,[\tau_b]$.*

(ii) *Suppose $a \in \mathcal{A}(I)$ and $b \in \mathcal{B}(I)$. Then the set $\mathcal{U}_{a,b}$ is the closure of $\mathcal{U}_{a,b} \cap \mathbb{B}(\mathcal{D}(\mathcal{A}), \mathcal{D}(\mathcal{B}))$ in $\mathcal{L}(\mathcal{D}_{\mathcal{A}}, \mathcal{D}_{\mathcal{B}}^+)\,[\tau_b]$.*

Proof. (i): By Theorem 3.3.16, (i), the topology τ_b on $\mathcal{L}(\mathcal{D}_{\mathcal{A}}, \mathcal{D}_{\mathcal{B}}^+)$ is generated by the directed family of seminorms $\{q_{c,d}\colon c \in \mathbb{B}(\mathcal{D}(\mathcal{A}))_+$ and $d \in \mathbb{B}(\mathcal{D}(\mathcal{B}))_+\}$. This implies that there are operators $c \in \mathbb{B}(\mathcal{D}(\mathcal{A}))_+$ and $d \in \mathbb{B}(\mathcal{D}(\mathcal{B}))_+$ such that $p \leqq q_{c,d}$. Let $c = \int_0^\infty \lambda \, de(\lambda)$ and $d = \int_0^\infty \lambda \, df(\lambda)$ be the spectral resolutions of c and d. Set $e := e((\varepsilon, +\infty))$ and $f := f((\varepsilon, +\infty))$, where $\varepsilon > 0$ will be chosen later. Since $c^{1/2} \in \mathbb{B}(\mathcal{D}(\mathcal{A}))_+$ and $d^{1/2} \in \mathbb{B}(\mathcal{D}(\mathcal{B}))_+$ by Corollary 3.1.5 and $\mathcal{R}$ is bounded in $\mathcal{L}(\mathcal{D}_{\mathcal{A}}, \mathcal{D}_{\mathcal{B}}^+)\,[\tau_b]$, $\alpha := \sup_{x \in \mathcal{R}} q_{c^{1/2},d^{1/2}}(x) = \sup_{x \in \mathcal{R}} \|\widehat{d^{1/2}}xc^{1/2}\| < \infty$. Further, $\|c^{1/2}(I - e)\| \leqq \varepsilon^{1/2}$ and $\|d^{1/2}(I - f)\| \leqq \varepsilon^{1/2}$ by the spectral theorem. Using these facts and Proposition 3.1.10, we get for $x \in \mathcal{R}$ and sufficiently small ε,

$$p(x - \hat{f}xe) \leqq q_{c,d}(x - \hat{f}xe) = \|\hat{d}(x - \hat{f}xe)\,c\| = \|\hat{d}xc(I - e) + \hat{d}(I - \hat{f})\,xce\|$$

$$= \sup_{\varphi,\psi \in \mathcal{U}_{\mathcal{H}}} |\langle \widehat{d^{1/2}}xc^{1/2}c^{1/2}(I - e)\,\varphi, d^{1/2}\psi\rangle + \langle \widehat{d^{1/2}}xc^{1/2}e\varphi, d^{1/2}(I - f)\,\psi\rangle|$$

$$\leqq \alpha\varepsilon^{1/2}\,\|d^{1/2}\| + \alpha\,\|c^{1/2}e\|\,\varepsilon^{1/2} \leqq 1.$$

If $\mathcal{A} = \mathcal{B}$, we can take $c = d$ by Theorem 3.3.16, (ii); then we obtain $e = f$. Corollary 3.3.6 shows that $\hat{f}\mathcal{R}e \subseteq \mathbf{B}(\mathcal{D}(\mathcal{A}), \mathcal{D}(\mathcal{B}))$. Since $z \to f \circ ze$ is a continuous map of $\mathcal{L}(\mathcal{D}_\mathcal{A}, \mathcal{D}_\mathcal{B}^+)[\tau_b]$ onto itself by Proposition 3.3.4, the image $f \circ \mathcal{R}e \equiv \hat{f}\mathcal{R}e \upharpoonright \mathcal{D}(\mathcal{A})$ of the bounded set $\mathcal{R}$ is again bounded in the topology τ_b.

(ii): Let $x \in \mathcal{U}_{a,b}$. By Proposition 3.2.3 there is an operator $y \in \mathbf{B}(\mathcal{H})$ such that $\|y\| \leqq 1$ and $x = b^+ \circ ya$. Applying (i) to the singleton $\mathcal{R} = \{y\}$, it follows that y is in the closure of the set $\{\hat{f}ye : e \in \mathbf{P}(\mathcal{D}(\mathcal{A}))$ and $f \in \mathbf{P}(\mathcal{D}(\mathcal{B}))\}$ in $\mathcal{L}(\mathcal{D}_\mathcal{A}, \mathcal{D}_\mathcal{B}^+)[\tau_b]$. By the continuity of the mapping $z \to b^+ \circ za$ in $\mathcal{L}(\mathcal{D}_\mathcal{A}, \mathcal{D}_\mathcal{B}^+)[\tau_b]$ (again by Proposition 3.3.4), $x = b^+ \circ ya$ belongs to the closure of $\{b^+ \circ \hat{f}yea : e \in \mathbf{P}(\mathcal{D}(\mathcal{A}))$ and $f \in \mathbf{P}(\mathcal{D}(\mathcal{B}))\}$ in $\mathcal{L}(\mathcal{D}_\mathcal{A}, \mathcal{D}_\mathcal{B}^+)[\tau_b]$. The proof of (ii) is complete once we have shown that the latter set is contained in $\mathcal{U}_{a,b} \cap \mathbf{B}(\mathcal{D}(\mathcal{A}), \mathcal{D}(\mathcal{B}))$. Suppose $e \in \mathbf{P}(\mathcal{D}(\mathcal{A}))$ and $f \in \mathbf{P}(\mathcal{D}(\mathcal{B}))$. Then $b^+ \circ \hat{f}yea \equiv b^+ fyea$ is in $\mathcal{U}_{a,b}$, since

$$|\langle b^+ fyea\varphi, \psi\rangle| = |\langle yea\varphi, fb\psi\rangle| \leqq \|a\varphi\| \, \|b\psi\| \quad \text{for} \quad \varphi \in \mathcal{D}(\mathcal{A}) \quad \text{and} \quad \psi \in \mathcal{D}(\mathcal{B}).$$

As noted above $fye \in \mathbf{B}(\mathcal{D}(\mathcal{A}), \mathcal{D}(\mathcal{B}))$. By Corollary 3.1.7, (i), we have $\overline{b^+ fyea} \in \mathbf{B}(\mathcal{D}(\mathcal{A}), \mathcal{D}(\mathcal{B}))$. □

Corollary 3.4.2. *Let $\mathcal{A}$ and $\mathcal{B}$ be as in Theorem 3.4.1. Then the algebra $\mathbf{B}(\mathcal{D}(\mathcal{A}), \mathcal{D}(\mathcal{B})) \equiv \mathcal{L}(\mathcal{D}_\mathcal{A}^+, \mathcal{D}_\mathcal{B}) \upharpoonright \mathcal{H}$ is dense in $\mathcal{L}(\mathcal{D}_\mathcal{A}, \mathcal{D}_\mathcal{B}^+)[\tau_b]$.*

Proof. By Proposition 3.1.10, $\mathbf{B}(\mathcal{D}(\mathcal{A}), \mathcal{D}(\mathcal{B})) = \mathcal{L}(\mathcal{D}_\mathcal{A}^+, \mathcal{D}_\mathcal{B}) \upharpoonright \mathcal{H}$. Since each $x \in \mathcal{L}(\mathcal{D}_\mathcal{A}, \mathcal{D}_\mathcal{B}^+)$ is contained in $\mathcal{U}_{a,b}$ for some $a \in \mathcal{A}$ and $b \in \mathcal{B}$, $\mathbf{B}(\mathcal{D}(\mathcal{A}), \mathcal{D}(\mathcal{B}))$ is dense in $\mathcal{L}(\mathcal{D}_\mathcal{A}, \mathcal{D}_\mathcal{B}^+)[\tau_b]$ by Theorem 3.4.1, (ii). □

Remark 1. Note that the first statement in Theorem 3.4.1 is actually much stronger than the density of $\mathbf{B}(\mathcal{D}(\mathcal{A}), \mathcal{D}(\mathcal{B}))$ in $\mathcal{L}(\mathcal{D}_\mathcal{A}, \mathcal{D}_\mathcal{B}^+)[\tau_b]$ means. It asserts that, given a bounded set $\mathcal{R}$ in $\mathcal{L}(\mathcal{D}_\mathcal{A}, \mathcal{D}_\mathcal{B}^+)[\tau_b]$, there exist nets $(e_i : i \in I)$ and $(f_i : i \in I)$ of projections in $\mathbf{P}(\mathcal{D}(\mathcal{A}))$ and $\mathbf{P}(\mathcal{D}(\mathcal{B}))$, respectively, such that $(\hat{f}_i x e_i : i \in I)$ converges to x in $\mathcal{L}(\mathcal{D}_\mathcal{A}, \mathcal{D}_\mathcal{B}^+)[\tau_b]$ uniformly on $\mathcal{R}$.

Remark 2. Roughly speaking, if more about the structure of the O*-algebras $\mathcal{A}$ and $\mathcal{B}$ is known, then it can be said more about the projections $e \in \mathbf{P}(\mathcal{D}(\mathcal{A}))$ and $f \in \mathbf{P}(\mathcal{D}(\mathcal{B}))$ which can be taken in Theorem 3.4.1, (i). We give a sample for this remark. Suppose that $\mathcal{A}$ is a commutatively dominated O*-algebra and $\mathcal{D}_\mathcal{A}$ is a Frechet space. We assume without loss of generality by Proposition 2.2.17 that $\mathcal{A}$ is as in Example 2.2.16. We also keep the notation used therein. *Then the projection $e \in \mathbf{P}(\mathcal{D}(\mathcal{A}))$ in Theorem 3.4.1, (i), can be chosen of the form $E(\mathfrak{Z})$, where $\mathfrak{Z}$ is a measurable subset of* $\mathbb{R}$.

Proof. As shown in Example 2.4.4, $\{h(A)\, \mathcal{U}_\mathcal{H} : h \in \mathfrak{F}_\infty\}$ is a fundamental system of bounded sets in $\mathcal{D}_\mathcal{A}$. Therefore, by Remark 5 in 3.3, we can assume in the proof of Theorem 3.4.1, (i), that $c = h(A)$ for some function $h \in \mathfrak{F}_\infty$. Since $A = \int \lambda \, dE(\lambda)$, the spectral projection $e = e((\varepsilon, +\infty))$ of $c = h(A)$ is $E(\mathfrak{Z})$, where $\mathfrak{Z} := \{t \in \mathbb{R} : \varepsilon < h(t) < \infty\}$. □

The same reasoning shows that if $\mathcal{A}$ satisfies the assumptions of Theorem 2.4.3, then we can take e in the von Neumann algebra $\mathcal{N}$. Similar assertions hold for the O*-algebra $\mathcal{B}$.

Remark 3. The following fact is needed in Remark 1 in Section 4.3. Set $\mathcal{A} = \mathcal{L}^+(\mathcal{D})$ and $\mathcal{D} = \mathcal{D}(\mathcal{A})$ in Example 2.2.16. *Then the identity map I is the limit in $\mathcal{L}(\mathcal{D}, \mathcal{D}^+)[\tau_\mathcal{D}]$ of a net $(E(\mathfrak{Z}_i) : i \in I)$ of projections in $\mathbf{P}(\mathcal{D})$, where $\mathfrak{Z}_i$, $i \in I$, are measurable subsets of* $\mathbb{R}$. Indeed, by Theorem 3.4.1 applied with $\mathcal{R} = \{I\}$ and $\mathcal{A} = \mathcal{B} = \mathcal{L}^+(\mathcal{D})$, I is the limit of a net $(\hat{e}_i I e_i \equiv e_i : i \in I)$, where $e_i \in \mathbf{P}(\mathcal{D})$ for $i \in I$. By the preceding remark, e_i can be chosen of the form $E(\mathfrak{Z}_i)$.

Remark 4. Suppose $\mathcal{A}$ is an O*-algebra such that $\mathcal{D}_{\mathcal{A}}$ is a QF-space. Since $\mathbb{B}(\mathcal{D}(\mathcal{A})) \subseteq \mathcal{L}^+(\mathcal{D}_{\mathcal{A}})$ (by the notational convention of Remark 4 in 3.2), Corollary 3.4.2 shows that in particular $\mathcal{L}^+(\mathcal{D}_{\mathcal{A}})$ is dense in $\mathcal{L}(\mathcal{D}_{\mathcal{A}}, \mathcal{D}_{\mathcal{A}}^+)\,[\tau_{\mathcal{D}}]$.

Corollary 3.4.3. *Suppose that $\mathcal{A}$ is an O*-algebra for which $\mathcal{D}_{\mathcal{A}}$ is a QF-space.*

(i) *The positive cone $\mathcal{L}(\mathcal{D}_{\mathcal{A}}, \mathcal{D}_{\mathcal{A}}^+)_+$ is the closure of the cone generated by $\mathbb{P}(\mathcal{D}(\mathcal{A}))$ (that is, the set of all linear combinations of operators in $\mathbb{P}(\mathcal{D}(\mathcal{A}))$ with positive coefficients) in $\mathcal{L}(\mathcal{D}_{\mathcal{A}}, \mathcal{D}_{\mathcal{A}}^+)\,[\tau_{\mathcal{D}}]$.*

(ii) *The real linear span of $\mathbb{P}(\mathcal{D}(\mathcal{A}))$ is dense in $\mathcal{L}(\mathcal{D}_{\mathcal{A}}, \mathcal{D}_{\mathcal{A}}^+)_{\mathrm{h}}\,[\tau_{\mathcal{D}}]$.*

(iii) *The complex linear span of $\mathbb{P}(\mathcal{D}(\mathcal{A}))$ is dense in $\mathcal{L}(\mathcal{D}_{\mathcal{A}}, \mathcal{D}_{\mathcal{A}}^+)\,[\tau_{\mathcal{D}}]$.*

Proof. (i): It is trivial that $\mathcal{L}(\mathcal{D}_{\mathcal{A}}, \mathcal{D}_{\mathcal{A}}^+)_+$ contains the closure of the cone generated by $\mathbb{P}(\mathcal{D}(\mathcal{A}))$. Conversely, suppose $x \in \mathcal{L}(\mathcal{D}_{\mathcal{A}}, \mathcal{D}_{\mathcal{A}}^+)_+$. By Theorem 3.4.1, x is in the closure of $\{\hat{e}xe : e \in \mathbb{P}(\mathcal{D}(\mathcal{A}))\}$ in $\mathcal{L}(\mathcal{D}_{\mathcal{A}}, \mathcal{D}_{\mathcal{A}}^+)\,[\tau_{\mathcal{D}}]$. Therefore, it suffices to show that each operator $\hat{e}xe$ belongs to the closure of the cone generated by $\mathbb{P}(\mathcal{D}(\mathcal{A}))$. Fix $e \in \mathbb{P}(\mathcal{D}(\mathcal{A}))$. Because $x \geqq 0$, $\hat{e}xe$ is a positive self-adjoint operator in $\mathbb{B}(\mathcal{H})$. Let $\hat{e}xe = \int_0^\infty \lambda \, \mathrm{d}f(\lambda)$ be the spectral resolution of this operator. Approximating the integral by Riemann-Stieltjes sums, it follows that $\hat{e}xe$ is the norm limit of bounded operators of the form $y = \sum_{n=1}^{k} \lambda_n(f(\lambda_{n+1}) - f(\lambda_n))$, where $\lambda_{n+1} > \lambda_n > 0$ for $n = 1, \ldots, k$. Since the convergence in the operator norm always implies the convergence in $\mathcal{L}(\mathcal{D}_{\mathcal{A}}, \mathcal{D}_{\mathcal{A}}^+)\,[\tau_{\mathcal{D}}]$, it only remains to check that the operators y belong to the cone generated by $\mathbb{P}(\mathcal{D}(\mathcal{A}))$. For this it is sufficient to know that $(f(\lambda_{n+1}) - f(\lambda_n)) \in \mathbb{P}(\mathcal{D}(\mathcal{A}))$ for $n = 1, \ldots, k$. But, since $\lambda_{n+1} > \lambda_n > 0$ and $\hat{e}xe \in \mathbb{B}(\mathcal{D}(\mathcal{A}))$ (as stated in Theorem 3.4.1), this follows immediately from Corollary 3.1.5, (iv).

(ii): By Corollary 3.2.5, each $x \in \mathcal{L}(\mathcal{D}_{\mathcal{A}}, \mathcal{D}_{\mathcal{A}}^+)_{\mathrm{h}}$ is of the form $x = x_1 - x_2$ with $x_1, x_2 \in \mathcal{L}(\mathcal{D}_{\mathcal{A}}, \mathcal{D}_{\mathcal{A}}^+)_+$; so the assertion follows at once from (i).

(iii) follows from (ii), since $\mathcal{L}(\mathcal{D}_{\mathcal{A}}, \mathcal{D}_{\mathcal{A}}^+) = \mathcal{L}(\mathcal{D}_{\mathcal{A}}, \mathcal{D}_{\mathcal{A}}^+)_{\mathrm{h}} + \mathrm{i}\mathcal{L}(\mathcal{D}_{\mathcal{A}}, \mathcal{D}_{\mathcal{A}}^+)_{\mathrm{h}}$. □

The next two theorems are concerned with the density of $\mathbb{F}(\mathcal{D}(\mathcal{A}), \mathcal{D}(\mathcal{B}))$, the set of finite rank operators in $\mathbb{B}(\mathcal{D}(\mathcal{A}), \mathcal{D}(\mathcal{B}))$, in $\mathcal{L}(\mathcal{D}_{\mathcal{A}}, \mathcal{D}_{\mathcal{B}}^+)$. We first prove an auxiliary lemma.

Lemma 3.4.4. *Suppose $\mathcal{A}$ and $\mathcal{B}$ are O-families in the Hilbert space $\mathcal{H}$.*

(i) *For any $a \in \mathcal{A}(I)$ and $b \in \mathcal{B}(I)$, $\mathbb{F}(\mathcal{D}(\mathcal{A}), \mathcal{D}(\mathcal{B}))$ is dense in the normed linear space $(\mathcal{F}(\mathcal{D}_{\mathcal{A}}, \mathcal{D}_{\mathcal{B}}^+)_{a,b}, \mathfrak{l}_{a,b})$.*

(ii) *$\mathcal{F}(\mathcal{D}_{\mathcal{A}}, \mathcal{D}_{\mathcal{B}}^+)$ is contained in the closure of $\mathbb{F}(\mathcal{D}(\mathcal{A}), \mathcal{D}(\mathcal{B}))$ in the locally convex space $\mathcal{L}(\mathcal{D}_{\mathcal{A}}, \mathcal{D}_{\mathcal{B}}^+)\,[\tau_{\mathrm{in}}]$.*

(iii) *$\mathbb{F}(\mathcal{D}(\mathcal{A}), \mathcal{D}(\mathcal{B}))$ is dense in $\mathcal{F}(\mathcal{D}_{\mathcal{A}}, \mathcal{D}_{\mathcal{B}}^+)\,[\tau_{\mathrm{b}}]$.*

Proof. (i): Let $z \in \mathcal{F}(\mathcal{D}_{\mathcal{A}}, \mathcal{D}_{\mathcal{B}}^+)_{a,b}$. As noted in 3.2, z is of the form $z = \sum_{n=1}^{k} \varphi_n^! \otimes \psi_n^!$. We can assume the $\varphi_n^!$ and the $\psi_n^!$ to be linearly independent. From elementary linear algebra we know that there are vectors $\varphi_1, \ldots, \varphi_k \in \mathcal{D}(\mathcal{A})$ such that $\langle \varphi_n, \varphi_m^! \rangle = \delta_{nm}$, $n, m = 1, \ldots, k$. From $|\langle z\varphi_n, \psi \rangle| = |\langle \psi_n^!, \psi \rangle| \leqq \mathfrak{l}_{a,b}(z) \, \|a\varphi_n\| \, \|b\psi\|$, $\psi \in \mathcal{D}(\mathcal{B})$, we conclude that $\psi_n^! \in \mathcal{H}^b$, $n = 1, \ldots, k$. Similarly, $\varphi_n^! \in \mathcal{H}^a$ for $n = 1, \ldots, k$. Therefore it suffices to show that each rank one operator $\varphi^{\prime} \otimes \psi^{\prime}$, where $\varphi^{\prime} \in \mathcal{H}^a$ and $\psi^{\prime} \in \mathcal{H}^b$, is in the closure of $\mathbb{F}(\mathcal{D}(\mathcal{A}), \mathcal{D}(\mathcal{B}))$ in $(\mathcal{F}(\mathcal{D}_{\mathcal{A}}, \mathcal{D}_{\mathcal{B}}^+)_{a,b}, \mathfrak{l}_{a,b})$. Fix $\varphi^{\prime} \in \mathcal{H}^a$ and $\psi^{\prime} \in \mathcal{H}^b$ and let $\varepsilon > 0$. By Lemma 2.3.4, there are vectors $\zeta \in \mathcal{D}(\mathcal{A})$ and $\eta \in \mathcal{D}(\mathcal{B})$ such that $\|\varphi^{\prime} - \zeta\|^a \leqq \varepsilon$ and $\|\psi^{\prime} - \eta\|^b \leqq \varepsilon$. Then $\|\zeta\|^a \leqq \|\varphi^{\prime}\|^a + \varepsilon$. If $\varphi \in \mathcal{D}(\mathcal{A})$ and $\psi \in \mathcal{D}(\mathcal{B})$, then

$$\begin{aligned} |\langle(\varphi^{\prime} \otimes \psi^{\prime} - \zeta \otimes \eta)\,\varphi, \psi\rangle| &= |\langle\varphi, \varphi^{\prime} - \zeta\rangle\,\langle\psi^{\prime}, \psi\rangle + \langle\varphi, \zeta\rangle\,\langle\psi^{\prime} - \eta, \psi\rangle| \\ &\leqq \|\varphi^{\prime} - \zeta\|^a\,\|\varphi\|_a\,\|\psi^{\prime}\|^b\,\|\psi\|_b + \|\zeta\|^a\,\|\varphi\|_a\,\|\psi^{\prime} - \eta\|^b\,\|\psi\|_b \\ &\leqq \varepsilon(\|\psi^{\prime}\|^b + \varepsilon + \|\varphi^{\prime}\|^a)\,\|a\varphi\|\,\|b\psi\|, \end{aligned}$$

i.e. $\mathfrak{l}_{a,b}(\varphi^{\prime} \otimes \psi^{\prime} - \zeta \otimes \eta) \leqq \text{const.}\,\varepsilon$.

Since $\zeta \otimes \eta \in \mathbb{F}(\mathcal{D}(\mathcal{A}), \mathcal{D}(\mathcal{B}))$, this yields the assertion.

(ii): We can assume without loss of generality by Proposition 2.2.13 that $\mathcal{A}$ and $\mathcal{B}$ are directed O-families. But then $\mathcal{L}(\mathcal{D}_{\mathcal{A}}, \mathcal{D}_{\mathcal{B}}^+)\,[\tau_{\text{in}}]$ is the inductive limit of the family of normed spaces $\{(\mathcal{L}(\mathcal{D}_{\mathcal{A}}, \mathcal{D}_{\mathcal{B}}^+)_{a,b}, \mathfrak{l}_{a,b}) : a \in \mathcal{A}(I) \text{ and } b \in \mathcal{B}(I)\}$, so that the assertion follows immediately from (i).

(iii) is an obvious consequence of (ii), since $\tau_{\text{b}} \subseteqq \tau_{\text{in}}$ on $\mathcal{L}(\mathcal{D}_{\mathcal{A}}, \mathcal{D}_{\mathcal{B}}^+)$. □

Theorem 3.4.5. *Let $\mathcal{A}$ and $\mathcal{B}$ be O-families in the Hilbert space $\mathcal{H}$. Suppose that at least one of the locally convex spaces $\mathcal{D}_{\mathcal{A}}$ and $\mathcal{D}_{\mathcal{B}}$ is a semi-Montel space. Then $\mathbb{F}(\mathcal{D}(\mathcal{A}), \mathcal{D}(\mathcal{B}))$ is dense in $\mathcal{L}(\mathcal{D}_{\mathcal{A}}, \mathcal{D}_{\mathcal{B}}^+)\,[\tau_{\text{b}}]$.*

Proof. By the continuity of the involution, $\mathbb{F}(\mathcal{D}(\mathcal{A}), \mathcal{D}(\mathcal{B}))$ is dense in $\mathcal{L}(\mathcal{D}_{\mathcal{A}}, \mathcal{D}_{\mathcal{B}}^+)\,[\tau_{\text{b}}]$ if and only if $\mathbb{F}(\mathcal{D}(\mathcal{B}), \mathcal{D}(\mathcal{A}))$ is dense in $\mathcal{L}(\mathcal{D}_{\mathcal{B}}, \mathcal{D}_{\mathcal{A}}^+)\,[\tau_{\text{b}}]$. Thus we can assume without loss of generality that $\mathcal{D}_{\mathcal{A}}$ is a semi-Montel space. By Corollary 2.3.2, (i), the space $\mathcal{D}_{\mathcal{A}}$ has the approximation property. From this it follows in particular (see e.g. SCHÄFER [1], III, 9.1) that the finite rank operators in $\mathfrak{L}(\mathcal{D}_{\mathcal{A}}, \mathcal{D}_{\mathcal{B}}^+[\beta])$ are dense in $\mathfrak{L}(\mathcal{D}_{\mathcal{A}}, \mathcal{D}_{\mathcal{B}}^+[\beta])$ in the topology of uniform convergence on precompact subsets of $\mathcal{D}_{\mathcal{A}}$. Since $\mathcal{D}_{\mathcal{A}}$ is a semi-Montel space, each bounded subset of $\mathcal{D}_{\mathcal{A}}$ is precompact. Moreover, as already noted in 3.2, $\mathcal{F}(\mathcal{D}_{\mathcal{A}}, \mathcal{D}_{\mathcal{B}}^+)$ is the set of finite rank operators in $\mathfrak{L}(\mathcal{D}_{\mathcal{A}}, \mathcal{D}_{\mathcal{B}}^+[\beta])$. Therefore, $\mathcal{F}(\mathcal{D}_{\mathcal{A}}, \mathcal{D}_{\mathcal{B}}^+)$ is dense in $\mathfrak{L}(\mathcal{D}_{\mathcal{A}}, \mathcal{D}_{\mathcal{B}}^+[\beta])\,[\iota_{\text{b}}]$. Since $\mathcal{L}(\mathcal{D}_{\mathcal{A}}, \mathcal{D}_{\mathcal{B}}^+) \subseteqq \mathfrak{L}(\mathcal{D}_{\mathcal{A}}, \mathcal{D}_{\mathcal{B}}^+[\beta])$ and $\mathbb{F}(\mathcal{D}(\mathcal{A}), \mathcal{D}(\mathcal{B}))$ is dense in $\mathcal{F}(\mathcal{D}_{\mathcal{A}}, \mathcal{D}_{\mathcal{B}}^+)\,[\tau_{\text{b}}]$ by Lemma 3.4.4, the assertion follows. □

Remark 5. If, in addition, $\mathcal{A}$ and $\mathcal{B}$ are O*-algebras and $\mathcal{D}_{\mathcal{A}}$ and $\mathcal{D}_{\mathcal{B}}$ are QF-spaces, then the assertion of Theorem 3.4.5 follows directly from Theorem 3.4.1 without appealing to the approximation property of $\mathcal{D}_{\mathcal{A}}$. We sketch this argument. Suppose again that $\mathcal{D}_{\mathcal{A}}$ is a semi-Montel space. If $e \in \mathbb{P}(\mathcal{D}(\mathcal{A}))$, then the bounded set $e\mathcal{U}_{\mathcal{H}}$ in $\mathcal{D}_{\mathcal{A}}$ (by Corollary 3.1.3) is relatively compact, so that e has finite rank. Therefore, for any $x \in \mathcal{L}(\mathcal{D}_{\mathcal{A}}, \mathcal{D}_{\mathcal{B}}^+)$ and $f \in \mathbb{P}(\mathcal{D}(\mathcal{B}))$, $\hat{f}xe$ is a finite rank operator in $\mathbb{B}(\mathcal{D}(\mathcal{A}), \mathcal{D}(\mathcal{B}))$ and hence contained in $\mathbb{F}(\mathcal{D}(\mathcal{A}), \mathcal{D}(\mathcal{B}))$. Thus Theorem 3.4.1, (i), implies that $\mathbb{F}(\mathcal{D}(\mathcal{A}), \mathcal{D}(\mathcal{B}))$ is dense in $\mathcal{L}(\mathcal{D}_{\mathcal{A}}, \mathcal{D}_{\mathcal{B}}^+)\,[\tau_{\text{b}}]$.

Theorem 3.4.6. *Let $\mathcal{A}$ and $\mathcal{B}$ be O-families in the Hilbert space $\mathcal{H}$, and let $\mathcal{L}$ be a linear subspace of $\mathcal{L}(\mathcal{D}_{\mathcal{A}}, \mathcal{D}_{\mathcal{B}}^+)$ which contains $\mathbb{F}(\mathcal{D}(\mathcal{A}), \mathcal{D}(\mathcal{B}))$. If at least one of the spaces $\mathcal{D}_{\mathcal{A}}$ and $\mathcal{D}_{\mathcal{B}}$ is a Schwartz space, then $\mathbb{F}(\mathcal{D}(\mathcal{A}), \mathcal{D}(\mathcal{B}))$ is dense in $\mathcal{L}[\tau_{\text{in}}]$.*

Proof. As in the proof of Theorem 3.4.5 it suffices to consider the case where $\mathcal{D}_{\mathcal{A}}$ is a Schwartz space. By Proposition 2.2.13, there is no loss of generality to assume that $\mathcal{A}$ and $\mathcal{B}$ are directed O*-vector spaces. Suppose $x \in \mathcal{L}$. Then $x \in \mathcal{U}_{a,b}$ for some $a \in \mathcal{A}(I)$ and $b \in \mathcal{B}(I)$. By Proposition 3.2.3, there is an operator $y \in \mathbb{B}(\mathcal{H})$, $\|y\| \leqq 1$, such that $\langle x\cdot, \cdot\rangle \equiv \langle ya\cdot, b\cdot\rangle$. From Proposition 2.3.14 it follows that there is an $a_1 \in \mathcal{A}(I)$ such that the embedding map of the Hilbert space $\mathcal{H}_{a_1}$ into the Hilbert space $\mathcal{H}_a$ is compact. We denote this map by $\mathfrak{j}$. Thus, if $\varepsilon > 0$, then there is a bounded finite rank operator $\mathfrak{j}_\varepsilon$ of $\mathcal{H}_{a_1}$ into $\mathcal{H}_a$ satisfying $\|(\mathfrak{j} - \mathfrak{j}_\varepsilon)\varphi\|_{\bar{a}} \leqq \varepsilon \|\varphi\|_{\bar{a}_1}$, $\varphi \in \mathcal{H}_{a_1}$. Writing $\mathfrak{j}_\varepsilon$ in the form

$$\mathfrak{j}_\varepsilon = \sum_{n=1}^{k} \langle\cdot, \varphi_n\rangle_{\bar{a}_1} \psi_n \quad \text{with} \quad \varphi_1, \dots, \varphi_k \in \mathcal{D}(\bar{a}_1) \quad \text{and} \quad \psi_1, \dots, \psi_k \in \mathcal{D}(\bar{a})$$

we have

$$\langle ya\mathfrak{j}_\varepsilon\varphi, b\psi\rangle = \sum_{n=1}^{k} \langle\varphi, \varphi_n\rangle_{\bar{a}_1} \langle y\bar{a}\psi_n, b\psi\rangle \quad \text{for} \quad \varphi \in \mathcal{D}(\mathcal{A}) \quad \text{and} \quad \psi \in \mathcal{D}(\mathcal{B}).$$

From this we see that there is an $x_\varepsilon \in \mathcal{F}(\mathcal{D}_\mathcal{A}, \mathcal{D}_\mathcal{B}^+)$ such that $\langle x_\varepsilon\cdot, \cdot\rangle \equiv \langle y\bar{a}\mathfrak{j}_\varepsilon\cdot, b\cdot\rangle$. Then

$$\begin{aligned}|\langle(x - x_\varepsilon)\varphi, \psi\rangle| &= |\langle y\bar{a}(\mathfrak{j} - \mathfrak{j}_\varepsilon)\varphi, b\psi\rangle| \leqq \|y\| \, \|\bar{a}(\mathfrak{j} - \mathfrak{j}_\varepsilon)\varphi\| \, \|b\psi\| \\ &\leqq \varepsilon \|a_1\varphi\| \, \|b\psi\| \quad \text{for} \quad \varphi \in \mathcal{D}(\mathcal{A}) \quad \text{and} \quad \psi \in \mathcal{D}(\mathcal{B}),\end{aligned}$$

i. e., $\mathfrak{l}_{a_1,b}(x - x_\varepsilon) \leqq \varepsilon$. This implies that $x = \lim_{\varepsilon\to+0} x_\varepsilon$ in $\mathcal{L}[\tau_{\text{in}}]$. Hence $\mathcal{F}(\mathcal{D}_\mathcal{A}, \mathcal{D}_\mathcal{B}^+)$ is dense in $\mathcal{L}[\tau_{\text{in}}]$. Combined with Lemma 3.4.4, (ii), the assertion follows. □

Some investigations in this chapter can be reformulated in terms of the notion of a topological quasi $*$-algebra. We briefly discuss this concept which has also independent interest.

Definition 3.4.7. A *topological quasi $*$-algebra* is a couple (X, A) of a locally convex space X and a $*$-algebra A which is a linear subspace of X such that:

(i) X is an A-bimodule (cf. Definition 3.2.8). The module operations $(a, x) \to a\cdot x$ and $(x, a) \to x\cdot a$ extend the multiplication of A, and they are separately continuous bilinear mappings of A $\times$ X resp. X $\times$ A into X, where A carries the induced topology of X.

(ii) There is a continuous involution $x \to x^+$ of X which extends the involution of A and satisfies $(a\cdot x)^+ = x^+\cdot a^+$ and $(x\cdot a)^+ = a^+\cdot x^+$ for all $a \in$ A and $x \in$ X.

(iii) A is dense in X.

One reason for introducing this concept is the following simple observation. If A is a topological $*$-algebra, then it is not possible in general to extend the algebraic operations of A to the completion $\tilde{\mathsf{A}}$ of A such that $\tilde{\mathsf{A}}$ is a topological $*$-algebra. (An example showing this is the $*$-algebra A $:= C[0, 1]$ equipped with the L^p-norm on $[0, 1]$ for some $p \in \mathbb{R}$, $1 \leqq p < +\infty$.) But it is easily seen that the multiplication and the involution of A extend by continuity to A $\times$ $\tilde{\mathsf{A}}$ and $\tilde{\mathsf{A}}$ $\times$ A resp. $\tilde{\mathsf{A}}$ such that the couple ($\tilde{\mathsf{A}}$, A) becomes a topological quasi $*$-algebra.

Now suppose that $\mathcal{A}$ is an O*-algebra, A is a $*$-subalgebra of $\mathcal{L}^+(\mathcal{D}_\mathcal{A})$ and X$: = \mathcal{L}(\mathcal{D}_\mathcal{A}, \mathcal{D}_\mathcal{A}^+)[\tau]$, where τ is either the topology $\tau_\mathcal{D}$ $(= \tau_\text{b})$ or the topology $\tau_\mathcal{O}$ $(= \tau_\text{in})$. We define $a\cdot x := a \circ x$ and $x\cdot a := xa$ for $a \in$ A and $x \in$ X. As involution of X we take the involution of the $*$-vector space $\mathcal{L}(\mathcal{D}_\mathcal{A}, \mathcal{D}_\mathcal{A}^+)$. Then the conditions (i) and (ii) in Definition 3.4.7 are satisfied. Indeed, the algebraic parts of these axioms follow from Lemma 3.2.7 and the continuity assertions are contained in Propositions 3.3.4 and 3.3.8. Therefore, if A is dense in X, then the couple (X, A) as just defined is a topological quasi $*$-algebra. The density results of this section (Corollary 3.4.3 and Theorems 3.4.5 and

3.4.6) show that (X, A) is a topological quasi $*$-algebra when one of the following three groups of assumptions are satisfied:

1. $\tau = \tau_{\mathcal{D}}$, $\mathcal{D}_{\mathcal{A}}$ is a QF-space and $\mathsf{A} \supseteq \mathbb{P}(\mathcal{D}(\mathcal{A}))$.
2. $\tau = \tau_{\mathcal{D}}$, $\mathcal{D}_{\mathcal{A}}$ is a semi-Montel space and $\mathsf{A} \supseteq \mathbb{F}(\mathcal{D}(\mathcal{A}))$.
3. $\tau = \tau_{\mathcal{O}}$, $\mathcal{D}_{\mathcal{A}}$ is a Schwartz space and $\mathsf{A} \supseteq \mathbb{F}(\mathcal{D}(\mathcal{A}))$.

In particular, the couple $(\mathcal{L}(\mathcal{D}_{\mathcal{A}}, \mathcal{D}_{\mathcal{A}}^+)[\tau_{\mathcal{D}}], \mathcal{L}^+(\mathcal{D}_{\mathcal{A}}))$ is a topological quasi $*$-algebra if $\mathcal{A}$ is an O*-algebra such that $\mathcal{D}_{\mathcal{A}}$ is a QF-space.

3.5. The Weak- and Strong-Operator Topologies and the Ultraweak and Ultrastrong Topologies

The Weak-Operator and the Ultraweak Topologies

Throughout the following, we assume that $\mathcal{A}$ and $\mathcal{B}$ are O-families in a Hilbert space $\mathcal{H}$ and $\mathcal{L}$ is a fixed linear subspace of $\mathcal{L}(\mathcal{D}_{\mathcal{A}}, \mathcal{D}_{\mathcal{B}}^+)$. The *weak-operator topology* on $\mathcal{L}$ is the locally convex topology on $\mathcal{L}$ which is defined by the family of seminorms

$$\mathcal{L} \ni x \to |\langle x\varphi, \psi\rangle|, \quad \varphi \in \mathcal{D}(\mathcal{A}) \quad \text{and } \psi \in \mathcal{D}(\mathcal{B}).$$

Of course, the weak-operator topology is a Hausdorff topology. It is the coarsest locally convex topology on $\mathcal{L}$ for which the map $\mathcal{L} \ni x \to x\varphi \in \mathcal{D}_{\mathcal{B}}^+$ is continuous for each $\varphi \in \mathcal{D}(\mathcal{A})$ if $\mathcal{D}_{\mathcal{B}}^+$ is endowed with the topology $\sigma(\mathcal{D}_{\mathcal{B}}^+, \mathcal{D}(\mathcal{A}))$.

For $\varphi \in \mathcal{D}(\mathcal{A})$ and $\psi \in \mathcal{D}(\mathcal{B})$, let $\omega_{\varphi,\psi}$ denote the linear functional on $\mathcal{L}$ defined by $\omega_{\varphi,\psi}(x) := \langle x\varphi, \psi\rangle$, $x \in \mathcal{L}$. Let $\mathcal{L}_\sim$ be the vector space of all linear functionals $\omega = \sum_{n=1}^{k} \omega_{\varphi_n,\psi_n}$ on $\mathcal{L}$, where $k \in \mathbb{N}$ and $\varphi_n \in \mathcal{D}(\mathcal{A})$ and $\psi_n \in \mathcal{D}(\mathcal{B})$ for $n = 1, \ldots, k$.

Proposition 3.5.1. *A linear functional ω on $\mathcal{L}$ is weak-operator continuous if and only if $\omega \in \mathcal{L}_\sim$, that is, there are vectors $\varphi_1, \ldots, \varphi_k \in \mathcal{D}(\mathcal{A})$ and $\psi_1, \ldots, \psi_k \in \mathcal{D}(\mathcal{B})$, $k \in \mathbb{N}$, such that $\omega = \sum_{n=1}^{k} \omega_{\varphi_n,\psi_n}$.*

Proof. Endowed with the bilinear form $(\omega, x) \to \omega(x)$ on $\mathcal{L}_\sim \times \mathcal{L}$, the vector spaces $\mathcal{L}_\sim$ and $\mathcal{L}$ form a dual pairing, and the weak-operator topology coincides with the topology $\sigma(\mathcal{L}, \mathcal{L}_\sim)$. Thus the assertion is a special case of the well-known fact (Schäfer [1], IV, 1.2) that $\mathcal{L}_\sim$ is precisely the set of $\sigma(\mathcal{L}, \mathcal{L}_\sim)$-continuous linear functionals on $\mathcal{L}$. □

The *ultraweak topology* (or σ-weak topology) on $\mathcal{L}$ is the locally convex topology on $\mathcal{L}$ which is determined by the family of seminorms

$$\mathcal{L} \ni x \to \left|\sum_{n=1}^{\infty} \langle x\varphi_n, \psi_n\rangle\right| \equiv \left|\sum_{n=1}^{\infty} \omega_{\varphi_n,\psi_n}(x)\right|, \tag{1}$$

where $(\varphi_n : n \in \mathbb{N})$ and $(\psi_n : n \in \mathbb{N})$ are sequences of vectors in $\mathcal{D}(\mathcal{A})$ and $\mathcal{D}(\mathcal{B})$, respectively, satisfying

$$\sum_{n=1}^{\infty} \|a\varphi_n\|^2 < \infty \quad \text{and} \quad \sum_{n=1}^{\infty} \|b\psi_n\|^2 < \infty \quad \text{for all } a \in \mathcal{A} \text{ and } b \in \mathcal{B}. \tag{2}$$

It is clear that (2) is fulfilled if and only if

$$\sum_{n=1}^{\infty} p(\varphi_n)^2 < \infty \quad \text{and} \quad \sum_{n=1}^{\infty} q(\psi_n)^2 < \infty \tag{3}$$

for arbitrary continuous seminorms p and q on $\mathcal{D}_{\mathcal{A}}$ and $\mathcal{D}_{\mathcal{B}}$, respectively. In particular, this shows that the ultraweak topology on $\mathcal{L}(\mathcal{D}_{\mathcal{A}}, \mathcal{D}_{\mathcal{B}}^{+})$ depends only on the graph topologies $\mathrm{t}_{\mathcal{A}}$ and $\mathrm{t}_{\mathcal{B}}$ rather than on $\mathcal{A}$ and $\mathcal{B}$.

We have to check that the series in (1) converges. Indeed, fix an $x \in \mathcal{L}$. Since $\mathcal{L} \subseteq \mathcal{L}(\mathcal{D}_{\mathcal{A}}, \mathcal{D}_{\mathcal{B}}^{+})$, there are continuous seminorms p and q on $\mathcal{D}_{\mathcal{A}}$ and $\mathcal{D}_{\mathcal{B}}$, respectively, such that $|\langle x\varphi, \psi\rangle| \leqq p(\varphi)\, q(\psi)$ for all $\varphi \in \mathcal{D}(\mathcal{A})$ and $\psi \in \mathcal{D}(\mathcal{B})$. By the Cauchy-Schwarz inequality and by (3), we have

$$\left|\sum_{n=1}^{\infty} \langle x\varphi_n, \psi_n\rangle\right|^2 \leqq \left(\sum_{n=1}^{\infty} p(\varphi_n)\, q(\psi_n)\right)^2 \leqq \left(\sum_{n=1}^{\infty} p(\varphi_n)^2\right)\left(\sum_{n=1}^{\infty} q(\psi_n)^2\right) < \infty.$$

Let $\mathcal{L}_*$ denote the vector space of all linear functionals $\omega = \sum_{n=1}^{\infty} \omega_{\varphi_n,\psi_n}$ on $\mathcal{L}$, where $(\varphi_n : n \in \mathbb{N})$ and $(\psi_n : n \in \mathbb{N})$ are sequences in $\mathcal{D}(\mathcal{A})$ and $\mathcal{D}(\mathcal{B})$, respectively, for which (2) hold. Replacing $\mathcal{L}_\sim$ by $\mathcal{L}_*$ in the proof of Proposition 3.5.1, we obtain

Proposition 3.5.2. *A linear functional ω on $\mathcal{L}$ is ultraweakly continuous if and only if $\omega \in \mathcal{L}_*$, that is, there are vectors $\varphi_n \in \mathcal{D}(\mathcal{A})$ and $\psi_n \in \mathcal{D}(\mathcal{B})$ for $n \in \mathbb{N}$ such that (2) is satisfied and $\omega = \sum_{n=1}^{\infty} \omega_{\varphi_n,\psi_n}$.*

Proposition 3.5.3. *Let p and q be continuous seminorms on $\mathcal{D}_{\mathcal{A}}$ and $\mathcal{D}_{\mathcal{B}}$, respectively. Then the weak-operator topology and the ultraweak topology coincide on $\mathcal{L} \cap \mathcal{U}_{p,q}$. In particular, both topologies coincide on $\mathcal{L} \cap \mathcal{U}_{a,b}$ for all $a \in \mathcal{A}$ and $b \in \mathcal{B}$.*

Proof. Fix $x \in \mathcal{L} \cap \mathcal{U}_{p,q}$. Let $\mathcal{W}(x)$ be a neighbourhood of x in the ultraweak topology on $\mathcal{L} \cap \mathcal{U}_{p,q}$. Then there exist sequences $(\varphi_{ln} : n \in \mathbb{N})$ in $\mathcal{D}(\mathcal{A})$ and $(\psi_{ln} : n \in \mathbb{N})$ in $\mathcal{D}(\mathcal{B})$, $l = 1, \ldots, k$ and $k \in \mathbb{N}$, satisfying (2) such that

$$\mathcal{W}(x) \supseteq \left\{y \in \mathcal{L} \cap \mathcal{U}_{p,q} : \left|\sum_{n=1}^{\infty} \langle (x-y)\, \varphi_{ln}, \psi_{ln}\rangle\right| \leqq 1 \quad \text{for} \quad l = 1, \ldots, k\right\}.$$

Since (2) implies (3), there is a $k \in \mathbb{N}$ such that $\sum_{n=k+1}^{\infty} p(\varphi_{ln})\, q(\psi_{ln}) \leqq \frac{1}{4}$ for $l = 1, \ldots, k$. Let $\mathcal{W}_1(x)$ be the neighbourhood of x in the weak-operator topology defined by $\mathcal{W}_1(x)$ $:= \left\{y \in \mathcal{L} \cap \mathcal{U}_{p,q} : \left|\sum_{n=1}^{k} \langle (x-y)\, \varphi_{ln}, \psi_{ln}\rangle\right| \leqq \frac{1}{2} \text{ for } l = 1, \ldots, k\right\}$. If $y \in \mathcal{W}_1(x)$, then $y - x \in 2\mathcal{U}_{p,q}$ and hence $y \in \mathcal{W}(x)$ by the preceding. Thus $\mathcal{W}_1(x) \subseteq \mathcal{W}(x)$. Since the weak-operator topology is trivially weaker than the ultraweak topology, both topologies coincide on $\mathcal{L} \cap \mathcal{U}_{p,q}$. □

Remark 1. If a given linear space $\mathcal{L}$ can be embedded into different spaces $\mathcal{L}(\mathcal{D}_{\mathcal{A}}, \mathcal{D}_{\mathcal{B}}^{+})$, then the corresponding weak-operator resp. ultraweak topologies on $\mathcal{L}$ are, of course, different in general; see Example 3.5.4 below. Therefore, if we speak about the weak-operator topology or the ultraweak topology, we always refer to a fixed underlying space $\mathcal{L}(\mathcal{D}_{\mathcal{A}}, \mathcal{D}_{\mathcal{B}}^{+})$; cf. Remark 7 in 3.3.

Example 3.5.4. For $n \in \mathbb{N}$, let φ_n be the function in $\mathcal{H} := L^2(\mathbb{R})$ defined by $\varphi_n(t) := \exp t^2$ if $t \in (n, n+1)$ and $\varphi_n(t) := 0$ otherwise. Let $\mathcal{D} := C_0^\infty(\mathbb{R})$. Then the sequence $(\varphi_n \otimes \varphi_n : n \in \mathbb{N})$ converges to zero in the weak-operator topology of $\mathcal{L}(\mathcal{D}, \mathcal{D}^+)$, but certainly not in the weak-operator topology of $\mathbb{B}(\mathcal{H})$. ○

For $a \in \mathcal{A}(I)$ and $b \in \mathcal{B}(I)$, we defined in Section 3.2 a mapping $R_{a,b}$ of $Q_b\mathbb{B}(\mathcal{H})\,Q_a$ into $\mathcal{L}(\mathcal{D}_\mathcal{A}, \mathcal{D}_\mathcal{B}^+)_{a,b}$ by $\langle R_{a,b}(y)\,\varphi, \psi\rangle = \langle ya\varphi, b\psi\rangle$, $\varphi \in \mathcal{D}(\mathcal{A})$ and $\psi \in \mathcal{D}(\mathcal{B})$. By Corollary 3.2.4, $R_{a,b}$ is bijective. Let $T_{a,b}$ denote the inverse of $R_{a,b}$.

Proposition 3.5.5. *Suppose $a \in \mathcal{A}(I)$ and $b \in \mathcal{B}(I)$.*

(i) *The mapping $R_{a,b}$ of $Q_b\mathbb{B}(\mathcal{H})\,Q_a$ onto $\mathcal{L}(\mathcal{D}_\mathcal{A}, \mathcal{D}_\mathcal{B}^+)_{a,b}$ is continuous in the corresponding weak-operator topologies resp. ultraweak topologies.*

(ii) *$T_{a,b}$ maps $\mathcal{U}_{a,b}$ into $Q_b\mathbb{B}(\mathcal{H})\,Q_a$ continuously in the corresponding weak-operator topologies resp. ultraweak topologies.*

Proof. The proof of (i) is straightforward; so we omit the details.

(ii): Suppose $(x_i : i \in I)$ is a net in $\mathcal{U}_{a,b}$ which converges to $x \in \mathcal{U}_{a,b}$ in the weak-operator topology (relative to $\mathcal{L}(\mathcal{D}_\mathcal{A}, \mathcal{D}_\mathcal{B}^+)$). Then, for $\varphi \in \mathcal{D}(\mathcal{A})$, $\psi \in \mathcal{D}(\mathcal{B})$, $\zeta \in (I - Q_a)\,\mathcal{H}$ and $\eta \in (I - Q_b)\,\mathcal{H}$, we have

$$\langle T_{a,b}(x_i)\,(a\varphi + \zeta), b\psi + \eta\rangle = \langle T_{a,b}(x_i)\,a\varphi, b\psi\rangle = \langle x_i\varphi, \psi\rangle$$

$$\to \langle x\varphi, \psi\rangle = \langle T_{a,b}(x)\,(a\varphi + \zeta), b\psi + \eta\rangle. \tag{4}$$

Since $x_i \in \mathcal{U}_{a,b}$, $\|T_{a,b}(x_i)\| \leqq 1$ for $i \in I$ by Corollary 3.2.4. Therefore, since $a\mathcal{D}(\mathcal{A}) + (I - Q_a)\,\mathcal{H}$ and $b\mathcal{D}(\mathcal{B}) + (I - Q_b)\,\mathcal{H}$ are both dense in $\mathcal{H}$, it follows from (4) that $(T_{a,b}(x_i) : i \in I)$ converges to $T_{a,b}(x)$ in the weak-operator topology relative to $\mathbb{B}(\mathcal{H})$. This proves the assertion of (ii) for the weak-operator topology. By Proposition 3.5.3, the weak-operator and the ultraweak topologies (of $\mathcal{L}(\mathcal{D}_\mathcal{A}, \mathcal{D}_\mathcal{B}^+)$ resp. $\mathbb{B}(\mathcal{H})$) coincide on $\mathcal{U}_{a,b}$ resp. on $\mathcal{U}_{\mathbb{B}(\mathcal{H})}$. Since $T_{a,b}$ maps $\mathcal{U}_{a,b}$ into $\mathcal{U}_{\mathbb{B}(\mathcal{H})}$, the assertion for the ultraweak topology follows. □

An immediate consequence of Proposition 3.5.5 is

Corollary 3.5.6. *Let $a \in \mathcal{A}(I)$ and $b \in \mathcal{B}(I)$. For each subset $\mathcal{R}$ of $\mathcal{U}_{a,b}$ the following assertions are equivalent:*

(i) *$\mathcal{R}$ is weak-operator closed in $\mathcal{L}(\mathcal{D}_\mathcal{A}, \mathcal{D}_\mathcal{B}^+)$.*

(ii) *$\mathcal{R}$ is ultraweakly closed in $\mathcal{L}(\mathcal{D}_\mathcal{A}, \mathcal{D}_\mathcal{B}^+)$.*

(iii) *$T_{a,b}(\mathcal{R})$ is weak-operator closed in $\mathbb{B}(\mathcal{H})$.*

We specialize to the space $\mathcal{L}(\mathcal{D}_\mathcal{A}, \mathcal{D}_\mathcal{A}^+)$.

A net $(x_i : i \in I)$ in $\mathcal{L}(\mathcal{D}_\mathcal{A}, \mathcal{D}_\mathcal{A}^+)_\mathrm{h}$ is called *monotone increasing* if $x_i \leqq x_j$ is equivalent to $i \leqq j$ for i, j in the directed index set I. Such a net $(x_i : i \in I)$ is said to be *bounded* (or more precisely bounded from above) if there is a $y \in \mathcal{L}(\mathcal{D}_\mathcal{A}, \mathcal{D}_\mathcal{A}^+)_+$ such that $x_i \leqq y$ for $i \in I$.

Lemma 3.5.7. *Each bounded monotone increasing net $(x_i : i \in I)$ in $\mathcal{L}(\mathcal{D}_\mathcal{A}, \mathcal{D}_\mathcal{A}^+)_\mathrm{h}$ has a least upper bound, denoted by $\sup_i x_i$, in the ordered vector space $\left(\mathcal{L}(\mathcal{D}_\mathcal{A}, \mathcal{D}_\mathcal{A}^+)_\mathrm{h}, \geqq\right)$. Moreover, $\sup_i x_i$ is the limit of the net $(x_i : i \in I)$ in the ultraweak topology in $\mathcal{L}(\mathcal{D}_\mathcal{A}, \mathcal{D}_\mathcal{A}^+)$.*

Proof. By Remark 6 in 3.2, there is no loss of generality to assume that $\mathcal{A}$ is a directed O-vector space. Let $y \in \mathcal{L}(\mathcal{D}_{\mathcal{A}}, \mathcal{D}_{\mathcal{A}}^+)_+$ be such that $x_i \leqq y$ for $i \in I$. Since the net $(x_i : i \in I)$ is monotone increasing and bounded, $\sup_i \langle x_i\varphi, \varphi\rangle = \lim_i \langle x_i\varphi, \varphi\rangle < \infty$ for $\varphi \in \mathcal{D}(\mathcal{A})$. From the polarization formula 3.2/(3) we conclude that $\lim_i \langle x_i\varphi, \psi\rangle$ exists for all $\varphi, \psi \in \mathcal{D}(\mathcal{A})$. Therefore, $\mathfrak{c}(\varphi, \psi) := \lim_i \langle x_i\varphi, \psi\rangle$, $\varphi, \psi \in \mathcal{D}(\mathcal{A})$, defines a sesquilinear form $\mathfrak{c}$ on $\mathcal{D}(\mathcal{A}) \times \mathcal{D}(\mathcal{A})$. We take an operator $a \in \mathcal{A}$ such that $y \in \mathcal{U}_a$ $(\equiv \mathcal{U}_{a,a})$. From the fact that $x_i \leqq y$ and the polarization formula we obtain that $x_i \in \mathcal{U}_{2a}$ for $i \in I$. Therefore, $|\mathfrak{c}(\varphi, \psi)| \leqq \|2a\varphi\| \, \|2a\psi\|$ for $\varphi, \psi \in \mathcal{D}(\mathcal{A})$. By Lemma 1.2.1, there is an x in $\mathcal{L}(\mathcal{D}_{\mathcal{A}}, \mathcal{D}_{\mathcal{A}}^+)$ such that $\mathfrak{c} = \mathfrak{c}_x$. From the construction it is clear that $x \in \mathcal{L}(\mathcal{D}_{\mathcal{A}}, \mathcal{D}_{\mathcal{A}}^+)_{\mathrm{h}}$ is the least upper bound of the set $\{x_i : i \in I\}$ and that $x = \lim_i x_i$ in the weak-operator topology of $\mathcal{L}(\mathcal{D}_{\mathcal{A}}, \mathcal{D}_{\mathcal{A}}^+)$. Since x and x_i, $i \in I$, are in $\mathcal{U}_{2a}$, Proposition 3.5.3 ensures that $x = \lim_i x_i$ in the ultraweak topology. □

A strongly positive linear functional f on an ultraweakly closed $*$-vector subspace $\mathcal{L}$ of $\mathcal{L}(\mathcal{D}_{\mathcal{A}}, \mathcal{D}_{\mathcal{A}}^+)$ is said to be *normal* if $\lim_i f(x_i) = f\left(\sup_i x_i\right)$ for each bounded monotone increasing net $(x_i : i \in I)$ in $\mathcal{L}_{\mathrm{h}}$. Note that $f\left(\sup_i x_i\right)$ is well-defined, since $\sup_i x_i$ is the ultraweak limit of $(x_i : i \in I)$ by Lemma 3.5.7, and $\mathcal{L}$ is ultraweakly closed on $\mathcal{L}(\mathcal{D}_{\mathcal{A}}, \mathcal{D}_{\mathcal{A}}^+)$. The final statement in Lemma 3.5.7 yields

Corollary 3.5.8. *If $\mathcal{L}$ is an ultraweakly closed $*$-vector subspace of $\mathcal{L}(\mathcal{D}_{\mathcal{A}}, \mathcal{D}_{\mathcal{A}}^+)$, then each ultraweakly continuous strongly positive linear functional on $\mathcal{L}$ is normal.*

Remark 2. Further results concerning ultraweakly continuous linear functionals will be obtained in Chapter 5, cf. Propositions 5.2.11 and 5.2.12.

The Strong-Operator and the Ultrastrong Topologies

In this subsection, $\mathcal{A}$ denotes an O-family in a Hilbert space $\mathcal{H}$ and $\mathcal{L}$ is a linear subspace of $\mathfrak{L}(\mathcal{D}_{\mathcal{A}}, \mathcal{H}) \equiv \mathcal{L}(\mathcal{D}_{\mathcal{A}}, \mathcal{H})$.

The *strong-operator topology* on $\mathcal{L}$ is the locally convex topology on $\mathcal{L}$ which is determined by the family of seminorms

$$\mathcal{L} \ni x \to \|x\varphi\|, \quad \varphi \in \mathcal{D}(\mathcal{A}).$$

It will be denoted by $\sigma^{\mathcal{D}}$.

Proposition 3.5.9. *A linear functional ω on $\mathcal{L}$ is continuous on $\mathcal{L}[\sigma^{\mathcal{D}}]$ if and only if there are vectors $\varphi_1, \ldots, \varphi_k \in \mathcal{D}(\mathcal{A})$ and $\psi_1, \ldots, \psi_k \in \mathcal{H}$, $k \in \mathbb{N}$, such that $\omega = \sum_{n=1}^{k} \omega_{\varphi_n, \psi_n}$.*

Proof. The sufficiency part is trivial. To prove the necessity, let ω be a continuous linear functional on $\mathcal{L}[\sigma^{\mathcal{D}}]$. Then there are vectors $\varphi_1, \ldots, \varphi_k \in \mathcal{D}(\mathcal{A})$ such that

$$|\omega(x)| \leqq (\|x\varphi_1\|^2 + \cdots + \|x\varphi_k\|^2)^{1/2} \quad \text{for all } x \in \mathcal{L}. \tag{5}$$

For $x \in \mathcal{L}$, let $\tilde{x}\varphi$ be the vector $(x\varphi_1, \ldots, x\varphi_k)$ in the Hilbert space $\mathcal{H}_k := \mathcal{H} \oplus \cdots \oplus \mathcal{H}$ (k times). (5) shows that the map $\tilde{x}\varphi \to \omega(x)$ defines a continuous linear functional on the linear subspace $\mathcal{D}_\varphi := \{\tilde{x}\varphi : x \in \mathcal{L}\}$ of $\mathcal{H}_k$. By the Riesz theorem there exists a vector

$\psi = (\psi_1, \ldots, \psi_k)$ contained in the closure of $\mathcal{D}_\varphi$ in $\mathcal{H}_k$ such that $\omega(x) = \langle \tilde{x}\varphi, \psi\rangle = \sum_{n=1}^{k} \langle x\varphi_n, \psi_n\rangle$ for $x \in \mathcal{L}$. Thus $\omega = \sum_{n=1}^{k} \omega_{\varphi_n, \psi_n}$. □

The *ultrastrong topology* (or σ-strong topology) is the locally convex topology on $\mathcal{L}$ defined by the family of seminorms

$$\mathcal{L} \ni x \to \|x\|_{(\varphi_n)} := \left(\sum_{n=1}^{\infty} \|x\varphi_n\|^2\right)^{1/2},$$

where $(\varphi_n : n \in \mathbb{N})$ is an arbitrary sequence in $\mathcal{D}(\mathcal{A})$ which satisfies $\|a\|_{(\varphi_n)} < \infty$ for all $a \in \mathcal{A}$. Since $\mathcal{L} \subseteq \mathfrak{L}(\mathcal{D}_\mathcal{A}, \mathcal{H})$, $\|\cdot\|_{(\varphi_n)}$ is finite and hence a seminorm on $\mathcal{L}$. Note that the family of all seminorms $\|\cdot\|_{(\varphi_n)}$ is directed.

Proposition 3.5.10. *A linear functional ω on $\mathcal{L}$ is ultrastrongly continuous if and only if there are vectors $\varphi_n \in \mathcal{D}(\mathcal{A})$ and $\psi_n \in \mathcal{H}$, $n \in \mathbb{N}$, satisfying $\|a\|_{(\varphi_n)} < \infty$ for all $a \in \mathcal{A}$ and $\sum_{n=1}^{\infty} \|\psi_n\|^2 < \infty$ such that $\omega = \sum_{n=1}^{\infty} \omega_{\varphi_n, \psi_n}$.*

Proof. The sufficiency follows immediately from the Cauchy-Schwarz inequality. We verify the necessity. Suppose that ω is an ultrastrongly continuous linear functional on $\mathcal{L}$. Since the family of seminorms $\{\|\cdot\|_{(\varphi_n)}\}$ is directed, there exists a sequence $(\varphi_n : n \in \mathbb{N})$ in $\mathcal{D}(\mathcal{A})$ such that $\|a\|_{(\varphi_n)} < \infty$ for $a \in \mathcal{A}$ and such that $|\omega(x)| \leqq \|x\|_{(\varphi_n)}$ for all $x \in \mathcal{L}$. We now slightly modify the proof of Proposition 3.5.9. Let $\mathcal{H}_\infty$ be the Hilbert space $\mathcal{H}_\infty := \sum_{n=1}^{\infty} \oplus \mathcal{H}$. Since $\|a\|_{(\varphi_n)} < \infty$ for $a \in \mathcal{A}$, $\tilde{x}\varphi := (x\varphi_n : n \in \mathbb{N})$ is a vector in $\mathcal{H}_\infty$ for each $x \in \mathcal{L}$. Then the map $\tilde{x}\varphi \to \omega(x)$ is a continuous linear functional on the linear subspace $\mathcal{D}_\varphi := \{\tilde{x}\varphi : x \in \mathcal{L}\}$ of $\mathcal{H}_\infty$. Again by the Riesz theorem there is a vector $\psi = (\psi_n)$ in $\mathcal{H}_\infty$ such that $\omega(x) = \langle \tilde{x}\varphi, \psi\rangle$ for $x \in \mathcal{L}$. From this the assertion follows. □

3.6. Continuity of $*$-Representations

Suppose E is a locally convex space and $\mathcal{A}$ is an O-family. A linear mapping π of E into $\mathcal{L}(\mathcal{D}_\mathcal{A}, \mathcal{D}_\mathcal{A}^+)$ is said to be *weakly continuous* if $\langle \pi(\cdot)\, \varphi, \psi\rangle$ is a continuous linear functional on E for arbitrary vectors $\varphi, \psi \in \mathcal{D}(\mathcal{A})$. In other words, π is weakly continuous if it is a continuous mapping of E into $\mathcal{L}(\mathcal{D}_\mathcal{A}, \mathcal{D}_\mathcal{A}^+)$ if the latter carries the weak-operator topology. It follows at once from the polarization formula 3.2/(3) that π is weakly continuous provided that all linear functionals $\omega_\varphi(\cdot) := \langle \pi(\cdot)\, \varphi, \varphi\rangle$, $\varphi \in \mathcal{D}(\mathcal{A})$, are continuous on E.

Assume that π is a $*$-representation of a topological $*$-algebra A with unit. We consider π as a mapping into $\mathcal{L}(\mathcal{D}_\mathcal{A}, \mathcal{D}_\mathcal{A}^+)$, where $\mathcal{A} := \pi(\mathsf{A})$. The results in this section are related, directly or indirectly, to the following basic question. Under what circumstances is π continuous as a mapping of A on $\pi(\mathsf{A})\,[\tau_\mathcal{D}]$ (or more generally on $\pi(\mathsf{A})\,[\tau]$, where τ denotes one of the topologies from Section 3.3)? We shall divide this problem into the following two subproblems.

(i) When is π weakly continuous?

(ii) Suppose that π is weakly continuous. When is π a continuous mapping of A on $\pi(\mathsf{A})\,[\tau_\mathcal{D}]$ (or on $\pi(\mathsf{A})\,[\tau]$)?

We briefly discuss problem (i). First note that the weak continuity of π is, of course, necessary for the continuity of the map $\pi\colon \mathsf{A} \to \pi(\mathsf{A})\,[\tau_{\mathcal{D}}]$, but it is not sufficient; cf. Theorem 6.2.7. Let $\varphi \in \mathcal{D}(\mathcal{A})$. Since π is a $*$-representation of A, $\omega_\varphi(a^+a) = \langle \pi(a^+a)\,\varphi, \varphi\rangle = \|\pi(a)\,\varphi\|^2 \geqq 0$ for $a \in \mathsf{A}$, i.e., ω_φ is a positive linear functional on A. That is, π is weakly continuous if and only if the positive linear functionals ω_φ, where $\varphi \in \mathcal{D}(\mathcal{A})$, are continuous on A. In particular, we see that π is weakly continuous provided that *all* positive linear functionals are continuous on A. Conversely, suppose that there exists a discontinuous positive linear functional, say ω, on A. Then the $*$-representation π_ω obtained from ω by the GNS construction (see Section 8.6) is not weakly continuous. (The latter follows from the formula $\omega(a) = \langle \pi_\omega(a)\,\varphi_\omega, \varphi_\omega\rangle$, $a \in \mathsf{A}$.) Summing up, this discussion shows that the continuity of positive linear functionals on A is the central question for problem (i).

Theorem 3.6.1. *Suppose that* A *is a Frechet topological $*$-algebra with unit element. Then each positive linear functional on* A *is continuous.*

Proof. Since A is a Frechet space, every separately continuous bilinear mapping of $\mathsf{A} \times \mathsf{A}$ into A is continuous (SCHÄFER [1], III, 5.1). Hence the multiplication is jointly continuous in A. Let ϑ be a metric which defines the topology of A. Assume to the contrary that there exists a positive linear functional ω on A which is not continuous. Since $\omega(1) = 0$ would imply $\omega \equiv 0$, we can suppose that $\omega(1) = 1$. Since ω is discontinuous, we can find a sequence $(a_n\colon n \in \mathbb{N})$ in A which converges to zero such that $(\omega(a_n)\colon n \in \mathbb{N})$ does not converge to zero. By passing to a subsequence if necessary, we can assume that $|\omega(a_n)| \geqq \varepsilon$, $n \in \mathbb{N}$, for some $\varepsilon > 0$. Then $\omega(a_n^+a_n) \geqq \varepsilon^2$ by the Cauchy-Schwarz inequality and by the fact that $\omega(1) = 1$. Set $b_n := \omega(a_n^+a_n)^{-1}\, a_n^+a_n$, $n \in \mathbb{N}$. From the continuity of the involution and the joint continuity of the multiplication in A it follows that $\lim_n b_n = 0$ in A. Moreover, $\omega(b_n) = 1$ for $n \in \mathbb{N}$.

We shall define inductively a subsequence $(c_n\colon n \in \mathbb{N})$ of $(b_n\colon n \in \mathbb{N})$. Let $c_1 := b_1$. Suppose $n > 1$ and $c_1, \ldots, c_{n-1}$ are chosen. Define $T_{kn}(x) := c_k + \big(c_{k+1}(\cdots + (c_{n-1} + x^2)^2)\cdots\big)^2$ for $k \in \mathbb{N}$, $1 \leqq k \leqq n-1$, and for $x \in \mathsf{A}$. Again by the joint continuity of the multiplication, each $T_{kn}(\cdot)$ is a continuous mapping of A into itself. Therefore, since $\lim_n b_n = 0$ and $T_{kn}(0) = T_{k,n-1}(c_{n-1})$ for $k = 1, \ldots, n-2$, we can choose a sufficiently large number $n' \in \mathbb{N}$ such that $c_n := b_{n'}$ satisfies

$$\vartheta\big(T_{kn}(c_n), T_{k,n-1}(c_{n-1})\big) \leqq 2^{-n} \quad \text{for} \quad k = 1, \ldots, n-2.$$

Let $k \in \mathbb{N}$. Then $(T_{k,k+n}(c_{k+n})\colon n \in \mathbb{N})$ is a Cauchy sequence in A, since

$$\vartheta\big(T_{k,k+n+m}(c_{k+n+m}), T_{k,k+n}(c_{k+n})\big) \leqq \sum_{l=1}^{m} \vartheta\big(T_{k,k+n+l}(c_{k+n+l}), T_{k,k+n+l-1}(c_{k+n+l-1})\big)$$

$$\leqq \sum_{l=1}^{m} 2^{-(k+n+l)} < 2^{-n}$$

for $n, m \in \mathbb{N}$. Since A is complete, there is an $x_k \in \mathsf{A}$ such that $x_k = \lim_n T_{k,k+n}(c_{k+n})$. By construction, $T_{k,k+n}(c_{k+n}) = c_k + \big(T_{k+1,k+1+n}(c_{k+1+n})\big)^2$ for $k, n \in \mathbb{N}$; so $x_k = c_k + x_{k+1}^2$. Because $x_k \in \mathsf{A}_{\mathrm{h}}$ and $\omega(c_k) = 1$, we obtain $\omega(x_k) = 1 + \omega(x_{k+1}^2) \geqq 1$ for $k \in \mathbb{N}$. By the

Cauchy-Schwarz inequality, this gives

$$\omega(x_1) = \omega(c_1) + \omega(x_2^2) \geqq 1 + \omega(x_2) = 1 + \omega(c_2) + \omega(x_3^2) \geqq \cdots \geqq k + \omega(x_{k+1}) > k$$

for any $k \in \mathbb{N}$. This is a contradiction. □

A seminorm p on an algebra A is called *submultiplicative* if $p(ab) \leqq p(a)\, p(b)$ for all $a, b \in$ A. By an *lmc *-algebra* we mean a topological *-algebra the topology of which can be given by a family of submultiplicative seminorms.

Proposition 3.6.2. *If* A *is a complete lmc *-algebra with unit element, then each positive linear functional* ω *on* A *is bounded (that is, the image* $\omega(\mathsf{R})$ *of every bounded subset* R *of* A *is bounded).*

Proof. If p is a continuous submultiplicative seminorm on A, then p^+ is, where $p^+(a) := \max\{p(a), p(a^+)\}$, $a \in$ A. From this it follows that the topology of the lmc *-algebra A can be generated by a family Γ of submultiplicative seminorms which are invariant under the involution. Let Γ_k denote the set of all $p \in \Gamma$ for which $p(x) \leqq k$ for all $x \in$ R. Set $\mathsf{A}_0 := \left\{a \in \mathsf{A} : \sup_{p \in \Gamma_k} p(a) < \infty \text{ for all } k \in \mathbb{N}\right\}$ and $p_k(a) := \sup_{p \in \Gamma_k} p(a)$ for $a \in \mathsf{A}_0$ and $k \in \mathbb{N}$. From the properties of the seminorms in Γ it follows easily that A_0 is a *-subalgebra of A and that each p_k, $k \in \mathbb{N}$, is a submultiplicative seminorm on A_0 which is invariant under the involution. We equip A_0 with the locally convex topology defined by the seminorms p_k, $k \in \mathbb{N}$. If $p \in \Gamma$, then $\lambda_p := \sup_{x \in \mathsf{R}} p(x) < \infty$, so that $p \in \Gamma_k$ and $p \leqq p_k$ on A_0 if $k > \lambda_p$, $k \in \mathbb{N}$. From this we see that the topology of A_0 is a Hausdorff topology and that it is stronger than the induced topology of A. Moreover, $1 \in \mathsf{A}_0$, since $p(1) \leqq 1$ for $p \in \Gamma$. Therefore, A_0 is a metrizable lmc *-algebra with unit element. Suppose for a moment we have shown that A_0 is complete. Then the positive linear functional $\omega_0 := \omega \restriction \mathsf{A}_0$ is continuous on A_0 by Theorem 3.6.1. From the construction it is clear that $\mathsf{R} \subseteqq \mathsf{A}_0$ and $p_k(x) \leqq k$ for $x \in$ R and $k \in \mathbb{N}$. Hence R is a bounded subset in A_0; so $\omega_0(\mathsf{R}) \equiv \omega(\mathsf{R})$ is bounded by the continuity of ω_0, and the proof would be complete.

It remains to show that A_0 is complete. We let $(x_n : n \in \mathbb{N})$ be a Cauchy sequence in A_0. Since the topology of A_0 is stronger than the induced topology of A, (x_n) is also a Cauchy sequence in A. Since we assumed A to be complete, there is an $x \in$ A such that $x = \lim_n x_n$ in A. Let $k \in \mathbb{N}$ and $\varepsilon > 0$ be given. Then there is an $n_0 \in \mathbb{N}$ such that $p_k(x_n - x_m) \leqq \varepsilon$ if $n, m > n_0$. Hence $p(x_n - x_m) \leqq \varepsilon$ for each $p \in \Gamma_k$, if $n, m > n_0$. Taking the limit in A, the latter gives $p(x_n - x) \leqq \varepsilon$ if $n > n_0$ and $p \in \Gamma_k$. Since $x_n \in \mathsf{A}_0$ for any $n \in \mathbb{N}$, this yields $x \in \mathsf{A}_0$. Moreover, we get $p_k(x_n - x) \leqq \varepsilon$ if $n > n_0$. Since $\{p_k : k \in \mathbb{N}\}$ is a directed family of seminorms defining the topology of A_0, this shows that $x = \lim_n x_n$ in A_0. □

Corollary 3.6.3. *Each positive linear functional on a complete bornological lmc *-algebra with unit element is continuous.*

Example 3.6.4. Let W denote the set of all ordinals less than the first uncountable ordinal, endowed with the order topology. (We refer to Gillman/Jerison [1], § 5, 12., for the topological facts used in this example.) We equip the *-algebra $\mathsf{A} := C(W)$ with the topology of uniform convergence on compact subsets of the topological space W. Then A is a complete lmc *-algebra with unit. Every function $f \in C(W)$ is constant on some set $W(\alpha_1) = \{\alpha \in W : \alpha \geqq \alpha_1\}$ with $\alpha_1 \in W$ depending upon f. Let $\omega(f)$ be that constant value. Then $\omega(\cdot)$ is a discontinuous positive linear functional on A. Note that ω is even a character, i.e., $\omega(fg) = \omega(f)\,\omega(g)$ for $f, g \in$ A and $\omega(1) = 1$. ○

We now turn to subproblem (ii).

Proposition 3.6.5. *Suppose that E is a barrelled locally convex space. Let $\mathcal{A}$ be an O-family, and let π be a weakly continuous linear mapping of E into $\mathcal{L}(\mathcal{D}_\mathcal{A}, \mathcal{D}_\mathcal{A}^+)$. Then π maps E continuously into $\mathcal{L}(\mathcal{D}_\mathcal{A}, \mathcal{D}_\mathcal{A}^+)[\tau_\mathcal{D}]$. If $\pi(E) \subseteqq \mathcal{L}(\mathcal{D}_\mathcal{A}, \mathcal{H})$, then π is a continuous mapping of E into $\mathcal{L}(\mathcal{D}_\mathcal{A}, \mathcal{H})[\tau^\mathcal{D}]$. If $\mathcal{A}$ is an O*-algebra and $\pi(E) \subseteqq \mathcal{L}^+(\mathcal{D}_\mathcal{A})$, then the mapping π of E into $\mathcal{L}^+(\mathcal{D}_\mathcal{A})[\tau^*]$ is continuous.*

Proof. Suppose $\mathcal{M}$ is a bounded subset of $\mathcal{D}_\mathcal{A}$. Then the set $W := \{x \in E : p_\mathcal{M}(\pi(x)) \leqq 1\}$ is obviously absolutely convex and absorbing in E. Since π is weakly continuous, each set $W_{\varphi,\psi} := \{x \in E : |\langle \pi(x)\,\varphi, \psi\rangle| \leqq 1\}$, where $\varphi, \psi \in \mathcal{D}(\mathcal{A})$, is closed in E. Therefore, $W = \bigcap_{\varphi,\psi \in \mathcal{M}} W_{\varphi,\psi}$ is closed and hence a barrel in E. Since E is assumed to be a barrelled space, W is a 0-neighbourhood in E. This proves that π is a continuous mapping of E into $\mathcal{L}(\mathcal{D}_\mathcal{A}, \mathcal{D}_\mathcal{A}^+)[\tau_\mathcal{D}]$.

The two other assertions will be proved by similar reasoning.

First suppose that $\pi(E) \subseteqq \mathcal{L}(\mathcal{D}_\mathcal{A}, \mathcal{H})$. Let $\mathcal{M}$ be a bounded set in $\mathcal{D}_\mathcal{A}$. Set $W := \{x \in E : p^\mathcal{M}(\pi(x)) \leqq 1\}$. From $W = \bigcap_{\varphi \in \mathcal{M}} \bigcap_{\psi \in \mathcal{U}_{\mathcal{D}(\mathcal{A})}} W_{\varphi,\psi}$ it follows that W is closed in E. Since W is absolutely convex and absorbing in E, it is a barrel and hence a 0-neighbourhood in E. Thus π maps E continuously into $\mathcal{L}(\mathcal{D}_\mathcal{A}, \mathcal{H})[\tau^\mathcal{D}]$.

Finally, suppose $\mathcal{A}$ is an O*-algebra and $\pi(E) \subseteqq \mathcal{L}^+(\mathcal{D}_\mathcal{A})$. Let $a \in \mathcal{A}$ and let $\mathcal{M}$ be a bounded subset of $\mathcal{D}_\mathcal{A}$. Now we define

$$W := \{x \in E : p^{a,\mathcal{M}}(\pi(x)) \leqq 1\} \quad \text{and} \quad W_+ = \{x \in E : p_+^{a,\mathcal{M}}(\pi(x)) \leqq 1\}.$$

We have

$$W = \bigcap_{\varphi \in \mathcal{M}} \bigcap_{\psi \in \mathcal{U}_{\mathcal{D}(\mathcal{A})}} \{x \in E : |\langle \pi(x)\,\varphi, a^+\psi\rangle| \leqq 1\}$$

and

$$W_+ = \bigcap_{\varphi \in \mathcal{M}} \bigcap_{\psi \in \mathcal{U}_{\mathcal{D}(\mathcal{A})}} \{x \in E : |\langle \varphi, \pi(x)\, a^+\psi\rangle| \leqq 1\}.$$

Using these formulas and the fact that $a^+\psi \in \mathcal{D}(\mathcal{A})$ for $\psi \in \mathcal{D}(\mathcal{A})$, the same argument as above shows that W and W_+ are 0-neighbourhoods. This proves the continuity of the map $\pi : E \to \mathcal{L}^+(\mathcal{D}_\mathcal{A})[\tau^*]$. □

An immediate consequence of Proposition 3.6.5 is the following corollary which gives a sufficient condition for an affirmative answer to question (ii) in case $\tau = \tau^*$.

Corollary 3.6.6. *Each weakly continuous *-representation π of a barrelled topological *-algebra A with unit is a continuous mapping of A onto $\pi(\mathsf{A})[\tau^*]$.*

Corollary 3.6.7. *If $\mathcal{A}$ is an O*-algebra and $\mathcal{A}[\tau_\mathcal{D}]$ is barrelled, then $\tau_\mathcal{D} = \tau^\mathcal{D} = \tau^*$ on $\mathcal{A}$.*

Proof. Letting π be the identity map in Corollary 3.6.6, we get $\tau^* \subseteqq \tau_\mathcal{D}$ on $\mathcal{A}$. Since always $\tau_\mathcal{D} \subseteqq \tau^\mathcal{D} \subseteqq \tau^*$ on $\mathcal{A}$, the assertion follows. □

The following theorem summarizes our main results concerning the question formulated at the beginning of this section.

Theorem 3.6.8. *Suppose that* A *is a Frechet topological* $*$*-algebra or that* A *is a complete bornological lmc* $*$*-algebra. Let* A *have a unit element. Then every* $*$*-representation* π *of* A *is a continuous mapping of* A *onto* $\pi(\mathsf{A})\,[\tau^*]$. *In particular, each* $*$*-representation* π *maps* A *continuously onto* $\pi(\mathsf{A})\,[\tau_{\mathcal{D}}]$.

Proof. From Theorem 3.6.1 and Corollary 3.6.3 it follows that each $*$-representation of A is weakly continuous. Frechet spaces and complete bornological spaces are both barrelled (SCHÄFER [1], II, 7 and 8). Thus Corollary 3.6.6 applies and yields the assertion. □

Proposition 3.6.9. *Suppose that* A *is a complete bornological topological* $*$*-algebra with unit. Let* $\bar{\mathcal{P}}$ *denote the closure of the wedge* $\mathcal{P}(\mathsf{A})$ *in* A. *Let* $\mathcal{A}$ *be an O-family. Suppose that* π *is a linear mapping of* A *into* $\mathcal{L}(\mathcal{D}_{\mathcal{A}}, \mathcal{D}_{\mathcal{A}}^+)$ *which satisfies* $\pi(\bar{\mathcal{P}}) \subseteq \mathcal{L}(\mathcal{D}_{\mathcal{A}}, \mathcal{D}_{\mathcal{A}}^+)_+$. *Then* $\pi(\mathsf{A})$ *is a* $*$*-vector subspace of* $\mathcal{L}(\mathcal{D}_{\mathcal{A}}, \mathcal{D}_{\mathcal{A}}^+)$ *and the mapping* π *of* A *onto* $\pi(\mathsf{A})\,[\tau_{\mathcal{O}}]$ *is continuous. If in addition* $\mathcal{A}$ *is an* O^**-algebra and* $\pi(\mathsf{A}) \subseteq \mathcal{L}^+(\mathcal{D}_{\mathcal{A}})$, *then* π *is a continuous mapping of* A *into* $\mathcal{L}^+(\mathcal{D}_{\mathcal{A}})\,[\tau^*]$.

In the proof we use the following simple lemma.

Lemma 3.6.10. *Suppose* A *is a barrelled topological* $*$*-algebra with unit. Then, for every bounded subset* R *of* A, *there are bounded sets* R_1, R_2, R_3, *and* R_4 *contained in* $\mathcal{P}(\mathsf{A})$ *such that* $\mathsf{R} \subseteq (\mathsf{R}_1 - \mathsf{R}_2) + i(\mathsf{R}_3 - \mathsf{R}_4)$.

Proof. First we show that the set $\mathsf{N}^2 \equiv \{xy : x, y \in \mathsf{N}\}$ is bounded provided that N is bounded in A. Let $T_y x := xy$, $x, y \in \mathsf{A}$. Since N is bounded, the subset $\{T_y : y \in \mathsf{N}\}$ of $\mathfrak{L}(\mathsf{A})$ is pointwise bounded and hence equicontinuous by the Banach-Steinhaus theorem. (Note that this theorem applies, since A is barrelled.) This implies that N^2 is bounded.

If R is a bounded set in A_h, then it follows from the identity $4x = (x+1)^2 - (x-1)^2$, $x \in \mathsf{A}_h$, that $\mathsf{R} \subseteq \mathsf{R}_1 - \mathsf{R}_2$, where $\mathsf{R}_1 := \{(x+1)^2 : x \in \mathsf{R}\}$ and $\mathsf{R}_2 := \{(x-1)^2 : x \in \mathsf{R}\}$. By the preceding, R_1 and R_2 are bounded sets. Obviously, $\mathsf{R}_1 \subseteq \mathcal{P}(\mathsf{A})$ and $\mathsf{R}_2 \subseteq \mathcal{P}(\mathsf{A})$. The assertion for a general set R follows at once from the continuity of the involution in A. □

Proof of Proposition 3.6.9

From $\mathsf{A}_h = \bar{\mathcal{P}} - \bar{\mathcal{P}}$ (by Lemma 3.6.10) and $\pi(\bar{\mathcal{P}}) \subseteq \mathcal{L}(\mathcal{D}_{\mathcal{A}}, \mathcal{D}_{\mathcal{A}}^+)_+$, we conclude that $\pi(a^+) = \pi(a)^+$ for $a \in \mathsf{A}$; so $\pi(\mathsf{A})$ is a $*$-vector subspace of $\mathcal{L}(\mathcal{D}_{\mathcal{A}}, \mathcal{D}_{\mathcal{A}}^+)$. We prove that π is a continuous mapping of A onto $\pi(\mathsf{A})\,[\tau_{\mathcal{O}}]$. Let $\mathcal{U}$ be an absolutely convex 0-neighbourhood in $\pi(\mathsf{A})\,[\tau_{\mathcal{O}}]$, and let $\mathsf{U} := \{a \in \mathsf{A} : \pi(a) \in \mathcal{U}\}$. Since A is bornological by assumption, it is sufficient to prove that U absorbs every bounded subset R in A. Assume to the contrary that there is a bounded set R in A which is not absorbed by U. Since complete bornological spaces are barrelled (SCHÄFER [1], II, 8), A is barrelled. Therefore, by Lemma 3.6.10, we can assume without loss of generality that $\mathsf{R} \subseteq \mathcal{P}(\mathsf{A})$. Since U does not absorb R, for each $k \in \mathbb{N}$ there is an element $x_k \in \mathsf{R}$ such that $k^{-3}x_k \notin \mathsf{U}$. Then the sequence $\left(y_n = \sum_{k=1}^{n} k^{-2}x_k : k \in \mathbb{N}\right)$ is a Cauchy sequence, since $\{x_k : k \in \mathbb{N}\}$ is a subset of the bounded set R. Because A is complete, there is a $y \in \mathsf{A}$ such that $y = \lim_n y_n$ in A. Let $k \in \mathbb{N}$. Since $l^{-2}x_l \in \mathcal{P}(\mathsf{A})$, $y_n - k^{-2}x_k \in \mathcal{P}(\mathsf{A})$ if $n \geq k$; so $y - k^{-2}x_k \in \bar{\mathcal{P}}$. Because $\pi(\bar{\mathcal{P}}) \subseteq \mathcal{L}(\mathcal{D}_{\mathcal{A}}, \mathcal{D}_{\mathcal{A}}^+)_+$ by assumption, this implies that $\pi(k^{-2}x_k)$ belongs to the order interval $[0, \pi(y)]$ in $\pi(\mathsf{A})_h$. But the 0-neighbourhood $\mathcal{U}$ for the topology $\tau_{\mathcal{O}}$ on $\pi(\mathsf{A})$

absorbs the order interval $[0, \pi(y)]$; hence there is an $m \in \mathbb{N}$ such that $[0, \pi(y)] \subseteq m\mathcal{U}$. This gives $\pi(m^{-1}k^{-2}x_k) \in \mathcal{U}$ and $m^{-1}x^{-2}x_k \in \mathsf{U}$ for $k \in \mathbb{N}$. In case $k = m$ we have a contradiction. Thus the first statement in the proposition is proved.

Now we assume in addition that $\mathcal{A}$ is an O*-algebra and that $\pi(\mathsf{A}) \subseteq \mathcal{L}^+(\mathcal{D}_\mathcal{A})$. Let ω be a $\bar{\mathcal{P}}$-positive linear functional on A. Applying the first statement in the one dimensional case (that is in case in which $\mathcal{D}(\mathcal{A}) = \mathbb{C}$ and $\omega = \pi$), it follows that ω is continuous on A. Since $\pi(\bar{\mathcal{P}}) \subseteq \mathcal{L}(\mathcal{D}_\mathcal{A}, \mathcal{D}_\mathcal{A}^+)_+$ by assumption, the linear functionals $\omega_\varphi(\cdot) = \langle \pi(\cdot)\, \varphi, \varphi \rangle$, $\varphi \in \mathcal{D}(\mathcal{A})$, on A are $\bar{\mathcal{P}}$-positive and hence continuous on A. This shows that π is weakly continuous. As already noted in this proof, A is barrelled. Therefore, by Proposition 3.6.5, π is a continuous mapping of A into $\mathcal{L}^+(\mathcal{D}_\mathcal{A})\,[\tau^*]$. □

Corollary 3.6.11. *If $\mathcal{A}$ is an O*-algebra and $\mathcal{A}[\tau_\mathcal{D}]$ is a complete bornological space, then we have $\tau_\mathcal{D} = \tau^\mathcal{D} = \tau^* = \tau_\mathcal{O}$ on $\mathcal{A}$.*

Proof. Applying Proposition 3.6.9 with π being the identity map, we get $\tau_\mathcal{O} \subseteq \tau_\mathcal{D}$ and $\tau^* \subseteq \tau_\mathcal{D}$. This yields the assertion, since $\tau_\mathcal{D} \subseteq \tau_\mathcal{O}$ and $\tau_\mathcal{D} \subseteq \tau^\mathcal{D} \subseteq \tau^*$ on any O*-algebra. □

Notes

3.1. In the case where $\mathcal{L}^+(\mathcal{D})$ is self-adjoint the ideal $\mathbb{B}(\mathcal{D})$ was investigated by Timmermann [2].
3.2. Continuous linear mappings of a Frechet domain $\mathcal{D}[t_+]$ into the conjugate space of its strong dual were studied by Lassner [6] and Kürsten [2]. The concepts of spaces $\mathcal{L}(\mathcal{D}^+_\mathcal{B}, \mathcal{D}_\mathcal{A})$ and $\mathcal{L}(\mathcal{D}_\mathcal{A}, \mathcal{D}^+_\mathcal{B})$ appear for the first time in this monograph.

Various kinds of "partial products" have been defined by Araki/Jurzak [1], Antoine/Karwowski [1], Lassner [8] and Kürsten [2], [5]. The latter paper contains a rather general concept which covers also the one we have used in the text.
Proposition 3.2.3 can be found in Kürsten [2], [5].
3.3. The topologization of unbounded operator algebras was initiated by Lassner [1] who introduced and studied the topologies $\tau_\mathcal{D}$, $\tau^\mathcal{D}$ and $\tau^*(\mathcal{A})$ (in our notation) on an O*-algebra $\mathcal{A}$; cf. also Lassner [4]. Later Jurzak [1] and Arnal/Jurzak [1] proposed the topologies $\tau_\mathcal{O}$ and $\tau^\mathcal{O}$ (which were called the ϱ- and λ-topologies in these papers) and established their basic properties. In the text we have given a more unified approach to these topologies which is based on the topologies τ_b and τ_{in}. Proposition 3.3.1 is due to Schmüdgen [4], and Proposition 3.3.7 can be found in Schmüdgen [2]. Propositions 3.3.13, (i), and 3.3.15 are due to Lassner [4]. Theorem 3.3.16 and Corollary 3.3.18 are from Kürsten [2], [5]. Proposition 3.3.19 appears here for the first time.
3.4. Theorem 3.4.1 and its subsequent corollaries are due to Kürsten [2], [5]. In a special case Theorem 3.4.1 was previously shown by Lassner [6]. Theorem 3.4.6 has been shown in the proof of Theorem 2 in Schmüdgen [5]. The concept of a topological quasi *-algebra was introduced by Lassner [8].
3.5. The basic properties of the weak-operator, strong-operator, ultraweak and ultrastrong topologies developed in the text can be found in Arnal/Jurzak [1] and in Araki/Jurzak [1].
3.6. Theorem 3.6.1 and its ingenious proof are due to Xia [1]. But Xia treated only the case of lmc *-algebras. Using his method Ng/Warner [1] obtained a rather general result which covers Theorem 3.6.1. Proposition 3.6.2 is from Dixon/Fremlin [1]. Proposition 3.6.5, its two corollaries and Theorem 3.6.8 are due to Lassner [1], [4]. In the proof of Proposition 3.6.9 we have combined the proof of a result on the continuity of positive linear functionals on ordered topological vector spaces (cf. Schäfer [1], V, 5.5) with some algebra technique.

Additional References:

3.3. Lassner [2], [3], [5], [6], [7], Kunze [1], [2], Schmüdgen [1], [3], [11], Junek [2].

4. Topologies for O-Families with Metrizable Graph Topologies

In the previous chapter some basic topologies on linear subspaces of $\mathcal{L}(\mathcal{D}_\mathcal{A}, \mathcal{D}_\mathcal{B}^+)$ were introduced and general properties of these topologies were established. Assuming throughout that the graph topologies of the O-families resp. O*-algebras are metrizable, the present chapter continues the study of the topologies $\tau_\mathcal{D}$, $\tau_\mathcal{N}$, $\tau_\mathcal{O}$, $\tau^\mathcal{D}$, $\tau^\mathcal{N}$ and $\tau^\mathcal{O}$. There are a number of results which can be obtained under this additional assumption.

In Section 4.1 we describe 0-neighbourhood bases for the topologies $\tau_\mathcal{D}$, $\tau_\mathcal{N}$, $\tau_\mathcal{O}$ and $\tau^\mathcal{D}$, $\tau^\mathcal{N}$, $\tau^\mathcal{O}$ which are convenient for many purposes and which will be used later on. Section 4.2 contains a few general results which are all based on a characterization of the bounded sets in the topologies τ_b and τ_in.

The remaining three sections in this chapter are related, directly or indirectly, to the following question. Under what condition do the topologies $\tau_\mathcal{D}$, $\tau_\mathcal{N}$ and $\tau_\mathcal{O}$ on a cofinal $*$-vector subspace $\mathcal{L}$ of $\mathcal{L}(\mathcal{D}_\mathcal{A}, \mathcal{D}_\mathcal{A}^+)$ coincide? Each positive result towards this end yields important information about these topologies. Thus the equalities $\tau_\mathcal{D} = \tau_\mathcal{O}$ and $\tau_\mathcal{N} = \tau_\mathcal{O}$ are valid if and only if the topologies $\tau_\mathcal{D}$ and $\tau_\mathcal{N}$, respectively, are bornological (cf. Corollary 4.2.3). From each of these two equalities it follows that the positive cone $\mathcal{L}_+$ is normal in the order topology $\tau_\mathcal{O}$. The latter implies that each linear functional on $\mathcal{L}$ which is bounded on order intervals (or equivalently, each $\tau_\mathcal{O}$-continuous linear functional) is a linear combination of strongly positive linear functionals on $\mathcal{L}$ (cf. Proposition 1.5.4). If $\tau_\mathcal{D} = \tau_\mathcal{N}$, then every strongly positive linear functional on $\mathcal{L}$ is continuous in the topology $\tau_\mathcal{D}$.

In Section 4.3 the above question is investigated for commutatively dominated closed O*-algebras. Section 4.4 provides some general results which give affirmative answers to this question under certain assumptions. In these two sections we restrict ourselves to the topologies $\tau_\mathcal{D}$, $\tau_\mathcal{N}$, $\tau_\mathcal{O}$, though similar results for the topologies $\tau^\mathcal{D}$, $\tau^\mathcal{N}$, $\tau^\mathcal{O}$ on O*-algebras could be obtained only by some slight modifications of the proofs. Section 4.5 contains some results about the topologies $\tau_\mathcal{D}$, $\tau_\mathcal{N}$, $\tau_\mathcal{O}$ and $\tau^\mathcal{D}$, $\tau^\mathcal{N}$, $\tau^\mathcal{O}$ on $*$-vector spaces $\mathcal{L}$ and O*-algebras $\mathcal{A}$, respectively, which have an at most countable basis. Finally, some illuminating examples are discussed.

4.1. 0-Neighbourhood Bases for the Topologies $\tau_\mathcal{D}$, $\tau_\mathcal{N}$, $\tau_\mathcal{O}$ and $\tau^\mathcal{D}$, $\tau^\mathcal{N}$, $\tau^\mathcal{O}$

In this section we assume that $\mathcal{A}$ is an O-family with metrizable graph topology $\mathrm{t}_\mathcal{A}$ and $(a_n : n \in \mathbb{N})$ is a sequence of operators in $\mathcal{A}$ such that the family of seminorms $\{\|\cdot\|_{a_n} : n \in \mathbb{N}\}$ generates the graph topology $\mathrm{t}_\mathcal{A}$. Later additional assumptions concerning $\mathcal{A}$ and $(a_n : n \in \mathbb{N})$ will be added.

Proposition 4.1.1. *If $\mathcal{L}$ is a linear subspace of $\mathcal{L}(\mathcal{D}_{\mathcal{A}}, \mathcal{D}_{\mathcal{A}}^+)$, then a 0-neighbourhood base for the topology $\tau_{\mathcal{D}}$ on $\mathcal{L}$ is given by the family of sets*

$$\mathcal{U}_{(\varepsilon_n)} := \{x \in \mathcal{L} : |\langle x\varphi, \varphi\rangle| \leqq \sum_{n=1}^{\infty} \varepsilon_n \|a_n\varphi\|^2 \quad \textit{for all } \varphi \in \mathcal{D}(\mathcal{A})\},$$

where $(\varepsilon_n : n \in \mathbb{N})$ varies over all positive sequences.

Proof. In this proof, we abbreviate $\mathfrak{h}_\varepsilon(\varphi) := \sum_{n=1}^{\infty} \varepsilon_n \|a_n\varphi\|^2$ for $\varphi \in \mathcal{D}(\mathcal{A})$ and $\varepsilon \equiv (\varepsilon_n : n \in \mathbb{N})$; see also the proof of Theorem 2.4.1. Suppose $\mathcal{M}$ is a bounded subset of $\mathcal{D}_{\mathcal{A}}$. We choose positive numbers ε_n such that $\varepsilon_n \left(\sup_{\varphi \in \mathcal{M}} \|a_n\varphi\|^2\right) \leqq 2^{-n}$ for all $n \in \mathbb{N}$. Letting $\varepsilon := (\varepsilon_n : n \in \mathbb{N})$, we then have $\mathfrak{h}_\varepsilon(\varphi) \leqq 1$ for all $\varphi \in \mathcal{M}$. If $x \in \mathcal{U}_{(\varepsilon_n)}$, then $|\langle x\varphi, \varphi\rangle| \leqq \mathfrak{h}_\varepsilon(\varphi) \leqq 1$ for $\varphi \in \mathcal{M}$, that is, $p'_{\mathcal{M}}(x) \leqq 1$. Hence $\mathcal{U}_{(\varepsilon_n)} \subseteq \{x \in \mathcal{L} : p'_{\mathcal{M}}(x) \leqq 1\}$.

Conversely, let $\varepsilon = (\varepsilon_n : n \in \mathbb{N})$ be a positive sequence. Define $\mathcal{M} := \{\mathfrak{h}_\varepsilon(\varphi)^{-1/2}\varphi : \varphi \in \mathcal{D}(\mathcal{A})\}$, where we set $(+\infty)^{-1/2} := 0$. Then we have $\sup_{\varphi \in \mathcal{M}} \|a_n\varphi\|^2 \leqq \varepsilon_n^{-1} \sup_{\varphi \in \mathcal{M}} \mathfrak{h}_\varepsilon(\varphi) \leqq \varepsilon_n^{-1}$ for $n \in \mathbb{N}$ which proves that $\mathcal{M}$ is bounded in $\mathcal{D}_{\mathcal{A}}$. If $x \in \mathcal{L}$ and $p'_{\mathcal{M}}(x) \leqq 1$, then $|\langle x\varphi, \varphi\rangle| \leqq \mathfrak{h}_\varepsilon(\varphi)$ by the definition of $\mathcal{M}$. Thus $\{x \in \mathcal{L} : p'_{\mathcal{M}}(x) \leqq 1\} \subseteq \mathcal{U}_{(\varepsilon_n)}$. □

If $\mathcal{L}$ is a $*$-vector space and $\mathcal{L}_+$ admits a countable order-dominating subset, then we have a similar result for the topologies $\tau_{\mathcal{N}}$ and $\tau_{\mathcal{O}}$.

Proposition 4.1.2. *Suppose $\mathcal{B}$ is an O-family and $\mathcal{L}$ is a $*$-vector subspace of $\mathcal{L}(\mathcal{D}_{\mathcal{B}}, \mathcal{D}_{\mathcal{B}}^+)$. Let $\{y_n : n \in \mathbb{N}\}$ be a subset of $\mathcal{L}_+$. For each positive sequence $(\varepsilon_n : n \in \mathbb{N})$, we define*

$$\mathcal{V}'_{(\varepsilon_n)} := \bigcup_{k \in \mathbb{N}} \left\{x \in \mathcal{L} : |\langle x\varphi, \varphi\rangle| \leqq \sum_{n=1}^{k} \varepsilon_n \langle y_n\varphi, \varphi\rangle \textit{ for } \varphi \in \mathcal{D}(\mathcal{B})\right\}$$

and

$$\mathcal{W}'_{(\varepsilon_n)} := \operatorname*{aco}_{k \in \mathbb{N}} \{x \in \mathcal{L} : |\langle x\varphi, \varphi\rangle| \leqq \varepsilon_k \langle y_k\varphi, \varphi\rangle \textit{ for } \varphi \in \mathcal{D}(\mathcal{B})\}.$$

(i) *Suppose that $\{y_1 + \cdots + y_n : n \in \mathbb{N}\}$ is an order-dominating set for the ordered $*$-vector space $\mathcal{L}$. Then the collection of all sets $\mathcal{V}'_{(\varepsilon_n)}$ is a 0-neighbourhood base for the topology $\tau_{\mathcal{N}}$ on $\mathcal{L}$.*

(ii) *Suppose that the set $\{y_n : n \in \mathbb{N}\}$ is order-dominating for the ordered $*$-vector space $\mathcal{L}$. Then the family of all sets $\mathcal{W}'_{(\varepsilon_n)}$ forms a 0-neighbourhood base for the topology $\tau_{\mathcal{O}}$ on $\mathcal{L}$.*

Proof. (i): We first show that each $\mathcal{V}'_{(\varepsilon_n)}$ is a 0-neighbourhood for $\tau_{\mathcal{N}}$. It is easily seen that $\mathcal{V}'_{(\varepsilon_n)}$ is absolutely convex and that $\mathcal{V}'_{(\varepsilon_n)} \cap \mathcal{L}_{\mathrm{h}}$ is $\mathcal{L}_+$-saturated. Suppose $x \in \mathcal{L}_{\mathrm{h}}$. Since the set $\{y_1 + \cdots + y_n : n \in \mathbb{N}\}$ is order-dominating for $\mathcal{L}$, there are numbers $k \in \mathbb{N}$ and $\lambda > 0$ such that $x \leqq \lambda(y_1 + \cdots + y_k)$ and $-x \leqq \lambda(y_1 + \cdots + y_k)$. Hence $|\langle x\varphi, \varphi\rangle| \leqq \lambda\langle (y_1 + \cdots + y_k)\varphi, \varphi\rangle$ for $\varphi \in \mathcal{D}(\mathcal{B})$. We choose $\alpha > 0$ such that $\lambda\alpha \leqq \varepsilon_n$ for $n = 1, \ldots, k$. Then $\alpha x \in \mathcal{V}'_{(\varepsilon_n)}$. This shows that $\mathcal{V}'_{(\varepsilon_n)} \cap \mathcal{L}_{\mathrm{h}}$ is absorbing in $\mathcal{L}_{\mathrm{h}}$. The preceding proves that $\mathcal{V}'_{(\varepsilon_n)}$ belongs to $\boldsymbol{U}_n$ (cf. p. 23); so it is a 0-neighbourhood for $\tau_{\mathcal{N}}$.

Conversely, suppose that $\mathcal{V}$ is a set from $\boldsymbol{U}_n$. Then $\mathcal{V}$ is absorbing for $\mathcal{L}$ by Lemma 1.5.1, so there are positive numbers δ_n such that $\delta_n y_n \in \mathcal{V}$, $n \in \mathbb{N}$. Put $\varepsilon_n := 2^{-(n+1)}\delta_n$, $n \in \mathbb{N}$. Since $\mathcal{V}$ is absolutely convex, $z_k := \sum_{n=1}^{k} 2\varepsilon_n y_n \in \mathcal{V}$ for each $k \in \mathbb{N}$. Suppose that $x \in \mathcal{L}$ is in $\mathcal{V}'_{(\varepsilon_n)}$. Then $|\langle x\varphi, \varphi\rangle| \leqq \frac{1}{2}\langle z_k\varphi, \varphi\rangle$ for some $k \in \mathbb{N}$ and for all $\varphi \in \mathcal{D}(\mathcal{B})$.

Writing x as $x = x_1 + ix_2$ with $x_1, x_2 \in \mathcal{L}_h$, the preceding gives $|\langle 2x_l\varphi, \varphi\rangle| \leqq \langle z_k\varphi, \varphi\rangle$ for all $\varphi \in \mathcal{D}(\mathcal{B})$, so that $2x_l \in [-z_k, z_k]$ for $l = 1, 2$. Because $\mathcal{V} \cap \mathcal{L}_h$ is $\mathcal{L}_+$-saturated, $2x_l \in \mathcal{V}$ for $l = 1, 2$. Using once more that $\mathcal{V}$ is absolutely convex, we get $x \in \mathcal{V}$. Thus we have proved that $\mathcal{V}'_{(\varepsilon_n)} \subseteqq \mathcal{V}$.

(ii): Arguing similarly as above, it follows that the absolutely convex set $\mathcal{W}'_{(\varepsilon_n)} \cap \mathcal{L}_h$ absorbs the order intervals of $(\mathcal{L}_h, \geqq)$. Hence $\mathcal{W}'_{(\varepsilon_n)}$ belongs to $\boldsymbol{U}_0$ and is a 0-neighbourhood for $\tau_\mathcal{O}$. Conversely, let $\mathcal{W}$ be a set in $\boldsymbol{U}_0$. Since $\mathcal{W} \cap \mathcal{L}_h$ absorbs order intervals, there are numbers $\varepsilon_n > 0$ such that $2\varepsilon_n [-y_n, y_n] \subseteqq \mathcal{W}$. We show that $\mathcal{W}'_{(\varepsilon_n)} \subseteqq \mathcal{W}$. Since $\mathcal{W}$ is absolutely convex, it suffices to check that each set $\mathcal{W}_n := \{x \in \mathcal{L} : |\langle x\varphi, \varphi\rangle| \leqq \varepsilon_n\langle y_n\varphi, \varphi\rangle$ for $\varphi \in \mathcal{D}(\mathcal{B})\}$ is a subset of $\mathcal{W}$. Fix $n \in \mathbb{N}$ and let $x \in \mathcal{W}_n$. We write $x = x_1 + ix_2$ with $x_1, x_2 \in \mathcal{L}_h$. From $x \in \mathcal{W}_n$ it follows that $x_l \in \varepsilon_n[-y_n, y_n]$ for $l = 1, 2$. Hence $2x_l \in \mathcal{W}$ for $l = 1, 2$ which yields $x \in \mathcal{W}$. This proves $\mathcal{W}'_{(\varepsilon_n)} \subseteqq \mathcal{W}$. □

Proposition 4.1.3. *Suppose that $\mathcal{A}$ is an O^*-algebra and $\mathcal{L}$ is a cofinal $*$-vector subspace of $\mathcal{L}(\mathcal{D}_\mathcal{A}, \mathcal{D}_\mathcal{A}^+)$. For every positive sequence $(\varepsilon_n : n \in \mathbb{N})$, we define*

$$\mathcal{V}_{(\varepsilon_n)} := \bigcup_{k\in\mathbb{N}} \left\{x \in \mathcal{L} : |\langle x\varphi, \varphi\rangle| \leqq \sum_{n=1}^{k} \varepsilon_n \|a_n\varphi\|^2 \text{ for } \varphi \in \mathcal{D}(\mathcal{A})\right\}$$

and

$$\mathcal{W}_{(\varepsilon_n)} := \underset{k\in\mathbb{N}}{\operatorname{aco}} \{x \in \mathcal{L} : |\langle x\varphi, \varphi\rangle| \leqq \varepsilon_k \|a_k\varphi\|^2 \text{ for } \varphi \in \mathcal{D}(\mathcal{A})\}.$$

(i) *The family of sets $\mathcal{V}_{(\varepsilon_n)}$ is a 0-neighbourhood base for the topology $\tau_\mathcal{N}$ on $\mathcal{L}$.*

(ii) *Suppose in addition to the above assumption that $\|a_n\varphi\| \leqq \|a_{n+1}\varphi\|$ for all $\varphi \in \mathcal{D}(\mathcal{A})$ and $n \in \mathbb{N}$. Then the sets $\mathcal{W}_{(\varepsilon_n)}$ form a 0-neighbourhood base for the topology $\tau_\mathcal{O}$ on $\mathcal{L}$.*

Proof. (i): Set $x_n := a_n^+a_n$ for $n \in \mathbb{N}$. Since $\mathcal{L}$ is assumed to be cofinal in $\mathcal{L}(\mathcal{D}_\mathcal{A}, \mathcal{D}_\mathcal{A}^+)$, for each $n \in \mathbb{N}$ there exists a $y_n \in \mathcal{L}_+$ such that $x_n \leqq y_n$. First we verify that the sets $\{y_1 + \cdots + y_n : n \in \mathbb{N}\}$ and $\{x_1 + \cdots + x_n : n \in \mathbb{N}\}$ are both order-dominating for $\mathcal{L}(\mathcal{D}_\mathcal{A}, \mathcal{D}_\mathcal{A}^+)$. Indeed, let $x \in \mathcal{L}(\mathcal{D}_\mathcal{A}, \mathcal{D}_\mathcal{A}^+)_h$. Since, by the above assumption, $t_\mathcal{A}$ is generated by the family of seminorms $\{\|\cdot\|_{a_n} : n \in \mathbb{N}\}$, there are numbers $n \in \mathbb{N}$ and $\lambda > 0$ such that $|\langle x\varphi, \varphi\rangle| \leqq \lambda(\|a_1\varphi\|^2 + \cdots + \|a_n\varphi\|^2)$ for all $\varphi \in \mathcal{D}(\mathcal{A})$. Then $x \leqq \lambda(a_1^+a_1 + \cdots + a_n^+a_n) = \lambda(x_1 + \cdots + x_n) \leqq \lambda(y_1 + \cdots + y_n)$ which proves that these sets are order-dominating for $\mathcal{L}(\mathcal{D}_\mathcal{A}, \mathcal{D}_\mathcal{A}^+)_h$. Since the set $\{y_1 + \cdots + y_n : n \in \mathbb{N}\}$ is in particular order-dominating for $\mathcal{L}$, Proposition 4.1.2, (i), applies (with $\mathcal{A} = \mathcal{B}$) and shows that the sets $\mathcal{V}'_{(\varepsilon_n)}$ form a 0-neighbourhood base for $\tau_\mathcal{N}$. (We retain the notation from Proposition 4.1.2.). Since $x_n \leqq y_n$ for $n \in \mathbb{N}$, it is obvious that $\mathcal{V}_{(\varepsilon_n)} \subseteqq \mathcal{V}'_{(\varepsilon_n)}$ for each positive sequence $(\varepsilon_n : n \in \mathbb{N})$. Let $(\varepsilon_n : n \in \mathbb{N})$ be a given positive sequence. Our proof is complete once we have shown that $\mathcal{V}'_{(\delta_n)} \subseteqq \mathcal{V}_{(\varepsilon_n)}$ for a certain positive sequence $(\delta_n : n \in \mathbb{N})$. Since $\{x_1 + \cdots + x_n : n \in \mathbb{N}\}$ is order-dominating for $\mathcal{L}(\mathcal{D}_\mathcal{A}, \mathcal{D}_\mathcal{A}^+)$, there are numbers $m_n \in \mathbb{N}$ and $\lambda_n > 0$ such that $y_n \leqq \lambda_n(x_1 + \cdots + x_{m_n})$ for $n \in \mathbb{N}$. There is no loss of generality to assume that $m_{n+1} > m_n$ for $n \in \mathbb{N}$. For $n \in \mathbb{N}$, we take positive numbers δ_n such that $\delta_n\lambda_n \leqq 2^{-n}\varepsilon_r$ for all $r \in \mathbb{N}$, $r \leqq m_n$. Put $m_0 := 0$. For $k \in \mathbb{N}$, we have

$$\sum_{n=1}^{k} \delta_n y_n \leqq \sum_{n=1}^{k} \delta_n\lambda_n(x_1 + \cdots + x_{m_n}) = \sum_{l=0}^{k-1} \sum_{r=1}^{m_{l+1}-m_l} \left(\sum_{n=l+1}^{k} \delta_n\lambda_n\right) x_{m_l+r}$$

$$\leqq \sum_{l=0}^{k-1} \sum_{r=1}^{m_{l+1}-m_l} \left(\sum_{n=l+1}^{k} 2^{-n}\varepsilon_{m_l+r}\right) x_{m_l+r} \leqq \sum_{n=1}^{m_k} \varepsilon_n x_n.$$

This implies that $\mathcal{V}'_{(\delta_n)} \subseteqq \mathcal{V}_{(\varepsilon_n)}$.

(ii): The proof is similar to the proof of (i). Because we have assumed in addition that $\|a_n\varphi\| \leqq \|a_{n+1}\varphi\|$ for $\varphi \in \mathcal{D}(\mathcal{A})$ and $n \in \mathbb{N}$, the sets $\{y_n : n \in \mathbb{N}\}$ and $\{x_n : n \in \mathbb{N}\}$ are both order-dominating for $\mathcal{L}(\mathcal{D}_{\mathcal{A}}, \mathcal{D}_{\mathcal{A}}^+)$. By Proposition 4.1.2, (ii), the sets $\mathcal{W}'_{(\varepsilon_n)}$ constitute a 0-neighbourhood base for $\tau_{\mathcal{O}}$ on $\mathcal{L}$. Since $x_n \leqq y_n$ for all $n \in \mathbb{N}$, we have always that $\mathcal{W}_{(\varepsilon_n)} \subseteqq \mathcal{W}'_{(\varepsilon_n)}$.

Suppose $(\varepsilon_n : n \in \mathbb{N})$ is a positive sequence. Since $\{x_n : n \in \mathbb{N}\}$ is order-dominating for $\mathcal{L}$, there are numbers $r_n \in \mathbb{N}$ and $\alpha_n > 0$ such that $y_n \leqq \alpha_n x_{r_n}$ for $n \in \mathbb{N}$. We choose a positive sequence $(\delta_n : n \in \mathbb{N})$ such that $\delta_n \alpha_n \leqq \varepsilon_{r_n}$ for $n \in \mathbb{N}$. Then $\mathcal{W}'_{(\delta_n)} \subseteqq \mathcal{W}_{(\varepsilon_n)}$. □

Remark 1. In the notation established above, we obviously have $\mathcal{W}_{(\varepsilon_n)} \subseteqq \mathcal{V}_{(\varepsilon_n)} \subseteqq \mathcal{U}_{(\varepsilon_n)}$ for each positive sequence $(\varepsilon_n : n \in \mathbb{N})$. In particular, we thus see again from Propositions 4.1.1 and 4.1.3 that $\tau_{\mathcal{D}} \subseteqq \tau_{\mathcal{N}} \subseteqq \tau_{\mathcal{O}}$ on $\mathcal{L}$ if $\mathcal{L}$ is a cofinal $*$-vector subspace of $\mathcal{L}(\mathcal{D}_{\mathcal{A}}, \mathcal{D}_{\mathcal{A}}^+)$ and $\mathcal{A}$ is an O*-algebra with metrizable graph topology. (Recall that this relation holds also without the latter assumptions, cf. 3.3/(2).)

Remark 2. Suppose $\mathcal{A}$ is an O*-algebra. For a positive sequence $(\varepsilon_n : n \in \mathbb{N})$, let

$$\tilde{\mathcal{V}}_{(\varepsilon_n)} := \bigcup_{k \in \mathbb{N}} \operatorname{aco}\left[-\sum_{n=1}^{k} \varepsilon_n a_n^+ a_n, \sum_{n=1}^{k} \varepsilon_n a_n^+ a_n\right]$$

and

$$\tilde{\mathcal{W}}_{(\varepsilon_n)} := \operatorname*{aco}_{k \in \mathbb{N}} [-\varepsilon_k a_k^+ a_k, \varepsilon_k a_k^+ a_k].$$

Clearly, $\tilde{\mathcal{V}}_{(\varepsilon_n)} \subseteqq \mathcal{V}_{(\varepsilon_n)} \subseteqq \tilde{\mathcal{V}}_{(2\varepsilon_n)}$ and $\tilde{\mathcal{W}}_{(\varepsilon_n)} \subseteqq \mathcal{W}_{(\varepsilon_n)} \subseteqq \tilde{\mathcal{W}}_{(2\varepsilon_n)}$. Therefore, under the assumptions of Proposition 4.1.3, (i) and (ii), the families $\{\tilde{\mathcal{V}}_{(\varepsilon_n)}\}$ and $\{\tilde{\mathcal{W}}_{(\varepsilon_n)}\}$ are 0-neighbourhood bases for the topologies $\tau_{\mathcal{N}}$ and $\tau_{\mathcal{O}}$ on $\mathcal{L}$, respectively.

The next proposition will be used in Sections 4.3 and 4.4.

Proposition 4.1.4. *Let $\mathcal{A}$ be an O*-algebra and let $\mathcal{L}$ be a cofinal $*$-vector subspace of $\mathcal{L}(\mathcal{D}_{\mathcal{A}}, \mathcal{D}_{\mathcal{A}}^+)$. Suppose $(a_n : n \in \mathbb{N})$ is a sequence of symmetric operators in $\mathcal{A}(I)$ such that $\|a_n^2\varphi\| \leqq \|a_{n+1}\varphi\|$ for all $\varphi \in \mathcal{D}(\mathcal{A})$ and $n \in \mathbb{N}$ and such that the topology $\mathrm{t}_{\mathcal{A}}$ is generated by the family of seminorms $\{\|\cdot\|_{a_n} : n \in \mathbb{N}\}$. Assume that for each positive sequence $(\alpha_n : n \in \mathbb{N})$ there is a positive sequence $(\delta_n : n \in \mathbb{N})$ with $\alpha_1 \leqq 2\delta_1$ such that for every $k \in \mathbb{N}$ the following is true: If $x \in \mathcal{L}$ satisfies*

$$|\langle x\varphi, \psi\rangle| \leqq \left\|\sum_{n=1}^{k} \delta_n a_n^2 \varphi\right\| \left\|\sum_{n=1}^{k} \delta_n a_n^2 \psi\right\| \quad \text{for all } \varphi, \psi \in \mathcal{D}(\mathcal{A}), \tag{1}$$

then there exist elements x_{lm} of $\mathcal{L}$ for $l, m = 1, \ldots, k$ such that

$$|\langle x_{lm}\varphi, \varphi\rangle| \leqq \alpha_l \alpha_m \|a_l^2\varphi\| \, \|a_m^2\varphi\| \quad \text{for } \varphi \in \mathcal{D}(\mathcal{A}) \quad \text{and} \quad l, m = 1, \ldots, k \tag{2}$$

and

$$x = \sum_{l,m=1}^{k} x_{lm}. \tag{3}$$

Then we have $\tau_{\mathcal{N}} = \tau_{\mathcal{O}}$ on $\mathcal{L}$.

Proof. Let $(\varepsilon_n : n \in \mathbb{N})$ be a given positive sequence. By induction we choose a positive sequence $(\alpha_n : n \in \mathbb{N})$ satisfying

$$2^{l+m+3} \alpha_1^{-1} \alpha_l \alpha_m \leqq \varepsilon_{l+m} \quad \text{for all} \quad l, m \in \mathbb{N}. \tag{4}$$

For this sequence $(\alpha_n : n \in \mathbb{N})$ we take a positive sequence $(\delta_n : n \in \mathbb{N})$ which has the property stated above. Our assumptions concerning $(a_n : n \in \mathbb{N})$ imply that $\|a_n\varphi\| \leqq \|a_{n+1}\varphi\|$ for $\varphi \in \mathcal{D}(\mathcal{A})$ and $n \in \mathbb{N}$, so that Proposition 4.1.3 applies. Since always $\tau_{\mathcal{N}} \subseteq \tau_{\mathcal{O}}$, it therefore suffices to show that $\mathcal{V}_{(\delta_n)} \subseteq \mathcal{W}_{(\varepsilon_n)}$.

Suppose $y \in \mathcal{V}_{(\delta_n)}$. Then there is a $k \in \mathbb{N}$ such that

$$|\langle y\varphi, \varphi\rangle| \leqq \langle b_k\varphi, \varphi\rangle \quad \text{for} \quad \varphi \in \mathcal{D}(\mathcal{A}), \tag{5}$$

where $b_k := \sum_{n=1}^{k} \delta_n a_n^2$. By $a_1 \in \mathcal{A}(I)$, $\delta_1 \|\varphi\|^2 \leqq \delta_1 \|a_1\varphi\|^2 \leqq \langle b_k\varphi, \varphi\rangle \leqq \|b_k\varphi\| \, \|\varphi\|$ and hence $\delta_1 \|\varphi\| \leqq \|b_k\varphi\|$ for $\varphi \in \mathcal{D}(\mathcal{A})$. Therefore, (5) implies that $|\langle \delta_1 y\varphi, \varphi\rangle| \leqq \|b_k\varphi\|^2$ for $\varphi \in \mathcal{D}(\mathcal{A})$. By the polarization formula 3.2/(3), $|\langle \delta_1 y\varphi, \psi\rangle| \leqq 4 \|b_k\varphi\| \, \|b_k\psi\|$ for $\varphi, \psi \in \mathcal{D}(\mathcal{A})$, i.e., $x := \delta_1 y/4$ satisfies (1). Let x_{lm} be the corresponding elements of $\mathcal{L}$. From $\alpha_1 \leqq 2\delta_1$, (2) and (4), we have $|\langle 2^{l+m+2}\delta_1^{-1} x_{lm}\varphi, \varphi\rangle| \leqq 2^{l+m+3}\alpha_1^{-1}\alpha_l\alpha_m \|a_l^2\varphi\| \, \|a_m^2\varphi\| \leqq \varepsilon_{l+m} \|a_{l+m}\varphi\|^2$ for $\varphi \in \mathcal{D}(\mathcal{A})$ and $l, m = 1, \ldots, k$. Combined with the equation $y = 4\delta_1^{-1} x = \sum_{l,m=1}^{k} 2^{-l-m}(2^{l+m+2}\delta_1^{-1} x_{lm})$ by (3), this proves that $y \in \mathcal{W}_{(\delta_n)}$. □

We now describe 0-neighbourhood bases for the topologies $\tau^{\mathcal{D}}$, $\tau^{\mathcal{N}}$ and $\tau^{\mathcal{O}}$. We retain the assumption that $(a_n : n \in \mathbb{N})$ is a sequence in $\mathcal{A}$ such that the seminorms $\|\cdot\|_{a_n}$, $n \in \mathbb{N}$, determine the graph topology $\mathrm{t}_{\mathcal{A}}$.

A similar reasoning as in the proof of Proposition 4.1.1 gives the following proposition.

Proposition 4.1.5. *Suppose that $\mathcal{A}$ is an O-vector space. Then the family of sets*

$$\mathcal{U}^{(\varepsilon_n)} := \left\{x \in \mathcal{A} : \|x\varphi\| \leqq \sum_{n=1}^{\infty} \varepsilon_n \|a_n\varphi\| \quad \text{for} \quad \varphi \in \mathcal{D}(\mathcal{A})\right\},$$

where $(\varepsilon_n : n \in \mathbb{N})$ is an arbitrary positive sequence, constitutes a 0-neighbourhood base for the topology $\tau^{\mathcal{D}}$ on $\mathcal{A}$.

Proposition 4.1.6. *Suppose $\mathcal{A}$ is an O*-algebra. For a positive sequence $(\varepsilon_n : n \in \mathbb{N})$, define*

$$\mathcal{V}^{(\varepsilon_n)} := \bigcup_{k \in \mathbb{N}} \left\{x \in \mathcal{A} : \|x\varphi\| \leqq \sum_{n=1}^{k} \varepsilon_n \|a_n\varphi\| \quad \text{for} \quad \varphi \in \mathcal{D}(\mathcal{A})\right\}$$

and

$$\mathcal{W}^{(\varepsilon_n)} := \operatorname*{aco}_{k \in \mathbb{N}} \{x \in \mathcal{A} : \|x\varphi\| \leqq \varepsilon_k \|a_k\varphi\| \quad \text{for} \quad \varphi \in \mathcal{D}(\mathcal{A})\}.$$

(i) *The collection of all sets $\mathcal{V}^{(\varepsilon_n)}$ is a 0-neighbourhood base for the topology $\tau^{\mathcal{N}}$ on $\mathcal{A}$.*

(ii) *Assume in addition that $\|a_n\varphi\| \leqq \|a_{n+1}\varphi\|$ for all $\varphi \in \mathcal{D}(\mathcal{A})$ and $n \in \mathbb{N}$. Then the family of all sets $\mathcal{W}^{(\varepsilon_n)}$ is a 0-neighbourhood base for the topology $\tau^{\mathcal{O}}$ on $\mathcal{A}$.*

Proof. (i): We set $\mathcal{L} := \mathcal{A}$ and $y_n := a_n^+ a_n$, $n \in \mathbb{N}$, in Proposition 4.1.2. As noted in the proof of Proposition 4.1.3, (i), the set $\{y_1 + \cdots + y_n : n \in \mathbb{N}\}$ is order-dominating for the ordered $*$-vector space $\mathcal{L} = \mathcal{A}$. Thus, by Proposition 4.1.2, the sets $\mathcal{V}'_{(\varepsilon_n)}$ form a 0-neighbourhood base for the topology $\tau_{\mathcal{N}}$ on $\mathcal{A}$. From the definition of $\tau^{\mathcal{N}}$ it is clear that a 0-neighbourhood base for $\tau^{\mathcal{N}}$ is given by the sets $\check{\mathcal{V}}_{(\varepsilon_n)} := \{x \in \mathcal{A} : x^+x \in \mathcal{V}'_{(\varepsilon_n)}\}$. Since obviously $\check{\mathcal{V}}_{(\varepsilon_n)} \subseteq \mathcal{V}^{(\varepsilon_n^{1/2})}$ and $\mathcal{V}^{(\varepsilon_n)} \subseteq \check{\mathcal{V}}_{(4^n\varepsilon_n^2)}$ for each positive sequence $(\varepsilon_n : n \in \mathbb{N})$, the assertion follows.

(ii): It is straightforward to check that each set $\mathscr{W}^{(\varepsilon_n)}$ absorbs all sets

$$R_a = \{x \in \mathcal{A} : x^+x \leqq a^+a\} = \{x \in \mathcal{A} : \|x\varphi\| \leqq \|a\varphi\| \quad \text{for} \quad \varphi \in \mathcal{D}(\mathcal{A})\},$$

$a \in \mathcal{A}$, and that each absolutely convex set in $\mathcal{A}$ which absorbs all R_{a_k}, $k \in \mathbb{N}$, contains some $\mathscr{W}^{(\varepsilon_n)}$. This gives the assertion. (One could also use Proposition 3.3.14, since the sets $\mathscr{W}^{(\varepsilon_n)}$ form a 0-neighbourhood base for the inductive limit topology of the family of normed spaces $\{(\mathcal{A}^{a_k}, \mathfrak{l}^{a_k}) : k \in \mathbb{N}\}$.) □

4.2. Bounded Sets for the Topologies τ_b and τ_{in}

Proposition 4.2.1. *Suppose that $\mathcal{A}$ and $\mathcal{B}$ are directed O-vector spaces in the Hilbert space $\mathcal{H}$. Suppose that the locally convex spaces $\mathcal{D}_{\mathcal{A}}$ and $\mathcal{D}_{\mathcal{B}}$ are metrizable. Let $(a_n : n \in \mathbb{N})$ resp. $(b_n : n \in \mathbb{N})$ be a sequence of operators in $\mathcal{A}$ resp. $\mathcal{B}$ such that $\|a_n\varphi\| \leqq \|a_{n+1}\varphi\|$, $\varphi \in \mathcal{D}(\mathcal{A})$, resp. $\|b_n\psi\| \leqq \|b_{n+1}\psi\|$, $\psi \in \mathcal{D}(\mathcal{B})$, for all $n \in \mathbb{N}$ and such that the family of seminorms $\{\|\cdot\|_{a_n} : n \in \mathbb{N}\}$ resp. $\{\|\cdot\|_{b_n} : n \in \mathbb{N}\}$ generates the graph topology $\mathfrak{t}_{\mathcal{A}}$ resp. $\mathfrak{t}_{\mathcal{B}}$. Let $\mathcal{L}$ be a linear subspace of $\mathcal{L}(\mathcal{D}_{\mathcal{A}}, \mathcal{D}_{\mathcal{B}}^+)$. For each subset $\mathcal{R}$ of $\mathcal{L}$ the following statements are equivalent:*

(i) *$\mathcal{R}$ is a bounded subset of $\mathcal{L}[\tau_b]$.*

(ii) *$\mathcal{R}$ is a bounded subset of $\mathcal{L}[\tau_{in}]$.*

(iii) *There are operators $a \in \mathcal{A}$ and $b \in \mathcal{B}$ such that $\mathcal{R} \subseteqq \mathcal{U}_{a,b}$.*

(iv) *There is an $n \in \mathbb{N}$ such that $\mathcal{R} \subseteqq \mathcal{U}_{na_n,b_n}$.*

If in addition the O-vector spaces $\mathcal{A}$ and $\mathcal{B}$ are closed (i.e., if $\mathcal{D}_{\mathcal{A}}$ and $\mathcal{D}_{\mathcal{B}}$ are Frechet spaces), then (i) *is also equivalent to*

(v) *$\mathcal{R}$ is bounded in the weak-operator topology, i.e., $\sup_{x \in \mathcal{R}} |\langle x\varphi, \psi\rangle| < \infty$ for arbitrary vectors $\varphi \in \mathcal{D}(\mathcal{A})$ and $\psi \in \mathcal{D}(\mathcal{B})$.*

Proof. (i) → (iv): Assume that (iv) is not true. Then for every $n \in \mathbb{N}$ there are $\varphi_n \in \mathcal{D}(\mathcal{A})$, $\psi_n \in \mathcal{D}(\mathcal{B})$ and $x_n \in \mathcal{R}$ such that $|\langle x_n\varphi_n, \psi_n\rangle| > n \|a_n\varphi_n\| \|b_n\psi_n\|$. Upon multiplying the vectors by some constants if necessary, we can assume that $\|a_n\varphi_n\| = \|b_n\psi_n\| = 1$ for $n \in \mathbb{N}$. Since $\|a_n \cdot\| \leqq \|a_{n+1} \cdot\|$, $n \in \mathbb{N}$, by assumption, we have $\sup_{n \in \mathbb{N}} \|a_k\varphi_n\| = \max(\|a_k\varphi_1\|, \ldots, \|a_k\varphi_{k-1}\|, 1) < \infty$ for any $k \in \mathbb{N}$. This shows that the set $\mathcal{M} := \{\varphi_n : n \in \mathbb{N}\}$ is bounded in $\mathcal{D}_{\mathcal{A}}$. Similarly, $\mathcal{N} := \{\psi_n : n \in \mathbb{N}\}$ is bounded in $\mathcal{D}_{\mathcal{B}}$. Since $p_{\mathcal{M},\mathcal{N}}(x_n) \geqq |\langle x_n\varphi_n, \psi_n\rangle| > n$ for $n \in \mathbb{N}$, we see that the set $\mathcal{R}$ is not bounded in $\mathcal{L}[\tau_b]$. That is, (i) is not satisfied.

(iii) → (ii): By (iii), $\mathcal{R}$ is a bounded subset of the normed space $(\mathcal{L}_{a,b}, \mathfrak{l}_{a,b})$. Since the embedding of this space into $\mathcal{L}[\tau_{in}]$ is continuous, $\mathcal{R}$ is bounded in $\mathcal{L}[\tau_{in}]$.

(ii) → (i) is obvious, since $\tau_b \subseteqq \tau_{in}$ on $\mathcal{L}$. (iv) → (iii) is trivial; so the equivalence of the first four conditions is proved.

Assume now that $\mathcal{A}$ and $\mathcal{B}$ are closed. (i) → (v) is trivial. We prove (v) → (iii). (v) means that the family of continuous bilinear mappings $\{\langle x\cdot, \cdot\rangle : x \in \mathcal{R}\}$ on $\mathcal{D}_{\mathcal{A}} \times \mathcal{D}_{\mathcal{B}}^-$ is weakly bounded. Since $\mathcal{D}_{\mathcal{A}}$ and $\mathcal{D}_{\mathcal{B}}$ are Frechet spaces, this family is equicontinuous (Schäfer [1], III, 5.1, Corollary 2). Because $\mathcal{A}$ and $\mathcal{B}$ are directed O-vector spaces, the latter implies that there are $a \in \mathcal{A}$ and $b \in \mathcal{B}$ such that $|\langle x\varphi, \psi\rangle| \leqq \|a\varphi\| \|b\psi\|$ for all $\varphi \in \mathcal{D}(\mathcal{A})$, $\psi \in \mathcal{D}(\mathcal{B})$ and $x \in \mathcal{R}$. Thus $\mathcal{R} \subseteqq \mathcal{U}_{a,b}$. □

There are several easy, but important consequences of this proposition. We state some of them as corollaries.

Corollary 4.2.2. *Let $\mathcal{A}$ and $\mathcal{B}$ be O-families in the Hilbert space $\mathcal{H}$, and let $\mathcal{L}$ be a linear subspace of $\mathcal{L}(\mathcal{D}_\mathcal{A}, \mathcal{D}_\mathcal{B}^+)$. Suppose that the graph topologies $t_\mathcal{A}$ and $t_\mathcal{B}$ are metrizable. Then the locally convex space $\mathcal{L}[\tau_b]$ has a fundamental sequence of bounded sets. The topologies τ_b and τ_{in} on $\mathcal{L}$ have the same families of bounded sets, and the topology τ_{in} is the bornological topology associated with τ_b.*

Proof. By Proposition 2.2.13, there is no loss of generality to assume that $\mathcal{A}$ and $\mathcal{B}$ are directed O-vector spaces. Then there are sequences (a_n) and (b_n) as in Proposition 4.2.1. By (i) $\leftrightarrow$ (iv) in this proposition, $\{\mathcal{U}_{na_n,b_n} \cap \mathcal{L} : n \in N\}$ is a fundamental sequence of bounded sets in $\mathcal{L}[\tau_b]$. For the last assertion it suffices to recall that $\mathcal{L}[\tau_{in}]$ is bornological as noted in Section 1.2. $\square$

Corollary 4.2.3. *Suppose $\mathcal{A}$ is an O-family with metrizable graph topology $t_\mathcal{A}$.*

(i) *If $\mathcal{L}$ is a cofinal $*$-vector subspace of $\mathcal{L}(\mathcal{D}_\mathcal{A}, \mathcal{D}_\mathcal{A}^+)$, then the bornological spaces associated with $\mathcal{L}[\tau_\mathcal{D}]$ and with $\mathcal{L}[\tau_\mathcal{N}]$ coincide with $\mathcal{L}[\tau_\mathcal{O}]$.*

(ii) *If in addition $\mathcal{A}$ is an O*-algebra, then the topology $\tau^\mathcal{O}$ on $\mathcal{A}$ coincides with the bornological topologies associated with $\tau^\mathcal{D}$ and with $\tau^\mathcal{N}$.*

Proof. (i): Since $\tau_\mathcal{D} = \tau_b$ by definition and $\tau_\mathcal{O} = \tau_{in}$ in $\mathcal{L}$ by Proposition 3.3.11, the assertion for $\tau_\mathcal{D}$ follows at once from Corollary 4.2.2 applied with $\mathcal{A} = \mathcal{B}$. Since $\tau_\mathcal{D} \subseteqq \tau_\mathcal{N} \subseteqq \tau_\mathcal{O}$ on $\mathcal{L}$, $\tau_\mathcal{O}$ is also the bornological topology associated with $\tau_\mathcal{N}$. (ii) follows quite similarly if we use Proposition 3.3.14 and Corollary 4.2.2 in case $\mathcal{B} = \mathbb{B}(\mathcal{H})$. $\square$

Remark 1. We mention another fact of similar nature which follows immediately from the Banach-Steinhaus theorem. Suppose $\mathcal{A}$ is an O-vector space such that the locally convex space $\mathcal{D}_\mathcal{A}$ is barrelled. If $\mathcal{R}$ is a subset of $\mathcal{A}$ which is bounded in the strong-operator topology, then R is bounded in $\mathcal{A}[\tau^\mathcal{D}]$.

The following example shows that the equivalence of (i) and (iii) in Proposition 4.2.1 is no longer true if the assumption that the graph topologies are metrizable is omitted.

Example 4.2.4. Let $\mathcal{D}$ be the domain of all finite sequences in the Hilbert space $\mathcal{H} := l^2(\mathbb{N})$. Let $x = (x_n : n \in \mathbb{N})$ be a complex sequence. We define $\alpha_k(x) := 1/k \operatorname{card}\{n \in \mathbb{N} : n \leqq k$ and $x_n \neq 0\}$, $k \in \mathbb{N}$. We also denote by x the diagonal operator on $\mathcal{D}$ defined by the sequence $x = (x_n : n \in \mathbb{N})$, i.e., $x(\varphi_n) := (x_n\varphi_n)$ for $(\varphi_n) \in \mathcal{D}$. Let $\mathcal{A}$ be the set of all operators $\lambda I + x$, where $\lambda \in \mathbb{C}$ and x is a complex sequence satisfying $\lim_{k\to\infty} \alpha_k(x) = 0$. Since $\alpha_k(x + y) \leqq \alpha_k(x) + \alpha_k(y)$ and $\alpha_k(xy) \leqq \alpha_k(x)$ for $k \in \mathbb{N}$ and arbitrary sequences x and y, $\mathcal{A}$ is an O*-algebra. It is easy to check that $\mathcal{A}$ is closed. Let $e_k := (\delta_{kn} : n \in \mathbb{N})$ for $k \in \mathbb{N}$ and $\mathcal{R} := \{ke_k : k \in \mathbb{N}\}$. Corollary 2.3.11 implies that each bounded set in $\mathcal{D}_\mathcal{A}$ is contained in some $\mathcal{D}_k := \{(\varphi_n) \in \mathcal{D} : \varphi_n = 0 \text{ for all } n \geqq k\}$, $k \in \mathbb{N}$. Therefore, $\mathcal{R}$ is bounded in $\mathcal{A}[\tau_\mathcal{D}]$. From the definition of $\mathcal{A}$ it is clear that $\mathcal{R}$ is not contained in one of the sets $\mathcal{U}_{a,b}$ with $a, b \in \mathcal{A}$. Moreover, we have $t_\mathcal{A} \neq t_+$ in this example. $\bigcirc$

For O*-algebras with metrizable graph topologies, a similar result as Proposition 3.3.13 is valid for the topologies $\tau_\mathcal{O}$ and $\tau^\mathcal{O}$ as well.

Proposition 4.2.5. *Suppose $\mathcal{A}$ is an O*-algebra and $\mathcal{D}_{\mathcal{A}}$ is metrizable. Then the multiplication is jointly continuous in $\mathcal{A}[\tau_{\mathcal{O}}]$ if and only if the topologies $\tau_{\mathcal{O}}$ and $\tau^{\mathcal{O}}$ on $\mathcal{A}$ coincide.*

Proof. Suppose first that the multiplication is jointly continuous in $\mathcal{A}[\tau_{\mathcal{O}}]$. Let $\mathcal{R}$ be a bounded subset of $\mathcal{A}[\tau_{\mathcal{O}}]$. Since the multiplication is jointly continuous and the involution is continuous in $\mathcal{A}[\tau_{\mathcal{O}}]$, the set $\mathcal{R}_1 := \{x^+x : x \in \mathcal{R}\}$ is also bounded in $\mathcal{A}[\tau_{\mathcal{O}}]$. Recall that $\tau_{\mathcal{O}}$ is the topology τ_{in} on $\mathcal{A}$ when $\mathcal{A}$ is considered as a linear subspace of $\mathcal{L}(\mathcal{D}_{\mathcal{A}}, \mathcal{D}_{\mathcal{A}}^+)$. Therefore, by Proposition 4.2.1, there is an $a \in \mathcal{A}$ such that $\mathcal{R}_1 \subseteq \mathcal{U}_{a,a}$. But $x^+x \in \mathcal{U}_{a,a}$ clearly implies that $x \in \mathcal{U}_{a,I}$. Hence $\mathcal{R} \subseteq \mathcal{U}_{a,I}$, and $\mathcal{R}$ is bounded in the topology $\tau^{\mathcal{O}}$. Because $\tau_{\mathcal{O}} \subseteq \tau^{\mathcal{O}}$ and $\mathcal{A}[\tau_{\mathcal{O}}]$ is a bornological space (by Proposition 1.5.2, (vi)), the preceding implies that $\tau_{\mathcal{O}} = \tau^{\mathcal{O}}$ on $\mathcal{A}$.

Now we prove the converse direction. Suppose that $\tau_{\mathcal{O}} = \tau^{\mathcal{O}}$ on $\mathcal{A}$. Then, it is sufficient to show that the bilinear mapping $T: (x, y) \to xy$ of $\mathcal{A}[\tau^{\mathcal{O}}] \times \mathcal{A}[\tau^{\mathcal{O}}]$ into $\mathcal{A}[\tau^{\mathcal{O}}]$ is continuous. The topology $\tau^{\mathcal{O}}$ is the topology τ_{in} on $\mathcal{A}$ if we consider $\mathcal{A}$ as a linear subspace of $\mathcal{L}(\mathcal{D}_{\mathcal{A}}, \mathcal{H})$. Thus, because $\mathcal{D}_{\mathcal{A}}$ is metrizable, the locally convex space $\mathcal{A}[\tau^{\mathcal{O}}]$ is the inductive limit of a sequence of normed spaces. Hence $\mathcal{A}[\tau^{\mathcal{O}}]$ is a DF-space (KÖTHE [1], § 29, 5., (4)). It therefore suffices to prove that T is hypocontinuous (KÖTHE [2], § 40, 2., (10)). That is, we have to show that for each bounded subset $\mathcal{R}$ of $\mathcal{A}[\tau^{\mathcal{O}}]$ the families of mappings $\{{}_xT: y \to xy; x \in \mathcal{R}\}$ and $\{T_x: y \to yx; x \in \mathcal{R}\}$ of $\mathcal{A}[\tau^{\mathcal{O}}]$ into $\mathcal{A}[\tau^{\mathcal{O}}]$ are equicontinuous. Since the involution is continuous in $\mathcal{A}[\tau^{\mathcal{O}}]$ (because of $\tau_{\mathcal{O}} = \tau^{\mathcal{O}}$) and T_x can be decomposed as $y \to y^+ \to x^+y^+ \to (x^+y^+)^+ = yx$, it is enough to prove this for the family $\{{}_xT: x \in \mathcal{R}\}$.

Because the set $\mathcal{R}$ is bounded in $\mathcal{A}[\tau^{\mathcal{O}}]$, there is a $b \in \mathcal{A}$ by Proposition 4.2.1, (ii) $\to$ (iv), such that $\mathcal{R} \subseteq \mathcal{U}_{b,I}$. Since $\mathcal{A}[\tau_{\mathcal{O}}]$ is a topological algebra by Proposition 3.3.10 and $\tau_{\mathcal{O}} = \tau^{\mathcal{O}}$ by assumption, the left multiplication is continuous in $\mathcal{A}[\tau^{\mathcal{O}}]$. Therefore, if $a \in \mathcal{A}$, then $\mathcal{U}_{a,I}$ and hence also $b\mathcal{U}_{a,I}$ is bounded in $\mathcal{A}[\tau^{\mathcal{O}}]$. Applying once more Proposition 4.2.1, there is an operator $a_1 \in \mathcal{A}$ such that $b\mathcal{U}_{a,I} \subseteq \mathcal{U}_{a_1,I}$. The latter and $\mathcal{R} \subseteq \mathcal{U}_{b,I}$ give $\|xy\varphi\| \leqq \|by\varphi\| \leqq \|a_1\varphi\|$ for all $\varphi \in \mathcal{D}(\mathcal{A})$, $y \in \mathcal{U}_{a,I}$ and $x \in \mathcal{R}$. Hence $\mathfrak{l}^{a_1}(xy) \leqq \mathfrak{l}^a(y)$ for $y \in \mathcal{A}^a$ and for all $x \in \mathcal{R}$. This shows that $\{{}_xT \restriction \mathcal{A}^a : x \in \mathcal{R}\}$ is an equicontinuous family of mappings of the normed space $(\mathcal{A}^a, \mathfrak{l}^a)$ into the normed space $(\mathcal{A}^{a_1}, \mathfrak{l}^{a_1})$. Recall that $\mathcal{A}[\tau^{\mathcal{O}}]$ was the inductive limit of the family of normed spaces $\{(\mathcal{A}^a, \mathfrak{l}^a): a \in \mathcal{A}\}$. Therefore, by the properties of the inductive limit, the preceding implies that $\{{}_xT: x \in \mathcal{R}\}$ is an equicontinuous family on $\mathcal{A}[\tau^{\mathcal{O}}]$ which completes the proof. □

Corollary 4.2.6. *Let $\mathcal{A}$ be an O*-algebra with metrizable graph topology. If the space $\mathcal{A}[\tau_{\mathcal{O}}]$ is complete, then we have $\tau_{\mathcal{O}} = \tau^{\mathcal{O}}$ on $\mathcal{A}$.*

Proof. Since $\mathrm{t}_{\mathcal{A}}$ is metrizable, $\mathcal{A}[\tau_{\mathcal{O}}]$ is the inductive limit of a sequence of normed spaces and hence a DF-space (KÖTHE [1], § 29, 5., (4)). Being complete and bornological, $\mathcal{A}[\tau_{\mathcal{O}}]$ is barrelled. The bilinear mapping $(x, y) \to xy$ of $\mathcal{A}[\tau_{\mathcal{O}}] \times \mathcal{A}[\tau_{\mathcal{O}}]$ in $\mathcal{A}[\tau_{\mathcal{O}}]$ is separately continuous (by Proposition 3.3.10); hence it is continuous, since $\mathcal{A}[\tau_{\mathcal{O}}]$ is a barrelled DF-space (KÖTHE [2], § 40, 2., (11)). By Proposition 4.2.5 this implies $\tau_{\mathcal{O}} = \tau^{\mathcal{O}}$ on $\mathcal{A}$. □

4.3. Commutatively Dominated Frechet Domains

By a *commutatively dominated Frechet domain* we mean a dense linear subspace $\mathcal{D}$ of a Hilbert space $\mathcal{H}$ such that $\mathcal{D}[\mathrm{t}_+]$ is a Frechet space and such that the O*-algebra $\mathcal{L}^+(\mathcal{D})$ is commutatively dominated.

In this section we assume that $\mathcal{D}$ is a commutatively dominated Frechet domain in the Hilbert space $\mathcal{H}$.

By applying Proposition 2.2.17 to $\mathcal{A} := \mathcal{L}^+(\mathcal{D})$ it follows that $\mathcal{D}$ is of the form set out in Example 2.2.16. This means that there exist a self-adjoint operator $A = \int \lambda \, dE(\lambda)$ on $\mathcal{H}$ and a sequence $(h_n : n \in \mathbb{N})$ of measurable a.e. finite real functions h_n on $\mathbb{R}$ satisfying

$$h_1(\cdot) \geqq 1 \text{ and } h_n(\cdot)^2 \leqq h_{n+1}(\cdot) \quad \text{a.e. on } \mathbb{R} \text{ for all } n \in \mathbb{N} \tag{1}$$

and $\mathcal{D} = \bigcap_{n \in \mathbb{N}} \mathcal{D}\big(h_n(A)\big)$. Then the operators $a_n := h_n(A) \upharpoonright \mathcal{D}$, $n \in \mathbb{N}$, belong to $\mathcal{L}^+(\mathcal{D})$, and the graph topology t_+ of $\mathcal{L}^+(\mathcal{D})$ is generated by the directed family of seminorms $\{\|\cdot\|_{a_n} : n \in \mathbb{N}\}$. We shall keep these notations and assumptions fixed throughout this section.

For a convenient formulation of our results, the following conditions concerning the sequence $(h_n : n \in \mathbb{N})$ are useful:

(*) For each positive sequence $\gamma = (\gamma_n : n \in \mathbb{N})$, there is a number $r = r_\gamma \in \mathbb{N}$ such that all functions h_n, $n \in \mathbb{N}$, are essentially bounded on the set $\mathfrak{X}(\gamma, r) := \{t \in \mathbb{R} : |h_k(t)| \leqq \gamma_k \text{ for } k = 1, \ldots, r\}$.

(**) For each positive sequence $\delta = (\delta_n : n \in \mathbb{N})$, there is a number $s = s_\delta \in \mathbb{N}$ such that all functions h_n, $n \in \mathbb{N}$, are essentially bounded on the set $\mathfrak{Y}(\delta, s) := \left\{t \in \mathbb{R} : \sum_{k=1}^{s} \delta_k \, |h_k(t)|^2 \leqq 1\right\}$.

Since obviously $\mathfrak{Y}(\delta, r) \subseteq \mathfrak{X}\big((\delta_n^{-1/2}), r\big)$ and $\mathfrak{X}(\gamma, r) \subseteq \mathfrak{Y}\big((2^{-n}\gamma_n^{-2}), r\big)$ for arbitrary $r \in \mathbb{N}$ and positive sequences $\delta = (\delta_n : n \in \mathbb{N})$ and $\gamma = (\gamma_n : n \in \mathbb{N})$, (*) and (**) are equivalent for each sequence $(h_n : n \in \mathbb{N})$.

Proposition 4.3.1. *Let $\mathcal{L}$ be a cofinal *-vector subspace of $\mathcal{L}(\mathcal{D}, \mathcal{D}^+)$. Suppose that the sequence $(h_n : n \in \mathbb{N})$ satisfies condition (*). Then $\tau_{\mathcal{D}} = \tau_{\mathcal{N}}$ on $\mathcal{L}$.*

Proof. In this proof we use the notation of Section 4.1. Let $(\varepsilon_n : n \in \mathbb{N})$ be a given positive sequence. By Propositions 4.1.1 and 4.1.3 it is sufficient to show that there is a positive sequence $(\delta_n : n \in \mathbb{N})$ such that $\mathcal{U}_{(\delta_n)} \subseteq \mathcal{V}_{(5\varepsilon_n)}$.

Fix a $k \in \mathbb{N}$ and consider the set $\mathfrak{Z}_{km} := \left\{t \in \mathbb{R} : \sum_{n=1}^{m} \varepsilon_n h_n(t)^2 \leqq 4^k h_k(t)^2\right\}$ for $m \in \mathbb{N}$, $m > k$. By (1), we have $h_{k+l}(t)\, h_k(t)^2 \leqq h_{k+l}(t)^2$ a.e. on $\mathbb{R}$ for $l \in \mathbb{N}$ and hence

$$\begin{aligned} \mathfrak{Z}_{km} &\subseteq \{t \in \mathbb{R} : \varepsilon_{k+l} h_{k+l}(t)^2 \leqq 4^k h_k(t)^2 \text{ for } l = 1, \ldots, m-k\} \\ &\subseteq \{t \in \mathbb{R} : \varepsilon_{k+l} h_{k+l}(t) \leqq 4^k \text{ for } l = 1, \ldots, m-k\} \subseteq \mathfrak{X}\big((\gamma_n), m\big), \end{aligned}$$

where $\gamma_n := 4^k \varepsilon_{k+1}^{-1}$ for $n \in \mathbb{N}$, $n \leqq k$, and $\gamma_n := 4^k \varepsilon_n^{-1}$ for $n \in \mathbb{N}$, $n \geqq k+1$. From condition (*), there is a number $r_k \in \mathbb{N}$ such that all functions h_n, $n \in \mathbb{N}$, are essentially bounded on $\mathfrak{Z}_{kr_k}$. There is no loss of generality to assume that $r_k \geqq k$. Let $\mathcal{H}_k := E(\mathfrak{Z}_{kr_k})\, \mathcal{H}$. Then $\mathcal{H}_k \subseteq \bigcap_{n \in \mathbb{N}} \mathcal{D}\big(h_n(A)\big) = \mathcal{D}$ and $a_n \upharpoonright \mathcal{H}_k \equiv h_n(A) \upharpoonright \mathcal{H}_k$ is bounded for each $n \in \mathbb{N}$.

We now choose a positive sequence $(\delta_n : n \in \mathbb{N})$ such that

$$\delta_n \leqq \varepsilon_n \quad \text{for} \quad n \in \mathbb{N} \quad \text{and} \quad \delta_n \, \|a_n \upharpoonright \mathcal{H}_k\|^2 \leqq \varepsilon_1 2^{-n} \quad \text{for} \quad n, k \in \mathbb{N},\, n > r_k. \tag{2}$$

(Since $r_k \geqq k$ for $k \in \mathbb{N}$, the latter is possible.) Our aim is to show that $\mathcal{U}_{(\delta_n)} \subseteqq \mathcal{V}_{(5\varepsilon_n)}$. For this suppose $x \in \mathcal{U}_{(\delta_n)}$.

Since the topology t_+ is determined by the directed family of seminorms $\{\|\cdot\|_{a_n} : n \in \mathbb{N}\}$, there are an $m \in \mathbb{N}$ and an $\alpha > 0$ such that $|\langle x\zeta, \eta\rangle| \leqq \alpha \|a_m\zeta\| \|a_m\eta\|$ for $\zeta, \eta \in \mathcal{D}$. We choose a $k \in \mathbb{N}$ such that $k \geqq m$ and $2^k \geqq \lambda\varepsilon_m^{-1/2}$. Then, for $\zeta, \eta \in \mathcal{D}$,

$$|\langle x\zeta, \eta\rangle| \leqq \left(\sum_{n=1}^{k} \varepsilon_n \|a_n\zeta\|^2\right)^{1/2} 2^k \|a_k\eta\|. \tag{3}$$

Suppose $\varphi \in \mathcal{D}$. Then $\varphi_1 := E(\mathfrak{Z}_{kr_k})\varphi \in \mathcal{H}_k$. Using $x \in \mathcal{U}_{(\delta_n)}$ and (2), we obtain

$$\begin{aligned} |\langle x\varphi_1, \varphi_1\rangle| &\leqq \sum_{n=1}^{\infty} \delta_n \|a_n\varphi_1\|^2 \leqq \sum_{n=1}^{r_k} \delta_n \|a_n\varphi_1\|^2 + \sum_{n>r_k} \delta_n \|a_n \restriction \mathcal{H}_k\|^2 \|\varphi_1\|^2 \\ &\leqq \sum_{n=1}^{r_k} \varepsilon_n \|a_n\varphi_1\|^2 + \sum_{n>r_k} 2^{-n}\varepsilon_1 \|\varphi_1\|^2 \leqq \sum_{n=1}^{r_k} 2\varepsilon_n \|a_n\varphi\|^2. \end{aligned} \tag{4}$$

Let $\varphi_2 := \varphi - \varphi_1$. Since $\varphi_2 = E(\mathbb{R} \setminus \mathfrak{Z}_{kr_k})\varphi$, it follows from (3) combined with the definition of $\mathfrak{Z}_{kr_k}$ that

$$\begin{aligned} |\langle x\varphi_1, \varphi_2\rangle| &\leqq \left(\sum_{n=1}^{k} \varepsilon_n \|a_n\varphi_1\|^2\right)^{1/2} \langle 4^k h_k(A)^2 \varphi_2, \varphi_2\rangle^{1/2} \\ &\leqq \left(\sum_{n=1}^{r_k} \varepsilon_n \|a_n\varphi_1\|^2\right)^{1/2} \left\langle \sum_{n=1}^{r_k} \varepsilon_n a_n^2 \varphi_2, \varphi_2\right\rangle^{1/2} \leqq \sum_{n=1}^{r_k} \varepsilon_n \|a_n\varphi\|^2. \end{aligned} \tag{5}$$

Similar inequalities hold for $|\langle x\varphi_2, \varphi_1\rangle|$ and for $|\langle x\varphi_2, \varphi_2\rangle|$. Since

$$\langle x\varphi, \varphi\rangle = \langle x\varphi_1, \varphi_1\rangle + \langle x\varphi_1, \varphi_2\rangle + \langle x\varphi_2, \varphi_1\rangle + \langle x\varphi_2, \varphi_2\rangle,$$

(4), (5) and these estimations give

$$|\langle x\varphi, \varphi\rangle| \leqq \sum_{n=1}^{r_k} 5\varepsilon_n \|a_n\varphi\|^2 \quad \text{for} \quad \varphi \in \mathcal{D}.$$

This proves that $x \in \mathcal{V}_{(5\varepsilon_n)}$. □

The next proposition gives some converse of the previous proposition.

Proposition 4.3.2. *Suppose that the sequence $(h_n : n \in \mathbb{N})$ does not fulfill condition $(*)$. Then there exists a strongly positive linear functional $f \not\equiv 0$ on $\mathcal{L}(\mathcal{D}, \mathcal{D}^+)$ such that $f \restriction \mathbf{B}(\mathcal{D}) \equiv 0$. If $\mathcal{L}$ is a $*$-vector subspace of $\mathcal{L}(\mathcal{D}, \mathcal{D}^+)$ which contains the operator $E(\mathfrak{Z})$ for each measurable subset $\mathfrak{Z}$ of $\mathbb{R}$, then $f \restriction \mathcal{L}$ is not continuous on $\mathcal{L}[\tau_\mathcal{D}]$. In particular, $\tau_\mathcal{D} \neq \tau_\mathcal{N}$ on $\mathcal{L}$.*

Proof. Since $(h_n : n \in \mathbb{N})$ does not satisfy $(*)$ by assumption, there exist a positive sequence $\gamma = (\gamma_n : n \in \mathbb{N})$ and a sequence $(m_k : k \in \mathbb{N})$ of natural numbers such that the function $h_{m_{k+1}}$ is not essentially bounded on $\mathfrak{X}(\gamma, k)$ for each $k \in \mathbb{N}$. For notational simplicity, let us assume that $m_k = k$ for $k \in \mathbb{N}$. Then, for arbitrary $k, n \in \mathbb{N}$, there are measurable subsets $\mathfrak{J}_{kn} \subseteqq \mathfrak{X}(\gamma, k)$ such that $E(\mathfrak{J}_{kn}) \neq 0$, $\mathfrak{J}_{k,n+1} \subseteqq \mathfrak{J}_{kn}$ and

$$h_{k+1}(t) \geqq n \quad \text{on} \quad \mathfrak{J}_{kn}. \tag{6}$$

Let φ_{kn} be a unit vector in $E(\mathfrak{J}_{kn})\,\mathcal{D}$. Of course, $\varphi_{kn} \in \mathcal{D}$. Since $\mathfrak{J}_{kn} \subseteq \mathfrak{X}(\gamma, k)$, we have $h_l(t) \leqq \gamma_l$ on $\mathfrak{J}_{kn}$ and hence

$$\|a_l \varphi_{kn}\| \leqq \gamma_l \quad \text{for all} \quad l, k, n \in \mathbb{N}, l \leqq k. \tag{7}$$

We take an ultrafilter $\mathbb{U}$ in $\mathbb{N} \times \mathbb{N}$ which contains all sets

$$\mathbb{N}_{l,\alpha} := \{(k, m) \in \mathbb{N} \times \mathbb{N} : k \geqq l \text{ and } m \geqq \alpha_k\}$$

for each positive sequence $\alpha = (\alpha_k : k \in \mathbb{N})$ and for each $l \in \mathbb{N}$. Suppose $x \in \mathcal{L}(\mathcal{D}, \mathcal{D}^+)$. Then there are numbers $l \in \mathbb{N}$ and $\lambda > 0$ such that $|\langle x\varphi, \varphi\rangle| \leqq \lambda \|a_l \varphi\|^2$ for $\varphi \in \mathcal{D}$. By (7), $|\langle x\varphi_{kn}, \varphi_{kn}\rangle| \leqq \lambda\gamma_l^2$ if $l, k, n \in \mathbb{N}$, $l \leqq k$. Therefore, $|\langle x\varphi_{kn}, \varphi_{kn}\rangle| \leqq \lambda\gamma_l^2$ if $(k, n) \in \mathbb{N}_{l,\alpha}$ for any positive sequence α. Hence $f_{\mathbb{U}}(x) := \lim_{\mathbb{U}} \langle x\varphi_{kn}, \varphi_{kn}\rangle$ is finite. Clearly, $f_{\mathbb{U}}(\cdot)$ is a strongly positive linear functional on $\mathcal{L}(\mathcal{D}, \mathcal{D}^+)$ and $f_{\mathbb{U}}(I) = 1$. We prove that $f_{\mathbb{U}} \restriction \mathbb{B}(\mathcal{D}) \equiv 0$. Suppose $c \in \mathbb{B}(\mathcal{D})$. Let $\varepsilon > 0$ be given. By Lemma 3.1.2, the operator ca_n is bounded for each $n \in \mathbb{N}$. Let $k \in \mathbb{N}$. We choose $n_k \in \mathbb{N}$ such that $\|ca_{k+1}\| \leqq \varepsilon n_k$. Recall that $h_{k+1}(A)$ has a bounded inverse (because $h_{k+1}(\cdot) \geqq 1$ a.e. by (1)) and $h_{k+1}(A)^{-1}\varphi_{kn} \in \mathcal{D}$. From (6), $\|h_{k+1}(A)^{-1}\varphi_{kn}\| \leqq n_k^{-1}$ and so $|\langle c\varphi_{kn}, \varphi_{kn}\rangle| \leqq \|c\varphi_{kn}\| = \|ca_{k+1}h_{k+1}(A)^{-1}\varphi_{kn}\| \leqq \|ca_{k+1}\|\, n_k^{-1} \leqq \varepsilon$ for all $n \geqq n_k$. Since $\varepsilon > 0$ was arbitrary and $\mathbb{N}_{1,(n_k)} \in \mathbb{U}$, this yields $f_{\mathbb{U}}(c) = 0$.

Now let $\mathcal{L}$ be as above. We show that $f_{\mathbb{U}} \restriction \mathcal{L}$ is not continuous on $\mathcal{L}[\tau_{\mathcal{D}}]$. Let $\mathcal{M}$ be a bounded subset of $\mathcal{D}[t_+]$. Then $\lambda_k := \sup_{\varphi \in \mathcal{M}} \|a_k\varphi\| < \infty$ for $k \in \mathbb{N}$. It suffices to show that $f_{\mathbb{U}}$ is unbounded on the 0-neighbourhood $\mathcal{U}'_{\mathcal{M}} := \{x \in \mathcal{L} : p'_{\mathcal{M}}(x) \leqq 1\}$ in $\mathcal{L}[\tau_{\mathcal{D}}]$. Let δ be a given positive number. For $k \in \mathbb{N}$, we take a number $n_k \in \mathbb{N}$ such that $\delta\lambda_{k+1}^2 n_k^{-2} \leqq 2^{-k}$. Put $\mathfrak{Z} := \bigcup_{k \in \mathbb{N}} \mathfrak{J}_{k,n_k}$ and $x := \delta E(\mathfrak{Z})$. By assumption, $x \in \mathcal{L}$. If $n \geqq n_k$ and $k, n \in \mathbb{N}$, then $\mathfrak{J}_{kn} \subseteq \mathfrak{J}_{k,n_k}$ and hence $\varphi_{kn} \in E(\mathfrak{J}_{kn})\,\mathcal{D} \subseteq E(\mathfrak{Z})\,\mathcal{D}$, so that $\langle x\varphi_{kn}, \varphi_{kn}\rangle = \delta$. Applying once more that $\mathbb{N}_{1,(n_k)} \in \mathbb{U}$, it follows that $f_{\mathbb{U}}(x) = \delta$. Our proof is complete once we have shown that $x \in \mathcal{U}'_{\mathcal{M}}$. By (6),

$$n_k \|E(\mathfrak{J}_{k,n_k})\,\varphi\| \leqq \|a_{k+1}E(\mathfrak{J}_{k,n_k})\,\varphi\| \leqq \|a_{k+1}\varphi\| \leqq \lambda_{k+1}$$

for $\varphi \in \mathcal{M}$. Therefore,

$$p'_{\mathcal{M}}(x) = \sup_{\varphi \in \mathcal{M}} \delta\, \|E(\mathfrak{Z})\,\varphi\|^2 \leqq \sup_{\varphi \in \mathcal{M}} \sum_{k=1}^{\infty} \delta\, \|E(\mathfrak{J}_{k,n_k})\,\varphi\|^2 \leqq \sum_{k=1}^{\infty} \delta n_k^{-2}\lambda_{k+1}^2 \leqq \sum_{k=1}^{\infty} 2^{-k} = 1$$

which means that $x \in \mathcal{U}'_{\mathcal{M}}$. Being a strongly positive linear functional on $\mathcal{L}$, $f_{\mathbb{U}} \restriction \mathcal{L}$ is continuous on $\mathcal{L}[\tau_{\mathcal{N}}]$. Therefore, $\tau_{\mathcal{D}} \neq \tau_{\mathcal{N}}$ on $\mathcal{L}$. □

Remark 1. There is a very short argument (which is, however, not so explicit as the one given above) which proves that the functional $f_{\mathbb{U}} \restriction \mathcal{L}$ in the previous proof is not continuous on $\mathcal{L}[\tau_{\mathcal{D}}]$. From Remark 3 in 3.4 we know that there exists a net $(E(\mathfrak{Z}_i) : i \in I)$ of projections in $\mathbb{P}(\mathcal{D})$, where $\mathfrak{Z}_i$, $i \in I$, are measurable subsets of $\mathbb{R}$, which converges to I in $\mathcal{L}[\tau_{\mathcal{D}}]$. Since $f_{\mathbb{U}}(E(\mathfrak{Z}_i)) = 0$ for $i \in I$ (by $f_{\mathbb{U}} \restriction \mathbb{B}(\mathcal{D}) \equiv 0$) and $f_{\mathbb{U}}(I) = 1$, $f_{\mathbb{U}} \restriction \mathcal{L}$ is not continuous on $\mathcal{L}[\tau_{\mathcal{D}}]$.

Proposition 4.3.3. *Suppose $\mathcal{L}$ is a cofinal $*$-vector subspace of $\mathcal{L}(\mathcal{D}, \mathcal{D}^+)$. If $E(\mathfrak{Y}) \circ xE(\mathfrak{Z}) \in \mathcal{L}$ for each $x \in \mathcal{L}$ and arbitrary measurable subsets $\mathfrak{Y}$ and $\mathfrak{Z}$ of $\mathbb{R}$, then $\tau_{\mathcal{N}} = \tau_{\mathcal{D}}$ on $\mathcal{L}$.*

Proof. We apply Proposition 4.1.4. Thus it is enough to verify the condition occuring therein. We let $(\alpha_n : n \in \mathbb{N})$ be a given positive sequence. Put $\delta_n := 2^{-n}\alpha_n$, $n \in \mathbb{N}$. Suppose that $x \in \mathcal{L}$ satisfies 4.1/(1) for some (fixed) $k \in \mathbb{N}$. We then choose mutually disjoint measurable subsets $\mathfrak{J}_1, \ldots, \mathfrak{J}_k$ of $\mathbb{R}$ such that $\mathbb{R} = \bigcup_{n=1}^{k} \mathfrak{J}_n$ and such that $\max_{n=1,\ldots,k} 2^n \delta_n h_n(t)^2 = 2^l \delta_l h_l(t)^2$ a.e. on $\mathfrak{J}_l$, $l = 1, \ldots, k$. Then

$$\sum_{n=1}^{k} \delta_n h_n(t)^2 \leqq \sum_{n=1}^{k} 2^{l-n} \delta_l h_l(t)^2 \leqq 2^l \delta_l h_l(t)^2 \quad \text{a.e. on } \mathfrak{J}_l. \tag{8}$$

By assumption, $x_{lm} := E(\mathfrak{J}_l) \circ x E(\mathfrak{J}_m) \in \mathcal{L}$ for $l, m = 1, \ldots, k$. From 4.1/(1) and (8) we conclude easily that $|\langle x_{lm}\varphi, \varphi\rangle| \leqq \alpha_l \alpha_m \|a_l^2 \varphi\| \|a_m^2 \varphi\|$ for $\varphi \in \mathcal{D}$ and $l, m = 1, \ldots, k$. Since obviously $x = \sum_{l,m=1}^{k} x_{lm}$, the assumptions of Proposition 4.1.4 are satisfied; so that $\tau_{\mathcal{N}} = \tau_{\mathcal{O}}$ on $\mathcal{L}$. □

Remark 2. The three propositions proved so far in this section apply (for instance) to $\mathcal{L} = \mathcal{L}(\mathcal{D}, \mathcal{D}^+)$, $\mathcal{L} = \mathcal{L}^+(\mathcal{D})$ and more generally to any O*-algebra $\mathcal{L}$ which contains all operators a_n, $n \in \mathbb{N}$, and $E(\mathfrak{Z})$, $\mathfrak{Z}$ a measurable subset of $\mathbb{R}$. Note also that the equality $\tau_{\mathcal{N}} = \tau_{\mathcal{O}}$ on $\mathcal{L}(\mathcal{D}, \mathcal{D}^+)$ holds for each (not necessarily commutatively dominated) Frechet domain $\mathcal{D}$; see Theorem 4.4.2.

The following theorem is the main result in this section. It summarizes some of the preceding investigations.

Theorem 4.3.4. *Suppose that $\mathcal{D}$ is a commutatively dominated Frechet domain. Let $(h_n : n \in \mathbb{N})$ be the sequence of functions set out at the beginning of this section. Then the following six assertions are equivalent:*

(i) *The sequence $(h_n : n \in \mathbb{N})$ satisfies condition $(*)$.*

(ii) $\tau_{\mathcal{D}} = \tau_{\mathcal{N}}$ *on* $\mathcal{L}(\mathcal{D}, \mathcal{D}^+)$.

(iii) $\tau_{\mathcal{D}} = \tau_{\mathcal{N}}$ *on* $\mathcal{L}^+(\mathcal{D})$.

(iv) *Each strongly positive linear functional on $\mathcal{L}(\mathcal{D}, \mathcal{D}^+)$ is continuous on $\mathcal{L}(\mathcal{D}, \mathcal{D}^+)\, [\tau_{\mathcal{D}}]$.*

(v) $\mathbb{B}(\mathcal{D})$ *is dense in* $\mathcal{L}(\mathcal{D}, \mathcal{D}^+)\, [\tau_{\mathcal{N}}]$.

(vi) *Each strongly positive linear functional on $\mathcal{L}(\mathcal{D}, \mathcal{D}^+)$ which vanishes on $\mathbb{B}(\mathcal{D})$ is identically zero.*

Proof. It suffices to prove the chains of implications (i) → (ii) → (v) → (vi) → (i), (i) ↔ (iii) and (ii) → (iv) → (vi). (i) → (ii) and (i) → (iii) have been shown in Proposition 4.3.1, (iii) → (i) and (vi) → (i) follow from Proposition 4.3.2. Since $\mathbb{B}(\mathcal{D})$ is dense in $\mathcal{L}(\mathcal{D}, \mathcal{D}^+)\, [\tau_{\mathcal{D}}]$ by Corollary 3.4.2, we have (ii) → (v) and (iv) → (vi). (ii) → (iv) and (v) → (vi) are immediate consequences of the fact that strongly positive linear functionals are always continuous in the topology $\tau_{\mathcal{N}}$. □

Remark 3. Since always $\tau_{\mathcal{N}} = \tau_{\mathcal{O}}$ on $\mathcal{L}(\mathcal{D}, \mathcal{D}^+)$ by Proposition 4.3.3, Theorem 4.3.4 remains true if we replace in (ii) and in (iv) the topology $\tau_{\mathcal{N}}$ by the topology $\tau_{\mathcal{O}}$. Moreover, the theorem is also valid with $\mathcal{L}^+(\mathcal{D})$ in place of $\mathcal{L}(\mathcal{D}, \mathcal{D}^+)$. Further equivalent statements will be given in Theorem 6.2.7.

We close this section by some examples. For this it is convenient to allow a more general (but equivalent) situation than the one described at the beginning of this section.

As above, let $\mathcal{D}$ be a dense linear subspace of a Hilbert space $\mathcal{H}$, and let A be a self-adjoint operator on $\mathcal{H}$. Suppose that $(g_n : n \in \mathbb{N})$ is a sequence of measurable functions (with respect to the spectral measure of A) on $\mathbb{R}$ such that

$$\mathcal{D} = \bigcap_{n \in \mathbb{N}} \mathcal{D}(g_n(A)) \quad \text{and} \quad b_n := g_n(A) \upharpoonright \mathcal{D} \quad \text{is in} \quad \mathcal{L}^+(\mathcal{D}) \quad \text{for} \quad n \in \mathbb{N}. \tag{9}$$

We define inductively a sequence $(h_n: n \in \mathbb{N})$ of functions on $\mathbb{R}$ by $h_1(t) := 1 + |g_1(t)|^2$ and $h_{n+1}(t) := 1 + h_n(t)^2 + |g_{n+1}(t)|^2$ for $n \in \mathbb{N}$. Then the sequence $(h_n: n \in \mathbb{N})$ fulfills the assumptions set out at the beginning of this section. Indeed, (1) is obvious; so it only remains to show that $\mathcal{D} = \bigcap_{n\in\mathbb{N}} \mathcal{D}(h_n(A))$. Since $b_n \in \mathcal{L}^+(\mathcal{D})$ by assumption, $|g_n(A)|^2 \upharpoonright \mathcal{D} = b_n^+ b_n$ is in $\mathcal{L}^+(\mathcal{D})$ for $n \in \mathbb{N}$. Using this, it follows easily by induction that $\mathcal{D} \subseteq \mathcal{D}(h_n(A))$ and $h_n(A)\,\mathcal{D} \subseteq \mathcal{D}$ for all $n \in \mathbb{N}$. Combined with $\mathcal{D}(h_n(A)) \subseteq \mathcal{D}(g_n(A))$ (by the definition of h_n) and (9), this yields $\mathcal{D} = \bigcap_{n\in\mathbb{N}} \mathcal{D}(h_n(A))$.

Retaining these assumptions and notations, we have

Lemma 4.3.5. *The sequence $(g_n: n \in \mathbb{N})$ satisfies condition $(*)$ (with g_n in place of h_n) if and only if the sequence $(h_n: n \in \mathbb{N})$ does it.*

Proof. We denote by $\mathfrak{X}(\gamma, r)_g$, $\mathfrak{X}(\gamma, r)_h$, $r_{\gamma,g}$ and $r_{\gamma,h}$ the corresponding subsets and numbers occuring in $(*)$, respectively. Assume that $(*)$ is true for $(h_n: n \in \mathbb{N})$. Let $\gamma = (\gamma_n: n \in \mathbb{N})$ be a given positive sequence. Define a sequence $\alpha = (\alpha_n: n \in \mathbb{N})$ by $\alpha_1 := 1 + \gamma_1^2$ and $\alpha_{n+1} := 1 + \alpha_n^2 + \gamma_{n+1}^2$ for $n \in \mathbb{N}$. From the definitions of h_n, $n \in \mathbb{N}$, we see that $\mathfrak{X}(\gamma, k)_g \subseteq \mathfrak{X}(\alpha, k)_h$ for $k \in \mathbb{N}$. Since $|g_n(\cdot)| \leqq h_n(\cdot)$ for $n \in \mathbb{N}$, this implies that all functions g_n, $n \in \mathbb{N}$, are essentially bounded on $\mathfrak{X}(\gamma, r)_g$ if $r := r_{\alpha,h}$.

Conversely, suppose $(g_n: n \in \mathbb{N})$ satisfies $(*)$. Since obviously $\mathfrak{X}(\gamma, k)_h \subseteq \mathfrak{X}(\gamma, k)_g$ for $k \in \mathbb{N}$, all functions g_n, $n \in \mathbb{N}$, are essentially bounded on $\mathfrak{X}(\gamma, r)_h$, where $r := r_{\gamma,g}$. From the definition of the functions h_n it follows immediately that each function h_n, $n \in \mathbb{N}$, is also essentially bounded on $\mathfrak{X}(\gamma, r)_h$. That is, $(h_n: n \in \mathbb{N})$ fulfills $(*)$. $\square$

Example 4.3.6. Let $g_n(t) := t^n$ for $t \in \mathbb{R}$ and $n \in \mathbb{N}$. Then $\mathcal{D} = \bigcap_{n\in\mathbb{N}} \mathcal{D}(A^n)$. In that case $(*)$ is obviously true for $(g_n: n \in \mathbb{N})$ and hence for $(h_n: n \in \mathbb{N})$ by Lemma 4.3.5. Therefore, by Proposition 4.3.1, we have $\tau_{\mathcal{D}} = \tau_{\mathcal{N}}$ on each cofinal $*$-vector subspace of $\mathcal{L}(\mathcal{D}, \mathcal{D}^+)$. Moreover, $\tau_{\mathcal{D}} = \tau_{\mathcal{N}} = \tau_{\mathcal{O}}$ on $\mathcal{L}^+(\mathcal{D})$ and on $\mathcal{L}(\mathcal{D}, \mathcal{D}^+)$ from Proposition 4.3.3. ○

Example 4.3.7. Let $(\alpha_n: n \in \mathbb{N})$ be a fixed complex sequence which will be specified later. Let $\mathfrak{x}_k$, $k \in \mathbb{N}$, be the $\mathbb{N} \times \mathbb{N}$ -matrix

$$\mathfrak{x}_k \equiv [x_{nm}^{(k)}]_{n,m\in\mathbb{N}} = [y_1^{(k)}, \ldots, y_k^{(k)}, e_{k+1}, e_{k+2}, \ldots],$$

where the column vectors $y_l^{(k)}$ and e_n are defined by $y_l^{(k)} := (1, 2^k, 3^k, \ldots)$ for $l = 1, \ldots, k$ and $e_n := (\alpha_n, \alpha_n, \alpha_n, \ldots)$ for $n \in \mathbb{N}$, $n \geqq k + 1$. (Here the infinite sequences will be written as columns.) Let j be a bijection of $\mathbb{N}$ onto $\mathbb{N} \times \mathbb{N}$. We denote by x_k, $k \in \mathbb{N}$, the diagonal operator on the Hilbert space $\mathcal{H} := l^2(\mathbb{N})$ which corresponds to the matrix $\mathfrak{x}_k$ via the bijection j, i.e., $x_k(\varphi_n) := (x_{j(n)}^{(k)} \varphi_n)$. Set $\mathcal{D} := \bigcap_{m\in\mathbb{N}} \bigcap_{k,n\in\mathbb{N}^m} \mathcal{D}(x_n^k)$, where $x_n^k := x_{n_1}^{k_1} \ldots x_{n_m}^{k_m}$ for arbitrary multiindices $k = (k_1, \ldots, k_m)$ and $n = (n_1, \ldots, n_m)$ in $\mathbb{N}^m$. Let A be the diagonal operator on $l^2(\mathbb{N})$ defined by $A(\varphi_n) := (n\varphi_n)$. We consider each x_n^k as a function $g_{n,k}$ of A by assigning to the r-th diagonal entry of A the corresponding entry of the diagonal operator x_n^k. Writing the functions $g_{n,k}$ as a sequence $(g_n: n \in \mathbb{N})$, (9) is fulfilled, so that $\mathcal{D}$ is of the form described above. We discuss two cases separately.

Case 1: $\lim_{n\to\infty} |\alpha_n| = +\infty$.

Then the sequence $(g_n: n \in \mathbb{N})$ satisfies condition $(*)$; so, by Lemma 4.3.5 and Proposition 4.3.1, $\tau_{\mathcal{D}} = \tau_{\mathcal{N}}$ on $\mathcal{L}^+(\mathcal{D})$ and also on $\mathcal{L}(\mathcal{D}, \mathcal{D}^+)$. Let $(*)'$ denote the condition which is obtained if we replace in (*) the set $\mathfrak{X}(\gamma, r)$ by the set $\mathfrak{X}(\gamma, r)' := \{t \in \mathbb{R} : |h_r(t)| \leqq \gamma_r\}$. It is easily seen that (*)' is not valid for the sequence $(g_n : n \in \mathbb{N})$ in this case. Hence (*)' is sufficient, but not necessary for the equality of the topologies $\tau_{\mathcal{D}}$ and $\tau_{\mathcal{N}}$ on $\mathcal{L}(\mathcal{D}, \mathcal{D}^+)$ or equivalently on $\mathcal{L}^+(\mathcal{D})$.

Case 2: The sequence $(\alpha_n : n \in \mathbb{N})$ is bounded.

Then condition (*) is not true for the sequence $(g_n : n \in \mathbb{N})$, so that the topologies $\tau_{\mathcal{D}}$ and $\tau_{\mathcal{N}}$ do not coincide on $\mathcal{L}(\mathcal{D}, \mathcal{D}^+)$ and on $\mathcal{L}^+(\mathcal{D})$. ○

4.4. General Results about the Topologies $\tau_{\mathcal{D}}$, $\tau_{\mathcal{N}}$, $\tau_{\mathcal{O}}$

Theorem 4.4.1. *Suppose that $\mathcal{A}$ is an O*-algebra and $\mathcal{L}$ is a cofinal *-vector subspace of $\mathcal{L}(\mathcal{D}_{\mathcal{A}}, \mathcal{D}_{\mathcal{A}}^+)$. If $\mathcal{D}_{\mathcal{A}}$ is a Frechet-Montel space, then we have $\tau_{\mathcal{D}} = \tau_{\mathcal{N}}$ on $\mathcal{L}$.*

Proof. Since $\mathcal{D}_{\mathcal{A}}$ is metrizable, we can find a sequence $(a_n : n \in \mathbb{N})$ in $\mathcal{A}$ such that the seminorms $\|\cdot\|_{a_n}$, $n \in \mathbb{N}$, determine the topology $t_{\mathcal{A}}$ on $\mathcal{D}(\mathcal{A})$. Let $(\varepsilon_n : n \in \mathbb{N})$ be a given positive sequence. Because of Propositions 4.1.1 and 4.1.3, it suffices to show that $\mathcal{U}_{(\varepsilon_n)} \subseteqq \mathcal{V}_{(2\varepsilon_n)}$. (Here and in the proofs of the following theorems in this section we freely use the notation established in Section 4.1.) Assume to the contrary that there exists an $x \in \mathcal{L}$ such that $x \in \mathcal{U}_{(\varepsilon_n)}$ and $x \notin \mathcal{V}_{(2\varepsilon_n)}$. That x is in $\mathcal{U}_{(\varepsilon_n)}$ means that

$$|\langle x\varphi, \varphi\rangle| \leqq \sum_{n=1}^{\infty} \varepsilon_n \|a_n\varphi\|^2 \quad \text{for all} \quad \varphi \in \mathcal{D}(\mathcal{A}). \tag{1}$$

Since x is not in $\mathcal{V}_{(2\varepsilon_n)}$, for each $k \in \mathbb{N}$ there exists a vector $\varphi_k \in \mathcal{D}(\mathcal{A})$ such that

$$|\langle x\varphi_k, \varphi_k\rangle| > \sum_{n=1}^{k} 2\varepsilon_n \|a_n\varphi_k\|^2 \quad \text{for} \quad k \in \mathbb{N}. \tag{2}$$

After norming the vectors we can assume that $|\langle x\varphi_k, \varphi_k\rangle| = 2$ for $k \in \mathbb{N}$. Then we conclude from (2) that for each $n \in \mathbb{N}$

$$\sup_{k \in \mathbb{N}} \|a_n\varphi_k\| \leqq \max \{\|a_n\varphi_1\|, \ldots, \|a_n\varphi_n\|, \varepsilon_n^{-1/2}\} < \infty.$$

This shows that $(\varphi_k : k \in \mathbb{N})$ is a bounded sequence in $\mathcal{D}_{\mathcal{A}}$. Because $\mathcal{D}_{\mathcal{A}}$ is a Frechet-Montel space, there is a subsequence $(\varphi_{k_r} : r \in \mathbb{N})$ of this sequence which converges to some vector $\varphi_0 \in \mathcal{D}(\mathcal{A})$ in $\mathcal{D}_{\mathcal{A}}$. Fix $m \in \mathbb{N}$. If $k_r \geqq m$, then $\sum_{n=1}^{m} \varepsilon_n \|a_n\varphi_{k_r}\|^2 < 1$ by (2). Letting $r \to \infty$, we get $\sum_{n=1}^{m} \varepsilon_n \|a_n\varphi_0\|^2 \leqq 1$. Since $m \in \mathbb{N}$ is arbitrary,

$$\sum_{n=1}^{\infty} \varepsilon_n \|a_n\varphi_0\|^2 \leqq 1. \tag{3}$$

On the other hand, the sesquilinear form $\langle x\cdot, \cdot\rangle$ is continuous on $\mathcal{D}_{\mathcal{A}} \times \mathcal{D}_{\mathcal{A}}$, since $x \in \mathcal{L}(\mathcal{D}_{\mathcal{A}}, \mathcal{D}_{\mathcal{A}}^+)$. Therefore, $|\langle x\varphi_k, \varphi_k\rangle| = 2$ for $k \in \mathbb{N}$ implies that $|\langle x\varphi_0, \varphi_0\rangle| = 2$. Setting $\varphi = \varphi_0$ in (1), the latter and (1) contradict (3). □

Remark 1. Let $\mathcal{A}$ be as in Theorem 4.4.1, and let $\mathcal{D} := \mathcal{D}(\mathcal{A})$. Then $\mathcal{L}(\mathcal{D}, \mathcal{D}^+) = \mathcal{L}(\mathcal{D}_{\mathcal{A}}, \mathcal{D}_{\mathcal{A}}^+)$ and $\mathcal{L}^+(\mathcal{D}) = \mathcal{L}^+(\mathcal{D}_{\mathcal{A}})$, since $\mathcal{D}_{\mathcal{A}}$ is a Frechet space. Each *-vector subspace of $\mathcal{L}(\mathcal{D}, \mathcal{D}^+)$ that contains $\mathcal{A}$ is cofinal in $\mathcal{L}(\mathcal{D}, \mathcal{D}^+)$; so Theorem 4.4.1 shows that $\tau_{\mathcal{D}} = \tau_{\mathcal{N}}$ on $\mathcal{A}$, $\mathcal{L}^+(\mathcal{D})$ and on $\mathcal{L}(\mathcal{D}, \mathcal{D}^+)$.

Theorem 4.4.2. *If $\mathcal{A}$ is an O^*-algebra with metrizable graph topology, then we have $\tau_{\mathcal{N}} = \tau_{\mathcal{O}}$ on $\mathcal{L}(\mathcal{D}_{\mathcal{A}}, \mathcal{D}_{\mathcal{A}}^+)$.*

Proof. By Lemma 2.2.7 we can choose a sequence $(a_n : n \in \mathbb{N})$ of symmetric operators in $\mathcal{A}(I)$ such that $\|a_n^2\varphi\| \leqq \|a_{n+1}\varphi\|$ for $\varphi \in \mathcal{D}(\mathcal{A})$, $n \in \mathbb{N}$, and such that the seminorms $\|\cdot\|_{a_n}$, $n \in \mathbb{N}$, generate the graph topology $t_{\mathcal{A}}$. We verify the conditions stated in Proposition 4.1.4 in case $\mathcal{L} := \mathcal{L}(\mathcal{D}_{\mathcal{A}}, \mathcal{D}_{\mathcal{A}}^+)$. Given a positive sequence $(\alpha_n : n \in \mathbb{N})$, we set $\delta_n := \alpha_n$ for $n \in \mathbb{N}$. Assume that $x \in \mathcal{L}(\mathcal{D}_{\mathcal{A}}, \mathcal{D}_{\mathcal{A}}^+)$ satisfies 4.1/(1) for some $k \in \mathbb{N}$. Put $b_k := \sum_{n=1}^{k} \delta_n a_n^2$. Then, by Proposition 3.2.3, there exists an operator $y \in \mathbb{B}(\mathcal{H})$ with $\|y\| \leqq 1$ such that $x = b_k \circ y b_k$. Defining $x_{lm} := \delta_l \delta_m a_l^2 \circ y a_m^2$ for $l, m = 1, \ldots, k$, we have $x_{lm} \in \mathcal{L}(\mathcal{D}_{\mathcal{A}}, \mathcal{D}_{\mathcal{A}}^+)$ because $a_n^2 \in \mathcal{A}$ for all $n \in \mathbb{N}$. Further,

$$|\langle x_{lm}\varphi, \varphi\rangle| = \delta_l \delta_m |\langle y a_m^2 \varphi, a_l^2 \varphi\rangle| \leqq \alpha_l \alpha_m \|a_m^2\varphi\| \|a_l^2\varphi\|$$

for $\varphi \in \mathcal{D}(\mathcal{A})$ and $x = \sum_{l,m=1}^{k} x_{lm}$. This shows that the assumptions of Proposition 4.1.4 are fulfilled; hence $\tau_{\mathcal{N}} = \tau_{\mathcal{O}}$ on $\mathcal{L}(\mathcal{D}_{\mathcal{A}}, \mathcal{D}_{\mathcal{A}}^+)$. □

Combining the two preceding theorems, we obtain the following

Corollary 4.4.3. *If $\mathcal{A}$ is an O^*-algebra such that $\mathcal{D}_{\mathcal{A}}$ is a Frechet-Montel space, then $\tau_{\mathcal{D}} = \tau_{\mathcal{N}} = \tau_{\mathcal{O}}$ on $\mathcal{L}(\mathcal{D}_{\mathcal{A}}, \mathcal{D}_{\mathcal{A}}^+)$.*

The following lemma is the key ingredient for our next result concerning the equality of the topologies $\tau_{\mathcal{N}}$ and $\tau_{\mathcal{O}}$. Since it will also be used in Section 7.4, we shall prove it in a more general version than needed here.

Lemma 4.4.4. *Suppose $\mathcal{C}$ is a $*$-subalgebra of $\mathbb{B}(\mathcal{H})$ (without unit element in general). Let c and d be positive operators contained in $\mathcal{C}$, and let $\gamma, \delta \in \mathbb{R}$, $0 < \gamma < 1$, $0 < \delta < 1$. Suppose that z is an operator in $\mathcal{C}$ which satisfies*

$$|\langle z\varphi, \psi\rangle|^2 \leqq \langle (c + \gamma)\varphi, \varphi\rangle \langle (d + \delta)\psi, \psi\rangle \quad \textit{for all} \quad \varphi, \psi \in \mathcal{H}. \tag{4}$$

Then there are operators $z_1, z_2 \in \mathcal{C}$ such that $z = z_1 + z_2$ and

$$|\langle z_1\varphi, \psi\rangle|^2 \leqq \langle c\varphi, \varphi\rangle \langle d\psi, \psi\rangle \tag{5}$$

and

$$|\langle z_2\varphi, \psi\rangle| \leqq 2((\gamma\delta)^{1/2} + (\gamma\|d\|)^{1/2} + (\delta\|c\|)^{1/2}) \|\varphi\| \|\psi\| \tag{6}$$

for all $\varphi, \psi \in \mathcal{H}$. Moreover, there is an operator $y_1 \in \mathcal{C}$ such that $z_1 = d y_1 c$.

Proof. Let $\lambda := (\max\{1, \|c\|, \|d\|\})^{-1}$. Upon replacing z, c, d, γ, δ by $\lambda z, \lambda c, \lambda d, \lambda\gamma, \lambda\delta$, respectively, we can assume without loss of generality that $\|c\| \leqq 1$ and $\|d\| \leqq 1$.

Fix $\alpha \in \mathbb{R}$, $0 < \alpha < 1$. Let f be the function on the interval $[0, 1]$ defined by $f(t) := \big(t(t + \alpha)\big)^{-1/2}$ if $t \in [\varepsilon, 1]$ and $f(t) := \big(\varepsilon(\varepsilon + \alpha)\big)^{-1/2}$ if $t \in [0, \varepsilon)$, where ε is a positive number satisfying $4\varepsilon \leqq \alpha^{1/2}$ and $\varepsilon \leqq \alpha$. We approximate the continuous real function $f(t) - \varepsilon$ on $[0, 1]$ by a real polynomial $p(t)$ such that $|p(t) - (f(t) - \varepsilon)| \leqq \varepsilon$ for all $t \in [0, 1]$. Put $q_\alpha(t) := tp(t)$. We check that for $t \in [0, 1]$

$$0 \leqq q_\alpha(t) \leqq t^{1/2}(t + \alpha)^{-1/2} \tag{7}$$

and

$$0 \leqq (t + \alpha)^{1/2} \big(1 - q_\alpha(t)\big) \leqq 2\alpha^{1/2}. \tag{8}$$

Since $2\varepsilon \leqq 1/2 \leqq (1+\alpha)^{-1/2}$, we have $f(t) - 2\varepsilon \geqq f(1) - 2\varepsilon \geqq 0$ and hence

$$0 \leqq t\big(f(t) - 2\varepsilon\big) \leqq tp(t) = q_\alpha(t) \leqq tf(t) \leqq t^{1/2}(t+\alpha)^{-1/2}$$

for $t \in [0, 1]$ which proves (7). Since obviously $f(t) \leqq t^{-1}$, we have $1 - q_\alpha(t) \geqq 1 - tf(t) \geqq 0$ on $[0, 1]$. If $t \in [\varepsilon, 1]$, then

$$(t+\alpha)^{1/2}\big(1 - q_\alpha(t)\big) \leqq (t+\alpha)^{1/2}\big(1 - t(f(t) - 2\varepsilon)\big)$$
$$= (t+\alpha)^{1/2} - t^{1/2} + 2\varepsilon t(t+\alpha)^{1/2} \leqq \alpha^{1/2} + 4\varepsilon \leqq 2\alpha^{1/2},$$

because $4\varepsilon \leqq \alpha^{1/2}$. If $t \in [0, \varepsilon)$, then

$$(t+\alpha)^{1/2}\big(1 - q_\alpha(t)\big) \leqq (t+\alpha)^{1/2} < (\varepsilon+\alpha)^{1/2} \leqq 2\alpha^{1/2},$$

since $\varepsilon \leqq \alpha$. This proves (8).

We now define $z_1 := q_\delta(d)\, zq_\gamma(c)$ and $z_2 := z - z_1$. Since q_δ and q_γ are polynomials with vanishing constant coefficients, $z_1 = dy_1c$ for some $y_1 \in \mathcal{C}$. Further, z_1 and z_2 are in $\mathcal{C}$. From (4) and (7) it follows that

$$|\langle z_1\varphi, \psi\rangle|^2 = |\langle zq_\gamma(c)\, \varphi, q_\delta(d)\, \psi\rangle|$$
$$\leqq \langle (c+\gamma)\, q_\gamma(c)\, \varphi, q_\gamma(c)\, \varphi\rangle \langle (d+\delta)\, q_\delta(d)\, \psi, q_\delta(d)\, \psi\rangle \leqq \langle c\varphi, \varphi\rangle \langle d\psi, \psi\rangle$$

for $\varphi, \psi \in \mathcal{H}$ which proves (5). Applying (4), (7) and (8), we get

$$|\langle z_2\varphi, \psi\rangle| \leqq |\langle zq_\gamma(c)\, \varphi, \big(I - q_\delta(d)\big)\, \psi\rangle| + |\langle z\big(I - q_\gamma(c)\big)\, \varphi, \psi\rangle|$$
$$\leqq \langle (c+\gamma)\, q_\gamma(c)\, \varphi, q_\gamma(c)\, \varphi\rangle^{1/2} \langle (d+\delta)\big(I - q_\delta(d)\big)\, \psi, \big(I - q_\delta(d)\big)\, \psi\rangle^{1/2}$$
$$+ \langle (c+\gamma)\big(I - q_\gamma(c)\big)\, \varphi, \big(I - q_\gamma(c)\big)\, \varphi\rangle^{1/2} \langle (d+\delta)\, \psi, \psi\rangle^{1/2}$$
$$\leqq \langle c\varphi, \varphi\rangle^{1/2}\, 2\delta^{1/2}\, \|\psi\| + 2\gamma^{1/2}\, \|\varphi\|\, (\|d\|^{1/2} + \delta^{1/2})\, \|\psi\|$$

for $\varphi, \psi \in \mathcal{H}$. From this (6) follows. □

Theorem 4.4.5. *Let $\mathcal{A}$ be an O*-algebra in the Hilbert space $\mathcal{H}$. Suppose that there exists a sequence $(a_n : n \in \mathbb{N})$ of operators in $\mathcal{A}(I)$ such that $a_n\mathcal{D}(\mathcal{A})$ is dense in $\mathcal{H}$ and $\|a_n \cdot\| \leqq \|a_{n+1} \cdot\|$ for each $n \in \mathbb{N}$ and such that the family of seminorms $\{\|\cdot\|_{a_n} : n \in \mathbb{N}\}$ generates the graph topology $t_\mathcal{A}$ on $\mathcal{D}(\mathcal{A})$. Suppose that $\mathcal{C}$ is a $*$-subalgebra of $\mathbf{B}(\mathcal{H})$ with $I \in \mathcal{C}$ and $\mathcal{L} := \bigcup_{n\in\mathbb{N}} a_n^+ \circ \mathcal{C}a_n$ is a linear subspace of $\mathcal{L}(\mathcal{D}_\mathcal{A}, \mathcal{D}_\mathcal{A}^+)$ such that $\mathcal{L}_{a_n} \subseteqq a_n^+ \circ \mathcal{C}a_n$ for every $n \in \mathbb{N}$. Then $\mathcal{L}$ is a cofinal $*$-vector subspace of $\mathcal{L}(\mathcal{D}_\mathcal{A}, \mathcal{D}_\mathcal{A}^+)$ and the topologies $\tau_\mathcal{N}$ and $\tau_\mathcal{O}$ of $\mathcal{L}$ coincide.*

Proof. From the equality $(a_n^+ \circ \mathcal{C}a_n)^+ = a_n^+ \circ \mathcal{C}a_n$ for $n \in \mathbb{N}$ we conclude that $\mathcal{L}$ is a $*$-vector subspace of $\mathcal{L}(\mathcal{D}_\mathcal{A}, \mathcal{D}_\mathcal{A}^+)$. Since $I \in \mathcal{C}$ by assumption, $a_n^+ \circ Ia_n = a_n^+a_n \in \mathcal{L}$ for $n \in \mathbb{N}$. By the assumptions concerning $(a_n : n \in \mathbb{N})$, the set $\{a_n^+a_n : n \in \mathbb{N}\}$ is order-dominating for $\mathcal{L}(\mathcal{D}_\mathcal{A}, \mathcal{D}_\mathcal{A}^+)$. Hence $\mathcal{L}$ is cofinal in $\mathcal{L}(\mathcal{D}_\mathcal{A}, \mathcal{D}_\mathcal{A}^+)$, and Proposition 4.1.3 applies. Let $(\varepsilon_n : n \in \mathbb{N})$ be a given positive sequence. We choose a positive sequence $(\delta_n : n \in \mathbb{N})$ satisfying $\sum_{n=1}^{\infty} \delta_n \leqq \frac{1}{4}$ and $2^{n+4}\delta_n^{1/2} \leqq \varepsilon_n$ for all $n \in \mathbb{N}$. In order to prove that $\tau_\mathcal{N} = \tau_\mathcal{O}$ on $\mathcal{L}$, it suffices to prove according to Proposition 4.1.3 that $\mathcal{V}_{(\delta_n)} \subseteqq \mathcal{W}_{(\varepsilon_n)}$.

Define $b_k := \sum_{n=1}^{k} \delta_n a_n^+ a_n$, $k \in \mathbb{N}$. Fix an element $x \in \mathcal{V}_{(\delta_n)}$. Then there is a $k \in \mathbb{N}$ such that $|\langle x\varphi, \varphi\rangle| \leqq \langle b_k\varphi, \varphi\rangle$ for all $\varphi \in \mathcal{D}(\mathcal{A})$. If $k = 1$, then obviously $x \in \mathcal{W}_{(\varepsilon_n)}$.

Suppose now that $k \geqq 2$. From 3.2/(3) we obtain

$$|\langle x\varphi, \psi\rangle|^2 \leqq 16\langle b_k\varphi, \varphi\rangle \langle b_k\psi, \psi\rangle \quad \text{for} \quad \varphi, \psi \in \mathcal{D}(\mathcal{A}). \tag{9}$$

Since $\|a_n\cdot\| \leqq \|a_k\cdot\|$ if $n \leqq k$ and $16(\delta_1 + \cdots + \delta_k)^2 \leqq 1$, (9) implies that x is in $\mathcal{U}_{a_k}$. The same argument with b_{k-1} in place of x shows that $b_{k-1} \in \mathcal{U}_{a_k}$. Moreover, $b_{k-1} \in \mathcal{L}$, since $a_n^+ a_n \in \mathcal{L}$ for $n \in \mathbb{N}$ as noted above. By the assumption $\mathcal{L}_{a_k} \subseteq a_k^+ \circ \mathcal{C} a_k$, there exist operators z and c_k in $\mathcal{C}$ such that $x = a_k^+ \circ z a_k$ and $b_{k-1} = a_k^+ \circ c_k a_k$. From the density of $a_k\mathcal{D}(\mathcal{A})$ in $\mathcal{H}$ and $0 \leqq b_{k-1} \leqq a_k^+ a_k$ we get $0 \leqq c_k \leqq I$. Putting $x = a_k^+ \circ z a_k$ and $b_k = b_{k-1} + \delta_k a_k^+ a_k = a_k^+ \circ (c_k + \delta_k I)\, a_k$ into (9) and using once more the density of $a_k\mathcal{D}(\mathcal{A})$ in $\mathcal{H}$, it follows that assumption (4) in Lemma 4.4.4 is fulfilled in case $c = d := 4c_k$ and $\gamma = \delta := 4\delta_k$. Therefore, by this lemma, there exist operators $z_1, z_2 \in \mathcal{C}$ such that $z = z_1 + z_2$,

$$|\langle z_1\varphi, \psi\rangle|^2 \leqq 16\langle c_k\varphi, \varphi\rangle \langle c_k\psi, \psi\rangle \tag{10}$$

and

$$|\langle z_2\varphi, \psi\rangle| \leqq 10\delta_k^{1/2} \|\varphi\| \|\psi\| \tag{11}$$

for $\varphi, \psi \in \mathcal{H}$, where we used that $\|c_k\| \leqq 1$. Define $x_{k-1} := a_k^+ \circ z_1 a_k$ and $y_k := 2^k a_k^+ \circ z_2 a_k$. By (11), $|\langle y_k\varphi, \varphi\rangle| \leqq 10 \cdot 2^k \delta_k^{1/2} \|a_k\varphi\|^2 \leqq \varepsilon_k \|a_k\varphi\|^2$ for $\varphi \in \mathcal{D}(\mathcal{A})$. That is, $y_k \in \mathcal{W}_k$, where

$$\mathcal{W}_n := \{y \in \mathcal{L} : |\langle y\varphi, \varphi\rangle| \leqq \varepsilon_n \|a_n\varphi\|^2 \quad \text{for all} \quad \varphi \in \mathcal{D}(\mathcal{A})\}, \quad n \in \mathbb{N}.$$

From (10) we obtain

$$|\langle x_{k-1}\varphi, \psi\rangle|^2 \leqq 16\langle c_k a_k\varphi, a_k\varphi\rangle \langle c_k a_k\psi, a_k\psi\rangle = 16\langle b_{k-1}\varphi, \varphi\rangle \langle b_{k-1}\psi, \psi\rangle$$

for $\varphi, \psi \in \mathcal{D}(\mathcal{A})$. This shows that (9) is valid with b_k replaced by b_{k-1} and x by x_{k-1} Moreover, $x = x_{k-1} + 2^{-k} y_k$. Proceeding by induction we find elements $y_1, \ldots, y_k \in \mathcal{L}$ such that $x = 2^{-1} y_1 + \cdots + 2^{-k} y_k$ and $y_n \in \mathcal{W}_n$ for $n = 1, \ldots, k$. Hence $x \in \operatorname{aco} \underset{n\in\mathbb{N}}{\mathcal{W}_n} \equiv \mathcal{W}_{(\varepsilon_n)}$. □

Corollary 4.4.6. *Let $\mathcal{A}$, $\mathcal{H}$ and $(a_n : n \in \mathbb{N})$ be as in Theorem 4.4.5. Suppose in addition that a_n^{-1} (which exists because of $a_n \in \mathcal{A}(I)$) belongs to $\mathcal{L}^+(\mathcal{D}(\mathcal{A}))$ for each $n \in \mathbb{N}$. Let $\mathcal{L}$ be a linear subspace of $\mathcal{L}(\mathcal{D}_{\mathcal{A}}, \mathcal{D}_{\mathcal{A}}^+)$ such that $\mathcal{C} := \mathcal{L} \cap \mathbf{B}(\mathcal{H})$ is a $*$-subalgebra of $\mathbf{B}(\mathcal{H})$ with $I \in \mathcal{C}$. Suppose that $a_n^+ \circ \mathcal{C} a_n \subseteq \mathcal{L}$ and $(a_n^{-1})^+ \circ \mathcal{L} a_n^{-1} \subseteq \mathcal{L}$ for all $n \in \mathbb{N}$. Then $\mathcal{L}$ is a cofinal $*$-vector space in $\mathcal{L}(\mathcal{D}_{\mathcal{A}}, \mathcal{D}_{\mathcal{A}}^+)$ and $\tau_{\mathcal{N}} = \tau_{\mathcal{O}}$ on $\mathcal{L}$.*

Proof. The assertion follows immediately from Theorem 4.4.5 once we have shown that $\mathcal{L} = \bigcup_{n\in\mathbb{N}} a_n^+ \circ \mathcal{C} a_n$ and that $\mathcal{L}_{a_n} \subseteq a_n^+ \circ \mathcal{C} a_n$. The inclusion $a_n^+ \circ \mathcal{C} a_n \subseteq \mathcal{L}$ is one of the above assumptions. Conversely, let $x \in \mathcal{L}$. Since $\{\|\cdot\|_{a_n} : n \in \mathbb{N}\}$ is a directed family of seminorms which generates the topology $\mathrm{t}_{\mathcal{A}}$, there exists $n \in \mathbb{N}$ such that $x \in \mathcal{L}_{a_n}$. By Proposition 3.2.3 there is an operator $y \in \mathbf{B}(\mathcal{H})$ such that $x = a_n^+ \circ y a_n$. Then $y = (a_n^{-1})^+ \circ x a_n^{-1}$ and hence $y \in \mathcal{L}$, since $(a_n^{-1})^+ \circ \mathcal{L} a_n^{-1} \subseteq \mathcal{L}$ by assumption. (Here we used again the notational convention from Remark 4 in 3.2.) Since $a_n \in \mathcal{A}(I)$, $\bar{a}_n^{-1} \in \mathbf{B}(\mathcal{H})$ and so $y = (\bar{a}_n^{-1})^* x(\bar{a}_n^{-1})$. Therefore, $y \in \mathcal{L} \cap \mathbf{B}(\mathcal{H}) = \mathcal{C}$. This proves that $\mathcal{L}_{a_n} \subseteq a_n^+ \circ \mathcal{C} a_n$ and $\mathcal{L} \subseteq \bigcup_{n\in\mathbb{N}} a_n^+ \circ \mathcal{C} a_n$. □

Corollary 4.4.7. *Let $\mathcal{A}$ be an O^*-algebra with metrizable graph topology $\mathrm{t}_{\mathcal{A}}$. If $\mathcal{A}$ is a symmetric $*$-algebra, then $\tau_{\mathcal{N}} = \tau_{\mathcal{O}}$ on $\mathcal{A}$.*

Proof. Since $t_{\mathcal{A}}$ is metrizable, there is a sequence $(b_n : n \in \mathbb{N})$ of symmetric operators in $\mathcal{A}(I)$ such that $2 \|b_n \cdot\| \leqq \|b_{n+1} \cdot\|$ for $n \in \mathbb{N}$ and the family of seminorms $\{\|\cdot\|_{b_n} : n \in \mathbb{N}\}$ generates the topology $t_{\mathcal{A}}$. Setting $a_n := b_n + \mathrm{i}$, we have $a_n^{-1} \in \mathcal{A} \subseteq \mathcal{L}^+(\mathcal{D}(\mathcal{A}))$, since $\mathcal{A}$ is symmetric. Hence the assumptions of Corollary 4.4.6 are satisfied when we set $\mathcal{L} := \mathcal{A}$, so $\tau_{\mathcal{N}} = \tau_{\mathcal{Q}}$ on $\mathcal{L} = \mathcal{A}$. □

Remark 2. Linear subspaces of $\mathcal{L}(\mathcal{D}_{\mathcal{A}}, \mathcal{D}_{\mathcal{A}}^+)$ which are closely related to the one occuring in Theorem 4.4.5 will be investigated in Section 7.4.

4.5. Topologies on Countably Generated O*-Algebras

Theorem 4.5.1. *Let $\mathcal{A}$ be an O*-algebra. Suppose that $\mathcal{L}$ is a cofinal *-vector subspace of $\mathcal{L}(\mathcal{D}_{\mathcal{A}}, \mathcal{D}_{\mathcal{A}}^+)$ which has an at most countable Hamel basis. Then we have $\tau_{\mathcal{D}} = \tau_{\mathcal{N}}$ on $\mathcal{L}$.*

Proof. Since the assertion is trivial when $\mathcal{L}$ is finite dimensional we can assume that the vector space $\mathcal{L}$ has a countable basis, say $\{x_n : n \in \mathbb{N}\}$. Then for each $n \in \mathbb{N}$ we can find an operator $a_n \in \mathcal{A}$ such that $|\langle x_n\varphi, \varphi\rangle| \leqq \|a_n\varphi\|^2$ for $\varphi \in \mathcal{D}(\mathcal{A})$. Suppose $a \in \mathcal{A}$. Since $\mathcal{L}$ is cofinal in $\mathcal{L}(\mathcal{D}_{\mathcal{A}}, \mathcal{D}_{\mathcal{A}}^+)$, there is an $x \in \mathcal{L}_+$ such that $a^+a \leqq x$. Writing x as a linear combination of $x_1, \ldots, x_k$, we get $\|a \cdot\| \leqq \lambda(\|a_1 \cdot\| + \cdots + \|a_k \cdot\|)$ with some $\lambda > 0$. This shows that the graph topology $t_{\mathcal{A}}$ is metrizable and generated by the family of seminorms $\{\|\cdot\|_{a_n} : n \in \mathbb{N}\}$.

We apply Propositions 4.1.1 and 4.1.3. Let $(\varepsilon_n : n \in \mathbb{N})$ be a given positive sequence. Since $\mathcal{L}$ has a countable basis, there is a countable subset $\mathcal{R} \equiv \{y_k : k \in \mathbb{N}\}$ of $\mathcal{L}$ which is dense in $\mathcal{L}[\tau_{\mathcal{N}}]$. (It suffices to take all elements of $\mathcal{L}$ whose coordinates w.r.t. the basis have rational real and imaginary parts.) Let $\mathbb{N}' := \{k \in \mathbb{N} : y_k \notin \mathcal{V}_{(\varepsilon_n)}\}$. If $k \in \mathbb{N}'$, then there is a vector $\varphi_k \in \mathcal{D}(\mathcal{A})$ such that $|\langle y_k\varphi_k, \varphi_k\rangle| > \sum_{n=1}^{k} \varepsilon_n \|a_n\varphi_k\|^2$. We choose a positive sequence $(\delta_n : n \in \mathbb{N})$ such that $\delta_n \leqq \varepsilon_n/2$ and $\delta_n \|a_n\varphi_k\|^2 \leqq 2^{-n} \sum_{l=1}^{k} \varepsilon_l \|a_l\varphi_k\|^2$ for all $n \in \mathbb{N}$, $k \in \mathbb{N}'$, $k < n$. If $k \in \mathbb{N}'$, then

$$\sum_{n=1}^{\infty} \delta_n \|a_n\varphi_k\|^2 \leqq \sum_{n=1}^{k} \frac{1}{2} \varepsilon_n \|a_n\varphi_k\|^2 + \sum_{n=k+1}^{\infty} 2^{-n} \sum_{l=1}^{k} \varepsilon_l \|a_l\varphi_k\|^2$$
$$\leqq \sum_{n=1}^{k} \varepsilon_n \|a_n\varphi_k\|^2 < |\langle y_k\varphi_k, \varphi_k\rangle| .$$

This proves that $y_k \notin \mathcal{U}_{(\delta_n)}$ if $y_k \notin \mathcal{V}_{(\varepsilon_n)}$ for $k \in \mathbb{N}$; so $\mathcal{R} \cap \mathcal{U}_{(\delta_n)} \subseteq \mathcal{R} \cap \mathcal{V}_{(\varepsilon_n)}$. Since $\tau_{\mathcal{D}} \subseteq \tau_{\mathcal{N}}$ on $\mathcal{L}$ and $\mathcal{R}$ is dense in $\mathcal{L}[\tau_{\mathcal{N}}]$, $\frac{1}{2} \mathcal{U}_{(\delta_n)} \subseteq (\mathcal{R} \cap \mathcal{U}_{(\delta_n)})^- \subseteq (\mathcal{R} \cap \mathcal{V}_{(\varepsilon_n)})^- \subseteq 2\mathcal{V}_{(\varepsilon_n)}$, where the bar denotes the closure in $\mathcal{L}[\tau_{\mathcal{N}}]$. Thus $\mathcal{U}_{(\delta_n)} \subseteq \mathcal{V}_{(4\varepsilon_n)}$, and the equality $\tau_{\mathcal{D}} = \tau_{\mathcal{N}}$ on $\mathcal{L}$ is proved. □

Remark 1. The main part of the preceding proof gives the following more general statement. If $\mathcal{A}$ is an O*-algebra with metrizable graph topology and $\mathcal{L}$ is a cofinal *-vector subspace of $\mathcal{L}(\mathcal{D}_{\mathcal{A}}, \mathcal{D}_{\mathcal{A}}^+)$ such that $\mathcal{L}[\tau_{\mathcal{N}}]$ is separable, then the topologies $\tau_{\mathcal{D}}$ and $\tau_{\mathcal{N}}$ of $\mathcal{L}$ coincide.

Corollary 4.5.2. *If an O*-algebra $\mathcal{A}$ is countably generated (as a *-algebra), then we have $\tau_{\mathcal{D}} = \tau_{\mathcal{N}}$ on $\mathcal{A}$.*

Proof. Apply Theorem 4.5.1 to $\mathcal{L} := \mathcal{A}$. □

Corollary 4.5.3. *Suppose $\mathcal{A}$ is an O*-algebra with metrizable graph topology and $\mathcal{L}$ is a cofinal $*$-vector subspace of $\mathcal{L}(\mathcal{D}_\mathcal{A}, \mathcal{D}_\mathcal{A}^+)$. Then the topologies $\tau_\mathcal{N}$ of $\mathcal{L}$ and $\tau_\mathcal{D}$ of $\mathcal{L}(\mathcal{D}_\mathcal{A}, \mathcal{D}_\mathcal{A}^+)$ induce the same topology on each $*$-vector subspace $\mathcal{L}_1$ of $\mathcal{L}$ which has an at most countable Hamel basis.*

Proof. We write $\tau_\mathcal{N}(\tilde{\mathcal{L}})$ for the topology $\tau_\mathcal{N}$ of a $*$-vector space $\tilde{\mathcal{L}} \subseteq \mathcal{L}(\mathcal{D}_\mathcal{A}, \mathcal{D}_\mathcal{A}^+)$. Take a sequence $(a_n : n \in \mathbb{N})$ in $\mathcal{A}$ such that the seminorms $\|\cdot\|_{a_n}$, $n \in \mathbb{N}$, generate the topology $t_\mathcal{A}$. Since $\mathcal{L}$ is cofinal in $\mathcal{L}(\mathcal{D}_\mathcal{A}, \mathcal{D}_\mathcal{A}^+)$, for each $n \in \mathbb{N}$ there is an $x_n \in \mathcal{L}_+$ such that $a_n^+ a_n \leqq x_n$. Then the linear span $\mathcal{L}_2$ of $\mathcal{L}_1$ and $\{x_n : n \in \mathbb{N}\}$ is a cofinal $*$-vector space in $\mathcal{L}(\mathcal{D}_\mathcal{A}, \mathcal{D}_\mathcal{A}^+)$ which has an at most countable Hamel basis. Hence $\tau_\mathcal{N}(\mathcal{L}_2) = \tau_\mathcal{D} \upharpoonright \mathcal{L}_2$ by Theorem 4.5.1. Since $\mathcal{L}$ and $\mathcal{L}_2$ are both cofinal in $\mathcal{L}(\mathcal{D}_\mathcal{A}, \mathcal{D}_\mathcal{A}^+)$, it follows from Proposition 4.1.3 that $\tau_\mathcal{N}(\mathcal{L}) \upharpoonright \mathcal{L}_2 = \tau_\mathcal{N}(\mathcal{L}_2)$. Thus $\tau_\mathcal{N}(\mathcal{L}) \upharpoonright \mathcal{L}_1 = \tau_\mathcal{D} \upharpoonright \mathcal{L}_1$. □

Remark 2. Let $\mathcal{A}$ and $\mathcal{L}$ be as in Corollary 4.5.3. This corollary shows that a *sequence* converges in $\mathcal{L}[\tau_\mathcal{D}]$ if and only if it converges in $\mathcal{L}[\tau_\mathcal{N}]$ (of course, to the same limit). From this we conclude that each strongly positive linear functional f on $\mathcal{L}$ is *sequentially* continuous on $\mathcal{L}[\tau_\mathcal{D}]$. Recall from Section 4.3 that in general the topologies $\tau_\mathcal{D}$ and $\tau_\mathcal{N}$ do not coincide on $\mathcal{L}$ and f is not continuous on $\mathcal{L}[\tau_\mathcal{D}]$.

In the following two theorems we characterize the countably generated O*-algebras $\mathcal{A}$ for which the topologies $\tau_\mathcal{D}$ or $\tau^\mathcal{D}$ coincide with the finest locally convex topology (always denoted by $\tau_{\rm st}$) on $\mathcal{A}$.

Theorem 4.5.4. *Suppose $\mathcal{A}$ is an O*-algebra and $\mathcal{L}$ is a cofinal $*$-vector subspace of $\mathcal{L}(\mathcal{D}_\mathcal{A}, \mathcal{D}_\mathcal{A}^+)$ which admits an at most countable Hamel basis. The following three statements are equivalent:*

(i) $\tau_\mathcal{D} = \tau_{\rm st}$ *on* $\mathcal{L}$.

(ii) $\tau_\mathcal{O} = \tau_{\rm st}$ *on* $\mathcal{L}$.

(iii) *For every $a \in \mathcal{A}$, the vector space*

$$\mathcal{L}_a \equiv \bigcup_{\lambda>0} \{x \in \mathcal{L} : |\langle x\varphi, \psi\rangle| \leqq \lambda \|a\varphi\| \|a\psi\| \quad \text{for all} \quad \varphi, \psi \in \mathcal{D}(\mathcal{A})\}$$

is finite dimensional.

Proof. (i) → (ii) is trivial, since $\tau_\mathcal{D} \subseteq \tau_\mathcal{O} \subseteq \tau_{\rm st}$.

(ii) → (iii): By Proposition 3.3.11 and by (ii), $\tau_{\rm in} = \tau_{\rm st}$ on $\mathcal{L}$. Hence the unit ball of the normed space $(\mathcal{L}_a, \mathfrak{l}_a)$ is bounded in $\mathcal{L}[\tau_{\rm st}]$. Of course, this is only possible if $\mathcal{L}_a$ is finite dimensional (Schäfer [1], II, Exercise 7, (b)).

(iii) → (i): As shown in the first paragraph of the proof of Theorem 4.5.1, the graph topology $t_\mathcal{A}$ is metrizable. Hence there exists a sequence $(a_n : n \in \mathbb{N})$ in $\mathcal{A}$ such that $\|a_n \cdot\| \leqq \|a_{n+1} \cdot\|$ for $n \in \mathbb{N}$ and such that the topology $t_\mathcal{A}$ is generated by the seminorms $\|\cdot\|_{a_n}$, $n \in \mathbb{N}$. Then we have $\mathcal{L}_{a_n} \subseteq \mathcal{L}_{a_{n+1}}$ for $n \in \mathbb{N}$ and $\bigcup_{n\in\mathbb{N}} \mathcal{L}_{a_n} = \mathcal{L}$.

From Theorem 4.5.1, $\tau_\mathcal{D} = \tau_\mathcal{N}$ on $\mathcal{L}$; hence it suffices to prove that $\tau_{\rm st} \subseteq \tau_\mathcal{N}$ on $\mathcal{L}$. Let $\mathcal{U}$ be an open 0-neighbourhood in $\mathcal{L}[\tau_{\rm st}]$. By Proposition 4.1.3 we have to show that there exists a positive sequence $(\varepsilon_n : n \in \mathbb{N})$ such that $\mathcal{V}_{(\varepsilon_n)} \subseteq \mathcal{U}$. Let $k \in \mathbb{N}$. Suppose that positive numbers $\varepsilon_1, \ldots, \varepsilon_k$ are chosen such that

$$\mathcal{V}_{(\varepsilon_1,\ldots,\varepsilon_k)} := \{x \in \mathcal{L} : |\langle x\varphi, \varphi\rangle| \leqq \sum_{n=1}^{k} \varepsilon_n \|a_n\varphi\|^2 \text{ for } \varphi \in \mathcal{D}(\mathcal{A})\}$$

is contained in $\mathcal{U}_k := \mathcal{U} \cap \mathcal{L}_{a_k}$. Set $\mathcal{X}_\delta := \mathcal{V}_{(\varepsilon_1,\ldots,\varepsilon_k,\delta)}$ for $\delta > 0$. Let $\mathcal{X}$ denote the one point compactification of the finite dimensional (by (iii)) normed space $(\mathcal{L}_{a_{k+1}}, \mathfrak{l}_{a_{k+1}})$. Since obviously $\bigcap_{\delta>0} \mathcal{X}_\delta = \mathcal{V}_{(\varepsilon_1,\ldots,\varepsilon_k)}$ and $\mathcal{V}_{(\varepsilon_1,\ldots,\varepsilon_k)} \subseteqq \mathcal{U}_k$, the sets $\mathcal{U}_k$ and $\mathcal{X} \setminus \mathcal{X}_\delta$, $\delta > 0$, form an open cover of $\mathcal{X}$. Since $\mathcal{X}$ is compact, there is a finite subcover, say $\{\mathcal{X} \setminus \mathcal{X}_{\delta_1}, \ldots, \mathcal{X} \setminus \mathcal{X}_{\delta_r}, \mathcal{U}_k\}$. Setting $\varepsilon_{k+1} := \min\{\delta_j : j = 1, \ldots r\}$, we have $\mathcal{X}_{\varepsilon_{k+1}} = \mathcal{V}_{(\varepsilon_1,\ldots,\varepsilon_{k+1})} \subseteqq \mathcal{U}_k \subseteqq \mathcal{U}_{k+1}$. It is clear that there exists a positive number ε_1 such that $\mathcal{V}_{(\varepsilon_1)} \subseteqq \mathcal{U}_1$. Therefore, by induction, we obtain a positive sequence (ε_n) such that $\mathcal{V}_{(\varepsilon_n)} = \bigcup_{k\in\mathbb{N}} \mathcal{V}_{(\varepsilon_1,\ldots,\varepsilon_k)} \subseteqq \bigcup_{k\in\mathbb{N}} \mathcal{U}_k = \mathcal{U}$. □

Remark 3. Theorem 4.5.4 applies in particular to $\mathcal{L} := \mathcal{A}$ if $\mathcal{A}$ is a countably generated O*-algebra.

Theorem 4.5.5. *Suppose $\mathcal{A}$ is a countably generated O*-algebra. Then $\tau^{\mathcal{D}} = \tau^{\mathcal{N}}$ on $\mathcal{A}$. Further, the following three assertions are equivalent:*

(i) $\tau^{\mathcal{D}} = \tau_{\mathrm{st}}$ *on $\mathcal{A}$.*

(ii) $\tau^{\mathcal{O}} = \tau_{\mathrm{st}}$ *on $\mathcal{A}$.*

(iii) *For every $a \in \mathcal{A}$, the vector space*

$$\mathcal{A}^a \equiv \bigcup_{\lambda>0} \{x \in \mathcal{A} : \|x\varphi\| \leqq \lambda \|a\varphi\| \text{ for } \varphi \in \mathcal{D}(\mathcal{A})\}$$

is finite dimensional.

Proof. The proof is similar to the proofs of Theorems 4.5.1 and 4.5.4 if we replace Propositions 4.1.1, 4.1.3, 3.3.11 by Propositions 4.1.5, 4.1.6, 3.3.14, respectively. □

Recall from Section 3.5 that $\sigma^{\mathcal{D}}$ denotes the strong-operator topology.

Remark 4. Actually a stronger result is valid. In Theorem 4.5.5, each of the conditions (i)—(iii) is equivalent to the following statement:

(iv) $\sigma^{\mathcal{D}} = \tau_{\mathrm{st}}$ on $\hat{\mathcal{A}}$.

We shall not prove this here and refer to the paper Schmüdgen [24] where a more general result is proved. We mention an obvious consequence of this strengthened version of Theorem 4.5.5: *Suppose $\mathcal{A}$ is a closed countably generated O*-algebra. If $\tau_{\mathcal{D}} = \tau_{\mathrm{st}}$ on $\mathcal{A}$, then, of course, $\tau^{\mathcal{D}} = \tau_{\mathrm{st}}$ and hence $\sigma^{\mathcal{D}} = \tau_{\mathrm{st}}$ by the implication* (i) → (iv).

Remark 5. We state (without giving proofs) two additional facts. First we note that the converse of the final statement in Remark 4 is not true in general. That is, there exists a closed countably generated O*-algebra $\mathcal{A}$ for which $\sigma^{\mathcal{D}} = \tau^{\mathcal{D}} = \tau_{\mathrm{st}}$, but $\tau_{\mathcal{D}} \neq \tau_{\mathrm{st}}$ on $\mathcal{A}$. Secondly, it may happen, again for closed countably generated O*-algebras $\mathcal{A}$, that the topology $\tau_{\mathcal{D}}$ ($= \tau_{\mathcal{N}}$ by Corollary 4.5.2) on $\mathcal{A}$ does not coincide with the order topology $\tau_{\mathcal{O}}$ of $\mathcal{A}$. Combined with Theorem 4.4.2 (which shows that $\tau_{\mathcal{D}} = \tau_{\mathcal{O}}$ on $\mathcal{L}(\mathcal{D}_{\mathcal{A}}, \mathcal{D}_{\mathcal{A}}^+)$), it follows from the latter that the topology $\tau_{\mathcal{O}}$ of $\mathcal{A}$ is different from the topology on $\mathcal{A}$ which is induced by the order topology $\tau_{\mathcal{O}}$ of $\mathcal{L}(\mathcal{D}_{\mathcal{A}}, \mathcal{D}_{\mathcal{A}}^+)$.

We close this section by a number of examples. The assertion $\tau_{\mathcal{D}} = \tau_{\mathrm{st}}$ in these examples follows always from Theorem 4.5.4 by verifying condition (iii) occuring therein. We omit some or all details of these proofs. Except from Example 4.5.9 which requires some more work it is easy to fill these gaps.

Example 4.5.6. Let x be a symmetric operator in some $\mathcal{L}^+(\mathcal{D})$, and let $\mathcal{A} := \mathbb{C}[x]$ be the O*-algebra of all polynomials in x. If the operator x is unbounded on $\mathcal{D}$, then $\tau_{\mathcal{D}} = \tau_{\mathrm{st}}$ on $\mathbb{C}[x]$.

Sketch of proof. First we note that for each $n \in \mathbb{N}$

$$\sup \{\|x^n\varphi\| : \varphi \in \mathcal{D} \text{ and } \|\varphi\|_n := \|\varphi\| + \|x\varphi\| + \cdots + \|x^{n-1}\varphi\| \leqq 1\} = \infty. \tag{1}$$

Otherwise, x would be a bounded operator on the normed space $(\mathcal{D}, \|\cdot\|_n)$. Since $\|\cdot\|_n$ is stronger than the Hilbert space norm and x is unbounded, this contradicts Proposition 2.1.11. By treating the cases n odd and n even separately, it follows from (1) that for $n \in \mathbb{N}$

$$\sup \{|\langle x^n\varphi, \varphi\rangle| : \varphi \in \mathcal{D} \text{ and } |\langle x^k\varphi, \varphi\rangle| \leqq 1 \text{ for } k \in \mathbb{N}_0,\ k < n\} = \infty. \tag{2}$$

Let $a \in \mathbb{C}[x]$ be a polynomial in x of degree n. From (2) we conclude that $\mathcal{A}_a$ consists only of polynomials of degree less or equal $2n$. Hence $\mathcal{A}_a$ is finite dimensional, and the assertion follows from Theorem 4.5.4, (iii) → (i). □ ○

Example 4.5.7. Suppose $n \in \mathbb{N}$. Let $\mathcal{A}$ be the O*-algebra $\mathbf{A}(p_1, q_1, \ldots, p_n, q_n)$ generated by the position operators q_k and momentum operators p_k, $k = 1, \ldots, n$, on the domain $\mathcal{D}(\mathcal{A}) := \mathcal{S}(\mathbb{R}^n)$; cf. Example 2.5.2. Then we have $\tau_{\mathcal{D}} = \tau_{\mathrm{st}}$ on $\mathcal{A}$.

Sketch of proof in case $n = 1$. Take a vector $\varphi \in \mathcal{D}(\mathcal{A})$ such that $\varphi \neq 0$ and $\operatorname{supp} \varphi \subseteqq [0, 1]$. Setting $\varphi_{\alpha,\beta}(t) := \beta\varphi\big(\beta(t - \alpha)\big)$ for $t, \alpha, \beta \in \mathbb{R}$, we have $\alpha^k\beta^l \|p^l\varphi\| \leqq \|q^k p^l \varphi_{\alpha,\beta}\| \leqq (\alpha + 1)^k \beta^l \|p^l\varphi\|$ for $k, l \in \mathbb{N}_0$ and for arbitrary $\alpha, \beta \in \mathbb{R}$. From this we conclude easily that each space $\mathcal{A}_a$, $a \in \mathcal{A}$, is finite dimensional. □ ○

Example 4.5.8. Suppose $n \in \mathbb{N}$. Let $\mathcal{A}$ be the O*-algebra $\mathbb{C}[x_1, \ldots, x_n]$ on the domain $\mathcal{D}(\mathcal{A}) := \{\varphi \in L^2(\mathbb{R}^n) : p(\cdot)\, \varphi(\cdot) \in L^2(\mathbb{R}^n)$ for all polynomials $p \in \mathbb{C}[\mathrm{x}_1, \ldots, \mathrm{x}_n]\}$ in the Hilbert space $L^2(\mathbb{R}^n)$, where the polynomials act as multiplication operators; cf. Example 2.6.11. Then $\tau_{\mathcal{D}} = \tau_{\mathrm{st}}$ on $\mathcal{A}$.

As an illustration we give an application of this result to the n-dimensional classical moment problem by proving the following statement:

For each complex multi-sequence $(\alpha_k : k \in \mathbb{N}_0^n)$ there exists a complex measure $\mu \in M(\mathbb{R}^n)$ such that $\alpha_k = \int_{\mathbb{R}^n} t^k \, \mathrm{d}\mu(t)$ for all $k \in \mathbb{N}_0^n$. (We use the notation from Example 2.6.11.)

Proof. Define a linear functional f on the O*-algebra $\mathcal{A} = \mathbb{C}[x_1, \ldots, x_n]$ by $f(x^k) := \alpha_k$, $k \in \mathbb{N}_0^n$. Since $\tau_{\mathcal{D}} = \tau_{\mathrm{st}}$ on $\mathcal{A}$, f is continuous on $\mathcal{A}[\tau_{\mathcal{D}}]$. By Proposition 3.3.7, the cone $\mathcal{A}_+$ is normal in $\mathcal{A}[\tau_{\mathcal{D}}]$. From Proposition 1.5.4, (ii), we conclude that there are strongly positive linear functionals f_1, f_2, f_3, f_4 on $\mathcal{A}$ such that $f = (f_1 - f_2) + \mathrm{i}(f_3 - f_4)$. By Statement 1 in Example 2.6.11, there are positive measures $\mu_l \in M_+(\mathbb{R}^n)$, $l = 1, \ldots, 4$, such that $f_l(p) = \int p(t)\, \mathrm{d}\mu_l(t)$ for $p \in \mathbb{C}[x_1, \ldots, x_n]$. Setting $\mu := (\mu_1 - \mu_2) + \mathrm{i}(\mu_3 - \mu_4)$, the proof is complete. □ ○

Example 4.5.9. Suppose G is a real Lie group with left Haar measure μ and Lie algebra $\mathfrak{g}$. Let $\mathcal{E}(\mathfrak{g})$ be the universal enveloping algebra of $\mathfrak{g}$. For $x \in \mathcal{E}(\mathfrak{g})$, let $\tilde{x}$ denote the associated right invariant differential operator on G. (See also Section 1.7.)

Let $\mathcal{A}$ be the O*-algebra on $\mathcal{D}(\mathcal{A}) := C_0^\infty(G)$ in the Hilbert space $L^2(G; \mu)$ which is formed by the operators $\tilde{x} \restriction C_0^\infty(G)$, where $x \in \mathcal{E}(\mathfrak{g})$. (In the notation of Example 10.1.8, $\mathcal{A}$ is the O*-algebra $dU_{lr}\big(\mathcal{E}(\mathfrak{g})\big) \restriction C_0^\infty(G)$.) Then we have $\tau_{\mathcal{D}} = \tau_{\mathrm{st}}$ on $\mathcal{A}$. (A proof is given in Schmüdgen [8].) ○

Remark 6. From the preceding and the assertion in Remark 4 we conclude that $\sigma^{\mathcal{D}} = \tau_{\mathrm{st}}$ on $\mathcal{A}$ in Examples 4.5.7 and 4.5.8 and $\sigma^{\mathcal{D}} = \tau_{\mathrm{st}}$ on $\hat{\mathcal{A}}$ in Examples 4.5.6 and 4.5.9.

Notes

4.1. The description of 0-neighbourhood bases for the topologies $\tau_{\mathcal{D}}$ and $\tau_{\mathcal{N}}$ given in Propositions 4.1.1 and 4.1.3, (i), are from KRÖGER [3].
4.2. Proposition 4.2.5 and the equivalence of (ii) and (iii) in Proposition 4.2.1 occur in ARNAL/JURZAK [1]. The equivalence of (i) and (iii) in Proposition 4.2.1 and Corollary 4.2.3 were observed in SCHMÜDGEN [9]. Proposition 4.2.5 and Corollary 4.2.6 are due to ARNAL/JURZAK [1].
4.3. Most of the results in this section are due to SCHMÜDGEN [9]. That conditions (v) and (vi) are equivalent to (i) in Theorem 4.3.4 was added by KÜRSTEN.
4.4. The assertion of Theorem 4.4.1 (without mentioning the topology $\tau_{\mathcal{N}}$ explicitly) was proved by SCHMÜDGEN [7]. Theorem 4.4.2 is due to KÜRSTEN [2], [3]. Theorem 4.4.5 generalizes a result of ARAKI/JURZAK [1].
4.5. Corollaries 4.5.2 and 4.5.3 are from KRÖGER [3]. Theorems 4.5.4 and 4.5.5 (also in the stronger version stated in Remark 4) have been proved by SCHMÜDGEN [10] without using Theorem 4.5.1 and the results from Section 4.1. The elegant compactness argument in the proof of Theorem 4.5.4 was found by J. FRIEDRICH. The result on the moment problem derived in Example 4.5.8 was proved by BOAS [1] for $n = 1$ and by SHERMAN [2] in general (of course, without using unbounded operator algebras).

It should be noted that some of the topological results occuring in Part I of this monograph are consequences or even special cases of general facts from the theory of locally convex spaces or from the theory of ordered vector spaces. For instance, some results in Section 2.3 (such as Propositions 2.3.1 and 2.3.10 and Corollary 2.3.2) could be mentioned in this respect. Assertion (i) in Proposition 4.1.3 is a general fact on ordered vector spaces stated here in a special case. Theorem 4.5.1 is closely related to Theorem 4 in GROTHENDIECK [3]. Further examples of this kind will be indicated in the notes after Chapters 5 and 6.

5. Ultraweakly Continuous Linear Functionals and Duality Theory

This chapter is devoted to a study of linear functionals on linear subspaces of $\mathcal{L}(\mathcal{D}_{\mathcal{A}}, \mathcal{D}_{\mathcal{B}}^{+})$ which are defined by means of a "generalized trace" and a "density matrix".

To be somewhat more precise, let $\mathcal{A}$ and $\mathcal{B}$ be directed O*-families in a Hilbert space $\mathcal{H}$ and let $\mathcal{L}$ be a linear subspace of $\mathcal{L}(\mathcal{D}_{\mathcal{A}}, \mathcal{D}_{\mathcal{B}}^{+})$. We are concerned with linear functionals on $\mathcal{L}$ of the form $f_t(x) := \operatorname{tr} \hat{t}x$, $x \in \mathcal{L}$. Here t belongs to a set $\mathbb{B}_1(\mathcal{B}, \mathcal{A})$ of trace class operators on $\mathcal{H}$ which have the property that for a in $\mathcal{A}$ and b in $\mathcal{B}$ the operator $\overline{atb}$ is also of trace class. This set $\mathbb{B}_1(\mathcal{B}, \mathcal{A})$ and its projective topology are investigated in Section 5.1. The symbol "tr" refers to a generalization of the usual trace of trace class operators on Hilbert space. This concept is developed in Section 5.2. If $\mathcal{A}$ and $\mathcal{B}$ are O*-algebras, then the functionals f_t, where $t \in \mathbb{B}_1(\mathcal{B}, \mathcal{A})$, are ultraweakly continuous.

The goal of Section 5.3 is to characterize the functionals $f_t(\cdot) \equiv \operatorname{tr} \hat{t}\,\cdot$. Among others it is shown that all strongly positive linear functionals on $\mathcal{L}(\mathcal{D}_{\mathcal{A}}, \mathcal{D}_{\mathcal{A}}^{+})$ are of the form f_t with $t \in \mathbb{B}_1(\mathcal{A})_+$ provided that $\mathcal{A}$ is a closed O*-algebra for which $\mathcal{D}_{\mathcal{A}}$ is a Frechet-Montel space or $\mathcal{D}_{\mathcal{A}}$ is a Schwartz space.

In Sections 5.4 and 5.5 we restrict ourselves to O*-algebras $\mathcal{A}$ and $\mathcal{B}$. Section 5.4 is devoted to a duality theorem which can be considered as a generalization of the classical fact that the norm dual of the space $\mathbb{B}_1(\mathcal{H})$ of trace class operators on $\mathcal{H}$ is the space $\mathbb{B}(\mathcal{H})$ of bounded operators on $\mathcal{H}$. This theorem states that if $\mathcal{D}_{\mathcal{A}}$ and $\mathcal{D}_{\mathcal{B}}$ are QF-spaces, then the space $\mathcal{L}(\mathcal{D}_{\mathcal{A}}, \mathcal{D}_{\mathcal{B}}^{+})$, equipped with the bounded topology, is the strong dual of the space $\mathbb{B}_1(\mathcal{B}, \mathcal{A})$ endowed with the projective topology. This is the reason we call the latter space the predual. In Section 5.5 we give a number of conditions which are equivalent to the Montel property of the space $\mathcal{D}_{\mathcal{A}}$.

5.1. The Predual

We begin with some terminology. Suppose $\mathcal{H}$ is a Hilbert space. Let $\mathbb{B}_1(\mathcal{H})$ denote the set of all trace class operators on $\mathcal{H}$. For $t \in \mathbb{B}_1(\mathcal{H})$, $\operatorname{Tr} t$ is the trace of t and $\nu(t) \equiv \operatorname{Tr} |t|$ is the trace norm of t. We sometimes write $\operatorname{Tr}_{\mathcal{H}} t$ when confusion is possible. In order to simplify the notation we shall write $\operatorname{Tr} t$ for $\operatorname{Tr} \bar{t}$ and $\nu(t)$ for $\nu(\bar{t})$ when t is a closable operator on $\mathcal{H}$ with $\bar{t} \in \mathbb{B}_1(\mathcal{H})$. By an *absolutely convergent series on* $\mathcal{H}$ we mean a series $\sum_{n=1}^{\infty} \psi_n \otimes \varphi_n$ such that φ_n and ψ_n are vectors in $\mathcal{H}$ for $n \in \mathbb{N}$ and $\sum_n \|\psi_n\| \, \|\varphi_n\| < \infty$. Such a series converges in particular in the operator norm on $\mathcal{H}$, so it defines a bounded operator t on $\mathcal{H}$. We shall say that t is represented by the series $\sum_n \psi_n \otimes \varphi_n$.

Let $t \in \mathbb{B}_1(\mathcal{H})$. Then t can be represented as $t = \sum_{n=1}^{\infty} \lambda_n \langle \cdot, \psi_n \rangle \varphi_n$, where $(\lambda_n : n \in \mathbb{N})$ is a complex sequence such that $\sum_{n=1}^{\infty} |\lambda_n| < \infty$ and $(\varphi_n : n \in \mathbb{N}')$ and $(\psi_n : n \in \mathbb{N}')$ are orthonormal sets in $\mathcal{H}$ with $\mathbb{N}' := \{n \in \mathbb{N} : \lambda_n \neq 0\}$. Moreover, $\nu(t) = \sum_n |\lambda_n|$. In case $t = t^*$ we can have in addition that $\lambda_n \in \mathbb{R}$ and $\varphi_n = \psi_n$ for all $n \in \mathbb{N}$. (For a proof of these facts, see e.g. Birman/Solomjak [1], ch. 11, § 1, or Köthe [2], § 42, 6., (1).) Further, we set $\varphi_n = \psi_n = 0$ for $n \in \mathbb{N} \setminus \mathbb{N}'$. If the preceding conditions are fulfilled, then we call the sum $\sum_n \lambda_n (\psi_n \otimes \varphi_n)$ a *canonical representation for t.*

We state two well-known facts from operator theory as a reference.

Lemma 5.1.1. *Suppose that $\sum_n \psi_n \otimes \varphi_n$ is an absolutely convergent series on $\mathcal{H}$. Then $t := \sum_n \psi_n \otimes \varphi_n$ is a trace class operator on $\mathcal{H}$ and $\operatorname{Tr} t = \sum_n \langle \varphi_n, \psi_n \rangle$. Further,*

$$\nu(t) = \inf \Big\{ \sum_n \|\psi_n\| \, \|\varphi_n\| \Big\},$$

where the infimum is taken over all absolutely convergent series $\sum_n \psi_n \otimes \varphi_n$ on $\mathcal{H}$ which represent the operator t.

Proof. Köthe [2], § 42, 5., (7), (8), and 7., (6); see also Weidmann [1], 7.12. □

Lemma 5.1.2. *An operator $t \in \mathbb{B}(\mathcal{H})$ is in $\mathbb{B}_1(\mathcal{H})$ if and only if $\sum_{i \in I} |\langle t\zeta_i, \eta_i \rangle| < \infty$ for arbitrary orthonormal sets $\{\zeta_i : i \in I\}$ and $\{\eta_i : i \in I\}$ in $\mathcal{H}$.*

Proof. Birman/Solomjak [1], ch. 11, § 2. □

The Algebra $\mathbb{B}_1(\mathcal{B}, \mathcal{A})$ for O*-Families

Throughout this subsection, $\mathcal{A}$ and $\mathcal{B}$ denote O*-families on the Hilbert space $\mathcal{H}$.

Definition 5.1.3. We define $\mathbb{B}_1(\mathcal{B}, \mathcal{A}) := \{t \in \mathbb{B}(\mathcal{H}) : t\mathcal{H} \subseteq \mathcal{D}(\mathcal{A}),\ t^*\mathcal{H} \subseteq \mathcal{D}(\mathcal{B})$ and $\overline{atb} \in \mathbb{B}_1(\mathcal{H})$ for all $a \in \mathcal{A}$ and $b \in \mathcal{B}\}$ and $\mathbb{B}_1(\mathcal{H}, \mathcal{A}) := \{t \in \mathbb{B}(\mathcal{H}) : t\mathcal{H} \subseteq \mathcal{D}(\mathcal{A})$ and $at \in \mathbb{B}_1(\mathcal{H})$ for all $a \in \mathcal{A}\}$. Set $\mathbb{B}_1(\mathcal{A}) := \mathbb{B}_1(\mathcal{A}, \mathcal{A})$ and $\mathbb{B}_1(\mathcal{A})_+ := \{t \in \mathbb{B}_1(\mathcal{A}) : t \geqq 0\}$. In case $\mathcal{A} = \mathcal{L}^+(\mathcal{D}_1)$, $\mathcal{D}(\mathcal{A}) = \mathcal{D}_1$ and $\mathcal{B} = \mathcal{L}^+(\mathcal{D}_2)$, $\mathcal{D}(\mathcal{B}) = \mathcal{D}_2$ we write $\mathbb{B}_1(\mathcal{D}_2, \mathcal{D}_1)$ for $\mathbb{B}_1(\mathcal{B}, \mathcal{A})$, $\mathbb{B}_1(\mathcal{D}_1)$ for $\mathbb{B}_1(\mathcal{A})$ and $\mathbb{B}_1(\mathcal{D}_1)_+$ for $\mathbb{B}_1(\mathcal{A})_+$.

Remark 1. The operator atb in Definition 5.1.3 is always closable, since $(atb)^* \supseteq b^+t^*a^+$ and this operator is densely defined because of $t^*\mathcal{H} \subseteq \mathcal{D}(\mathcal{B})$.

Remark 2. Setting $b = I$ or $a = I$ we see that $at \in \mathbb{B}_1(\mathcal{H})$ and $\overline{tb} \in \mathbb{B}_1(\mathcal{H})$ for $t \in \mathbb{B}_1(\mathcal{B}, \mathcal{A})$, $a \in \mathcal{A}$ and $b \in \mathcal{B}$. In particular, $\mathbb{B}_1(\mathcal{B}, \mathcal{A}) \subseteq \mathbb{B}_1(\mathcal{H})$. It is not difficult to check that $\mathbb{B}_1(\mathcal{H}, \mathcal{A})$ is the space $\mathbb{B}_1(\mathcal{B}, \mathcal{A})$, where $\mathcal{B}$ is the O*-family $\mathbb{B}(\mathcal{H})$ on $\mathcal{D}(\mathcal{B}) = \mathcal{H}$.

Remark 3. Let $\sum_n \lambda_n(\psi_n \otimes \varphi_n)$ be a canonical representation for $t \in \mathbb{B}_1(\mathcal{B}, \mathcal{A})$. Then $\varphi_n \in \mathcal{D}(\mathcal{A})$ and $\psi_n \in \mathcal{D}(\mathcal{B})$ for all $n \in \mathbb{N}$. Indeed, if $\lambda_n \neq 0$, then $\varphi_n = \lambda_n^{-1} t\psi_n \in \mathcal{D}(\mathcal{A})$ and $\psi_n = \overline{\lambda_n}^{-1} t^* \varphi_n \in \mathcal{D}(\mathcal{B})$. If $\lambda_n = 0$, then $\varphi_n = \psi_n = 0$ by the definition of a canonical representation.

Lemma 5.1.4. (i) $\mathbb{B}_1(\mathcal{B}, \mathcal{A})^* = \mathbb{B}_1(\mathcal{A}, \mathcal{B})$.

(ii) *$\mathbb{B}_1(\mathcal{B}, \mathcal{A})$ is a subalgebra of $\mathbb{B}(\mathcal{H})$.*

(iii) *$\mathbb{B}_1(\mathcal{A})$ is a $*$-subalgebra of $\mathbb{B}(\mathcal{H})$.*

Proof. (i): Let $t \in \mathbb{B}_1(\mathcal{B}, \mathcal{A})$, $a \in \mathcal{A}$ and $b \in \mathcal{B}$. Then $\overline{a^+tb^+} \in \mathbb{B}_1(\mathcal{H})$ and so $(\overline{a^+tb^+})^* \in \mathbb{B}_1(\mathcal{H})$. Since $(\overline{a^+tb^+})^* \supseteq bt^*a$, this yields $\overline{bt^*a} \in \mathbb{B}_1(\mathcal{H})$, so $\mathbb{B}_1(\mathcal{B}, \mathcal{A})^* \subseteq \mathbb{B}_1(\mathcal{A}, \mathcal{B})$, and the assertion follows by symmetry. For (ii) it suffices to note that $\overline{at_1t_2b} = at_1\overline{t_2b}$ and $at_1, \overline{t_2b} \in \mathbb{B}_1(\mathcal{H})$ (cf. Remark 2) for $t_1, t_2 \in \mathbb{B}_1(\mathcal{B}, \mathcal{A})$, $a \in \mathcal{A}$ and $b \in \mathcal{B}$. (iii) follows at once from (i) and (ii). □

Definition 5.1.5. An O*-family $\mathcal{A}$ is said to be *self-adjoint* if $\mathcal{D}(\mathcal{A}) = \mathcal{D}^*(\mathcal{A}) := \bigcap_{a \in \mathcal{A}} \mathcal{D}(a^*)$.

Note that $\mathcal{D}(\mathcal{A})$ is always contained in $\mathcal{D}^*(\mathcal{A})$, since $\mathcal{A}$ is an O*-family.

For self-adjoint O*-families $\mathcal{A}$ and $\mathcal{B}$ the next proposition gives a characterization of $\mathbb{B}_1(\mathcal{B}, \mathcal{A})$ where the domain conditions $t\mathcal{H} \subseteq \mathcal{D}(\mathcal{A})$ and $t^*\mathcal{H} \subseteq \mathcal{D}(\mathcal{B})$ do not occur. For this we need a lemma.

Lemma 5.1.6. *Suppose $t \in \mathbb{B}(\mathcal{H})$. Then $\overline{ta} \in \mathbb{B}_1(\mathcal{H})$ for all $a \in \mathcal{A}$ if and only if $t^*\mathcal{H} \subseteq \mathcal{D}^*(\mathcal{A})$ and $a^*t^* \in \mathbb{B}_1(\mathcal{H})$ for all $a \in \mathcal{A}$.*

Proof. Assume that $\overline{ta} \in \mathbb{B}_1(\mathcal{H})$ for all $a \in \mathcal{A}$. Since, in particular, ta is bounded, we have $\langle a\varphi, t^*\psi \rangle = \langle ta\varphi, \psi \rangle = \langle \varphi, (ta)^* \psi \rangle$ for $\psi \in \mathcal{H}$ and $\varphi \in \mathcal{D}(\mathcal{A})$. This yields $t^*\psi \in \mathcal{D}(a^*)$ and so $t^*\mathcal{H} \subseteq \bigcap_{a \in \mathcal{A}} \mathcal{D}(a^*) \equiv \mathcal{D}^*(\mathcal{A})$. Since $\overline{ta} \in \mathbb{B}_1(\mathcal{H})$, $(\overline{ta})^* \in \mathbb{B}_1(\mathcal{H})$. Because $(\overline{ta})^* \supseteq a^*t^*$, we get $a^*t^* \in \mathbb{B}_1(\mathcal{H})$. Now we verify the opposite direction. Let $a \in \mathcal{A}$. Since $a^*t^* \in \mathbb{B}_1(\mathcal{H})$, $(a^*t^*)^* \in \mathbb{B}_1(\mathcal{H})$. But $ta \subseteq (a^*t^*)^*$, so that $\overline{ta} \in \mathbb{B}_1(\mathcal{H})$. □

Proposition 5.1.7. *The O*-families $\mathcal{A}$ and $\mathcal{B}$ are both self-adjoint if and only if $\mathbb{B}_1(\mathcal{B}, \mathcal{A}) = \{t \in \mathbb{B}(\mathcal{H})\colon (tb)^* a^*$ is closable and $\overline{(tb)^* a^*} \in \mathbb{B}_1(\mathcal{H})$ for all $a \in \mathcal{A}$ and $b \in \mathcal{B}\}$.*

Proof. Throughout this proof let ${}_1\mathbb{B}(\mathcal{B}, \mathcal{A})$ denote the set on the right-hand side of the equality sign. First we note that $\mathbb{B}_1(\mathcal{B}, \mathcal{A}) \subseteq {}_1\mathbb{B}(\mathcal{B}, \mathcal{A})$. Suppose $t \in \mathbb{B}_1(\mathcal{B}, \mathcal{A})$. Then $(\overline{atb})^* \in \mathbb{B}_1(\mathcal{H})$ and so, since $(\overline{atb})^* \supseteq (tb)^* a^*$, $\overline{(tb)^* a^*} \in \mathbb{B}_1(\mathcal{H})$ for $a \in \mathcal{A}$ and $b \in \mathcal{B}$, i.e., $t \in {}_1\mathbb{B}(\mathcal{B}, \mathcal{A})$.

Suppose that $\mathcal{A}$ and $\mathcal{B}$ are self-adjoint. Let $t \in {}_1\mathbb{B}(\mathcal{B}, \mathcal{A})$. Setting $b = I$ we conclude that $\overline{t^*a^*} = \overline{t^*a^+} \in \mathbb{B}_1(\mathcal{H})$ for all $a \in \mathcal{A}$. Therefore, by Lemma 5.1.6 (applied to t^*), $(t^*)^* \mathcal{H} \equiv t\mathcal{H} \subseteq \mathcal{D}^*(\mathcal{A}) = \mathcal{D}(\mathcal{A})$. Letting $a = I$ we see that $(tb)^*$ and so $\overline{tb} = (tb)^{**}$ is in $\mathbb{B}_1(\mathcal{H})$ for $b \in \mathcal{B}$. Hence $t^*\mathcal{H} \subseteq \mathcal{D}^*(\mathcal{B}) = \mathcal{D}(\mathcal{B})$ again by Lemma 5.1.6. Let $a \in \mathcal{A}$ and $b \in \mathcal{B}$. Since the closure of $(tb^+)^* (a^+)^*$ is in $\mathbb{B}_1(\mathcal{H})$ by the definition of ${}_1\mathbb{B}(\mathcal{B}, \mathcal{A})$ and $bt^*a \subseteq (tb^+)^* (a^+)^*$, we obtain that $\overline{bt^*a} \in \mathbb{B}_1(\mathcal{H})$. The preceding facts together prove that $t^* \in \mathbb{B}_1(\mathcal{A}, \mathcal{B})$, so $t \in \mathbb{B}_1(\mathcal{B}, \mathcal{A})$ by Lemma 5.1.4, (i). Since always $\mathbb{B}_1(\mathcal{B}, \mathcal{A}) \subseteq {}_1\mathbb{B}(\mathcal{B}, \mathcal{A})$, we have shown that $\mathbb{B}_1(\mathcal{B}, \mathcal{A}) = {}_1\mathbb{B}(\mathcal{B}, \mathcal{A})$.

Conversely, assume that $\mathbb{B}_1(\mathcal{B}, \mathcal{A}) = {}_1\mathbb{B}(\mathcal{B}, \mathcal{A})$. Let $\zeta \in \mathcal{D}^*(\mathcal{A})$ and $\eta \in \mathcal{D}^*(\mathcal{B})$ be unit vectors. Put $t := \eta \otimes \zeta$. It is not difficult to check that $(tb)^* a^*\varphi = \langle \varphi, (a^+)^* \zeta \rangle b^*\eta$ for $\varphi \in \mathcal{D}(\mathcal{A})$ and hence $\overline{(tb)^* a^*} = (a^+)^* \zeta \otimes b^*\eta \in \mathbb{B}_1(\mathcal{H})$ for $a \in \mathcal{A}$ and $b \in \mathcal{B}$; so $t \in {}_1\mathbb{B}(\mathcal{B}, \mathcal{A})$. By the equality $\mathbb{B}_1(\mathcal{B}, \mathcal{A}) = {}_1\mathbb{B}(\mathcal{B}, \mathcal{A})$, $t \in \mathbb{B}_1(\mathcal{B}, \mathcal{A})$ and hence $\zeta = t\eta \in \mathcal{D}(\mathcal{A})$ and $\eta = t^*\zeta \in \mathcal{D}(\mathcal{B})$. This implies that $\mathcal{D}^*(\mathcal{A}) \subseteq \mathcal{D}(\mathcal{A})$ and $\mathcal{D}^*(\mathcal{B}) \subseteq \mathcal{D}(\mathcal{B})$. Since the converse inclusions are always true, this shows that the O*-families $\mathcal{A}$ and $\mathcal{B}$ are self-adjoint. □

Lemma 5.1.8. *If $\mathcal{A}, \mathcal{A}_0, \mathcal{B}, \mathcal{B}_0$ are directed O*-families on domains $\mathcal{D}(\mathcal{A}_0) = \mathcal{D}(\mathcal{A})$ resp. $\mathcal{D}(\mathcal{B}_0) = \mathcal{D}(\mathcal{B})$ such that $\mathrm{t}_{\mathcal{A}_0} = \mathrm{t}_{\mathcal{A}}$ and $\mathrm{t}_{\mathcal{B}_0} = \mathrm{t}_{\mathcal{B}}$, then $\mathbb{B}_1(\mathcal{B}, \mathcal{A}) = \mathbb{B}_1(\mathcal{B}_0, \mathcal{A}_0)$.*

Proof. Suppose $t \in \mathbb{B}_1(\mathcal{B}_0, \mathcal{A}_0)$, $a \in \mathcal{A}$ and $b \in \mathcal{B}$. Since $t_{\mathcal{A}_0} = t_{\mathcal{A}}$ and $t_{\mathcal{B}_0} = t_{\mathcal{B}}$, it follows that there are operators $a_0 \in \mathcal{A}_0$, $b_0 \in \mathcal{B}_0$ and $x, y \in \mathbb{B}(\mathcal{H})$ such that $a = xa_0$ and $b^+ = yb_0$. Then $b \subseteq (b^+)^* = b_0^* y^*$ and so $atb \subseteq x(a_0tb_0^*)\, y^*$. Since $\overline{a_0tb_0^+} \in \mathbb{B}_1(\mathcal{H})$ because of $t \in \mathbb{B}_1(\mathcal{B}, \mathcal{A})$ and $a_0tb_0^+ \subseteq a_0tb_0^*$, we have $\overline{a_0tb_0^+} = \overline{a_0tb_0^*}$ and $\overline{atb} = xa_0tb_0^*y^* = x\overline{a_0tb_0^+}y^* \in \mathbb{B}_1(\mathcal{H})$, so $t \in \mathbb{B}_1(\mathcal{B}, \mathcal{A})$. By symmetry, $\mathbb{B}_1(\mathcal{B}, \mathcal{A}) = \mathbb{B}_1(\mathcal{B}_0, \mathcal{A}_0)$. □

The following terminology will be frequently used. Let $(\varphi_n : n \in \mathbb{N})$ and $(\psi_n : n \in \mathbb{N})$ be sequences of vectors in $\mathcal{D}(\mathcal{A})$ and $\mathcal{D}(\mathcal{B})$, respectively. Suppose $a \in \mathcal{A}(I)$ and $b \in \mathcal{B}(I)$. We shall say that the series $\sum_n \psi_n \otimes \varphi_n$ *converges absolutely with respect to a and b* if $\sum_n \|b\psi_n\|\, \|a\varphi_n\| < \infty$. Because of $a \in \mathcal{A}(I)$ and $b \in \mathcal{B}(I)$ this implies that the series $\sum_n \psi_n \otimes \varphi_n$ converges absolutely on $\mathcal{H}$, so it defines an operator t in $\mathbb{B}_1(\mathcal{H})$. We say that the series $\sum_n \psi_n \otimes \varphi_n$ *converges absolutely with respect to $\mathcal{A}$ and $\mathcal{B}$* if $\sum_n \|b\psi_n\|\, \|a\varphi_n\| < \infty$ for all $a \in \mathcal{A}$ and $b \in \mathcal{B}$. It is clear that the latter notion depends only on the graph topologies $t_{\mathcal{A}}$ and $t_{\mathcal{B}}$ rather than on $\mathcal{A}$ and $\mathcal{B}$.

Proposition 5.1.9. *Suppose that the O*-families $\mathcal{A}$ and $\mathcal{B}$ are directed, and let $t \in \mathbb{B}(\mathcal{H})$. Consider the following condition:*

(*) *For arbitrary $a \in \mathcal{A}(I)$ and $b \in \mathcal{B}(I)$ there exist a sequence $(\varphi_n : n \in \mathbb{N})$ in $\mathcal{D}(\mathcal{A})$ and a sequence $(\psi_n : n \in \mathbb{N})$ in $\mathcal{D}(\mathcal{B})$ such that the series $\sum_n \psi_n \otimes \varphi_n$ converges absolutely with respect to a and b and represents the operator t.*

(i) *If $\mathcal{A}$ and $\mathcal{B}$ are closed and t satisfies (*), then $t \in \mathbb{B}_1(\mathcal{B}, \mathcal{A})$. Moreover, then $\overline{atb^+} = \sum_n b\psi_n \otimes a\varphi_n$ and $\operatorname{Tr} atb^+ = \sum_n \langle a\varphi_n, b\psi_n\rangle$.*

(ii) If $t \in \mathbb{B}_1(\mathcal{B}, \mathcal{A})$, *then (*) is fulfilled.*

Proof. (i): Suppose $a \in \mathcal{A}(I)$ and $b \in \mathcal{B}(I)$. Let $\sum_n \psi_n \otimes \varphi_n$ be a series which exists by (*). Since $\sum_n \|b\psi_n\|\, \|a\varphi_n\| < \infty$, Lemma 5.1.1 shows that the operator $z := \sum_n b\psi_n \otimes a\varphi_n$ is in $\mathbb{B}_1(\mathcal{H})$ and $\operatorname{Tr} z = \sum_n \langle a\varphi_n, b\psi_n\rangle$. Suppose for a moment we know already that $t\mathcal{H} \subseteq \mathcal{D}(\mathcal{A})$. Then we have

$$\begin{aligned}\langle z\psi, \varphi\rangle &= \sum_n \langle \psi, b\psi_n\rangle \langle a\varphi_n, \varphi\rangle = \sum_n \langle b^+\psi, \psi_n\rangle \langle \varphi_n, a^+\varphi\rangle \\ &= \langle tb^+\psi, a^+\varphi\rangle = \langle atb^+\psi, \varphi\rangle\end{aligned}$$

for $\varphi \in \mathcal{D}(\mathcal{A})$ and $\psi \in \mathcal{D}(\mathcal{B})$, so $atb^+ = z \upharpoonright \mathcal{D}(\mathcal{B})$. Therefore, $\overline{atb^+} = z \in \mathbb{B}_1(\mathcal{H})$ and the above formulas for the operator atb^+ follow.

We show that $t\mathcal{H} \subseteq \mathcal{D}(\mathcal{A})$. Let $\varphi \in \mathcal{H}$. We retain the notation from the preceding paragraph. Since $\sum_n \|\psi_n\|\, \|a\varphi_n\| < \infty$, the series $\sum_n \langle \varphi, \psi_n\rangle \varphi_n$ converges in the Hilbert space $\mathcal{H}_a \equiv \big(\mathcal{D}(\bar{a}), \|\cdot\|_{\bar{a}}\big)$. It converges to $t\varphi$ in $\mathcal{H}$, so $t\varphi \in \mathcal{D}(\bar{a})$ for all $a \in \mathcal{A}(I)$. Since $\mathcal{A}$ is directed and closed, Proposition 2.2.12 yields $\mathcal{D}(\mathcal{A}) = \bigcap_{a \in \mathcal{A}(I)} \mathcal{D}(\bar{a})$. Hence $t\varphi \in \mathcal{D}(\mathcal{A})$ and $t\mathcal{H} \subseteq \mathcal{D}(\mathcal{A})$. A similar argument proves that $t^*\mathcal{H} \subseteq \mathcal{D}(\mathcal{B})$. Thus we have shown that $t \in \mathbb{B}_1(\mathcal{B}_0, \mathcal{A}_0)$, where $\mathcal{A}_0 := \mathcal{A}(I)$ and $\mathcal{B}_0 := \mathcal{B}(I)$. Since $\mathcal{A}$ and $\mathcal{B}$ are directed, $\mathcal{A}$, $\mathcal{A}_0$, $\mathcal{B}$ and $\mathcal{B}_0$ satisfy the assumptions of Lemma 5.1.8; hence $t \in \mathbb{B}_1(\mathcal{B}, \mathcal{A})$.

(ii): Fix $a \in \mathcal{A}(\mathrm{I})$ and $b \in \mathcal{B}(\mathrm{I})$. Since the operators $\overline{atb^+}$, $\overline{tb^+}$, $\overline{bt^*a^+}$ and $\overline{t^*a^+}$ are bounded (because of $t \in \mathbb{B}_1(\mathcal{B}, \mathcal{A})$), we conclude easily that $\overline{atb^+} = \bar{a}\overline{tb^+}$ and $\overline{bt^*a^+} = \bar{b}\overline{t^*a^+}$. Since $t \in \mathbb{B}_1(\mathcal{B}, \mathcal{A})$, $\overline{atb^+} \in \mathbb{B}_1(\mathcal{H})$. Let $\sum_n \lambda_n(\gamma_n \otimes \delta_n)$ be a canonical representation for $\overline{atb^+}$. Set $\mathbb{N}' := \{n \in \mathbb{N}: \lambda_n \neq 0\}$. Define $\zeta_n := \lambda_n^{-1}\, \overline{tb^+}\, \gamma_n$ and $\eta_n := \overline{\lambda_n^{-1}}\, \overline{t^*a^+}\, \delta_n$ for $n \in \mathbb{N}'$ and $\zeta_n = \eta_n := 0$ for $n \in \mathbb{N} \setminus \mathbb{N}'$. Since $\overline{atb^+} = \bar{a}\overline{tb^+}$, $\zeta_n \in \mathcal{D}(\bar{a})$ and $\bar{a}\zeta_n = \lambda_n^{-1}\overline{atb^+}\gamma_n = \delta_n$ for $n \in \mathbb{N}'$. Similarly, $\eta_n \in \mathcal{D}(\bar{b})$ and $\bar{b}\eta_n = \gamma_n$ for $n \in \mathbb{N}'$. Also, $\bar{a}\zeta_n = \delta_n = 0$ and $\bar{b}\eta_n = \gamma_n = 0$ for $n \in \mathbb{N} \setminus \mathbb{N}'$. Thus $\bar{a}\overline{tb^+} = \overline{atb^+} = \sum_n \lambda_n(\gamma_n \otimes \delta_n) = \sum_n \gamma_n \otimes \bar{a}(\lambda_n\zeta_n)$. Since $a \in \mathcal{A}(I)$, the latter implies that $\overline{tb^+} = \sum_n \gamma_n \otimes (\lambda_n\zeta_n)$, and this series converges absolutely on $\mathcal{H}$. Hence $bt^* = (\overline{tb^+})^* = \sum_n (\lambda_n\zeta_n) \otimes \gamma_n = \sum_n (\lambda_n\zeta_n) \otimes \bar{b}\eta_n$. By $b \in \mathcal{B}(I)$, this gives $t^* = \sum_n (\lambda_n\zeta_n) \otimes \eta_n$ and so $t = \sum_n \eta_n \otimes (\lambda_n\zeta_n)$. Since $\bar{b}\eta_n \otimes \bar{a}(\lambda_n\zeta_n) = \lambda_n(\gamma_n \otimes \delta_n)$ by construction, $\sum_n \|\bar{b}\eta_n\|\, \|\bar{a}(\lambda_n\zeta_n)\| < \infty$.

In order to get the desired representation for t (with vectors $\varphi_n \in \mathcal{D}(\mathcal{A})$ and $\psi_n \in \mathcal{D}(\mathcal{B})$), we proceed as follows. Let $n \in \mathbb{N}$. Since $\mathcal{D}(\mathcal{A})$ is dense in the Hilbert space $\mathcal{H}_a$, there is a sequence $(\zeta_{n,k}: k \in \mathbb{N})$ in $\mathcal{D}(\mathcal{A})$ such that $\zeta_{n,1} = 0$ and $\|\zeta_n - \zeta_{n,k}\|_{\bar{a}} \leqq 2^{-n-k}\|\zeta_n\|_{\bar{a}}$ for $k \in \mathbb{N}$, $k \geqq 2$. Likewise, there exists a sequence $(\eta_{n,k}: k \in \mathbb{N})$ in $\mathcal{D}(\mathcal{B})$ such that $\eta_{n,1} = 0$ and $\|\eta_n - \eta_{n,k}\|_{\bar{b}} \leqq 2^{-n-k}\|\eta_n\|_{\bar{b}}$ for $k \in \mathbb{N}$, $k \geqq 2$. Then $\zeta_n = \sum_k (\zeta_{n,k+1} - \zeta_{n,k})$ in $\mathcal{H}_a$ and $\eta_n = \sum_k (\eta_{n,k+1} - \eta_{n,k})$ in $\mathcal{H}_b$. From the preceding it follows easily that

$$\sum_{n,k,l} \|\eta_{n,k+1} - \eta_{n,k}\|_b\, \|\lambda_n(\zeta_{n,l+1} - \zeta_{n,l})\|_a < \infty$$

and

$$t = \sum_n \eta_n \otimes (\lambda_n\zeta_n) = \sum_n \sum_{k,l} (\eta_{n,k+1} - \eta_{n,k}) \otimes \big(\lambda_n(\zeta_{n,l+1} - \zeta_{n,l})\big).$$

Writing the last threefold sum as one sum, we obtain the required series. □

Remark 4. If $\sum_n \lambda_n(\psi_n \otimes \varphi_n)$ is a canonical representation of an operator $t \in \mathbb{B}_1(\mathcal{B}, \mathcal{A})$, then one might think that the series $\sum_n \psi_n \otimes (\lambda_n\varphi_n)$ converges absolutely w.r.t. $\mathcal{A}$ and $\mathcal{B}$. This is indeed true for O*-algebras $\mathcal{A}$ and $\mathcal{B}$ (cf. Proposition 5.1.12), but not for general directed O*-families even not if $\mathcal{A} = \mathcal{B}$ and $t = t^*$. However, if $t \in \mathbb{B}_1(\mathcal{A})_+$ and $\sum_n \lambda_n(\varphi_n \otimes \varphi_n)$ is a canonical representation for t, then the series $\sum_n \varphi_n \otimes (\lambda_n\varphi_n)$ converges absolutely w.r.t. $\mathcal{A}$ and $\mathcal{A}$. This follows immediately from Lemma 5.2.9 below.

The Algebra $\mathbb{B}_1(\mathcal{B}, \mathcal{A})$ for O*-Algebras

In this subsection $\mathcal{A}$ and $\mathcal{B}$ denote O*-algebras on the Hilbert space $\mathcal{H}$.

Lemma 5.1.10. *Suppose that t is an operator of $\mathbb{B}_1(\mathcal{H})$ such that $t\mathcal{H} \subseteq \mathcal{D}(\mathcal{A})$ and $t^*\mathcal{H} \subseteq \mathcal{D}(\mathcal{B})$. Let $\sum_n \lambda_n(\psi_n \otimes \varphi_n)$ be a canonical representation for t. If $at \in \mathbb{B}_1(\mathcal{H})$ for all $a \in \mathcal{A}$ and $bt^* \in \mathbb{B}_1(\mathcal{H})$ for all $b \in \mathcal{B}$, then the series $\sum_n \psi_n \otimes (\lambda_n\varphi_n)$ converges absolutely with respect to $\mathcal{A}$ and $\mathcal{B}$.*

Proof. First note that $\lambda_n\varphi_n \in \mathcal{D}(\mathcal{A})$ and $\psi_n \in \mathcal{D}(\mathcal{B})$ for $n \in \mathbb{N}$ by the definition of a canonical representation. Let $a \in \mathcal{A}$ and $b \in \mathcal{B}$. Set $\mathbb{N}' := \{n \in \mathbb{N}: \lambda_n \neq 0\}$. Recall that $\{\varphi_n : n \in \mathbb{N}'\}$ and $\{\psi_n : n \in \mathbb{N}'\}$ are orthonormal sets in $\mathcal{H}$. Therefore, since a^+at and b^+bt^* are in $\mathbb{B}_1(\mathcal{H})$ by assumption, Lemma 5.1.2 yields

$$\sum_{n\in\mathbb{N}'} |\langle a^+at\psi_n, \varphi_n\rangle| = \sum_{n\in\mathbb{N}'} |\langle a^+a(\lambda_n\varphi_n), \varphi_n\rangle| = \sum_{n\in\mathbb{N}} |\lambda_n| \, \|a\varphi_n\|^2 < \infty \tag{1}$$

and

$$\sum_{n\in\mathbb{N}'} |\langle b^+bt^*\varphi_n, \psi_n\rangle| = \sum_{n\in\mathbb{N}'} |\langle b^+b(\overline{\lambda_n}\psi_n), \psi_n\rangle| = \sum_{n\in\mathbb{N}} |\lambda_n| \, \|b\psi_n\|^2 < \infty. \tag{2}$$

By the Cauchy-Schwarz inequality,

$$\sum_n \|b\psi_n\| \, \|a(\lambda_n\varphi_n)\| \leqq \Big(\sum_n |\lambda_n| \, \|b\psi_n\|^2\Big)^{1/2} \Big(\sum_n |\lambda_n| \, \|a\varphi_n\|^2\Big)^{1/2} < \infty. \ \square$$

Lemma 5.1.11. *Let $(\varphi_n : n \in \mathbb{N})$ and $(\psi_n : n \in \mathbb{N})$ be sequences of vectors in $\mathcal{D}(\mathcal{A})$ and $\mathcal{D}(\mathcal{B})$, respectively, such that the series $\sum_n \psi_n \otimes \varphi_n$ converges absolutely w.r.t. $\mathcal{A}$ and $\mathcal{B}$. Suppose that the operator $t := \sum_n \psi_n \otimes \varphi_n$ maps $\mathcal{H}$ into $\mathcal{D}(\mathcal{A})$. Then, for all $a \in \mathcal{A}$ and $b \in \mathcal{B}$, $\overline{atb^+}$ is in $\mathbb{B}_1(\mathcal{H})$ and* $\operatorname{Tr} \overline{atb^+} = \sum_n \langle a\varphi_n, b\psi_n\rangle$.

Proof. Since $t\mathcal{H} \subseteqq \mathcal{D}(\mathcal{A})$ by assumption, the assertion follows by the same arguments as used in the first paragraph of the proof of Proposition 5.1.9. $\square$

Proposition 5.1.12. (i) *Let $t \in \mathbb{B}_1(\mathcal{B}, \mathcal{A})$ and let $\sum_n \lambda_n(\psi_n \otimes \varphi_n)$ be a canonical representation for t. Then the series $\sum_n \psi_n \otimes (\lambda_n\varphi_n)$ converges absolutely w.r.t. $\mathcal{A}$ and $\mathcal{B}$.*

(ii) *Assume that the spaces $\mathcal{D}_\mathcal{A}$ and $\mathcal{D}_\mathcal{B}$ are sequentially complete. Let $(\varphi_n : n \in \mathbb{N})$ and $(\psi_n : n \in \mathbb{N})$ be sequences in $\mathcal{D}(\mathcal{A})$ and $\mathcal{D}(\mathcal{B})$, respectively, such that the series $\sum_n \psi_n \otimes \varphi_n$ converges absolutely w.r.t. $\mathcal{A}$ and $\mathcal{B}$. Then the operator $t := \sum_n \psi_n \otimes \varphi_n$ belongs to $\mathbb{B}_1(\mathcal{B}, \mathcal{A})$ and $\overline{atb} \in \mathbb{B}_1(\mathcal{B}, \mathcal{A})$ for all $a \in \mathcal{A}$ and $b \in \mathcal{B}$.*

Proof. (i): Since $t \in \mathbb{B}_1(\mathcal{B}, \mathcal{A})$, $at \in \mathbb{B}_1(\mathcal{H})$ and $bt^* = (\overline{tb^+})^* \in \mathbb{B}_1(\mathcal{H})$ for $a \in \mathcal{A}$ and $b \in \mathcal{B}$, so the assertion follows from Lemma 5.1.10.

(ii): We first check that $t\mathcal{H} \subseteqq \mathcal{D}(\mathcal{A})$. Let $\varphi \in \mathcal{H}$. Since $\sum_n \|\psi_n\| \, \|a\varphi_n\| < \infty$ by assumption, the series $\sum_n \langle \varphi, \psi_n\rangle \, \varphi_n$ converges in the sequentially complete locally convex space $\mathcal{D}_\mathcal{A}$. Since its sum in $\mathcal{H}$ is $t\varphi$, this gives $t\varphi \in \mathcal{D}(\mathcal{A})$ and so $t\mathcal{H} \subseteqq \mathcal{D}(\mathcal{A})$. Similarly, $t^*\mathcal{H} \subseteqq \mathcal{D}(\mathcal{B})$. The operator t satisfies the assumptions of Lemma 5.1.11, hence $\overline{atb} \in \mathbb{B}_1(\mathcal{H})$ for $a \in \mathcal{A}$ and $b \in \mathcal{B}$. This shows that $t \in \mathbb{B}_1(\mathcal{B}, \mathcal{A})$. Let $a \in \mathcal{A}$ and $b \in \mathcal{B}$. Since $\mathcal{A}$ and $\mathcal{B}$ are O*-algebras, the series $\sum_n b^+\psi_n \otimes a\varphi_n$ also converges absolutely w.r.t. $\mathcal{A}$ and $\mathcal{B}$. It represents the operator $\overline{atb}$. Therefore, applying the preceding with $\overline{atb}$ in place of t, we obtain $\overline{atb} \in \mathbb{B}_1(\mathcal{B}, \mathcal{A})$. $\square$

We derive a number of corollaries.

Corollary 5.1.13. (i): $\mathbb{B}_1(\mathcal{B}, \mathcal{A})$
$= \{t \in \mathbb{B}(\mathcal{H}) : t\mathcal{H} \subseteqq \mathcal{D}(\mathcal{A}),\ t^*\mathcal{H} \subseteqq \mathcal{D}(\mathcal{B}),\ at \in \mathbb{B}_1(\mathcal{H})$ *and* $bt^* \in \mathbb{B}_1(\mathcal{H})$ *for* $a \in \mathcal{A}$ *and* $b \in \mathcal{B}\}$
$= \{t \in \mathbb{B}(\mathcal{H}) : t\mathcal{H} \subseteqq \mathcal{D}(\mathcal{A}),\ t^*\mathcal{H} \subseteqq \mathcal{D}(\mathcal{B}),\ \overline{tb} \in \mathbb{B}_1(\mathcal{H})$ *and* $\overline{t^*a} \in \mathbb{B}_1(\mathcal{H})$ *for* $a \in \mathcal{A}$ *and* $b \in \mathcal{B}\}$.

(ii): *Suppose that the O*-algebras $\mathcal{A}$ and $\mathcal{B}$ are self-adjoint. Then we have* $\mathbb{B}_1(\mathcal{B}, \mathcal{A}) = \{t \in \mathbb{B}(\mathcal{H}) : tb$ *and* t^*a *are closable,* $\overline{tb} \in \mathbb{B}_1(\mathcal{H})$ *and* $\overline{t^*a} \in \mathbb{B}_1(\mathcal{H})$ *for* $a \in \mathcal{A}$ *and* $b \in \mathcal{B}\}$.

Proof. (i): We have already noted that $t \in \mathbf{B}_1(\mathcal{B}, \mathcal{A})$ implies that $at \in \mathbf{B}_1(\mathcal{H})$ and $bt^* = (\overline{tb^+})^* \in \mathbf{B}_1(\mathcal{H})$ for $a \in \mathcal{A}$ and $b \in \mathcal{B}$. Conversely, let $t \in \mathbf{B}(\mathcal{H})$ be such that $t\mathcal{H} \subseteq \mathcal{D}(\mathcal{A})$, $t^*\mathcal{H} \subseteq \mathcal{D}(\mathcal{B})$, $at \in \mathbf{B}_1(\mathcal{H})$ and $bt^* \in \mathbf{B}_1(\mathcal{H})$ for $a \in \mathcal{A}$ and $b \in \mathcal{B}$. Then $t \in \mathbf{B}_1(\mathcal{H})$, and t satisfies the assumptions of Lemma 5.1.10 and so of Lemma 5.1.11. Therefore, $\overline{atb} \in \mathbf{B}_1(\mathcal{H})$ for $a \in \mathcal{A}$ and $b \in \mathcal{B}$; hence $t \in \mathbf{B}_1(\mathcal{B}, \mathcal{A})$. The second equality follows by applying the adjoint operation.

(ii) follows by combining (i) with Lemma 5.1.6. □

Corollary 5.1.14. (i) *If* $t \in \mathbf{B}_1(\mathcal{A})$ *and* $a, b \in \mathcal{A}$, *then* $\operatorname{Tr} atb = \operatorname{Tr} tba = \operatorname{Tr} bat$.

(ii) *Each operator* $t \in \mathbf{B}_1(\mathcal{A})$ *can be written as* $t = (t_1 - t_2) + \mathrm{i}(t_3 - t_4)$ *with* $t_1, t_2, t_3, t_4 \in \mathbf{B}_1(\mathcal{A})_+$.

Proof. (i): Proposition 5.1.12, (i), shows that t satisfies the assumptions of Lemma 5.1.11; so the assertion follows immediately from the last formula in Lemma 5.1.11.

(ii): Since $\mathbf{B}_1(\mathcal{A})$ is $*$-invariant (cf. Lemma 5.1.4), it suffices to assume that $t = t^* \in \mathbf{B}_1(\mathcal{A})$. Then t has a canonical representation $\sum_n \lambda_n(\varphi_n \otimes \varphi_n)$ with $\lambda_n \in \mathbb{R}$ for $n \in \mathbb{N}$. Define $t_1 := \sum_{n \in \mathbf{N}_+} \lambda_n(\varphi_n \otimes \varphi_n)$, where $\mathbf{N}_+ := \{n \in \mathbf{N} : \lambda_n > 0\}$. Letting e be the projection whose range is spanned by the set $\{\varphi_n : n \in \mathbf{N}_+\}$, we have $t_1\mathcal{H} \subseteq te\mathcal{H} \subseteq \mathcal{D}(\mathcal{A})$. Since the series $\sum_n \varphi_n \otimes (\lambda_n \varphi_n)$ (by Proposition 5.1.12,(i)) and so $\sum_{n \in \mathbf{N}_+} \varphi_n \otimes (\lambda_n \varphi_n)$ converges absolutely w.r.t. $\mathcal{A}$ and $\mathcal{A}$, Lemma 5.1.11 shows that $\overline{at_1b} \in \mathbf{B}_1(\mathcal{H})$ for all $a, b \in \mathcal{A}$. Hence $t_1 \in \mathbf{B}_1(\mathcal{A})_+$. Obviously, $t_2 := t_1 - t \geqq 0$. Since t and t_1 are in $\mathbf{B}_1(\mathcal{A})$, $t_2 \in \mathbf{B}_1(\mathcal{A})_+$. □

Corollary 5.1.15. *Suppose that* $\mathcal{D}_\mathcal{A}$ *and* $\mathcal{D}_\mathcal{B}$ *are sequentially complete.*

(i) *If* $t \in \mathbf{B}_1(\mathcal{B}, \mathcal{A})$, $a \in \mathcal{L}^+(\mathcal{D}_\mathcal{A})$ *and* $b \in \mathcal{L}^+(\mathcal{D}_\mathcal{B})$, *then* $\overline{atb} \in \mathbf{B}_1(\mathcal{B}, \mathcal{A})$ *and* $\overline{bt^*a} \in \mathbf{B}_1(\mathcal{A}, \mathcal{B})$.

(ii) $\mathbf{B}_1(\mathcal{A})$ *is a two-sided* $*$-*ideal in the* $*$-*algebra* $\mathcal{L}^+(\mathcal{D}_\mathcal{A})$.

Proof. (i): Set $\mathcal{A}_1 := \mathcal{L}^+(\mathcal{D}_\mathcal{A})$ and $\mathcal{B}_1 := \mathcal{L}^+(\mathcal{D}_\mathcal{B})$. Since $\mathrm{t}_{\mathcal{A}_1} = \mathrm{t}_\mathcal{A}$ and $\mathrm{t}_{\mathcal{B}_1} = \mathrm{t}_\mathcal{B}$ and hence $\mathbf{B}_1(\mathcal{B}, \mathcal{A}) = \mathbf{B}_1(\mathcal{B}_1, \mathcal{A}_1)$ by Lemma 5.1.8, we can assume without loss of generality that $\mathcal{A} = \mathcal{L}^+(\mathcal{D}_\mathcal{A})$ and $\mathcal{B} = \mathcal{L}^+(\mathcal{D}_\mathcal{B})$. But then the first assertion is stated in Proposition 5.1.12, (ii); the second one follows from Lemma 5.1.4, (i).
(ii) follows at once from (i). □

The Projective Topology on $\mathbf{B}_1(\mathcal{B}, \mathcal{A})$

In this subsection, $\mathcal{A}$ and $\mathcal{B}$ are directed O*-families on the Hilbert space $\mathcal{H}$.

First we introduce some seminorms. For $a \in \mathcal{A}$ and $b \in \mathcal{B}$, we define

$$\nu_{a,b}(t) := \nu(atb), \quad t \in \mathbf{B}_1(\mathcal{B}, \mathcal{A}).$$

Further, we define for $a \in \mathcal{A}(I)$, $b \in \mathcal{B}(I)$ and $t \in \mathbf{B}_1(\mathcal{B}, \mathcal{A})$

$$\|\cdot\|_b \widehat{\otimes}_\pi \|\cdot\|_a\,(t) := \inf \left\{ \sum_{n=1}^{\infty} \|\psi_n\|_b \, \|\varphi_n\|_a \right\}, \tag{3}$$

where the infimum is extended over all absolutely convergent series $\sum_n \psi_n \otimes \varphi_n$ with respect to a and b which represent the operator t.

Let τ_π denote the locally convex topology on $\mathbb{B}_1(\mathcal{B}, \mathcal{A})$ which is generated by the family of seminorms $\{\nu_{a,b}: a \in \mathcal{A} \text{ and } b \in \mathcal{B}\}$. We call τ_π the *projective topology* of $\mathbb{B}_1(\mathcal{B}, \mathcal{A})$.

The vector space of all finite rank operators in $\mathbb{B}_1(\mathcal{B}, \mathcal{A})$ is equal to $\mathbb{F}(\mathcal{D}(\mathcal{B}), \mathcal{D}(\mathcal{A}))$. Clearly, each operator in $\mathbb{F}(\mathcal{D}(\mathcal{B}), \mathcal{D}(\mathcal{A}))$ is of the form $\sum_{n=1}^{k} \psi_n \otimes \varphi_n$, where $\varphi_1, \ldots, \varphi_k \in \mathcal{D}(\mathcal{A})$, $\psi_1, \ldots, \psi_k \in \mathcal{D}(\mathcal{B})$ and $k \in \mathbb{N}$. To be somewhat more precise, this means that we have identified the element $z = \sum_{n=1}^{k} \psi_n \otimes \varphi_n$ in the algebraic tensor product $\mathcal{D}_{\mathcal{B}}^- \otimes \mathcal{D}_{\mathcal{A}}$ with the operator $\chi(z) := \sum_{n=1}^{k} \langle \cdot, \psi_n \rangle \varphi_n$ on $\mathcal{H}$. That is, in our notation the vector spaces $\mathcal{D}_{\mathcal{B}}^- \otimes \mathcal{D}_{\mathcal{A}}$ and $\mathbb{F}(\mathcal{D}(\mathcal{B}), \mathcal{D}(\mathcal{A}))$ coincide via the identifying map χ. The projective tensor topology on $\mathcal{D}_{\mathcal{B}}^- \otimes \mathcal{D}_{\mathcal{A}}$ is generated by the family of seminorms $\{\|\cdot\|_b \otimes_\pi \|\cdot\|_a : a \in \mathcal{A} \text{ and } b \in \mathcal{B}\}$; cf. p. 15. Recall that by definition

$$\|\cdot\|_b \otimes_\pi \|\cdot\|_a(t) = \inf \left\{ \sum_{n=1}^{k} \|\psi_n\|_b \|\varphi_n\|_a \right\}, \tag{4}$$

where the infimum is taken over all representations of the operator $t \in \mathcal{D}_{\mathcal{B}}^- \otimes \mathcal{D}_{\mathcal{A}} \equiv \mathbb{F}(\mathcal{D}(\mathcal{B}), \mathcal{D}(\mathcal{A}))$ as a finite sum $\sum_{n=1}^{k} \psi_n \otimes \varphi_n$ with $\varphi_1, \ldots, \varphi_k \in \mathcal{D}(\mathcal{A})$ and $\psi_1, \ldots, \psi_k \in \mathcal{D}(\mathcal{B})$.

Lemma 5.1.16. *Suppose that $a \in \mathcal{A}(I)$ and $b \in \mathcal{B}(I)$. Then we have*

$$\nu_{a,b^+}(t) = \|\cdot\|_b \mathbin{\widehat{\otimes}_\pi} \|\cdot\|_a(t) \quad \textit{for} \quad t \in \mathbb{B}_1(\mathcal{B}, \mathcal{A}) \tag{5}$$

and

$$\nu_{a,b^+}(t) = \|\cdot\|_b \otimes_\pi \|\cdot\|_a(t) \quad \textit{for} \quad t \in \mathbb{F}(\mathcal{D}(\mathcal{B}), \mathcal{D}(\mathcal{A})). \tag{6}$$

Proof. We first verify that for $t \in \mathbb{B}_1(\mathcal{B}, \mathcal{A})$

$$\nu_{a,b^+}(t) \leqq \|\cdot\|_b \mathbin{\widehat{\otimes}_\pi} \|\cdot\|_a(t). \tag{7}$$

Let $\varepsilon > 0$. Then there exists a representation of t as an absolutely convergent series $\sum_n \psi_n \otimes \varphi_n$ w.r.t. a and b such that $\sum_n \|\psi_n\|_b \|\varphi_n\|_a \leqq \|\cdot\|_b \mathbin{\widehat{\otimes}_\pi} \|\cdot\|_a(t) + \varepsilon$. Since $t\mathcal{H} \subseteq \mathcal{D}(\mathcal{A})$, the absolutely convergent series $\sum_n b\psi_n \otimes a\varphi_n$ on $\mathcal{H}$ represents the operator atb^+; see the proof of Proposition 5.1.9, (i). Therefore, by Lemma 5.1.1, $\nu_{a,b^+}(t) = \nu(atb^+) \leqq \sum_n \|b\psi_n\| \|a\varphi_n\|$ and so $\nu_{a,b^+}(t) \leqq \|\cdot\|_b \mathbin{\widehat{\otimes}_\pi} \|\cdot\|_a(t) + \varepsilon$. Since $\varepsilon > 0$ is arbitrary, (7) follows.

Next we show that for $t \in \mathbb{F}(\mathcal{D}(\mathcal{B}), \mathcal{D}(\mathcal{A}))$

$$\|\cdot\|_b \otimes_\pi \|\cdot\|_a (t) \leqq \nu_{a,b^+}(t). \tag{8}$$

We argue similarly as in the proof of Proposition 5.1.9, (ii). Let $\sum_n \lambda_n(\gamma_n \otimes \delta_n)$ be a canonical representation for the operator $\overline{atb^+} \in \mathbb{B}_1(\mathcal{H})$. Since $t \in \mathbb{F}(\mathcal{D}(\mathcal{B}), \mathcal{D}(\mathcal{A}))$, $\overline{atb^+} \in \mathbb{F}(\mathcal{H})$, so that the set $\mathbb{N}' := \{n \in \mathbb{N}: \lambda_n \neq 0\}$ is finite, and also $\overline{atb^+} = atb^+$ and $\overline{bt^*a^+} = bt^*a$. Hence $\varphi_n := \overline{tb^+}\gamma_n \in \mathcal{D}(\mathcal{A})$ and $\psi_n := \overline{\lambda_n^{-1} t^*a^+}\delta_n \in \mathcal{D}(\mathcal{B})$ for $n \in \mathbb{N}'$. Then $a\varphi_n = \lambda_n\delta_n$ and $b\psi_n = \gamma_n$ for $n \in \mathbb{N}$, so $\overline{atb^+} = atb^+ = \sum_{n \in \mathbb{N}'} \gamma_n \otimes a\varphi_n$. Since $\ker a = \{0\}$, this yields $\overline{tb^+} = \sum_{n \in \mathbb{N}'} \gamma_n \otimes \varphi_n$; hence $bt^* = (\overline{tb^+})^* = \sum_{n \in \mathbb{N}'} \varphi_n \otimes \gamma_n = \sum_{n \in \mathbb{N}'} \varphi_n \otimes b\psi_n$.

By $\ker b = \{0\}$, $t = \sum_{n\in\mathbb{N}'} \psi_n \otimes \varphi_n$. From the latter and (4), we obtain

$$\|\cdot\|_b \otimes_\pi \|\cdot\|_a (t) \leqq \sum_{n\in\mathbb{N}'} \|\psi_n\|_b \|\varphi_n\|_a = \sum_{n\in\mathbb{N}'} \|\gamma_n\| \|\lambda_n \delta_n\| = \sum_n |\lambda_n| = \nu(atb^+) = \nu_{a,b^+}(t)$$

which proves (8).

Since trivially $\|\cdot\|_b \widehat{\otimes}_\pi \|\cdot\|_a \leqq \|\cdot\|_b \otimes_\pi \|\cdot\|_a$ on $\mathbb{F}(\mathcal{D}(\mathcal{B}), \mathcal{D}(\mathcal{A}))$, (7) and (8) imply (6).

Now suppose $t \in \mathbb{B}_1(\mathcal{B}, \mathcal{A})$. Let $\sum_n \psi_n \otimes \varphi_n$ be an absolutely convergent series w.r.t. a and b which represents t. Let $\varepsilon > 0$. There exists a $k \in \mathbb{N}$ such that $\sum_{n\geqq k+1} \|\psi_n\|_b \|\varphi_n\|_a < \varepsilon$. Set $t_k := \sum_{n=1}^{k} \psi_n \otimes \varphi_n$. Then

$$\|\cdot\|_b \widehat{\otimes}_\pi \|\cdot\|_a (t - t_k) \leqq \sum_{n\geqq k+1} \|\psi_n\|_b \|\varphi_n\|_a < \varepsilon. \tag{9}$$

Since $t_k \in \mathbb{F}(\mathcal{D}(\mathcal{B}), \mathcal{D}(\mathcal{A}))$, we have $\nu_{a,b^+}(t_k) = \|\cdot\|_b \otimes_\pi \|\cdot\|_a (t_k)$ by (6). Therefore,

$$\begin{aligned}\|\cdot\|_b \widehat{\otimes}_\pi \|\cdot\|_a (t) &\leqq \|\cdot\|_b \widehat{\otimes}_\pi \|\cdot\|_a (t - t_k) + \|\cdot\|_b \widehat{\otimes}_\pi \|\cdot\|_a (t_k)\\ &\leqq \varepsilon + \nu_{a,b^+}(t_k) \leqq \varepsilon + \nu_{a,b^+}(t) + \nu_{a,b^+}(t - t_k)\\ &\leqq \varepsilon + \nu_{a,b^+}(t) + \|\cdot\|_b \widehat{\otimes}_\pi \|\cdot\|_a (t - t_k) \leqq 2\varepsilon + \nu_{a,b^+}(t),\end{aligned}$$

where we used once more (7). Letting $\varepsilon \downarrow 0$, we get $\|\cdot\|_b \otimes_\pi \|\cdot\|_a (t) \leqq \nu_{a,b^+}(t)$. Together with (7), this gives (5). □

Corollary 5.1.17. *Let $\mathcal{A}_0 \subseteqq \mathcal{A}(I)$ and $\mathcal{B}_0 \subseteqq \mathcal{B}(I)$ be such that $\{\|\cdot\|_a : a \in \mathcal{A}_0\}$ and $\{\|\cdot\|_b : b \in \mathcal{B}_0\}$ are directed families of seminorms which generate the graph topologies $t_\mathcal{A}$ and $t_\mathcal{B}$, respectively. Then the family of seminorms $\{\|\cdot\|_b \widehat{\otimes}_\pi \|\cdot\|_a : a \in \mathcal{A}_0$ and $b \in \mathcal{B}_0\}$ is directed and determines the projective topology τ_π on $\mathbb{B}_1(\mathcal{B}, \mathcal{A})$.*

Proof. From the definition it is obvious that $\|\cdot\|_{b_1} \widehat{\otimes}_\pi \|\cdot\|_{a_1} \leqq \|\cdot\|_{b_2} \widehat{\otimes}_\pi \|\cdot\|_{a_2}$ when $\|\cdot\|_{b_1} \leqq \|\cdot\|_{b_2}$ and $\|\cdot\|_{a_1} \leqq \|\cdot\|_{a_2}$. Hence the family $\{\|\cdot\|_b \widehat{\otimes}_\pi \|\cdot\|_a : a \in \mathcal{A}_0$ and $b \in \mathcal{B}_0\}$ is directed. By (5) these seminorms are continuous on $\mathbb{B}_1(\mathcal{B}, \mathcal{A})\,[\tau_\pi]$. Let $a \in \mathcal{A}$ and $b \in \mathcal{B}$. The above assumptions imply that there are operators $a_0 \in \mathcal{A}_0$, $b_0 \in \mathcal{B}_0$ and $x, y \in \mathbb{B}(\mathcal{H})$ such that $a = xa_0$ and $b^+ = yb_0$. As shown in the proof of Lemma 5.1.8 we then have $\overline{atb} = \overline{xa_0tb_0^+y^*}$ for $t \in \mathbb{B}_1(\mathcal{B}, \mathcal{A})$. Hence $\nu_{a,b} \leqq \|x\| \|y^*\| \nu_{a_0,b_0^+} = \|x\| \|y^*\| (\|\cdot\|_{b_0} \widehat{\otimes}_\pi \|\cdot\|_{a_0})$ by (5). Thus the above family generates the topology τ_π on $\mathbb{B}_1(\mathcal{B}, \mathcal{A})$. □

The first assertion of the next corollary is the reason we call the topology τ_π on $\mathbb{B}_1(\mathcal{B},\mathcal{A})$ the projective topology.

Corollary 5.1.18. (i) *The projective tensor topology on $\mathcal{D}_\mathcal{B}^- \otimes \mathcal{D}_\mathcal{A} = \mathbb{F}(\mathcal{D}(\mathcal{B}), \mathcal{D}(\mathcal{A}))$ coincides with the induced topology of τ_π.*

(ii) *$\mathbb{F}(\mathcal{D}(\mathcal{B}), \mathcal{D}(\mathcal{A}))$ is dense in $\mathbb{B}_1(\mathcal{B}, \mathcal{A})\,[\tau_\pi]$.*

Proof. (i) follows easily from Corollary 5.1.17 and Lemma 5.1.16. We verify (ii). By Corollary 5.1.17, the topology τ_π is generated by the directed family of seminorms $\{\|\cdot\|_b \widehat{\otimes}_\pi \|\cdot\|_a : a \in \mathcal{A}(I)$ and $b \in \mathcal{B}(I)\}$. (Recall that we assumed that the O*-families $\mathcal{A}$ and $\mathcal{B}$ are directed!) Hence the formula (9) above shows that $\mathbb{F}(\mathcal{D}(\mathcal{B}), \mathcal{D}(\mathcal{A}))$ is dense in $\mathbb{B}_1(\mathcal{B}, \mathcal{A})\,[\tau_\pi]$. □

The following lemma is an auxiliary result which is also used in the proof of Corollary 5.4.7.

Lemma 5.1.19. *Let $\mathcal{E}$ and $\mathcal{G}$ be subspaces of the locally convex spaces $\mathcal{D}_{\mathcal{A}}$ and $\mathcal{D}_{\mathcal{B}}$, respectively, and let $\mathbb{B}_1(\mathcal{B}, \mathcal{A}; \mathcal{G}, \mathcal{E}) := \{t \in \mathbb{B}_1(\mathcal{B}, \mathcal{A}) : t\mathcal{H} \subseteq \mathcal{E}$ and $t^*\mathcal{H} \subseteq \mathcal{G}\}$. Suppose that $\mathcal{E}$ and $\mathcal{G}$ are complete (in the corresponding graph topologies). Then $\mathbb{B}_1(\mathcal{B}, \mathcal{A}; \mathcal{G}, \mathcal{E})\,[\tau_\pi]$ is complete. If $\mathcal{E}$ and $\mathcal{G}$ are Frechet spaces, then $\mathbb{B}_1(\mathcal{B}, \mathcal{A}; \mathcal{G}, \mathcal{E})\,[\tau_\pi]$ is also a Frechet space.*

Proof. First we prove that $\mathbb{B}_1(\mathcal{B},\mathcal{A}; \mathcal{G},\mathcal{E})\,[\tau_\pi]$ is complete. Let $(t_i : i \in I)$ be a Cauchy net in $\mathbb{B}_1(\mathcal{B}, \mathcal{A}; \mathcal{G}, \mathcal{E})\,[\tau_\pi]$. Suppose $a \in \mathcal{A}$ and $b \in \mathcal{B}$. Then $(\overline{at_ib} : i \in I)$ is a Cauchy net in the Banach space $(\mathbb{B}_1(\mathcal{H}), \nu)$. Hence there exists an operator $t_{a,b} \in \mathbb{B}_1(\mathcal{H})$ such that $\lim_i \nu(at_ib - t_{a,b}) = 0$. Let $t := t_{I,I}$.

We show that $t\mathcal{H} \subseteq \mathcal{E}$ and $t^*\mathcal{H} \subseteq \mathcal{G}$. Let $\varphi \in \mathcal{H}$. For $i, j \in I$, $a \in \mathcal{A}$ and $b \in \mathcal{B}$, we have

$$\|a(t_i\varphi - t_j\varphi)\| \leqq \|a(t_i - t_j)\| \, \|\varphi\| \leqq \nu_{a,I}(t_i - t_j) \, \|\varphi\|$$

and

$$\|b(t_i^*\varphi - t_j^*\varphi)\| \leqq \|b(t_i^* - t_j^*)\| \, \|\varphi\| = \|\overline{(t_i - t_j)b^+}\| \, \|\varphi\| \, \nu_{I,b^+}(t_i - t_j) \, \|\varphi\|.$$

Since $t_i \in \mathbb{B}_1(\mathcal{B}, \mathcal{A}; \mathcal{G}, \mathcal{E})$ and hence $t_i\mathcal{H} \subseteq \mathcal{E}$ and $t_i^*\mathcal{H} \subseteq \mathcal{G}$, the preceding implies that $(t_i\varphi : i \in I)$ and $(t_i^*\varphi : i \in I)$ are Cauchy nets in the complete locally convex spaces $\mathcal{E}$ and $\mathcal{G}$, respectively. Therefore, $t\varphi = \lim_i t_i\varphi$ in $\mathcal{E}$ and $t^*\varphi = \lim_i t_i^*\varphi$ in $\mathcal{G}$. Hence $t\varphi \in \mathcal{E}$ and $t^*\varphi \in \mathcal{G}$, so that $t\mathcal{H} \subseteq \mathcal{E}$ and $t^*\mathcal{H} \subseteq \mathcal{G}$. Suppose $a \in \mathcal{A}$ and $b \in \mathcal{B}$. Since also $\lim_i \|\overline{at_ib} - t_{a,b}\| = 0$, we conclude from $\langle at_ib\psi, \varphi\rangle = \langle t_ib\psi, a^+\varphi\rangle$ that $\langle t_{a,b}\psi, \varphi\rangle = \langle tb\psi, a^+\varphi\rangle = \langle atb\psi, \varphi\rangle$ for $\varphi \in \mathcal{D}(\mathcal{A})$ and $\psi \in \mathcal{D}(\mathcal{B})$, where we used that $t\mathcal{H} \subseteq \mathcal{D}(\mathcal{A})$. Consequently, $\overline{atb} = t_{a,b} \in \mathbb{B}_1(\mathcal{H})$. Thus $t \in \mathbb{B}_1(\mathcal{B}, \mathcal{A})$. From $t\mathcal{H} \subseteq \mathcal{E}$ and $t^*\mathcal{H} \subseteq \mathcal{G}$, $t \in \mathbb{B}_1(\mathcal{B}, \mathcal{A}; \mathcal{G}, \mathcal{E})$. From $\lim_i \nu(at_ib - t_{a,b}) = \lim_i \nu\big(a(t_i - t)\,b\big) = 0$ for $a \in \mathcal{A}$ and $b \in \mathcal{B}$ we see that $t = \lim_i t_i$ in $\mathbb{B}_1(\mathcal{B}, \mathcal{A}; \mathcal{G}, \mathcal{E})\,[\tau_\pi]$. Thus we have shown that $\mathbb{B}_1(\mathcal{B}, \mathcal{A}; \mathcal{G}, \mathcal{E})\,[\tau_\pi]$ is complete.

Assume now that $\mathcal{E}$ and $\mathcal{G}$ are Frechet spaces. Then the topologies of $\mathcal{E}$ and $\mathcal{G}$ are generated by directed families of seminorms $\{\|\cdot\|_{a_n} : n \in \mathbb{N}\}$ and $\{\|\cdot\|_{b_n} : n \in \mathbb{N}\}$, respectively, where $a_n \in \mathcal{A}$ and $b_n \in \mathcal{B}$ for $n \in \mathbb{N}$. Fix $a \in \mathcal{A}$ and $b \in \mathcal{B}$. There are numbers $n \in \mathbb{N}$ and $\lambda > 0$ such that $\|a\varphi\| \leqq \lambda \, \|a_n\varphi\|$, $\varphi \in \mathcal{E}$. Hence we can find an operator $x \in \mathbb{B}(\mathcal{H})$ such that $a\varphi = xa_n\varphi$, $\varphi \in \mathcal{E}$. Let $t \in \mathbb{B}_1(\mathcal{B}, \mathcal{A}; \mathcal{G}, \mathcal{E})$. Since $t\mathcal{H} \subseteq \mathcal{E}$, $atb = xa_ntb$, so that $\overline{atb} = x\overline{a_ntb}$. Similarly, there exist an $m \in \mathbb{N}$ and a $y \in \mathbb{B}(\mathcal{H})$ such that $\overline{b^+t^*a_n^+} = y\overline{b_mt^*a_n^+}$ for all $t \in \mathbb{B}_1(\mathcal{B}, \mathcal{A}; \mathcal{G}, \mathcal{E})$. Then

$$\begin{aligned} \nu_{a,b}(t) = \nu(atb) &\leqq \|x\| \, \nu(a_ntb) = \|x\| \, \nu\big((a_ntb)^*\big) = \|x\| \, \nu(b^+t^*a_n^+) \\ &\leqq \|x\| \, \|y\| \, \nu(b_mt^*a_n^+) = \|x\| \, \|y\| \, \nu(a_ntb_m^+) = \|x\| \, \|y\| \, \nu_{a_n,b_m^+}(t). \end{aligned}$$

This proves that the family of seminorms $\{\nu_{a_n,b_m^+} : n, m \in \mathbb{N}\}$ determines the topology τ_π on $\mathbb{B}_1(\mathcal{B}, \mathcal{A}; \mathcal{G}, \mathcal{E})$. Hence $\mathbb{B}_1(\mathcal{B}, \mathcal{A}; \mathcal{G}, \mathcal{E})\,[\tau_\pi]$ is metrizable. Since it is complete as just shown, $\mathbb{B}_1(\mathcal{B}, \mathcal{A}; \mathcal{G}, \mathcal{E})\,[\tau_\pi]$ is a Frechet space. □

Proposition 5.1.20. *Suppose $\mathcal{A}$ and $\mathcal{B}$ are closed (directed) O*-families on the Hilbert space $\mathcal{H}$. Then the locally convex space $\mathbb{B}_1(\mathcal{B}, \mathcal{A})\,[\tau_\pi]$ is complete. The identifying map χ of $\mathcal{D}_{\mathcal{B}}^- \otimes_\pi \mathcal{D}_{\mathcal{A}}$ and $\mathbb{F}\big(\mathcal{D}(\mathcal{B}), \mathcal{D}(\mathcal{A})\big)\,[\tau_\pi]$ extends by continuity to a topological isomorphism of the completed projective tensor product $\mathcal{D}_{\mathcal{B}}^- \widehat{\otimes}_\pi \mathcal{D}_{\mathcal{A}}$ and $\mathbb{B}_1(\mathcal{B}, \mathcal{A})\,[\tau_\pi]$. If $\mathcal{D}_{\mathcal{A}}$ and $\mathcal{D}_{\mathcal{B}}$ are Frechet spaces, then $\mathbb{B}_1(\mathcal{B}, \mathcal{A})\,[\tau_\pi]$ is also a Frechet space.*

Proof. The special case $\mathcal{E} = \mathcal{D}_{\mathcal{A}}, \mathcal{G} = \mathcal{D}_{\mathcal{B}}$ in Lemma 5.1.19 gives the first and the third assertion of the proposition. We verify the second statement. In terms of the identifying map χ Corollary 5.1.18, (ii), means that χ is a topological isomorphism of $\mathcal{D}_{\overline{\mathcal{B}}} \otimes_\pi \mathcal{D}_{\mathcal{A}}$ on $\mathbb{F}(\mathcal{D}(\mathcal{B}), \mathcal{D}(\mathcal{A}))\,[\tau_\pi]$. Therefore, χ extends by continuity to the completions of both spaces. By definition, the completion of $\mathcal{D}_{\overline{\mathcal{B}}} \otimes_\pi \mathcal{D}_{\mathcal{A}}$ is $\mathcal{D}_{\overline{\mathcal{B}}} \widehat{\otimes}_\pi \mathcal{D}_{\mathcal{A}}$. By Corollary 5.1.18, (ii), $\mathbb{F}(\mathcal{D}(\mathcal{B}), \mathcal{D}(\mathcal{A}))$ is dense in $\mathbb{B}_1(\mathcal{B}, \mathcal{A})\,[\tau_\pi]$. Since $\mathbb{B}_1(\mathcal{B}, \mathcal{A})\,[\tau_\pi]$ is also complete, it is a completion of $\mathbb{F}(\mathcal{D}(\mathcal{B}), \mathcal{D}(\mathcal{A}))\,[\tau_\pi]$. □

5.2. The Generalized Trace

Throughout this section $\mathcal{A}$ and $\mathcal{B}$ denote directed O*-families on the Hilbert space $\mathcal{H}$.

Let a be a fixed operator in $\mathcal{A}(I)$. We let $N(a)$ denote the set of all operators z on $\mathcal{D}(\mathcal{A})$ which have the following property: For each $b \in \mathcal{A}(I)$ there exist a sequence $(\varphi_n^{\mathsf{l}} : n \in \mathbb{N})$ in $\mathcal{H}^a$ and a sequence $(\varphi_n : n \in \mathbb{N})$ in $\mathcal{D}(\mathcal{A})$ such that

$$\sum_{n=1}^{\infty} \|\varphi_n^{\mathsf{l}}\|^a \, \|\varphi_n\|_b < \infty \tag{1}$$

and

$$z\varphi = \sum_{n=1}^{\infty} \langle \varphi, \varphi_n^{\mathsf{l}} \rangle \, \varphi_n \quad \text{for} \quad \varphi \in \mathcal{D}(\mathcal{A}). \tag{2}$$

Lemma 5.2.1. *Suppose that $z \in N(a)$. Then z is a continuous linear mapping of $\mathcal{D}_a \equiv (\mathcal{D}(\mathcal{A}), \|\cdot\|_a)$ into $\mathcal{D}_{\hat{\mathcal{A}}}$. Let z_a denote its continuous extension to a mapping of $\mathcal{H}_a \equiv (\mathcal{D}(\bar{a}), \|\cdot\|_{\bar{a}})$ into $\mathcal{D}_{\hat{\mathcal{A}}}$. Then z_a is a trace class operator on the Hilbert space $\mathcal{H}_a$. If (φ_n^{l}) and (φ_n) are sequences as above which satisfy* (1) *in case $a = b$ and* (2), *then*

$$\operatorname{Tr}_{\mathcal{H}_a} z_a = \sum_{n=1}^{\infty} \langle \varphi_n, \varphi_n^{\mathsf{l}} \rangle. \tag{3}$$

Proof. Suppose $b \in \mathcal{A}(I)$, and let (φ_n^{l}) and (φ_n) be as above. Let $\varphi \in \mathcal{D}(\mathcal{A})$. We have $\|\langle \varphi, \varphi_n^{\mathsf{l}} \rangle \varphi_n\|_b \leqq \|\varphi\|_a \|\varphi_n^{\mathsf{l}}\|^a \|\varphi_n\|_b$ for $n \in \mathbb{N}$. Combining this with (1) it follows that the series in (2) converges in the Hilbert space $\mathcal{H}_b$. Hence $z\varphi \in \mathcal{H}_b = \mathcal{D}(\bar{b})$ for all $b \in \mathcal{A}(I)$. By Proposition 2.2.12, $\hat{\mathcal{D}}(\mathcal{A}) = \bigcap_{b \in \mathcal{A}(I)} \mathcal{D}(\bar{b})$; so $z\varphi \in \hat{\mathcal{D}}(A)$. Applying the preceding inequality once more, we obtain

$$\|z\varphi\|_{\bar{b}} \leqq \sum_n \|\langle \varphi, \varphi_n^{\mathsf{l}} \rangle \varphi_n\|_b \leqq \Big(\sum_n \|\varphi_n^{\mathsf{l}}\|^a \|\varphi_n\|_b\Big) \|\varphi\|_a$$

for $\varphi \in \mathcal{D}(A)$. Since the sum in the parentheses is finite by (1), this estimate shows that z maps $\mathcal{D}_a$ continuously into $\mathcal{D}_{\hat{\mathcal{A}}}$. Since $\hat{\mathcal{D}}(\mathcal{A}) \subseteq \mathcal{D}(\bar{a})$, it is obvious that z_a maps $\mathcal{H}_a$ into itself. Now let (φ_n^{l}) and (φ_n) be sequences as above which satisfy (1) with $a = b$ and (2). Since the mapping $\xi \to \langle \cdot, \xi \rangle_{\bar{a}}$ is an isometric isomorphism of $\mathcal{H}_a$ onto $\mathcal{H}^a$ (see the discussion before Lemma 2.3.4), there exist vectors $\xi_n \in \mathcal{H}_a$, $n \in \mathbb{N}$, such that $\|\xi_n\|_{\bar{a}} = \|\varphi_n^{\mathsf{l}}\|^a$ and $\langle \cdot, \varphi_n^{\mathsf{l}} \rangle \equiv \langle \cdot, \xi_n \rangle_{\bar{a}}$ on $\mathcal{D}(\mathcal{A})$. Then $\sum_n \|\xi_n\|_{\bar{a}} \|\varphi_n\|_{\bar{a}} < \infty$ by (1). Therefore, we conclude from Lemma 5.1.1 that the operator y defined by $y\varphi := \sum_n \langle \varphi, \xi_n \rangle_{\bar{a}} \varphi_n$, $\varphi \in \mathcal{H}_a$, belongs to $\mathbb{B}_1(\mathcal{H}_a)$ and that $\operatorname{Tr} y = \sum_n \langle \varphi_n, \xi_n \rangle_{\bar{a}} = \sum_n \langle \varphi_n, \eta_n^{\mathsf{l}} \rangle$. Since $\langle \cdot, \xi_n \rangle_{\bar{a}} \equiv \langle \cdot, \varphi_n^{\mathsf{l}} \rangle$ by definition, we have $z\varphi = y\varphi$ for $\varphi \in \mathcal{D}(\mathcal{A})$. Hence $z_a = y$. □

Lemma 5.2.2. *Suppose that $a, c \in \mathcal{A}(I)$ and $\|\cdot\|_a \leqq \|\cdot\|_c$. Then $N(a) \subseteqq N(c)$ and $\mathrm{Tr}_{\mathcal{H}_a} z_a = \mathrm{Tr}_{\mathcal{H}_c} z_c$ for $z \in N(a)$.*

Proof. Since $\|\cdot\|_a \leqq \|\cdot\|_c$, $\mathcal{H}^a \subseteqq \mathcal{H}^c$ and $\|\cdot\|^c \leqq \|\cdot\|^a$. This yields $N(a) \subseteqq N(c)$. Let $z \in N(a)$. Then there exist a sequence $(\varphi_n^!)$ in $\mathcal{H}^a$ and a sequence (φ_n) in $\mathcal{D}(\mathcal{A})$ such that $\sum_n \|\varphi_n^!\|^a \|\varphi_n\|_c < \infty$ and (2) is fulfilled. Because $\|\cdot\|_a \leqq \|\cdot\|_c$ and $\|\cdot\|^c \leqq \|\cdot\|^a$, this gives $\sum_n \|\varphi_n^!\|^a \|\varphi_n\|_a < \infty$ and $\sum_n \|\varphi_n^!\|^c \|\varphi_n\|_c < \infty$. Therefore, by (3), $\mathrm{Tr}_{\mathcal{H}_a} z_a$ and $\mathrm{Tr}_{\mathcal{H}_c} z_c$ are both equal to $\sum_n \langle \varphi_n, \varphi_n^! \rangle$. □

Let $N(\mathcal{A})$ denote the union of all $N(a)$, where $a \in \mathcal{A}(I)$. Obviously, each $N(a)$ is a vector space. Since the O*-family $\mathcal{A}$ is assumed to be directed, Lemma 5.2.2 shows that $\{N(a) : a \in \mathcal{A}(I)\}$ is a directed family of vector spaces. In particular, we see that $N(\mathcal{A})$ is a vector space.

Now we define a "generalized trace" on $N(\mathcal{A})$. Suppose $z \in N(\mathcal{A})$. Then $z \in N(a)$ for some $a \in \mathcal{A}(I)$, and we define

$$\mathrm{tr}_{\mathcal{A}}\, z := \mathrm{Tr}_{\mathcal{H}_a} z_a. \tag{4}$$

In other words, if $(\varphi_n^! : n \in \mathbb{N})$ and $(\varphi_n : n \in \mathbb{N})$ are sequences in $\mathcal{H}^a$ and $\mathcal{D}(\mathcal{A})$, respectively, such that (1) in case $a = b$ and (2) are fulfilled, then, by (3), we have defined

$$\mathrm{tr}_{\mathcal{A}}\, z = \sum_{n=1}^{\infty} \langle \varphi_n, \varphi_n^! \rangle. \tag{5}$$

We have to check that the number $\mathrm{tr}_{\mathcal{A}} z$ depends only on the operator z, but not on a. Indeed, let $\tilde{a}$ be another operator in $\mathcal{A}(I)$ such that $z \in N(\tilde{a})$. Since $\mathcal{A}$ is directed, there exists $c \in \mathcal{A}(I)$ such that $\|\cdot\|_a \leqq \|\cdot\|_c$ and $\|\cdot\|_{\tilde{a}} \leqq \|\cdot\|_c$, so $\mathrm{Tr}_{\mathcal{H}_a} z_a = \mathrm{Tr}_{\mathcal{H}_c} z_c = \mathrm{Tr}_{\mathcal{H}_{\tilde{a}}} z_{\tilde{a}}$ by Lemma 5.2.2.

We call the number $\mathrm{tr}_{\mathcal{A}} z$ defined by (5) the *generalized trace* of the operator z in $N(\mathcal{A})$. When no confusion is possible we write simply $\mathrm{tr}\, z$ instead of $\mathrm{tr}_{\mathcal{A}} z$.

Remark 1. If $\mathcal{A} = \mathbf{B}(\mathcal{H})$, then $N(\mathcal{A}) = \mathbf{B}_1(\mathcal{H})$ and $\mathrm{tr}_{\mathcal{A}}\, z = \mathrm{Tr}_{\mathcal{H}}\, z$ for $z \in N(\mathcal{A}) = \mathbf{B}_1(\mathcal{H})$.

Lemma 5.2.3. *Suppose that $t \in \mathbf{B}_1(\mathcal{B}, \mathcal{A})$. Then there exists a unique linear mapping $\hat{t}$ of $\mathcal{D}_{\mathcal{B}}^+$ into $\hat{\mathcal{D}}(\mathcal{A})$ such that $\hat{t}$ is an extension of t and for arbitrary $a \in \mathcal{A}(I)$ and $b \in \mathcal{B}(I)$, $\hat{t}$ maps $\mathcal{H}^b$ continuously into $\mathcal{H}_a$. More precisely, if $a \in \mathcal{A}(I)$, $b \in \mathcal{B}(I)$ and if $\sum_n \psi_n \otimes \varphi_n$ is an absolutely convergent series with respect to a and b which represents the operator t, then*

$$\hat{t}\psi^! = \sum_{n=1}^{\infty} \langle \psi^!, \psi_n \rangle \varphi_n \quad \text{for all} \quad \psi^! \in \mathcal{H}^b. \tag{6}$$

Proof. Let us assume for a moment we have shown that for arbitrary $a \in \mathcal{A}(I)$ and $b \in \mathcal{B}(I)$ there exists a continuous linear mapping $t_{a,b}$ of $\mathcal{H}^b$ into $\mathcal{H}_a$ such that $t_{a,b} \restriction \mathcal{H} = t$. Let $a, a_1 \in \mathcal{A}(I)$ and $b, b_1 \in \mathcal{B}(I)$ be such that $\|\cdot\|_a \leqq \|\cdot\|_{a_1}$ and $\|\cdot\|_b \leqq \|\cdot\|_{b_1}$. Since $\mathcal{H}^b \subseteqq \mathcal{H}^{b_1}$ and $\mathcal{H}_{a_1} \subseteqq \mathcal{H}_a$ and the corresponding inclusions are continuous, we conclude that the restriction $t_{a_1,b_1} \restriction \mathcal{H}^b$ maps $\mathcal{H}^b$ continuously into $\mathcal{H}_a$. Since t_{a_1,b_1} and $t_{a,b}$ are both extensions of t and $\mathcal{H}$ is dense in $\mathcal{H}^b$ by Lemma 2.3.4, (iii), this implies that $t_{a_1,b_1} \restriction \mathcal{H}^b = t_{a,b}$. Now let $\psi^! \in \mathcal{D}_{\mathcal{B}}^+$. Since $\mathcal{B}$ is directed, there is a $b \in \mathcal{B}(I)$ such that $\psi^! \in \mathcal{H}^b$. Define $\hat{t}\psi^! := t_{a,b}\psi^!$, where a is some element of $\mathcal{A}(I)$. Since the O*-families $\mathcal{A}$ and $\mathcal{B}$ are directed, it follows from the preceding discussion that this definition is independent of the particular choices of a and b. Further, since $t_{a,b}$ maps $\mathcal{H}^b$ into $\mathcal{H}_a$, we have $\hat{t}\psi^\dagger \in \bigcap_{a\in\mathcal{A}(I)} \mathcal{H}_a \equiv \bigcap_{a\in\mathcal{A}(I)} \mathcal{D}(\bar{a})$. By Proposition 2.2.12, the latter is equal to $\hat{\mathcal{D}}(\mathcal{A})$, so $\hat{t}\psi^\dagger \in \hat{\mathcal{D}}(\mathcal{A})$ and the map $\hat{t}$ has the required properties. The uniqueness of $\hat{t}$ follows again from the density of $\mathcal{H}$ in the Hilbert space $\mathcal{H}^b$.

To complete the proof, it suffices to define mappings $t_{a,b}$ which have the above properties. Suppose $a \in \mathcal{A}(I)$ and $b \in \mathcal{B}(I)$. By Proposition 5.1.9, (i), there exists a series $\sum_n \psi_n \otimes \varphi_n$ as stated above. We define $t_{a,b}\psi^\dagger$, $\psi^\dagger \in \mathcal{H}^b$, by (6). Since $\sum_n \|\psi_n\|_b \|\varphi_n\|_a < \infty$ by assumption and $\|\langle\psi^\dagger, \psi_n\rangle \varphi_n\|_a \leqq \|\psi^\dagger\|^b \|\psi_n\|_b \|\varphi_n\|_a$ for $n \in \mathbb{N}$, we conclude that the series in (6) converges in the Hilbert space $\mathcal{H}_a$; so $t_{a,b}\psi^\dagger \in \mathcal{H}_a$ for all $\psi^\dagger \in \mathcal{H}^b$. The same inequalities show that the mapping $t_{a,b}$ defined in this way is continuous from $\mathcal{H}^b$ into $\mathcal{H}_a$. Since the series $\sum_n \psi_n \otimes \varphi_n$ represents the operator t, we have $t_{a,b} \restriction \mathcal{H} = t$ by (6). □

Proposition 5.2.4. *Suppose that $t \in \mathbb{B}_1(\mathcal{B}, \mathcal{A})$. Then for all x in $\mathcal{L}(\mathcal{D}_\mathcal{A}, \mathcal{D}_\mathcal{B}^+)$ the operator $\hat{t}x$, i.e., the composition of the mappings $\hat{t}$ (defined by Lemma 5.2.3) and x, belongs to $N(\mathcal{A})$. If $a \in \mathcal{A}(I)$ and $b \in \mathcal{B}(I)$ and if $\sum_n \psi_n \otimes \varphi_n$ is an absolutely convergent series with respect to a and b which represents the operator t, then*

$$\mathrm{tr}_\mathcal{A}\, \hat{t}x = \sum_{n=1}^{\infty} \langle x\varphi_n, \psi_n\rangle \quad \textit{for all} \quad x \in \mathcal{L}(\mathcal{D}_\mathcal{A}, \mathcal{D}_\mathcal{B}^+)_{a,b}. \tag{7}$$

Proof. Fix $x \in \mathcal{L}(\mathcal{D}_\mathcal{A}, \mathcal{D}_\mathcal{B}^+)_{a,b}$. Let $c \in \mathcal{A}(I)$. For a moment let $\sum_n \psi_n \otimes \varphi_n$ denote an absolutely convergent series w.r.t. c and b which represents t. (Such a series exists by Proposition 5.1.9.) Put $\varphi_n^\dagger := x^+\psi_n$, $n \in \mathbb{N}$. For $\varphi \in \mathcal{D}(\mathcal{A})$, $|\langle\varphi, \varphi_n^\dagger\rangle| = |\langle x\varphi, \psi_n\rangle| \leqq \mathfrak{l}_{a,b}(x) \|\varphi\|_a \|\psi_n\|_b$ and hence $\varphi_n^\dagger \in \mathcal{H}^a$ and $\|\varphi_n^\dagger\|^a \leqq \mathfrak{l}_{a,b}(x) \|\psi_n\|_b$ for $n \in \mathbb{N}$. Since $\sum_n \|\psi_n\|_b \|\varphi_n\|_c < \infty$, (1) follows. By (6),

$$\hat{t}x\varphi = \sum_n \langle x\varphi, \psi_n\rangle \varphi_n = \sum_n \langle \varphi, \varphi_n^\dagger\rangle \varphi_n \quad \text{for} \quad \varphi \in \mathcal{D}(\mathcal{A}).$$

This shows that $\hat{t}x \in N(a)$. Setting $a = c$ in the preceding, (7) follows from (5). □

Corollary 5.2.5. *Let $t \in \mathbb{B}_1(\mathcal{B}, \mathcal{A})$ and let $\mathcal{L}$ be a linear subspace of $\mathcal{L}(\mathcal{D}_\mathcal{A}, \mathcal{D}_\mathcal{B}^+)$. Define $f_t(x) := \mathrm{tr}_\mathcal{A}\, \hat{t}x$ for $x \in \mathcal{L}$. Then f_t is a continuous linear functional on $\mathcal{L}[\tau_{\mathrm{in}}]$.*

Proof. It is obvious that f_t is linear. If a, b and $\sum_n \psi_n \otimes \varphi_n$ are as in Proposition 5.2.4, then, by (7) and 3.2/(5),

$$|f_t(x)| \leqq \sum_n |\langle x\varphi_n, \psi_n\rangle| \leqq \Big(\sum_n \|\varphi_n\|_a \|\psi_n\|_b\Big) \mathfrak{l}_{a,b}(x)$$

for all $x \in \mathcal{L}_{a,b}$. Since the sum in the parentheses is finite, this proves that $f_t \restriction \mathcal{L}_{a,b}$ is continuous on $(\mathcal{L}_{a,b}, \mathfrak{l}_{a,b})$. Hence f_t is continuous on $\mathcal{L}[\tau_{\mathrm{in}}]$. □

Corollary 5.2.6. *Let $t \in \mathbb{B}_1(\mathcal{B}, \mathcal{A})$ and let $x \in \mathcal{L}(\mathcal{D}_\mathcal{A}, \mathcal{D}_\mathcal{B}^+)$. Suppose that $\langle x\cdot, \cdot\rangle \equiv \langle ya\cdot, b\cdot\rangle$ (cf. Proposition 3.2.3), where $a \in \mathcal{A}(I)$, $b \in \mathcal{B}(I)$ and $y \in \mathbb{B}(\mathcal{H})$. Then $\mathrm{tr}_\mathcal{A}\, \hat{t}x = \mathrm{Tr}\, y\overline{atb^+}$.*

Proof. First note that $y\overline{atb^+} \in \mathbb{B}_1(\mathcal{H})$, since $t \in \mathbb{B}_1(\mathcal{B}, \mathcal{A})$ and so $\overline{atb^+} \in \mathbb{B}_1(\mathcal{H})$. By Proposition 5.1.9, (ii), there exists a series $\sum_n \psi_n \otimes \varphi_n$ which converges absolutely w.r.t. a and b and which represents the operator t. Then $y\overline{atb^+}$ is represented by the absolutely convergent series $\sum_n b\psi_n \otimes ya\varphi_n$ on $\mathcal{H}$. Clearly, $x \in \mathcal{L}(\mathcal{D}_\mathcal{A}, \mathcal{D}_\mathcal{B}^+)_{a,b}$. Therefore, by Lemma 5.1.1 and (7), we have $\mathrm{Tr}\, y\overline{atb^+} = \sum_n \langle ya\varphi_n, b\psi_n\rangle = \sum_n \langle x\varphi_n, \psi_n\rangle = \mathrm{tr}_\mathcal{A}\, \hat{t}x$. □

Corollary 5.2.7. *If $t \in \mathbf{B}_1(\mathcal{H}, \mathcal{A})$ and $x \in \mathfrak{L}(\mathcal{D}_\mathcal{A}, \mathcal{H})$, then $xt \in \mathbf{B}_1(\mathcal{H})$ and $\operatorname{tr}_\mathcal{A} \hat{t}x = \operatorname{Tr} xt$.*

Proof. Recall that $\mathfrak{L}(\mathcal{D}_\mathcal{A}, \mathcal{H}) = \mathcal{L}(\mathcal{D}_\mathcal{A}, \mathcal{D}_\mathcal{B}^+)$ and $\mathbf{B}_1(\mathcal{H}, \mathcal{A}) = \mathbf{B}_1(\mathcal{B}, \mathcal{A})$ when $\mathcal{B} := \mathbf{B}(\mathcal{H})$ on $\mathcal{D}(\mathcal{B}) = \mathcal{H}$. Since each $x \in \mathfrak{L}(\mathcal{D}_\mathcal{A}, \mathcal{H})$ is of the form $x = ya$ for some $a \in \mathcal{A}(I)$ and $y \in \mathbf{B}(\mathcal{H})$, the assertion follows from Corollary 5.2.6 applied in case $b = I$, $\mathcal{B} = \mathbf{B}(\mathcal{H})$. □

Remark 2. In the notation of Corollary 5.2.7, the operator tx on $\mathcal{D}(\mathcal{A})$ is not closable in general. An example showing this can be obtained by setting $x = a$ and $t = \zeta \otimes \eta$, where a, ζ and η are chosen as in Remark 5 of 3.2.

Before we state the next corollary, we prove two auxiliary lemmas.

Lemma 5.2.8. *Let $\mathcal{D}$ be a dense linear subspace of an infinite dimensional Hilbert space $\mathcal{H}$ and let $\mathcal{K}$ be a separable closed linear subspace of $\mathcal{H}$. Then there exists an orthonormal sequence in $\mathcal{D}$ such that their closed linear span contains $\mathcal{K}$.*

Proof. Let $\mathcal{R}$ be a countable dense subset of $\mathcal{K}$. Since $\mathcal{D}$ is dense in $\mathcal{H}$, there is a countable subset $\mathcal{R}_1$ of $\mathcal{D}$ such that $\mathcal{R}$ is contained in the closure of $\mathcal{R}_1$ in $\mathcal{H}$. We write $\mathcal{R}_1$ as a sequence and apply the Gram-Schmidt procedure. □

Lemma 5.2.9. *Suppose $t \in \mathbf{B}_1(\mathcal{A})_+$, and let $\sum_n \lambda_n(\zeta_n \otimes \zeta_n)$ be a canonical representation for t. Then we have*

$$\sum_n \lambda_n \|a\zeta_n\|^2 < \infty \text{ for all } a \in \mathcal{A}. \tag{8}$$

Proof. We can assume that $\mathcal{H}$ is infinite dimensional. Fix $a \in \mathcal{A}$. Since $t \in \mathbf{B}_1(\mathcal{A})_+$, $s := \overline{ata^+} \in \mathbf{B}_1(\mathcal{H})$ and $s \geqq 0$. By Lemma 5.2.8. there exists an orthonormal sequence $(\psi_k : k \in \mathbb{N})$ of vectors in $\mathcal{D}(\mathcal{A})$ such that their closed linear span contains all vectors $a\zeta_n$, $n \in \mathbb{N}$. Then

$$\begin{aligned}\|s^{1/2}\psi_k\|^2 &= \langle s\psi_k, \psi_k\rangle = \langle ata^+\psi_k, \psi_k\rangle = \langle ta^+\psi_k, a^+\psi_k\rangle \\ &= \sum_n \lambda_n \langle a^+\psi_k, \zeta_n\rangle \langle \zeta_n, a^+\psi_k\rangle = \sum_n |\langle \psi_k, \lambda_n^{1/2} a\zeta_n\rangle|^2 \quad \text{for } k \in \mathbb{N}.\end{aligned}$$

Recall that $\lambda_n \geqq 0$ for $n \in \mathbb{N}$, because $t \geqq 0$. Since $s \in \mathbf{B}_1(\mathcal{H})$, $s^{1/2}$ is a Hilbert-Schmidt operator on $\mathcal{H}$ and so $\sum_k \|s^{1/2}\psi_k\|^2 < \infty$ (cf. Birman/Solomjak [1], ch. 11, § 3). By the preceding and the Parseval identity, we have

$$\sum_k \|s^{1/2}\psi_k\|^2 = \sum_n \sum_k |\langle \psi_k, \lambda_n^{1/2} a\zeta_n\rangle|^2 = \sum_n \|\lambda_n^{1/2} a\zeta_n\|^2 < \infty$$

which gives (8). □

Corollary 5.2.10. (i) *If $t = t^* \in \mathbf{B}_1(\mathcal{A})$, then $f_t(\cdot) \equiv \operatorname{tr}_\mathcal{A} \hat{t} \cdot$ is a hermitian linear functional on $\mathcal{L}(\mathcal{D}_\mathcal{A}, \mathcal{D}_\mathcal{A}^+)$.*

(ii) *If $t \in \mathbf{B}_1(\mathcal{A})_+$, then the linear functional $f_t(\cdot) \equiv \operatorname{tr}_\mathcal{A} \hat{t} \cdot$ on $\mathcal{L}(\mathcal{D}_\mathcal{A}, \mathcal{D}_\mathcal{A}^+)$ is strongly positive.*

Proof. (i): Let $x \in \mathcal{L}(\mathcal{D}_\mathcal{A}, \mathcal{D}_\mathcal{A}^+)$. Since $\mathcal{A}$ is directed, $x \in \mathcal{L}(\mathcal{D}_\mathcal{A}, \mathcal{D}_\mathcal{A}^+)_a$ for some $a \in \mathcal{A}(I)$. By Proposition 5.1.9, (ii), t is represented by a series $\sum_n \psi_n \otimes \varphi_n$ which converges absolutely w.r.t. a and a. Since $t = t^*$, the series $\sum_n \varphi_n \otimes \psi_n$ has the same property. Therefore, by (7), $f_t(x) = \operatorname{tr}_\mathcal{A} \hat{t}x = \sum_n \langle x\varphi_n, \psi_n\rangle = \sum_n \overline{\langle x^+\psi_n, \varphi_n\rangle} = \overline{\operatorname{tr}_\mathcal{A} \hat{t}x^+} = \overline{f_t(x^+)}$. Hence $f_t = (f_t)^+$.

(ii): If $t \in \mathbb{B}_1(\mathcal{A})_+$ and $\sum_n \lambda_n(\zeta_n \otimes \zeta_n)$ is a canonical representation for t, then Lemma 5.2.9 shows that the series $\sum_n \zeta_n \otimes (\lambda_n\zeta_n)$ converges absolutely w.r.t. $\mathcal{A}$ and $\mathcal{A}$. Further, $\lambda_n \geqq 0$ for all $n \in \mathbb{N}$, so the assertion follows from (7). □

We close this section by characterizing the ultraweakly continuous linear functionals in terms of the generalized trace; see also Remark 3 in 5.3.

Proposition 5.2.11. (i) *If the O*-families $\mathcal{A}$ and $\mathcal{B}$ are closed, and f is an ultraweakly continuous linear functional on $\mathcal{L}(\mathcal{D}_\mathcal{A}, \mathcal{D}_\mathcal{B}^+)$, then there is a $t \in \mathbb{B}_1(\mathcal{B}, \mathcal{A})$ such that $f(x) = \operatorname{tr}_\mathcal{A} \hat{t}x$ for all $x \in \mathcal{L}(\mathcal{D}_\mathcal{A}, \mathcal{D}_\mathcal{B}^+)$.*

(ii) *If $\mathcal{A}$ and $\mathcal{B}$ are O*-algebras and $t \in \mathbb{B}_1(\mathcal{B}, \mathcal{A})$, then the linear functional $f_t(\cdot) \equiv \operatorname{tr}_\mathcal{A} \hat{t}\,\cdot$ on $\mathcal{L}(\mathcal{D}_\mathcal{A}, \mathcal{D}_\mathcal{B}^+)$ is ultraweakly continuous.*

(iii) *If $t \in \mathbb{B}_1(\mathcal{A})_+$, then $f_t(\cdot) \equiv \operatorname{tr}_\mathcal{A} \hat{t}\,\cdot$ is an ultraweakly continuous linear functional on $\mathcal{L}(\mathcal{D}_\mathcal{A}, \mathcal{D}_\mathcal{A}^+)$.*

Proof. (i): By Proposition 3.5.2, there exist a sequence $(\varphi_n : n \in \mathbb{N})$ in $\mathcal{D}(\mathcal{A})$ and a sequence $(\psi_n : n \in \mathbb{N})$ in $\mathcal{D}(\mathcal{B})$ such that 3.5/(2) is satisfied and $f = \sum_n \omega_{\varphi_n,\psi_n}$. Combining 3.5/(2) with the Cauchy-Schwarz inequality we conclude that the series $\sum_n \psi_n \otimes \varphi_n$ converges absolutely w.r.t. a and b for all $a \in \mathcal{A}(I)$ and $b \in \mathcal{B}(I)$. Therefore, since $\mathcal{A}$ and $\mathcal{B}$ are closed (and directed by the assumption stated at the beginning of this section), Proposition 5.1.9, (i), ensures that this series represents an operator $t \in \mathbb{B}_1(\mathcal{B}, \mathcal{A})$. For $x \in \mathcal{L}(\mathcal{D}_\mathcal{A}, \mathcal{D}_\mathcal{B}^+)$, $f(x) = \sum_n \omega_{\varphi_n,\psi_n}(x) = \sum_n \langle x\varphi_n, \psi_n\rangle = \operatorname{tr}_\mathcal{A} \hat{t}x$ by (7).

(ii): Let $\sum_n \lambda_n(\eta_n \otimes \zeta_n)$ be a canonical representation for the operator $t \in \mathbb{B}_1(\mathcal{B}, \mathcal{A})$. For $n \in \mathbb{N}$, we set $\varphi_n := \lambda_n|\lambda_n|^{-1/2}\zeta_n$, $\psi_n := |\lambda_n|^{1/2}\eta_n$ if $\lambda_n \neq 0$ and $\varphi_n = \psi_n = 0$ otherwise. The inequalities 5.1/(1) and 5.1/(2) show that the sequences (φ_n) and (ψ_n) satisfy 3.5/(2). Further, $t = \sum_n \psi_n \otimes \varphi_n$ and hence $\operatorname{tr}_\mathcal{A} \hat{t}x = \sum_n \langle x\varphi_n, \psi_n\rangle = \sum_n \omega_{\varphi_n,\psi_n}(x)$, so f_t is ultraweakly continuous by Proposition 3.5.2.

(iii) follows in the same way as (ii) when we use the inequality (8) instead of 5.1./(1) and 5.1/(2). □

Proposition 5.2.12. *If $\mathcal{A}$ is a closed (directed) O*-family, then the following three conditions on a strongly positive linear functional f on $\mathcal{L}(\mathcal{D}_\mathcal{A}, \mathcal{D}_\mathcal{A}^+)$ are equivalent:*

(i) *There is a $t \in \mathbb{B}_1(\mathcal{A})_+$ such that $f(x) = \operatorname{tr}_\mathcal{A} \hat{t}x$ for $x \in \mathcal{L}(\mathcal{D}_\mathcal{A}, \mathcal{D}_\mathcal{A}^+)$.*

(ii) *f is ultraweakly continuous.*

(iii) *f is normal.*

Proof. The implications (i) → (ii) → (iii) are already shown by Proposition 5.2.11, (iii), and Corollary 3.5.8. We prove that (iii) implies (i).

Fix $a \in \mathcal{A}(I)$. Define $g_a(y) := f\big(R_a(Q_a y Q_a)\big)$, $y \in \mathbb{B}(\mathcal{H})$. Recall from Section 3.2 that Q_a is the projection onto the closure of $a\mathcal{D}(\mathcal{A})$ and $\langle R_a(Q_a y Q_a)\cdot, \cdot\rangle \equiv \langle ya\cdot, a\cdot\rangle$ by definition. We show that the linear functional g_a on $\mathbb{B}(\mathcal{H})$ is normal. Suppose $(y_i : i \in I)$ is a bounded monotone increasing net in $\mathbb{B}(\mathcal{H})_\mathrm{h}$. Let $y := \sup_i y_i$. Then $\big(R_a(Q_a y_i Q_a); i \in I\big)$ is clearly a bounded monotone increasing net in $\mathcal{L}(\mathcal{D}_\mathcal{A}, \mathcal{D}_\mathcal{A}^+)_\mathrm{h}$. Since y is the ultraweak limit of $(y_i : i \in I)$ in $\mathbb{B}(\mathcal{H})$, $R_a(Q_a y Q_a)$ is the ultraweak limit of $\big(R_a(Q_a y_i Q_a) : i \in I\big)$ by

Proposition 3.5.5. Hence $R_a(Q_a y Q_a) = \sup_i R_a(Q_a y_i Q_a)$ by Lemma 3.5.7. Since f is normal by (iii), this yields $g_a(y) = \lim_i g_a(y_i)$; so g_a is a normal positive linear functional on $\mathbb{B}(\mathcal{H})$. Thus there is an operator $s_a \in \mathbb{B}_1(\mathcal{H})_+$ such that $g_a(y) = \operatorname{Tr} s_a y$ (see e.g. KADISON/RINGROSE [2], 7.1.12).

Set $t := s_I$. Suppose $a \in \mathcal{A}(I)$. We define an operator $c_a \in \mathbb{B}(\mathcal{H})$ by $c_a(\bar{a}\varphi) := \varphi$ if $\varphi \in \mathcal{D}(\bar{a})$ and $c_a \psi := 0$ if $\psi \in \left(\bar{a}\mathcal{D}(\bar{a})\right)^\perp = (I - Q_a)\mathcal{H}$. It is not difficult to check that $c_a^* y c_a \in Q_a \mathbb{B}(\mathcal{H}) Q_a$ and $R_a(c_a^* y c_a) = y$ for $y \in \mathbb{B}(\mathcal{H})$. This gives $g_a(c_a^* y c_a) = f(y) = g_I(y)$ and so $\operatorname{Tr} s_a(c_a^* y c_a) = \operatorname{Tr} ty$ for all $y \in \mathbb{B}(\mathcal{H})$. Consequently, $c_a s_a c_a^* = t$ for all $a \in \mathcal{A}(I)$. Since $c_a\mathcal{H} = \mathcal{D}(\bar{a})$ for $a \in \mathcal{A}(I)$, the latter yields $t\mathcal{H} \subseteq \bigcap_{a \in \mathcal{A}(I)} \mathcal{D}(\bar{a}) = \mathcal{D}(\mathcal{A})$, where the last equality follows from Proposition 2.2.12 and the assumptions that $\mathcal{A}$ is closed and directed. From the relation $\operatorname{Tr} s_a y(I - Q_a) = g_a\left(y(I - Q_a)\right) = f\left(R_a\left(Q_a y(I - Q_a) Q_a\right)\right) = 0$ for $y \in \mathbb{B}(\mathcal{H})$ we see that $(I - Q_a) s_a = 0$; so $s_a\mathcal{H} \subseteq Q_a\mathcal{H}$. Combined with $c_a s_a c_a^* = t$ and $t\mathcal{H} \subseteq \mathcal{D}(\mathcal{A})$, this gives $s_a c_a^* = at$. Therefore, $c_a s_a = (s_a c_a^*)^* = (at)^* \supseteq t^* a^+ = ta^+$. Using once more that $s_a\mathcal{H} \subseteq Q_a\mathcal{H}$, this yields $s_a = \overline{ata^+}$. Since $s_a \in \mathbb{B}_1(\mathcal{H})$, it follows that $\overline{ata^+} = s_a \in \mathbb{B}_1(\mathcal{H})$. Now let a_1 and a_2 be arbitrary elements of $\mathcal{A}$. Since $\mathcal{A}$ is directed, there are operators $a \in \mathcal{A}(I)$ and $x, y \in \mathbb{B}(\mathcal{H})$ such that $a_1 = xa$ and $a_2^+ = ya$. Repeating the arguments from the proof of Lemma 5.1.8 we conclude from the latter and the fact $\overline{ata^+} \in \mathbb{B}_1(\mathcal{H})$ shown above that $\overline{a_1 t a_2} = x\, \overline{ata^+}\, y^* \in \mathbb{B}_1(\mathcal{H})$. Since $t\mathcal{H} \subseteq \mathcal{D}(\mathcal{A})$ and $t \geqq 0$, this proves that $t \in \mathbb{B}_1(\mathcal{A})_+$.

Now let $x \in \mathcal{L}(\mathcal{D}_\mathcal{A}, \mathcal{D}_\mathcal{A}^+)$. Then $x \in \mathcal{U}_a$ for some $a \in \mathcal{A}(I)$. By Proposition 3.2.3, there is a $y \in \mathbb{B}(\mathcal{H})$ such that $\langle x\cdot, \cdot\rangle \equiv \langle ya\cdot, a\cdot\rangle$ on $\mathcal{D}(\mathcal{A}) \times \mathcal{D}(\mathcal{A})$, i.e., $R_a(Q_a y Q_a) = x$. We have $f(x) = f\left(R_a(Q_a y Q_a)\right) = g_a(y) = \operatorname{Tr} s_a y = \operatorname{Tr} y s_a = \operatorname{Tr} yata^+ = \operatorname{tr}_\mathcal{A} \hat{t} x$, where the last equality is true by Corollary 5.2.6. □

5.3. Representation of Linear Functionals by Density Matrices

Throughout this section we suppose that $\mathcal{A}$ and $\mathcal{B}$ are directed O*-families acting on the Hilbert space $\mathcal{H}$.

Proposition 5.3.1. *Suppose that the O*-families $\mathcal{A}$ and $\mathcal{B}$ are closed. Let $\mathcal{L}$ be a linear subspace of $\mathcal{L}(\mathcal{D}_\mathcal{A}, \mathcal{D}_\mathcal{B}^+)$ that contains $\mathbb{F}(\mathcal{H})$. Let $\overline{\mathbb{F}}$ denote the closure of $\mathbb{F}(\mathcal{H})$ in $\mathcal{L}[\tau_{in}]$. Suppose f is a continuous linear functional on $\mathcal{L}[\tau_{in}]$. Then there exists a unique operator $t \in \mathbb{B}_1(\mathcal{B}, \mathcal{A})$ such that $f(x) = \operatorname{tr}_\mathcal{A} \hat{t} x$ for all $x \in \overline{\mathbb{F}}$. In particular, $f(x) = \operatorname{tr}_\mathcal{A} \hat{t} x$ for all $x \in \mathcal{L} \cap \mathcal{F}(\mathcal{D}_\mathcal{A}, \mathcal{D}_\mathcal{B}^+)$.*

Before proving the proposition, we derive two corollaries.

Corollary 5.3.2. *Let $\mathcal{A}$ and $\mathcal{B}$ be as in Proposition 5.3.1, and let $\mathcal{L}$ be an arbitrary linear subspace of $\mathcal{L}(\mathcal{D}_\mathcal{A}, \mathcal{D}_\mathcal{B}^+)$. If f is a continuous linear functional on $\mathcal{L}[\tau_b]$, then there is an operator $t \in \mathbb{B}_1(\mathcal{B}, \mathcal{A})$ such that $f(x) = \operatorname{tr}_\mathcal{A} \hat{t} x$ for all $x \in \mathcal{L} \cap \mathcal{F}(\mathcal{D}_\mathcal{A}, \mathcal{D}_\mathcal{B}^+)$.*

Proof. The Hahn-Banach theorem allows us to extend f to a continuous linear functional $\tilde{f}$ on $\mathcal{L}(\mathcal{D}_\mathcal{A}, \mathcal{D}_\mathcal{B}^+)\,[\tau_b]$. Since $\tau_b \subseteq \tau_{in}$, $\tilde{f}$ is continuous on $\mathcal{L}(\mathcal{D}_\mathcal{A}, \mathcal{D}_\mathcal{B}^+)\,[\tau_{in}]$, and the last statement in Proposition 5.3.1 applies to $\tilde{f}$. □

Corollary 5.3.3. *Suppose that $\mathcal{A}$ is a closed O*-algebra and $\mathcal{L}$ is a cofinal $*$-vector subspace of $\mathcal{L}(\mathcal{D}_\mathcal{A}, \mathcal{D}_\mathcal{A}^+)$. For each strongly positive linear functional f there exists an operator $t \in \mathbb{B}_1(\mathcal{A})_+$ such that $f(x) = \operatorname{tr}_\mathcal{A} \hat{t} x$ for $x \in \mathcal{L} \cap \mathcal{F}(\mathcal{D}_\mathcal{A}, \mathcal{D}_\mathcal{A}^+)$.*

Proof. By Lemma 1.3.2, f can be extended to a strongly positive linear functional $\tilde{f}$ on $\mathcal{L}(\mathcal{D}_{\mathcal{A}}, \mathcal{D}_{\mathcal{A}}^{+})$. Since $\tau_{\mathcal{O}} = \tau_{\text{in}}$ on $\mathcal{L}(\mathcal{D}_{\mathcal{A}}, \mathcal{D}_{\mathcal{A}}^{+})$ by Proposition 3.3.11, $\tilde{f}$ is continuous on $\mathcal{L}(\mathcal{D}_{\mathcal{A}}, \mathcal{D}_{\mathcal{A}}^{+})\,[\tau_{\text{in}}]$. From Proposition 5.3.1, $f(x) = \text{tr}_{\mathcal{A}}\, \hat{t}x$, $x \in \mathcal{L} \cap \mathcal{F}(\mathcal{D}_{\mathcal{A}}, \mathcal{D}_{\mathcal{A}}^{+})$, for some $t \in \mathbb{B}_1(\mathcal{A})$. Since $f(\varphi \otimes \varphi) = \text{tr}\, \hat{t}(\varphi \otimes \varphi) = \langle t\varphi, \varphi\rangle \geqq 0$ for all $\varphi \in \mathcal{H}$, $t \geqq 0$ and hence $t \in \mathbb{B}_1(\mathcal{A})_+$. □

Proof of Proposition 5.3.1. Since f is continuous on $\mathcal{L}[\tau_{\text{in}}]$, its restriction to each normed space $(\mathcal{L}_{a,b}, \mathfrak{l}_{a,b})$ is continuous. That is, for arbitrary $a \in \mathcal{A}$ and $b \in \mathcal{B}$ there is a constant $\lambda_{a,b} > 0$ such that

$$|f(x)| \leqq \lambda_{a,b}\mathfrak{l}_{a,b}(x) \quad \text{for all } x \in \mathcal{L}_{a,b}. \tag{1}$$

Suppose $a \in \mathcal{A}$, $b \in \mathcal{B}$, $\eta \in \mathcal{D}(a^*)$ and $\zeta \in \mathcal{D}(b^*)$. For $\varphi \in \mathcal{D}(\mathcal{A})$ and $\psi \in \mathcal{D}(\mathcal{B})$, we have

$$|\langle (a^*\eta \otimes b^*\zeta)\, \varphi, \psi\rangle| = |\langle \varphi, a^*\eta\rangle \langle b^*\zeta, \psi\rangle| \leqq \|\eta\|\, \|\zeta\|\, \|a\varphi\|\, \|b\psi\|,$$

i.e.,

$$\mathfrak{l}_{a,b}(a^*\eta \otimes b^*\zeta) \leqq \|\eta\|\, \|\zeta\|.$$

Hence, by (1),

$$|f(a^*\eta \otimes b^*\zeta)| \leqq \lambda_{a,b}\|\eta\|\, \|\zeta\|. \tag{2}$$

In case $a = I$, $b = I$ the preceding shows that $(\varphi, \psi) \to f(\psi \otimes \varphi)$ is a continuous sesquilinear form on $\mathcal{H} \times \mathcal{H}$, so there exists an operator $t \in \mathbb{B}(\mathcal{H})$ such that

$$f(\psi \otimes \varphi) = \langle t\varphi, \psi\rangle \quad \text{for all } \varphi, \psi \in \mathcal{H}. \tag{3}$$

We prove that $t\mathcal{H} \subseteq \mathcal{D}(\mathcal{A})$. Let $\zeta \in \mathcal{H}$ and $a \in \mathcal{A}$. Applying (2) in case $b = I$ and (1), we obtain $|\langle t\zeta, a^*\eta\rangle| = |f(a^*\eta \otimes \zeta)| \leqq \lambda_{a,I}\|\eta\|\, \|\zeta\|$ for all $\eta \in \mathcal{D}(a^*)$. Therefore, $t\zeta \in \mathcal{D}((a^*)^*) = \mathcal{D}(\bar{a})$. By assumption, $\mathcal{A}$ is closed and directed, so that $\mathcal{D}(\mathcal{A}) = \bigcap_{a \in \mathcal{A}} \mathcal{D}(\bar{a})$ by Proposition 2.2.12. Thus $t\zeta \in \mathcal{D}(\mathcal{A})$ and $t\mathcal{H} \subseteq \mathcal{D}(\mathcal{A})$. A similar argument shows that $t^*\mathcal{H} \subseteq \mathcal{D}(\mathcal{B})$.

Fix $a \in \mathcal{A}$ and $b \in \mathcal{B}$. By (3) and (2),

$$|\langle atb^+\zeta, \eta\rangle| = |\langle tb^*\zeta, a^*\eta\rangle| = |f(a^*\eta \otimes b^*\zeta)| \leqq \lambda_{a,b}\|\eta\|\, \|\zeta\|$$

for $\eta \in \mathcal{D}(\mathcal{A})$ and $\zeta \in \mathcal{D}(\mathcal{B})$. This shows that the operator atb^+ is bounded on $\mathcal{D}(\mathcal{B})$. We prove that $\overline{atb^+} \in \mathbb{B}_1(\mathcal{H})$. Let $\{\zeta_i : i \in I\}$ and $\{\eta_i : i \in I\}$ be orthonormal sets in $\mathcal{H}$. Suppose $\{i_1, \ldots, i_k\}$ is a finite subset of I. Since the domains $\mathcal{D}(\mathcal{A})$ and $\mathcal{D}(\mathcal{B})$ are dense in $\mathcal{H}$, for each $n \in \{1, \ldots, k\}$ there exist vectors $\gamma_n \in \mathcal{D}(\mathcal{A})$ and $\delta_n \in \mathcal{D}(\mathcal{B})$ such that

$$(1 + \|\overline{atb^+}\|)\,(\|\gamma_n - \eta_{i_n}\| + \|\delta_n - \zeta_{i_n}\|) \leqq 2^{-n}. \tag{4}$$

Further, we take a number $\alpha_n \in \mathbb{C}$, $|\alpha_n| = 1$, such that $|\langle tb^+\delta_n, a^+\gamma_n\rangle| = \langle t(\alpha_n b^+\delta_n), a^+\gamma_n\rangle$. Define $x := \sum_{n=1}^{k} a^+\gamma_n \otimes (\alpha_n b^+\delta_n)$. By (4) and the Bessel inequality, we have for $\varphi \in \mathcal{D}(\mathcal{A})$ and $\psi \in \mathcal{D}(\mathcal{B})$

$$\begin{aligned} |\langle x\varphi, \psi\rangle| &\leqq \sum_{n=1}^{k} |\langle a\varphi, \gamma_n\rangle \langle \delta_n, b\psi\rangle| \\ &= \sum_{n=1}^{k} |\langle a\varphi, \gamma_n - \eta_{i_n}\rangle \langle \delta_n, b\psi\rangle + \langle a\varphi, \eta_{i_n}\rangle \langle \delta_n - \zeta_{i_n}, b\psi\rangle + \langle a\varphi, \eta_{i_n}\rangle \langle \zeta_{i_n}, b\psi\rangle| \\ &\leqq \sum_{n=1}^{k} (\|a\varphi\|\, 2^{-n} \cdot 2\|b\psi\| + \|a\varphi\|\, 2^{-n}\|b\psi\|) + \|a\varphi\|\, \|b\psi\| \leqq 3\|a\varphi\|\, \|b\psi\|, \end{aligned}$$

i.e., $\mathfrak{l}_{a,b}(x) \leqq 3$ and so $|f(x)| \leqq 3\lambda_{a,b}$ by (1). Using (4) and (2), we obtain

$$\sum_{n=1}^{k} |\langle \overline{atb^+ \zeta_{i_n}, \eta_{i_n}} \rangle|$$

$$= \sum_{n=1}^{k} |\langle \overline{atb^+}(\zeta_{i_n} - \delta_n), \eta_{i_n}\rangle + \langle \delta_n, (atb^+)^* (\eta_{i_n} - \gamma_n)\rangle + \langle atb^+\delta_n, \gamma_n\rangle|$$

$$\leqq \sum_{n=1}^{k} (2^{-n} + 2 \cdot 2^{-n} + |\langle tb^+\delta_n, a^+\gamma_n\rangle|) \leqq 3 + \sum_{n=1}^{k} \langle t(\alpha_n b^+\delta_n), a^+\gamma_n\rangle$$

$$= 3 + \sum_{n=1}^{k} f\big(a^+\gamma_n \otimes (\alpha_n b^+ \delta_n)\big) = 3 + f(x) \leqq 3 + 3\lambda_{a,b}.$$

Therefore, $\sum_{i \in I} |\langle \overline{atb^+} \zeta_i, \eta_i\rangle| \leqq 3 + 3\lambda_{a,b}$. By Lemma 5.1.2 this yields $\overline{atb^+} \in \mathbb{B}_1(\mathcal{H})$. The preceding proves that $t \in \mathbb{B}_1(\mathcal{B}, \mathcal{A})$.

From (3) we see that $f(x) = \operatorname{tr}_{\mathcal{A}} \hat{t}x$ ($\equiv \operatorname{Tr} tx$) for all rank one operators x in $\mathbb{F}(\mathcal{H})$. By linearity this holds for all x in $\mathbb{F}(\mathcal{H})$. From Corollary 5.2.5, $f_t(\cdot) \equiv \operatorname{tr}_{\mathcal{A}} \hat{t}\,\cdot$ is a continuous linear functional on $\mathcal{L}[\tau_{\text{in}}]$. Since f is continuous on $\mathcal{L}[\tau_{\text{in}}]$ by assumption, the equality $f = f_t$ on $\mathbb{F}(\mathcal{H})$ extends by continuity to the closure $\overline{\mathbb{F}}$ of $\mathbb{F}(\mathcal{H})$ in $\mathcal{L}[\tau_{\text{in}}]$. That is, $f(x) = \operatorname{tr}_{\mathcal{A}} \hat{t}x$ for all $x \in \overline{\mathbb{F}}$. By Lemma 3.4.4, each $x \in \mathcal{L} \cap \mathcal{F}(\mathcal{D}_{\mathcal{A}}, \mathcal{D}_{\mathcal{B}}^+)$ belongs to $\overline{\mathbb{F}}$.

It only remains to verify the uniqueness assertion. If s is another operator in $\mathbb{B}_1(\mathcal{B}, \mathcal{A})$ such that $f(x) = \operatorname{tr}_{\mathcal{A}} \hat{s}x$ on $\overline{\mathbb{F}}$, then, in particular, $f(\psi \otimes \varphi) = \operatorname{tr}_{\mathcal{A}} \hat{s}(\psi \otimes \varphi) = \langle s\varphi, \psi\rangle$ for all $\varphi, \psi \in \mathcal{H}$. Combined with (3), this gives $s = t$. □

Our next objective is to characterize (under certain assumptions) the linear functionals of the form $f_t(\cdot) \equiv \operatorname{tr} \hat{t}\,\cdot$, $t \in \mathbb{B}_1(\mathcal{B}, \mathcal{A})$, as those linear functionals which are continuous in the topology τ_c defined now. The *precompact topology* τ_c is the locally convex topology on $\mathcal{L}(\mathcal{D}_{\mathcal{A}}, \mathcal{D}_{\mathcal{B}}^+)$ which is determined by the family of seminorms

$$p_{\mathcal{M},\mathcal{N}}(x) \equiv \sup_{\varphi \in \mathcal{M}} \sup_{\psi \in \mathcal{N}} |\langle x\varphi, \psi\rangle|, \quad x \in \mathcal{L}(\mathcal{D}_{\mathcal{A}}, \mathcal{D}_{\mathcal{B}}^+),$$

where $\mathcal{M}$ and $\mathcal{N}$ range over the precompact subsets of $\mathcal{D}_{\mathcal{A}}$ and $\mathcal{D}_{\mathcal{B}}$, respectively. Note that the family of these seminorms $p_{\mathcal{M},\mathcal{N}}$ is directed, since the union of finitely many precompact sets is again precompact.

Remark 1. In a complete semi-Montel space a set is precompact if and only if it is bounded. Therefore, if the O*-families $\mathcal{A}$ and $\mathcal{B}$ are closed and $\mathcal{D}_{\mathcal{A}}$ and $\mathcal{D}_{\mathcal{B}}$ are semi-Montel spaces, then the precompact topology τ_c on $\mathcal{L}(\mathcal{D}_{\mathcal{A}}, \mathcal{D}_{\mathcal{B}}^+)$ coincides with the bounded topology τ_b.

Proposition 5.3.4. *Suppose that the (directed) O*-families $\mathcal{A}$ and $\mathcal{B}$ are closed and $\mathcal{L}$ is a linear subspace of $\mathcal{L}(\mathcal{D}_{\mathcal{A}}, \mathcal{D}_{\mathcal{B}}^+)$. If f is a continuous linear functional on $\mathcal{L}[\tau_c]$, then there exists an operator $t \in \mathbb{B}_1(\mathcal{B}, \mathcal{A})$ such that $f(x) = \operatorname{tr}_{\mathcal{A}} \hat{t}x$ for all $x \in \mathcal{L}$.*

Proof. By the Hahn-Banach theorem, f extends to a continuous linear functional on $\mathcal{L}(\mathcal{D}_{\mathcal{A}}, \mathcal{D}_{\mathcal{B}}^+)[\tau_c]$. Therefore, it is sufficient to prove the assertion in case $\mathcal{L} = \mathcal{L}(\mathcal{D}_{\mathcal{A}}, \mathcal{D}_{\mathcal{B}}^+)$. Since the family of seminorms $p_{\mathcal{M},\mathcal{N}}$ of the above form is directed, there are precompact subsets $\mathcal{M}$ and $\mathcal{N}$ of $\mathcal{D}_{\mathcal{A}}$ and $\mathcal{D}_{\mathcal{B}}$, respectively, such that

$$|f(x)| \leqq p_{\mathcal{M},\mathcal{N}}(x) \equiv \sup \{|\langle x\varphi, \psi\rangle| : \varphi \in \mathcal{M} \text{ and } \psi \in \mathcal{N}\} \text{ for } x \in \mathcal{L}. \tag{5}$$

Without loss of generality we assume that $\mathcal{M}$ and $\mathcal{N}$ are closed in $\mathcal{D}_{\mathcal{A}}$ and $\mathcal{D}_{\mathcal{B}}$, respectively. Then $\mathcal{M}$ and $\mathcal{N}$ are compact in the corresponding graph topologies, since $\mathcal{D}_\mathcal{A}$ and $\mathcal{D}_\mathcal{B}$ are complete by assumption. For $x \in \mathcal{L}$, let h_x denote the continuous function on the compact Hausdorff space $\mathcal{M} \times \mathcal{N}$ which is defined by $h_x(\varphi, \psi) := \langle x\varphi, \psi\rangle$, $(\varphi, \psi) \in \mathcal{M} \times \mathcal{N}$. From (5) we see that the mapping $h_x \to f(x)$ is a continuous linear functional on the linear subspace $\{h_x : x \in \mathcal{L}\}$ of the Banach space $C(\mathcal{M} \times \mathcal{N})$ relative to the supremum norm. Let g denote a Hahn-Banach extension of this functional to $C(\mathcal{M} \times \mathcal{N})$. We can write g as $g = (g_1 - g_2) + \mathrm{i}(g_3 - g_4)$, where g_1, g_2, g_3 and g_4 are positive linear functionals on the C^*-algebra $C(\mathcal{M} \times \mathcal{N})$. Let $k \in \{1, 2, 3, 4\}$. Define a linear functional f_k on $\mathcal{L}$ by $f_k(x) := g_k(h_x)$, $x \in \mathcal{L}$. Since the positive linear functional g_k on $C(\mathcal{M} \times \mathcal{N})$ is continuous, we have $|f_k(\cdot)| \leqq \lambda_k p_{\mathcal{M},\mathcal{N}}(\cdot)$ for some constant $\lambda_k > 0$. Hence f_k is continuous on $\mathcal{L}[\tau_b]$. From Corollary 5.3.2, there is an operator $t_k \in \mathbb{B}_1(\mathcal{B}, \mathcal{A})$ such that

$$f_k(x) = \operatorname{tr} \hat{t}_k x \quad \text{for all} \quad x \in \mathcal{L} \cap \mathcal{F}(\mathcal{D}_\mathcal{A}, \mathcal{D}_\mathcal{B}^+) \equiv \mathcal{F}(\mathcal{D}_\mathcal{A}, \mathcal{D}_\mathcal{B}^+). \tag{6}$$

We shall prove below that $f_k(x) = \operatorname{tr} \hat{t}_k x$ for all x in $\mathcal{L}$. Suppose for a moment that this is done. Setting $t := (t_1 - t_2) + \mathrm{i}(t_3 - t_4)$ and using that $f = (f_1 - f_2) + \mathrm{i}(f_3 - f_4)$ by construction, we then have $f(x) = \operatorname{tr} \hat{t}x$ for all $x \in \mathcal{L}$, and the proof is complete. Now we fix $x \in \mathcal{L}$ and $k \in \{1, 2, 3, 4\}$ and we prove $f_k(x) = \operatorname{tr} \hat{t}_k x$.

By the Riesz representation theorem there exists a positive regular Borel measure μ_k on $\mathcal{M} \times \mathcal{N}$ such that

$$g_k(h) = \int_{\mathcal{M} \times \mathcal{N}} h(\varphi, \psi) \, \mathrm{d}\mu_k(\varphi, \psi) \quad \text{for } h \in C(\mathcal{M} \times \mathcal{N}).$$

In case $h = h_z$ this gives

$$f_k(z) = g_k(h_z) = \int \langle z\varphi, \psi\rangle \, \mathrm{d}\mu_k(\varphi, \psi) \quad \text{for all } z \in \mathcal{L}. \tag{7}$$

Since $x \in \mathcal{L}(\mathcal{D}_\mathcal{A}, \mathcal{D}_\mathcal{B}^+)$, there are $a \in \mathcal{A}(I)$ and $b \in \mathcal{B}(I)$ such that $x \in \mathcal{L}(\mathcal{D}_\mathcal{A}, \mathcal{D}_\mathcal{B}^+)_{a,b}$. Let $\bar{x}$ be the extension of $x \in \mathfrak{L}(\mathcal{D}_a, \mathcal{H}^b)$ (by Lemma 3.2.6) to a bounded operator of $\mathcal{H}_a$ into $\mathcal{H}^b$. Further, there exists a bounded operator y of $\mathcal{H}_b$ into $\mathcal{H}_a$ such that

$$\langle \bar{x}\varphi, \psi\rangle = \langle \varphi, y\psi\rangle_{\bar{a}} \quad \text{for} \quad \varphi \in \mathcal{D}(\bar{a}) \quad \text{and} \quad \psi \in \mathcal{D}(\mathcal{B}). \tag{8}$$

From the proof of Proposition 5.2.4 we know that $\hat{t}_k x \in N(a)$. Obviously, $(\hat{t}_k x)_a = \hat{t}_k \bar{x}$. By 5.2/(3), we have $\operatorname{tr} \hat{t}_k x = \operatorname{Tr}_{\mathcal{H}_a} (\hat{t}_k x)_a$. Since $\mathcal{M}$ is compact in $\mathcal{D}_\mathcal{A}$, $\mathcal{M}$ is also compact in the Hilbert space $\mathcal{H}_a$. By the same reason $\mathcal{N}$ and so $y(\mathcal{N})$ is compact in $\mathcal{H}_a$. Hence there is a separable subspace $\mathcal{K}$ of the Hilbert space $\mathcal{H}_a$ which contains $(t_k x)_a \mathcal{H}_a$, $\mathcal{M}$ and $y(\mathcal{N})$. Further, $\mathcal{D}(\bar{a}^*\bar{a}) \equiv \mathcal{D}(|\bar{a}|^2)$ is dense in $(\mathcal{D}(|\bar{a}|), \|\cdot\|_{|\bar{a}|}) = (\mathcal{D}(\bar{a}), \|\cdot\|_{\bar{a}}) = \mathcal{H}_a$. Therefore, by Lemma 5.2.8, we can find an orthonormal sequence $(\varphi_n : n \in \mathbb{N})$ in the Hilbert space $\mathcal{H}_a$ of vectors $\varphi_n \in \mathcal{D}(\bar{a}^*\bar{a})$ such that their closed linear span contains $\mathcal{K}$. (Of course, we exclude the trivial case where $\mathcal{H}$ is finite dimensional.) Using the previous facts, the definition of the trace, (6) and (7), the Parseval identity and finally (8) and again (7), we get

$$\begin{aligned}
\operatorname{tr} \hat{t}_k x &= \operatorname{Tr}_{\mathcal{H}_a} (\hat{t}_k x)_a = \sum_n \langle \hat{t}_k \bar{x}\varphi_n, \varphi_n\rangle_{\bar{a}} = \sum_n \langle \hat{t}_k \bar{x}\varphi_n, \bar{a}^*\bar{a}\varphi_n\rangle \\
&= \sum_n \operatorname{tr} \hat{t}_k (\bar{a}^*\bar{a}\varphi_n \otimes \bar{x}\varphi_n) = \sum_n \int \langle (\bar{a}^*\bar{a}\varphi_n \otimes \bar{x}\varphi_n) \varphi, \psi\rangle \, \mathrm{d}\mu_k(\varphi, \psi) \\
&= \int \sum_n \langle \varphi, \bar{a}^*\bar{a}\varphi_n\rangle \langle \bar{x}\varphi_n, \psi\rangle \, \mathrm{d}\mu_k(\varphi, \psi) \\
&= \int \sum_n \langle \varphi, \varphi_n\rangle_{\bar{a}} \langle \varphi_n, y\psi\rangle_{\bar{a}} \, \mathrm{d}\mu_k(\varphi, \psi) = \int \langle \varphi, y\psi\rangle_{\bar{a}} \, \mathrm{d}\mu_k(\varphi, \psi) = f_k(x).
\end{aligned}$$

The interchange of the summation and integration is justified, since

$$\sum_n |\langle \varphi, \varphi_n \rangle_{\bar{a}} \langle \varphi_n, y\psi \rangle_{\bar{a}}| \leqq \|\varphi\|_a \|y\psi\|_{\bar{a}} \leqq \|y\| \|\varphi\|_a \|\psi\|_b =: h(\varphi, \psi)$$

for $(\varphi, \psi) \in \mathcal{M} \times \mathcal{N}$ and the function h is continuous on $\mathcal{M} \times \mathcal{N}$. □

Proposition 5.3.5. *Suppose that the locally convex spaces $\mathcal{D}_\mathcal{A}$ and $\mathcal{D}_\mathcal{B}$ are metrizable. If $t \in \mathbf{B}_1(\mathcal{B}, \mathcal{A})$, then the linear functional $f_t(\cdot) \equiv \operatorname{tr}_\mathcal{A} \hat{t} \cdot$ is continuous on $\mathcal{L}(\mathcal{D}_\mathcal{A}, \mathcal{D}_\mathcal{B}^+)$ $[\tau_c]$.*

Proof. From Corollary 5.1.18 we conclude that the operator t of $\mathbf{B}_1(\mathcal{B}, \mathcal{A})$ belongs to the completion of the projective tensor product $\mathcal{D}_{\bar{\mathcal{B}}} \otimes_\pi \mathcal{D}_\mathcal{A}$. Therefore, since $\mathcal{D}_\mathcal{A}$ and $\mathcal{D}_\mathcal{B}$ are metrizable, a classical result of Grothendieck (see e.g. Köthe [2], § 41, 4., (6)) says that t admits a representation $t = \sum_n \lambda_n(\eta_n \otimes \zeta_n)$, where $(\lambda_n : n \in \mathbb{N})$ is a sequence in $l_1(\mathbb{N})$, $(\zeta_n : n \in \mathbb{N})$ is a null sequence in $\mathcal{D}_\mathcal{A}$ and $(\eta_n : n \in \mathbb{N})$ is a null sequence in $\mathcal{D}_\mathcal{B}$. Then $\sum_n \eta_n \otimes (\lambda_n \zeta_n)$ is an absolutely convergent series w.r.t. $\mathcal{A}$ and $\mathcal{B}$ which represents t, so that for $x \in \mathcal{L}(\mathcal{D}_\mathcal{A}, \mathcal{D}_\mathcal{B}^+)$

$$|f_t(x)| = |\operatorname{tr}_\mathcal{A} \hat{t}x| = \left|\sum_n \langle x(\lambda_n \zeta_n), \eta_n \rangle\right| \leqq \left(\sum_n |\lambda_n|\right) p_{\mathcal{M},\mathcal{N}}(x),$$

where $\mathcal{M} := \{\zeta_n : n \in \mathbb{N}\}$ and $\mathcal{N} := \{\eta_n : n \in \mathbb{N}\}$. Since $\mathcal{M}$ and $\mathcal{N}$ are obviously precompact in $\mathcal{D}_\mathcal{A}$ and $\mathcal{D}_\mathcal{B}^+$, respectively, this shows that the functional f_t is continuous on $\mathcal{L}(\mathcal{D}_\mathcal{A}, \mathcal{D}_\mathcal{B}^+)$ $[\tau_c]$. □

Remark 2. Proposition 5.3.5 remains valid if the metrizability assumption is replaced by a weaker requirement. It is sufficient to assume that every bounded subset of the space $\mathcal{D}_\mathcal{A}$ resp. $\mathcal{D}_\mathcal{B}$ is contained in a metrizable linear subspace.

Remark 3. Under the assumptions of Proposition 5.3.5, the functional f_t is ultraweakly continuous on $\mathcal{L}(\mathcal{D}_\mathcal{A}, \mathcal{D}_\mathcal{B}^+)$. Indeed, we define sequences (φ_n) and (ψ_n) as in the proof of Proposition 5.2.11, (ii). They satisfy 3.5/(2) and we have $f_t = \sum_n \omega_{\varphi_n, \psi_n}$; so the ultraweak continuity follows from Proposition 3.5.2.

Combining Propositions 5.3.4 and 5.3.5 we obtain the following theorem.

Theorem 5.3.6. *Suppose that $\mathcal{A}$ and $\mathcal{B}$ are (directed) O*-families on the Hilbert space $\mathcal{H}$ such that $\mathcal{D}_\mathcal{A}$ and $\mathcal{D}_\mathcal{B}$ are Frechet spaces. Let $\mathcal{L}$ be a linear subspace of $\mathcal{L}(\mathcal{D}_\mathcal{A}, \mathcal{D}_\mathcal{B}^+)$, and let f be a linear functional on $\mathcal{L}$. Then f is continuous on $\mathcal{L}[\tau_c]$ if and only if there is an operator $t \in \mathbf{B}_1(\mathcal{B}, \mathcal{A})$ such that $f(x) = \operatorname{tr}_\mathcal{A} \hat{t}x$ for all $x \in \mathcal{L}$.*

Remark 4. Assume in Theorem 5.3.6 that in addition $\mathcal{D}_\mathcal{A}$ and $\mathcal{D}_\mathcal{B}$ are Montel spaces. Then we have $\tau_b = \tau_c$ on $\mathcal{L}(\mathcal{D}_\mathcal{A}, \mathcal{D}_\mathcal{B}^+)$; cf. Remark 1. Therefore, in this case the continuous linear functionals on $\mathcal{L}[\tau_b]$ are precisely the functionals of the form $f_t(\cdot) = \operatorname{tr}_\mathcal{A} \hat{t} \cdot$ with $t \in \mathbf{B}_1(\mathcal{B}, \mathcal{A})$.

We give some applications of the results obtained so far.

Theorem 5.3.7. *Suppose that $\mathcal{A}$ and $\mathcal{B}$ are closed (directed) O*-families on the Hilbert space $\mathcal{H}$, and f is a linear functional on a linear subspace $\mathcal{L}$ of $\mathcal{L}(\mathcal{D}_\mathcal{A}, \mathcal{D}_\mathcal{B}^+)$. Suppose that at least one of the following three groups of assumptions is satisfied:*

(i) *$\mathcal{D}_\mathcal{A}$ or $\mathcal{D}_\mathcal{B}$ is a Schwartz space, $\mathbb{F}(\mathcal{H}) \subseteqq \mathcal{L}$, and f is continuous on $\mathcal{L}[\tau_{\text{in}}]$.*

(ii) *$\mathcal{D}_\mathcal{A}$ and $\mathcal{D}_\mathcal{B}$ are semi-Montel spaces and f is continuous on $\mathcal{L}[\tau_b]$.*

(iii) *$\mathcal{D}_\mathcal{A}$ or $\mathcal{D}_\mathcal{B}$ is a semi-Montel space, $\mathcal{D}_\mathcal{A}$ and $\mathcal{D}_\mathcal{B}$ are both metrizable, and f is continuous on $\mathcal{L}[\tau_b]$.*

Then there exists a $t \in \mathbf{B}_1(\mathcal{B}, \mathcal{A})$ such that $f(x) = \operatorname{tr}_\mathcal{A} \hat{t}x$ for $x \in \mathcal{L}$.

Proof. If (i) is fulfilled, then Theorem 3.4.6 shows that $\overline{\mathbb{F}} = \mathcal{L}$, so that Proposition 5.3.1 gives the assertion. If (ii) is valid, then $\tau_c = \tau_b$ by Remark 1, and the assertion follows from Proposition 5.3.4. Assume finally that (iii) is satisfied. Upon extending f to a continuous linear functional on $\mathcal{L}(\mathcal{D}_{\mathcal{A}}, \mathcal{D}_{\mathcal{B}}^+) [\tau_b]$ by the Hahn-Banach theorem, we can assume that $\mathcal{L} = \mathcal{L}(\mathcal{D}_{\mathcal{A}}, \mathcal{D}_{\mathcal{B}}^+)$. Then, by Corollary 5.3.2, there is a $t \in \mathbb{B}_1(\mathcal{B}, \mathcal{A})$ such that $f(x) = \operatorname{tr} \hat{t}x$ for all $x \in \mathcal{F}(\mathcal{D}_{\mathcal{A}}, \mathcal{D}_{\mathcal{B}}^+)$. Because at least one of the spaces $\mathcal{D}_{\mathcal{A}}$ and $\mathcal{D}_{\mathcal{B}}$ is a semi-Montel space, $\mathcal{F}(\mathcal{D}_{\mathcal{A}}, \mathcal{D}_{\mathcal{B}}^+)$ is dense in $\mathcal{L}(\mathcal{D}_{\mathcal{A}}, \mathcal{D}_{\mathcal{B}}^+) [\tau_b]$ by Theorem 3.4.5. Since $\mathcal{D}_{\mathcal{A}}$ and $\mathcal{D}_{\mathcal{B}}$ are metrizable, Proposition 5.3.5 shows that the functional $\operatorname{tr} \hat{t}\,\cdot$ is continuous on $\mathcal{L}(\mathcal{D}_{\mathcal{A}}, \mathcal{D}_{\mathcal{B}}^+) [\tau_b]$. Hence the equality $f(\cdot) = \operatorname{tr} \hat{t}\,\cdot$ extends from $\mathcal{F}(\mathcal{D}_{\mathcal{A}}, \mathcal{D}_{\mathcal{B}}^+)$ to the whole $\mathcal{L}(\mathcal{D}_{\mathcal{A}}, \mathcal{D}_{\mathcal{B}}^+)$. $\square$

Theorem 5.3.8. *Let $\mathcal{A}$ be a closed O*-algebra, and let $\mathcal{L}$ be a cofinal *-vector subspace of $\mathcal{L}(\mathcal{D}_{\mathcal{A}}, \mathcal{D}_{\mathcal{A}}^+)$. Suppose that $\mathcal{D}_{\mathcal{A}}$ is a Frechet-Montel space or that $\mathcal{D}_{\mathcal{A}}$ is a Schwartz space. If f is a strongly positive linear functional on $\mathcal{L}$, then there is an operator $t \in \mathbb{B}_1(\mathcal{A})_+$ such that $f(x) = \operatorname{tr}_{\mathcal{A}} \hat{t}x$ for all $x \in \mathcal{L}$.*

Proof. The proof repeats some arguments from the proof of Corollary 5.3.3. There is no loss of generality to assume that $\mathcal{L} = \mathcal{L}(\mathcal{D}_{\mathcal{A}}, \mathcal{D}_{\mathcal{A}}^+)$, since f can be extended to a strongly positive linear functional on $\mathcal{L}(\mathcal{D}_{\mathcal{A}}, \mathcal{D}_{\mathcal{A}}^+)$ by Lemma 1.3.2. Since $\tau_{\mathcal{O}} = \tau_{in}$ on $\mathcal{L}(\mathcal{D}_{\mathcal{A}}, \mathcal{D}_{\mathcal{A}}^+)$ by Proposition 3.3.11, it follows that the strongly positive linear functional f is continuous on $\mathcal{L}(\mathcal{D}_{\mathcal{A}}, \mathcal{D}_{\mathcal{A}}^+)[\tau_{in}]$. If $\mathcal{D}_{\mathcal{A}}$ is a Frechet-Montel space, then Theorem 4.4.1 shows that $\tau_{\mathcal{D}} = \tau_{\mathcal{N}}$ on $\mathcal{L}(\mathcal{D}_{\mathcal{A}}, \mathcal{D}_{\mathcal{A}}^+)$, so that f is continuous on $\mathcal{L}(\mathcal{D}_{\mathcal{A}}, \mathcal{D}_{\mathcal{A}}^+) [\tau_{\mathcal{D}}]$ in this case. Hence the assumptions of Theorem 5.3.7 ((i) if $\mathcal{D}_{\mathcal{A}}$ is a Schwartz space and (ii) if $\mathcal{D}_{\mathcal{A}}$ is a Frechet-Montel space) are fulfilled. Therefore, $f(\cdot) = \operatorname{tr} \hat{t}\,\cdot$ for some $t \in \mathbb{B}_1(\mathcal{A})$. From $f(\varphi \otimes \varphi) = \operatorname{tr} \hat{t}(\varphi \otimes \varphi) = \langle t\varphi, \varphi \rangle \geqq 0$ for $\varphi \in \mathcal{H}$ we conclude that $t \geqq 0$; so $t \in \mathbb{B}_1(\mathcal{A})_+$. $\square$

Example 5.3.9. Let $\mathcal{A}$ be the O*-algebra $\mathbb{A}(p_1, q_1, \ldots, p_n, q_n)$ of Example 2.5.2. Since $t_{\mathcal{A}}$ is the usual topology of the space $\mathcal{S}(\mathbb{R}^n)$, $\mathcal{D}_{\mathcal{A}}$ is a Frechet-Montel space. Further, we have $t_{\mathcal{A}} = t_+$ on $\mathcal{D} := \mathcal{D}(\mathcal{A})$, so that $\mathcal{L}(\mathcal{D}_{\mathcal{A}}, \mathcal{D}_{\mathcal{A}}^+) = \mathcal{L}(\mathcal{D}, \mathcal{D}^+)$ and $\mathbb{B}_1(\mathcal{A}) = \mathbb{B}_1(\mathcal{D})$. Suppose $\mathcal{L}$ is a *-vector subspace of $\mathcal{L}(\mathcal{D}, \mathcal{D}^+)$ which contains $\mathcal{A}$. Then $\mathcal{L}$ is cofinal in $\mathcal{L}(\mathcal{D}, \mathcal{D}^+)$. By Theorem 5.3.7, case (ii), the dual of $\mathcal{L}[\tau_{\mathcal{D}}]$ is precisely the vector space of all functionals $f_t(\cdot) = \operatorname{tr} \hat{t}\,\cdot$, $t \in \mathbb{B}_1(\mathcal{D})$, on $\mathcal{L}$. If f is a strongly positive linear functional on $\mathcal{L}$, then Theorem 5.3.8 shows that f is of the form $f(\cdot) = \operatorname{tr} \hat{t}\,\cdot$ with $t \in \mathbb{B}_1(\mathcal{D})_+$. Finally, we consider the special case $\mathcal{L} = \mathcal{A}$. Then, by Example 4.5.7, the topology $\tau_{\mathcal{D}}$ is the finest locally convex topology τ_{st} on $\mathcal{A}$. Therefore, every linear functional f on $\mathcal{A}$ is continuous on $\mathcal{A}[\tau_{\mathcal{D}}]$ and hence of the form $f(\cdot) = \operatorname{tr} \hat{t}\,\cdot$ with $t \in \mathbb{B}_1(\mathcal{D})$. $\bigcirc$

5.4. The Duality Theorem

In this section $\mathcal{A}$ and $\mathcal{B}$ are O*-algebras acting on the Hilbert space $\mathcal{H}$.

For $x \in \mathcal{L}(\mathcal{D}_{\mathcal{A}}, \mathcal{D}_{\mathcal{B}}^+)$, let g_x denote the linear functional on $\mathbb{B}_1(\mathcal{B}, \mathcal{A})$ defined by $g_x(t) = \operatorname{tr}_{\mathcal{A}} \hat{t}x$, $t \in \mathbb{B}_1(\mathcal{B}, \mathcal{A})$.

Proposition 5.4.1. *The map $\mathsf{I}: x \to g_x$ is a bijective linear mapping of $\mathcal{L}(\mathcal{D}_{\mathcal{A}}, \mathcal{D}_{\mathcal{B}}^+)$ onto the dual space of $\mathbb{B}_1(\mathcal{B}, \mathcal{A}) [\tau_\pi]$. The inverse of I maps the strong dual of $\mathbb{B}_1(\mathcal{B}, \mathcal{A}) [\tau_\pi]$ continuously on $\mathcal{L}(\mathcal{D}_{\mathcal{A}}, \mathcal{D}_{\mathcal{B}}^+) [\tau_b]$.*

Proof. First we show that each functional g_x is continuous on $\mathbb{B}_1(\mathcal{B}, \mathcal{A})[\tau_\pi]$. Fix $x \in \mathcal{L}(\mathcal{D}_\mathcal{A}, \mathcal{D}_\mathcal{B}^+)$. There are operators $a \in \mathcal{A}(I)$ and $b \in \mathcal{B}(I)$ such that x is in $\mathcal{U}_{a,b}$. Suppose $t \in \mathbb{B}_1(\mathcal{B}, \mathcal{A})$. Let $\sum_n \psi_n \otimes \varphi_n$ be an arbitrary absolutely convergent series w.r.t. a and b which represents the operator t. By 5.2/(7),

$$|g_x(t)| = |\mathrm{tr}_\mathcal{A}\, \hat{t}x| = \Big|\sum_n \langle x\varphi_n, \psi_n\rangle\Big| \leqq \sum_n \|\varphi_n\|_a \|\psi_n\|_b .$$

This implies $|g_x(t)| \leqq \|\cdot\|_b \mathbin{\widehat{\otimes}_\pi} \|\cdot\|_a (t)$. Since $\|\cdot\|_b \mathbin{\widehat{\otimes}_\pi} \|\cdot\|_a$ is a continuous seminorm on $\mathbb{B}_1(\mathcal{B}, \mathcal{A})[\tau_\pi]$ by 5.1/(5), this proves the continuity of g_x.

It is clear that l is linear. If $\varphi \in \mathcal{D}(\mathcal{A})$ and $\psi \in \mathcal{D}(\mathcal{B})$, then $\psi \otimes \varphi \in \mathbb{B}_1(\mathcal{B}, \mathcal{A})$ and so $g_x(\psi \otimes \varphi) = \langle x\varphi, \psi\rangle$ by 5.2/(7) for $x \in \mathcal{L}(\mathcal{D}_\mathcal{A}, \mathcal{D}_\mathcal{B}^+)$. From this we see that l is injective. To prove that l is surjective, let $f \in \mathbb{B}_1(\mathcal{B}, \mathcal{A})[\tau_\pi]'$. From Corollary 5.1.17 and Lemma 5.1.16, the family of seminorms $\{\nu_{a,b}: a \in \mathcal{A}$ and $b \in \mathcal{B}\}$ is directed. Hence we can find operators $a \in \mathcal{A}$ and $b \in \mathcal{B}$ such that $|f(t)| \leqq \nu_{a,b}(t)$ for all $t \in \mathbb{B}_1(\mathcal{B}, \mathcal{A})$. Let $\varphi \in \mathcal{D}(\mathcal{A})$ and $\psi \in \mathcal{D}(\mathcal{B})$. Setting $t = \psi \otimes \varphi$, we obtain $|f(\psi \otimes \varphi)| \leqq \nu_{a,b}(\psi \otimes \varphi) = \|b^+\psi\|\, \|a\varphi\|$. This shows that the mapping $(\varphi, \psi) \to f(\psi \otimes \varphi)$ is a continuous sesquilinear form on $\mathcal{D}_\mathcal{A} \times \mathcal{D}_\mathcal{B}^+$. From Lemma 1.2.1, there is an $x \in \mathcal{L}(\mathcal{D}_\mathcal{A}, \mathcal{D}_\mathcal{B}^+)$ such that $f(\psi \otimes \varphi) = \langle x\varphi, \psi\rangle$ for all $\varphi \in \mathcal{D}(\mathcal{A})$ and $\psi \in \mathcal{D}(\mathcal{B})$. Now let $t \in \mathbb{B}_1(\mathcal{B}, \mathcal{A})$ and let $\sum_n \psi_n \otimes \varphi_n$ be an absolutely convergent series w.r.t. $\mathcal{A}$ and $\mathcal{B}$ which represents t. From Corollary 5.1.17 it follows immediately that this series converges to t in the locally convex space $\mathbb{B}_1(\mathcal{B}, \mathcal{A})[\tau_\pi]$. Since f is continuous on $\mathbb{B}_1(\mathcal{B}, \mathcal{A})[\tau_\pi]$, we obtain $f(t) = \sum_n f(\psi_n \otimes \varphi_n) = \sum_n \langle x\varphi_n, \psi_n\rangle$ $= \mathrm{tr}_\mathcal{A}\, \hat{t}x = g_x(t)$, where we used once more formula 5.2/(7). Thus $f = g_x$ and l is surjective.

It remains to prove the continuity of l^{-1}. Let $\mathcal{M}$ and $\mathcal{N}$ be bounded subsets of $\mathcal{D}_\mathcal{A}$ and $\mathcal{D}_\mathcal{B}$, respectively. Then $\mathcal{R} := \{\psi \otimes \varphi: \varphi \in \mathcal{M}$ and $\psi \in \mathcal{N}\}$ is obviously a bounded subset of $\mathbb{B}_1(\mathcal{B}, \mathcal{A})[\tau_\pi]$ and we have

$$\sup_{t \in \mathcal{R}} |g_x(t)| = \sup_{\varphi \in \mathcal{M}} \sup_{\psi \in \mathcal{N}} |\langle x\varphi, \psi\rangle| = p_{\mathcal{M},\mathcal{N}}(x)$$

for $x \in \mathcal{L}(\mathcal{D}_\mathcal{A}, \mathcal{D}_\mathcal{B}^+)$. This shows that l^{-1} is a continuous mapping of the strong dual of $\mathbb{B}_1(\mathcal{B}, \mathcal{A})[\tau_\pi]$ on $\mathcal{L}(\mathcal{D}_\mathcal{A}, \mathcal{D}_\mathcal{B}^+)[\tau_\mathrm{b}]$. □

The main result in this section is

Theorem 5.4.2. *Let $\mathcal{A}$ and $\mathcal{B}$ be O*-algebras in the Hilbert space $\mathcal{H}$. Suppose that $\mathcal{D}_\mathcal{A}$ and $\mathcal{D}_\mathcal{B}$ are QF-spaces. Then the mapping $\mathsf{l}: x \to g_x$ is a topological isomorphism of the locally convex space $\mathcal{L}(\mathcal{D}_\mathcal{A}, \mathcal{D}_\mathcal{B}^+)[\tau_\mathrm{b}]$ onto the strong dual of the locally convex space $\mathbb{B}_1(\mathcal{B}, \mathcal{A})[\tau_\pi]$.*

We state two important special cases of this theorem separately as

Corollary 5.4.3. *Suppose $\mathcal{A}$ is an O*-algebra on the Hilbert space $\mathcal{H}$ such that $\mathcal{D}_\mathcal{A}$ is a QF-space.*

(i) *The map $\mathsf{l}: x \to g_x$ is a topological isomorphism of $\mathcal{L}(\mathcal{D}_\mathcal{A}, \mathcal{D}_\mathcal{A}^+)[\tau_\mathcal{D}]$ onto the strong dual of $\mathbb{B}_1(\mathcal{A})[\tau_\pi]$.*

(ii) *The map $\mathsf{l}: x \to g_x(\cdot) \equiv \mathrm{Tr}\, x\cdot$ is a topological isomorphism of $\mathfrak{L}(\mathcal{D}_\mathcal{A}, \mathcal{H})[\tau^\mathcal{D}]$ onto the strong dual of $\mathbb{B}_1(\mathcal{H}, \mathcal{A})[\tau_\pi]$.*

Proof. (i) is the special case $\mathcal{A} = \mathcal{B}$ of Theorem 5.4.2. (ii): By Corollary 5.2.7, $g_x(t)$ $= \mathrm{tr}\, \hat{t}x = \mathrm{Tr}\, xt$ for $t \in \mathbb{B}_1(\mathcal{H}, \mathcal{A})$ and $x \in \mathfrak{L}(\mathcal{D}_\mathcal{A}, \mathcal{H})$. The other assertions follow from

Theorem 5.4.2 applied in case $\mathcal{B} = \mathbb{B}(\mathcal{H})$. Recall that $\mathfrak{L}(\mathcal{D}_{\mathcal{A}}, \mathcal{H}) = \mathcal{L}(\mathcal{D}_{\mathcal{A}}, \mathcal{D}_{\mathcal{B}}^+)$ and $\mathbb{B}_1(\mathcal{B}, \mathcal{A}) = \mathbb{B}_1(\mathcal{H}, \mathcal{A})$ in this case. □

The crucial step in the proof of the theorem is the following lemma.

Lemma 5.4.4. *Keep the assumptions of Theorem 5.4.2. Let $\mathcal{R}$ be a bounded subset of the locally convex space $\mathbb{B}_1(\mathcal{B}, \mathcal{A})[\tau_n]$. Then there are operators $c \in \mathbb{B}(\mathcal{D}(\mathcal{A}))_+$ and $d \in \mathbb{B}(\mathcal{D}(\mathcal{B}))_+$ such that $\mathcal{R} \subseteq c\mathcal{U}_{\mathbb{B}_1(\mathcal{H})}d$, where $\mathcal{U}_{\mathbb{B}_1(\mathcal{H})} := \{t \in \mathbb{B}_1(\mathcal{H}) : \nu(t) \leqq 1\}$.*

Proof. For $a \in \mathcal{A}$, $b \in \mathcal{B}$ and $\varphi \in \mathcal{U}_{\mathcal{H}}$ we have

$$\sup_{t \in \mathcal{R}} \|at\varphi\| \leqq \sup_{t \in \mathcal{R}} \|at\| \leqq \sup_{t \in \mathcal{R}} \nu_{a,I}(t) < \infty$$

and

$$\sup_{t \in \mathcal{R}} \|bt^*\varphi\| \leqq \sup_{t \in \mathcal{R}} \|(bt^*)^*\| = \sup_{t \in \mathcal{R}} \|\overline{tb^+}\| \leqq \sup_{t \in \mathcal{R}} \nu_{I,b}(t) < \infty.$$

This shows that $\mathcal{M} := \bigcup_{t \in \mathcal{R}} t\mathcal{U}_{\mathcal{H}}$ and $\mathcal{N} := \bigcup_{t \in \mathcal{R}} t^*\mathcal{U}_{\mathcal{H}}$ are bounded subsets of $\mathcal{D}_{\mathcal{A}}$ and $\mathcal{D}_{\mathcal{B}}$, respectively. Since $\mathcal{D}_{\mathcal{A}}$ and $\mathcal{D}_{\mathcal{B}}$ are QF-spaces, there are Frechet subspaces $\mathcal{E}$ and $\mathcal{G}$ of $\mathcal{D}_{\mathcal{A}}$ and $\mathcal{D}_{\mathcal{B}}$, respectively, such that $\mathcal{M} \subseteq \mathcal{E}$ and $\mathcal{N} \subseteq \mathcal{G}$. Then there exists a sequence $(a_n : n \in \mathbb{N})$ resp. $(b_n : n \in \mathbb{N})$ of symmetric operators in $\mathcal{A}$ resp. $\mathcal{B}$ such that the topology $t_{\mathcal{A}}$ on $\mathcal{E}$ resp. $t_{\mathcal{B}}$ on $\mathcal{G}$ is generated by the seminorms $\{\|\cdot\|_{a_n} : n \in \mathbb{N}\}$ resp. $\{\|\cdot\|_{b_n} : n \in \mathbb{N}\}$. Of course, we can assume that $a_1 = I$ and $b_1 = I$. Since $\mathcal{R}$ is bounded in $\mathbb{B}_1(\mathcal{B}, \mathcal{A})$ $[\tau_n]$, $\alpha_{m,n} := \sup_{t \in \mathcal{R}} \nu(a_m^2 t b_n^2) < \infty$ for all $m, n \in \mathbb{N}$. By induction, we choose a positive sequence $(\delta_n : n \in \mathbb{N})$ satisfying

$$\delta_m \delta_n \alpha_{m,n} \leqq 2^{-(m+n)} \quad \text{for } m, n \in \mathbb{N}. \tag{1}$$

(Indeed, let $\delta_1 := \frac{1}{2} (\max\{1, \alpha_{1,1}\})^{-1/2}$. If the positive numbers $\delta_1, \ldots, \delta_n$ are chosen, it suffices to take $\delta_{n+1} > 0$ such that $\delta_{n+1}\delta_m(\alpha_{m,n+1} + \alpha_{n+1,m}) \leqq 2^{-(m+n+1)}$ for $m = 1, \ldots, n + 1$.) For $t \in \mathcal{R}$, we define an operator t_1 on $\mathcal{H}$ by

$$t_1\varphi = \sum_{m,n=1}^{\infty} \delta_m \delta_n \overline{a_m^2 t b_n^2} \varphi, \quad \varphi \in \mathcal{H}. \tag{2}$$

From $\sum_{m,n=1}^{\infty} \delta_m\delta_n \|\overline{a_m^2 t b_n^2}\| \leqq \sum_{m,n=1}^{\infty} \delta_m \delta_n \nu(a_m^2 t b_n^2) \leqq \sum_{m,n=1}^{\infty} \delta_m \delta_n \alpha_{m,n} \leqq \sum_{m,n=1}^{\infty} 2^{-(k+n)} = 1$ we conclude that t_1 is a well-defined bounded operator on $\mathcal{H}$ and that $t_1 \in \mathcal{U}_{\mathbb{B}_1(\mathcal{H})}$.

Next we apply Lemma 2.4.2 to $\mathcal{A}$, $\mathcal{E}$, $(a_n : n \in \mathbb{N})$ and to $\mathcal{B}$, $\mathcal{G}$, $(b_n : n \in \mathbb{N})$ with the sequence $(\delta_n : n \in \mathbb{N})$. Let $\tilde{c}$ and $\tilde{d}$ be the corresponding operators of $\mathbb{B}(\mathcal{D}(\mathcal{A}))_+$ and $\mathbb{B}(\mathcal{D}(\mathcal{B}))_+$, respectively. Suppose $t \in \mathcal{R}$ and $\zeta \in \mathcal{H}$. Using (1) and $b_1 = I$, we have for $n \in \mathbb{N}$

$$\delta_n \|b_n^2 t^* \zeta\| \leqq \delta_n \|b_n^2 t^*\| \, \|\zeta\| = \delta_n \|\overline{tb_n^2}\| \, \|\zeta\| \leqq \delta_n \nu(\overline{tb_n^2}) \, \|\zeta\| \leqq \delta_n \alpha_{1,n} \|\xi\| \leqq 2^{-(n+1)} \delta_1^{-1} \|\zeta\|$$

and

$$\delta_n \|b_n t^* \zeta\|^2 = \delta_n \langle b_n^2 t^* \zeta, t^* \zeta \rangle \leqq \delta_n \|b_n^2 t^* \zeta\| \, \|t^* \zeta\| \leqq 2^{-(n+1)} \delta_1^{-1} \|\zeta\| \, \|t^* \zeta\|.$$

Therefore, $\sum_{n=1}^{\infty} \delta_n \|b_n t^* \zeta\|^2 < \infty$ and the series $\sum_{n=1}^{\infty} \delta_n b_n^2 t^* \zeta$ converges in $\mathcal{H}$. Moreover, by construction, $t^*\zeta \in \mathcal{G}$. This shows that the vector $t^*\zeta$ satisfies the assumptions of Lemma

2.4.2, (iii), in case of $(b_n : n \in \mathbb{N})$ and $\tilde{d}$. Replacing b_n by a_n and t^* by t, a similar reasoning shows that each vector $t\zeta$, where $t \in \mathcal{R}$ and $\zeta \in \mathcal{H}$, fulfills the assumptions of Lemma 2.4.2, (iii), in case $(a_n : n \in \mathbb{N})$, $\tilde{c}$. Applying (2) and Lemma 2.4.2, (iii), we get

$$\langle \tilde{c}^2 t_1 \tilde{d}^2 \varphi, \psi \rangle = \sum_{n=1}^{\infty} \delta_n \left\langle \tilde{c}^2 \sum_{m=1}^{\infty} \delta_m a_m^2 (t b_n^2 \tilde{d}^2 \varphi), \psi \right\rangle$$

$$= \sum_{n=1}^{\infty} \delta_n \langle t b_n^2 \tilde{d}^2 \varphi, \psi \rangle = \left\langle \varphi, \tilde{d}^2 \sum_{n=1}^{\infty} \delta_n b_n^2 t^* \varphi, \psi \right\rangle = \langle \varphi, t^* \psi \rangle = \langle t\varphi, \psi \rangle$$

for all $t \in \mathcal{R}$ and $\varphi, \psi \in \mathcal{H}$. This proves that $\tilde{c}^2 t_1 \tilde{d}^2 = t$ for each $t \in \mathcal{R}$. That is, $\mathcal{R} \subseteq \tilde{c}^2 \mathcal{U}_{\mathbb{B}_1(\mathcal{H})} \tilde{d}^2$. The proof of the lemma is complete if we set $c := \tilde{c}^2$ and $d := \tilde{d}^2$. □

The following corollary is of interest in itself.

Corollary 5.4.5. *Let $\mathcal{A}$ and $\mathcal{B}$ be as in Theorem 5.4.2. Then $\mathbb{B}_1(\mathcal{B}, \mathcal{A})$ is precisely the set of operators ct_1d, where $c \in \mathbb{B}(\mathcal{D}(\mathcal{A}))_+$, $d \in \mathbb{B}(\mathcal{D}(\mathcal{B}))_+$ and $t_1 \in \mathbb{B}_1(\mathcal{H})$. The sets $c\mathcal{U}_{\mathbb{B}_1(\mathcal{H})}d$, where $c \in \mathbb{B}(\mathcal{D}(\mathcal{A}))_+$ and $d \in \mathbb{B}(\mathcal{D}(\mathcal{B}))_+$, form a fundamental system of bounded sets of the locally convex space $\mathbb{B}_1(\mathcal{B}, \mathcal{A})\,[\tau_\pi]$.*

Proof. If $t \in \mathbb{B}_1(\mathcal{B}, \mathcal{A})$, then Lemma 5.4.4 applied to the singleton $\mathcal{R} = \{t\}$ shows that $t = ct_1d$ for some $c \in \mathbb{B}(\mathcal{D}(\mathcal{A}))_+$ and $d \in \mathbb{B}(\mathcal{D}(\mathcal{B}))_+$. Conversely, let $t = ct_1d$ with $c \in \mathbb{B}(\mathcal{D}(\mathcal{A}))_+$, $d \in \mathbb{B}(\mathcal{D}(\mathcal{B}))_+$ and $t \in \mathbb{B}_1(\mathcal{H})$. Then $t\mathcal{H} \subseteq c\mathcal{H} \subseteq \mathcal{D}(\mathcal{A})$ and $t^*\mathcal{H} \subseteq d\mathcal{H} \subseteq \mathcal{D}(\mathcal{B})$. If $a \in \mathcal{A}$ and $b \in \mathcal{B}$, then the operators ac and bd are bounded by Lemma 3.1.2 and hence $at = (ac)\, t_1 d$ and $bt^* = (bd)\, t_1^* c$ are in $\mathbb{B}_1(\mathcal{H})$. This proves $t \in \mathbb{B}_1(\mathcal{B}, \mathcal{A})$. Let $\mathcal{R} := c\mathcal{U}_{\mathbb{B}_1(\mathcal{H})}d$, where $c \in \mathbb{B}(\mathcal{D}(\mathcal{A}))_+$ and $d \in \mathbb{B}(\mathcal{D}(\mathcal{B}))_+$. Suppose $a \in \mathcal{A}$ and $b \in \mathcal{B}$. Since ac and $\overline{db}$ are bounded again by Lemma 3.1.2, $\nu_{a,b}(t_1) = \nu(act_1db) \leqq \|ac\|\, \nu(t_1)\, \|\overline{db}\| \leqq \|ac\|\, \|\overline{db}\|$ for each $t_1 \in \mathcal{U}_{\mathbb{B}_1(\mathcal{H})}$. Hence $\mathcal{R}$ is bounded in $\mathbb{B}_1(\mathcal{B}, \mathcal{A})\,[\tau_\pi]$. By Lemma 5.4.4, each bounded subset of $\mathbb{B}_1(\mathcal{B}, \mathcal{A})\,[\tau_\pi]$ is contained in a set $\mathcal{R}$ of that kind. Thus these sets form a fundamental system of bounded sets of $\mathbb{B}_1(\mathcal{B}, \mathcal{A})\,[\tau_\pi]$. □

Remark 1. Keep the assumptions of Theorem 5.4.2. We mention two additional facts concerning the final assertion in Corollary 5.4.5.

(i) *The family $\{c\mathcal{U}_{\mathbb{B}_1(\mathcal{H})}c : c \in \mathbb{B}(\mathcal{D}(\mathcal{A}))_+\}$ is a fundamental system of bounded sets in $\mathbb{B}_1(\mathcal{A})\,[\tau_\pi]$.* (To prove this, it suffices to show that in case $\mathcal{A} = \mathcal{B}$ we can take $c = d$ in Lemma 5.4.4. Replacing in the proof of this lemma $\mathcal{M}$ and $\mathcal{N}$ by $\mathcal{M} \cup \mathcal{N}$ and letting $\mathcal{E} = \mathcal{S}$, $a_n = b_n$ for $n \in \mathbb{N}$, we obtain $c = d$.)

(ii) Suppose $\mathcal{D}_\mathcal{A}$ and $\mathcal{D}_\mathcal{B}$ are Frechet spaces. If the O*-algebras $\mathcal{A}$ and $\mathcal{B}$ satisfy the assumptions of Theorem 2.4.3 with von Neumann algebras $\mathcal{M}$ and $\mathcal{N}$, respectively, then $\{c\mathcal{U}_{\mathbb{B}_1(\mathcal{H})}d : c \in \mathcal{M} \cap \mathbb{B}(\mathcal{D}(\mathcal{A}))_+$ and $d \in \mathcal{N} \cap \mathbb{B}(\mathcal{D}(\mathcal{B}))_+\}$ *is a fundamental system of bounded sets in $\mathbb{B}_1(\mathcal{B}, \mathcal{A})\,[\tau_\pi]$.* (Indeed, it was shown in the proof of Theorem 2.4.3 that the operators c and d occuring in the proof of Lemma 5.4.4 can be chosen in $\mathcal{M}$ and $\mathcal{N}$, respectively.)

The following lemma is a second step in the proof of Theorem 5.4.2; it holds for arbitrary O*-algebras $\mathcal{A}$ and $\mathcal{B}$ on $\mathcal{H}$.

Lemma 5.4.6. *For $c \in \mathbb{B}(\mathcal{D}(\mathcal{A}))_+$, $d \in \mathbb{B}(\mathcal{D}(\mathcal{B}))_+$ and $x \in \mathcal{L}(\mathcal{D}_\mathcal{A}, \mathcal{D}_\mathcal{B}^+)$, we have*

$$\sup_{t \in c\mathcal{U}_{\mathbb{B}_1(\mathcal{H})}d} |g_x(t)| = \|\hat{d}xc\|. \tag{3}$$

Proof. Fix $x \in \mathcal{L}(\mathcal{D}_\mathcal{A}, \mathcal{D}_\mathcal{B}^+)$. Let $t_1 \in \mathcal{U}_{\mathbb{B}_1(\mathcal{H})}$, and let $\sum_n \lambda_n (\psi_n \otimes \varphi_n)$ be a canonical representation for t. Then $t := ct_1d$ is in $\mathbb{B}(\mathcal{D}(\mathcal{B}), \mathcal{D}(\mathcal{A}))$. Since ac and bd are bounded for

$a \in \mathcal{A}$ and $b \in \mathcal{B}$, the series $\sum_n d\psi_n \otimes (\lambda_n c\varphi_n)$ converges absolutely w.r.t. $\mathcal{A}$ and $\mathcal{B}$. Obviously, this series represents the operator t; hence $atb \in \mathbf{B}_1(\mathcal{H})$ for all $a \in \mathcal{A}$ and $b \in \mathcal{B}$ by Lemma 5.1.11. Thus $t \in \mathbf{B}_1(\mathcal{B}, \mathcal{A})$. Further, $\hat{d}xct_1$ is represented by the absolutely convergent series $\sum_n \psi_n \otimes (\lambda_n \hat{d}xc\varphi_n)$ on $\mathcal{H}$. By Proposition 5.2.4 and Lemma 5.1.1,

$$g_x(t) := \operatorname{tr}_{\mathcal{A}} \hat{t}x = \sum_n \langle x(\lambda_n c\varphi_n), d\psi_n\rangle = \sum_n \langle \lambda_n \hat{d}xc\varphi_n, \psi_n\rangle = \operatorname{Tr} \hat{d}xct_1 .$$

Since $\sup_{t_1 \in \mathcal{U}_{\mathbf{B}_1(\mathcal{H})}} |\operatorname{Tr} \hat{d}xct_1| = \|\hat{d}xc\|$, (3) follows. □

Proof of Theorem 5.4.2. Because of Proposition 5.4.1, it is sufficient to prove that $\mathfrak{l}$ is continuous. But this follows immediately from Lemmas 5.4.4 and 5.4.6 if we take into account that (by Corollary 3.3.6) each seminorm $x \to \|\hat{d}xc\|$, where $c \in \mathbf{B}(\mathcal{D}(\mathcal{A}))_+$ and $d \in \mathbf{B}(\mathcal{D}(\mathcal{B}))_+$, is continuous on $\mathcal{L}(\mathcal{D}_{\mathcal{A}}, \mathcal{D}_{\mathcal{B}}^+)[\tau_b]$. □

We give some applications of the preceding results.

Corollary 5.4.7. *If $\mathcal{A}$ and $\mathcal{B}$ are as in Theorem 5.4.2, then $\mathbf{B}_1(\mathcal{B}, \mathcal{A})[\tau_\pi]$ is a QF-space.*

Proof. Let $\mathcal{R}$ be a bounded subset of $\mathbf{B}_1(\mathcal{B}, \mathcal{A})[\tau_\pi]$. As shown in the proof of Lemma 5.4.4 there are Frechet subspaces $\mathcal{E}$ and $\mathcal{G}$ of $\mathcal{D}_{\mathcal{A}}$ and $\mathcal{D}_{\mathcal{B}}$, respectively, such that $t\mathcal{H} \subseteq \mathcal{E}$ and $t^*\mathcal{H} \subseteq \mathcal{G}$ for all $t \in \mathcal{R}$. In the terminology of Lemma 5.1.19, $\mathcal{R} \subseteq \mathbf{B}_1(\mathcal{B}, \mathcal{A}; \mathcal{G}, \mathcal{E})$. From Lemma 5.1.19, $\mathbf{B}_1(\mathcal{B}, \mathcal{A}; \mathcal{G}, \mathcal{E})[\tau_\pi]$ is a Frechet space. This proves that $\mathbf{B}_1(\mathcal{B}, \mathcal{A})[\tau_\pi]$ is a QF-space. □

By a *Frechet domain* in the Hilbert space $\mathcal{H}$ we mean a dense linear subspace $\mathcal{D}$ of $\mathcal{H}$ for which the locally convex space $\mathcal{D}[t_+]$ is a Frechet space.

In the remaining part of this section we consider only Frechet domains. Let $\mathcal{D}_1$ and $\mathcal{D}_2$ be Frechet domains in a Hilbert space $\mathcal{H}$. Recall that, by definition, $\mathbf{B}_1(\mathcal{D}_2, \mathcal{D}_1) = \mathbf{B}_1(\mathcal{B}, \mathcal{A})$ and $\mathcal{L}(\mathcal{D}_1, \mathcal{D}_2^+) = \mathcal{L}(\mathcal{D}_{\mathcal{A}}, \mathcal{D}_{\mathcal{B}}^+)$ for $\mathcal{A} := \mathcal{L}^+(\mathcal{D}_1)$ and $\mathcal{B} := \mathcal{L}^+(\mathcal{D}_2)$. Thus Theorem 5.4.2 states that $\mathfrak{l}$ is a topological isomorphism of $\mathcal{L}(\mathcal{D}_1, \mathcal{D}_2^+)[\tau_b]$ on the strong dual of $\mathbf{B}_1(\mathcal{D}_2, \mathcal{D}_1)[\tau_\pi]$.

Corollary 5.4.8. *Suppose that $\mathcal{D}_1$ and $\mathcal{D}_2$ are Frechet domains in the Hilbert space $\mathcal{H}$. Then $\mathcal{L}(\mathcal{D}_1, \mathcal{D}_2^+)[\tau_b]$ is a complete DF-space. If $\mathcal{L}$ is a linear subspace of $\mathcal{L}(\mathcal{D}_1, \mathcal{D}_2^+)$ which contains $\mathbf{B}(\mathcal{D}_1, \mathcal{D}_2)$, then $\mathcal{L}[\tau_b]$ is a DF-space. In particular, $\mathbf{B}(\mathcal{D}_1, \mathcal{D}_2)[\tau_b]$ is a DF-space.*

Proof. By Proposition 5.1.20, applied with $\mathcal{A} = \mathcal{L}^+(\mathcal{D}_1)$ and $\mathcal{B} = \mathcal{L}^+(\mathcal{D}_2)$, $\mathbf{B}_1(\mathcal{D}_2, \mathcal{D}_1)[\tau_\pi]$ is a Frechet space. Being topologically isomorphic to the strong dual of the Frechet space $\mathbf{B}(\mathcal{D}_2, \mathcal{D}_1)[\tau_\pi]$, $\mathcal{L}(\mathcal{D}_1, \mathcal{D}_2][\tau_b]$ is a complete DF-space (Jarchow [1], 12.4.5).

Now we prove the second assertion. Let $(\mathcal{U}_n : n \in \mathbb{N})$ be a sequence of closed absolutely convex 0-neighbourhoods in $\mathcal{L}[\tau_b]$ such that $\mathcal{U} := \bigcap_{n=1}^{\infty} \mathcal{U}_n$ absorbs all bounded sets in $\mathcal{L}[\tau_b]$. We have to show that $\mathcal{U}$ is a 0-neighbourhood in $\mathcal{L}[\tau_b]$. Let $\overline{\mathcal{U}_n}$, $n \in \mathbb{N}$, denote the closure of $\mathcal{U}_n$ in $\mathcal{L}(\mathcal{D}_1, \mathcal{D}_2^+)[\tau_b]$, and let $\tilde{\mathcal{U}} := \bigcap_{n=1}^{\infty} \overline{\mathcal{U}_n}$. Since $\mathbf{B}(\mathcal{D}_1, \mathcal{D}_2) \subseteq \mathcal{L}$ by assumption, Corollary 3.4.2 implies that $\mathcal{L}$ is dense in $\mathcal{L}(\mathcal{D}_1, \mathcal{D}_2)[\tau_b]$. From this it follows that $\overline{\mathcal{U}_n}$, the closure of a 0-neighbourhood in $\mathcal{L}[\tau_b]$, is an (absolutely convex) 0-neighbourhood in $\mathcal{L}(\mathcal{D}_1 \mathcal{D}_2^+)[\tau_b]$. Let $a \in \mathcal{L}^+(\mathcal{D}_1)$ and $b \in \mathcal{L}^+(\mathcal{D}_2)$. Since $\mathcal{U}$ absorbs bounded sets, there is a $\delta > 0$ such that $\delta'(\mathcal{U}_{a,b} \cap \mathcal{L}) \subseteq \mathcal{U} \subseteq \tilde{\mathcal{U}}$ when $0 < \delta' \leqq \delta$. From Theorem 3.4.1, $\mathcal{U}_{a,b}$ is the closure of $\mathcal{U}_{a,b} \cap \mathcal{L}$ in $\mathcal{L}(\mathcal{D}_1, \mathcal{D}_2^+)[\tau_b]$. Because $\tilde{\mathcal{U}}$ is closed in $\mathcal{L}(\mathcal{D}_1, \mathcal{D}_2^+)[\tau_b]$, we conclude that $\delta' \mathcal{U}_{a,b} \subseteq \tilde{\mathcal{U}}$. By Proposition 4.2.1, $\{\mathcal{U}_{a,b}: a \in \mathcal{L}^+(\mathcal{D}_1)$ and $b \in \mathcal{L}^+(\mathcal{D}_2)\}$ is a fundamental system of bounded sets in $\mathcal{L}(\mathcal{D}_1, \mathcal{D}_2^+)$ $[\tau_b]$. Therefore, the preceding shows that $\tilde{\mathcal{U}}$ absorbs each bounded subset of $\mathcal{L}(\mathcal{D}_1, \mathcal{D}_2^+)$ $[\tau_b]$. Since $\mathcal{L}(\mathcal{D}_1, \mathcal{D}_2)[\tau_b]$ is a DF-space as already proved, it follows from these properties of $\tilde{\mathcal{U}}$ and $\overline{\mathcal{U}_n}$ that $\tilde{\mathcal{U}}$ is a 0-neighbourhood in $\mathcal{L}(\mathcal{D}_1, \mathcal{D}_2)$ $[\tau_b]$. Because $\mathcal{U}_n$, $n \in \mathbb{N}$, is closed in $\mathcal{L}[\tau_b]$, $\tilde{\mathcal{U}} \cap \mathcal{L} = \mathcal{U}$, so that $\mathcal{U}$ is a 0-neighbourhood in $\mathcal{L}[\tau_b]$. Combined with the fact that $\mathcal{L}[\tau_b]$ has a countable fundamental system of bounded sets by Corollary 4.2.2 this proves that $\mathcal{L}[\tau_b]$ is a DF-space. □

Corollary 5.4.9. *If $\mathcal{D}$ is a Frechet domain in a Hilbert space, then $\mathcal{L}^+(\mathcal{D})$ $[\tau_{\mathcal{D}}]$ is a DF-space.*

Proof. Since $\mathbb{B}(\mathcal{D}) \subseteq \mathcal{L}^+(\mathcal{D})$, Corollary 5.4.8 applies with $\mathcal{L} = \mathcal{L}^+(\mathcal{D})$ and $\mathcal{D} = \mathcal{D}_1 = \mathcal{D}_2$. □

Proposition 5.4.10. *Suppose $\mathcal{D}_1$ and $\mathcal{D}_2$ are Frechet domains in the same Hilbert space. If $\mathcal{M}$ is a convex subset of $\mathcal{L}(\mathcal{D}_1, \mathcal{D}_2^+)$, then the following three assertions are equivalent:*

(i) *$\mathcal{M}$ is ultraweakly closed in $\mathcal{L}(\mathcal{D}_1, \mathcal{D}_2^+)$.*

(ii) *For each $a \in \mathcal{L}^+(\mathcal{D}_1)$ and $b \in \mathcal{L}^+(\mathcal{D}_2)$, $\mathcal{M} \cap \mathcal{U}_{a,b}$ is ultraweakly closed in $\mathcal{L}(\mathcal{D}_1, \mathcal{D}_2^+)$.*

(iii) *For arbitrary $a \in \mathcal{L}^+(\mathcal{D}_1)$ and $b \in \mathcal{L}^+(\mathcal{D}_2)$, $\mathcal{M} \cap \mathcal{U}_{a,b}$ is weak-operator closed in $\mathcal{L}(\mathcal{D}_1, \mathcal{D}_2^+)$.*

Proof. Since $\mathcal{U}_{a,b}$ is always ultraweakly closed, (i) → (ii) is obvious. Conditions (ii) and (iii) are equivalent, because the ultraweak topology and the weak-operator topology coincide on $\mathcal{U}_{a,b}$. It suffices to prove the implication (ii) → (i). From Proposition 5.1.20, $E := \mathbb{B}_1(\mathcal{D}_2, \mathcal{D}_1)$ $[\tau_\pi]$ is a Frechet space, since $\mathcal{D}_1[t_+]$ and $\mathcal{D}_2[t_+]$ are Frechet spaces. Our proof is essentially based on the Krein-Šmulian theorem applied to this space E. Let us identify $\mathcal{L}(\mathcal{D}_1, \mathcal{D}_2^+)[\tau_b]$ with the strong dual $E'[\beta]$ of E by means of the topological isomorphism I. By Proposition 5.2.11 the ultraweakly continuous linear functionals on $\mathcal{L}(\mathcal{D}_1, \mathcal{D}_2^+)$ are precisely the functionals $f_t(\cdot) \equiv \operatorname{tr} \hat{t} \cdot$, $t \in \mathbb{B}_1(\mathcal{D}_2, \mathcal{D}_1)$. Therefore, the ultraweak topology on $E' \equiv \mathcal{L}(\mathcal{D}_1, \mathcal{D}_2^+)$ equals the weak topology σ' on E'. Let U be a 0-neighbourhood in E. Then the polar U^0 of U in E' is bounded in $E'[\beta] \equiv L(\mathcal{D}_1, \mathcal{D}_2^+)$ $[\tau_b]$. By Proposition 4.2.1 there are operators $a \in \mathcal{L}^+(\mathcal{D}_1)$ and $b \in \mathcal{L}^+(\mathcal{D}_2)$ such that $U^0 \subseteq \mathcal{U}_{a,b}$. Therefore, by (ii), $\mathcal{M} \cap U^0$ is closed in $E'[\sigma']$. The preceding facts show that the assumptions of the Krein-Šmulian theorem (Schäfer [1], IV, 6.4) are satisfied; so $\mathcal{M}$ is closed in $E'[\sigma']$. But this is only a reformulation of (i). □

Corollary 5.4.11. *Let $\mathcal{D}_1$ and $\mathcal{D}_2$ be Frechet domains in the same Hilbert space. If $\mathcal{L}$ is an ultraweakly closed linear subspace of $\mathcal{L}(\mathcal{D}_1, \mathcal{D}_2^+)$ and f is a linear functional on $\mathcal{L}$, then the following statements are equivalent:*

(i) *f is ultraweakly continuous on $\mathcal{L}$.*

(ii) *The restriction to $\mathcal{L} \cap \mathcal{U}_{a,b}$ of f is ultraweakly continuous for each $a \in \mathcal{L}^+(\mathcal{D}_1)$ and $b \in \mathcal{L}^+(\mathcal{D}_2)$.*

(iii) *The restriction to $\mathcal{L} \cap \mathcal{U}_{a,b}$ of f is weak-operator continuous for all $a \in \mathcal{L}^+(\mathcal{D}_1)$ and $b \in \mathcal{L}^+(\mathcal{D}_2)$.*

Proof. (i) → (ii) is trivial. Proposition 3.5.3 gives (ii) → (iii). We prove (iii) → (i). Set $\mathcal{M} := \ker f$. Since $\mathcal{L}$ is ultraweakly closed in $\mathcal{L}(\mathcal{D}_1, \mathcal{D}_2^+)$, it follows from (iii) and Proposition 5.4.10 that $\mathcal{M}$ is ultraweakly closed in $\mathcal{L}(\mathcal{D}_1, \mathcal{D}_2^+)$. Hence f is ultraweakly continuous. □

Remark 2. In Sections 5.4 and 5.5 only O*-algebras are considered. But the duality theorem and so some of its applications hold more generally. KÜRSTEN [6] proved that the conclusion of Theorem 5.4.2 is true for directed O*-families $\mathcal{A}$ and $\mathcal{B}$ on the same Hilbert space such that $\mathcal{D}_\mathcal{A}$ and $\mathcal{D}_\mathcal{B}$ are QF-spaces.

5.5. Characterizations of Montel Domains

Proposition 5.5.1. *Suppose that $\mathcal{A}$ is a closed O*-algebra on the Hilbert space $\mathcal{H}$ and $\mathcal{L}$ is a *-vector subspace of $\mathcal{L}(\mathcal{D}_\mathcal{A}, \mathcal{D}_\mathcal{A}^+)$ which contains $\mathcal{A}$ and $\mathbb{F}(\mathcal{D}(\mathcal{A}))$. Then the following statements are equivalent:*

(i) *$\mathcal{D}_\mathcal{A}$ is a semi-Montel space.*

(ii) *Each continuous linear functional f on $\mathcal{L}[\tau_\mathcal{D}]$ is of the form $f(x) = \operatorname{tr}_\mathcal{A} \hat{t}x$, $x \in \mathcal{L}$, for some $t \in \mathbb{B}_1(\mathcal{A})$.*

(iii) *Each continuous strongly positive linear functional f on $\mathcal{L}[\tau_\mathcal{D}]$ is of the form $f(x) = \operatorname{tr}_\mathcal{A} \hat{t}x$, $x \in \mathcal{L}$, for some $t \in \mathbb{B}_1(\mathcal{A})$.*

Proof. (i) → (ii) is a special case of Theorem 5.3.7. (ii) → (iii) is trivial. We prove the implication (iii) → (i).

To prove (i), it suffices to show that each closed bounded subset $\mathcal{M}$ of $\mathcal{D}_\mathcal{A}$ is compact in the graph topology $t_\mathcal{A}$. Fix such a set $\mathcal{M}$, and let $\mathbb{U}$ be an ultrafilter on $\mathcal{M}$. There is a $\lambda > 0$ such that $\mathcal{M} \subseteq \lambda \mathcal{U}_\mathcal{H}$. Since $\mathbb{U}$ is an ultrafilter basis in $\lambda \mathcal{U}_\mathcal{H}$ and $\lambda \mathcal{U}_\mathcal{H}$ is compact in the weak topology of the Hilbert space $\mathcal{H}$, there exists a vector $\zeta_0 \in \mathcal{H}$ such that

$$\lim_{\zeta, \mathbb{U}} \langle \zeta, \eta \rangle = \langle \zeta_0, \eta \rangle \quad \text{for all} \quad \eta \in \mathcal{H}. \tag{1}$$

(We refer to BOURBAKI [1], I, § 6 and 7, for the facts and the notation concerning ultrafilter limits we use.) Define $f(x) := \lim_{\zeta, \mathbb{U}} \langle x\zeta, \zeta \rangle$. Then we have $|f(x)| \leq p_\mathcal{M}(x)$ for all $x \in \mathcal{L}$. From this we see that $f(x)$ is finite for $x \in \mathcal{L}$. Hence f is a strongly positive linear functional on $\mathcal{L}$. Further, this inequality shows that f is continuous on $\mathcal{L}[\tau_\mathcal{D}]$. By (iii), there exists an operator $t \in \mathbb{B}_1(\mathcal{A})$ such that $f(x) = \operatorname{tr} \hat{t}x$, $x \in \mathcal{L}$. Letting $x = \psi \otimes \varphi$ with $\varphi, \psi \in \mathcal{D}(\mathcal{A})$, we obtain

$$\langle t\varphi, \psi \rangle = \operatorname{tr} \hat{t}(\psi \otimes \varphi) = f(\psi \otimes \varphi) = \lim_{\zeta, \mathbb{U}} \langle \zeta, \psi \rangle \langle \varphi, \zeta \rangle = \langle \zeta_0, \psi \rangle \langle \varphi, \zeta_0 \rangle.$$

Because $\mathcal{D}(\mathcal{A})$ is dense in $\mathcal{H}$, this gives $t = \zeta_0 \otimes \zeta_0$. From $t \in \mathbb{B}_1(\mathcal{A})$ we conclude that $\zeta_0 \in \mathcal{D}(\mathcal{A})$.

Suppose $a \in \mathcal{A}$. We next prove that $\lim_{\zeta, \mathbb{U}} \|a(\zeta - \zeta_0)\|^2 = 0$. By (1), $\lim_{\zeta, \mathbb{U}} \langle a^+ a\zeta_0, \zeta \rangle = \langle a^+ a\zeta_0, \zeta_0 \rangle$. Moreover,

$$\lim_{\zeta, \mathbb{U}} \langle a^+ a\zeta, \zeta \rangle = f(a^+ a) = \operatorname{tr} \widehat{(\zeta_0 \otimes \zeta_0)}\, a^+ a = \langle a^+ a\zeta_0, \zeta_0 \rangle.$$

Using both facts, we get

$$\begin{aligned} \lim_{\zeta, \mathbb{U}} \|a(\zeta - \zeta_0)\|^2 &= \lim_{\zeta, \mathbb{U}} \langle a^+ a\zeta, \zeta \rangle - 2 \operatorname{Re} \lim_{\zeta, \mathbb{U}} \langle a^+ a\zeta_0, \zeta \rangle + \|a\zeta_0\|^2 \\ &= \langle a^+ a\zeta_0, \zeta_0 \rangle - 2 \operatorname{Re} \langle a^+ a\zeta_0, \zeta_0 \rangle + \|a\zeta_0\|^2 = 0. \end{aligned}$$

From $\lim_{\zeta,\mathbb{U}} \|a(\zeta - \zeta_0)\|^2 = 0$ for all $a \in \mathcal{A}$ it follows that $\zeta_0 \in \mathcal{M}$ (because $\mathcal{M}$ is closed in $\mathcal{D}_\mathcal{A}$) and that the ultrafilter $\mathbb{U}$ on $\mathcal{M}$ converges to ζ_0. This proves that $\mathcal{M}$ is a compact subset of $\mathcal{D}_\mathcal{A}$. □

Corollary 5.5.2. *Suppose $\mathcal{A}$ is a closed O*-algebra such that $\mathcal{D}_\mathcal{A}$ is a QF-space. Then $\mathcal{D}_\mathcal{A}$ is a semi-Montel space if and only if $\mathbb{B}_1(\mathcal{A})[\tau_\pi]$ is semireflexive.*

Proof. Because of Corollary 5.4.3, (i), the semireflexivity of $\mathbb{B}_1(\mathcal{A})[\tau_\pi]$ means that each continuous linear functional on $\mathcal{L}(\mathcal{D}_\mathcal{A}, \mathcal{D}_\mathcal{A}^+)[\tau_\mathcal{D}]$ is of the form $f_t(\cdot) \equiv \mathrm{tr}_\mathcal{A}\, \hat{t}\,\cdot$ for some $t \in \mathbb{B}_1(\mathcal{A})$. By Proposition 5.5.1, (i) $\leftrightarrow$ (ii), this is the case if and only if $\mathcal{D}_\mathcal{A}$ is a semi-Montel space. □

Corollary 5.5.3. *If $\mathcal{D}$ is a Frechet domain in a Hilbert space, then the following assertions are equivalent:*

(i) *$\mathcal{D}[t_+]$ is a Montel space.*

(ii) *$\mathbb{B}_1(\mathcal{D})[\tau_\pi]$ is reflexive.*

(iii) *$\mathcal{L}(\mathcal{D}, \mathcal{D}^+)[\tau_\mathcal{D}]$ is reflexive.*

Proof. First recall that $\mathbb{B}_1(\mathcal{D})[\tau_\pi]$ is a Frechet space by Proposition 5.1.20, since $\mathcal{D}[t_+]$ is a Frechet space by assumption. Further, note that a Frechet space is semireflexive [resp. a semi-Montel space] if and only if it is reflexive [resp. a Montel space]. Therefore, Corollary 5.5.2 (applied with $\mathcal{A} = \mathcal{L}^+(\mathcal{D})$) yields the equivalence of (i) and (ii). Since $\mathcal{L}(\mathcal{D}, \mathcal{D}^+)[\tau_\mathcal{D}]$ is topologically isomorphic to the strong dual of $\mathbb{B}_1(\mathcal{D})[\tau_\pi]$, (ii) $\leftrightarrow$ (iii) follows at once from the fact that a Frechet space is reflexive if and only if its strong dual has this property (Schäfer [1] IV, 5.6). □

Proposition 5.5.4. *If $\mathcal{A}$ is an O*-algebra on the Hilbert space $\mathcal{H}$ such that $\mathcal{D}_\mathcal{A}$ is a QF-space, then the following assertions are equivalent:*

(i) *$\mathcal{D}_\mathcal{A}$ is a semi-Montel space.*

(ii) *Each closed linear subspace of $\mathcal{H}$ which is contained in $\mathcal{D}(\mathcal{A})$ is finite dimensional.*

(ii)′ *Each projection in $\mathbb{P}(\mathcal{D}(\mathcal{A}))$ has finite rank.*

iii) *Each operator in $\mathbb{B}(\mathcal{D}(\mathcal{A}))$ is compact.*

(iv) *$\mathbb{F}(\mathcal{D}(\mathcal{A}))$ is dense in $\mathcal{L}(\mathcal{D}_\mathcal{A}, \mathcal{D}_\mathcal{A}^+)[\tau_\mathcal{D}]$.*

(iv)′ *$\mathbb{F}(\mathcal{D}(\mathcal{A}))$ is dense in $\mathcal{L}^+(\mathcal{D}_\mathcal{A})[\tau_\mathcal{D}]$.*

(iv)″ *$\mathcal{F}(\mathcal{D}_\mathcal{A}, \mathcal{D}_\mathcal{A}^+)$ is dense in $\mathcal{L}(\mathcal{D}_\mathcal{A}, \mathcal{D}_\mathcal{A}^+)[\tau_\mathcal{D}]$.*

We first prove an auxiliary lemma.

Lemma 5.5.5. *Let $\mathcal{A}$ be as in Proposition 5.5.4 and let g be a linear functional on $\mathbb{F}(\mathcal{H})$. If g is continuous in the topology $\tau_\mathcal{D}$ of $\mathcal{L}(\mathcal{D}_\mathcal{A}, \mathcal{D}_\mathcal{A}^+)$, then there exists a $t \in \mathbb{B}_1(\mathcal{A})$ such that $g(x) = \mathrm{Tr}\, tx$ for $x \in \mathbb{F}(\mathcal{H})$.*

Proof. We consider $\mathbb{F}(\mathcal{H})$ as a linear subspace of $\mathcal{L}(\mathcal{D}_{\hat{\mathcal{A}}}, \mathcal{D}_{\hat{\mathcal{A}}}^+)$. Since g is obviously continuous in the bounded topology of $\mathcal{L}(\mathcal{D}_{\hat{\mathcal{A}}}, \mathcal{D}_{\hat{\mathcal{A}}}^+)$, it follows from Corollary 5.3.2 (applied to $\mathcal{L} = \mathbb{F}(\mathcal{H})$ and $\hat{\mathcal{A}}$) that there is an operator $t \in \mathbb{B}_1(\hat{\mathcal{A}})$ such that $g(x) = \mathrm{tr}_{\hat{\mathcal{A}}}\, \hat{t}x$, $x \in \mathbb{F}(\mathcal{H})$. This gives $g(x) = \mathrm{Tr}\, tx$ for $x \in \mathbb{F}(\mathcal{H})$. It remains to prove that $t \in \mathbb{B}_1(\mathcal{A})$. By the continuity assumption there exists a bounded subset $\mathcal{M}$ of $\mathcal{D}_\mathcal{A}$ such that $|g(x)|$

$\leq p_{\mathcal{M}}(x)$ for all $x \in \mathbb{F}(\mathcal{H})$. Letting $x = \psi \otimes \varphi$, we get

$$|\langle t\varphi, \psi\rangle| = |\mathrm{Tr}\, t(\psi \otimes \varphi)| = |g(\psi \otimes \varphi)| \leq p_{\mathcal{M}}(\psi \otimes \varphi) = r_{\mathcal{M}}(\varphi)\, r_{\mathcal{M}}(\psi) \tag{2}$$

for $\varphi, \psi \in \mathcal{H}$, where $r_{\mathcal{M}}(\cdot) \equiv \sup_{\eta \in \mathcal{M}} |\langle \cdot, \eta\rangle|$. Because $\mathcal{D}_{\mathcal{A}}$ is a QF-space, $\mathcal{M}$ is contained in a Frechet linear subspace $\mathcal{E}$ of $\mathcal{D}_{\mathcal{A}}$. We prove that $t\mathcal{H} \subseteq \mathcal{E}$. Assume the contrary, that is, $t\varphi \notin \mathcal{E}$ for some $\varphi \in \mathcal{H}$. Since $t\varphi \in \mathcal{D}(\hat{\mathcal{A}})$ because of $t \in \mathbb{B}_1(\hat{\mathcal{A}})$ and $\mathcal{E}$ is a closed linear subspace of $\mathcal{D}_{\hat{\mathcal{A}}}$, the separation theorems for convex sets ensure the existence of a linear functional $h \in \mathcal{D}'_{\hat{\mathcal{A}}}$ satisfying $h(t\varphi) = 1$ and $h(\psi) = 0$ for $\psi \in \mathcal{E}$. By Proposition 2.3.5, there are $a \in \mathcal{A}$ and $\xi \in \mathcal{H}_{\hat{a}}$ such that $h(\cdot) \equiv \langle \bar{a}\,\cdot, \bar{a}\xi\rangle$ on $\mathcal{D}_{\hat{\mathcal{A}}}$. Let $\varepsilon > 0$. We choose a vector $\xi_\varepsilon \in \mathcal{D}(\mathcal{A})$ such that $\|\bar{a}(\xi_\varepsilon - \xi)\| < \varepsilon$. Then

$$|\langle \bar{a}t\varphi, \bar{a}\xi_\varepsilon\rangle| = |\langle \bar{a}t\varphi, \bar{a}(\xi_\varepsilon - \xi)\rangle + h(t\varphi)| \geq 1 - \varepsilon\, \|\bar{a}t\varphi\|.$$

On the other hand, by (2) and $h(\cdot) \equiv \langle \bar{a}\,\cdot, \bar{a}\xi\rangle = 0$ on $\mathcal{M}$, we have

$$\begin{aligned} |\langle \bar{a}t\varphi, \bar{a}\xi_\varepsilon\rangle| &= |\langle t\varphi, a^+a\xi_\varepsilon\rangle| \leq r_{\mathcal{M}}(\varphi)\, r_{\mathcal{M}}(a^+a\xi_\varepsilon) = r_{\mathcal{M}}(\varphi) \sup_{\eta\in\mathcal{M}} |\langle a\eta, a\xi_\varepsilon\rangle| \\ &= r_{\mathcal{M}}(\varphi) \sup_{\eta\in\mathcal{M}} |\langle a\eta, \bar{a}(\xi_\varepsilon - \xi)\rangle| \leq \varepsilon r_{\mathcal{M}}(\varphi) \sup_{\eta\in\mathcal{M}} \|a\eta\|. \end{aligned}$$

Since $\varepsilon > 0$ was arbitrary, we arrived at a contradiction. Thus we have shown that $t\mathcal{H} \subseteq \mathcal{E} \subseteq \mathcal{D}(\mathcal{A})$. A similar reasoning yields $t^*\mathcal{H} \subseteq \mathcal{E} \subseteq \mathcal{D}(\mathcal{A})$. Combined with $t \in \mathbb{B}_1(\hat{\mathcal{A}})$, this proves that $t \in \mathbb{B}_1(\mathcal{A})$. □

Proof of Proposition 5.5.4.

We prove that (i) → (ii)′ → (iv) → (iv)″ → (i), (iii) ↔ (ii)′ ↔ (ii) and (iv) ↔ (iv)′. (iii) → (ii)′ ↔ (ii) and (iv) → (iv)″ are trivial. Since $\mathcal{L}^+(\mathcal{D}_{\mathcal{A}})$ is dense in $\mathcal{L}(\mathcal{D}_{\mathcal{A}}, \mathcal{D}_{\mathcal{A}}^+)\,[\tau_{\mathcal{D}}]$ as noted in Remark 4 in 3.4, we have (iv) ↔ (iv)′. (ii)′ → (iv) follows immediately from Corollary 3.4.3, (iii).

(i) → (ii)′: Suppose $e \in \mathbb{P}(\mathcal{D}(\mathcal{A}))$. By Corollary 3.1.3, $e\mathcal{U}_{\mathcal{H}}$ is a bounded set in $\mathcal{D}_{\mathcal{A}}$. By (i), this set is relatively compact in $\mathcal{D}_{\mathcal{A}}$. But this is only possible if the projection e has finite rank.

(ii)′ → (iii): Since $\mathbb{B}(\mathcal{D}(\mathcal{A}))$ is a $*$-vector space, it suffices to show that self-adjoint operators in $\mathbb{B}(\mathcal{D}(\mathcal{A}))$ are compact. Let $c = c^* \in \mathbb{B}(\mathcal{D}(\mathcal{A}))$. Let e_ε be the spectral projection of c associated with the set $(-\infty, -\varepsilon) \cup (\varepsilon, +\infty)$, where $\varepsilon > 0$. From $e_\varepsilon\mathcal{H} \subseteq c\mathcal{H} \subseteq \mathcal{D}(\mathcal{A})$, $e_\varepsilon \in \mathbb{P}(\mathcal{D}(\mathcal{A}))$. By (ii)′, e_ε has finite rank for every $\varepsilon > 0$. This implies that c is compact.

(iv)″ → (i): We slightly modify the argument used in the proof of implication (iii) → (i) in Proposition 5.5.1. Note that the assumption that $\mathcal{A}$ is closed was not needed in this proof. Let $\mathcal{M}$, $\mathbb{U}$ and ζ_0 be as in this proof. We define a continuous linear functional f on $\mathcal{L}(\mathcal{D}_{\mathcal{A}}, \mathcal{D}_{\mathcal{A}}^+)\,[\tau_{\mathcal{D}}]$ by $f(x) = \lim_{\zeta,\mathbb{U}} \langle x\zeta, \zeta\rangle$, $x \in \mathcal{L}(\mathcal{D}_{\mathcal{A}}, \mathcal{D}_{\mathcal{A}}^+)$. Applying Lemma 5.5.5 to $g := f \upharpoonright \mathbb{F}(\mathcal{H})$, there is an operator $t \in \mathbb{B}_1(\mathcal{A})$ such that $f(x) = g(x) = \mathrm{Tr}\, tx$ for all $x \in \mathbb{F}(\mathcal{H})$. In the same way as in the proof of Proposition 5.5.1 we obtain $t = \zeta_0 \otimes \zeta_0$ and $\zeta_0 \in \mathcal{D}(\mathcal{A})$. Therefore, $f(x) = \mathrm{Tr}\, tx = \langle x\zeta_0, \zeta_0\rangle$ for $x \in \mathbb{F}(\mathcal{H})$. Since $\mathbb{F}(\mathcal{H})$ is dense in $\mathcal{F}(\mathcal{D}_{\mathcal{A}}, \mathcal{D}_{\mathcal{A}}^+)\,[\tau_{\mathcal{D}}]$ by Lemma 3.4.4 and $\mathcal{F}(\mathcal{D}_{\mathcal{A}}, \mathcal{D}_{\mathcal{A}}^+)$ is dense in $\mathcal{L}(\mathcal{D}_{\mathcal{A}}, \mathcal{D}_{\mathcal{A}}^+)\,[\tau_{\mathcal{D}}]$ by (iv)″, the latter implies that $f(x) = \langle x\zeta_0, \zeta_0\rangle$ for all $x \in \mathcal{L}(\mathcal{D}_{\mathcal{A}}, \mathcal{D}_{\mathcal{A}}^+)$. Arguing now as in the proof of Proposition 5.5.1 it follows that the ultrafilter $\mathbb{U}$ converges to ζ_0 and that $\mathcal{M}$ is compact in the graph topology $\mathrm{t}_{\mathcal{A}}$. This proves (i). □

Proposition 5.5.6. *Let $\mathcal{A}$ and $\mathcal{B}$ be O^*-algebras on the Hilbert space $\mathcal{H}$, and let $\mathcal{L}$ be a linear subspace of $\mathcal{L}(\mathcal{D}_\mathcal{A}, \mathcal{D}_\mathcal{B}^+)$ which contains $\mathcal{F}(\mathcal{D}_\mathcal{A}, \mathcal{D}_\mathcal{B}^+)$. Suppose that $\mathcal{D}_\mathcal{A}$ and $\mathcal{D}_\mathcal{B}$ are QF-spaces. Then the following assertions are equivalent:*

(i) *$\mathcal{D}_\mathcal{A}$ and $\mathcal{D}_\mathcal{B}$ are semi-Montel spaces.*

(ii) *$\mathbf{B}_1(\mathcal{B}, \mathcal{A})[\tau_\pi]$ is a semi-Montel space.*

(iii) *Each bounded subset of $\mathcal{L}[\tau_b]$ is precompact.*

Proof. (i) → (ii): We have to show that each bounded subset of $\mathbf{B}_1(\mathcal{B}, \mathcal{A})[\tau_\pi]$ is relatively compact. Since $\mathbf{B}_1(\mathcal{B}, \mathcal{A})[\tau_\pi]$ is again a QF-space by Corollary 5.4.7, the closure of a bounded set is complete. Thus it is enough to show that bounded subsets of $\mathbf{B}_1(\mathcal{B}, \mathcal{A})[\tau_\pi]$ are precompact. By Corollary 5.4.5, it suffices to prove this for bounded sets $\mathcal{R}$ of the form $\mathcal{R} = c\mathcal{U}_{\mathbf{B}_1(\mathcal{H})}d$, where $c \in \mathbf{B}(\mathcal{D}(\mathcal{A}))_+$ and $d \in \mathbf{B}(\mathcal{D}(\mathcal{B}))_+$. Fix c and d. If $\varepsilon > 0$, let e_ε and f_ε denote the spectral projections of c and d, respectively, associated with the interval $[\varepsilon, +\infty)$. Put $\mathcal{R}_\varepsilon := e_\varepsilon \mathcal{R} f_\varepsilon$. Since $\mathcal{D}_\mathcal{A}$ and $\mathcal{D}_\mathcal{B}$ are semi-Montel spaces by (i), e_ε and f_ε have finite rank by Proposition 5.5.4, (i) ↔ (iii). We have $\mathcal{R}_\varepsilon = e_\varepsilon c\mathcal{U}_{\mathbf{B}_1(\mathcal{H})}df_\varepsilon$ with $e_\varepsilon c \in \mathbf{B}(\mathcal{D}(\mathcal{A}))_+$ and $df_\varepsilon \in \mathbf{B}(\mathcal{D}(\mathcal{B}))_+$. Therefore, $\mathcal{R}_\varepsilon$ is a bounded subset (by Corollary 5.4.5) of a finite dimensional subspace of $\mathbf{B}_1(\mathcal{B}, \mathcal{A})[\tau_\pi]$. Suppose $a \in \mathcal{A}$ and $b \in \mathcal{B}$. Let $t_1 \in \mathcal{U}_{\mathbf{B}_1(\mathcal{H})}$. Set $t = ct_1d$ and $t_\varepsilon = e_\varepsilon ct_1df_\varepsilon$. Then we have

$$\begin{aligned}\nu_{a,b}(t - t_\varepsilon) &= \nu(act_1\overline{db} - ace_\varepsilon t_1 f_\varepsilon \overline{db})\\ &\leq \nu(ac^{1/2}c^{1/2}(I - e_\varepsilon)\, t_1\overline{db} + \nu(ace_\varepsilon t_1(I - f_\varepsilon)\, d^{1/2}\overline{d^{1/2}b})\\ &\leq \varepsilon^{1/2}(\|ac^{1/2}\|\, \|\overline{db}\| + \|ac\|\, \|\overline{d^{1/2}b}\|).\end{aligned}$$

Here we used that $c^{1/2} \in \mathbf{B}(\mathcal{D}(\mathcal{A}))$ and $d^{1/2} \in \mathbf{B}(\mathcal{D}(\mathcal{B}))$ by Corollary 3.1.5 and that the operators $ac^{1/2}$, db, ac and $d^{1/2}b$ are bounded by Lemma 3.1.2. Since the topology τ_π is generated by the directed family (by Lemma 5.1.16 and Corollary 5.1.17) of seminorms $\{\nu_{a,b}: a \in \mathcal{A} \text{ and } b \in \mathcal{B}\}$, it follows from the preceding estimate and Lemma 1.1.1 that $\mathcal{R}$ is precompact in $\mathbf{B}_1(\mathcal{B}, \mathcal{A})[\tau_\pi]$.

(i) → (iii): Suppose $\mathcal{R}$ is a bounded subset of $\mathcal{L}[\tau_b]$. Let p be a continuous seminorm on $\mathcal{L}(\mathcal{D}_\mathcal{A}, \mathcal{D}_\mathcal{B}^+)[\tau_b]$. From Theorem 3.4.1, there are projections $e \in \mathbf{P}(\mathcal{D}(\mathcal{A}))$ and $f \in \mathbf{P}(\mathcal{D}(\mathcal{B}))$ such that $p(x - \hat{f}xe) \leq 1$ for all $x \in \mathcal{R}$. By (i) and Proposition 5.5.4, e and f have finite rank. This implies that $\hat{f}Re$ is a bounded subset of a finite dimensional subspace of $\mathcal{F}(\mathcal{D}_\mathcal{A}, \mathcal{D}_\mathcal{B}^+)[\tau_b]$ and hence of $\mathcal{L}[\tau_b]$, since $\mathcal{F}(\mathcal{D}_\mathcal{A}, \mathcal{D}_\mathcal{B}^+) \subseteq \mathcal{L}$ by assumption. By Lemma 1.1.1 this proves that $\mathcal{R}$ is precompact in $\mathcal{L}[\tau_b]$.

(ii) → (i): Fix a non-zero vector $\zeta \in \mathcal{D}(\mathcal{B})$ and define $T\varphi = \zeta \otimes \varphi$ for $\varphi \in \mathcal{D}(\mathcal{A})$. From $\nu_{a,b}(\zeta \otimes \varphi) = \|a\varphi\|\, \|b^+\zeta\|$ for $a \in \mathcal{A}$, $b \in \mathcal{B}$ and $\varphi \in \mathcal{D}(\mathcal{A})$ we see that T is a topological isomorphism of $\mathcal{D}_\mathcal{A}$ onto a subspace of $\mathbf{B}_1(\mathcal{B}, \mathcal{A})[\tau_\pi]$. Let $\mathcal{M}$ be a bounded set in $\mathcal{D}_\mathcal{A}$. By (ii), the bounded subset $T(\mathcal{M})$ of $\mathbf{B}_1(\mathcal{B}, \mathcal{A})[\tau_\pi]$ is relatively compact and hence precompact. Therefore, $\mathcal{M}$ is precompact in $\mathcal{D}_\mathcal{A}$. Since $\mathcal{D}_\mathcal{A}$ is a QF-space, the closure of $\mathcal{M}$ in $\mathcal{D}_\mathcal{A}$ is complete, so that $\mathcal{M}$ is relatively compact in $\mathcal{D}_\mathcal{A}$. This proves that $\mathcal{D}_\mathcal{A}$ is a semi-Montel space. The proof for $\mathcal{D}_\mathcal{B}$ is similar.

(iii) → (i): The proof is based on a similar idea as the previous proof. Suppose $e \in \mathbf{P}(\mathcal{D}(\mathcal{A}))$. Take a fixed unit vector $\zeta \in \mathcal{D}(\mathcal{B})$ and define $T\varphi = \zeta \otimes \varphi$ for $\varphi \in e\mathcal{H}$. Then $T(e\mathcal{H}) \subseteq \mathcal{F}(\mathcal{D}_\mathcal{A}, \mathcal{D}_\mathcal{B}^+) \subseteq \mathcal{L}$. If $\varphi \in e\mathcal{H}$, $c \in \mathbf{B}(\mathcal{D}(\mathcal{A}))_+$ and $d \in \mathbf{B}(\mathcal{D}(\mathcal{B}))_+$, then we have

$$q_{c,d}(\zeta \otimes \varphi) = \|d(\zeta \otimes \varphi)\, c\| \leq \|d\|\, \|c\zeta\|\, \|\varphi\| \quad \text{and}$$

$$\|\varphi\| = \|e(\zeta \otimes \varphi)\, (\zeta \otimes \zeta)\| = q_{\zeta \otimes \zeta, e}(\zeta \otimes \varphi).$$

By Theorem 3.3.16 this shows that $e\mathcal{H}$ (endowed with the norm topology of $\mathcal{H}$) and $T(e\mathcal{H})$ (equipped with the topology τ_b) are homeomorphic. Therefore, the set $T(e\mathcal{U}_{\mathcal{H}})$ is a bounded and hence a precompact subset of $\mathcal{L}[\tau_b]$ by (iii). Thus $e\mathcal{U}_{\mathcal{H}}$ is precompact in the norm topology which implies that e is of finite rank. From Proposition 5.5.4, (i) $\leftrightarrow$ (ii)$'$, $\mathcal{D}_{\mathcal{A}}$ is a semi-Montel space. A similar reasoning proves that $\mathcal{D}_{\mathcal{B}}$ is a semi-Montel space. □

Corollary 5.5.7. *If $\mathcal{D}_1$ and $\mathcal{D}_2$ are Frechet domains in the same Hilbert space, then the following assertions are equivalent:*

(i) *$\mathcal{D}_1[t_+]$ and $\mathcal{D}_2[t_+]$ are Montel spaces.*
(ii) *$\mathbb{B}_1(\mathcal{D}_2, \mathcal{D}_1)[\tau_\pi]$ is a Montel space.*
(iii) *$\mathcal{L}(\mathcal{D}_1, \mathcal{D}_2^+)[\tau_b]$ is a Montel space.*

Proof. First we recall that $\mathbb{B}_1(\mathcal{D}_2, \mathcal{D}_1)[\tau_\pi]$ is a Frechet space by Proposition 5.1.20. Note also that a semi-Montel space which is a Frechet space is Montel space. Therefore, Proposition 5.5.6 applied with $\mathcal{A} = \mathcal{L}^+(\mathcal{D}_1)$, $\mathcal{B} = \mathcal{L}^+(\mathcal{D}_2)$ and $\mathcal{L} = \mathcal{L}(\mathcal{D}_{\mathcal{A}}, \mathcal{D}_{\mathcal{B}}^+)$ gives (iii) $\rightarrow$ (ii) $\leftrightarrow$ (i). Since $\mathcal{L}(\mathcal{D}_1, \mathcal{D}_2^+)[\tau_b]$ is topologically isomorphic to the strong dual of $\mathbb{B}_1(\mathcal{D}_2, \mathcal{D}_1)[\tau_\pi]$ by Theorem 5.4.2, (ii) $\rightarrow$ (iii) follows at once from the fact that the strong dual of a Montel space is again a Montel space (Schäfer [1], IV, 5.9). □

Notes

5.1. The space $\mathbb{B}_1(\mathcal{D})$ was introduced by Lassner and Timmermann [1]. It was further investigated in Schmüdgen [5]. The material developed in the third subsection is mostly taken from Kürsten [2], [5]. The case of general O*-families is treated here for the first time.
5.2. The equivalence of (ii) and (iii) in Proposition 5.2.12 is from Araki/Jurzak [1].
5.3. The starting point for the investigations in this section was the following question: under what conditions to an O*-algebra $\mathcal{A}$ is every strongly positive linear function f on $\mathcal{A}$ of the form $f = f_t \equiv \operatorname{Tr} t\cdot$ with some $t \in \mathbb{B}_1(\mathcal{A})_+$? This problem was first studied by Sherman [1] who gave an affirmative answer for a countably generated O*-algebra which contains the restriction to $\mathcal{D}(\mathcal{A})$ of the inverse of some compact operator. Woronowicz [1], [2] proved this for the O*-algebras $\mathbb{A}(p_1, q_1)$ and $\mathcal{L}^+(\mathcal{S}(\mathbb{R}))$. His idea of proof combined with Corollary 5.3.3 was used by Schmüdgen [5] to show that this is true for any self-adjoint O*-algebra which contains the restriction of the inverse of a compact operator. (This result is included in the second half of Theorem 5.3.8). Lassner and Timmermann [1] studied the continuity of the functionals f_t in the topology $\tau_{\mathcal{D}}$. In Schmüdgen [5], [7] it was proved, that the above question has an affirmative answer if $\mathcal{D}_{\mathcal{A}}$ is a Frechet-Montel space. (This result is contained in the first half of Theorem 5.3.8.) The linear functionals f_t with general t in $\mathbb{B}_1(\mathcal{A})$ were characterized by Schmüdgen [5]; the corresponding result is covered by Theorem 5.3.7.
5.4. Let E and F be locally convex Hausdorff spaces. It is wellknown that the dual of the completed projective tensor product $E \mathbin{\widehat{\otimes}_\pi} F$ is (canonically isomorphic to) the space $\mathcal{B}(E, F)$ of continuous sesquilinear forms on $E \times F$ and that the strong topology on $(E \mathbin{\widehat{\otimes}_\pi} F)'$ is finer than the bi-bounded topology on $\mathcal{B}(E, F)$. (Proposition 5.4.1 expresses this general fact in a concrete setting.) A natural question is: under what conditions do these two topologies coincide? The question whether or not this is true for Frechet spaces E and F was first raised by Grothendieck [1], ch. I, § 1, pp. 33–34. It is equivalent to his "problème de topologies"; cf. Grothendieck [1], questions non résolues, 2. This problem was open for many years; it was solved by Taskinen [1] who gave a counter-example. Theorem 5.4.2 says, in particular, that the answer to the above question is affirmative in case where $E = \mathcal{D}_{\mathcal{A}}$ and $F = \mathcal{D}_{\mathcal{B}}$ when $\mathcal{A}$ and $\mathcal{B}$ satisfy the assumptions of the

theorem. Further affirmative results concerning this question can be found in KÜRSTEN [6] and TASKINEN [2].
The central result of this section, Theorem 5.4.2, and its important consequences, Corollaries 5.4.5, 5.4.7 and 5.4.8, are due to KÜRSTEN [2], [5]. KÜRSTEN considered only the case $\mathscr{A} = \mathscr{B}$, but the general case uses the same idea of proof.
5.5. Proposition 5.5.1 and Proposition 5.5.4, (i)$\leftrightarrow$(ii), under some additional assumptions are from SCHMÜDGEN [5]. Proposition 5.5.6 and the other statements of Proposition 5.5.4 are due to KÜRSTEN [2].

Additional References:

LASSNER/LASSNER [1], LÖFFLER/TIMMERMANN [2], [3], TIMMERMANN [1].

6. The Generalized Calkin Algebra and the $*$-Algebra $\mathcal{L}^+(\mathcal{D})$

In this chapter we develop various results concerning completely continuous operators in $\mathcal{L}(\mathcal{D}_\mathcal{A}, \mathcal{D}_\mathcal{B}^+)$ and in $\mathcal{L}^+(\mathcal{D}_\mathcal{A})$, the generalized Calkin algebra of $\mathcal{D}_\mathcal{A}$ and the maximal O*-algebra $\mathcal{L}^+(\mathcal{D})$. Some of them (but not all) can be considered as generalizations of classical facts about compact operators in Hilbert space, the Calkin algebra and the $*$-algebra $\mathbb{B}(\mathcal{H})$, respectively.

In Section 6.1 we study the vector space $\mathcal{V}(\mathcal{D}_\mathcal{A}, \mathcal{D}_\mathcal{B}^+)$, which is defined as the closure of the finite rank mappings $\mathcal{F}(\mathcal{D}_\mathcal{A}, \mathcal{D}_\mathcal{B}^+)$ in $\mathcal{L}(\mathcal{D}_\mathcal{A}, \mathcal{D}_\mathcal{B}^+)\,[\tau_b]$, and the closed two-sided $*$-ideal $\mathcal{V}(\mathcal{D}_\mathcal{A})$ of $\mathcal{L}^+(\mathcal{D}_\mathcal{A})$. If $\mathcal{A}$ and $\mathcal{B}$ are O*-algebras for which $\mathcal{D}_\mathcal{A}$ and $\mathcal{D}_\mathcal{B}$ are Frechet spaces, then the strong dual of $\mathcal{V}(\mathcal{D}_\mathcal{A}, \mathcal{D}_\mathcal{B}^+)\,[\tau_b]$ is topologically isomorphic in canonical way to the space $\mathbb{B}_1(\mathcal{B}, \mathcal{A})\,[\tau_\pi]$ considered in the previous chapter. If $\mathcal{A}$ is an O*-algebra and $\mathcal{D}_\mathcal{A}$ is a quasi-Frechet space, then the quotient $*$-algebra $\mathcal{L}^+(\mathcal{D}_\mathcal{A})/\mathcal{V}(\mathcal{D}_\mathcal{A})$, endowed with the quotient topology of $\tau_\mathcal{D}$, is called the (generalized) Calkin algebra of $\mathcal{D}_\mathcal{A}$. In Section 6.2 for this topological $*$-algebra a class of faithful $*$-representations with continuous inverse is constructed and the problem of the existence of continuous faithful $*$-representations is investigated. In Section 6.3 it is shown that $*$-automorphisms and derivations of the $*$-algebra $\mathcal{L}^+(\mathcal{D})$ are always inner. In Section 6.4 two classes of $*$-algebras, called atomic $*$-algebras and maximal atomic $*$-algebras, are analyzed, and their structure is described up to $*$-isomorphisms. The maximal atomic $*$-algebras are unbounded generalizations in some sense of atomic W^*-algebras.

6.1. Completely Continuous Linear Mappings

The Vector Space $\mathcal{V}(\mathcal{D}_\mathcal{A}, \mathcal{D}_\mathcal{B}^+)$

Suppose $\mathcal{A}$ and $\mathcal{B}$ are O-families in the Hilbert space $\mathcal{H}$.

Definition 6.1.1. Let $\mathcal{V}(\mathcal{D}_\mathcal{A}, \mathcal{D}_\mathcal{B}^+)$ be the closure of $\mathcal{F}(\mathcal{D}_\mathcal{A}, \mathcal{D}_\mathcal{B}^+)$ in $\mathcal{L}(\mathcal{D}_\mathcal{A}, \mathcal{D}_\mathcal{B}^+)\,[\tau_b]$. If $\mathcal{A} = \mathcal{L}^+(\mathcal{D}_1)$, $\mathcal{D}_1 = \mathcal{D}(\mathcal{A})$ and $\mathcal{B} = \mathcal{L}^+(\mathcal{D}_2)$, $\mathcal{D}_2 = \mathcal{D}(\mathcal{B})$, then we write $\mathcal{V}(\mathcal{D}_1, \mathcal{D}_2^+)$ for $\mathcal{V}(\mathcal{D}_\mathcal{A}, \mathcal{D}_\mathcal{B}^+)$.

Remark 1. An element x of $\mathcal{L}(\mathcal{D}_\mathcal{A}, \mathcal{D}_\mathcal{B}^+)$ is in $\mathcal{V}(\mathcal{D}_\mathcal{A}, \mathcal{D}_\mathcal{B}^+)$ if and only if x^+ is in $\mathcal{V}(\mathcal{D}_\mathcal{B}, \mathcal{D}_\mathcal{A}^+)$. This is an immediate consequence of the fact that $x \to x^+$ is a homeomorphism of $\mathcal{L}(\mathcal{D}_\mathcal{A}, \mathcal{D}_\mathcal{B}^+)\,[\tau_b]$ onto $\mathcal{L}(\mathcal{D}_\mathcal{B}, \mathcal{D}_\mathcal{A}^+)\,[\tau_b]$ which maps $\mathcal{F}(\mathcal{D}_\mathcal{A}, \mathcal{D}_\mathcal{B}^+)$ onto $\mathcal{F}(\mathcal{D}_\mathcal{B}^+, \mathcal{D}_\mathcal{A})$.

Remark 2. If at least one of the spaces $\mathcal{D}_\mathcal{A}$ and $\mathcal{D}_\mathcal{B}$ is a semi-Montel space, then $\mathcal{F}(\mathcal{D}_\mathcal{A}, \mathcal{D}_\mathcal{B}^+)$ is dense in $\mathcal{L}(\mathcal{D}_\mathcal{A}, \mathcal{D}_\mathcal{B}^+)\,[\tau_b]$ by Theorem 3.4.5 and hence $\mathcal{V}(\mathcal{D}_\mathcal{A}, \mathcal{D}_\mathcal{B}^+) = \mathcal{L}(\mathcal{D}_\mathcal{A}, \mathcal{D}_\mathcal{B}^+)$.

The following simple fact is used in the proofs of Propositions 6.1.3 and 6.1.10 and Theorem 6.2.4.

Lemma 6.1.2. *Let $\mathscr{A}$ be an O-family in the Hilbert space $\mathscr{H}$. Suppose $(\psi_n : n \in \mathbb{N})$ is a sequence in $\mathscr{H}$ which converges weakly in the Hilbert space $\mathscr{H}$ to $\psi \in \mathscr{H}$. If $c \in \mathbb{B}(\mathscr{D}(\mathscr{A}))$, then $(c\psi_n : n \in \mathbb{N})$ converges weakly in $\mathscr{D}_\mathscr{A}$ to $c\psi$.*

Proof. Since $c \in \mathfrak{L}(\mathscr{H}, \mathscr{D}_\mathscr{A})$ by Corollary 3.1.3, the assumption implies that $c\psi_n \to c\psi$ weakly in $\mathscr{D}_\mathscr{A}$. □

Proposition 6.1.3. *Suppose $\mathscr{A}$ and $\mathscr{B}$ are O*-algebras in the Hilbert space $\mathscr{H}$ such that $\mathscr{D}_\mathscr{A}$ and $\mathscr{D}_\mathscr{B}$ are QF-spaces. Then for each x in $\mathscr{L}(\mathscr{D}_\mathscr{A}, \mathscr{D}_\mathscr{B}^+)$ the following three statements are equivalent:*

(i) $x \in \mathcal{V}(\mathscr{D}_\mathscr{A}, \mathscr{D}_\mathscr{B}^+)$.
(ii) *x maps every bounded subset of $\mathscr{D}_\mathscr{A}$ into a relatively compact subset of $\mathscr{D}_\mathscr{B}^+[\beta]$.*
(iii) *x maps every weak null sequence in $\mathscr{D}_\mathscr{A}$ into a null sequence in $\mathscr{D}_\mathscr{B}^+[\beta]$.*

Proof. (i) $\to$ (ii): Fix a bounded set $\mathscr{M}$ in $\mathscr{D}_\mathscr{A}$. There is no loss of generality to assume that $\mathscr{M}$ is closed in $\mathscr{D}(\mathscr{A})\,[\sigma]$. Let $\mathscr{N}$ be a bounded subset of $\mathscr{D}_\mathscr{B}$. By the definition of $\mathcal{V}(\mathscr{D}_\mathscr{A}, \mathscr{D}_\mathscr{B}^+)$ there is a $y \in \mathscr{F}(\mathscr{D}_\mathscr{A}, \mathscr{D}_\mathscr{B}^+)$ such that $p_{\mathscr{M},\mathscr{N}}(x - y) \equiv \sup_{\varphi \in \mathscr{M}} r_\mathscr{N}(x\varphi - y\varphi) \leqq 1$. Since $y \in \mathscr{F}(\mathscr{D}_\mathscr{A}, \mathscr{D}_\mathscr{B}^+)$ and $y \in \mathscr{L}(\mathscr{D}_\mathscr{A}, \mathscr{D}_\mathscr{B}^+)\,[\beta]$, $y(\mathscr{M})$ is a bounded subset of a finite dimensional linear subspace of $\mathscr{D}_\mathscr{B}^+[\beta]$. Therefore, by Lemma 1.1.1, $x(\mathscr{M})$ is precompact in $\mathscr{D}_\mathscr{B}^+[\beta]$. To prove that $x(\mathscr{M})$ is relatively compact in $\mathscr{D}_\mathscr{B}^+[\beta]$, it suffices to show that $x(\mathscr{M})$ is β-complete. Since $\mathscr{M}$ is closed in $\mathscr{D}(\mathscr{A})\,[\sigma]$ and the QF-space $\mathscr{D}_\mathscr{A}$ is semireflexive by Proposition 2.3.12, $\mathscr{M}$ is σ-compact (Schäfer [1], IV, 5.5). Because $\mathscr{L}(\mathscr{D}_\mathscr{A}, \mathscr{D}_\mathscr{B}^+) \subseteq \mathfrak{L}(\mathscr{D}(\mathscr{A})\,[\sigma], \mathscr{D}_\mathscr{B}^+[\sigma'])$, $x(\mathscr{M})$ is σ'-compact and hence σ'-complete. Since the topology β on $\mathscr{D}_\mathscr{B}^+$ has a 0-neighbourhood basis of σ'-closed sets, $x(\mathscr{M})$ is β-complete (Jarchow [1], 3.2.4).

(ii) $\to$ (iii): Let $(\varphi_n : n \in \mathbb{N})$ be a null sequence in $\mathscr{D}(\mathscr{A})\,[\sigma]$. Since $x \in \mathfrak{L}(\mathscr{D}(\mathscr{A})\,[\sigma], \mathscr{D}_\mathscr{B}^+[\sigma'])$, $(x\varphi_n : n \in \mathbb{N})$ is a null sequence in $\mathscr{D}_\mathscr{B}^+[\sigma']$. On the other hand, since $\{\varphi_n : n \in \mathbb{N}\}$ is a bounded set in $\mathscr{D}_\mathscr{A}$, $\{x\varphi_n : n \in \mathbb{N}\}$ is relatively compact in $\mathscr{D}_\mathscr{B}^+[\beta]$ by (ii). These two facts imply that $(x\varphi_n : n \in \mathbb{N})$ converges to 0 in $\mathscr{D}_\mathscr{B}^+[\beta]$.

(iii) $\to$ (i): Let $\mathcal{U}$ be a given open neighbourhood of x in $\mathscr{L}(\mathscr{D}_\mathscr{A}, \mathscr{D}_\mathscr{B}^+)\,[\tau_b]$. By Theorem 3.4.1, (i), there exist projections $e \in \mathbb{P}(\mathscr{D}(\mathscr{A}))$ and $f \in \mathbb{P}(\mathscr{D}(\mathscr{B}))$ such that $\hat{f}xe \in \mathcal{U}$. We check that the operator $c := \hat{f}xe$ of $\mathbb{B}(\mathscr{H})$ is compact. Let $(\zeta_n : n \in \mathbb{N})$ be a weak null sequence in the Hilbert space $\mathscr{H}$. Combining Lemma 6.1.2 with (iii), it follows that $(xe\zeta_n : n \in \mathbb{N})$ is a null sequence in $\mathscr{D}_\mathscr{B}^+[\beta]$. Since $f\mathcal{U}_\mathscr{H}$ is bounded in $\mathscr{D}_\mathscr{B}$ by Corollary 3.1.3, this gives

$$\lim_n \|c\zeta_n\| = \lim_n \sup_{\eta \in \mathcal{U}_\mathscr{H}} |\langle \hat{f}xe\zeta_n, \eta\rangle| = \lim_n r_{f\mathcal{U}_\mathscr{H}}(xe\zeta_n) = 0.$$

Hence the operator c is compact, so c is the limit in the operator norm of a sequence $(c_n : n \in \mathbb{N})$ of operators $c_n \in \mathbb{F}(\mathscr{H})$. Then, of course, $c = \lim_n c_n$ in $\mathscr{L}(\mathscr{D}_\mathscr{A}, \mathscr{D}_\mathscr{B}^+)\,[\tau_b]$. Since $c \in \mathcal{U}$ and $\mathcal{U}$ is open, $c_n \in \mathcal{U}$ for sufficiently large n. Since $c_n \in \mathscr{F}(\mathscr{D}_\mathscr{A}, \mathscr{D}_\mathscr{B}^+)$, this shows that x belongs to the closure of $\mathscr{F}(\mathscr{D}_\mathscr{A}, \mathscr{D}_\mathscr{B}^+)$ in $\mathscr{L}(\mathscr{D}_\mathscr{A}, \mathscr{D}_\mathscr{B}^+)\,[\tau_b]$, i.e., $x \in \mathcal{V}(\mathscr{D}_\mathscr{A}, \mathscr{D}_\mathscr{B}^+)$. □

Remark 3. As the preceding proof shows, the implications (i) $\to$ (ii) $\to$ (iii) are already valid if $\mathscr{A}$ and $\mathscr{B}$ are arbitrary O-families in $\mathscr{H}$ and if the space $\mathscr{D}_\mathscr{A}$ is semireflexive.

Remark 4. Let E and F be locally convex spaces. Let us say that a continuous linear mapping of E into F is *completely continuous* if it maps a weak null sequence in E into a null sequence of F. In this terminology, condition (iii) of Proposition 6.1.3 means that $x \in \mathcal{L}(\mathcal{D}_\mathcal{A}, \mathcal{D}_\mathcal{B}^+)$ $(\subseteq \mathfrak{L}(\mathcal{D}_\mathcal{A}, \mathcal{D}_\mathcal{B}^+[\beta]))$ is a completely continuous mapping of $\mathcal{D}_\mathcal{A}$ into $\mathcal{D}_\mathcal{B}^+[\beta]$.

Recall from Section 3.2 that the algebraic tensor product $\mathcal{D}_\mathcal{A}^| \otimes \mathcal{D}_\mathcal{B}^+$ was identified with the vector space $\mathcal{F}(\mathcal{D}_\mathcal{A}, \mathcal{D}_\mathcal{B}^+)$ and the identifying map χ was defined by $\chi(z) = \sum_{n=1}^{k} \langle \cdot, \varphi_n^| \rangle \psi_n^|$ for $z = \sum_{n=1}^{k} \varphi_n^| \otimes \psi_n^| \in \mathcal{D}_\mathcal{A}^| \otimes \mathcal{D}_\mathcal{B}^+$.

Proposition 6.1.4. *Let $\mathcal{A}$ and $\mathcal{B}$ be O-families in the Hilbert space $\mathcal{H}$. If the locally convex spaces $\mathcal{D}_\mathcal{A}$ and $\mathcal{D}_\mathcal{B}$ are semireflexive, then the identifying map χ is a topological isomorphism of the injective tensor product $\mathcal{D}_\mathcal{A}^|[\beta] \otimes_\varepsilon \mathcal{D}_\mathcal{B}^+[\beta]$ and $\mathcal{F}(\mathcal{D}_\mathcal{A}, \mathcal{D}_\mathcal{B}^+)[\tau_b]$. If $\mathcal{D}_\mathcal{A}$ and $\mathcal{D}_\mathcal{B}$ are Frechet spaces, then the map χ has a continuous extension to a topological isomorphism of the completed injective tensor product $\mathcal{D}_\mathcal{A}^|[\beta] \widehat{\otimes}_\varepsilon \mathcal{D}_\mathcal{B}^+[\beta]$ onto $\mathcal{V}(\mathcal{D}_\mathcal{A}, \mathcal{D}_\mathcal{B}^+)[\tau_b]$.*

Proof. We prove the first assertion. Since $\mathcal{D}_\mathcal{A}$ and $\mathcal{D}_\mathcal{B}$ are semireflexive, $(\mathcal{D}_\mathcal{A}^|[\beta])' = \{\langle \cdot, \varphi \rangle : \varphi \in \mathcal{D}(\mathcal{A})\}$ and $(\mathcal{D}_\mathcal{B}^+[\beta])' = \{\langle \psi, \cdot \rangle : \psi \in \mathcal{D}(\mathcal{B})\}$. Further, the equicontinuous subsets of $(\mathcal{D}_\mathcal{A}^|[\beta])'$ and $(\mathcal{D}_\mathcal{B}^+[\beta])'$ correspond to the bounded subsets of $\mathcal{D}_\mathcal{A}$ and $\mathcal{D}_\mathcal{B}$, respectively. Therefore, the injective tensor topology on $\mathcal{D}_\mathcal{A}^|[\beta] \widehat{\otimes}_\varepsilon \mathcal{D}_\mathcal{B}^+[\beta]$ is defined by the family of seminorms

$$\varepsilon_{\mathcal{M},\mathcal{N}}(z) = \sup_{\varphi \in \mathcal{M}} \sup_{\psi \in \mathcal{N}} \left| \sum_{n=1}^{k} \langle \varphi, \varphi_n^| \rangle \langle \psi_n^|, \psi \rangle \right| = p_{\mathcal{M},\mathcal{N}}(\chi(z)) \tag{1}$$

for $z = \sum_{n=1}^{k} \varphi_n^| \otimes \psi_n^| \in \mathcal{D}_\mathcal{A}^| \otimes \mathcal{D}_\mathcal{B}^+$; so (1) gives the first assertion.

Now suppose that $\mathcal{D}_\mathcal{A}$ and $\mathcal{D}_\mathcal{B}$ are Frechet spaces. Then $\mathcal{D}_\mathcal{A}$ and $\mathcal{D}_\mathcal{B}$ are semireflexive by Corollary 2.3.2, so that the preceding applies. The homeomorphism χ extends by continuity to the completions of $\mathcal{D}_\mathcal{A}^|[\beta] \otimes_\varepsilon \mathcal{D}_\mathcal{B}^+[\beta]$ and $\mathcal{F}(\mathcal{D}_\mathcal{A}, \mathcal{D}_\mathcal{B}^+)[\tau_b]$. By Lemma 3.3.3, $\mathcal{L}(\mathcal{D}_\mathcal{A}, \mathcal{D}_\mathcal{B}^+)[\tau_b]$ is complete. Hence $\mathcal{V}(\mathcal{D}_\mathcal{A}, \mathcal{D}_\mathcal{B}^+)[\tau_b]$ is complete and so a completion of $\mathcal{F}(\mathcal{D}_\mathcal{A}, \mathcal{D}_\mathcal{B}^+)[\tau_b]$. □

Remark 5. Suppose $\mathcal{D}_1$ and $\mathcal{D}_2$ are Frechet domains in the same Hilbert space both endowed with the graph topologies t_+. Then $\mathcal{D}_1^|[\beta]$ and $\mathcal{D}_2^+[\beta]$ are complete locally convex spaces which have the approximation property (by Corollary 3.3.18). From this it follows that $\mathcal{D}_1^|[\beta] \widehat{\otimes}_\varepsilon \mathcal{D}_2^+[\beta]$ coincides with L. Schwartz' ε-product $\mathcal{D}_1^|[\beta] \,\varepsilon\, \mathcal{D}_2^+[\beta]$ (Köthe [2], § 43, 3., (7); see also Jarchow [1], 18.1.8). Thus, by Proposition 6.1.4, $\mathcal{D}_1^|[\beta] \,\varepsilon\, \mathcal{D}_2^+[\beta]$ is topologically isomorphic to $\mathcal{V}(\mathcal{D}_1, \mathcal{D}_2^+)[\tau_b]$; so the equivalence of conditions (i) and (ii) in Proposition 6.1.3 is a well-known property of ε-products (see Köthe [2], § 43, 3., (2)).

The next proposition generalizes the classical result that the Banach space of trace class operators on a Hilbert space is the norm dual of the Banach space of compact operators on the space.

Proposition 6.1.5. *Let $\mathcal{D}_1$ and $\mathcal{D}_2$ be Frechet domains (cf. p. 147) in a Hilbert space $\mathcal{H}$. For $t \in \mathbb{B}_1(\mathcal{D}_2, \mathcal{D}_1)$, let f_t be the linear functional on $\mathcal{V}(\mathcal{D}_1, \mathcal{D}_2^+)$ which is defined by $f_t(x) := \operatorname{tr} \hat{t}x$, $x \in \mathcal{V}(\mathcal{D}_1, \mathcal{D}_2^+)$. Then the mapping $\mathrm{J}: t \to f_t$ is a topological isomorphism of the Frechet space $\mathbb{B}_1(\mathcal{D}_2, \mathcal{D}_1)[\tau_\pi]$ onto the strong dual of $\mathcal{V}(\mathcal{D}_1, \mathcal{D}_2^+)[\tau_b]$.*

Proof. First recall that $\mathbb{B}_1(\mathcal{D}_2, \mathcal{D}_1)[\tau_\pi]$ is a Frechet space by Proposition 5.1.20. By Proposition 5.3.5, each functional f_t is continuous on $\mathcal{V}(\mathcal{D}_1, \mathcal{D}_2^+)[\tau_c]$. Therefore, f_t

$\in \mathcal{V}(\mathcal{D}_1, \mathcal{D}_2^+)[\tau_b]'$, since $\tau_c \subseteqq \tau_b$. It is obvious that J is injective. To show that J is surjective, let $f \in \mathcal{V}(\mathcal{D}_1, \mathcal{D}_2^+)[\tau_b]'$. By Corollary 5.3.2, there is a $t \in \mathbb{B}_1(\mathcal{D}_2, \mathcal{D}_1)$ such that $f(x) = \operatorname{tr} \hat{t}x$ for all $x \in \mathcal{F}(\mathcal{D}_1, \mathcal{D}_2^+)$. That is, $f = f_t$ on $\mathcal{F}(\mathcal{D}_1, \mathcal{D}_2^+)$. Since both f and f_t are continuous on $\mathcal{V}(\mathcal{D}_1, \mathcal{D}_2^+)[\tau_b]$ and $\mathcal{F}(\mathcal{D}_1, \mathcal{D}_2^+)$ is dense in $\mathcal{V}(\mathcal{D}_1, \mathcal{D}_2^+)[\tau_b]$, the latter implies that $f = f_t$ on $\mathcal{V}(\mathcal{D}_1, \mathcal{D}_2^+)$. Thus we have shown that J is a bijective mapping of $\mathbb{B}_1(\mathcal{D}_2, \mathcal{D}_1)$ onto $\mathcal{V}(\mathcal{D}_1, \mathcal{D}_2)[\tau_b]'$.

Next we prove that J and J^{-1} are cotinuous. In view of Corollary 6.1.6 below, we begin with more generality than is needed for this. Let $\mathcal{L}$ be a subset of $\mathcal{V}(\mathcal{D}_1, \mathcal{D}_2^+)$ which contains $\mathbb{F}(\mathcal{D}_1, \mathcal{D}_2)$. Suppose $a \in \mathcal{L}^+(\mathcal{D}_1)$ and $b \in \mathcal{L}^+(\mathcal{D}_2)$. From Proposition 3.2.3, each $x \in \mathcal{U}_{a,b}$ is of the form $x = b^+ \circ ya$ with $y \in \mathcal{U}_{\mathbb{B}(\mathcal{H})}$. Recall that $\mathcal{U}_{\mathbb{B}(\mathcal{H})}$ denotes the unit ball of $\mathbb{B}(\mathcal{H})$ in the operator norm. If $y \in \mathbb{F}(\mathcal{D}_1, \mathcal{D}_2) \cap \mathcal{U}_{\mathbb{B}(\mathcal{H})}$, then obviously $b^+ \circ ya \in \mathbb{F}(\mathcal{D}_1, \mathcal{D}_2) \cap \mathcal{U}_{a,b}$. From these two facts it follows that there exists a subset $\mathcal{R}$ of $\mathcal{U}_{\mathbb{B}(\mathcal{H})}$ such that $\mathcal{L} \cap \mathcal{U}_{a,b} = b^+ \circ \mathcal{R}a$ and $\mathbb{F}(\mathcal{D}_1, \mathcal{D}_2) \cap \mathcal{U}_{\mathbb{B}(\mathcal{H})} \subseteqq \mathcal{R}$. If $t \in \mathbb{B}_1(\mathcal{D}_2, \mathcal{D}_1)$, we then have

$$\sup_{x \in \mathcal{U}_{a,b} \cap \mathcal{L}} |f_t(x)| = \sup_{y \in \mathcal{R}} |\operatorname{tr} \hat{t}(b^+ \circ ya)| = \sup_{y \in \mathcal{R}} |\operatorname{Tr} \overline{yatb^+}| = \sup_{y \in \mathcal{U}_{\mathbb{B}(\mathcal{H})}} |\operatorname{Tr} \overline{yatb^+}|$$
$$= \nu(\overline{atb^+}) = \nu_{a,b^+}(t). \tag{2}$$

Here the second equality follows from Corollary 5.2.6, and the third equality is true since $\mathbb{F}(\mathcal{H}) \cap \mathcal{U}_{\mathbb{B}(\mathcal{H})}$ is ultraweakly dense in $\mathcal{U}_{\mathbb{B}(\mathcal{H})}$ by the Kaplansky density theorem. Now we specialize to the case $\mathcal{L} = \mathcal{V}(\mathcal{D}_1, \mathcal{D}_2^+)$. By Proposition 4.2.1, $\{\mathcal{U}_{a,b} \cap \mathcal{V}(\mathcal{D}_1, \mathcal{D}_2^+)$: $a \in \mathcal{L}^+(\mathcal{D}_1)$ and $b \in \mathcal{L}^+(\mathcal{D}_2)\}$ is a fundamental system of bounded sets in $\mathcal{V}(\mathcal{D}_1, \mathcal{D}_2^+)[\tau_b]$. We therefore conclude from (2) that J is a topological isomorphism of $\mathbb{B}_1(\mathcal{D}_2, \mathcal{D}_1)[\tau_\pi]$ onto the strong dual of $\mathcal{V}(\mathcal{D}_1, \mathcal{D}_2^+)[\tau_b]$. □

Corollary 6.1.6. *Keep the assumptions of Proposition* 6.1.5. *If $a \in \mathcal{L}^+(\mathcal{D}_1)$ and $b \in \mathcal{L}^+(\mathcal{D}_2)$, then $\mathcal{U}_{a,b} \cap \mathbb{F}(\mathcal{D}_1, \mathcal{D}_2)$ is dense in $\mathcal{U}_{a,b} \cap \mathcal{V}(\mathcal{D}_1, \mathcal{D}_2^+)$ in the bounded topology τ_b.*

Proof. From (2), applied in case $\mathcal{L} = \mathbb{F}(\mathcal{D}_1, \mathcal{D}_2)$ and in case $\mathcal{L} = \mathcal{V}(\mathcal{D}_1, \mathcal{D}_2^+)$, we see that the absolutely convex sets $\mathcal{U}_{a,b} \cap \mathbb{F}(\mathcal{D}_1, \mathcal{D}_2)$ and $\mathcal{U}_{a,b} \cap \mathcal{V}(\mathcal{D}_1, \mathcal{D}_2^+)$ have the same polar (namely, $\{f_t : t \in \mathbb{B}_1(\mathcal{D}_2, \mathcal{D}_1)$ and $\nu_{a,b^+}(t) \leqq 1\}$) in $\mathcal{V}(\mathcal{D}_1, \mathcal{D}_2^+)[\tau_b]'$ and hence the same bipolar in $\mathcal{V}(\mathcal{D}_1, \mathcal{D}_2^+)$. By the bipolar theorem (see e.g. SCHÄFER [1], IV, 1.5), the bipolar is equal to the closure of each of these two sets in $\mathcal{V}(\mathcal{D}_1, \mathcal{D}_2^+)[\tau_b]$. □

Corollary 6.1.7. *Let $\mathcal{D}_1$ and $\mathcal{D}_2$ be as in Proposition* 6.1.5. *Then $\mathcal{V}(\mathcal{D}_1, \mathcal{D}_2^+)[\tau_b]$ is a complete barrelled DF-space. If $\mathcal{L}$ is a linear subspace of $\mathcal{V}(\mathcal{D}_1, \mathcal{D}_2^+)$ which contains $\mathbb{F}(\mathcal{D}_1, \mathcal{D}_2)$, then $\mathcal{L}[\tau_b]$ is a DF-space.*

Proof. The space $\mathcal{V}(\mathcal{D}_1, \mathcal{D}_2^+)[\tau_b]$ is complete, since $\mathcal{L}(\mathcal{D}_1, \mathcal{D}_2^+)[\tau_b]$ is complete by Lemma 3.3.3. We prove that $\mathcal{V}(\mathcal{D}_1, \mathcal{D}_2^+)[\tau_b]$ is barrelled. Let $\mathcal{W}$ be a barrel in $\mathcal{V}(\mathcal{D}_1, \mathcal{D}_2^+)[\tau_b]$, that is, $\mathcal{W}$ is a closed, absorbing and absolutely convex subset of $\mathcal{V}(\mathcal{D}_1, \mathcal{D}_2^+)[\tau_b]$. We have to show that $\mathcal{W}$ is a 0-neighbourhood in $\mathcal{V}(\mathcal{D}_1, \mathcal{D}_2^+)[\tau_b]$. We denote by $\mathcal{R}^0$ the polar and by $\mathcal{R}^{00}$ the bipolar of a set $\mathcal{R}$ taken in the dual pairing $(\mathcal{V}(\mathcal{D}_1, \mathcal{D}_2^+), \mathbb{B}_1(\mathcal{D}_2, \mathcal{D}_1))$ with respect to the bilinear form $(x, t) \to f_t(x) \equiv \operatorname{tr} \hat{t}x$. Since $\mathcal{V}(\mathcal{D}_1, \mathcal{D}_2^+)[\tau_b]'$ $= \{f_t : t \in \mathbb{B}_1(\mathcal{D}_2, \mathcal{D}_1)\}$ by Proposition 6.1.5 and $\mathcal{W}$ is absolutely convex and closed, it follows from the bipolar theorem that $\mathcal{W} = \mathcal{W}^{00}$.

Suppose $a \in \mathcal{L}^+(\mathcal{D}_1)$ and $b \in \mathcal{L}^+(\mathcal{D}_2)$. The set $\mathcal{U}_{a,b} \cap \mathcal{V}(\mathcal{D}_1, \mathcal{D}_2^+)$ is bounded in $\mathcal{V}(\mathcal{D}_1, \mathcal{D}_2^+)[\tau_b]$ and τ_b-complete, since $\mathcal{V}(\mathcal{D}_1, \mathcal{D}_2^+)[\tau_b]$ is complete. By the Banach-Mackey theorem (SCHÄFER [1], II, 8.5), the barrel $\mathcal{W}$ absorbs $\mathcal{U}_{a,b} \cap \mathcal{V}(\mathcal{D}_1, \mathcal{D}_2^+)$. Thus there is a $\delta > 0$ such that $\delta(\mathcal{U}_{a,b} \cap \mathcal{V}(\mathcal{D}_1, \mathcal{D}_2^+)) \subseteq \mathcal{W}$. Therefore, if $t \in \mathcal{W}^0$, then $\delta t \in (\mathcal{U}_{a,b} \cap \mathcal{V}(\mathcal{D}_1, \mathcal{D}_2^+))^0$ and hence $\nu_{a,b+}(t) = \sup\{|f_t(x)| : x \in \mathcal{U}_{a,b} \cap \mathcal{V}(\mathcal{D}_1, \mathcal{D}_2^+)\} \leqq \delta^{-1}$ by (2). This proves that $\mathcal{W}^0$ is bounded in $\mathbb{B}_1(\mathcal{D}_2, \mathcal{D}_1)[\tau_\pi]$. By Lemma 5.4.4, there are $c \in \mathbb{B}(\mathcal{D}_1)_+$ and $d \in \mathbb{B}(\mathcal{D}_2)_+$ such that $\mathcal{W}^0 \subseteq c\mathcal{U}_{\mathbb{B}_1(\mathcal{H})}d$. Hence $\mathcal{W} = \mathcal{W}^{00} \supseteq (c\mathcal{U}_{\mathbb{B}_1(\mathcal{H})}d)^0$. Lemma 5.4.6 (note that $g_x(t) \equiv f_t(x)$) shows that $(c\mathcal{U}_{\mathbb{B}_1(\mathcal{H})}d)^0 = \{x \in \mathcal{V}(\mathcal{D}_1, \mathcal{D}_2^+) : \|\hat{d}xc\| \leqq 1\}$. By Corollary 3.3.6, the latter set is a 0-neighbourhood in $\mathcal{V}(\mathcal{D}_1, \mathcal{D}_2^+)[\tau_b]$ and hence is $\mathcal{W}$. Thus we have proved that $\mathcal{V}(\mathcal{D}_1, \mathcal{D}_2^+)[\tau_b]$ is barrelled.

Being a barrelled space which has a fundamental sequence of bounded sets (by Corollary 4.2.2), $\mathcal{V}(\mathcal{D}_1, \mathcal{D}_2^+)[\tau_b]$ is a DF-space (Köthe [1], § 29, 3).

That $\mathcal{L}[\tau_b]$ is a DF-space can be proved by the same arguments as used in the proof of the second assertion in Corollary 5.4.8. We replace only $\mathcal{L}(\mathcal{D}_1, \mathcal{D}_2^+)$ by $\mathcal{V}(\mathcal{D}_1, \mathcal{D}_2^+)$ and apply Corollary 6.1.6 in place of Theorem 3.4.1. □

The Ideal $\mathcal{V}(\mathcal{D}_\mathcal{A})$

In this subsection $\mathcal{A}$ is an O*-algebra in a Hilbert space $\mathcal{H} \neq \{0\}$.

Definition 6.1.8. Let $\mathcal{V}(\mathcal{D}_\mathcal{A})$ be the closure of $\mathbb{F}(\mathcal{D}(\mathcal{A}))$ in $\mathcal{L}^+(\mathcal{D}_\mathcal{A})[\tau_\mathcal{D}]$.

Remark 6. Suppose that $\mathcal{D}_\mathcal{A}$ is a QF-space. Then Proposition 5.5.4, (i) $\leftrightarrow$ (iv)′, states that $\mathcal{V}(\mathcal{D}_\mathcal{A}) = \mathcal{L}^+(\mathcal{D}_\mathcal{A})$ if and only if $\mathcal{D}_\mathcal{A}$ is a semi-Montel space.

Remark 7. We have $\mathcal{V}(\mathcal{D}_\mathcal{A}) = \mathcal{V}(\mathcal{D}_\mathcal{A}, \mathcal{D}_\mathcal{A}^+) \cap \mathcal{L}^+(\mathcal{D}_\mathcal{A})$. Indeed, Lemma 3.4.4, (iii), implies that the closures of $\mathbb{F}(\mathcal{D}(\mathcal{A}))$ and $\mathcal{F}(\mathcal{D}_\mathcal{A}, \mathcal{D}_\mathcal{A}^+)$ in, $\mathcal{L}(\mathcal{D}_\mathcal{A}, \mathcal{D}_\mathcal{A}^+)[\tau_\mathcal{D}]$ coincide. Intersecting these closures with $\mathcal{L}^+(\mathcal{D}_\mathcal{A})$, we obtain $\mathcal{V}(\mathcal{D}_\mathcal{A}) = \mathcal{V}(\mathcal{D}_\mathcal{A}, \mathcal{D}_\mathcal{A}^+) \cap \mathcal{L}^+(\mathcal{D}_\mathcal{A})$.

Lemma 6.1.9. (i) $\mathbb{F}(\mathcal{D}(\mathcal{A}))$ *and* $\mathcal{V}(\mathcal{D}_\mathcal{A})$ *are two-sided *-ideals in the *-algebra* $\mathcal{L}^+(\mathcal{D}_\mathcal{A})$.

(ii) $\mathbb{F}(\mathcal{D}(\mathcal{A}))$ *is the smallest non-zero two-sided ideal in* $\mathcal{L}^+(\mathcal{D}_\mathcal{A})$, *and* $\mathcal{V}(\mathcal{D}_\mathcal{A})$ *is the smallest non-zero closed two-sided ideal in* $\mathcal{L}^+(\mathcal{D}_\mathcal{A})[\tau_\mathcal{D}]$.

Proof. (i): It is obvious that $\mathbb{F}(\mathcal{D}(\mathcal{A}))$ is a two-sided *-ideal in $\mathcal{L}^+(\mathcal{D}_\mathcal{A})$. Since $\mathcal{L}^+(\mathcal{D}_\mathcal{A})[\tau_\mathcal{D}]$ is a topological *-algebra, its closure is again a two-sided *-ideal. (ii): Let $\mathcal{J}$ be a non-zero two-sided ideal in $\mathcal{L}^+(\mathcal{D}_\mathcal{A})$. Let $x \in \mathcal{J}$, $x \neq 0$. Then there are vectors $\zeta, \eta \in \mathcal{D}(\mathcal{A})$ such that $\langle x\zeta, \eta\rangle \neq 0$. For $\varphi, \psi \in \mathcal{D}(\mathcal{A})$, we have $\psi \otimes \varphi = \langle x\zeta, \eta\rangle^{-1} (\eta \otimes \varphi)\, x(\psi \otimes \zeta)$, so it follows that $\psi \otimes \varphi \in \mathcal{J}$. This yields $\mathcal{J} \supseteq \mathbb{F}(\mathcal{D}(\mathcal{A}))$. The assertion concerning $\mathcal{V}(\mathcal{D}_\mathcal{A})$ follows immediately from the latter. □

Proposition 6.1.10. *Suppose* $\mathcal{A}$ *is an O*-algebra such that* $\mathcal{D}_\mathcal{A}$ *is a QF-space. For each* $x \in \mathcal{L}^+(\mathcal{D}_\mathcal{A})$ *the following three statements are equivalent:*

(i) $x \in \mathcal{V}(\mathcal{D}_\mathcal{A})$.

(ii) x *maps each weak null sequence in* $\mathcal{D}_\mathcal{A}$ *into a null sequence in* $\mathcal{D}_\mathcal{A}$.

(iii) x *maps every bounded set in* $\mathcal{D}_\mathcal{A}$ *into a relatively compact set in* $\mathcal{D}_\mathcal{A}$.

Proof. (i) $\to$ (ii): Suppose $(\varphi_n : n \in \mathbb{N})$ is a weak null sequence in $\mathcal{D}_\mathcal{A}$. Then $\lim_n \|(\eta \otimes \zeta)\varphi_n\| = \lim_n \|\langle\varphi_n, \eta\rangle \zeta\| = 0$ for arbitrary vectors $\eta, \zeta \in \mathcal{D}(\mathcal{A})$. Since these operators $\eta \otimes \zeta$ span $\mathbb{F}(\mathcal{D}(\mathcal{A}))$, $\lim_n \|y\varphi_n\| = 0$ for all $y \in \mathbb{F}(\mathcal{D}(\mathcal{A}))$.

Suppose $a \in \mathcal{A}$. The set $\{\varphi_n : n \in \mathbb{N}\}$ is bounded in $\mathcal{D}_\mathcal{A}$. Since $x \in \mathcal{L}^+(\mathcal{D}_\mathcal{A})$, the set $\mathcal{M} := \{a^+ax\varphi_n, \varphi_n : n \in \mathbb{N}\}$ is also bounded in $\mathcal{D}_\mathcal{A}$. Let $\varepsilon > 0$. By (i), x belongs to the closure of $\mathbb{F}(\mathcal{D}(\mathcal{A}))$ in $\mathcal{L}^+(\mathcal{D}_\mathcal{A})[\tau_\mathcal{D}]$. Thus there is a $y \in \mathbb{F}(\mathcal{D}(\mathcal{A}))$ such that $p_\mathcal{M}(x - y) \leqq \varepsilon$. For $n \in \mathbb{N}$, we have $\|ax\varphi_n\|^2 = \langle (x - y)\, \varphi_n, a^+ax\varphi_n\rangle + \langle y\varphi_n, a^+ax\varphi_n\rangle \leqq \varepsilon + \lambda \|y\varphi_n\|$, where $\lambda := \sup_{\psi \in \mathcal{M}} \|\psi\|$. Since $\lim_n \|y\varphi_n\| = 0$ as noted above, this proves that $\lim_n \|ax\varphi_n\| = 0$. Hence $\lim_n x\varphi_n = 0$ in $\mathcal{D}_\mathcal{A}$.

(ii) $\to$ (iii): Let $\mathcal{M}$ be a bounded subset of $\mathcal{D}_\mathcal{A}$. Since $\mathcal{D}_\mathcal{A}$ is a QF-space by assumption, Theorem 2.4.1 applies and shows that there is a $c \in \mathbb{B}(\mathcal{D}(\mathcal{A}))_+$ such that $\mathcal{M} \subseteq c\mathcal{U}_\mathcal{H}$. Further, the set $c\mathcal{U}_\mathcal{H}$ is bounded in $\mathcal{D}_\mathcal{A}$ by Corollary 3.1.3 and hence is contained in a Frechet subspace of $\mathcal{D}_\mathcal{A}$. Therefore, it suffices to prove that each sequence $(xc\zeta_n : n \in \mathbb{N})$, where $\zeta_n \in \mathcal{U}_\mathcal{H}$ for $n \in \mathbb{N}$, possesses a convergent subsequence in $\mathcal{D}_\mathcal{A}$. Fix such a sequence. We choose a subsequence $(\zeta_{n_k} : n_k \in \mathbb{N})$ of $(\zeta_n : n \in \mathbb{N})$ which converges weakly in the Hilbert space $\mathcal{H}$ to some vector $\zeta \in \mathcal{H}$. By Lemma 6.1.2, $(c(\zeta_{n_k} - \zeta) : k \in \mathbb{N})$ is a weak null sequence in $\mathcal{D}_\mathcal{A}$. Thus, by (ii), $(xc(\zeta_{n_k} - \zeta) : n \in \mathbb{N})$ is a null sequence in $\mathcal{D}_\mathcal{A}$, i.e., $xc\zeta = \lim_k xc\zeta_{n_k}$ in $\mathcal{D}_\mathcal{A}$.

(iii) $\to$ (i): Let $\mathcal{M}$ be a bounded set in $\mathcal{D}_\mathcal{A}$. By (iii), the closure of $x(\mathcal{M})$ in $\mathcal{D}_\mathcal{A}$ is a compact subset of $\mathcal{D}_\mathcal{A}$ and hence also of $\mathcal{D}_\mathcal{A}^+[\beta]$, since on $\mathcal{D}(\mathcal{A})$ the strong topology of $\mathcal{D}_\mathcal{A}^+$ is weaker than the graph topology $t_\mathcal{A}$. That is, $x(\mathcal{M})$ is relatively compact in $\mathcal{D}_\mathcal{A}^+[\beta]$. Therefore, by Proposition 6.1.3, (i) $\leftrightarrow$ (ii), $x \in \mathcal{V}(\mathcal{D}_\mathcal{A}, \mathcal{D}_\mathcal{A}^+)$. Combined with Remark 7, this gives $x \in \mathcal{V}(\mathcal{D}_\mathcal{A})$. □

Remark 8. In the terminology of Remark 4, condition (ii) in Proposition 6.1.10 says that x is a completely continuous linear mapping of the locally convex space $\mathcal{D}_\mathcal{A}$ into itself.

The following lemma contains the main part of the proof of the next proposition.

Lemma 6.1.11. *Suppose $\mathcal{A}$ is an O^*-algebra in the Hilbert space $\mathcal{H}$ and $\mathcal{D}_\mathcal{A}$ is a QF-space. Then each two-sided ideal $\mathcal{J}$ of $\mathcal{L}^+(\mathcal{D}_\mathcal{A})$ is contained in the closure of the left ideal generated by $\mathcal{J} \cap \mathbb{P}(\mathcal{D}(\mathcal{A}))$ in $\mathcal{L}^+(\mathcal{D}_\mathcal{A})[\tau_\mathcal{D}]$.*

Proof. Suppose $x \in \mathcal{J}$. Let p be a continuous seminorm on $\mathcal{L}(\mathcal{D}_\mathcal{A}, \mathcal{D}_\mathcal{A}^+)[\tau_\mathcal{D}]$. By Theorem 3.4.1, there is a projection $e \in \mathbb{P}(\mathcal{D}(\mathcal{A}))$ such that $p(x - exe) \leqq 1$. As noted therein (and is easy to verify), $c := exe \in \mathbb{B}(\mathcal{D}(\mathcal{A}))$. Let $c = u\,|c|$ be the polar decomposition of c. By Corollary 3.1.5, $|c| \in \mathbb{B}(\mathcal{D}(\mathcal{A}))$. Recall that u is a partial isometry with initial space $\overline{|c|\,\mathcal{H}}$ and range $\overline{c\mathcal{H}}$. Since $c^*c\mathcal{H} \subseteq e\mathcal{H}$ and $|c| = (c^*c)^{1/2}$ is a norm limit of polynomials without constant terms in c^*c, $\overline{|c|\,\mathcal{H}} \subseteq e\mathcal{H}$. Combined with $\overline{c\mathcal{H}} \subseteq e\mathcal{H}$, this yields $u \in \mathbb{B}(\mathcal{D}(\mathcal{A}))$. Let $|c| = \int_0^\infty \lambda \, de(\lambda)$ be the spectral decomposition of $|c|$. For $\varepsilon > 0$, define $c_\varepsilon := \int_\varepsilon^\infty \lambda^{-1} \, de(\lambda)$ and $e_\varepsilon := e([\varepsilon, +\infty))$. Again by Corollary 3.1.5, c_ε and e_ε are in $\mathbb{B}(\mathcal{D}(\mathcal{A}))$. The operators e, u^* and c_ε are in $\mathbb{B}(\mathcal{D}(\mathcal{A}))$ and hence in $\mathcal{L}^+(\mathcal{D}_\mathcal{A})$. Therefore, since x belongs to the ideal $\mathcal{J}$, $e_\varepsilon = c_\varepsilon\,|c| = c_\varepsilon u^*c = c_\varepsilon u^*exe$ belongs to $\mathcal{J} \cap \mathbb{P}(\mathcal{D}(\mathcal{A}))$ for any $\varepsilon > 0$. Since $c = u\,|c| = \lim_{\varepsilon \to +0} u\,|c|\,e_\varepsilon$ in the operator norm on $\mathcal{H}$, there is an $\varepsilon > 0$ such that $p(c - u\,|c|\,e_\varepsilon) \leqq 1$. By $p(x - c) = p(x - exe) \leqq 1$, $p(x - u\,|c|\,e_\varepsilon) \leqq 2$. From $u, |c| \in \mathbb{B}(\mathcal{D}(\mathcal{A})) \subseteq \mathcal{L}^+(\mathcal{D}_\mathcal{A})$ and $e_\varepsilon \in \mathcal{J} \cap \mathbb{P}(\mathcal{D}(\mathcal{A}))$ it follows that $u\,|c|e_\varepsilon$ belongs to the left ideal in $\mathcal{L}^+(\mathcal{D}_\mathcal{A})$ generated by $\mathcal{J} \cap \mathbb{P}(\mathcal{D}(\mathcal{A}))$. This gives the assertion. □

Proposition 6.1.12. *Suppose that $\mathcal{A}$ is an O^*-algebra in the Hilbert space $\mathcal{H}$ such that $\mathcal{D}_\mathcal{A}$ is a QF-space. Suppose that the Hilbert space $e\mathcal{H}$ is separable for every projection*

$e \in \mathbb{P}(\mathcal{D}(\mathcal{A}))$. *Then* $\{0\}$, $\mathcal{V}(\mathcal{D}_\mathcal{A})$ *and* $\mathcal{L}^+(\mathcal{D}_\mathcal{A})$ *are the only closed two-sided ideals in* $\mathcal{L}^+(\mathcal{D}_\mathcal{A})[\tau_\mathcal{D}]$.

Proof. Suppose $\mathcal{J}$ is a closed two-sided ideal in $\mathcal{L}^+(\mathcal{D}_\mathcal{A})[\tau_\mathcal{D}]$ which is different from $\{0\}$ and $\mathcal{V}(\mathcal{D}_\mathcal{A})$. We want to prove that $\mathcal{J} = \mathcal{L}^+(\mathcal{D}_\mathcal{A})$. By Lemma 6.1.9, $\mathcal{V}(\mathcal{D}_\mathcal{A}) \subseteq \mathcal{J}$. Because of Corollary 3.4.3, (iii), it suffices to show that all projections of $\mathbb{P}(\mathcal{D}(\mathcal{A}))$ are in $\mathcal{J}$. Fix $e \in \mathbb{P}(\mathcal{D}(\mathcal{A}))$. If e is a finite rank projection, then $e \in \mathbb{F}(\mathcal{D}(\mathcal{A})) \subseteq \mathcal{J}$; so we can assume that $e\mathcal{H}$ is infinite dimensional. From Lemma 6.1.11 it follows that there exists an $f \in \mathcal{J} \cap \mathbb{P}(\mathcal{D}(\mathcal{A}))$ with infinite dimensional range, since otherwise $\mathcal{J} \subseteq \mathcal{V}(\mathcal{D}_\mathcal{A})$. By assumption, $e\mathcal{H}$ and $f\mathcal{H}$ are separable. Thus there exists a partial isometry u on $\mathcal{H}$ with initial space $e\mathcal{H}$ and range $f\mathcal{H}$. Since e and f are in $\mathbb{P}(\mathcal{D}(\mathcal{A}))$, u is in $\mathbb{B}(\mathcal{D}(\mathcal{A}))$ and hence in $\mathcal{L}^+(\mathcal{D}_\mathcal{A})$. From $f \in \mathcal{J}$ and $e = u^*fu$, $e \in \mathcal{J}$. □

Remark 9. The separability assumption in Proposition 6.1.12 is of course fulfilled if the Hilbert space $\mathcal{H}$ is separable, but there are also O*-algebras in non-separable Hilbert spaces which satisfy the assumptions of Proposition 6.1.12.

6.2. Faithful $*$-Representations of the Generalized Calkin Algebra

Throughout this section we assume that $\mathcal{A}$ is an O*-algebra in a Hilbert space $\mathcal{H} \neq \{0\}$ and that $\mathcal{D}_\mathcal{A}$ is a QF-space.

Let $\mathcal{Q}(\mathcal{D}_\mathcal{A}) := \mathcal{L}^+(\mathcal{D}_\mathcal{A})/\mathcal{V}(\mathcal{D}_\mathcal{A})$ be the quotient $*$-algebra, and let $\iota\colon \mathcal{L}^+(\mathcal{D}_\mathcal{A}) \to \mathcal{Q}(\mathcal{D}_\mathcal{A})$ be the quotient mapping. Let $\check{\tau}$ denote the quotient topology of $\mathcal{L}^+(\mathcal{D}_\mathcal{A})[\tau_\mathcal{D}]$ on $\mathcal{Q}(\mathcal{D}_\mathcal{A})$. Since $\mathcal{V}(\mathcal{D}_\mathcal{A})$ is closed in $\mathcal{L}^+(\mathcal{D}_\mathcal{A})[\tau_\mathcal{D}]$, $\mathcal{Q}(\mathcal{D}_\mathcal{A})[\check{\tau}]$ is a locally convex Hausdorff space. The topology $\check{\tau}$ on $\mathcal{Q}(\mathcal{D}_\mathcal{A})$ is determined by the directed family of seminorms

$$\check{p}_\mathcal{M}(\iota(x)) := \inf_{y \in \mathcal{V}(\mathcal{D}_\mathcal{A})} p_\mathcal{M}(x+y), \quad x \in \mathcal{L}^+(\mathcal{D}_\mathcal{A}),$$

where $\mathcal{M}$ runs through the bounded subsets of $\mathcal{D}_\mathcal{A}$. Since $\mathcal{L}^+(\mathcal{D}_\mathcal{A})[\tau_\mathcal{D}]$ is a topological $*$-algebra, $\mathcal{Q}(\mathcal{D}_\mathcal{A})[\check{\tau}]$ is a topological $*$-algebra as well.

Definition 6.2.1. The topological $*$-algebra $\mathcal{Q}(\mathcal{D}_\mathcal{A})[\check{\tau}]$ is called the *Calkin algebra* of $\mathcal{D}_\mathcal{A}$.

In the case where $\mathcal{A} = \mathcal{L}^+(\mathcal{D})$ and $\mathcal{D}[t_+]$ is a QF-space we omit the subscript $\mathcal{A}$ and we call $\mathcal{Q}(\mathcal{D})[\check{\tau}]$ the Calkin algebra of the domain $\mathcal{D}$. Note that if $\mathcal{D}_\mathcal{A}$ is a semi-Montel space, then $\mathcal{V}(\mathcal{D}_\mathcal{A}) = \mathcal{L}^+(\mathcal{D}_\mathcal{A})$ and hence $\mathcal{Q}(\mathcal{D}_\mathcal{A}) = \{0\}$.

Our next objective is to define the generalized Calkin representations π_θ of the $*$-algebra $\mathcal{Q}(\mathcal{D}_\mathcal{A})$.

Suppose that θ is a singular state of the W^*-algebra $l^\infty \equiv l^\infty(\mathbb{N})$. This means that θ is a positive linear functional on the $*$-algebra l^∞ satisfying $\theta(1) = 1$ which annihilates the vector space c_0 of all null sequences. A typical example is the following one: If $\mathbb{U}$ is a free ultrafilter on $\mathbb{N}$, then $\theta_\mathbb{U}((x_n)) := \lim_\mathbb{U} x_n$, $(x_n) \in l^\infty$, defines a singular state $\theta_\mathbb{U}$ on l^∞. (Recall that an ultrafilter on $\mathbb{N}$ is said to be *free* if the intersection of all its members is empty.)

Let $\mathcal{D}_\infty$ be the set of all weak null sequences of the locally convex space $\mathcal{D}_\mathcal{A}$, and let $\mathcal{H}_\infty$ denote the set of all weak null sequences of the Hilbert space $\mathcal{H}$. With pointwise addition and scalar multiplication of sequences, $\mathcal{D}_\infty$ and $\mathcal{H}_\infty$ are vector spaces. Let $\mathcal{N}_\theta$ be the vector space of all $(\varphi_n) \in \mathcal{H}_\infty$ for which $\theta((\|\varphi_n\|)) = 0$. Define the quotient spaces

$\mathcal{D}_\theta := \mathcal{D}_\infty/(\mathcal{D}_\infty \cap \mathcal{N}_\theta)$ and $\mathcal{H}_\theta := \mathcal{H}_\infty/\mathcal{N}_\theta$. Since $\mathcal{D}_\infty \subseteqq \mathcal{H}_\infty$, $\mathcal{D}_\theta$ is a linear subspace of $\mathcal{H}_\theta$ in a canonical way. The image of a sequence (φ_n) in $\mathcal{D}_\infty$ or in $\mathcal{H}_\infty$ under the quotient map will be denoted by $(\varphi_n)_\theta$. For (φ_n) and (ψ_n) in $\mathcal{H}_\infty$, we define $\langle(\varphi_n)_\theta, (\psi_n)_\theta\rangle := \theta\big((\langle\varphi_n, \psi_n\rangle)\big)$. It is straightforward to check that $\langle\cdot,\cdot\rangle$ is a scalar product on $\mathcal{H}_\theta$. (We verify (for instance) the positive definiteness. Suppose that $\langle(\varphi_n)_\theta, (\varphi_n)_\theta\rangle = \theta\big((\|\varphi_n\|^2)\big) = 0$ for some $(\varphi_n) \in \mathcal{H}_\infty$. By the Cauchy-Schwarz inequality, $\theta\big((\|\varphi_n\|)\big)^2 \leqq \theta\big((\|\varphi_n\|^2)\big) = 0$, so that $(\varphi_n) \in \mathcal{N}_\theta$ and $(\varphi_n)_\theta = 0$.) Endowed with the scalar product $\langle\cdot,\cdot\rangle$, $\mathcal{H}_\theta$ and $\mathcal{D}_\theta$ are unitary spaces.

Remark 1. In general, $\mathcal{D}_\theta$ is not dense in $\mathcal{H}_\theta$. For instance, if $\mathcal{D}_\mathcal{A}$ is a semi-Montel space, then $\mathcal{D}_\theta = \{0\}$, but $\mathcal{H}_\theta \neq \{0\}$ if $\mathcal{H}$ is infinite dimensional.

Lemma 6.2.2. *Suppose* $x \in \mathcal{L}^+(\mathcal{D}_\mathcal{A})$.

(i) *If* $(\varphi_n) \in \mathcal{D}_\infty$, *then* $(x\varphi_n) \in \mathcal{D}_\infty$.

(ii) *If* $(\varphi_n) \in \mathcal{D}_\infty \cap \mathcal{N}_\theta$, *then* $(x\varphi_n) \in \mathcal{D}_\infty \cap \mathcal{N}_\theta$.

(iii) $x \in \mathcal{V}(\mathcal{D}_\mathcal{A})$ *if and only if* $(x\varphi_n) \in \mathcal{N}_\theta$ *for all* $(\varphi_n) \in \mathcal{D}_\infty$.

Proof. (i): Since $x \in \mathcal{L}^+(\mathcal{D}_\mathcal{A})$, $x \in \mathfrak{L}(\mathcal{D}_\mathcal{A})$. Hence x maps the weak null sequences of $\mathcal{D}_\mathcal{A}$ into itself.

(ii): Let $(\varphi_n) \in \mathcal{D}_\infty \cap \mathcal{N}_\theta$. Since a weakly convergent sequence of a locally convex space is always bounded, $\lambda := \sup_{n\in\mathbb{N}} \|x^+x\varphi_n\| < \infty$. Thus $\|x\varphi_n\|^2 = \langle x^+x\varphi_n, \varphi_n\rangle \leqq \lambda\,\|\varphi_n\|$ for $n \in \mathbb{N}$. Using this and the Cauchy-Schwarz inequality for θ, we get $\theta\big((\|x\varphi_n\|)\big)^2 \leqq \theta\big((\|x\,\varphi_n\|^2)\big) \leqq \lambda\theta\big((\|\varphi_n\|)\big) = 0$, i.e., $(x\varphi_n) \in \mathcal{N}_\theta$. By (i), $(x\varphi_n) \in \mathcal{D}_\infty \cap \mathcal{N}_\theta$.

(iii): First suppose $x \in \mathcal{V}(\mathcal{D}_\mathcal{A})$. Let $(\varphi_n) \in \mathcal{D}_\infty$. By Proposition 6.1.10, (i) $\leftrightarrow$ (ii), $\lim_n \|x\varphi_n\| = 0$. Thus $\theta\big((\|x\varphi_n\|)\big) = 0$, since $\theta(c_0) = \{0\}$. That is, $(x\varphi_n) \in \mathcal{N}_\theta$.

Now suppose that $x \notin \mathcal{V}(\mathcal{D}_\mathcal{A})$. Applying once more Proposition 6.1.10, there exists a weak null sequence (ψ_n) in $\mathcal{D}_\mathcal{A}$ such that the sequence $(x\psi_n)$ does not converge to zero in $\mathcal{D}_\mathcal{A}$. Then there is an $a \in \mathcal{A}$ such that $(ax\psi_n)$ is not convergent to zero in the Hilbert space $\mathcal{H}$. Hence we can find a subsequence (φ_n) of (ψ_n) satisfying $\delta := \inf_{n\in\mathbb{N}} \|ax\varphi_n\| > 0$. Obviously, $(\varphi_n) \in \mathcal{D}_\infty$. We have $\lambda := \sup_{n\in N} \|a^+ax\varphi_n\| < \infty$, since $\{\varphi_n : n \in \mathbb{N}\}$ is bounded in $\mathcal{D}_\mathcal{A}$. From $\delta^2 \leqq \|ax\varphi_n\|^2 = \langle a^+ax\varphi_n, x\varphi_n\rangle \leqq \lambda\,\|x\varphi_n\|$ for all $n \in \mathbb{N}$ we conclude that $\delta^2 \leqq \lambda\theta\big((\|x\varphi_n\|)\big)$. This implies that $(x\varphi_n) \notin \mathcal{N}_\theta$. $\square$

Let $x \in \mathcal{L}^+(\mathcal{D}_\mathcal{A})$. We define $\varrho_\theta(x)\,(\varphi_n)_\theta = (x\varphi_n)_\theta$ for $(\varphi_n)_\theta \in \mathcal{D}_\theta$. From Lemma 6.2.2, (i) and (ii), we see that $\varrho_\theta(x)$ is a well-defined linear operator on $\mathcal{D}_\theta$ that maps $\mathcal{D}_\theta$ into itself.

Lemma 6.2.3. *The mapping $x \to \varrho_\theta(x)$ is a weakly continuous $*$-representation of the topological $*$-algebra $\mathcal{L}^+(\mathcal{D}_\mathcal{A})\,[\tau_\mathcal{D}]$ on the unitary space $\mathcal{D}(\varrho_\theta) := \mathcal{D}_\theta$. The kernel of ϱ_θ is the ideal $\mathcal{V}(\mathcal{D}_\mathcal{A})$. The map $\varrho_\theta\colon \mathcal{L}^+(\mathcal{D}_\mathcal{A}) \to \mathcal{L}^+(\mathcal{D}_\theta)$ is strongly positive, i.e., if $x \in \mathcal{L}^+(\mathcal{D}_\mathcal{A})$ and $x \geqq 0$, then $\varrho_\theta(x) \geqq 0$.*

Proof. For $x \in \mathcal{L}^+(\mathcal{D}_\mathcal{A})$ and $(\varphi_n), (\psi_n) \in \mathcal{D}_\infty$, we have

$$\langle\varrho_\theta(x)\,(\varphi_n)_\theta, (\psi_n)_\theta\rangle = \theta\big((\langle x\varphi_n, \psi_n\rangle)\big) = \theta\big((\langle\varphi_n, x^+\psi_n\rangle)\big) = \langle(\varphi_n)_\theta, \varrho_\theta(x^+)\,(\psi_n)_\theta\rangle.$$

From this it follows that ϱ_θ is a $*$-preserving map in $\mathcal{L}^+(\mathcal{D}_\mathcal{A})$ into $\mathcal{L}^+(\mathcal{D}_\theta)$. Since ϱ_θ is obviously a homomorphism of the algebra $\mathcal{L}^+(\mathcal{D}_\mathcal{A})$ in $\mathcal{L}^+(\mathcal{D}_\theta)$, it is a $*$-representation of $\mathcal{L}^+(\mathcal{D}_\mathcal{A})$. The weak continuity of ϱ_θ follows from

$$|\langle\varrho_\theta(x)\,(\varphi_n)_\theta, (\psi_n)_\theta\rangle| = \big|\theta\big((\langle x\varphi_n, \psi_n\rangle)\big)\big| \leqq \sup_{n\in\mathbb{N}} |\langle x\varphi_n, \psi_n\rangle| \leqq p_\mathcal{M}(x)$$

for $x \in \mathcal{L}^+(\mathcal{D}_{\mathcal{A}})$ and $(\varphi_n), (\psi_n) \in \mathcal{D}_\infty$, where $\mathcal{M}$ denotes the bounded set $\{\varphi_n, \psi_n : n \in \mathbb{N}\}$ in $\mathcal{D}_{\mathcal{A}}$. The second assertion is only a reformulation of Lemma 6.2.2, (iii), and the final assertion is obvious. □

Since $\ker \varrho_\theta = \mathcal{V}(\mathcal{D}_{\mathcal{A}})$, there exists a unique faithful $*$-representation π_θ of the $*$-algebra $\mathcal{Q}(\mathcal{D}_{\mathcal{A}})$ on $\mathcal{D}(\pi_\theta) := \mathcal{D}_\theta$ such that $\varrho_\theta = \pi_\theta \circ \iota$, that is, $\pi_\theta(\iota(x)) = \varrho_\theta(x)$ for all $x \in \mathcal{L}^+(\mathcal{D}_{\mathcal{A}})$. The $*$-representation π_θ of $\mathcal{Q}(\mathcal{D}_{\mathcal{A}})[\check{\tau}]$ is weakly continuous, because ϱ_θ is weakly continuous on $\mathcal{L}^+(\mathcal{D}_{\mathcal{A}})[\tau_{\mathcal{D}}]$.

Theorem 6.2.4. *Suppose $\mathcal{A}$ is an O^*-algebra and $\mathcal{D}_{\mathcal{A}}$ is a QF-space. Let θ be a singular state on the W^*-algebra $l^\infty(\mathbb{N})$. Then the inverse of the faithful $*$-representation π_θ is a continuous mapping of $\pi_\theta(\mathcal{Q}(\mathcal{D}_{\mathcal{A}}))[\tau_{\mathcal{D}}]$ onto $\mathcal{Q}(\mathcal{D}_{\mathcal{A}})[\check{\tau}]$. If the topologies $\tau_{\mathcal{D}}$ and $\tau_{\mathcal{N}}$ on $\mathcal{L}^+(\mathcal{D}_{\mathcal{A}})$ coincide, then π_θ is a topological $*$-isomorphism of the Calkin algebra $\mathcal{Q}(\mathcal{D}_{\mathcal{A}})[\check{\tau}]$ onto $\pi_\theta(\mathcal{Q}(\mathcal{D}_{\mathcal{A}}))[\tau_{\mathcal{D}}]$.*

Proof. Suppose $\mathcal{M}$ is a bounded subset of $\mathcal{D}_{\mathcal{A}}$. To prove the continuity of the inverse of π_θ, we have to show that there is a set $\mathcal{N} \subseteq \mathcal{D}_\theta$ which is bounded in the graph topology of $\pi_\theta(\mathcal{Q}(\mathcal{D}_{\mathcal{A}}))$ such that $\check{p}_{\mathcal{M}}(z) \leqq p_{\mathcal{N}}(\pi_\theta(z))$ for all $z \in \mathcal{Q}(\mathcal{D}_{\mathcal{A}})$ or, equivalently,

$$\check{p}_{\mathcal{M}}(\iota(x)) \leqq p_{\mathcal{N}}(\varrho_\theta(x)) \quad \text{for all} \quad x \in \mathcal{L}^+(\mathcal{D}_{\mathcal{A}}). \tag{1}$$

Since we assumed that $\mathcal{D}_{\mathcal{A}}$ is a QF-space, there exists $c \in \mathbb{B}(\mathcal{D}(\mathcal{A}))_+$ by Theorem 2.4.1 such that $\mathcal{M} \subseteq c\mathcal{U}_{\mathcal{H}}$. Using the density of $\mathbb{F}(\mathcal{D}(\mathcal{A}))$ in $\mathcal{V}(\mathcal{D}_{\mathcal{A}})[\tau_{\mathcal{D}}]$, we obtain

$$\begin{aligned}\check{p}_{\mathcal{M}}(\iota(x)) &= \inf_{y \in \mathbb{F}(\mathcal{D}(\mathcal{A}))} p_{\mathcal{M}}(x+y) \leqq \inf_{y \in \mathbb{F}(\mathcal{D}(\mathcal{A}))} \sup_{\varphi,\psi \in \mathcal{U}_{\mathcal{H}}} |\langle (x+y)\, c\varphi, c\psi\rangle| \\ &= \inf_{y \in \mathbb{F}(\mathcal{D}(\mathcal{A}))} \|cxc + cyc\| \quad \text{for} \quad x \in \mathcal{L}^+(\mathcal{D}_{\mathcal{A}}).\end{aligned}$$

If $\mathcal{S}$ is the closure of $c\mathcal{D}(\mathcal{A})$ in $\mathcal{H}$, then $\{cyc : y \in \mathbb{F}(\mathcal{D}(\mathcal{A}))\} = \mathbb{F}(c\mathcal{D}(\mathcal{A}))$ is norm dense in $\mathbb{F}(\mathcal{S})$. Therefore,

$$\check{p}_{\mathcal{M}}(\iota(x)) \leqq \inf_{y \in \mathbb{F}(\mathcal{S})} \|cxc + y\| = \inf_{y \in \mathbb{F}(\mathcal{H})} \|cxc + y\| \quad \text{for} \quad x \in \mathcal{L}^+(\mathcal{D}_{\mathcal{A}}). \tag{2}$$

Next we apply some results of the preceding discussion with $\mathcal{L}^+(\mathcal{D}_{\mathcal{A}})$ replaced by $\mathbb{B}(\mathcal{H})$. The equation $\omega_\theta(b)\,(\varphi_n)_\theta := (b\varphi_n)_\theta$, $b \in \mathbb{B}(\mathcal{H})$ and $(\varphi_n)_\theta \in \mathcal{H}_\theta$, defines a $*$-representation ω_θ of $\mathbb{B}(\mathcal{H})$ on the unitary space $\mathcal{H}_\theta$. Since obviously $\|\omega_\theta(b)\| \leqq \|b\|$ for $b \in \mathbb{B}(\mathcal{H})$, ω_θ extends to a $*$-representation $\hat{\omega}_\theta$ of $\mathbb{B}(\mathcal{H})$ on the completion $\hat{\mathcal{H}}_\theta$ of $\mathcal{H}_\theta$. Since $\ker \omega_\theta \equiv \ker \hat{\omega}_\theta$ is the ideal $\mathbb{K}(\mathcal{H})$ $(= \mathcal{V}(\mathcal{H})$ in the above terminology) of compact operators on $\mathcal{H}$, there is a faithful $*$-representation ν_θ of the quotient C^*-algebra $\mathbb{B}(\mathcal{H})/\mathbb{K}(\mathcal{H})$ on $\hat{\mathcal{H}}_\theta$ such that $\hat{\omega}_\theta = \nu_\theta \circ \iota$. Since ν_θ is faithful, it is isometric. In particular, this yields

$$\inf_{y \in \mathbb{F}(\mathcal{H})} \|cxc + y\| = \inf_{y \in \mathbb{K}(\mathcal{H})} \|cxc + y\| = \|\nu_\theta(\iota(cxc))\| = \|\hat{\omega}_\theta(cxc)\| \quad \text{for} \quad x \in \mathcal{L}^+(\mathcal{D}_{\mathcal{A}}). \tag{3}$$

Recall that $cxc \in \mathbb{B}(\mathcal{H})$ by Lemma 3.1.2, since $c \in \mathbb{B}(\mathcal{D}(\mathcal{A}))_+$.

Define $\mathcal{N} := \omega_\theta(c)\, \mathcal{U}_{\mathcal{H}_\theta}$. Suppose $(\varphi_n) \in \mathcal{H}_\infty$ and $(\varphi_n)_\theta \in \mathcal{U}_{\mathcal{H}_\theta}$. By Lemma 6.1.2, $(c\varphi_n) \in \mathcal{D}_\infty$, so that $\omega_\theta(c)\,(\varphi_n)_\theta = (c\varphi_n)_\theta \in \mathcal{D}_\theta$. This proves that $\mathcal{N} \subseteq \mathcal{D}_\theta$. Suppose $x \in \mathcal{L}^+(\mathcal{D}_{\mathcal{A}})$. Recall that $xc \in \mathbb{B}(\mathcal{H})$ by Lemma 3.1.2. By the definition of ϱ_θ and ω_θ, we have

$$\varrho_\theta(x)\, \omega_\theta(c)\, (\varphi_n)_\theta = (xc\varphi_n)_\theta = \omega_\theta(xc)\, (\varphi_n)_\theta. \tag{4}$$

From (4) and

$$\|(xc\varphi_n)_\theta\|^2 = \theta\big((\|xc\varphi_n\|^2)\big) \leqq \|xc\|^2\, \theta\big((\|\varphi_n\|^2)\big) = \|xc\|^2\, \|(\varphi_n)_\theta\|^2 \leqq \|xc\|^2$$

we conclude that $\mathcal{N}$ is bounded in the graph topology of $\varrho_\theta\big(\mathcal{L}^+(\mathcal{D}_\mathcal{A})\big) \equiv \pi_\theta\big(\mathcal{Q}(\mathcal{D}_\mathcal{A})\big)$.

From (2), (3) and (4), we obtain

$$p_\mathcal{M}\big(\iota(x)\big) \leqq \|\omega_\theta(cxc)\| = \sup_{\varphi,\psi\in\mathcal{U}_{\mathcal{H}_\theta}} |\langle\omega_\theta(xc)\,\varphi, \omega_\theta(c)\,\psi\rangle|$$

$$= \sup_{\varphi,\psi\in\mathcal{U}_{\mathcal{H}_\theta}} |\langle\varrho_\theta(x)\,\omega_\theta(c)\,\varphi, \omega_\theta(c)\,\psi\rangle| = p_\mathcal{N}\big(\varrho_\theta(x)\big) \quad \text{for} \quad x \in \mathcal{L}^+(\mathcal{D}_\mathcal{A})$$

which proves (1).

Now assume that $\tau_\mathcal{D} = \tau_\mathcal{N}$ on $\mathcal{L}^+(\mathcal{D}_\mathcal{A})$. As stated in Lemma 6.2.3, ϱ_θ is strongly positive. Therefore, by Lemma 6.2.5 below applied with $\mathcal{B} := \mathcal{L}^+(\mathcal{D}_\mathcal{A})$, ϱ_θ is a continuous mapping of $\mathcal{L}^+(\mathcal{D}_\mathcal{A})\,[\tau_\mathcal{D}]$ onto $\varrho_\theta\big(\mathcal{L}^+(\mathcal{D}_\mathcal{A})\big)\,[\tau_\mathcal{D}]$. Because $\varrho_\theta = \pi_\theta \circ \iota$ and $\check{\tau}$ is the quotient topology of $\mathcal{L}^+(\mathcal{D}_\mathcal{A})\,[\tau_\mathcal{D}]$ on $\mathcal{Q}(\mathcal{D}_\mathcal{A})$, this means that π_θ maps $\mathcal{Q}(\mathcal{D}_\mathcal{A})\,[\check{\tau}]$ continuously onto $\pi_\theta\big(\mathcal{Q}(\mathcal{D}_\mathcal{A})\big)\,[\tau_\mathcal{D}]$. Together with the preceding, this proves that π_θ is a topological isomorphism. $\square$

Lemma 6.2.5. *Suppose $\mathcal{B}$ is an O*-algebra for which $\tau_\mathcal{D} = \tau_\mathcal{N}$ on $\mathcal{B}$. If ϱ is a strongly positive *-representation of $\mathcal{B}$, then ϱ is a continuous mapping of $\mathcal{B}[\tau_\mathcal{D}]$ onto $\varrho(\mathcal{B})\,[\tau_\mathcal{D}]$.*

Proof. Since ϱ is strongly positive, it is clear that ϱ is a continuous mapping of $\mathcal{B}[\tau_\mathcal{N}]$ onto $\varrho(\mathcal{B})\,[\tau_\mathcal{N}]$ and hence on $\varrho(\mathcal{B})\,[\tau_\mathcal{D}]$, because $\tau_\mathcal{D} \subseteqq \tau_\mathcal{N}$ on $\varrho(\mathcal{B})$. Combined with the equality $\tau_\mathcal{D} = \tau_\mathcal{N}$ on $\mathcal{B}$, the assertion follows. $\square$

Lemma 6.2.6. *Suppose $\mathcal{A}$ is an O*-algebra such that $\mathcal{D}_\mathcal{A}$ is a QF-space. If ϱ is a weakly continuous *-representation of $\mathcal{L}^+(\mathcal{D}_\mathcal{A})\,[\tau_\mathcal{D}]$, then ϱ is strongly positive.*

Proof. Suppose $x \in \mathcal{L}^+(\mathcal{D}_\mathcal{A})_+$. Let $\varphi \in \mathcal{D}(\varrho)$. By Corollary 3.4.3, (i), there is a net $(x_i : i \in I)$ of operators in $\mathbb{B}\big(\mathcal{D}(\mathcal{A})\big)_+$ that converges to x in $\mathcal{L}^+(\mathcal{D}_\mathcal{A})\,[\tau_\mathcal{D}]$. Since ϱ is weakly continuous by assumption, $\langle\varrho(x)\,\varphi, \varphi\rangle = \lim_i \langle\varrho(x_i)\,\varphi, \varphi\rangle$. Because $x_i^{1/2} \in \mathbb{B}\big(\mathcal{D}(\mathcal{A})\big)_+ \subseteqq \mathcal{L}^+(\mathcal{D}_\mathcal{A})_+$ by Corollary 3.1.5, we have $\langle\varrho(x_i)\,\varphi, \varphi\rangle = \|\varrho(x_i^{1/2})\,\varphi\|^2 \geqq 0$ for any $i \in I$ and so $\langle\varrho(x)\,\varphi, \varphi\rangle \geqq 0$. Thus $\varrho(x) \geqq 0$. $\square$

Let us say that a *-representation π of a topological *-algebra A with unit is *continuous* if π is a continuous mapping of A onto $\pi(\mathsf{A})\,[\tau_\mathcal{D}]$. The next theorem completely characterizes those commutatively dominated Frechet domains for which the generalized Calkin algebra has a continuous faithful *-representation. Among others, it also contains a converse to the final assertion in Theorem 6.2.4.

Theorem 6.2.7. *Suppose that $\mathcal{D}$ is a commutatively dominated Frechet domain* (*cf.* p. 108). *Then the following six statements are equivalent:*

(i) $\tau_\mathcal{D} = \tau_\mathcal{N}$ *on* $\mathcal{L}^+(\mathcal{D})$.

(ii) *There exists a faithful *-representation π of $\mathcal{Q}(\mathcal{D})$ which is a topological isomorphism of $\mathcal{Q}(\mathcal{D})\,[\check{\tau}]$ onto $\pi\big(\mathcal{Q}(\mathcal{D})\big)\,[\tau_\mathcal{D}]$.*

(iii) *There exists a continuous faithful *-representation of $\mathcal{Q}(\mathcal{D})\,[\check{\tau}]$.*

(iv) *There exists a continuous *-representation π of $\mathcal{L}^+(\mathcal{D})\,[\tau_\mathcal{D}]$ such that* $\ker \pi = \mathcal{V}(\mathcal{D})$.

(v) *Each weakly continuous *-representation of $\mathcal{L}^+(\mathcal{D})\,[\tau_\mathcal{D}]$ is continuous.*

(vi) *Each weakly continuous strongly positive *-representation of $\mathcal{L}^+(\mathcal{D})\,[\tau_\mathcal{D}]$ is continuous.*

Proof. We prove the implications (i) → (ii) → (iii) → (iv) → (i) and (i) → (v) → (vi) → (iv). Take a singular state on $l^\infty(\mathbb{N})$. Theorem 6.2.4 shows that (i) → (ii) by setting $\pi := \pi_\theta$. By Lemma 6.2.3, ϱ_θ is a weakly continuous strongly positive *-representation of $\mathcal{L}^+(\mathcal{D})[\tau_\mathcal{D}]$ such that $\ker \varrho_\theta = \mathcal{V}(\mathcal{D})$; so if (vi) is satisfied, then (iv) follows by setting $\pi := \varrho_\theta$. (ii) → (iii) and (v) → (vi) are trivial, and (iii) → (iv) follows at once from the definition of $\mathcal{Q}(\mathcal{D})[\dot{\tau}]$. (i) → (v) is a consequence of Lemmas 6.2.5 and 6.2.6. Thus the proof of the theorem will be complete once we have shown that (iv) implies (i).

Suppose that there exists a continuous *-representation π of $\mathcal{L}^+(\mathcal{D})[\tau_\mathcal{D}]$ such that $\ker \pi = \mathcal{V}(\mathcal{D})$. To prove (i), suppose, on the contrary, that $\tau_\mathcal{D} \neq \tau_\mathcal{N}$ on $\mathcal{L}^+(\mathcal{D})$. The commutatively dominated Frechet domain $\mathcal{D}$ must be of the form described at the beginning of Section 4.3. We use the notation established therein. By Theorem 4.3.4, condition (*) is not satisfied, since $\tau_\mathcal{D} \neq \tau_\mathcal{N}$ on $\mathcal{L}^+(\mathcal{D})$. Then, similarly as in the proof of Proposition 4.3.2, we may assume (without loss of generality) that there exist a sequence $\gamma = (\gamma_n : n \in \mathbb{N})$ satisfying $\gamma_{n+1} > \gamma_n \geqq n$ for $n \in \mathbb{N}$ and measurable subsets $\mathfrak{J}_{kn}$, $k, n \in \mathbb{N}$, of $\mathfrak{X}(\gamma, k)$ such that $E(\mathfrak{J}_{kn}) \neq 0$ and

$$h_{k+1}(\cdot) \geqq \gamma_n \text{ on } \mathfrak{J}_{kn} \quad \text{for all} \quad k, n \in \mathbb{N}. \tag{5}$$

Take a unit vector φ_{kn} of $E(\mathfrak{J}_{kn})\mathcal{D}$, $k, n \in \mathbb{N}$. Since $\mathfrak{J}_{kn} \subseteqq \mathfrak{X}(\gamma, k)$, we have

$$h_l(\cdot) \leqq \gamma_l \text{ on } \mathfrak{J}_{kn} \quad \text{for all} \quad l, k, n \in \mathbb{N}, l \leqq k. \tag{6}$$

We denote by Γ the collection of all sequences $n = (n_k : k \in \mathbb{N})$ of natural numbers satisfying $n_k \geqq k + 2$ for $k \in \mathbb{N}$. Fix $n \in \Gamma$. We verify that

$$\left\|\pi(a_l)\, \pi\Big(E\Big(\bigcup_{k \geqq l+1} \mathfrak{J}_{k,n_k}\Big)\Big) \varphi\right\| \leqq \gamma_l \|\varphi\| \tag{7}$$

for $l \in \mathbb{N}$ and $\varphi \in \mathcal{D}(\pi)$. Let χ_l be the characteristic function of the set $\mathfrak{Z}_l := \bigcup_{k \geqq l+1} \mathfrak{J}_{k,n_k}$. By (6), we have $h_l(\cdot)\, \chi_l(\cdot) \leqq \gamma_l$ a.e. on $\mathbb{R}$. Define a function f_l on $\mathbb{R}$ by $f_l := (\gamma_l^2 - h_l^2 \chi_l)^{1/2}$. Clearly, $y_l := f_l(A) \restriction \mathcal{D} \in \mathcal{L}^+(\mathcal{D})$. If $\varphi \in \mathcal{D}(\pi)$, then

$$\langle \pi(y_l^2)\, \varphi, \varphi \rangle = \|\pi(y_l)\, \varphi\|^2 \geqq 0$$

and hence

$$\gamma_l^2 \|\varphi\|^2 = \langle \pi(\gamma_l^2 I)\, \varphi, \varphi \rangle \geqq \langle \pi\big(a_l^2 \chi_l(A)\big)\, \varphi, \varphi \rangle = \big\|\pi(a_l)\, \pi\big(E(\mathfrak{Z}_l)\big)\, \varphi\big\|^2$$

which proves (7).

Let c_n denote the orthogonal projection onto the closure of $\mathcal{D}_n := \text{l.h.}\ \{\varphi_{k,n_k} : k \in \mathbb{N}\}$. We show that $c_n \in \mathbb{P}(\mathcal{D})$. Fix a number $l \in \mathbb{N}$. Each vector $\varphi \in \mathcal{D}_n$ can be written as a finite sum $\varphi = \sum_{k=1}^{s} \lambda_k \varphi_{k,n_k}$ with $\lambda_1, \ldots, \lambda_s \in \mathbb{C}$ and $s \in \mathbb{N}$, $s > l$. Let $k, m \in \mathbb{N}$, where $m > k$. Since $h_{k+1}(\cdot) \leqq \gamma_{k+1}$ on $\mathfrak{J}_{m,n_m}$, by (6), and $h_{k+1}(\cdot) \geqq \gamma_{n_k} \geqq \gamma_{k+2} > \gamma_{k+1}$ on $\mathfrak{J}_{k,n_k}$, by (5), it follows that $E(\mathfrak{J}_{m,n_m})\, E(\mathfrak{J}_{k,n_k}) = 0$, so that

$$\varphi_{m,n_m} \perp \varphi_{k,n_k} \quad \text{and} \quad a_l \varphi_{m,n_m} \perp a_l \varphi_{k,n_k} \quad \text{for} \quad k, m = 1, \ldots, s,\ k \neq m.$$

Therefore,

$$\|a_l \varphi\|^2 = \sum_{k=1}^{l} |\lambda_k|^2 \|a_l \varphi_{k,n_k}\|^2 + \sum_{k=l+1}^{s} |\lambda_k|^2 \|a_l \varphi_{k,n_k}\|^2 \leqq \alpha_l \sum_{k=1}^{s} |\lambda_k|^2 = \alpha_l \|\varphi\|^2,$$

where $\alpha_l := \max\{\|a_l\varphi_{1,n_1}\|, \ldots, \|a_l\varphi_{l,n_l}\|, \gamma_l\}$. Since this holds for all $\varphi \in \mathcal{D}_n$, we conclude that $c_n\mathcal{H} \subseteq \mathcal{D}(\overline{a_l}) \equiv \mathcal{D}(h_l(A))$. By $\mathcal{D} = \bigcap_{l\in\mathbb{N}} \mathcal{D}(h_l(A))$, $c_n\mathcal{H} \subseteq \mathcal{D}$, that is, $c_n \in \mathbb{P}(\mathcal{D})$.

Next we prove that $\mathcal{N} := \bigcup_{n\in\Gamma} \pi(c_n)\mathcal{U}_{\mathcal{D}(\pi)}$ is a bounded subset of $\mathcal{D}(\pi)$ in the graph topology of $\pi(\mathcal{L}^+(\mathcal{D}))$. We suppose $l \in \mathbb{N}$ and $n \in \Gamma$. Let $c_{l,n}$ be the projection on $\mathcal{H}$ with range l.h. $\{\varphi_{k,n_k} : k = 1, \ldots, l\}$. Clearly, $c_{l,n} \in \mathcal{L}^+(\mathcal{D})$ and $c_{l,n} \in \mathcal{V}(\mathcal{D}) = \ker \pi$. Moreover, we have $c_n - c_{l,n} = E\Big(\bigcup_{k\geqq l+1} \mathfrak{J}_{k,n_k}\Big)(c_n - c_{l,n})$. Using these facts and (7), we get

$$\begin{aligned}\|\pi(a_l)\,\pi(c_n)\,\varphi\| &= \|\pi(a_l)\,\pi(c_n - c_{l,n})\,\varphi\| \\ &= \Big\|\pi(a_l)\,\pi\Big(E\Big(\bigcup_{k\geqq l+1}\mathfrak{J}_{k,n_k}\Big)\Big)\,\pi(c_n - c_{l,n})\,\varphi\Big\| \leqq \gamma_l\,\|\pi(c_n - c_{l,n})\,\varphi\| \leqq \gamma_l\end{aligned}$$

for all $\varphi \in \mathcal{U}_{\mathcal{D}(\pi)}$. By Corollary 2.6.8, the family of seminorms $\{\|\cdot\|_{\pi(a_l)} : l \in \mathbb{N}\}$ generates the graph topology of $\pi(\mathcal{L}^+(\mathcal{D}))$. Therefore, the preceding estimates prove the boundedness of the set $\mathcal{N}$.

Since π is a continuous $*$-representation of $\mathcal{L}^+(\mathcal{D})\,[\tau_{\mathcal{D}}]$, there is a bounded subset $\mathcal{M}$ of $\mathcal{D}[t_+]$ such that

$$p_{\mathcal{N}}(\pi(x)) \leqq p_{\mathcal{M}}(x) \quad \text{for all} \quad x \in \mathcal{L}^+(\mathcal{D}). \tag{8}$$

Because $\mathcal{M}$ is bounded, $\alpha_k := \sup_{\varphi\in\mathcal{M}} \|a_k\varphi\| < \infty$ for $k \in \mathbb{N}$. We choose natural numbers n_k such that $n_k \geqq k + 2$ and $\gamma_{n_k} \geqq \alpha_{k+1}2^k$ for $k \in \mathbb{N}$. This is possible because $\gamma_n \geqq n$ for $n \in \mathbb{N}$. Define $\mathfrak{Z} := \bigcup_{k\in\mathbb{N}} \mathfrak{J}_{k,n_k}$ and $z := E(\mathfrak{Z})$. If $\varphi \in \mathcal{M}$ and $k \in \mathbb{N}$, then by (5)

$$\gamma_{n_k}\,\|E(\mathfrak{J}_{k,n_k})\,\varphi\| \leqq \|h_{k+1}(A)\,E(\mathfrak{J}_{k,n_k})\,\varphi\| \leqq \|a_{k+1}\,\varphi\| \leqq \alpha_{k+1}.$$

Therefore,

$$p_{\mathcal{M}}(z) = \sup_{\varphi,\psi\in\mathcal{M}} |\langle E(\mathfrak{Z})\,\varphi, \psi\rangle| \leqq \sup_{\varphi\in\mathcal{M}} \sum_{k=1}^{\infty} \|E(\mathfrak{J}_{k,n_k})\,\varphi\|^2 \leqq \sum_{k=1}^{\infty} \alpha_{k+1}^2\,\gamma_{n_k}^{-2} \leqq \sum_{k=1}^{\infty} 2^{-2k} < 1. \tag{9}$$

On the other hand, as noted above, $(\varphi_{k,n_k} : k \in \mathbb{N})$ is an orthonormal sequence of vectors contained in the range of the projection c_n of $\mathbb{P}(\mathcal{D})$. This clearly implies that $c_n \notin \mathcal{V}(\mathcal{D})$; so $c_n \notin \ker \pi$. Consequently, the closure of $\pi(c_n)$ is a non-zero projection. This gives

$$1 = \sup_{\varphi\in\mathcal{U}_{\mathcal{D}(\pi)}} \|\pi(c_n)\,\varphi\|^2 = \sup_{\varphi\in\mathcal{U}_{\mathcal{D}(\pi)}} |\langle\pi(z)\,\pi(c_n)\,\varphi, \pi(c_n)\,\varphi\rangle| \leqq p_{\mathcal{N}}(\pi(z)). \tag{10}$$

Comparing (9) and (10) with (8), we obtain the desired contradiction. □

Remark 2. The preceding proof should be compared with the final part in the proof of Proposition 4.3.2.

6.3. Derivations and $*$-Automorphisms of $\mathcal{L}^+(\mathcal{D})$

Definition 6.3.1. A *derivation* on an algebra A is a linear mapping δ of A into itself such that

$$\delta(ab) = a\delta(b) + \delta(a)\,b \quad \text{for all} \quad a, b \in \mathsf{A}. \tag{1}$$

A $*$-*derivation* on a $*$-algebra A is a derivation δ on A which satisfies $\delta(a^+) = \delta(a)^+$ for all $a \in \mathsf{A}$.

Suppose A is an algebra and $x \in$ A. Define $\delta_x(a) := xa - ax$ for $a \in$ A. Then δ_x is a derivation on A. Each derivation of this form is called an *inner derivation* on A. If A is a $*$-algebra and if $x^+ = -x$, then δ_x is obviously a $*$-derivation on A.

Proposition 6.3.2. *Suppose that $\mathcal{D}$ is a unitary space and $\mathcal{A}$ is a subalgebra of $L(\mathcal{D})$ which contains $\mathbb{F}(\mathcal{D}) \upharpoonright \mathcal{D}$. Suppose δ is a derivation on $\mathcal{A}$. Then there exists an operator $x \in L(\mathcal{D})$ such that $\delta(a) = xa - ax$ for all $a \in \mathcal{A}$. If $\mathcal{A}$ is contained in $\mathcal{L}^+(\mathcal{D})$, then x can be chosen in $\mathcal{L}^+(\mathcal{D})$. If $\mathcal{A} \subseteq \mathcal{L}^+(\mathcal{D})$ and if $\delta(a^+) = \delta(a)^+$ for all $a \in \mathbb{F}(\mathcal{D})$, then we can choose the operator x such that $x \in \mathcal{L}^+(\mathcal{D})$ and $x^+ = -x$.*

Proof. Clearly, we can assume that $\mathcal{D} \neq \{0\}$. Take a fixed unit vector ξ of $\mathcal{D}$. We define a linear mapping of $\mathcal{D}$ into $\mathcal{D}$ by $x\varphi := \delta(\xi \otimes \varphi)\,\xi$, $\varphi \in \mathcal{D}$. This definition makes sense, since $\mathbb{F}(\mathcal{D}) \upharpoonright \mathcal{D} \subseteq \mathcal{A}$. By (1), we have that

$$\begin{aligned} xa\varphi &= \delta(\xi \otimes a\varphi)\,\xi = \delta\big(a(\xi \otimes \varphi)\big)\,\xi = a\delta(\xi \otimes \varphi)\,\xi + \delta(a)\,(\xi \otimes \varphi)\,\xi \\ &= ax\varphi + \delta(a)\,\varphi \end{aligned}$$

for $a \in \mathcal{A}$ and $\varphi \in \mathcal{D}$. That is,

$$\delta(a) = xa - ax \quad \text{for} \quad a \in \mathcal{A}. \tag{2}$$

Now suppose in addition that $\mathcal{A} \subseteq \mathcal{L}^+(\mathcal{D})$. We prove that $x \in \mathcal{L}^+(\mathcal{D})$. The definition of x and (2) applied with $a := \xi \otimes \xi$ yield

$$\langle x\xi, \xi\rangle = \langle \delta(\xi \otimes \xi)\,\xi, \xi\rangle = \langle x(\xi \otimes \xi)\,\xi, \xi\rangle - \langle (\xi \otimes \xi)\,x\xi, \xi\rangle = 0.$$

Since $\mathcal{A} \subseteq \mathcal{L}^+(\mathcal{D})$, the equation $y\varphi := \big(\delta(\varphi \otimes \xi)\big)^+\,\xi$, $\varphi \in \mathcal{D}$, defines a linear mapping y of $\mathcal{D}$ into $\mathcal{D}$. Suppose $\varphi, \psi \in \mathcal{D}$. From (2) and since $\langle x\xi, \xi\rangle = 0$, we have

$$\begin{aligned} \langle y\varphi, \psi\rangle &= \langle \big(\delta(\varphi \otimes \xi)\big)^+\,\xi, \psi\rangle = \langle \xi, \delta(\varphi \otimes \xi)\,\psi\rangle \\ &= \langle \xi, x(\varphi \otimes \xi)\,\psi\rangle - \langle \xi, (\varphi \otimes \xi)\,x\psi\rangle \\ &= \langle \xi, x\xi\rangle\,\langle \varphi, \psi\rangle - \langle \xi, \xi\rangle\,\langle \varphi, x\psi\rangle = -\langle \varphi, x\psi\rangle. \end{aligned}$$

Since $x\mathcal{D} \subseteq \mathcal{D}$ and $y\mathcal{D} \subseteq \mathcal{D}$, this shows that $x \in \mathcal{L}^+(\mathcal{D})$ and $x^+ = -y$.

Finally, suppose that $\mathcal{A} \subseteq \mathcal{L}^+(\mathcal{D})$ and $\delta(a^+) = \delta(a)^+$ for $a \in \mathbb{F}(\mathcal{D})$. Then $y\varphi = \big(\delta(\varphi \otimes \xi)\big)^+\,\xi = \delta\big((\varphi \otimes \xi)^+\big)\,\xi = \delta(\xi \otimes \varphi)\,\xi = x\varphi$ for $\varphi \in \mathcal{D}$, so that $x^+ = -x$. □

Corollary 6.3.3. *If $\mathcal{D}$ is a unitary space, then each derivation of the algebra $\mathcal{L}^+(\mathcal{D})$ is inner.*

Proof. Apply Proposition 6.3.2 to $\mathcal{A} := \mathcal{L}^+(\mathcal{D})$. □

Definition 6.3.4. Let $\mathcal{D}_1$ and $\mathcal{D}_2$ be dense linear subspaces of Hilbert spaces $\mathcal{H}_1$ and $\mathcal{H}_2$, and let $\mathcal{A}_1$ and $\mathcal{A}_2$ be $*$-subalgebras of $\mathcal{L}^+(\mathcal{D}_1)$ and $\mathcal{L}^+(\mathcal{D}_2)$, respectively. A $*$-isomorphism π of $\mathcal{A}_1$ onto $\mathcal{A}_2$ is called *spatial* if there exists an isometry U of $\mathcal{H}_1$ onto $\mathcal{H}_2$ such that $U\mathcal{D}_1 = \mathcal{D}_2$ and $\pi(a)\,\varphi = UaU^{-1}\varphi$, $\varphi \in \mathcal{D}_2$, for all $a \in \mathcal{A}_1$. Then we say that π is *implemented* by U. A $*$-automorphism of $\mathcal{A}_1$ is said to be *inner* if it is spatial and it can be implemented by a unitary operator U on $\mathcal{H}_1$ such that $U \upharpoonright \mathcal{D}_1$ is in $\mathcal{A}_1$.

We shall prove that $*$-automorphisms of O*-algebras $\mathcal{L}^+(\mathcal{D})$ are always inner. This will be obtained as a corollary from a more general result (Theorem 6.3.6) which will be used in the next section as well. For this we need some preliminaries.

Suppose A is an abstract $*$-algebra with or without unit element. By a *projection* in A we mean a hermitian idempotent of A. If e_1 and e_2 are projections in A, we write $e_1 \leqq e_2$

if and only if $e_1e_2 = e_1$. It is easy to check that "$\leqq$" is a reflexive, antisymmetric and transitive relation in the set of all projections of A. (We verify, for instance, the transitivity. If $e_1 \leqq e_2$ and $e_2 \leqq e_3$, then $e_1e_2 = e_1$ and $e_2e_3 = e_2$ and hence $e_1e_3 = (e_1e_2)\, e_3 = e_1(e_2e_3) = e_1e_2 = e_1$, so that $e_1 \leqq e_3$.) This terminology is justified by the following fact. If there exists a $*$-isomorphism π of A onto a $*$-subalgebra of some $\mathcal{L}^+(\mathcal{D})$, and if e_1 and e_2 are projections in A, then $\overline{\pi(e_1)}$ and $\overline{\pi(e_2)}$ are Hilbert space projections and the relation $e_1 \leqq e_2$ is equivalent to the usual relation $\overline{\pi(e_1)} \leqq \overline{\pi(e_2)}$ for the projections $\overline{\pi(e_1)}$ and $\overline{\pi(e_2)}$.

A projection $e \neq 0$ in A is said to be *minimal* if the relation $e_1 \leqq e$ for a projection $e_1 \neq 0$ in A always implies that $e_1 = e$. We denote the set of all minimal projections in A by M(A). For $e_1, e_2 \in$ M(A), we write $e_1 \approx e_2$ if $e_1 \mathsf{A} e_2 \neq \{0\}$. Of course, it may happen that the set M(A) is empty.

Suppose that I is an index set. For every $i \in I$, let $\mathcal{D}_i$ be a dense linear subspace of a Hilbert space $\mathcal{H}_i$. Let $\mathcal{H}_I$ be the Hilbert space $\sum_{i \in I} \oplus \mathcal{H}_i$, and let $\mathcal{D}_I$ denote the dense linear subspace of $\mathcal{H}_I$ formed by the vectors (φ_i) which have only a finite number of non-zero components $\varphi_i \in \mathcal{D}_i$. We consider each $\mathcal{H}_i$ as a subspace of $\mathcal{H}_I$ in the obvious way. Each element (a_i) of the product $\prod_{i \in I} \mathcal{L}^+(\mathcal{D}_i)$ acts as an operator on $\mathcal{D}_I$ by the definition $(a_i)\,(\varphi_i) := (a_i\varphi_i)$, $(\varphi_i) \in \mathcal{D}_I$. The set of all these operators (a_i) forms an O*-algebra on the domain $\mathcal{D}_I$ which we will denote by the symbol $\mathcal{L}^+(\mathcal{D}_i : i \in I)$.

Retaining this notation, we have

Lemma 6.3.5. *Suppose $\mathcal{A}$ is a $*$-subalgebra of $\mathcal{L}^+(\mathcal{D}_I)$.*

(i) *$\mathsf{M}\big(\mathcal{L}^+(\mathcal{D}_i : i \in I)\big)$ is the set of all rank one projections of the form $\varphi_i \otimes \varphi_i$, where $i \in I$, $\varphi_i \in \mathcal{D}_i$ and $\|\varphi_i\| = 1$. If $\varphi_i \otimes \varphi_i$ and $\psi_{i'} \otimes \psi_{i'}$ are two such projections, then $\varphi_i \otimes \varphi_i \approx \psi_{i'} \otimes \psi_{i'}$ if and only if $i = i'$.*

(ii) *$\mathsf{M}(\mathcal{A}) = \mathsf{M}\big(\mathcal{L}^+(\mathcal{D}_i : i \in I)\big)$ if and only if $\mathcal{A} \subseteqq \mathcal{L}^+(\mathcal{D}_i : i \in I)$ and $\mathbb{F}(\mathcal{D}_i) \subseteqq \mathcal{A}$ for all $i \in I$.*

(iii) *If $\mathsf{M}(\mathcal{A}) = \mathsf{M}\big(\mathcal{L}^+(\mathcal{D}_i : i \in I)\big)$, then the relation "$\approx$" relative to the $*$-algebra $\mathcal{A}$ coincides with the relation "$\approx$" relative to the $*$-algebra $\mathcal{L}^+(\mathcal{D}_i : i \in I)$ on the set $\mathsf{M}(\mathcal{A}) \equiv \mathsf{M}\big(\mathcal{L}^+(\mathcal{D}_i : i \in I)\big)$.*

Proof. The proofs of (i) and of the if part in (ii) are straightforward, so we omit the details. We prove the only if part of (ii). Suppose that $\mathsf{M}(\mathcal{A}) = \mathsf{M}\big(\mathcal{L}^+(\mathcal{D}_i : i \in I)\big)$.

First we show that $\mathcal{A} \subseteqq \mathcal{L}^+(\mathcal{D}_i : i \in I)$. Let $i \in I$ and $\varphi \in \mathcal{D}_i$. It clearly suffices to verify that $a\varphi \in \mathcal{D}_i$ for each $a \in \mathcal{A}$. Without loss of generality we assume that $\|\varphi\| = 1$ and $a\varphi \neq 0$. Set $\psi := \|a\varphi\|^{-1}\, \varphi$. By (i), $\varphi \otimes \varphi \in \mathsf{M}\big(\mathcal{L}^+(\mathcal{D}_i : i \in I)\big) = \mathsf{M}(\mathcal{A})$. Since $\mathcal{A}$ is a $*$-algebra, the rank one projection $\|a\varphi\|^{-2}\, a(\varphi \otimes \varphi)\, a^+ \equiv a\psi \otimes a\psi$ is in $\mathcal{A}$ and so in $\mathsf{M}(\mathcal{A})$. By $\mathsf{M}(\mathcal{A}) = \mathsf{M}\big(\mathcal{L}^+(\mathcal{D}_i : i \in I)\big)$ and (i), $a\psi$ and hence $a\varphi$ is in $\mathcal{D}_{i'}$ for some $i' \in I$. Assume that $i \neq i'$. Then $(a + \varphi \otimes \varphi)\, \varphi \neq 0$. Since $a + \varphi \otimes \varphi \in \mathcal{A}$, the preceding argument applied with a replaced by $a + \varphi \otimes \varphi$ shows that $a\varphi + \varphi \equiv (a + \varphi \otimes \varphi)\, \varphi \in \mathcal{D}_{i''}$ for some $i'' \in I$. Since $\varphi \in \mathcal{D}_i$ and $a\varphi \in \mathcal{D}_{i'}$, this is impossible. Thus $i = i'$ and $a\varphi \in \mathcal{D}_i$; so we have proved that $\mathcal{A} \subseteqq \mathcal{L}^+(\mathcal{D}_i : i \in I)$.

Let $i \in I$. We show that $\mathbb{F}(\mathcal{D}_i) \subseteqq \mathcal{A}$. It is sufficient to check that $\varphi \otimes \psi \in \mathcal{A}$ for arbitrary unit vectors φ, ψ in $\mathcal{D}_i$. Set $\xi := 2^{-1/2}(\varphi + \psi)$ if $\varphi \perp \psi$ and $\xi := \varphi$ otherwise. Again by $\mathsf{M}(\mathcal{A}) = \mathsf{M}\big(\mathcal{L}^+(\mathcal{D}_i : i \in I)\big)$, the operators $\varphi \otimes \varphi$, $\psi \otimes \psi$ and $\xi \otimes \xi$ are in $\mathcal{A}$

and so

$$\varphi \otimes \psi = \langle \varphi, \xi \rangle^{-1} \langle \psi, \xi \rangle^{-1} (\psi \otimes \psi) (\xi \otimes \xi) (\varphi \otimes \varphi) \in \mathcal{A}.$$

Finally, we prove (iii). Suppose $\mathsf{M}(\mathcal{A}) = \mathsf{M}(\mathcal{L}^+(\mathcal{D}_i : i \in I))$. Then $\mathcal{A} \subseteq \mathcal{L}^+(\mathcal{D}_i : i \in I)$ by (ii). Hence, by (i), it suffices to show that $(\psi \otimes \psi) \mathcal{A}(\varphi \otimes \varphi) \neq \{0\}$ for arbitrary unit vectors $\varphi, \psi \in \mathcal{D}_i$. But this follows from

$$(\psi \otimes \psi) (\xi \otimes \xi) (\varphi \otimes \varphi) = \langle \varphi, \xi \rangle \langle \psi, \xi \rangle \varphi \otimes \psi \neq 0$$

and $\xi \otimes \xi \in \mathcal{A}$, where ξ is as in the preceding proof. □

Theorem 6.3.6. *Let $\{\mathcal{D}_i : i \in I\}$ and $\{\mathcal{D}_j : j \in J\}$ be indexed families of non-zero unitary spaces. Suppose $\mathcal{A}$ and $\mathcal{B}$ are $*$-subalgebras of $\mathcal{L}^+(\mathcal{D}_I)$ and $\mathcal{L}^+(\mathcal{D}_J)$, respectively, satisfying $\mathsf{M}(\mathcal{A}) = \mathsf{M}(\mathcal{L}^+(\mathcal{D}_i : i \in I))$ and $\mathsf{M}(\mathcal{B}) = \mathsf{M}(\mathcal{L}^+(\mathcal{D}_j : j \in J))$. Suppose that there exists a $*$-isomorphism π of $\mathcal{A}$ onto $\mathcal{B}$. Then π is a spatial $*$-isomorphism. More precisely, there exist a bijective map $\varkappa$ of I onto J and an isometry U of $\mathcal{H}_I$ onto $\mathcal{H}_J$ such that U implements π and $U\mathcal{D}_i = \mathcal{D}_{\varkappa(i)}$ for $i \in I$.*

Proof. From the definitions it is clear that the set $\mathsf{M}(\mathcal{A})$ and the relation "$\approx$" are preserved under $*$-isomorphisms. Therefore, $\pi(\mathsf{M}(\mathcal{A})) = \mathsf{M}(\mathcal{B})$. Combined with the assumptions, this yields that

$$\pi(\mathsf{M}(\mathcal{L}^+(\mathcal{D}_i : i \in I))) = \mathsf{M}(\mathcal{L}^+(\mathcal{D}_j : j \in J)). \tag{3}$$

For every $i \in I$ we take a fixed unit vector φ_i of $\mathcal{D}_i$. This is possible, since $\mathcal{D}_i \neq \{0\}$ by assumption. By Lemma 6.3.5, (i), $\varphi_i \otimes \varphi_i \in \mathsf{M}(\mathcal{L}^+(\mathcal{D}_i : i \in I))$, and hence, by (3), there exist an index $\varkappa(i) \in J$ and a unit vector $\psi_{\varkappa(i)} \in \mathcal{D}_{\varkappa(i)}$ such that $\pi(\varphi_i \otimes \varphi_i) = \psi_{\varkappa(i)} \otimes \psi_{\varkappa(i)}$. From (3) and Lemma 6.3.5, (i) and (iii), we conclude easily that the map $i \to \varkappa(i)$ is a bijection of I onto J.

Suppose that $\varphi \in \mathcal{D}_i$ and $\psi \in \mathcal{D}_j$ are unit vectors such that $\pi(\varphi \otimes \varphi) = \psi \otimes \psi$. We show that

$$\|x\varphi\| = \|\pi(x)\, \psi\| \quad \text{for all } x \in \mathcal{A}. \tag{4}$$

Fix $x \in \mathcal{A}$. Lemma 6.3.5, (ii), yields $x\varphi \in \mathcal{D}_i$ and so $x\varphi \otimes x\varphi \in \mathcal{A}$. We have $\pi(x\varphi \otimes x\varphi) = \pi(x(\varphi \otimes \varphi)\, x^+) = \pi(x)\, (\psi \otimes \psi)\, \pi(x)^+ = \pi(x)\, \psi \otimes \pi(x)\, \psi$. In case where $\pi(x)\, \psi = 0$ this implies already (4). Taking the square in the preceding equality, we obtain

$$\pi(x\varphi \otimes x\varphi)^2 = \pi((x\varphi \otimes x\varphi)^2) = \pi(\|x\varphi\|^2\, x\varphi \otimes x\varphi) = \|x\varphi\|^2\, \pi(x)\, \psi \otimes \pi(x)\, \psi$$

at the left-hand side and

$$\|\pi(x)\, \psi\|^2\, \pi(x)\, \psi \otimes \pi(x)\, \psi$$

at the right-hand side. This gives (4) also in case $\pi(x)\, \psi \neq 0$.

Let $i \in I$. From (4) we conclude that the equation $U_i(x\varphi_i) := \pi(x)\, \psi_{\varkappa(i)}$, $x \in \mathcal{A}$, defines, unambiguously, a norm-preserving linear mapping of $\mathcal{A}\varphi_i$ onto $\pi(\mathcal{A})\, \psi_{\varkappa(i)} = \mathcal{B}\psi_{\varkappa(i)}$. Since $\mathbb{F}(\mathcal{D}_i) \subseteq \mathcal{A} \restriction \mathcal{D}_i \subseteq \mathcal{L}^+(\mathcal{D}_i)$ by Lemma 6.3.5, (ii), we have $\mathcal{A}\varphi_i = \mathcal{D}_i$. Similarly, $\mathcal{B}\psi_{\varkappa(i)} = \mathcal{D}_{\varkappa(i)}$. That is, U_i maps $\mathcal{D}_i$ onto $\mathcal{D}_{\varkappa(i)}$. Consequently, there is a unique isometry U of $\mathcal{H}_I$ onto $\mathcal{H}_J$ such that $U \restriction \mathcal{D}_i = U_i \restriction \mathcal{D}_i$ for $i \in I$. By construction, $U\mathcal{D}_i = \mathcal{D}_{\varkappa(i)}$ for $i \in I$ and $U\mathcal{D}_I = \mathcal{D}_J$. Suppose $a \in \mathcal{A}$. From the preceding definitions, we have $\pi(a)\, (\pi(x)\, \psi_{\varkappa(i)}) = \pi(ax)\, \psi_{\varkappa(i)} = Uax\varphi_i = UaU^{-1}(\pi(x)\, \psi_{\varkappa(i)})$ for all $x \in \mathcal{A}$ and $i \in I$. From this it follows that $\pi(a)\, \varphi = UaU^{-1}\varphi$ for all $\varphi \in \mathcal{D}_J$; so π is spatial and implemented by U. □

Corollary 6.3.7. *Let $\{\mathcal{D}_i : i \in I\}$ and $\{\mathcal{D}_j : j \in J\}$ be indexed families of unitary spaces. If π is a $*$-isomorphism of $\mathcal{L}^+(\mathcal{D}_i : i \in I)$ onto a $*$-subalgebra of $\mathcal{L}^+(\mathcal{D}_J)$ such that $\mathsf{M}(\pi(\mathcal{L}^+(\mathcal{D}_i : i \in I))) = \mathsf{M}(\mathcal{L}^+(\mathcal{D}_j : j \in J))$, then π is a spatial $*$-isomorphism of $\mathcal{L}^+(\mathcal{D}_i : i \in I)$ onto $\mathcal{L}^+(\mathcal{D}_j : j \in J)$.*

Proof. There is no loss of generality to assume that the unitary spaces are non-zero. Then the assumptions of Theorem 6.3.6 are satisfied when we set $\mathcal{A} := \mathcal{L}^+(\mathcal{D}_i : i \in I)$ and $\mathcal{B} := \pi(\mathcal{A})$; so π is spatial by Theorem 6.3.6. From the properties of the isometry U in Theorem 6.3.6 it is clear that $a \to UaU^{-1}$ maps $\mathcal{L}^+(\mathcal{D}_i : i \in I)$ onto $\mathcal{L}^+(\mathcal{D}_j : j \in J)$. Since $\pi(a) = UaU^{-1}$ for $a \in \mathcal{A}$, this gives $\pi(\mathcal{A}) = \mathcal{L}^+(\mathcal{D}_j : j \in J)$. □

Corollary 6.3.8. *Suppose that $\mathcal{D}$ is a unitary space. Then each $*$-automorphism of $\mathcal{L}^+(\mathcal{D})$ is inner.*

Proof. Obviously, we can assume that $\mathcal{D} \neq \{0\}$. We apply Theorem 6.3.6 in case where both families are the singleton $\{\mathcal{D}\}$ and $\mathcal{A} = \mathcal{B} := \mathcal{L}^+(\mathcal{D})$. Thus every $*$-automorphism π of $\mathcal{L}^+(\mathcal{D})$ is spatial. If π is implemented by U, then $U\mathcal{D} = \mathcal{D}$ and hence $U^*\mathcal{D} = \mathcal{D}$, so that $U \upharpoonright \mathcal{D}$ is in $\mathcal{L}^+(\mathcal{D})$ and π is inner. □

6.4. Atomic $*$-Algebras

Throughout this section A denotes an abstract $*$-algebra such that $\mathsf{A} \neq \{0\}$. We do not assume that A has a unit element.

A left ideal of A is called *minimal* if it is different from $\{0\}$ and if it does not contain properly any other non-zero left ideal.

Definition 6.4.1. The $*$-algebra A is called *$*$-semisimple* if it is $*$-isomorphic to a $*$-subalgebra of some $\mathcal{L}^+(\mathcal{D})$. We say that A is *atomic* if A is $*$-semisimple and every non-zero left ideal of A contains a minimal left ideal. We say that A is *maximal atomic* if A is atomic and if each atomic $*$-algebra B which contains A as a $*$-subalgebra and satisfies $\mathsf{M}(\mathsf{B}) = \mathsf{M}(\mathsf{A})$ is equal to A.

Remark 1. Clearly, each $*$-subalgebra of $\mathcal{L}^+(\mathcal{D})$ and hence each $*$-semisimple $*$-algebra A possesses the following property: if $\sum_{n=1}^{k} a_n^+ a_n = 0$ for some $a_1, \ldots, a_k \in \mathsf{A}$ and $k \in \mathbb{N}$, then $a_1 = \cdots = a_k = 0$. In particular, $a^+a = 0$ always implies $a = 0$ for $a \in \mathsf{A}$. The latter fact will be frequently used in the sequel.

Remark 2. The three notions defined in Definition 6.4.1 are, of course, preserved under $*$-isomorphisms.

The following two theorems describe the structure of atomic and maximal atomic $*$-algebras up to $*$-isomorphisms.

Theorem 6.4.2. *A $*$-algebra A is atomic if and only if there exist a family $\{\mathcal{D}_i : i \in I\}$ of unitary spaces and a $*$-ismorphism of A onto a $*$-subalgebra $\mathcal{A}$ of $\mathcal{L}^+(\mathcal{D}_i : i \in I)$ such that $\mathsf{M}(\mathcal{A}) = \mathsf{M}(\mathcal{L}^+(\mathcal{D}_i : i \in I))$. If A is atomic, then the $*$-algebras $\mathcal{A}$ and $\mathcal{L}^+(\mathcal{D}_i : i \in I)$ are both uniquely determined up to spatial $*$-isomorphisms by the above properties.*

Theorem 6.4.3. *A $*$-algebra A is maximal atomic if and only if there exists a family $\{\mathcal{D}_i : i \in I\}$ of unitary spaces such that A is $*$-isomorphic to $\mathcal{L}^+(\mathcal{D}_i : i \in I)$. The O*-algebra $\mathcal{L}^+(\mathcal{D}_i : i \in I)$ is then uniquely determined up to spatial $*$-isomorphisms by A.*

Remark 3. The $*$-subalgebras $\mathcal{A}$ of $\mathcal{L}^+(\mathcal{D}_i : i \in I)$ that satisfy $\mathsf{M}(\mathcal{A}) = \mathsf{M}(\mathcal{L}^+(\mathcal{D}_i : i \in I))$ are characterized in Lemma 6.3.5, (ii).

Remark 4. From Theorem 6.4.2 and Lemma 6.3.5 or from Corollary 6.4.8 it can be seen that a W^*-algebra is atomic in the sense of classification theory of W^*-algebras (see e.g. TAKESAKI [1], III, Definition 5.9) if and only if it is atomic according to Definition 6.4.1. This is the reason we used the name atomic. From Theorem 6.4.3 it is clear that the maximal atomic $*$-algebras can be considered as unbounded generalizations of atomic W^*-algebras. Let us note that an atomic W^*-algebra (for instance, $l^\infty(\mathbb{N})$) is, in general, not maximal atomic.

Remark 5. In this and the preceding section we did not assume that the $*$-algebras have unit elements. However, a maximal atomic $*$-algebra has always a unit as Theorem 6.4.3 shows.

The proof of Theorem 6.4.2 will be completed at the end of this section. First we derive Theorem 6.4.3 from Theorem 6.4.2.

Proof of Theorem 6.4.3 (granted Theorem 6.4.2). Suppose A is maximal atomic. By Theorem 6.4.2, there is a $*$-isomorphism π of A on a $*$-subalgebra $\mathcal{A}$ of some $\mathcal{L}^+(\mathcal{D}_i : i \in I)$ such that $\mathsf{M}(\mathcal{A}) = \mathsf{M}(\mathcal{L}^+(\mathcal{D}_i : i \in I))$. Since the $*$-algebras $\mathcal{A} = \pi(\mathsf{A})$ and $\mathcal{B} := \mathcal{L}^+(\mathcal{D}_i : i \in I)$ (by Theorem 6.4.2) are atomic and $\mathsf{M}(\mathcal{A}) = \mathsf{M}(\mathcal{B})$, the maximality of A yields $\mathcal{A} = \mathcal{B}$. Hence π is a $*$-isomorphism of A on $\mathcal{L}^+(\mathcal{D}_i : i \in I)$. The uniqueness statement follows directly from the corresponding statement in Theorem 6.4.2.

To prove the converse direction, it suffices to show that $\mathsf{A} := \mathcal{L}^+(\mathcal{D}_i : i \in I)$ is maximal atomic. By Theorem 6.4.2, this $*$-algebra is atomic. Let B be an atomic $*$-algebra which contains A and satisfies $\mathsf{M}(\mathsf{B}) = \mathsf{M}(\mathsf{A})$. Applying once more Theorem 6.4.2, there is a $*$-isomorphism π of B on a $*$-subalgebra of some $\mathcal{L}^+(\mathcal{D}_j : j \in J)$ such that $\mathsf{M}(\pi(\mathsf{B})) = \mathsf{M}(\mathcal{L}^+(\mathcal{D}_j : j \in J))$. Then $\mathsf{M}(\pi(\mathsf{A})) = \pi(\mathsf{M}(\mathsf{A})) = \pi(\mathsf{M}(\mathsf{B})) = \mathsf{M}(\pi(\mathsf{B})) = \mathsf{M}(\mathcal{L}^+(\mathcal{D}_j : j \in J))$, so that $\pi(\mathsf{A}) = \mathcal{L}^+(\mathcal{D}_j : j \in J)$ by Corollary 6.3.7. This obviously implies that $\mathsf{A} = \mathsf{B}$. □

A consequence of Theorem 6.4.3 is the following inner characterization of $\mathcal{L}^+(\mathcal{D})$.

Corollary 6.4.4. *For any $*$-algebra A, the following three conditions are equivalent:*

(i) *There is a unitary space $\mathcal{D}$ such that A is $*$-isomorphic to $\mathcal{L}^+(\mathcal{D})$.*

(ii) *A is maximal atomic and $e\mathsf{A}f \neq \{0\}$ for all $e, f \in \mathsf{M}(\mathsf{A})$.*

(iii) *A is maximal atomic and the centre of A consists of scalar multiples of the unit of A (which exists by Theorem 6.4.3).*

Proof. It is easy to check (by Theorem 6.4.3 and Lemma 6.3.5, (i)) that $\mathcal{L}^+(\mathcal{D})$ and so each $*$-isomorphic $*$-algebra A satisfies (ii) and (iii). For the implications (ii) → (i) and (iii) → (i), we apply Theorem 6.4.3 and note that the additional requirements imply that the family $\{\mathcal{D}_i : i \in I\}$ reduces to a singleton. □

Now we begin with the preliminaries of the proof of Theorem 6.4.2. Some of these investigations are of interest in itself, and not all results are needed in full strength to prove Theorem 6.4.2.

Lemma 6.4.5. *Suppose that A is $*$-semisimple.*

(i) *Suppose J is a minimal left ideal of A. Then there is a unique projection e in A such that $\mathsf{J} = \mathsf{A}e$. Further, $e \in \mathsf{M}(\mathsf{A})$ and $e\mathsf{A}e = \mathbb{C} \cdot e$.*

(ii) *If $e \neq 0$ is a projection in A such that $e\mathsf{A}e = \mathbb{C} \cdot e$, then $\mathsf{A}e$ is a minimal left ideal of A.*

Proof. (i): Since $\mathsf{J} \neq \{0\}$, there is an element $x \neq 0$ in J. Since A is $*$-semisimple, $y := x^+x \neq 0$ and $y^2 \neq 0$. Because $y^2 \in \mathsf{J}y$, $\mathsf{J}y$ is a non-zero left ideal contained in J, so that $\mathsf{J}y = \mathsf{J}$ by the minimality of J. We next check that $z \in \mathsf{J}$ and $zy = 0$ imply $z = 0$. Otherwise, $\mathsf{A}z \neq \{0\}$ (because of $z^+z \neq 0$) and so $\mathsf{A}z$ is a non-zero left ideal contained in J; hence $\mathsf{J} = \mathsf{A}z$ and $\mathsf{J} = \mathsf{J}y = \mathsf{A}zy = \{0\}$ which is the desired contradiction. From $\mathsf{J} = \mathsf{J}y$, there is $u \in \mathsf{J}$ such that $uy = y$. This gives $(u^2 - u)\, y = 0$. Since $u^2 - u \in \mathsf{J}$, we have $u^2 = u = 0$ by the preceding. Since $u^2 = u \neq 0$ (by $uy = y \neq 0$), the minimality of J yields $\mathsf{J} = \mathsf{J}u$. From $y \in \mathsf{J}$ and the latter, there is $v \in \mathsf{J}$ such that $y = vu$. Thus $yu = vu^2 = vu = y$. By $y = y^+$, this gives $u^+y = y$. Set $e := u^+u$. We have $e \in \mathsf{J}$ and $ey = u^+uy = u^+y = y$, so that $(e^2 - e)\, y = 0$. As shown above, this implies $e^2 - e = 0$. Hence e is a projection. Since $e^2 = e \neq 0$ and $e \in \mathsf{J}$ it follows again from the minimality of J that $\mathsf{J} = \mathsf{A}e$.

We prove the uniqueness of e. Suppose $\tilde{e}$ is another projection in A such that $\mathsf{J} = \mathsf{A}\tilde{e}$. Then $\tilde{e} = (\tilde{e})^2 \in \mathsf{J}$ and there are $a, \tilde{a} \in \mathsf{J}$ such that $e = a\tilde{e}$ and $\tilde{e} = \tilde{a}e$. Hence $e\tilde{e} = a\tilde{e}\tilde{e} = a\tilde{e} = e$ and $e\tilde{e} = (\tilde{e}e)^+ = (\tilde{a}ee)^+ = (\tilde{a}e)^+ = (\tilde{e})^+ = \tilde{e}$, so that $e = \tilde{e}$.

We show that $e \in \mathsf{M}(\mathsf{A})$. We let $f \neq 0$ be a projection in A such that $f \leqq e$, i.e., $fe = f$. Using once more the minimality of J, this gives $\mathsf{J} = \mathsf{A}f$. Therefore, by the uniqueness assertion just shown, $f = e$ and so $e \in \mathsf{M}(\mathsf{A})$.

We verify that $e\mathsf{A}e$ is a division algebra with unit. Since $e^2 = e$, e is the unit element of the algebra $e\mathsf{A}e$. Let $x \in \mathsf{A}$ be such that $exe \neq 0$. Then $\mathsf{A}exe$ is a non-zero left ideal of A contained in $\mathsf{J} = \mathsf{A}e$, hence $\mathsf{A}exe = \mathsf{A}e$ and there is $y \in \mathsf{A}$ such that $yexe = ee$. Then $(eye)\,(exe) = e$ which shows that exe has a left inverse, so $e\mathsf{A}e$ is a division algebra (see 2.1). Since A and hence $e\mathsf{A}e$ is $*$-semisimple, we conclude from Proposition 2.1.12 that $e\mathsf{A}e = \mathbb{C} \cdot e$.

(ii): Since $e \neq 0$, $\mathsf{A}e$ is a non-zero left ideal of A. Let J be another non-zero left ideal of A contained in $\mathsf{A}e$. Then there is $x \in \mathsf{A}$ such that $xe \in \mathsf{J}$ and $xe \neq 0$. Since A is $*$-semisimple, $(xe)^+\, xe \equiv ex^+xe \neq 0$. From $e\mathsf{A}e = \mathbb{C} \cdot e$, $ex^+xe = \lambda e$ for some $\lambda \in \mathbb{C}$, $\lambda \neq 0$. Since $xe \in \mathsf{J}$, $e = \lambda^{-1}ex^+xe \in \mathsf{J}$. This yields $\mathsf{J} = \mathsf{A}e$ and proves that $\mathsf{A}e$ is minimal. □

Corollary 6.4.6. *Suppose* A *is atomic. If e is a projection in* A, *then* $e \in \mathsf{M}(\mathsf{A})$ *if and only if* $\mathsf{A}e$ *is a minimal left ideal (or equivalently, if $e \neq 0$ and* $e\mathsf{A}e = \mathbb{C} \cdot e$).

Proof. Because of Lemma 6.4.5, it suffices to show that the left ideal $\mathsf{A}e$ is minimal when $e \in \mathsf{M}(\mathsf{A})$. Since $e \neq 0$ by $e \in \mathsf{M}(\mathsf{A})$, $\mathsf{A}e \neq \{0\}$. Let J be a non-zero left ideal of A with $\mathsf{J} \subseteqq \mathsf{A}e$. Since A is atomic, there is a minimal left ideal J_0 such that $\mathsf{J}_0 \subseteqq \mathsf{J}$. By Lemma 6.4.5, (i), $\mathsf{J}_0 = \mathsf{A}f$ for some projection $f \neq 0$ in A. Then $\mathsf{A}f \subseteqq \mathsf{A}e$ and there exists $a \in \mathsf{A}$ such that $f \equiv ff = ae$. Then $fe = (ae)\, e = ae = f$, so $f \leqq e$ and hence $f = e$, since $e \in \mathsf{M}(\mathsf{A})$. Consequently, $\mathsf{A}f = \mathsf{J}_0 = \mathsf{J} = \mathsf{A}e$ which shows that $\mathsf{A}e$ is minimal. □

Lemma 6.4.7. *Suppose that* A *is atomic and* $e \in \mathsf{M}(\mathsf{A})$.

(i) *There exists a unique positive linear functional g_e on* A *such that $exe = g_e(x)\, e$ for all* $x \in \mathsf{A}$. *For* $x, y \in \mathsf{A}$, *we define* $\langle xe, ye\rangle_e := g_e(y^+x)$ *and* $\varrho_e(x)\, ye := xye$. *Then* $\mathcal{D}_e := (\mathsf{A}e, \langle\cdot, \cdot\rangle_e)$ *is a unitary space, and ϱ_e is a $*$-homomorphism of* A *into* $\mathcal{L}^+(\mathcal{D}_e)$.

(ii) *If* $a \in \mathsf{A}$ *and* $\|ae\|_e = 1$, *then* $aea^+ \in \mathsf{M}(\mathsf{A})$, $aea^+ \approx e$ *and* $\varrho_e(aea^+) = ae \otimes ae$. *Conversely, if* $f \in \mathsf{M}(\mathsf{A})$ *and* $f \approx e$, *then there exists* $a \in \mathsf{A}$ *such that* $\|ae\|_e = 1$ *and* $f = aea^+$. *If* $f \in \mathsf{M}(\mathsf{A})$ *and* $f \not\approx e$, *then* $\varrho_e(f) = 0$.

Proof. (i): By Corollary 6.4.6, $e\mathsf{A}e = \mathbb{C} \cdot e$. Since $e \neq 0$, this means that for each $x \in \mathsf{A}$ there is a unique complex number $g_e(x)$ such that $exe = g_e(x)\, e$. Obviously, $g_e(\cdot)$ is a hermitian linear functional on A. If there were an $x \in \mathsf{A}$ such that $g_e(x^+x) < 0$, then

$(xe)^+xe + (\lambda e)^2 = 0$ with $\lambda := \big(-g_e(x^+x)\big)^{1/2}$. Since $\lambda e \neq 0$, this contradicts the $*$-semisimplicity of A. Hence g_e is a positive linear functional on A.

Since $e\big((ye)^+xe\big)e = ey^+xe$, we obtain $g_e\big((ye)^+xe\big) = g_e(y^+x)$ for $x, y \in \mathsf{A}$. From this we see that $\langle\cdot,\cdot\rangle_e$ is well-defined on Ae. Clearly, $\langle xe, xe\rangle_e = g_e(x^+x) \geqq 0$ for $x \in \mathsf{A}$. If $\langle xe, xe\rangle_e = 0$, then $(xe)^+xe = g_e(x^+x)\, e = \langle xe, xe\rangle_e\, e = 0$ and so $xe = 0$. Therefore, $\langle\cdot,\cdot\rangle_e$ is a scalar product on Ae, and $\mathcal{D}_e$ is a unitary space. Clearly, ϱ_e is a homomorphism of A into $L(\mathcal{D}_e)$. That ϱ_e is a $*$-homomorphism of A into $\mathcal{L}^+(\mathcal{D}_e)$ follows from $\langle \varrho_e(a)\, xe, ye\rangle_e = \langle axe, ye\rangle_e = g_e(y^+ax) = g_e\big((a^+y)^+x\big) = \langle xe, a^+ye\rangle_e = \langle xe, \varrho_e(a^+)\, ye\rangle_e$, $a, x, y \in \mathsf{A}$.

(ii): Suppose $a \in \mathsf{A}$ and $\|ae\|_e = 1$. Since $g_e(a^+a) \equiv \langle ae, ae\rangle_e = 1$, $aea^+aea^+ = a(ea^+ae)\, a^+ = g_e(a^+a)\, aea^+ = aea^+$, so that aea^+ is a projection. We have $aea^+xaea^+ = a(ea^+xae)\, a^+ = g_e(a^+xa)\, aea^+$ for $x \in \mathsf{A}$. Therefore, $aea^+\mathsf{A}aea^+ = \mathbb{C}\cdot aea^+$ and so $aea^+ \in \mathsf{M}(\mathsf{A})$ by Corollary 6.4.6. Since $(aea^+)ae = g_e(a^+a)\, ae \neq 0$, $aea^+ \approx e$. From $\langle \varrho_e(aea^+)\, xe, ye\rangle_e = \langle aea^+xe, ye\rangle_e = g_e(a^+x)\, \langle ae, ye\rangle_e = \langle xe, ae\rangle_e \langle ae, ye\rangle_e$ for $x, y \in \mathsf{A}$ we conclude that $\varrho_e(aea^+) = ae \otimes ae$.

Conversely, suppose that $f \in \mathsf{M}(\mathsf{A})$ and $f \approx e$. Then $fbe \neq 0$ for some $b \in \mathsf{A}$ and so $fbe(fbe)^+ = fbeb^+f = g_f(beb^+)\, f \neq 0$, since A is $*$-semisimple. Thus $g_f(beb^+) = g_e\big(be(be)^+\big) > 0$. Setting $a := g_e(beb^+)^{-1/2}\, fb$, we have $f = aea^+$. From $f = ff = aea^+aea^+ = g_e(a^+a)\, aea^+ = \|ae\|_e^2\, f$ we get $\|ae\|_e = 1$.

Finally, if $f \in \mathsf{M}(\mathsf{A})$ and $f \not\approx e$, then $\varrho_e(f)\, ae = fae = 0$ for $a \in \mathsf{A}$ and so $\varrho_e(f) = 0$. □

Remark 6. In Lemma 6.4.7 we also have that $\varrho_e(\mathsf{A})\, e = \mathcal{D}_e$ and $g_e(\cdot) = \langle \varrho_e(\cdot)\, e, e\rangle_e$ on A. From this and Theorem 8.6.2 it follows that if A has a unit, then ϱ_e is (unitarily equivalent to) the $*$-representation $\tilde{\pi}_{g_e}$ of A obtained from the positive linear functional g_e by the GNS construction; cf. Section 8.6.

Corollary 6.4.8. *Suppose that* A *is an atomic $*$-algebra. Then* $ea^+ae = 0$ *(or equivalently,* $g_e(a^+a) \equiv \|ae\|_e^2 = 0$*) for all* $e \in \mathsf{M}(\mathsf{A})$ *implies* $a = 0$ *for arbitrary* a *in* A.

Proof. Suppose $a \in \mathsf{A}$, $a \neq 0$. Since the $*$-algebra A is atomic, the non-zero left ideal Aa contains a minimal left ideal J. By Lemma 6.4.5, (i), $\mathsf{J} = \mathsf{A}e$ for some $e \in \mathsf{M}(\mathsf{A})$. Thus $\mathsf{A}a \supseteq \mathsf{A}e$, and there exists $b \in \mathsf{A}$ such that $ba = ee$. Then $bae = eee = e \neq 0$ and so $ae \neq 0$. Since A is $*$-semisimple, $(ae)^+\, ae \equiv ea^+ae \neq 0$. □

Corollary 6.4.9. *If the $*$-algebra* A *is atomic, then the relation "$\approx$" is an equivalence relation in* M(A).

Proof. Reflexivity and symmetry are obvious. We prove the transitivity. Suppose that $e_1 \approx e_2$ and $e_2 \approx e_3$ for $e_1, e_2, e_3 \in \mathsf{M}(\mathsf{A})$. Then there are $a, b \in \mathsf{A}$ such that $e_1ae_2 \neq 0$ and $e_2be_3 \neq 0$. Since A is $*$-semisimple, $e_1ae_2(e_1ae_2)^+ = e_1ae_2a^+e_1 = g_{e_1}(ae_2a^+)\, e_1 \neq 0$ and so $g_{e_1}(ae_2a^+) \neq 0$. Similarly, $g_{e_2}(be_3b^+) \neq 0$. Thus $e_1ae_2be_3(e_1ae_2be_3)^+ = e_1ae_2be_3b^+e_2a^+e_1 = g_{e_2}(be_3b^+)\, e_1ae_2a^+e_1 = g_{e_2}(be_3b^+)\, g_{e_1}(ae_2a^+)\, e_1 \neq 0$. Therefore, $e_1ae_2be_3 \neq 0$, so that $e_1 \approx e_3$. □

Proof of Theorem 6.4.2. First suppose A is atomic. The equivalence relation "$\approx$" partitions the non-empty set M(A) (by Corollaries 6.4.8. and 6.4.9) into equivalence classes. Let $\{e_i : i \in I\}$ be an indexed subset of M(A) obtained by choosing precisely one element from each of these equivalence classes. Setting $\mathcal{D}_i := \mathcal{D}_{e_i}$ for $i \in I$, the family $\{\mathcal{D}_i : i \in I\}$ of unitary spaces has the desired properties. Define $\pi(x)\, (\varphi_i) := (\varrho_{e_i}(x)\, \varphi_i)$ for $x \in \mathsf{A}$ and $(\varphi_i) \in \mathcal{D}_I$. Since each ϱ_{e_i} is a $*$-homomorphism of A into $\mathcal{L}^+(\mathcal{D}_i)$ by Lemma 6.4.7, (i), π is a $*$-homomorphism of A into $\mathcal{L}^+(\mathcal{D}_I)$. We verify that π is injective. We suppose $a \in \mathsf{A}$, $a \neq 0$. By Corollary 6.4.8. there is a projection $e \in \mathsf{M}(\mathsf{A})$ such that $g_e(a^+a) \equiv \|ae\|_e^2 \neq 0$. Putting $b := \|ae\|_e^{-1}ae$, we have $\|beb\|_e = 1$ and hence $beb^+ \in \mathsf{M}(\mathsf{A})$ by Lemma 6.4.7, (ii). There exists $i \in I$ such that $beb^+ \approx e_i$. Lemma 6.4.7, (ii), shows that $\varrho_{e_i}(beb^+) \neq 0$. Since $\varrho_{e_i}(beb^+) = \|ae\|_e^{-2}\, \varrho_{e_i}(a)\, \varrho_{e_i}(ea^+)$, this gives $\varrho_{e_i}(a) \neq 0$, so that $\pi(a) \neq 0$. Therefore, π is a $*$-isomorphism of A on $\mathcal{A} := \pi(\mathsf{A})$. It remains to show that $\mathsf{M}(\mathcal{A}) = \mathsf{M}(\mathcal{L}^+(\mathcal{D}_i : i \in I))$. Let $f \in \mathsf{M}(\mathsf{A})$. We can find an index $i \in I$ such that $f \approx e_i$. By Lemma 6.4.7, (ii), $\varrho_{e_i}(f) = \varphi_i \otimes \varphi_i$ for some unit vector $\varphi_i \in \mathcal{D}_i$ and $\varrho_{e_{i'}}(f) = 0$ if $i' \in I$, $i \neq i'$; so $\pi(f) = \varphi_i \otimes \varphi_i$. The first assertion in Lemma 6.4.7, (ii), implies that each rank one projection $\varphi_i \otimes \varphi_i$ with $\varphi_i \in \mathcal{D}_i$ is of the form $\pi(f)$ for some $f \in \mathsf{M}(\mathsf{A})$. Combined with Lemma 6.3.5, (i), the preceding shows that $\mathsf{M}(\pi(\mathsf{A})) \equiv \mathsf{M}(\mathcal{A}) = \pi(\mathsf{M}(\mathsf{A})) = \mathsf{M}(\mathcal{L}^+(\mathcal{D}_i : i \in I))$.

Conversely, suppose $\mathcal{A}$ is a $*$-subalgebra of some $\mathcal{L}^+(\mathcal{D}_i : i \in I)$ satisfying $\mathsf{M}(\mathcal{A}) = \mathsf{M}(\mathcal{L}^+(\mathcal{D}_i : i \in I))$. Suppose J is a non-zero left ideal of $\mathcal{A}$. Let $a \in \mathsf{J}$, $a \neq 0$. There exists a vector $\varphi_i \in \mathcal{D}_i$ such that $\|a^+\varphi_i\| = 1$. Since $\varphi_i \otimes \varphi_i \in \mathcal{A}$ by Lemma 6.3.5, (ii), $e := a^+(\varphi_i \otimes \varphi_i)\, a \in \mathsf{J}$. Clearly, $e = a^+\varphi_i \otimes a^+\varphi_i$ is a rank one projection and $e\mathcal{A}e = \mathbb{C} \cdot e$, so $\mathcal{A}e$ is a minimal left ideal of $\mathcal{A}$ by Lemma 6.4.5, (ii). Since $e \in \mathsf{J}$, $\mathcal{A}e \subseteq \mathsf{J}$. This proves that $\mathcal{A}$ is atomic. Hence each image of $\mathcal{A}$ under $*$-isomorphisms is atomic.

The uniqueness assertion follows immediately from Theorem 6.3.6. □

Notes

6.1. Operator ideals associated with bounded operator ideals have been systematically investigated by Timmermann [1], [2]. The ideals $\mathbb{B}(\mathcal{D})$, $\mathbb{B}_1(\mathcal{D})$ and $\mathcal{V}(\mathcal{D}_{\mathcal{A}})$ occuring in this monograph are examples of such operator ideals. Propositions 6.1.3 and 6.1.5 and Corollary 6.1.7 can be found in Kürsten [2], [4]. However, parts of these assertions reformulate known results from the theory of locally convex spaces; see e.g. Remark 5 in 6.1. Further, the algebraic part of Proposition 6.1.5 also follows from Theorem 2.1 in Collins/Ruess [1]. Propositions 6.1.10 and 6.1.12 appeared in Kürsten [2], [4].

6.2. The main results in this section were proved by Schmüdgen [20]. We have given a somewhat more general version of Theorem 6.2.4 than in Schmüdgen [20] by incorporating some modifications of the construction given by Kürsten [2]. For a special class of domains (covered by our Theorem 6.2.4) a realization of the generalized Calkin algebra was obtained independently and simultaneously by Löffler/Timmermann [1].

6.3. Corollary 6.3.3 is due to Kröger [1]. Our proof follows Uhlmann [3]. Theorem 6.3.6 and its two corollaries were obtained by Uhlmann [3]. In contrast to Uhlmann [3], our proof of this theorem avoids the use of the Wigner theorem.

6.4. The main ideas and results occuring in this section are due to Uhlmann [3]. We have chosen a slightly different approach which is essentially based on Proposition 2.1.12. Note also that our terminology differs from the one in Uhlmann [3].

Additional References:

6.1. Junek [1].
6.2. Kürsten/Milde [1].
6.3. Inoue/Ota [1].

7. Commutants

In a broad sense this chapter deals with the commutativity of both single unbounded operators and families of operators. In particular, various notions of commutants of O*-algebras are investigated. Section 7.1 contains some general results on strongly commuting self-adjoint operators. Apart from being of interest in itself, they are used in Sections 7.3, 9.1 and 10.2. In Section 7.2 we define six (in general different) concepts of unbounded and bounded commutants for an O*-algebra, and we discuss the relations between them. The self-adjointness of the O*-algebra implies that the weak and the strong unbounded commutants coincide, but it is not sufficient to ensure a close connection between unbounded and bounded commutants. For this further restrictions are needed. Such a class of O*-algebras which we call strictly self-adjoint O*-algebras is considered in Section 7.3.

Commutativity for unbounded operators is a rather delicate matter. As a consequence, the attempt to generalize results which are based on commutation properties from the bounded to the unbounded case often meets serious difficulties. We illustrate this for the bicommutant theorem by a simple example: Let $\mathcal{A}$ be the self-adjoint O*-algebra generated by the multiplication operator by the independent variable on the Hilbert space $\mathcal{H} = L^2(\mathbb{R})$. Then the strong-operator topology of $\mathcal{A}$ coincides with the finest locally convex topology on $\mathcal{A}$. (See Remark 4 in 4.5.) Hence $\mathcal{A}$ is strong-operator closed in $\mathfrak{L}(\mathcal{D}_{\mathcal{A}}, \mathcal{H})$. But the bicommutant of $\mathcal{A}$ (in any reasonable definition) certainly contains all multiplication operators by L^∞-functions, so $\mathcal{A}$ is different from its bicommutant.

In Section 7.4 we study a class of subspaces of the space $\mathcal{L}(\mathcal{D}_{\mathcal{A}}, \mathcal{D}_{\mathcal{B}}^+)$ which are built around a $*$-algebra of bounded operators (in a way defined therein), and we prove some results which can be interpreted as generalizations of the bicommutant theorem and Kaplansky density theorem, respectively.

7.1. Some Results on Strongly Commuting Self-Adjoint Operators

In this section $\mathcal{D}$ is a dense linear subspace of the Hilbert space $\mathcal{H}$.

Lemma 7.1.1. *Suppose a_1 and a_2 are symmetric operators in $\mathcal{L}^+(\mathcal{D})$ and $a = a_1 + ia_2$. The operator a is formally normal if and only if $a_1a_2 = a_2a_1$.*

Proof. That a is formally normal means that $\mathcal{D}(a) \subseteq \mathcal{D}(a^*)$ and $\|a\varphi\| = \|a^*\varphi\|$ for $\varphi \in \mathcal{D}(a)$. Thus the assertion follows at once from the identity $\|a\varphi\|^2 - \|a^*\varphi\|^2 = 2i\langle (a_1a_2 - a_2a_1)\varphi, \varphi\rangle$, $\varphi \in \mathcal{D}$. □

Lemma 7.1.2. *If $a \in \mathcal{L}^+(\mathcal{D})$ and aa^+ is essentially self-adjoint, then $\overline{a^+} = a^*$.*

Proof. It is clear that $\overline{a^+} \subseteq a^*$. Therefore, it is sufficient to show that each element in the graph of a^* which is orthogonal to the graph of a^+ is zero. Suppose that $(\zeta, a^*\zeta)$ is orthogonal in $\mathcal{H} \oplus \mathcal{H}$ to the graph of a^+. Then $\langle \zeta, \varphi \rangle + \langle a^*\zeta, a^+\varphi \rangle = 0$ for all $\varphi \in \mathcal{D}$. Since $a^+\varphi \in \mathcal{D}$ for $\varphi \in \mathcal{D}$, this gives $\langle \zeta, (I + aa^+)\varphi \rangle = 0$ for $\varphi \in \mathcal{D}$. Since aa^+ is essentially self-adjoint, $(I + aa^+)\mathcal{D}$ is dense in $\mathcal{H}$, so that $\zeta = 0$. □

Recall that the strong commutativity of two normal (in particular, self-adjoint) operators means by definition that the spectral projections of both operators mutually commute.

Proposition 7.1.3. *Suppose that a_1 and a_2 are symmetric operators in $\mathcal{L}^+(\mathcal{D})$ such that $a_1 a_2 = a_2 a_1$. Set $a = a_1 + ia_2$.*

(i) *The operator $\bar{a}$ is normal if and only if $\overline{a^+} = a^*$. If this is true, then $\overline{a_1}$ and $\overline{a_2}$ are strongly commuting self-adjoint operators.*

(ii) *If the operator a^+a is essentially self-adjoint, then $\bar{a}$ is normal and $\overline{a^+a} = \bar{a}\bar{a}^* = \bar{a}^*\bar{a}$.*

Proof. (i): By Lemma 7.1.1, a is formally normal. Hence $\|a\varphi\| = \|a^+\varphi\|$ for $\varphi \in \mathcal{D}$. This implies that $\mathcal{D}(\bar{a}) = \mathcal{D}(\overline{a^+})$. Further, it follows that $\bar{a}$ is formally normal.

First suppose that $\bar{a}$ is normal. Then $\mathcal{D}(\bar{a}) = \mathcal{D}((\bar{a})^*) = \mathcal{D}(a^*)$. Since $\overline{a^+} \subseteq a^*$ and $\mathcal{D}(\bar{a}) = \mathcal{D}(\overline{a^+})$ as just shown, we obtain that $\overline{a^+} = a^*$. Further, since $\bar{a}$ is normal, $A_1 := \frac{1}{2}\overline{(\bar{a} + (\bar{a})^*)}$ and $A_2 := \frac{1}{2i}\overline{(\bar{a} - (\bar{a})^*)}$ are strongly commuting self-adjoint operators (Dunford/Schwartz [2], XII, 9.11). Fix $l \in \{1, 2\}$. We show that $\overline{a_l} = A_l$. Obviously, $a_l \subseteq A_l \upharpoonright \mathcal{D}$ and so $\overline{a_l} \subseteq A_l$. From the inequality $\|a\varphi\|^2 = \|a_1\varphi\|^2 + \|a_2\varphi\|^2 \geqq \|a_l\varphi\|^2$ for $\varphi \in \mathcal{D}$ it follows that $\mathcal{D}(\bar{a}) \subseteq \mathcal{D}(\overline{a_l})$. By definition $\mathcal{D}(\bar{a}) = \mathcal{D}((\bar{a})^*)$ is a core for A_l, so that $\overline{A_l \upharpoonright \mathcal{D}(\bar{a})} = A_l$. Combined with the preceding, this yields $\overline{A_l \upharpoonright \mathcal{D}(\overline{a_l})} = A_l$. Since $\overline{a_l} \subseteq A_l$, the latter gives $A_l = \overline{a_l}$.

Conversely, suppose that $\overline{a^+} = a^*$. As noted above, $\mathcal{D}(\bar{a}) = \mathcal{D}(\overline{a^+})$. Therefore, $\mathcal{D}(\bar{a}) = \mathcal{D}((\bar{a})^*)$. Because $\bar{a}$ is formally normal, this means that $\bar{a}$ is normal.

(ii): Suppose that a^+a is essentially self-adjoint. Since $a_1a_2 = a_2a_1$ by assumption, $a^+a = aa^+$ and a is formally normal by Lemma 7.1.1. By Lemma 7.1.2 and (i), $\bar{a}$ is normal; so $\bar{a}^*\bar{a} = \bar{a}\bar{a}^*$. Since $\bar{a}^*\bar{a}$ is a symmetric extension of the self-adjoint operator $\overline{a^+a}$, $\overline{a^+a} = \bar{a}^*\bar{a}$. □

Remark 1. By applying the adjoint operation it follows that the equality $\overline{a^+} = a^*$ is equivalent to $\bar{a} = (a^+)^*$ for any $a \in \mathcal{L}^+(\mathcal{D})$. Therefore, by Lemma 7.1.1 and Proposition 7.1.3, (i), for arbitrary $a \in \mathcal{L}^+(\mathcal{D})$ the operator $\bar{a}$ is normal if and only if $\overline{a^+}$ is.

For a and c in $\mathcal{L}^+(\mathcal{D})$, the operator c is said to be *a-bounded* if there exists a constant $\lambda > 0$ such that $\|c\varphi\| \leqq \lambda(\|\varphi\| + \|a\varphi\|)$ for all $\varphi \in \mathcal{D}$.

Proposition 7.1.4. *Let a be an operator in $\mathcal{L}^+(\mathcal{D})$ such that $\|a \cdot\| \geqq \|\cdot\|$. Suppose that the operator $\bar{a}$ is normal. Let c_1 and c_2 be a-bounded symmetric operators in $\mathcal{L}^+(\mathcal{D})$ such that $ac_1 = c_1a$, $ac_2 = c_2a$ and $c_1c_2 = c_2c_1$. Then $\overline{c_1}$ and $\overline{c_2}$ are strongly commuting self-adjoint operators.*

The key step in the proof of Proposition 7.1.4 is contained in the following lemma.

Lemma 7.1.5. *Let a be as in Proposition 7.1.4 and let $l \in \{1, 2\}$. Suppose that c_l is an a-bounded symmetric operator in $\mathcal{L}^+(\mathcal{D})$ such that $ac_l = c_l a$. Then $\overline{c_l}$ is a self-adjoint operator which commutes strongly with the normal operator $\bar{a}$.*

Proof. Since c_l is a-bounded and $\|\cdot\| \leqq \|a\cdot\|$, $\|c_l\cdot\| \leqq \lambda\|a\cdot\|$ for some $\lambda > 0$. Hence there exists an operator $x_l \in \mathbb{B}(\mathcal{H})$ such that $c_l = x_l a$. Then $c_l^* = a^* x_l^*$. Let $\bar{a} = u|\bar{a}|$ be the polar decomposition of $\bar{a}$. Set $y_l := u^* x_l^*$. From $a^* = |\bar{a}|\, u^*$ (by the properties of the polar decomposition) and $c_l^* = a^* x_l^*$ we get $c_l^* = |\bar{a}|\, y_l$. Since $c_l = c_l^+$ by assumption, the latter gives that $c_l\varphi = |\bar{a}|\, y_l\varphi$ for $\varphi \in \mathcal{D}$. Therefore, by $ac_l = c_l a$, we have

$$c_l a\varphi = |\bar{a}|\, y_l a\varphi = ac_l\varphi = a|\bar{a}|\, y_l\varphi = |\bar{a}|\, \bar{a} y_l\varphi \quad \text{for all } \varphi \in \mathcal{D}, \tag{1}$$

where the relation $a|\bar{a}| \subseteq |\bar{a}|\, \bar{a}$ follows from the normality of the operator $\bar{a}$. Since $\|a\cdot\| \geqq \|\cdot\|$ by assumption, $|\bar{a}| \geqq I$. Thus (1) yields $y_l a\varphi = \bar{a} y_l\varphi$ for $\varphi \in \mathcal{D}$. This implies that $y_l\bar{a} \subseteq \bar{a} y_l$.

Next we prove that $y_l|\bar{a}| \subseteq |\bar{a}|\, y_l$. Since $\|a\cdot\| \geqq \|\cdot\|$, $\bar{a}\mathcal{D}(\bar{a})$ is closed in $\mathcal{H}$. We show that $\bar{a}\mathcal{D}(\bar{a}) = \mathcal{H}$. Indeed, if $\psi \in \mathcal{H}$ is orthogonal to $\bar{a}\mathcal{D}(\bar{a})$, then $\bar{a}^*\psi = 0$ and hence $|\bar{a}|^2\, \psi = \bar{a}^*\bar{a}\psi = \bar{a}\bar{a}^*\psi = 0$. By $|\bar{a}| \geqq I$, this gives $\psi = 0$, so that $\bar{a}\mathcal{D}(\bar{a}) = \mathcal{H}$. Since $\|a\cdot\| \geqq \|\cdot\|$ and $\bar{a}$ is normal, it follows easily that $\bar{a}^{-1}$ is a bounded normal operator on $\mathcal{H}$. From $y_l\bar{a} \subseteq \bar{a} y_l$ we obtain $y_l\bar{a}^{-1} = \bar{a}^{-1}y_l$. By Fuglede's theorem (see e.g. Douglas [1], 4.76), y_l commutes with the operator $(\bar{a}^{-1})^* = (\bar{a}^*)^{-1}$. This in turn yields $y_l\bar{a}^* \subseteq \bar{a}^* y_l$; so $y_l|\bar{a}|^2 = y_l\bar{a}^*\bar{a} \subseteq \bar{a}^*\bar{a} y_l = |\bar{a}|^2\, y_l$ and hence $y_l|\bar{a}| \subseteq |\bar{a}|\, y_l$.

We prove that $y_l = y_l^*$. If $\varphi \in \mathcal{D}$, then, by the preceding, $c_l\varphi = |\bar{a}|\, y_l\varphi = y_l|\bar{a}|\, \varphi = x_l a\varphi = x_l u|\bar{a}|\, \varphi = y_l^*|\bar{a}|\, \varphi$. Since $|\bar{a}| \geqq I$ and $\mathcal{D}$ is a core for $\bar{a}$ and hence for $|\bar{a}|$, $|\bar{a}|\, \mathcal{D}$ is dense in $\mathcal{H}$, and so $y_l = y_l^*$.

We show that $\overline{c_l}$ is self-adjoint. We let $z \in \mathbb{C} \setminus \mathbb{R}$ and suppose that $\varphi \in \mathcal{D}(c_l^*)$ satisfies $c_l^*\varphi = z\varphi$. Since $c_l^* = |\bar{a}|\, y_l$ and $|\bar{a}|^{-1} \in \mathbb{B}(\mathcal{H})$, we have $y_l\varphi = z|\bar{a}|^{-1}\, \varphi$, so that $\langle y_l\varphi, \varphi\rangle = z\, \||\bar{a}|^{-1/2}\varphi\|^2$. By $y_l = y_l^*$, $\langle y_l\varphi, \varphi\rangle$ is real. Consequently, $|\bar{a}|^{-1/2}\varphi = 0$ and so $\varphi = 0$. This shows c_l has zero deficiency indices. Hence $\overline{c_l}$ is self-adjoint. Since $\overline{c_l} = c_l^*$ as just shown, we have $\overline{c_l} = |\bar{a}|\, y_l$. Using again the normality of $\bar{a}$ and the fact that y_l commutes with $\bar{a}^{-1}$, we conclude that $\bar{a}^{-1}\overline{c_l} = \bar{a}^{-1}|\bar{a}|\, y_l \subseteq |\bar{a}|\, \bar{a}^{-1}y_l = |\bar{a}|\, y_l\bar{a}^{-1} = \overline{c_l}\bar{a}^{-1}$. By Lemma 1.6.2, $\overline{c_l}$ and $\bar{a}$ strongly commute. □

Remark 2. Suppose in Lemma 7.1.5 that in addition $\bar{a}$ is self-adjoint and $\bar{a} \geqq I$. Then the assertion that $\overline{c_l}$ is self-adjoint follows also from the commutator theorem (see e.g. Reed/Simon [2], Theorem X.37).

Proof of Proposition 7.1.4. Because of Lemma 7.1.5, the operators $\overline{c_1}$ and $\overline{c_2}$ are self-adjoint. It remains to show that these operators commute strongly. In order to prove this, we use some notation and some facts from the proof of Lemma 7.1.5. Since $y_1\, |\bar{a}| \subseteq |\bar{a}|\, y_1$ it follows that $c_1c_2\varphi = |\bar{a}|\, y_1\, |\bar{a}|\, y_2\varphi = |\bar{a}|^2\, y_1y_2\varphi$ for $\varphi \in \mathcal{D}$. Similarly, $c_2c_1\varphi = |\bar{a}|^2\, y_2y_1\varphi$ for $\varphi \in \mathcal{D}$. Since $c_1c_2 = c_2c_1$ by assumption and $|\bar{a}| \geqq I$, we obtain $y_1y_2 = y_2y_1$ on $\mathcal{D}$ and hence on $\mathcal{H}$. Therefore, $y_1\overline{c_2} = y_1\, |\bar{a}|\, y_2 \subseteq |\bar{a}|\, y_1y_2 = |\bar{a}|\, y_2y_1 = \overline{c_2}y_1$. Since $\overline{c_2}$ is a self-adjoint operator, y_1 commutes with $(\overline{c_2} + \mathrm{i})^{-1}$. From Lemma 7.1.5, $\overline{c_2}$ and $\bar{a}$ and hence $\overline{c_2}$ and $|\bar{a}|$ strongly commute; so $(\overline{c_2} + \mathrm{i})^{-1}\, |\bar{a}| \subseteq |\bar{a}|\, (\overline{c_2} + \mathrm{i})^{-1}$ by Lemma 1.6.2. From these two facts we obtain $(\overline{c_2} + \mathrm{i})^{-1}\, \overline{c_1} = (\overline{c_2} + \mathrm{i})^{-1}\, |\bar{a}|\, y_1 \subseteq |\bar{a}|\, (\overline{c_2} + \mathrm{i})^{-1}\, y_1 = |\bar{a}|\, y_1(\overline{c_2} + \mathrm{i})^{-1} = \overline{c_1}(\overline{c_2} + \mathrm{i})^{-1}$. By Lemma 1.6.2, $\overline{c_1}$ and $\overline{c_2}$ strongly commute. □

The next proposition is of similar nature as Proposition 7.1.4. The main difference is that the assumption $ac_2 = c_2a$ of Proposition 7.1.4 is omitted and a stronger assumption concerning a is required.

Proposition 7.1.6. *Let a be a formally normal operator in $\mathcal{L}^+(\mathcal{D})$ such that $\|\cdot\| \leqq \|a\cdot\|$. Suppose that the operator a^+a is essentially self-adjoint. Let c_1 and c_2 be a-bounded operators in $\mathcal{L}^+(\mathcal{D})$ satisfying $ac_1 = c_1a$ and $c_1c_2 = c_2c_1$. Suppose that c_1 is a symmetric operator. Then $\bar{c}_1$ is a self-adjoint operator and $(\bar{c}_1 + \mathrm{i})^{-1}\,\bar{c}_2 \subseteq \bar{c}_2(\bar{c}_1 + \mathrm{i})^{-1}$. In particular, if in addition $\bar{c}_2$ is self-adjoint, then $\bar{c}_1$ and $\bar{c}_2$ strongly commute.*

Proof. By Proposition 7.1.3, (ii), the above assumptions concerning a imply that $\bar{a}$ is a normal operator. Therefore, by Lemma 7.1.5, $\bar{c}_1$ is self-adjoint, and $\bar{c}_1$ and $\bar{a}$ strongly commute. Since c_2 is a-bounded and $\|\cdot\| \leqq \|a\cdot\|$, there is an operator $x_2 \in \mathbb{B}(\mathcal{H})$ such that $c_2 = x_2a$.

Suppose that $\varphi \in \mathcal{D}(\bar{a}^*\bar{a})$. Since a^+a is essentially self-adjoint, $\overline{a^+a} = \bar{a}^*\bar{a}$ by Proposition 7.1.3, (ii). Hence there exists a sequence $(\varphi_n : n \in \mathbb{N})$ in $\mathcal{D}$ such that $\overline{a^+a}\varphi = \lim a^+a\varphi_n$ and $\varphi = \lim \varphi_n$ in $\mathcal{H}$. By the assumptions $c_1c_2 = c_2c_1$ and $ac_1 = c_1a$, we have for $n \in \mathbb{N}$

$$c_2(c_1 + \mathrm{i})\,\varphi_n = x_2a(c_1 + \mathrm{i})\,\varphi_n = x_2(c_1 + \mathrm{i})\,a\varphi_n = (c_1 + \mathrm{i})\,c_2\varphi_n = (c_1 + \mathrm{i})\,x_2a\varphi_n\,. \tag{2}$$

Since c_1 is a-bounded and $\|\cdot\| \leqq \|a\cdot\| \equiv \|a^+\cdot\|$, we have $\|a\cdot\| \leqq \|a^+a\cdot\|$ and $\|(c_1 + \mathrm{i})\,a\cdot\| \leqq \lambda\,\|a^+a\cdot\|$ with some constant $\lambda > 0$. From this it follows that the sequences $(a\varphi_n)$, $\big((c_1 + \mathrm{i})\,a\varphi_n\big)$, $\big(x_2(c_1 + \mathrm{i})\,a\varphi_n\big)$ and $(x_2a\varphi_n)$ converge in $\mathcal{H}$. Therefore, letting $n \to \infty$ in (2), we obtain $x_2(\bar{c}_1 + \mathrm{i})\,\bar{a}\varphi = (\bar{c}_1 + \mathrm{i})\,x_2\bar{a}\varphi$ for $\varphi \in \mathcal{D}(\bar{a}^*\bar{a})$. Since $\bar{a}^*\bar{a}\,\mathcal{D}(\bar{a}^*\bar{a}) \equiv |\bar{a}|^2\,\mathcal{D}(|\bar{a}|^2) = \mathcal{H}$ because of $|\bar{a}| \geqq I$, we have $\bar{a}\mathcal{D}(\bar{a}^*\bar{a}) = \mathcal{D}(\bar{a}^*)$. From this fact and the preceding we get $x_2(\bar{c}_1 + \mathrm{i})\,\psi = (\bar{c}_1 + \mathrm{i})\,x_2\psi$ for all $\psi \in \mathcal{D}(\bar{a}^*)$. As noted above, c_1 is essentially self-adjoint, so that $(\bar{c}_1 + \mathrm{i})\,\mathcal{D}(\bar{a}^*)$ $\big(\supseteq (c_1 + \mathrm{i})\,\mathcal{D}\big)$ is dense in $\mathcal{H}$ and the latter gives that $(\bar{c}_1 + \mathrm{i})^{-1}\,x_2 = x_2(\bar{c}_1 + \mathrm{i})^{-1}$. Hence $(\bar{c}_1 + \mathrm{i})^{-1}\,c_2 = (\bar{c}_1 + \mathrm{i})^{-1}\,x_2a$ $= x_2(\bar{c}_1 + \mathrm{i})^{-1}\,a \subseteq x_2\bar{a}(\bar{c}_1 + \mathrm{i})^{-1} \subseteq \bar{c}_2(\bar{c}_1 + \mathrm{i})^{-1}$, where we used the strong commutativity of $\bar{c}_1$ and $\bar{a}$. Therefore, $(\bar{c}_1 + \mathrm{i})^{-1}\,\bar{c}_2 \subseteq \bar{c}_2(\bar{c}_1 + \mathrm{i})^{-1}$. If in addition $\bar{c}_2$ is self-adjoint, then it follows from the latter and Lemma 1.6.2 that $\bar{c}_1$ and $\bar{c}_2$ strongly commute. □

7.2. Unbounded and Bounded Commutants of O*-Algebras

First we recall the notion of a self-adjoint O*-family which occured already in Section 5.1. Let $\mathcal{A}$ be an O*-family in the Hilbert space $\mathcal{H}$. Set $\mathcal{D}^*(\mathcal{A}) := \bigcap_{a \in \mathcal{A}} \mathcal{D}(a^*)$. Since $\mathcal{A}$ is an O*-family, $\mathcal{D}^*(\mathcal{A}) \supseteq \mathcal{D}(\mathcal{A})$, so that $\mathcal{D}^*(\mathcal{A})$ is dense in $\mathcal{H}$. Thus $\mathcal{A}^* := \{a^* \upharpoonright \mathcal{D}^*(\mathcal{A}) : a \in \mathcal{A}\}$ is an O-family with domain $\mathcal{D}(\mathcal{A}^*) := \mathcal{D}^*(\mathcal{A})$ in the Hilbert space $\mathcal{H}$ which is called the *adjoint O-family to* $\mathcal{A}$. It is easily seen that $\mathcal{A}^* \upharpoonright \mathcal{D}(\mathcal{A}) = \mathcal{A}$ and that $\mathcal{A}^*$ is a closed O-family. According to Definition 5.1.5, the O*-family $\mathcal{A}$ is said to be self-adjoint if $\mathcal{D}(\mathcal{A}) = \mathcal{D}^*(\mathcal{A})$ or equivalently if $\mathcal{A}$ is equal to its adjoint O-family $\mathcal{A}^*$.

Remark 1. This definition is quite similar to the definition of a self-adjoint operator. But is should be noted that the self-adjointness (and also the closedness) of an O*-family is a "collective notion" which cannot be reduced to the self-adjointness of single operators in general.

We now define unbounded commutants of O*-algebras.

Definition 7.2.1. Suppose $\mathcal{A}$ is an O*-algebra in the Hilbert space $\mathcal{H}$. The set $\mathcal{A}_f^c := \{x \in \mathcal{L}(\mathcal{D}_\mathcal{A}, \mathcal{D}_\mathcal{A}^+) : xa = a \circ x \text{ for all } a \in \mathcal{A}\}$ is called the *form commutant* of $\mathcal{A}$. The *weak unbounded commutant* of $\mathcal{A}$ is the set $\mathcal{A}_w^c := \{x \in \mathfrak{L}(\mathcal{D}_\mathcal{A}, \mathcal{H}) : \langle xa\varphi, \psi\rangle = \langle x\varphi, a^+\psi\rangle$ for all $a \in \mathcal{A}$ and $\varphi, \psi \in \mathcal{D}(\mathcal{A})\}$, and the *strong unbounded commutant* of $\mathcal{A}$ is the set $\mathcal{A}_s^c := \{x \in \mathfrak{L}(\mathcal{D}_\mathcal{A}, \mathcal{H}) : x(\mathcal{D}(\mathcal{A})) \subseteq \mathcal{D}(\mathcal{A})$ and $xa\varphi = ax\varphi$ for all $a \in \mathcal{A}$ and $\varphi \in \mathcal{D}(\mathcal{A})\}$.

Remark 2. Note that all expressions in the preceding definition are well-defined, since $\mathcal{A}$ is an O*-algebra. If one tries to generalize these definitions to general O*-families, additional domain problems occur, since then $a\varphi$ is not in $\mathcal{D}(\mathcal{A})$ in general. However, these problems do not appear if we take only bounded operators for x; see Definition 7.2.7 below.

The name "form commutant" stems from assertion (i) in the following proposition.

Proposition 7.2.2. *Suppose $\mathcal{A}$ is an O*-algebra in the Hilbert space $\mathcal{H}$.*

(i) *The bijection $x \to \mathfrak{c}_x$ of $\mathcal{L}(\mathcal{D}_\mathcal{A}, \mathcal{D}_\mathcal{A}^+)$ onto $\mathcal{B}(\mathcal{D}_\mathcal{A}, \mathcal{D}_\mathcal{A})$ maps $\mathcal{A}_f^c$ onto the set of all continuous sesquilinear forms $\mathfrak{c}$ on $\mathcal{D}_\mathcal{A} \times \mathcal{D}_\mathcal{A}$ which satisfy $\mathfrak{c}(a\varphi, \psi) = \mathfrak{c}(\varphi, a^+\psi)$ for all $a \in \mathcal{A}$ and $\varphi, \psi \in \mathcal{D}(\mathcal{A})$.*

(ii) $\mathcal{A}_w^c = \{x \in \mathfrak{L}(\mathcal{D}_\mathcal{A}, \mathcal{H}) : x(\mathcal{D}(\mathcal{A})) \subseteq \mathcal{D}^*(\mathcal{A})$ *and* $xa\varphi = (a^+)^* x\varphi$ *for* $a \in \mathcal{A}$ *and* $\varphi \in \mathcal{D}(\mathcal{A})\}$.

(iii) $\mathcal{A}_w^c = \mathcal{A}_f^c \cap \mathfrak{L}(\mathcal{D}_\mathcal{A}, \mathcal{H})$ *and* $\mathcal{A}_s^c = \mathcal{A}_w^c \cap L(\mathcal{D}(\mathcal{A}))$.

(iv) *If $\mathcal{A}$ is self-adjoint, then* $\mathcal{A}_w^c = \mathcal{A}_s^c$.

Proof. (i): If $x \in \mathcal{A}_f^c$, then $\mathfrak{c}_x(a\varphi, \psi) = \langle xa\varphi, \psi\rangle = \langle a \circ x\varphi, \psi\rangle = \langle x\varphi, a^+\psi\rangle = \mathfrak{c}_x(\varphi, a^+\psi)$ for $a \in \mathcal{A}$ and $\varphi, \psi \in \mathcal{D}(\mathcal{A})$. Conversely, suppose that $\mathfrak{c} \in \mathcal{B}(\mathcal{D}_\mathcal{A}, \mathcal{D}_\mathcal{A})$ has the property stated above. By Lemma 1.2.1, $\mathfrak{c} = \mathfrak{c}_x$ for some $x \in \mathcal{L}(\mathcal{D}_\mathcal{A}, \mathcal{D}_\mathcal{A}^+)$. If $a \in \mathcal{A}$ and $\varphi, \psi \in \mathcal{D}(\mathcal{A})$, then $\langle xa\varphi, \psi\rangle = \mathfrak{c}(a\varphi, \psi) = \mathfrak{c}(\varphi, a^+\psi) = \langle x\varphi, a^+\psi\rangle = \langle a \circ x\varphi, \psi\rangle$ by 3.2/(6) and hence $xa\varphi = a \circ x\varphi$, so that $x \in \mathcal{A}_f^c$.

(ii): Suppose $x \in \mathcal{A}_w^c$. Let $a \in \mathcal{A}$ and $\varphi \in \mathcal{D}(\mathcal{A})$. From $\langle xa\varphi, \psi\rangle = \langle x\varphi, a^+\psi\rangle$ for all $\psi \in \mathcal{D}(\mathcal{A})$ we conclude that $x\varphi \in \mathcal{D}((a^+)^*)$ and $xa\varphi = (a^+)^* x\varphi$. Hence we get $x\varphi \in \bigcap_{a \in \mathcal{A}} \mathcal{D}((a^+)^*) = \mathcal{D}^*(\mathcal{A})$; so $x(\mathcal{D}(\mathcal{A})) \subseteq \mathcal{D}^*(\mathcal{A})$. The converse direction is straight-forward.

(iii): Using again 3.2/(6), the equality $\mathcal{A}_w^c = \mathcal{A}_f^c \cap \mathfrak{L}(\mathcal{D}_\mathcal{A}, \mathcal{H})$ follows directly from the definitions. $\mathcal{A}_s^c = \mathcal{A}_w^c \cap L(\mathcal{D}(\mathcal{A}))$ follows from (ii), since $(a^+)^* \restriction \mathcal{D}(\mathcal{A}) = a$ for $a \in \mathcal{A}$.

(iv) is an immediate consequence of (ii). $\square$

Proposition 7.2.3. *Let $\mathcal{A}$ be an O*-algebra in the Hilbert space $\mathcal{H}$.*

(i) $\mathcal{A}_f^c$ *is a weak-operator closed $*$-vector subspace of* $\mathcal{L}(\mathcal{D}_\mathcal{A}, \mathcal{D}_\mathcal{A}^+)$.

(ii) $\mathcal{A}_w^c$ *is a weak-operator closed linear subspace of* $\mathfrak{L}(\mathcal{D}_\mathcal{A}, \mathcal{H})$. $\mathcal{A}_w^c$ *is contained in* $\mathfrak{L}(\mathcal{D}_\mathcal{A}, \mathcal{D}_{\mathcal{A}^*})$.

(iii) $\mathcal{A}_s^c$ *is a subalgebra of* $\mathfrak{L}(\mathcal{D}_\mathcal{A})$.

(iv) $\hat{\mathcal{A}}_w^c \restriction \mathcal{D}(\mathcal{A}) = \mathcal{A}_w^c$ *and* $\hat{\mathcal{A}}_s^c \restriction \mathcal{D}(\mathcal{A}) \supseteq \mathcal{A}_s^c$.

Proof. (i): We verify the invariance of $\mathcal{A}_f^c$ under the involution. The other statements in (i) are obvious. Let $x \in \mathcal{A}_f^c$. Then $\langle x^+a\varphi, \psi\rangle = \langle a\varphi, x\psi\rangle = \langle \varphi, a^+ \circ x\psi\rangle = \langle \varphi, xa^+\psi\rangle = \langle x^+\varphi, a^+\psi\rangle = \langle a \circ x^+\varphi, \psi\rangle$ for $a \in \mathcal{A}$ and $\varphi, \psi \in \mathcal{D}(\mathcal{A})$. Hence $x^+ \in \mathcal{A}_f^c$.

(ii) and (iii): We prove that $\mathcal{A}_w^c \subseteq \mathfrak{L}(\mathcal{D}_\mathcal{A}, \mathcal{D}_{\mathcal{A}^*})$. Let $x \in \mathcal{A}_w^c$. Since $x \in \mathfrak{L}(\mathcal{D}_\mathcal{A}, \mathcal{H})$, there is an operator $a \in \mathcal{A}$ such that $\|x\cdot\| \le \|a\cdot\|$ on $\mathcal{D}(\mathcal{A})$. If $b \in \mathcal{A}$, then $\|b^*x\varphi\| = \|xb^+\varphi\| \le \|ab^+\varphi\|$, $\varphi \in \mathcal{D}(\mathcal{A})$, by Proposition 7.2.2, (ii). Since $ab^+ \in \mathcal{A}$ because $\mathcal{A}$ is an O*-

algebra, this proves that $x \in \mathfrak{L}(\mathcal{D}_{\mathcal{A}}, \mathcal{D}_{\mathcal{A}^*})$. A similar reasoning shows that $\mathcal{A}_s^c \subseteqq \mathfrak{L}(\mathcal{D}_{\mathcal{A}})$. The remaining statements in (ii) and (iii) are clear.

(iv): $\hat{\mathcal{A}}_w^c \upharpoonright \mathcal{D}(\mathcal{A}) \subseteqq \mathcal{A}_w^c$ is trivial. If $x \in \mathcal{A}_w^c$, then it is easily seen that the continuous extension of $x \in \mathfrak{L}(\mathcal{D}_{\mathcal{A}}, \mathcal{H})$ to an operator in $\mathfrak{L}(\mathcal{D}_{\hat{\mathcal{A}}}, \mathcal{H})$ belongs to $\hat{\mathcal{A}}_w^c$; so $\hat{\mathcal{A}}_w^c \upharpoonright \mathcal{D}(\mathcal{A}) = \mathcal{A}_w^c$. Let $x \in \mathcal{A}_s^c$. By (iii), $x \in \mathfrak{L}(\mathcal{D}_{\mathcal{A}})$, so that x has a continuous extension to a mapping of $\mathfrak{L}(\mathcal{D}_{\hat{\mathcal{A}}})$. By continuity, this operator is in $\hat{\mathcal{A}}_s^c$. □

If $\mathcal{A}_f^c = \mathcal{A}_w^c = \mathcal{A}_s^c$, then we write $\mathcal{A}^c$ for $\mathcal{A}_f^c = \mathcal{A}_w^c = \mathcal{A}_s^c$.

Corollary 7.2.4. *Suppose $\mathcal{A}$ is an O*-algebra such that $\mathcal{A}_f^c = \mathcal{A}_w^c = \mathcal{A}_s^c$. Then $\mathcal{A}^c$ is also an O*-algebra on the domain $\mathcal{D}(\mathcal{A})$.*

Proof. Let $x \in \mathcal{A}^c$. Because $\mathcal{A}^c = \mathcal{A}_f^c$, $x^+ \in \mathcal{A}^c$ by Proposition 7.2.3, (i). Since $\mathcal{A}^c = \mathcal{A}_s^c$, x and x^+ are Hilbert space operators. Therefore, in the formula $\langle x\varphi, \psi\rangle = \langle \varphi, x^+\psi\rangle$, $\varphi, \psi \in \mathcal{D}(\mathcal{A})$, the expression $\langle \cdot, \cdot \rangle$ refers on both sides to the scalar product of the Hilbert space. Consequently, x is closable and $x^+ = x^* \upharpoonright \mathcal{D}(\mathcal{A})$. This shows that $\mathcal{A}^c$ is an O*-family on $\mathcal{D}(\mathcal{A})$. Since $\mathcal{A}^c = \mathcal{A}_s^c$, $\mathcal{A}^c$ is a subalgebra of $L(\mathcal{D}_{\mathcal{A}})$ and hence an O*-algebra. □

Remark 3. In general, the sets $\mathcal{A}_w^c$ and $\mathcal{A}_s^c$ are not invariant under the involution of $\mathcal{L}(\mathcal{D}_{\mathcal{A}}, \mathcal{D}_{\mathcal{A}}^+)$ (or equivalently, under the restriction to $\mathcal{D}(\mathcal{A})$ of the adjoint operation in the Hilbert space) even not if $\mathcal{A}$ is a self-adjoint O*-algebra. A counter-example is provided by Example 9.4.6. Moreover, the sets $\mathcal{A}_w^c$ and $\mathcal{A}_s^c$ contain also nonclosable operators in general.

The three commutants defined above satisfy the relation $\mathcal{A}_s^c \subseteqq \mathcal{A}_w^c \subseteqq \mathcal{A}_f^c$, where the inclusions are proper in general; cf. Examples 7.2.14 and 9.4.6. It is quite natural to ask for necessary and/or sufficient conditions for the equality of two of these commutants. By Proposition 7.2.2, (iv), the self-adjointness of the O*-algebra $\mathcal{A}$ is sufficient to ensure that $\mathcal{A}_w^c = \mathcal{A}_s^c$. Example 7.2.15 below shows that the self-adjointness of $\mathcal{A}$ is not necessary for the equality $\mathcal{A}_w^c = \mathcal{A}_s^c$. By Example 9.4.6, the self-adjointness of $\mathcal{A}$ does not imply that $\mathcal{A}_f^c = \mathcal{A}_w^c$. A simple sufficient condition for $\mathcal{A}_f^c = \mathcal{A}_w^c$ (which also applies to certain non-self-adjoint O*-algebras; cf. Example 7.2.6) is given by

Proposition 7.2.5. *Let $\mathcal{A}$ be an O*-algebra in the Hilbert space $\mathcal{H}$. Suppose that there exists a subset $\{a_j : j \in J\}$ of operators in $\mathcal{A}(I)$ such that $a_j^2\mathcal{D}(\mathcal{A})$ is dense in $\mathcal{H}$ for each $j \in J$ and such that $\{\|\cdot\|_{a_j} : j \in J\}$ is a directed family of seminorms generating the graph topology $\mathfrak{t}_{\mathcal{A}}$. Then $\mathcal{A}_f^c = \mathcal{A}_w^c$.*

Proof. Since always $\mathcal{A}_w^c \subseteqq \mathcal{A}_f^c$, it remains to show that $\mathcal{A}_f^c \subseteqq \mathcal{A}_w^c$. We suppose $x \in \mathcal{A}_f^c$. Since $x \in \mathcal{L}(\mathcal{D}_{\mathcal{A}}, \mathcal{D}_{\mathcal{A}}^+)$, there are an index $j \in J$ and a positive constant λ such that $|\langle x\varphi, \psi\rangle| \leqq \lambda \|a_j\varphi\| \|a_j\psi\|$ for $\varphi, \psi \in \mathcal{D}(\mathcal{A})$. From Proposition 3.2.3 it follows that there is an operator $y \in \mathbb{B}(\mathcal{H})$ such that $\langle x\cdot, \cdot\rangle \equiv \langle ya_j\cdot, a_j\cdot\rangle$. Since $a_j^+ \in \mathcal{A}$ and $x \in \mathcal{A}_f^c$, we have for $\varphi, \psi \in \mathcal{D}(\mathcal{A})$

$$\langle ya_ja_j^+\varphi, a_j\psi\rangle = \langle xa_j^+\varphi, \psi\rangle = \langle x\varphi, a_j\psi\rangle = \langle ya_j\varphi, a_j^2\psi\rangle. \tag{1}$$

From $a_j \in \mathcal{A}(I)$ it follows that there is an operator $z \in \mathbb{B}(\mathcal{H})$ such that $z(a_j\eta) = \eta$ for $\eta \in \mathcal{D}(\mathcal{A})$. Setting $\zeta = a_j^2\psi$ in (1), we obtain $\langle ya_ja_j^+\varphi, z\zeta\rangle = \langle ya_j\varphi, \zeta\rangle$ for all $\zeta \in a_j^2\mathcal{D}(\mathcal{A})$. Since $a_j^2\mathcal{D}(\mathcal{A})$ is assumed to be dense in $\mathcal{H}$, we get $z^*ya_ja_j^+\varphi = ya_j\varphi$ for $\varphi \in \mathcal{D}(\mathcal{A})$. Thus $\langle x\varphi, \psi\rangle = \langle ya_j\varphi, a_j\psi\rangle = \langle z^*ya_ja_j^+\varphi, a_j\psi\rangle = \langle ya_ja_j^+\varphi, za_j\psi\rangle = \langle ya_ja_j^+\varphi, \psi\rangle$ for all $\varphi, \psi \in \mathcal{D}(\mathcal{A})$ which gives $x = ya_ja_j^+$. Since the Hilbert space operator $ya_ja_j^+$ is obviously contained in $\mathfrak{L}(\mathcal{D}_{\mathcal{A}}, \mathcal{H})$, we have $x \in \mathcal{A}_f^c \cap \mathfrak{L}(\mathcal{D}_{\mathcal{A}}, \mathcal{H}) = \mathcal{A}_w^c$. □

Example 7.2.6. Suppose that A is a closed symmetric operator in the Hilbert space $\mathcal{H}$ such that $\ker(A^* - \mathrm{i}) = \{0\}$. Then the Cayley transform U of A is an isometry on $\mathcal{H}$, and $\mathcal{D}(A^n) = (I - U)^n \mathcal{H}$ for $n \in \mathbb{N}$. By Proposition 1.6.1, $\mathcal{D} := \mathcal{D}^\infty(A)$ is dense in $\mathcal{H}$. Let $\mathcal{A}$ be an O*-algebra on $\mathcal{D}$ which contains the operator $a := A \restriction \mathcal{D}$. Then $\mathcal{D}_\mathcal{A}$ is a Frechet space, and the graph topology $\mathrm{t}_\mathcal{A}$ is generated by the directed family of seminorms $\{\|\cdot\|_{a_n} : n \in \mathbb{N}\}$, where $a_n := (a + \mathrm{i})^n$. From Proposition 1.6.1, $\mathcal{D} \equiv \mathcal{D}^\infty(A)$ is a core for each operator A^k, $k \in \mathbb{N}$. This implies that $a_n^2 \mathcal{D} \equiv (A + \mathrm{i})^{2n} \mathcal{D}$ is dense in $(A + \mathrm{i})^{2n} \mathcal{D}(A^{2n}) \equiv (A + \mathrm{i})^{2n}(I - U)^{2n} \mathcal{H} \equiv \mathcal{H}$ for every $n \in \mathbb{N}$, so the assumptions of the preceding proposition are satisfied and we have $\mathcal{A}_\mathrm{I}^\mathrm{c} = \mathcal{A}_\mathrm{w}^\mathrm{c}$. Note that if $\mathcal{A} = \mathbb{C}[a]$ and if the operator A is not self-adjoint, then the O*-algebra $\mathcal{A}$ is not self-adjoint. In fact, $\ker(A^* + \mathrm{i}) \subseteq \mathcal{D}^*(\mathcal{A})$ in this case. ○

Next we turn to bounded commutants. They will be defined for more general sets of operators.

Definition 7.2.7. Suppose that $\mathcal{A}$ is a set of closable operators in the Hilbert space $\mathcal{H}$. Let $\mathcal{A}'_{\mathrm{ss}} := \{x \in \mathbf{B}(\mathcal{H}) : x\bar{a} \subseteq \bar{a}x \text{ and } x^*\bar{a} \subseteq \bar{a}x^* \text{ for all } a \in \mathcal{A}\}$. If $\mathcal{A}$ is an O-family, then $\mathcal{A}'_\mathrm{s} := \{x \in \mathbf{B}(\mathcal{H}) : x(\mathcal{D}(\mathcal{A})) \subseteq \mathcal{D}(\mathcal{A}) \text{ and } xa\varphi = ax\varphi \text{ for } a \in \mathcal{A} \text{ and } \varphi \in \mathcal{D}(\mathcal{A})\}$ is the *strong commutant* of $\mathcal{A}$. For an O*-family $\mathcal{A}$, $\mathcal{A}'_\mathrm{w} := \{x \in \mathbf{B}(\mathcal{H}) : \langle xa\varphi, \psi\rangle = \langle x\varphi, a^+\psi\rangle$ for $a \in \mathcal{A}$ and $\varphi \in \mathcal{D}(\mathcal{A})\}$ is called the *weak commutant* of $\mathcal{A}$.

Often it is convenient to work with the corresponding sets for single operators. The set $(a)'_\mathrm{s} := \{x \in \mathbf{B}(\mathcal{H}) : xa \subseteq ax\}$ is called the *strong commutant* of an operator a in the Hilbert space $\mathcal{H}$. If the operator a is symmetric, then the set $(a)'_\mathrm{w} := \{x \in \mathbf{B}(\mathcal{H}) : \langle xa\varphi, \psi\rangle = \langle x\varphi, a\psi\rangle \text{ for } \varphi, \psi \in \mathcal{D}(a)\}$ is said to be the *weak commutant* of a. It is easy to check (cf. Proposition 7.2.10, (i)) that $(a)'_\mathrm{w} = \{x \in \mathbf{B}(\mathcal{H}) : xa \subseteq a^*x\}$. Therefore, if the operator a is self-adjoint, then $(a)'_\mathrm{s} = (a)'_\mathrm{w}$. In the latter case we shall write simply $(a)'$ for the set $(a)'_\mathrm{s} \equiv (a)'_\mathrm{w}$.

Lemma 7.2.8. *Suppose a is a closable linear operator in the Hilbert space $\mathcal{H}$. Then the weak-operator closure of $(a)'_\mathrm{s}$ in $\mathbf{B}(\mathcal{H})$ is contained in $(\bar{a})'_\mathrm{s}$. In particular, $(a)'_\mathrm{s}$ is weak-operator closed in $\mathbf{B}(\mathcal{H})$ if a is a closed operator.*

Proof. First we show that $(a)'_\mathrm{s} \subseteq (\bar{a})'_\mathrm{s}$. We let $x \in (a)'_\mathrm{s}$ and $\varphi \in \mathcal{D}(\bar{a})$. There is a sequence $(\varphi_n : n \in \mathbb{N})$ in $\mathcal{D}(a)$ such that $\varphi = \lim \varphi_n$ and $\bar{a}\varphi = \lim a\varphi_n$ in $\mathcal{H}$. Then $x\varphi = \lim x\varphi_n$ and $\lim ax\varphi_n = \lim xa\varphi_n = x\bar{a}\varphi$. From the latter we conclude that $x\varphi \in \mathcal{D}(\bar{a})$ and $x\bar{a}\varphi = \bar{a}x\varphi$. Thus $x \in (\bar{a})'_\mathrm{s}$ and $(a)'_\mathrm{s} \subseteq (\bar{a})'_\mathrm{s}$.

To complete the proof, it suffices to verify that $(\bar{a})'_\mathrm{s}$ is weak-operator closed in $\mathbf{B}(\mathcal{H})$. Since $(\bar{a})'_\mathrm{s}$ is a convex subset of $\mathbf{B}(\mathcal{H})$, its weak-operator closure and its strong-operator closure in $\mathbf{B}(\mathcal{H})$ coincide (see e.g. Kadison/Ringrose [1], 5.1.2). Let $\mathcal{N}$ denote this set, and let $x \in \mathcal{N}$. Then there exists a net $(x_i : i \in I)$ in $(\bar{a})'_\mathrm{s}$ such that $x\psi = \lim x_i\psi$ in $\mathcal{H}$ for any $\psi \in \mathcal{H}$. If $\varphi \in D(\bar{a})$, then it follows from $x\varphi = \lim x_i\varphi$ and $x\bar{a}\varphi = \lim x_i\bar{a}\varphi = \lim \bar{a}x_i\varphi$ that $x\varphi \in \mathcal{D}(\bar{a})$ and $x\bar{a}\varphi = \bar{a}x\varphi$. This yields $x \in (\bar{a})'_\mathrm{s}$, so $(\bar{a})'_\mathrm{s}$ is weak-operator closed in $\mathbf{B}(\mathcal{H})$. □

Proposition 7.2.9. *Suppose $\mathcal{A}$ is a family of closable linear operators in the Hilbert space $\mathcal{H}$.*

(i) *$\mathcal{A}'_{\mathrm{ss}}$ is a von Neumann algebra in $\mathcal{H}$. For every $a \in \mathcal{A}$, the operator $\bar{a}$ is affiliated with $(\mathcal{A}'_{\mathrm{ss}})'$, and $(\mathcal{A}'_{\mathrm{ss}})'$ is the smallest von Neumann algebra in $\mathcal{H}$ which admits this property.*

(ii) *Suppose $\mathcal{A}$ is an O-family. Then $\mathcal{A}'_\mathrm{s}$ is a subalgebra of $\mathbf{B}(\mathcal{H})$, and $\mathcal{A}'_\mathrm{s} \cap (\mathcal{A}'_\mathrm{s})^* \subseteq \mathcal{A}'_{\mathrm{ss}}$. If the O-family $\mathcal{A}$ is directed, then $\mathcal{A}'_\mathrm{s} \subseteq \hat{\mathcal{A}}'_\mathrm{s}$. If $\mathcal{A}$ is closed and directed, then $\mathcal{A}'_\mathrm{s}$ is weak-operator closed in $\mathbf{B}(\mathcal{H})$ and $\mathcal{A}'_{\mathrm{ss}} = \mathcal{A}'_\mathrm{s} \cap (\mathcal{A}'_\mathrm{s})^*$.*

(iii) *Suppose $\mathcal{A}$ is an O*-family. Then $\mathcal{A}'_w$ is a $*$-invariant linear subspace of $\mathbb{B}(\mathcal{H})$ which is closed in the weak-operator topology of $\mathbb{B}(\mathcal{H})$ and spanned by its positive elements.*

Proof. (i): By definition $\mathcal{A}'_{ss}$ is the intersection of the sets $(\bar{a})'_s \cap ((\bar{a})'_s)^*$, where $a \in \mathcal{A}$. By Lemma 7.2.8, these sets are weak-operator closed in $\mathbb{B}(\mathcal{H})$. Since obviously $\mathcal{A}'_{ss}$ is a $*$-algebra and $I \in \mathcal{A}'_{ss}$, $\mathcal{A}'_{ss}$ is a von Neumann algebra. From the definition of $\mathcal{A}'_{ss}$ it is clear that the operators $\bar{a}$, where $a \in \mathcal{A}$, are affiliated with $(\mathcal{A}'_{ss})'$. Let $\mathcal{N}$ be another von Neumann algebra which has this property. Suppose $y \in \mathcal{N}'$. Then $y\bar{a} \subseteq \bar{a}y$ for $a \in \mathcal{A}$, since $\bar{a}$ is affiliated with $\mathcal{N}$. From $y^* \in \mathcal{N}'$ we obtain $y^*\bar{a} \subseteq \bar{a}y^*$ for $a \in \mathcal{A}$; so $y \in \mathcal{A}'_{ss}$. Hence $\mathcal{N}' \subseteq \mathcal{A}'_{ss}$ and $(\mathcal{A}'_{ss})' \subseteq \mathcal{N}'' = \mathcal{N}$.
(ii): It is clear that $\mathcal{A}'_s$ is a subalgebra of $\mathbb{B}(\mathcal{H})$. Let $x \in \mathcal{A}'_s \cap (\mathcal{A}'_s)^*$. Then, by definition, x and x^* are in $(a)'_s$ and hence in $(\bar{a})'_s$ by Lemma 7.2.8 for each $a \in \mathcal{A}$. Thus $x \in \mathcal{A}'_{ss}$. Suppose that $\mathcal{A}$ is directed. Let x be in the weak-operator closure of $\mathcal{A}'_s$ in $\mathbb{B}(\mathcal{H})$. Suppose $a \in \mathcal{A}$. Since $\mathcal{A}'_s \subseteq (a)'_s$, we have $x \in (\bar{a})'_s$ by Lemma 7.2.8. Thus $x(\hat{\mathcal{D}}(\mathcal{A})) \subseteq \bigcap_{a\in\mathcal{A}} \mathcal{D}(\bar{a})$. Because $\mathcal{A}$ is directed, Proposition 2.2.12 shows that the latter set is equal to $\hat{\mathcal{D}}(\mathcal{A})$, so that $x(\hat{\mathcal{D}}(\mathcal{A})) \subseteq \hat{\mathcal{D}}(\mathcal{A})$. Therefore, $x \in \hat{\mathcal{A}}'_s$. In particular this proves that $\mathcal{A}'_s \subseteq \hat{\mathcal{A}}'_s$. Suppose now that $\mathcal{A}$ is closed and directed. Then the preceding argument shows that $\mathcal{A}'_s$ is weak-operator closed in $\mathbb{B}(\mathcal{H})$. Suppose $x \in \mathcal{A}'_{ss}$. Then $x \in (\bar{a})'_s$ and $x^* \in (\bar{a})'_s$ for every $a \in \mathcal{A}$. Combined with the equality $\hat{\mathcal{D}}(\mathcal{A}) = \bigcap_{a\in\mathcal{A}} \mathcal{D}(\bar{a})$, this gives $x \in \mathcal{A}'_s$ and $x^* \in \mathcal{A}'_s$. Thus $\mathcal{A}'_{ss} \subseteq \mathcal{A}'_s \cap (\mathcal{A}'_s)^*$. Since the reversed inclusion is already proved, we get $\mathcal{A}'_{ss} = \mathcal{A}'_s \cap (\mathcal{A}'_s)^*$.
(iii): To verify the $*$-invariance of $\mathcal{A}'_w$, we repeat the argument of Proposition 7.2.3, (i). Suppose $x \in \mathcal{A}'_w$. Then $\langle x^*a\varphi, \psi\rangle = \langle a\varphi, x\psi\rangle = \langle \varphi, xa^+\psi\rangle = \langle x^*\varphi, a^+\psi\rangle$ for $a \in \mathcal{A}$ and $\varphi, \psi \in \mathcal{D}(\mathcal{A})$, where the second equality holds because of $x \in \mathcal{A}'_w$. This yields $x^* \in \mathcal{A}'_w$. Since $I \in \mathcal{A}'_w$, $x + \|x\| \cdot I$ and $\|x\| \cdot I - x$ are in $\mathcal{A}'_w$ for any $x = x^* \in \mathcal{A}'_w$. Hence $\mathcal{A}'_w$ is spanned by its positive elements. The remaining statements in (iii) are clear. □

Proposition 7.2.10. *Suppose that $\mathcal{A}$ is an O*-family in the Hilbert space $\mathcal{H}$.*
(i) $\mathcal{A}'_w = \{x \in \mathbb{B}(\mathcal{H}) : x(\mathcal{D}(\mathcal{A})) \subseteq \mathcal{D}^*(\mathcal{A})$ *and* $xa \subseteq (a^+)^* x$ *for all* $a \in \mathcal{A}\}$.
(ii) $\mathcal{A}'_s = \{x \in \mathcal{A}'_w : x(\mathcal{D}(\mathcal{A})) \subseteq \mathcal{D}(\mathcal{A})\}$.
(iii) *If $\mathcal{A}$ is self-adjoint, then* $\mathcal{A}'_w = \mathcal{A}'_s$.
(iv) *If* $\mathcal{A}'_w = \mathcal{A}'_s$, *then* $\mathcal{A}'_w \equiv \mathcal{A}'_s = \mathcal{A}'_{ss}$, *and this set is a von Neumann algebra.*
(v) $\hat{\mathcal{A}}'_w = \mathcal{A}'_w$, *and* $\mathcal{A}'_{ss} \subseteq \mathcal{A}'_w$.
(vi) *If $\mathcal{A}$ is an O*-algebra, then* $\mathcal{A}'_s = \{x \in \mathbb{B}(\mathcal{H}) : x \restriction \mathcal{D}(\mathcal{A}) \in \mathcal{A}^c_s\}$ *and* $\mathcal{A}'_w = \{x \in \mathbb{B}(\mathcal{H}) : x \restriction \mathcal{D}(\mathcal{A}) \in \mathcal{A}^c_w\}$.

Proof. The proof of (i) is the same as the proof of Proposition 7.2.2, (ii). Since $(a^+)^* \supseteq \bar{a}$ for $a \in \mathcal{A}$, (ii), (iii) and $\mathcal{A}'_{ss} \subseteq \mathcal{A}'_w$ follow immediately from (i). (vi) and the equality $\hat{\mathcal{A}}'_w = \mathcal{A}'_w$ follow easily from the corresponding definitions. We verify (iv). Suppose $\mathcal{A}'_w = \mathcal{A}'_s$. Let $x \in \mathcal{A}'_w \equiv \mathcal{A}'_s$. Since $\mathcal{A}'_w$ is $*$-invariant, $x^* \in \mathcal{A}'_w \equiv \mathcal{A}'_s$. Thus $x \in \mathcal{A}'_s \cap (\mathcal{A}'_s)^*$. By Proposition 7.2.9, (ii), $x \in \mathcal{A}'_{ss}$. This proves that $\mathcal{A}'_w \subseteq \mathcal{A}'_{ss}$. Since $\mathcal{A}'_{ss} \subseteq \mathcal{A}'_w$ by (v), we get $\mathcal{A}'_w = \mathcal{A}'_{ss}$. As stated in Proposition 7.2.9, (i), $\mathcal{A}'_{ss}$ is a von Neumann algebra. □

If $\mathcal{A}$ is an O*-family such that $\mathcal{A}'_w = \mathcal{A}'_s$ (in particular, if $\mathcal{A}$ is a self-adjoint O*-family), then we write simply $\mathcal{A}'$ for $\mathcal{A}'_w = \mathcal{A}'_s$ and we call $\mathcal{A}'$ the *commutant* of $\mathcal{A}$. By Propositions

7.2.9, (i), and 7.2.10, (iv), in this case $\mathcal{A}'$ is a von Neumann algebra, and for any $a \in \mathcal{A}$, the operator $\bar{a}$ is affiliated with the von Neumann algebra $\mathcal{A}''$.

Now we consider the relations between bounded and unbounded commutants of O*-algebras.

Proposition 7.2.11. *Suppose that $\mathcal{A}$ is a closed O*-algebra in the Hilbert space $\mathcal{H}$, and $\mathcal{N}$ is a von Neumann algebra contained in the strong commutant $\mathcal{A}'_s$. Let x be a closed linear operator on $\mathcal{H}$ such that $\mathcal{D}(\mathcal{A}) \subseteq \mathcal{D}(x)$ and $x \restriction \mathcal{D}(\mathcal{A})$ is in $\mathfrak{L}(\mathcal{D}_{\mathcal{A}}, \mathcal{H})$. If x is affiliated with $\mathcal{N}$, then $x \restriction \mathcal{D}(\mathcal{A})$ belongs to $\mathcal{A}^c_s$.*

Proof. Let $x = u\,|x|$ be the polar decomposition of x, and let $|x| = \int_0^\infty \lambda\, de(\lambda)$ be the spectral decomposition of the positive self-adjoint operator $|x|$. Since x is affiliated with $\mathcal{N}$, $u \in \mathcal{N}$ and $|x|$ is also affiliated with $\mathcal{N}$. Therefore, $|x|\, e\big((0, n)\big) \in \mathcal{N}$ and hence $x_n := xe\big((0, n)\big) = u\,|x|\, e\big((0, n)\big) \in \mathcal{N}$ for any $n \in \mathbb{N}$. Since $\mathcal{N} \subseteq \mathcal{A}'_s$, we have $x_n\big(\mathcal{D}(\mathcal{A})\big) \subseteq \mathcal{D}(\mathcal{A})$ and $x_n a\varphi = a x_n \varphi$ for $a \in \mathcal{A}$, $\varphi \in \mathcal{D}(\mathcal{A})$ and $n \in \mathbb{N}$. Suppose $\varphi \in \mathcal{D}(\mathcal{A})$ and $a \in \mathcal{A}$. From $x\varphi = u\,|x|\,\varphi = \lim_n u\,|x|\, e\big((0, n)\big)\, \varphi$ and $xa\varphi = \lim_n a x_n \varphi$ in $\mathcal{H}$ we conclude that $x\varphi \in \mathcal{D}(\bar{a})$ and $xa\varphi = \bar{a} x\varphi$. Because $\mathcal{A}$ is a closed O*-algebra, we have $\mathcal{D}(\mathcal{A}) = \bigcap_{a \in \mathcal{A}} \mathcal{D}(\bar{a})$. Therefore, the preceding implies $x\varphi \in \mathcal{D}(\mathcal{A})$ and $xa\varphi = ax\varphi$. Since $\varphi \in \mathcal{D}(\mathcal{A})$ and $a \in \mathcal{A}$ are arbitrary, this proves that $x \restriction \mathcal{D}(\mathcal{A})$ is in $\mathcal{A}^c_s$. □

Proposition 7.2.12. *Suppose that $\mathcal{A}$ is an O*-algebra in the Hilbert space $\mathcal{H}$ and $\mathcal{N}$ is a von Neumann algebra contained in the weak commutant $\mathcal{A}'_w$. Suppose that x is a closed linear operator on $\mathcal{H}$ which satisfies $\mathcal{D}(\mathcal{A}) \subseteq \mathcal{D}(x)$ and $x \restriction \mathcal{D}(\mathcal{A}) \in \mathfrak{L}(\mathcal{D}_{\mathcal{A}}, \mathcal{H})$. If x is affiliated with $\mathcal{N}$, then $x \restriction \mathcal{D}(\mathcal{A})$ is in $\mathcal{A}^c_w$.*

Proof. Up to the following modification, the proof follows the lines of the preceding proof. From $x_n \in \mathcal{N} \subseteq \mathcal{A}'_w$ we have $x_n a\varphi = (a^+)^*\, x_n\varphi$ for $a \in \mathcal{A}$, $\varphi \in \mathcal{D}(\mathcal{A})$ and $n \in \mathbb{N}$. Since $\mathcal{D}^*(\mathcal{A}) = \bigcap_{a \in \mathcal{A}} \mathcal{D}(a^*)$, it follows that $x\big(\mathcal{D}(\mathcal{A})\big) \subseteq \mathcal{D}^*(\mathcal{A})$. □

A consequence of each of the previous propositions is the following corollary.

Corollary 7.2.13. *Let $\mathcal{A}$ be a self-adjoint O*-algebra in the Hilbert space $\mathcal{H}$, and let x be a closed linear operator on $\mathcal{H}$ such that $\mathcal{D}(\mathcal{A}) \subseteq \mathcal{D}(x)$ and $x \restriction \mathcal{D}(\mathcal{A}) \in \mathfrak{L}(\mathcal{D}_{\mathcal{A}}, \mathcal{H})$. If the operator x is affiliated with the von Neumann algebra $\mathcal{A}'$, then $x \restriction \mathcal{D}(\mathcal{A})$ belongs to $\mathcal{A}^c_s \equiv \mathcal{A}^c_w$.*

Remark 4. If the locally convex space $\mathcal{D}_{\mathcal{A}}$ is barrelled, then $\mathrm{t}_{\mathcal{A}} = \mathrm{t}_c$ by Proposition 2.3.9; so the assumption $x \restriction \mathcal{D}(\mathcal{A}) \in \mathfrak{L}(\mathcal{D}_{\mathcal{A}}, \mathcal{H})$ in Propositions 7.2.11 and 7.2.12 is automatically fulfilled in this case.

Remark 5. We state a remarkable property of the strong commutant. (At the end of Example 7.2.14 we shall show that a similar statement for the weak commutant is not true in general.) *If $\mathcal{A}$ is an O*-algebra and x is a positive operator in $\mathcal{A}'_s$, then $xa \geqq 0$ for any $a \in \mathcal{A}_+$.*

Proof. By Proposition 7.2.9, x belongs to the von Neumann algebra $\mathcal{A}'_{ss}$. Hence $x^{1/2} \in \mathcal{A}'_{ss}$ which gives $\langle xa\varphi, \varphi\rangle = \langle x^{1/2} a\varphi, x^{1/2}\varphi\rangle = \langle \bar{a} x^{1/2}\varphi, x^{1/2}\varphi\rangle \geqq 0$ for $\varphi \in \mathcal{D}(\mathcal{A})$. □

We next discuss two examples more in detail.

Example 7.2.14. Let S be the shift operator on the Hardy space $\mathcal{H} := H^2(\mathbb{T})$, and let A be the closed symmetric operator on $\mathcal{H}$ the Cayley transform of which is S. That is,

$\mathcal{D}(A) = (I - S)\,\mathcal{H}$ and A is defined by $(A\psi)(z) = \mathrm{i}(1 + z)\,\varphi(z)$ for $\psi(z) := (1 - z)\varphi(z)$, where $\varphi \in \mathcal{H}$ and $z \in \mathbb{T}$. Since $\ker(A^* - \mathrm{i}) = \{0\}$, we are in a special case of Example 7.2.6. For $v \in L^\infty(\mathbb{T})$, let T_v denote the Toeplitz operator on $H^2(\mathbb{T})$ with symbol v. (All operator-theoretic notions and facts used in this example can be found, for instance, in HALMOS [2].)

Statement 1: $(A)'_{\mathrm{w}} = \{T_v : v \in L^\infty(\mathbb{T})\}$ *and* $(A)'_{\mathrm{s}} = \{T_v : v \in H^\infty(\mathbb{T})\}$.

Proof. Suppose $x \in \mathbf{B}(\mathcal{H})$. Then $x \in (A)'_{\mathrm{w}}$ if and only if $\langle x(A + \mathrm{i})\,\varphi, \psi\rangle = \langle x\varphi, (A - \mathrm{i})\psi\rangle$ for all $\varphi, \psi \in \mathcal{D}(A)$. Setting $\varphi = (I - S)\,\zeta$ and $\psi = (I - S)\,\eta$ with $\zeta, \eta \in \mathcal{H}$, it follows that the latter is equivalent to $x = S^*xS$. But this relation is true if and only if x is a Toeplitz operator with symbol in $L^\infty(\mathbb{T})$ (HALMOS [2], ch. 20).

We verify the assertion concerning $(A)'_{\mathrm{s}}$. It is plain that $T_v \in (A)'_{\mathrm{s}}$ if $v \in H^\infty(\mathbb{T})$. Conversely, let $x \in (A)'_{\mathrm{s}}$. Then $x \in (A)'_{\mathrm{w}}$ and hence $x = T_v$ with $v \in L^\infty(\mathbb{T})$. Since $x \in (A)'_{\mathrm{s}}$, $x(A - \mathrm{i})\,\mathcal{D}(A) \equiv xS\mathcal{H} \subseteq (A - \mathrm{i})\,\mathcal{D}(A) \equiv S\mathcal{H}$. This implies that the negative Fourier coefficients of v are zero, i.e., $v \in H^\infty(\mathbb{T})$. $\square$

Let $\mathcal{A}$ be the O*-algebra of all polynomials in $a := A \restriction \mathcal{D}$ on the domain $\mathcal{D} := \mathcal{D}^\infty(A)$ in $\mathcal{H}$. It is easily seen that $\mathcal{A}'_{\mathrm{w}} = (A)'_{\mathrm{w}}$ and $\mathcal{A}'_{\mathrm{s}} = (A)'_{\mathrm{s}}$, so that Statement 1 gives an explicit description of these commutants. In particular we see that $\mathcal{A}'_{\mathrm{w}}$ is not an algebra and that $\mathcal{A}'_{\mathrm{s}}$ is not $*$-invariant.

We check that $\mathcal{A}'_{\mathrm{ss}} = \mathbb{C} \cdot I$. Suppose $x = x^* \in \mathcal{A}'_{\mathrm{ss}}$. Then $x \in \mathcal{A}'_{\mathrm{s}}$. By Statement 1, $x = T_v$ with $v \in H^\infty(\mathbb{T})$. But each hermitian Toeplitz operator with analytic symbol is a multiple of the identity; so $x = \lambda I$ for some $\lambda \in \mathbb{R}$. Since $\mathcal{A}'_{\mathrm{ss}}$ is a von Neumann algebra, this yields $\mathcal{A}'_{\mathrm{ss}} = \mathbb{C} \cdot I$.

We now determine the unbounded commutants of $\mathcal{A}$.

Statement 2: $\mathcal{A}^{\mathrm{c}}_{\mathrm{f}} = \mathcal{A}^{\mathrm{c}}_{\mathrm{w}} = \{T_v b : v \in L^\infty(\mathbb{T})$ *and* $b \in \mathcal{A}\}$ *and* $\mathcal{A}^{\mathrm{c}}_{\mathrm{s}} = \{T_v b : v \in H^\infty(\mathbb{T})$ *and* $b \in \mathcal{A}\}$.

Proof. As already noted in Example 7.2.6, $\mathcal{A}^{\mathrm{c}}_{\mathrm{f}} = \mathcal{A}^{\mathrm{c}}_{\mathrm{w}}$. Suppose $c \in \mathcal{A}^{\mathrm{c}}_{\mathrm{w}}$. Since $c \in \mathfrak{L}(\mathcal{D}_{\mathcal{A}}, \mathcal{H})$, there are an $n \in \mathbb{N}_0$ and an operator $x \in \mathbf{B}(\mathcal{H})$ such that $c = x(a + \mathrm{i})^n$. From $c \in \mathcal{A}^{\mathrm{c}}_{\mathrm{w}}$ we obtain $\langle c(a + \mathrm{i})\,\varphi, \psi\rangle = \langle c\varphi, (a - \mathrm{i})\,\psi\rangle$ for $\varphi, \psi \in \mathcal{D}$. Hence $\langle x(A + \mathrm{i})^{n+1}\,\varphi, \psi\rangle = \langle x(A + \mathrm{i})^n\,\varphi, (A - \mathrm{i})\,\psi\rangle$ for all $\varphi, \psi \in \mathcal{D}$. Since $\mathcal{D} = \mathcal{D}^\infty(A)$ is a core for any power A^k, $k \in \mathbb{N}$, the latter is true for all $\varphi \in \mathcal{D}(A^{n+1})$ and for all $\psi \in \mathcal{D}(A)$. Setting $\varphi = (I - S)^{n+1}\,\zeta$ and $\psi = (I - S)\,\eta$ with $\zeta, \eta \in \mathcal{H}$ it follows that $x = S^*xS$. Therefore, as in the proof of Statement 1, $x = T_v$ for some $v \in L^\infty(\mathbb{T})$.

Now suppose that $c \in \mathcal{A}^{\mathrm{c}}_{\mathrm{s}}$. Then $c \in \mathcal{A}^{\mathrm{c}}_{\mathrm{w}}$, so that $c = T_v(a + \mathrm{i})^n$ for some $n \in \mathbb{N}$ and $v \in L^\infty(\mathbb{T})$ by the preceding. By $c \in \mathcal{A}^{\mathrm{c}}_{\mathrm{s}}$, $c(a - \mathrm{i})\,\varphi = (a - \mathrm{i})\,c\varphi$ and hence $T_v(A + \mathrm{i})^n\,(A - \mathrm{i})\,\varphi = (A - \mathrm{i})\,T_v(A + \mathrm{i})^n\,\varphi$ for $\varphi \in \mathcal{D}$. Using once more the core property of $\mathcal{D}^\infty(A)$, it follows that the last equality holds for all $\varphi \in \mathcal{D}(A^{n+1})$. Hence $T_v(A + \mathrm{i})^n\,(A - \mathrm{i})\,\mathcal{D}(A^{n+1}) \equiv T_vS\mathcal{H} \subseteq (A - \mathrm{i})\,\mathcal{D}(A) \equiv S\mathcal{H}$. As in the proof of Statement 1 we obtain $v \in H^\infty(\mathbb{T})$. The converse inclusions in Statement 2 follow by straightforward computations. $\square$

The next statement shows that the commutant of $\mathcal{A}$ within $\mathcal{L}^+(\mathcal{D})$ is very small. It consists only of $\mathcal{A}$ itself.

Statement 3: $\mathcal{A}^{\mathrm{c}}_{\mathrm{s}} \cap \mathcal{L}^+(\mathcal{D}) = \mathcal{A}$.

Proof. Suppose $c \in \mathcal{A}^{\mathrm{c}}_{\mathrm{s}} \cap \mathcal{L}^+(\mathcal{D})$. Since $\mathcal{D}_{\mathcal{A}}$ is a Frechet space, $c^+ \in \mathfrak{L}(\mathcal{D}_{\mathcal{A}}, \mathcal{H})$. Further, $c^+b = (b^+c)^+ = (cb^+)^+ = bc^+$ for $b \in \mathcal{A}$. Hence $c^+ \in \mathcal{A}^{\mathrm{c}}_{\mathrm{s}}$. There exists $n \in \mathbb{N}_0$ such that

c and c^+ are both $(a + \mathrm{i})^n$-bounded, so that, by the proof of Statement 2, $c = T_v(a+\mathrm{i})^n$ and $c^+ = T_\theta(a + \mathrm{i})^n$ for some $v, \theta \in H^\infty(\mathbb{T})$. For $\varphi, \psi \in \mathcal{D}$, we have

$$\langle T_v(a + \mathrm{i})^n \varphi, \psi\rangle = \langle c\varphi, \psi\rangle = \langle \varphi, c^+\psi\rangle = \langle \varphi, T_\theta(a + \mathrm{i})^n \psi\rangle = \langle \varphi, (a + \mathrm{i})^n T_\theta\psi\rangle$$
$$= \langle (a - \mathrm{i})^n \varphi, T_\theta\psi\rangle = \langle T_{\bar\theta}S^n(a + \mathrm{i})^n \varphi, \psi\rangle,$$

where we used that $T_\theta \in \mathcal{A}'_s$. Since $\mathcal{D}$ is a core for A^n, $(a + \mathrm{i})^n \mathcal{D}$ is dense in $(A+\mathrm{i})^n \mathcal{D}(A^n) = \mathcal{H}$; so the preceding yields $T_v = T_{\bar\theta}S^n$. Hence $v(z) = \overline{\theta(z)}\, z^n$, $z \in \mathbb{T}$. Since v and θ are both in $H^\infty(\mathbb{T})$, v must be a polynomial in z of degree at most n. Writing $v(z)$ as $\sum\limits_{k=0}^{n} \alpha_k(1 - z)^k$, we have

$$c\varphi = (a + \mathrm{i})^n v\varphi = \sum_{k=0}^{n} (2\mathrm{i})^k \alpha_k(a + \mathrm{i})^{n-k} \varphi \quad \text{for} \quad \varphi \in \mathcal{D}.$$

Thus $c \in \mathcal{A}$. The reversed inclusion is trivial. □

Let $\mathcal{B}$ be the O*-algebra on $\mathcal{D}$ generated by a^2. Let e denote the rank one projection onto the space of constant functions on $\mathbb{T}$, i.e., $e = z^0 \otimes z^0$. Since $(A - \mathrm{i})\, \mathcal{D}(A) = S\mathcal{H} \perp z^0$, $e(a^2 + I) = e(A - \mathrm{i})\, (A + \mathrm{i}) \upharpoonright \mathcal{D} = 0$. From this we see easily that $e \in \mathcal{B}'_w$ and that $ea^2 = -e \upharpoonright \mathcal{D}$ is not positive, though $e \geqq 0$ and $a^2 \in \mathcal{B}_+$. ○

Example 7.2.15. Suppose $\mathcal{H}$ is the Hilbert space $L^2\big((0, 1), t^{-1}\, \mathrm{d}t\big)$. Let A be the symmetric operator $-\mathrm{i}t\dfrac{\mathrm{d}}{\mathrm{d}t}$ on $\mathcal{H}$ with boundary condition $\varphi(1) = 0$ for φ in $\mathcal{D}(A)$, and let B denote the multiplication operator by the independent variable t on $\mathcal{H}$. Then A has deficiency indices $(0, 1)$, so that we are again in a special case of Example 7.2.6 when we set $\mathcal{D} := \mathcal{D}^\infty(A)$ and $a := A \upharpoonright \mathcal{D}$. Obviously, $b := B \upharpoonright \mathcal{D}$ is in $\mathcal{L}^+(\mathcal{D})$. Let $\mathcal{A}$ be the O*-algebra on $\mathcal{D}$ generated by a and b. The function $\xi_+(t) := t\big(\in \ker (A^* + \mathrm{i})\big)$ is contained in $\mathcal{D}^*(\mathcal{A})$, but not in $\mathcal{D}(\mathcal{A})$. Hence $\mathcal{A}$ is not self-adjoint. Nevertheless we have $\mathcal{A}^c_f = \mathcal{A}^c_w = \mathcal{A}^c_s$ and $\mathcal{A}'_w = \mathcal{A}'_s$ as the following statement shows.

Statement: $\mathcal{A}^c_f = \mathbb{C}\cdot I$.

Proof. Suppose $c \in \mathcal{A}^c_f$. As shown in Example 7.2.6, $\mathcal{A}^c_f = \mathcal{A}^c_w$; so $c \in \mathcal{A}^c_w$. The graph topology $t_\mathcal{A}$ is generated by the directed family of seminorms $\{\|\cdot\|_{(a+\mathrm{i})^n} : n \in \mathbb{N}\}$. By $c \in \mathfrak{L}(\mathcal{D}_\mathcal{A}, \mathcal{H})$, there exist an $n \in \mathbb{N}_0$ and an operator $x \in \mathbf{B}(\mathcal{H})$ such that $c = x(a + \mathrm{i})^n$. Set $C := x(A + \mathrm{i})^n$. From $c \in \mathcal{A}^c_w$ we obtain $ca \subseteq a^*c$ and $cb \subseteq b^*c$. Since $\mathcal{D}$ is a core for any power A^k, $k \in \mathbb{N}$, and B is bounded, it follows from the latter that

$$CA\varphi = A^*C\varphi \quad \text{for all} \quad \varphi \in \mathcal{D}(A^{n+1}) \tag{2}$$

and

$$CB\varphi = BC\varphi \quad \text{for all} \quad \varphi \in \mathcal{D}(A^n). \tag{3}$$

Set $\zeta(t) := (1 - t)^n$, $t \in (0, 1)$, and $\eta := C\zeta$. It is clear that A^* acts as $-\mathrm{i}t\dfrac{\mathrm{d}}{\mathrm{d}t}$ in the distributional sense. Since $\zeta(t) \in \mathcal{D}(A^n)$ and $(1 - t)\, \zeta(t) \in \mathcal{D}(A^{n+1})$, we obtain from (2) and (3) that

$$C\big(A(1 - t)\,\zeta\big) = C\big(\mathrm{i}t(n + 1)\, \zeta\big) = \mathrm{i}(n + 1)\, t\, C\zeta = \mathrm{i}(n + 1)\, t\eta = A^*\big(C(1 - t)\, \zeta\big)$$
$$= A^*\big((1 - t)\, \eta\big) = +\mathrm{i}t\eta - \mathrm{i}t(1 - t)\, \eta'$$

and hence

$$(1 - t)\,\eta'(t) = -n\eta(t) \quad \text{for} \quad t \in (0, 1).$$

By the uniqueness of the solution of the differential equation $(1 - t)\,f'(t) = -nf(t)$, there is a $\lambda \in \mathbb{C}$ such that $\eta(t) \equiv (C\zeta)(t) = \lambda\zeta(t)$ on $(0, 1)$. By (3), this gives $C\varphi = \lambda\varphi$ for all $\varphi \in \mathcal{D}_0 := \text{l.h.}\,\{t^k\zeta(t) : k \in \mathbb{N}_0\}$.

We show that $\mathcal{D}_0$ is a core for $(A + \mathrm{i})^n$. It is easily seen that $C_0^\infty(0, 1)$ is a core for any A^k, $k \in \mathbb{N}$. Hence it suffices to approximate a given function $\psi \in C_0^\infty(0, 1)$ in the graph norm $\|(A + \mathrm{i})^n \cdot\|$. We have $\psi\zeta^{-1} \in C_0^\infty(0, 1)$. Thus there exists a sequence $(p_k : k \in \mathbb{N})$ of polynomials in t such that $(\psi\zeta^{-1})^{(l)}(t) = \lim\limits_{k\to\infty} p_k^{(l)}(t)$ uniformly on $(0, 1)$ for $l = 0, \ldots, n$. Then

$$(A + \mathrm{i})^n\,(p_k\zeta - \psi) \equiv (A + \mathrm{i})^n\,\big((p_k - \psi\zeta^{-1})\,\zeta\big) \to 0 \quad \text{in } \mathcal{H} \text{ as } k \to \infty$$

which proves that $\mathcal{D}_0$ is a core for $(A + \mathrm{i})^n$. Therefore, the equality $C\varphi \equiv x(A + \mathrm{i})^n\,\varphi = \lambda\varphi$ is valid for all $\varphi \in \mathcal{D}(A^n)$, so that $c \equiv C \upharpoonright \mathcal{D} = \lambda I$. □

We give another perspective on the O*-algebra $\mathcal{A}$. Let $\mathfrak{g}$ be the Lie algebra of the affine group of the real line. There is a basis $\{x_1, x_2\}$ of $\mathfrak{g}$ satisfying the relation $[x_1, x_2] = x_2$. Since $ba - ab = \mathrm{i}b$, there exists a unique $*$-representation π of the universal enveloping algebra $\mathcal{E}(\mathfrak{g})$ of $\mathfrak{g}$ such that $\pi(x_1) = \mathrm{i}a$ and $\pi(x_2) = \mathrm{i}b$; cf. Sections 1.7 and 10.1. Then $\mathcal{A}$ is the image of $\mathcal{E}(\mathfrak{g})$ under π. ○

For the next proposition we recall a notation from Section 1.6. If $\mathcal{H}$ is a closed linear subspace of a Hilbert space $\mathcal{H}_1$ and $x \in \mathbf{B}(\mathcal{H}_1)$, then the operator $P_{\mathcal{H}}x \upharpoonright \mathcal{H}$ is denoted by $\mathrm{pr}_{\mathcal{H}}\,x$ or simply by $\mathrm{pr}\,x$.

Proposition 7.2.16. *Let $\mathcal{B}$ be an O*-family in a Hilbert space $\mathcal{H}_1$, and let $\mathcal{H}$ be a closed linear subspace of $\mathcal{H}_1$. Suppose that $\mathcal{A}$ is an O*-family in the Hilbert space $\mathcal{H}$ such that $\mathcal{D}(\mathcal{A}) \subseteq \mathcal{D}(\mathcal{B})$ and $\mathcal{B} \upharpoonright \mathcal{D}(\mathcal{A}) = \mathcal{A}$. Then we have $\mathrm{pr}_{\mathcal{H}}\,y^*x \in \mathcal{A}'_{\mathrm{w}}$ for all $x \in \mathcal{B}'_{\mathrm{w}}$ and $y \in \mathcal{B}'_{\mathrm{s}}$. In particular, $\mathrm{pr}_{\mathcal{H}}\,x \in \mathcal{A}'_{\mathrm{w}}$ when $x \in \mathcal{B}'_{\mathrm{w}}$.*

Proof. Suppose $a \in \mathcal{A}$. By assumption, there is a $b \in \mathcal{B}$ such that $a = b \upharpoonright \mathcal{D}(\mathcal{A})$. For $\varphi, \psi \in \mathcal{D}(\mathcal{A})$, we have

$$\langle b\varphi, \psi\rangle = \langle a\varphi, \psi\rangle = \langle \varphi, b^+\psi\rangle = \langle \varphi, a^+\psi\rangle.$$

Hence $P_{\mathcal{H}}b^+ \upharpoonright \mathcal{D}(\mathcal{A}) = a^+$. Since $b^+ \upharpoonright \mathcal{D}(\mathcal{A})$ is in $\mathcal{A}$, b^+ maps $\mathcal{D}(\mathcal{A})$ into $\mathcal{H}$ and so $(I - P_{\mathcal{H}})\,b^+ = 0$. Therefore, $b^+ \upharpoonright \mathcal{D}(\mathcal{A}) = a^+$. Using this and the assumptions, we obtain

$$\begin{aligned}\langle(\mathrm{pr}\,y^*x)\,a\varphi, \psi\rangle &= \langle y^*xa\varphi, \psi\rangle = \langle xb\varphi, y\psi\rangle = \langle x\varphi, b^+y\psi\rangle = \langle x\varphi, yb^+\psi\rangle \\ &= \langle y^*x\varphi, a^+\psi\rangle = \langle(\mathrm{pr}\,y^*x)\,\varphi, a^+\psi\rangle\end{aligned}$$

for $x \in \mathcal{B}'_{\mathrm{w}}$, $y \in \mathcal{B}'_{\mathrm{s}}$ and $\varphi, \psi \in \mathcal{D}(\mathcal{A})$. This shows that $\mathrm{pr}\,y^*x \in \mathcal{A}'_{\mathrm{w}}$. □

Remark 6. Exactly the same argument as above, yields the corresponding result for single operators. Suppose that a and b are symmetric operators in Hilbert spaces $\mathcal{H}$ and $\mathcal{H}_1$, respectively, such that $\mathcal{H}$ is a subspace of $\mathcal{H}_1$ and $a \subseteq b$. Then we have $\mathrm{pr}_{\mathcal{H}}\,y^*x \in (a)'_{\mathrm{w}}$ for $x \in (b)'_{\mathrm{w}}$ and $y \in (b)'_{\mathrm{s}}$. If in addition $\mathcal{H} = \mathcal{H}_1$, then $y^*x \in (a)'_{\mathrm{w}}$ for $x \in (b)'_{\mathrm{w}}$ and $y \in (b)'_{\mathrm{s}}$.

Remark 7. Suppose that b is a self-adjoint extension of the symmetric operator a in the same Hilbert space. Then all bounded measurable functions of b are obviously in $(b)'_{\mathrm{w}}$ and hence in $(a)'_{\mathrm{w}}$ by Remark 6, but they are not in $(a)'_{\mathrm{s}}$ in general.

7.3. Commutants of Strictly Self-Adjoint O*-Algebras

In the first subsection we develop some auxiliary results which are of some interest in its own right. They will be applied in the proof of the main theorem in the second subsection.

Preliminary Results on Operators Affiliated with von Neumann Algebras

We introduce some notation which will be used throughout this section. Suppose a and b are closable linear operators in a Hilbert space $\mathcal{H}$. We write $a \lesssim b$ if $\mathcal{D}(b) \subseteq \mathcal{D}(a)$ and $\|a\varphi\| \leqq \|b\varphi\|$ for all $\varphi \in \mathcal{D}(b)$. (This corresponds to the notation of Section 2.3.) If a and b are symmetric operators on $\mathcal{H}$ such that $\mathcal{D}(b) \subseteq \mathcal{D}(a)$ and $\langle a\varphi, \varphi\rangle \leqq \langle b\varphi, \varphi\rangle$ for $\varphi \in \mathcal{D}(b)$, then we say that $a \leqq b$. If a and b are positive self-adjoint operators on $\mathcal{H}$, then $a \lesssim b$ means that $\mathcal{D}(b^{1/2}) \subseteq \mathcal{D}(a^{1/2})$ and $\|a^{1/2}\varphi\| \leqq \|b^{1/2}\varphi\|$ for all $\varphi \in \mathcal{D}(b^{1/2})$. (Note that "$\lesssim$" is the order relation induced by the associated sesquilinear forms; see e.g. KATO [1], VI, § 2, 6.)

Lemma 7.3.1. *If a and b are densely defined closed operators in a Hilbert space $\mathcal{H}$, then the following conditions are equivalent:*
(i) $a \lesssim b$, (ii) $|a| \lesssim |b|$, (iii) $|a|^2 \lesssim |b|^2$.

Proof. (i) $\leftrightarrow$ (ii) follows from the fact that $\mathcal{D}(c) = \mathcal{D}(|c|)$ and $\|c\cdot\| = \||c|\cdot\|$ for any densely defined closed operator c. (ii) $\leftrightarrow$ (iii) is only a reformulation of the definitions. □

Recall that $\mathbb{A}(\mathcal{N})$ denotes the set of all densely defined closed operators which are affiliated with a von Neumann algebra $\mathcal{N}$.

Lemma 7.3.2. *If $\mathcal{N}$ is an abelian von Neumann algebra and if $a, b \in \mathbb{A}(\mathcal{N})$, then the following four statements are equivalent:*
(i) $a \lesssim b$, (ii) $a^k \lesssim b^k$ *for all* $k \in \mathbb{N}$, (iii) $|a| \lesssim |b|$, (iv) $|a| \leqq |b|$.

Proof. (i) $\to$ (ii) and (i) $\to$ (iii): Let e_n, $n \in \mathbb{N}$, be the spectral projection of the positive self-adjoint operator b associated with the interval $[0, n]$. Fix $n \in \mathbb{N}$. Since $a \lesssim b$, $|a| \lesssim |b|$ by Lemma 7.3.1. This implies that $a_n := |a|\, e_n$ and $b_n := |b|\, e_n$ are bounded operators on $\mathcal{H}$ satisfying $a_n \lesssim b_n$. From $a, b \in \mathbb{A}(\mathcal{N})$ we obtain $e_n, a_n, b_n \in \mathcal{N}$. Hence $a_n = a_n e_n = e_n a_n = e_n\, |a|\, e_n \geqq 0$ and similarly $b_n \geqq 0$. By $a_n \lesssim b_n$, $a_n^2 \leqq b_n^2$. By realizing the commutative von Neumann algebra $\mathcal{N}$ as a $*$-algebra of continuous functions on a compact Hausdorff space, it follows that $a_n^\alpha \leqq b_n^\alpha$ for any $\alpha > 0$. In order to prove (ii), we set $\alpha = 2k$, where $k \in \mathbb{N}$. Since $\mathcal{N}$ is abelian, the operators in $\mathbb{A}(\mathcal{N})$ are normal. (In this proof we freely use the properties of $\mathbb{A}(\mathcal{N})$ for abelian $\mathcal{N}$ stated in Lemma 1.6.3.) This yields $(a^*a)^k = (a^*)^k a^k$. Therefore, $a_n^{2k} = (|a|\, e_n)^{2k} = |a|^{2k} e_n = (a^*a)^k e_n = (a^*)^k a^k e_n$ and so $a_n^{2k} = e_n a_n^{2k} = e_n (a^*)^k a^k e_n$. Similarly, $b_n^{2k} = e_n (b^*)^k b^k e_n$. Combined with $a_n^\alpha \leqq b_n^\alpha$, the latter gives $\|a^k e_n \varphi\| \leqq \|b^k e_n \varphi\|$ for $\varphi \in \mathcal{H}$. Suppose $\varphi \in \mathcal{D}(a^k) \cap \mathcal{D}(b^k)$. Then $\|e_n a^k \varphi\| = \|a^k e_n \varphi\| \leqq \|b^k e_n \varphi\| = \|e_n b^k \varphi\|$. Letting $n \to \infty$, we get $\|a^k\varphi\| \leqq \|b^k\varphi\|$. Since $\mathcal{D}(a^k) \cap \mathcal{D}(b^k)$ is a core for b^k, this implies that $a^k \lesssim b^k$, thus proving (ii). Now we set $\alpha = 1$. Since $e_n\, |a|\, e_n \equiv a_n \leqq b_n \equiv e_n\, |b|\, e_n$, it follows that $\|\, |a|^{1/2} e_n \varphi\| \leqq \|\, |b|^{1/2} e_n \varphi\|$ for $\varphi \in \mathcal{H}$. Proceeding as in the proof of (ii) just given, we conclude that $|a|^{1/2} \lesssim |b|^{1/2}$. By Lemma 7.3.1, $|a| \lesssim |b|$ and (iii) is proved.

(ii) $\to$ (i) is trivial.

(iii) $\to$ (i): From $|a| \lesssim |b|$, $|a|^{1/2} \lesssim |b|^{1/2}$. Therefore, by the implication (i) $\to$ (ii), $|a| \lesssim |b|$, so that $a \lesssim b$.

(iii) $\to$ (iv): Since (iii) $\to$ (i) as just shown, $|a| \lesssim |b|$ implies that $\mathcal{D}(|b|) \subseteq \mathcal{D}(|a|)$. By the assumption $|a| \lesssim |b|$ we have

$$\langle |a| \varphi, \varphi \rangle = \| \, |a|^{1/2} \varphi\|^2 \leqq \| \, |b|^{1/2} \varphi\|^2 = \langle |b| \varphi, \varphi \rangle \quad \text{for}$$

$\varphi \in \mathcal{D}(|b|) \left(\subseteq \mathcal{D}(|b|^{1/2}) \cap \mathcal{D}(|a|^{1/2})\right)$, i.e., $|a| \leqq |b|$.

(iv) $\to$ (iii): From $|a| \leqq |b|$, $\| |a|^{1/2} \varphi\| \leqq \| \, |b|^{1/2} \varphi\|$ for $\varphi \in \mathcal{D}(|b|) \left(\subseteq \mathcal{D}(|a|^{1/2}) \cap \mathcal{D}(|b|^{1/2})\right)$. Since $\mathcal{D}(|b|)$ is a core for $|b|^{1/2}$, this implies that $|a|^{1/2} \lesssim |b|^{1/2}$, that is, $|a| \lesssim |b|$. $\square$

Proposition 7.3.3. *Let $\mathcal{N}$ be a von Neumann algebra with center $\mathcal{Z}$ acting on a Hilbert space $\mathcal{H}$. Suppose that a is a positive self-adjoint operator affiliated with $\mathcal{N}$.*

(i) *There is a largest (relative to the relation "$\lesssim$") positive self-adjoint operator b affiliated with $\mathcal{Z}$ such that $b \lesssim a$. This operator is uniquely determined by a and denoted by $[a]_{\mathcal{Z}}$ or simply by $[a]$ if no confusion can arise.*

(ii) *If c is a positive self-adjoint operator affiliated with $\mathcal{N}'$ such that $c \lesssim a$, then $c \lesssim [a]_{\mathcal{Z}}$.*

(iii) $[a^2]_{\mathcal{Z}} = [a]^2_{\mathcal{Z}}$.

Proof. (i): Fix $n \in \mathbb{N}$. Let e_n denote the spectral projection of a associated with the interval $[0, n]$. Since $a \in \mathbf{A}(\mathcal{N})$, $e_n \in \mathcal{N}$ and $ae_n \in \mathcal{N}$. Let $\mathcal{Z}_n := \{x \in \mathcal{Z}_+ : xe_n \leqq ae_n\}$. Suppose $x_1, x_2 \in \mathcal{Z}_n$. Since $\mathcal{Z}$ is an abelian von Neumann algebra, there exists a projection $e \in \mathcal{Z}$ such that $x_1 e \leqq x_2 e$ and $x_1(I - e) \geqq x_2(I - e)$. Then $x_3 := x_1(I - e) + x_2 e \in \mathcal{Z}_n$, $x_3 \geqq x_1$ and $x_3 \geqq x_2$ which proves that $\mathcal{Z}_n$ is upward directed. Thus $\mathcal{Z}_n e_n$ is a non-empty upward directed subset of $\mathcal{Z} e_n$ which is closed in the weak-operator topology. Hence there exists an operator $x_n \in \mathcal{Z}_n$ such that $x_n e_n = \sup \mathcal{Z}_n e_n$.

We define a positive sesquilinear form $\mathfrak{h}$ with domain $\mathcal{D}_{\mathfrak{h}} := \left\{\varphi \in \mathcal{H} : \sup_{n \in \mathbb{N}} \langle x_n e_n \varphi, \varphi \rangle < \infty\right\}$ by $\mathfrak{h}(\varphi, \psi) := \lim_n \langle x_n e_n \varphi, \psi \rangle$ for $\varphi, \psi \in \mathcal{D}_{\mathfrak{h}}$. Since $\mathcal{Z}_n e_n \subseteq \mathcal{Z}_{n+1} e_{n+1}$ and hence $0 \leqq x_n e_n \leqq x_{n+1} e_{n+1}$ for $n \in \mathbb{N}$, $\mathfrak{h}(\varphi, \varphi) = \lim_n \langle x_n e_n \varphi, \varphi \rangle = \sup_{n \in \mathbb{N}} \langle x_n e_n \varphi, \varphi \rangle$ exists and is finite for all $\varphi \in \mathcal{D}_{\mathfrak{h}}$. From the polarization formula it follows that $\mathfrak{h}(\varphi, \psi)$ exists for all φ and ψ in $\mathcal{D}_{\mathfrak{h}}$. It is straightforward to check that $\mathfrak{h}$ is closed. By the representation theorem of forms (cf. KATO [1], VI, § 2, Theorem 2.23), there exists a positive self-adjoint operator $[a]$ on $\mathcal{H}$ such that $\mathcal{D}_{\mathfrak{h}} = \mathcal{D}([a]^{1/2})$ and $\mathfrak{h}(\varphi, \psi) = \langle [a]^{1/2} \varphi, [a]^{1/2} \psi \rangle$ for all $\varphi, \psi \in \mathcal{D}_{\mathfrak{h}}$. Suppose $\varphi \in \mathcal{D}(a^{1/2})$. From $\sup_{n \in \mathbb{N}} \langle x_n e_n \varphi, \varphi \rangle \leqq \sup_{n \in \mathbb{N}} \langle a e_n \varphi, \varphi \rangle \leqq \|a^{1/2} \varphi\|^2$ it follows that $\varphi \in \mathcal{D}_{\mathfrak{h}} \equiv \mathcal{D}([a]^{1/2})$ and $\|[a]^{1/2} \varphi\|^2 = \mathfrak{h}(\varphi, \varphi) \leqq \|a^{1/2} \varphi\|^2$. This proves that $[a] \lesssim a$.

We show that $[a]$ is affiliated with $\mathcal{Z}$. Let $k \in \mathbb{N}$ and $x \in \mathcal{N}' \cup e_k \mathcal{N} e_k$. If $x \in e_k \mathcal{N} e_k$, then x commutes with $x_n e_n$ if $n \geqq k$, since $x_n \in \mathcal{Z}$ and $e_k e_n = e_k$ for $n \geqq k$. If $x \in \mathcal{N}'$, then x obviously commutes with $x_n e_n$. Combined with the definition of $\mathfrak{h}$, this implies that x and also x^* map $\mathcal{D}_{\mathfrak{h}}$ into itself and $\mathfrak{h}(x\varphi, \psi) = \mathfrak{h}(\varphi, x^* \psi)$ for $\varphi, \psi \in \mathcal{D}_{\mathfrak{h}}$. Let $\psi \in \mathcal{D}([a])$. Then $\psi \in \mathcal{D}([a]^{1/2}) \equiv \mathcal{D}_{\mathfrak{h}}$ and $\langle \varphi, x^*[a] \psi \rangle = \langle [a]^{1/2} x\varphi, [a]^{1/2} \psi \rangle = \mathfrak{h}(x\varphi, \psi) = \mathfrak{h}(\varphi, x^*\psi) = \langle [a]^{1/2} \varphi, [a]^{1/2} x^* \psi \rangle$ for all $\varphi \in \mathcal{D}([a]^{1/2})$. From this we conclude that $[a]^{1/2} x^* \psi \in \mathcal{D}\left(([a]^{1/2})^*\right) \equiv \mathcal{D}([a]^{1/2})$ and $x^*[a] \psi = [a]^{1/2} [a]^{1/2} x^* \psi$, i.e., $x^*[a] \psi = [a] x^* \psi$. Hence $x^* \in ([a])'_s$ and $\mathcal{N}' \cup e_k \mathcal{N} e_k \subseteq ([a])'_s$. Since the operator $[a]$ is self-adjoint, $([a])'_s$ is a von Neumann algebra. Therefore, by letting $k \to \infty$, we get $\mathcal{N}' \cup \mathcal{N} \subseteq ([a])'_s$ and so $(\mathcal{N}' \cup \mathcal{N})'' = \mathcal{Z}' \subseteq ([a])'_s$. The latter means that $[a]$ is affiliated with $\mathcal{Z}$.

Let b be another positive self-adjoint operator on $\mathscr{H}$ such that $b \in \mathbb{A}(\mathscr{Z})$ and $b \preceq a$. Let $k \in \mathbb{N}$ and $n \in \mathbb{N}$. Let f_k denote the spectral projection of b associated with $[0, k]$. Then $bf_k \in \mathscr{Z}$ and $f_k \in \mathscr{Z}$. If $\varphi \in \mathscr{H}$, then we have $f_k e_n \varphi = e_n f_k \varphi \in \mathscr{D}(a) \cap \mathscr{D}(b)$ and

$$\langle bf_k e_n \varphi, \varphi \rangle = \langle bf_k e_n \varphi, f_k e_n \varphi \rangle \leqq \langle af_k e_n \varphi, f_k e_n \varphi \rangle = \langle f_k a e_n \varphi, f_k e_n \varphi \rangle$$
$$= \|f_k (ae_n)^{1/2} \varphi\|^2 \leqq \|(ae_n)^{1/2} \varphi\|^2 = \langle ae_n \varphi, \varphi \rangle,$$

where the first inequality follows from the relation $b \preceq a$. That is, $bf_k e_n \leqq ae_n$ and $bf_k \in \mathscr{Z}_n$. This leads to $e_n bf_k \equiv bf_k e_n \leqq x_n e_n$. If $\varphi \in \mathscr{D}([a]^{1/2})$, then

$$\|b^{1/2} f_k \varphi\|^2 = \lim_n \langle e_n bf_k \varphi, \varphi \rangle \leqq \lim_n \langle x_n e_n \varphi, \varphi \rangle$$
$$= \mathfrak{h}(\varphi, \varphi) = \|[a]^{1/2} \varphi\|^2 \quad \text{for all } k \in \mathbb{N}.$$

Therefore, $\varphi \in \mathscr{D}(b^{1/2})$ and $\|b^{1/2}\varphi\| \leqq \|[a]^{1/2} \varphi\|$, i.e., $b \preceq [a]$.

The uniqueness of the operator $[a]$ follows from the antisymmetry of the relation "$\preceq$".

(ii): Let $\mathscr{N}_1$, $\mathscr{N}_2$ and $\mathscr{N}_3$ denote the von Neumann algebras generated by the spectral projections of a and $[a]$, c and $[a]$ and a and c, respectively. Since $a \in \mathbb{A}(\mathscr{N})$, $[a] \in \mathbb{A}(\mathscr{Z})$ and $c \in \mathbb{A}(\mathscr{N}')$, these three algebras are abelian. Moreover, $a \in \mathbb{A}(\mathscr{N}_1)$, $[a] \in \mathbb{A}(\mathscr{N}_1)$, $c \in \mathbb{A}(\mathscr{N}_2)$, $[a] \in \mathbb{A}(\mathscr{N}_2)$, $a \in \mathbb{A}(\mathscr{N}_3)$ and $c \in \mathbb{A}(\mathscr{N}_3)$. In the rest of this proof we freely use the properties of operators which are affiliated with an abelian von Neumann algebra as stated in Lemma 1.6.3. Further, we use Lemma 7.3.2 without mention. From Lemma 1.6.3, the operators $\overline{a - [a]}$ and $\overline{[a] - c}$ are self-adjoint. Since $[a] \preceq a$, we have $[a] \leqq a$ which implies that $\overline{a - [a]} \geqq 0$. Let ε and δ be positive numbers such that $\varepsilon < \delta$. Let v_ε and u_δ be the spectral projections of $\overline{a - [a]}$ and $\overline{[a] - c}$ associated with the intervals $[0, \varepsilon]$ and $(-\infty, -\delta]$, respectively. We show that the central carrier $z(v_\varepsilon)$ of v_ε is I. From the inequalities $\overline{a - [a]} \geqq \varepsilon(I - v_\varepsilon) \geqq \varepsilon\big(I - z(v_\varepsilon)\big)$ we see that the positive self-adjoint operator $b := [a] + \varepsilon\big(I - z(v_\varepsilon)\big)$ in $\mathbb{A}(\mathscr{Z})$ satisfies $b \leqq a$. Hence $b \preceq a$. Therefore, by (i), $b \preceq [a]$ and so $b \leqq [a]$. But this is only possible if $z(v_\varepsilon) = I$.

Recall that $c \preceq a$ by assumption. Hence $c \leqq a$. By construction we have $v_\varepsilon \in \mathscr{N}_1 \subseteq \mathscr{N}$ and $u_\delta \in \mathscr{N}_2 \subseteq \mathscr{N}'$, so that $v_\varepsilon u_\delta = u_\delta v_\varepsilon$. Suppose that $\varphi \in \mathscr{D}(a)$. Since $\mathscr{D}([a]) \supseteq \mathscr{D}(a)$, $\varphi \in \mathscr{D}([a])$. Since $[a] \in \mathbb{A}(\mathscr{Z})$, $v_\varepsilon \in \mathscr{N}$ and $u_\delta \in \mathscr{N}'$, this gives $v_\varepsilon u_\delta \varphi \in \mathscr{D}([a])$. By the definition of v_ε, $v_\varepsilon u_\delta \varphi \in \mathscr{D}(\overline{a - [a]})$. Combined with the relation $\mathscr{D}(\overline{a - [a]}) \cap \mathscr{D}([a]) \subseteq \mathscr{D}(a)$, the preceding yields $v_\varepsilon u_\delta \varphi \in \mathscr{D}(a)$. Since $c \leqq a$ and hence $\mathscr{D}(c) \supseteq \mathscr{D}(a)$, $v_\varepsilon u_\delta \varphi \in \mathscr{D}(c)$. Using these facts, the spectral theorem and the relation $c \leqq a$, we have

$$\langle av_\varepsilon u_\delta \varphi, v_\varepsilon u_\delta \varphi \rangle = \langle (a - [a])\, v_\varepsilon u_\delta \varphi, v_\varepsilon u_\delta \varphi \rangle + \langle ([a] - c)\, u_\delta v_\varepsilon \varphi, u_\delta v_\varepsilon \varphi \rangle$$
$$+ \langle cv_\varepsilon u_\delta \varphi, v_\varepsilon u_\delta \varphi \rangle \leqq \varepsilon \|v_\varepsilon u_\delta \varphi\|^2 - \delta \|v_\varepsilon u_\delta \varphi\|^2 + \langle av_\varepsilon u_\delta \varphi, v_\varepsilon u_\delta \varphi \rangle.$$

Since $\varepsilon < \delta$, this yields $v_\varepsilon u_\delta = 0$ on $\mathscr{D}(a)$ and so on $\mathscr{H}$. Since $v_\varepsilon \in \mathscr{N}$ and $u_\delta \in \mathscr{N}'$, it follows from the equality $v_\varepsilon u_\delta = 0$ that $z(v_\varepsilon)\, z(u_\delta) = 0$ (Kadison/Ringrose [1], 5.5.4). As shown above, $z(v_\varepsilon) = I$. Therefore, $z(u_\delta) = 0$ and hence $u_\delta = 0$. This shows that the operator $\overline{[a] - c}$ is positive. Therefore, if $\varphi \in \mathscr{D}([a]) \cap \mathscr{D}(c)$, then

$$\langle [a]\, \varphi, \varphi \rangle = \|[a]^{1/2} \varphi\|^2 \geqq \|c^{1/2} \varphi\|^2 = \langle c\varphi, \varphi \rangle.$$

Since $\mathscr{D}([a]) \cap \mathscr{D}(c)$ is a core for $[a]^{1/2}$, the latter implies that $c \preceq [a]$.

(iii): By (i), $[a] \lesssim a$. Recall that a and $[a]$ are affiliated with the abelian von Neumann algebra $\mathcal{N}_1$. Therefore, by Lemma 7.3.2, $[a] \lesssim a$, $[a]^2 \lesssim a^2$ and so $[a]^2 \lesssim a^2$. By the characterization given in (i) this implies $[a]^2 \lesssim [a^2]$. By Lemma 7.3.1, $[a^2] \lesssim a^2$ yields $[a^2]^{1/2} \lesssim a$. Hence $[a^2]^{1/2} \lesssim a$ by Lemma 7.3.2 and $[a^2]^{1/2} \lesssim [a]$ by (i). Applying Lemma 7.3.2 once more, $[a^2] \lesssim [a]^2$. By the antisymmetry of the relation "$\lesssim$", $[a]^2 = [a^2]$. □

Remark 1. If $\mathcal{N}$ is a factor, then, of course, $[a] = \lambda I$ for some $\lambda \geqq 0$ for any positive self-adjoint operator $a \in \mathbb{A}(\mathcal{N})$.

Corollary 7.3.4. *Let $\mathcal{N}$ be as in Proposition 7.3.3. Suppose a, a_1, a_2 and c are closed operators on $\mathcal{H}$ such that a, a_1 and a_2 are affiliated with $\mathcal{N}$ and c is affiliated with $\mathcal{N}'$.*

(i) $[|a|] \lesssim a$.

(ii) *If* $c \lesssim a$, *then* $c \lesssim [|a|]$.

(iii) *If* $a_1 \lesssim a_2$, *then* $[|a_1|] \lesssim [|a_2|]$.

Proof. (i): By Proposition 7.3.3, (i), $[|a|] \lesssim |a|$. Since $|a|$ and $[|a|]$ are affiliated with a common abelian von Neumann algebra, Lemma 7.3.2 gives $[|a|] \lesssim a$.

(ii): Since $a \in \mathbb{A}(\mathcal{N})$ and $c \in \mathbb{A}(\mathcal{N}')$, a and c are affiliated with a common abelian von Neumann algebra and Lemma 7.3.2 applies. Since $c \lesssim a$, $|c| \lesssim |a|$ and hence $|c| \lesssim [|a|]$ by Proposition 7.3.3, (ii). Thus $c \lesssim [|a|]$.

(iii): From $a_1 \lesssim a_2$ we get $|a_1|^2 \lesssim |a_2|^2$ by Lemma 7.3.1. This implies $[|a_1|^2] \lesssim [|a_2|^2]$. Combined with Proposition 7.3.3, (iii), the latter yields $[|a_1|^2] \lesssim [|a_2|^2]$, so that $[|a_1|] \lesssim [|a_2|]$ again by Lemma 7.3.1. □

Commutants of Strictly Self-Adjoint O*-Algebras

Definition 7.3.5. A closed O*-algebra $\mathcal{A}$ is said to be *strictly self-adjoint* if there exists a subset $\{a_i : i \in I\}$ of $\mathcal{A}$ such that:

(i) For every $i \in I$ the operator a_i is formally normal and the operator $a_i^+ a_i$ is essentially self-adjoint.

(ii) The family of seminorms $\{\|\cdot\|_{a_i} : i \in I\}$ is directed and generates the graph topology $t_{\mathcal{A}}$.

Remark 2. Of course, condition (i) is fulfilled if $a_i = b_i + \delta_i I$, where $\delta_i \in \mathbb{C}$ and b_i is a symmetric operator in $\mathcal{A}$ such that b_i^2 is essentially self-adjoint. From this it follows in particular that a closed O*-algebra $\mathcal{A}$ is strictly self-adjoint provided that there exists an O*-subalgebra $\mathcal{A}_0$ of $\mathcal{A}$ with $t_{\mathcal{A}} = t_{\mathcal{A}_0}$ such that each symmetric operator in $\mathcal{A}_0$ is essentially self-adjoint.

Remark 3. *Each closed commutatively dominated O*-algebra is strictly self-adjoint.*

Proof. By Lemma 2.2.15, the O-family $\mathcal{A}_0$ in Definition 2.2.14 can be chosen to be an O*-algebra. Since the closures of symmetric operators in $\mathcal{A}_0$ are affiliated with an abelian von Neumann algebra, they are self-adjoint, and the O*-algebra is strictly self-adjoint by Remark 2. □

Thus in particular the O*-algebras in Examples 2.2.16 and 2.5.2 are strictly self-adjoint. All examples of strictly self-adjoint O*-algebras occuring in the monograph are commutatively dominated O*-algebras.

Remark 4. In this remark we use the terminology and some results of Chapters 9 and 10. If π is a G-integrable $*$-representation of the enveloping algebra $\mathscr{E}(\mathfrak{g})$ of the Lie algebra $\mathfrak{g}$ of a Lie group G

(see Chapter 10), then the O*-algebra $\pi(\mathscr{E}(\mathfrak{g}))$ is strictly self-adjoint. This follows immediately from the Corollaries 10.2.3, 10.2.4 and 10.2.5. Further, if A is a commutative $*$-algebra with unit and π is a $*$-representation of A, then the O*-algebra $\pi(\mathsf{A})$ is strictly self-adjoint if and only if π is integrable. This will be stated in Corollary 9.1.3. In this way a large class of examples of strictly self-adjoint O*-algebras will be obtained by taking integrable representations for π. On the other hand, this shows that the non-integrable self-adjoint $*$-representations π of the polynomial algebra $\mathbb{C}[\mathsf{x}_1, \mathsf{x}_2]$ constructed in Section 9.4 give rise to examples of self-adjoint O*-algebras $\pi(\mathbb{C}[\mathsf{x}_1, \mathsf{x}_2])$ which are not strictly self-adjoint.

Remark 5. Suppose that a_i is an operator as in condition (i). Then, by Proposition 7.1.3, $\overline{a_i}$ is a normal operator. In particular, if a_i is symmetric, then $\overline{a_i}$ is self-adjoint.

The following theorem is our first main result in this section.

Theorem 7.3.6. *Suppose that $\mathcal{A}$ is a strictly self-adjoint O*-algebra.*

(i) *$\mathcal{A}$ is a self-adjoint O*-algebra which satisfies $\mathcal{A}_\mathrm{f}^\mathrm{c} = \mathcal{A}_\mathrm{w}^\mathrm{c} = \mathcal{A}_\mathrm{s}^\mathrm{c}$ and $\mathcal{A}'_\mathrm{w} = \mathcal{A}'_\mathrm{s} = \mathcal{A}'_\mathrm{ss}$. $\mathcal{A}^\mathrm{c}$ is an O*-algebra on $\mathcal{D}(\mathcal{A})$ and $\mathcal{A}'$ is a von Neumann algebra.*

(ii) *Suppose c is a symmetric operator in $\mathcal{A}^\mathrm{c}$. Then $\bar{c}$ is a self-adjoint operator which is affiliated with the von Neumann algebra $\mathcal{A}'$. If x is an essentially self-adjoint operator in $\mathcal{L}^+(\mathcal{D}_\mathcal{A})$ such that $xc\varphi = cx\varphi$ for $\varphi \in \mathcal{D}(\mathcal{A})$, then the self-adjoint operators $\bar{x}$ and $\bar{c}$ strongly commute.*

Proof. Throughout this proof, let $\{a_i : i \in I\}$ be as in Definition 7.3.5.

(i): For $i \in I$, $\overline{a_i}$ is a normal operator by Remark 5 and hence $\mathcal{D}(\overline{a_i}) = \mathcal{D}(a_i^*)$. Therefore, $\mathcal{D}^*(\mathcal{A}) \subseteq \bigcap_{i\in I} \mathcal{D}(a_i^*) = \bigcap_{i\in I} \mathcal{D}(\overline{a_i}) = \hat{\mathcal{D}}(\mathcal{A})$, where the last equality follows from condition (ii) in Definition 7.3.5 and Proposition 2.2.12. Since $\mathcal{A}$ is closed, $\mathcal{D}^*(\mathcal{A}) \subseteq \mathcal{D}(\mathcal{A})$, and $\mathcal{A}$ is self-adjoint. From Definition 7.3.5, (ii), there are $i_0 \in I$ and $\lambda > 0$ such that $\|\cdot\| \leqq \lambda \|a_{i_0}\cdot\|$. Upon replacing $\{a_i : i \in I\}$ by the set $\{\lambda a_i : i \in I \text{ and } \|a_{i_0}\cdot\| \leqq \|a_i\cdot\|\}$, we can assume without loss of generality that all operators a_i, $i \in I$, are contained in $\mathcal{A}(I)$. Fix $i \in I$. By $a_i \in \mathcal{A}(I)$, $a_i^+ a_i \geqq I$. Hence $a_i^+ a_i \mathcal{D}$ is dense in $\mathcal{H}$, since $a_i^+ a_i$ is essentially self-adjoint by Definition 7.3.5, (i). From the functional calculus for the normal operator $\overline{a_i}$ we conclude that $w_i := \overline{a_i}^2 |\overline{a_i}|^{-2}$ is a unitary operator on $\mathcal{H}$. Therefore, $w_i(a_i^+ a_i \mathcal{D}) = \overline{a_i}^2 \mathcal{D}$ is dense in $\mathcal{H}$. Thus the assumptions of Proposition 7.2.5 are satisfied and hence $\mathcal{A}_\mathrm{f}^\mathrm{c} = \mathcal{A}_\mathrm{w}^\mathrm{c}$.

Using the self-adjointness of $\mathcal{A}$ and the equality $\mathcal{A}_\mathrm{f}^\mathrm{c} = \mathcal{A}_\mathrm{w}^\mathrm{c}$, the other statements of (i) are contained in Propositions 7.2.2 and 7.2.10 and in Corollary 7.2.4.

(ii): Suppose $x \in \mathcal{A}$. From $c \in \mathcal{A}^\mathrm{c} \subseteq \mathfrak{L}(\mathcal{D}_\mathcal{A}, \mathcal{H})$ and condition (ii) in Definition 7.3.5 it follows that there exists an $i \in I$ such that x and c are both a_i-bounded. Therefore, Proposition 7.1.6 applies (with $a := a_i$, $c_1 := c$, $c_2 := x$) and shows that $\bar{c}$ is self-adjoint and that $(\bar{c} + \mathrm{i})^{-1} \bar{x} \subseteq \bar{x}(\bar{c} + \mathrm{i})^{-1}$. Hence $(\bar{c} + \mathrm{i})^{-1} \bar{x} \subseteq (x^+)^* (\bar{c} + \mathrm{i})^{-1}$ for all $x \in \mathcal{A}$ which yields $(\bar{c} + \mathrm{i})^{-1} \in \mathcal{A}'_\mathrm{w} = \mathcal{A}'$, so $\bar{c}$ is affiliated with $\mathcal{A}'$.

Suppose that $x \in \mathcal{L}^+(\mathcal{D}_\mathcal{A})$ and $xc = cx$. Then the same proof gives $(\bar{c} + \mathrm{i})^{-1} \bar{x} \subseteq \bar{x}(\bar{c} + \mathrm{i})^{-1}$. Therefore, if $\bar{x}$ is self-adjoint, then $\bar{x}$ and $\bar{c}$ strongly commute by Lemma 1.6.2. □

Now we use the results obtained in the preceding subsection in order to give a much more explicit description of the unbounded commutants $\mathcal{A}^\mathrm{c}$ and $(\widehat{\mathcal{A}^\mathrm{c}})^\mathrm{c}$.

Assume that $\mathcal{A}$ is a strictly self-adjoint O*-algebra. Let $\{a_i : i \in I\}$ be the corresponding set in Definition 7.3.5. Since $\mathcal{A}' = \mathcal{A}'_{ss}$ by Theorem 7.3.6, each operator $\overline{a}_i$, $i \in I$, is affiliated with the von Neumann algebra $\mathcal{A}''$; cf. Proposition 7.2.9. Therefore, by Proposition 7.3.3, the operators $[|\overline{a}_i|]_{\mathcal{Z}}$ are well-defined if we let $\mathcal{N} := \mathcal{A}''$ and if $\mathcal{Z}$ denotes the center of $\mathcal{A}''$. In what follows we omit the subscript $\mathcal{Z}$. Set $c_i := [|\overline{a}_i|] \upharpoonright \mathcal{D}(\mathcal{A})$ and $\hat{c}_i := [|\overline{a}_i|] \upharpoonright \hat{\mathcal{D}}(\mathcal{A}^c)$ for $i \in I$. (Note that $\hat{\mathcal{D}}(\mathcal{A}^c) \subseteq \mathcal{D}([|\overline{a}_i|])$, since $c_i \in \mathcal{A}^c$ as we shall show below.)

Under these assumptions and notations, we have

Theorem 7.3.7. (i) $\mathcal{A}^c = \{xc_i : x \in \mathcal{A}' \text{ and } i \in I\}$.

(ii) $(\widehat{\mathcal{A}^c})^c = \{x\hat{c}_i : x \in \mathcal{A}'' \text{ and } i \in I\}$ *and* $(\widehat{\mathcal{A}^c})' = \mathcal{A}''$.

(iii) $\widehat{\mathcal{A}^c}$ *and* $(\widehat{\mathcal{A}^c})^c$ *are strictly self-adjoint O*-algebras on the same domain* $\hat{\mathcal{D}}(\mathcal{A}^c) = \bigcap_{i \in I} \mathcal{D}([|\overline{a}_i|])$.

More precisely, $\{\hat{c}_i : i \in I\}$ *is a subset of* $\widehat{\mathcal{A}^c}$ *and of* $(\widehat{\mathcal{A}^c})^c$ *which satisfies conditions* (i) *and* (ii) *in Definition* 7.3.5 *for both O*-algebras* $\widehat{\mathcal{A}^c}$ *and* $(\widehat{\mathcal{A}^c})^c$.

Proof. First we prove (i). Let $i \in I$. Since $[|\overline{a}_i|] \lesssim \overline{a}_i$ by Corollary 7.3.4, (i), $c_i \in \mathfrak{L}(\mathcal{D}_{\mathcal{A}}, \mathcal{H})$. By construction, $[|\overline{a}_i|] \in \mathbf{A}(\mathcal{Z})$ and hence $[|\overline{a}_i|] \in \mathbf{A}(\mathcal{A}')$. From Proposition 7.2.11 it therefore follows that $c_i \equiv [|\overline{a}_i|] \upharpoonright \mathcal{D}(\mathcal{A})$ is in $\mathcal{A}^c$. Since $\mathcal{A}^c$ is an O*-algebra and $\mathcal{A}' \upharpoonright \mathcal{D}(\mathcal{A}) \subseteq \mathcal{A}^c$, $xc_i \in \mathcal{A}^c$ for each $x \in \mathcal{A}'$. Conversely, let $c \in \mathcal{A}^c$. Since $c \in \mathfrak{L}(\mathcal{D}_{\mathcal{A}}, \mathcal{H})$, it follows from Definition 7.3.5, (ii), that there exists an $i \in I$ such that $a_i \in \mathcal{A}(I)$ and $\|c\cdot\| \leqq \lambda\|a_i\cdot\|$ for some $\lambda > 0$. Thus $\lambda^{-1}\overline{c} \lesssim \overline{a}_i$. By Corollary 7.3.4, (ii), $\lambda^{-1}\overline{c} \lesssim [|\overline{a}_i|]$. Since $I \lesssim \overline{a}_i$ by $a_i \in \mathcal{A}(I)$, Corollary 7.3.4, (iii), yields $I \lesssim [|\overline{a}_i|]$; so $[|\overline{a}_i|]^{-1} \in \mathbf{B}(\mathcal{H})$. From $\lambda^{-1}\overline{c} \lesssim [|\overline{a}_i|]$ we conclude that $x := \overline{c}[|\overline{a}_i|]^{-1}$ is a bounded operator on $\mathcal{H}$. Since $[|\overline{a}_i|] \in \mathbf{A}(\mathcal{Z})$, $[|\overline{a}_i|]^{-1} \in \mathcal{Z} \subseteq \mathcal{A}'$. Consequently, $[|\overline{a}_i|]^{-1} \upharpoonright \mathcal{D}(\mathcal{A})$ is in $\mathcal{A}^c$. Since $\mathcal{A}^c$ is an algebra and $c \in \mathcal{A}^c$, the latter implies that $x \upharpoonright \mathcal{D}(\mathcal{A})$ is in $\mathcal{A}^c$. Thus $x \in \mathcal{A}'$. Clearly, $c = x([|\overline{a}_i|] \upharpoonright \mathcal{D}(\mathcal{A})) \equiv xc_i$. This completes the proof of (i).

Next we show that the closure $\widehat{\mathcal{A}^c}$ of the O*-algebra $\mathcal{A}^c$ is strictly self-adjoint. From $\mathcal{A}^c = \{xc_i : x \in \mathcal{A}' \text{ and } i \in I\}$ we obviously get $\widehat{\mathcal{A}^c} = \{x\hat{c}_i : x \in \mathcal{A}' \text{ and } i \in I\}$. Hence the graph topology of $\widehat{\mathcal{A}^c}$ is generated by the family of seminorms $\{\|\cdot\|_{\hat{c}_i} : i \in I\}$. In order to prove that this family is directed, let $i, i' \in I$ and $\|a_i\cdot\| \leqq \|a_{i'}\cdot\|$. Then $\overline{a}_i \lesssim \overline{a}_{i'}$, so that $\|\hat{c}_i\cdot\| \leqq \|\hat{c}_{i'}\cdot\|$ by Corollary 7.3.4, (iii). Suppose $i \in I$. As noted above, $c_i \in \mathcal{A}^c$. Hence $c_i^2 \in \mathcal{A}^c$. By Theorem 7.3.6, (ii), $\widehat{c_i^2} \equiv \hat{c}_i^2$ is essentially self-adjoint. This shows that Definition 7.3.5 is satisfied with $\widehat{\mathcal{A}^c}$ and $\hat{c}_i$ in place of $\mathcal{A}$ and a_i, respectively; so $\widehat{\mathcal{A}^c}$ is strictly self-adjoint. Further, since $\overline{c}_i$ is self-adjoint (again by Theorem 7.3.6, (ii)), we have $\overline{c}_i = [|\overline{a}_i|]$ for $i \in I$. Therefore, by Proposition 2.2.12,

$$\hat{\mathcal{D}}(\mathcal{A}^c) = \bigcap_{i \in I} \mathcal{D}(\overline{c}_i) = \bigcap_{i \in I} \mathcal{D}([|\overline{a}_i|]).$$

Now we verify that $(\widehat{\mathcal{A}^c})' = \mathcal{A}''$. First note that the notation $(\widehat{\mathcal{A}^c})'$ makes sense, since $\widehat{\mathcal{A}^c}$ is self-adjoint. Suppose $y \in \mathcal{A}''$. Since $[|\overline{a}_i|] \in \mathbf{A}(\mathcal{Z})$, $y[|\overline{a}_i|] \subseteq [|\overline{a}_i|]\, y$ for $i \in I$. Thus, if $x \in \mathcal{A}'$ and $i \in I$, we have $y(xc_i) \subseteq yx[|\overline{a}_i|] = xy[|\overline{a}_i|] \subseteq x[|\overline{a}_i|]\, y \subseteq (\overline{xc_i})\, y$. Since the operators $x\hat{c}_i$ exhaust $\widehat{\mathcal{A}^c}$, it follows that $y \in (\widehat{\mathcal{A}^c})'_{\mathrm{w}} = (\widehat{\mathcal{A}^c})'$. Conversely, if $y \in (\widehat{\mathcal{A}^c})'$, then y commutes with the subset $\mathcal{A}' \upharpoonright \hat{\mathcal{D}}(\mathcal{A}^c)$ of $\widehat{\mathcal{A}^c}$; hence $y \in \mathcal{A}''$. Thus $(\widehat{\mathcal{A}^c})' = \mathcal{A}''$.

Since $\overline{c}_i = [|\overline{a}_i|]$ for $i \in I$, it is obvious from the characterization of $[|\overline{a}_i|]$ given in Proposition 7.3.3, (i), that $[|\overline{c}_i|] = [|\overline{a}_i|]$. Therefore, since $\widehat{\mathcal{A}^c}$ is strictly self-adjoint and

$(\widehat{\mathcal{A}^c})' = \mathcal{A}''$, the remaining assertion in the theorem follows if we replace in the preceding $\mathcal{A}$ by $\widehat{\mathcal{A}^c}$ and a_i by $\widehat{c}_i$. □

Corollary 7.3.8. *If $\mathcal{A}$ is a strictly self-adjoint O*-algebra, then $\widehat{\mathcal{A}^c}$ and $(\widehat{\mathcal{A}^c})^c$ are commutatively dominated O*-algebras.*

Proof. Let $\mathcal{A}_0$ be the O-family on $\mathcal{D}(\mathcal{A}_0) := \hat{\mathcal{D}}(\mathcal{A}^c)$ formed by the operators I and $\widehat{c}_i$, $i \in I$. By the preceding proof, this O-family is directed and the operators $\widehat{\overline{c}_i} \equiv [|\overline{a}_i|]$, $i \in I$, are affiliated with the abelian von Neumann algebra $\mathcal{Z}$, so Definition 2.2.14 is satisfied. □

Corollary 7.3.9. *If $\mathcal{A}$ is a strictly self-adjoint O*-algebra for which the von Neumann algebra $\mathcal{A}'$ is a factor, then $\mathcal{A}^c = \mathcal{A}' \upharpoonright \mathcal{D}(\mathcal{A})$, i.e., $\mathcal{A}^c$ only consists of bounded operators.*

Proof. In this case $\mathcal{Z}$ is trivial, so each $[|\overline{a}_i|] \in \mathbf{A}(\mathcal{Z})$ and hence c_i is a multiple of the identity. Theorem 7.3.7, (i), gives the assertion. □

An obvious consequence of Corollary 7.3.9 is

Corollary 7.3.10. *If $\mathcal{A}$ is a strictly self-adjoint O*-algebra such that $\mathcal{A}' = \mathbb{C} \cdot I$, then $\mathcal{A}^c = \mathbb{C} \cdot I$.*

7.4. A Class of Subspaces of $\mathcal{L}(\mathcal{D}_\mathcal{A}, \mathcal{D}_\mathcal{B}^+)$

Throughout this section we assume that $\mathcal{A}$ and $\mathcal{B}$ are closed O*-algebras in a Hilbert space $\mathcal{H}$, $\mathcal{N}$ is a von Neumann algebra acting on $\mathcal{H}$ and $\mathcal{C}$ is a non-degenerate $*$-subalgebra of $\mathcal{N}$ such that $\mathcal{C}'' = \mathcal{N}$. (Recall that $\mathcal{C}$ is said to be non-degenerate if the linear span of vectors $c\varphi$, where $c \in \mathcal{C}$ and $\varphi \in \mathcal{H}$, is dense in $\mathcal{H}$.)

In order to formulate our first result, we need some more types of commutants. For a subset $\mathcal{L}$ of $\mathcal{L}(\mathcal{D}_\mathcal{A}, \mathcal{D}_\mathcal{A}^+)$ and a subset $\mathcal{R}$ of $\mathcal{L}^+(\mathcal{D}_\mathcal{A})$, we define $\mathcal{L}^\supset := \{a \in \mathcal{L}^+(\mathcal{D}_\mathcal{A}) : a \circ x = xa$ for all $x \in \mathcal{L}\}$, $\mathcal{R}_\mathrm{f}^\mathrm{c} := \{x \in \mathcal{L}(\mathcal{D}_\mathcal{A}, \mathcal{D}_\mathcal{A}^+) : a \circ x = xa$ for all $a \in \mathcal{R}\}$ and $\mathcal{R}_\mathrm{w}^\mathrm{c} := \{x \in \mathfrak{L}(\mathcal{D}_\mathcal{A}, \mathcal{H}) : a \circ x = xa$ for all $a \in \mathcal{R}\}$. Further, let $\mathcal{L}_\mathrm{b}^\supset$ denote the set of all bounded operators in $\mathcal{L}^\supset$.

Remark 1. If $\mathcal{L} \subseteq \mathcal{L}^+(\mathcal{D}_\mathcal{A})$, then $\mathcal{L}^\supset$ is simply the commutant of $\mathcal{L}$ within the algebra $\mathcal{L}^+(\mathcal{D}_\mathcal{A})_\mathrm{f}^\mathrm{c}$ If $\mathcal{R}$ is an O*-algebra on $\mathcal{D}(\mathcal{R}) \equiv \mathcal{D}(\mathcal{A})$ such that $\mathrm{t}_\mathcal{R} = \mathrm{t}_\mathcal{A}$, then $\mathcal{R}_\mathrm{f}^\mathrm{c}$ is the form commutant $\mathcal{R}$. and $\mathcal{R}_\mathrm{w}^\mathrm{c}$ is the weak unbounded commutant $\mathcal{R}_\mathrm{w}^\mathrm{c}$; cf. Definition 7.2.1.

Proposition 7.4.1. *Suppose that there exists an indexed subset $\{a_j : j \in J\}$ of $\mathcal{A}(I)$ such that $a_j\mathcal{D}(\mathcal{A})$ is dense in $\mathcal{H}$ for every $j \in J$ and such that $\{\|\cdot\|_{a_j} : j \in J\}$ is a directed family of seminorms which generates the graph topology $\mathrm{t}_\mathcal{A}$. Suppose that $\mathcal{N}$ contains the operators $\overline{a}_j^{-1}$, $j \in J$. Let $\mathcal{L}$ be the linear span of $a_j^+ \circ \mathcal{C}a_j$, $j \in J$, in $\mathcal{L}(\mathcal{D}_\mathcal{A}, \mathcal{D}_\mathcal{A}^+)$.*

Then $(\mathcal{L}^\supset)_\mathrm{f}^\mathrm{c} = (\mathcal{L}_\mathrm{b}^\supset)_\mathrm{f}^\mathrm{c} = \bigcup_{j \in J} a_j^+ \circ \mathcal{N} a_j$, and this vector space is equal to the ultraweak closure of $\mathcal{L}$ in $\mathcal{L}(\mathcal{D}_\mathcal{A}, \mathcal{D}_\mathcal{A}^+)$.

Lemma 7.4.2. *Let a and b be operators of $\mathcal{A}$ such that $a\mathcal{D}(\mathcal{A})$ and $b\mathcal{D}(\mathcal{A})$ are dense in $\mathcal{H}$. Let $c \in \mathbb{B}(\mathcal{H})$. Suppose that c (that is, $c \upharpoonright \mathcal{D}(\mathcal{A})$) is in $\mathcal{L}^+(\mathcal{D}_\mathcal{A})$, $ac\varphi = ca\varphi$ and $bc^*\varphi = c^*b\varphi$ for $\varphi \in \mathcal{D}(\mathcal{A})$. Let $z := b^+ \circ xa$, where $x \in \mathbb{B}(\mathcal{H})$. Then $c \circ z = zc$ if and only if $cx = xc$.*

Proof. For $\varphi, \psi \in \mathcal{D}(\mathcal{A})$, we have by definition

$$\langle c \circ z\varphi, \psi\rangle = \langle z\varphi, c^*\psi\rangle = \langle xa\varphi, bc^*\psi\rangle = \langle a\varphi, x^*c^*b\psi\rangle$$

and

$$\langle zc\varphi, \psi\rangle = \langle xac\varphi, b\psi\rangle = \langle a\varphi, c^*x^*b\psi\rangle.$$

Here we essentially used the commutativity assumptions concerning a, c and b, c^*. Since $a\mathcal{D}(\mathcal{A})$ and $b\mathcal{D}(\mathcal{A})$ are assumed to be dense in $\mathcal{H}$, we conclude from the preceding equalities that $c \circ z = zc$ is equivalent to $x^*c^* = c^*x^*$ and so to $cx = xc$. □

Proof of Proposition 7.4.1. First we check that $\mathcal{C}' \subseteq \mathcal{L}^+(\mathcal{D}_\mathcal{A})$. Fix $c \in \mathcal{C}'$. Since $\bar{a}_j^{-1} \in \mathcal{N} \equiv \mathcal{C}''$ by assumption, we have $c\bar{a}_j^{-1} = \bar{a}_j^{-1}c$ and hence $c\bar{a}_j \subseteq \bar{a}_jc$ for $j \in J$. From the assumptions and Proposition 2.2.12 we obtain that $\mathcal{D}(\mathcal{A}) = \bigcap_{j\in J} \mathcal{D}(\bar{a}_j)$. Therefore, the preceding yields $c\mathcal{D}(\mathcal{A}) \subseteq \mathcal{D}(\mathcal{A})$. Since the topology $\mathfrak{t}_\mathcal{A}$ is generated by the seminorms $\|\cdot\|_{a_j}$, $j \in J$, the preceding also shows that $c \upharpoonright \mathcal{D}(\mathcal{A})$ is in $\mathfrak{L}(\mathcal{D}_\mathcal{A})$. Since $\mathcal{C}'$ is a $*$-algebra, we can replace c by c^* and obtain that $c^* \upharpoonright \mathcal{D}(\mathcal{A})$ is in $\mathfrak{L}(\mathcal{D}_\mathcal{A})$. Thus $c \upharpoonright \mathcal{D}(\mathcal{A})$ is in $\mathcal{L}^+(\mathcal{D}_\mathcal{A})$. By the convention formulated in Remark 4 of Section 3.2, $c \in \mathcal{L}^+(\mathcal{D}_\mathcal{A})$ and so $\mathcal{C}' \subseteq \mathcal{L}^+(\mathcal{D}_\mathcal{A})$.

Suppose $c \in \mathcal{C}'$. As just noted, c and c^* commute both with $\bar{a}_j$ and so with a_j on $\mathcal{D}(\mathcal{A})$ for each $j \in J$, since $c \in \mathcal{L}^+(\mathcal{D}_\mathcal{A})$. Therefore, by Lemma 7.4.2, we have $c \circ z = zc$ for every element $z \in \mathcal{L}$ of the form $z = a_j^+ \circ xa_j$, where $x \in \mathcal{C}$ and $j \in J$. Combined with $\mathcal{C}' \subseteq \mathcal{L}^+(\mathcal{D}_\mathcal{A})$, this gives $\mathcal{C}' \subseteq \mathcal{L}_\mathrm{b}^\circ$.

Now suppose that $z \in (\mathcal{L}_\mathrm{b}^\circ)_\mathrm{f}^\mathrm{c}$. By definition, $z \in \mathcal{L}(\mathcal{D}_\mathcal{A}, \mathcal{D}_\mathcal{A}^+)$. From the assumptions and Proposition 3.2.3 (cf. Remark 8 in 3.2) there are an index $j \in J$ and an operator $x \in \mathbf{B}(\mathcal{H})$ such that $z = a_j^+ \circ xa_j$. Since $z \in (\mathcal{L}_\mathrm{b}^\circ)_\mathrm{f}^\mathrm{c}$, $c \circ z = zc$ for each $c \in \mathcal{L}_\mathrm{b}^\circ$ and so in particular for each $c \in \mathcal{C}'$. Lemma 7.4.2, now applied in reversed order, yields $cx = xc$ for all $c \in \mathcal{C}'$, i.e., $x \in \mathcal{C}'' \equiv \mathcal{N}$.

Let x be an arbitrary element of $\mathcal{N}$. Since $\mathcal{C}$ is a non-degenerate $*$-subalgebra of $\mathbf{B}(\mathcal{H})$, the von Neumann density theorem (see e.g. TAKESAKI [1], II, Theorem 3.9) applies and there exists a net $(x_i : i \in I)$ of operators in $\mathcal{C}$ which converges to the operator x of $\mathcal{C}'' \equiv \mathcal{N}$ in the ultraweak topology of $\mathbf{B}(\mathcal{H})$. Then the net $(a_j^+ \circ x_ia_j : i \in I)$ converges to $a_j^+ \circ xa_j$ in the ultraweak topology of $\mathcal{L}(\mathcal{D}_\mathcal{A}, \mathcal{D}_\mathcal{A}^+)$. Since $a_j^+ \circ x_ia_j \in \mathcal{L}$ for $i \in I$, $a_j^+ \circ xa_j$ is in the ultraweak closure $\bar{\mathcal{L}}^{\mathrm{uw}}$ of $\mathcal{L}$ within $\mathcal{L}(\mathcal{D}_\mathcal{A}, \mathcal{D}_\mathcal{A}^+)$. Thus we have shown that $(\mathcal{L}_\mathrm{b}^\circ)_\mathrm{f}^\mathrm{c} \subseteq \bigcup_{j\in J} a_j^+ \circ \mathcal{N} a_j \subseteq \bar{\mathcal{L}}^{\mathrm{uw}}$. Since $\mathcal{L} \subseteq (\mathcal{L}^\circ)_\mathrm{f}^\mathrm{c}$ and $(\mathcal{L}^\circ)_\mathrm{f}^\mathrm{c}$ is obviously ultraweakly closed in $\mathcal{L}(\mathcal{D}_\mathcal{A}, \mathcal{D}_\mathcal{A}^+)$, we have $\bar{\mathcal{L}}^{\mathrm{uw}} \subseteq (\mathcal{L}^\circ)_\mathrm{f}^\mathrm{c}$. Clearly, $(\mathcal{L}^\circ)_\mathrm{f}^\mathrm{c} \subseteq (\mathcal{L}_\mathrm{b}^\circ)_\mathrm{f}^\mathrm{c}$. Combining these relations, we get $(\mathcal{L}_\mathrm{b}^\circ)_\mathrm{f}^\mathrm{c} = (\mathcal{L}^\circ)_\mathrm{f}^\mathrm{c} = \bigcup_{j\in J} a_j^+ \circ \mathcal{N} a_j = \bar{\mathcal{L}}^{\mathrm{uw}}$. □

The next proposition gives a similar result for the ultrastrong topology.

Proposition 7.4.3. *Let* $\{a_j : j \in J\}$ *and* $\mathcal{N}$ *satisfy the assumptions of Proposition* 7.4.1, *and let* $\mathcal{L}$ *be the linear span of the spaces* $\mathcal{C}a_j$, $j \in J$.
Then $(\mathcal{L}_\mathrm{b}^\circ)_\mathrm{w}^\mathrm{c} = (\mathcal{L}^\circ)_\mathrm{w}^\mathrm{c} = \bigcup_{j\in J} \mathcal{N} a_j$, *and this vector space is the ultrastrong closure of* $\mathcal{L}$ *within* $\mathfrak{L}(\mathcal{D}_\mathcal{A}, \mathcal{H})$.

Proof. The proof is similar to the previous proof, so we sketch only the necessary modifications. As shown in the first paragraph of the proof of Proposition 7.4.1, $\mathcal{C}' \subseteq \mathcal{L}^+(\mathcal{D}_\mathcal{A})$ and $ca_j\varphi = a_jc\varphi$ for $c \in \mathcal{C}'$, $j \in J$ and $\varphi \in \mathcal{D}(\mathcal{A})$. Lemma 7.4.2 (applied with $a = a_j$ and $b = I$) yields $\mathcal{C}' \subseteq \mathcal{L}_\mathrm{b}^\circ$. Suppose that $z \in (\mathcal{L}_\mathrm{b}^\circ)_\mathrm{w}^\mathrm{c}$. Since $z \in \mathfrak{L}(\mathcal{D}_\mathcal{A}, \mathcal{H})$, there are $j \in J$ and $x \in \mathbf{B}(\mathcal{H})$ such that $z = xa_j$. From Lemma 7.4.2 and the relation $\mathcal{C}' \subseteq \mathcal{L}_\mathrm{b}^\circ$

we obtain that $x \in \mathcal{C}''$. By the von Neumann density theorem each operator $x \in \mathcal{C}''$ is also in the ultrastrong closure of $\mathcal{C}$. The rest follows similarly as above. □

It is clear that the two preceding propositions can be considered as generalizations of the von Neumann bicommutant theorem. Our next proposition and also Proposition 7.4.9 below could be interpreted as generalized versions of the Kaplansky density theorem. (In order to see this, it suffices to recall that in case $\mathcal{A} = \mathcal{B} = \mathbb{B}(\mathcal{H})$ the space $\mathcal{L}(\mathcal{D}_{\mathcal{A}}, \mathcal{D}_{\mathcal{B}}^+)$ is equal to $\mathbb{B}(\mathcal{H})$ and $\mathcal{U}_{I,I}$ is the unit ball of $\mathbb{B}(\mathcal{H})$.)

Let J denote a fixed index set. For a convenient formulation of the results, the following two conditions are useful:

(I) There exist a subset $\{a_j : j \in J\}$ of $\mathcal{A}(I)$ and a subset $\{b_j : j \in J\}$ of $\mathcal{B}(I)$ such that for each $j \in J$ the operators $\overline{a}_j$ and $\overline{b}_j$ are normal and their inverses $\overline{a}_j^{-1}$ and $\overline{b}_j^{-1}$ belong to $\mathcal{N}$.

(II) The families of seminorms $\{\|\cdot\|_{a_j} : j \in J\}$ and $\{\|\cdot\|_{b_j} : j \in J\}$ are directed, and they generate the graph topologies of $\mathcal{A}$ and $\mathcal{B}$, respectively.

Remark 2. Since $a_j \in \mathcal{A}(I)$, $b_j \in \mathcal{B}(I)$ and $\overline{a}_j$ and $\overline{b}_j$ are normal, the operators $\overline{a}_j^{-1}$ and $\overline{b}_j^{-1}$ are bounded and everywhere defined on $\mathcal{H}$ (by Lemma 7.4.5), so the requirements $\overline{a}_j^{-1} \in \mathcal{N}$ and $\overline{b}_j^{-1} \in \mathcal{N}$ in (I) make sense.

Proposition 7.4.4. *Suppose that the conditions* (I) *and* (II) *are satisfied. Let $\mathcal{L}$ be the vector space spanned by $b_j^+ \circ \mathcal{C} a_j$, $j \in J$, and let $\mathcal{L}_1$ be another linear subspace of $\mathcal{L}(\mathcal{D}_{\mathcal{A}}, \mathcal{D}_{\mathcal{B}}^+)$ which contains $\mathcal{L}$. If $\mathcal{L}$ is weak-operator dense in $\mathcal{L}_1$, then $\mathcal{L} \cap \mathcal{U}_{a_j,b_j}$ is ultraweakly dense in $\mathcal{L}_1 \cap \mathcal{U}_{a_j,b_j}$ for every $j \in J$.*

Before proving this proposition, we require two auxiliary lemmas.

Lemma 7.4.5. *Suppose that a is a closable operator on $\mathcal{H}$ such that $\|a\cdot\| \geqq \|\cdot\|$ and $\overline{a}$ is normal. Then $\overline{a}^{-1}$ is a bounded everywhere defined (normal) operator on $\mathcal{H}$ and*

$$\|\overline{a}^{-2}\varphi\|^2 \leqq \varepsilon^2\|\varphi\|^2 + \varepsilon^{-1}\|\overline{a}^{-3}\varphi\|^2 \quad \textit{for all } \varphi \in \mathcal{H} \textit{ and } \varepsilon > 0.$$

Proof. The first assertion was already shown in the second paragraph of the proof of Lemma 7.1.5. To prove the second assertion, we fix an $\varepsilon > 0$ and let e denote the spectral projection of the normal operator a associated with the set $\{\lambda \in \mathbb{C} : |\lambda|^2 \geqq \varepsilon^{-1}\}$. From the spectral theorem we have for $\varphi \in \mathcal{H}$

$$\|\overline{a}^{-2}\varphi\|^2 = \|\overline{a}^{-2}e\varphi\|^2 + \|\overline{a}^{-2}(I - e)\,\varphi\|^2$$

$$\leqq \varepsilon^2\|e\varphi\|^2 + \varepsilon^{-1}\|\overline{a}^{-3}(I - e)\,\varphi\|^2 \leqq \varepsilon^2\|\varphi\|^2 + \varepsilon^{-1}\|\overline{a}^{-3}\varphi\|^2. \; \square$$

Lemma 7.4.6. *Assume that condition* (I) *is satisfied. Let $\tilde{\mathcal{L}}$ denote the linear subspace of $\mathcal{L}(\mathcal{D}_{\mathcal{A}}, \mathcal{D}_{\mathcal{B}}^+)$ generated by the spaces $b_j^+ \circ \mathcal{N} a_j$, $j \in J$. Then $\mathcal{N}$ is a dense subset of $\tilde{\mathcal{L}}[\tau_{\mathrm{in}}]$.*

Proof. If $x \in \mathcal{N}$ and $j \in J$, then $(\overline{b}_j^{-1})^* \, x\overline{a}_j^{-1} \in \mathcal{N}$ and so $x = b_j^+ \circ \left((\overline{b}_j^{-1})^* \, x\overline{a}_j^{-1}\right) a_j \in \tilde{\mathcal{L}}$. Thus $\mathcal{N}$ is indeed a subset of $\tilde{\mathcal{L}}$. We fix an index j in J and an operator $x \neq 0$ in $\mathcal{N}$. Set $y := b_j^+ \circ xa_j$. Since $\tilde{\mathcal{L}}$ is the linear span of such elements y, it suffices to show that y is in the closure of $\mathcal{N}$ in $\tilde{\mathcal{L}}[\tau_{\mathrm{in}}]$. For simplicity we omit the index j throughout the rest of this proof. Let ε be a positive number such that $\varepsilon^2\|x\| < 1$. From Lemma 7.4.5 applied to $\overline{a}$ and to $\overline{b}$ we have that

$$|\langle(\overline{b}^{-2})^* \, x\overline{a}^{-2}\varphi, \psi\rangle|^2 \leqq \|x\|^2 \, \|\overline{a}^{-2}\varphi\|^2 \, \|\overline{b}^{-2}\psi\|^2$$

$$\leqq \|x\|^2 \, (\varepsilon^2\|\varphi\|^2 + \varepsilon^{-1}\|\overline{a}^{-3}\varphi\|^2) \, (\varepsilon^2\|\psi\|^2 + \varepsilon^{-1}\|\overline{b}^{-3}\psi\|^2)$$

for $\varphi, \psi \in \mathcal{H}$. This shows that the assumptions of Lemma 4.4.4 are fulfilled when we set $z := (\bar{b}^{-2})^* x\bar{a}^{-2}$, $c := \|x\| \, \varepsilon^{-1}(\bar{a}^{-3})^* \bar{a}^{-3}$, $d := \|x\| \, \varepsilon^{-1}(\bar{b}^{-3})^* \bar{b}^{-3}$ and $\gamma = \delta := \varepsilon^2\|x\|$. From Lemma 4.4.4 it follows that there exist operators z_1, z_2 and y_1 in $\mathcal{N}$ such that $z_1 = (\bar{b}^{-3})^* y_1\bar{a}^{-3}$ and

$$|\langle z_2\varphi, \psi\rangle| \leqq \alpha\varepsilon^{1/2}\|\varphi\| \, \|\psi\| \quad \text{for} \quad \varphi, \psi \in \mathcal{H}, \tag{1}$$

where α is a certain constant depending only on the norms of x, $\bar{a}^{-1}$ and $\bar{b}^{-1}$. (The inequality (5) in Lemma 4.4.4 is not needed here.) Setting $y_2 := (b^+)^3 \circ z_2a^3$, we have

$$y_1 + y_2 = (b^+)^3 \circ \left((\bar{b}^{-3})^* y_1\bar{a}^{-3}\right) a^3 + y_2 = (b^+)^3 \circ z_1a^3 + (b^+)^3 \circ z_2a^3$$
$$= (b^+)^3 \circ za^3 = (b^+)^3 \circ \left((\bar{b}^{-2})^* x\bar{a}^{-2}\right) a^3 = b^+ \circ xa = y.$$

From this and (1),

$$|\langle (y - y_1) \, \varphi, \psi\rangle| = |\langle y_2\varphi, \psi\rangle| = |\langle z_2a^3\varphi, b^3\psi\rangle| \leqq \alpha\varepsilon^{1/2}\|a^3\varphi\| \, \|b^3\psi\|$$

for all $\varphi \in \mathcal{D}(\mathcal{A})$ and $\psi \in \mathcal{D}(\mathcal{B})$. Since $y_1 \in \mathcal{N}$ by construction and α depends only on x, a and b, this implies that y belongs to the closure of $\mathcal{N}$ in $\tilde{\mathcal{I}}[\tau_{\text{in}}]$. □

Remark 3. The assumption that $\mathcal{N}$ is a von Neumann algebra was not used in the preceding proof. In fact, Lemma 7.4.6 is valid if condition (I) is fulfilled and if $\mathcal{N}$ is a $*$-subalgebra of $\mathbb{B}(\mathcal{H})$ which contains the operators $\bar{a}_j^{-1}$ and $\bar{b}_j^{-1}$, $j \in J$.

Proof of Proposition 7.4.4. Suppose that $j \in J$. By condition (I) and the first assertion of Lemma 7.4.5, $\bar{a}_j\mathcal{D}(\bar{a}_j) = \mathcal{H}$. Hence $a_j\mathcal{D}(a_j) \equiv a_j\mathcal{D}(\mathcal{A})$ is dense in $\mathcal{H}$. For the same reason it follows that $b_j\mathcal{D}(\mathcal{B})$ is dense in $\mathcal{H}$. Let $\mathcal{U}_1$ be the unit ball of $\mathbb{B}(\mathcal{H})$. By Proposition 3.2.3 (cf. Remark 8 in 3.2.), for each $x \in (\mathcal{L}_1)_{a_j,b_j}$ there exists an operator $y \in \mathbb{B}(\mathcal{H})$ such that $x = b_j^+ \circ ya_j$. Let $\mathcal{C}_j$ denote the set of all such operators y if x runs through $(\mathcal{L}_1)_{a_j,b_j}$. From the density of the spaces $a_j\mathcal{D}(\mathcal{A})$ and $b_j\mathcal{D}(\mathcal{B})$ in $\mathcal{H}$ we conclude easily that $\mathcal{C} \subseteqq \mathcal{C}_j$ and $b_j^+ \circ (\mathcal{C}_j \cap \mathcal{U}_1) \, a_j = \mathcal{L}_1 \cap \mathcal{U}_{a_j,b_j}$.

Next we prove that $\mathcal{C}_j \subseteqq \mathcal{N}$. We let $y \in \mathcal{C}_j$. By Lemma 7.4.6, $\mathcal{N}$ is dense in $\tilde{\mathcal{I}}[\tau_{\text{in}}]$. Hence $\mathcal{N}$ and so $\mathcal{C}$ is dense in $\mathcal{L}$ in the weak-operator topology of $\mathcal{L}(\mathcal{D}_{\mathcal{A}}, \mathcal{D}_{\mathcal{B}}^+)$. Since $\mathcal{L}$ is weak-operator dense in $\mathcal{L}_1$ by assumption, $\mathcal{C}$ is weak-operator dense in $\mathcal{L}_1$. Hence there exists a net $(y_i : i \in I)$ from $\mathcal{C}$ which converges to the element $b_j^+ \circ ya_j$ of $\mathcal{L}_1$ in the weak-operator topology. Suppose that $c \in \mathcal{C}'$. Arguing in the same way as in the first paragraph of the proof of Proposition 7.4.1 (recall that condition (II) is valid), it follows that $c\mathcal{D}(\mathcal{A}) \subseteqq \mathcal{D}(\mathcal{A})$, $ca_j\varphi = a_jc\varphi$ for $\varphi \in \mathcal{D}(\mathcal{A})$, $c^*\mathcal{D}(\mathcal{B}) \subseteqq \mathcal{D}(\mathcal{B})$ and $c^*b_j\psi = b_jc^*\psi$ for $\psi \in \mathcal{D}(\mathcal{B})$. Further, $cy_i = y_ic$ for $i \in I$, since $y_i \in \mathcal{C}$ and $c \in \mathcal{C}'$. From these facts we have

$$\langle yca_j\varphi, b_j\psi\rangle = \langle ya_jc\varphi, b_j\psi\rangle = \langle (b_j^+ \circ ya_j) \, c\varphi, \psi\rangle$$
$$= \lim_i \langle y_ic\varphi, \psi\rangle = \lim_i \langle y_i\varphi, c^*\psi\rangle = \langle (b_j^+ \circ ya_j) \, \varphi, c^*\psi\rangle$$
$$= \langle ya_j\varphi, b_jc^*\psi\rangle = \langle cya_j\varphi, b_j\psi\rangle \text{ for all } \varphi \in \mathcal{D}(\mathcal{A}) \text{ and } \psi \in \mathcal{D}(\mathcal{B}).$$

Since $a_j\mathcal{D}(\mathcal{A})$ and $b_j\mathcal{D}(\mathcal{B})$ are dense in $\mathcal{H}$, the latter implies that $yc = cy$. Since $c \in \mathcal{C}'$ is arbitrary, this shows that $y \in \mathcal{C}''$ and so $\mathcal{C}_j \subseteqq \mathcal{C}'' = \mathcal{N}$.

The Kaplansky density theorem (see e.g. KADISON/RINGROSE [1], 5.3.5.) states that $\mathcal{C} \cap \mathcal{U}_1$ is weak-operator and so ultraweakly dense in $\mathcal{N} \cap \mathcal{U}_1$. Since $\mathcal{C} \subseteqq \mathcal{C}_j \subseteqq \mathcal{N}$ as just shown, $\mathcal{C} \cap \mathcal{U}_1$ is ultraweakly dense in $\mathcal{C}_j \cap \mathcal{U}_1$. From this it follows that $b_j^+ \circ (\mathcal{C} \cap \mathcal{U}_1) \, a_j$ is ultraweakly dense (i.e., dense in the ultraweak topology of $\mathcal{L}(\mathcal{D}_{\mathcal{A}}, \mathcal{D}_{\mathcal{B}}^+)$) in

$b_j^+ \circ (\mathcal{C}_j \cap \mathcal{U}_1)\, a_j$. Since obviously $b_j^+ \circ (\mathcal{C} \cap \mathcal{U}_1)\, a_j \subseteq \mathcal{L} \cap \mathcal{U}_{a_j,b_j}$ and $b_j^+ \circ (\mathcal{C}_j \cap \mathcal{U}_1)\, a_j = \mathcal{L}_1 \cap \mathcal{U}_{a_j,b_j}$ as noted above, this gives the assertion. □

A by-product of the preceding proof is

Corollary 7.4.7. *Let $\mathcal{N}$, $\mathcal{L}$, $\{a_j : j \in J\}$ and $\{b_j : j \in J\}$ be as in Proposition 7.4.4. Then the weak-operator closure and the ultraweak closure of $\mathcal{L}$ within $\mathcal{L}(\mathcal{D}_{\mathcal{A}}, \mathcal{D}^+_{\mathcal{B}})$ coincide, and they are equal to $\mathcal{L}_0 := \bigcup_{j \in J} b_j^+ \circ \mathcal{N} a_j$. Moreover, $\mathcal{L}_0 \cap \mathcal{U}_{a_j,b_j} = b_j^+ \circ (\mathcal{N} \cap \mathcal{U}_1)\, a_j$ for $j \in J$, where $\mathcal{U}_1$ is the unit ball of $\mathbf{B}(\mathcal{H})$.*

Proof. Let $\mathcal{L}_1$ denote the weak-operator closure of $\mathcal{L}_0$ in $\mathcal{L}(\mathcal{D}_{\mathcal{A}}, \mathcal{D}^+_{\mathcal{B}})$, and let $\mathcal{N}_j$, $j \in J$, be the corresponding subsets of $\mathbf{B}(\mathcal{H})$ for $\mathcal{L}_1$ as defined in the preceding proof. The proof of Proposition 7.4.4 (with $\mathcal{L}$ and $\mathcal{C}$ replaced by $\mathcal{L}_0$ and $\mathcal{N}$, respectively) showed that $\mathcal{N} \subseteq \mathcal{N}_j \subseteq \mathcal{N}''$ for $j \in J$, so that $\mathcal{N}_j = \mathcal{N}$. Therefore, $\mathcal{L}_1 = \mathcal{L}_0$, and $\mathcal{L}_0$ is weak-operator and so ultraweakly closed in $\mathcal{L}(\mathcal{D}_{\mathcal{A}}, \mathcal{D}^+_{\mathcal{B}})$. Since $b_j^+ \circ (\mathcal{N}_j \cap \mathcal{U}_1)\, a_j = \mathcal{L}_1 \cap \mathcal{U}_{a_j,b_j}$ (by the above proof), $\mathcal{N}_j = \mathcal{N}$ and $\mathcal{L}_1 = \mathcal{L}_0$, we obtain the final assertion. Since $\mathcal{C}$ is ultraweakly dense in $\mathcal{N}$ by the von Neumann density theorem, it follows that $\mathcal{L}$ is ultraweakly and so weak-operator dense in $\mathcal{L}_0$. Combined with the preceding, this yields the first assertion. □

An immediate consequence of Corollary 7.4.7 is

Corollary 7.4.8. *Under the assumptions and the notation of Proposition 7.4.4., the following three conditions are equivalent:*

(i) *$\mathcal{L}$ is weak-operator closed in $\mathcal{L}(\mathcal{D}_{\mathcal{A}}, \mathcal{D}^+_{\mathcal{B}})$.*

(ii) *$\mathcal{L}$ is ultraweakly closed in $\mathcal{L}(\mathcal{D}_{\mathcal{A}}, \mathcal{D}^+_{\mathcal{B}})$.*

(iii) *$\mathcal{C}$ is a von Neumann algebra on $\mathcal{H}$.*

There are similar results for the ultrastrong topology on $\mathfrak{L}(\mathcal{D}_{\mathcal{A}}, \mathcal{H})$. Let us recall that the weak-operator topology on $\mathfrak{L}(\mathcal{D}_{\mathcal{A}}, \mathcal{H})$ is defined by the family of seminorms $|\langle \cdot \varphi, \psi \rangle|$, where $\varphi \in \mathcal{D}(\mathcal{A})$ and $\psi \in \mathcal{H}$; cf. Remark 1 in 3.5.

Proposition 7.4.9. *Suppose that the O*-algebra $\mathcal{A}$ satisfies the parts of conditions (I) and (II) that apply to $\mathcal{A}$. Let $\mathcal{L}$ be the linear subspace of $\mathfrak{L}(\mathcal{D}_{\mathcal{A}}, \mathcal{H})$ spanned by the operators ca_j, where $c \in \mathcal{C}$ and $j \in J$, and let $\mathcal{L}_1$ be another linear subspace of $\mathfrak{L}(\mathcal{D}, \mathcal{H})$ such that $\mathcal{L} \subseteq \mathcal{L}_1$. If $\mathcal{L}$ is dense in $\mathcal{L}_1$ in the weak-operator topology of $\mathfrak{L}(\mathcal{D}_{\mathcal{A}}, \mathcal{H})$, then for each $j \in J$ the set $\mathcal{L} \cap \mathcal{U}^{a_j}$ is dense in $\mathcal{L}_1 \cap \mathcal{U}^{a_j}$ in the ultrastrong topology of $\mathfrak{L}(\mathcal{D}_{\mathcal{A}}, \mathcal{H})$.*

Proof. The proof is similar to the proof of Proposition 7.4.4 when we set $\mathcal{B} := \mathbf{B}(\mathcal{H})$, $\mathcal{D}(\mathcal{B}) := \mathcal{H}$ and $b_j := I$. It suffices to replace the ultraweak density of $\mathcal{C} \cap \mathcal{U}_1$ in $\mathcal{C}_j \cap \mathcal{U}_1$ by the ultrastrong density. □

The following two corollaries can be derived in a similar way as Corollaries 7.4.7 and 7.4.8. We retain the assumptions and notations of Proposition 7.4.9.

Corollary 7.4.10. *The closure of $\mathcal{L}$ in any one of the weak-operator, ultraweak, strong-operator or ultrastrong topologies within $\mathfrak{L}(\mathcal{D}_{\mathcal{A}}, \mathcal{H})$ coincides with $\bigcup_{j \in J} \mathcal{N} a_j$.*

Corollary 7.4.11. *The following three statements are equivalent:*

(i) *$\mathcal{L}$ is weak-operator closed in $\mathfrak{L}(\mathcal{D}_{\mathcal{A}}, \mathcal{H})$.*

(ii) *$\mathcal{L}$ is ultraweakly closed in $\mathfrak{L}(\mathcal{D}_{\mathcal{A}}, \mathcal{H})$.*

(iii) *$\mathcal{C}$ is a von Neumann algebra on $\mathcal{H}$.*

Remark 4. The assumption $\bar{a}_j^{-1} \in \mathcal{N}$ in Propositions 7.4.1 and 7.4.3 and in condition (I) is obviously equivalent to the requirement $\bar{a}_j \in \mathbb{A}(\mathcal{N})$.

Remark 5. Concrete examples of spaces $\mathcal{L}$ satisfying the above assumptions are easily obtained by means of operators which are affiliated with a fixed von Neumann algebra $\mathcal{N}$. For instance, let $\mathcal{A}$, $(a_n : n \in \mathbb{N})$ and $E(\cdot)$ be as in Example 2.2.16. Set $\mathcal{B} := \mathcal{A}$ and $b_n := a_n$ for $n \in \mathbb{N}$, and let $\mathcal{N}$ be a von Neumann algebra such that $E(\lambda) \in \mathcal{N}$ for $\lambda \in \mathbb{R}$. Then $\mathcal{A}$, $\mathcal{B}$, $(a_n : n \in \mathbb{N})$, $(b_n : n \in \mathbb{N})$ and $\mathcal{N}$ satisfy the conditions (I) and (II).

Remark 6. Suppose that $\mathcal{A}$ is a closed O*-algebra which is a symmetric *-algebra. Then $\mathcal{A}$ is selfadjoint, and all results of this section apply to $\mathcal{L} := \mathcal{A}$ (with $\mathcal{A} = \mathcal{B}$ in Proposition 7.4.4 and Corollary 7.4.7) and to each von Neumann algebra $\mathcal{N}$ which contains $\mathcal{A}''$; cf. the proof of Corollary 4.4.7.

Notes

7.1. Lemma 7.1.2 is from NELSON/STINESPRING [1]. Proposition 7.1.3, (ii), is Corollary 9.2 in NELSON [1]. In the case where $\bar{a}$ is selfadjoint Proposition 7.1.4 was proved by POULSEN [2] who used an analytic domination result of NELSON [1]. The operator-theoretic proof in the text is taken from SCHMÜDGEN [22]. Proposition 7.1.6 is also from SCHMÜDGEN [22].
7.2. The various types of commutants have different sources. The strong commutant expresses the way in which commutativity of a bounded and an unbounded operator is defined in the standard text books on functional analysis (see e.g. RIESZ/SZ.-NAGY [1], Nr. 116). The weak commutant first appeared in papers on quantum field theory; cf. RUELLE [1], p. 162. Intertwining sesquilinear forms (and so in fact form commutants) have been studied in representation theory of Lie groups by BRUHAT [1] and POULSEN [1]. In the context of unbounded operator algebras or *-representations weak commutants were first studied by VASILIEV [1], POWERS [1] and UHLMANN [2], and form commutants first appeared in ARAKI/JURZAK [1].

The unbounded commutants $\mathcal{A}_{\mathrm{f}}^{\mathrm{c}}$, $\mathcal{A}_{\mathrm{w}}^{\mathrm{c}}$ and $\mathcal{A}_{\mathrm{s}}^{\mathrm{c}}$ occuring in the text were introduced and studied by SCHMÜDGEN [22]. Other types of unbounded commutants can be found e.g. in GUDDER/HUDSON [1], INOUE [3], ANTOINE/KARWOWSKI [1] and in MATHOT [1]. The idea of the proof of Proposition 7.2.5 has been adapted from POULSEN [1], p. 98; cf. ARAKI/JURZAK [1]. Propositions 7.2.11 and 7.2.12 are taken from SCHMÜDGEN [22]. Examples 7.2.14 and 7.2.15 are in SCHMÜDGEN [21].
7.3. The main reference for this section is ARAKI/JURZAK [1], though our proofs are different in many respects. The central result of this section, Theorem 7.3.7, can be found in ARAKI/JURZAK [1] under more restrictive assumptions. Our main intention for introducing strictly self-adjoint O*-algebras was to find a rather general class of O*-algebras for which the assertions of Theorems 7.3.6 and 7.3.7 can be proved.
7.4. This section follows the paper of SCHMÜDGEN [23].

Additional References:

7.1. FRÖHLICH [1], NUSSBAUM [3], [4].
7.2. BORCHERS/YNGVASON [1], INOUE/UEDA/YAMAUCHI [1], VORONIN/SUSHKO/HORUZHY [1], ANTOINE/MATHOT/TRAPANI [1], NGUYEN [1], VAN DAELE/KASPAREK [1].
7.3. BHATT [1].

Part II.
*-Representations

The main theme of the second part of this monograph are $*$-representations of general $*$-algebras by unbounded operators on Hilbert space.

Part II is organized as follows. Chapter 8 provides a detailed study of general $*$-representations. In Chapters 9 and 10 we specialize to particular classes of $*$-representations and $*$-algebras. In Chapter 9 integrable representations of commutative $*$-algebras are investigated and non-integrable self-adjoint representations of the polynomial algebra $\mathbb{C}[x_1, x_2]$ are constructed. Chapter 10 deals mainly with integrable representations of enveloping algebras. Especially, the infinitesimal representation dU associated with a unitary representation U of a Lie group and the exponentiation problem for $*$-representations of enveloping algebras are studied. Chapters 11 and 12 are devoted to two special topics. In Chapter 11 n-positive and completely positive $*$-representations and mappings of $*$-algebras are considered. Chapter 12 is concerned with the decomposition theory of closed operators, $*$-representations and states.

As already mentioned in the preface, this part is to a large extent independent of Part I. There are only two earlier sections from which concepts and facts are frequently used. These are Section 2.2 with notions like graph topology, closed O-algebras and closure of an O-algebra, and Section 7.2 with the three types of bounded commutants $\mathcal{A}'_s$, $\mathcal{A}'_w$ and $\mathcal{A}'_{ss}$. Sometimes only a single result is applied (for instance, Proposition 2.3.3 or Proposition 7.1.3 in the proof of Theorem 12.3.5. or Theorem 9.1.2). Often results or remarks which use terminology or facts from Part I indicate links to earlier sections, but they are not needed later (for instance, Corollaries 9.1.3 and 9.1.10).

8. Basics of $*$-Representations

In this chapter we develop fundamental concepts and constructs of $*$-representations of general $*$-algebras. Suppose A is a $*$-algebra with unit element. A representation of A is a homomorphism π of the algebra A onto an O-algebra which maps the unit of A into the identity map. If the image $\pi(A)$ is an O*-algebra and π preserves the involution, then π is called a $*$-representation of A. Though our main intention is the study of $*$-representations, we need to consider also representations, since, for instance, the adjoint of a $*$-representation is a representation, but not a $*$-representation in general.

In Section 8.1 representations, $*$-representations and special subclasses such as closed, adjoint and self-adjoint representations are defined and some of their basic properties are established. In Section 8.2 we consider the space $\mathbb{I}(\pi_1, \pi_2)$ of intertwining operators for two representations π_1 and π_2 of A. The strong commutant $\pi(A)'_s$ and the weak commutant $\pi(A)'_w$ appear as the special cases $\mathbb{I}(\pi, \pi)$ and $\mathbb{I}(\pi, \pi^*)$, respectively, of these spaces. Section 8.3 is concerned with various basic notions in representation theory like direct sums, subrepresentations, invariant or reducing subspaces, irreducibility and cyclic vectors of representations. It is shown that the self-adjoint subrepresentations of a given self-adjoint representation π of A are in one-to-one correspondence with the projections in the commutant $\pi(A)'$. In Section 8.4 we deal with the similarity, unitary equivalence and disjointness of representations. In Section 8.5 we investigate a general procedure of constructing extensions of a $*$-representation π of A in a possibly larger Hilbert space by means of certain subsets of the weak commutant $\pi(A)'_w$.

The main subject of Section 8.6 is the so-called Gelfand-Neumark-Segal construction which allows to produce a cyclic $*$-representation π_ω from a positive linear functional ω on A. This procedure is also an extremely useful tool to study properties of positive linear functionals. For instance, the order relation, the purity and the orthogonality of positive linear functionals can be characterized in terms of the $*$-representations π_ω.

8.1. Representations and $*$-Representations

Representations

Suppose A is an algebra with unit.

Definition 8.1.1. Let $\mathcal{D}$ be a dense linear subspace of a Hilbert space $\mathcal{H}$. A *representation* of A on $\mathcal{D}$ is a mapping π of A into the set of linear operators defined on $\mathcal{D}$ such that:

(i) $\pi(\alpha_1 a_1 + \alpha_2 a_2)\,\varphi = \alpha_1 \pi(a_1)\,\varphi + \alpha_2 \pi(a_2)\,\varphi$ and $\pi(1)\,\varphi = \varphi$,

(ii) $\pi(a_2)\varphi \in \mathcal{D}$ and $\pi(a_1 a_2)\varphi = \pi(a_1)\pi(a_2)\varphi$,

(iii) $\pi(a)$ is a closable operator on $\mathcal{D}$

for all $a, a_1, a_2 \in \mathsf{A}$, $\alpha_1, \alpha_2 \in \mathbb{C}$ and $\varphi \in \mathcal{D}$. We call $\mathcal{D}$ the *domain* of π and we write $\mathcal{D}(\pi) := \mathcal{D}$ and $\mathcal{H}(\pi) := \mathcal{H}$.

In other words, a representation of A is an identity preserving homomorphism of A into an O-algebra.

Suppose π_1 and π_2 are two representations of A. We say π_2 is an *extension* of π_1 and π_1 is a *subrepresentation* of π_2 and write $\pi_1 \subseteq \pi_2$ if $\mathcal{H}(\pi_1) \subseteq \mathcal{H}(\pi_2)$, $\mathcal{D}(\pi_1) \subseteq \mathcal{D}(\pi_2)$ and $\pi_1(a) = \pi_2(a) \upharpoonright \mathcal{D}(\pi_1)$ for all $a \in \mathsf{A}$. By the relation $\mathcal{H}(\pi_1) \subseteq \mathcal{H}(\pi_2)$ we always mean that $\mathcal{H}(\pi_1)$ is a closed subspace of the Hilbert space $\mathcal{H}(\pi_2)$; that is, this also means that the scalar product of $\mathcal{H}(\pi_1)$ is the restriction to $\mathcal{H}(\pi_1)$ of the scalar product of $\mathcal{H}(\pi_2)$.

Let π be a representation of A. Then $\pi(\mathsf{A})$ is an O-algebra on $\mathcal{D}(\pi)$. We recall some notions and facts from Section 2.2 and we reformulate them in the present context. The graph topology $\mathfrak{t}_{\pi(\mathsf{A})}$ of $\pi(\mathsf{A})$ is the locally convex topology on $\mathcal{D}(\pi)$ which is generated by the family of seminorms $\{\|\cdot\|_{\pi(a)} : a \in \mathsf{A}\}$. If no confusion is possible, we shall write $\mathfrak{t}_\pi$ for $\mathfrak{t}_{\pi(\mathsf{A})}$. Further, $\hat{\mathcal{D}}(\pi(\mathsf{A}))$ is the domain of the closure $\widehat{\pi(\mathsf{A})}$ of the O-algebra $\pi(\mathsf{A})$, and $\widehat{\pi(\mathsf{A})}$ consists of the operators $\widehat{\pi(a)} := \overline{\pi(a)} \upharpoonright \hat{\mathcal{D}}(\pi(\mathsf{A}))$, $a \in \mathsf{A}$. By Proposition 2.2.11, $\widehat{\pi(\mathsf{A})}$ is an O-algebra and $\pi(a) \to \widehat{\pi(a)}$ is a homomorphism of $\pi(\mathsf{A})$ onto $\widehat{\pi(\mathsf{A})}$. Therefore, $\hat{\pi}(a) := \widehat{\pi(a)}$, $a \in \mathsf{A}$, defines a representation of A on $\mathcal{D}(\hat{\pi}) := \hat{\mathcal{D}}(\pi(\mathsf{A}))$.

Definition 8.1.2. The representation $\hat{\pi}$ is called the *closure* of π. π is said to be *closed* if $\pi = \hat{\pi}$.

Remark 1. We mention some simple facts which follow immediately from the preceding definitions and from the results in Section 2.2. The representation $\hat{\pi}$ is always closed; it is the smallest (relative to the relation "$\subseteq$" defined above) closed extension of π. A representation π of A is closed if and only if the O-algebra $\pi(\mathsf{A})$ is closed or equivalently if the locally convex space $\mathcal{D}(\pi)[\mathfrak{t}_\pi]$ is complete.

From now on we assume that A is a $*$-algebra with unit.

Proposition 8.1.3. *Suppose π is a representation of the $*$-algebra A. Let $\mathcal{D}(\pi^*) := \bigcap_{a \in \mathsf{A}} \mathcal{D}(\pi(a)^*)$, and let $\mathcal{H}(\pi^*)$ be the closure of $\mathcal{D}(\pi^*)$ in $\mathcal{H}(\pi)$. Define $\pi^*(a) := \pi(a^+)^* \upharpoonright \mathcal{D}(\pi^*)$ for $a \in \mathsf{A}$.*

(i) *π^* is a closed representation of A on $\mathcal{D}(\pi^*)$ in the Hilbert space $\mathcal{H}(\pi^*)$. We have $(\hat{\pi})^* = \pi^*$.*

(ii) *π^* is the largest among the representations π_1 of A on the Hilbert space $\mathcal{H}(\pi_1) = \mathcal{H}(\pi^*)$ which satisfy $\langle \pi(a)\varphi, \psi\rangle = \langle \varphi, \pi_1(a^+)\psi\rangle$ for all $a \in \mathsf{A}$, $\varphi \in \mathcal{D}(\pi)$ and $\psi \in \mathcal{D}(\pi_1)$.*

(iii) *If π_0 is another representation of A on $\mathcal{H}(\pi_0) = \mathcal{H}(\pi)$ such that $\pi \subseteq \pi_0$, then $\pi_0^* \subseteq \pi^*$.*

Proof. (ii) and (iii) follow in a rather straightforward way from the corresponding definitions. We carry out the proof of (i). Suppose $a_1, a_2 \in \mathsf{A}$ and $\alpha_1, \alpha_2 \in \mathbb{C}$. Let $\varphi \in \mathcal{D}(\pi)$ and let $\psi \in \mathcal{D}(\pi^*)$. From

$$\begin{aligned}\langle \pi^*(\alpha_1 a_1 + \alpha_2 a_2)\psi, \varphi\rangle &= \langle \pi(\overline{\alpha}_1 a_1^+ + \overline{\alpha}_2 a_2^+)^* \psi, \varphi\rangle \\ &= \alpha_1\langle \psi, \pi(a_1^+)\varphi\rangle + \alpha_2\langle \psi, \pi(a_2^+)\varphi\rangle \\ &= \langle (\alpha_1\pi(a_1^+)^* + \alpha_2\pi(a_2^+)^*)\psi, \varphi\rangle = \big\langle \big(\alpha_1\pi^*(a_1) + \alpha_2\pi^*(a_2)\big)\psi, \varphi\big\rangle\end{aligned}$$

we conclude that π^* is linear. It is trivial that $\pi^*(1)\,\psi = \psi$. Since $\pi(a_1^+)\,\varphi \in \mathcal{D}(\pi)$, we have

$$\langle \pi(a_1^+)\,\varphi, \pi^*(a_2)\,\psi\rangle = \langle \pi(a_1^+)\,\varphi, \pi(a_2^+)^*\,\psi\rangle = \langle \pi(a_2^+)\,\pi(a_1^+)\,\varphi, \psi\rangle$$

$$= \langle \pi\big((a_1a_2)^+\big)\,\varphi, \psi\rangle = \langle \varphi, \pi\big((a_1a_2)^+\big)^*\,\psi\rangle = \langle \varphi, \pi^*(a_1a_2)\,\psi\rangle.$$

Since $\varphi \in \mathcal{D}(\pi)$ is arbitrary, this gives $\pi^*(a_2)\,\psi \in \mathcal{D}(\pi(a_1^+)^*)$ and $\pi(a_1^+)^*\,\pi^*(a_2)\,\psi = \pi^*(a_1a_2)\,\psi$. Because $a_1 \in \mathsf{A}$ is arbitrary, it follows that $\pi^*(a_2)\,\psi \in \mathcal{D}(\pi^*)$; so, by the definition of π^*, $\pi^*(a_1)\,\pi^*(a_2)\,\psi = \pi^*(a_1a_2)\,\psi$. Since $\pi^*(a) \subseteq \pi(a^+)^*$ by definition, each operator $\pi^*(a)$, $a \in \mathsf{A}$, is closable. Further, $\mathcal{D}(\pi^*)$ is a dense linear subspace of the Hilbert space $\mathcal{H}(\pi^*)$. All this together proves that π^* is a representation of A on $\mathcal{D}(\pi^*)$ in the Hilbert space $\mathcal{H}(\pi^*)$. It is obvious that $\mathcal{D}(\pi^*) = \bigcap_{a\in\mathsf{A}} \mathcal{D}\big(\overline{\pi^*(a)}\big)$, where the bar means the closure of the operator in the Hilbert space $\mathcal{H}(\pi^*)$. Therefore, by Lemma 2.2.9, the O-family $\pi^*(\mathsf{A})$ is closed. Hence π^* is closed. From $\pi(a) \subseteq \hat{\pi}(a) \subseteq \overline{\pi(a)}$ for $a \in \mathsf{A}$ it follows immediately that $(\hat{\pi})^* = \pi^*$. $\square$

Definition 8.1.4. Let π be a representation of A. We call π^* the *adjoint representation* to π, and $\pi^{**} := (\pi^*)^*$ the *biadjoint representation* to π. π is said to be *adjointable* if $\mathcal{H}(\pi) = \mathcal{H}(\pi^*)$. π is called *biclosed* if $\pi = \pi^{**}$.

Remark 2. Since an adjoint representation is always closed, each biclosed representation is closed. The converse is not true; see Example 8.1.14 below.

Proposition 8.1.5. *Suppose π is an adjointable representation of A.*

(i) $\pi \subseteq \hat{\pi} \subseteq \pi^{**}$ *and* $\pi^* = \pi^{***}$.

(ii) π^{**} *is biclosed. It is the smallest biclosed extension of* π.

Proof. (i): By the definition of π^*, we have

$$\langle \pi(a)\,\varphi, \psi\rangle = \langle \varphi, \pi^*(a^+)\,\psi\rangle \quad \text{for } a \in \mathsf{A}, \varphi \in \mathcal{D}(\pi) \text{ and } \psi \in \mathcal{D}(\pi^*). \tag{1}$$

Since $\mathcal{H}(\pi) = \mathcal{H}(\pi^*)$ by assumption, this shows that $\mathcal{D}(\pi) \subseteq \mathcal{D}(\pi^*(a^+)^*)$ for $a \in \mathsf{A}$. Consequently, $\mathcal{D}(\pi) \subseteq \mathcal{D}\big((\pi^*)^*\big) \equiv \mathcal{D}(\pi^{**})$. Further, we conclude from (1) that $\pi(a) = \pi^*(a^+)^* \upharpoonright \mathcal{D}(\pi) = \pi^{**}(a) \upharpoonright \mathcal{D}(\pi)$. Thus $\pi \subseteq \pi^{**}$. Since π^{**} is closed, $\hat{\pi} \subseteq \pi^{**}$.

We verify that $\pi^* = \pi^{***}$. (Note that π^{***} and π^{****} are only abbreviations for $(\pi^{**})^*$ and $(\pi^{***})^*$, respectively.) By Proposition 8.1.3, (iii), $\pi \subseteq \pi^{**}$ yields $\pi^* \supseteq \pi^{***}$. From $\pi \subseteq \pi^{**}$ we see that π^* is again an adjointable representation. Therefore, replacing π by π^* in $\pi \subseteq \pi^{**}$, we obtain $\pi^* \subseteq \pi^{***}$. Hence $\pi^* = \pi^{***}$.

(ii): Replacing π by π^* in $\pi^* = \pi^{***}$, we get $\pi^{**} = \pi^{****}$. This means that π^{**} is biclosed. Let π_1 be a biclosed extension of π. Applying Proposition 8.1.3, (iii), twice, we obtain $\pi^{**} \subseteq \pi_1^{**} = \pi_1$. $\square$

The following simple results are useful for an explicit determination of $\mathcal{D}(\pi^*)$ and of π^*.

Lemma 8.1.6. *Suppose that the $*$-algebra A is the linear span of a certain set $\{a_{j_1} \cdots a_{j_r} : (j_1, \ldots, j_r) \in J\}$, where J is an index set and a_j are elements of A. If π is a representation of A, then* $\mathcal{D}(\pi^*) = \bigcap_{(j_1,\ldots,j_r)\in J} \mathcal{D}\big(\pi(a_{j_r})^* \cdots \pi(a_{j_1})^*\big)$.

Proof. Let $\mathcal{D}_1$ denote the set on the right-hand side of the equality sign. Suppose $\psi \in \mathcal{D}(\pi^*)$ and $(j_1, \ldots, j_r) \in J$. Then $\psi \in \mathcal{D}(\pi(a_{j_1})^*)$. By Proposition 8.1.3, (i), $\pi(a_{j_1})^* \psi$ is again in $\mathcal{D}(\pi^*)$ and hence in $\mathcal{D}(\pi(a_{j_2})^*)$. Continuing this reasoning we get $\psi \in \mathcal{D}\big(\pi(a_{j_r})^* \cdots \pi(a_{j_1})^*\big)$. Hence $\psi \in \mathcal{D}_1$ and $\mathcal{D}(\pi^*) \subseteq \mathcal{D}_1$. Let $a \in \mathsf{A}$. From the assumption it follows that $\pi(a)$ is equal to a certain finite sum $\sum \lambda_{(j_1,\ldots,j_r)}\pi(a_{j_1}) \cdots \pi(a_{j_r})$ with complex coefficients $\lambda_{(j_1,\ldots,j_r)}$. Then $\pi(a)^* \supseteq \sum \lambda_{(j_1,\ldots,j_r)}\pi(a_{j_r})^* \cdots \pi(a_{j_1})^*$. This gives $\mathcal{D}_1 \subseteq \mathcal{D}(\pi^*)$. □

An immediate consequence of this lemma is

Corollary 8.1.7. *Suppose that there exist d elements $a_1, \ldots, a_d$ in A such that* $\mathsf{A} =$ l.h. $\{a_1^{n_1} \cdots a_d^{n_d} : (n_1, \ldots, n_d) \in \mathbb{N}_0^d\}$, where $d \in \mathbb{N}$. *Then for any representation π of A we have*

$$\mathcal{D}(\pi^*) = \bigcap_{(n_1,\ldots,n_d)\in\mathbb{N}_0^d} \mathcal{D}\Big(\big(\pi(a_d)^*\big)^{n_d} \cdots \big(\pi(a_1)^*\big)^{n_1}\Big).$$

Remark 3. The Weyl algebra $\mathbb{A}(p_1, q_1, \ldots, p_n, q_n)$, the polynomial algebra $\mathbb{C}[x_1, \ldots, x_n]$, and more generally the enveloping algebra $\mathcal{E}(\mathfrak{g})$ of a finite dimensional Lie algebra $\mathfrak{g}$ satisfy the assumption of Corollary 8.1.7.

Corollary 8.1.8. *Suppose that B is a subset of A such that $\mathsf{B} \cup \{1\}$ generates A as an algebra. Suppose π_1 and π_2 are representations of the $*$-algebra A in the same Hilbert space $\mathcal{H}(\pi_1) = \mathcal{H}(\pi_2)$. If $\overline{\pi_1(b)} = \overline{\pi_2(b)}$ for all $b \in \mathsf{B}$, then $\pi_1^* = \pi_2^*$ and $\pi_1^{**} = \pi_2^{**}$.*

Proof. Since $\overline{\pi_1(b)} = \overline{\pi_2(b)}$ by assumption, we have $\pi_1(b)^* = \pi_2(b)^*$ for all $b \in \mathsf{B}$. By Lemma 8.1.6, this gives $\mathcal{D}(\pi_1^*) = \mathcal{D}(\pi_2^*)$. Further, we have $\pi_1^*(b^+) = \pi_1(b)^* \upharpoonright \mathcal{D}(\pi_1^*) = \pi_2(b)^* \upharpoonright \mathcal{D}(\pi_2^*) = \pi_2^*(b^+)$, $b \in \mathsf{B}$. Since π_1^* and π_2^* are representations of A and $\mathsf{B}^+ \cup \{1\}$ also generates the algebra A, this implies that $\pi_1^* = \pi_2^*$. Hence $\pi_1^{**} = \pi_2^{**}$. □

$*$-Representations

In this subsection we assume that A is a $*$-algebra with unit.

Definition 8.1.9. Suppose $\mathcal{D}$ is a dense linear subspace of a Hilbert space $\mathcal{H}$. A mapping π of A into the set of linear operators defined on $\mathcal{D}$ is said to be a *$*$-representation* of A on $\mathcal{D}$ if the following conditions are fulfilled:

(i) $\pi(\alpha_1 a_1 + \alpha_2 a_2)\,\varphi = \alpha_1 \pi(a_1)\,\varphi + \alpha_2 \pi(a_2)\,\varphi$ and $\pi(1)\,\varphi = \varphi$,

(ii) $\pi(a_2)\,\varphi \in \mathcal{D}$ and $\pi(a_1 a_2)\,\varphi = \pi(a_1)\,\pi(a_2)\,\varphi$,

(iii) $\langle \pi(a)\,\varphi, \psi\rangle = \langle \varphi, \pi(a^+)\,\psi\rangle$

for all $a, a_1, a_2 \in \mathsf{A}$, $\alpha_1, \alpha_2 \in \mathbb{C}$ and $\varphi, \psi \in \mathcal{D}$.

It is clear that an equivalent definition is obtained if (iii) is replaced by

(iii)′ $\pi(a) \in \mathcal{L}^+(\mathcal{D})$ and $\pi(a^+) = \pi(a)^+$.

Therefore, by another slight reformulation of Definition 8.1.9, a $*$-representation of A on $\mathcal{D}$ is a $*$-homomorphism π of A into the O*-algebra $\mathcal{L}^+(\mathcal{D})$ which satisfies $\pi(1) = I$. In that way $*$-representations were defined in Definition 2.1.13; that is, Definitions 2.1.13 and 8.1.9 are equivalent.

Each $*$-representation π of A is, of course, an adjointable representation of A, since condition (iii)′ above implies that $\pi(a)$, $a \in \mathsf{A}$, is closable and $\mathcal{D}(\pi) \subseteq \mathcal{D}(\pi^*)$. Thus the terminology and all results of the preceding subsection apply in particular for $*$-representations.

Of course, concepts like closed representations or adjoint representations are suggested by the corresponding notions in single operator theory. Some more concepts in a similar spirit are contained in the next definition.

Definition 8.1.10. Let π be a representation of the $*$-algebra A. We say that π is *self-adjoint* if $\pi = \pi^*$. π is called *essentially self-adjoint* if $\bar{\pi}$ is self-adjoint, i. e., if $\bar{\pi} = \pi^*$. We say that π is *hermitian* if π is a $*$-representation of A.

Lemma 8.1.11. *Suppose π is a representation of* A.

(i) *π is hermitian if and only if $\pi \subseteq \pi^*$.*

(ii) *π is self-adjoint if and only if π is biclosed and π^* is self-adjoint.*

Proof. (i): Suppose π is hermitian. Then, by condition (iii)′ above, $\mathcal{D}(\pi) \subseteq \mathcal{D}(\pi^*)$ and $\pi^*(a) \restriction \mathcal{D}(\pi) = \pi(a^+)^* \restriction \mathcal{D}(\pi) = \pi(a^+)^+ = \pi(a)$ for $a \in$ A. That is, $\pi \subseteq \pi^*$. Conversely, $\pi \subseteq \pi^*$ obviously implies that Definition 8.1.9, (iii), is satisfied, so that π is hermitian.

(ii): If $\pi = \pi^*$, then $\pi = \pi^* = \pi^{**}$, so that π is biclosed and π^* is self-adjoint. Conversely, if π is biclosed and π^* is self-adjoint, then $\pi = \pi^{**}$ and $\pi^* = \pi^{**}$; hence $\pi = \pi^*$. $\square$

Some basic properties of $*$-representations are collected in

Proposition 8.1.12. *Suppose that π is a $*$-representation of the $*$-algebra* A.

(i) *$\bar{\pi}$ and π^{**} are $*$-representations, and $\pi \subseteq \bar{\pi} \subseteq \pi^{**} \subseteq \pi^*$. Moreover, $\mathcal{D}(\bar{\pi}) = \bigcap_{a \in A} \mathcal{D}(\overline{\pi(a)})$.*

(ii) *π is self-adjoint if and only if $\mathcal{D}(\pi^*) \subseteq \mathcal{D}(\pi)$.*

(iii) *π^* is self-adjoint if and only if π^* is hermitian.*

(iv) *If π is self-adjoint and π_1 is a hermitian extension of π in the Hilbert space $\mathcal{H}(\pi) = \mathcal{H}(\pi_1)$, then $\pi_1 = \pi$.*

(v) *Suppose that* B *is a subset of* A *such that* B $\cup$ {1} *generates* A *as an algebra. If $\overline{\pi(b^+)} = \pi(b)^*$ for all $b \in$ B, then π^* is self-adjoint.*

Proof. First note that $\pi \subseteq \pi^*$ by Lemma 8.1.11, (i), since π is a $*$-representation.

(i): Since π^* is closed, $\pi \subseteq \pi^*$ yields $\bar{\pi} \subseteq \pi^* = (\bar{\pi})^*$. But $\bar{\pi} \subseteq (\bar{\pi})^*$ means (again by Lemma 8.1.11) that $\bar{\pi}$ is a $*$-representation. Applying Proposition 8.1.3, (iii), twice to $\pi \subseteq \pi^*$ we get $\pi^{**} \subseteq \pi^{***}$ which shows that π^{**} is hermitian. Combined with $\pi^* = \pi^{***}$, the latter gives $\pi^{**} \subseteq \pi^*$. The other inclusions have been already stated in Proposition 8.1.5. Since π is a $*$-representation, π(A) is an O*-algebra, and the equality $\mathcal{D}(\bar{\pi}) = \hat{\mathcal{D}}(\pi(A)) = \bigcap_{a \in A} \mathcal{D}(\overline{\pi(a)})$ follows therefore from Proposition 2.2.12.

(ii): Since $\pi \subseteq \pi^*$, the equality $\pi = \pi^*$ is obviously equivalent to $\mathcal{D}(\pi^*) \subseteq \mathcal{D}(\pi)$.

(iii): Since $\pi^{**} \subseteq \pi^*$, both statements are equivalent to $\pi^* \subseteq \pi^{**}$.

(iv): From $\pi \subseteq \pi_1$, $\pi_1^* \subseteq \pi^*$. Since π is self-adjoint and π_1 is hermitian, $\pi^* = \pi$ and $\pi_1 \subseteq \pi_1^*$, so that $\pi_1 \subseteq \pi$. Hence $\pi_1 = \pi$.

(v): Let $b_1, \ldots, b_n \in \mathsf{B}$. For $\varphi, \psi \in \mathcal{D}(\pi^*)$, we have

$$\begin{aligned}\langle \pi^*\big((b_1 \cdots b_n)^+\big) \varphi, \psi\rangle &= \langle \pi^*(b_n^+) \cdots \pi^*(b_1^+) \varphi, \psi\rangle \\ &= \langle \pi(b_n)^* \cdots \pi(b_1)^* \varphi, \psi\rangle = \langle \overline{\pi(b_n^+)} \cdots \overline{\pi(b_1^+)} \varphi, \psi\rangle \\ &= \langle \varphi, \pi(b_1^+)^* \cdots \pi(b_n^+)^* \psi\rangle = \langle \varphi, \pi^*(b_1) \cdots \pi^*(b_n) \psi\rangle \\ &= \langle \varphi, \pi^*(b_1 \cdots b_n) \psi\rangle .\end{aligned}$$

Since an arbitrary $a \in \mathsf{A}$ is a linear combination of $\mathbf{1}$ and of elements of the form $b_1 \cdots b_n$ where $b_1, \ldots, b_n \in \mathsf{B}$, this gives $\langle \pi^*(a^+) \varphi, \psi\rangle = \langle \varphi, \pi^*(a) \psi\rangle$ for $a \in \mathsf{A}$ and $\varphi, \psi \in \mathcal{D}(\pi^*)$. Therefore, the representation π^* is hermitian. By (iii), this shows that π^* is self-adjoint. □

An important special case of Proposition 8.1.12, (v), will be stated separately as

Corollary 8.1.13. *Let $\{x_1, \ldots, x_d\}$, where $d \in \mathbb{N}$, be a basis of a Lie algebra $\mathfrak{g}$. If π is a $*$-representation of the enveloping algebra $\mathcal{E}(\mathfrak{g})$ of $\mathfrak{g}$ such that all operators $\pi(\mathrm{i}x_k)$, $k = 1, \ldots, d$, are essentially self-adjoint, then π^* is a self-adjoint representation.*

Remark 4. Let π be a $*$-representation of a $*$-algebra A. Then we clearly have that $\pi^*(\mathsf{A}) = \pi(\mathsf{A})^*$, where $\pi^*(\mathsf{A})$ is the image of A under the adjoint representation to π, and $\pi(\mathsf{A})^*$ is the adjoint O-family of the O*-algebra $\pi(\mathsf{A})$ as defined in Section 7.2. From this it follows that π is self-adjoint if and only if the O*-algebra $\pi(\mathsf{A})$ is self-adjoint in the sense of Definition 5.1.5.

Among others the following example shows that if π is a $*$-representation, then all inclusions in $\pi \subseteq \bar{\pi} \subseteq \pi^{**} \subseteq \pi^*$ are proper in general.

Example 8.1.14. Let $\mathsf{A} := \mathbb{C}[\mathsf{x}]$. We define two $*$-representations π_1 and π_2 of A in the Hilbert space $\mathcal{H}(\pi_1) = \mathcal{H}(\pi_2) := L^2(0, 1)$ by $\pi_l\big(p(\mathsf{x})\big) = p\left(-\mathrm{i}\,\frac{\mathrm{d}}{\mathrm{d}t}\right) \upharpoonright \mathcal{D}(\pi_l)$ for $l = 1, 2$ and $p(\mathsf{x}) \in \mathbb{C}[\mathsf{x}]$, where

$$\mathcal{D}(\pi_1) := \{\varphi \in C^\infty[0, 1] : \varphi(0) = \varphi(1) \text{ and } \varphi^{(n)}(0) = \varphi^{(n)}(1) = 0 \text{ for } n \in \mathbb{N}\}$$

and

$$\mathcal{D}(\pi_2) := \{\varphi \in C^\infty[0, 1] : \operatorname{supp} \varphi \subseteq (0, 1)\} .$$

Then it is easily seen that $\pi_1 = \bar{\pi}_1 \neq \pi_1^{**} = \pi_1^*$ and $\pi_2 \neq \bar{\pi}_2 = \pi_2^{**} \neq \pi_2^*$. Hence π_2^* is not hermitian, and π_1^* is hermitian and so self-adjoint. The operator $\overline{\pi_1(\mathsf{x})}$ is self-adjoint, but $\overline{\pi_1(\mathsf{x}^2 + \alpha\mathsf{x})}$ is not self-adjoint for all $\alpha \in \mathbb{R}$. The latter fact will be used in Examples 8.1.18 and 9.1.15. ○

We discuss the concepts introduced above in case of the polynomial algebra in one variable. Let π be a representation of $\mathsf{A} = \mathbb{C}[\mathsf{x}]$. By Corollary 8.1.7, $\mathcal{D}(\pi^*) = \mathcal{D}^\infty\big(\pi(\mathsf{x})^*\big)$; so π is adjointable if and only if $\mathcal{D}^\infty\big(\pi(\mathsf{x})^*\big)$ is dense in $\mathcal{H}(\pi)$. It is obvious that π is hermitian if and only if the operator $\pi(\mathsf{x})$ is symmetric.

Proposition 8.1.15. *Suppose π is a $*$-representation of $\mathsf{A} = \mathbb{C}[\mathsf{x}]$.*

(i) *$\mathcal{D}(\pi^{**}) = \mathcal{D}^\infty\big(\overline{\pi(\mathsf{x})}\big)$ and $\pi^{**}(\mathsf{x}) = \overline{\pi(\mathsf{x})} \upharpoonright \mathcal{D}(\pi^{**})$.*

(ii) *π is biclosed if and only if $\mathcal{D}(\pi) = \mathcal{D}^\infty\big(\overline{\pi(\mathsf{x})}\big)$.*

(iii) *π^* is self-adjoint if and only if $\pi(\mathsf{x})$ is essentially self-adjoint.*

(iv) *π is essentially self-adjoint if and only if $\pi(\mathsf{x})^n$ is essentially self-adjoint for all $n \in \mathbb{N}$.*

(v) *π is self-adjoint if and only if $\pi(\mathsf{x})$ is essentially self-adjoint and $\mathcal{D}(\pi) = \mathcal{D}^\infty\big(\overline{\pi(\mathsf{x})}\big)$.*

In the proof we need the following simple lemma.

Lemma 8.1.16. *Let π_1 and π_2 be $*$-representations of $\mathsf{A} = \mathbb{C}[\mathsf{x}]$ acting on the same Hilbert space such that $\overline{\pi_1(\mathsf{x})} \subseteq \overline{\pi_2(\mathsf{x})}$. Then $\overline{\pi_1(\mathsf{x})} = \overline{\pi_2(\mathsf{x})}$ if and only if $\pi_1^* = \pi_2^*$.*

Proof. First suppose that $\overline{\pi_1(\mathsf{x})} = \overline{\pi_2(\mathsf{x})}$. Then $\pi_1(\mathsf{x})^* = \pi_2(\mathsf{x})^*$ and so $\mathcal{D}(\pi_1^*) = \mathcal{D}^\infty(\pi_1(\mathsf{x})^*) = \mathcal{D}^\infty(\pi_2(\mathsf{x})^*) = \mathcal{D}(\pi_2^*)$ by Corollary 8.1.7. Since $\pi_1^*(\mathsf{x})$ and $\pi_2^*(\mathsf{x})$ are both restriction of $\pi_1(\mathsf{x})^* \equiv \pi_2(\mathsf{x})^*$, this yields $\pi_1^* = \pi_2^*$.

Now suppose that $\overline{\pi_1(\mathsf{x})} \neq \overline{\pi_2(\mathsf{x})}$. Then $\mathcal{H}_+^2 + \mathcal{H}_-^2 \subsetneqq \mathcal{H}_+^1 + \mathcal{H}_-^1$, where $\mathcal{H}_\pm^1$ and $\mathcal{H}_\pm^2$ are the deficiency spaces of the closed symmetric operators $\overline{\pi_1(\mathsf{x})}$ and $\overline{\pi_2(\mathsf{x})}$, respectively. If $\varphi \in (\mathcal{H}_+^1 + \mathcal{H}_-^1) \setminus (\mathcal{H}_+^2 + \mathcal{H}_-^2)$, then obviously $\varphi \in \mathcal{D}^\infty(\pi_1(\mathsf{x})^*) = \mathcal{D}(\pi_1^*)$, but $\varphi \notin \mathcal{D}(\pi_2(\mathsf{x})^*)$, since $\pi_2(\mathsf{x})^* \subseteq \pi_1(\mathsf{x})^*$. Therefore, $\mathcal{D}(\pi_1^*) \neq \mathcal{D}(\pi_2^*)$. □

Proof of Proposition 8.1.15

(i): Let π_1 be the $*$-representation of $\mathsf{A} = \mathbb{C}[\mathsf{x}]$ defined by $\mathcal{D}(\pi_1) := \mathcal{D}^\infty(\overline{\pi(\mathsf{x})})$ and $\pi_1(\mathsf{x}) := \overline{\pi(\mathsf{x})} \restriction \mathcal{D}(\pi_1)$. Since obviously $\overline{\pi(\mathsf{x})} = \overline{\pi_1(\mathsf{x})}$, Lemma 8.1.16 gives $\pi^* = \pi_1^*$. Hence $\pi^{**} = \pi_1^{**}$. Further, we have $\overline{\pi(\mathsf{x})} \subseteq \overline{\pi^{**}(\mathsf{x})}$ and $\pi^* = (\pi^{**})^*$. Therefore, Lemma 8.1.16, applied in reversed order with $\pi_1 = \pi$ and $\pi_2 = \pi^{**}$, yields $\overline{\pi(\mathsf{x})} = \overline{\pi^{**}(\mathsf{x})}$. Because π^{**} is closed, this gives

$$\mathcal{D}(\pi^{**}) = \bigcap_{n\in\mathbb{N}} \mathcal{D}\left(\overline{\pi^{**}(\mathsf{x})^n}\right) \subseteq \bigcap_{n\in\mathbb{N}} \mathcal{D}\left(\left(\overline{\pi^{**}(\mathsf{x})}\right)^n\right) = \mathcal{D}^\infty\left(\overline{\pi(\mathsf{x})}\right) = \mathcal{D}(\pi_1).$$

Thus

$$\mathcal{D}(\pi_1^{**}) = \mathcal{D}(\pi^{**}) \subseteq \mathcal{D}(\pi_1) \subseteq \mathcal{D}(\pi_1^{**})$$

which implies that $\pi_1 = \pi_1^{**}$. Combined with $\pi^{**} = \pi_1^{**}$, we obtain $\pi^{**} = \pi_1$.

(ii) follows from (i) and the inclusion $\pi \subseteq \pi^{**}$.

(iii): If $\pi(\mathsf{x})$ is essentially self-adjoint, then the operator $\pi^*(\mathsf{x})$ $\left(\subseteq \pi(\mathsf{x})^* = \overline{\pi(\mathsf{x})}\right)$ is symmetric; so π^* is hermitian and hence self-adjoint. Conversely, suppose that $\pi(\mathsf{x})$ is not essentially self-adjoint. Then both deficiency spaces for $\pi(\mathsf{x})$ are contained in $\mathcal{D}^\infty(\pi(\mathsf{x})^*) = \mathcal{D}(\pi^*)$, so the operator $\pi^*(\mathsf{x})$ is not symmetric, and π^* is not self-adjoint.

(iv): If $\pi(\mathsf{x})^n$ is essentially self-adjoint for all $n \in \mathbb{N}$, then we have

$$\mathcal{D}(\hat{\pi}) = \bigcap_{n\in\mathbb{N}} \mathcal{D}\left(\overline{\pi(\mathsf{x})^n}\right) = \bigcap_{n\in\mathbb{N}} \mathcal{D}\left(\left(\pi(\mathsf{x})^n\right)^*\right) = \mathcal{D}(\pi^*)$$

which gives $\hat{\pi} = \pi^*$. Conversely, assume that $\hat{\pi} = \pi^*$. Then π^* is hermitian and hence self-adjoint. By (iii), $\pi(\mathsf{x})$ is essentially self-adjoint. Therefore,

$$\mathcal{D}(\hat{\pi}) = \mathcal{D}(\pi^*) = \mathcal{D}^\infty\left(\pi(\mathsf{x})^*\right) = \mathcal{D}^\infty\left(\overline{\pi(\mathsf{x})}\right).$$

By definition $\mathcal{D}(\hat{\pi})$ is a core for $\overline{\pi(\mathsf{x})^n}$. Since $\overline{\pi(\mathsf{x})}$ is self-adjoint, $\mathcal{D}^\infty\left(\overline{\pi(\mathsf{x})}\right)$ is a core for $\left(\overline{\pi(\mathsf{x})}\right)^n$. Therefore, $\overline{\pi(\mathsf{x})^n} = \left(\overline{\pi(\mathsf{x})}\right)^n$ and, this operator is self-adjoint, so that $\pi(\mathsf{x})^n$ is essentially self-adjoint.

(v) follows at once from (ii), (iii) and Lemma 8.1.11, (ii). □

Now let π be a $*$-representation of A and let B be a subset of the $*$-algebra A such that $\mathsf{B} \cup \{1\}$ generates A as an algebra. We consider the family of all closed $*$-representations

ϱ of A on $\mathcal{H}(\pi) = \mathcal{H}(\varrho)$ which are extensions of π and which have the property that $\mathcal{D}(\pi)$ is a core for each operator $\varrho(b)$, $b \in \mathsf{B}$. In case $\mathsf{B} = \mathsf{A}$ there is only one $*$-representation of this kind, the closure of π. In general there are many different representations in this family. For instance in the case where $\mathsf{A} = \mathbb{C}[\mathsf{x}]$ and $\mathsf{B} = \{\mathsf{x}\}$ we have $\overline{\hat{\pi}(b)} = \overline{\pi^{**}(b)}$ for $b \in \mathsf{B}$ (by Proposition 8.1.15, (i)), but $\hat{\pi} \neq \pi^{**}$ in general (cf. Example 8.1.14). However, this family contains always a *largest* representation which we describe now.

Proposition 8.1.17. *Suppose that π is a $*$-representation of the $*$-algebra* A *and* B *is a subset of* A *such that* $\mathsf{B} \cup \{1\}$ *generates* A *as an algebra.*

(i) *Define*

$$\mathcal{D}_0 := \bigcap_{k \in \mathbb{N}} \bigcap_{b_1, \ldots, b_k \in \mathsf{B}} \mathcal{D}\big(\overline{\pi(b_1)} \cdots \overline{\pi(b_k)}\big).$$

Then $\mathcal{D}_0$ is a linear subspace of $\mathcal{D}(\pi^*)$ which is invariant under $\pi^*(a)$, $a \in \mathsf{A}$, and $\pi_0 := \pi^* \upharpoonright \mathcal{D}_0$ is a closed $*$-representation of A *which extends π. Moreover, $\pi_0^* = \pi^*$ and $\pi_0(b) = \overline{\pi(b)} \upharpoonright \mathcal{D}_0$ for $b \in \mathsf{B}$.*

(ii) *π_0 is the largest among the closed $*$-representations ϱ of* A *on $\mathcal{H}(\varrho) \equiv \mathcal{H}(\pi)$ which satisfy $\pi \subseteq \varrho$ and $\overline{\pi(b)} = \overline{\varrho(b)}$ for all $b \in \mathsf{B}$.*

(iii) *If $\overline{\pi(b)} = \pi(b^+)^*$ for all $b \in \mathsf{B}$, then π_0 is self-adjoint and $\pi_0 = \pi^*$.*

Proof. (i): From the inclusions

$$\overline{\pi(b_1)} \cdots \overline{\pi(b_k)} \subseteq \pi(b_1^+)^* \cdots \pi(b_k^+)^* \subseteq \big(\pi(b_k^+) \cdots \pi(b_1^+)\big)^* = \pi(b_k^+ \cdots b_1^+)^*$$

for $b_1, \ldots, b_k \in \mathsf{B}$ we conclude easily that $\mathcal{D}_0 \subseteq \mathcal{D}(\pi^*)$. It is plain from the definition of $\mathcal{D}_0$ that $\mathcal{D}_0$ is invariant under $\overline{\pi(b)}$, $b \in \mathsf{B}$. Since $\overline{\pi(b)} \subseteq \pi(b^+)^*$ and hence $\overline{\pi(b)} \upharpoonright \mathcal{D}_0 = \overline{\pi^*(b)} \upharpoonright \mathcal{D}_0$, $\mathcal{D}_0$ is invariant under $\pi^*(b)$ for all $b \in \mathsf{B}$. Because $\mathsf{B} \cup \{1\}$ generates the algebra A and π^* is a homomorphism, the latter is true for all $b \in \mathsf{A}$. Being the restriction of the representation π^* to the invariant domain $\mathcal{D}_0$, π_0 is a representation of A. We show that π_0 preserves the involution. It suffices to prove this for the elements b^+, where $b \in \mathsf{B}$. Fix $b \in \mathsf{B}$ and let $\psi \in \mathcal{D}_0$. Since $\mathcal{D}_0 \subseteq \mathcal{D}(\pi^*)$, we have

$$\langle \pi(b)\, \varphi, \psi \rangle = \langle \varphi, \pi(b)^* \psi \rangle = \langle \varphi, \pi^*(b^+)\, \psi \rangle \quad \text{for } \varphi \in \mathcal{D}(\pi)$$

and hence

$$\langle \overline{\pi(b)}\, \varphi, \psi \rangle = \langle \varphi, \pi^*(b^+)\, \psi \rangle \quad \text{for } \varphi \in \mathcal{D}_0.$$

Because $\overline{\pi(b)} \upharpoonright \mathcal{D}_0 = \pi^*(b) \upharpoonright \mathcal{D}_0 = \pi_0(b)$ and $\pi^*(b^+) \upharpoonright \mathcal{D}_0 = \pi_0(b^+)$, this shows that $\pi_0(b)^+ = \pi_0(b^+)$. Thus π_0 is a $*$-representation of A. The equality $\pi_0(b) = \overline{\pi(b)} \upharpoonright \mathcal{D}_0$, $b \in \mathsf{B}$, was just mentioned. It implies $\overline{\pi_0(b)} = \overline{\pi(b)}$ for $b \in \mathsf{B}$. Therefore, by Corollary 8.1.8, $\pi_0^* = \pi^*$.

Next we prove that π_0 is closed. Let $(\varphi_i : i \in I)$ be a Cauchy net in $\mathcal{D}_0[\mathrm{t}_{\pi_0}]$. Since $\pi_0 \subseteq \pi^*$, $(\varphi_i : i \in I)$ is also a Cauchy net in $\mathcal{D}(\pi^*)$ relative to the graph topology of $\pi^*(\mathsf{A})$. Because π^* is closed, this net has a limit, say φ, in $\mathcal{D}(\pi^*)\,[\mathrm{t}_{\pi^*}]$. The proof that π_0 is closed is complete once we have shown that $\varphi \in \mathcal{D}_0$. In order to prove this we verify by induction on k that $\varphi \in \mathcal{D}\big(\overline{\pi(b_1)} \cdots \overline{\pi(b_k)}\big)$ and $\pi^*(b_1 \cdots b_k)\, \varphi = \overline{\pi(b_1)} \cdots \overline{\pi(b_k)}\, \varphi$ for arbitrary elements $b_1, \ldots, b_k$ in B. In case $k = 1$ this has been already noted above. Suppose that this is shown for $k \in \mathbb{N}$. Take $b_1, \ldots, b_{k+1} \in$ B. Since $\varphi = \lim_i \varphi_i$ in the graph topology of $\pi^*(\mathsf{A})$, the nets $(\pi^*(b_1 \cdots b_{k+1}) \varphi_i : i \in I)$ and $(\pi^*(b_2 \cdots b_{k+1}) \varphi_i : i \in I)$ converge to $\pi^*(b_1 \cdots b_{k+1}) \varphi$ and $\pi^*(b_2 \cdots b_{k+1}) \varphi$ in $\mathcal{H}(\pi)$, respectively. By $\pi^*(b_1) \upharpoonright \mathcal{D}_0 = \pi(b_1) \upharpoonright \mathcal{D}_0$, we have

$$\pi^*(b_1 \cdots b_{k+1}) \varphi_i = \pi^*(b_1) \pi^*(b_2 \cdots b_{k+1}) \varphi_i = \overline{\pi(b_1)} \pi^*(b_2 \cdots b_{k+1}) \varphi_i, \quad i \in I.$$

Therefore, it follows that $\pi^*(b_2 \cdots b_{k+1}) \varphi \in \mathcal{D}(\overline{\pi(b_1)})$ and $\overline{\pi(b_1)} \pi^*(b_2 \cdots b_{k+1}) \varphi = \pi^*(b_1 \cdots b_{k+1}) \varphi$. From the induction hypothesis, $\pi^*(b_2 \cdots b_{k+1}) \varphi = \overline{\pi(b_2)} \cdots \overline{\pi(b_{k+1})} \varphi$, so $\varphi \in \mathcal{D}(\overline{\pi(b_1)} \cdots \overline{\pi(b_{k+1})})$ and $\pi^*(b_1 \cdots b_{k+1}) \varphi = \overline{\pi(b_1)} \cdots \overline{\pi(b_{k+1})} \varphi$ which completes the induction proof.

(ii): It is clear that $\pi \subseteq \pi_0$ and $\overline{\pi_0(b)} = \overline{\pi(b)}$ for $b \in$ B. Let ϱ be a $*$-representation of A on $\mathcal{H}(\varrho) \equiv \mathcal{H}(\pi)$ such that $\pi \subseteq \varrho$ and $\overline{\pi(b)} = \overline{\varrho(b)}$ for $b \in$ B. Suppose $\varphi \in \mathcal{D}(\varrho)$, and let $b_1, \ldots, b_k \in$ B. We have $\varphi \in \mathcal{D}(\overline{\varrho(b_k)}) = \mathcal{D}(\overline{\pi(b_k)})$ and $\overline{\pi(b_k)} \varphi = \overline{\varrho(b_k)} \varphi = \varrho(b_k) \varphi \in \mathcal{D}(\varrho)$. Replacing φ by $\varrho(b_k) \varphi$ and b_k by b_{k-1}, we get $\varphi \in \mathcal{D}(\overline{\pi(b_{k-1})}\, \overline{\pi(b_k)})$ and $\overline{\pi(b_{k-1})}\, \overline{\pi(b_k)} \varphi = \varrho(b_{k-1}) \varrho(b_k) \varphi = \varrho(b_{k-1} b_k) \varphi \in \mathcal{D}(\varrho)$. Proceeding along this line, we obtain $\varphi \in \mathcal{D}(\overline{\pi(b_1)} \cdots \overline{\pi(b_k)})$ and $\overline{\pi(b_1)} \cdots \overline{\pi(b_k)} \varphi = \varrho(b_1 \cdots b_k) \varphi$. Hence $\varphi \in \mathcal{D}_0$ and $\varrho \subseteq \pi_0$, since $\pi_0(b) = \overline{\pi(b)} \upharpoonright \mathcal{D}_0$ for $b \in$ B.

(iii): Suppose $\overline{\pi(b)} = \pi(b^+)^*$ for $b \in$ B. From the definition of $\mathcal{D}_0$ and the formula for $\mathcal{D}(\pi^*)$ in Lemma 8.1.6 we see that $\mathcal{D}_0 \equiv \mathcal{D}(\pi_0) = \mathcal{D}(\pi^*)$. Thus $\pi_0 = \pi^*$, since $\pi_0 \subseteq \pi^*$ by definition. Assertion (v) (or (iii)) of Proposition 8.1.12 shows that $\pi^* \equiv \pi_0$ is self-adjoint. □

The $*$-representation π_0 in the preceding proposition satisfies $\pi_0^* = \pi^*$ and hence $\pi_0 \subseteq \pi_0^{**} = \pi^{**}$. From the following example we see that π_0 is not biclosed and hence different from π^{**} in general. Moreover, this example shows that π_0 really depends on the set B.

Example 8.1.18. Let $\mathsf{A} := \mathbb{C}[\mathsf{x}]$ and let π be the $*$-representation π_1 from Example 8.1.14. Since $\pi(\mathsf{x}) \equiv \pi_1(\mathsf{x})$ is essentially self-adjoint, π^* is self-adjoint by Proposition 8.1.15, (iii). Applying Proposition 8.1.15, (iv), to π^*, it follows that $\pi^*(\mathsf{x}^2)$ is essentially self-adjoint. Setting $\mathsf{B} := \{\mathsf{x}, \mathsf{x}^2\}$, we have $\pi_0(\mathsf{x}^2) = \overline{\pi(\mathsf{x}^2)} \equiv \overline{\pi_1(\mathsf{x}^2)}$. By Example 8.1.14, this operator is not self-adjoint. Hence $\pi_0 \neq \pi^* = \pi^{**} = \pi_0^{**}$. However, in case $\mathsf{B} := \{\mathsf{x}\}$ we clearly have $\pi_0 = \pi^*$. ○

The next proposition is only a reformulation of a well-known criterion for the essential self-adjointness of a symmetric operator in the context of $*$-representations.

Proposition 8.1.19. *Suppose π is a $*$-representation of the $*$-algebra* A *and a is a hermitian element of* A. *Let α_1 and α_2 be complex numbers with $\operatorname{Im} \alpha_1 > 0$ and $\operatorname{Im} \alpha_2 < 0$. Suppose that there are linear operators x_1 and x_2 defined on $\mathcal{D}(\pi)$ and leaving $\mathcal{D}(\pi)$ invariant such that $(\pi(a) - \alpha_1) x_1 \varphi = (\pi(a) - \alpha_2) x_2 \varphi = \varphi$ for $\varphi \in \mathcal{D}(\pi)$. (In particular, the latter is fulfilled if there are elements $b_1, b_2 \in$ A such that $(a - \alpha_1) b_1 = (a - \alpha_2) b_2 = 1$.)*

Then the operator $\overline{\pi(a)}$ is self-adjoint. Moreover, the operators x_1 and x_2 are bounded and we have that $\overline{x_1} = (\overline{\pi(a)} - \alpha_1)^{-1}$ and $\overline{x_2} = (\overline{\pi(a)} - \alpha_2)^{-1}$.

Proof. Since $(\pi(a) - \alpha_k)\,\mathcal{D}(\pi) \supseteq (\pi(a) - \alpha_k)\,x_k\mathcal{D}(\pi) = \mathcal{D}(\pi)$, $(\pi(a) - \alpha_k)\,\mathcal{D}(\pi)$ is dense in $\mathcal{H}(\pi)$ for $k = 1, 2$, and the operator $\pi(a)$ is essentially self-adjoint (cf. p. 29). From the relations $(\overline{\pi(a)} - \alpha_k)\,x_k\varphi = \varphi = (\overline{\pi(a)} - \alpha_k)\,(\overline{\pi(a)} - \alpha_k)^{-1}\varphi$ and $\ker(\overline{\pi(a)} - \alpha_k) = \{0\}$ (because of $\operatorname{Im} \alpha_k \neq 0$) we obtain $x_k\varphi = (\overline{\pi(a)} - \alpha_k)^{-1}\,\varphi$ for $k = 1, 2$ and $\varphi \in \mathcal{D}(\pi)$. This yields the second assertion. □

Corollary 8.1.20. *If* A *is a symmetric* $*$*-algebra and* π *is a* $*$*-representation of* A*, then* $\overline{\pi(a)}$ *is self-adjoint and* $\overline{\pi\big((a - \alpha)^{-1}\big)} = \big(\overline{\pi(a)} - \alpha\big)^{-1}$ *for all* $a = a^+ \in$ A *and* $\alpha \in \mathbb{C} \setminus \mathbb{R}$.

Proof. Apply Proposition 8.1.19 with $x_1 := \pi\big((a - \alpha)^{-1}\big)$ and $x_2 := \pi\big((a - \bar{\alpha})^{-1}\big)$ when $\operatorname{Im} \alpha > 0$; otherwise we interchange x_1 and x_2. Recall that $a - \alpha$ is invertible in A, since A is a symmetric $*$-algebra. □

8.2. Intertwining Operators

In this section A will denote an algebra with unit.

Definition 8.2.1. Let π_1 and π_2 be representations of A. A bounded operator x from $\mathcal{H}(\pi_1)$ into $\mathcal{H}(\pi_2)$ is called an *intertwining operator* for π_1 and π_2 if $x\big(\mathcal{D}(\pi_1)\big) \subseteq \mathcal{D}(\pi_2)$ and $x\pi_1(a)\,\varphi = \pi_2(a)\,x\varphi$ for $a \in$ A and $\varphi \in \mathcal{D}(\pi_1)$. The vector space of these operators x is called the *intertwining space* for π_1 and π_2 and denoted by $\mathbb{I}(\pi_1, \pi_2)$.

The intertwining space of two representations is an important tool in representation theory. Concepts like unitary equivalence, similarity and disjointness of representations will be defined in terms of this space; see Section 8.4. There are two special cases of these spaces which are of particular interest. For any representation π of A, the intertwining space $\mathbb{I}(\pi, \pi)$ is equal to the *strong commutant* $\pi(\mathsf{A})'_{\mathrm{s}}$ of the O-algebra $\pi(\mathsf{A})$. If A is a $*$-algebra and π is a $*$-representation of A, then $\mathbb{I}(\pi, \pi^*)$ coincides with the *weak commutant* $\pi(\mathsf{A})'_{\mathrm{w}}$ of the O*-algebra $\pi(\mathsf{A})$. The first of these two statements follows at once from Definition 7.2.7, and the second one from Proposition 7.2.10, (i), combined with the definition of π^*. (Recall that $\mathcal{D}^*(\pi(\mathsf{A})) = \mathcal{D}(\pi^*)$ and $(\pi(a)^+)^* \upharpoonright \mathcal{D}(\pi^*) = \pi(a^+)^* \upharpoonright \mathcal{D}(\pi^*) = \pi^*(a)$ for $a \in$ A by definition.)

Some simple properties of the intertwining spaces are collected in the following propositions. They will be often used in the sequel.

Proposition 8.2.2. *Suppose* π_1, π_2 *and* π_3 *are representations of* A.

(i) *Each operator* x *of* $\mathbb{I}(\pi_1, \pi_2)$ *is a continuous mapping of* $\mathcal{D}(\pi_1)\,[\mathrm{t}_{\pi_1}]$ *into* $\mathcal{D}(\pi_2)\,[\mathrm{t}_{\pi_2}]$.

(ii) *If* $\pi_3 \subseteq \pi_1$ *and* $\mathcal{H}(\pi_3) = \mathcal{H}(\pi_1)$, *then* $\mathbb{I}(\pi_1, \pi_2) \subseteq \mathbb{I}(\pi_3, \pi_2)$. *If* $\pi_2 \subseteq \pi_3$ *and* $\mathcal{H}(\pi_2) = \mathcal{H}(\pi_3)$, *then* $\mathbb{I}(\pi_1, \pi_2) \subseteq \mathbb{I}(\pi_1, \pi_3)$.

(iii) *If* $x_1 \in \mathbb{I}(\pi_1, \pi_2)$ *and* $x_2 \in \mathbb{I}(\pi_2, \pi_3)$, *then* $x_2x_1 \in \mathbb{I}(\pi_1, \pi_3)$.

(iv) $\mathbb{I}(\pi_1, \pi_2) \subseteq \mathbb{I}\big(\widehat{\pi}_1, \widehat{\pi}_2\big)$.

(v) *The closure of* $\mathbb{I}(\pi_1, \pi_2)$ *in the weak-operator topology in* $\mathbb{B}\big(\mathcal{H}(\pi_1), \mathcal{H}(\pi_2)\big)$ *is contained in* $\mathbb{I}\big(\pi_1, \widehat{\pi}_2\big)$. *If* π_2 *is a closed representation, then* $\mathbb{I}(\pi_1, \pi_2)$ *is weak-operator closed in* $\mathbb{B}\big(\mathcal{H}(\pi_1), \mathcal{H}(\pi_2)\big)$.

Proof. (i), (ii) and (iii) follow immediately from the definition.

(iv): Suppose $x \in \mathbb{I}(\pi_1, \pi_2)$. Let $\varphi \in \mathcal{D}(\widehat{\pi}_1)$. Then there is a net $(\varphi_i : i \in I)$ in $\mathcal{D}(\pi_1)$ such that $\varphi = \lim \varphi_i$ in the graph topology of $\widehat{\pi}_1(\mathsf{A})$. By (i), $(x\varphi_i : i \in I)$ is a Cauchy net in $\mathcal{D}(\pi_2)\,[\mathrm{t}_{\pi_2}]$. Since x is bounded, this implies that $x\varphi = \lim x\varphi_i$ in the locally convex space $\mathcal{D}(\widehat{\pi}_2)\,[\mathrm{t}_{\widehat{\pi}_2}]$. Thus $x\varphi \in \mathcal{D}(\widehat{\pi}_2)$. From $x\pi_1(a)\,\varphi_i = \pi_2(a)\,x\varphi_i$ it follows that $x\widehat{\pi}_1(a)\,\varphi = \widehat{\pi}_2(a)\,x\varphi$ for $a \in \mathsf{A}$. Hence $x \in \mathbb{I}(\widehat{\pi}_1, \widehat{\pi}_2)$.

(v): Set $\tilde{\mathcal{H}} := \mathcal{H}(\pi_1) \oplus \mathcal{H}(\pi_2)$. For $x \in \mathbb{B}(\mathcal{H}(\pi_1), \mathcal{H}(\pi_2))$, let $\tilde{x}$ be the operator in in $\mathbb{B}(\tilde{\mathcal{H}})$ defined by $\tilde{x}(\varphi_1, \varphi_2) := (0, x\varphi_1)$, $\varphi_1 \in \mathcal{H}(\pi_1)$ and $\varphi_2 \in \mathcal{H}(\pi_2)$. Let x be in the weak-operator closure of $\mathbb{I}(\pi_1, \pi_2)$ in $\mathbb{B}(\mathcal{H}(\pi_1), \mathcal{H}(\pi_2))$. Since $\tilde{\mathbb{I}} := \{\tilde{z} : z \in \mathbb{I}(\pi_1, \pi_2)\}$ is a linear subspace of $\mathbb{B}(\mathcal{H})$, $\tilde{x}$ belongs to the strong-operator closure of $\tilde{\mathbb{I}}$ in $\mathbb{B}(\tilde{\mathcal{H}})$. Then there is a net $(x_i : i \in I)$ in $\mathbb{I}(\pi_1, \pi_2)$ such that $\tilde{x} = \lim \tilde{x}_i$ in the strong-operator topology. Fix $\varphi \in \mathcal{D}(\pi_1)$. For each $a \in \mathsf{A}$, $(x_i\pi_1(a)\,\varphi \equiv \pi_2(a)\,x_i\varphi : i \in I)$ is a net in $\mathcal{H}$ which converges to $x\pi_1(a)\,\varphi$. From this we conclude that $(x_i\varphi : i \in I)$ is a Cauchy net in $\mathcal{D}(\pi_2)\,[\mathrm{t}_{\pi_2}]$, $x\varphi = \lim x_i\varphi$ in the graph topology of $\widehat{\pi}_2(A)$ and $x\pi_1(a)\,\varphi = \widehat{\pi}_2(a)\,x\varphi$. Therefore, $x \in \mathbb{I}(\pi_1, \widehat{\pi}_2)$. (The two preceding proofs are based on similar arguments as the proof of Lemma 7.2.8.) □

Proposition 8.2.3. *Suppose that* A *is a* $*$*-algebra and* π_1 *and* π_2 *are adjointable representations of* A.

(i) $\mathbb{I}(\pi_1, \pi_2)^* \subseteq \mathbb{I}(\pi_2^*, \pi_1^*)$.

(ii) $\mathbb{I}(\widehat{\pi}_1, \widehat{\pi}_2) \subseteq \mathbb{I}(\pi_1^{**}, \pi_2^{**})$.

(iii) $\mathbb{I}(\pi_1, \pi_2^*) = \mathbb{I}(\widehat{\pi}_1, \pi_2^*) = \mathbb{I}(\pi_1^{**}, \pi_2^*)$.

(iv) $\mathbb{I}(\pi_1, \pi_2^*)^* = \mathbb{I}(\pi_2, \pi_1^*)$.

Here $\mathbb{I}(\pi_1, \pi_2)^*$ *and* $\mathbb{I}(\pi_1, \pi_2^*)^*$ *denote the sets of all operators* x^*, *where* $x \in \mathbb{I}(\pi_1, \pi_2)$ *and* $x \in \mathbb{I}(\pi_1, \pi_2^*)$, *respectively.*

Proof. First note that $\mathcal{H}(\pi_1) = \mathcal{H}(\pi_1^*) = \mathcal{H}(\pi_1^{**})$ and $\mathcal{H}(\pi_2) = \mathcal{H}(\pi_2^*) = \mathcal{H}(\pi_2^{**})$, since π_1 and π_2 are assumed to be adjointable.

(i): Suppose $x \in \mathbb{I}(\pi_1, \pi_2)$. Let $\varphi_2 \in \mathcal{D}(\pi_2^*)$. Then we have $\langle x^*\varphi_2, \pi_1(a)\,\varphi_1\rangle = \langle \varphi_2, x\pi_1(a)\varphi_1\rangle = \langle \varphi_2, \pi_2(a)\,x\varphi_1\rangle = \langle x^*\pi_2^*(a^+)\,\varphi_2, \varphi_1\rangle$ for all $\varphi_1 \in \mathcal{D}(\pi_1)$ and $a \in \mathsf{A}$. This implies that $x^*\varphi_2 \in \bigcap_{a \in \mathsf{A}} \mathcal{D}(\pi_1(a)^*) = \mathcal{D}(\pi_1^*)$ and $\pi_1^*(a^+)\,x^*\varphi_2 = \pi_1(a)^*\,x^*\varphi_2 = x^*\pi_2^*(a^+)\,\varphi_2$ for $a \in \mathsf{A}$. This proves that $x^* \in \mathbb{I}(\pi_2^*, \pi_1^*)$.

(ii): Applying (i) twice gives $\mathbb{I}(\pi_1, \pi_2) \subseteq \mathbb{I}(\pi_1^{**}, \pi_2^{**})$. Replacing π_l by $\widehat{\pi}_l$ and using that $\pi_l^{**} = (\widehat{\pi}_l)^{**}$ for $l = 1, 2$, we obtain $\mathbb{I}(\widehat{\pi}_1, \widehat{\pi}_2) \subseteq \mathbb{I}(\pi_1^{**}, \pi_2^{**})$.

(iii): Applying again (i) twice it follows that $\mathbb{I}(\pi_1, \pi_2^*) \subseteq \mathbb{I}(\pi_1^{**}, \pi_2^{***}) = \mathbb{I}(\pi_1^{**}, \pi_2^*)$. On the other hand, since $\pi_1 \subseteq \widehat{\pi}_1 \subseteq \pi_1^{**}$ by Proposition 8.1.5, we have $\mathbb{I}(\pi_1^{**}, \pi_2^*) \subseteq \mathbb{I}(\widehat{\pi}_1, \pi_2^*) \subseteq \mathbb{I}(\pi_1, \pi_2^*)$. Both together give the assertion.

(iv): From (i) we obtain $\mathbb{I}(\pi_1, \pi_2^*)^* \subseteq \mathbb{I}(\pi_2^{**}, \pi_1^*) \subseteq \mathbb{I}(\pi_2, \pi_1^*)$. By symmetry, $\mathbb{I}(\pi_2, \pi_1^*)^* \subseteq \mathbb{I}(\pi_1, \pi_2^*)$ and so $\mathbb{I}(\pi_1, \pi_2^*)^* = \mathbb{I}(\pi_2, \pi_1^*)$. □

Corollary 8.2.4. *Let* π_1, π_2 *and* A *be as in Proposition 8.2.3. Suppose that* $x \in \mathbb{I}(\pi_1, \pi_2)$.

(i) *If* π_2 *is a* $*$*-representation of* A, *then* $x^*x \in \mathbb{I}(\pi_1, \pi_1^*)$.

(ii) *If* π_1 *is a self-adjoint representation, then* $xx^* \in \mathbb{I}(\pi_2^*, \pi_2)$.

Proof. We freely use the properties established in the two previous propositions. First note that $x^* \in \mathbb{I}(\pi_2^*, \pi_1^*)$, since $x \in \mathbb{I}(\pi_1, \pi_2)$.

(i): Since π_2 is a $*$-representation, $\pi_2 \subseteq \pi_2^*$ by Lemma 8.1.11 and so $x^* \in \mathbb{I}(\pi_2, \pi_1^*)$. Hence $x^*x \in \mathbb{I}(\pi_1, \pi_1^*)$.

(ii): Since $\pi_1 = \pi_1^*$ by assumption, $x^* \in \mathbb{I}(\pi_2^*, \pi_1)$. Thus $xx^* \in \mathbb{I}(\pi_2^*, \pi_2)$. $\square$

Corollary 8.2.5. *Suppose that* A *is a* $*$*-algebra.*

(i) *If* π *is an adjointable representation of* A, *then* $\pi(\mathsf{A})'_{\mathrm{s}} \subseteq \pi^{**}(\mathsf{A})'_{\mathrm{s}}$.

(ii) *If* π *is a* $*$*-representation of* A, *then* $\pi(\mathsf{A})'_{\mathrm{w}} = \pi^{**}(\mathsf{A})'_{\mathrm{w}}$.

Proof. (i): $\pi(\mathsf{A})'_{\mathrm{s}} \equiv \mathbb{I}(\pi, \pi) \subseteq \mathbb{I}(\pi^{**}, \pi^{**}) \equiv \pi^{**}(\mathsf{A})'_{\mathrm{s}}$ by Proposition 8.2.3, (ii).

(ii): Since $\pi^* = \pi^{***}$, Proposition 8.2.3, (iii), gives $\pi(\mathsf{A})'_{\mathrm{w}} = \mathbb{I}(\pi, \pi^*) = \mathbb{I}(\pi^{**}, \pi^{***}) = \pi^{**}(\mathsf{A})'_{\mathrm{w}}$. $\square$

Of course, the notion of an intertwining space can also be defined for single operators. Suppose that a and b are closable linear operators in Hilbert spaces $\mathcal{H}$ and $\mathcal{K}$, respectively. Then the vector space $\mathbb{I}(a, b) := \{x \in \mathbb{B}(\mathcal{H}, \mathcal{K}) : xa \subseteq bx\}$ is called the *intertwining space* for a and b. Obviously, $\mathbb{I}(a, a) = (a)'_{\mathrm{s}}$. If the operator a is symmetric, then $\mathbb{I}(a, a^*) = (a)'_{\mathrm{w}}$. The following properties are proved quite similarly as in case of representations. We omit the details.

Lemma 8.2.6. (i) $\mathbb{I}(a, b) \subseteq \mathbb{I}(\bar{a}, \bar{b})$.

(ii) $\mathbb{I}(a, b)^* \subseteq \mathbb{I}(b^*, a^*)$.

(iii) *If the operator* b *is closed, then* $\mathbb{I}(a, b)$ *is a closed vector space of* $\mathbb{B}(\mathcal{H}, \mathcal{K})$ *in the weak-operator topology.*

Proposition 8.2.7. *Let* B *be a subset of the* $*$*-algebra* A *such that* $\mathsf{B} \cup \{1\}$ *generates* A *as an algebra. Suppose* π_1 *and* π_2 *are representations of* A *and* π_2 *is adjointable. Then we have* $\mathbb{I}(\pi_1, \pi_2^*) = \bigcap_{b \in \mathsf{B}} \mathbb{I}(\overline{\pi_1(b)}, \pi_2(b^+)^*)$.

Proof. If $x \in \mathbb{I}(\pi_1, \pi_2^*)$, then

$$x \in \mathbb{I}(\pi_1(b), \pi_2^*(b)) \subseteq \mathbb{I}(\pi_1(b), \pi_2(b^+)^*) \subseteq \mathbb{I}(\overline{\pi_1(b)}, \pi_2(b^+)^*) \quad \text{for} \quad b \in \mathsf{B},$$

where the last inclusion follows from Lemma 8.2.6, (i). Conversely, suppose that x is in $\mathbb{I}(\overline{\pi_1(b)}, \pi_2(b^+)^*)$ for all $b \in \mathsf{B}$. Let $\varphi \in \mathcal{D}(\pi_1)$. By induction on n it follows easily that $x\varphi \in \mathcal{D}(\pi_2(b_1^+)^* \dots \pi_2(b_n^+)^*)$ and $x\pi_1(b_1 \dots b_n)\, \varphi \equiv x\pi_1(b_1) \dots \pi_1(b_n)\, \varphi = \pi_2(b_1^+)^* \dots \pi_2(b_n^+)^*\, x\varphi$ for arbitrary $b_1, \dots, b_n \in \mathsf{B}$ and $n \in \mathbb{N}$. Since A is the linear span of 1 and of elements of the form $b_1^+ \dots b_n^+$, where $b_1, \dots, b_n \in \mathsf{B}$, Lemma 8.1.6 yields $x\varphi \in \mathcal{D}(\pi_2^*)$. Then the preceding gives $x\pi_1(b_1 \dots b_n)\, \varphi = \pi_2^*(b_1) \dots \pi_2^*(b_n)\, x\varphi = \pi_2^*(b_1 \dots b_n)\, x\varphi$ for $\varphi \in \mathcal{D}(\pi_1)$ and $b_1, \dots, b_n \in \mathsf{B}$. By linearity, $x\pi_1(a)\, \varphi = \pi_2^*(a)\, x\varphi$ for all $a \in \mathsf{A}$, so $x \in \mathbb{I}(\pi_1, \pi_2^*)$. $\square$

An immediate consequence of this proposition is

Corollary 8.2.8. *Suppose that* B *is a subset of* A_{h} *such that* $\mathsf{B} \cup \{1\}$ *generates* A *as an algebra. For any* $*$*-representation* π *of* A, *we have* $\pi(\mathsf{A})'_{\mathrm{w}} = \bigcap_{b \in \mathsf{B}} (\overline{\pi(b)})'_{\mathrm{w}}$.

Remark 1. The assertion of the preceding corollary is not true in general if we replace the weak commutants by the strong commutants. In order to see this, let π be the representation π_1 of $A = \mathbb{C}[x]$ defined in Example 8.1.14 and let $\mathcal{B} := \{x\}$. If $u(t)$ denotes the left translation in $L^2(0, 1)$ by t modulo 1, we clearly have $u(t) \in (\overline{\pi(x)})'_s$ and $u(t) \notin \pi(A)'_s$ for all t in $(0, 1)$.

8.3. Invariant and Reducing Subspaces

In this section A denotes an algebra with unit. When we speak about $*$-representations of A, we always assume that A is a $*$-algebra with unit.

First we define the direct sum of representations. Suppose that $\{\pi_i : i \in I\}$ is a family of representations of A. Let $\mathcal{H}(\pi) := \sum_{i \in I} \oplus \mathcal{H}(\pi_i)$ be the direct sum of the family of Hilbert spaces $\{\mathcal{H}(\pi_i) : i \in I\}$. Let $\mathcal{D}(\pi)$ denote the set of all vectors $\varphi \equiv (\varphi_i)$ in $\mathcal{H}(\pi)$ for which $\varphi_i \in \mathcal{D}(\pi_i)$ for all $i \in I$ and $\pi(a)\,\varphi := (\pi_i(a)\,\varphi_i)$ is a vector in $\mathcal{H}(\pi)$ for all $a \in A$. Of course, $\mathcal{D}(\pi)$ is a dense linear subspace of $\mathcal{H}(\pi)$. It is easily seen that π is a representation of A on the domain $\mathcal{D}(\pi)$ in the Hilbert space $\mathcal{H}(\pi)$. We call π the *direct sum of the family* $\{\pi_i : i \in I\}$ *of representations of* A and write $\pi = \sum_{i \in I} \oplus \pi_i$.

We mention some properties of direct sums. The (easy) proofs of these assertions will be omitted. Suppose that $\pi = \sum_{i \in I} \oplus \pi_i$. Then $\hat{\pi} = \sum_{i \in I} \oplus \hat{\pi}_i$. Therefore, π is closed if and only if all π_i, $i \in I$, are closed. For each $i \in I$, the projection $P_{\mathcal{H}(\pi_i)}$ of $\mathcal{H}(\pi)$ onto its subspace $\mathcal{H}(\pi_i)$ belongs to the strong commutant $\pi(A)'_s$ and satisfies $P_{\mathcal{H}(\pi_i)}\mathcal{D}(\pi) = \mathcal{D}(\pi_i)$. Assume now that A is a $*$-algebra. Then $\pi^* = \sum_{i \in I} \oplus \pi_i^*$. The representation π is adjointable resp. biclosed, hermitian, self-adjoint if and only if π_i is adjointable resp. biclosed, hermitian, self-adjoint for each $i \in I$.

Next we consider subrepresentations of representations. We shall use the following notation. If $\mathcal{E}$ is a linear subspace of a Hilbert space $\mathcal{H}$, then $\overline{\mathcal{E}}$ means the closure of $\mathcal{E}$ in $\mathcal{H}$ relative to the Hilbert space norm.

Suppose that π is a representation of A.

Definition 8.3.1. A linear subspace $\mathcal{E}$ of $\mathcal{D}(\pi)$ is said to be *invariant for* π if $\pi(a)\,\varphi \in \mathcal{E}$ for all $a \in A$ and $\varphi \in \mathcal{E}$. A closed linear subspace $\mathcal{K}$ of $\mathcal{H}(\pi)$ is called *invariant for* π if there exists a linear subspace $\mathcal{E}$ of $\mathcal{D}(\pi)$ which is dense in $\mathcal{K}$ and invariant for π.

Remark 1. Suppose $\mathcal{K}$ is a closed linear subspace of $\mathcal{H}(\pi)$ which is contained in $\mathcal{D}(\pi)$. We check that in this case the above definition is not ambiguous, that is, both parts of this definition are equivalent. Suppose that $\mathcal{K}$ satisfies the second part of Definition 8.3.1. Then there is a dense linear subspace $\mathcal{E}$ of $\mathcal{K}$ such that $\mathcal{E} \subseteq \mathcal{D}(\pi)$ and $\pi(a)\,\mathcal{E} \subseteq \mathcal{E}$ for all $a \in A$. Since each operator $\pi(a)$, $a \in A$, is closable and $\mathcal{K}$ is closed in the Hilbert space norm in $\mathcal{H}(\pi)$, the closed graph theorem shows that $\pi(a) \restriction \mathcal{K}$ is a bounded operator of $\mathcal{K}$ into $\mathcal{H}(\pi)$ for every $a \in A$. Therefore, $\pi(a)\,\mathcal{E} \subseteq \mathcal{E}$ implies that $\pi(a)\,\mathcal{K} \subseteq \mathcal{K}$ for $a \in A$. That is, $\mathcal{K}$ satisfies the first part of Definition 8.3.1. The opposite direction is trivial.

Let $\mathcal{E}$ be a linear subspace of $\mathcal{D}(\pi)$ which is invariant for π. Then the mapping $a \to \pi(a) \restriction \mathcal{E}$ defines a representation of A on $\mathcal{E}$ in the Hilbert space $\overline{\mathcal{E}}$. We denote this representation by $\pi \restriction \mathcal{E}$. Moreover, the closed linear subspace $\overline{\mathcal{E}}$ of $\mathcal{H}(\pi)$ is invariant for π in the sense of the second part of Definition 8.3.1.

Now let $\mathcal{K}$ be a closed linear subspace of $\mathcal{H}(\pi)$ which is invariant for π. We denote by $\mathcal{D}(\pi)_{\mathcal{K}}$ the set of all vectors $\varphi \in \mathcal{D}(\pi) \cap \mathcal{K}$ for which $\pi(a)\varphi \in \mathcal{K}$ for all $a \in \mathsf{A}$. It is not difficult to see that $\mathcal{D}(\pi)_{\mathcal{K}}$ is the largest linear subspace of $\mathcal{D}(\pi) \cap \mathcal{K}$ which is invariant for π. Since $\mathcal{K}$ is assumed to be invariant for π, Definition 8.3.1 ensures that $\mathcal{D}(\pi)_{\mathcal{K}}$ is dense in $\mathcal{K}$. Therefore, $\pi_{\mathcal{K}} := \pi \upharpoonright \mathcal{D}(\pi)_{\mathcal{K}}$ is a representation of A on $\mathcal{D}(\pi)_{\mathcal{K}}$ in the Hilbert space $\mathcal{K}$.

In general, the orthogonal complement $\mathcal{K}^{\perp}$ of $\mathcal{K}$ in $\mathcal{H}(\pi)$ is not invariant for π, and $\mathcal{D}(\pi) \cap \mathcal{K}$ is not invariant for π and hence different from $\mathcal{D}(\pi)_{\mathcal{K}}$. A counter-example where π is even a self-adjoint representation is provided by Example 8.3.8 below. Moreover, in this example the projection $P_{\mathcal{K}}$ onto the invariant closed linear subspace $\mathcal{K}$ for π is not contained in the weak commutant of the O*-algebra $\pi(\mathsf{A})$.

The pathologies just mentioned do not occur if the subspace $\mathcal{K}$ is reducing in the sense of the following definition.

Definition 8.3.2. Let $\mathcal{E}$ be a linear subspace of $\mathcal{D}(\pi)$ and let $\mathcal{K}$ be a closed linear subspace of $\mathcal{H}(\pi)$. We say that $\mathcal{E}$ [resp. $\mathcal{K}$] is *reducing for* π if there exist representations π_1 and π_2 of the algebra A such that $\pi = \pi_1 \oplus \pi_2$ and $\mathcal{E} = \mathcal{D}(\pi_1)$ [resp. $\mathcal{K} = \mathcal{H}(\pi_1)$].

Remark 2. It is obvious that for a closed linear subspace $\mathcal{K}$ of $\mathcal{H}(\pi)$ contained in $\mathcal{D}(\pi)$ both parts of Definition 8.3.2 are equivalent, cf. Remark 1 above.

We note some immediate consequences of the above definitions. Let $\mathcal{E}$ be a linear subspace of $\mathcal{D}(\pi)$. Then $\mathcal{E}$ is reducing for π if and only if $\mathcal{D}(\pi) \cap \overline{\mathcal{E}} = \mathcal{E}$ and if the closed linear subspace $\overline{\mathcal{E}}$ of $\mathcal{H}(\pi)$ is reducing for π. If $\mathcal{E}$ is reducing for π, then $\mathcal{E}$ is invariant for π and $\mathcal{E} = \mathcal{D}(\pi)_{\overline{\mathcal{E}}}$. If a closed linear subspace $\mathcal{K}$ of $\mathcal{H}(\pi)$ is reducing for π, then $\mathcal{K}$ is invariant for π, $\mathcal{D}(\pi_{\mathcal{K}}) = P_{\mathcal{K}}\mathcal{D}(\pi) = \mathcal{D}(\pi) \cap \mathcal{K}$ and this space is reducing for π.

Lemma 8.3.3. *For each closed linear subspace $\mathcal{K}$ of $\mathcal{H}(\pi)$, the following conditions are equivalent:*

(i) *$\mathcal{K}$ is reducing for π.*

(ii) *The linear subspaces $\mathcal{D}(\pi) \cap \mathcal{K}$ and $\mathcal{D}(\pi) \cap \mathcal{K}^{\perp}$ of $\mathcal{D}(\pi)$ are invariant for π and $P_{\mathcal{K}}\mathcal{D}(\pi) \subseteq \mathcal{D}(\pi)$.*

(iii) *$P_{\mathcal{K}} \in \pi(\mathsf{A})'_{\mathrm{s}}$.*

Proof. The proof is straightforward. We sketch e.g. the proof of the implication (ii) $\to$ (i). Since $\mathcal{D}(\pi) \cap \mathcal{K}$ is invariant for π and $P_{\mathcal{K}}\mathcal{D}(\pi) \subseteq \mathcal{D}(\pi)$, we have $P_{\mathcal{K}}\mathcal{D}(\pi) = \mathcal{D}(\pi_{\mathcal{K}})$. Similarly, $P_{\mathcal{K}^{\perp}}\mathcal{D}(\pi) = \mathcal{D}(\pi_{\mathcal{K}^{\perp}})$. This implies $\pi = \pi_{\mathcal{K}} \oplus \pi_{\mathcal{K}^{\perp}}$ and $\mathcal{K} = \mathcal{H}(\pi_{\mathcal{K}})$. □

Definition 8.3.4. A representation π of A is called *irreducible* if the only linear subspaces of $\mathcal{D}(\pi)$ which are reducing for π are $\{0\}$ and $\mathcal{D}(\pi)$ itself.

Lemma 8.3.5. *For every representation π of A, the following statements are equivalent:*

(i) *π is irreducible.*

(ii) *Each decomposition $\pi = \pi_1 \oplus \pi_2$ of π as a direct sum of representations π_1 and π_2 of A implies that $\mathcal{H}(\pi_1) = \{0\}$ or $\mathcal{H}(\pi_2) = \{0\}$.*

(iii) *The only closed linear subspaces of $\mathcal{H}(\pi)$ which are reducing for π are $\{0\}$ and $\mathcal{H}(\pi)$.*

(iv) *The only projections contained in* $\pi(\mathsf{A})'_{\mathrm{s}}$ *are* 0 *and* I.

If A *is a* $*$*-algebra and* π *is a closed* $*$*-representation, then* (i) *is also equivalent to* (v) $\pi(\mathsf{A})'_{\mathrm{ss}} = \mathbb{C}\cdot I$.

Proof. The equivalence of (i)—(iv) follows immediately from Lemma 8.3.3 combined with the corresponding definitions. Suppose that A is a $*$-algebra and π is a closed $*$-representation of A. Then $\pi(\mathsf{A})$ is a closed O*-algebra. By Proposition 7.2.10, (ii), we have $\pi(\mathsf{A})'_{\mathrm{ss}} = \pi(\mathsf{A})'_{\mathrm{s}} \cap (\pi(\mathsf{A})'_{\mathrm{s}})^*$, and this set is a von Neumann algebra on $\mathcal{H}(\pi)$. From this the equivalence of (iv) and (v) follows. □

Remark 3. The irreducibility in the sense of the above definition means that the representation is not decomposable as a direct sum of two representations in a non-trivial way. Probably it would be better to call these representations "indecomposable". There exist several other possible (in general much stronger) definitions of irreducibility for $*$-representations. One could define irreducibility by the requirement that the whole strong commutant $\pi(\mathsf{A})'_{\mathrm{s}}$, the weak commutant $\pi(\mathsf{A})'_{\mathrm{w}}$ or some of the unbounded commutants of $\pi(\mathsf{A})$ are trivial. We briefly discuss the relations between these concepts.

Suppose π is a closed $*$-representation of a $*$-algebra A. If π is irreducible (in the sense of Definition 8.3.4), then it follows from Lemma 8.3.5., (i) → (v), that the hermitian part of $\pi(\mathsf{A})'_{\mathrm{s}}$ is trivial; the whole strong commutant $\pi(\mathsf{A})'_{\mathrm{s}}$ is not trivial in general as Example 8.3.6 shows. However, if π is self-adjoint, then $\pi(\mathsf{A})' \equiv \pi(\mathsf{A})'_{\mathrm{s}} \equiv \pi(\mathsf{A})'_{\mathrm{w}}$ is a von Neumann algebra, so that π is irreducible if and only if $\pi(\mathsf{A})'$ is trivial. This justifies to some extent, at least for self-adjoint representations, the above definition of irreducibility. If π is self-adjoint and irreducible, then we cannot conclude in general that $\pi(\mathsf{A})^{\mathrm{c}}_{\mathrm{s}}$ is trivial; see Example 9.4.6.

Suppose now that π is a $*$-representation of a $*$-algebra A such that $\pi(\mathsf{A})$ is a strictly self-adjoint O*-algebra (cf. Definition 7.3.5). Then π is self-adjoint by Theorem 7.3.6. Corollary 7.3.10 (applied with $\mathcal{A} := \pi(\mathsf{A})$) states that if π is irreducible, then $\pi(\mathsf{A})^{\mathrm{c}}$ $(= \pi(\mathsf{A})^{\mathrm{c}}_{\mathrm{s}} = \pi(\mathsf{A})^{\mathrm{c}}_{\mathrm{w}} = \pi(\mathsf{A})^{\mathrm{c}}_{\mathrm{f}}$ by Theorem 7.3.6) is trivial.

We illustrate the preceding by four examples.

Example 8.3.6. Let a, $\mathcal{D}$ and $\mathcal{A}$ be as in Example 7.2.14. Define a $*$-representation π of $\mathsf{A} := \mathbb{C}[\mathsf{x}]$ on $\mathcal{D}(\pi) := \mathcal{D}$ by $\pi(p) := p(a)$, $p(\mathsf{x}) \in \mathbb{C}[\mathsf{x}]$. π is closed, but not self-adjoint. As shown in Example 7.2.14, $\pi(\mathsf{A})'_{\mathrm{ss}} \equiv \mathcal{A}'_{\mathrm{ss}} = \mathbb{C}\cdot I$. Therefore, by Lemma 8.3.5, π is irreducible. As discussed in Example 7.2.14, the strong commutant $\pi(\mathsf{A})'_{\mathrm{s}} \equiv \mathcal{A}'_{\mathrm{s}}$ consists of all Toeplitz operators with symbols in $\mathcal{H}^\infty(\mathbb{T})$. In particular, $\pi(\mathsf{A})'_{\mathrm{s}} \neq \mathbb{C}\cdot I$. ○

Example 8.3.7. Let π be the $*$-representation of the Weyl algebra $\mathsf{A} := \mathbb{A}(\mathsf{p}_1, \mathsf{q}_1, \ldots, \mathsf{p}_n, \mathsf{q}_n)$ on $\mathcal{D}(\pi) = \mathcal{S}(\mathbb{R}^n)$ defined in Example 2.5.2. Recall that the operators $p_l = \pi(\mathsf{p}_l)$ and $q_l = \pi(\mathsf{q}_l)$, $l = 1, \ldots, n$, form the Schrödinger representation of the canonical commutation relations 2.5/(2). That is, $(p_l\varphi)(t) = -\mathrm{i}\,\dfrac{\partial\varphi}{\partial t_l}(t)$ and $(q_l\varphi)(t) = t_l\varphi(t)$ for $\varphi \in \mathcal{D}(\pi)$, $l = 1, \ldots, n$ and $t = (t_1, \ldots, t_n) \in \mathbb{R}^n$. The operators $\overline{p_l}$ and $\overline{q_l}$, $l = 1, \ldots, n$, are self-adjoint, and we have

$$\pi(\mathsf{A})'_{\mathrm{s}} = \bigcap_{l=1}^{n} (p_l)'_{\mathrm{s}} \cap (q_l)'_{\mathrm{s}} \subseteq \bigcap_{l=1}^{n} (\overline{p_l})'_{\mathrm{s}} \cap (\overline{q_l})'_{\mathrm{s}}$$

$$\subseteq \{\exp \mathrm{i}\lambda\overline{p_l}, \exp \mathrm{i}\lambda\overline{q_l} : \lambda \in \mathbb{R} \text{ and } l = 1, \ldots, n\}'.$$

It is well-known that the latter is trivial (cf. Barut/Raczka [1], ch. 20, § 2). Hence π is irreducible. Let $a := I + p_1^2 + q_1^2 + \cdots + p_n^2 + q_n^2$. Since all powers a^m, $m \in \mathbb{N}$, of a are essentially self-adjoint and the graph topology of $\pi(\mathsf{A}) \equiv \mathbb{A}(p_1, q_1, \dots, p_n, q_n)$ is generated by the directed family of seminorms $\{\|\cdot\|_{a^m} : m \in \mathbb{N}_0\}$, it follows that the (closed) O*-algebra $\pi(\mathsf{A})$ is strictly self-adjoint (cf. Remark 2 in 7.3). Thus, by Corollary 7.3.10, $\pi(\mathsf{A})^c = \mathbb{C} \cdot I$.

Now we set $n = 1$. Let $\mathcal{K} := \{\varphi \in \mathcal{H}(\pi) \equiv L^2(\mathbb{R}) : \varphi(t) = 0 \text{ a.e. on } (0, 1)\}$. Clearly, $\mathcal{K}$ and $\mathcal{K}^\perp$ are closed linear subspaces of $\mathcal{H}(\pi)$ which are invariant for π. But $\mathcal{K}$ is not reducing for π, since $P_{\mathcal{K}}\varphi \notin \mathcal{D}(\pi)$ if $\varphi \in \mathcal{D}(\pi)$ and $\varphi(1) \neq 0$. In fact, we have

$$\mathcal{D}(\pi_{\mathcal{K}} \oplus \pi_{\mathcal{K}^\perp}) = \{\varphi \in \mathcal{D}(\pi) : \varphi^{(m)}(0) = \varphi^{(m)}(1) = 0 \text{ for } m \in \mathbb{N}_0\} \subsetneqq \mathcal{D}(\pi).$$

Since $\pi(\mathsf{A})'_{\mathrm{w}} \equiv \pi(\mathsf{A})'_{\mathrm{s}} = \mathbb{C} \cdot I$ as noted above, $P_{\mathcal{K}} \notin \pi(\mathsf{A})'_{\mathrm{w}}$. ○

Example 8.3.8. Suppose that A is an unbounded self-adjoint operator on a Hilbert space $\mathcal{H}$. We define a *-representation of the *-algebra $\mathsf{A} := \mathbb{C}[\mathsf{x}]$ on $\mathcal{D}(\pi) := \mathcal{D}^\infty(A)$ in the Hilbert space $\mathcal{H}(\pi) := \mathcal{H}$ by $\pi(\mathsf{x}) := A \upharpoonright \mathcal{D}(\pi)$. By Proposition 8.1.15, π is self-adjoint. Since the operator A is unbounded, we can find a vector $\xi \in \mathcal{H}$ with $\xi \notin \mathcal{D}(A)$. Let $\mathcal{K} := \{\varphi \in \mathcal{H} : \varphi \perp \xi\}$ and $\mathcal{K}_1 := \{\varphi \in \mathcal{K} : \varphi \perp U^*\xi\}$, where $U := (A - \mathrm{i})(A + \mathrm{i})^{-1}$ denotes the Cayley transform of A. It is not difficult to check that U maps $\mathcal{K}_1$ into $\mathcal{K}$ and that $(I - U)\mathcal{K}_1$ is dense in $\mathcal{K}$. From this it follows that $A_1 := A \upharpoonright (I - U)\mathcal{K}_1$ is a densely defined closed symmetric operator in the Hilbert space $\mathcal{K}$ with deficiency indices $(1, 1)$. Therefore, by Proposition 1.6.1, $\mathcal{E} := \mathcal{D}^\infty(A_1)$ is dense in $\mathcal{K}$. If $\varphi \in \mathcal{E}$, then $\pi(\mathsf{x})\varphi = A\varphi = A_1\varphi \in \mathcal{E}$. Hence the linear subspace $\mathcal{E}$ of $\mathcal{D}(\pi)$ and the closed linear subspace $\mathcal{K}$ of $\mathcal{H}(\pi)$ are invariant for π. Since $\xi \notin \mathcal{D}(\pi)$, we have $P_{\mathcal{K}} \notin \pi(\mathsf{A})'_{\mathrm{s}} \equiv \pi(\mathsf{A})'_{\mathrm{w}}$ and $\mathcal{K}^\perp \cap \mathcal{D}(\pi) = \{0\}$; so $\mathcal{K}^\perp$ is not an invariant subspace for π. Further, $\mathcal{D}(\pi) \cap \mathcal{K} \neq \mathcal{D}(\pi)_{\mathcal{K}}$. We prove the latter. Let $e(\lambda)$, $\lambda \in \mathbb{R}$, be the spectral projections of A. We choose numbers $\gamma, \delta \in \mathbb{R}$ such that $\gamma + 2 \leqq \delta$, $e\big((\gamma, \gamma + 1)\big)\xi \neq 0$ and $e\big((\delta - 1, \delta)\big)\xi \neq 0$. From the spectral theorem we easily conclude that there is a vector $\varphi \in e\big((\gamma, \delta)\big)\mathcal{H}$ which is orthogonal to ξ such that $A\varphi$ is not orthogonal to ξ. Then $\varphi \in \mathcal{D}(\pi) \cap \mathcal{K}$ and $\varphi \notin \mathcal{D}(\pi)_{\mathcal{K}}$, since $\pi(\mathsf{x})\varphi = A\varphi \notin \mathcal{K}$. ○

Example 8.3.9. As in Example 8.3.8, we let A be an unbounded self-adjoint operator in a Hilbert space with spectral projections $e(\lambda)$, $\lambda \in \mathbb{R}$. We take a number $\alpha \in \mathbb{R}$ and non-zero vectors ξ_1 and ξ_2 in $\mathcal{H}$ such that $\xi_1 \in e\big((\alpha, \alpha + 1)\big)\mathcal{H}$, $e\big((\alpha, \alpha + 1)\big)\xi_2 = 0$ and $\xi_2 \notin \mathcal{D}(A)$. Set $\xi := \xi_1 + \xi_2$ and $\mathcal{H}_1 := \{\varphi \in \mathcal{H} : \varphi \perp \xi\}$. Since $\xi \notin \mathcal{D}(A)$, it follows that $A_1 := A \upharpoonright (A + \mathrm{i})^{-1}\mathcal{H}_1$ is a densely defined closed symmetric operator on $\mathcal{H}$ with deficiency indices $(1, 1)$. By Proposition 1.6.1, $\mathcal{D}^\infty(A_1)$ is dense in $\mathcal{H}$. We define *-representations π and π_1 of $\mathsf{A} := \mathbb{C}[\mathsf{x}]$ by $\pi(\mathsf{x}) := A \upharpoonright \mathcal{D}^\infty(A)$, $\mathcal{D}(\pi) := \mathcal{D}^\infty(A)$ and $\pi_1(\mathsf{x}) := A_1 \upharpoonright \mathcal{D}^\infty(A_1)$, $\mathcal{D}(\pi_1) := \mathcal{D}^\infty(A_1)$ in the Hilbert space $\mathcal{H}$. Set $\mathcal{K} := e\big((\alpha, \alpha + 1)\big)\mathcal{H}$. We have $P_{\mathcal{K}} \in \pi_1(A)'_{\mathrm{w}}$, since $\pi_1 \subseteq \pi$ and so $P_{\mathcal{K}} \in \pi(\mathsf{A})'_{\mathrm{s}} \subseteq \pi_1(\mathsf{A})'_{\mathrm{w}}$. But the closed linear subspace $\mathcal{K}$ of $\mathcal{H}$ is not invariant for π_1. In order to prove this we show that $\mathcal{D}(\pi_1) \cap \mathcal{K}$ is not dense in $\mathcal{K}$. More precisely, we prove that the non-zero vector $(A - \mathrm{i})\xi_1 \equiv e\big((\alpha, \alpha + 1)\big)(A - \mathrm{i})\xi_1$ of $\mathcal{K}$ is orthogonal to $\mathcal{D}(\pi_1) \cap \mathcal{K}$. We let $\varphi \in \mathcal{D}(\pi_1) \cap \mathcal{K}$. Since $\varphi \in \mathcal{D}(\pi_1)$, $\varphi = (A + \mathrm{i})^{-1}\psi$ with $\psi \perp \xi$. Since $\varphi \in \mathcal{K}$, $\psi \in e\big((\alpha, \alpha + 1)\big)\mathcal{H}$. Therefore,

$$\langle \varphi, (A - \mathrm{i})\xi_1 \rangle = \langle (A + \mathrm{i})^{-1}\psi, (A - \mathrm{i})\xi_1 \rangle = \langle \psi, \xi_1 \rangle = \langle \psi, \xi_1 + \xi_2 \rangle = 0$$

where we used the assumption $e\big((\alpha, \alpha + 1)\big)\xi_2 = 0$. ○

As we have seen in the preceding examples invariant subspaces for self-adjoint representations are not reducing in general. The next proposition shows that there is a one-to-one correspondence between self-adjoint subrepresentations and reducing subspaces of self-adjoint representations.

Proposition 8.3.10. *Suppose π is a $*$-representation of A and $\mathscr{E}$ is a linear subspace of $\mathscr{D}(\pi)$.*

(i) *If $\mathscr{E}$ is invariant for π and $\pi \restriction \mathscr{E}$ is self-adjoint, then $\mathscr{E}$ is reducing for π. Moreover, $\mathscr{E} = P_{\bar{\mathscr{E}}}\mathscr{D}(\pi)$, $\pi \restriction \mathscr{E} = \pi_{\bar{\mathscr{E}}}$ and $P_{\bar{\mathscr{E}}} \in \pi(\mathsf{A})'_{\mathrm{s}}$.*

(ii) *If $\mathscr{E}$ is reducing for π and π is self-adjoint, then $\pi \restriction \mathscr{E}$ is self-adjoint.*

Proof. (i): Since $\mathscr{E}$ is invariant for π, $\mathscr{K} := \bar{\mathscr{E}}$ is invariant for π and $\pi \restriction \mathscr{E} \subseteq \pi_{\mathscr{K}}$. Hence $\pi \restriction \mathscr{E} = \pi_{\mathscr{K}}$, since $\pi_{\mathscr{K}}$ is a hermitian extension of the self-adjoint representation $\pi \restriction \mathscr{E}$ in the Hilbert space $\mathscr{K}$. Obviously, $P_{\mathscr{K}} \restriction \mathscr{K} \in \mathbb{I}(\pi_{\mathscr{K}}, \pi)$. Because $\pi_{\mathscr{K}} \equiv \pi \restriction \mathscr{E}$ is self-adjoint, Corollary 8.2.4, (ii), yields $P_{\mathscr{K}} \equiv (P_{\mathscr{K}} \restriction \mathscr{K})(P_{\mathscr{K}} \restriction \mathscr{K})^* \in \mathbb{I}(\pi^*, \pi) \subseteq \mathbb{I}(\pi, \pi) = \pi(\mathsf{A})'_{\mathrm{s}}$. Therefore, by Lemma 8.3.3, $\mathscr{K}$ is reducing for π. Because of $\pi \restriction \mathscr{E} = \pi_{\mathscr{K}}$, $\mathscr{E}$ is reducing for π and hence $\mathscr{E} = P_{\mathscr{K}}\mathscr{D}(\pi)$.

(ii): Again let $\mathscr{K} := \bar{\mathscr{E}}$. By $\tilde{P}_{\mathscr{K}}$ we mean $P_{\mathscr{K}}$ considered as an operator of $\mathscr{H}$ into $\mathscr{K}$. Since $\mathscr{E}$ is reducing by assumption, $\mathscr{K}$ is reducing for π. Hence $\pi \restriction \mathscr{E} = \pi_{\mathscr{K}}$ and $\pi = \pi_{\mathscr{K}} \oplus \pi_{\mathscr{K}^\perp}$. Consequently, $\tilde{P}_{\mathscr{K}} \in \mathbb{I}(\pi, \pi_{\mathscr{K}})$. Since π is self-adjoint, Corollary 8.2.4, (ii), implies that $I_{\mathscr{K}} = \tilde{P}_{\mathscr{K}}(\tilde{P}_{\mathscr{K}})^* \in \mathbb{I}(\pi^*_{\mathscr{K}}, \pi_{\mathscr{K}})$ which gives $\pi^*_{\mathscr{K}} \subseteq \pi_{\mathscr{K}}$. Thus $\pi_{\mathscr{K}} \equiv \pi \restriction \mathscr{E}$ is self-adjoint. □

Remark 4. A shorter proof of part (ii) in the preceding proposition which avoids the use of intertwining spaces goes as follows. Since $\mathscr{K}$ is reducing for π, $\pi = \pi_{\mathscr{K}} \oplus \pi_{\mathscr{K}^\perp}$. By $\pi = \pi^*$, it follows that $\pi_{\mathscr{K}} \oplus \pi_{\mathscr{K}^\perp} = (\pi_{\mathscr{K}})^* \oplus (\pi_{\mathscr{K}^\perp})^*$ and hence $\pi_{\mathscr{K}} = (\pi_{\mathscr{K}})^*$.

We give a reformulation of Proposition 8.3.10 in terms of the strong commutant which is more convenient for later applications.

Proposition 8.3.11. *Let π be a $*$-representation of the $*$-algebra A.*

(i) *If π_1 is a self-adjoint subrepresentation of π, then the projection $P \equiv P_{\mathscr{H}(\pi_1)}$ of $\mathscr{H}(\pi)$ onto $\mathscr{H}(\pi_1)$ is in $\pi(\mathsf{A})'_{\mathrm{s}}$ and $\pi_1 = \pi \restriction P\mathscr{D}(\pi)$.*

(ii) *If π is self-adjoint and P is a projection contained in $\pi(\mathsf{A})' \equiv \pi(\mathsf{A})'_{\mathrm{s}}$, then $\pi \restriction P\mathscr{D}(\pi)$ is a self-adjoint subrepresentation of π.*

Proof. (i): Apply Proposition 8.3.10, (i), with $\mathscr{E} := \mathscr{D}(\pi_1)$.

(ii): Apply Proposition 8.3.10, (ii), with $\mathscr{E} := P\mathscr{D}(\pi)$. □

We mention two interesting corollaries which follow immediately from Proposition 8.3.10. By the trivial subrepresentations of a representation π we mean the representation π itself and the restriction of π to $\{0\}$.

Corollary 8.3.12. *A self-adjoint representation π of A is irreducible if and only if the only self-adjoint subrepresentations of π are the trivial subrepresentations.*

Corollary 8.3.13. *Suppose π is a self-adjoint representation of A. If π_1 is a $*$-representation of A in a possibly larger Hilbert space such that $\pi \subseteq \pi_1$, then there exists a $*$-representation π_0 of A on the Hilbert space $\mathscr{H}(\pi_1) \ominus \mathscr{H}(\pi)$ such that $\pi_1 = \pi \oplus \pi_0$.*

We give another application of Proposition 8.3.11. First however we introduce some more terminology.

Definition 8.3.14. Let π be a *-representation of A and $\mathcal{M}$ a subset of $\mathcal{D}(\pi)$. We say that $\mathcal{M}$ is *generating* or *cyclic* [resp. *weakly generating*] *for* π if $\pi(\mathsf{A})\,\mathcal{M} :=$ l.h. $\{\pi(a)\,\varphi : a \in \mathsf{A}$ and $\varphi \in \mathcal{M}\}$ is dense in $\mathcal{D}(\pi)\,[\mathrm{t}_\pi]$ [resp. $\mathcal{H}(\pi)$]. A vector $\varphi \in \mathcal{D}(\pi)$ is said to be *cyclic* [resp. *weakly cyclic*] *for* π if $\{\varphi\}$ is generating [resp. *weakly generating*] for π. The set $\mathcal{M}$ is called *separating for a linear subspace* $\mathcal{B}$ of $\pi(\mathsf{A})^{\mathrm{c}}_{\mathrm{f}}$ if $b \in \mathcal{B}$ is equal to 0 when $b\varphi = 0$ for all $\varphi \in \mathcal{M}$.

Lemma 8.3.15. *Suppose π is a *-representation of* A *and* $\mathcal{M} \subseteq \mathcal{D}(\pi)$.

(i) *If $\mathcal{M}$ is weakly generating for π, then $\mathcal{M}$ is separating for* $\pi(\mathsf{A})'_{\mathrm{w}}$.

(ii) *If $\mathcal{M}$ is generating for π, then $\mathcal{M}$ is separating for* $\pi(\mathsf{A})^{\mathrm{c}}_{\mathrm{f}}$.

Proof. (i): Suppose $x \in \pi(\mathsf{A})'_{\mathrm{w}}$ satisfies $x\varphi = 0$ for all $\varphi \in \mathcal{M}$. Then $x\pi(a)\,\varphi = \pi(a^+)^*\,x\varphi = 0$ for all $a \in \mathsf{A}$ and $\varphi \in \mathcal{M}$. Hence $x \upharpoonright \pi(\mathsf{A})\,\mathcal{M} = 0$. Since $\pi(\mathsf{A})\,\mathcal{M}$ is dense in $\mathcal{H}(\pi)$ and x is bounded, $x = 0$.

(ii): The proof is similar to the proof of (i). Let $x \in \pi(\mathsf{A})^{\mathrm{c}}_{\mathrm{f}}$ be such that $x\varphi = 0$ for $\varphi \in \mathcal{M}$. For $a \in \mathsf{A}$, $\varphi \in \mathcal{M}$ and $\psi \in \mathcal{D}(\pi)$, we have $\langle x\pi(a)\,\varphi, \psi\rangle = \langle(\pi(a) \circ x)\,\varphi, \psi\rangle = \langle x\varphi, \pi(a)^+\,\psi\rangle = 0$, so that $\langle x\zeta, \psi\rangle = 0$ for $\zeta \in \pi(\mathsf{A})\,\mathcal{M}$ and $\psi \in \mathcal{D}(\pi)$. Since $\langle x\cdot, \psi\rangle$, $\psi \in \mathcal{D}(\pi)$, is continuous on $\mathcal{D}(\pi)\,[\mathrm{t}_\pi]$ and $\pi(\mathsf{A})\,\mathcal{M}$ is dense in $\mathcal{D}(\pi)\,[\mathrm{t}_\pi]$, the latter implies that $x = 0$. □

The converses of the assertions in Lemma 8.3.15 are not true in general; see Example 8.3.17 below. However, we have

Proposition 8.3.16. *Let π be a *-representation of* A. *If $\mathcal{M}$ is a subset of $\mathcal{D}(\pi)$ such that the representation $\pi \upharpoonright \pi(\mathsf{A})\,\mathcal{M}$ is essentially self-adjoint, then the following statements are equivalent:*

(i) *$\mathcal{M}$ is generating for π.*

(ii) *$\mathcal{M}$ is weakly generating for π.*

(iii) *$\mathcal{M}$ is separating for* $\pi(\mathsf{A})^{\mathrm{c}}_{\mathrm{f}}$.

(iv) *$\mathcal{M}$ is separating for* $\bar{\pi}(\mathsf{A})'_{\mathrm{s}}$.

Proof. First note that $\bar{\pi}(\mathsf{A})'_{\mathrm{s}} \subseteq \bar{\pi}(\mathsf{A})'_{\mathrm{w}} = \pi(\mathsf{A})'_{\mathrm{w}} \subseteq \pi(\mathsf{A})^{\mathrm{c}}_{\mathrm{f}}$. Thus (i) → (ii) and (iii) → (iv) are trivial, and (i) → (iii) and (ii) → (iv) follow from Lemma 8.3.15. Therefore, our proof will be complete once we have shown that (iv) implies (i). Letting $\mathcal{E}$ be the closure of $\pi(\mathsf{A})\mathcal{M}$ in $\mathcal{D}(\bar{\pi})\,[\mathrm{t}_{\bar{\pi}}]$, then $\pi_1 := \bar{\pi} \upharpoonright \mathcal{E}$ is the closure of the *-representation $\pi \upharpoonright \pi(\mathsf{A})\,\mathcal{M}$. By assumption, π_1 is self-adjoint. Therefore, $P_{\bar{\mathcal{E}}} \in \bar{\pi}(\mathsf{A})'_{\mathrm{s}}$ and $\pi_1 = \bar{\pi} \upharpoonright P_{\bar{\mathcal{E}}}\mathcal{D}(\bar{\pi})$ by Proposition 8.3.11, (i). From $\bar{\pi}(1) = I$ we obtain that $\mathcal{M} \subseteq \bar{\mathcal{E}}$ and hence $(I - P_{\bar{\mathcal{E}}})\,\varphi = 0$ for all $\varphi \in \mathcal{M}$. Since $\mathcal{M}$ is separating for $\bar{\pi}(\mathsf{A})'_{\mathrm{s}}$ by (iv), we get $I - P_{\bar{\mathcal{E}}} = 0$. Thus $\pi_1 = \bar{\pi}$, that is, $\mathcal{E} = \mathcal{D}(\bar{\pi})$ which gives (i). □

Example 8.3.17. Let A and π be as in Example 8.3.7. Since $\pi(\mathsf{A})^{\mathrm{c}}_{\mathrm{f}} = \mathbb{C}\cdot I$ as shown therein, each non-zero vector $\varphi \in \mathcal{D}(\pi) \equiv \mathcal{S}(\mathbb{R}^n)$ is separating for $\pi(\mathsf{A})^{\mathrm{c}}_{\mathrm{f}}$ and also for $\pi(\mathsf{A})'_{\mathrm{w}}$. But if φ has compact support, then φ is certainly not weakly cyclic for π and hence not cyclic. ○

The next example illustrates the difference between weakly cyclic vectors and cyclic vectors. The terminology and the results from the theory of the moment problem used in this example can be found in the monograph of Akhiezer [1].

Example 8.3.18. Suppose that μ is an N-extremal indeterminate measure of $M_+(\mathbb{R})$ (cf. Example 2.2.16). (The existence of such measures is well-known in the theory of the moment problem.) We define a (self-adjoint) $*$-representation π of the $*$-algebra $\mathsf{A} := \mathbb{C}[\mathsf{x}]$ on $\mathcal{D}(\pi) := \{\varphi \in L^2(\mathbb{R};\mu) : t^k\varphi(t) \in L^2(\mathbb{R};\mu)$ for all $k \in \mathbb{N}\}$ in the Hilbert space $\mathcal{H}(\pi) := L^2(\mathbb{R};\mu)$ by $\big(\pi(p(\mathsf{x}))\,\varphi\big)(t) := p(t)\,\varphi(t), p \in \mathbb{C}[\mathsf{x}]$ and $\varphi \in \mathcal{D}(\pi)$. Let φ_0 be the function in $\mathcal{D}(\pi)$ which is constant equal to 1. Then $\pi(\mathsf{A})\,\varphi_0$ are the polynomials in the independent variable t considered as a subspace of $\mathcal{D}(\pi)$. Therefore, since μ is N-extremal, $\pi(\mathsf{A})\,\varphi_0$ is dense in $\mathcal{H}(\pi) = L^2(\mathbb{R};\mu)$, and the vector $\varphi_0 \in \mathcal{D}(\pi)$ is weakly cyclic for π. Because μ is indeterminate, the restriction to $\pi(\mathsf{A})\,\varphi_0$ of $\pi(\mathsf{x})$ is not essentially self-adjoint. Since $\pi(\mathsf{x})$ is essentially self-adjoint, $\pi(\mathsf{A})\,\varphi_0$ cannot be dense in $\mathcal{D}(\pi)$ $[\mathsf{t}_\pi]$; so φ_0 is not cyclic for π. ○

8.4. Similarity, Unitary Equivalence and Disjointness of Representations

Definition 8.4.1. Suppose A is an algebra with unit and π_1 and π_2 are representations of A.

(i) π_1 and π_2 are said to be *similar* if there exists an operator $T \in \mathbb{I}(\pi_1, \pi_2)$ with bounded inverse T^{-1} contained in $\mathbb{I}(\pi_2, \pi_1)$. We then write $\pi_1 \sim \pi_2$.

(ii) π_1 and π_2 are *unitarily equivalent* and we write $\pi_1 \cong \pi_2$ if there exists an isometry U of $\mathcal{H}(\pi_1)$ onto $\mathcal{H}(\pi_2)$ such that $U \in \mathbb{I}(\pi_1, \pi_2)$ and $U^{-1} \in \mathbb{I}(\pi_2, \pi_1)$.

(iii) We write $\pi_1 \leqq \pi_2$ if there exists a subrepresentation π_{20} of π_2 such that $\pi_1 \cong \pi_{20}$.

(iv) π_1 and π_2 are said to be *disjoint* if $\mathbb{I}(\pi_1, \pi_2) = \{0\}$ and $\mathbb{I}(\pi_2, \pi_1) = \{0\}$. We denote this fact by $\pi_1 \between \pi_2$.

By a slight reformulation of the preceding definition, π_1 and π_2 are unitarily equivalent if and only if there is a unitary operator U of $\mathcal{H}(\pi_1)$ onto $\mathcal{H}(\pi_2)$ such that $U\mathcal{D}(\pi_1) = \mathcal{D}(\pi_2)$ and $U^{-1}\pi_2(a)\,U\varphi = \pi_1(a)\,\varphi$ for all $a \in \mathsf{A}$ and $\varphi \in \mathcal{D}(\pi_1)$.

It is clear that "$\sim$" and "$\cong$" are both equivalence relations.

Definition 8.4.2. Suppose A is a $*$-algebra with unit and π is a self-adjoint representation of A. π is called a *factor representation* if the von Neumann algebra $\pi(\mathsf{A})'$ is a factor, and π is said to be *multiplicity-free* if the von Neumann algebra $\pi(\mathsf{A})'$ is commutative. By the *type of* π we mean the type of the von Neumann algebra $\pi(\mathsf{A})''$.

Proposition 8.4.3. *Suppose π_1 and π_2 are representations of the $*$-algebra A with unit. If π_1 is self-adjoint and if π_1 and π_2 are similar, then π_1 and π_2 are unitarily equivalent and π_2 is self-adjoint.*

Proof. Since $\pi_1 \sim \pi_2$, there is a $T \in \mathbb{I}(\pi_1, \pi_2)$ such that $T^{-1} \in \mathbb{I}(\pi_2, \pi_1)$. Let $T = U\,|T|$ be the polar decomposition of T. Since T has a bounded inverse defined on the whole $\mathcal{H}(\pi_2)$, U is an isometry of $\mathcal{H}(\pi_1)$ onto $\mathcal{H}(\pi_2)$ and $|T|$ has a bounded inverse in $\mathbb{B}\big(\mathcal{H}(\pi_1)\big)$. By Corollary 8.2.4, (i), $|T|^2 = T^*T \in \mathbb{I}(\pi_1, \pi_1^*) = \pi_1(\mathsf{A})'_{\mathrm{w}}$. Because π_1 is self-adjoint, $\pi_1(\mathsf{A})'_{\mathrm{w}} = \pi_1(\mathsf{A})'_{\mathrm{s}}$ is a von Neumann algebra and hence $|T|^{-1} \in \pi_1(\mathsf{A})'_{\mathrm{s}} = \mathbb{I}(\pi_1, \pi_1)$. From Proposition 8.2.2, (iii), $U = T\,|T|^{-1} \in \mathbb{I}(\pi_1, \pi_2)$ and $U^{-1} = |T|\,T^{-1} \in \mathbb{I}(\pi_2, \pi_1)$, so that $\pi_1 \cong \pi_2$. Since the unitary equivalence obviously preserves self-adjointness, π_2 is self-adjoint. □

Proposition 8.4.3 and also Proposition 8.3.10 show how results from the representation theory by bounded operators carry over to the unbounded case if we assume the self-adjointness of the corresponding representation or subrepresentation. The next two propositions are in the same spirit. The proofs of the results use the same technique as in the bounded case (see e.g. DIXMIER [2], ch. 5), of course with the necessary modifications for the unbounded situation. We collect some basic properties of the above notions in the following proposition.

Proposition 8.4.4. *Suppose that π_1 and π_2 are self-adjoint representations of the $*$-algebra* A *with unit such that $\mathcal{H}(\pi_1) \neq \{0\}$ and $\mathcal{H}(\pi_2) \neq \{0\}$.*

(i) *If $\pi_1 \leqq \pi_2$ and $\pi_2 \leqq \pi_1$, then $\pi_1 \cong \pi_2$.*

(ii) *$\pi_1 \between \pi_2$ if and only if there are no self-adjoint subrepresentations π_{10} and π_{20} of π_1 and π_2, respectively, with $\mathcal{H}(\pi_{10}) \neq \{0\}$ and $\pi_{10} \cong \pi_{20}$.*

(iii) *If there is an operator $T \in \mathbb{I}(\pi_1, \pi_2)$ such that* $\ker T = \{0\}$ *and $T(\mathcal{H}(\pi_1))$ is dense in $\mathcal{H}(\pi_2)$, then $\pi_1 \cong \pi_2$.*

(iv) *If π_1 and π_2 are factor representations, then one of the following relations hold: $\pi_1 \between \pi_2$, $\pi_1 \leqq \pi_2$ or $\pi_2 \leqq \pi_1$.*

(v) *If π_1 and π_2 are irreducible, then either $\pi_1 \cong \pi_2$ or $\pi_1 \between \pi_2$.*

Proof. Let $\pi := \pi_1 \oplus \pi_2$ and $\mathcal{H} := \mathcal{H}(\pi_1) \oplus \mathcal{H}(\pi_2)$. Then π is a self-adjoint representation of A and $\mathcal{N} := \pi(\mathsf{A})'$ is a von Neumann algebra on $\mathcal{H}$. Let $\mathcal{Z}$ be the center of $\mathcal{N}$, and let $z(e)$ denote the central carrier of a projection e in $\mathcal{N}$. In this proof $e \sim f$ and $e \precsim f$ denote the equivalence and ordering, respectively, of projections e and f in $\mathcal{N}$ (see e.g. KADISON/RINGROSE [2], ch. 6). For $l = 1, 2$, P_l is the projection of $\mathcal{H}$ onto $\mathcal{H}(\pi_l)$.

Statement 1: $\mathbb{I}(\pi_1, \pi_2) = \{T \in \mathbb{B}(\mathcal{H}(\pi_1), \mathcal{H}(\pi_2)) : TP_1 \in \mathcal{N}\}$.

Proof. If $T \in \mathbb{I}(\pi_1, \pi_2)$, then $TP_1\pi(a)\varphi = T\pi_1(a) P_1\varphi = \pi_2(a) TP_1\varphi = \pi(a) TP_1\varphi$ for $a \in \mathsf{A}$ and $\varphi \in \mathcal{D}(\pi)$. Hence $TP_1 \in \pi(\mathsf{A})' = \mathcal{N}$. Conversely, if $TP_1 \in \mathcal{N}$, then $T\pi_1(a)\varphi = TP_1\pi(a)\varphi = \pi(a) TP_1\varphi = \pi_2(a) T\varphi$ for $a \in \mathsf{A}$ and $\varphi \in \mathcal{D}(\pi_1)$, so that $T \in \mathbb{I}(\pi_1, \pi_2)$. □

Let $l \in \{1, 2\}$ and let e_l be a projection in $\mathcal{N}$ such that $e_l \leqq P_l$. Then $e_l \upharpoonright \mathcal{H}(\pi_l)$ is in $\pi_l(\mathsf{A})'$. By Proposition 8.3.11, (ii), $\pi_l \upharpoonright e_l\mathcal{D}(\pi_l)$ is a self-adjoint subrepresentation of π_l.

Statement 2: *$\pi_1 \upharpoonright e_1\mathcal{D}(\pi_1) \cong \pi_2 \upharpoonright e_2\mathcal{D}(\pi_2)$ if and only if $e_1 \sim e_2$.*

Proof. We abbreviate $\tilde{\pi}_l := \pi_l \upharpoonright e_l\mathcal{D}(\pi_l)$, $l = 1, 2$. Suppose $\tilde{\pi}_1 \cong \tilde{\pi}_2$. Let U be an isometry of $e_1\mathcal{H}(\pi_1)$ onto $e_2\mathcal{H}(\pi_2)$ that establishes this unitary equivalence. Setting $V := 0$ on $(I - e_1)\mathcal{H}$ and $V := U$ on $e_1\mathcal{H}$, V becomes a partial isometry on $\mathcal{H}$ with initial space $e_1\mathcal{H}$ and range $e_2\mathcal{H}$. Using that $e_1 \upharpoonright \mathcal{H}(\pi_1)$ is in $\pi_1(\mathsf{A})_s'$, we have $V\pi_1(a)\varphi = Ue_1\pi_1(a)\varphi = U\pi_1(a) e_1\varphi = \pi_2(a) Ue_1\varphi = \pi_2(a) V\varphi$ for $a \in \mathsf{A}$ and $\varphi \in \mathcal{D}(\pi_1)$. Therefore, $V \upharpoonright \mathcal{H}(\pi_1) \in \mathbb{I}(\pi_1, \pi_2)$. By Statement 1, $V = VP_1 \in \mathcal{N}$. Thus $e_1 \sim e_2$.

Conversely, assume that $e_1 \sim e_2$. Then there is a partial isometry V in $\mathcal{N}$ with initial space $e_1\mathcal{H}$ and range $e_2\mathcal{H}$. Statement 1 yields that $V \upharpoonright \mathcal{H}(\pi_1) \in \mathbb{I}(\pi_1, \pi_2)$. Combined with $e_1 \upharpoonright \mathcal{H}(\pi_1) \in \pi_1(\mathsf{A})_s'$ this implies that the isometry $U := V \upharpoonright e_1\mathcal{H}$ of $e_1\mathcal{H}$ onto $e_2\mathcal{H}$ belongs to $\mathbb{I}(\tilde{\pi}_1, \tilde{\pi}_2)$. From this it follows that $U^{-1}\tilde{\pi}_2(\cdot)\, U$ is a $*$-representation of A on the Hilbert space $e_1\mathcal{H}$ which is an extension of the self-adjoint representation $\tilde{\pi}_1$

on $e_1\mathcal{H}$. Therefore, $U^{-1}\tilde{\pi}_2(\cdot)\,U = \pi_1(\cdot)$ and hence U establishes the unitary equivalence of $\tilde{\pi}_1$ and $\tilde{\pi}_2$. □

Statement 3: The relations $\pi_1 \cong \pi_2$, $\pi_1 \subseteq \pi_2$ and $\pi_2 \subseteq \pi_1$ are equivalent to $P_1 \sim P_2$, $P_1 \precsim P_2$ and $P_2 \precsim P_1$, respectively.

Proof. We show that $\pi_1 \subseteq \pi_2$ implies $P_1 \precsim P_2$. The other assertions follow similarly or directly from Statement 2. Suppose $\pi_1 \subseteq \pi_2$. Then there is a subrepresentation π_{20} of π_2 such that $\pi_1 \cong \pi_{20}$. Since $\pi_2 \subseteq \pi$ and π_1 is self-adjoint, π_{20} is a self-adjoint subrepresentation of π. Therefore, by Proposition 8.3.11, there is a projection $e_2 \in \pi(\mathsf{A})'_s = \mathcal{N}$ such that $\pi_{20} = \pi \upharpoonright e_2\mathcal{D}(\pi) \equiv \pi \upharpoonright e_2\mathcal{D}(\pi_2)$. By $\pi_{20} \subseteq \pi_2$, $e_2 \leqq P_2$. Since $\pi_1 \equiv \pi \upharpoonright P_1\mathcal{D}(\pi_1)$ $\cong \pi_2 \upharpoonright e_2\mathcal{D}(\pi_2)$, Statement 2 gives $P_1 \sim e_2$. Thus $P_1 \precsim P_2$. □

After these preliminaries we turn to the proof of the assertions stated in Proposition 8.4.4. (i) follows from Statement 3 combined with the fact that for any von Neumann algebra the relations $e \precsim f$ and $f \precsim e$ imply that $e \sim f$.

We prove (ii). For later applications (in the proofs of (iii) and of Proposition 8.4.5 below) we prove in addition that $\pi_1 \between \pi_2$ is equivalent to $z(P_1)\,z(P_2) = 0$.

First we show the necessity part of (ii). Suppose that $\pi_1 \between \pi_2$. Let π_{10} and π_{20} be unitarily equivalent self-adjoint subrepresentations of π_1 and π_2, respectively, and let U be an isometry which gives the unitary equivalence. As noted already above, π_{l0} $= \pi_l \upharpoonright e_l\mathcal{D}(\pi_l)$ for some projection $e_l \in \mathcal{N}$, $l = 1, 2$. Let V be as in the proof of Statement 2. Then $V \upharpoonright \mathcal{H}(\pi_1) \in \mathbb{I}(\pi_1, \pi_2) = \{0\}$, since $\pi_1 \between \pi_2$. This yields $\mathcal{H}(\pi_{10}) = \{0\}$.

Next we show that the condition formulated in (ii) implies that $z(P_1)\,z(P_2) = 0$. Assume to the contrary that $z(P_1)\,z(P_2) \neq 0$. Then there exist non-zero projections e_1 and e_2 in $\mathcal{N}$ such that $e_1 \leqq P_1$, $e_2 \leqq P_2$ and $e_1 \sim e_2$ (Kadison/Ringrose [2], 6.1.8). By Statement 2 and the remark before, $\pi_1 \upharpoonright e_1\mathcal{D}(\pi_1)$ and $\pi_2 \upharpoonright e_2\mathcal{D}(\pi_2)$ are unitarily equivalent self-adjoint subrepresentations of π_1 and π_2, respectively. Since $e_1 \neq 0$, this contradicts the condition in (ii).

Finally, we prove that $z(P_1)\,z(P_2) = 0$ implies $\pi_1 \between \pi_2$. We let $T \in \mathbb{I}(\pi_1, \pi_2)$. By Statement 1, $TP_1 \in \mathcal{N}$. Let e_1 and e_2 be the projections onto the closures of $(TP_1)^*\,\mathcal{H}$ and $TP_1\mathcal{H}$, respectively. Then $e_1, e_2 \in \mathcal{N}$ and $e_1 \sim e_2$ (Kadison/Ringrose [2], 6.1.6). Hence $z(e_1) = z(e_2)$. Since obviously $e_1 \leqq P_1$ and $e_2 \leqq P_2$, we have $e_1 \leqq z(e_1) = z(e_1)\,z(e_2)$ $\leqq z(P_1)\,z(P_2) = 0$. Thus $e_1 = 0$ which yields $TP_1 = 0$ and $T = 0$ on $\mathcal{H}(\pi_1)$. Hence $\mathbb{I}(\pi_1, \pi_2)$ $= \{0\}$. By symmetry, $\mathbb{I}(\pi_2, \pi_1) = \{0\}$, so that $\pi_1 \between \pi_2$. This completes the proof of (ii).

We verify (iii). Retaining the notation of the final part of the preceding proof of (ii), the assumptions concerning $T \in \mathbb{I}(\pi_1, \pi_2)$ imply that $e_1 = P_1$ and $e_2 = P_2$. Since $e_1 \sim e_2$, Statement 2 gives $\pi_1 \cong \pi_2$.

Now we prove (iv). By assumption, the von Neumann algebras $\pi_1(\mathsf{A})'$ and $\pi_2(\mathsf{A})'$ are factors. We first show that either $z(P_1)\,z(P_2) = 0$ or $z(P_1) = z(P_2)$. Assume that $e := z(P_1)\,z(P_2) \neq 0$. Since $\mathcal{N}'_{P_1} = \pi_1(\mathsf{A})''$ is a factor isomorphic to $\mathcal{N}'_{z(P_1)}$, $\mathcal{N}'_{z(P_1)}$ is a factor with centre $\mathcal{Z}_{z(P_1)}$. Therefore, $e = z(P_1)$, since e is a non-zero projection in $\mathcal{Z}$. Changing the role of P_1 and P_2, we get $e = z(P_2)$, so that $z(P_1) = z(P_2)$.

If $z(P_1)\,z(P_2) = 0$, then $\pi_1 \between \pi_2$ by the above proof of (ii). If $z(P_1)\,z(P_2) \neq 0$, then $z(P_1) = z(P_2)$ as just shown. Since $\mathcal{N}'_{z(P_1)}$ is a factor, $\mathcal{N}_{z(P_1)} = (\mathcal{N}'_{z(P_1)})'$ is also a factor. Therefore, the projections P_1 and P_2 satisfy $P_1 \precsim P_2$ or $P_2 \precsim P_1$ (Kadison/Ringrose [2], 6.2.6). Hence, in virtue of Statement 3, $\pi_1 \subseteq \pi_2$ or $\pi_2 \subseteq \pi_1$. This completes the proof of (iv).

(v) follows easily from (iv) and (i). □

Proposition 8.4.5. *Suppose π is a self-adjoint representation of the *-algebra A with unit.*

(i) *π is a factor representation (i.e., $\pi(\mathsf{A})'$ is a factor) if and only if π cannot be decomposed into a direct sum of two nontrivial disjoint subrepresentations of π.*

(ii) *π is multiplicity-free (i.e., $\pi(\mathsf{A})'$ is commutative) if and only if for each decomposition $\pi = \pi_1 \oplus \pi_2$ of π as a direct sum of subrepresentations, π_1 and π_2 are disjoint.*

Proof. We first prove both necessity parts. Suppose that $\pi = \pi_1 \oplus \pi_2$, where π_1 and π_2 are representations of A. Since π is self-adjoint by assumption, π_1 and π_2 are self-adjoint, so that we are in the setup of the proof of Proposition 8.4.4. As shown therein, $\pi_1 \circ \pi_2$ is equivalent to $z(P_1)\, z(P_2) = 0$. Since $P_2 = I - P_1$, the latter is equivalent to $P_1 \in \mathcal{Z}$; so $\pi_1 \circ \pi_2$ if and only if $P_1 \in \mathcal{Z}$. Therefore, if $\mathcal{N} = \pi(\mathsf{A})'$ is a factor, then $P_1 = 0$ or $P_1 = I$ which means that π_1 and π_2 are trivial subrepresentations of π. If $\mathcal{N}$ is commutative, then always $P_1 \in \mathcal{Z} \equiv \mathcal{N}$ and hence $\pi_1 \circ \pi_2$. Now we verify the sufficiency parts. Let e be a projection in the von Neumann algebra $\mathcal{N} = \pi(\mathsf{A})'$. By Proposition 8.3.11, $\pi_1 := \pi \restriction e\mathcal{D}(\pi)$ and $\pi_2 := \pi \restriction (I - e)\, \mathcal{D}(\pi)$ are self-adjoint subrepresentations of π. It is obvious that $\pi = \pi_1 \oplus \pi_2$, so that we are again in the situation of the proof of Proposition 8.4.4. Recall that $\pi_1 \circ \pi_2$ is equivalent to $P_1 \equiv e \in \mathcal{Z}$. Therefore, if the condition of (i) is fulfilled, and if we take $e \in \mathcal{Z}$, then $e = 0$ or $e = I$; that is, $\mathcal{Z}$ is trivial, and $\mathcal{N}$ is a factor. In case of (ii) we have $\pi_1 \circ \pi_2$ and hence $e \in \mathcal{Z}$. This shows that $\mathcal{Z}$ and $\mathcal{N}$ have the same projections; so $\mathcal{N}$ is commutative. □

8.5. Induced Extensions

In this section A denotes a *-algebra with unit.

We begin with two examples from single operator theory which contain the basic idea and serve as a motivation for the construction given below.

Example 8.5.1. Suppose that A is a (densely defined) closed symmetric operator on a Hilbert space $\mathcal{H}$. Let Q_+ and Q_- denote the projections of $\mathcal{H}$ onto the deficiency spaces $\ker(A^* - \mathrm{i})$ and $\ker(A^* + \mathrm{i})$, respectively. Recall that the Caley transform U of A is defined by $U(A + \mathrm{i})\,\varphi = (A - \mathrm{i})\,\varphi$, $\varphi \in \mathcal{D}(A)$. By a slight abuse of notation, we let U also denote the partial isometry on $\mathcal{H}$ which acts as U on $(I - Q_+)\,\mathcal{H} = (A + \mathrm{i})\mathcal{D}(A)$ and which is zero on $Q_+\mathcal{H}$.

Being a contraction, the operator U has a minimal unitary dilation V (see Sz.-Nagy/Foais [1], I, § 4). That is, V is a unitary operator on a Hilbert space $\mathcal{H}_1$ which contains $\mathcal{H}$ as a subspace such that $\mathrm{pr}_{\mathcal{H}}\, V^n = U^n$ for $n \in \mathbb{N}_0$ and $\mathcal{H}_1 = \text{c.l.h.}\,\{V^n\mathcal{H} : n \in \mathbb{Z}\}$. We first check that $\ker(V - I) = \{0\}$. Let $\varphi \in \ker(V - I)$. Then

$$\begin{aligned} \langle \varphi, V^n(V - I)\,(I - Q_+)\,\psi\rangle &= \langle V^{-(n+1)}\varphi - V^{-n}\varphi, (I - Q_+)\,\psi\rangle \\ &= \langle \varphi - \varphi, (I - Q_+)\,\psi\rangle = 0 \end{aligned}$$

for all $\psi \in \mathcal{H}$ and $n \in \mathbb{Z}$. Since $\mathcal{D}(A) = (U - I)\,(I - Q_+)\,\mathcal{H} = (V - I)\,(I - Q_+)\,\mathcal{H}$ is dense in $\mathcal{H}$, this yields $\varphi \perp V^n\mathcal{H}$ for $n \in \mathbb{Z}$. Because $\mathcal{H}_1 = \text{c.l.h.}\,\{V^n\mathcal{H} : n \in \mathbb{Z}\}$, this implies $\varphi = 0$.

Since $\ker(V - I) = \{0\}$ as just shown, $B := \mathrm{i}(V + I)\,(I - V)^{-1}$ is a well-defined self-adjoint operator on the Hilbert space $\mathcal{H}_1$. It is clear that $A \subseteq B$, since $\mathcal{D}(A) = (U - I)\,(I - Q_+)\,\mathcal{H}$ and $U \restriction (I - Q_+)\,\mathcal{H} = V \restriction (I - Q_+)\,\mathcal{H}$. Let $\mathcal{D}_1 := \text{l.h.}\,\{V^n\mathcal{D}(A):$

$n \in \mathbb{Z}\}$. Since $\mathcal{D}(A) \subseteq \mathcal{D}(B)$, $\mathcal{D}_1$ is contained in $\mathcal{D}(B)$. We prove that $\mathcal{D}_1$ is a core for the self-adjoint operator B.

First we note that $(I - Q_+)\,\mathcal{H} + (I - Q_-)\,\mathcal{H}$ is dense in $\mathcal{H}$, Indeed, if $\zeta \in \mathcal{H}$ is orthogonal to $(I - Q_+)\,\mathcal{H}$ and to $(I - Q_-)\,\mathcal{H}$, then $\zeta \in Q_+\mathcal{H} \cap Q_-\mathcal{H}$; hence $A^*\zeta = \mathrm{i}\zeta = -\mathrm{i}\zeta$ and so $\zeta = 0$. Now let $\psi \in \mathcal{H}_1$ be such that $\psi \perp (B + \mathrm{i})\,\mathcal{D}_1$. Then

$$0 = \langle \psi, (B + \mathrm{i})\, V^n\varphi\rangle = \langle \psi, V^n(A + \mathrm{i})\,\varphi\rangle = \left\langle \psi, V^{n-1}\big(U(A + \mathrm{i})\,\varphi\big)\right\rangle$$

for all $\varphi \in \mathcal{D}(A)$ and $n \in \mathbb{Z}$. Since $(A + \mathrm{i})\,\mathcal{D}(A) = (I - Q_+)\,\mathcal{H}$ and $U(A + \mathrm{i})\,\mathcal{D}(A) = (I - Q_-)\,\mathcal{H}$, this shows that $\psi \perp V^n(I - Q_+)\,\mathcal{H}$ and $\psi \perp V^n(I - Q_-)\,\mathcal{H}$ for all $n \in \mathbb{Z}$. But $(I - Q_+)\,\mathcal{H} + (I - Q_-)\,\mathcal{H}$ is dense in $\mathcal{H}$. Thus $\psi \perp V^n\mathcal{H}$ for $n \in \mathbb{Z}$ which implies $\psi = 0$. Hence $(B + \mathrm{i})\,\mathcal{D}_1$ is dense in $\mathcal{H}_1$. Since B is a self-adjoint operator, it follows that $B \restriction \mathcal{D}_1$ is essentially self-adjoint, so that $\mathcal{D}_1$ is a core for B.

In view of Definition 8.5.3 and the investigations below, we state the following facts which follow easily from the preceding. The operator B on $\mathcal{H}_1$ is a self-adjoint extension of the closed symmetric operator A on $\mathcal{H}$ and $\mathcal{D}_1 \equiv \text{l.h.}\,\{V^n\mathcal{D}(A): n \in \mathbb{Z}\}$ is dense in $\mathcal{D}(B)$ relative to the norm $\|\cdot\| + \|\cdot\|_B$. We have $V^n \in (B)_s'$ and hence $U^n \equiv \mathrm{pr}_{\mathcal{H}}\, V^n \in (A)_w'$ for $n \in \mathbb{N}$ by Remark 6 in 7.2. Moreover, the operators $V^n \restriction \mathcal{H}$, $n \in \mathbb{Z}$, belong to the intertwining space $\mathbb{I}(A, B)$. (This follows immediately from $BV^n\varphi = V^nB\varphi = V^nA\varphi$, $\varphi \in \mathcal{D}(A)$.) ○

Example 8.5.2. Suppose that A is a (densely defined) closed symmetric operator in the Hilbert space $\mathcal{H}$ with equal deficiency indices. Let B be a self-adjoint extension of A in the same Hilbert space $\mathcal{H}$. Then the subspace $\mathcal{D}_1 := (B + \mathrm{i})^{-1}\,\mathcal{D}(A)$ of $\mathcal{D}(B)$ is a core for the operator B, since $(B + \mathrm{i})\,\mathcal{D}_1 = \mathcal{D}(A)$ is dense in $\mathcal{H}$. Therefore, $\mathcal{D}_1$ is dense in $\mathcal{D}(B)$ relative to the norm $\|\cdot\| + \|\cdot\|_B$. Further, $(B + \mathrm{i})^{-1} \in (B)_s' \subseteq (A)_w'$ and $(B+\mathrm{i})^{-1} \in \mathbb{I}(A, B)$. ○

Definition 8.5.3. Let π be a $*$-representation of A. An *induced extension* of π is a pair $(\pi_1, \mathcal{M})$, where π_1 is a $*$-representation of A and $\mathcal{M}$ is a subset of $\pi_1(\mathsf{A})_s'$ such that $\pi \subseteq \pi_1$ and $\mathcal{M}\mathcal{D}(\pi) \equiv \text{l.h.}\,\{x\varphi: x \in \mathcal{M} \text{ and } \varphi \in \mathcal{D}(\pi)\}$ is a dense linear subspace of $\mathcal{D}(\pi_1)$ $[\mathfrak{t}_{\pi_1}]$.

Let π_1 be an extension of a $*$-representation π of A. We call π_1 an *induced extension* of π if there is a set $\mathcal{M}$ such that the pair $(\pi_1, \mathcal{M})$ is an induced extension of π in the sense of Definition 8.5.3.

Remark 1. If $(\pi_1, \mathcal{M})$ is an induced extension of π and $\mathcal{M}_1$ is the algebra generated by $\mathcal{M}$ and I, then $(\pi_1, \mathcal{M}_1)$ is, of course, again an induced extension of π. That is, *we can assume without loss of generality in Definition 8.5.3 that $\mathcal{M}$ is an algebra which contains I.*

In the two above examples we have seen how (essentially self-adjoint) extensions of a closed symmetric operator can be defined with the aid of certain elements (U^n and $(B + \mathrm{i})^{-1}$) in the weak commutant of the operator. We now describe a similar extension procedure for $*$-representations.

Proposition 8.5.4. *Let π be a $*$-representation of A, and let $\mathcal{H}_1$ be a Hilbert space which contains $\mathcal{H}(\pi)$ as a subspace. Suppose that $\mathcal{M}$ is a subset of $\mathbf{B}(\mathcal{H}_1)$ with $I \in \mathcal{M}$ for which $\mathcal{M}\mathcal{D}(\pi)$ is dense in $\mathcal{H}_1$ and $\mathrm{pr}_{\mathcal{H}(\pi)}\, y^*x \in \pi(\mathsf{A})_w'$ for all $x, y \in \mathcal{M}$.*

(i) *Then there exists a closed $*$-representation $\pi_{\mathcal{M}}$ of A on $\mathcal{H}(\pi_{\mathcal{M}}) := \mathcal{H}_1$ such that $\pi \subseteq \pi_{\mathcal{M}}$, $\mathcal{M} \restriction \mathcal{H}(\pi) \subseteq \mathbb{I}(\pi, \pi_{\mathcal{M}})$ and $\mathcal{M}\mathcal{D}(\pi)$ is a dense linear subspace of $\mathcal{D}(\pi_{\mathcal{M}})$ relative to* *the graph topology of $\pi_{\mathcal{M}}(\mathsf{A})$. These conditions determine the (closed) $*$-representation $\pi_{\mathcal{M}}$ uniquely.*

(ii) *An operator z in $\mathbb{B}(\mathcal{H}_1)$ belongs to $\pi_{\mathcal{M}}(\mathsf{A})'_{\mathrm{w}}$ if and only if $\mathrm{pr}_{\mathcal{H}(\pi)}\, y^*zx \in \pi(\mathsf{A})'_{\mathrm{w}}$ for all $x, y \in \mathcal{M}$.*

(iii) *If xy is in $\mathcal{M}$ when x and y are in $\mathcal{M}$, then $\mathcal{M} \subseteq \pi_{\mathcal{M}}(\mathsf{A})'_{\mathrm{s}}$ and so $(\pi_{\mathcal{M}}, \mathcal{M})$ is an induced extension of π.*

Proof. (i): Suppose that φ and ψ are in $\mathcal{M}\mathcal{D}(\pi)$. We can write φ and ψ as $\varphi = x_1\varphi_1 + \dots + x_n\varphi_n$ and $\psi = y_1\psi_1 + \dots + y_m\psi_m$ with $x_k, y_l \in \mathcal{M}$ and $\varphi_k, \psi_l \in \mathcal{D}(\pi)$ for $k = 1, \dots, n$ and $l = 1, \dots, m$. Suppose $a \in \mathsf{A}$.
Then we have

$$\begin{aligned}\Big\langle \sum_k x_k\pi(a)\,\varphi_k, \psi\Big\rangle &= \sum_{k,l} \langle x_k\pi(a)\,\varphi_k, y_l\psi_l\rangle = \sum_{k,l} \langle (\mathrm{pr}\; y_l^* x_k)\,\pi(a)\,\varphi_k, \psi_l\rangle \\ &= \sum_{k,l} \langle (\mathrm{pr}\; y_l^* x_k)\,\varphi_k, \pi(a^+)\,\psi_l\rangle = \sum_{k,l} \langle x_k\varphi_k, y_l\pi(a^+)\,\psi_l\rangle \\ &= \Big\langle \varphi, \sum_l y_l\pi(a^+)\,\psi_l\Big\rangle, \end{aligned} \tag{1}$$

where we used that $\mathrm{pr}\; y_l^* x_k \in \pi(\mathsf{A})'_{\mathrm{w}}$ by assumption. We define

$$\pi_0(a)\,\varphi \equiv \pi_0(a)\Big(\sum_k x_k\varphi_k\Big) := \sum_k x_k\pi(a)\,\varphi_k. \tag{2}$$

Since $\mathcal{M}\mathcal{D}(\pi)$ is dense in $\mathcal{H}_1$, we conclude from (1) that $\pi_0(a)$ is a well-defined linear operator on $\mathcal{D}(\pi_0) := \mathcal{M}\mathcal{D}(\pi)$, i.e., $\varphi = 0$ implies that $\pi_0(a)\,\varphi = 0$. Further, since $\pi(a)$ maps $\mathcal{D}(\pi)$ into itself, $\pi_0(a)$ leaves $\mathcal{D}(\pi_0)$ invariant. It is not difficult to check that $a \to \pi_0(a)$ is a homomorphism of A into $L(\mathcal{D}(\pi_0))$. Putting the definition (2) into (1) we obtain that $\langle \pi_0(a)\,\varphi, \psi\rangle = \langle \varphi, \pi_0(a^+)\,\psi\rangle$ for $a \in \mathsf{A}$ and $\varphi, \psi \in \mathcal{D}(\pi_0)$. Therefore the preceding shows that π_0 is a $*$-representation of A on the Hilbert space $\mathcal{H}(\pi_0) := \mathcal{H}_1$. Setting $x_1 = I$ (recall that $I \in \mathcal{M}$ by assumption) and $n = 1$ in (2) we see that $\pi \subseteq \pi_0$. Letting $\pi_{\mathcal{M}} := \widehat{\pi_0}$, it is clear that $\pi \subseteq \pi_{\mathcal{M}}$ and $\mathcal{M}\mathcal{D}(\pi) \equiv \mathcal{D}(\pi_0)$ is dense in $\mathcal{D}(\pi_{\mathcal{M}})\,[\mathrm{t}_{\pi_{\mathcal{M}}}]$. From (2) we conclude that $\mathcal{M} \restriction \mathcal{H}(\pi)$ is contained in $\mathbb{I}(\pi, \pi_0)$. Proposition 8.2.2, (ii), yields $\mathcal{M} \restriction \mathcal{H}(\pi) \subseteq \mathbb{I}(\pi, \pi_{\mathcal{M}})$.

We prove the uniqueness assertion. Let π_1 be another closed $*$-representation of A on $\mathcal{H}_1$ having the properties ascribed to $\pi_{\mathcal{M}}$ in (i). If $a \in \mathsf{A}$ and $\varphi \in \mathcal{D}(\pi_0)$ is as above, then we have

$$\pi_{\mathcal{M}}(a)\,\varphi = \pi_0(a)\,\varphi = \sum_k x_k\pi(a)\,\varphi_k = \sum_k \pi_1(a)\,x_k\varphi_k = \pi_1(a)\,\varphi,$$

where we first used (2) and then $x_k \restriction \mathcal{H}(\pi) \in \mathbb{I}(\pi, \pi_1)$. Thus $\pi_{\mathcal{M}} \restriction \mathcal{D}(\pi_0) = \pi_1 \restriction \mathcal{D}(\pi_0)$. Since $\pi_{\mathcal{M}}$ and π_1 are closed and $\mathcal{D}(\pi_0) \equiv \mathcal{M}\mathcal{D}(\pi)$ is dense in $\mathcal{D}(\pi_{\mathcal{M}})\,[\mathrm{t}_{\pi_{\mathcal{M}}}]$ and in $\mathcal{D}(\pi_1)\,[\mathrm{t}_{\pi_1}]$, this implies that $\pi_{\mathcal{M}} = \pi_1$.

(ii): Let $z \in \mathbb{B}(\mathcal{H}_1)$. Suppose $a \in \mathsf{A}$. Letting φ and ψ be as in the proof of (i), we have

$$\langle z\pi_0(a)\,\varphi, \psi\rangle = \sum_{k,l} \langle zx_k\pi(a)\,\varphi_k, y_l\psi_l\rangle = \sum_{k,l} \langle (\mathrm{pr}\; y_l^* z x_k)\,\pi(a)\,\varphi_k, \psi_l\rangle$$

and similarly

$$\langle z\varphi, \pi_0(a^+)\,\psi\rangle = \sum_{k,l} \langle (\mathrm{pr}\; y_l^* z x_k)\,\varphi_k, \pi(a^+)\,\psi_l\rangle.$$

Comparing these two formulas, we conclude that $z \in \pi_0(\mathsf{A})'_{\mathrm{w}}$ if and only if pr $y^*zx \in \pi(\mathsf{A})'_{\mathrm{w}}$ for all $x, y \in \mathcal{M}$. Since $\pi_{\mathcal{M}} = \widehat{\pi_0}$ and hence $\pi_0(\mathsf{A})'_{\mathrm{w}} = \pi_{\mathcal{M}}(\mathsf{A})'_{\mathrm{w}}$, this gives the assertion.

(iii): Suppose that $xy \in \mathcal{M}$ for all $x, y \in \mathcal{M}$. Suppose $a \in \mathsf{A}$ and $x \in \mathcal{M}$. Letting φ be as in the proof of (i), we have $x\varphi = \sum_k xx_k\varphi_k \in \mathcal{D}(\pi_0)$, since $xx_k \in \mathcal{M}$ for $k = 1, \ldots, n$. By (2),

$$\pi_0(a)\, x\varphi = \sum_k xx_k\pi(a)\, \varphi_k = \sum_k x\pi_0(a)\, x_k\varphi_k = x\pi_0(a)\, \varphi .$$

Hence $x \in \pi_0(\mathsf{A})'_{\mathrm{s}}$. Proposition 7.2.9, (ii), gives $x \in \hat{\pi}_0(\mathsf{A})'_{\mathrm{s}} \equiv \pi_{\mathcal{M}}(\mathsf{A})'_{\mathrm{s}}$; so $\mathcal{M} \subseteq \pi_{\mathcal{M}}(\mathsf{A})'_{\mathrm{s}}$. Together with (i), this shows that $(\pi_{\mathcal{M}}, \mathcal{M})$ is an induced extension of π. □

Remark 2. A slight reformulation of the uniqueness assertion in Proposition 8.5.4, (i), says that $\pi_{\mathcal{M}}$ is the smallest among the closed $*$-representations π_1 of A such that $\pi \subseteq \pi_1$ and $\mathcal{M} \upharpoonright \mathcal{H}(\pi) \subseteq \mathbb{I}(\pi, \pi_1)$.

Proposition 8.5.5. *Let π be a $*$-representation of A. Suppose that $\mathcal{H}(\pi)$ is a subspace of a possibly larger Hilbert space $\mathcal{H}_1$ and $\mathcal{M}$ is a subalgebra of $\mathbf{B}(\mathcal{H}_1)$ with $I \in \mathcal{M}$ such that $\mathcal{M}\mathcal{D}(\pi)$ is dense in $\mathcal{H}_1$. Then there exists a $*$-representation π_1 on $\mathcal{H}(\pi_1) := \mathcal{H}_1$ such that $(\pi_1, \mathcal{M})$ is an induced extension of π if and only if $\mathrm{pr}_{\mathcal{H}(\pi)}\, y^*x$ is in $\pi(\mathsf{A})'_{\mathrm{w}}$ for all x and y in $\mathcal{M}$. If the latter is satisfied, then we can take $\pi_{\mathcal{M}}$ for π_1.*

Proof. The if part is stated in Proposition 8.5.4, (iii). Conversely, suppose that $(\pi_1, \mathcal{M})$ is an induced extension. Then $\mathcal{M} \subseteq \pi_1(\mathsf{A})'_{\mathrm{s}} \subseteq \pi_1(\mathsf{A})'_{\mathrm{w}}$ and $\pi(\mathsf{A}) = \pi_1(\mathsf{A}) \upharpoonright \mathcal{D}(\pi)$; so Proposition 7.2.16 (applied with $\mathcal{A} := \pi(\mathsf{A})$ and $\mathcal{B} := \pi_1(\mathsf{A})$) gives pr $y^*x \in \pi(\mathsf{A})'_{\mathrm{w}}$ for $x, y \in \mathcal{M}$. □

We derive two corollaries in which the Hilbert space $\mathcal{H}(\pi_{\mathcal{M}})$ coincides with $\mathcal{H}(\pi)$.

Corollary 8.5.6. *Suppose that π is a $*$-representation of A and $\mathcal{M}$ is a $*$-algebra with $I \in \mathcal{M}$ contained in $\pi(\mathsf{A})'_{\mathrm{w}}$. Then the $*$-representation $\pi_{\mathcal{M}}$ is well-defined and $(\pi_{\mathcal{M}}, \mathcal{M})$ is an induced extension of π. Moreover, $\pi_{\mathcal{M}}$ is the smallest among the closed $*$-representations π_1 of A which are extensions of π and satisfy $\mathcal{M} \subseteq \pi_1(\mathsf{A})'_{\mathrm{s}}$.*

Proof. We have $y^*x \in \pi(\mathsf{A})'_{\mathrm{w}}$ for $x, y \in \mathcal{M}$, since $\mathcal{M}$ is a $*$-algebra and $\mathcal{M} \subseteq \pi(\mathsf{A})'_{\mathrm{w}}$. Thus the first assertion follows from Proposition 8.5.5 applied with $\mathcal{H}_1 = \mathcal{H}(\pi)$. Since $\pi \subseteq \pi_1$ and $\mathcal{M} \subseteq \pi_1(\mathsf{A})'_{\mathrm{s}}$ obviously imply that $\mathcal{M} \subseteq \mathbb{I}(\pi, \pi_1)$, Remark 2 above yields the second assertion. □

Corollary 8.5.7. *Suppose that π is a $*$-representation of A for which $\mathcal{M} := \pi(\mathsf{A})'_{\mathrm{w}}$ is an algebra. Then we have $\pi(\mathsf{A})'_{\mathrm{w}} = \pi_{\mathcal{M}}(\mathsf{A})'_{\mathrm{s}} = \pi_{\mathcal{M}}(\mathsf{A})'_{\mathrm{w}}$ and $\pi_{\mathcal{M}}$ is the smallest of the closed $*$-representations π_1 of A which satisfy $\pi \subseteq \pi_1$ and $\pi(\mathsf{A})'_{\mathrm{w}} \subseteq \pi_1(\mathsf{A})'_{\mathrm{s}}$.*

Proof. Since $\pi(\mathsf{A})'_{\mathrm{w}}$ is always $*$-invariant and $\mathcal{M} \equiv \pi(\mathsf{A})'_{\mathrm{w}}$ is an algebra, $\mathcal{M}$ is a $*$-algebra with $I \in \mathcal{M}$; so Corollary 8.5.6 applies. Therefore, $\mathcal{M} \equiv \pi(\mathsf{A})'_{\mathrm{w}} \subseteq \pi_{\mathcal{M}}(\mathsf{A})'_{\mathrm{s}}$. On the other hand, by $\pi \subseteq \pi_{\mathcal{M}}$, we have $\pi_{\mathcal{M}}(\mathsf{A})'_{\mathrm{s}} \subseteq \pi_{\mathcal{M}}(\mathsf{A})'_{\mathrm{w}} \subseteq \pi(\mathsf{A})'_{\mathrm{w}}$; hence $\pi(\mathsf{A})'_{\mathrm{w}} = \pi_{\mathcal{M}}(\mathsf{A})'_{\mathrm{s}} = \pi_{\mathcal{M}}(\mathsf{A})'_{\mathrm{w}}$. The final assertion is a special case of the corresponding statement in Corollary 8.5.6. □

Remark 3. Suppose P is a projection contained in $\pi(\mathsf{A})'_{\mathrm{w}}$ for a $*$-representation π of A. Then Corollary 8.5.6 applies in case $\mathcal{M} := \mathrm{l.h.}\,\{I, P\}$. Hence $P \in \pi_{\mathcal{M}}(\mathsf{A})'_{\mathrm{s}}$, and P provides a decomposition of $\pi_{\mathcal{M}}$ into a direct sum according to Lemma 8.3.3.

We now discuss some examples. In the first two examples we use some standard constructions from dilation theory (see, e.g., the Appendix of RIESZ/SZ.-NAGY [1]). Throughout these examples we assume that π is a $*$-representation of A.

Example 8.5.8. Suppose that u is an operator in $\pi(\mathsf{A})'_{\mathrm{w}}$ such that $\|u\| \leqq 1$ and $u^n \in \pi(\mathsf{A})'_{\mathrm{w}}$ for all $n \in \mathbb{N}$. (E.g., if π and A are as in Example 8.3.6 and θ is a function in $H^\infty(\mathbb{T})$ of norm less or equal to one in $H^\infty(\mathbb{T})$, then the Toeplitz operator $u := T_\theta$ with symbol θ has these properties. In this case we even have $u^n \in \pi(\mathsf{A})'_{\mathrm{s}}$ for $n \in \mathbb{N}$). Let v be the minimal unitary dilation of the contraction u on the Hilbert space $\mathcal{H}_1$ and let $\mathcal{M} := \{v^n : n \in \mathbb{Z}\}$. Since $\mathcal{H}_1 = \text{c.l.h.}\ \{v^n \mathcal{H}(\pi) : n \in \mathbb{Z}\}$ and $\operatorname{pr} v^n = u^n$ for $n \in \mathbb{N}_0$, it follows easily that the assumptions of Proposition 8.5.4 are fulfilled. Hence $(\pi_{\mathcal{M}}, \mathcal{M})$ is an induced extension of π. ○

Example 8.5.9. Suppose that u is a self-adjoint operator in $\pi(\mathsf{A})'_{\mathrm{w}}$ with $0 \leqq u \leqq I$. Let x be the self-adjoint operator on $\mathcal{H}_0 := \mathcal{H}(\pi) \oplus \mathcal{H}(\pi)$ which is given by the matrix

$$x = \begin{bmatrix} u & w \\ w & I - u \end{bmatrix}, \quad \text{where } w := (u - u^2)^{1/2}.$$

It is easy to calculate that $x^2 = x$; so x is a projection on $\mathcal{H}_0$. Let $\mathcal{H}_1$ be the closure of $\mathcal{H}(\pi) + x\mathcal{H}(\pi)$ in $\mathcal{H}_0$, where $\mathcal{H}(\pi)$ is identified with the linear subspace $\mathcal{H}(\pi) \oplus \{0\}$ of $\mathcal{H}_0$. Then $v := x \restriction \mathcal{H}_1$ is a projection on the Hilbert space $\mathcal{H}_1$ and $\operatorname{pr} v = u$. Setting $\mathcal{M} := \{I, v\}$, the assumptions of Proposition 8.5.4 are again satisfied; so $(\pi_{\mathcal{M}}, \mathcal{M})$ is an induced extension of π. In particular, the projection v is in $\pi_{\mathcal{M}}(\mathsf{A})'_{\mathrm{s}}$.

Let $a \in \mathsf{A}$. For $\varphi, \psi \in \mathcal{D}(\pi)$, we have

$$\begin{aligned} &\langle \pi_{\mathcal{M}}(a) \big(v\varphi + (I - v)\,\psi\big), v\varphi + (I - v)\,\psi \rangle \\ &= \langle v\pi(a)\,\varphi + (I - v)\,\pi(a)\,\psi, v\varphi + (I - v)\,\psi \rangle \\ &= \langle v\pi(a)\,\varphi, \varphi \rangle + \langle (I - v)\,\pi(a)\,\psi, \psi \rangle \\ &= \langle u\pi(a)\,\varphi, \varphi \rangle + \langle (I - u)\,\pi(a)\,\psi, \psi \rangle. \end{aligned}$$

This implies that $\pi_{\mathcal{M}}(a) \geqq 0$ if and only if $u\pi(a) \geqq 0$ and $(I - u)\,\pi(a) \geqq 0$. ○

Example 8.5.10. Suppose that u is an operator on $\mathcal{H}(\pi)$ such that u and u^*u are both in $\pi(\mathsf{A})'_{\mathrm{w}}$. We define an operator v on $\mathcal{H}_1 := \mathcal{H}(\pi) \oplus \mathcal{H}(\pi)$ by the matrix

$$v := \begin{bmatrix} u & 0 \\ I & 0 \end{bmatrix}.$$

Set $\mathcal{M} := \{I, v\}$. Then $\mathcal{M}\mathcal{D}(\pi)$ is dense in $\mathcal{H}_1$, and $\operatorname{pr} v = u$ and $\operatorname{pr} v^*v = u^*u + I$ are in $\pi(\mathsf{A})'_{\mathrm{w}}$; so the $*$-representation $\pi_{\mathcal{M}}$ is well-defined by Proposition 8.5.4, (i). Since $\operatorname{pr} v^*v^2 = u^*u^2 + u$ and $\operatorname{pr} v^2 = u^2$, Proposition 8.5.4, (ii), shows that $v \in \pi_{\mathcal{M}}(\mathsf{A})'_{\mathrm{w}}$ if and only if u^*u^2 and u^2 are in $\pi(\mathsf{A})'_{\mathrm{w}}$.

In order to describe an example where these conditions are satisfied, let π and A be as in Example 8.3.6. Suppose that θ is a function in $H^\infty(\mathbb{T})$ and let $u := T_\theta$ be the Toeplitz operator with symbol θ. The operators u^*u, u^*u^2 and u^2 are Toeplitz operators with symbols $\bar\theta\theta$, $\bar\theta\theta^2$ and θ^2, respectively, and so contained in $\pi(\mathsf{A})'_{\mathrm{w}}$. Hence we have $v \in \pi_{\mathcal{M}}(\mathsf{A})'_{\mathrm{w}}$ in this case. ○

8.6. The Gelfand-Neumark-Segal Construction

Throughout this section A denotes a $*$-algebra with unit element.

Suppose π is a $*$-representation of A and φ is a vector in $\mathcal{D}(\pi)$. Then $\omega(a) := \langle\pi(a)\varphi, \varphi\rangle$, $a \in \mathsf{A}$, defines a positive linear functional on A. (Indeed, $\omega(a^+a) = \langle\pi(a^+a)\,\varphi, \varphi\rangle = \langle\pi(a)^+\,\pi(a)\,\varphi, \varphi\rangle = \|\pi(a)\,\varphi\|^2 \geqq 0$ for $a \in \mathsf{A}$.) Theorem 8.6.2 shows that each positive linear functional on A arises in this way. Further, since $\omega(1) = \|\varphi\|^2$, ω is a state on A if and only if φ is a unit vector in $\mathcal{D}(\pi)$. In this case, ω is called a *vector state* in the $*$-representation π. Theorem 8.6.2 says that each state on A is a vector state in a certain $*$-representation of A.

Definition 8.6.1. Let π be a representation of A. We say that π is *algebraically cyclic* if there exists a vector $\varphi \in \mathcal{D}(\pi)$ such that $\mathcal{D}(\pi) = \pi(\mathsf{A})\,\varphi$. In this case, φ is called an *algebraically cyclic vector* for π. The representation π is said to be *cyclic* if there exists a vector $\varphi \in \mathcal{D}(\pi)$ which is cyclic for π, i.e., $\pi(\mathsf{A})\,\varphi$ is dense in $\mathcal{D}(\pi)$ $[\mathrm{t}_\pi]$.

Theorem 8.6.2. *Suppose that ω is a positive linear functional on A. There exists an algebraically cyclic $*$-representation $\tilde{\pi}_\omega$ and an algebraically cyclic vector φ_ω for $\tilde{\pi}_\omega$ such that $\omega(a) = \langle\tilde{\pi}_\omega(a)\,\varphi_\omega, \varphi_\omega\rangle$ for all $a \in \mathsf{A}$. If π is another algebraically cyclic $*$-representation of A with algebraically cyclic vector φ such that $\omega(a) = \langle\pi(a)\,\varphi, \varphi\rangle$, $a \in \mathsf{A}$, then π is unitarily equivalent to $\tilde{\pi}_\omega$.*

We first prove an auxiliary lemma.

Lemma 8.6.3. *If ω is a positive linear functional on A, then $\mathsf{N}_\omega := \{x \in \mathsf{A} : \omega(x^+x) = 0\}$ is a left ideal in the algebra A.*

Proof. The proof is based on the Cauchy-Schwarz inequality and the fact that $\omega(a^+) = \overline{\omega(a)}$ for $a \in \mathsf{A}$; cf. Lemma 1.4.1. Let $x, y \in \mathsf{N}_\omega$. Then

$$\omega\big((x+y)^+(x+y)\big) = \omega(x^+x + x^+y + y^+x + y^+y) = 2\operatorname{Re}\omega(x^+y)$$
$$\leqq 2\omega(x^+x)^{1/2}\,\omega(y^+y)^{1/2} = 0;$$

so $x + y \in \mathsf{N}_\omega$. It is obvious that $\lambda x \in \mathsf{N}_\omega$ for $\lambda \in \mathbb{C}$. If $a \in \mathsf{A}$, then

$$\omega\big((ax)^+\,ax\big) = \omega\big(x^+(a^+ax)\big) \leqq \omega(x^+x)^{1/2}\,\omega\big((a^+ax)^+\,a^+ax\big) = 0.$$

Hence $ax \in \mathsf{N}_\omega$. This shows that N_ω is a left ideal in A. $\square$

Proof of Theorem 8.6.2. For $x, y \in \mathsf{A}$, we define $\langle x, y\rangle_1 := \omega(y^+x)$. Since the linear functional ω is positive, $\langle\cdot,\cdot\rangle_1$ is a semi-definite inner product on A. Let $\mathcal{D}_\omega := \mathsf{A}/\mathsf{N}_\omega$ be the quotient vector space and let ι denote the quotient map of A into $\mathcal{D}_\omega = \mathsf{A}/\mathsf{N}_\omega$. Then the equation $\langle\iota(x), \iota(y)\rangle := \omega(y^+x)$, $x, y \in \mathsf{A}$, defines a scalar product on $\mathcal{D}_\omega$.

Since N_ω is a left ideal in A by Lemma 8.6.3, $\iota(x_1) = \iota(x_2)$ always implies $\iota(ax_1) = \iota(ax_2)$ for $a, x_1, x_2 \in \mathsf{A}$. Therefore, if $a \in \mathsf{A}$, then $\tilde{\pi}_\omega(a)\,\iota(x) := \iota(ax)$, $x \in \mathsf{A}$, defines, unambiguously, a linear operator on $\mathcal{D}_\omega$ which maps $\mathcal{D}_\omega$ into itself. If a, b, x, and y are in A, then

$$\tilde{\pi}_\omega(ab)\,\iota(x) = \iota(abx) = \tilde{\pi}_\omega(a)\,\iota(bx) = \tilde{\pi}_\omega(a)\,\tilde{\pi}_\omega(b)\,\iota(x)$$

and

$$\langle\tilde{\pi}_\omega(a)\,\iota(x), \iota(y)\rangle = \langle\iota(ax), \iota(y)\rangle = \omega(y^+ax) = \omega\big((a^+y)^+\,x\big)$$
$$= \langle\iota(x), \iota(a^+y)\rangle = \langle\iota(x), \tilde{\pi}_\omega(a^+)\,\iota(y)\rangle.$$

It is obvious that the map $a \to \tilde{\pi}_\omega(a)$ of A into $L(\mathcal{D}_\omega)$ is linear and that $\tilde{\pi}_\omega(1) = I$. Thus we have shown that $\tilde{\pi}_\omega$ is a $*$-representation of A on $\mathcal{D}(\tilde{\pi}_\omega) := \mathcal{D}_\omega$ in the Hilbert space $\mathcal{H}(\tilde{\pi}_\omega) \equiv \mathcal{H}_\omega$ which is, by definition, the completion of the unitary space $\mathcal{D}_\omega$.

Set $\varphi_\omega := \iota(1)$. Clearly, $\tilde{\pi}_\omega(\mathsf{A})\,\varphi_\omega = \mathcal{D}_\omega$. Hence φ_ω is an algebraically cyclic vector for $\tilde{\pi}_\omega$. We have $\langle \tilde{\pi}_\omega(a)\,\varphi_\omega, \varphi_\omega\rangle = \langle \iota(a \cdot 1), \iota(1)\rangle = \omega(a)$ for $a \in \mathsf{A}$.

Now we prove the uniqueness assertion. For $a \in \mathsf{A}$, we have

$$\|\pi(a)\,\varphi\|^2 = \langle \pi(a)\,\varphi, \pi(a)\,\varphi\rangle = \langle \pi(a^+a)\,\varphi, \varphi\rangle = \omega(a^+a)$$
$$= \langle \tilde{\pi}_\omega(a^+a)\,\varphi_\omega, \varphi_\omega\rangle = \langle \tilde{\pi}_\omega(a)\,\varphi_\omega, \tilde{\pi}_\omega(a)\,\varphi_\omega\rangle = \|\tilde{\pi}_\omega(a)\,\varphi_\omega\|^2.$$

From this we conclude that the equation $U(\pi(a)\,\varphi) = \tilde{\pi}_\omega(a)\,\varphi_\omega$, $a \in \mathsf{A}$, defines a norm-preserving linear map of $\pi(\mathsf{A})\,\varphi$ onto $\tilde{\pi}_\omega(\mathsf{A})\,\varphi_\omega$. Since φ and φ_ω are algebraically cyclic vectors for π and $\tilde{\pi}_\omega$, respectively, $\mathcal{D}(\pi) = \pi(\mathsf{A})\,\varphi$ and $\mathcal{D}(\tilde{\pi}_\omega) = \tilde{\pi}_\omega(\mathsf{A})\,\varphi_\omega$. Therefore, U extends by continuity to a unitary operator, again denoted by U, of $\mathcal{H}(\pi)$ onto $\mathcal{H}(\tilde{\pi}_\omega)$. By construction, U maps $\mathcal{D}(\pi)$ onto $\mathcal{D}(\tilde{\pi}_\omega)$. For $a, b \in \mathsf{A}$, we have

$$U^{-1}\tilde{\pi}_\omega(a)\,U(\pi(b)\,\varphi) = U^{-1}\tilde{\pi}_\omega(a)\,\tilde{\pi}_\omega(b)\,\varphi_\omega = U^{-1}\tilde{\pi}_\omega(ab)\,\varphi_\omega = \pi(ab)\,\varphi$$
$$= \pi(a)\,(\pi(b)\,\varphi).$$

Hence $\pi \cong \tilde{\pi}_\omega$. $\square$

Theorem 8.6.4. *Let ω be a positive linear functional on* A. *Then there exists a closed cyclic $*$-representation π_ω of* A *and a cyclic vector φ_ω for π_ω such that $\omega(a) = \langle \pi_\omega(a)\,\varphi_\omega, \varphi_\omega\rangle$ for $a \in \mathsf{A}$. If π is a closed cyclic $*$-representation of* A *with a cyclic vector φ such that $\omega(a) = \langle \pi(a)\,\varphi, \varphi\rangle$ for $a \in \mathsf{A}$, then π is unitarily equivalent to π_ω.*

Proof. Let $\tilde{\pi}_\omega$ and φ_ω be as in Theorem 8.6.2 and let π_ω be the closure of $\tilde{\pi}_\omega$. From Theorem 8.6.2 it follows that π_ω has the stated properties. We verify the uniqueness assertion. Set $\pi_0 := \pi \upharpoonright \pi(\mathsf{A})\,\varphi$. By the uniqueness part of Theorem 8.6.2, $\pi_0 \cong \tilde{\pi}_\omega$; so there is an isometry U of $\mathcal{H}(\pi)$ onto $\mathcal{H}(\pi_\omega)$ such that $U \in \mathbb{I}(\pi_0, \tilde{\pi}_\omega)$ and $U^{-1} \in \mathbb{I}(\tilde{\pi}_\omega, \pi_0)$. Since $\pi = \widehat{\pi_0}$, Proposition 8.2.2, (iv), gives $U \in \mathbb{I}(\pi, \pi_\omega)$ and $U^{-1} \in \mathbb{I}(\pi_\omega, \pi)$ which proves that $\pi \cong \pi_\omega$. $\square$

The method used in the preceding proofs is called the *Gelfand-Neumark-Segal construction* or briefly the *GNS construction*. It is one of the fundamental tools in representation theory of $*$-algebras. We shall retain the notation introduced in the GNS construction and we shall use it sometimes without comment. That is, if ω is a positive linear functional on A, then $\tilde{\pi}_\omega$, π_ω, $\mathcal{D}_\omega$, $\mathcal{H}_\omega$ and φ_ω have the meaning attached to them in the preceding proof.

Remark 1. Let π be a $*$-representation of A. As above, let $\omega(a) := \langle \pi(a)\,\varphi, \varphi\rangle$, $a \in \mathsf{A}$, when $\varphi \in \mathcal{D}(\pi)$. From the uniqueness assertion in Theorem 8.6.2 if follows immediately that the $*$-representation $\pi \upharpoonright \pi(\mathsf{A})\,\varphi$ is unitarily equivalent to $\tilde{\pi}_\omega$. Moreover, the proof given above shows that the unitary equivalence is implemented by an isometry U which satisfies $U(\pi(a)\,\varphi) = \tilde{\pi}_\omega(a)\,\varphi_\omega$, $a \in \mathsf{A}$.

Remark 2. Recall that $\mathcal{Z}(\mathsf{A})$ denotes the states of A. The $*$-representation $\pi_{\mathrm{unr}} := \sum_{\omega \in \mathcal{Z}(A)} \oplus\, \pi_\omega$ is called the *universal representation* of the $*$-algebra A. It has the important property that each state on A is a vector state of the $*$-representation π_{unr}. It is easily seen that π_{unr} is faithful if and only if $\mathcal{Z}(\mathsf{A})$ separates the elements of A, i.e., given $a \neq 0$ in A there is a state ω on A such that $\omega(a) \neq 0$.

In what follows we use the GNS construction as a tool in studying positive linear functionals on the $*$-algebra A.

We need some more notation. Recall that $\mathcal{P}(\mathsf{A})^*$ is the set of all positive linear functionals on A. For $\omega, \nu \in \mathcal{P}(A)^*$, we define

$$\nu \lesssim \omega \text{ if and only if } \nu(a^+a) \leqq \omega(a^+a) \text{ for all } a \in \mathsf{A}. \tag{1}$$

Let $[0, \omega]$ denote the set $\{\nu \in \mathcal{P}(\mathsf{A})^*: \nu \lesssim \omega\}$ equipped with the order relation "$\lesssim$". (This notation stems from the following fact. If we equip the real vector space A_h^* of all hermitian linear functionals on A with the order relation defined by (1), then $[0, \omega]$ is an order interval in the ordered vector space $(\mathsf{A}_h^*, \gtrsim)$.) By $[0, I]$ we mean the set $\{x \in \mathbf{B}(\mathcal{H}): 0 \leqq x \leqq I\}$ endowed with the usual order relation of self-adjoint operators on the Hilbert space $\mathcal{H}$.

Definition 8.6.5. A positive linear functional ω on A is said to be *pure* if it is an extremal point of the wedge $\mathcal{P}(\mathsf{A})^*$, i.e., if $[0, \omega] = \{\lambda\omega: 0 \leqq \lambda \leqq 1\}$.

Proposition 8.6.6. *Suppose that ω is a positive linear functional on* A. *If $x \in \pi_\omega(\mathsf{A})'_w \cap [0, I]$, then $\omega_x(a) := \langle x\pi_\omega(a)\, \varphi_\omega, \varphi_\omega\rangle$, $a \in \mathsf{A}$, defines a positive linear functional on* A *which satisfies $\omega_x \lesssim \omega$. The mapping $x \to \omega_x$ is an order isomorphism of $\pi_\omega(\mathsf{A})'_w \cap [0, I]$ onto $[0, \omega]$, i.e., the map $x \to \omega_x$ is bijective, and $x \leqq y$ is equivalent to $\omega_x \lesssim \omega_y$ for arbitrary $x, y \in \pi_\omega(\mathsf{A})'_w \cap [0, I]$.*

Proof. Suppose $x \in \pi_\omega(\mathsf{A})'_w \cap [0, I]$. For $a \in \mathsf{A}$, we have

$$\omega_x(a^+a) = \langle x\pi_\omega(a^+)\, \pi_\omega(a)\, \varphi_\omega, \varphi_\omega\rangle = \langle x\pi_\omega(a)\, \varphi_\omega, \pi_\omega(a)\, \varphi_\omega\rangle \geqq 0. \tag{2}$$

Hence $\omega_x \in \mathcal{P}(\mathsf{A})^*$. Since $x \leqq I$, (2) gives that

$$\omega_x(a^+a) \leqq \langle \pi_\omega(a)\, \varphi_\omega, \pi_\omega(a)\, \varphi_\omega\rangle = \omega(a^+a) \quad \text{for } a \in \mathsf{A};$$

so $\omega_x \lesssim \omega$. Further, because $\mathcal{D}_\omega \equiv \pi_\omega(\mathsf{A})\, \varphi_\omega$ is dense in $\mathcal{H}(\pi_\omega)$, it follows from (2) that $x \leqq y$ if and only if $\omega_x \lesssim \omega_y$ for $x, y \in \pi_\omega(\mathsf{A})'_w \cap [0, I]$. It is clear that the mapping $x \to \omega_x$ is injective. We prove that this mapping is surjective. Suppose $\nu \in \mathcal{P}(\mathsf{A})^*$ and $\nu \lesssim \omega$. We show that

$$\langle \pi_\omega(a)\, \varphi_\omega, \pi_\omega(b)\, \varphi_\omega\rangle_\nu := \nu(b^+a), \qquad a, b \in \mathsf{A}, \tag{3}$$

defines a bounded sesquilinear form on the unitary space $\mathcal{D}_\omega \equiv \pi_\omega(\mathsf{A})\, \varphi_\omega$. First we check that the definition (3) is correct, that is, the definition of $\langle \varphi, \psi\rangle_\nu$ does not depend on the representations $\varphi = \pi_\omega(a)\, \varphi_\omega$ and $\psi = \pi_\omega(b)\, \varphi_\omega$ for $\varphi, \psi \in \mathcal{D}_\omega$. Suppose that $\pi_\omega(a)\, \varphi_\omega = \pi_\omega(a_1)\, \varphi_\omega$ with $a, a_1 \in \mathsf{A}$. Then $\omega\big((a - a_1)^+ (a - a_1)\big) = \|\pi_\omega(a - a_1)\, \varphi_\omega\|^2 = 0$ and hence $\nu\big((a - a_1)^+ (a - a_1)\big) = 0$, since $\nu \lesssim \omega$. The Cauchy-Schwarz inequality gives $\nu\big(b^+(a - a_1)\big) = 0$ for each $b \in \mathsf{A}$, so that $\nu(b^+a) = \nu(b^+a_1)$. The same argument works for the second variable in $\langle \cdot, \cdot\rangle_\nu$ as well. This shows that the definition (3) makes sense. From

$$\|\pi_\omega(a)\, \varphi_\omega\|_\nu^2 = \nu(a^+a) \leqq \omega(a^+a) = \|\pi_\omega(a)\, \varphi_\omega\|^2, \qquad a \in \mathsf{A},$$

we see that the sesquilinear form $\langle \cdot, \cdot\rangle_\nu$ is bounded on $\mathcal{D}_\omega$. Hence there exists an operator $x \in \mathbf{B}\big(\mathcal{H}(\pi_\omega)\big)$ such that for $a, b \in \mathsf{A}$

$$\nu(b^+a) \equiv \langle \pi_\omega(a)\, \varphi_\omega, \pi_\omega(b)\, \varphi_\omega\rangle_\nu = \langle x\pi_\omega(a)\, \varphi_\omega, \pi_\omega(b)\, \varphi_\omega\rangle. \tag{4}$$

Since $\nu \in \mathcal{P}(\mathsf{A})^*$ and $\nu \lesssim \omega$, it follows from (4) that $0 \leqq x \leqq I$. If $a, b, c \in \mathsf{A}$, then

$$\begin{aligned}\langle x\bar{\pi}_\omega(a)\, \pi_\omega(b)\, \varphi_\omega, \pi_\omega(c)\, \varphi_\omega\rangle &= \nu(c^+ab)\\ &= \nu(a^+c)^+b = \langle x\pi_\omega(b)\, \varphi_\omega, \bar{\pi}_\omega(a^+)\, \pi_\omega(c)\, \varphi_\omega\rangle\end{aligned}$$

by (4). This shows that $x \in \bar{\pi}_\omega(\mathsf{A})'_{\mathrm{w}}$. Since π_ω is the closure of $\bar{\pi}_\omega$, $x \in \pi_\omega(\mathsf{A})'_{\mathrm{w}} \cap [0, I]$ and the mapping $x \to \omega_x$ is onto. □

Corollary 8.6.7. *If ω is a positive linear functional on A, then ω is pure if and only if the weak commutant of $\pi_\omega(\mathsf{A})$ is trivial, i.e., $\pi_\omega(\mathsf{A})'_{\mathrm{w}} = \mathbb{C} \cdot I$.*

Proof. Recall that ω is pure means that $[0, \omega] = \{\lambda\omega : 0 \leqq \lambda \leqq 1\}$. Therefore, by Proposition 8.6.6, ω is pure if and only if $\pi_\omega(\mathsf{A})'_{\mathrm{w}} \cap [0, I] = \{\lambda I; 0 \leqq \lambda \leqq 1\}$. Since $\pi_\omega(\mathsf{A})'_{\mathrm{w}}$ is a *-invariant vector subspace of $\mathbb{B}(\mathcal{H}(\pi_\omega))$, the latter is clearly equivalent to $\pi_\omega(\mathsf{A})'_{\mathrm{w}} = \mathbb{C} \cdot I$. □

Since $\bar{\pi}_\omega(\mathsf{A})'_{\mathrm{s}} \subseteq \pi_\omega(\mathsf{A})'_{\mathrm{s}} \subseteq \pi_\omega(\mathsf{A})'_{\mathrm{w}} = \bar{\pi}_\omega(\mathsf{A})'_{\mathrm{w}}$, Corollary 8.6.7 and Lemma 8.3.5 imply

Corollary 8.6.8. *If ω is a pure positive linear functional on A, then the representations π_ω and $\bar{\pi}_\omega$ are irreducible.*

Remark 3. Since $\bar{\pi}_\omega(\mathsf{A})'_{\mathrm{w}} = \pi_\omega(\mathsf{A})'_{\mathrm{w}}$, Proposition 8.6.6 and Corollary 8.6.7 remain valid if we replace π_ω by $\bar{\pi}_\omega$.

Remark 4. If π_ω or $\bar{\pi}_\omega$ is irreducible, then ω is not pure in general.

If ν and ω are positive linear functionals on A, we say that ν *is dominated by* ω if there is a $\lambda > 0$ such that $\nu \lesssim \lambda\omega$, i.e., $\nu(a^+a) \leqq \lambda\omega(a^+a)$ for all $a \in \mathsf{A}$. We next investigate the relation between π_ν and π_ω if ν is dominated by ω.

Let $\pi_\omega(\mathsf{A})'_{\mathrm{w},+}$ denote the set of positive self-adjoint operators in $\pi_\omega(\mathsf{A})'_{\mathrm{w}}$. If $x \in \pi_\omega(\mathsf{A})'_{\mathrm{w},+}$, then $\omega_x(a) := \langle x\pi_\omega(a)\, \varphi_\omega, \varphi_\omega\rangle$, $a \in \mathsf{A}$, is a positive linear functional on A, and $x \to \omega_x$ is a bijective mapping of $\pi_\omega(\mathsf{A})'_{\mathrm{w},+}$ onto the positive linear functionals on A which are dominated by ω. (This follows at once from Proposition 8.6.6; it suffices to replace x by some multiple of x.)

Suppose ν and ω are positive linear functionals on A and ν is dominated by ω. Then $\nu \lesssim \lambda\omega$ for some $\lambda > 0$. If $a \in \mathsf{A}$, we have

$$\|\pi_\nu(a)\, \varphi_\nu\|^2 = \nu(a^+a) \leqq \lambda\omega(a^+a) = \lambda\|\pi_\omega(a)\, \varphi_\omega\|^2.$$

Therefore, the equation $K_{\omega,\nu}(\pi_\omega(a)\, \varphi_\omega) := \pi_\nu(a)\, \varphi_\nu$, $a \in \mathsf{A}$, defines a bounded linear mapping of the unitary space $\mathcal{D}_\omega$ onto the unitary space $\mathcal{D}_\nu$. Let $K_{\omega,\nu}$ also denote the continuous extension of this mapping to a bounded operator of $\mathcal{H}_\omega$ into $\mathcal{H}_\nu$. The operator $K_{\omega,\nu}$ is a useful tool in order to compare π_ν and π_ω. Some simple properties of this operator are collected in

Lemma 8.6.9. (i) $K_{\omega,\nu} \in \mathbb{I}(\bar{\pi}_\omega, \bar{\pi}_\nu)$, $K_{\omega,\nu} \in \mathbb{I}(\pi_\omega, \pi_\nu)$ *and* $K_{\omega,\nu}\varphi_\omega = \varphi_\nu$.
(ii) $x := (K_{\omega,\nu})^*\, K_{\omega,\nu} \in \pi_\omega(\mathsf{A})'_{\mathrm{w}}$ *and* $\nu = \omega_x$.
(iii) *If* $\nu \lesssim \omega$, *then* $(K_{\omega,\nu})^*\, K_{\omega,\nu} + (K_{\omega,\omega-\nu})^*\, K_{\omega,\omega-\nu} = I$.
(iv) *If ω is also dominated by ν, then* $K_{\omega,\nu} = K_{\nu,\omega}^{-1}$.

Proof. (i): By definition, $K_{\omega,\nu}$ maps $\mathcal{D}(\bar{\pi}_\omega)$ onto $\mathcal{D}(\bar{\pi}_\nu)$. If $a, b \in \mathsf{A}$, then

$$\begin{aligned}K_{\omega,\nu}\bar{\pi}_\omega(a)\left(\pi_\omega(b)\, \varphi_\omega\right) &= K_{\omega,\nu}\pi_\omega(ab)\, \varphi_\omega = \pi_\nu(ab)\, \varphi_\nu\\ &= \pi_\nu(a)\, \pi_\nu(b)\, \varphi_\nu = \bar{\pi}_\nu(a)\, K_{\omega,\nu}\left(\pi_\omega(b)\, \varphi_\omega\right);\end{aligned}$$

so $K_{\omega,\nu} \in \mathbb{I}(\tilde{\pi}_\omega, \tilde{\pi}_\nu)$. Since π_ω and π_ν are the closures of $\tilde{\pi}_\omega$ and $\tilde{\pi}_\nu$, respectively, Proposition 8.2.2, (iv), yields $K_{\omega,\nu} \in \mathbb{I}(\pi_\omega, \pi_\nu)$.

(ii): From Propositions 8.2.3, (i), and 8.2.2, (iii), $x \equiv (K_{\omega,\nu})^* K_{\omega,\nu} \in \mathbb{I}(\pi_\omega, \pi_\omega^*) = \pi_\omega(\mathsf{A})'_{\mathrm{w}}$. If $a \in \mathsf{A}$, then

$$\omega_x(a) = \langle (K_{\omega,\nu})^* K_{\omega,\nu}\pi_\omega(a)\, \varphi_\omega, \varphi_\omega\rangle = \langle K_{\omega,\nu}\pi_\omega(a)\, \varphi_\omega, K_{\omega,\nu}\varphi_\omega\rangle$$
$$= \langle \pi_\nu(a)\, \varphi_\nu, \varphi_\nu\rangle = \nu(a).$$

(iii): Suppose $\nu \lesssim \omega$. Set $x := (K_{\omega,\nu})^* K_{\omega,\nu}$ and $y := (K_{\omega,\omega-\nu})^* K_{\omega,\omega-\nu}$. By (ii), $\nu = \omega_x$ and $\omega - \nu = \omega_y$. Hence $\omega = \omega_x + \omega_y$ which implies $\omega_I = \omega_{x+y}$ and $I = x + y$.

(iv) follows immediately from the definitions of $K_{\omega,\nu}$ and $K_{\nu,\omega}$. □

Corollary 8.6.10. *Let ν and ω be positive linear functionals on A. Then ν is dominated by ω if and only if there exists an operator $x \in \mathbb{I}(\pi_\omega, \pi_\nu)$ such that $x\varphi_\omega = \varphi_\nu$.*

Proof. If ν is dominated by ω, then Lemma 8.6.9, (i), shows that $x := K_{\omega,\nu}$ has the desired properties. Conversely, assume that there exists such an operator x. Then

$$\nu(a^+a) = \|\pi_\nu(a)\, \varphi_\nu\|^2 = \|\pi_\nu(a)\, x\varphi_\omega\|^2 = \|x\pi_\omega(a)\, \varphi_\omega\|^2$$
$$\leqq \|x\|^2\, \|\pi_\omega(a)\, \varphi_\omega\|^2 = \|x\|^2\, \omega(a^+a)$$

for $a \in \mathsf{A}$. Hence $\nu \lesssim \|x\|^2\, \omega$. □

Proposition 8.6.11. *Suppose that ω is a positive linear functional on A such that $\pi_\omega(\mathsf{A})'_{\mathrm{s}} = \pi_\omega(\mathsf{A})'_{\mathrm{w}}$. Then $\pi_\nu \leqq \pi_\omega$ for all positive linear functionals ν on A which are dominated by ω. More precisely, π_ν is unitarily equivalent to the closure of the subrepresentation $\pi_\omega \upharpoonright \pi_\omega(\mathsf{A})\, |K_{\omega,\nu}|\, \varphi_\omega$ of π_ω.*

Proof. Suppose that $\nu \in \mathcal{P}(\mathsf{A})^*$ is dominated by ω. By Lemma 8.6.9, (ii), $x = (K_{\omega,\nu})^* K_{\omega,\nu} \in \pi_\omega(\mathsf{A})'_{\mathrm{w}}$. Since $\pi_\omega(\mathsf{A})'_{\mathrm{s}} = \pi_\omega(\mathsf{A})'_{\mathrm{w}} =: \mathcal{N}$ by assumption, $\mathcal{N}$ is a von Neumann algebra on $\mathcal{H}_\omega$. Therefore, $|K_{\omega,\nu}| = x^{1/2} \in \mathcal{N} = \pi_\omega(\mathsf{A})'_{\mathrm{s}}$. Hence $\psi_{\omega,\nu} := |K_{\omega,\nu}|\, \varphi_\omega \in \mathcal{D}(\pi_\omega)$. Let $\varrho_{\omega,\nu}$ denote the closure of $\pi_\omega \upharpoonright \pi_\omega(\mathsf{A})\, \psi_{\omega,\nu}$. If $a \in \mathsf{A}$, then

$$\langle \varrho_{\omega,\nu}(a)\, \psi_{\omega,\nu}, \psi_{\omega,\nu}\rangle = \langle \pi_\omega(a)\, |K_{\omega,\nu}|\, \varphi_\omega, |K_{\omega,\nu}|\, \varphi_\omega\rangle$$
$$= \langle |K_{\omega,\nu}|^2\, \pi_\omega(a)\, \varphi_\omega, \varphi_\omega\rangle = \omega_x(a) = \nu(a),$$

where we used the fact that $|K_{\omega,\nu}| \in \pi_\omega(\mathsf{A})'_{\mathrm{s}}$ and Lemma 8.6.9, (ii). Since $\varrho_{\omega,\nu}$ is a closed cyclic $*$-representation of A with cyclic vector $\psi_{\omega,\nu}$, the uniqueness assertion of Theorem 8.6.4 yields $\pi_\nu \simeq \varrho_{\omega,\nu}$. Since $\varrho_{\omega,\nu} \subseteq \pi_\omega$, this gives $\pi_\nu \leqq \pi_\omega$. □

In the case where A is commutative we have the following characterization of the equality $\pi_\omega(\mathsf{A})'_{\mathrm{s}} = \pi_\omega(\mathsf{A})'_{\mathrm{w}}$.

Proposition 8.6.12. *Suppose that the $*$-algebra A is commutative. For each positive linear functional ω on A, the following three statements are equivalent:*

(i) $\pi_\omega(\mathsf{A})'_{\mathrm{s}} = \pi_\omega(\mathsf{A})'_{\mathrm{w}}$.

(ii) *If $\nu \in \mathcal{P}(\mathsf{A})^*$ and $\nu \lesssim \omega$, then $\pi_\nu \leqq \pi_\omega$.*

(iii) *If $\nu \in \mathcal{P}(\mathsf{A})^*$ and $\nu \lesssim \omega$, then there exists a vector $\varphi \in \mathcal{D}(\pi_\omega)$ such that $\nu(a) = \langle \pi_\omega(a)\varphi, \varphi\rangle$ for all $a \in \mathsf{A}$.*

Proof. Proposition 8.6.11 shows that (i) implies (ii). (ii) → (iii) is clear. We now prove that (iii) implies (i). We let x be in $\pi_\omega(\mathsf{A})'_{\mathrm{w}} \cap [0, I]$. Since $\pi_\omega(\mathsf{A})'_{\mathrm{w}}$ is the linear span of operators x of this form, our proof is complete once we have shown that x is in $\pi_\omega(\mathsf{A})'_{\mathrm{s}}$.

By Proposition 8.6.6, $\omega_x(\cdot) \equiv \langle x\pi_\omega(\cdot)\,\varphi_\omega, \varphi_\omega\rangle$ is a positive linear functional on A which satisfies $\omega_x \precsim \omega$. Thus, by (iii), there is a vector $\varphi \in \mathcal{D}(\pi_\omega)$ such that $\omega_x(a) = \langle \pi_\omega(a)\,\varphi, \varphi\rangle$ for $a \in \mathsf{A}$. Since $\|\pi_\omega(a)\,\varphi\|^2 = \omega_x(a^+a) \leqq \omega(a^+a) = \|\pi_\omega(a)\,\varphi_\omega\|^2$ for $a \in \mathsf{A}$, the equation $R(\pi_\omega(a)\,\varphi_\omega) = \pi_\omega(a)\,\varphi$, $a \in \mathsf{A}$, defines a bounded linear mapping of $\mathcal{D}_\omega$ into $\mathcal{D}(\pi_\omega)$. Let R also denote the continuous extension of this mapping to an operator of $\mathbf{B}(\mathcal{H}_\omega)$. It is straightforward (see the proof of Lemma 8.6.9, (i)) to check that $R \in \mathbf{I}(\bar{\pi}_\omega, \pi_\omega)$. By Proposition 8.2.2, (iv), $R \in \mathbf{I}(\pi_\omega, \pi_\omega) \equiv \pi_\omega(\mathsf{A})'_{\mathrm{s}}$. If $a, b \in \mathsf{A}$, then

$$\begin{aligned}\langle x\pi_\omega(a)\,\varphi_\omega, \pi_\omega(b)\,\varphi_\omega\rangle &= \langle x\pi_\omega(b^+a)\,\varphi_\omega, \varphi_\omega\rangle = \omega_x(b^+a)\\ &= \langle \pi_\omega(b^+a)\,\varphi, \varphi\rangle = \langle \pi_\omega(a)\,\varphi, \pi_\omega(b)\,\varphi\rangle\\ &= \langle R\pi_\omega(a)\,\varphi_\omega, R\pi_\omega(b)\,\varphi_\omega\rangle = \langle R^*R\pi_\omega(a)\,\varphi_\omega, \pi_\omega(b)\,\varphi_\omega\rangle .\end{aligned}$$

Since $\mathcal{D}_\omega$ is dense in $\mathcal{H}_\omega$, this gives $x = R^*R$.

The main step of the proof is to show that R^* is in $\pi_\omega(\mathsf{A})'_{\mathrm{s}}$. Since the vector π_ω is cyclic for π_ω, there exists a net $(\pi_\omega(a_i)\,\varphi_\omega : i \in I)$, where $a_i \in \mathsf{A}$ for $i \in I$, which converges to φ in the graph topology of $\pi_\omega(\mathsf{A})$. Since A is commutative, we have

$$\begin{aligned}\|\pi_\omega(a_i)\,\varphi_\omega - \pi_\omega(a_j)\varphi_\omega\|^2 &= \omega\big((a_i - a_j)^+\,(a_i - a_j)\big)\\ &= \omega\big((a_i^+ - a_j^+)^+\,(a_i^+ - a_j^+)\big)\\ &= \|\pi_\omega(a_i^+)\,\varphi_\omega - \pi_\omega(a_j^+)\,\varphi_\omega\|^2 \quad \text{for} \quad i, j \in I .\end{aligned}$$

From this we see that $(\pi_\omega(a_i^+)\,\varphi_\omega : i \in I)$ is a Cauchy net in the locally convex space $\mathcal{D}(\pi_\omega)\,[\mathrm{t}_{\pi_\omega}]$. Since π_ω is closed, the latter space is complete and hence this net has a limit $\varphi^+ \in \mathcal{D}(\pi_\omega)$. Using once more that A is commutative, we obtain for $a, b \in \mathsf{A}$

$$\begin{aligned}\langle R\pi_\omega(a)\,\varphi_\omega, \pi_\omega(b)\,\varphi_\omega\rangle &= \langle \pi_\omega(a)\,\varphi, \pi_\omega(b)\,\varphi_\omega\rangle = \lim_i\, \langle \pi_\omega(a)\,\pi_\omega(a_i)\,\varphi_\omega, \pi_\omega(b)\,\varphi_\omega\rangle\\ &= \lim_i\, \langle \pi_\omega(a)\,\varphi_\omega, \pi_\omega(b)\,\pi_\omega(a_i^+)\,\varphi_\omega\rangle = \langle \pi_\omega(a)\,\varphi_\omega, \pi_\omega(b)\,\varphi^+\rangle .\end{aligned}$$

Hence $R^*\pi_\omega(b)\,\varphi_\omega = \pi_\omega(b)\,\varphi^+$ for $b \in \mathsf{A}$. Similarly as above (with R^* and φ^+ in place of R and φ, respectively) this implies that $R^* \in \pi_\omega(\mathsf{A})'_{\mathrm{s}}$. Because $R \in \pi_\omega(\mathsf{A})'_{\mathrm{s}}$ and $\pi_\omega(\mathsf{A})'_{\mathrm{s}}$ is an algebra, we get $x = R^*R \in \pi_\omega(\mathsf{A})'_{\mathrm{s}}$. □

We turn to another application of Propositions 8.6.6 and 8.6.11.

Definition 8.6.13. If ω_1 and ω_2 are positive linear functionals on A, we say that ω_1 and ω_2 are *orthogonal* and write $\omega_1 \perp \omega_2$ if for each $v \in \mathcal{P}(\mathsf{A})^*$ the relations $v \precsim \omega_1$ and $v \precsim \omega_2$ imply that $v = 0$.

Proposition 8.6.14. *Suppose that* v *and* ω *are positive linear functionals on* A *such that* $v \precsim \omega$ *and* $\pi_\omega(\mathsf{A})'_{\mathrm{s}} = \pi_\omega(\mathsf{A})'_{\mathrm{w}}$. *Then* v *is orthogonal to* $\omega - v$ *if and only if* $|K_{\omega,v}|$ *is a projection. In this case we have* $\pi_v \oplus \pi_{\omega-v} \cong \pi_\omega$.

Proof. In this proof we freely use the notation and the facts established in the proofs of Lemma 8.6.9 and Proposition 8.6.11.

Suppose first $|K_{\omega,v}|$ is a projection. Then $x \equiv (K_{\omega,v})^*\,K_{\omega,v} \equiv |K_{\omega,v}|^2$ and $y \equiv I - x$

are both projections. Let $v \in \mathcal{P}(\mathsf{A})^*$ be such that $v \lesssim \nu$ and $v \lesssim \omega - \nu$. Proposition 8.6.6 ensures that there is an operator $z \in \pi_\omega(\mathsf{A})'_{\mathrm{w}} \cap [0, I]$ such that $v = \omega_z$. From $\omega_z = v \lesssim \nu = \omega_x$ and $\omega_z = v \lesssim \omega - \nu = \omega_y$ we conclude that $z \leqq x$ and $z \leqq y = I - x$. Since $x = x^2$ and $y = y^2$, the latter implies that $\|z^{1/2}x\varphi\|^2 = \langle zx\varphi, x\varphi\rangle \leqq \langle (I - x)\, x\varphi, x\varphi\rangle = 0$ and similarly $\|z^{1/2}y\varphi\|^2 = 0$ for $\varphi \in \mathcal{H}(\pi_\omega)$. Thus $z^{1/2}x = z^{1/2}y = 0$ which leads to $z = 0$ and $v = 0$. This proves that $\nu \perp \omega - \nu$. Next we show that $\pi_\nu \oplus \pi_{\omega-\nu} \cong \pi_\omega$. Since $\pi_\omega(\mathsf{A})'_{\mathrm{s}} = \pi_\omega(\mathsf{A})'_{\mathrm{w}}$ by assumption, the projection $x \equiv |K_{\omega,\nu}|^2$ belongs to the strong commutant $\pi_\omega(\mathsf{A})'_{\mathrm{s}}$. Therefore, we have $\pi_\omega = (\pi_\omega)_{\mathcal{K}} \oplus (\pi_\omega)_{\mathcal{K}^\perp}$ (in the notation of Section 8.3), where $\mathcal{K} := x(\mathcal{H}(\pi_\omega))$. Since φ_ω is a cyclic vector for π_ω, $x\pi_\omega(\mathsf{A})\,\varphi_\omega \equiv \pi_\omega(\mathsf{A})\, x\varphi_\omega$ is dense in $x(\mathcal{D}(\pi_\omega)) \equiv \mathcal{D}((\pi_\omega)_{\mathcal{K}})$ relative to the graph topology for $(\pi_\omega)_{\mathcal{K}}$. Hence $\psi_{\omega,\nu} = x\varphi_\omega$ is a cyclic vector for $(\pi_\omega)_{\mathcal{K}}$. This implies that $(\pi_\omega)_{\mathcal{K}} = \varrho_{\omega,\nu}$. Similarly, $\psi_{\omega,\omega-\nu} = (I - x)\,\varphi_\omega = y\varphi_\omega$ is cyclic for $(\pi_\omega)_{\mathcal{K}^\perp}$, so that $(\pi_\omega)_{\mathcal{K}^\perp} = \varrho_{\omega,\omega-\nu}$. As shown in the proof of Proposition 8.6.11 (applied to ν and to $\omega - \nu$) we have $\pi_\nu \cong \varrho_{\omega,\nu}$ and $\pi_{\omega-\nu} \cong \varrho_{\omega,\omega-\nu}$, so $\pi_\nu \oplus \pi_{\omega-\nu} \cong \varrho_{\omega,\nu} \oplus \varrho_{\omega,\omega-\nu} = (\pi_\omega)_{\mathcal{K}} \oplus (\pi_\omega)_{\mathcal{K}^\perp} = \pi_\omega$.

Now assume that $\nu \perp \omega - \nu$. Set $z := x(I - x)$. From $x \in [0, I]$ we have that $0 \leqq z \leqq x$ and $0 \leqq z \leqq y = I - x$. Since $\pi_\omega(\mathsf{A})'_{\mathrm{s}} = \pi_\omega(\mathsf{A})'_{\mathrm{w}}$, $\pi_\omega(\mathsf{A})'_{\mathrm{w}}$ is an algebra. Therefore, since $x \in \pi_\omega(\mathsf{A})'_{\mathrm{w}}$, $z \in \pi_\omega(\mathsf{A})'_{\mathrm{w}}$; so $z \in \pi_\omega(\mathsf{A})'_{\mathrm{w}} \cap [0, I]$. By Proposition 8.6.6, ω_z is a positive linear functional on A which satisfies $\omega_z \lesssim \omega_x \equiv \nu$ and $\omega_z \lesssim \omega_y \equiv \omega - \nu$. Therefore, by $\nu \perp \omega - \nu$, $\omega_z = 0$ and hence $z = x(I - x) = 0$. Therefore, $x \equiv |K_{\omega,\nu}|^2$ and so $|K_{\omega,\nu}|$ is a projection. □

Remark 5. The assumption $\pi_\omega(\mathsf{A})'_{\mathrm{s}} = \pi_\omega(\mathsf{A})'_{\mathrm{w}}$ was not used in the proof of the if part of Proposition 8.6.14. Further, some simple operator-theoretic arguments show that $|K_{\omega,\nu}|$ is a projection if and only if $K_{\omega,\nu}$ is a partial isometry of $\mathcal{H}_\omega$ into $\mathcal{H}_\nu$.

Remark 6. A slight reformulation of the previous proposition is as follows. Suppose ω is a positive linear functional on A such that $\pi_\omega(\mathsf{A})'_{\mathrm{s}} = \pi_\omega(\mathsf{A})'_{\mathrm{w}}$. If $x \in \pi_\omega(\mathsf{A})'_{\mathrm{w}} \cap [0, I]$, then $\omega_x \perp \omega - \omega_x$ if and only if x is a projection. In order to prove this, we set $\tilde{x} := (K_{\omega,\omega_x})^* K_{\omega,\omega_x}$. By Lemma 8.6.9, (ii), $\omega_x = \omega_{\tilde{x}}$; hence $x = \tilde{x} \equiv |K_{\omega,\omega_x}|^2$ and the assertion follows from Proposition 8.6.14.

We close this section with the following example.

Example 8.6.15. Let A be the $*$-algebra $\mathbb{A}(\mathsf{p}_1, \mathsf{q}_1)$ of Example 2.5.2 and let π be the $*$-representation of A defined there. Recall that $p_1 \equiv \pi(\mathsf{p}_1)$ is the differential operator $-\mathrm{i}\,\dfrac{\mathrm{d}}{dt}$ and $q_1 \equiv \pi(\mathsf{q}_1)$ is the multiplication operator by the independent variable t on the domain $\mathcal{D}(\pi) = \mathcal{S}(\mathbb{R})$ in the Hilbert space $L^2(\mathbb{R})$. Set $\varphi_0(t) := \exp(-t^2/2)$, $t \in \mathbb{R}$, and $\omega_0(a) := \langle \pi(a)\, \varphi_0, \varphi_0\rangle$, $a \in \mathsf{A}$. It is obvious that $\pi(\mathsf{A})\, \varphi_0$ is equal to the linear span of the Hermite functions. Since the Hermite functions form a basis of the space $\mathcal{S}(\mathbb{R})$ in its "usual" topology (Reed/Simon [1], Theorem V.13) and this topology coincides with the graph topology t_π (cf. Example 2.5.2), we conclude that φ_0 is a cyclic vector for π. Therefore, by the uniqueness part of Theorem 8.6.4 (cf. Remark 1), π_{ω_0} is unitarily equivalent to π. Because $\pi(\mathsf{A})'_{\mathrm{w}} = \mathbb{C} \cdot I$ as noted in Example 8.3.7, the latter implies that ω_0 is pure by Corollary 8.6.7.

Now suppose φ is a fixed function from $C^\infty(\mathbb{R})$ such that $\operatorname{supp} \varphi \subseteqq [0, 2]$ and $\varphi(t) \neq 0$ for all $t \in (0, 1) \cup (1, 2)$. Define $\omega(a) := \langle \pi(a)\, \varphi, \varphi\rangle$, $a \in \mathsf{A}$. Then π_ω is unitarily equivalent to the closure of the $*$-representation $\pi \restriction \pi(\mathsf{A})\, \varphi$ of A. For notational simplicity we shall identify π_ω with the latter throughout the following discussion. Then $\mathcal{H}(\pi_\omega)$ is the Hilbert space $L^2(0, 2)$ considered as a subspace of $L^2(\mathbb{R})$ in the obvious way. Let e be the multiplication operator on $L^2(0, 2)$ by the characteristic function of the interval $(0, 1)$.

Case 1: $\varphi^{(k)}(1) = 0$ for all $k \in \mathbb{N}$.

In this case we have $\pi_\omega(\mathsf{A})'_{\mathrm{w}} = \{\alpha \cdot e + \lambda \cdot I : \alpha, \lambda \in \mathbb{C}\}$. Indeed, suppose $x \in \pi_\omega(\mathsf{A})'_{\mathrm{w}}$. Then x commutes with $\pi_\omega(q_1)$ and hence with $\overline{\pi_\omega(q_1)}$ which is the multiplication operator by t on $L^2(0, 2)$. Therefore, x is the multiplication operator by some L^∞-function η on $(0, 2)$. Further, let a denote the symmetric operator $-\mathrm{i}\frac{\mathrm{d}}{\mathrm{d}t}$ with domain $\mathcal{D}(a)$ $:= C_0^\infty(0, 1) + C_0^\infty(1, 2)$ in the Hilbert space $L^2(0, 2)$. We check that $\mathcal{D}(\pi_\omega)$ is a core for $\bar{a}$. We let $\psi \in \mathcal{D}(a)$. Then $\psi\varphi_0^{-1} \in \mathcal{D}(a)$, and we can find a sequence of polynomials $(r_n : n \in \mathbb{N})$ in t such that $(\psi\varphi_0^{-1})^{(l)}(t) = \lim_n r_n^{(l)}(t)$ uniformly on $(0, 2)$ for $l = 0, 1$. Then $\psi - r_n\varphi_0$ $\equiv (\psi\varphi_0^{-1} - r_n)\,\varphi_0 \to 0$ and $a(\psi - r_n\varphi_0) \equiv -\mathrm{i}(\psi\varphi_0^{-1} - r_n)'\,\varphi_0 - \mathrm{i}(\psi\varphi_0^{-1} - r_n)\,\varphi_0' \to 0$ in $L^2(0, 2)$ as $n \to \infty$. Since $r_n\varphi_0 \in \pi(\mathsf{A})\,\varphi_0 \subseteq \mathcal{D}(\pi_\omega)$ for $n \in \mathbb{N}$, this shows that $\mathcal{D}(\pi_\omega)$ is a core for $\bar{a}$. Therefore, $\big(\pi_\omega(p_1)\big)'_{\mathrm{w}} \subseteq (a)'_{\mathrm{w}}$, so that $x \in (a)'_{\mathrm{w}}$, i.e., $xa \subseteq a^*x$. Since a^* acts as $-\mathrm{i}\frac{\mathrm{d}}{\mathrm{d}t}$ in the distribution sense on $(0, 1) \cap (1, 2)$, the latter implies that the function η is constant on $(0, 1)$ and on $(1, 2)$ and so of the form $\alpha \cdot e + \lambda \cdot I$ for some $\alpha, \lambda \in \mathbb{C}$. Conversely, since $\varphi^{(k)}(1) = 0$, e is in $\big(\pi_\omega(p_1)\big)'_{\mathrm{w}} \cap \big(\pi_\omega(q_1)\big)'_{\mathrm{w}}$. Hence $e \in \pi_\omega(\mathsf{A})'_{\mathrm{w}}$ by Corollary 8.2.8, and the above description of $\pi_\omega(\mathsf{A})'_{\mathrm{w}}$ is proved.

In particular, ω is not pure, since $\pi_\omega(\mathsf{A})'_{\mathrm{w}} \neq \mathbb{C} \cdot I$. Because e is a projection, we have $\omega_e \perp \omega - \omega_e$ by Remark 6.

Case 2: $\varphi^{(k)}(1) \neq 0$ for some $k \in \mathbb{N}$.

A similar reasoning as in case 1 shows that $\pi_\omega(\mathsf{A})'_{\mathrm{w}} = \mathbb{C} \cdot I$, so ω is pure in this case. ○

Notes

*-Representations of *-algebras by unbounded operators first appeared in representation theory of Lie algebras and in quantum field theory. Some history in the former case is discussed in the notes after Chapter 10. The pioneering papers in the latter case are Borchers [1] and Uhlmann [1]. After an algebraic reformulation of the Wightman axioms had been given by these papers tensor algebras and their representations have gained some interest. They were studied by Lassner/Uhlmann [1], Wyss [1] and Borchers [2] and later in many other papers.

A systematic investigation of (unbounded) *-representations of general *-algebras was initiated independently and almost simultaneously by Vasiliev [1], [2], Powers [1], [2] and Uhlmann [2]. A major step towards to a general theory were the two papers Powers [1], [2] which contain both new concepts (i.e., standard representations and completely strongly positive maps) and important non-trivial results; cf. Sections 9.1 and 9.2 and Chapter 11.

8.1. Most of the basic notions and properties of *-representations discussed here are from the pioneering papers Vasiliev [1], [2], Powers [1] and Uhlmann [2] and from Gudder/Scruggs [1]. Lemma 8.1.6 and its subsequent applications (e.g., Proposition 8.1.12, (v)) are (in a special case) from Schmüdgen [13]. The assertion (v) of Proposition 8.1.15 was obtained by Borisov/Reichert [1], and the assertions (iii) and (iv) of this proposition are in Richter [1]. Proposition 8.1.17 appears to be new.

8.2. Propositions 8.2.2 and 8.2.3 are from Richter [1].

8.3. The most useful result in this section is Proposition 8.3.11 which was discovered by Powers [1]. It should be noted that (in contrast to our Definition 8.3.4) Powers and other authors define the irreducibility of a $*$-representation by the requirement that the weak (bounded) commutant is trivial; cf. Remark 3 in 8.3.

8.4. Proposition 8.4.3 was proved (independently) by OTA [1] and VORONIN/SUSHKO/HORUZHY [1]. The rest of Section 8.4 (and also parts of Sections 8.2 and 8.3) follow the paper SCHMÜDGEN [12].

8.5. Induced extensions (with another definition!) were studied by BORCHERS/YNGAVSON [1] in their approach to the decomposition theory. Corollary 8.5.6 is due to SCHMÜDGEN [21]. The special case stated in Corollary 8.5.7 was obtained independently by INOUE/UEDA/YAMAUCHI [1]. Propositions 8.5.4 and 8.5.5 and the examples appear here for the first time.

8.6. The GNS construction for normed $*$-algebras and bounded $*$-representations is known since the fourties by the work of GELFAND, NEUMARK and SEGAL. It has been adapted for tensor algebras by BORCHERS [1] and UHLMANN [1]. For general $*$-algebras this construction and also Proposition 8.6.6 appeared (again independently and almost simultaneously) in POWERS [1], VASILIEV [2] and UHLMANN [2].

The embedding map $K_{\omega,\nu}$ occurs in INOUE [8] and in TODOROV [1]. A result like Proposition 8.6.11 is in TODOROV [1]. Proposition 8.6.12 seems to be new. Proposition 8.6.14 is the unbounded version of a known result for C^*-algebras.

Additional References:

DIXON [1], INOUE [7], [9], INOUE/TAKESUE [1], JORGENSEN [1], [3], LASSNER [2], SCHMÜDGEN [4].
8.2. VORONIN/SUSHKO/HORUZHY [1].
8.3. BHATT [2].
8.4. SCHMÜDGEN [15].
8.6. GUDDER [1], GUDDER/HUDSON [1], INOUE [8], [10], TAKESUE [1], VORONIN/SUSHKO/HORUZHY [1].

9. Self-Adjoint Representations of Commutative $*$-Algebras

The results obtained in Section 8.4 have shown that a part of the representation theory of C^*-algebras can be generalized to unbounded $*$-representations if, roughly speaking, the self-adjointness of certain $*$-representations is assumed. Thus self-adjoint representations are basic objects in the theory of $*$-representations of general $*$-algebras. In this chapter we are concerned with self-adjoint representations of *commutative* $*$-algebras.

In Section 9.1 we investigate a class of well-behaved self-adjoint representations of a commutative $*$-algebra A which we call *integrable* (or standard) representations. By one of several characterizations, they are precisely those self-adjoint representations π of A for which the von Neumann algebra $\pi(\mathsf{A})''$ is abelian. In Section 9.2 we investigate cyclic integrable representations and we show that an integrable representation with metrizable graph topology can be decomposed as a direct sum of cyclic representations.

The remaining two sections in this chapter are devoted to the construction of non-integrable self-adjoint representations of the polynomial algebra $\mathbb{C}[x_1, x_2]$. In Section 9.3 we study two classes of pairs of self-adjoint operators which give rise to (certain) self-adjoint representations of $\mathbb{C}[x_1, x_2]$. They are used in Section 9.4 to construct non-integrable self-adjoint representations of $\mathbb{C}[x_1, x_2]$ which have some additional properties. To mention the most striking result, we prove that for each properly infinite von Neumann algebra $\mathcal{N}$ in a separable Hilbert space there exists a self-adjoint (of course, non-integrable) representation π of $\mathbb{C}[x_1, x_2]$ such that $\pi(\mathbb{C}[x_1, x_2])'' = \mathcal{N}$.

9.1. Integrable Representations of Commutative $*$-Algebras

Throughout this section, A will denote a commutative $*$-algebra with unit.

Definition 9.1.1. A representation π of A is called *integrable* (or *standard*) if π is closed and $\overline{\pi(a^+)} = \pi(a)^*$ for all $a \in \mathsf{A}$.

Remark 1. We shall prefer the word "integrable" rather than "standard". The reason for the name "integrable" stems from the terminology used in representation theory of enveloping algebras, cf. Section 10.1. It is motivated and justified to some extent by the following fact. Let us identify the $*$-algebra $\mathsf{A} := \mathbb{C}[x_1, \ldots, x_n]$, $n \in \mathbb{N}$, with the enveloping algebra of the complexified Lie algebra of the Lie group $\mathbb{R}^n$ in the usual way. Then a representation of A is integrable in the sense of Definition 9.1.1 (applied to the commutative $*$-algebra A) if and only if it is integrable according to Definition 10.1.7 (applied to the enveloping algebra A). Since in both cases integrability implies self-adjointness (by Remark 2 or by Corollary 10.2.3), this assertion follows if we compare Theorem 9.1.2, (i) $\leftrightarrow$ (iv), with Corollary 10.2.10 and Theorem 10.5.8, (iii)″ $\to$ (i).

Remark 2. Each integrable representation π of A is self-adjoint. Indeed, the second condition in Definition 9.1.1 shows that π is a *-representation. Therefore, by Proposition 2.2.12, $\mathcal{D}(\pi) = \mathcal{D}(\hat{\pi}) = \bigcap_{a\in A} \mathcal{D}\big(\overline{\pi(a)}\big) = \bigcap_{a\in A} \mathcal{D}\big(\pi(a^+)^*\big) = \mathcal{D}(\pi^*)$, so π is self-adjoint.

Our main objective in this section is to characterize those *-representations π of A for which $\hat{\pi}$ or π^* is integrable. If we assume in addition that π is closed or self-adjoint, then these results give us criteria for the integrability of π itself.

Theorem 9.1.2. *For every *-representation π of* A*, the following statements are equivalent:*

(i) *$\hat{\pi}$ is integrable.*

(ii) *$\overline{\pi(a)}$ is a normal operator for each $a \in$ A.*

(iii) *$\overline{\pi(a_1)}$ and $\overline{\pi(a_2)}$ are strongly commuting self-adjoint operators for arbitrary a_1 and a_2 in* A_h.

(iv) *$\overline{\pi(a)}$ is a self-adjoint operator for each $a \in A_h$.*

Proof. There is no loss of generality to assume that π is closed, since $\overline{\pi(a)} = \overline{\hat{\pi}(a)}$, $a \in$ A, for any representation π. Let $a \in$ A. We write $a = a_1 + ia_2$ with $a_1, a_2 \in A_h$, and we apply Proposition 7.1.3 to the operators $\pi(a)$, $\pi(a_1)$ and $\pi(a_2)$ in $\mathcal{L}^+(\mathcal{D}(\pi))$. Proposition 7.1.3, (i), gives (i) $\leftrightarrow$ (ii) $\to$ (iii), and Proposition 7.1.3, (ii), shows that (iv) $\to$ (i). (iii) $\to$ (iv) is trivial. □

Corollary 9.1.3. *A *-representation π of* A *is integrable if and only if the O^*-algebra π(A) is strictly self-adjoint (in the sense of Definition* 7.3.5).

Proof. If π is integrable, then, by Theorem 9.1.2, each operator $\overline{\pi(a)}$, $a \in A_h$, is self-adjoint, so $\pi(A_h)$ can be taken for the set $\{a_i : i \in I\}$ in Definition 7.3.5. Conversely, suppose π(A) is strictly self-adjoint, and let $a \in A_h$. Since obviously $\pi(a) \in \pi(A)^c$ (recall that A is commutative), we conclude from Theorem 7.3.6., (ii), that $\overline{\pi(a)}$ is self-adjoint. (This can be also derived from Proposition 7.1.6 (or from Lemma 7.1.5) which was built into the proof of Theorem 7.3.6.) Hence π is integrable by Theorem 9.1.2. □

Corollary 9.1.4. *If* A *is a symmetric *-algebra, then each closed *-representation of* A *is integrable.*

Proof. By Corollary 8.1.20, condition (iv) in Theorem 9.1.2 is valid. □

Corollary 9.1.5. *If π is an integrable representation of* A, *then $\big(\overline{\pi(a)} - \alpha\big)^{-1} \in \pi(A)'$ for any a in A_h and α in the resolvent set of $\overline{\pi(a)}$.*

Proof. Suppose $b \in A_h$. By Theorem 9.1.2, $\overline{\pi(a)}$ and $\overline{\pi(b)}$ are strongly commuting self-adjoint operators, so $A_\alpha := \big(\overline{\pi(a)} - \alpha\big)^{-1}$ commutes with $\overline{\pi(b)}$ by Lemma 1.6.2. From this it follows that A_α maps $\mathcal{D}(\pi)$ into $\bigcap_{b\in A_h} \mathcal{D}\big(\overline{\pi(b)}\big) = \mathcal{D}(\hat{\pi}) = \mathcal{D}(\pi)$ and $A_\alpha \in \pi(A)'_s$. □

Our next theorem contains some characterizations of integrable representations in terms of commutants. First, however, we prove an auxiliary lemma.

Lemma 9.1.6. *Suppose* B *is a subset of* A_h *such that* $\mathsf{B} \cup \{1\}$ *generates the* $*$*-algebra* A. *Suppose that* π *is a* $*$*-representation of* A *such that* $\overline{\pi(b_1)}$ *and* $\overline{\pi(b_2)}$ *are strongly commuting self-adjoint operators for all* $b_1, b_2 \in \mathsf{B}$. *Then* $\pi(\mathsf{A})'_w$ *is a von Neumann algebra with abelian commutant.*

Proof. Let $e(\lambda; b)$, $\lambda \in \mathbb{R}$, be the spectral projections of the self-adjoint operator $\overline{\pi(b)}$, $b \in \mathsf{B}$. Since B is a subset of A_h, A is also generated, as an algebra, by $\mathsf{B} \cup \{1\}$. Therefore, by Corollary 8.2.8, $\pi(\mathsf{A})'_w = \bigcap_{b \in \mathsf{B}} (\pi(b))'_w$. Because $\overline{\pi(b)}$ is a self-adjoint operator, $(\pi(b))'_w = \{e(\lambda; b) : \lambda \in \mathbb{R}\}'$. Thus $\pi(\mathsf{A})'_w = \{e(\lambda; b) : \lambda \in \mathbb{R} \text{ and } b \in \mathsf{B}\}'$, and this set is, of course, a von Neumann algebra. Since $\overline{\pi(b_1)}$ and $\overline{\pi(b_2)}$ strongly commute, $e(\lambda_1; b_1)$ and $e(\lambda_2; b_2)$ commute for all $\lambda_1, \lambda_2 \in \mathbb{R}$ and $b_1, b_2 \in \mathsf{B}$, so that $\pi(\mathsf{A})''_w$ is commutative. □

Theorem 9.1.7. *For any* $*$*-representation* π *of* A, *the following six statements are equivalent:*

(i) $\hat{\pi}$ *is integrable.*

(ii) $\hat{\pi}$ *is self-adjoint, and the von Neumann algebra* $\pi(\mathsf{A})''_w$ *is abelian.*

(iii) $\pi(\mathsf{A})'_w = \hat{\pi}(\mathsf{A})'_s$, *and the von Neumann algebra* $\pi(\mathsf{A})''_w$ *is abelian.*

(iv) *The von Neumann algebra* $(\pi(\mathsf{A})'_{ss})'$ *is abelian.*

(v) *There is an abelian von Neumann algebra* $\mathcal{N}$ *such that* $\overline{\pi(a)}$ *is affiliated with* $\mathcal{N}$ *for all* $a \in \mathsf{A}$.

(vi) *There is an abelian von Neumann algebra* $\mathcal{N}$ *such that* $\overline{\pi(a)}$ *is affiliated with* $\mathcal{N}$ *for all* $a \in \mathsf{A}_h$.

Proof. First note that $\pi(\mathsf{A})''_w$ is always a von Neumann algebra, since $\pi(\mathsf{A})'_w$ is $*$-invariant.

(i) $\to$ (ii): From Theorem 9.1.2, (i) $\to$ (iii), $\hat{\pi}$ satisfies the assumptions of Lemma 9.1.6 with $\mathsf{B} := \mathsf{A}_h$. Hence $\hat{\pi}(\mathsf{A})''_w \equiv \pi(\mathsf{A})''_w$ is abelian. By Remark 2, $\hat{\pi}$ is self-adjoint.

(ii) $\to$ (iii): Since $\hat{\pi}$ is self-adjoint, $\pi(\mathsf{A})'_w = \hat{\pi}(\mathsf{A})'_w = \hat{\pi}(\mathsf{A})'_s$.

(iii) $\to$ (iv) follows from Proposition 7.2.10, (iv), combined with $\pi(\mathsf{A})'_w = \hat{\pi}(\mathsf{A})'_w$ and $\pi(\mathsf{A})'_{ss} = \hat{\pi}(\mathsf{A})'_{ss}$.

(iv) $\to$ (v): By Proposition 7.2.9, (i), it suffices to set $\mathcal{N} := (\pi(\mathsf{A})'_{ss})'$.

(v) $\to$ (vi) is trivial.

(vi) $\to$ (i) follows from Theorem 9.1.2, (iv) $\to$ (i), and Lemma 1.6.3, (i). □

The next two corollaries are nothing but special cases of the preceding theorem.

Corollary 9.1.8. *A closed* $*$*-representation* π *of* A *is integrable if and only if the von Neumann algebra* $(\pi(\mathsf{A})'_{ss})'$ *is abelian.*

Corollary 9.1.9. *A self-adjoint representation* π *of* A *is integrable if and only if the von Neumann algebra* $\pi(\mathsf{A})''_w$ *is abelian.*

Corollary 9.1.10. *A closed* $*$*-representation* π *of* A *is integrable if and only if the O*-algebra* $\pi(\mathsf{A})$ *is commutatively dominated (in the sense of Definition 2.2.14).*

Proof. Suppose $\pi(\mathsf{A})$ is commutatively dominated. Since π is closed, Remark 3 in 7.3 shows that the O*-algebra $\pi(\mathsf{A})$ is then strictly self-adjoint, so π is integrable by Corollary 9.1.3. The opposite inclusion follows at once from Theorem 9.1.7, (i) $\to$ (v). □

Corollary 9.1.11. *If π is an irreducible integrable representation of* A *on a Hilbert space $\mathcal{H}(\pi) \neq \{0\}$, then $\mathcal{H}(\pi)$ is one-dimensional.*

Proof. Since π is integrable, $\pi(\mathsf{A})''$ is abelian by Theorem 9.1.7. Because π is irreducible and self-adjoint, $\pi(\mathsf{A})' = \mathbb{C} \cdot I$ by Lemma 8.3.5, (i) $\rightarrow$ (iv), and so $\pi(\mathsf{A})'' = \mathbb{B}(\mathcal{H}(\pi))$. Hence dim $\mathcal{H}(\pi) = 1$. □

Remark 3. Let π be an integrable representation of A. By Theorem 9.1.7, there is an abelian von Neumann algebra $\mathcal{N}$ such that each operator $\overline{\pi(a)}$, $a \in \mathsf{A}$, is affiliated with $\mathcal{N}$. Keep $\mathcal{N}$ fixed throughout this remark. Recall from Lemma 1.6.3 that the family $\mathbb{A}(\mathcal{N})$ of operators affiliated with $\mathcal{N}$ forms a commutative *-algebra with unit element I under the operations $x \mathbin{\hat{+}} y := \overline{x + y}$ for addition, $x \mathbin{\hat{\cdot}} y := \overline{xy}$ for multiplication, and $x \rightarrow x^*$ for involution.

*Then the map θ defined by $\theta(a) := \overline{\pi(a)}$, $a \in \mathsf{A}$, is a *-homomorphism of the *-algebra* A *into the *-algebra $\mathbb{A}(\mathcal{N})$.*

Proof. Let $a, b \in \mathsf{A}$. From

$$\begin{aligned}\theta(a) + \theta(b) &= \overline{\pi(a)} + \overline{\pi(b)} = \pi(a^+)^* + \pi(b^+)^* \subseteq (\pi(a^+) + \pi(b^+))^* = \pi((a+b)^+)^* \\ &= \overline{\pi(a+b)} = \theta(a+b)\end{aligned}$$

and

$$\theta(a)\,\theta(b) = \overline{\pi(a)}\,\overline{\pi(b)} = \pi(a^+)^*\,\pi(b^+)^* \subseteq (\pi(b^+)\,\pi(a^+))^* = \pi((ab)^+)^* = \overline{\pi(ab)} = \theta(ab)$$

it follows that

$$\theta(a) \mathbin{\hat{+}} \theta(b) \subseteq \theta(a+b) \quad \text{and} \quad \theta(a) \mathbin{\hat{\cdot}} \theta(b) \subseteq \theta(ab).$$

Since the reversed inclusions are obviously true, we have

$$\theta(a+b) = \theta(a) \mathbin{\hat{+}} \theta(b) \quad \text{and} \quad \theta(ab) = \theta(a) \mathbin{\hat{\cdot}} \theta(b).$$

Further, $\theta(a^+) = \overline{\pi(a^+)} = \pi(a)^* = \theta(a)^*$. Of course, $\theta(\lambda a) = \lambda\theta(a)$ for $\lambda \in \mathbb{C}$. □

This *-homomorphism θ could be a useful tool for a detailed study of the integrable representation π, because the *-algebra $\mathbb{A}(\mathcal{N})$ has many nice properties (see Kadison/Ringrose [1], Section 5.6, or Kadison [1]). For instance, $\mathbb{A}(\mathcal{N})$ is *-isomorphic to a *-algebra of functions (in general, not bounded and not every-where defined) on an extremely disconnected compact Hausdorff space, the spectrum of $\mathcal{N}$.

Proposition 9.1.12. *Let π be a *-representation of* A. *There exists an integrable extension π_1 of π acting in the same Hilbert space as π if and only if there is a *-algebra (or equivalently, a von Neumann algebra) $\mathcal{M}$ contained in $\pi(\mathsf{A})'_{\mathrm{w}}$ with abelian commutant $\mathcal{M}'$. If this is true, then the *-representation $\pi_{\mathcal{M}}$ (as defined by Proposition* 8.5.4) *can be taken for π_1.*

Proof. First suppose that there exists an integrable extension π_1 of π on $\mathcal{H}(\pi_1) = \mathcal{H}(\pi)$. Set $\mathcal{M} := \pi_1(\mathsf{A})'$. Since π_1 is self-adjoint, $\mathcal{M}$ is a von Neumann algebra. From Theorem 9.1.7, (i) $\rightarrow$ (ii), $\mathcal{M}'$ is abelian. Moreover, $\mathcal{M} = \mathbb{I}(\pi_1, \pi_1) \subseteq \mathbb{I}(\pi, \pi^*) = \pi(\mathsf{A})'_{\mathrm{w}}$.

Conversely, assume that there is a *-subalgebra $\mathcal{M}$ of $\mathbb{B}(\mathcal{H}(\pi))$ contained in $\pi(\mathsf{A})'_{\mathrm{w}}$ for which $\mathcal{M}'$ is abelian. There is no loss of generality to assume that $\mathcal{M}$ contains the identity map I. By Corollary 8.5.6, $\pi_{\mathcal{M}}$ is a closed *-representation of A, and we have that $\pi \subseteq \pi_{\mathcal{M}}$ and $\mathcal{M} \subseteq \pi_{\mathcal{M}}(\mathsf{A})'_{\mathrm{s}}$. Since $\mathcal{M}$ is *-invariant, $\mathcal{M} \subseteq \pi_{\mathcal{M}}(\mathsf{A})'_{\mathrm{s}} \cap (\pi_{\mathcal{M}}(\mathsf{A})'_{\mathrm{s}})^* = \pi_{\mathcal{M}}(\mathsf{A})'_{\mathrm{ss}}$, so $(\pi_{\mathcal{M}}(\mathsf{A})'_{\mathrm{ss}})' \subseteq \mathcal{M}'$ and $(\pi_{\mathcal{M}}(\mathsf{A})'_{\mathrm{ss}})'$ is abelian. Because $\pi_{\mathcal{M}}$ is closed, we conclude from Theorem 9.1.7, (iv) $\rightarrow$ (i), that $\pi_{\mathcal{M}}$ is integrable. □

Remark 4. Proposition 9.1.12 gives a necessary and sufficient condition for a $*$-representation to have an integrable extension acting in the same Hilbert space. Though being certainly of theoretical importance, this condition is not very explicit. It seems to be of some interest to have more useful necessary and/or sufficient criteria (in case $\mathsf{A} = \mathbb{C}[\mathsf{x}_1, \mathsf{x}_2]$, for instance, in terms of the Cayley transforms of the closed symmetric operators $\overline{\pi(x_1)}$ and $\overline{\pi(x_2)}$).

In order to verify that a concrete $*$-representation is integrable, it is often better to have conditions in terms of generating subsets of the $*$-algebra A.

Theorem 9.1.13. *Let B be a subset of A_h such that $\mathsf{B} \cup \{1\}$ generates the $*$-algebra A. Suppose π is a $*$-representation of A such that $\overline{\pi(b_1)}$ and $\overline{\pi(b_2)}$ are strongly commuting self-adjoint operators for arbitrary b_1 and b_2 in B. Then π^* is an integrable representation of A, and $\mathcal{D}(\pi^*) = \bigcap_{b \in \mathsf{B}} \bigcap_{n \in \mathbb{N}} \mathcal{D}\big((\overline{\pi(b)})^n\big)$. Further, we have that $\pi^* = \pi_{\mathcal{M}}$ with $\mathcal{M} := \pi(\mathsf{A})'_{\mathrm{w}}$ and $\pi^*(\mathsf{A})' = \bigcap_{b \in \mathsf{B}} \big(\overline{\pi(b)}\big)'$.*

Proof. From Lemma 9.1.6, $\mathcal{M} := \pi(\mathsf{A})'_{\mathrm{w}}$ is a von Neumann algebra with abelian commutant. Therefore, by Proposition 9.1.12, $\pi_{\mathcal{M}}$ is integrable. We show that $\pi^* = \pi_{\mathcal{M}}$. Because $\mathsf{B} \subseteq \mathsf{A}_h$, A is generated, as an algebra, by $\mathsf{B} \cup \{1\}$. By assumption, $\overline{\pi(b)}$ is self-adjoint for every $b \in \mathsf{B}$. From these two facts and Proposition 8.1.12, (v), we conclude that π^* is self-adjoint. Since $\pi \subseteq \pi_{\mathcal{M}}$, we have $(\pi_{\mathcal{M}})^* \subseteq \pi^*$. Since $\pi_{\mathcal{M}}$ is integrable and hence self-adjoint, the preceding implies that $\pi^* = \pi_{\mathcal{M}}$ by Proposition 8.1.12, (iv). For $b \in \mathsf{B}$, the symmetric operator $\overline{\pi^*(b)}$ is an extension of the self-adjoint operator $\overline{\pi(b)}$. Hence $\overline{\pi^*(b)} = \overline{\pi(b)}$ and $\big(\overline{\pi^*(b)}\big)'_{\mathrm{w}} = \big(\overline{\pi(b)}\big)'$ for $b \in \mathsf{B}$, so that the above description of $\pi^*(\mathsf{A})'$ follows immediately from Corollary 8.2.8 applied to the self-adjoint representation π^*.

It remains to prove the formula for $\mathcal{D}(\pi^*)$. First note that

$$\mathcal{D}(\pi^*) = \bigcap_{m \in \mathbb{N}} \bigcap_{b_1, \dots, b_m \in \mathsf{B}} \mathcal{D}\big(\pi(b_1)^* \dots \pi(b_m)^*\big)$$

by Lemma 8.1.6. Fix $m \in \mathbb{N}$ and $b_1, \dots, b_m \in \mathsf{B}$. By assumption, $\overline{\pi(b_k)}$ and $\overline{\pi(b_l)}$ are strongly commuting self-adjoint operators for $k, l = 1, \dots, m$. Hence the self-adjoint operators $\overline{\pi(b_1)}, \dots, \overline{\pi(b_m)}$ have a common spectral resolution. From the corresponding functional calculus we conclude that

$$\mathcal{D}\big(\pi(b_1)^* \dots \pi(b_m)^*\big) \equiv \mathcal{D}\big(\overline{\pi(b_1)} \dots \overline{\pi(b_m)}\big) \supseteq \mathcal{D}\Big(\big(\overline{\pi(b_1)}\big)^n\Big) \cap \dots \cap \mathcal{D}\Big(\big(\overline{\pi(b_m)}\big)^n\Big)$$

provided that $n \in \mathbb{N}$ is sufficiently large. Thus

$$\mathcal{D}(\pi^*) \supseteq \bigcap_{b \in \mathsf{B}} \bigcap_{n \in \mathbb{N}} \mathcal{D}\Big(\big(\overline{\pi(b)}\big)^n\Big).$$

The opposite inclusion follows immediately from $\mathcal{D}(\pi^*) \subseteq \mathcal{D}\big((\pi(b)^*)^n\big) = \mathcal{D}\Big(\big(\overline{\pi(b)}\big)^n\Big)$, $b \in \mathsf{B}$ and $n \in \mathbb{N}$. $\square$

Remark 5. Let $\mathsf{B} \subseteq \mathsf{A}_h$ be such that $\mathsf{B} \cup \{1\}$ generates the $*$-algebra A. Suppose π is a $*$-representation of A such that $\overline{\pi(b_1 - ib_2)} = \pi(b_1 + ib_2)^*$ (or equivalently, by Proposition 7.1.3, $\overline{\pi(b_1 - ib_2)}$ is a normal operator) for all $b_1, b_2 \in \mathsf{B}$. Then π^* is integrable. Indeed, from Proposition 7.1.3, (i), $\overline{\pi(b_1)}$ and $\overline{\pi(b_2)}$ are strongly commuting self-adjoint operators for $b_1, b_2 \in \mathsf{B}$, and so Theorem 9.1.13 applies.

Corollary 9.1.14. *Let* B *and* A *be as in Theorem 9.1.13. For any self-adjoint representation* π *of* A, *the following three conditions are equivalent:*

(i) π *is integrable.*

(ii) $\overline{\pi(b_1 - ib_2)} = \pi(b_1 + ib_2)^*$ *for all* $b_1, b_2 \in$ B.

(iii) $\overline{\pi(b_1)}$ *and* $\overline{\pi(b_2)}$ *are strongly commuting self-adjoint operators for all* $b_1, b_2 \in$ B.

Proof. (i) → (ii) is clear by the definition of integrability. Proposition 7.1.3, (i), shows that (ii) → (iii). (iii) → (i) follows from Theorem 9.1.13 if we take into account that $\pi = \pi^*$ by assumption. □

Remark 6. We state some of the previous results separately in the case where $\mathsf{A} = \mathbb{C}[\mathsf{x}_1, \dots, \mathsf{x}_n]$ and $\mathsf{B} = \{\mathsf{x}_1, \dots, \mathsf{x}_n\}$, $n \in \mathbb{N}$. Let π be a *-representation of A. First suppose $n = 1$. Then π is integrable if and only if π is self-adjoint. From Theorem 9.1.13 (or from Proposition 8.1.15) this is the case if and only if the operator $\overline{\pi(x_1)}$ is self-adjoint and $\mathcal{D}(\pi) = \mathcal{D}^\infty\big(\overline{\pi(\mathsf{x}_1)}\big)$. Now let $n \in \mathbb{N}$ be arbitrary. Then π^* is integrable if $\overline{\pi(\mathsf{x}_k - \mathrm{i}\mathsf{x}_l)} = \pi(\mathsf{x}_k + \mathrm{i}\mathsf{x}_l)^*$ (or equivalently, if the operator $\overline{\pi(\mathsf{x}_k - \mathrm{i}\mathsf{x}_l)}$ is normal) for all $k, l = 1, \dots, n$. (If $n \geqq 2$, it suffices to assume this for all $k, l = 1, \dots, n$, $k \neq l$.) If the representation π is self-adjoint or if $\mathcal{D}(\pi) = \bigcap_{k=1}^{n} \mathcal{D}^\infty\big(\overline{\pi(\mathsf{x}_k)}\big)$, then π is integrable if and only if $\overline{\pi(\mathsf{x}_k - \mathrm{i}\mathsf{x}_l)} = \pi(\mathsf{x}_k + \mathrm{i}\mathsf{x}_l)^*$ for all $k, l = 1, \dots, n$.

Remark 7. By a (slight) reformulation of Definition 9.1.1, a *-representation π of A is integrable if and only if the O*-algebra $\pi(\mathsf{A})$ is self-adjoint and $\overline{x^+} = x^*$ for all $x \in \pi(\mathsf{A})$. From this we see that the integrability of π depends only on the O*-algebra $\pi(\mathsf{A})$ rather than on π and A. That is, if π_1 and π_2 are *-representations of commutative *-algebras A_1 and A_2 with units, respectively, such that $\pi_1(\mathsf{A}_1) = \pi_2(\mathsf{A}_2)$, then π_1 is integrable if and only if π_2 is.

Remark 8. Let π be a *-representation of $\mathsf{A} := \mathbb{C}[\mathsf{x}_1, \dots, \mathsf{x}_n]$, $n \in \mathbb{N}$, such that π^* is integrable. In case $n = 1$ this implies that the operator $\overline{\pi(\mathsf{x}_1)}$ is self-adjoint (see Proposition 8.1.15, (iii)). Example 9.1.15 below shows that a similar assertion is no longer true if $n \geqq 2$. Moreover, this example also shows that the sufficient condition in Theorem 9.1.13 for the integrability of π^* is not a necessary one.

Example 9.1.15. Let π_1 be a *-representation of $\mathbb{C}[\mathsf{x}]$ such that the operator $\overline{\pi_1(\mathsf{x})}$ is self-adjoint and such that the operators $\overline{\pi_1(\mathsf{x}^2 + \mathsf{x})}$ and $\overline{\pi_1(\mathsf{x}^2)}$ are both not self-adjoint. (The *-representation π_1 in Example 8.1.14 has these properties.) Define a *-representation π of $\mathsf{A} = \mathbb{C}[\mathsf{x}_1, \mathsf{x}_2]$ by $\pi(\mathsf{x}_1) := \pi_1(\mathsf{x}^2 + \mathsf{x})$ and $\pi(\mathsf{x}_2) := \pi_1(\mathsf{x}^2)$. Since $\overline{\pi_1(\mathsf{x}_1)}$ is self-adjoint, π_1^* is integrable by Theorem 9.1.13. From this and $\pi(\mathbb{C}[\mathsf{x}_1, \mathsf{x}_2]) = \pi_1(\mathbb{C}[\mathsf{x}])$ we conclude that π^* is a *-representation and $\pi^*(\mathbb{C}[\mathsf{x}_1, \mathsf{x}_2]) = \pi_1^*(\mathbb{C}[\mathsf{x}])$. By Remark 8, π^* is an integrable representation of $\mathbb{C}[\mathsf{x}_1, \mathsf{x}_2]$, though the operators $\overline{\pi(\mathsf{x}_1)}$ and $\overline{\pi(\mathsf{x}_2)}$ are both not self-adjoint. ○

Example 9.1.16. Let μ be a positive regular Borel measure on $\mathbb{R}^n$. Define $\mathcal{D} := \{\varphi \in L^2(\mathbb{R}^n; \mu) : p(t)\, \varphi(t) \in L^2(\mathbb{R}^n; \mu)$ for all $p \in \mathbb{C}[\mathsf{x}_1, \dots, \mathsf{x}_n]\}$. Let A_0 be the *-algebra of all polynomially bounded measurable functions on $\mathbb{R}^n$ with the usual pointwise algebraic operations, and let A be a *-subalgebra of A_0 which contains all polynomials. For $a \in \mathsf{A}$ and $\varphi \in \mathcal{D}$, we define $\pi(a)\, \varphi := a\varphi$. Then π is an integrable representation of A on $\mathcal{D}(\pi) = \mathcal{D}$ in the Hilbert space $L^2(\mathbb{R}^n; \mu)$. ○

Proposition 9.1.17. (i) *Let π be a representation of* A *which is the direct sum of a family $\{\pi_i : i \in I\}$ of representations of* A. *Then π is integrable if and only if all π_i, $i \in I$, are integrable.*

(ii) *Every self-adjoint subrepresentation π of an integrable representation π_0 is itself integrable.*

Proof. The proof of (i) is straightforward, so we omit the details. (ii): Since π is self-adjoint, it follows from Corollary 8.3.13 that there exists a $*$-representation π_1 of A such that $\pi_0 = \pi \oplus \pi_1$. By (i), π is integrable, since π_0 is integrable. □

9.2. Decomposition of Integrable Representations as Direct Sums of Cyclic Representations

In this section A is a commutative $*$-algebra with unit.

Our first theorem contains the main step in the proof of the decomposition theorem, but it is also of interest in itself.

Theorem 9.2.1. *Suppose that π is an integrable representation of* A *such that the graph topology of $\pi(\mathsf{A})$ is metrizable. Then the following three assertions are equivalent:*

(i) *π is cyclic.*

(ii) *π admits a weakly cyclic vector.*

(iii) *The von Neumann algebra $\pi(\mathsf{A})''$ has a cyclic vector.*

Proof. (i) → (ii) is trivial. We prove (ii) → (iii). Let $\varphi_0 \in \mathcal{D}(\pi)$ be a weakly cyclic vector for π. By Lemma 8.3.15, (i), φ_0 is a separating vector for the von Neumann algebra $\pi(\mathsf{A})'_{\mathrm{w}} \equiv \pi(\mathsf{A})'$. Hence the vector φ_0 is cyclic for $\pi(\mathsf{A})''$.

In the rest of this proof we show that (iii) implies (i). Since the graph topology of $\pi(\mathsf{A})$ is metrizable, we know from Lemma 2.2.7 that there is a sequence $(a_n : n \in \mathbb{N})$ in A_{h} such that the family of seminorms $\{\|\cdot\|_{\pi(a_n)} : n \in \mathbb{N}\}$ is directed and generates the graph topology of $\pi(\mathsf{A})$. Set $A_n := \overline{\pi(a_n)}$, $n \in \mathbb{N}$. Because π is integrable, it follows from Theorem 9.1.2 that A_k and A_m are strongly commuting self-adjoint operators for all k, $m \in \mathbb{N}$. Put $T_n(\lambda) := \exp(-\lambda A_n^2)$ for $\lambda > 0$ and $n \in \mathbb{N}$. By (iii), there exists a vector $\varphi_0 \in \mathcal{H}(\pi)$ which is cyclic for the von Neumann algebra $\pi(\mathsf{A})''$. Without loss of generality we assume that $\|\varphi_0\| = 1$. Since $T_n(\lambda)$ converges strongly to I as $\lambda \to +0$, there is a number $\lambda_n > 0$ such that

$$\left\|\left(T_n(\lambda_n) - I\right) \varphi_0\right\| \leqq 2^{-n} \quad \text{for } n \in \mathbb{N}. \tag{1}$$

Set $R_n := T_1(\lambda_1) \dots T_n(\lambda_n)$, $n \in \mathbb{N}$. Since the operators A_k and A_m, $k, m \in \mathbb{N}$, strongly commute, $(R_n : n \in \mathbb{N})$ is a decreasing sequence of positive bounded operators on $\mathcal{H}(\pi)$, so it converges in the strong-operator topology to some operator R. Define $\psi_0 := R\varphi_0$.

We show that ψ_0 is in $\mathcal{D}(\pi)$. We have

$$\lim_n T_{k+1}(\lambda_{k+1}) \dots T_{k+n}(\lambda_{k+n}) \varphi_0 = \lim_n R_k^{-1} R_{k+n} \varphi_0 = R_k^{-1} R \varphi_0 = R_k^{-1} \psi_0 \quad \text{for } k \in \mathbb{N}.$$

Thus $\psi_0 \in \mathcal{D}(\exp \lambda_k A_k^2) \subseteq \mathcal{D}(A_k)$ for $k \in \mathbb{N}$, so $\psi_0 \in \bigcap_{k \in \mathbb{N}} \mathcal{D}(A_k)$. From Proposition 2.2.12

and the properties of the seminorms $\|\cdot\|_{\pi(a_n)}$, we have

$$\mathcal{D}(\hat{\pi}) = \bigcap_{k\in\mathbb{N}} \mathcal{D}(\overline{\pi(a_k)}) \equiv \bigcap_{k\in\mathbb{N}} \mathcal{D}(A_k).$$

Hence $\psi_0 \in \mathcal{D}(\pi)$, since π is closed.

Let $\mathcal{H}_0$ be the closure of $\pi(\mathsf{A})\,\psi_0$ in $\mathcal{H}(\pi)$, and let π_0 be the closure of the $*$-representation $\pi \upharpoonright \pi(\mathsf{A})\,\psi_0$ in the Hilbert space $\mathcal{H}(\pi_0) := \mathcal{H}_0$. We next prove that π_0 is self-adjoint. Let $a \in \mathsf{A}$, and let $m \in \mathbb{N}$. Since the family of seminorms $\{\|\cdot\|_{\pi(a_n)} : n \in \mathbb{N}\}$ is directed and generates the graph topology of $\pi(\mathsf{A})$, there are numbers $k \in \mathbb{N}$ and $\alpha > 0$ such that $\|\pi(a)\,\varphi\| \leqq \alpha\|\pi(a_k)\,\varphi\|$ and $\|\pi(a_m)\,\varphi\| \leqq \|\pi(a_k)\,\varphi\|$ for all $\varphi \in \mathcal{D}(\pi)$. The second estimate implies that $\mathcal{D}(A_k) \subseteq \mathcal{D}(A_m)$ and $\|A_m\varphi\| \leqq \|A_k\varphi\|$ for all $\varphi \in \mathcal{D}(A_k)$. Since the strongly commuting self-adjoint operators A_k and A_m are affiliated with a common abelian von Neumann algebra, it follows from the latter and Lemma 7.3.2 that

$$\|\pi(a_m)^n\varphi\| = \|A_m^n\varphi\| \leqq \|A_k^n\varphi\| = \|\pi(a_k)^n\varphi\|$$

for all $\varphi \in \mathcal{D}(\pi)$ and $n \in \mathbb{N}$. Therefore,

$$\begin{aligned}\|\pi_0(a_m)^n\,\pi_0(a)\,\psi_0\| &\leqq \|\pi(a_k)^n\,\pi(a)\,\psi_0\| = \|\pi(a)\,\pi(a_k)^n\,\psi_0\| \leqq \alpha\|\pi(a_k)^{n+1}\,\psi_0\|\\ &= \alpha\|A_k^{n+1}\,\psi_0\| = \alpha\|A_k^{n+1}R_kR_k^{-1}\psi_0\| \leqq \alpha\|A_k^{n+1}T_k(\lambda_k)\|\,\|R_k^{-1}\psi_0\|\\ &\leqq \alpha\|R_k^{-1}\psi_0\|\sup\{|t^{n+1}\exp(-\lambda_k t^2)| : t \in \mathbb{R}\} \leqq M_k^n n!\end{aligned}$$

for all $n \in \mathbb{N}$ with some constant $M_k > 0$. This shows that all vectors in $\pi_0(\mathsf{A})\,\psi_0$ are analytic vectors for the symmetric operator $\pi_0(a_m)$ in the Hilbert space $\mathcal{H}_0$. By Nelson's lemma (cf. Proposition 10.3.4), $\pi_0(a_m)$ is essentially self-adjoint. Hence $\mathcal{D}(\pi_0) = \bigcap_{m\in\mathbb{N}} \mathcal{D}(\overline{\pi_0(a_m)}) = \bigcap_{m\in\mathbb{N}} \mathcal{D}(\pi_0(a_m)^*) \supseteq \mathcal{D}(\pi_0^*)$, where the first equality follows again from Proposition 2.2.12. Therefore, π_0 is self-adjoint.

Since π_0 is self-adjoint and $\pi_0 \subseteq \pi$, it follows from Proposition 8.3.11 that $P_0 \in \pi(\mathsf{A})'$ and $\mathcal{D}(\pi_0) = P_0\mathcal{D}(\pi)$, where P_0 is the projection of $\mathcal{H}(\pi)$ onto $\mathcal{H}_0$. By definition π_0 is a cyclic $*$-representation of A. Therefore our proof is complete once we have shown that $\pi_0 = \pi$. Suppose $\varepsilon > 0$. Using (1), we obtain

$$\begin{aligned}\|R_k^{-1}\psi_0 - \varphi_0\| &= \lim_{n\to\infty} \|T_{k+1}(\lambda_{k+1}) \dots T_{k+n}(\lambda_{k+n})\,\varphi_0 - \varphi_0\|\\ &\leqq \lim_{n\to\infty} \sum_{j=1}^{n-1} \left\|T_{k+1}(\lambda_{k+1}) \dots T_{k+j}(\lambda_{k+j})\left(T_{k+j+1}(\lambda_{k+j+1}) - I\right)\varphi_0\right\|\\ &\quad + \left\|\left(T_{k+1}(\lambda_{k+1}) - I\right)\varphi_0\right\|\\ &\leqq \sum_{j=1}^{\infty} \left\|\left(T_{k+j}(\lambda_{k+j}) - I\right)\varphi_0\right\| \leqq \sum_{j=1}^{\infty} 2^{-(k+j)} = 2^{-k}\end{aligned}$$

for $k \in \mathbb{N}$. Hence $\|R_k^{-1}\psi_0 - \varphi_0\| < \varepsilon$ for some $k \in \mathbb{N}$. Let E_δ be the spectral projection of the positive self-adjoint operator R_k associated with the interval $[0, \delta]$. There is a $\delta > 0$ such that $\|R_k^{-1}\psi_0 - R_k^{-1}E_\delta\psi_0\| < \varepsilon$. Then we have $\|\varphi_0 - R_k^{-1}E_\delta\psi_0\| < 2\varepsilon$. Since π is self-adjoint, $A_m = \overline{\pi(a_m)}$ is affiliated with $\pi(\mathsf{A})'' \left(\equiv (\pi(\mathsf{A})'_{\mathrm{ss}})'\right)$ for $m \in \mathbb{N}$. Hence $R_k \in \pi(\mathsf{A})''$ and so $R_k^{-1}E_\delta \in \pi(\mathsf{A})''$. Thus we have proved that φ_0 is in the closure of the set $\pi(\mathsf{A})''\,\psi_0$ in $\mathcal{H}(\pi)$. Because $P_0 \in \pi(\mathsf{A})'$ and $\psi_0 \in P_0\mathcal{H}(\pi)$, we obtain $\pi(\mathsf{A})''\,\psi_0 \subseteq P_0\mathcal{H}(\pi)$. Therefore, $\varphi_0 \in P_0\mathcal{H}(\pi)$ and so $\pi(\mathsf{A})''\,\varphi_0 \subseteq P_0\mathcal{H}(\pi)$. By construction, φ_0 is cyclic for the von Neumann algebra $\pi(\mathsf{A})''$. Hence the latter implies $P_0 = I$, so that $\pi_0 = \pi$. □

The main assertion in Theorem 9.2.1 (that is, the implication (iii) $\rightarrow$ (i)) is no longer true in general if the graph topology of $\pi(\mathsf{A})$ is not metrizable. This is shown by the following

Example 9.2.2. Let $\mathsf{A} = C(\mathbb{R})$ and let $\mathcal{D}(\pi)$ be the linear subspace of all functions in $\mathcal{H}(\pi) := L^2(\mathbb{R})$ with compact support. Define $\pi(a)\,\varphi = a\varphi$, $a \in \mathsf{A}$ and $\varphi \in \mathcal{D}(\pi)$. Then π is an integrable representation of A which has obviously no cyclic vector. Clearly, $\pi(\mathsf{A})'' = L^\infty(\mathbb{R})$, where the functions of $L^\infty(\mathbb{R})$ act as multiplication operators on $L^2(\mathbb{R})$. The vector $\varphi_0(t) := \exp(-t^2)$, $t \in \mathbb{R}$, is cyclic for the von Neumann algebra $\pi(\mathsf{A})''$. ○

Theorem 9.2.3. *Suppose that π is an integrable representation of A such that the graph topology of $\pi(\mathsf{A})$ is metrizable. Then π is a direct sum of cyclic integrable representations of A.*

Proof. The identity representation of the von Neumann algebra $\pi(\mathsf{A})''$ can be expressed as a direct sum of cyclic representations of $\pi(\mathsf{A})''$. Hence there exists a set $\{\varphi_i : i \in I\}$ of vectors from $\mathcal{H}(\pi)$ such that $\mathcal{H}_i$ is orthogonal to $\mathcal{H}_j$ for $i, j \in I$, $i \neq j$, and $\mathcal{H}(\pi) = \sum_{i \in I} \oplus\, \mathcal{H}_i$, where $\mathcal{H}_i$ denotes the closure of $\pi(\mathsf{A})''\,\varphi_i$ in $\mathcal{H}(\pi)$. Let $i \in I$, and let P_i be the projection of $\mathcal{H}(\pi)$ onto $\mathcal{H}_i$. Since $\mathcal{H}_i$ reduces $\pi(\mathsf{A})''$, $P_i \in \pi(\mathsf{A})'$. By Proposition 8.3.11, $\pi_i := \pi \upharpoonright P_i\mathcal{D}(\pi)$ is a self-adjoint representation of A. It is straightforward to check that $\pi = \sum_{i \in I} \oplus\, \pi_i$. Since π is integrable, each π_i is integrable as well. Since $\pi_i(\mathsf{A})'' = \pi(\mathsf{A})'' \upharpoonright \mathcal{H}_i$, $\pi_i(\mathsf{A})''\,\varphi_i \equiv \pi(\mathsf{A})''\,\varphi_i$ is dense in $\mathcal{H}_i$ and so the von Neumann algebra $\pi_i(\mathsf{A})''$ admits a cyclic vector. By Theorem 9.2.1, π_i is cyclic for each $i \in I$. □

Remark 1. As shown later (cf. Corollary 11.6.8), there exists a closed $*$-representation of the $*$-algebra $\mathbb{C}[\mathsf{x}_1, \mathsf{x}_2]$ which cannot be decomposed as a direct sum of cyclic representations.

9.3. Two Classes of Couples of Self-Adjoint Operators

In this section we develop some technical tools which are used for the construction of non-integrable self-adjoint representations of the polynomial algebra $\mathbb{C}[\mathsf{x}_1, \mathsf{x}_2]$. Concrete applications will be considered in the next section.

Definitions and Basic Properties of the Classes N_∞ and N_∞^∞

Throughout this subsection A and B denote self-adjoint operators in the Hilbert space $\mathcal{H}$.

Definition 9.3.1. We say that the couple $\{A, B\}$ belongs to the class N_∞ if there exists a linear subspace $\mathcal{D}$ of $\mathcal{H}$ such that:

(i) $\mathcal{D} \subseteq \mathcal{D}(A) \cap \mathcal{D}(B)$, $A\mathcal{D} \subseteq \mathcal{D}$ and $B\mathcal{D} \subseteq \mathcal{D}$,

(ii) $AB\varphi = BA\varphi$ for $\varphi \in \mathcal{D}$,

(iii)$_1$ $A \upharpoonright \mathcal{D}$ and $B \upharpoonright \mathcal{D}$ are both essentially self-adjoint.

We say that $\{A, B\}$ is in the class N_∞^∞ if there is a linear subspace $\mathcal{D}$ of $\mathcal{H}$ satisfying (i), (ii) and

iii)$_2$ $A^n \upharpoonright \mathcal{D}$ and $B^n \upharpoonright \mathcal{D}$ are essentially self-adjoint for all $n \in \mathbb{N}$.

Our first objective is to give sufficient conditions in terms of the resolvents of A and B for a couple $\{A, B\}$ to be in $\boldsymbol{N}_\infty$ or in $\boldsymbol{N}_\infty^\infty$. For this we need some preliminaries.

First we fix some notation which will be kept throughout this subsection. We let α and β be complex numbers in the resolvent sets $\mathbb{C} \setminus \sigma(A)$ and $\mathbb{C} \setminus \sigma(B)$, respectively, and we set $X_\alpha := (A - \alpha)^{-1}$ and $Y_\beta := (B - \beta)^{-1}$. Suppose $r, s \in \mathbb{N}$. Let $\mathcal{Q}_{r,s}(\alpha, \beta)$ be the linear span of the ranges of the commutators $[X_\alpha^n, Y_\beta^m] \equiv X_\alpha^n Y_\beta^m - Y_\beta^m X_\alpha^n$, $n = 1, \ldots, r$ and $m = 1, \ldots, s$. Let $Q_{r,s}(\alpha, \beta)$ denote the projection onto the closure of $\mathcal{Q}_{r,s}(\alpha, \beta)$ in $\mathcal{H}$.

Lemma 9.3.2. *Suppose $n, m, r, s \in \mathbb{N}$, $n \leqq r$ and $m \leqq s$.*

(i) $\mathcal{Q}_{r,s}(\alpha, \beta) = \text{l.h.}\,\{X_\alpha^k Y_\beta^l [X_\alpha, Y_\beta]\,\mathcal{H} : k = 0, \ldots, r-1$ *and* $l = 0, \ldots, s-1\}$
$= \text{l.h.}\,\{Y_\beta^l X_\alpha^k [X_\alpha, Y_\beta]\,\mathcal{H} : k = 0, \ldots, r-1$ *and* $l = 0, \ldots, s-1\}$.

(ii) $X_{\bar\alpha}^n Y_{\bar\beta}^m \varphi = Y_{\bar\beta}^m X_{\bar\alpha}^n \varphi$ *for all* $\varphi \in \big(I - Q_{r,s}(\alpha, \beta)\big)\,\mathcal{H}$.

Proof. (i): We prove the first equality. The second one follows by symmetry. Since

$$X_\alpha^k Y_\beta^l [X_\alpha, Y_\beta] = [X_\alpha^{k+1}, Y_\beta^{l+1}] - [X_\alpha^k, Y_\beta^{l+1}]\,X_\alpha - [X_\alpha^{k+1}, Y_\beta^l]\,Y_\beta + [X_\alpha^k, Y_\beta^l]\,X_\alpha Y_\beta,$$

we have

$$\mathcal{Q}_{r,s}(\alpha, \beta) \supseteqq \text{l.h.}\,\{X_\alpha^k Y_\beta^l [X_\alpha, Y_\beta]\,\mathcal{H};\ k = 0, \ldots, r-1 \text{ and } l = 0, \ldots, s-1\}.$$

The opposite inclusion follows from the identity

$$[X_\alpha^k, Y_\beta^l] = \sum_{n=0}^{k-1} \sum_{m=0}^{l-1} X_\alpha^{k-1-n} Y_\beta^{l-1-m} [X_\alpha, Y_\beta]\, Y_\beta^m X_\alpha^n, \quad k, l \in \mathbb{N}.$$

(ii): By definition,

$$[X_\alpha^n, Y_\beta^m]\,\mathcal{H} \subseteqq \big(I - Q_{r,s}(\alpha, \beta)\big)\,\mathcal{H}.$$

Hence

$$T := \big(I - Q_{r,s}(\alpha, \beta)\big)\,[X_\alpha^n, Y_\beta^m] \equiv 0$$

and so

$$T^* = -[X_{\bar\alpha}^n, Y_{\bar\beta}^m]\,\big(I - Q_{r,s}(\alpha, \beta)\big) \equiv 0$$

which gives the assertion. □

For $r, s \in \mathbb{N}$, define

$$\mathcal{D}_{r,s} \equiv \mathcal{D}_{r,s}(A, B) := \{\varphi \in \mathcal{H} : \varphi \in \mathcal{D}(A^n B^m) \cap \mathcal{D}(B^m A^n) \text{ and } A^n B^m \varphi = B^m A^n \varphi$$
$$\text{for all } n = 1, \ldots, r \text{ and } s = 1, \ldots, m\}.$$

Set $\mathcal{D}_\infty(A, B) := \bigcap_{r,s \in \mathbb{N}} \mathcal{D}_{r,s}(A, B)$.

Lemma 9.3.3. (i) *Suppose $r, s \in \mathbb{N}$. If $n_1, \ldots, n_k, m_1, \ldots, m_k$ are non-negative integers such that $n := n_1 + \cdots + n_k \leqq r$ and $m := m_1 + \cdots + m_k \leqq s$, then $\mathcal{D}_{r,s} \subseteqq \mathcal{D}(A^{n_1} B^{m_1} \ldots A^{n_k} B^{m_k})$ and $A^{n_1} B^{m_1} \ldots A^{n_k} B^{m_k} \varphi = B^n A^m \varphi$ for all $\varphi \in \mathcal{D}_{r,s}$.*

(ii) *$A\mathcal{D}_\infty(A, B) \subseteqq \mathcal{D}_\infty(A, B)$ and $B\mathcal{D}_\infty(A, B) \subseteqq \mathcal{D}_\infty(A, B)$. Thus $\mathcal{D} := \mathcal{D}_\infty(A, B)$ is the largest linear subspace of $\mathcal{H}$ which satisfies the conditions* (i) *and* (ii) *in Definition* 9.3.1.

Proof. Using the definition of $\mathcal{D}_{r,s}$, (i) follows easily by induction on k. To verify (ii), let $\varphi \in \mathcal{D}_\infty(A, B)$ and $r, s \in \mathbb{N}$. Then, by definition, $\varphi \in \mathcal{D}_{r+1,s}$. Thus, by (i), $\varphi \in \mathcal{D}(A^n B^m A) \cap \mathcal{D}(B^m A^n A)$ and $A^n B^m A\varphi = B^m A^n A\varphi$ for all $n, m \in \mathbb{N}$, $n \leqq r$ and $m \leqq s$. This shows that $A\varphi \in \mathcal{D}_{r,s}$. Hence $A\varphi \in \mathcal{D}_\infty(A, B)$. Similarly, $B\mathcal{D}_\infty(A, B) \subseteqq \mathcal{D}_\infty(A, B)$. $\square$

Lemma 9.3.4. *For $r, s \in \mathbb{N}$, $\mathcal{D}_{r,s} = X^r_{\bar\alpha} Y^s_{\bar\beta}\big(I - Q_{r,s}(\alpha, \beta)\big)\,\mathcal{H}$.*

Proof. Suppose that $\varphi \in \mathcal{D}_{r,s}$. Let $n, m \in \mathbb{N}$, $n \leqq r$ and $m \leqq s$. Lemma 9.3.3, (i), implies that

$$\varphi \in \mathcal{D}\big((A - \bar\alpha)^n (B - \bar\beta)^m (A - \bar\alpha)^{r-n} (B - \bar\beta)^{s-m}\big) \cap \mathcal{D}\big((B - \bar\beta)^m (A - \bar\alpha^r) (B - \bar\beta)^{s-m}\big)$$

and

$$(A - \bar\alpha)^n (B - \bar\beta)^m (A - \bar\alpha)^{r-n} (B - \bar\beta)^{s-m} \varphi$$
$$= (B - \bar\beta)^m (A - \bar\alpha)^r (B - \alpha)^{s-m} \varphi = (B - \bar\beta)^s (A - \bar\alpha)^r \varphi =: \xi.$$

Thus

$$\varphi = Y^{s-m}_{\bar\beta} X^{r-n}_{\bar\alpha} Y^m_{\bar\beta} X^n_{\bar\alpha}\xi = Y^{s-m}_{\bar\beta} X^r_{\bar\alpha} Y^m_{\bar\beta}\xi = X^r_{\bar\alpha} Y^s_{\bar\beta}\xi.$$

Since $\ker X_{\bar\alpha} = \ker Y_{\bar\beta} = \{0\}$, $[X^n_{\bar\alpha}, Y^m_{\bar\beta}]\,\xi = 0$, so that $\xi \perp [X^n_\alpha, Y^m_\beta]\,\mathcal{H}$. Consequently, $\xi \perp \mathcal{Q}_{r,s}(\alpha, \beta)$ and hence

$$\varphi = X^r_{\bar\alpha} Y^s_{\bar\beta}\xi = X^r_{\bar\alpha} Y^s_{\bar\beta}\big(I - Q_{r,s}(\alpha, \beta)\big)\,\xi,$$

which proves that $\mathcal{D}_{r,s} \subseteqq X^r_{\bar\alpha} Y^s_{\bar\beta}\big(I - Q_{r,s}(\alpha, \beta)\big)\,\mathcal{H}$.

We now prove the reversed inclusion. Let $\varphi \in X^r_{\bar\alpha} Y^s_{\bar\beta}\big(I - Q_{r,s}(\alpha, \beta)\big)\,\mathcal{H}$, i.e., $\varphi = X^r_{\bar\alpha} Y^s_{\bar\beta}\xi$ with $\xi \in \big(I - Q_{r,s}(\alpha, \beta)\big)\,\mathcal{H}$. Suppose $n, m \in \mathbb{N}$, $n \leqq r$ and $m \leqq s$. Since $\xi \perp \mathcal{Q}_{r,s}(\alpha, \beta)$, we have that $\xi \perp Y^{s-m}_{\bar\beta} Y^l_\beta X^k_\alpha [X_\alpha, Y_\beta]\,\mathcal{H}$ by Lemma 9.3.2, (i), and so $Y^{s-m}_{\bar\beta}\xi \perp Y^l_\beta X^k_\alpha [X_\alpha, Y_\beta]\,\mathcal{H}$ for all $k, l \in \mathbb{N}_0$, $k \leqq r - 1$ and $l \leqq m - 1$. Employing Lemma 9.3.2, (i), once more, this gives $Y^{s-m}_{\bar\beta}\xi \in \big(I - Q_{r,s}(\alpha, \beta)\big)\,\mathcal{H}$. Hence, by Lemma 9.3.2, (ii),

$$\varphi = X^n_{\bar\alpha} X^{r-n}_{\bar\alpha} Y^m_{\bar\beta} Y^{s-m}_{\bar\beta}\xi = X^n_{\bar\alpha} Y^m_{\bar\beta} X^{r-n}_{\bar\alpha} Y^{s-m}_{\bar\beta}\xi = X^n_{\bar\alpha} Y^m_{\bar\beta} Y^{s-m}_{\bar\beta} X^{r-n}_{\bar\alpha}\xi,$$

where we used again $\xi \in \big(I - Q_{r,s}(\alpha, \beta)\big)\,\mathcal{H}$. A similar argument shows that

$$\varphi = Y^m_{\bar\beta} X^n_{\bar\alpha} Y^{s-m}_{\bar\beta} X^{r-n}_{\bar\alpha}\xi.$$

Thus

$$\varphi \in \mathcal{D}\big((A - \bar\alpha)^n (B - \bar\beta)^m\big) \cap \mathcal{D}\big((B - \bar\beta)^m (A - \bar\alpha)^n\big)$$

and

$$Y^{s-m}_{\bar\beta} X^{r-n}_{\bar\alpha}\xi = (A - \bar\alpha)^n (B - \bar\beta)^m \varphi = (B - \bar\beta)^m (A - \bar\alpha)^n \varphi.$$

Since A^n and B^m are polynomials in $(A - \bar\alpha)$ resp. $(B - \bar\beta)$ with degree n resp. m, it follows that $\varphi \in \mathcal{D}_{r,s}$. $\square$

For a convenient formulation of the next results we introduce some conditions denoted by $(\mathrm{I})^{\alpha,\beta}_{n,m}$ and $(\mathrm{II})^{\alpha,\beta}_{n,m}$. First we extend the definitions of $Q_{r,s}(\alpha, \beta)$ and $\mathcal{D}_{r,s}$ to the cases $r = 0$ and $s = 0$ by setting $Q_{r,0}(\alpha, \beta) = Q_{0,s}(\alpha, \beta) = 0$, $\mathcal{D}_{r,0} = \mathcal{D}(A^r)$ and $\mathcal{D}_{0,s} = \mathcal{D}(B^s)$ when r and s in $\mathbb{N}_0$. For $\alpha \in \mathbb{C} \setminus \sigma(A)$, $\beta \in \mathbb{C} \setminus \sigma(B)$ and $n, m \in \mathbb{N}$, we consider the following conditions:

$(\mathrm{I})^{\alpha,\beta}_{n,m}$ If $X_\alpha\varphi \in Q_{n,m}(\alpha, \beta)\,\mathcal{H}$ for some $\varphi \in \mathcal{H}$, then $\varphi \in Q_{n-1,m}(\alpha, \beta)\,\mathcal{H}$.

$(\mathrm{II})^{\alpha,\beta}_{n,m}$ If $Y_\beta\varphi \in Q_{n,m}(\alpha, \beta)\,\mathcal{H}$ for some $\varphi \in \mathcal{H}$, then $\varphi \in Q_{n,m-1}(\alpha, \beta)\,\mathcal{H}$.

Lemma 9.3.5. *Suppose $n, m, r, s \in \mathbb{N}$, $n \leqq r$ and $m \leqq s$.*

(i) *$A^n \upharpoonright \mathcal{D}_{r,s}$ is essentially self-adjoint if and only if $Y^s_\beta X^{r-n}_\alpha \mathcal{H} \cap Q_{r,s}(\alpha, \beta)\, \mathcal{H} = \{0\}$. This is true if for some $\alpha \in \mathbb{C} \setminus \sigma(A)$ and $\beta \in \mathbb{C} \setminus \sigma(B)$ the conditions $(\mathrm{II})^{\alpha,\beta}_{r,l}$ are satisfied for $l = 1, \ldots, s$.*

(ii) *$B^m \upharpoonright \mathcal{D}_{r,s}$ is essentially self-adjoint if and only if $X^r_\alpha Y^{s-m}_\beta \mathcal{H} \cap Q_{r,s}(\alpha, \beta)\, \mathcal{H} = \{0\}$. This is the case if for some $\alpha \in \mathbb{C} \setminus \sigma(A)$ and $\beta \in \mathbb{C} \setminus \sigma(B)$ the conditions $(\mathrm{I})^{\alpha,\beta}_{k,s}$ are fulfilled for $k = 1, \ldots, r$.*

Proof. Again, by symmetry, it suffices to prove the assertion of (i). First note that $\mathcal{D}_{r,s} = Y^r_{\bar\alpha} Y^s_{\bar\beta}(I - Q_{r,s}(\alpha, \beta))\, \mathcal{H}$ by Lemma 9.3.4. From the spectral theorem for self-adjoint operators it follows that the operator $T := (A^n - \mathrm{i})\,(A - \bar\alpha)^{-n}$ is an isomorphism of the Hilbert space $\mathcal{H}$. Since A^n is self-adjoint, $A^n \upharpoonright \mathcal{D}_{r,s}$ is essentially self-adjoint if and only if

$$\begin{aligned}(A^n - \mathrm{i})\, \mathcal{D}_{r,s} &\equiv (A^n - \mathrm{i})\,(A - \bar\alpha)^{-n} X^{r-n}_{\bar\alpha} Y^s_{\bar\beta}(I - Q_{r,s}(\alpha, \beta))\, \mathcal{H} \\ &\equiv T X^{r-n}_{\bar\alpha} Y^s_{\bar\beta}(I - Q_{r,s}(\alpha, \beta))\, \mathcal{H}\end{aligned}$$

is dense in $\mathcal{H}$ or equivalently, if $X^{r-n}_{\bar\alpha} Y^s_{\bar\beta}(I - Q_{r,s}(\alpha, \beta))\, \mathcal{H}$ is dense in $\mathcal{H}$. But this is equivalent to

$$\ker \left(X^{r-n}_{\bar\alpha} Y^s_{\bar\beta}(I - Q_{r,s}(\alpha, \beta))\right)^* = \{0\}$$

and so to

$$Y^s_\beta X^{r-n}_\alpha \mathcal{H} \cap Q_{r,s}(\alpha, \beta)\, \mathcal{H} = \{0\}.$$

To prove the second assertion of (i), assume that $(\mathrm{II})_{r,l}$ is true for $l = 1, \ldots, s$. Suppose that $Y^s_\beta X^{r-n}_\alpha \varphi \in Q_{r,s}(\alpha, \beta)\, \mathcal{H}$ for some $\varphi \in \mathcal{H}$. A repeated application of $(\mathrm{II})_{r,l}$, $l = 1, \ldots, s$, yields $X^{r-n}_\alpha \varphi \in Q_{r,0}(\alpha, \beta)\, \mathcal{H}$. By definition, the latter is $\{0\}$; so $Y^s_\beta X^{r-n}_\alpha \mathcal{H} \cap Q_{r,s}(\alpha, \beta)\, \mathcal{H} = \{0\}$. □

Lemma 9.3.6. *Suppose $n, m, r, s \in \mathbb{N}_0$, $n \leqq r$ and $m \leqq s$. If for some $\alpha \in \mathbb{C} \setminus \sigma(A)$ and $\beta \in \mathbb{C} \setminus \sigma(B)$ the conditions $(\mathrm{I})^{\alpha,\beta}_{r+k,s+k}$ and $(\mathrm{II})^{\alpha,\beta}_{r+k-1,s+k}$ are satisfied for all $k \in \mathbb{N}$, then $\mathcal{D}_\infty(A, B)$ is a core for the (not necessarily densely defined) operator $A^n B^m \upharpoonright \mathcal{D}_{r,s}$.*

Proof. We fix n, m, r, s, α and β. For $k, l \in \mathbb{N}_0$, let $\mathcal{H}_{k,l}$ be the linear space $\mathcal{D}_{r+k,s+l}$ equipped with the inner product $\langle \cdot, \cdot \rangle_{k,l} := \langle (B - \bar\beta)^{s+l} (A - \bar\alpha)^{r+k} \cdot, (B - \bar\beta)^{s+l} (A - \bar\alpha)^{r+k} \cdot \rangle$. Let $\|\cdot\|_{k,l}$ be the corresponding norm. Since

$$\mathcal{D}_{r+k,s+l} = X^{r+k}_{\bar\alpha} Y^{s+l}_{\bar\beta}(I - Q_{r+k,s+l}(\alpha, \beta))\, \mathcal{H}$$

by Lemma 9.3.4, $\mathcal{H}_{k,l}$ is a Hilbert space.

Fix $k \in \mathbb{N}$. We prove that $(\mathrm{I})_{r+k,s+k}$ implies that $\mathcal{H}_{k,k}$ is dense in the Hilbert space $\mathcal{H}_{k-1,k}$. We suppose that $\psi \in \mathcal{H}_{k-1,k}$ is orthogonal to $\mathcal{H}_{k,k}$ in the Hilbert space $\mathcal{H}_{k-1,k}$. We can write ψ as $\psi = X^{r+k-1}_{\bar\alpha} Y^{s+k}_{\bar\beta}\, \xi$ with $\xi \in (I - Q_{r+k-1,s+k}(\alpha, \beta))\, \mathcal{H}$. From Lemmas 9.3.4 and 9.3.2, (ii),

$$\mathcal{D}_{r+k,s+k} = X^{r+k-1}_{\bar\alpha} Y^{s+k}_{\bar\beta} X_{\bar\alpha}(I - Q_{r+k,s+k}(\alpha, \beta))\, \mathcal{H}.$$

Hence if $\varphi \in (I - Q_{r+k,s+k}(\alpha, \beta))\, \mathcal{H}$,

$$0 = \langle \psi, X^{r+k-1}_{\bar\alpha} Y^{s+k}_{\bar\beta} X_{\bar\alpha} \varphi \rangle_{k-1,k} = \langle \xi, X_{\bar\alpha} \varphi \rangle = \langle X_\alpha \xi, \varphi \rangle,$$

i.e., $X_\alpha \xi \in Q_{r+k,s+k}(\alpha,\beta)\,\mathcal{H}$. By $(\mathrm{I})^{\alpha,\beta}_{r+k,s+k}$, $\xi \in Q_{r+k-1,s+k}(\alpha,\beta)\,\mathcal{H}$. Since, by construction, $\xi \in \big(I - Q_{r+k-1,s+k}(\alpha,\beta)\big)\,\mathcal{H}$, we get $\xi = 0$ and so $\psi = 0$. This proves that $\mathcal{H}_{k,k}$ is dense in $\mathcal{H}_{k-1,k}$. By a similar reasoning it follows from $(\mathrm{II})^{\alpha,\beta}_{r+k-1,s+k}$ that $\mathcal{H}_{k-1,k}$ is dense in the Hilbert space $\mathcal{H}_{k-1,k-1}$.

We consider the following chain of Hilbert spaces:

$$\mathcal{H}_{0,0} \supseteqq \mathcal{H}_{0,1} \supseteqq \mathcal{H}_{1,1} \supseteqq \mathcal{H}_{1,2} \supseteqq \mathcal{H}_{2,2} \supseteqq \cdots.$$

Obviously, each Hilbert space is continuously embedded in the preceding one. Further, as just shown, each space is dense in its predecessor. Thus Lemma 1.1.2 applies and shows that $\bigcap_{k\in\mathbb{N}_0} (\mathcal{H}_{k,k} \cap \mathcal{H}_{k,k+1}) = \bigcap_{k\in\mathbb{N}} \mathcal{D}_{r+k,s+k} = \mathcal{D}_\infty(A,B)$ is $\|\cdot\|_{0,0}$-dense in $\mathcal{H}_{0,0} \equiv \mathcal{D}_{r,s}$. Since $\bar\alpha \notin \sigma(A)$ and $\bar\beta \notin \sigma(B)$, we have $\|A^nB^m\cdot\| + \|\cdot\| \leqq \text{const.}\ \|(B-\bar\beta)^r\,(A-\bar\alpha)^s\cdot\| \equiv \text{const.}\ \|\cdot\|_{0,0}$ on $\mathcal{D}_{r,s}$, so $\mathcal{D}_\infty(A,B)$ is a core for $A^nB^m \restriction \mathcal{D}_{r,s}$. □

Proposition 9.3.7. *Suppose $n, m \in \mathbb{N}_0$.*

(i) *$A^n \restriction \mathcal{D}_\infty(A,B)$ is essentially self-adjoint if there exist $r, s \in \mathbb{N}$, $n \leqq r$, $\alpha \in \mathbb{C} \setminus \sigma(A)$ and $\beta \in \mathbb{C}\setminus\sigma(B)$ such that for all $k \in \mathbb{N}$ and $l = 1, \ldots, s$ the conditions $(\mathrm{I})^{\alpha,\beta}_{r+k,s+k}$, $(\mathrm{II})^{\alpha,\beta}_{r+k-1,s+k}$ and $(\mathrm{II})^{\alpha,\beta}_{r,l}$ are satisfied.*

(ii) *$B^m \restriction \mathcal{D}_\infty(A,B)$ is essentially self-adjoint if there are $r, s \in \mathbb{N}$, $m \leqq s$, $\alpha \in \mathbb{C}\setminus\sigma(A)$ and $\beta \in \mathbb{C}\setminus\sigma(B)$ such that for all $k \in \mathbb{N}$ and $l = 1, \ldots, r$ the conditions $(\mathrm{I})^{\alpha,\beta}_{r+k,s+k-1}$, $(\mathrm{II})^{\alpha,\beta}_{r+k,s+k}$ and $(\mathrm{I})^{\alpha,\beta}_{l,s}$ are satisfied.*

Proof. Since (ii) follows from (i) by symmetry, it is sufficient to prove (i). Because we assumed $(\mathrm{II})^{\alpha,\beta}_{r,l}$ for $l = 1, \ldots, s$, Lemma 9.3.5, (i), ensures that $A^n \restriction \mathcal{D}_{r,s}$ is essentially self-adjoint. Applying Lemma 9.3.6 in case $m = 0$, it follows that $\mathcal{D}_\infty(A,B)$ is a core for $A^n \restriction \mathcal{D}_{r,s}$. Combining both statements, we obtain the assertion. □

An immediate consequence of Proposition 9.3.7 is

Corollary 9.3.8. *If there are numbers $\alpha \in \mathbb{C}\setminus\sigma(A)$ and $\beta \in \mathbb{C}\setminus\sigma(B)$ such that $(\mathrm{I})^{\alpha,\beta}_{n,m}$ and $(\mathrm{II})^{\alpha,\beta}_{n,m}$ are valid for all $n, m \in \mathbb{N}$, then $\{A, B\} \in N^\infty_\infty$.*

Corollary 9.3.9. *Let $\alpha \in \mathbb{C}\setminus\sigma(A)$ and $\beta \in \mathbb{C}\setminus\sigma(B)$ be such that the conditions $(\mathrm{I})^{\alpha,\beta}_{1,1}$, $(\mathrm{II})^{\alpha,\beta}_{1,1}$, $(\mathrm{I})^{\alpha,\beta}_{n,n}$, $(\mathrm{I})^{\alpha,\beta}_{n,n-1}$, $(\mathrm{II})^{\alpha,\beta}_{n,n}$ and $(\mathrm{II})^{\alpha,\beta}_{n-1,n}$ are satisfied for arbitrary $n \in \mathbb{N}$, $n \geqq 2$. Then $\{A, B\} \in N_\infty$.*

Proof. Apply Proposition 9.3.7 with $n = m = r = s = 1$. □

For the applications given in the next section it is more convenient to work with the Cayley transforms of A and B. Recall that the Cayley transforms of A and B are defined by $U := (A - \mathrm{i})\,(A + \mathrm{i})^{-1}$ and $V := (B - \mathrm{i})\,(B + \mathrm{i})^{-1}$, respectively. We abbreviate $Q_{n,m} := Q_{n,m}(-\mathrm{i}, -\mathrm{i})$, $n, m \in \mathbb{N}_0$. Suppose that n and m are in $\mathbb{N}$. Since $U = I - 2\mathrm{i}X_{-\mathrm{i}}$ and $V = I - 2\mathrm{i}Y_{-\mathrm{i}}$, it follows at once from Lemma 9.3.2, (i), that

$$\begin{aligned} Q_{n,m}\mathcal{H} &= \text{c.l.h.}\ \{U^kV^l[U,V]\,\mathcal{H} : k = 0, \ldots, n-1 \text{ and } l = 0, \ldots, m-1\} \\ &= \text{c.l.h.}\ \{V^lU^k[U,V]\,\mathcal{H} : k = 0, \ldots, n-1 \text{ and } l = 0, \ldots, m-1\}. \end{aligned}$$

Further, condition $(\mathrm{I})^{-\mathrm{i},-\mathrm{i}}_{n,m}$ resp. $(\mathrm{II})^{-\mathrm{i},-\mathrm{i}}_{n,m}$ is equivalent to the following condition $(\mathrm{I})_{n,m}$ resp. $(\mathrm{II})_{n,m}$:

$(\mathrm{I})_{n,m}$ If $(I - U)\,\varphi \in Q_{n,m}\mathcal{H}$ for some $\varphi \in \mathcal{H}$, then $\varphi \in Q_{n-1,m}\mathcal{H}$.

$(\mathrm{II})_{n,m}$ If $(I - V)\,\varphi \in Q_{n,m}\mathcal{H}$ for some $\varphi \in \mathcal{H}$, then $\varphi \in Q_{n,m-1}\mathcal{H}$.

The special case $\alpha = \beta = -\mathrm{i}$ of Corollary 9.3.8 is

Corollary 9.3.10. *If* $(\mathrm{I})_{n,m}$ *and* $(\mathrm{II})_{n,m}$ *are fulfilled for all n and m in* $\mathbb{N}$, *then* $\{A, B\} \in \boldsymbol{N}_\infty^\infty$.

Remark 1. If the self-adjoint operators A and B strongly commute, then, of course, $\{A, B\} \in \boldsymbol{N}_\infty^\infty$. This follows at once from Corollary 9.3.10 (note that $Q_{n,m} = 0$ for $n, m \in \mathbb{N}_0$ in this case) or also from the functional calculus based on the joint spectral resolution of A and B.

Lemma 9.3.11. *If* $\alpha, \alpha' \in \mathbb{C} \setminus \sigma(A)$ *and* $\beta, \beta' \in \mathbb{C} \setminus \sigma(B)$, *then* $\dim Q_{1,1}(\alpha, \beta) \mathcal{H} = \dim Q_{1,1}(\alpha', \beta') \mathcal{H}$.

Proof. By Lemma 9.3.4,

$$\mathcal{D}_{1,1} = X_{\bar{\alpha}} Y_{\bar{\beta}} \big(I - Q_{1,1}(\alpha, \beta)\big) \mathcal{H} = X_{\bar{\alpha}} Y_{\bar{\beta}'} \big(I - Q_{1,1}(\alpha, \beta')\big) \mathcal{H}.$$

Since $\ker X_\alpha = \{0\}$,

$$(B - \overline{\beta'}) \, Y_{\bar{\beta}} \big(I - Q_{1,1}(\alpha, \beta)\big) \mathcal{H} = \big(I - Q_{1,1}(\alpha, \beta')\big) \mathcal{H}.$$

Since $(B - \overline{\beta'}) \, Y_{\bar{\beta}}$ is an isomorphism of $\mathcal{H}$ (by the spectral theorem), the preceding gives

$$\operatorname{codim} \big(I - Q_{1,1}(\alpha, \beta)\big) \mathcal{H} = \operatorname{codim} \big(I - Q_{1,1}(\alpha, \beta')\big) \mathcal{H},$$

i.e.,

$$\dim Q_{1,1}(\alpha, \beta) \mathcal{H} = \dim Q_{1,1}(\alpha, \beta') \mathcal{H}.$$

Since also

$$\mathcal{D}_{1,1} = Y_{\bar{\beta}'} X_{\bar{\alpha}} \big(I - Q_{1,1}(\alpha, \beta')\big) \mathcal{H} = Y_{\bar{\beta}'} X_{\bar{\alpha}'} \big(I - Q_{1,1}(\alpha', \beta')\big) \mathcal{H},$$

the same reasoning shows that $\dim Q_{1,1}(\alpha, \beta') \mathcal{H} = \dim Q_{1,1}(\alpha', \beta') \mathcal{H}$. □

Since, by Lemma 9.3.11, $\dim Q_{1,1}(\alpha, \beta) \mathcal{H}$ does not depend on the numbers $\alpha \in \mathbb{C} \setminus \sigma(A)$. and $\beta \in \mathbb{C} \setminus \sigma(B)$, the following definition is justified.

Definition 9.3.12. The dimension of the space $Q_{1,1}(\alpha, \beta) \mathcal{H}$, where $\alpha \in \mathbb{C} \setminus \sigma(A)$ and $\beta \in \mathbb{C} \setminus \sigma(B)$, is called the *defect number* of the couple $\{A, B\}$ and denoted by $d(A, B)$

Remark 2. The following fact indicates that $d(A, B)$ measures the distance to the strong commutativity in some sense: A and B strongly commute if and only if $d(A, B) = 0$. (Indeed, the latter is equivalent to $X_\alpha Y_\beta = Y_\beta X_\alpha$ for all $\alpha \in \mathbb{C} \setminus \sigma(A)$ and $\beta \in \mathbb{C} \setminus \sigma(B)$.)

The next proposition establishes a one-to-one correspondence between couples in $\boldsymbol{N}_\infty$ and certain self-adjoint representations of the polynomial algebra $\mathbb{C}[\mathrm{x}_1, \mathrm{x}_2]$.

Proposition 9.3.13. (i) *Suppose that* $\{A, B\} \in \boldsymbol{N}_\infty$. *Then* $\pi(\mathrm{x}_1) := A \upharpoonright \mathcal{D}_\infty(A, B)$ *and* $\pi(\mathrm{x}_2) := B \upharpoonright \mathcal{D}_\infty(A, B)$ *defines a self-adjoint representation of* $\mathbb{C}[\mathrm{x}_1, \mathrm{x}_2]$ *on* $\mathcal{D}(\pi) := \mathcal{D}_\infty(A, B)$ *for which the operators* $\overline{\pi(\mathrm{x}_1)}$ *and* $\overline{\pi(\mathrm{x}_2)}$ *are self-adjoint. The representation* π *is integrable if and only if* $d(A, B) = 0$. *Moreover, we have that* $\pi(\mathbb{C}[\mathrm{x}_1, \mathrm{x}_2])' = (A)' \cap (B)'$.

(ii) *Suppose* π *is a self-adjoint representation of* $\mathbb{C}[\mathrm{x}_1, \mathrm{x}_2]$ *such that* $\overline{\pi(\mathrm{x}_1)}$ *and* $\overline{\pi(\mathrm{x}_2)}$ are *self-adjoint operators. Then* $\{\overline{\pi(\mathrm{x}_1)}, \overline{\pi(\mathrm{x}_2)}\} \in \boldsymbol{N}_\infty$, *and* $\mathcal{D}(\pi) = \mathcal{D}_\infty\big(\overline{\pi(\mathrm{x}_1)}, \overline{\pi(\mathrm{x}_2)}\big)$.

Proof. (i): From Lemma 9.3.3, (ii), we see immediately that π defines indeed a $*$-representation of $\mathbb{C}[\mathrm{x}_1, \mathrm{x}_2]$. The operators $\overline{\pi(\mathrm{x}_1)}$ and $\overline{\pi(\mathrm{x}_2)}$ are self-adjoint, since $\{A, B\} \in \boldsymbol{N}_\infty$ and hence $\overline{\pi(\mathrm{x}_1)} = A$ and $\overline{\pi(\mathrm{x}_2)} = B$ by Definition 9.3.1, (iii). We show that π is self-adjoint. From $\pi(x_1)^* = A$ and $\pi(x_2)^* = B$ we conclude that the domain $\mathcal{D} := \mathcal{D}(\pi^*)$ satisfies the conditions (i) and (ii) in Definition 9.3.1. Therefore, by Lemma 9.3.3, (ii), $\mathcal{D}(\pi^*) \subseteq \mathcal{D}_\infty(A, B) = \mathcal{D}(\pi)$, so π is self-adjoint. From Corollary 9.1.14, π is integrable if and only if the self-adjoint operators $\overline{\pi(\mathsf{x}_1)} = A$ and $\overline{\pi(\mathsf{x}_2)} = B$ strongly commute. As noted in Remark 2, the latter is equivalent to the equality $d(A, B) = 0$. Since $(\pi(\mathsf{x}_1))'_{\mathrm{w}} = (\overline{\pi(\mathsf{x}_1)})'_{\mathrm{w}} = (A)'$ and also $(\pi(\mathsf{x}_2))'_{\mathrm{w}} = (B)'$, Corollary 8.2.8 yields $\pi(\mathbb{C}[\mathsf{x}_1, \mathsf{x}_2])' = (A)' \cap (B)'$.

(ii): Letting $\mathcal{D} := \mathcal{D}(\pi)$, $A := \overline{\pi(\mathsf{x}_1)}$ and $B := \overline{\pi(\mathsf{x}_2)}$ in Definition 9.3.1, we see that $\{A, B\} \in \boldsymbol{N}_\infty$. Using once more the characterization of $\mathcal{D}_\infty(A, B)$ given in Lemma 9.3.3, (ii), we obtain that $\mathcal{D}(\pi) \subseteq \mathcal{D}_\infty(A, B)$. On the other hand, $\pi_0(\mathsf{x}_1) := A \restriction \mathcal{D}_\infty(A, B)$ and $\pi_0(\mathsf{x}_2) := B \restriction \mathcal{D}_\infty(A, B)$ define a $*$-representation of $\mathbb{C}[\mathsf{x}_1, \mathsf{x}_2]$ with $\pi \subseteq \pi_0$. Since π is self-adjoint by assumption, this implies that $\pi = \pi_0$ and so $\mathcal{D}(\pi) = \mathcal{D}_\infty(A, B)$. □

We illustrate the preceding investigations by an example.

Example 9.3.14. Let S be the unilateral shift on the Hardy space $\mathcal{H} \equiv H^2(\mathbb{T})$. That is, $(S\varphi)(z) = z\varphi(z)$ for $\varphi \in H^2(\mathbb{T})$. Put $X_0 := S + S^*$ and $Y_0 := -\mathrm{i}(S - S^*)$. Since $\ker X_0 = \ker Y_0 = \{0\}$, $A := X_0^{-1}$ and $B := Y_0^{-1}$ are well-defined self-adjoint operators on $\mathcal{H}$. Clearly, $0 \notin \sigma(A)$ and $0 \notin \sigma(B)$. It is easy to check that $Q_{n,m}(0, 0)\, \mathcal{H} = \mathrm{l.h.} \{z^k : k = 0, \ldots, n + m - 2\}$ for $n, m \in \mathbb{N}$ and that the assumptions of Corollary 9.3.9 are fulfilled in case $\alpha = \beta = 0$. Hence $\{A, B\} \in \boldsymbol{N}_\infty$. Since $[X_0, Y_0]\, \mathcal{H}$ is one-dimensional, $d(A,B) = 1$. By Proposition 9.3.13, $\pi(\mathsf{x}_1) := A \restriction \mathcal{D}_\infty(A,B)$ and $\pi(\mathsf{x}_2) := B \restriction \mathcal{D}_\infty(A,B)$ defines a non-integrable self-adjoint representation π of $\mathbb{C}[\mathsf{x}_1, \mathsf{x}_2]$. Since $z \in Y_0\mathcal{H} \cap Q_{2,1}(0,0)\, \mathcal{H}$, Lemma 9.3.5, (i), shows that $A^2 \restriction \mathcal{D}_{2,1}$ and hence $\pi(\mathsf{x}_1)^2 \equiv A^2 \restriction \mathcal{D}_\infty(A,B)$ is not essentially self-adjoint. Similarly, $\pi(\mathsf{x}_2)^2$ is not essentially self-adjoint. ○

An Auxiliary Construction

A useful method for the construction of couples $\{A, B\}$ in $\boldsymbol{N}_\infty$ with non-zero defect numbers and so of non-integrable self-adjoint representations of $\mathbb{C}[\mathsf{x}_1, \mathsf{x}_2]$ is obtained by the following general setup. We let the Cayley transforms of A and B be the vector-valued bilateral shift operator and a diagonal operator with unitary diagonal entries, respectively. By specifying these entries, we can produce couples in $\boldsymbol{N}_\infty$ which have special additional properties. The proof of Theorem 9.4.1 will be based on this method. In this subsection we develop some preliminaries for the proof of this theorem.

Let $\mathcal{H}$ be a Hilbert space. Set $\mathcal{H}_{\mathbb{Z}} := \sum_{n \in \mathbb{Z}} \oplus \mathcal{H}_n$, where each $\mathcal{H}_n$ is $\mathcal{H}$. Vectors of $\mathcal{H}_{\mathbb{Z}}$ will be written as sequences (φ_n) or as $(\ldots, \varphi_0, \underline{\varphi_1}, \varphi_2, \ldots)$, where the component with index 1 is underlined. For $\varphi \in \mathcal{H}_{\mathbb{Z}}$ and $n \in \mathbb{Z}$, φ_n denotes the n-th component of φ. If $\mathcal{M}_2, \ldots, \mathcal{M}_k$ are subsets of $\mathcal{H}$, then $(\ldots, \underline{0}, \mathcal{M}_2, \ldots, \mathcal{M}_k, 0, \ldots)$ means the set of all $\varphi \in \mathcal{H}_{\mathbb{Z}}$ such that $\varphi_n \in \mathcal{M}_n$ if $n = 2, \ldots, k$ and $\varphi_n = 0$ otherwise. The symbol $(\ldots, \underline{0}, 0, \mathcal{M}_3, \ldots, \mathcal{M}_k, 0, \ldots)$ has a similar meaning. Let U be the vector-valued bilateral shift on $\mathcal{H}_{\mathbb{Z}}$, i.e., U is defined by $U(\ldots, \varphi_0, \underline{\varphi_1}, \varphi_2, \ldots) = (\ldots, \underline{\varphi_0}, \varphi_1, \varphi_2, \ldots)$. Let v_1, v_2 and v_3 be unitary operators on $\mathcal{H}$ which will be specified later. Suppose that $\ker(v_n - I) = \{0\}$ for $n = 1, 2, 3$. Set $v_n = v_1$ if $n \in \mathbb{Z}$, $n < 1$, and $v_n = v_3$ if $n \in \mathbb{Z}$, $n > 3$. Define a unitary operator V on $\mathcal{H}_{\mathbb{Z}}$ by $V(\varphi_n) = (v_n\varphi_n)$ for $(\varphi_n) \in \mathcal{H}_{\mathbb{Z}}$. Then $\ker(U - I) = \ker(V - I) = \{0\}$, so that $A := \mathrm{i}(U + I)(U - I)^{-1}$ and $B := \mathrm{i}(V + I)(V - I)^{-1}$

define self-adjoint operators on $\mathcal{H}_{\mathbb{Z}}$. Let $\mathcal{H}_{12}$ and $\mathcal{H}_{23}$ denote the closures of $(v_1 - v_2)\,\mathcal{H}$ and $(v_2 - v_3)\,\mathcal{H}$ in $\mathcal{H}$, respectively. For $m \in \mathbb{N}$, $z \in \mathbf{B}(\mathcal{H})$ and $\mathcal{M} \subseteq \mathcal{H}$, $\mathcal{S}_m(z; \mathcal{M})$ denotes the linear span of $z^n\mathcal{M}$, $n = 0, \ldots, m-1$.

Lemma 9.3.15. *Suppose that the operators v_1, v_2 and v_3 satisfy the following conditions:*

(i) *The linear spaces $\mathcal{S}_m(v_2; \mathcal{H}_{12})$, $\mathcal{S}_m(v_3; \mathcal{H}_{23})$ and $\mathcal{S}_m(v_3; \mathcal{H}_{12} + \mathcal{H}_{23})$ are closed in $\mathcal{H}$ for each $m \in \mathbb{N}$.*

(ii) $(I - v_2)\,\mathcal{H} \cap \mathcal{H}_{12} = (I - v_3)\,\mathcal{H} \cap (\mathcal{H}_{12} + \mathcal{H}_{23}) = \{0\}$.

(iii) $\mathcal{S}_m(v_2; \mathcal{H}_{12}) \cap \mathcal{S}_m(v_3; \mathcal{H}_{23}) = \{0\}$ *for* $m \in \mathbb{N}$.

Then, $\{A, B\} \in \mathcal{N}_{\infty}^{\infty}$.

Proof. From (i) and from the concrete form of the operators U and V we obtain for n and m in $\mathbb{N}$

$$Q_{n,m}\mathcal{H} = \big(\ldots, \underset{-}{0}, \mathcal{S}_m(v_2; \mathcal{H}_{12}), \mathcal{S}_m(v_3; \mathcal{H}_{12} + \mathcal{H}_{23}), \ldots, \mathcal{S}_m(v_3; \mathcal{H}_{12} + \mathcal{H}_{23}),$$
$$\mathcal{S}_m(v_3; \mathcal{H}_{23}), 0, \ldots\big), \tag{1}$$

where the space $\mathcal{S}_m(v_3; \mathcal{H}_{23})$ stands at the place with index $n + 2$. By Corollary 9.3.10, it suffices to show that the conditions $(\mathrm{I})_{n,m}$ and $(\mathrm{II})_{n,m}$ are fulfilled for all $n, m \in \mathbb{N}$. Fix $n \in \mathbb{N}$ and $m \in \mathbb{N}$. Suppose that $(I - U)\,\varphi \in Q_{n,m}\mathcal{H}$ for some $\varphi \in \mathcal{H}$. First let $n = 1$. From (1) it then follows that $\varphi_2 \in \mathcal{S}_m(v_2; \mathcal{H}_{12}) \cap \mathcal{S}_m(v_3; \mathcal{H}_{23})$ and $\varphi_k = 0$ if $k \in \mathbb{Z}, k \neq 2$. By (iii), $\varphi = 0$ and $(\mathrm{I})_{1,m}$ is proved. Now suppose that $n \geq 2$. Then, by (1), $\varphi_2 \in \mathcal{S}_m(v_2; \mathcal{H}_{12})$, $\varphi_k - \varphi_{k-1} \in \mathcal{S}_m(v_3; \mathcal{H}_{12} + \mathcal{H}_{23})$ if $k = 3, \ldots, n+1$ and $\varphi_k = 0$ if $k \leq 1$ and if $k \geq n + 2$. Hence $\varphi_k \in \mathcal{S}_m(v_3; \mathcal{H}_{12} + \mathcal{H}_{23})$ if $k \in \mathbb{N}$, $2 \leq k \leq n$, so that, again by (1), $\varphi \in Q_{n-1,m}\mathcal{H}$. This proves $(\mathrm{I})_{n,m}$.

Now suppose $(I - V)\,\varphi \in Q_{n,m}\mathcal{H}$ for $\varphi \in \mathcal{H}$. We treat only the case where $m \geq 2$. A slight modification of this argument also applies in case $m = 1$. By (1), $(I - v_2)\,\varphi_2 \in \mathcal{S}_m(v_2; \mathcal{H}_{12})$, $(I - v_3)\,\varphi_k \in \mathcal{S}_m(v_3; \mathcal{H}_{12} + \mathcal{H}_{23})$ if $k \in \mathbb{N}$, $3 \leq k \leq n + 1$, and $(I - v_3)\,\varphi_{n+2} \in \mathcal{S}_m(v_3; \mathcal{H}_{23})$. Further, $\varphi_k = 0$ if $k \leq 1$ and if $k \geq n + 3$. Let $k \in \mathbb{N}$, $3 \leq k \leq n + 1$. Since $m \geq 2$, it follows from the definitions of the spaces $\mathcal{S}_m(\cdot\,;\cdot)$ that $(I - v_3)\,\varphi_k$ can be written as $(I - v_3)\,\psi_k + \zeta_k$ with $\psi_k \in \mathcal{S}_{m-1}(v_3; \mathcal{H}_{12} + \mathcal{H}_{23})$ and $\zeta_k \in \mathcal{H}_{12} + \mathcal{H}_{23}$. Thus $(I - v_3)\,(\varphi_k - \psi_k) = \zeta_k \in \mathcal{H}_{12} + \mathcal{H}_{23}$. By (ii), $\varphi_k - \psi_k = 0$, so $\varphi_k \in \mathcal{S}_{m-1}(v_3; \mathcal{H}_{12} + \mathcal{H}_{23})$. A similar reasoning shows that $\varphi_2 \in \mathcal{S}_{m-1}(v_2; \mathcal{H}_{12})$ and $\varphi_{n+2} \in \mathcal{S}_{m-1}(v_3; \mathcal{H}_{23})$. Therefore, again by (1), $\varphi \in Q_{n,m-1}\mathcal{H}$ which proves $(\mathrm{II})_{n,m}$. □

The following example illustrates how the preceding lemma can be used for the construction of couples $\{A, B\}$ in $\mathcal{N}_{\infty}^{\infty}$ with non-zero $d(A, B)$.

Example 9.3.16. Let v_1 be a unitary on $\mathcal{H}$, and let e_1 and e_2 be projections on $\mathcal{H}$ such that $\ker(v_1 - I) = \{0\}$ and $(e_1\mathcal{H} + e_2\mathcal{H}) \cap (I - v_1)\,\mathcal{H} = \{0\}$. Suppose that for each $m \in \mathbb{N}$ $\mathcal{S}_m(v_1; e_1\mathcal{H})$ and $\mathcal{S}_m(v_2; e_2\mathcal{H})$ are closed linear subspaces of $\mathcal{H}$ which intersect only in $\{0\}$. (For instance, all these assumptions are certainly fulfilled if v_1 is the Cayley transform of an unbounded self-adjoint operator a, $e_1 = 0$ and e_2 is a finite rank projection such that $e_2\mathcal{H} \cap \mathcal{D}(a) = \{0\}$.) Set $v_2 := (I - 2e_1)\,v_1$ and $v_3 := (I - 2e_2)\,(I - 2e_1)\,v_1$. Then the assumptions of Lemma 9.3.15 are satisfied, so that $\{A, B\} \in \mathcal{N}_{\infty}^{\infty}$. Clearly, $d(A, B) \neq 0$ if $e_1 \neq 0$ or if $e_2 \neq 0$. ○

9.4. Construction of Non-Integrable Self-Adjoint Representations of $\mathbb{C}[x_1, x_2]$

Self-Adjoint Representations of Types II_∞ and III

If π is a self-adjoint representation of a commutative *-algebra A with unit such that the von Neumann algebra $\pi(A)''$ is finite, then π is integrable and hence $\pi(A)''$ is abelian. (Indeed, since π is self-adjoint, each operator $\overline{\pi(a)}$, $a \in A$, is affiliated with $\pi(A)'' \equiv (\pi(A)'_{ss})'$. Since the von Neumann algebra $\pi(A)''$ is finite, this implies that $\overline{\pi(a)}$ is a self-adjoint operator for each $a \in A_h$ (KADISON/RINGROSE [2], 6.9.53). By Theorem 9.1.2, π is integrable.)

In sharp contrast to this fact we now prove that each properly infinite von Neumann algebra on a separable Hilbert space is equal to the bicommutant $\pi(\mathbb{C}[x_1, x_2])''$ for some self-adjoint (of course, non-integrable) representation π of the polynomial algebra $\mathbb{C}[x_1, x_2]$. Since the type of π was defined to be the type of the von Neumann algebra $\pi(\mathbb{C}[x_1, x_2])''$ (cf. Definition 8.4.2), this shows that the polynomial algebra $\mathbb{C}[x_1, x_2]$ has self-adjoint representations of types I_∞, II_∞ and III. (In case of I_∞ much simpler examples can be constructed, see Example 9.4.6 below.)

Theorem 9.4.1. *Suppose that $\mathcal{N}$ is a properly infinite von Neumann algebra on a separable Hilbert space $\mathcal{H}$. Then there exists a self-adjoint representation π of the *-algebra $\mathbb{C}[x_1, x_2]$ such that $\pi(\mathbb{C}[x_1, x_2])'' = \mathcal{N}$ and such that the operators $\pi(x_1)^n$ and $\pi(x_2)^n$ are essentially self-adjoint for all $n \in \mathbb{N}$.*

Throughout this subsection, we retain the assumptions of Theorem 9.4.1. For an index set I, we set $\mathcal{H}_I := \sum_{i \in I} \oplus \mathcal{H}_i$, where $\mathcal{H}_i := \mathcal{H}$ for $i \in I$, and we let $M_I(\mathcal{N})$ denote the von Neumann algebra of all matrices $[a_{nm}]_{n,m \in I}$ over $\mathcal{N}$ which act boundedly on $\mathcal{H}_I$. In the proof of Theorem 9.4.1 we require some auxiliary lemmas.

Lemma 9.4.2. *If the index set I is finite or countable, then the von Neumann algebras $\mathcal{N}$ and $M_I(\mathcal{N})$ are spatially isomorphic.*

Proof. TOPPING [1], § 7, Corollary 14. □

Lemma 9.4.3. *The properly infinite von Neumann algebra $\mathcal{N}$ on the separable Hilbert space $\mathcal{H}$ is generated (as a von Neumann algebra) by a self-adjoint operator a and a projection q, i.e., $\{a, q\}'' = \mathcal{N}$.*

Proof. Since the von Neumann algebras $\mathcal{N}$ and $M_{\mathbb{N}}(\mathcal{N})$ are spatially isomorphic by Lemma 9.4.2, it is sufficient to prove the assertion with $M_{\mathbb{N}}(\mathcal{N})$ in place of $\mathcal{N}$. By assumption the Hilbert space $\mathcal{H}$ is separable. Hence there exists a countable subset $\{a_n : n \in \mathbb{N}\}$ of $\mathcal{N}$ which generates $\mathcal{N}$ as a von Neumann algebra. Obviously, we can assume without loss of generality that the operators a_n are self-adjoint and satisfy $I \leqq a_n \leqq 2I$. We define a and q by the infinite matrices

$$a := \begin{bmatrix} 0 & a_1 & 0 & \dots \\ a_1 & 0 & a_2 & \dots \\ 0 & a_2 & 0 & \dots \\ \dots & \dots & \dots & \dots \end{bmatrix} \quad \text{and} \quad q := \begin{bmatrix} I & 0 & 0 & \dots \\ 0 & 0 & 0 & \dots \\ 0 & 0 & 0 & \dots \\ \dots & \dots & \dots & \dots \end{bmatrix}.$$

For $r \in \mathbb{N}$, let q_r denote the matrix $[\delta_{nr}\delta_{mr}]_{n,m\in\mathbb{N}}$. Let $\mathcal{M}$ be the von Neumann algebra $\{a, q\}''$. We prove by induction on r that $q_r \in \mathcal{M}$ for every $r \in \mathbb{N}$. In case $r = 1$ this is clear, since $q_1 = q$. Suppose that $q_1, \ldots, q_r \in \mathcal{M}$ for some $r \in \mathbb{N}$. Then we have $b_r := (I - q_1 - \cdots - q_r)\, aq_r \in \mathcal{M}$. The only non-vanishing matrix entry of b_r is a_r in the $(r+1, r)$-th position. Hence $b_r b_r^*$ has a_r^2 in the $(r+1, r+1)$-th position and zeros elsewhere. Take a sequence $(p_k : k \in \mathbb{N})$ of polynomials in one variable such that $p_k(t) \to t^{-1}$ uniformly on $[1, 2]$. Then $a_r^2 p_k(a_r^2) \to I$ on $\mathcal{H}$ and so $b_r b_r^* p_k(b_r b_r^*) \to q_{r+1}$ on $\mathcal{H}_{\mathbb{N}}$ as $k \to \infty$ in the corresponding operator norms. Since $b_r \in \mathcal{M}$ and so $b_r b_r^* p_k(b_r b_r^*) \in \mathcal{M}$, this gives $q_{r+1} \in \mathcal{M}$, and the induction proof is complete.

Now let $c = c^* \in \mathcal{M}' \equiv \{a, q\}'$. We write c as a matrix $[c_{nm}]_{n,m\in\mathbb{N}}$ over $\mathbb{B}(\mathcal{H})$. Since $q_r \in \mathcal{M}$ as just shown, we have $cq_r = q_r c$ for all $r \in \mathbb{N}$. This yields $c_{nm} = 0$ for $n, m \in \mathbb{N}$, $n \neq m$. Set $c_n := c_{nn}$ for $n \in \mathbb{N}$. From the equality $ca = ca$ we obtain $c_n a_n = a_n c_{n+1}$ for $n \in \mathbb{N}$. Fix $n \in \mathbb{N}$. Recall that the operators a_n (by construction) and c_n, c_{n+1} (because of $c = c^*$) are self-adjoint. Taking the adjoints in $c_n a_n = a_n c_{n+1}$, we get $a_n c_n = c_{n+1} a_n$. Hence $c_n a_n^2 = a_n c_{n+1} a_n = a_n^2 c_n$. Since $a_n \geqq I$, it follows that c_n also commutes with $a_n \equiv (a_n^2)^{1/2}$ and so $a_n c_{n+1} = c_n a_n = a_n c_n$ which yields $c_{n+1} = c_n$. Thus $c_1 = c_n$ for all $n \in \mathbb{N}$. Using the equality $c_n a_n = a_n c_{n+1}$ once more, we obtain $c_1 \in \{a_n : n \in \mathbb{N}\}' \equiv \mathcal{N}'$. Therefore, $c \in M_{\mathbb{N}}(\mathcal{N})'$ and so $\mathcal{M}' \subseteq M_{\mathbb{N}}(\mathcal{N})'$. Since $a, q \in M_{\mathbb{N}}(\mathcal{N})$ by construction, $M_{\mathbb{N}}(\mathcal{N})' \subseteq \{a, q\}' \equiv \mathcal{M}'$. Thus $\mathcal{M} \equiv \{a, q\}'' = M_{\mathbb{N}}(\mathcal{N})$. □

Lemma 9.4.4. *There exist a unitary operator $w \in \mathcal{N}$ and a projection $e \in \mathcal{N}$ such that $\mathcal{H}_w \cap e\mathcal{H} = \{0\}$, $\mathcal{H}_w + e\mathcal{H}_w = \mathcal{H}$ and $\{w, e\}'' = \mathcal{N}$, where $\mathcal{H}_w$ denotes the closure of $(w - I)\,\mathcal{H}$ in $\mathcal{H}$.*

Proof. For the index set $I := \{1, 2\}$, we set $\mathcal{H}_2 := \mathcal{H}_I$ and $M_2(\mathcal{N}) := M_I(\mathcal{N})$. By Lemma 9.4.2, $\mathcal{N}$ is spatially isomorphic to the von Neumann algebra $M_2(\mathcal{N})$ of all 2×2 matrices over $\mathcal{N}$ acting on $\mathcal{H}_2 \equiv \mathcal{H} \oplus \mathcal{H}$. Thus it suffices to prove the assertion for the von Neumann algebra $M_2(\mathcal{N})$ on $\mathcal{H}_2$. By Lemma 9.4.3, there are a self-adjoint operator $a \in \mathcal{N}$ and a projection $q \in \mathcal{N}$ such that $\{a, q\}'' = \mathcal{N}$. Obviously, we can assume that $\frac{1}{2} \cdot I \leqq a \leqq \frac{3}{4} \cdot I$. We define the unitary operator $w \in M_2(\mathcal{N})$ and the projection $e \in M_2(\mathcal{N})$ by the matrices

$$w := \begin{bmatrix} \mathrm{i}(I - 2q) & 0 \\ 0 & I \end{bmatrix} \quad \text{and} \quad e := \begin{bmatrix} a & b \\ b & I - a \end{bmatrix},$$

where $b := (a - a^2)^{1/2}$. Obviously, $(\mathcal{H}_2)_w = \mathcal{H} \oplus \{0\}$ and $e\mathcal{H}_2 = \{(\varphi, ba^{-1}\varphi) : \varphi \in \mathcal{H}\}$, so $(\mathcal{H}_2)_w \cap e\mathcal{H}_2 = \{0\}$ and $(\mathcal{H}_2)_w + e(\mathcal{H}_2)_w = \mathcal{H}_2$.

It remains to prove that $\{w, e\}'' = M_2(\mathcal{N})$. Suppose $c \in \{w, e\}'$. We write c as a 2×2 matrix $[c_{nm}]_{n,m=1,2}$ over $\mathbb{B}(\mathcal{H})$. The equality $wc = cw$ gives $qc_{11} = c_{11}q$, $\mathrm{i}(I - 2q)\, c_{12} = c_{12}$ and $c_{21} = \mathrm{i}(I - 2q)\, c_{21}$. Since $\mathrm{i}(I - 2q)$ has a bounded inverse on $\mathcal{H}$, $c_{12} = c_{21} = 0$. The relation $ec = ce$ yields $ac_{11} = c_{11}a$ and $bc_{22} = c_{11}b$. Therefore, $c_{11} \in \{a, q\}'$. Since $\{a, q\}'' = \mathcal{N}$ by Lemma 9.4.3, we have $c_{11} \in \mathcal{N}'$. Thus $bc_{22} = c_{11}b = bc_{11}$. Since b has a bounded inverse on $\mathcal{H}$, $c_{22} = c_{11}$ and hence $c \in M_2(\mathcal{N})'$. This shows that $\{w, e\}' \subseteq M_2(\mathcal{N})'$. The opposite inclusion is trivial, so $\{w, e\}'' = M_2(\mathcal{N})$. □

For Lemma 9.4.5 and for the proof of Theorem 9.4.1 we shall retain the notation introduced in the second subsection of 9.3.

Lemma 9.4.5. (i) *If* l.h. $\{v_1^k\mathcal{H}_{12}, v_3^k\mathcal{H}_{12}: k \in \mathbb{Z}\}$ *is dense in* $\mathcal{H}$ *and* $\{v_1, v_2, v_3\}'' = \mathcal{N}$, *then* $\{U, V\}'' = \boldsymbol{M}_{\mathbb{Z}}(\mathcal{N})$.
(ii) *Let* x_1 *and* x_2 *be operators from* $\mathcal{N}$ *such that* $x_1\mathcal{H} + x_2\mathcal{H}$ *is dense in* $\mathcal{H}$. *Define operators* y_1 *and* y_2 *on* $\mathcal{H}_{\mathbb{Z}}$ *by* $y_1(\varphi_n) = (x_1\delta_{1n}\varphi_n)$ *and* $y_2(\varphi_n) = (x_2\delta_{2n}\varphi_n)$ *for* $(\varphi_n) \in \mathcal{H}_{\mathbb{Z}}$. *If the von Neumann algebra* $\mathcal{N}$ *is generated by* x_1 *and* x_2, *then* $\boldsymbol{M}_{\mathbb{Z}}(\mathcal{N})$ *is generated by* U, y_1 *and* y_2.

Proof. (i): Suppose $c = c^* \in \{U, V\}'$. As above, we write c as an infinite matrix $[c_{nm}]_{n,m\in\mathbb{Z}}$ over $\mathbf{B}(\mathcal{H})$. Since $U, V \in \boldsymbol{M}_{\mathbb{Z}}(\mathcal{N})$ and so $\{U, V\}'' \subseteq \boldsymbol{M}_{\mathbb{Z}}(\mathcal{N})$, it is sufficient to show that $c \in \boldsymbol{M}_{\mathbb{Z}}(\mathcal{N})'$, that is, c is diagonal, and the entries c_{nn} do not depend on $n \in \mathbb{Z}$ and belong to $\mathcal{N}'$.

Because of $Uc = cU$, we have $c_{nm} = c_{n-1,m-1}$ for all $n, m \in \mathbb{Z}$. The relation $Vc = cV$ yields $c_{nm}v_m = v_nc_{nm}$ for all $n, m \in \mathbb{Z}$.

Fix $n, m \in \mathbb{Z}$. We first check that $c_{nm} \in \{v_1, v_3\}'$. We choose $r \in \mathbb{N}$ such that $n - r \leqq 1$ and $m - r \leqq 1$. Then $v_1 = v_{n-r} = v_{m-r}$ and hence

$$c_{nm}v_1 = c_{n-r,m-r}v_{m-r} = v_{n-r}c_{n-r,m-r} = v_1c_{nm}.$$

Similarly, $c_{nm}v_3 = v_3c_{nm}$. Now suppose $n, m \in \mathbb{Z}$, $n < m$. Then

$$c_{nm}v_2 = c_{n-(m-2),2}v_2 = v_{n-(m-2)}c_{n-(m-2),2} = v_1c_{nm} = c_{nm}v_1.$$

This yields $c_{nm}(v_1 - v_2) = 0$ and so $c_{nm}\mathcal{H}_{12} = 0$. Since c_{nm} commutes with v_1 and v_3 as just shown, it follows from our first assumption that $c_{nm} = 0$. Since $c = c^*$, $c_{mn} = c_{nm}^*$. Thus $c_{mn} = 0$. This proves that c is diagonal. Since $c_{nn} = c_{n-1,n-1}$ as noted above, the diagonal entries do not depend on $n \in \mathbb{Z}$. As mentioned above, $c_{22}v_2 = v_2c_{22}$. Therefore, $c_{22} \in \{v_1, v_2, v_3\}'$. By assumption, the latter is equal to $\mathcal{N}'$. Hence $c_{22} \in \mathcal{N}'$ and $c \in \boldsymbol{M}_{\mathbb{Z}}(\mathcal{N})'$.

(ii): The proof is similar to the proof of (i). Take a $c \equiv [c_{nm}]$ from $\{U, y_1, y_2, y_1^*, y_2^*\}'$. As in (i), $Uc = cU$ yields $c_{nm} = c_{n-1,m-1}$ for $n, m \in \mathbb{Z}$. Combined with $y_kc = cy_k$, this implies that $c_{nm}x_k = 0$ for $k = 1, 2$ and $n, m \in \mathbb{Z}$, $n \neq m$. Since $x_1\mathcal{H} + x_2\mathcal{H}$ is dense in $\mathcal{H}$, $c_{nm} = 0$ for all $n, m \in \mathbb{Z}$, $n \neq m$. Hence c is diagonal with diagnonals not depending on n. Since $c \in \{y_1, y_2, y_1^*, y_2^*\}'$, we obtain $c_{11} \in \{x_1, x_2, x_1^*, x_2^*\}'$. By assumption the latter is $\mathcal{N}'$, so $c \in \boldsymbol{M}_{\mathbb{Z}}(\mathcal{N})'$. □

Proof of Theorem 9.4.1. Since $\mathcal{N}$ and $\boldsymbol{M}_{\mathbb{Z}}(\mathcal{N})$ are spatially isomorphic by Lemma 9.4.2, it suffices to prove the theorem for the von Neumann algebra $\boldsymbol{M}_{\mathbb{Z}}(\mathcal{N})$ on $\mathcal{H}_{\mathbb{Z}}$. By Proposition 9.3.13,(i), the proof is complete once we have shown that there exists a couple $\{\tilde{A}, \tilde{B}\} \in \boldsymbol{N}_\infty^\infty$ on the Hilbert space $\mathcal{H}_{\mathbb{Z}}$ such that $\{(\tilde{A})' \cap (\tilde{B})'\}' = \boldsymbol{M}_{\mathbb{Z}}(\mathcal{N})$. (Recall that we prove the theorem with $\mathcal{N}$ replaced by $\boldsymbol{M}_{\mathbb{Z}}(\mathcal{N})$.) To do this we construct $\tilde{A}$ and $\tilde{B}$ of the form $\tilde{A} = A$ and $\tilde{B} = B$, where A and B are as in the second subsection of 9.3. We also use the notation introduced therein. Obviously, $(A)' = (U)'$ and $(B)' = (V)'$. Therefore, it is sufficient to show that there are unitaries $v_1, v_2, v_3 \in \mathcal{N}$ such that $\ker(v_k - I) = \{0\}$ for $k = 1, 2, 3$ and such that the assumptions of Lemma 9.3.15 and of Lemma 9.4.5, (i), are satisfied. Using once more the fact that $\mathcal{N}$ and $\boldsymbol{M}_{\mathbb{Z}}(\mathcal{N})$ are spatially isomorphic, we conclude that it suffices to prove the latter assertion with $\mathcal{N}$ replaced by $\boldsymbol{M}_{\mathbb{Z}}(\mathcal{N})$ and $\mathcal{H}$ by $\mathcal{H}_{\mathbb{Z}}$. Let $w \in \mathcal{N}$ and $e \in \mathcal{N}$ be as in Lemma 9.4.4. We define $v_1 := U$,

$$v_2(\ldots, \varphi_0, \underline{\varphi_1}, \varphi_2, \ldots) := (\ldots, \underline{\varphi_0}, w_1\varphi_1, \varphi_2, \ldots)$$

and

$$v_3(\ldots, \varphi_0, \underline{\varphi_1}, \varphi_2, \ldots) := (\ldots, \varphi_0, \underline{w_1\varphi_1}, w_2\varphi_2, \varphi_3, \ldots)$$

for $(\ldots, \varphi_0, \underline{\varphi_1}, \varphi_2, \ldots) \in \mathcal{H}_{\mathbb{Z}}$, where $w_1 := w$ and $w_2 := I - 2e$. Then, obviously, $\ker(v_k - I) = \{0\}$ for $k = 1, 2, 3$. Further, we have $\mathcal{H}_{12} = (\ldots, \underline{0}, \mathcal{H}_w, 0, \ldots)$ an d$\mathcal{H}_{23} = (\ldots, \underline{0}, 0, e\mathcal{H}, 0, \ldots)$. Since $\mathcal{H}_w + e\mathcal{H}_w = \mathcal{H}$ by Lemma 9.4.4, we have for $m \in \mathbb{N}$

$$\mathcal{S}_m(v_2; \mathcal{H}_{12}) = (\ldots, \underline{0}, \mathcal{H}_w, \mathcal{H}_w, \ldots, \mathcal{H}_w, 0, \ldots),$$

$$\mathcal{S}_m(v_3; \mathcal{H}_{23}) = (\ldots, \underline{0}, 0, e\mathcal{H}, e\mathcal{H}, \ldots, e\mathcal{H}, 0, \ldots) \text{ and}$$

$$\mathcal{S}_m(v_3; \mathcal{H}_{12} + \mathcal{H}_{23}) = (\ldots, \underline{0}, \mathcal{H}_w, \mathcal{H}, \ldots, \mathcal{H}, e\mathcal{H}, 0, \ldots),$$

where in case $m = 1$ the last formula has to be interpreted as $(\ldots, \underline{0}, \mathcal{H}_w, e\mathcal{H}, 0, \ldots)$. From these formulas, assumption (i) of Lemma 9.3.15 is obvious, and (ii) and (iii) follow easily from the fact that $\mathcal{H}_w \cap e\mathcal{H} = \{0\}$ by Lemma 9.4.4. We verify the assumptions of Lemma 9.4.5, (i). Since $\mathcal{H}_w + e\mathcal{H} = \mathcal{H}$, $\mathcal{H}_{12} + v_1^{-1}v_3\mathcal{H}_{12} = (\ldots, \underline{0}, \mathcal{H}, 0, \ldots)$, so that l.h. $\{v_1^k\mathcal{H}_{12}, v_3^k\mathcal{H}_{12} : k \in \mathbb{Z}\}$ is dense in $\mathcal{H}_{\mathbb{Z}}$. (Recall that we have to replace $\mathcal{H}$ by $\mathcal{H}_{\mathbb{Z}}$ and $\mathcal{N}$ by $M_{\mathbb{Z}}(\mathcal{N})$ in the assumptions of Lemma 9.4.5, (i).) In order to prove that $\{v_1, v_2, v_3\}'' = M_{\mathbb{Z}}(\mathcal{N})$, we apply Lemma 9.4.5, (ii). Letting $x_1 := w_1 - I$ and $x_2 := w_2 - I$, $v_1^{-1}v_2 - I$ and $v_1^{-1}v_3 - I$ are precisely the operators y_1 and y_2, respectively, as defined in Lemma 9.4.5, (ii). We have $x_1\mathcal{H} + x_2\mathcal{H} = (w - I)\,\mathcal{H} + e\mathcal{H} \supseteq (w - I)\,\mathcal{H} + e\mathcal{H}_w$. Hence $x_1\mathcal{H} + x_2\mathcal{H}$ is dense in $\mathcal{H}$, since $\mathcal{H}_w + e\mathcal{H}_w = \mathcal{H}$. Since $\mathcal{N}$ is generated by w and e and so by x_1 and x_2, Lemma 9.4.5, (ii), applies and shows that $M_{\mathbb{Z}}(\mathcal{N})$ is generated by U, y_1 and y_2 and so by v_1, v_2 and v_3. Thus the proof of the theorem is complete. □

Further Examples

The following example has been already quoted in Section 7.2.

Example 9.4.6. Let π be the (non-integrable) self-adjoint representation of the $*$-algebra $\mathbb{C}[x_1, x_2]$ defined in Example 9.3.14. Then $\mathcal{A} := \pi(\mathbb{C}[x_1, x_2])$ is a self-adjoint O*-algebra on $\mathcal{D}(\pi)$. The main objective of this example is to prove the following

Statement: There exists an operator $x \in \mathcal{A}_{\mathrm{w}}^{\mathrm{c}}$ such that $\mathcal{D}(\pi)$ is not contained in $\mathcal{D}(x^*)$. In particular, $\mathcal{A}_{\mathrm{w}}^{\mathrm{c}} \neq \mathcal{A}_{\mathrm{f}}^{\mathrm{c}}$.

Proof. We freely use the notation from the preceding section. We write X, Y and $Q_{n,m}$ for X_0, Y_0 and $Q_{n,m}(0, 0)$, respectively. Recall that $\overline{\pi(x_1)} = A = X^{-1} = (S + S^*)^{-1}$, $\overline{\pi(x_2)} = B = Y^{-1} = \mathrm{i}(S - S^*)^{-1}$ and $Q_{n,m}\mathcal{H} = \text{l.h.}\,\{z^k : k = 0, \ldots, n + m - 2\}$. Define $x := (S^*)^2 AB \upharpoonright \mathcal{D}(\pi)$.

We first show that $x \in \mathcal{A}_{\mathrm{w}}^{\mathrm{c}}$. It is obvious that $x \in \mathfrak{L}(\mathcal{D}_{\mathcal{A}}, \mathcal{H})$. To prove that $x \in \mathcal{A}_{\mathrm{w}}^{\mathrm{c}}$, it clearly suffices to show that for $k = 1,2$

$$\langle x\pi(x_k)\, \varphi, \psi \rangle = \langle x\varphi, \pi(x_k)\, \psi \rangle, \quad \varphi, \psi \in \mathcal{D}(\pi). \tag{1}$$

We let $\varphi, \psi \in \mathcal{D}(\pi)$. Since $\mathcal{D}(\pi) \subseteq \mathcal{D}_{2,1}$, there is a $\zeta \in (I - Q_{2,1})\,\mathcal{H}$ such that $\varphi = X^2Y\zeta$. Further, $\psi = X\xi$ for some $\xi \in \mathcal{H}$. From $Q_{2,1}\mathcal{H} = \text{l.h.}\,\{z^0, z\}$, $\varphi = XYX\zeta = YX^2\zeta$

and $(S^*)^2 X\zeta = X(S^*)^2 \zeta$. Therefore,

$$\begin{aligned}\langle x\pi(\mathsf{x}_1)\varphi, \psi\rangle &= \langle (S^*)^2 ABA\varphi, \psi\rangle = \langle (S^*)^2 ABAXYX\zeta, X\xi\rangle = \langle X(S^*)^2\zeta, \xi\rangle \\ &= \langle (S^*)^2 X\zeta, \xi\rangle = \langle (S^*)^2 ABYX^2\zeta, \xi\rangle = \langle (S^*)^2 AB\varphi, A\psi\rangle \\ &= \langle x\varphi, \pi(\mathsf{x}_1)\psi\rangle .\end{aligned}$$

A similar reasoning proves (1) in case $k = 2$.

Next we prove that $\mathcal{D}(\pi)$ is not contained in $\mathcal{D}(x^*)$. Assume to the contrary that $\mathcal{D}(\pi) \subseteq \mathcal{D}(x^*)$. Since the operator $y := x^* \upharpoonright \mathcal{D}(\pi)$ is closable and $\mathcal{D}_{\mathcal{A}} \equiv \mathcal{D}(\pi)[t_\pi]$ is a Frechet space, it follows from the closed graph theorem that y maps $\mathcal{D}_{\mathcal{A}}$ continuously into $\mathcal{H}$. We have $\|A^r B^s\varphi\| \leqq \|X\|\, \|A^{r+1}B^s\varphi\|$ and $\|A^r B^s\varphi\| \leqq \|Y\|\, \|A^r B^{s+1}\varphi\|$ for $\varphi \in \mathcal{D}(\pi)$ and $r, s \in \mathbb{N}_0$. Hence there are $\lambda > 0$ and $n \in \mathbb{N}$ such that $\|y\cdot\| \leqq \lambda\, \|A^n B^n\cdot\|$ on $\mathcal{D}(\pi)$, so that there exists a bounded operator Z on $\mathcal{H}$ satisfying $y = ZA^nB^n \upharpoonright \mathcal{D}(\pi)$. From $y = x^* \upharpoonright \mathcal{D}(\pi)$ we have

$$\langle (S^*)^2 AB\varphi, \psi\rangle = \langle \varphi, ZA^nB^n\psi\rangle \quad \text{for} \quad \varphi, \psi \in \mathcal{D}(\pi). \tag{2}$$

From Lemma 9.3.6 it follows that, $\mathcal{D}(\pi) \equiv \mathcal{D}_\infty(A, B)$ is a core for $AB \upharpoonright \mathcal{D}_{1,1}$ and also for $A^nB^n \upharpoonright \mathcal{D}_{n,n}$. Hence (2) is valid for arbitrary $\varphi \in \mathcal{D}_{1,1}$ and $\psi \in \mathcal{D}_{n,n}$. Since $\mathcal{D}_{1,1} = YX(I - Q_{1,1})\,\mathcal{H}$ and $\mathcal{D}_{n,n} = X^nY^n(I - Q_{n,n})\,\mathcal{H}$, this gives

$$\langle (S^*)^2 (I - Q_{1,1})\,\zeta, X^nY^n(I - Q_{n,n})\,\xi\rangle = \langle YX(I - Q_{1,1})\,\zeta, Z(I - Q_{n,n})\,\xi\rangle$$

for all $\zeta, \xi \in \mathcal{H}$, i.e., $(I - Q_{1,1})\,(S^2X^nY^n - XYZ)\,(I - Q_{n,n}) = 0$.

In particular the latter yields $S^2X^nY^n(I - Q_{n,n})\,\mathcal{H} \subseteq XY\mathcal{H} + Q_{1,1}\mathcal{H}$. For $k, l = 1, \dots, n$, X^k and Y^l commute on $(I - Q_{n,n})\,\mathcal{H}$ by Lemma 9.3.2, (ii), hence also S^k and $(S^*)^l$. Therefore, by the preceding,

$$\begin{aligned}&S^2(S + S^*)\,(S - S^*)\,\big(S^2 - (S^*)^2\big)^{n-1}\,(I - Q_{n,n})\,\mathcal{H} \\ &\subseteq (S + S^*)\,(S - S^*)\,\mathcal{H} + Q_{1,1}\mathcal{H}.\end{aligned} \tag{3}$$

We have

$$(S + S^*)\,\mathcal{H} \cap Q_{1,1}\mathcal{H} = \{0\} \text{ (by (I)}_{1,1}^{0,0}\text{)}, \ \ker\,(S + S^*) = \{0\}$$

and

$$S^2(S + S^*)\,(S - S^*) = (S + S^*)\,(S - S^*)\,(S^2 + Q_{1,1}) + (S + S^*)\,Q_{1,1}S^*.$$

From these facts and (3) we get

$$Q_{1,1}S^*\big(S^2 - (S^*)^2\big)^{n-1}\,(I - Q_{n,n})\,\mathcal{H} \subseteq (S - S^*)\,\mathcal{H} = Y\mathcal{H}.$$

But

$$Q_{1,1}S^*\big(S^2 - (S^*)^2\big)^{n-1}\,(I - Q_{n,n})\, z^{2n-1} = (-1)^{n-1}\, z^0 \notin Y\mathcal{H}$$

by $(\mathrm{I})_{1,1}^{0,0}$.

This is a contradiction, so we have proved that $\mathcal{D}(\pi) \nsubseteq \mathcal{D}(x^*)$.

We verify that $\mathcal{A}_\mathrm{w}^\mathrm{c} \neq \mathcal{A}_\mathrm{f}^\mathrm{c}$. Since $\mathcal{A}_\mathrm{f}^\mathrm{c}$ is $*$-invariant, $x \in \mathcal{A}_\mathrm{w}^\mathrm{c} \subseteq \mathcal{A}_\mathrm{f}^\mathrm{c}$ implies $x^+ \in \mathcal{A}_\mathrm{f}^\mathrm{c}$. If x^+ were in $\mathcal{A}_\mathrm{w}^\mathrm{c}$, then the Hilbert space operator x^+ would be a restriction of x^* and so $\mathcal{D}(\pi) \subseteq \mathcal{D}(x^*)$. Since the latter is not true, $x^+ \notin \mathcal{A}_\mathrm{w}^\mathrm{c}$ and hence $\mathcal{A}_\mathrm{w}^\mathrm{c} \neq \mathcal{A}_\mathrm{f}^\mathrm{c}$. □

Since the shift operator S is irreducible, it follows from Proposition 9.3.13 that $\mathcal{A}' \equiv \pi(\mathbb{C}[\mathsf{x}_1, \mathsf{x}_2])'$ is trivial; so the self-adjoint representation π is of type I_∞. Moreover, the operator $\pi(x_1)$ is obviously in $\mathcal{A}^c$, but $\overline{\pi(x_1)} = A$ is not affiliated with $\mathcal{A}'$; compare also with Corollary 7.2.13 and Theorem 7.3.6, (ii). ○

Example 9.4.7. Let $\mathfrak{R}$ be the C^∞-manifold with boundary obtained by cutting $\mathbb{R}^2 \setminus \{(0,0)\}$ along the positive y-axis and adding two copies $\mathfrak{Y}_+$ and $\mathfrak{Y}_-$ of the positive y-axis as the boundary of $\mathfrak{R}$. The points of $\mathfrak{Y}_+$ and $\mathfrak{Y}_-$ are written as $(+0, y)$ and $(-0, y)$, respectively, with $y > 0$. Let α be a complex number with $|\alpha| = 1$ and $\alpha \neq 1$ which will be fixed for the moment. Let $\mathcal{D}(\pi)$ be the set of all functions $\varphi \in C^\infty(\mathfrak{R})$ satisfying $\frac{\partial^n}{\partial x^n} \frac{\partial^m}{\partial y^m} \varphi \in L^2(\mathbb{R}^2)$ for $n, m \in \mathbb{N}_0$ and

$$\frac{\partial^n \varphi}{\partial x^n}(+0, y) = \alpha \frac{\partial^n \varphi}{\partial x^n}(-0, y) \quad \text{for all } n \in \mathbb{N}_0 \text{ and } y > 0. \tag{4}$$

Then $\mathcal{D}(\pi)$ is a dense linear subspace of the Hilbert space $\mathcal{H}(\pi) := L^2(\mathbb{R}^2)$. We define

$$\pi(x_1)\varphi := -i\frac{\partial \varphi}{\partial x} \quad \text{and} \quad \pi(x_2)\varphi := -i\frac{\partial \varphi}{\partial y}, \quad \varphi \in \mathcal{D}(\pi).$$

Since $\pi(x_1)$ and $\pi(x_2)$ are symmetric operators which leave $\mathcal{D}(\pi)$ invariant and which commute pointwise on $\mathcal{D}(\pi)$, π defines a $*$-representation of the $*$-algebra $\mathbb{C}[x_1, x_2]$.

Statement 1: π is a self-adjoint representation.

Proof. Suppose $\psi \in \mathcal{D}(\pi^*)$. Since $\psi \in \mathcal{D}\big((\pi(x_1)^*)^n (\pi(x_2)^*)^m\big)$ for all $n, m \in \mathbb{N}_0$, ψ has distributive derivatives in $L^2(\mathbb{R}^2)$ of arbitrary high order. Therefore, by the Sobolev lemma (see e.g. WLOKA [1]), $\psi \in C^\infty(\mathfrak{R})$. To prove that $\psi \in \mathcal{D}(\pi)$, it suffices to verify the boundary conditions (4). Since

$$\pi^*(x_1^n)\psi = (-i)^n \frac{\partial^n \psi}{\partial x^n} \in \mathcal{D}(\pi^*) \quad \text{for} \quad n \in \mathbb{N},$$

it is sufficient to treat the case $n = 0$. Using integration by parts and condition (4) for $\varphi \in \mathcal{D}(\pi)$, we obtain

$$0 = \langle \pi(x_1)\varphi, \psi\rangle - \langle \varphi, \pi(x_1)^*\psi\rangle = -i\int_0^\infty \varphi(-0, y)\,\big(\overline{\psi(-0, y) - \alpha\psi(+0, y)}\big)\,dy$$

for arbitrary $\varphi \in \mathcal{D}(\pi)$ with compact support. Hence $\psi(+0, y) = \alpha\psi(-0, y)$ for $y > 0$ and so $\psi \in \mathcal{D}(\pi)$ which shows that π is self-adjoint. □

We define two strongly continuous one-parameter unitary groups $U_1(\cdot)$ and $U_2(\cdot)$ on $\mathcal{H}(\pi)$ by

$$\big(U_1(t)\varphi\big)(x, y) = \begin{cases} \bar{\alpha}\varphi(x + t, y) & \text{if} \quad y > 0 \quad \text{and} \quad -t < x < 0, \\ \alpha\varphi(x + t, y) & \text{if} \quad y > 0 \quad \text{and} \quad 0 < x < -t, \\ \varphi(x + t, y) & \text{if} \quad y \leqq 0 \quad \text{or} \quad x(x + t) \geqq 0 \end{cases}$$

and

$$\big(U_2(t)\varphi\big)(x, y) = \varphi(x, y + t) \quad \text{for} \quad \varphi \in \mathcal{H}(\pi) \quad \text{and} \quad t \in \mathbb{R}.$$

Statement 2: $\pi(x_k)^n$ is essentially self-adjoint and $U_k(t) = \exp it\overline{\pi(x_k)}$ for $k = 1, 2, n \in \mathbb{N}$ and $t \in \mathbb{R}$.

Proof. Let A_k be the infinitesimal generator of $U_k(\cdot)$, $k = 1, 2$. From the definition of U_k it is clear that $\pi(x_k) \subseteq -iA_k$. Let $\mathcal{D}_2$ be the set of all $\varphi \in \mathcal{D}(\pi)$ which vanish in some neighbourhood of the y-axis (more precisely, of $\mathfrak{Y}_+ \cup \mathfrak{Y}_- \cup \{(x, y) \in \mathbb{R}^2 : x = 0, y \leqq 0\}$). Clearly, we have $U_2(t)\, \mathcal{D}_2 \subseteq \mathcal{D}_2$ for $t \in \mathbb{R}$, and $\mathcal{D}_2$ is dense in $\mathcal{H}(\pi)$. Therefore, by Corollary 10.1.15, $\mathcal{D}_2$ and so the larger set $\mathcal{D}(\pi)$ is a core for each power A_2^n, $n \in \mathbb{N}$. This implies that $\pi(x_2)^n$ is essentially self-adjoint. The proof in case $k = 1$ is similar. $\square$

Statement 3: *π is irreducible.*

Proof. Let $t_1 \geqq 0$ and $t_2 \geqq 0$. From the definitions of U_1 and U_2 we conclude that

$$W(t_1, t_2) := I - U_1(-t_1)\, U_2(-t_2)\, U_1(t_1)\, U_2(t_2) = (1 - \alpha)\, \chi_{t_1,t_2} \tag{5}$$

where χ_{t_1,t_2} denotes the multiplication operator by the characteristic function of the rectangle $\{(x, y) \in \mathbb{R}^2 : 0 \leqq x \leqq t_1,\ 0 \leqq y \leqq t_2\}$. Similar formulas are true in the other cases for t_1 and t_2. Suppose $z \in \pi(\mathbb{C}[x_1, x_2])'$. Then z commutes with $\overline{\pi(x_k)}$ and hence with $U_k(t)$ for $k = 1, 2$ and $t \in \mathbb{R}$. Consequently, z commutes with $W(t_1, t_2)$ for all $t_1, t_2 \in \mathbb{R}$. Since $\alpha \neq 1$, it follows from the formulas for $W(t_1, t_2)$ that z commutes with the whole maximal abelian von Neumann algebra $L^\infty(\mathbb{R}^2)$ on $\mathcal{H}(\pi)$. (Here the functions of $L^\infty(\mathbb{R}^2)$ act as multiplication operators on $L^2(\mathbb{R}^2)$.) Hence there is a $\psi \in L^\infty(\mathbb{R}^2)$ such that $z\varphi = \psi \cdot \varphi$, $\varphi \in L^2(\mathbb{R}^2)$. Since z commutes with $U_k(t)$ for $k = 1, 2$ and all $t \in \mathbb{R}$, the latter implies that ψ is constant a.e. on $\mathbb{R}^2$. Thus $z = \lambda \cdot I$ for some $\lambda \in \mathbb{C}$, and π is irreducible by Lemma 8.3.5. $\square$

From Statement 2 we see that the operators $\overline{\pi(x_1)}$ and $\overline{\pi(x_2)}$ are self-adjoint and that the couple $\{\overline{\pi(x_1)}, \overline{\pi(x_2)}\}$ belongs to N_∞^∞. A little computation shows that the commutator $[(\overline{\pi(x_1)} - i)^{-1}, (\overline{\pi(x_2)} - i)^{-1}]$ is a rank one operator with range spanned by the function $\chi(x, y)\, e^{-x-y}$, where χ is the characteristic function of $\{(x, y) \in \mathbb{R}^2 : x \geqq 0, y \geqq 0\}$. Thus $d(\overline{\pi(x_1)}, \overline{\pi(x_2)}) = 1$, and the self-adjoint representation π is not integrable. (The latter fact can be also seen as follows. If π were integrable, then $\overline{\pi(x_1)}$ and $\overline{\pi(x_2)}$ would strongly commute. But then the unitary groups U_1 and U_2 would commute which contradicts (5).)

Finally, we consider the dependence of π on the number α. Two different numbers α_1 and α_2 of the set $\{\alpha \in \mathbb{C} : |\alpha| = 1 \text{ and } \alpha \neq 1\}$ give rise to inequivalent representations. (Indeed, otherwise the corresponding operators $W(\cdot, \cdot)$ would be unitarily equivalent. By (5), this is only possible if $\alpha_1 = \alpha_2$.) Thus, even this rather simple example produces a continuum of inequivalent irreducible non-integrable self-adjoint representations of the polynomial algebra $\mathbb{C}[x_1, x_2]$.

Notes

9.1. Integrable representations of commutative $*$-algebras have been introduced by Powers [1] who called them standard representations. The characterizations given in Theorem 9.1.2 and in Corollary 9.1.9 are due to Powers [1]. Some assertions stated in Remark 6 are due to Inoue/Takesue [1]. Several results in this section such as Proposition 9.1.12 and Theorem 9.1.13 seem to be new.

9.2. Theorem 9.2.1 and Theorem 9.2.3 are both due to Powers [1].

9.3. Couples of self-adjoint operators which commute on a common core for both operators are extensively studied by SCHMÜDGEN [16], [17, [18], [19] and by SCHMÜDGEN/FRIEDRICH [1]. The resolvent approach used in the text was invented by SCHMÜDGEN [16] and developed further by SCHMÜDGEN/FRIEDRICH [1]. The first subsection of 9.3 mainly follows the latter paper. Proposition 9.3.13 is from SCHMÜDGEN [18].

9.4. Theorem 9.4.1 is due to SCHMÜDGEN [19]. Lemma 9.4.3 was proved by BEHNCKE [1]. It strengthens a theorem of WOGEN [1] which states that properly infinite von Neumann algebras on separable Hilbert spaces are singly generated.

Example 9.4.7 has a longer history. NELSON [1] discovered the first example of two self-adjoint operators which commute on a common core and for which the spectral projections do not commute. Another interesting example of this kind was published by FUGLEDE [1]. A somewhat simpler example (also due to NELSON) can be found in REED/SIMON [1], VIII. 6. Our Example 9.4.7 (which is reproduced from SCHMÜDGEN [17]) is very much in the spirit of Nelson's example and the example in POWERS [1]. The elegant proof of Statement 3 is from POWERS [1]. Example 9.4.6 is in SCHMÜDGEN [22].

Additional References:

9.1. FUGLEDE [3], INOUE [6], [7], KADISON [1], SAMOILENKO [1], [2], SLINKER [1], TAKESUE [2].
9.3. FRIEDRICH [2].
9.4. FRIEDRICH [3], FUGLEDE [2], JORGENSEN/MOORE [1], ch. 11, NGUYEN [1].

10. Integrable Representations of Enveloping Algebras

This chapter deals with $*$-representations of enveloping algebras. Though some of the considerations and of the main results (e.g., Theorem 10.4.4) are valid for general $*$-representations, we aim to present a detailed study of integrable representations. To be more precise, let G be a Lie group with Lie algebra $\mathfrak{g}$, and let $\mathscr{E}(\mathfrak{g})$ be the universal enveloping algebra of the complexification of $\mathfrak{g}$. A representation of the $*$-algebra $\mathscr{E}(\mathfrak{g})$ is said to be *G-integrable* if it is equal to the infinitesimal representation dU of some unitary representation U of G. When G is connected and simply connected, the G-integrable representations are called simply *integrable*.

Sections 10.1 and 10.2 provide a systematic study of the infinitesimal representation dU associated with a unitary representation U of the Lie group G. The representation dU is defined on the space $\mathscr{D}^\infty(U)$ of C^∞-vectors for U which is the principal tool in these two sections. Several characterizations of C^∞-vectors are given. The basic properties of these notions are developed in Section 10.1. It is shown that any dense linear subspace of $\mathscr{D}^\infty(U)$ which is invariant under the action of U is a core for each operator $dU(x)$, $x \in \mathscr{E}(\mathfrak{g})$. Section 10.2 is concerned with conditions on a hermitian element a of $\mathscr{E}(\mathfrak{g})$ which ensure that the operator $dU(a)$ is essentially self-adjoint. Among others, we prove that hermitian elements which commute with an elliptic element of $\mathscr{E}(\mathfrak{g})$ have this property. As an application, the continuous group invariant sesquilinear forms on $\mathscr{D}^\infty(U) \times \mathscr{D}^\infty(U)$ are characterized.

The main technical tool in the remaining four sections of this chapter are analytic vectors. Section 10.3 deals with analytic and semi-analytic vectors for symmetric operators in Hilbert space and with the analytic domination of families of operators. In Section 10.4 analytic vectors for $*$-representations of the enveloping algebra $\mathscr{E}(\mathfrak{g})$, for unitary representations of the Lie group G and for the image of single elements of the Lie algebra $\mathfrak{g}$ under $*$-representations are studied in detail. The main result (Theorem 10.4.4) states that, for each $*$-representation π of $\mathscr{E}(\mathfrak{g})$, the space of analytic vectors for π is precisely the space of semi-analytic vectors for the operator $\pi(1 - \Delta)$, where Δ is the Nelson Laplacian relative to a basis of $\mathfrak{g}$. Section 10.5 is concerned with the following question: When is a $*$-representation of the enveloping algebra exponentiable? Here we say that a $*$-representation π of $\mathscr{E}(\mathfrak{g})$ is exponentiable if there exists a basis $\{x_1, \ldots, x_d\}$ for $\mathfrak{g}$ and a unitary representation U of the universal covering group $\tilde{G}$ of G such that $\overline{\pi(x_k)} = \overline{dU(x_k)}$, $k = 1, \ldots, d$. The two main results in this section (Theorems 10.5.4 and 10.5.6) establish criteria for a $*$-representation to be exponentiable. The first one (due to Flato, Simon, Snellman and Sternheimer) shows that it suffices that there exists a dense linear subspace consisting of analytic vectors for the operators $\pi(x_k)$, $k = 1, \ldots, d$.

The second result (due to NELSON) assumes that the image of the Nelson Laplacian relative to some basis is essentially self-adjoint. These results are used to characterize the integrable representations by various equivalent conditions. In Section 10.6 it is shown that each G-integrable representation is a direct sum of cyclic G-integrable representations when the Lie group G is connected.

Throughout this chapter we assume that G is a real Lie group with Lie algebra $\mathfrak{g}$ and $\mathscr{E}(\mathfrak{g})$ is the universal enveloping algebra of $\mathfrak{g}$. Further, we shall use the notation and the facts collected in Section 1.7.

10.1. The Infinitesimal Representation of a Unitary Representation

In this section we assume that U is a unitary representation of the Lie group G in the Hilbert space $\mathscr{H}(U)$.

Definition 10.1.1. A vector φ in $\mathscr{H}(U)$ is called a *C^∞-vector* for U if the mapping $g \to U(g)\varphi$ from the C^∞-manifold G into the Hilbert space $\mathscr{H}(U)$ is a C^∞-mapping.

We denote the set of C^∞-vectors for U by $\mathscr{D}^\infty(U)$. Obviously, $\mathscr{D}^\infty(U)$ is a linear subspace of $\mathscr{H}(U)$. Since translations by group elements are C^∞-mappings of G, $\mathscr{D}^\infty(U)$ is invariant under $U(g)$, $g \in G$.

The next proposition is the heart of the "scalar" characterization of C^∞-vectors given in Corollary 10.1.3 below.

Proposition 10.1.2. *Suppose that $\mathfrak{D}$ is an open subset of $\mathbb{R}^d$ and $\varphi(\cdot)$ is a mapping of $\mathfrak{D}$ into a Hilbert space $\mathscr{H}$. Define $f_\psi(t) := \langle \psi, \varphi(t) \rangle$ for $\psi \in \mathscr{H}$ and $t \in \mathfrak{D}$.*

(i) *If $f_\psi \in C^2(\mathfrak{D})$ for each $\psi \in \mathscr{H}$, then φ is a C^1-mapping of $\mathfrak{D}$ into $\mathscr{H}$.*

(ii) *If $f_\psi \in C^\infty(\mathfrak{D})$ for each $\psi \in \mathscr{H}$, then φ is a C^∞-mapping of $\mathfrak{D}$ into $\mathscr{H}$.*

Proof. (i): Let $\{a_1, \ldots, a_d\}$ be a basis of $\mathbb{R}^d$. We write D_k for the directional derivative in the direction a_k, $k = 1, \ldots, d$. Fix $k \in \{1, \ldots, d\}$. Let $t \in \mathfrak{D}$. The continuous linear functionals $\psi \to \left\langle \psi, \lambda^{-1}\big(\varphi(t + \lambda a_k) - \varphi(t)\big) \right\rangle$ on $\mathscr{H}$ converge pointwise to the linear functional $\psi \to D_k f_\psi(t)$ on $\mathscr{H}$ as $\lambda \to 0$. By the Banach-Steinhaus theorem, $\psi \to D_k f_\psi(t)$ is a continuous linear functional on $\mathscr{H}$. Hence there is a vector $\zeta_k(t) \in \mathscr{H}$ such that

$$D_k f_\psi(t) = \langle \psi, \zeta_k(t) \rangle \quad \text{for } \psi \in \mathscr{H} \text{ and } t \in \mathfrak{D}. \tag{1}$$

Fix $t \in \mathfrak{D}$. We next show that the map $s \to \varphi(s)$ of $\mathfrak{D}$ into $\mathscr{H}$ is continuous at t. Take a compact convex neighbourhood $\mathfrak{K}$ of t in $\mathfrak{D}$. By assumption, $D_k f_\psi(\cdot)$ is continuous on $\mathfrak{D}$; so the map $s \to \zeta_k(s)$ of $\mathfrak{D}$ into $\mathscr{H}$ is continuous relative to the weak topology on $\mathscr{H}$. Therefore, $\zeta_k(\mathfrak{K})$ is weakly compact and hence norm bounded in $\mathscr{H}$. Thus there exists a $\gamma > 0$ such that $\|\zeta_k(s)\| \leqq \gamma$ for all $s \in \mathfrak{K}$ and $k = 1, \ldots, d$.

There is a number $\varepsilon > 0$ such that $t + b \in \mathfrak{K}$ for all $b = \sum_{k=1}^{d} \lambda_k a_k$ with $(\lambda_1, \ldots, \lambda_d) \in \mathbb{R}^d$, $|\lambda_1| \leqq \varepsilon, \ldots, |\lambda_d| \leqq \varepsilon$. Fix such a vector b. Put $b_1 = 0$ and $b_n = \sum_{k=1}^{n-1} \lambda_k a_k$ for

$n = 2, \ldots, d$. Then

$$|\langle \psi, \varphi(t+b) - \varphi(t) \rangle| = |f_\psi(t+b) - f_\psi(t)| \leqq \sum_{n=1}^{d} |f_\psi(t + b_n + \lambda_n a_n) - f_\psi(t + b_n)|$$
$$\leqq \sum_{n=1}^{d} |\lambda_n| \sup \{|D_n f_\psi(t + b_n + \alpha a_n)| : |\alpha| \leqq |\lambda_n|\}$$
$$\leqq \sum_{n=1}^{d} |\lambda_n| \, \gamma \|\psi\| \quad \text{for } \psi \in \mathcal{H},$$

where we used the mean value theorem and (1). This implies that

$$\|\varphi(t+b) - \varphi(t)\| \leqq \sum_{n=1}^{d} |\lambda_n| \, \gamma,$$

so φ is continuous at t.

Since $f_\psi \in C^2(\mathfrak{D})$ for each $\psi \in \mathcal{H}$, the same argument applies to the map $t \to \zeta_k(t)$ of $\mathfrak{D}$ into $\mathcal{H}$ and shows that this map is continuous for $k = 1, \ldots, d$. Thus the proof of (i) is complete once we have shown that $\zeta_k(t) = D_k\varphi(t)$ for $t \in \mathfrak{D}$ and $k = 1, \ldots, d$. Using (1) once more, we have for $\psi \in \mathcal{H}$ and sufficiently small $|\lambda| \neq 0$

$$\left|\left\langle \psi, \lambda^{-1}\big(\varphi(t + \lambda a_k) - \varphi(t)\big) - \zeta_k(t) \right\rangle\right| = \left|\lambda^{-1}\big(f_\psi(t + \lambda a_k) - f_\psi(t)\big) - D_k f_\psi(t)\right|$$
$$= \left|\lambda^{-1} \int_0^\lambda \big(D_k f_\psi(t + \alpha a_k) - D_k f_\psi(t)\big) \, d\alpha\right|$$
$$\leqq \|\psi\| \sup \{\|\zeta_k(t + \alpha a_k) - \zeta_k(t)\| : |\alpha| \leqq |\lambda|\}.$$

By the continuity of $\zeta_k(\cdot)$ it follows that $\lim_{\lambda \to 0} \lambda^{-1}\big(\varphi(t + \lambda a_k) - \varphi(t)\big) = \zeta_k(t)$ in $\mathcal{H}$, that is, $\zeta_k(t) = D_k\varphi(t)$.

(ii): Using induction with respect to the order of the partial derivatives, the following assertion can be immediately derived from (i):

If $f_\psi \in C^{n+1}(\mathfrak{D})$ for each $\psi \in \mathcal{H}$, then φ is a C^n-mapping of $\mathfrak{D}$ into $\mathcal{H}$ for $n \in \mathbb{N}$. This gives (ii). □

Corollary 10.1.3. *For each vector φ in $\mathcal{H}(U)$ the following conditions are equivalent:*

(i) $\varphi \in \mathcal{D}^\infty(U)$.

(ii) *The function $g \to \langle U(g)\,\varphi, \psi \rangle$ is in $C^\infty(G)$ for each $\psi \in \mathcal{H}(U)$.*

(iii) *The function $g \to \langle U(g)\,\psi, \varphi \rangle$ is in $C^\infty(G)$ for each $\psi \in \mathcal{H}(U)$.*

Proof. (i) → (ii) is obvious. (ii) ↔ (iii) follows from $\langle U(g)\,\varphi, \psi \rangle = \overline{\langle U(g^{-1})\,\psi, \varphi \rangle}$ and the fact that $g \to g^{-1}$ is a C^∞-map of G. To prove (ii) → (i), we choose a diffeomorphism $t \to g(t)$ of an open subset $\mathfrak{D}$ of $\mathbb{R}^d$ onto a neighbourhood of a given point $g_0 \in G$ and we apply Proposition 10.1.2, (ii), to the map φ defined by $\varphi(t) = U\big(g(t)\big)\,\varphi$, $t \in \mathfrak{D}$. □

For $f \in C_0^\infty(G)$ and $\varphi \in \mathcal{H}(U)$, we define $U_f\varphi = \int_G f(g)\, U(g)\, \varphi \, d\mu(g)$, where the integral is to be understood as an $\mathcal{H}(U)$-valued Bochner integral. (Note that the integrand is a continuous mapping of G into $\mathcal{H}(U)$.) The linear span $\mathcal{D}_G(U)$ of the vectors $U_f\varphi$, where $f \in C_0^\infty(G)$ and $\varphi \in \mathcal{H}(U)$, is called the *Gårding subspace* of $\mathcal{D}(U)$ for U. Some simple properties of this space are collected in

Lemma 10.1.4. (i) *$\mathcal{D}_G(U)$ is dense in $\mathcal{H}(U)$.*

(ii) *For $g \in G$, $f \in C_0^\infty(G)$ and $\varphi \in \mathcal{H}(U)$, $U(g)\, U_f\varphi = U_{f(g^{-1}\cdot)}\varphi$.*

(iii) $\mathcal{D}_G(U)$ *is invariant under* $U(g)$ *for* $g \in G$.

(iv) $\mathcal{D}_G(U) \subseteq \mathcal{D}^\infty(U)$.

Proof. (i): Suppose $\varphi \in \mathcal{H}(U)$. Let f be a non-negative function of $C_0^\infty(G)$ such that $\int_G f(g)\, \mathrm{d}\mu(g) = 1$. Then

$$\|U_f\varphi - \varphi\| = \left\|\int_G f(g)\left(U(g) - U(e)\right)\varphi\, \mathrm{d}\mu(g)\right\| \leq \sup_{g \in \operatorname{supp} f} \left\|\left(U(g) - U(e)\right)\varphi\right\|.$$

Therefore, if supp f shrinks to $\{e\}$, then $U_f\varphi$ tends to φ in $\mathcal{H}(U)$ by the continuity of U. Thus φ is in the closure of $\mathcal{D}_G(U)$.

(ii) follows immediately from the left-invariance of the Haar measure μ and (iii) is a consequence of (ii).

(iv): Let $\varphi \in \mathcal{H}(U)$ and $f \in C_0^\infty(G)$. The function $g \to \langle U(g)\, U_f\varphi, \psi\rangle$ is in $C^\infty(G)$ for all $\psi \in \mathcal{H}(U)$, since $\langle U(g)\, U_f\varphi, \psi\rangle = \int_G f(g^{-1}h) \langle U(h)\, \varphi, \psi\rangle\, \mathrm{d}\mu(h)$ by (ii). From Corollary 10.1.3, $U_f\varphi \in \mathcal{D}^\infty(U)$, so $\mathcal{D}_G(U) \subseteq \mathcal{D}^\infty(U)$. □

Definition 10.1.5. Let $\mathcal{D}$ be a dense linear subspace of a Hilbert space $\mathcal{H}$. A *$*$-representation of the Lie algebra* $\mathfrak{g}$ on $\mathcal{D}$ is a mapping π of $\mathfrak{g}$ into $L(\mathcal{D})$ such that

(i) $\pi(\alpha x + \beta y) = \alpha\pi(x) + \beta\pi(y)$,

(ii) $\pi([x, y]) = \pi(x)\, \pi(y) - \pi(y)\, \pi(x)$,

(iii) $\langle \pi(x)\, \varphi, \psi\rangle = -\langle \varphi, \pi(x)\, \psi\rangle$,

whenever $x, y \in \mathfrak{g}$, $\alpha, \beta \in \mathbb{R}$ and $\varphi, \psi \in \mathcal{D}$.

We call $\mathcal{D}$ the *domain* of π and we write $\mathcal{D}(\pi) := \mathcal{D}$. Condition (iii) means that the operator $\pi(x)$ is skew-symmetric for each x in $\mathfrak{g}$. Since also $\pi(x) \in L(\mathcal{D})$, (iii) implies that $\pi(x) \in \mathcal{L}^+(\mathcal{D})$ for $x \in \mathfrak{g}$. By a slight reformulation of the preceding definition, a $*$-representation of the Lie algebra $\mathfrak{g}$ on $\mathcal{D}$ is a homomorphism π of $\mathfrak{g}$ into the algebra $\mathcal{L}^+(\mathcal{D})$ satisfying $\pi(x)^+ = -\pi(x)$ for all x in $\mathfrak{g}$.

For x in $\mathfrak{g}$, we define an operator $\mathrm{d}U(x)$ with domain $\mathcal{D}^\infty(U)$ by

$$\mathrm{d}U(x)\, \varphi = \frac{\mathrm{d}}{\mathrm{d}t}\, U(\exp tx)\, \varphi|_{t=0} = \lim_{t \to 0} t^{-1}\left(U(\exp tx) - I\right)\varphi, \quad \varphi \in \mathcal{D}^\infty(U).$$

Proposition 10.1.6. *The map* $x \to \mathrm{d}U(x)$ *is a $*$-representation of the Lie algebra* $\mathfrak{g}$ *on the dense linear subspace* $\mathcal{D}(\mathrm{d}U) := \mathcal{D}^\infty(U)$ *of the Hilbert space* $\mathcal{H}(U)$.

Proof. Since $\mathcal{D}_G(U) \subseteq \mathcal{D}^\infty(U)$ and $\mathcal{D}_G(U)$ is dense in $\mathcal{H}(U)$ by Lemma 10.1.4, $\mathcal{D}^\infty(U)$ is dense in $\mathcal{H}(U)$. The vector $\mathrm{d}U(x)\, \varphi$ is, by definition, the value of the derivative in the direction of x of the function $g \to U(g)\, \varphi$ at e. Therefore, since $\varphi \in \mathcal{D}^\infty(U)$, $\mathrm{d}U(x)\, \varphi \in \mathcal{D}^\infty(U)$ for $x \in \mathfrak{g}$. It is obvious that the map $x \to \mathrm{d}U(x)$ is (real-) linear. We show that $\mathrm{d}U(\cdot)$ preserves the Lie bracket. We suppose $x, y \in \mathfrak{g}$ and $y \in \mathcal{D}^\infty(U)$. For $\psi \in \mathcal{H}(U)$, we have

$$\begin{aligned}
&\langle (\mathrm{d}U(x)\, \mathrm{d}U(y) - \mathrm{d}U(y)\, \mathrm{d}U(x))\, \varphi, \psi\rangle \\
&= \frac{\mathrm{d}}{\mathrm{d}t}\left(\frac{\mathrm{d}}{\mathrm{d}s}\left\langle U(\exp(-tx)\exp(-sy))\, \varphi, \psi\right\rangle\Big|_{s=0}\right)\Big|_{t=0} - \\
&\quad \frac{\mathrm{d}}{\mathrm{d}s}\left(\frac{\mathrm{d}}{\mathrm{d}t}\left\langle U(\exp(-sy)\exp(-tx))\, \varphi, \psi\right\rangle\Big|_{t=0}\right)\Big|_{s=0}
\end{aligned}$$

$$= \big((\tilde{y}\tilde{x} - \tilde{x}\tilde{y})\,\langle U(\cdot)\,\varphi, \psi\rangle\big)\,(e) = \big(\widetilde{[y, x]}\,\langle U(\cdot)\,\varphi, \psi\rangle\big)\,(e)$$

$$= \frac{\mathrm{d}}{\mathrm{d}t}\big\langle U\big(\exp\,(-t[y, x])\big)\,\varphi, \psi\big\rangle\big|_{t=0} = \langle \mathrm{d}U(-[y, x])\,\varphi, \psi\rangle$$

$$= \langle \mathrm{d}U([x, y])\,\varphi, \psi\rangle,$$

where we used the formulas 1.7/(1) and 1.7/(2). Thus $\mathrm{d}U([x, y]) = \mathrm{d}U(x)\,\mathrm{d}U(y) - \mathrm{d}U(y)\,\mathrm{d}U(x)$ for $x, y \in \mathfrak{g}$ which proves condition (ii) in Definition 10.1.5. Condition (iii) rests on the assumption that the representation U is unitary. If $\varphi, \psi \in \mathcal{D}^\infty(U)$ and $x \in \mathfrak{g}$,

$$\langle \mathrm{d}U(x)\,\varphi, \psi\rangle = \frac{\mathrm{d}}{\mathrm{d}t}\,\langle U(\exp tx)\varphi, \psi\rangle|_{t=0} = \frac{\mathrm{d}}{\mathrm{d}t}\,\big\langle U\big(\exp\,(-tx)\big)^{-1}\varphi, \psi\big\rangle|_{t=0}$$

$$= \frac{\mathrm{d}}{\mathrm{d}t}\,\big\langle \varphi, U\big(\exp\,(-tx)\big)\,\psi\big\rangle\big|_{t=0} = -\langle \varphi, \mathrm{d}U(x)\,\psi\rangle. \; \square$$

From the universal property of the enveloping algebra $\mathcal{E}(\mathfrak{g})$ it follows that the $*$-representation $\mathrm{d}U$ of the Lie algebra $\mathfrak{g}$ on $\mathcal{D}^\infty(U)$ has a unique extension to an identity preserving $*$-homomorphism, also denoted by $\mathrm{d}U$, of the $*$-algebra $\mathcal{E}(\mathfrak{g})$ into the $*$-algebra $\mathcal{L}^+\big(\mathcal{D}^\infty(U)\big)$. Then $\mathrm{d}U$ is a $*$-representation of the $*$-algebra $\mathcal{E}(\mathfrak{g})$ on $\mathcal{D}(\mathrm{d}U) := \mathcal{D}^\infty(U)$ in the sense of Definition 8.1.9.

Definition 10.1.7. The $*$-representation $\mathrm{d}U$ of $\mathcal{E}(\mathfrak{g})$ (or of $\mathfrak{g}$) on $\mathcal{D}^\infty(U)$ is called the *infinitesimal representation* or the *differential* of the unitary representation U of G. A representation π of the $*$-algebra $\mathcal{E}(\mathfrak{g})$ is called *G-integrable* if there exists a unitary representation U of the Lie group G on the Hilbert space $\mathcal{H}(\pi)$ such that $\pi = \mathrm{d}U$. We say that π is *integrable* if π is $\tilde{G}$-integrable.

Recall that $\tilde{G}$ is the connected and simply connected Lie group which has $\mathfrak{g}$ as its Lie algebra. Note that the equality $\pi = \mathrm{d}U$ means that $\pi(x) \subseteq \mathrm{d}U(x)$ for all $x \in \mathcal{E}(\mathfrak{g})$ (or equivalently, for all $x \in \mathfrak{g}$) and that $\mathcal{D}(\pi) = \mathcal{D}(\mathrm{d}U) \equiv \mathcal{D}^\infty(U)$.

Example 10.1.8. For $g \in G$, let $U_{lr}(g)$ denote the operator in the Hilbert space $\mathcal{H}(U_{lr}) := L^2(G; \mu)$ defined by $\big(U_{lr}(g)\,\varphi\big)\,(h) = \varphi(g^{-1}h)$, $\varphi \in L^2(G; \mu)$ and $h \in G$. Then the mapping $g \to U_{lr}(g)$ is a unitary representation of G, the *left regular representation* of G. By the definition of U_{lr}, we have

$$\mathrm{d}U_{lr}(x)\,\varphi = \frac{\mathrm{d}}{\mathrm{d}t}\,U_{lr}(\exp tx)\,\varphi|_{t=0} = \frac{\mathrm{d}}{\mathrm{d}t}\,\varphi\big(\exp\,(-tx)\,\cdot\big)\big|_{t=0}$$

for $x \in \mathfrak{g}$ an $\mathrm{d}\varphi \in \mathcal{D}^\infty(U)$. Thus $C_0^\infty(G) \subseteq \mathcal{D}^\infty(U)$ and $\mathrm{d}U_{lr}(x)\,\varphi = \tilde{x}\varphi$ for all $x \in \mathcal{E}(\mathfrak{g})$ and $\varphi \in C_0^\infty(G)$. Recall that $\tilde{x}$ is the right-invariant differential operator on G associated with $x \in \mathcal{E}(\mathfrak{g})$. It is well-known that the map $x \to \tilde{x} \restriction C_0^\infty(G)$ is an isomorphism. Hence $\mathrm{d}U_{lr}$ is faithful. In particular, this shows that $\mathcal{E}(\mathfrak{g})$ is $*$-isomorphic to an O*-algebra. $\bigcirc$

Next we describe the space $\mathcal{D}^\infty(U)$ of C^∞-vectors in terms of domains of certain operators. Another result in this direction is proved in Section 10.2, cf. Corollary 10.2.4.

Suppose $x \in \mathfrak{g}$. Let $\partial U(x)$ denote the infinitesimal generator of the strongly continuous one-parameter unitary group $t \to U(\exp tx)$ on $\mathcal{H}(U)$. Then $\mathrm{i}\partial U(x)$ is a self-adjoint operator on $\mathcal{H}(U)$ and $U(\exp tx) = \exp t\,\partial U(x)$, $t \in \mathbb{R}$. The domain of $\partial U(x)$

consists of all vectors φ in $\mathscr{H}(U)$ for which limit $\lim_{t\to 0} t^{-1}(U(\exp tx) - I)\,\varphi$ exists in $\mathscr{H}(U)$ and $U(x)\,\varphi = \lim_{t\to 0} t^{-1}(U(\exp tx) - I)\,\varphi$ for $\varphi \in \mathscr{D}(\partial U(x))$. (These well-known facts can be found, e.g., in REED/SIMON [1], VIII. 4.) In particular, the latter implies that $\mathscr{D}^\infty(U) \subseteq \mathscr{D}(\partial U(x))$ and $\mathrm{d}U(x) \subseteq \partial U(x)$. (We show by Corollary 10.2.11 that $\overline{\mathrm{d}U(x)} = \partial U(x)$.) Since $\mathrm{d}U(x)$ leaves $\mathscr{D}^\infty(U)$ invariant, $\mathscr{D}^\infty(U) \subseteq \mathscr{D}(\partial U(x)^n)$ for all $n \in \mathbb{N}$.

Theorem 10.1.9. *If $\{x_1, \ldots, x_d\}$ is a basis of the Lie algebra* $\mathfrak{g}$, *then* $\mathscr{D}^\infty(U) = \bigcap_{k=1}^{d} \mathscr{D}^\infty(\partial U(x_k))$.

Proof. One inclusion has been already mentioned above. To prove the non-trivial part, let $\varphi \in \mathscr{D}(\partial U(x_k)^n)$ for all $k = 1, \ldots, d$ and $n \in \mathbb{N}$. For $x \in \mathfrak{g}$, let $l(x)$ denote the left-invariant vector field on G defined by $(l(x)\,f)\,(g) = \frac{\mathrm{d}}{\mathrm{d}t} f(g \exp tx)|_{t=0}$, $f \in C^\infty(G)$. Further, let μ_r be the right-invariant Haar measure on G. Fix $\psi \in \mathscr{H}(U)$. Let $k \in \{1, \ldots, d\}$, $n \in \mathbb{N}$ and $f \in C_0^\infty(G)$. We have

$$\begin{aligned}
&\int_G f(g)\,\langle U(g)\,\mathrm{d}U(x_k)^n\,\varphi, \psi\rangle\,\mathrm{d}\mu_r(g)\\
&= \left(\frac{\mathrm{d}}{\mathrm{d}t}\right)^n \left(\int_G f(g)\,\langle U(g \exp tx_k)\,\varphi, \psi\rangle\,\mathrm{d}\mu_r(g)\right)\Big|_{t=0}\\
&= \left(\frac{\mathrm{d}}{\mathrm{d}t}\right)^n \left(\int_G f(g \exp(-tx_k))\,\langle U(g)\,\varphi, \psi\rangle\,\mathrm{d}\mu_r(g)\right)\Big|_{t=0}\\
&= (-1)^n \int_G (l(x_k)^n f)\,(g)\,\langle U(g)\,\varphi, \psi\rangle\,\mathrm{d}\mu_r(g).
\end{aligned}$$

Consider the differential operator $L_m := l(x_1)^{2m} + \cdots + l(x_d)^{2m}$, $m \in \mathbb{N}$, on G. The above formula shows that the function $h(g) := \langle U(g)\,\varphi, \psi\rangle$ is a distribution solution to the equation $L_m h = h_m$, where $h_m(g) := \sum_{k=1}^{d} \langle U(g)\,\mathrm{d}U(x_k)^{2m}\,\varphi, \psi\rangle$. Since U is assumed to be strongly continuous, the function $h_m(g)$ is continuous on G. The differential operator L_m on G is an elliptic operator of order $2m$ with C^∞-coefficients relative to local coordinates on G. By the local regularity theorem for weak solutions of elliptic equations (see e.g. BERS/JOHN/SCHECHTER [1], p. 190), h has derivatives of order $\leqq 2m$ which are locally in $L^2(G; \mu_r)$. This is true for all $m \in \mathbb{N}$, so that, by the classical Sobolev lemma (see e.g. WLOKA [1], p. 115), $h(g) \equiv \langle U(g)\,\varphi, \psi\rangle$ is in $C^\infty(G)$. Since this holds for all $\psi \in \mathscr{H}(U)$, we conclude from Corollary 10.1.3 that $\varphi \in \mathscr{D}^\infty(U)$. □

From Theorem 10.1.9 we obtain a corollary which sharpens Corollary 10.1.3.

Corollary 10.1.10. *Let $\{x_1, \ldots, x_d\}$ be a basis of* $\mathfrak{g}$. *A vector $\varphi \in \mathscr{H}(U)$ is in $\mathscr{D}^\infty(U)$ if and only if for each $\psi \in \mathscr{H}(U)$ and $k = 1, \ldots, d$, the function $t \to \langle U(\exp tx_k)\,\varphi, \psi\rangle$ is in $C^\infty(\mathbb{R})$.*

Proof. The necessity is obvious. We verify the sufficiency. Suppose that the above condition is satisfied. From Corollary 10.1.3 (applied to the unitary representation $t \to U(\exp tx_k)$ of the Lie group $\mathbb{R}$) it follows that the map $t \to U(\exp tx_k)\,\varphi$ of $\mathbb{R}$ into $\mathscr{H}(U)$ is C^∞ for $k = 1, \ldots, d$. Hence $\varphi \in \bigcap_{k=1}^{d} \mathscr{D}^\infty(\partial U(x_k))$. By Theorem 10.1.9, $\varphi \in \mathscr{D}^\infty(U)$. □

Proposition 10.1.11. *For any vector $\varphi \in \mathcal{D}^\infty(U)$, $g \to U(g)\,\varphi$ is a C^∞-mapping of G into the locally convex space $\mathcal{D}^\infty(U)\,[\mathfrak{t}_{\mathrm{d}U}]$.*

Proof. Fix $\varphi \in \mathcal{D}^\infty(U)$. Let $\{x_1, \ldots, x_d\}$ be a basis of $\mathfrak{g}$, and set $g(t) := \exp t_1x_1 \ldots \exp t_dx_d$ for $t = (t_1, \ldots, t_d) \in \mathbb{R}^d$. The map $g(t) \to t$ is an analytic coordinate system in a certain neighbourhood of e in G. Therefore, being the composition of the two C^∞-mappings $g \to U(g)\,\varphi$ of G into $\mathcal{H}(U)$ and $(s, t) \to g(s)\,g(t)$ of $\mathbb{R}^{2d}$ into G, $(s, t) \to U\big(g(s)\,g(t)\big)\,\varphi$ is a C^∞-mapping of $\mathbb{R}^{2d}$ into $\mathcal{H}(U)$. If $t = (t_1, \ldots, t_d) \in \mathbb{R}^d$ and $n = (n_1, \ldots, n_d) \in \mathbb{N}_0^d$, we write D_t^n for $\left(\frac{\partial}{\partial t_1}\right)^{n_1} \cdots \left(\frac{\partial}{\partial t_d}\right)^{n_d}$. We have

$$\mathrm{d}U(x^n)\,U\big(g(t)\big)\,\varphi = D_s^nU\big(g(s)\big)\,U\big(g(t)\big)\,\varphi|_{s=0} = D_s^nU\big(g(s)\,g(t)\big)\,\varphi|_{s=0}$$

for $n \in \mathbb{N}_0^d$ and $s, t \in \mathbb{R}^d$. Since the x^n, $n \in \mathbb{N}_0^d$, span $\mathcal{E}(\mathfrak{g})$, this shows that the map $t \to \mathrm{d}U(x)\,U\big(g(t)\big)\,\varphi$ of $\mathbb{R}^d$ into $\mathcal{H}(U)$ is C^∞ for each $x \in \mathcal{E}(\mathfrak{g})$. Because the operators $\mathrm{d}U(x)$, $x \in \mathcal{E}(\mathfrak{g})$, are closable, this implies that $D_t^mU\big(g(t)\big)\,\varphi$ relative to the Hilbert space norm is equal to $D_t^mU\big(g(t)\big)\,\varphi$ relative to the graph topology $\mathfrak{t}_{\mathrm{d}U}$. Hence the map $g \to U(g)\,\varphi$ of G into $\mathcal{D}^\infty(U)\,[\mathfrak{t}_{\mathrm{d}U}]$ is C^∞ in a neighbourhood of e. Replacing φ by $U(g)\,\varphi$, $g \in G$, we see that it is C^∞ on the whole G. □

Lemma 10.1.12. *For $x \in \mathcal{E}(\mathfrak{g})$, $g \in G$, $\varphi \in \mathcal{D}^\infty(U)$, $\psi \in \mathcal{H}(U)$ and $f \in C_0^\infty(G)$, we have*

$$\mathrm{d}U\big(Ad\; g(x)\big)\,\varphi = U(g)\,\mathrm{d}U(x)\,U(g^{-1})\,\varphi \tag{2}$$

and

$$\mathrm{d}U(x)\,U_f\psi = U_{\tilde{x}f}\psi. \tag{3}$$

Proof. The mappings $x \to \mathrm{d}U(x)$, $x \to Ad\;g(x)$ and $x \to \tilde{x}$ are homomorphisms of the algebra $\mathcal{E}(\mathfrak{g})$ into $\mathcal{L}^+\big(\mathcal{D}^\infty(U)\big)$, $\mathcal{E}(\mathfrak{g})$ and $\mathfrak{D}(G)$, respectively. From the Poincaré-Birkhoff-Witt theorem we therefore conclude that it suffices to prove both formulas in the case where x is in $\mathfrak{g}$. Fix $x \in \mathfrak{g}$. By formula 1.7/(3),

$$U\big(\exp tAd\;g(x)\big)\,\varphi = U(g \exp tx\; g^{-1})\,\varphi = U(g)\,U(\exp tx)\,U(g^{-1})\,\varphi.$$

Differentiation of this identity at $t = 0$ yields (2). From Lemma 10.1.4, $U(\exp tx)\,U_f\psi = U_{f(\exp(-tx)\cdot)}\psi$, $t \in \mathbb{R}$. Differentiating at $t = 0$, (3) follows. □

Corollary 10.1.13. *Each operator $U(g)$, $g \in G$, maps $\mathcal{D}^\infty(U)\,[\mathfrak{t}_{\mathrm{d}U}]$ continuously into itself.*

Proof. By (2), $\mathrm{d}U(x)\,U(g)\,\varphi = U(g)\,\mathrm{d}U\big(Ad\,g^{-1}(x)\big)\,\varphi$ and so $\|\mathrm{d}U(x)\,U(g)\,\varphi\| = \|\mathrm{d}U\big(Ad\,g^{-1}(x)\big)\,\varphi\|$ for $x \in \mathcal{E}(\mathfrak{g})$, $g \in G$ and $\varphi \in \mathcal{D}^\infty(U)$. □

Theorem 10.1.14. *Let $\mathcal{D}$ be a dense linear subspace of $\mathcal{H}(U)$, which is contained in $\mathcal{D}^\infty(U)$ and invariant under $U(g)$ for all g in the connected component G_0 of the unit element of G. Then $\mathcal{D}$ is dense in $\mathcal{D}^\infty(U)\,[\mathfrak{t}_{\mathrm{d}U}]$ and $\mathcal{D}$ is a core for each operator $\mathrm{d}U(x)$, $x \in \mathcal{E}(\mathfrak{g})$.*

Proof. By Corollary 10.1.13, each $U(g)$, $g \in G_0$, is a continuous mapping of $\mathcal{D}^\infty(U)\,[\mathfrak{t}_{\mathrm{d}U}]$ into itself. Thus we can assume without loss of generality that $\mathcal{D}$ is $\mathfrak{t}_{\mathrm{d}U}$-closed in $\mathcal{D}^\infty(U)$. Let $\varphi \in \mathcal{D}$ and $f \in C_0^\infty(G_0)$. Since $\mathrm{d}U(x)$, $x \in \mathcal{E}(\mathfrak{g})$, is closable in $\mathcal{H}(U)$ and continuous on $\mathcal{D}^\infty(U)\,[\mathfrak{t}_{\mathrm{d}U}]$, we have $\mathrm{d}U(x)\,U_f\varphi = \int f(g)\,\mathrm{d}U(x)\,U(g)\,\varphi\,\mathrm{d}\mu(g)$. This implies that $U_f\varphi$ is the $\mathfrak{t}_{\mathrm{d}U}$-limit of Riemann sums for the integral $\int f(g)\,U(g)\,\varphi\,\mathrm{d}\mu(g)$. Since $U(g)\,\varphi \in \mathcal{D}$ for $g \in G_0$ and since $\mathcal{D}$ is $\mathfrak{t}_{\mathrm{d}U}$-closed in $\mathcal{D}^\infty(U)$, this yields $U_f\varphi \in \mathcal{D}$.

Suppose that $\psi \in \mathcal{D}^\infty(U)$. We next check that $U_f\psi \in \mathcal{D}$ for $f \in C_0^\infty(G_0)$. Since $\mathcal{D}$ is dense in $\mathcal{H}(U)$, there is a sequence $(\psi_n : n \in \mathbb{N})$ in $\mathcal{D}$ such that $\psi = \lim_n \psi_n$ in $\mathcal{H}(U)$. If $x \in \mathcal{E}(\mathfrak{g})$, then, by (3),

$$\lim_n \mathrm{d}U(x)\, U_f\psi_n = \lim_n U_{\tilde{x}f}\psi_n = U_{\tilde{x}f}\psi = \mathrm{d}U(x)\, U_f\psi \quad \text{in } \mathcal{D}(U);$$

so

$$\lim_n U_f\psi_n = U_f\psi \quad \text{in} \quad \mathcal{D}^\infty(U)\,[\mathfrak{t}_{\mathrm{d}U}].$$

Since $U_f\psi_n \in \mathcal{D}$ as proved above, $U_f\psi \in \mathcal{D}$.

Now we prove that $\psi \in \mathcal{D}^\infty(U)$ is the $\mathfrak{t}_{\mathrm{d}U}$-limit of vectors $U_f\psi$, $f \in C_0^\infty(G_0)$. Take a sequence $(f_n : n \in \mathbb{N})$ of non-negative functions of $C_0^\infty(G_0)$ such that $\int f_n(g)\, \mathrm{d}\mu(g) = 1$ for $n \in \mathbb{N}$ and such that $\operatorname{supp} f_n$ shrinks to $\{e\}$ as $n \to \infty$. For $x \in \mathcal{E}(\mathfrak{g})$ and $n \in \mathbb{N}$, we have

$$\|\mathrm{d}U(x)\,(U_{f_n}\psi - \psi)\| = \left\|\int f_n(g)\, \mathrm{d}U(x) \big(U(g) - I\big)\, \psi\, \mathrm{d}\mu(g)\right\|$$
$$\leq \sup_{g \in \operatorname{supp} f_n} \left\|\mathrm{d}U(x)\, \big(U(g) - I\big)\, \psi\right\| \equiv \sup_{g \in \operatorname{supp} f_n} \left\|\big(U(g) - I\big)\, \psi\right\|_{\mathrm{d}U(x)}.$$

Using once more that $U(g)$, $g \in G_0$, is continuous relative to the graph topology $\mathfrak{t}_{\mathrm{d}U}$, it follows from the latter that $\lim_n \mathrm{d}U(x)\, U_{f_n}\psi = \mathrm{d}U(x)\,\psi$ in $\mathcal{H}(U)$, i.e., $\lim_n U_{f_n}\psi = \psi$ in $\mathcal{D}^\infty(U)\,[\mathfrak{t}_{\mathrm{d}U}]$. Since $U_{f_n}\psi \in \mathcal{D}$ as shown above, this proves that $\mathcal{D}$ is dense in $\mathcal{D}^\infty(U)\,[\mathfrak{t}_{\mathrm{d}U}]$. By the definition of the graph topology $\mathfrak{t}_{\mathrm{d}U}$, this means that $\mathcal{D}$ is a core for $\mathrm{d}U(x)$, $x \in \mathcal{E}(\mathfrak{g})$. □

The special case of Theorem 10.1.14 where U is a one-parameter unitary group is stated separately as

Corollary 10.1.15. *Let A be a self-adjoint operator in a Hilbert space $\mathcal{H}$ and let $U(t) := \mathrm{e}^{\mathrm{i}tA}$, $t \in \mathbb{R}$. Suppose $\mathcal{D}$ is a dense linear subspace of $\mathcal{H}$ contained in $\mathcal{D}^\infty(A)$. If $\mathcal{D}$ is invariant under $U(t)$ for all $t \in \mathbb{R}$, then $\mathcal{D}$ is a core for each operator A^n, $n \in \mathbb{N}$.*

In the last part of the above proof of Theorem 10.1.14 the following corollary was shown. (It is also a direct consequence of the theorem, because $\mathcal{D}_G(U)$ is dense in $\mathcal{H}(U)$ and invariant under $U(g)$, $g \in G_0$, by Lemma 10.1.4.)

Corollary 10.1.16. *The Gårding subspace $\mathcal{D}_G(U)$ of $\mathcal{H}(U)$ for U is dense in $\mathcal{D}^\infty(U)\,[\mathfrak{t}_{\mathrm{d}U}]$ and hence a core for each operator $\mathrm{d}U(x)$, $x \in \mathcal{E}(\mathfrak{g})$.*

Remark 1. In fact a much stronger result is true. It was proved by DIXMIER/MALLIAVIN [1], p. 313, Theorem 3.3, that the Gårding space $\mathcal{D}_G(U)$ is *equal* to $\mathcal{D}^\infty(U)$, i.e., each vector in $\mathcal{D}^\infty(U)$ can be represented as a finite sum of vectors $U_f\psi$, where $f \in G_0^\infty(G)$ and $\psi \in \mathcal{H}(U)$. Moreover, the functions f can be chosen such that their supports are contained in a given neighbourhood of the identity in G.

10.2. Elliptic Elements in the Enveloping Algebra

Throughout this section, U denotes a unitary representation of the Lie group G on the Hilbert space $\mathcal{H}(U)$.

Definition 10.2.1. An element a in $\mathcal{E}(\mathfrak{g})$ is called *elliptic* if $\tilde{a}$ is an elliptic partial differential operator on G and if $a \neq \lambda \cdot 1$ for all $\lambda \in \mathbb{C}$.

Remark 1. The last requirement in Definition 10.2.1 is only included for a convenient formulation of the results. Some results such as Lemma 10.2.2 and Theorem 10.2.6 are certainly not true in general when $a = \lambda \cdot 1$, $\lambda \in \mathbb{C}$.

Remark 2. Let $\{x_1, \ldots, x_d\}$ be a basis for $\mathfrak{g}$. Recall that, by the Poincaré-Birkhoff-Witt theorem, each element $a \in \mathcal{E}(\mathfrak{g})$ can be written as

$$a = \sum_{k=0}^{m} \sum_{\substack{n \in \mathbb{N}_0^d \\ |n|=k}} \alpha_n x^n \tag{1}$$

with $m \in \mathbb{N}_0$ and complex coefficients α_n. Here we set $|n| := n_1 + \cdots + n_d$ for $n = (n_1, \ldots, n_d) \in \mathbb{N}_0^d$. If $a \in \mathcal{E}(\mathfrak{g})$ is of the form (1), then a is an elliptic element if $m \neq 0$ and if $\sum_{|n|=m} \alpha_n t^n \neq 0$ for all non-zero vectors $t \in \mathbb{R}^d$. Important examples of elliptic elements in $\mathcal{E}(\mathfrak{g})$ are the Nelson Laplacian $\Delta = x_1^2 + \cdots + x_d^2$ relative to the basis $\{x_1, \ldots, x_d\}$ of $\mathfrak{g}$ and $(1 - \Delta)^k$ for every $k \in \mathbb{N}$.

The following preliminary lemma is the key for most of the results in this section.

Lemma 10.2.2. *If a is an elliptic element of $\mathcal{E}(\mathfrak{g})$, then $\bigcap_{n \in \mathbb{N}} \mathcal{D}\big((\mathrm{d}U(a)^n)^*\big) \subseteq \mathcal{D}^\infty(U)$.*

Proof. Suppose that $\varphi \in \bigcap_{n \in \mathbb{N}} \mathcal{D}\big((\mathrm{d}U(a)^n)^*\big)$. Let $\psi \in \mathcal{H}(U)$. By Lemma 10.1.12, 10.1/(3), we have for each $f \in C_0^\infty(G)$ and $n \in \mathbb{N}$

$$\int_G \big((\tilde{a})^n f\big)(g) \langle U(g)\,\psi, \varphi\rangle \,\mathrm{d}\mu(g)$$

$$= \langle \mathrm{d}U(a)^n\, U_f\psi, \varphi\rangle = \big\langle U_f\psi, (\mathrm{d}U(a)^n)^*\,\varphi\big\rangle = \int_G f(g) \big\langle U(g)\,\psi, (\mathrm{d}U(a)^n)^*\,\varphi\big\rangle \,\mathrm{d}\mu(g).$$

This shows that the function $h(g) := \langle U(g)\,\psi, \varphi\rangle$ on G is a weak solution of the elliptic equation $(\tilde{a})^n\, h = h_n$ on G, where h_n is defined by $h_n(g) = \big\langle U(g)\,\psi, (\mathrm{d}U(a)^n)^*\,\varphi\big\rangle$, $g \in G$. Arguing in a similar way as in the proof of Theorem 10.1.9 it follows from the elliptic regularity theorem and from the Sobolev lemma that $h(\cdot) \equiv \langle U(\cdot)\,\psi, \varphi\rangle$ is in $C^\infty(G)$. Since $\psi \in \mathcal{H}(U)$ is arbitrary, $\varphi \in \mathcal{D}^\infty(U)$ by Corollary 10.1.3. □

Corollary 10.2.3. *The representation $\mathrm{d}U$ is self-adjoint. Thus each G-integrable representation of $\mathcal{E}(\mathfrak{g})$ is self-adjoint.*

Proof. Let a be any elliptic element of $\mathcal{E}(\mathfrak{g})$; see e.g. Remark 2. By definition, $\mathcal{D}\big((\mathrm{d}U)^*\big) \subseteq \bigcap_{n \in \mathbb{N}} \mathcal{D}\big((\mathrm{d}U(a)^n)^*\big)$, so $\mathcal{D}\big((\mathrm{d}U)^*\big) \subseteq \mathcal{D}^\infty(U) \equiv \mathcal{D}(\mathrm{d}U)$ by Lemma 10.2.2. Since $\mathrm{d}U$ is a $*$-representation, $\mathrm{d}U$ is self-adjoint. □

Remark 3. Since self-adjoint representations are always closed (cf. 8.1), $\mathrm{d}U$ is closed and hence $\mathcal{D}^\infty(U)\,[\mathrm{t}_{\mathrm{d}U}]$ is complete. The graph topology $\mathrm{t}_{\mathrm{d}U}$ is generated by a countable family of seminorms, so $\mathcal{D}^\infty(U)\,[\mathrm{t}_{\mathrm{d}U}]$ is a Frechet space. This fact could be also derived from Theorem 10.1.9.

Corollary 10.2.4. *Let a be an elliptic element of $\mathcal{E}(\mathfrak{g})$. Then $\mathcal{D}^\infty(U) = \mathcal{D}^\infty\big(\overline{\mathrm{d}U(a)}\big)$ and the graph topology $\mathrm{t}_{\mathrm{d}U}$ on $\mathcal{D}^\infty(U)$ is generated by the family of seminorms $\|\cdot\|_{\mathrm{d}U(a)^n}$, $n \in \mathbb{N}_0$.*

Proof. If a is elliptic, then so is a^+. Therefore, by Lemma 10.2.2,

$$\bigcap_{n \in \mathbb{N}} \mathcal{D}\big((\mathrm{d}U(a^+)^n)^*\big) \subseteq \mathcal{D}^\infty(U).$$

Since $\mathrm{d}U(a) \subseteq \mathrm{d}U(a^+)^*$, we have

$$\left(\overline{\mathrm{d}U(a)}\right)^n \subseteq \left(\mathrm{d}U(a^+)^*\right)^n \subseteq \left(\mathrm{d}U(a^+)^n\right)^*$$

for $n \in \mathbb{N}$. Hence

$$\mathcal{D}^\infty\left(\overline{\mathrm{d}U(a)}\right) \subseteq \bigcap_{n\in\mathbb{N}} \mathcal{D}\left(\left(\mathrm{d}U(a^+)^n\right)^*\right) \subseteq \mathcal{D}^\infty(U).$$

Since trivially $\mathcal{D}^\infty(U) \subseteq \mathcal{D}^\infty\left(\overline{\mathrm{d}U(a)}\right)$, we get $\mathcal{D}^\infty(U) = \mathcal{D}^\infty\left(\overline{\mathrm{d}U(a)}\right)$. Let $\mathfrak{k}$ denote the locally convex topology on $\mathcal{D}^\infty(U)$ which is generated by the seminorms $\|\cdot\|_{\mathrm{d}U(a)^n}$, $n \in \mathbb{N}_0$. Because of $\mathcal{D}^\infty(U) = \mathcal{D}^\infty\left(\overline{\mathrm{d}U(a)}\right)$, $\mathfrak{k}$ is a Frechet topology. Each $\mathrm{d}U(x)$, $x \in \mathscr{E}(\mathfrak{g})$, considered as an operator of $\mathcal{D}^\infty(U)\,[\mathfrak{k}]$ into $\mathcal{H}(U)$ is closed and hence continuous by the closed graph theorem. This yields $\mathfrak{t}_{\mathrm{d}U} \subseteq \mathfrak{k}$. Since obviously $\mathfrak{k} \subseteq \mathfrak{t}_{\mathrm{d}U}$, $\mathfrak{k} = \mathfrak{t}_{\mathrm{d}U}$. □

Corollary 10.2.5. *For each hermitian elliptic element a of $\mathscr{E}(\mathfrak{g})$, the operator $\mathrm{d}U(a)$ is essentially self-adjoint.*

Proof. Let $\xi \in \ker\left(\mathrm{d}U(a)^* - \alpha\right)$ for some $\alpha \in \mathbb{C} \diagup \mathbb{R}$. Then $\xi \in \mathcal{D}\left((\mathrm{d}U(a)^*)^n\right) \subseteq \mathcal{D}\left((\mathrm{d}U(a)^n)^*\right)$ for $n \in \mathbb{N}$, so that $\xi \in \mathcal{D}^\infty(U)$ by Lemma 10.2.2. From $a = a^+$, $\mathrm{d}U(a)^*\xi = \mathrm{d}U(a)\,\xi = \alpha\xi$. Since $\mathrm{d}U(a)$ is a symmetric operator, $\xi = 0$. □

Remark 4. Let a be a hermitian elliptic element of $\mathscr{E}(\mathfrak{g})$ and let π be the $*$-representation of the polynomial algebra $\mathbb{C}[\mathrm{x}]$ defined by $\pi(\mathrm{x}) = \mathrm{d}U(a)$. Lemma 10.2.2 shows that π is self-adjoint. Therefore, the assertions $\mathcal{D}^\infty(U) = \mathcal{D}^\infty\left(\overline{\mathrm{d}U(a)}\right)$ in Corollary 10.2.4 (in the case where a is hermitian) and of Corollary 10.2.5 follow also from Proposition 8.1.15, (v).

Corollary 10.2.5 is the starting point for a number of results which give (among others) sufficient conditions for the image $\mathrm{d}U(x)$ of a hermitian element x of $\mathscr{E}(\mathfrak{g})$ to be essentially self-adjoint. Our main result in this direction is

Theorem 10.2.6. *Let a be an elliptic element of $\mathscr{E}(\mathfrak{g})$. If T is an operator of $\mathcal{L}^+\left(\mathcal{D}^\infty(U)\right)$ such that TT^+ commutes with $\mathrm{d}U(a)$ on $\mathcal{D}^\infty(U)$, then TT^+ is essentially self-adjoint and $\overline{T^+} = T^*$. In particular, each symmetric operator on $\mathcal{D}^\infty(U)$ which leaves $\mathcal{D}^\infty(U)$ invariant and which commutes with $\mathrm{d}U(a)$ is essentially self-adjoint.*

Proof. From the closed graph theorem it follows that the operator TT^+ maps the Frechet space $\mathcal{D}^\infty(U)\,[\mathfrak{t}_{\mathrm{d}U}]$ continuously into the Hilbert space $\mathcal{H}(U)$. Since a is elliptic, so is $b := a^+a + 1$. By Corollary 10.2.4, the graph topology $\mathfrak{t}_{\mathrm{d}U}$ is generated by the seminorms $\|\cdot\|_{\mathrm{d}U(b)^n}$, $n \in \mathbb{N}_0$. Moreover, this family of seminorms is directed. Hence there are numbers $n \in \mathbb{N}_0$ and $\lambda > 0$ such that $\|TT^+\varphi\| \leqq \lambda\|\mathrm{d}U(b)^n\varphi\|$ for $\varphi \in \mathcal{D}^\infty(U)$. Since the symmetric operator TT^+ commutes with $\mathrm{d}U(a)$ on $\mathcal{D}^\infty(U)$, it commutes with $\mathrm{d}U(a)^+$ and so with $\mathrm{d}U(b)^n \equiv \left(\mathrm{d}U(a)^+\,\mathrm{d}U(a) + I\right)^n$. Since b^n is elliptic and hermitian, Corollary 10.2.5 says that $\mathrm{d}U(b^n) \equiv \mathrm{d}U(b)^n$ is essentially self-adjoint. Thus we have shown that the assumptions of Lemma 7.1.5 are satisfied in case $c_l := TT^+$, $a := \mathrm{d}U(b)^n$. Therefore, by this lemma, TT^+ is essentially self-adjoint. Lemma 7.1.2 gives $\overline{T^+} = T^*$. □

Now we derive some corollaries from Theorem 10.2.6. The first one generalizes Corollary 10.2.5 to general elliptic elements.

Corollary 10.2.7. *For each elliptic element a of $\mathscr{E}(\mathfrak{g})$, $\overline{\mathrm{d}U(a^+)} = \mathrm{d}U(a)^*$.*

Proof. Apply Theorem 10.2.6 to $T := \mathrm{d}U(a)$ and the elliptic element aa^+. □

Corollary 10.2.8. *Let a be an elliptic element of $\mathscr{E}(\mathfrak{g})$, and let x be an element of $\mathscr{E}(\mathfrak{g})$ which satisfies* $\mathrm{d}U(x)\,\mathrm{d}U(a) = \mathrm{d}U(a)\,\mathrm{d}U(x)$ *and* $\mathrm{d}U(x)\,\mathrm{d}U(a)^+ = \mathrm{d}U(a)^+\,\mathrm{d}U(x)$. *Then* $\overline{\mathrm{d}U(x^+)} = \mathrm{d}U(x)^*$.

Proof. Apply Theorem 10.2.6 to $T := \mathrm{d}U(x)$ and the elliptic element a. □

Corollary 10.2.9. *Let $\mathscr{Z}$ be the center of $\mathscr{E}(\mathfrak{g})$. For each $z \in \mathscr{Z}$, $\overline{\mathrm{d}U(z^+)} = \mathrm{d}U(z)^*$. If z_1 and z_2 are hermitian elements of $\mathscr{Z}$, then $\overline{\mathrm{d}U(z_1)}$ and $\overline{\mathrm{d}U(z_2)}$ are strongly commuting self-adjoint operators.*

Proof. Let a be any elliptic element of $\mathscr{E}(\mathfrak{g})$. Applying Corollary 10.2.8 in case $x := z$, we get $\overline{\mathrm{d}U(z^+)} = \mathrm{d}U(z)^*$. Letting $z := z_1 + \mathrm{i}z_2$, this yields $\overline{\mathrm{d}U(z_1 - \mathrm{i}z_2)} = \mathrm{d}U(z_1 + \mathrm{i}z_2)^*$. From Proposition 7.1.3, (i), applied with $a_1 := \mathrm{d}U(z_1)$, $a_2 := \mathrm{d}U(z_2)$, the second assertion follows. □

Corollary 10.2.10. *Suppose that the Lie group G is abelian or compact. Then $\overline{\mathrm{d}U(x^+)} = \mathrm{d}U(x)^*$ for all x in $\mathscr{E}(\mathfrak{g})$.*

Proof. By Corollary 10.2.8 it suffices to check that the center of $\mathscr{E}(\mathfrak{g})$ contains a hermitian elliptic element. In the case where G is abelian this is trivial, since then $\mathscr{E}(\mathfrak{g})$ is abelian. Suppose now G is compact. Then G is the direct product of an abelian Lie group G_1 and a semi-simple Lie group G_2 (Barut/Raczka [1], ch. 3, § 8). Let $\mathfrak{g}_k$ be the Lie algebra of G_k, $k = 1, 2$. Let Δ_2 be the Nelson Laplacian relative to an orthonormal basis with respect to the Killing form of $\mathfrak{g}_2$, and let Δ_1 be the Nelson Laplacian relative to a basis of $\mathfrak{g}_1$. Then Δ_2 is in the center of $\mathscr{E}(\mathfrak{g}_2)$ (Varadarajan [1], 3.11.1), so $\Delta := \Delta_1 + \Delta_2$ is obviously a hermitian elliptic element in the center of $\mathscr{E}(\mathfrak{g})$. □

Corollary 10.2.11. *Let x be an element of $\mathfrak{g}$. If p is a complex polynomial, then $\overline{\mathrm{d}U\big(p(\mathrm{i}x)^+\big)} = \mathrm{d}U\big(p(\mathrm{i}x)\big)^*$. If p is a polynomial with real coefficients, then $\mathrm{d}U\big(p(\mathrm{i}x)\big)$ is essentially self-adjoint. In particular, $\mathrm{d}U(\mathrm{i}x)^n$ is essentially self-adjoint and $\overline{\mathrm{d}U(x)^n} = \partial U(x)^n$ for every $n \in \mathbb{N}$.*

Proof. Define a unitary representation U_1 of the Lie group $G_1 := \mathbb{R}$ by $U_1(t) := U(\exp tx)$, $t \in \mathbb{R}$. Then $\partial U_1(s) = \partial U(x)$, $\mathscr{D}^\infty(U_1) \supseteqq \mathscr{D}^\infty(U)$ and $\mathrm{d}U_1\big(q(s)\big) \supseteqq \mathrm{d}U\big(q(x)\big)$ for any polynomial q, where s is a basis element of the Lie algebra of $\mathbb{R}$. Corollary 10.2.10 applied to the representation U_1 of $\mathbb{R}$ yields $\overline{\mathrm{d}U_1\big(p(\mathrm{i}s)^+\big)} = \mathrm{d}U_1\big(p(\mathrm{i}s)\big)^*$. From the equality $U_1(t)\,U_f\varphi = U_{f(\exp(-tx)\cdot)}\varphi$ for $t \in \mathbb{R}$, $f \in C_0^\infty(G)$ and $\varphi \in \mathscr{H}(U)$ we see that U_1 leaves the Gårding domain $\mathscr{D}_G(U)$ invariant. Moreover, $\mathscr{D}_G(U) \subseteqq \mathscr{D}^\infty(U_1)$. Therefore, by Theorem 10.1.14, $\mathscr{D}_G(U)$ and so $\mathscr{D}^\infty(U)$ is a core for $\mathrm{d}U_1\big(q(\mathrm{i}s)\big)$, that is,

$$\overline{\mathrm{d}U\big(q(\mathrm{i}x)\big)} = \overline{\mathrm{d}U_1\big(q(\mathrm{i}s)\big) \upharpoonright \mathscr{D}^\infty(U)} = \overline{\mathrm{d}U_1\big(q(\mathrm{i}s)\big)}$$

for each polynomial q. Combined with the preceding, we get $\overline{\mathrm{d}U\big(p(\mathrm{i}x)^+\big)} = \mathrm{d}U\big(p(\mathrm{i}x)\big)^*$. The next two assertions are only reformulations of the first one. We verify the last statement. Let $n \in \mathbb{N}$. The operator $\mathrm{i}\partial U(x)$ is self-adjoint, and $\big(\mathrm{i}\partial U(x)\big)^n \supseteqq \big(\mathrm{i}\mathrm{d}U(x)\big)^n = \mathrm{d}U(\mathrm{i}x)^n$. Since $\mathrm{d}U(\mathrm{i}x)^n$ is essentially self-adjoint, this gives $\big(\mathrm{i}\partial U(x)\big)^n = \overline{\big(\mathrm{i}\mathrm{d}U(x)\big)^n}$. □

Combining the last assertion of Corollary 10.2.11 with Theorem 10.1.9, we obtain

Corollary 10.2.12. *If* $\{x_1, \ldots, x_d\}$ *is a basis for* $\mathfrak{g}$, *then* $\mathcal{D}^\infty(U) = \bigcap_{k=1}^{d} \mathcal{D}^\infty\big(\overline{\mathrm{d}U(x_k)}\big)$.

Corollary 10.2.13. *For each* x *in* $\mathfrak{g}$, $\overline{\mathrm{d}U(x)}$ *is the infinitesimal generator of the one-parameter unitary group* $t \to U(\exp tx)$, *i.e.,* $U(\exp tx) = \exp t\,\overline{\mathrm{d}U(x)}$ *for* $t \in \mathbb{R}$.

Proof. Combine the definition of $\partial U(x)$ with the equality $\overline{\mathrm{d}U(x)} = \partial U(x)$. □

Example 10.2.14. Let G be the Heisenberg group, that is, the three dimensional Lie group of all matrices

$$g(a, b, c) = \begin{bmatrix} 1 & a & c \\ 0 & 1 & b \\ 0 & 0 & 1 \end{bmatrix}, \quad a, b, c \in \mathbb{R}.$$

The Lie algebra $\mathfrak{g}$ of G is spanned by basis elements x, y, z satisfying the relations $[x, y] = z$, $[x, z] = [y, z] = 0$. The corresponding one-parameter groups in G are given by $\exp tx = g(t, 0, 0)$, $\exp ty = g(0, t, 0)$ and $\exp tz = g(0, 0, t)$, $t \in \mathbb{R}$.

For each $\lambda \in \mathbb{R} \setminus \{0\}$, the formula

$$\big(U_\lambda(g(a, b, c))\,\varphi\big)(t) := \exp\big(\mathrm{i}t\,\lambda(tb + c)\big)\,\varphi(t + a), \quad t \in \mathbb{R} \text{ and } \varphi \in L^2(\mathbb{R}),$$

defines an irreducible unitary representation U_λ of G on the Hilbert space $\mathcal{H}(U_\lambda) \equiv L^2(\mathbb{R})$. By differentiation we obtain that $\partial U_\lambda(x) = \frac{\mathrm{d}}{\mathrm{d}t}$, $\partial U_\lambda(y) = \mathrm{i}\lambda t$ and $\partial U_\lambda(z) = \mathrm{i}\lambda$. Therefore, it follows from Theorem 10.1.9 that $\mathcal{D}^\infty(U)$ is equal to the Schwartz space $\mathcal{S}(\mathbb{R})$. In fact, Theorem 10.1.9 gives an appearently weaker (but equivalent) condition: A function $\varphi \in C^\infty(\mathbb{R})$ is in $\mathcal{S}(\mathbb{R})$ if (and only if) for all $n \in \mathbb{N}_0$ and all polynomials $p \in \mathbb{C}[\mathrm{x}]$ the functions $\varphi^{(n)}(t)$ and $p(t)\,\varphi(t)$ are in $L^2(\mathbb{R})$. Moreover, it is obvious that $\mathrm{d}U(\mathcal{E}(\mathfrak{g}))$ coincides with the O*-algebra $\mathbf{A}(p_1, q_1)$ of Example 2.5.2.

Set $\Delta := x^2 + y_z + z^2$. By Corollary 10.2.5, $\mathrm{d}U_\lambda(-\Delta) = -\left(\frac{\mathrm{d}}{\mathrm{d}t}\right)^2 + \lambda^2 t^2 + \lambda^2$ is an essentially self-adjoint operator on $\mathcal{S}(\mathbb{R})$. Combined with Theorem 10.1.14 it follows that its restriction to $C_0^\infty(\mathbb{R})$ is essentially self-adjoint. (Both facts are well-known in quantum physics.) On the other hand, the image $T := \mathrm{d}U_1(\mathrm{i}yxy) = -t^2 \frac{\mathrm{d}}{\mathrm{d}t} - \mathrm{i}t$ of the hermitian element $\mathrm{i}yxy$ of $\mathcal{E}(\mathfrak{g})$ is not essentially self-adjoint. The symmetric operator T has deficiency indices (1, 1). (In fact, $\ker(T^* + \mathrm{i})$ is spanned by the function φ_+ and $\ker(T^* - \mathrm{i})$ by φ_-, where $\varphi_+(t) = t^{-1}\exp(-t^{-1})$ if $t > 0$, $\varphi_+(t) = 0$ if $t \leqq 0$, $\varphi_-(t) = t^{-1}\exp t^{-1}$ if $t < 0$ and $\varphi_-(t) = 0$ if $t \geqq 0$.) ○

Example 10.2.15. Let G be the affine group of the real line, that is, $G = \{(a,b) : a > 0, b \in \mathbb{R}\}$ with the multiplication rule $(a_1, b_1)(a_2, b_2) = (a_1a_2, a_1b_2 + b_1)$. The Lie algebra $\mathfrak{g}$ of G has a basis $\{x, y\}$ which satisfies the relation $[x, y] = y$. We have $\exp tx = (\mathrm{e}^t, 0)$ and $\exp ty = (1, t)$ for $t \in \mathbb{R}$. The formula $\big(U(a, b)\,\varphi\big)(t) = \exp(\mathrm{i}\mathrm{e}^t b)\,\varphi(t + \log a)$, $\varphi \in L^2(\mathbb{R})$, defines an irreducible unitary representation of G on $\mathcal{H}(U) \equiv L^2(\mathbb{R})$. Clearly, $\partial U(x) = \frac{\mathrm{d}}{\mathrm{d}t}$ and $\partial U(y) = \mathrm{i}\mathrm{e}^t$. By Theorem 10.1.9, $\mathcal{D}^\infty(U)$ consists of the C^∞-functions on $\mathbb{R}$ for which $\varphi^{(n)}(t)$ and $\mathrm{e}^{nt}\varphi(t)$ are in $L^2(\mathbb{R})$ for all $n \in \mathbb{N}_0$. From Corollary 10.2.5 and

Theorem 10.1.14, the restriction of the operator $\mathrm{d}U(-x^2-y^2) = -\left(\frac{\mathrm{d}}{\mathrm{d}t}\right)^2 + e^{2t}$ to $C_0^\infty(\mathbb{R})$ is essentially self-adjoint. The image $\mathrm{d}U(xy+yx) = 2ie^t\frac{\mathrm{d}}{\mathrm{d}t} + ie^t$ of the hermitian element $xy+yx$ of $\mathscr{E}(\mathfrak{g})$ has deficiency indices (0, 1). ○

Next we consider group invariant continuous sesquilinear forms. Let $\mathfrak{c}$ be a sesquilinear form on $\mathscr{D}^\infty(U) \times \mathscr{D}^\infty(U)$. We say $\mathfrak{c}$ is group invariant if $\mathfrak{c}(U(g)\varphi, U(g)\psi) = \mathfrak{c}(\varphi, \psi)$ for all $\varphi, \psi \in \mathscr{D}^\infty(U)$ and $g \in G_0$. Note that this definition makes sense since $U(g)$ leaves $\mathscr{D}^\infty(U)$ invariant. (The connected component G_0 of the unit in G is used only for a convenient formulation of the results.) Let $\mathscr{B}(\mathscr{D}^\infty(U))$ denote the vector space of all continuous sesquilinear forms on $\mathscr{D}^\infty(U) \times \mathscr{D}^\infty(U)$ relative to the graph topology $\mathfrak{t}_{\mathrm{d}U}$ on $\mathscr{D}^\infty(U)$.

We summarize our results concerning group invariant sesquilinear forms in the following theorem. In the proof of this theorem we shall use Theorem 7.3.6.

Theorem 10.2.16. *Let $\mathfrak{c}$ be a sesquilinear form of $\mathscr{B}(\mathscr{D}^\infty(U))$.*
The following are equivalent:

(i) $\mathfrak{c}$ *is group invariant.*

(ii) *There exists a linear operator T on $\mathscr{D}^\infty(U)$ such that $\mathfrak{c}(\cdot,\cdot) \equiv \langle T\cdot,\cdot\rangle$ and $U(g)\,T \subseteq TU(g)$ for all g in G_0.*

(iii) $\mathfrak{c}(\mathrm{d}U(x)\,\varphi, \psi) = \mathfrak{c}(\varphi, \mathrm{d}U(x)^+\,\psi)$ *for all φ and ψ in $\mathscr{D}^\infty(U)$ and x in $\mathscr{E}(\mathfrak{g})$.*

(iv) *There exists a linear operator T on $\mathscr{D}^\infty(U)$ such that $\mathfrak{c}(\cdot,\cdot) \equiv \langle T\cdot,\cdot\rangle$, $T\mathscr{D}^\infty(U) \subseteq \mathscr{D}^\infty(U)$ and $T\,\mathrm{d}U(x)\,\varphi = \mathrm{d}U(x)\,T\varphi$ for all φ in $\mathscr{D}^\infty(U)$ and x in $\mathscr{E}(\mathfrak{g})$.*

Further, if T is a linear operator on $\mathscr{D}^\infty(U)$ as in (ii) *or in* (iv), *then $T \in \mathscr{L}^+(\mathscr{D}^\infty(U))$ and $\overline{T^+} = T^*$.*

Remark 5. Theorem 10.2.16 remains valid if we only take x from $\mathfrak{g}$ in (iii) and in (iv).

Proof of Theorem 10.2.16:

(i) → (iii): Suppose $\varphi, \psi \in \mathscr{D}^\infty(U)$ and $x \in \mathfrak{g}$. From the group invariance of $\mathfrak{c}$, we have that $\mathfrak{c}(U(\exp tx)\,\varphi, \psi) = \mathfrak{c}(\varphi, U(\exp(-tx))\,\psi) =: f(t)$ for $t \in \mathbb{R}$. Since $\mathfrak{c}$ is continuous relative to the graph topology $\mathfrak{t}_{\mathrm{d}U}$ and since the map $t \to U(\exp tx)\,\varphi$ of $\mathbb{R}$ into $\mathscr{D}^\infty(U)[\mathfrak{t}_{\mathrm{d}U}]$ is C^∞ by Proposition 10.1.11, f is a complex-valued differentiable function on $\mathbb{R}$ and we have

$$f'(0) = \mathfrak{c}(\mathrm{d}U(x)\,\varphi, \psi) = \mathfrak{c}(\varphi, -\mathrm{d}U(x)\,\psi) = \mathfrak{c}(\varphi, \mathrm{d}U(x)^+\,\psi)$$

which proves (iii) in case where $x \in \mathfrak{g}$. Because of the Poincaré-Birkhoff-Witt theorem, a repeated application of the last equation yields (iii) for general elements x in $\mathscr{E}(\mathfrak{g})$.

(iii) → (i): Fix φ and ψ in $\mathscr{D}^\infty(U)$ and x in $\mathfrak{g}$. Define $f(t, s) := \mathfrak{c}(U(\exp tx)\,\varphi, U(\exp sx)\,\psi)$, $t, s \in \mathbb{R}$. Similarly as in the preceding proof of (i) → (iii), we conclude that f is differentiable on $\mathbb{R}^2$. By the chain rule,

$$\begin{aligned}\frac{\mathrm{d}}{\mathrm{d}t} f(t,t) &= \mathfrak{c}(\mathrm{d}U(x)\,U(\exp tx)\,\varphi, U(\exp tx)\,\psi) + \mathfrak{c}(U(\exp tx)\,\varphi, \mathrm{d}U(x)\,U(\exp tx)\,\psi) \\ &= \mathfrak{c}(U(\exp tx)\,\varphi, -\mathrm{d}U(x)\,U(\exp tx)\,\psi) \\ &\quad + \mathfrak{c}(U(\exp tx)\,\varphi, \mathrm{d}U(x)\,U(\exp tx)\,\psi) = 0\end{aligned}$$

for all $t \in \mathbb{R}$, where we used (iii) and that fact that $U(\exp tx)\,\mathcal{D}^\infty(U) \subseteq \mathcal{D}^\infty(U)$. Therefore, $f(t,t)$ is constant on $\mathbb{R}$. Hence

$$\mathfrak{c}\big(U(\exp tx)\,\varphi, U(\exp tx)\,\psi\big) = f(t,t) = f(0,0) = \mathfrak{c}\,(\varphi,\psi)$$

for $t \in \mathbb{R}$. Since each $g \in G_0$ is a product of elements $\exp x$, $x \in \mathfrak{g}$, this yields the group invariance of $\mathfrak{c}$.

(iii) $\to$ (iv): Let Δ be the Nelson Laplacian relative to a basis of $\mathfrak{g}$. By Corollary 10.2.5, the operator $(\mathrm{d}U(1-\Delta)^n)^2 \equiv \mathrm{d}U((1-\Delta)^{2n})$ is essentially self-adjoint for each $n \in \mathbb{N}$. From Corollary 10.2.4, the graph topology $\mathrm{t}_{\mathrm{d}U}$ is generated by the (directed) family of seminorms $\{\|\cdot\|_{\mathrm{d}U(1-\Delta)^n} : n \in \mathbb{N}_0\}$. Further, the $*$-representation $\mathrm{d}U$ and so the O*-algebra $\mathrm{d}U(\mathcal{E}(\mathfrak{g}))$ is closed. These facts show that the O*-algebra $\mathcal{A} := \mathrm{d}U(\mathcal{E}(\mathfrak{g}))$ is strictly self-adjoint (cf. Definition 7.3.5). By Proposition 7.2.2, (i), it follows from condition (iii) that there is a $T \in \mathcal{A}_{\mathrm{f}}^{\mathrm{c}}$ such that $\mathfrak{c}(\cdot,\cdot) \equiv \langle T\cdot,\cdot\rangle$. By Theorem 7.3.6, (i), $\mathcal{A}_{\mathrm{f}}^{\mathrm{c}} = \mathcal{A}^{\mathrm{c}}$, and $\mathcal{A}^{\mathrm{c}}$ is an O*-algebra on $\mathcal{D}^\infty(U)$. Hence $T \in \mathcal{A}^{\mathrm{c}} \subseteq \mathcal{L}^+(\mathcal{D}^\infty(U))$. Since $T \in \mathcal{A}_{\mathrm{f}}^{\mathrm{c}} = \mathcal{A}^{\mathrm{c}}$, $TT^+ \in \mathcal{A}^{\mathrm{c}}$. Therefore, by Theorem 7.3.6, (ii), TT^+ is essentially self-adjoint. From Lemma 7.1.2, $\overline{T^+} = T^*$.

(i) $\to$ (ii): Since (i) $\to$ (iii) as shown above, it follows from the preceding proof that there is an operator $T \in \mathcal{L}^+(\mathcal{D}^\infty(U))$ such that $\mathfrak{c}(\cdot,\cdot) \equiv \langle T\cdot,\cdot\rangle$. Let $g \in G_0$ and $\varphi \in \mathcal{D}^\infty(U)$. By the group invariance of $\mathfrak{c}$, $\langle T\varphi,\psi\rangle = \langle TU(g)\,\varphi, U(g)\,\psi\rangle$ for all $\psi \in \mathcal{D}^\infty(U)$. Hence $T\varphi = U(g)^*\,TU(g)\,\varphi$ which yields $U(g)\,T \subseteq TU(g)$.

(ii) $\to$ (i): Since $U(g)\,T \subseteq TU(g)$, we have $\mathfrak{c}\big(U(g)\,\varphi, U(g)\,\psi\big) = \langle TU(g)\,\varphi, U(g)\,\psi\rangle$ $= \langle U(g)\,T\varphi, U(g)\,\psi\rangle = \langle T\varphi,\psi\rangle = \mathfrak{c}(\varphi,\psi)$ for $\varphi,\psi \in \mathcal{D}^\infty(U)$ and $g \in G_0$.

A similar reasoning proves (iv) $\to$ (iii). Thus the four statements are equivalent.

Finally, suppose that T is as in (ii) or in (iv). Since (ii) $\to$ (iii) and (iv) $\to$ (iii), we have, by the above proof of the implication (iii) $\to$ (iv), $T \in \mathcal{L}^+(\mathcal{D}^\infty(U))$ and $\overline{T^+} = T^*$. □

Corollary 10.2.17. *If T is a formally normal operator on $\mathcal{D}^\infty(U)$ such that $U(g)\,T \subseteq TU(g)$ for all g in G_0, then $\overline{T}$ is normal.*

Proof. It follows from the closed graph theorem that T maps $\mathcal{D}^\infty(U)\,[\mathrm{t}_{\mathrm{d}U}]$ continuously into $\mathcal{H}(U)$. Hence the sesquilinear form $\mathfrak{c}(\cdot,\cdot) := \langle T\cdot,\cdot\rangle$ is in $\mathcal{B}(\mathcal{D}^\infty(U))$, and the result follows from the last assertion in Theorem 10.2.16 and Proposition 7.1.3, (i). □

Corollary 10.2.18. $\mathrm{d}U(\mathcal{E}(\mathfrak{g}))' = \{U(g) : g \in G_0\}'$.

Proof. For each $C \in \mathbf{B}(\mathcal{H}(U))$, the sesquilinear form $\mathfrak{c}(\cdot,\cdot) := \langle C\cdot,\cdot\rangle$ is, of course, in $\mathcal{B}(\mathcal{D}^\infty(U))$; so the equivalence of (ii) and (iv) in Theorem 10.2.16 gives the assertion. □

We sketch a direct proof of Corollary10.2.18 which does not use Theorem 10.2.16.

Second proof of Corollary 10.2.18. Let $x \in \mathfrak{g}$. An operator $C \in \mathbf{B}(\mathcal{H}(U))$ commutes with the self-adjoint operator $\mathrm{i}\overline{\mathrm{d}U(x)}$ if and only if it commutes with $U(\exp tx)$ $\equiv \exp t\,\overline{\mathrm{d}U(x)}$ for all $t \in \mathbb{R}$. (Here we used Corollaries 10.2.11 and 10.2.13.) Since $U(G_0)'$ $= \{U(\exp x) : x \in \mathfrak{g}\}'$, this implies $\mathrm{d}U(\mathcal{E}(\mathfrak{g}))' \subseteq U(G_0)'$. Since the algebra $\mathcal{E}(\mathfrak{g})$ is generated by $\mathfrak{g} \cup \{1\}$, the opposite inclusion also follows from the above fact once we have shown that each operator $C \in U(G_0)'$ leaves $\mathcal{D}^\infty(U)$ invariant. But $C \in U(G_0)'$ commutes with $\overline{\mathrm{d}U(x)}$, $x \in \mathfrak{g}$, so that C leaves $\mathcal{D}^\infty(\overline{\mathrm{d}U(x)})$ invariant. By Corollary 10.2.12, we obtain $T(\mathcal{D}^\infty(U)) \subseteq \mathcal{D}^\infty(U)$. □

Proposition 10.2.19. *If the Lie group G is connected, then each self-adjoint subrepresentation of a G-integrable representation is again G-integrable.*

Proof. Let π_0 be a self-adjoint representation of $\mathcal{E}(\mathfrak{g})$ such that $\pi_0 \subseteq dU$, and let P be the projection of $\mathcal{H}(U)$ onto $\mathcal{H}(\pi_0)$. Since π_0 is self-adjoint, Proposition 8.3.11 yields $P \in dU(\mathcal{E}(\mathfrak{g}))'$. By Corollary 10.2.18, $dU(\mathcal{E}(\mathfrak{g}))' = U(G_0)'$. Since G is connected, $G = G_0$ and hence $P \in U(G)'$. Therefore, the map $g \to U_0(g) := U(g) \upharpoonright \mathcal{H}(\pi_0)$ is a unitary representation of G on $\mathcal{H}(\pi_0)$. From $\pi_0 \subseteq dU$ it follows that $\pi_0 \subseteq dU_0$, so that the $*$-representation dU_0 is an extension of the self-adjoint representation π_0 in the same Hilbert space $\mathcal{H}(\pi_0)$. This implies that $\pi_0 = dU_0$. □

10.3. Analytic Vectors and Analytic Domination of Families of Operators

Suppose that E is a linear space equipped with a seminorm $\|\cdot\|$. The word "operator" and the notation $\mathcal{D}(A)$ and $\mathcal{D}^\infty(A) \equiv \bigcap_{n\in\mathbb{N}} \mathcal{D}(A^n)$ will be used in the same way as in the case where $(E, \|\cdot\|)$ is a Hilbert space, cf. 1.6.

Definition 10.3.1. Let A be a linear operator in E. A vector φ in E is called an *analytic vector* [resp. *semi-analytic vector*] *for* A if $\varphi \in \mathcal{D}(A^n)$ for all $n \in \mathbb{N}$ and if there exists a constant M (depending on φ) such that $\|A^n\varphi\| \leq M^n n!$ [resp. $\|A^n\varphi\| \leq M^n(2n)!$] for all $n \in \mathbb{N}$.

We let $\mathcal{D}^\omega(A)$ and $\mathcal{D}^{s\omega}(A)$ denote the sets of all analytic vectors and semi-analytic vectors for A, respectively. Obviously, $\mathcal{D}^\omega(A)$ and $\mathcal{D}^{s\omega}(A)$ are linear subspaces of $\mathcal{D}^\infty(A)$, and $\mathcal{D}^\omega(A) \subseteq \mathcal{D}^{s\omega}(A)$.

We introduce some quantities which measure the growth of the sequence $(\|A^n\varphi\| : n \in \mathbb{N})$ for a vector $\varphi \in \mathcal{D}^\infty(A)$. If $t > 0$ and $\varphi \in \mathcal{D}^\infty(A)$, we define

$$e_t^A(\varphi) = \sum_{n=0}^{\infty} \frac{t^n}{n!} \|A^n\varphi\| \tag{1}$$

and

$$\mathfrak{s}_t^A(\varphi) = \sum_{n=0}^{\infty} \frac{t^{2n}}{(2n)} \|A^n\varphi\|. \tag{2}$$

Let $\mathcal{D}_t^\omega(A)$ and $\mathcal{D}_t^{s\omega}(A)$ be the linear subspaces of $\mathcal{D}^\infty(A)$ defined by $\mathcal{D}_t^\omega(A) = \{\varphi \in \mathcal{D}^\infty(A) : e_t^A(\varphi) < \infty\}$ and $\mathcal{D}_t^{s\omega}(A) = \{\varphi \in \mathcal{D}^\infty(A) : \mathfrak{s}_t^A(\varphi) < \infty\}$ and equipped with the seminorms $e_t^A(\cdot)$ and $\mathfrak{s}_t^A(\cdot)$, respectively. From the above definitions it is easily seen that $\mathcal{D}^\omega(A) = \bigcup_{t>0} \mathcal{D}_t^\omega(A)$ and $\mathcal{D}^{s\omega}(A) = \bigcup_{t>0} \mathcal{D}_t^{s\omega}(A)$. I. e. a vector $\varphi \in \mathcal{D}^\infty(A)$ is an analytic vector [resp. semi-analytic vector] for A if and only if there is a $t > 0$ such that the power series in (1) [resp. (2)] converges.

Definition 10.3.2. Let $\mathcal{X}$ be a set of linear mappings of E into itself. A vector φ in E is called an *analytic vector for the family* $\mathcal{X}$ if there exists a constant M such that $\|X_1 \dots X_n\varphi\| \leq M^n n!$ for arbitrary elements $X_1, \dots, X_n \in \mathcal{X}$ and for all $n \in \mathbb{N}$.

Let $\mathcal{D}^\omega(\mathcal{X})$ be the set of analytic vectors for $\mathcal{X}$. Clearly, $\mathcal{D}^\omega(\mathcal{X})$ is a linear subspace of E. We now define similar quantities and spaces as in case of a single operator. For $n \in \mathbb{N}$, let $\nu_n^{\mathcal{X}}(\cdot)$ be defined by

$$\nu_n^{\mathcal{X}}(\varphi) = \sup \{\|X_1 \dots X_n\varphi\| : X_1, \dots, X_n \in \mathcal{X}\}, \quad \varphi \in E.$$

Put $\nu_0^{\mathscr{X}}(\cdot) := \|\cdot\|$. Further, if $t > 0$ and $\varphi \in E$, set

$$\mathrm{e}_t^{\mathscr{X}}(\varphi) = \sum_{n=0}^{\infty} \frac{t^n}{n!} \nu_n^{\mathscr{X}}(\varphi). \tag{3}$$

Let $\mathcal{D}_t^\omega(\mathscr{X})$ be the linear subspace of E defined by $\mathcal{D}_t^\omega(\mathscr{X}) = \{\varphi \in E : \mathrm{e}_t^{\mathscr{X}}(\varphi) < \infty\}$ and endowed with the seminorm $\mathrm{e}_t^{\mathscr{X}}(\cdot)$. Then a vector $\varphi \in E$ is an analytic vector for $\mathscr{X}$ if and only if there is a constant M such that $\nu_n^{\mathscr{X}}(\varphi) \leqq M^n n!$ for all $n \in \mathbb{N}_0$ or equivalently if there is a $t > 0$ such that the series in (3) converges. Thus we have that $\mathcal{D}^\omega(\mathscr{X}) = \bigcup_{t>0} \mathcal{D}_t^\omega(\mathscr{X})$.

Remark 1. The above quantities and the notion of an analytic vector depend, of course, on the seminorm $\|\cdot\|$. If confusion can arise, we speak about analytic vectors relative to $\|\cdot\|$.

Analytic Vectors and Semi-Analytic Vectors for Symmetric Operators in Hilbert Space

In this subsection $\mathcal{H}$ is a Hilbert space with norm $\|\cdot\|$.

Lemma 10.3.3. *Suppose that A is a self-adjoint operator on $\mathcal{H}$ and $\varphi \in \mathcal{D}_t^\omega(A)$ for some $t > 0$. Then $\varphi \in \mathcal{D}(\mathrm{e}^{zA})$ and*

$$\mathrm{e}^{zA}\varphi = \sum_{n=0}^{\infty} \frac{z^n}{n!} A^n\varphi \tag{4}$$

for $z \in \mathbb{C}$, $|z| \leqq t$, where the series in (4) converges absolutely. The map $z \to \mathrm{e}^{\mathrm{i}zA}\varphi$ is a holomorphic function in the strip $\{z \in \mathbb{C} : |\mathrm{Im}\, z| < t\}$ with values in $\mathcal{H}$.

Proof. Let $A = \int \lambda \, \mathrm{d}E(\lambda)$ be the spectral decomposition of A. Fix $z \in \mathbb{C}$, $|z| \leqq t$. From the properties of the spectral decomposition, we have for $k \in \mathbb{N}$,

$$\left(\int_{-k}^{k} |\mathrm{e}^{z\lambda}|^2 \, \mathrm{d}\, \|E(\lambda)\, \varphi\|^2 \right)^{1/2} = \left\| \int_{-k}^{k} \mathrm{e}^{z\lambda} \, \mathrm{d}E(\lambda)\, \varphi \right\| = \left\| \int_{-k}^{k} \sum_{n=0}^{\infty} \frac{(z\lambda)^n}{n!} \, \mathrm{d}E(\lambda)\, \varphi \right\|$$

$$\leqq \sum_{n=0}^{\infty} \frac{|z|^n}{n!} \left\| \int_{-k}^{k} \lambda^n \, \mathrm{d}E(\lambda)\, \varphi \right\| \leqq \sum_{n=0}^{\infty} \frac{t^n}{n!} \|A^n\varphi\| < \infty.$$

Letting $k \to \infty$, this shows that $\varphi \in \mathcal{D}(\mathrm{e}^{zA})$.

Because of $\mathrm{e}_t^A(\varphi) < \infty$, we have

$$\left\| \int \sum_{n=m+1}^{\infty} \frac{(z\lambda)^n}{n!} \, \mathrm{d}E(\lambda)\, \varphi \right\| = \lim_{k\to\infty} \left\| \int_{-k}^{k} \sum_{n=m+1}^{\infty} \frac{(z\lambda)^n}{n!} \, \mathrm{d}E(\lambda)\, \varphi \right\|$$

$$\leqq \lim_{k\to\infty} \sum_{n=m+1} \frac{|z|^n}{n!} \left\| \int_{-k}^{k} \lambda^n \, \mathrm{d}E(\lambda)\, \varphi \right\|$$

$$\leqq \sum_{n=m+1}^{\infty} \frac{t^n}{n!} \|A^n\varphi\| \to 0 \quad \text{as} \quad m \to \infty,$$

so that

$$e^{zA}\varphi = \int e^{z\lambda}\, dE(\lambda)\, \varphi = \sum_{n=0}^{m} \frac{z^n}{n!} \int \lambda^n\, dE(\lambda)\, \varphi + \int \sum_{n=m+1}^{\infty} \frac{(z\lambda)^n}{n!}\, dE(\lambda)\, \varphi$$

$$= \lim_{m\to\infty} \sum_{n=0}^{m} \frac{z^n}{n!} A^n\varphi = \sum_{n=0}^{\infty} \frac{z^n}{n!} A^n\varphi.$$

This proves (4).

Let $s \in \mathbb{R}$. From $\|A^n e^{isA}\varphi\| = \|A^n\varphi\|$ we see that $e^{isA}\varphi \in \mathcal{D}_t^\omega(A)$. Applying (4) with z replaced by $i(z-s)$ and φ by $e^{izA}\varphi$, we have $e^{izA}\varphi = e^{i(z-s)A}e^{isA}\varphi = \sum_{n=0}^{\infty} \frac{(i(z-s))^n}{n!} A^n e^{isA}\varphi$ for all $z \in \mathbb{C}$, $|z-s| < t$. This implies that the map $z \to e^{izA}\varphi$ is holomorphic in the strip $\{z \in \mathbb{C}: |\mathrm{Im}\, z| < t\}$. □

Proposition 10.3.4. *Suppose that T is a symmetric linear operator on $\mathcal{H}$ such that $\mathcal{D}^\omega(T)$ is dense in $\mathcal{H}$. Then T is essentially self-adjoint.*

Proof. We first prove the assertion under the additional assumption that T has equal deficiency indices. Then there exists a self-adjoint extension, say A, of T on the Hilbert space $\mathcal{H}$. Fix a vector ξ of $\ker (T^* - i)$. Let $\varphi \in \mathcal{D}^\omega(T)$. Then $\varphi \in \mathcal{D}_t^\omega(T)$ for some $t > 0$. Since $T \subseteq A$, $\varphi \in \mathcal{D}_t^\omega(A)$. From Lemma 10.3.3 it follows that $f(z) := \langle e^{izA}\varphi, \xi\rangle$ defines a holomorphic function in the strip $\{z \in \mathbb{C}: |\mathrm{Im}\, z| < t\}$, and

$$f(z) = \sum_{n=0}^{\infty} \frac{(iz)^n}{n!} \langle A^n\varphi, \xi\rangle \quad \text{for} \quad z \in \mathbb{C},\ |z| < t.$$

From the latter and from $\xi \in \ker (T^* - i)$, we obtain that

$$f^{(n)}(0) = i^n\langle A^n\varphi, \xi\rangle = i^n\langle T^n\varphi, \xi\rangle = i^n\langle \varphi, (T^*)^n\, \xi\rangle$$
$$= i^n\langle \varphi, i^n\xi\rangle = \langle \varphi, \xi\rangle \quad \text{for} \quad n \in \mathbb{N}.$$

Moreover, $f(0) = \langle \varphi, \xi\rangle$. By the uniqueness theorem for holomorphic functions, we have

$$f(z) = \sum_{n=0}^{\infty} \frac{z^n}{n!} \langle \varphi, \xi\rangle = \langle \varphi, \xi\rangle\, e^z \quad \text{for all} \quad z \in \mathbb{C},\ |\mathrm{Im}\, z| < t. \tag{5}$$

On the other hand, if z is real, then e^{izA} is unitary and hence $|f(z)| \leqq \|\varphi\|\, \|\xi\|$. That is, f is bounded on $\mathbb{R}$. Combined with (5), this yields $\langle \varphi, \xi\rangle = 0$. Since this holds for all φ in the dense set $\mathcal{D}^\omega(T)$ in $\mathcal{H}$, $\xi = 0$. Thus $\ker (T^* - i) = \{0\}$. Similarly, $\ker (T^* + i) = \{0\}$. Hence T is essentially self-adjoint.

To prove the assertion in the general case, consider the symmetric operator $T_1 := T \oplus (-T)$ in the Hilbert space $\mathcal{H}_1 := \mathcal{H} \oplus \mathcal{H}$. Then $\mathcal{D}^\omega(T_1) = \mathcal{D}^\omega(T) \oplus \mathcal{D}^\omega(T)$ is dense in $\mathcal{H}_1$ and T_1 has equal deficiency indices. Therefore, by the preceding, T_1 and so T is essentially self-adjoint. □

Corollary 10.3.5. *A closed symmetric linear operator T on a Hilbert space $\mathcal{H}$ is self-adjoint if and only if $\mathcal{D}^\omega(T)$ is dense in $\mathcal{H}$.*

Proof. The sufficiency follows from Proposition 10.3.4. Suppose that T is self-adjoint. Let $E(\lambda)$, $\lambda \in \mathbb{R}$, be the spectral projections of T. If $\varphi \in E((-k, k))\, \mathcal{H}$, then $\|T^n\varphi\| \leqq k^n\|\varphi\|$

for $n \in \mathbb{N}$, so that $\mathcal{D}^b(T) := \bigcup_{k \in \mathbb{N}} E((-k, k))\,\mathcal{H} \subseteq \mathcal{D}^\omega(T)$. The spectral theorem shows that $\mathcal{D}^b(T)$ is dense in $\mathcal{H}$. $\square$

Remark 2. Each vector $\varphi \in \mathcal{D}^b(T)$ satisfies a much stronger growth condition than is needed to prove that $\varphi \in \mathcal{D}^\omega(T)$: There is a constant M such that $\|T^n\varphi\| \leqq M^n$ for all $n \in \mathbb{N}$. Such vectors are called *bounded vectors* for T.

For non-negative self-adjoint operators A there is a strong link between the spaces $\mathcal{D}_t^\omega(A)$ and the domains $\mathcal{D}(e^{tA})$ and between $\mathcal{D}_t^{s\omega}(A)$ and $\mathcal{D}_t^\omega(A^{1/2})$, $t \in \mathbb{R}$.

Proposition 10.3.6. *Suppose A is a non-negative self-adjoint operator on $\mathcal{H}$. Let $B := A^{1/2}$ and let $t, t' \in \mathbb{R}$ be such that $0 < t' < t$. Then*

(i) $\mathcal{D}_t^\omega(A) \subseteq \mathcal{D}(e^{tA}) \subseteq \mathcal{D}_{t'}^\omega(A)$,

(ii) $\mathcal{D}_t^\omega(B) \subseteq \mathcal{D}_t^{s\omega}(A) \subseteq \mathcal{D}_{t'}^\omega(B)$.

The embedding maps in (i) *and* (ii) *are continuous if $\mathcal{D}_t^\omega(A)$, $\mathcal{D}(e^{tA})$, $\mathcal{D}_{t'}^\omega(A)$, $\mathcal{D}_t^\omega(B)$, $\mathcal{D}_t^{s\omega}(A)$ and $\mathcal{D}_{t'}^\omega(B)$ carry the norms $e_t^A(\cdot)$, $\|e^{tA}\cdot\|$, $e_{t'}^A(\cdot)$, $e_t^B(\cdot)$, $\mathfrak{s}_t^A(\cdot)$ and $e_{t'}^B(\cdot)$, respectively.*

Proof. (i): By Lemma 10.3.3, $\mathcal{D}_t^\omega(A) \subseteq \mathcal{D}(e^{tA})$ and

$$\|e^{tA}\varphi\| \leqq \sum_{n=0}^{\infty} \frac{t^n}{n!} \|A^n\varphi\| = e_t^A(\varphi) \quad \text{for} \quad \varphi \in \mathcal{D}_t^\omega(A).$$

To prove that $\mathcal{D}(e^{tA}) \subseteq \mathcal{D}_{t'}^\omega(A)$, we make use of the assumption $A \geqq 0$. For $\varphi \in \mathcal{D}(e^{tA})$ and $n \in \mathbb{N}$, we have

$$\begin{aligned}\|A^n\varphi\| &= \|A^n e^{-tA} e^{tA}\varphi\| \leqq \|e^{tA}\varphi\| \sup\{\lambda^n e^{-t\lambda} : \lambda \geqq 0\}\\ &= \|e^{tA}\varphi\|\, n^n e^{-n} t^{-n} \leqq \|e^{tA}\varphi\| t^{-n} n!,\end{aligned}$$

so that

$$e_{t'}^A(\varphi) = \sum_{n=0}^{\infty} \frac{t'^n}{n!} \|A^n\varphi\| \leqq \sum_{n=0}^{\infty} (t't^{-1})^n \|e^{tA}\varphi\| = t(t - t')^{-1} \|e^{tA}\varphi\|.$$

(ii): From the definitions it is obvious that $\mathfrak{s}_t^A(\cdot) \leqq e_t^B(\cdot)$ and $\mathcal{D}_t^\omega(B) \subseteq \mathcal{D}_t^{s\omega}(A)$. From the spectral theorem, we have

$$\|B^{2n+1}\varphi\| \leqq \|B^{2n}\varphi\| + \|B^{2n+2}\varphi\| = \|A^n\varphi\| + \|A^{n+1}\varphi\| \tag{6}$$

for $n \in \mathbb{N}$ and $\varphi \in \mathcal{D}^\infty(A) \equiv \mathcal{D}^\infty(B)$. Put $\delta := t't^{-1}$. Since $\delta < 1$, $\alpha := \sup\{n\delta^n : n \in \mathbb{N}\} < \infty$. From (6),

$$\begin{aligned} e_{t'}^B(\varphi) &= \sum_{n=0}^{\infty} \frac{t'^{2n}}{(2n)!} \|B^{2n}\varphi\| + \sum_{n=0}^{\infty} \frac{t'^{2n+1}}{(2n+1)!} \|B^{2n+1}\varphi\| \\ &\leqq \mathfrak{s}_t^A(\varphi) + \sum_{n=0}^{\infty} \left(\frac{t^{2n}}{(2n)!} \frac{t'\delta^{2n}}{2n+1} \|A^n\varphi\| + \frac{t^{2n+2}}{(2n+2)!} \frac{(2n+2)\,\delta^{2n+2}}{t'} \|A^{n+1}\varphi\| \right) \\ &\leqq (1 + t' + \alpha t'^{-1})\, \mathfrak{s}_t^A(\varphi) \quad \text{for} \quad \varphi \in \mathcal{D}^\infty(A). \end{aligned}$$

Hence $\mathcal{D}_t^{s\omega}(A) \subseteq \mathcal{D}_{t'}^\omega(B)$. $\square$

Corollary 10.3.7. *For any self-adjoint operator A on $\mathcal{H}$ we have $\mathcal{D}^\omega(A) = \bigcup_{t>0} \mathcal{D}(e^{t|A|})$ and $\mathcal{D}^{s\omega}(A) = \bigcup_{t>0} \mathcal{D}(e^{t|A|^{1/2}})$.*

Proof. Since A is self-adjoint, $\mathcal{D}^\omega(A) = \mathcal{D}^\omega(|A|)$ and $\mathcal{D}^{s\omega}(A) = \mathcal{D}^{s\omega}(|A|)$, so the assertions follow from Proposition 10.3.6 applied to $|A|$. □

Proposition 10.3.8. *Let T be a non-negative symmetric linear operator on $\mathcal{H}$. If $\mathcal{D}^{s\omega}(T)$ is dense in $\mathcal{H}$, then T is essentially self-adjoint.*

Proof. The proof is similar to the proof of Proposition 10.3.4. Being non-negative and symmetric, T has a non-negative self-adjoint extension A in $\mathcal{H}$. Let $B := A^{1/2}$. Suppose $\xi \in \ker(T^* - \mathrm{i})$. Let $\varphi \in \mathcal{D}^{s\omega}(T)$. Then there is a $t > 0$ such that $\varphi \in \mathcal{D}_t^{s\omega}(T)$. Fix $t' \in \mathbb{R}$, $0 < t' < t$. Since $T \subseteq A$, we have $\varphi \in \mathcal{D}_t^{s\omega}(A)$; so $\varphi \in \mathcal{D}_{t'}^\omega(B)$ by Proposition 10.3.6, (ii). From Lemma 10.3.3 we conclude that for $z \in \mathbb{C}$, $|\mathrm{Im}\, z| < t'$, $\varphi \in \mathcal{D}(\mathrm{e}^{\mathrm{i}zB} + \mathrm{e}^{-\mathrm{i}zB}) \subseteq \mathcal{D}(\cos zB)$ and that the function

$$f(z) := \frac{1}{2} \langle (\mathrm{e}^{\mathrm{i}zB} + \mathrm{e}^{-\mathrm{i}zB})\, \varphi, \xi\rangle \equiv \langle \cos zB\, \varphi, \xi\rangle$$

is holomorphic in the strip $\{z \in \mathbb{C} : |\mathrm{Im}\, z| < t'\}$. Formula (4) in Lemma 10.3.3 yields

$$f(z) = \sum_{n=0}^{\infty} \frac{(-1)^n z^{2n}}{(2n)!} \langle B^{2n}\varphi, \xi\rangle .$$

Hence

$$\begin{aligned} f^{(2n)}(0) &= (-1)^n \langle B^{2n}\varphi, \xi\rangle = (-1)^n \langle A^n\varphi, \xi\rangle \\ &= (-1)^n \langle T^n\varphi, \xi\rangle = (-1)^n \langle \varphi, (T^*)^n \xi\rangle = \mathrm{i}^n \langle \varphi, \xi\rangle \quad \text{for} \quad n \in \mathbb{N}. \end{aligned}$$

Also $f(0) = \langle \varphi, \xi\rangle$. Thus

$$f(z) = \sum_{n=0}^{\infty} \frac{z^{2n}}{(2n)!} \mathrm{i}^n \langle \varphi, \xi\rangle = \langle \varphi, \xi\rangle \cos \frac{1}{\sqrt{2}} (1 - \mathrm{i})\, z \tag{7}$$

for $z \in \mathbb{C}$, $|z| < t'$. The uniqueness theorem for holomorphic functions shows that (7) holds for all $z \in \mathbb{C}$, $|\mathrm{Im}\, z| < t'$, and so in particular on $\mathbb{R}$. But since $f(z) = \langle \cos zB\, \varphi, \xi\rangle$ is obviously bounded on $\mathbb{R}$, this is only possible if $\langle \varphi, \xi\rangle = 0$. Because $\mathcal{D}^{s\omega}(T)$ is dense in $\mathcal{H}$, $\xi = 0$ and so $\ker(T^* - \mathrm{i}) = \{0\}$. The same reasoning shows that $\ker(T^* + \mathrm{i}) = \{0\}$. Thus T is essentially self-adjoint. □

Analytic Domination of Families of Operators

As at the beginning of this section, we assume in this subsection that E is a linear space endowed with a (fixed) seminorm $\|\cdot\|$.

Let $A \in L(E)$ and let $\mathcal{X} \subseteq L(E)$. (Recall that $L(E)$ denotes the algebra of all linear mappings of E into itself.) We say that A *analytically dominates* $\mathcal{X}$ if $\mathcal{D}^\omega(A) \subseteq \mathcal{D}^\omega(\mathcal{X})$, that is, if every analytic vector for the operator A is an analytic vector for the family $\mathcal{X}$.

The purpose of this subsection is to prove two general results about analytic domination of an operator family. The second one (Proposition 10.3.11) is an essential step in the proof of Theorem 10.4.4. First we verify a preliminary lemma.

If $X, Y \in L(E)$, we shall write $\mathrm{ad}\, X(Y)$ for the commutator $XY - YX$.

Lemma 10.3.9. *For $n \in \mathbb{N}$ and $k = 0, 1, \ldots, n$, let $P_{n,k}$ denote the set of all $\binom{n}{k}$ permutations v of $1, \ldots, n$ such that*

$$v(n) > v(n-1) > \cdots > v(k+1) \quad \text{and} \quad v(k) > v(k-1) > \cdots > v(1)$$

(with the obvious interpretation that the first resp. the second inequalities are always true if $k = n$ resp. $k = 0$). Let $A, X_1, \ldots, X_n$ and X be in $L(E)$. Then

$$X_n \ldots X_1 A = \sum_{k=0}^{n} \sum_{v \in P_{n,k}} \left(\text{ad } X_{v(k)} \ldots \text{ad } X_{v(1)}(A)\right) X_{v(n)} \ldots X_{v(k+1)}.$$

(The summands for $k = 0$ and $k = n$ are interpreted as $AX_{v(n)} \ldots X_{v(1)}$ and $\text{ad } X_{v(n)} \ldots \text{ad } X_{v(1)}(A)$, respectively.) In particular, we have

$$X^n A = \sum_{k=0}^{n} \binom{n}{k} (\text{ad } X)^k (A) X^{n-k}.$$

Proof. We proceed by induction on n. For $n = 1$ the assertion says $X_1 A = AX_1 + \text{ad } X_1(A)$, so it is true by definition. Suppose that the assertion holds for $n \in \mathbb{N}$, and let $X_{n+1} \in L(E)$. Then, by the induction assumption,

$$\begin{aligned} X_{n+1} X_n \ldots X_1 A &= \sum_{k=0}^{n} \sum_{v \in P_{n,k}} X_{n+1}\left(\text{ad } X_{v(k)} \ldots \text{ad } X_{v(1)}(A)\right) X_{v(n)} \ldots X_{v(k+1)} \\ &= \sum_{k=0}^{n} \sum_{v \in P_{n,k}} \left\{\left(\text{ad } X_{v(k)} \ldots \text{ad } X_{v(1)}(A)\right) X_{n+1} X_{v(n)} \ldots X_{v(k+1)} \right. \\ &\quad \left. + \left(\text{ad } X_{n+1} \text{ ad } X_{v(k)} \ldots \text{ad } X_{v(1)}(A)\right) X_{v(n)} \ldots X_{v(k+1)}\right\}. \end{aligned}$$

Let $k \in \{1, \ldots, n\}$ and let v be a permutation in $P_{n+1,k}$. We consider the term $\left(\text{ad } X_{v(k)} \ldots \text{ad } X_{v(1)}(A)\right) X_{v(n+1)} \ldots X_{v(k+1)}$. From the definition of $P_{n+1,k}$ it follows that either $v(n+1) = n+1$ or $v(k) = n+1$. In the first case, the term occurs in the sum before the $+$ sign and it corresponds to a permutation in $P_{n,k}$. In the second case, it appears in the sum after the $+$ sign and it corresponds to a permutation in $P_{n,k-1}$. Since $\binom{n+1}{k} = \binom{n}{k} + \binom{n}{k-1}$, the correspondence between the terms $\left(\text{ad } X_{v(k)} \ldots \text{ad } X_{v(1)}(A)\right) X_{v(n+1)} \ldots X_{v(k+1)}$, $v \in P_{n+1,k}$, and the corresponding terms in the above sum is one-to-one. This is also true for $k = 0$ and $k = n+1$, so that the assertion for $n+1$ follows. □

Proposition 10.3.10. *Let A be an operator of $L(E)$ and let $\mathcal{X}$ be a subset of $L(E)$. Suppose that*

$$\|X\varphi\| \leqq \|A\varphi\| \tag{8}$$

and

$$\|\text{ad } X_1 \ldots \text{ad } X_n(A)\, \varphi\| \leqq n!\, \|A\varphi\| \tag{9}$$

for arbitrary $X, X_1, \ldots, X_n$ of $\mathcal{X}$, $n \in \mathbb{N}$ and $\varphi \in E$.

Then A analytically dominates $\mathcal{X}$. More precisely, for every $t > 0$, there exists an $s(t) > 0$ such that $\mathcal{D}_t^\omega(A) \subseteq \mathcal{D}_{s(t)}^\omega(\mathcal{X})$, the inclusion being continuous in the corresponding seminorms.

Proof. Let $\varphi \in \mathcal{D}^\omega(A)$. Then there exists a constant M such that

$$\|A^n\varphi\| \leqq M^n n! \quad \text{for all} \quad n \in \mathbb{N}_0. \tag{10}$$

For $n \in \mathbb{N}$ and $m \in \mathbb{N}_0$, define $\alpha_{n,m} := \sup\{\|AX_n \ldots X_1 A^m \varphi\| : X_1, \ldots, X_n \in \mathcal{X}\}$ and

$\alpha_{0,m} := \|A^{m+1}\varphi\|$. We verify the recursive inequalities

$$\alpha_{n+1,m} \leqq \alpha_{n,m+1} + \sum_{k=0}^{n} (n+1)! \frac{\alpha_{k,m}}{k!} \quad \text{for} \quad n, m \in \mathbb{N}_0 \tag{11}$$

and

$$\alpha_{0,m} \leqq M^{m+1}(m+1)! \quad \text{for} \quad m \in \mathbb{N}_0 . \tag{12}$$

(12) is nothing but (10). To prove (11), let $n, m \in \mathbb{N}_0$ and let $X_1, \ldots, X_{n+1} \in \mathscr{X}$. From Lemma 10.3.9,

$$X_{n+1} \ldots X_1 A^{m+1}\varphi = (X_{n+1} \ldots X_1 A) A^m\varphi$$

$$= AX_{n+1} \ldots X_1 A^m\varphi + \sum_{k=1}^{n+1} \sum_{v \in P_{n+1,k}} \big(\operatorname{ad} X_{v(k)} \ldots \operatorname{ad} X_{v(1)}(A)\big) X_{v(n+1)} \ldots X_{v(k+1)} A^m \varphi .$$

Because of the assumptions (8) and (9), we therefore obtain

$$\|AX_{n+1} \ldots X_1 A^m\varphi\|$$

$$\leqq \|X_{n+1}X_n \ldots X_1 A^{m+1}\varphi\| + \sum_{k=1}^{n+1} \sum_{v \in P_{n+1,k}} k!\, \|AX_{v(n+1)} \ldots X_{v(k+1)} A^m \varphi\|$$

$$\leqq \|AX_n \ldots X_1 A^{m+1}\varphi\| + \sum_{k=1}^{n+1} \sum_{v \in P_{n+1,k}} k!\, \alpha_{n-k+1,m}$$

$$\leqq \alpha_{n,m+1} + \sum_{k=1}^{n+1} k! \binom{n+1}{k} \alpha_{n-k+1,m} ,$$

where we used that $P_{n+1,k}$ consists of $\binom{n+1}{k}$ permutations. This gives (11).

On the other hand, if u and v are in a sufficiently small neighbourhood W of zero in $\mathbb{R}$, then the function

$$f(u, v) := M(1-u) \left[(1-2u) \left(1 - Mv - \frac{1}{2} Mu + M \log (1-2v) \right)^{1/4} \right)^{2} \right]^{-1} \tag{13}$$

has a power series expansion

$$f(u, v) = \sum_{n,m=0}^{\infty} \frac{\beta_{n,m}}{n!m!} u^n v^m$$

which converges absolutely in W. In particular,

$$f(u, 0) = \sum_{n=0}^{\infty} \frac{\beta_{n,0}}{n!} u^n$$

converges in a neighbourhood of zero; so there is a constant $M_1 \geqq 1$ (depending on M only) such that

$$\beta_{n,0} \leqq M_1^n n! \quad \text{for} \quad n \in \mathbb{N}_0 . \tag{14}$$

A direct calculation shows that

$$f_u(u, v) = f_v(u, v) + \big(u(1-u)^{-1} f(u, v)\big)_u \tag{15}$$

and

$$f(0, v) = M(1 - Mv)^{-2} \tag{16}$$

for u and v near zero, where subscripts denote partial differentiation with respect to the indicated variable. Putting the power series expansion of f into (15) and (16) and comparing coefficients, we obtain

$$\beta_{n+1,m} = \beta_{n,m+1} + \sum_{k=0}^{n} (n+1)! \frac{\beta_{k,m}}{k!} \quad \text{for} \quad n, m \in \mathbb{N}_0 \tag{11'}$$

and

$$\beta_{0,m} = M^{m+1}(m+1)! \quad \text{for} \quad m \in \mathbb{N}_0. \tag{12'}$$

That is, the numbers $\beta_{n,m}$ are recursively defined by replacing the inequalities in (11) and (12) by equalities. Consequently, $\alpha_{n,m} \leqq \beta_{n,m}$ for $n, m \in \mathbb{N}_0$. Combining the latter with (14), (8) and (9), we get $\|X_{n+1}X_n \dots X_1\varphi\| \leqq \|AX_n \dots X_1\varphi\| \leqq \alpha_{n,0} \leqq \beta_{n,0} \leqq M_1^n n!$ $\leqq M_1^{n+1}(n+1)!$ and similarly $\|X_1\varphi\| \leqq M_1$ for arbitrary elements $X_1, \dots, X_{n+1}$ of $\mathcal{X}$ and $n \in \mathbb{N}_0$. Hence $v_n^{\mathcal{X}}(\varphi) \leqq M_1^n n!$ for $n \in \mathbb{N}$ and φ is an analytic vector for $\mathcal{X}$.

Given $t > 0$, we let $M := t^{-1}$ in the preceding. We assert that

$$e_{s(t)}^{\mathcal{X}}(\varphi) \leqq \big(1 - s(t)\, M_1\big)^{-1} e_t^A(\varphi) \quad \text{for all} \quad \varphi \in \mathcal{D}_t^{\omega}(A) \tag{17}$$

if $s(t)$ is any positive number such that $s(t)\, M_1 < 1$. Upon multiplying φ by a constant if necessary, it suffices to prove this in case where $e_t^A(\varphi) \leqq 1$. But then $\|A^n\varphi\| \leqq M^n n!$ for $n \in \mathbb{N}_0$ and hence by $v_0^{\mathcal{X}}(\varphi) \leqq 1$ and the above estimate for $v_n^{\mathcal{X}}(\varphi)$, we have

$$e_{s(t)}^{\mathcal{X}}(\varphi) = \sum_{n=0}^{\infty} \frac{s(t)^n}{n!} v_n^{\mathcal{X}}(\varphi) \leqq \sum_{n=0}^{\infty} \frac{s(t)^n}{n!} M_1^n n! = \big(1 - s(t)\, M_1\big)^{-1}$$

which proves the second assertion of the proposition. □

Remark 3. The constant M_1 occuring in the preceding proof depends only on the function f defined by (13) and so only on the constant M satisfying (10).

Proposition 10.3.11. *Let $A \in L(E)$ and let $\mathcal{X} \subseteqq L(E)$. Suppose that for $X, Y, X_1, \dots, X_n \in \mathcal{X}$, $n \in \mathbb{N}$ and $\varphi \in E$*

$$\|X\varphi\| \leqq \|A\varphi\| \quad \text{and} \quad \|YX\varphi\| \leqq \|A\varphi\| \tag{18}$$

and

$$\|\text{ad}\, X_1 \dots \text{ad}\, X_n(A)\, \varphi\| \leqq n!\, \|A\varphi\|. \tag{19}$$

Then every semi-analytic vector for A is an analytic vector for $\mathcal{X}$. More precisely, for each $t > 0$, there is an $s(t) > 0$ such that $\mathcal{D}_t^{s\omega}(A) \subseteqq \mathcal{D}_{s(t)}^{\omega}(\mathcal{X})$ and the inclusion is continuous in the corresponding seminorms.

Proof. Let $\tilde{E}$ be the linear space $E \oplus E$ (direct sum) endowed with the seminorm $\|(\varphi, \psi)\|^{\sim} := \sup_{X \in \mathcal{X}} \|X\varphi\| + \|\psi\|$, $\varphi, \psi \in E$. Because of (18), $\|\cdot\|^{\sim}$ is finite on $\tilde{E}$. Define operators $\tilde{A}$ and $\tilde{X}$, $X \in \mathcal{X}$, of $L(\tilde{E})$ by the matrices $\tilde{A} = \begin{bmatrix} 0 & 1 \\ A & 0 \end{bmatrix}$ and $\tilde{X} = \begin{bmatrix} X & 0 \\ 0 & X \end{bmatrix}$. From (18), we have for $\varphi, \psi \in E$

$$\|\tilde{X}(\varphi, \psi)\|^{\sim} = \sup_{Y \in \mathcal{X}} \|YX\varphi\| + \|X\psi\| \leqq \|A\varphi\| + \|X\psi\| \leqq \|(\psi, A\varphi)\|^{\sim} = \|\tilde{A}(\varphi, \psi)\|^{\sim}.$$

From (19) and

$$\text{ad}\,\tilde{X}_1 \dots \text{ad}\,\tilde{X}_n(\tilde{A}) = \begin{bmatrix} 0 & 0 \\ \text{ad}\,X_1 \dots \text{ad}\,X_n(A) & 0 \end{bmatrix}$$

we obtain that

$$\|\text{ad}\,\tilde{X}_1 \dots \text{ad}\,\tilde{X}_n(\tilde{A})\,(\varphi, \psi)\|^\sim = \|\text{ad}\,X_1 \dots \text{ad}\,X_n(A)\,\varphi\| \leqq n!\,\|A\varphi\| \leqq n!\,\|\tilde{A}(\varphi, \psi)\|^\sim$$

for all $X_1, \dots, X_n \in \mathcal{X}$, $n \in \mathbb{N}$ and $\varphi, \psi \in E$. This shows that the operator $\tilde{A}$ and the set $\tilde{\mathcal{X}} := \{\tilde{X} : X \in \mathcal{X}\}$ satisfy the assumptions of Proposition 10.3.10 with $\tilde{E}$ in place of E. For $n \in \mathbb{N}$,

$$(\tilde{A})^{2n} = \begin{bmatrix} A^n & 0 \\ 0 & A^n \end{bmatrix} \quad \text{and} \quad (\tilde{A})^{2n+1} = \begin{bmatrix} 0 & A^n \\ A^{n+1} & 0 \end{bmatrix},$$

so that

$$\|(\tilde{A})^{2n}\,(\varphi, 0)\|^\sim = \sup_{X \in \mathcal{X}} \|XA^n\varphi\| \leqq \|A^{n+1}\varphi\| \quad \text{and} \quad \|(\tilde{A})^{2n+1}(\varphi, 0)\|^\sim = \|A^{n+1}\varphi\|. \tag{20}$$

From (20), if φ is a semi-analytic vector for A, then $(\varphi, 0)$ is an analytic vector for $\tilde{A}$ and so for $\tilde{\mathcal{X}}$ by Proposition 10.3.10. This implies that φ is an analytic vector for $\mathcal{X}$.

Let $t > 0$ be given. Take $t' \in \mathbb{R}$, $0 < t' < t$. By Proposition 10.3.10 applied to $\tilde{A}$, $\tilde{\mathcal{X}}$ and t', there are $s > 0$ and $\tilde{\lambda} > 0$ such that

$$e_s^{\tilde{\mathcal{X}}}\big((\varphi, 0)\big) \leqq \tilde{\lambda} e_{t'}^{\tilde{A}}\big((\varphi, 0)\big) \quad \text{for} \quad (\varphi, 0) \in \mathcal{D}_{t'}^{\omega}(\tilde{A}). \tag{21}$$

Put $s(t) := s$. Using $e_s^{\tilde{\mathcal{X}}}\big((\varphi, 0)\big) = e_s^{\mathcal{X}}(\varphi)$, (20) and (21), a simple calculation shows that $e_{s(t)}^{\mathcal{X}}(\varphi) \leqq \lambda \mathfrak{s}_t^A(\varphi)$ for some constant λ and for all $\varphi \in \mathcal{D}_t^{s\omega}(A)$. □

10.4. Analytic Vectors for *-Representations of Enveloping Algebras

Analytic Vectors for General *-Representations of Enveloping Algebras

Suppose that π is a representation of the enveloping algebra $\mathcal{E}(\mathfrak{g})$.

Definition 10.4.1. Let $\{x_1, \dots, x_d\}$ be a basis of $\mathfrak{g}$. A vector φ in $\mathcal{D}(\pi)$ is called an *analytic vector for* π if φ is an analytic vector for the family of operators $\mathcal{X} := \{\pi(x_1), \dots, \pi(x_d)\}$ of $L(\mathcal{D}(\pi))$ relative to the Hilbert space norm of $\mathcal{H}(\pi)$.

We denote the set of analytic vectors for π by $\mathcal{D}^\omega(\pi)$. According to the above definition, a vector $\varphi \in \mathcal{D}(\pi)$ is in $\mathcal{D}^\omega(\pi)$ if and only if there is a constant M such that $\|\pi(x_{k_1}) \dots \pi(x_{k_n})\,\varphi\| \leqq M^n n!$ for arbitrary indices $k_1, \dots, k_n$ from $\{1, \dots, d\}$ and for all $n \in \mathbb{N}$.

Keep the notation of Definition 10.4.1. We shall write ν_n^π, $\mathcal{D}_t^\omega(\pi)$ and $e_t^\pi(\cdot)$ for $\nu_n^{\mathcal{X}}$, $\mathcal{D}_t^\omega(\mathcal{X})$ and $e_t^{\mathcal{X}}(\cdot)$, respectively. Of course, then the seminorms ν_n^π and the normed linear spaces $\big(\mathcal{D}_t^\omega(\pi), e_t^\pi(\cdot)\big)$ depend on the basis $\{x_1, \dots, x_d\}$ of $\mathfrak{g}$. However, by Lemma 10.4.2 below, the linear space $\mathcal{D}^\omega(\pi)$ as defined above is independent of the special basis for $\mathfrak{g}$.

Let $|\cdot|$ denote the l_1-norm on $\mathfrak{g}$ relative to the basis $\{x_1, \dots, x_d\}$ of $\mathfrak{g}$. It follows immediately from the triangle inequality that

$$\nu_m^{\mathfrak{X}}\big(\pi(y_1)\dots\pi(y_n)\,\varphi\big) \leqq |y_1|\dots|y_n|\,\nu_{n+m}^{\mathfrak{X}}(\varphi) \tag{1}$$

for arbitrary elements $y_1, \dots, y_n \in \mathfrak{g}$, $n \in \mathbb{N}$, $m \in \mathbb{N}_0$ and $\varphi \in \mathcal{D}(\pi)$. The next two lemmas are easy consequences of the inequality (1).

Lemma 10.4.2. *Let $\{x_1, \dots, x_d\}$ and $\{\tilde{x}_1, \dots, \tilde{x}_d\}$ be bases of $\mathfrak{g}$. Then a vector $\varphi \in \mathcal{D}(\pi)$ is an analytic vector for the family $\mathfrak{X} := \{\pi(x_1), \dots, \pi(x_d)\}$ if and only if it is an analytic vector for the family $\tilde{\mathfrak{X}} := \{\pi(\tilde{x}_1), \dots, \pi(\tilde{x}_d)\}$. More precisely, there are positive constants α, β (independent of π) such that $\mathcal{D}_{\alpha t}^{\omega}(\mathfrak{X}) \subseteq \mathcal{D}_t^{\omega}(\tilde{\mathfrak{X}})$ and $\mathcal{D}_{\beta t}^{\omega}(\tilde{\mathfrak{X}}) \subseteq \mathcal{D}_t^{\omega}(\mathfrak{X})$ for all $t > 0$, and the embedding maps are continuous in the corresponding norms.*

Proof. Put $\alpha := \max\,\{1, |x_k| : k = 1, \dots, d\}$. Let $\varphi \in \mathcal{D}(\pi)$. Applying (1) in case $m = 0$, we get $\|\pi(\tilde{x}_{k_1}) \dots \pi(\tilde{x}_{k_n})\,\varphi\| \leqq \alpha^n \nu_n^{\mathfrak{X}}(\varphi)\|$ for arbitrary indices $k_1, \dots, k_n$ of $\{1, \dots, d\}$ and $n \in \mathbb{N}$. Hence $\nu_n^{\tilde{\mathfrak{X}}}(\varphi) \leqq \alpha^n \nu_n^{\mathfrak{X}}(\varphi)$ for $n \in \mathbb{N}$ which gives $e_t^{\tilde{\mathfrak{X}}}(\varphi) \leqq e_{\alpha t}^{\mathfrak{X}}(\varphi)$. Thus $\mathcal{D}_{\alpha t}^{\omega}(\mathfrak{X}) \subseteq \mathcal{D}_t^{\omega}(\tilde{\mathfrak{X}})$. The other assertions follow by symmetry. $\square$

Lemma 10.4.3. *For arbitrary elements x in $\mathfrak{g}$ and y in $\mathcal{E}(\mathfrak{g})$ we have $\mathcal{D}_{|x|t}^{\omega}(\pi) \subseteq \mathcal{D}_t^{\omega}\big(\pi(x)\big)$ and $\pi(y)\,\mathcal{D}_t^{\omega}(\pi) \subseteq \mathcal{D}_{t'}^{\omega}(\pi)$ if $t, t' \in \mathbb{R}$, $0 < t' < t$. In particular, $\mathcal{D}^{\omega}(\pi) \subseteq \mathcal{D}^{\omega}\big(\pi(x)\big)$ and $\pi(y)\,\mathcal{D}^{\omega}(\pi) \subseteq \mathcal{D}^{\omega}(\pi)$.*

Proof. From (1), $\|\pi(x)^n\,\varphi\| \leqq |x|^n\,\nu_n^{\mathfrak{X}}(\varphi)$ for $n \in \mathbb{N}$ and $\varphi \in \mathcal{D}(\pi)$ which implies that $\mathcal{D}_{|x|t}^{\omega}(\pi) \subseteq \mathcal{D}_t^{\omega}\big(\pi(x)\big)$. Since $\mathcal{E}(\mathfrak{g})$ is generated, as an algebra, by $\mathfrak{g} \cup \{1\}$, it suffices to prove the second assertion for elements y in $\mathfrak{g}$. But then the assertion follows easily from the inequality $\nu_n^{\mathfrak{X}}\big(\pi(y)\,\varphi\big) \leqq |y|\,\nu_{n+1}^{\mathfrak{X}}(\varphi)$ which holds by (1). $\square$

The following theorem is the main result in this section. It gives a precise description of the space $\mathcal{D}^{\omega}(\pi)$ in terms of *one* operator.

Theorem 10.4.4. *Let $\{x_1, \dots, x_d\}$ be a basis of the Lie algebra $\mathfrak{g}$ and let $\Delta := x_1^2 + \dots + x_d^2$ be the Nelson Laplacian relative to this basis. Suppose that π is a $*$-representation of the enveloping algebra $\mathcal{E}(\mathfrak{g})$.*

Then $\mathcal{D}^{s\omega}\big(\pi(1 - \Delta)\big) = \mathcal{D}^{\omega}(\pi)$. For every $t > 0$ there, exist positive numbers $s_1 = s_1(t)$ and $s_2 = s_2(t)$ such that $\mathcal{D}_t^{\omega}(\pi) \subseteq \mathcal{D}_{s_1}^{s\omega}\big(\pi(1 - \Delta)\big)$ and $\mathcal{D}_t^{s\omega}\big(\pi(1 - \Delta)\big) \subseteq \mathcal{D}_{s_2}^{\omega}(\pi)$ and the inclusion maps are continuous in the corresponding norms.

The proof of Theorem 10.4.4 essentially rests on Proposition 10.3.11. In order to show that the assumptions of this proposition are fulfilled, we prove two preliminary lemmas. In the rest of this subsection we keep the assumptions and the notation of Theorem 10.4.4. Further, we abbreviate $A := \pi(1 - \Delta)$.

Lemma 10.4.5. *For each element $x \in \mathcal{E}_2(\mathfrak{g})$ there exists a number $\lambda_x > 0$ (independent of π) such that $\|\pi(x)\,\varphi\| \leqq \lambda_x \|A\varphi\|$ for all $\varphi \in \mathcal{D}(\pi)$.*

Proof. It suffices to prove the assertion for the elements 1, x_n and $x_n x_m$, $n, m = 1, \dots, d$, because these elements span $\mathcal{E}_2(\mathfrak{g})$. Since $x_n^+ = -x_n$ for $n = 1, \dots, d$ and π is a $*$-representation, $\pi(\Delta) \leqq 0$. This in turn implies that for $\varphi \in \mathcal{D}(\pi)$

$$\|\pi(1)\,\varphi\| = \|\varphi\| \leqq \|A\varphi\| \quad \text{and} \quad \|\pi(\Delta)\,\varphi\| \leqq \|A\varphi\|. \tag{2}$$

Using this, we have for $\varphi \in \mathcal{D}(\pi)$

$$\|\pi(x_n)\,\varphi\|^2 = \langle \pi(x_n^+ x_n)\,\varphi, \varphi\rangle \leqq \sum_{k=1}^{d} \langle \pi(x_k^+ x_k)\,\varphi, \varphi\rangle = \langle \pi(-\Delta)\,\varphi, \varphi\rangle \leqq \|A\varphi\|^2. \tag{3}$$

Thus we have proved the assertion for the elements 1 and x_n, $n = 1, \dots, d$.

For $n, m \in \{1, \ldots, d\}$, let $y_{nm} := x_n^2 x_m^2 + x_m^2 x_n^2$. By the commutation relations of the Lie algebra, we can write y_{nm} in the form

$$y_{nm} = x_n x_m^2 x_n + x_m x_n^2 x_m + z_{nm}, \tag{4}$$

where $z_{nm} \in \mathscr{E}_3$. Here $\mathscr{E}_3$ denotes the real linear span of x^n, $n \in \mathbb{N}_0^d$ and $0 < |n| \leqq 3$. Let $\mathscr{F}$ be the real span of $x_{kl} := x_k x_l + x_l x_k$, $k, l = 1, \ldots, d$, and let $\mathscr{G}$ be the real span of x_k and $x_{jkl} := x_j x_k x_l + x_j x_l x_k + x_k x_j x_l + x_k x_l x_j + x_l x_j x_k + x_l x_k x_j$, where $j, k, l = 1, \ldots, d$. Since obviously $\mathscr{E}_3 = \mathscr{F} + \mathscr{G}$, we can write $z_{nm} = u_{nm} + v_{nm}$ with $u_{nm} \in \mathscr{F}$ and $v_{nm} \in \mathscr{G}$. We have $z_{nm} = z_{nm}^+$, since this is true for the other terms in (4). Because all elements of $\mathscr{F}$ are hermitian, $u_{nm} = u_{nm}^+$. Hence $v_{nm} = v_{nm}^+$. But the element v_{nm} of $\mathscr{G}$ is skew-hermitian, since x_k and x_{jkl} are also. Thus, $v_{nm} = 0$, so that $z_{nm} \in \mathscr{F}$. Since $\mathscr{F}$ is spanned by the elements x_{kl}, it follows from (3) that there is a $\lambda_{nm} > 0$ such that

$$|\langle \pi(z_{nm}) \varphi, \varphi \rangle| \leqq \lambda_{nm} \|A\varphi\|^2, \quad \varphi \in \mathscr{D}(\pi) \quad \text{and} \quad n, m = 1, \ldots, d. \tag{5}$$

We have for $\varphi \in \mathscr{D}(\pi)$

$$\begin{aligned}\|\pi(\Delta)\varphi\|^2 &= \langle \pi(\Delta^2)\varphi, \varphi\rangle = \sum_{n,m=1}^{d} \langle \pi(x_n^2 x_m^2)\varphi, \varphi\rangle \\ &= \sum_{n,m} \frac{1}{2} \langle \pi(x_n x_m^2 x_n + x_m x_n^2 x_m + z_{nm})\varphi, \varphi\rangle \\ &\geqq \sum_{n,m} \left(\frac{1}{2} \|\pi(x_m x_n)\varphi\|^2 + \|\pi(x_n x_m)\varphi\|^2 - |\langle \pi(z_{nm})\varphi, \varphi\rangle| \right).\end{aligned}$$

Combined with (5) and (2), this gives

$$\|\pi(x_n x_m)\varphi\|^2 \leqq \Big(1 + \sum_{n,m} \lambda_{nm}\Big) \|A\varphi\|^2 \quad \text{for} \quad \varphi \in \mathscr{D}(\pi) \quad \text{and} \quad n, m = 1, \ldots, d. \quad \square$$

Lemma 10.4.6. *There exists a positive number α such that*

$$\|\pi(x_k)\varphi\| \leqq \alpha \|A\varphi\| \quad \text{and} \quad \|\pi(x_k)\pi(x_m)\varphi\| \leqq \alpha^2 \|A\varphi\| \tag{6}$$

and

$$\|\operatorname{ad} \pi(x_{k_1}) \ldots \operatorname{ad} \pi(x_{k_n}) (A) \varphi\| \leqq \alpha^n \|A\varphi\| \tag{7}$$

for all indices $k, m, k_1, \ldots, k_d$ from $\{1, \ldots, d\}$, $n \in \mathbb{N}$ and $\varphi \in \mathscr{D}(\pi)$.

Proof. For $x \in \mathscr{E}_2(\mathfrak{g})$, define $|||x||| := \sup\{\|\pi(x)\varphi\| : \varphi \in \mathscr{D}(\pi) \text{ and } \|A\varphi\| \leqq 1\}$. If λ_x is the constant from Lemma 10.4.5, then $|||x||| \leqq \lambda_x$ for all $x \in \mathscr{E}_2(\mathfrak{g})$. Thus $|||\cdot|||$ is finite and hence a seminorm on $\mathscr{E}_2(\mathfrak{g})$. If $x \in \mathfrak{g}$, $y \to \operatorname{ad} x(y)$ is a linear mapping of the finite dimensional vector space $\mathscr{E}_2(\mathfrak{g})$ into itself, so it is continuous with respect to any seminorm on $\mathscr{E}_2(\mathfrak{g})$. Hence there is a number $\alpha > 0$ such that $|||\operatorname{ad} x_n(y)||| \leqq \alpha\, |||y|||$ for all $y \in \mathscr{E}_2(\mathfrak{g})$ and $n = 1, \ldots, d$. Therefore $|||\operatorname{ad} x_{k_1} \ldots \operatorname{ad} x_{k_n}(y)||| \leqq \alpha^n\, |||y|||$ for all $k_1, \ldots, k_n \in \{1, \ldots, d\}$, $n \in \mathbb{N}$ and $y \in \mathscr{E}_2(\mathfrak{g})$. The preceding gives in terms of the Hilbert space norm on $\mathscr{H}(\pi)$

$$\big\|\pi(\operatorname{ad} x_{k_1} \ldots \operatorname{ad} x_{k_n}(I - \Delta))\varphi\big\| = \|\operatorname{ad} \pi(x_{k_1}) \ldots \operatorname{ad} \pi(x_{k_n})(A)\varphi\| \leqq \alpha^n \|A\varphi\|$$

for all $k_1, \ldots, k_n \in \{1, \ldots, d\}$, $n \in \mathbb{N}$ and $\varphi \in \mathscr{D}(\pi)$. Here we used also that π is a representation. This proves (7). Without loss of generality we can assume that $\alpha \geqq |||x_k|||$ and $\alpha^2 \geqq |||x_k x_m|||$ for $k, m = 1, \ldots, d$. This in turn implies (6). $\square$

For a later application given in Section 10.5 (in the proof of Lemma 10.5.7) we state a corollary which follows immediately from the formulas (6) and (7) in Lemma 10.4.6. Recall that $A = \pi(I - \Delta)$ and Δ is the Nelson Laplacian relative to the basis $\{x_1, \ldots, x_d\}$ of $\mathfrak{g}$.

Corollary 10.4.7. *For arbitrary numbers* $k, k_1, \ldots, k_n$ *of* $\{1, \ldots, d\}$ *and* $n \in \mathbb{N}$, *we have* $\mathcal{D}(\bar{A}) \subseteq \mathcal{D}(\overline{\pi(x_k)})$ *and* $\mathcal{D}(\bar{A}) \subseteq \mathcal{D}(\overline{\operatorname{ad} \pi(x_{k_1}) \ldots \operatorname{ad} \pi(x_{k_n}) (A)})$.

Proof of Theorem 10.4.4. First suppose $\varphi \in \mathcal{D}^\omega(\pi)$. Then there is an $M \geqq 1$ such that $\|\pi(x_{k_1}) \ldots \pi(x_{k_n}) \varphi\| \leqq M^n n!$ for arbitrary indices $k_1, \ldots, k_n$ from $\{1, \ldots, d\}$ and $n \in \mathbb{N}$. Since $A^n = \pi((I - \Delta)^n)$ is a sum of $(d+1)^n$ terms of the form $\pm\pi(x_{k_1}) \ldots \pi(x_{k_m})$ with $k_1, \ldots, k_m \in \{1, \ldots, d\}$ and $m \leqq 2n$, it follows that

$$\|A^n\varphi\| \leqq (d+1)^n M^{2n}(2n)! \leqq ((d+1) M)^{2n} (2n)! \quad \text{for} \quad n \in \mathbb{N}. \tag{8}$$

Thus $\varphi \in \mathcal{D}^{s\omega}(A)$.

If $t > 0$ is given, put $M := \max\{1, t^{-1}\}$ and take an $s_1 > 0$ such that $s_1 M(d+1) < 1$. We verify that

$$\mathfrak{s}^A_{s_1}(\varphi) \leqq (1 - s_1 M(d+1))^{-1} \mathfrak{e}^\pi_t(\varphi) \quad \text{for all} \quad \varphi \in \mathcal{D}^\omega_t(\pi). \tag{9}$$

To prove this, we assume without loss of generality that $\mathfrak{e}^\pi_t(\varphi) \leqq 1$. Then $\nu^\pi_n(\varphi) \leqq M^n n!$ for $n \in \mathbb{N}$, so that, by (8), $\mathfrak{s}^A_{s_1}(\varphi) \leqq \sum_{n=0}^{\infty} (s_1(d+1) M)^{2n} \leqq (1 - s_1 M(d+1))^{-1}$. This proves (9). By (9), $\mathcal{D}^\omega_t(\pi)$ is continuously embedded in $\mathcal{D}^{s}_{s_1}(A)$.

Now we turn to the opposite inclusion. From Lemma 10.4.6 we see that the assumptions of Proposition 10.3.11 are fulfilled in case $E = \mathcal{D}(\pi)$, $\mathcal{X} = \{\pi(\alpha^{-1}x_1), \ldots, \pi(\alpha^{-1}x_d)\}$. Therefore, by Proposition 10.3.11, $\mathcal{D}^{s\omega}(A) \subseteq \mathcal{D}^\omega(\mathcal{X}) = \mathcal{D}^\omega(\pi)$.

Let $t > 0$ be given. Take $s(t)$ as in Proposition 10.3.11 and put $s_2 := s(t)\,\alpha^{-1}$. From Proposition 10.3.11, there is a $\lambda > 0$ such that $\mathfrak{e}^\pi_{s_2}(\varphi) = \mathfrak{e}^{\mathcal{X}}_{s(t)}(\varphi) \leqq \lambda \mathfrak{s}^A_t(\varphi)$ for all $\varphi \in \mathcal{D}^{s\omega}_t(A)$. This shows that $\mathcal{D}^{s\omega}_t(A) \subseteq \mathcal{D}^\omega_{s_2}(\pi)$, and the embedding map is continuous in the respective norms. $\square$

Remark 1. Since $\mathcal{D}^\omega(\pi(I - \Delta)) \subseteq \mathcal{D}^{s\omega}(\pi(I - \Delta))$, Theorem 10.4.4 shows in particular that each analytic vector for the operator $\pi(I - \Delta)$ is an analytic vector for the $*$-representation π.

Analytic Vectors for Unitary Representations of Lie groups

In this subsection, U denotes a unitary representation of the Lie group G in the Hilbert space $\mathcal{H}(U)$.

In the previous Sections 10.1 and 10.2 we only needed the C^∞-structure of the Lie group G. Now we essentially use the (real) analytic structure of G.

A map u of the Lie group G into the Hilbert space $\mathcal{H}$ is said to be *analytic at a point* $g_0 \in G$ if there exists a neighbourhood V of g_0, an analytic coordinate system $t_1(g), \ldots, t_d(g)$ on V such that $t_1(g) = \cdots = t_d(g) = 0$ and coefficients $\psi_n \in \mathcal{H}$, $n \in \mathbb{N}_0^d$, such that $\sum_{n \in \mathbb{N}_0^d} \|\psi_n\|\, |t^n(g)| < \infty$ and $u(g) = \sum_{n \in \mathbb{N}_0^d} \psi_n\, t^n(g)$ for all $g \in V$. Here, $t^n(g) := t_1(g)^{n_1} \ldots t_d(g)^{n_d}$ with the interpretation $0^0 = 1$ if $t = (t_1, \ldots, t_d) \in \mathbb{R}^d$ and $n = (n_1, \ldots, n_d) \in \mathbb{N}_0^d$. The map u is said to be *analytic on* G if u is analytic at each point g_0 in G.

Definition 10.4.8. A vector φ in $\mathcal{H}(U)$ is called an *analytic vector for U* if the map $g \to U(g)\varphi$ of G into $\mathcal{H}(U)$ is analytic on G in the sense just defined.

Let $\mathcal{D}^\omega(U)$ denote the set of analytic vectors for U. Since translations by group elements are analytic isomorphisms of the Lie group G, the linear space $\mathcal{D}^\omega(U)$ is invariant under $U(g)$ for $g \in G$. For the same reason it follows that a vector $\varphi \in \mathcal{H}(U)$ is analytic for U if the map $g \to U(g)\,\varphi$ is analytic at the identity element e of G.

Lemma 10.4.9. $\mathcal{D}^\omega(\mathrm{d}U) \subseteq \mathcal{D}^\omega(U)$.

Proof. Fix a basis $\{x_1, \ldots, x_d\}$ for $\mathfrak{g}$ and let $|\cdot|$ be the l_1-norm on $\mathfrak{g}$ relative to this basis. For $t = (t_1, \ldots, t_d) \in \mathbb{R}^d$, put $x(t) := t_1x_1 + \cdots + t_dx_d$ and $g(t) := \exp x(t)$. The mapping $g(t) \to t$ is an analytic coordinate system in a neighbourhood of e in G. Suppose that $\varphi \in \mathcal{D}^\omega(\mathrm{d}U)$. Then there is an $s > 0$ such that $\varphi \in \mathcal{D}_s^\omega(\mathrm{d}U)$ with respect to the basis $\{x_1, \ldots, x_d\}$. Let $t = (t_1, \ldots, t_d)$ be a vector of $\mathbb{R}^d$ such that $|t_l| \leqq s2^{-d}$ for $l = 1, \ldots, d$. Then $|x(t)| \leqq s$ and hence $\varphi \in \mathcal{D}_1^\omega(\mathrm{d}U(x(t)))$ by Lemma 10.4.3. From Corollary 10.2.11, the operator $-\mathrm{i}\overline{\mathrm{d}U(x(t))}$ is self-adjoint. Since $\varphi \in \mathcal{D}_1^\omega(-\mathrm{i}\overline{\mathrm{d}U(x(t))})$, Lemma 10.3.3 shows that

$$U(g(t))\,\varphi = U(\exp x(t))\,\varphi = \exp \mathrm{i}(-\mathrm{i}\overline{\mathrm{d}U(x(t))})\,\varphi = \sum_{k=0}^{\infty} \frac{\mathrm{d}U(x(t))^k\,\varphi}{k!}.$$

We write $\frac{1}{k!}\,\mathrm{d}U(x(t))^k\,\varphi \equiv \frac{1}{k!}\,\mathrm{d}U((t_1x_1 + \cdots + t_dx_d)^k)\,\varphi$ as $\sum_{|n|=k} \varphi_n t^n$ with vectors $\varphi_n \in \mathcal{H}(U)$, where $|n| := n_1 + \cdots + n_d$ for $n = (n_1, \ldots, n_d) \in \mathbb{N}_0^d$. Then $U(g(t))\,\varphi = \sum_{n \in \mathbb{N}_0^d} \varphi_n t^n$. We show that this series converges absolutely. Let $n = (n_1, \ldots, n_d) \in \mathbb{N}_0^d$ and $|n| = k$. From the definition it follows that φ_n is a sum of $\frac{k!}{n_1! \ldots n_d!}$ terms of the form $\frac{1}{k!}\,\mathrm{d}U(x_{m_1}) \ldots \mathrm{d}U(x_{m_k})\,\varphi$, where $m_1, \ldots, m_k$ are (certain) numbers of $\{1, \ldots, d\}$. Hence $\|\varphi_n\| \leqq \frac{1}{n_1! \ldots n_d!}\,\nu_k^{\mathrm{d}U}(\varphi)$. Since $\varphi \in \mathcal{D}_s^\omega(\mathrm{d}U)$, $\nu_k^{\mathrm{d}U}(\varphi) \leqq \lambda s^{-k}k!$ for some $\lambda > 0$ and all $k \in \mathbb{N}$. Therefore, $\|\varphi_n\|\,|t^n| \leqq \frac{k!}{n_1! \ldots n_d!}\,\lambda s^{-k}\,|t^n|$. Since $|t_l| \leqq s2^{-d}$ for $l = 1, \ldots, d$, $s^{-k}\,|t^n| \leqq 2^{-kd}$. Hence the preceding estimate implies that the series $\sum \varphi_n\,t^n$ converges absolutely. This proves that the map $g \to U(g)\,\varphi$ is analytic at the point e. Thus $\varphi \in \mathcal{D}^\omega(U)$. □

Remark 2. The reversed inclusion $\mathcal{D}^\omega(U) \subseteq \mathcal{D}^\omega(\mathrm{d}U)$ is also true, but the proof of this fact is longer; cf. Nelson [1], p. 590, Lemma 7.1.

In the case where the $*$-representation π is G-integrable, Theorem 10.4.4 allows a more elegant formulation. Let $\{x_1, \ldots, x_d\}$ be a basis for $\mathfrak{g}$, and let $\Delta = x_1^2 + \cdots + x_d^2$ be the associated Nelson Laplacian. From Corollary 10.2.5, $A := \overline{\mathrm{d}U(1 - \Delta)}$ is a self-adjoint operator in $\mathcal{H}(U)$. Obviously, $A \geqq 0$. Set $B := A^{1/2}$.

Theorem 10.4.10. *Keep the above notation. Then* $\mathcal{D}^\omega(\mathrm{d}U) = \mathcal{D}^\omega(B) = \bigcup_{t>0} \mathcal{D}(\mathrm{e}^{tB})$. *For every* $t > 0$ *there exist positive numbers* $r_1 = r_1(t)$ *and* $r_2 = r_2(t)$ *such that* $\mathcal{D}_t^\omega(\mathrm{d}U) \subseteq \mathcal{D}_{r_1}^\omega(B) \subseteq \mathcal{D}(\mathrm{e}^{r_1B})$ *and* $\mathcal{D}(\mathrm{e}^{t'B}) \subseteq \mathcal{D}_t^\omega(B) \subseteq \mathcal{D}_{r_2}^\omega(\mathrm{d}U)$, *if* $t' \in \mathbb{R}$, $0 < t < t'$, *where the inclusion maps are continuous in the corresponding norms* $\mathrm{e}_t^{\mathrm{d}U}(\cdot)$, $\mathrm{e}_{r_1}^B(\cdot)$, $\|\mathrm{e}^{r_1B}\cdot\|$, $\|\mathrm{e}^{t'B}\cdot\|$, $\mathrm{e}_t^B(\cdot)$ *and* $\mathrm{e}_{r_2}^{\mathrm{d}U}(\cdot)$, *respectively.*

Proof. By Corollary 10.2.4, $\mathcal{D}(\mathrm{d}U) = \mathcal{D}^\infty(U) = \mathcal{D}^\infty(A)$. Hence $\mathcal{D}(\mathrm{d}U) = \mathcal{D}^\infty(B)$ and $\mathcal{D}_t^{\mathrm{s}\omega}(A) = \mathcal{D}_t^{\mathrm{s}\omega}(\mathrm{d}U(I - \Delta))$ for all $t > 0$. The assertions now follow by combining Theorem 10.4.4 with Proposition 10.3.6. □

Corollary 10.4.11. *There exists a positive number t such that* $\mathcal{D}_t^\omega(\mathrm{d}U)$ *is dense in* $\mathcal{D}^\infty(U)[\mathrm{t}_{\mathrm{d}U}]$.

Proof. We retain the above notation. Let $E(\lambda), \lambda \in \mathbb{R}$, be the spectral projections of the positive self-adjoint operator A, and let t denote the locally convex topology on $\mathcal{D}^\infty(A)$ defined by the seminorms $\|\cdot\|_{A^n}$, $n \in \mathbb{N}_0$. From the spectral theorem we conclude that $\mathcal{D}^b := \bigcup_{n\in\mathbb{N}} E([0, n]) \mathcal{H}(U)$ is dense in $\mathcal{D}^\infty(A)$ [t]. By Corollary 10.2.4, $\mathcal{D}^\infty(A) = \mathcal{D}^\infty(U)$ and $\mathrm{t} = \mathrm{t}_{\mathrm{d}U}$. Thus $\mathcal{D}^b$ is dense in $\mathcal{D}^\infty(U)$ $[\mathrm{t}_{\mathrm{d}U}]$. Since obviously $\mathcal{D}^b \subseteq \bigcap_{s>0} \mathcal{D}(e^{sB})$, Theorem 10.4.10 shows that $\mathcal{D}_b \subseteq \mathcal{D}_t^\omega(\mathrm{d}U)$ for some $t > 0$. □

Corollary 10.4.12. *The linear space* $\mathcal{D}^\omega(U)$ *is dense in the Hilbert space* $\mathcal{H}(U)$.

Proof. By Lemma 10.4.9, $\mathcal{D}^\omega(\mathrm{d}U) \subseteq \mathcal{D}^\omega(U)$; hence the space $\mathcal{D}_t^\omega(\mathrm{d}U)$ of Corollary 10.4.11 is contained in $\mathcal{D}^\omega(U)$. □

We close this subsection with a result which shows the usefulness of the concept of analytic vectors. Suppose that $\mathcal{D}$ is a linear subspace of $\mathcal{D}^\infty(U)$ which is invariant under $\mathrm{d}U(x)$ for all $x \in \mathcal{E}(\mathfrak{g})$. Then the closure $\overline{\mathcal{D}}$ of $\mathcal{D}$ in $\mathcal{H}(U)$ is not invariant under $U(g)$ for $g \in G_0$ in general. However, if $\mathcal{D} \subseteq \mathcal{D}^\omega(\mathrm{d}U)$, then we have $U(g)\, \overline{\mathcal{D}} \subseteq \overline{\mathcal{D}}$ for $g \in G_0$. These two facts follow from Example 10.4.13 and Proposition 10.4.14.

Example 10.4.13. Let U be the unitary representation of $G := \mathbb{R}$ defined by $(U(t)\, \varphi)\,(s) = \varphi(t + s)$, $t, s \in \mathbb{R}$, on the Hilbert space $\mathcal{H}(U) := L^2(\mathbb{R})$. The infinitesimal generator of $U(\cdot)$ is the differential operator $A := \frac{\mathrm{d}}{\mathrm{d}t}$. Let $x := A \upharpoonright \mathcal{D}^\infty(U)$. Then $\mathcal{D} := C_0^\infty(0, 1) \subseteq \mathcal{D}^\infty(U)$ is invariant under $\mathrm{d}U(\mathcal{E}(\mathfrak{g})) \equiv \mathbb{C}[x]$, but $\mathcal{D}$ is not invariant under $U(t)$, $t \in \mathbb{R}$, $t \neq 0$. ○

Proposition 10.4.14. *Let* $\{x_1, \ldots, x_d\}$ *be a basis for* $\mathfrak{g}$. *Suppose that* $\mathcal{D}$ *is a linear subspace of* $\bigcap_{k=1}^{d} \mathcal{D}^\omega(\mathrm{d}U(x_k))$ *which is invariant under* $\mathrm{d}U(x)$ *for* $x \in \mathcal{E}(\mathfrak{g})$. *Then the closure of* $\mathcal{D}$ *in* $\mathcal{H}(U)$ *is invariant under* $U(g)$ *for all* $g \in G_0$.

Proof. Suppose $\varphi \in \mathcal{D}$ and $k \in \{1, \ldots, d\}$. Put $\varphi_{k,m}(t) := \sum_{n=0}^{m} \frac{t^n}{n!}\, \mathrm{d}U(x_k)^n\, \varphi$ for $t \in \mathbb{R}$ and $m \in \mathbb{N}$. By Corollary 10.2.11, $\overline{-\mathrm{i}\mathrm{d}U(x_k)}$ is a self-adjoint operator. Since $\varphi \in \mathcal{D}^\omega(\mathrm{d}U(x_k))$, there is an $s > 0$ such that $\varphi \in \mathcal{D}_s^\omega(\overline{-\mathrm{i}\mathrm{d}U(x_k)})$. Let $t \in \mathbb{R}$, $|t| \leq s$. From Lemma 10.3.3 it follows that $\varphi_{k,m}(t)$ converges to $U(\exp tx_k)\, \varphi = \exp \mathrm{i}t(\overline{-\mathrm{i}\mathrm{d}U(x_k)})\, \varphi$ as $m \to \infty$. Since $\varphi_{k,m}(t) \in \mathcal{D}$ for $m \in \mathbb{N}$, $U(\exp tx_k)\, \varphi \in \overline{\mathcal{D}}$. Hence we have that $U(\exp tx_k)\, \overline{\mathcal{D}} \subseteq \overline{\mathcal{D}}$ for $t \in \mathbb{R}$, $|t| \leq s$, and so for all real t. Each element g in the connected component G_0 of e in G is a finite product of elements of the form $\exp tx_k$, where $t \in \mathbb{R}$ and $k \in \{1, \ldots, d\}$. Thus $U(g)\, \overline{\mathcal{D}} \subseteq \overline{\mathcal{D}}$ for $g \in G_0$. □

Analytic Vectors for Single Elements of the Lie Algebra

In the two preceding subsections the space $\mathcal{D}^\omega(\pi)$ of analytic vectors for a *-representation π of $\mathcal{E}(\mathfrak{g})$ was investigated. In this subsection we are concerned with the space $\mathcal{D}^\omega(\pi(x))$ of analytic vectors for the single operator $\pi(x)$, where x is a fixed element of $\mathfrak{g}$.

Proposition 10.4.15. *Suppose that π is a $*$-representation of $\mathscr{E}(\mathfrak{g})$. Let x be an element of $\mathfrak{g}$. Then the space $\mathcal{D}^\omega(\pi(x))$ is invariant under $\pi(y)$ for all y in $\mathscr{E}(\mathfrak{g})$.*

First we verify a simple lemma which is also used in the proof of Theorem 10.5.4.

Lemma 10.4.16. *For x and y in $\mathfrak{g}$ and φ in $\mathcal{D}(\pi)$, we have*

$$\pi(\mathrm{Ad}\exp x(y))\,\varphi = \sum_{n=0}^{\infty} \frac{1}{n!}\,(\mathrm{ad}\,\pi(x))^n\,(\pi(y))\,\varphi,$$

where the series converges absolutely in $\mathscr{H}(\pi)$.

Proof. By 1.7/(4), $\mathrm{Ad}\exp x(y) = \sum_{n=0}^{\infty} \frac{1}{n!}(\mathrm{ad}\,x)^n\,(y)$, and the series converges in any locally convex topology on the finite dimensional real vector space $\mathfrak{g}$. The convergence relative to the seminorm $\|\pi(\cdot)\,\varphi\|$ on $\mathfrak{g}$ means that the series $\sum_{n=0}^{\infty} \frac{1}{n!}\,\pi((\mathrm{ad}\,x)^n\,(y))\,\varphi$ converges absolutely in $\mathscr{H}(\pi)$ and its sum is $\pi(\mathrm{Ad}\exp x(y))\,\varphi$. Since π is a homomorphism of $\mathscr{E}(\mathfrak{g})$ into $L(\mathcal{D}(\pi))$, $\pi((\mathrm{ad}\,x)^n\,(y)) = (\mathrm{ad}\,\pi(x))^n\,(\pi(y))$ for $n\in\mathbb{N}$, and the assertion follows. $\square$

Proof of Proposition 10.4.15. Since π is a homomorphism of $\mathscr{E}(\mathfrak{g})$ into $L(\mathcal{D}(\pi))$, it suffices to prove the assertion for y in $\mathfrak{g}$. Fix $x\in\mathfrak{g}$ and $\varphi\in\mathcal{D}^\omega(\pi(x))$. By the last formula in Lemma 10.3.9, we have

$$\pi(x)^{2n}\,\pi(y) = \sum_{k=0}^{2n}\binom{2n}{k}(\mathrm{ad}\,\pi(x))^k\,(\pi(y))\,\pi(x)^{2n-k} \quad \text{for} \quad n\in\mathbb{N}.$$

Hence

$$\begin{aligned}\|\pi(x)^n\,\pi(y)\,\varphi\|^2 &= (-1)^n\,\langle\pi(x)^{2n}\,\pi(y)\,\varphi,\ \pi(y)\,\varphi\rangle\\ &= (-1)^n\sum_{k=0}^{2n}\binom{2n}{k}\langle\pi(x)^{2n-k}\,\varphi,\ -(\mathrm{ad}\,\pi(x))^k\,(\pi(y))\,\pi(y)\,\varphi\rangle \qquad (10)\end{aligned}$$

for $n\in\mathbb{N}$, where we used that π is a $*$-representation. By Lemma 10.4.16, the series $\sum_{k=0}^{\infty}\frac{1}{k!}(\mathrm{ad}\,\pi(x))^k\,(\pi(y))\,\pi(y)\,\varphi$ converges absolutely in $\mathscr{H}(U)$, so that there exists a constant $\lambda>0$ such that

$$\|(\mathrm{ad}\,\pi(x))^k\,(\pi(y))\,\pi(y)\,\varphi\| \leqq \lambda^k k! \quad \text{for} \quad k\in\mathbb{N}. \qquad (11)$$

Since $\varphi\in\mathcal{D}^\omega(\pi(x))$ by assumption, there is a constant $M\geqq\lambda$ such that

$$\|\pi(x)^k\,\varphi\| \leqq M^k k! \quad \text{for} \quad k\in\mathbb{N}. \qquad (12)$$

Putting (11) and (12) into (10), we obtain for $n\in\mathbb{N}$

$$\|\pi(x)^n\,\pi(y)\,\varphi\|^2 \leqq \sum_{k=0}^{2n}\binom{2n}{k}M^{2n-k}(2n-k)!\,\lambda^k k! \leqq M^{2n}(2n+1)!\,.$$

Using the Stirling formula it follows that there is an $M_1>0$ such that $\|\pi(x)^n\,\pi(y)\,\varphi\|^2 \leqq M_1^{2n}(n!)^2$ for $n\in\mathbb{N}$, that is, $\pi(y)\,\varphi\in\mathcal{D}^\omega(\pi(x))$. $\square$

Let π be a $*$-representation of $\mathscr{E}(\mathfrak{g})$ and let x be in $\mathfrak{g}$. Since $\mathcal{D}^\omega(\pi(x))$ is invariant under $\pi(y)$, $y\in\mathscr{E}(\mathfrak{g})$, by Proposition 10.4.15, the restriction of π to $\mathcal{D}^\omega(\pi(x))$ is a $*$-representation of $\mathscr{E}(\mathfrak{g})$. We shall denote this *-representation by θ_x. Further, if we assume that $\mathcal{D}^\omega(\pi(x))$ is dense in $\mathcal{H}(\pi)$, then the symmetric operator $i\pi(x)$ is essentially self-adjoint, so $\overline{\pi(x)}$ is the infinitesimal generator of a strongly continuous one-parameter unitary group $t \to U_x(t) := \exp t\overline{\pi(x)}$, $t \in \mathbb{R}$.

The following proposition is needed in proving Theorem 10.5.4 in the next section, but it is also of interest in itself.

Proposition 10.4.17. *Let π be a *-representation of $\mathscr{E}(\mathfrak{g})$. Suppose that x is an element of $\mathfrak{g}$ such that $\mathcal{D}^\omega(\pi(x))$ is dense in $\mathcal{H}(\pi)$. Then the unitary group $t \to U_x(t)$ maps $\mathcal{D}(\pi)$ into $\mathcal{D}(\theta_x^*)$ and we have*

$$\pi(\mathrm{Ad}\exp tx(y))\,\varphi = U_x(t)\,\theta_x^*(y)\,U_x(-t)\,\varphi, \quad \varphi \in \mathcal{D}(\pi), \tag{13}$$

for all y in $\mathscr{E}(\mathfrak{g})$ and t in $\mathbb{R}$.

Proof. Let $\varphi \in \mathcal{D}(\pi)$, $\psi \in \mathcal{D}^\omega(\pi(x))$ and $y \in \mathscr{E}(\mathfrak{g})$. We consider the functions $f(t) := \langle U_x(t)\,\pi(y^+)\,\psi, \varphi\rangle$ and $g(t) := \langle U_x(t)\,\psi, \pi(\mathrm{Ad}\exp tx(y))\,\varphi\rangle$ on $\mathbb{R}$. Since ψ and $\pi(y^+)\,\psi$ are in $\mathcal{D}^\omega(\pi(x))$, there is an $s > 0$ such that both vectors are in $\mathcal{D}_s^\omega(i\pi(x))$. From Lemma 10.3.3 applied with $A := i\overline{\pi(x)}$ it follows that the mappings $t \to U_x(t)\,\pi(y^+)\,\psi$ and $t \to U_x(t)\,\psi$ of $\mathbb{R}$ into $\mathcal{H}(\pi)$ are restrictions to $\mathbb{R}$ of $\mathcal{H}(\pi)$-valued holomorphic functions in the strip $\mathbb{R}_s := \{z \in \mathbb{C} : |\mathrm{Im}\, z| < s\}$. Lemma 10.4.16 applied with tx in place of x shows that the map $t \to \pi(\mathrm{Ad}\exp tx(y))\,\varphi$ is also the restriction to $\mathbb{R}$ of a $\mathcal{H}(\pi)$-valued holomorphic function in $\mathbb{R}_s$. Hence f and g have holomorphic extensions to the strip $\mathbb{R}_s$. For $n \in \mathbb{N}$, we have

$$f^{(n)}(0) = \langle \pi(x)^n\,\pi(y^+)\,\psi, \varphi\rangle$$

and

$$g^{(n)}(0) = \sum_{k=0}^{n} \binom{n}{k} \langle \pi(x)^{n-k}\,\psi, (\mathrm{ad}\,\pi(x))^k\,(\pi(y))\,\varphi\rangle$$

$$= \sum_{k=0}^{n} \binom{n}{k} \langle (\mathrm{ad}\,\pi(x))^k\,(\pi(y^+))\,\pi(x)^{n-k}\,\psi, \varphi\rangle,$$

where we used again Lemma 10.4.16 and the formula

$$((\mathrm{ad}\,\pi(x))^k\,(\pi(y)))^+ = (\mathrm{ad}\,\pi(x))^k\,(\pi(y^+)).$$

From the last formula in Lemma 10.3.9 we see that $f^{(n)}(0) = g^{(n)}(0)$ for $n \in \mathbb{N}$. Obviously, $f(0) = g(0)$. Therefore, the analytic functions f and g coincide on the whole real line. Hence

$$\begin{aligned}\langle \theta_x(y^+)\,\psi, U_x(-t)\,\varphi\rangle &= \langle U_x(t)\,\pi(y^+)\,\psi, \varphi\rangle = f(t) = g(t)\\ &= \langle U_x(t)\,\psi, \pi(\mathrm{Ad}\exp tx(y))\,\varphi\rangle\\ &= \langle \psi, U_x(-t)\,\pi(\mathrm{Ad}\exp tx(y))\,\varphi\rangle \quad \text{for} \quad t \in \mathbb{R}.\end{aligned}$$

Since this is true for all $y \in \mathscr{E}(\mathfrak{g})$ and $\psi \in \mathcal{D}^\omega(\pi(x)) = \mathcal{D}(\theta_x)$, we have

$$U_x(-t)\,\varphi \in \bigcap_{y \in \mathscr{E}(\mathfrak{g})} \mathcal{D}(\theta_x(y^+)^*) = \mathcal{D}(\theta_x^*)$$

and

$$\theta_x^*(y)\, U_x(-t)\, \varphi = U_x(-t)\, \pi\big(\mathrm{Ad}\exp tx(y)\big)\, \varphi$$

which gives (13). □

Corollary 10.4.18. *Keep the assumptions of Proposition* 10.4.17. *If, in addition, the unitary group $t \to U_x(t)$ leaves $\mathcal{D}(\pi)$ invariant, then*

$$\pi\big(\mathrm{Ad}\exp tx(y)\big)\, \varphi = U_x(t)\, \pi(y)\, U_x(-t)\, \varphi, \quad \varphi \in \mathcal{D}(\pi), \tag{14}$$

for all y in $\mathcal{E}(\mathfrak{g})$ and t in $\mathbb{R}$.

Proof. From $\theta_x \subseteq \pi$, $\theta_x^* \supseteq \pi^* \supseteq \pi$. Since $U_x(-t)\, \varphi \in \mathcal{D}(\pi)$ by assumption, $\theta_x^*(y)\, U_x(-t)\varphi = \pi(y)\, U_x(-t)\, \varphi$, and (14) follows from (13). □

Remark 3. If U is a unitary representation of G and $\pi = \mathrm{d}U$, then the assumptions of Corollary 10.4.18 are fulfilled for each x in g. In this case $U_x(t) = U\,(\exp tx)$ for $x \in g$ and $t \in \mathbb{R}$, and (14) is already known from Lemma 10.1.12.

10.5. Exponentiation of *-Representations of Enveloping Algebras

Let π be a given *-representation of the enveloping algebra $\mathcal{E}(\mathfrak{g})$. An important and natural question is: When is π integrable? In other words, when does there exist a unitary representation U of the connected and simply connected Lie group $\tilde{G}$ which has g as its Lie algebra such that $\pi = \mathrm{d}U$?

By definition the equality $\pi = \mathrm{d}U$ requires also that the domains $\mathcal{D}(\pi)$ and $\mathcal{D}(\mathrm{d}U) \equiv \mathcal{D}^\infty(U)$ are equal, that is, the domain $\mathcal{D}(\pi)$ has to be maximal in some sense. In concrete applications this is often too strong. For a convenient formulation of the main results in this section we introduce the following notion which is weaker than the concept of integrable representations.

Definition 10.5.1. A representation π of the enveloping algebra $\mathcal{E}(\mathfrak{g})$ is called *exponentiable* if there exist a unitary representation U of the Lie group $\tilde{G}$ on $\mathcal{H}(\pi)$ and a basis $\{x_1, \ldots, x_d\}$ for $\mathfrak{g}$ such that $\overline{\pi(x_k)} = \overline{\mathrm{d}U(x_k)}$, $k = 1, \ldots, d$.

Each exponentiable representation is a *-representation, because dU is. Integrable representations are always exponentiable, the converse is not true in general. But the concept of exponentiable representations is still sufficient to ensure a strong connection between π and $\mathrm{d}U$ as the next proposition shows.

Proposition 10.5.2. *Suppose that π is an exponentiable representation of $\mathcal{E}(\mathfrak{g})$, that is, there are a unitary representation U of $\tilde{G}$ in $\mathcal{H}(\pi)$ and a basis $\{x_1, \ldots, x_d\}$ for $\mathfrak{g}$ such that $\overline{\pi(x_k)} = \overline{\mathrm{d}U(x_k)}$, $k = 1, \ldots, d$. Then the unitary representation U is uniquely determined by this property and we have $\pi \subseteq \mathrm{d}U$ and $\pi^* = \mathrm{d}U$. If, in addition, π is self-adjoint (or equivalently, if $\mathcal{D}(\pi) = \mathcal{D}^\infty(U)$), then $\pi = \mathrm{d}U$ and so π is integrable.*

Proof. Let U_1 be another unitary representation of $\tilde{G}$ in $\mathcal{H}(\pi)$ such that $\overline{\pi(x_k)} = \overline{\mathrm{d}U_1(x_k)}$, $k = 1, \ldots, d$. Then $\overline{\mathrm{d}U(x_k)} = \overline{\mathrm{d}U_1(x_k)}$ and we obtain $U(\exp tx_k) = \exp t\, \overline{\mathrm{d}U(x_k)} = \exp t\, \overline{\mathrm{d}U_1(x_k)} = U_1(\exp tx_k)$ for $k = 1, \ldots, d$ and $t \in \mathbb{R}$ by Corollary 10.2.13. Since $\tilde{G}$ is connected, this implies that $U = U_1$ on the whole group $\tilde{G}$.

From $\overline{\pi(x_k)} = \overline{\mathrm{d}U(x_k)}$ for $k = 1, \dots, d$, we have $\mathcal{D}(\pi) \subseteq \bigcap_{k=1}^{d} \mathcal{D}^\infty\left(\overline{\pi(x_k)}\right) = \bigcap_{k=1}^{d} \mathcal{D}^\infty\left(\overline{\mathrm{d}U(x_k)}\right)$ $= \mathcal{D}^\infty(U) = \mathcal{D}(\mathrm{d}U)$ by Corollary 10.2.12. Hence $\pi(x_k) \subseteq \mathrm{d}U(x_k)$ for $k = 1, \dots, d$. Since the algebra $\mathcal{E}(\mathfrak{g})$ is generated by $\{x_1, \dots, x_d, 1\}$, $\pi \subseteq \mathrm{d}U$. By Corollary 10.2.11, the operators $\mathrm{i}\overline{\mathrm{d}U(x_k)} = \overline{\pi(\mathrm{i}x_k)}$ are self-adjoint, so π^* is self-adjoint by Proposition 8.1.12, (v). The relation $\pi \subseteq \mathrm{d}U$ leads to $(\mathrm{d}U)^* \subseteq \pi^*$. Since $\mathrm{d}U$ and π^* are both self-adjoint, the latter yields $\mathrm{d}U = \pi^*$. If π is self-adjoint, then $\pi = \pi^* = \mathrm{d}U$, that is, π is integrable. □

The following simple lemma is essentially used in the proofs of the two main theorems of this section.

Lemma 10.5.3. *Suppose π is a $*$-representation of $\mathcal{E}(\mathfrak{g})$. Let $\{x_1, \dots, x_d\}$ be a basis of $\mathfrak{g}$ and let $\mathcal{D}_0$ denote the intersection of the domains $\mathcal{D}\left(\overline{\pi(x_{k_1})} \dots \overline{\pi(x_{k_n})}\right)$ for arbitrary indices $k_1, \dots, k_n$ of $\{1, \dots, d\}$ and $n \in \mathbb{N}$. Then there is a (unique) $*$-representation π_0 of $\mathcal{E}(\mathfrak{g})$ on $\mathcal{D}_0$ such that $\pi_0(x_k) = \overline{\pi(x_k)} \upharpoonright \mathcal{D}_0$ for $k = 1, \dots, d$ and $\pi \subseteq \pi_0$.*

Remark 1. Lemma 10.5.3 follows at once from Proposition 8.1.17. We give another proof of Lemma 10.5.3 which is more transparent in this special case.

Proof of Lemma 10.5.3. Define $\pi_0\left(\sum_{k=1}^{d} \alpha_k x_k\right) := \sum_{k=1}^{d} \alpha_k \overline{\pi(x_k)} \upharpoonright \mathcal{D}_0$ for $\alpha_1, \dots, \alpha_d \in \mathbb{R}$. From the definition it is obvious that $\mathcal{D}_0$ is invariant under the operators $\overline{\pi(x_k)}$; so π_0 is a linear mapping of $\mathfrak{g}$ into $L(\mathcal{D}_0)$. Since the $\overline{\pi(x_k)}$ are skew-symmetric, each $\pi_0(x)$, $x \in \mathfrak{g}$, is skew-symmetric. Let $x = \sum_{k=1}^{d} \alpha_k x_k$ and $y = \sum_{k=1}^{d} \beta_k x_k$ be elements of $\mathfrak{g}$. Using the skew-symmetry of $\overline{\pi(x_k)}$, $k = 1, \dots, d$, and of $\pi_0([x, y])$, it follows that for $\varphi \in \mathcal{D}_0$ and $\psi \in \mathcal{D}(\pi)$

$$
\begin{aligned}
&\langle (\pi_0(x)\, \pi_0(y) - \pi_0(y)\, \pi_0(x))\, \varphi, \psi \rangle \\
&= \sum_{k,l=1}^{d} \alpha_k \beta_l \langle (\overline{\pi(x_k)}\, \overline{\pi(x_l)} - \overline{\pi(x_l)}\, \overline{\pi(x_k)})\, \varphi, \psi \rangle \\
&= \sum_{k,l=1}^{d} \alpha_k \beta_l \langle \varphi, (\pi(x_l)\, \pi(x_k) - \pi(x_k)\, \pi(x_l))\, \psi \rangle \\
&= \langle \varphi, (\pi(y)\, \pi(x) - \pi(x)\, \pi(y))\, \psi \rangle = \langle \varphi, -\pi([x, y])\, \psi \rangle = \langle \varphi, -\pi_0([x, y])\, \psi \rangle \\
&= \langle \pi_0([x, y])\, \varphi, \psi \rangle, \text{ i.e. } [\pi_0(x), \pi_0(y)] = \pi_0([x, y]).
\end{aligned}
$$

This shows that π_0 is a $*$-representation of the Lie algebra $\mathfrak{g}$. By the universal property of $\mathcal{E}(\mathfrak{g})$, π_0 extends to a $*$-representation, again denoted by π_0, of $\mathcal{E}(\mathfrak{g})$. From the construction it is clear that $\pi \subseteq \pi_0$. □

Our first main result in this section is

Theorem 10.5.4. *Let $\{x_1, \dots, x_d\}$ be a basis for the Lie algebra $\mathfrak{g}$ and let π be a $*$-representation of the enveloping algebra $\mathcal{E}(\mathfrak{g})$. Suppose that there exists a subset $\mathcal{D}_1$ of $\mathcal{D}(\pi)$ consisting of analytic vectors for every operator $\pi(x_k)$, $k = 1, \dots, d$, such that the subspace $\pi(\mathcal{E}(\mathfrak{g}))\, \mathcal{D}_1 \equiv$ l.h. $\{\pi(x)\, \varphi : x \in \mathcal{E}(\mathfrak{g})$ and $\varphi \in \mathcal{D}_1\}$ is dense in $\mathcal{H}(\pi)$. Then π is exponentiable. If in addition π is self-adjoint, then π is integrable.*

An immediate consequence of Theorem 10.5.4 is

Corollary 10.5.5. *If π is a $*$-representation of $\mathcal{E}(\mathfrak{g})$ such that $\mathcal{D}^\omega(\pi)$ is dense in $\mathcal{D}(\pi)$, then π is exponentiable.*

Proof. Put $\mathcal{D}_1 := \mathcal{D}^\omega(\pi)$ and use that $\mathcal{D}^\omega(\pi) \subseteq \mathcal{D}^\omega(\pi(x_k))$, $k = 1, \ldots, d$. □

Proof of Theorem 10.5.4. Let π_0 be the $*$-representation of $\mathcal{E}(\mathfrak{g})$ which is associated with π according to Lemma 10.5.3. Since $\mathcal{D}_1 \subseteq \bigcap_{k=1}^{d} \mathcal{D}^\omega(\pi(x_k))$ by assumption, it follows from Proposition 10.4.15 that $\mathcal{D}_a := \pi(\mathcal{E}(\mathfrak{g}))\, \mathcal{D}_1 \subseteq \bigcap_{k=1}^{d} \mathcal{D}^\omega(\pi(x_k))$. Since $\pi \subseteq \pi_0$, $\mathcal{D}_a \subseteq \bigcap_{k=1}^{d} \mathcal{D}^\omega(\pi_0(x_k))$. Suppose $n \in \{1, \ldots, d\}$. Let θ_n denote the restriction to $\mathcal{D}^\omega(\pi_0(x_n))$ of π_0. We first show that $\mathcal{D}(\theta_n^*) \subseteq \mathcal{D}(\pi_0)$. From Corollary 8.1.7 we know that $\mathcal{D}(\theta_n^*)$ is the intersection of all domains $\mathcal{D}(\theta_n(x_{k_1})^* \ldots \theta_n(x_{k_m})^*)$ with $k_1, \ldots, k_m \in \{1, \ldots, d\}$ and $m \in \mathbb{N}$. Since $\theta_n \subseteq \pi_0$ and $\mathcal{D}_a \subseteq \bigcap_{k=1}^{d} \mathcal{D}^\omega(\pi_0(x_k))$, we have that $\mathcal{D}_a \subseteq \bigcap_{k=1}^{d} \mathcal{D}^\omega(i\theta_n(x_k))$. Therefore, since $\mathcal{D}_a$ is dense in $\mathcal{H}(\pi)$, Proposition 10.3.4 shows that the symmetric operators $i\theta_n(x_k)$, $k = 1, \ldots, d$, are essentially self-adjoint. Hence $\theta_n(x_k)^* = -\overline{\theta_n(x_k)} \subseteq -\overline{\pi_0(x_k)} = -\overline{\pi(x_k)}$ for $k = 1, \ldots, d$. By the definition of $\mathcal{D}_0$, this implies that $\mathcal{D}(\theta_n^*) \subseteq \mathcal{D}_0 = \mathcal{D}(\pi_0)$. Since $\mathcal{D}_a \subseteq \mathcal{D}^\omega(i\pi_0(x_k))$ and $\mathcal{D}_a$ is dense in $\mathcal{H}(\pi)$, Proposition 10.3.4 also shows that $i\pi_0(x_n)$ is essentially self-adjoint. Thus $\overline{\pi_0(x_n)}$ is the infinitesimal generator of a one-parameter unitary group $t \to U_n(t) := \exp t\overline{\pi_0(x_n)}$. By Proposition 10.4.17 applied to the $*$-representation π_0, $U_n(\cdot)$ maps $\mathcal{D}(\pi_0)$ into $\mathcal{D}(\theta_{x_n}^*)$ and hence into $\mathcal{D}(\pi_0)$. Therefore, by Corollary 10.4.18, we have

$$\pi_0(\mathrm{Ad}\exp tx_n(y))\, \varphi = U_n(t)\, \pi_0(y)\, U_n(-t)\, \varphi \tag{1}$$

for $n = 1, \ldots, d$, $d \in \mathcal{E}(\mathfrak{g})$, $t \in \mathbb{R}$ and $\varphi \in \mathcal{D}(\pi_0)$.

In order to continue the proof, we need some general facts from the theory of Lie groups. We can choose an open neighbourhood W of e in $\tilde{G}$ such that the map $s = (s_1, \ldots, s_d) \to g(s) = \exp s_1 x_1 \ldots \exp s_d x_d$ is an analytic diffeomorphism of some open neighbourhood V of the origin in $\mathbb{R}^d$ onto W. (The numbers $s_1, \ldots, s_d$ are then the canonical coordinates of the second kind of $g(s)$.) Further, we choose a $\delta > 0$ and a neighbourhood W' of e in $\tilde{G}$ such that $\exp s_1 x_1 \ldots \exp s_d x_d \cdot g \in W$ if $|s_k| < \delta$ for $k = 1, \ldots, d$ and if $g \in W'$.

For $g(s) = \exp s_1 x_1 \ldots \exp s_d x_d$ with $s = (s_1, \ldots, s_d) \in V$ we define

$$U(g(s)) := U_1(s_1) \ldots U_d(s_d). \tag{2}$$

Our aim is to show that U extends to a unitary representation of $\tilde{G}$. Suppose that $n \in \{1, \ldots, d\}$ and $g \in W$ such that $\exp tx_n \cdot g \in W$ for $t \in (-\delta, \delta)$. The next important step is to prove that

$$U(\exp tx_n \cdot g) = U(\exp tx_n)\, U(g) \quad \text{for all} \quad t \in (-\delta, \delta). \tag{3}$$

Because $\exp tx_n \cdot g \in W$ if $|t| < \delta$, there are analytic functions $\alpha_1(t), \ldots, \alpha_d(t)$ on $(-\delta, \delta)$ such that

$$\exp tx_n \cdot g = \exp \alpha_1(t)\, x_1 \ldots \exp \alpha_d(t)\, x_d, \quad t \in (-\delta, \delta). \tag{4}$$

By Ado's theorem we can assume that all elements of $\mathfrak{g}$ and W are matrices. Then, differentiation of (4) yields

$$x_n \exp tx_n \cdot g$$

$$= \sum_{k=1}^{d} \left(\exp \alpha_1(t)\, x_1 \ldots \exp \alpha_{k-1}(t)\, x_{k-1}\right) \alpha_k'(t)\, x_k \exp \alpha_k(t)\, x_k \ldots \exp \alpha_d(t)\, x_d$$

$$= \sum_{k=1}^{d} \alpha_k'(t) \left(\exp \alpha_1(t)\, x_1 \ldots \exp \alpha_{k-1}(t)\, x_{k-1}\right) x_k \exp\left(-\alpha_{k-1}(t)\, x_{k-1}\right) \cdots$$

$$\ldots \exp\left(-\alpha_1(t)\, x_1\right) \exp tx_n \cdot g$$

$$= \sum_{k=1}^{n} \alpha_k'(t) \operatorname{Ad} \exp \alpha_1(t)\, x_1 \ldots \operatorname{Ad} \exp \alpha_{k-1}(t)\, x_{k-1}(x_n) \exp tx_n \cdot g$$

and hence

$$x_n = \sum_{k=1}^{d} \alpha_k'(t) \operatorname{Ad} \exp \alpha_1(t)\, x_1 \ldots \operatorname{Ad} \exp \alpha_{k-1}(t)\, x_{k-1}(x_n) \tag{5}$$

with the obvious interpretations of the term for $k = 1$. In (5), this term is $\alpha_1'(t)\, x_1$ by definition.

Recall that $U_k(t) = \exp t\overline{\pi_0(x_k)}$ for $k = 1, \ldots, d$. Therefore, if $\varphi \in \mathcal{D}_0$, the mapping $t \to U_k(t)\, \varphi$ of $\mathbb{R}$ into $\mathcal{H}(\pi)$ is differentiable. Fix $\varphi \in \mathcal{D}_0$. Since $U_k(t)\, \mathcal{D}_0 \subseteq \mathcal{D}_0$ for $k = 1, \ldots, d$ and $t \in \mathbb{R}$, the $\mathcal{H}(\pi)$-valued function

$$f(t) := U(\exp tx_n)^{-1}\, U(\exp tx_n \cdot g)\, \varphi \equiv U_n(-t)\, U_1(\alpha_1(t)) \ldots U_d(\alpha_d(t))\, \varphi,$$

$t \in (-\delta, \delta)$, is differentiable. Applying the product rule and using the formulas (1) and (5), we obtain for $t \in (-\delta, \delta)$

$$\frac{\mathrm{d}}{\mathrm{d}t} f(t) = U_n(-t) \left(-\pi_0(x_n)\right) U_1(\alpha_1(t)) \ldots U_d(\alpha_d(t))\, \varphi$$

$$+ \sum_{k=1}^{d} U_n(-t)\, U_1(\alpha_1(t)) \ldots U_{k-1}(\alpha_{k-1}(t))\, \alpha_k'(t)\, \pi_0(x_k)\, U_k(\alpha_k(t)) \ldots U_d(\alpha_d(t))\, \varphi$$

$$= U_n(-t) \left(-\pi_0(x_n)\right) U(\exp tx_n \cdot g)\, \varphi$$

$$+ U_n(-t) \left\{ \sum_{k=1}^{d} \alpha_k'(t)\, U_1(\alpha_1(t)) \ldots U_{k-1}(\alpha_{k-1}(t))\, \pi_0(x_k)\, U_{k-1}(-\alpha_{k-1}(t)) \cdots \right.$$

$$\left. \ldots U_1(-\alpha_1(t)) \right\} U(\exp tx_n \cdot g)\, \varphi$$

$$\underset{(1)}{=} U_n(-t) \left\{ -\pi_0(x_n) \right.$$

$$\left. + \sum_{k=1}^{d} \alpha_k'(t)\, \pi_0\left(\operatorname{Ad} \exp \alpha_1(t)\, x_1 \ldots \operatorname{Ad} \exp \alpha_{k-1}(t)\, x_{k-1}(x_k)\right) \right\} U(\exp tx_n \cdot g)\, \varphi$$

$$\underset{(5)}{=} U_n(-t)\, \{0\}\, U(\exp tx_n \cdot g)\, \varphi = 0.$$

(In case $k = 1$ we interpret terms like $U(\alpha_1(t)) \ldots U_{k-1}(\alpha_{k-1}(t))$ as to be the identity.) Thus the function $f(t)$ is constant on the interval $(-\delta, \delta)$. Since obviously $f(0) = U(g)\, \varphi$, we have $U(\exp tx_n)^{-1}\, U(\exp tx_n \cdot g)\, \varphi = U(g)\, \varphi$ on $(-\delta, \delta)$ for all $\varphi \in \mathcal{D}_0$ and hence for all $\varphi \in \mathcal{H}(\pi)$. Consequently, $U(\exp tx_n \cdot g) = U(\exp tx_n)\, U(g)$ for $t \in (-\delta, \delta)$, and (3) is proved.

Now let $s_n \in (-\delta, \delta)$ for $n = 1, \ldots, d$ and let $h \in W'$. By the above assumptions, the elements $\exp t_k x_k \ldots \exp t_d x_d \cdot h$, where $k = 1, \ldots, d$ and $t_k, \ldots, t_d \in (-\delta, \delta)$, are all in W. Hence (3) applies with $t = s_k$, $n = k$ and $g = \exp s_{k+1} x_{k+1} \ldots \exp s_d x_d \cdot h$. If $k = d$, we set $g = h$. Applying (3) d times and using (2), we get

$$\begin{aligned} U(\exp s_1 x_1 \ldots \exp s_d x_d \cdot h) &= U(\exp s_1 x_1)\, U(\exp s_2 x_2 \ldots \exp s_d x_d \cdot h) \\ &= \cdots = U(\exp s_1 x_1)\, U(\exp s_2 x_2) \ldots U(\exp s_d x_d)\, U(h) \\ &= U_1(s_1) \ldots U_d(s_d)\, U(h) \\ &= U(\exp s_1 x_1 \ldots \exp s_d x_d)\, U(h). \end{aligned}$$

This shows that U is a local homomorphism of a neighbourhood of the identity in $\tilde{G}$ into the group of unitaries on $\mathcal{H}(\pi)$. From the definition (2) it is clear that $\|U(\exp x)\varphi - \varphi\| \to 0$ as $x \to 0$ in $\mathfrak{g}$ for each vector $\varphi \in \mathcal{H}(\pi)$; so the map $g \to U(g)$ is strongly continuous at the identity of $\tilde{G}$. Since $\tilde{G}$ is connected and simply connected, there is a unique extension of U to a unitary representation, again denoted by U, of $\tilde{G}$ on $\mathcal{H}(\pi)$.

Let $k \in \{1, \ldots, d\}$. By Lemma 10.5.3, $\overline{\pi(x_k)} = \overline{\pi_0(x_k)}$. As noted above, $\overline{\pi_0(x_k)}$ is the infinitesimal generator of the unitary group $t \to U_k(t) \equiv U(\exp tx_k)$, $t \in \mathbb{R}$. Therefore, $\overline{\pi(x_k)} = \overline{\mathrm{d}U(x_k)}$ which proves that π is exponentiable. If π is self-adjoint, then π is integrable by Proposition 10.5.2. □

The second main result in this section is

Theorem 10.5.6. *Let $\{x_1, \ldots, x_d\}$ be a basis for the Lie algebra $\mathfrak{g}$, and let $\Delta := x_1^2 + \ldots + x_d^2$ be the corresponding Nelson Laplacian. Suppose π is a $*$-representation of $\mathcal{E}(\mathfrak{g})$ such that the operator $\pi(\Delta)$ is essentially self-adjoint.*

Then the representation π is exponentiable. If in addition π is self-adjoint, then π is integrable.

The proof of Theorem 10.5.6 requires a lemma.

Lemma 10.5.7. *Keep the assumptions and the notation of Theorem 10.5.6. Let $\mathcal{D}_0$ be the domain defined in Lemma 10.5.3. Then $\mathcal{D}^\infty\big(\overline{\pi(1 - \Delta)}\big) \subseteq \mathcal{D}_0$.*

Proof. In this proof we abbreviate $A := \pi(1 - \Delta)$ and $X_k := \pi(x_k)$, $k = 1, \ldots, d$. By Corollary 10.4.7, we have for $k, k_1, \ldots, k_n \in \{1, \ldots, d\}$ and $n \in \mathbb{N}$

$$\mathcal{D}(\bar{A}) \subseteq \mathcal{D}(\overline{X_k}) \tag{6}$$

and

$$\mathcal{D}(\bar{A}) \subseteq \mathcal{D}\big(\overline{\operatorname{ad} X_{k_1} \ldots \operatorname{ad} X_{k_n}(A)}\big). \tag{7}$$

We prove by induction on n that for arbitrary numbers $k_1, \ldots, k_{n-1} \in \{1, \ldots, d\}$

$$\mathcal{D}\big((\bar{A})^n\big) \subseteq \mathcal{D}(\bar{A}\, \overline{X_{k_{n-1}}} \ldots \overline{X_{k_1}}) \tag{8}$$

with the interpretation that in case $n = 1$, (8) means that $\mathcal{D}(\bar{A}) \subseteq \mathcal{D}(\bar{A})$. Combined with (6), (8) leads to $\mathcal{D}^\infty(\bar{A}) \subseteq \mathcal{D}(\bar{A}\, \overline{X_{k_{n-1}}} \ldots \overline{X_{k_1}}) \subseteq \mathcal{D}(\overline{X_{k_n}}\, \overline{X_{k_{n-1}}} \ldots \overline{X_{k_1}})$ which gives the assertion.

Let $n \in \mathbb{N}$. Assume that (8) is true for arbitrary numbers $k_1, \ldots, k_{n'-1} \in \{1, \ldots, d\}$

and all $n' \in \mathbb{N}$, $n' \leqq n$. Fix $k_1, \dots, k_n \in \{1, \dots, d\}$ and $\varphi \in \mathcal{D}((\bar{A})^{n+1})$. The operators X_k, $k = 1, \dots, d$, and $\mathrm{ad}\, X_{l_m} \dots \mathrm{ad}\, X_{l_1}(A)$, $l_1, \dots, l_m \in \{1, \dots, d\}$, are skew-symmetric and symmetric, respectively. Therefore, applying the involution to the first formula in Lemma 10.3.9, we see that $X_{k_n} \dots X_{k_1} A$ is a finite sum of $AX_{k_1} \dots X_{k_n}$ (the term $k = 0$ in the sum) and of terms of the form YZ, where $Y = X_{l_n} \dots X_{l_{k+1}}$ and $Z = \pm\, \mathrm{ad}\, X_{l_k} \dots \mathrm{ad}\, X_{l_1}(A)$ for some $l_1, \dots, l_n \in \{1, \dots, d\}$ and $k \in \{1, \dots, n\}$. (In case $k = n$ we set $Y = I$.) Suppose $\psi \in \mathcal{D}(\pi)$. From the induction hypothesis and (6), we have

$$\bar{A}\varphi \in \mathcal{D}((\bar{A})^n) \subseteq \mathcal{D}(\bar{A}\, \overline{X_{k_{n-1}}} \dots \overline{X_{k_1}}) \subseteq \mathcal{D}(\overline{X_{k_n}}\, \overline{X_{k_{n-1}}} \dots \overline{X_{k_1}})$$

and so

$$\langle AX_{k_1} \dots X_{k_n}\psi, \varphi\rangle = \langle X_{k_1} \dots X_{k_n}\psi, \bar{A}\varphi\rangle = (-1)^n \langle \psi, \overline{X_{k_n}} \dots \overline{X_{k_1}}\, \bar{A}\varphi\rangle, \tag{9}$$

where we again used the skew-symmetry of $\overline{X_k}$ and the symmetry of $\bar{A}$. Applying once more the induction hypothesis and (7), we get $\overline{X_{l_{k+1}}} \dots \overline{X_{l_n}}\varphi \in \mathcal{D}(\bar{A}) \subseteq \mathcal{D}(\bar{Z})$. Thus, by the symmetry of $\bar{Z}$,

$$\langle YZ\psi, \varphi\rangle = (-1)^{n-k} \langle Z\psi, \overline{X_{l_{k+1}}} \dots \overline{X_{l_n}}\varphi\rangle = (-1)^{n-k} \langle \psi, \bar{Z}\, \overline{X_{l_{k+1}}} \dots \overline{X_{l_n}}\varphi\rangle. \tag{10}$$

From (9) and (10) it follows that the linear functional $\psi \to \langle X_{k_n} \dots X_{k_1} A\psi, \varphi\rangle$ is continuous on $(\mathcal{D}(\pi), \|\cdot\|)$. Similarly as above, we have $\varphi \in \mathcal{D}((\bar{A})^n) \subseteq \mathcal{D}(\overline{X_{k_1}} \dots \overline{X_{k_n}})$ and hence $\langle X_{k_n} \dots X_{k_1} A\psi, \varphi\rangle = (-1)^n \langle A\psi, \overline{X_{k_1}} \dots \overline{X_{k_n}}\varphi\rangle$. Therefore, $\overline{X_{k_1}} \dots \overline{X_{k_n}}\varphi \in \mathcal{D}(A^*)$. Since $\pi(\Delta)$ and hence $A = \pi(1 - \Delta)$ is essentially self-adjoint by assumption, $\bar{A} = A^*$, so that $\varphi \in \mathcal{D}(\bar{A}\, \overline{X_{k_1}} \dots \overline{X_{k_n}})$. This proves (8) in case $n + 1$. □

Proof of Theorem 10.5.6: Let π_0 be the *-representation from Lemma 10.5.3. We first show that $\mathcal{D}^\omega(\pi_0)$ is dense in $\mathcal{H}(\pi_0) \equiv \mathcal{H}(\pi)$. Since $\pi \subseteq \pi_0$, $A := \pi(1 - \Delta) \subseteq \pi_0(1 - \Delta)$. By assumption, the operator $\bar{A}$ is self-adjoint. Hence $\bar{A} = \overline{\pi_0(1 - \Delta)}$. Let $E(\lambda)$, $\lambda \in \mathbb{R}$, be the spectral projections of the positive self-adjoint operator $\bar{A}$, and let $\mathcal{D}^b := \bigcup_{n \in \mathbb{N}} E([0, n])\, \mathcal{H}(\pi)$. From Lemma 10.5.7, $\mathcal{D}^\infty(\bar{A}) \subseteq \mathcal{D}_0$, so that $\mathcal{D}^b \subseteq \mathcal{D}_0$. Since the vectors in $\mathcal{D}^b$ are, of course, semi-analytic vectors for the operator $A \upharpoonright \mathcal{D}_0 = \pi_0(1 - \Delta)$, Theorem 10.4.4 shows that $\mathcal{D}^b \subseteq \mathcal{D}^\omega(\pi_0)$. Since $\mathcal{D}^b$ is dense in $\mathcal{H}(\pi)$ by the spectral theorem, $\mathcal{D}^\omega(\pi_0)$ is dense in $\mathcal{H}(\pi)$. By Corollary 10.5.5, π_0 is exponentiable. Since $\overline{\pi(x_k)} = \overline{\pi_0(x_k)}$ for $k = 1, \dots, d$ by the definition of π_0, this implies that the representation π is exponentiable. □

Remark 2. The preceding proof of Theorem 10.5.6 consists of two independent parts. The first one is to prove that $\mathcal{D}^\omega(\pi_0)$ is dense in $\mathcal{H}(\pi)$. This is done by combining Theorem 10.4.4 and Lemma 10.5.7. The second part uses Corollary 10.5.5 which was derived from Theorem 10.5.4. However, we have not used the full generality of Theorem 10.5.4. Moreover, the proof of Theorem 10.5.4 was rather long. Thus it seems to be worth to indicate an alternative proof of Theorem 10.5.6 which avoids Theorem 10.5.4. From the analytic domination theorem 10.4.4 and the technical Lemma 10.5.7 it follows as in the above proof of Theorem 10.5.6 that $\mathcal{D}^\omega_s(\pi_0)$ is dense in $\mathcal{H}(\pi)$ for some $s > 0$. (Indeed, since $\mathcal{D}^b \subseteq \mathcal{D}^{s\omega}_t(\pi_0(1 - \Delta))$ for all $t > 0$, $\mathcal{D}^b \subseteq \mathcal{D}^\omega_s(\pi_0)$ for some $s > 0$ by Theorem 10.4.4.) Therefore, the Campbell-Hausdorff formula can be used instead of Corollary 10.5.5; see Nelson [1], p. 601–602, Goodman [1], p. 60, or Warner [1], p. 289–299, for details. In this approach, Corollary 10.5.5 then follows from Theorem 10.5.6. (Indeed, assume that $\mathcal{D}^\omega(\pi)$ is dense in $\mathcal{H}(\pi)$. Since $\mathcal{D}^\omega(\pi) = \mathcal{D}^{s\omega}(\pi(1 - \Delta))$ by Theorem 10.4.4, $\mathcal{D}^{s\omega}(\pi(1 - \Delta))$ is dense in

$\mathcal{H}(\pi)$. From Proposition 10.3.8, $\pi(1-\Delta)$ and hence $\pi(\Delta)$ is essentially self-adjoint, so that the assumptions of Theorem 10.5.6 are fulfilled.)

The next theorem summarizes some of the results obtained so far in this chapter.

Theorem 10.5.8. *Let $\Delta = x_1^2 + \dots + x_d^2$ be the Nelson Laplacian relative to a basis $\{x_1, \dots, x_d\}$ for the Lie algebra $\mathfrak{g}$. For any $*$-representation π of the enveloping algebra $\mathcal{E}(\mathfrak{g})$, the following statements are equivalent:*

(i) *π is integrable.*

(ii) *$\mathcal{D}^\omega(\pi)$ is dense in $\mathcal{H}(\pi)$, and $\mathcal{D}(\pi) = \bigcap_{k=1}^{d} \mathcal{D}^\infty\big(\overline{\pi(x_k)}\big)$.*

(ii)′ *$\mathcal{D}^\omega(\pi)$ is dense in $\mathcal{H}(\pi)$, and $\mathcal{D}(\pi) = \mathcal{D}^\infty\big(\overline{\pi(\Delta)}\big)$.*

(ii)″ *$\mathcal{D}^\omega(\pi)$ is dense in $\mathcal{H}(\pi)$, and π is self-adjoint.*

(iii) *$\pi(\Delta)$ is essentially self-adjoint, and $\mathcal{D}(\pi) = \bigcap_{k=1}^{d} \mathcal{D}^\infty\big(\overline{\pi(x_k)}\big)$.*

(iii)′ *$\pi(\Delta)$ is essentially self-adjoint, and $\mathcal{D}(\pi) = \mathcal{D}^\infty\big(\overline{\pi(\Delta)}\big)$.*

(iii)″ *$\pi(\Delta)$ is essentially self-adjoint, and π is self-adjoint.*

Proof. The implications (i) → (ii), (i) → (ii)′ and (i) → (ii)″ follow from the Corollaries 10.4.12, 10.2.12, 10.2.4 and 10.2.3, respectively. Suppose that $\mathcal{D}^\omega(\pi)$ is dense in $\mathcal{H}(\pi)$. Since $\mathcal{D}^\omega(\pi) = \mathcal{D}^{s\omega}\big(\pi(1-\Delta)\big)$ by Theorem 10.4.4, this implies that $\pi(1-\Delta)$ and hence $\pi(\Delta)$ is essentially self-adjoint by Proposition 10.3.8. This proves that (ii) → (iii), (ii)′ → (iii)′ and (ii)″ → (iii)″. If $\pi(\Delta)$ is essentially self-adjoint, then

$$\mathcal{D}^\infty\big(\overline{\pi(1-\Delta)}\big) \subseteq \mathcal{D}_0 \subseteq \bigcap_{k=1}^{d} \mathcal{D}^\infty\big(\overline{\pi(x_k)}\big)$$

by Lemma 10.5.7. Since $\mathcal{D}^\infty\big(\overline{\pi(1-\Delta)}\big) = \mathcal{D}^\infty\big(\overline{\pi(\Delta)}\big)$, this shows that (iii) → (iii)′. If (iii)′ is satisfied, then

$$\mathcal{D}(\pi^*) \subseteq \bigcap_{n=1}^{\infty} \mathcal{D}\big((\pi(\Delta)^*)^n\big) = \bigcap_{n=1}^{\infty} \mathcal{D}\big((\overline{\pi(\Delta)})^n\big) = \mathcal{D}(\pi),$$

so that π is self-adjoint. Hence (iii)′ → (iii)″. If (iii)″ is true, then Theorem 10.5.6 shows that π is exponentiable. Since π is self-adjoint, π is integrable by Proposition 10.5.2. □

10.6. Decomposition of G-Integrable Representations as Direct Sums of Cyclic Representations

In this section, U denotes a unitary representation of the Lie group G in the Hilbert space $\mathcal{H}(U)$.

The two theorems proved in this section are analogous to those obtained in Section 9.2 for integrable representations of commutative $*$-algebras. We begin with an auxiliary result. It should be compared with Proposition 10.4.14.

Proposition 10.6.1. *Suppose that $\mathcal{D}$ is a linear subspace of $\mathcal{D}^\omega(\mathrm{d}U)$ which is invariant under $\mathrm{d}U(x)$ for all x in $\mathcal{E}(\mathfrak{g})$. Let $\hat{\mathcal{D}}$ denote the closure of $\mathcal{D}$ in $\mathcal{D}^\infty(U)\,[\mathfrak{t}_{\mathrm{d}U}]$. Then $U(g)\hat{\mathcal{D}} \subseteq \hat{\mathcal{D}}$ for all g in G_0.*

Proof. Let $\varphi \in \mathcal{D}$. Fix a basis $\{x_1, \ldots, x_d\}$ for $\mathfrak{g}$. Since $\mathcal{D} \subseteq \mathcal{D}^\omega(\mathrm{d}U)$, there exists an $s > 0$ such that $\varphi \in \mathcal{D}^\omega_s(\mathrm{d}U)$ relative to the basis $\{x_1, \ldots, x_d\}$. Suppose $k \in \{1, \ldots, d\}$ and $t \in \mathbb{R}$, $|t| < s$. Define $\varphi_{k,m}(t) = \sum\limits_{n=0}^{m} \frac{t^n}{n!}\, \mathrm{d}U(x_k)^n\, \varphi$, $m \in \mathbb{N}$. From Lemma 10.3.3, applied to the self-adjoint operator $-\mathrm{i}\overline{\mathrm{d}U(x_k)}$, we conclude that $\varphi_{k,m}(t)$ converges to $U(\exp tx_k)\, \varphi \equiv \exp it\big(-\mathrm{i}\overline{\mathrm{d}U(x_k)}\big)\, \varphi$ in $\mathcal{H}(U)$ as $m \to \infty$. For $l = \{1, \ldots, d\}$ and $r \in \mathbb{N}_0$, we have

$$\sum_{n=0}^{\infty} \left\| \mathrm{d}U(x_l)^r \left(\frac{t^n}{n!}\, \mathrm{d}U(x_k)^n\, \varphi \right) \right\| \leqq \sum_{n=0}^{\infty} \frac{|t|^n}{n!}\, \nu^{\mathrm{d}U}_{n+r}(\varphi). \tag{1}$$

The power series $\sum\limits_n \frac{z^n}{n!}\, \nu^{\mathrm{d}U}_{n+r}(\varphi)$ has the same radius of convergence as $\sum\limits_n \frac{z^n}{n!}\, \nu^{\mathrm{d}U}_n(\varphi)$. Since $\varphi \in \mathcal{D}^\omega_s(\mathrm{d}U)$, the latter converges for $z = s$. Therefore, since $|t| < s$, the series in (1) converges. Because the seminorms $\|\cdot\|_{\mathrm{d}U(x_l)^r}$, $l = 1, \ldots, d$ and $r \in \mathbb{N}_0$, generate the graph topology $\mathfrak{t}_{\mathrm{d}U}$, this shows that the series $\sum\limits_n \frac{t^n}{n!}\, \mathrm{d}U(x_k)^n\, \varphi$ converges absolutely in the locally convex space $\mathcal{D}^\infty(U)\,[\mathfrak{t}_{\mathrm{d}U}]$. Hence

$$U(\exp tx_k)\, \varphi = \lim_{m \to \infty} \varphi_{k,m}(t) \quad \text{in} \quad \mathcal{D}^\infty(U)\,[\mathfrak{t}_{\mathrm{d}U}].$$

Since $\mathcal{D}$ is invariant under $\mathrm{d}U(x)$ for x in $\mathcal{E}(\mathfrak{g})$, $\varphi_{k,m}(t) \in \mathcal{D}$ for $m \in \mathbb{N}$ and hence $U(\exp tx_k)\, \varphi \in \hat{\mathcal{D}}$. Thus $U(\exp tx_k)\, \mathcal{D} \subseteq \hat{\mathcal{D}}$. By Corollary 10.1.13, the operator $U(\exp tx_k)$ maps $\mathcal{D}^\infty(U)\,[\mathfrak{t}_{\mathrm{d}U}]$ continuously into itself. Therefore, the preceding implies that $U(\exp tx_k)\, \hat{\mathcal{D}} \subseteq \hat{\mathcal{D}}$ for all $t \in \mathbb{R}$, $|t| < s$, and $k = 1, \ldots, d$. Every element in G_0 is a finite product of such elements $\exp tx_k$. Thus $U(g)\, \hat{\mathcal{D}} \subseteq \hat{\mathcal{D}}$ for all g in G_0. □

Theorem 10.6.2. *The following three conditions are equivalent:*

(i) $\mathrm{d}U$ *is cyclic.*

(ii) $\mathrm{d}U$ *is weakly cyclic.*

(iii) *The von Neumann algebra $\mathrm{d}U\big(\mathcal{E}(\mathfrak{g})\big)''$ has a cyclic vector.*

Proof. (i) → (ii) is trivial. The proof of (ii) → (iii) is precisely the same as the proof of the corresponding assertion in Theorem 9.2.1. We prove that (iii) implies (i).

Suppose that φ_0 is a cyclic vector for the von Neumann algebra $\mathrm{d}U\big(\mathcal{E}(\mathfrak{g})\big)''$. We choose a basis $\{x_1, \ldots, x_d\}$ for the Lie algebra $\mathfrak{g}$. Let $\Delta = x_1^2 + \cdots + x_d^2$ be the corresponding Nelson Laplacian. By Corollary 10.2.5, the operator $A := \overline{\mathrm{d}U(1 - \Delta)}$ is self-adjoint. Define $\psi_0 := \exp(-A^2)\, \varphi_0$. Let $\hat{\mathcal{D}}_0$ be the closure of $\mathcal{D}_0 := \mathrm{d}U\big(\mathcal{E}(\mathfrak{g})\big)\, \psi_0$ in $\mathcal{D}^\infty(U)\,[\mathfrak{t}_{\mathrm{d}U}]$, and let $\mathcal{H}_0$ be the closure of $\mathcal{D}_0$ in $\mathcal{H}(U)$. Since, of course, $\psi_0 \in \mathcal{D}(\exp A^{1/2})$, Theorem 10.4.10 shows that $\psi_0 \in \mathcal{D}^\omega(\mathrm{d}U)$. From Lemma 10.4.3, $\mathcal{D}_0 = \mathrm{d}U\big(\mathcal{E}(\mathfrak{g})\big)\, \psi_0 \in \mathcal{D}^\omega(\mathrm{d}U)$. Therefore, by Proposition 10.6.1, we have $U(g)\, \hat{\mathcal{D}}_0 \subseteq \hat{\mathcal{D}}_0$ for all g in G_0. In particular, this implies that $U(g)\, \mathcal{H}_0 \subseteq \mathcal{H}_0$ for g in G_0. On the other hand, since the sequence $\left(\sum\limits_{n=0}^{k} \frac{t^{2n}}{n!} \exp(-t^2);\, k \in \mathbb{N} \right)$ converges monotonically to 1 for all real t, we conclude from the spectral theorem that φ_0 is in the closure of the set $\mathcal{D}_1 := \{p(A)\exp(-A^2)\varphi_0 : p \in \mathbb{C}[x]\}$ in $\mathcal{H}(U)$. But $\mathcal{D}_1$ is contained in $\mathcal{D}_0$, so that $\varphi_0 \in \mathcal{H}_0$. Since $\mathcal{H}_0$ is invariant under $U(g)$, $g \in G_0$, we have $U(G_0)\,\mathcal{H}_0 \subseteq \mathcal{H}_0$. Hence $U(G_0)''\,\mathcal{H}_0 \subseteq \mathcal{H}_0$. By Corollary 10.2.18, $\mathrm{d}U(\mathcal{E}(\mathfrak{g}))' = U(G_0)'$; so $\mathrm{d}U(\mathcal{E}(\mathfrak{g}))''\,\mathcal{H}_0 \subseteq \mathcal{H}_0$. Since the cyclic vector φ_0 for $\mathrm{d}U(\mathcal{E}(\mathfrak{g}))''$ is in $\mathcal{H}_0$, the latter implies that $\mathcal{H}_0 = \mathcal{H}(U)$. This means that $\hat{\mathcal{D}}_0$ is dense in $\mathcal{H}(U)$. Being invariant under $U(g)$ for all $g \in G_0$ as noted above, $\hat{\mathcal{D}}_0$ is dense in $\mathcal{D}^\infty(U)\,[t_{\mathrm{d}U}]$ by Theorem 10.1.14. Since, by definition, $\hat{\mathcal{D}}_0$ is $t_{\mathrm{d}U}$-closed in $\mathcal{D}^\infty(U)$, we get $\hat{\mathcal{D}}_0 = \mathcal{D}^\infty(U)$ which proves that φ_0 is a cyclic vector for the representation $\mathrm{d}U$. □

Theorem 10.6.3. *Suppose that the Lie group G is connected. Then each G-integrable representation of $\mathcal{E}(\mathfrak{g})$ is a direct sum of cyclic G-integrable representations.*

Proof. Let π be a G-integrable representation of $\mathcal{E}(\mathfrak{g})$. Arguing precisely in the same way as in the proof of Theorem 9.2.3 with $\mathcal{E}(\mathfrak{g})$ in place of A, we conclude that there is a family $\{\pi_i : i \in I\}$ of self-adjoint representations π_i of $\mathcal{E}(\mathfrak{g})$ such that $\pi = \sum_{i \in I} \oplus\, \pi_i$ and such that for each $i \in I$ the von Neumann algebra $\pi_i(\mathcal{E}(\mathfrak{g}))''$ has a cyclic vector. By Proposition 10.2.19, π_i is G-integrable for $i \in I$, that is, there exists a unitary representation U_i of G with $\pi_i = \mathrm{d}U_i$. Applying Theorem 10.6.2 to U_i, it follows that $\pi_i = \mathrm{d}U_i$ is cyclic. □

Notes

10.1. A corner-stone for the theory of infinitesimal representations was the paper [1] of Gårding who showed that the operators $\partial U(x)$, $x \in \mathfrak{g}$, have a common dense invariant domain of definition, the so-called Gårding subspace. Proposition 10.1.2 is the Hilbert space version of a result due to Grothendieck [2], p. 134, which is valid for general quasi-complete locally convex spaces. Theorem 10.1.9 is from Goodman [1], and Proposition 10.1.11 and Theorem 10.1.14 are due to Poulsen [1]. Most of the results in this section have generalizations to Banach space representations of the Lie group G; cf. Poulsen [1].

10.2. The importance of elliptic elements in the enveloping algebra was pointed out by Nelson/Stinespring [1] who proved the fundamental result stated as Corollary 10.2.8. We have given an alternative proof based on Lemma 7.1.5. Some of the applications (e.g., the essential self-adjointness of $\mathrm{i}\,\mathrm{d}U(x)$, $x \in \mathfrak{g}$) were much earlier known by the pioneering work of Segal [1], [2]. Theorems 10.2.6 and 10.2.16 and Corollaries 10.2.4 and 10.2.18 are due to Poulsen [1]. Proposition 10.2.19 is in Powers [2].

10.3. Analytic vectors were introduced by Harish-Chandra [1] who called them well-behaved vectors. Proposition 10.3.4 and Corollary 10.3.5 are now classical results obtained by Nelson [1]. The concept of semi-analytic vectors is due to Simon [1] who proved Proposition 10.3.8. The notion of analytic domination of operator families and Proposition 10.3.10 are from Nelson [1]. Proposition 10.3.11 can be found in Goodman [2].

10.4. Corollary 10.4.12 is the Hilbert space version of a general result which states that any (strongly continuous) representation of the Lie group G in Banach space has a dense set of analytic vectors. This theorem was proved in special cases by Harish-Chandra [1], Cartier/Dixmier [1], and in full generality by Nelson [1]; see also Gårding [2]. In case of a unitary representation U of G in Hilbert space Nelson showed that every analytic vector for the (self-adjoint) operator $\mathrm{d}U(\Delta)$ is an analytic vector for $\mathrm{d}U$ and so for U. The more precise description of the space $\mathcal{D}^\omega(\mathrm{d}U)$ given in Theorem 10.4.10 is due to Goodman [2], but the proof relies heavily on the fundamental work of Nelson [1]. The analytic domination theorem 10.4.4 for general $*$-representations of enveloping algebras appears here for the first time. Proposition 10.4.14 is a slight generalization of a result in HARISH-CHANDRA [1], and Proposition 10.4.15 was proved by FLATO/SIMON [1].

10.5. The second basic exponentiation theorem 10.5.6 was discovered by NELSON [1]. The first exponentiation theorem 10.5.4 is due to FLATO/SIMON/SNELLMAN/STERNHEIMER [1] combined with a result from FLATO/SIMON [1]. As shown by SIMON [1], Theorem 10.5.4 remains valid if we replace the basis of $\mathfrak{g}$ by a set of Lie generators for the Lie algebra $\mathfrak{g}$.

10.6. The results in this section are due to the author.

Additional References:

BARUT/RACZKA [1], JORGENSEN/MOORE [1], KIRILLOV [1], WARNER [1], JORGENSEN [3], KNAPP [1].
10.1. and **10.2.** BRUHAT [1], GOODMAN [3], SEGAL [3], [4].
10.3. CHERNOFF [1], NUSSBAUM [4].
10.5. FRÖHLICH [1], JORGENSEN [1], [2].

11. n-Positivity and Complete Positivity of $*$-Representations

This chapter is concerned with $*$-representations of a $*$-algebra A (or more generally with linear maps of A into the space of sesquilinear forms on a vector space) which map a distinguished wedge of "positive" matrices over A into the positive matrices of operators (or sesquilinear forms). The general study of such order properties leads to applications which are all formulated according to the following pattern: The $*$-representation admits an extension to a "well-behaved" $*$-representation in a larger Hilbert space if and only if it satisfies a certain additional positivity condition of the above form. In order to explain this idea by a simple pertinent example, let ω be a positive linear functional on the polynomial algebra $\mathbb{C}[\mathsf{x}_1, \mathsf{x}_2]$. Then ω is non-negative on non-negative polynomials if and only if it can be represented by a positive measure (see Example 2.6.11), or equivalently, if π_ω has an integrable extension.

Section 11.1 deals with n-positive and completely positive maps of general matrix ordered vector spaces. An extension theorem for completely positive mappings is proved in this rather general setting. In Section 11.2 we specialize to $*$-algebras, and we prove a generalized Stinespring dilation theorem for completely positive maps. By combining these two results, we obtain an extension theorem for $*$-representations from which all three applications are derived.

Sections 11.3, 11.4 and 11.5 are devoted to applications of the general theory. In Section 11.3 we characterize the $*$-representations of a commutative $*$-algebra with unit which have an integrable extension in a larger Hilbert space as those which are completely positive with respect to a certain cone of matrices. Section 11.4 contains a similar result for enveloping algebras. In Section 11.6 we prove the existence of a $*$-representation of the polynomial algebra $\mathbb{C}[\mathsf{x}_1, \mathsf{x}_2]$ which is 1-positive, but not 2-positive with respect to the corresponding cones of matrices. This shows that in the unbounded case matrix ordering is indespensable even for $*$-representations.

11.1. n-Positive and Completely Positive Maps of Matrix Ordered Spaces

We begin with some notation which will be frequently used in this chapter.

Suppose that E is a $*$-vector space with involution $x \to x^+$. We let $M_{n,m}(E)$, $n, m \in \mathbb{N}$, denote the vector space of all $n \times m$ matrices with entries in E. Set $M_n(E) := M_{n,n}(E)$. By a *finite matrix* over E we mean a matrix $[x_{kl}]_{k,l\in\mathbb{N}}$ over E whose entries x_{kl} are all zero but a finite number. Let $M(E)$ denote the vector space of all finite matrices over E. Matrices in $M_{n,m}(E)$ or in $M(E)$ will be often written as $[x_{kl}]$. We equip $M(E)$ with the involution defined by $[x_{kl}]^+ := [x_{lk}^+]$, so $M(E)$ becomes a $*$-vector space. If $\Lambda \equiv [\lambda_{kl}] \in M(\mathbb{C})$ and $X \equiv [x_{kl}] \in M(E)$, then $X\Lambda$ denotes the matrix in $M(E)$ with (k, l)-th entry $\sum_j \lambda_{jl} x_{kj}$. (Note that the sum is finite and $X\Lambda \in M(E)$, since Λ and X are finite matrices.) The product ΛX is defined in the usual way. In order to simplify the notation we shall use the following conventions. We identify a matrix $[x_{kl}]_{k=1,\ldots,n;l=1,\ldots,m}$ of $M_{n,m}(E)$ with the matrix $[x_{kl}]_{k,l\in\mathbb{N}}$ of $M(E)$ obtained by setting $x_{kl} = 0$ when $k > n$ or $l > m$. Further, we identify an element x of E with the matrix of $M(E)$ which has x in the $(1, 1)$-th position and zeros elsewhere. In this way, each space $M_{n,m}(E)$, $n, m \in \mathbb{N}$, and the space E itself are linear subspaces of $M(E)$. Moreover, we then have $E \equiv M_1(E)$ and $M(E) = \bigcup_{n\in\mathbb{N}} M_n(E)$. Let Φ be a mapping of E into another $*$-vector space F. For $n \in \mathbb{N}$, let $\Phi_{(n)}$ denote the mapping of $M_n(E)$ into $M_n(F)$ defined by $\Phi_{(n)}([x_{kl}]) := [\Phi(x_{kl})]$. Likewise, let $\Phi_{(\infty)}$ be the mapping of $M(E)$ into $M(F)$ defined by the same formula.

Definition 11.1.1. Suppose E is a $*$-vector space. An *admissible wedge* in $M(E)$ is a wedge K in $M(E)_h$ (i.e., a subset K of $M(E)_h$ for which $\lambda_1 X_1 + \lambda_2 X_2 \in K$ when $X_1, X_2 \in K$ and $\lambda_1 \geqq 0$, $\lambda_2 \geqq 0$) such that $\Lambda^+ X \Lambda \in K$ for all $X \in K$ and $\Lambda \in M(\mathbb{C})$. A *matrix ordered space* E is a $*$-vector space E together with an admissible wedge in $M(E)$.

Suppose E is a matrix ordered space. We denote the corresponding admissible wedge by $K(E)$. For each $n \in \mathbb{N}$, $K_n(E) := K(E) \cap M_n(E)$ is a wedge in $M_n(E)_h$. Moreover, we obviously have $K(E) = \bigcup_{n\in\mathbb{N}} K_n(E)$ and

$$\Lambda^+ X \Lambda \in K_m(E) \quad \text{for all} \quad X \in K_n(E),\ \Lambda \in M_{n,m}(\mathbb{C}) \quad \text{and} \quad m, n \in \mathbb{N}. \tag{1}$$

Conversely, if E is a $*$-vector space and $(K_n : n \in \mathbb{N})$ is a sequence of wedges K_n in $M_n(E)_h$ satisfying (1), then $K := \bigcup_{n\in\mathbb{N}} K_n$ is an admissible wedge in $M(E)$.

Let E_0 be a $*$-invariant linear subspace of a matrix ordered space E. If not stated explicitly otherwise, we also consider E_0 as a matrix ordered vector space by letting $K(E_0) := K(E) \cap M(E_0)$ be the admissible wedge in $M(E_0)$.

We now describe our standard example of a matrix ordered space.

Example 11.1.2. Suppose that $\mathfrak{X}$ is a vector space. Recall that the vector space $E := B(\mathfrak{X})$ of all sesquilinear forms on $\mathfrak{X} \times \mathfrak{X}$ is a $*$-vector space with involution $\mathfrak{x} \to \mathfrak{x}^+$, where $\mathfrak{x}^+$ is defined by $\mathfrak{x}^+(\varphi, \psi) := \overline{\mathfrak{x}(\psi, \varphi)}$, $\varphi, \psi \in \mathfrak{X}$. For $n \in \mathbb{N}$, let $K_n \equiv M_n(B(\mathfrak{X}))_+$ be the wedge of all matrices $[\mathfrak{x}_{kl}]$ in $M_n(B(\mathfrak{X}))_h$ which satisfy

$$\sum_{k,l=1}^{n} \mathfrak{x}_{kl}(\varphi_l, \varphi_k) \geqq 0 \tag{2}$$

for arbitrary vectors $\varphi_1, \ldots, \varphi_n$ in $\mathfrak{X}$.

We show that the compatibility condition (1) is fulfilled. Let $X \equiv [\mathfrak{x}_{kl}] \in K_n$, $\Lambda \equiv [\lambda_{rs}] \in M_{n,m}(\mathbb{C})$ and $\varphi_1, \ldots, \varphi_m \in \mathfrak{X}$ be given. Then the matrix $[\mathfrak{y}_{kl}] := \Lambda^+ X \Lambda$ has the entries $\mathfrak{y}_{kl} = \sum_{r,s=1}^{n} \overline{\lambda_{rk}} \lambda_{sl} \mathfrak{x}_{rs}$, hence

$$\sum_{k,l=1}^{m} \mathfrak{y}_{kl}(\varphi_l, \varphi_k) = \sum_{r,s=1}^{n} \mathfrak{x}_{rs}\left(\sum_{l=1}^{m} \lambda_{sl}\varphi_l, \sum_{k=1}^{m} \lambda_{rk}\varphi_k\right).$$

The last expression is non-negative by (2), since $X \in K_n$. Thus (1) is proved, so

$K \equiv M(B(\mathfrak{X}))_+ := \bigcup_{n\in\mathbb{N}} K_n$ is an admissible wedge in $M(B(\mathfrak{X}))$. Endowed with this wedge K, the $*$-vector space $E \equiv B(\mathfrak{X})$ becomes a matrix ordered space. ○

Definition 11.1.3. Suppose E and F are matrix ordered spaces and Φ is a linear mapping of E into F. Let $n \in \mathbb{N}$. We say that Φ is *n-positive* if $\Phi_{(n)}$ maps $K_n(E)$ into $K_n(F)$. The map Φ is called *completely positive* if $\Phi_{(\infty)}$ maps $K(E)$ into $K(F)$ or equivalently if Φ is n-positive for all $n \in \mathbb{N}$.

Definition 11.1.4. Suppose E is a matrix ordered space and $\mathfrak{X}$ is a vector space. Let $n \in \mathbb{N}$. An *n-positive* [resp. *completely positive*] *mapping* of E on $\mathfrak{X}$ is an n-positive [resp. completely positive] mapping of the matrix ordered vector space E into the matrix ordered vector space $B(\mathfrak{X})$ of Example 11.1.2.

That is, an n-positive map of E on $\mathfrak{X}$ is a linear map Φ of E into $B(\mathfrak{X})$ which has the property that

$$\sum_{k,l=1}^{n} \Phi(\mathfrak{x}_{kl})\,(\varphi_l, \varphi_k) \geqq 0$$

for all matrices $[\mathfrak{x}_{kl}] \in K_n(E)$ and for all vectors $\varphi_1, \ldots, \varphi_n \in \mathfrak{X}$. If the latter is true for all $n \in \mathbb{N}$, then Φ is completely positive.

Though the concept of complete positivity as defined above is rather general, it allows to prove an extension theorem for completely positive mappings which generalizes Arveson's theorem on extensions of completely positive maps on C^*-algebras.

Theorem 11.1.5. *Suppose E is a matrix ordered vector space. Let E_0 be a $*$-invariant linear subspace of E which is cofinal in E with respect to the wedge $K_1(E)$ (i.e., for every $x \in E_{\rm h}$ there is a $y \in (E_0)_{\rm h}$ such that $y \in K_1(E)$ and $y - x \in K_1(E)$). Suppose that Φ_0 is a completely positive mapping of the matrix ordered space E_0 on a vector space $\mathfrak{X}$. Then there exists a completely positive linear mapping Φ of E on $\mathfrak{X}$ such that $\Phi \restriction E_0 = \Phi_0$.*

The proof of this theorem requires the following lemma.

Lemma 11.1.6. *If E_0 and E are as in Theorem 11.1.5, then $M(E_0)$ is cofinal in $M(E)$ with respect to the wedge $K(E)$.*

Proof. We have to show that for each $X \equiv [x_{kl}] \in M(E)_{\rm h}$ there is a $Y \in M(E_0)_{\rm h}$ such that $Y \in K(E)$ and $Y - X \in K(E)$. Every matrix in $M(E)_{\rm h}$ is a finite sum of matrices in $M(E)_{\rm h}$ which have only vanishing entries except possibly at (r, r), (r, s), (s, r) and (s, s) for some $r, s \in \mathbb{N}$. Because $K(E)$ is a wedge it suffices to prove the assertion for these matrices. For notational simplicity, let $r = 1$ and $s = 2$. We write $x_{12} = x_1 + {\rm i}x_2$ with $x_1, x_2 \in E_{\rm h}$. Take an $n \in \mathbb{N}$, $n \geqq 2$. Let $\Lambda \equiv [\lambda_{kl}]$ be the matrix in $M(\mathbb{C})$ with entries $\lambda_{11} = \lambda_{12} = \lambda_{21} = \lambda_{n,1} = \lambda_{n+1,2} = 1$, $\lambda_{22} = -{\rm i}$ and $\lambda_{kl} = 0$ otherwise. For $m \in \mathbb{N}$, set $\Lambda_m := [\delta_{k1}\delta_{lm}]_{k,l\in\mathbb{N}}$. Putting $x_n := x_{11} - x_1 - x_2$ and $x_{n+1} := x_{22} - x_1 - x_2$, we then have

$$X = \Lambda^+(\Lambda_1^+ x_1 \Lambda_1 + \Lambda_2^+ x_2 \Lambda_2 + \Lambda_n^+ x_n \Lambda_n + \Lambda_{n+1}^+ x_{n+1} \Lambda_{n+1})\,\Lambda,$$

where we used the identification of $x \in E$ with the matrix $[x\delta_{1k}\delta_{1l}] \in M(E)$. Since E_0 is cofinal in E w.r.t. $K_1(E)$, there are elements $y_m \in (E_0)_{\rm h}$ such that $y_m \in K_1(E)$ and $y_m - x_m \in K_1(E)$ for $m = 1, 2, n, n+1$. Then the matrix

$$Y := \Lambda^+(\Lambda_1^+ y_1 \Lambda_1 + \Lambda_2^+ y_2 \Lambda_2 + \Lambda_n^+ y_n \Lambda_n + \Lambda_{n+1}^+ y_{n+1} \Lambda_{n+1})\,\Lambda$$

is in $K(E)$ (because $K(E)$ is an admissible wedge in $M(E)$) and it has the desired properties. □

Proof of Theorem 11.1.5. Without loss of generality we assume that $\Phi_0 \not\equiv 0$. Because of Zorn's lemma it is sufficient to prove the theorem in case where E_0 has codimension 1 in E. We shall assume this and take an element $x_0 = x_0^+ \in E$ which is not in E_0. Then E is the linear span of E_0 and x_0 and it suffices to find a form $\Phi(x_0) \in B(\mathfrak{X})$ such that

$$\sum_{k,l=1}^{n} \left(\alpha_{kl}\Phi(x_0)(\varphi_l, \varphi_k) + \Phi_0(x_{kl})(\varphi_l, \varphi_k)\right) \geqq 0 \tag{3}$$

for all vectors $\varphi_1, \ldots, \varphi_n \in \mathfrak{X}$, matrices $[\alpha_{kl}x_0 + x_{kl}] \in \boldsymbol{K}_n(E)$ with $\alpha_{kl} \in \mathbb{C}$ and $x_{kl} \in E_0$ for $k, l = 1, \ldots, n$ and all $n \in \mathbb{N}$. Note that $[\alpha_{kl}x_0 + x_{kl}] \in \boldsymbol{K}_n(E)$ implies that $\alpha_{kl} = \overline{\alpha_{lk}}$ for $k, l = 1, \ldots, n$, since $x_0 = x_0^+ \notin E_0$.

The existence of $\Phi(x_0)$ with these properties will be derived from the separation theorem for convex sets. To apply this theorem, we still need some preliminaries. The vector space $\mathfrak{X} \otimes \mathfrak{X}^-$ becomes a $*$-vector space by the definition $(\varphi \otimes \psi)^+ := \psi \otimes \varphi$, $\varphi \in \mathfrak{X}$ and $\psi \in \mathfrak{X}^-$. Its hermitian part $(\mathfrak{X} \otimes \mathfrak{X}^-)_{\mathrm{h}}$ is a real vector space. Let C denote the convex hull of all elements

$$(y, \lambda) \equiv \left(\sum_{k,l=1}^{n} \alpha_{kl}\varphi_l \otimes \varphi_k, \sum_{k,l=1}^{n} \Phi_0(x_{kl})(\varphi_l, \varphi_k)\right) \tag{4}$$

in the real vector space $G := (\mathfrak{X} \otimes \mathfrak{X}^-)_{\mathrm{h}} \oplus \mathbb{R}$, where $\varphi_1, \ldots, \varphi_n \in \mathfrak{X}$, $[\alpha_{kl}x_0 + x_{kl}] \in \boldsymbol{K}_n(E)$ with $\alpha_{kl} \in \mathbb{C}$ and $x_{kl} \in E_0$, $n \in \mathbb{N}$. (Note that the above assumptions imply indeed that $y \in (\mathfrak{X} \otimes \mathfrak{X}^-)_{\mathrm{h}}$ and $\lambda \in \mathbb{R}$.)

We first show that $(0, 1)$ is an internal point of C. Since $\Phi_0 \not\equiv 0$ and E_0 is cofinal in E w.r.t. $\boldsymbol{K}_1(E)$, there are $x \in \boldsymbol{K}_1(E) \cap E_0$ and $\varphi \in \mathfrak{X}$ such that $\Phi_0(x)(\varphi, \varphi) > 0$. This implies that $(0, \gamma) \in C$ for $\gamma > 0$. Now let $(y, \lambda) \in G$. Then $y \in (\mathfrak{X} \otimes \mathfrak{X}^-)_{\mathrm{h}}$ is of the form $y = \sum_{k,l=1}^{n} \alpha_{kl}\varphi_l \otimes \varphi_k$ for some vectors $\varphi_1, \ldots, \varphi_n \in \mathfrak{X}$, a hermitian matrix $[\alpha_{kl}] \in \boldsymbol{M}_n(\mathbb{C})$ and $n \in \mathbb{N}$. By Lemma 11.1.6 there exists a matrix $[x_{kl}] \in \boldsymbol{M}_n(E_0)$ such that $[\alpha_{kl}x_0 + x_{kl}] \in \boldsymbol{K}_n(E)$. Set $\lambda_1 := \sum_{k,l=1}^{n} \Phi_0(x_{kl})(\varphi_l, \varphi_k)$ and $\gamma_1 := \frac{1}{2}(|\lambda - \lambda_1| + 1)^{-1}$. Suppose $0 < \gamma < \gamma_1$. Then $\delta := \left(\gamma(\lambda - \lambda_1) + 1\right)(1 - \gamma)^{-1} > 0$ and $\gamma(y, \lambda) + (0, 1) = \gamma(y, \lambda_1) + (1 - \gamma)(0, \delta)$. From $(y, \lambda_1) \in C$ and $(0, \delta) \in C$ it follows that $\gamma(y, \lambda) + (0, 1) \in C$ which proves that $(0, 1)$ is an internal point of C.

Next we prove that $(0, 0)$ is not an internal point of C. Assume the contrary. Then we have $(0, -\varepsilon) \in C$ for some $\varepsilon > 0$. Hence there are vectors $\varphi_{kj} \in \mathfrak{X}$, $k = 1, \ldots, n_j$ and $j = 1, \ldots, m$, hermitian matrices $[\gamma_{kl}^{(j)}] \in \boldsymbol{M}_{n_j}(\mathbb{C})$, $[x_{kl}^{(j)}] \in \boldsymbol{M}_{n_j}(E_0)$ and numbers $\lambda_j \in [0, 1]$, $j = 1, \ldots, m$, such that

$$[\gamma_{kl}^{(j)}x_0 + x_{kl}^{(j)}] \in \boldsymbol{K}_{n_j}(E), \tag{5}$$

$$\sum_{j=1}^{m} \lambda_j \sum_{k,l=1}^{n_j} \gamma_{kl}^{(j)}\varphi_{lj} \otimes \varphi_{kj} = 0 \tag{6}$$

and

$$\sum_{j=1}^{m} \lambda_j \sum_{k,l=1}^{n_j} \Phi_0(x_{kl}^{(j)})(\varphi_{lj}, \varphi_{kj}) = -\varepsilon. \tag{7}$$

Put $\gamma_{(kj)(lr)} := \lambda_j\delta_{jr}\gamma_{kl}^{(j)}$ and $x_{(kj)(lr)} := \lambda_j\delta_{jr}x_{kl}^{(j)}$. After combining pairs of indices (kj), (lr) to single indices, say $\mathfrak{k}, \mathfrak{l}$, we obtain hermitian matrices $\Gamma \equiv [\gamma_{\mathfrak{k}\mathfrak{l}}] \in \boldsymbol{M}_n(\mathbb{C})$ and $\tilde{X} \equiv [x_{\mathfrak{k}\mathfrak{l}}] \in \boldsymbol{M}_n(E_0)$ for a certain $n \in \mathbb{N}$. Let $\psi_1, \ldots, \psi_d$ be a basis of the linear span of all vectors $\varphi_{\mathfrak{k}}$, $\mathfrak{k} = 1, \ldots, n$. Then there is a matrix $\Lambda \equiv [\lambda_{\mathfrak{k}\mathfrak{r}}] \in \boldsymbol{M}_{n,d}(\mathbb{C})$ such that $\varphi_{\mathfrak{k}} = \sum_{\mathfrak{r}=1}^{d} \lambda_{\mathfrak{k}\mathfrak{r}}\psi_{\mathfrak{r}}$ for $\mathfrak{k} = 1, \ldots, n$. From (6) we obtain $\sum_{\mathfrak{k},\mathfrak{l}=1}^{n} \overline{\lambda_{\mathfrak{l}\mathfrak{r}}}\gamma_{\mathfrak{k}\mathfrak{l}}\lambda_{\mathfrak{k}\mathfrak{s}} = 0$ for $\mathfrak{r}, \mathfrak{s} = 1, \ldots, d$. Hence $\Lambda^+\Gamma\Lambda = 0$. By (3), $[\gamma_{\mathfrak{k}\mathfrak{l}}x_0 + x_{\mathfrak{k}\mathfrak{l}}] \in \boldsymbol{K}_n(E)$. Therefore, since $\boldsymbol{K}(E)$ is an admissible wedge in $\boldsymbol{M}(E)$ and $\Lambda^+\Gamma\Lambda = 0$, $[z_{\mathfrak{k}\mathfrak{l}}] := \Lambda^+\tilde{X}\Lambda = \Lambda^+[\gamma_{\mathfrak{k}\mathfrak{l}}x_0 + x_{\mathfrak{k}\mathfrak{l}}]\,\Lambda \in \boldsymbol{K}_d(E) \cap \boldsymbol{M}_d(E_0) \equiv \boldsymbol{K}_d(E_0)$. But

$$\sum_{\mathfrak{k},\mathfrak{l}=1}^{d} \Phi_0(z_{\mathfrak{k}\mathfrak{l}})\,(\psi_{\mathfrak{l}}, \psi_{\mathfrak{k}}) = \sum_{\mathfrak{k},\mathfrak{l}=1}^{d} \sum_{\mathfrak{r},\mathfrak{s}=1}^{n} \overline{\lambda_{\mathfrak{r}\mathfrak{k}}}\lambda_{\mathfrak{s}\mathfrak{l}}\Phi_0(x_{\mathfrak{r}\mathfrak{s}})\,(\psi_{\mathfrak{l}}, \psi_{\mathfrak{k}}) = \sum_{\mathfrak{r},\mathfrak{s}=1}^{n} \Phi_0(x_{\mathfrak{r}\mathfrak{s}})\,(\varphi_{\mathfrak{s}}, \varphi_{\mathfrak{r}}) = -\varepsilon$$

by (7), and this contradicts the complete positivity of the mapping Φ_0. Thus $(0, 0)$ is not an internal point of C.

Since $(0, 0)$ is not an internal point of the convex set C in G and C admits an internal point, we conclude from the separation theorem for convex sets (see e.g. KÖTHE [1], § 17, 1., (3)) that there exists a real linear functional $F \not\equiv 0$ on the real vector space G such that $0 = F\big((0, 0)\big) \leqq \inf\,\{F(x) : x \in C\}$. By $F \not\equiv 0$, there is a $z \in G$ such that $F(z) \neq 0$. Since $(0, 1)$ is an internal point of C, there exists $\alpha_1 > 0$ such that $\alpha z + (0, 1) \in C$ whenever $\alpha \in (-\alpha_1, \alpha_1)$. Thus $\alpha F(z) + F\big((0, 1)\big) \geqq 0$ if $\alpha \in (-\alpha_1, \alpha_1)$. Because $F(z) \neq 0$, this gives $F\big((0, 1)\big) > 0$. Upon multiplying F by some positive constant, we can assume that $F\big((0, 1)\big) = 1$. Set $f_{\mathrm{h}}(x) := F\big((x, 0)\big)$ for $x \in (\mathfrak{X} \otimes \mathfrak{X}^-)_{\mathrm{h}}$. We extend f_{h} to a complex linear functional f on $\mathfrak{X} \otimes \mathfrak{X}^-$ (by Lemma 1.3.1) and define the sesquilinear form $\Phi(x_0) \in B(\mathfrak{X})$ by $\Phi(x_0)\,(\varphi, \psi) := f(\varphi \otimes \psi)$, $\varphi, \psi \in \mathfrak{X}$. If (y, λ) is as in (4), then $f(y) + \lambda = F\big((y, 0)\big) + \lambda F\big((0, 1)\big) = F\big((y, \lambda)\big) \geqq 0$. Therefore, by the definition of $\Phi(x_0)$, (3) is satisfied. □

The following proposition shows that complete positivity and n-positivity of mappings are the same if the vector space $\mathfrak{X}$ has finite dimension n.

Proposition 11.1.7. *Suppose that $\mathfrak{X}$ is a finite dimensional vector space and $\{\psi_1, \ldots, \psi_n\}$ is a basis of $\mathfrak{X}$. Let Φ be a linear mapping of a matrix ordered space E into $B(\mathfrak{X})$. Define a linear functional f on $\boldsymbol{M}_n(E)$ by $f([x_{kl}]) := \sum_{k,l=1}^{n} \Phi(x_{kl})\,(\psi_l, \psi_k)$, $[x_{kl}] \in \boldsymbol{M}_n(E)$. Then the following three conditions are equivalent:*

(i) *Φ is completely positive.*

(ii) *Φ is n-positive.*

(iii) *f is non-negative on $\boldsymbol{K}_n(E)$.*

Proof. The implications (i) → (ii) and (ii) → (iii) are trivial. To prove (iii) → (i), we use the same calculation as in Example 11.1.2, but in reversed order. Suppose $m \in \mathbb{N}$, $X = [x_{kl}] \in \boldsymbol{K}_m(E)$ and $\varphi_1, \ldots, \varphi_m \in \mathfrak{X}$. Then there are complex numbers λ_{rs}, $r = 1, \ldots, m$ and $s = 1, \ldots, n$, such that $\varphi_r = \sum_{k=1}^{n} \lambda_{rk}\psi_k$. Set $\Lambda := [\lambda_{rs}]$ and $y_{kl} := \sum_{r,s=1}^{m} \overline{\lambda_{rk}}\lambda_{sl}x_{rs}$ for $k, l = 1, \ldots, n$. Then we have $[y_{kl}] = \Lambda^+X\Lambda \in \boldsymbol{K}_n(E)$ by (1). Therefore, by (iii), we have

$$\sum_{r,s=1}^{m} \Phi(x_{rs})\,(\varphi_s, \varphi_r) = \sum_{k,l=1}^{n} \Phi(y_{kl})\,(\psi_l, \psi_k) = f([y_{kl}]) \geqq 0.$$

This proves that Φ is m-positive for each $m \in \mathbb{N}$. □

Remark 1. In the case where the vector space $\mathfrak{X}$ is finite dimensional, Theorem 11.1.5 can be easily derived from Lemma 1.3.2. Indeed, let f_0 be the linear functional defined in Proposition 11.1.7 in case of Φ_0 and E_0. By Lemma 1.3.2, f_0 can be extended to a linear functional f on $\boldsymbol{M}_n(E)$ which is non-negative on $\boldsymbol{K}_n(E)$. Define Φ by $\Phi(x)\,(\psi_s, \psi_r) := f([x\delta_{kr}\delta_{ls}])$, $x \in E$ and $r, s = 1, \ldots, n$. Then Φ is completely positive by Proposition 11.1.7, and $\Phi_0 \subseteq \Phi$.

11.2. n-Positive and Completely Positive Maps of $*$-Algebras

In this section A is a $*$-algebra with unit element.

With the usual algebraic operations of (finite) matrices over A, $M(\mathsf{A})$ is a $*$-algebra. Recall that $\mathcal{P}(M(\mathsf{A}))$ is the wedge in $M(\mathsf{A})_{\mathrm{h}}$ of all finite sums of elements X^+X, where $X \in M(\mathsf{A})$. By carrying out the matrix multiplication we see that $\mathcal{P}(M(\mathsf{A}))$ coincides with the set of all finite sums of matrices $[a_{kl}]$ in $M(\mathsf{A})$ of the form $a_{kl} = a_k^+ a_l$ for $k, l \in \mathbb{N}$, where $(a_k : k \in \mathbb{N})$ is a finite sequence in A. From the identity $\Lambda^+(X^+X)\Lambda = (X\Lambda)^+(X\Lambda)$ for $X \in M(\mathsf{A})$ and $\Lambda \in M(\mathbb{C})$ we conclude that $\mathcal{P}(M(\mathsf{A}))$ is an admissible wedge in $M(\mathsf{A})$. With this wedge, A becomes a matrix ordered space in the sense of Definition 11.1.1. By an *n-positive* or a *completely positive* mapping of A on a vector space $\mathfrak{X}$ we mean the corresponding notions for this matrix ordered space according to Definition 11.1.4. From the description of the wedge $\mathcal{P}(M(\mathsf{A}))$ given above it follows that a linear map Φ of A into $B(\mathfrak{X})$ is n-positive if and only if

$$\sum_{k,l=1}^{n} \Phi(a_k^+ a_l)\,(\varphi_l, \varphi_k) \geqq 0 \tag{1}$$

for arbitrary elements $a_1, \ldots, a_n \in \mathsf{A}$ and vectors $\varphi_1, \ldots, \varphi_n \in \mathfrak{X}$. If this holds for all $n \in \mathbb{N}$, then Φ is completely positive.

Example 11.2.1. Let $\mathfrak{X}$ be a vector space. Suppose that π is a $*$-representation of A and V is a linear mapping of $\mathfrak{X}$ into $\mathcal{D}(\pi)$. Define $\Phi(a)\,(\varphi, \psi) := \langle \pi(a)\, V\varphi, V\psi\rangle$ for $a \in \mathsf{A}$ and $\varphi, \psi \in \mathfrak{X}$. *Then Φ is a completely positive linear mapping of* A *on* $\mathfrak{X}$. Indeed, it is clear that $\Phi\,(a) \in B(\mathfrak{X})$ for $a \in \mathsf{A}$ and that the map Φ of A into $B(\mathfrak{X})$ is linear. For $a_1, \ldots, a_n \in \mathsf{A}$ and $\varphi_1, \ldots, \varphi_n \in \mathfrak{X}$ we have

$$\sum_{k,l=1}^{n} \Phi(a_k^+ a_l)\,(\varphi_l, \varphi_k) = \sum_{k,l=1}^{n} \langle \pi(a_k^+ a_l)\, V\varphi_l, V\varphi_k\rangle$$
$$= \left\langle \sum_{l=1}^{n} \pi(a_l)\, V\varphi_l, \sum_{k=1}^{n} \pi(a_k)\, V\varphi_k \right\rangle \geqq 0.$$

Therefore, Φ is n-positive for every $n \in \mathbb{N}$ and so completely positive. ○

The following theorem shows that all completely positive mappings of A on a vector space $\mathfrak{X}$ are of the form set out in Example 11.2.1.

Theorem 11.2.2. *Suppose that Φ is a completely positive linear map of* A *on a vector space $\mathfrak{X}$. Then there exists a closed $*$-representation π of* A *and a linear map V of $\mathfrak{X}$ into $\mathcal{D}(\pi)$ such that:*

(i) $\Phi(a)\,(\varphi, \psi) = \langle \pi(a)\, V\varphi, V\psi\rangle$ *for all $a \in \mathsf{A}$ and $\varphi, \psi \in \mathfrak{X}$,*

(ii) $\pi(\mathsf{A})\, V(\mathfrak{X}) \equiv \text{l.h.}\,\{\pi(a)\, V\varphi : a \in \mathsf{A}$ *and* $\varphi \in \mathfrak{X}\}$ *is dense in* $\mathcal{D}(\pi)\,[\mathsf{t}_\pi]$.

The couple $\{\pi, V\}$ is uniquely determined by the above requirements up to unitary equivalence, i.e., if $\{\tilde{\pi}, \tilde{V}\}$ is another such couple satisfying (i) *and* (ii), *then there exists a unitary operator U of $\mathcal{H}(\pi)$ onto $\mathcal{H}(\tilde{\pi})$ such that $UV = \tilde{V}$, $U(\mathcal{D}(\pi)) = \mathcal{D}(\tilde{\pi})$ and $\pi(a) = U^{-1}\tilde{\pi}(a)\, U$ for $a \in \mathsf{A}$.*

Proof. We define a sesquilinear form $\langle \cdot, \cdot \rangle_1$ on the vector space tensor product $\mathsf{A} \otimes \mathfrak{X}$ as follows: if $\eta = \sum_{k=1}^{n} a_k \otimes \varphi_k$ and $\zeta = \sum_{l=1}^{m} b_l \otimes \psi_l$ with $a_k, b_l \in \mathsf{A}$ and $\varphi_k, \psi_l \in \mathfrak{X}$, then we set $\langle \zeta, \eta \rangle_1 = \sum_{k=1}^{n} \sum_{l=1}^{m} \Phi(a_k^+ b_l)\,(\psi_l, \varphi_k)$.

Since Φ is a completely positive map on $\mathfrak{X}$, we conclude from (1) that $\langle\cdot,\cdot\rangle_1$ is a semidefinite inner product on $\mathsf{A}\otimes\mathfrak{X}$. Set $N := \{\eta \in \mathsf{A}\otimes\mathfrak{X} : \langle\eta,\eta\rangle_1 = 0\}$. For $a \in \mathsf{A}$, we define a linear mapping $\varrho(a)$ of $\mathsf{A}\otimes\mathfrak{X}$ into itself by $\varrho(a)\left(\sum_{k=1}^n a_k\otimes\varphi_k\right) = \sum_{k=1}^n aa_k\otimes\varphi_k$. It is clear that ϱ is a homomorphism of the algebra A in $L(\mathsf{A}\otimes\mathfrak{X})$ which satisfies $\varrho(1)\,\eta = \eta$ and $\langle\varrho(a)\,\eta,\zeta\rangle_1 = \langle\eta,\varrho(a^+)\,\zeta\rangle_1$ for $a \in \mathsf{A}$ and $\eta,\zeta \in \mathsf{A}\otimes\mathfrak{X}$. Suppose $\eta \in N$. The complete positivity of Φ implies that $g_\eta(a) := \langle\varrho(a)\,\eta,\eta\rangle_1$, $a \in \mathsf{A}$, is a positive linear functional on A. Since $|g_\eta(a)|^2 \leqq g_\eta(1)\,g_\eta(a^+a)$ for $a \in \mathsf{A}$ by the Cauchy-Schwarz inequality and $g_\eta(1) = \langle\eta,\eta\rangle_1 = 0$ by $\eta \in N$, we get $g_\eta \equiv 0$. Therefore, $g_\eta(a^+a) = \langle\varrho(a)\,\eta,\varrho(a)\,\eta\rangle_1 = 0$ and hence $\varrho(a)\,\eta \in N$ for $a \in \mathsf{A}$; so N is invariant under $\varrho(a)$.

Let $\mathcal{D}$ be the quotient vector space $(\mathsf{A}\otimes\mathfrak{X})/N$ and let ι be the corresponding quotient map. The equation $\langle\iota(\zeta),\iota(\eta)\rangle := \langle\zeta,\eta\rangle_1$, $\zeta,\eta \in \mathsf{A}\otimes\mathfrak{X}$, defines a scalar product on $\mathcal{D}$. Let $a \in \mathsf{A}$. Since $\varrho(a)\,N \subseteq N$, $\pi_0(a)\,\iota(\eta) := \iota(\varrho(a)\,\eta)$, $\eta \in \mathsf{A}\otimes\mathfrak{X}$, is a well-defined linear mapping of $\mathcal{D}$ into itself. From the properties of ϱ stated above we conclude immediately that π_0 is a $*$-representation of A on $\mathcal{D}(\pi_0) := \mathcal{D}$. Let π be the closure of π_0. We define a linear mapping V of $\mathfrak{X}$ into $\mathcal{D} \subseteq \mathcal{D}(\pi)$ by $V\varphi := \iota(1\otimes\varphi)$, $\varphi \in \mathfrak{X}$. From $\pi(a)V\varphi = \iota(a\otimes\varphi)$ for $a \in \mathsf{A}$ and $\varphi \in \mathfrak{X}$ we see that $\pi(\mathsf{A})\,V(\mathfrak{X}) = \mathcal{D}$, and this set is dense in $\mathcal{D}(\pi)\,[\mathrm{t}_\pi]$, since π is the closure of $\pi_0 \equiv \pi \upharpoonright \mathcal{D}$. For $a \in \mathsf{A}$ and $\varphi,\psi \in \mathfrak{X}$, we have $\langle\pi(a)\,V\varphi, V\psi\rangle = \langle\iota(a\otimes\varphi),\iota(1\otimes\psi)\rangle = \langle a\otimes\varphi, 1\otimes\psi\rangle_1 = \Phi(a)\,(\varphi,\psi)$. The proof of the first part of the theorem is complete.

Now we prove the uniqueness. For this let $\{\tilde{\pi},\tilde{V}\}$ be another such couple. We define a linear mapping of $\mathcal{D} = \pi(\mathsf{A})\,V(\mathfrak{X})$ onto $\tilde{\mathcal{D}} := \tilde{\pi}(\mathsf{A})\,\tilde{V}(\mathfrak{X})$ by $U\left(\sum_{k=1}^n \pi(a_k)\,V\varphi_k\right) = \sum_{k=1}^n \tilde{\pi}(a_k)\,\tilde{V}\varphi_k$, where $a_k \in \mathsf{A}$ and $\varphi_k \in \mathfrak{X}$. Applying (i) twice, we then have

$$\begin{aligned}\left\|U\left(\sum_{k=1}^n \pi(a_k)\,V\varphi_k\right)\right\|^2 &= \left\|\sum_{k=1}^n \tilde{\pi}(a_k)\,\tilde{V}\varphi_k\right\|^2 = \sum_{k,l=1}^n \langle\tilde{\pi}(a_k^+a_l)\,\tilde{V}\varphi_l,\tilde{V}\varphi_k\rangle\\ &= \sum_{k,l=1}^n \Phi(a_k^+a_l)\,(\varphi_l,\varphi_k) = \sum_{k,l=1}^n \langle\pi(a_k^+a_l)\,V\varphi_l, V\varphi_k\rangle\\ &= \left\|\sum_{k=1}^n \pi(a_k)\,V\varphi_k\right\|^2.\end{aligned}$$

From this we see that U is well-defined and isometric. Since $\mathcal{D}$ and $\tilde{\mathcal{D}}$ are dense in the Hilbert spaces $\mathcal{H}(\pi)$ and $\mathcal{H}(\tilde{\pi})$, respectively, by (ii), U has a unique extension to a unitary operator, again denoted by U, of $\mathcal{H}(\pi)$ onto $\mathcal{H}(\tilde{\pi})$. From the definition of U it is clear that $U\pi(a)\,\eta = \tilde{\pi}(a)\,U\eta$ for $a \in \mathsf{A}$ and $\eta \in \mathcal{D}$, i.e., $U \in \mathbb{I}(\pi \upharpoonright \mathcal{D}, \tilde{\pi} \upharpoonright \tilde{\mathcal{D}})$. By (ii), π and $\tilde{\pi}$ are the closures of $\pi \upharpoonright \mathcal{D}$ and $\tilde{\pi} \upharpoonright \tilde{\mathcal{D}}$, respectively. Hence $U \in \mathbb{I}(\pi,\tilde{\pi})$ by Proposition 8.2.2, (iv). Similarly, $U^{-1} \in \mathbb{I}(\tilde{\pi},\pi)$; so $\pi(a) = U^{-1}\tilde{\pi}(a)\,U$ for $a \in \mathsf{A}$ and U implements the unitary equivalence of $\tilde{\pi}$ and π. By definition of U we have $UV\varphi = U\pi(1)\,V\varphi = \tilde{\pi}(1)\,\tilde{V}\varphi = \tilde{V}\varphi$ for $\varphi \in X$; hence $UV = \tilde{V}$. □

We denote by $\{\pi_\Phi, V_\Phi\}$ the couple $\{\pi, V\}$ of Theorem 11.2.2. The $*$-representation π_Φ of A (or more precisely, the couple $\{\pi_\Phi, V_\Phi\}$) is called the *Stinespring dilation* of the completely positive map Φ.

Remark 1. Theorem 8.6.4 can be considered as the one dimensional version of Theorem 11.2.2. Indeed, for a linear functional ω on A, let Φ_ω be the linear map of A into $B(\mathbb{C})$ defined by $\Phi_\omega(a)\,(\varphi,\psi) = \omega(a)\,\varphi\overline{\psi}$, $a \in \mathsf{A}$ and $\varphi,\psi \in \mathbb{C}$. If ω is a positive linear functional on A, then Φ_ω is a completely positive map of A on $\mathbb{C}$ by Proposition 11.1.7. In this case the Stinespring dilation π_{Φ_ω} is nothing but the $*$-representation π_ω produced from ω by the GNS construction.

In the applications given in the next four sections other (that is, larger) wedges than $\mathcal{P}(\boldsymbol{M}(\mathsf{A}))$ play a central role. It is therefore necessary to extend our definitions to general wedges.

Definition 11.2.3. Let $n \in \mathbb{N}$. Suppose $\boldsymbol{K}_n$ and $\boldsymbol{K}$ are wedges in $\boldsymbol{M}_n(\mathsf{A})$ and $\boldsymbol{M}(\mathsf{A})$, respectively. Let $\mathfrak{X}$ be a vector space and let Φ be a linear mapping of A into $B(\mathfrak{X})$. We say that Φ is *n-positive with respect to $\boldsymbol{K}_n$* if $\Phi_{(n)}$ maps $\boldsymbol{K}_n$ into $\boldsymbol{M}_n(B(\mathfrak{X}))_+$ and that Φ is *completely positive with respect to $\boldsymbol{K}$* if $\Phi_{(\infty)}$ maps $\boldsymbol{K}$ into $\boldsymbol{M}(B(\mathfrak{X}))_+$.

Remark 2. If π is a $*$-representation of A, we always consider π as a mapping of A into $B(\mathcal{D}(\pi))$ by identifying $\pi(a)$ with the sesquilinear form $\langle \pi(a)\cdot, \cdot\rangle$ on $\mathcal{D}(\pi)\times \mathcal{D}(\pi)$ for $a \in \mathsf{A}$. Thus Definition 11.2.3 and the preceding investigations apply, in particular, to $*$-representations of A. For instance, Example 11.2.1 (with $\mathfrak{X} = \mathcal{D}(\pi)$ and V the identity map) shows that every $*$-representation π of A is completely positive (w.r.t. $\mathcal{P}(\boldsymbol{M}(\mathsf{A}))$).

Remark 3. Suppose that $\boldsymbol{K}$ is an m-admissible wedge in the $*$-algebra $\boldsymbol{M}(\mathsf{A})$. Recall from 1.4 that this means we have $\mathcal{P}(\boldsymbol{M}(\mathsf{A})) \subseteq \boldsymbol{K} \subseteq \boldsymbol{M}(\mathsf{A})_{\mathrm{h}}$ and $A^+XA \in \boldsymbol{K}$ for all $X \in \boldsymbol{K}$ and $A \in \boldsymbol{M}(\mathsf{A})$. Then $\boldsymbol{K}$ is an admissible wedge in $\boldsymbol{M}(\mathsf{A})$ in the sense of Definition 11.1.1. (Indeed, if $\Lambda \equiv [\lambda_{kl}] \in \boldsymbol{M}(\mathbb{C})$ and $X \in \boldsymbol{K}$, then $A := [\lambda_{kl}\cdot 1] \in \boldsymbol{M}(\mathsf{A})$ and hence $\Lambda^+X\Lambda \equiv A^+XA \in \boldsymbol{K}$.) Further, if Φ is a linear mapping of A into $B(\mathfrak{X})$ which is completely positive with respect to $\boldsymbol{K}$, then Φ is completely positive (because $\mathcal{P}(\boldsymbol{M}(\mathsf{A})) \subseteq \boldsymbol{K}$) and so the Stinespring dilation π_Φ is well-defined.

The following easy calculations are essential for the proofs of the next two propositions. Suppose π is a $*$-representation of A. Let $m, n \in \mathbb{N}$, $X \equiv [x_{kl}] \in \boldsymbol{M}_m(\mathsf{A})$, $A \equiv [a_{kl}] \in \boldsymbol{M}_{m,n}(\mathsf{A})$ and $\gamma_1, \ldots, \gamma_n \in \mathcal{D}(\pi)$. We define vectors $\varphi_1, \ldots, \varphi_m \in \mathcal{D}(\pi)$ by

$$\varphi_l := \pi(a_{l1})\,\gamma_1 + \cdots + \pi(a_{ln})\,\gamma_n, \quad l = 1, \ldots, m. \tag{2}$$

Letting $B \equiv [b_{kl}] := A^+XA$, we then have

$$\sum_{k,l=1}^{m} \langle \pi(x_{kl})\,\varphi_l, \varphi_k\rangle = \sum_{k,l=1}^{m} \sum_{r,s=1}^{n} \langle \pi(x_{kl})\,\pi(a_{ls})\,\gamma_s, \pi(a_{kr})\,\gamma_r\rangle$$
$$= \sum_{r,s=1}^{n} \left\langle \pi\left(\sum_{k,l=1}^{m} a_{kr}^+ x_{kl} a_{ls}\right)\gamma_s, \gamma_r\right\rangle = \sum_{r,s=1}^{n} \langle \pi(b_{rs})\,\gamma_s, \gamma_r\rangle. \tag{3}$$

Proposition 11.2.4. *Let $\mathfrak{X}$ be a vector space and let Φ be a linear mapping of A into $B(\mathfrak{X})$. Suppose $\boldsymbol{K}$ is an m-admissible wedge in $\boldsymbol{M}(\mathsf{A})$. Then the Stinespring dilation π_Φ is completely positive with respect to $\boldsymbol{K}$ if and only if Φ is completely positive with respect to $\boldsymbol{K}$.*

Proof. The only if part follows at once from formula (i) in Theorem 11.2.2 (without using that $\boldsymbol{K}$ is m-admissible). We prove the if part. We let $m \in \mathbb{N}$ and $X \equiv [x_{kl}] \in \boldsymbol{K} \cap \boldsymbol{M}_m(\mathsf{A})$. We have to show that

$$\sum_{k,l=1}^{m} \langle \pi_\Phi(x_{kl})\,\varphi_l, \varphi_k\rangle \geqq 0 \tag{4}$$

for arbitrary vectors $\varphi_1, \ldots, \varphi_m \in \mathcal{D}(\pi_\Phi)$. Since $\pi_\Phi(A)\,V_\Phi\mathfrak{X}$ is dense in $\mathcal{D}(\pi_\Phi)$ $[\mathrm{t}_{\pi_\Phi}]$, it suffices to prove this for vectors $\varphi_1, \ldots, \varphi_m$ in $\pi_\Phi(\mathsf{A})\,V_\Phi\mathfrak{X}$. But then the vectors $\varphi_1, \ldots, \varphi_m$ are of the form (2) with $\gamma_1, \ldots, \gamma_n \in V_\Phi\mathfrak{X}$ and $n \in \mathbb{N}$. We can write γ_r as $\gamma_r = V_\Phi\psi_r$ with $\psi_r \in \mathfrak{X}$, $r = 1, \ldots, n$. By (3), we have

$$\sum_{k,l=1}^{m} \langle \pi_\Phi(x_{kl})\,\varphi_l, \varphi_k\rangle = \sum_{r,s=1}^{n} \langle \pi_\Phi(b_{rs})\,V_\Phi\psi_s, V_\Phi\psi_r\rangle = \sum_{r,s=1}^{n} \Phi(b_{rs})\,(\psi_s, \psi_r). \tag{5}$$

Since $\boldsymbol{K}$ is m-admissible, $B \equiv [b_{kl}] = A^+XA$ is in $\boldsymbol{K}$. Therefore, the right-hand side in (5) is non-negative, since Φ is completely positive with respect to $\boldsymbol{K}$. This proves (4). □

A $*$-representation π of A is said to be *n-cyclic* if there exists a subset $\Gamma = \{\gamma_1, \ldots, \gamma_n\}$ of $\mathcal{D}(\pi)$ which is cyclic for π (cf. Definition 8.3.14), i.e., $\pi(\mathsf{A})\,\Gamma$ is dense in $\mathcal{D}(\pi)$ $[\mathrm{t}_\pi]$.

Remark 4. The 1-cyclic $*$-representations are precisely the cyclic $*$-representations in the sense of Definition 8.6.1.

Remark 5. Suppose Φ is a completely positive mapping of A on a vector space $\mathfrak{X}$ which has finite dimension n. Then the Stinespring dilation π_Φ is n-cyclic. When $\{\psi_1, \ldots, \psi_n\}$ is a basis of $\mathfrak{X}$, then the set $\Gamma := \{V_\Phi\psi_1, \ldots, V_\Phi\psi_n\}$ is, of course, cyclic for π_Φ.

Next we briefly discuss the concepts introduced above in case of n-cyclic $*$-representations. In order to state the results, we need further notations.

If π is a $*$-representation of A and $\Gamma = \{\gamma_1, \ldots, \gamma_n\}$ is a subset of $\mathcal{D}(\pi)$, we define a linear functional $f_{\pi,\Gamma}$ on $\boldsymbol{M}_n(\mathsf{A})$ by $f_{\pi,\Gamma}([x_{kl}]) := \sum_{k,l=1}^{n} \langle \pi(x_{kl})\,\gamma_l, \gamma_k\rangle$, $[x_{kl}] \in \boldsymbol{M}_n(\mathsf{A})$.

Let $m, n \in \mathbb{N}$. If $\boldsymbol{K}_m$ is a wedge in $\boldsymbol{M}_m(\mathsf{A})$, let $\boldsymbol{K}_n(m)$ denote the set of all finite sums of matrices A^+XA, where $X \in \boldsymbol{K}_m$ and $A \in \boldsymbol{M}_{m,n}(\mathsf{A})$. Clearly, $\boldsymbol{K}_n(m)$ is again a wedge in $\boldsymbol{M}_n(\mathsf{A})$.

Proposition 11.2.5. *Suppose $n \in \mathbb{N}$ and $m \in \mathbb{N}$. Let $\boldsymbol{K}_m$ be a wedge in $\boldsymbol{M}_m(\mathsf{A})$. Suppose that π is an n-cyclic $*$-representation of A and $\Gamma = \{\gamma_1, \ldots, \gamma_n\}$ is a cyclic set for π. Then π is m-positive with respect to $\boldsymbol{K}_m$ if and only if the functional $f_{\pi,\Gamma}$ is non-negative on $\boldsymbol{K}_n(m)$.*

Proof. Suppose $f_{\pi,\Gamma}$ is non-negative on $\boldsymbol{K}_n(m)$. Since Γ is a cyclic set for π, it suffices to prove (4) for all $X \equiv [x_{kl}] \in \boldsymbol{K}_m$ and all $\varphi_1, \ldots, \varphi_m \in \pi(\mathsf{A})\,\Gamma$. Then the vectors $\varphi_1, \ldots, \varphi_m$ are of the form (2) with some $A = [a_{kl}] \in \boldsymbol{M}_{m,n}(\mathsf{A})$. Because $B \equiv A^+XA \in \boldsymbol{K}_n(m)$ by definition, we have $f_{\pi,\Gamma}(B) \geqq 0$. Since $f_{\pi,\Gamma}(B)$ is equal to the right-hand side in (3), we see from (3) that (4) is valid. Hence π is m-positive w.r.t. $\boldsymbol{K}_m$. The opposite direction follows by a similar reasoning in reversed order. □

Corollary 11.2.6. *Suppose $\boldsymbol{K}$ is an m-admissible wedge in $\boldsymbol{M}(\mathsf{A})$ and π is an n-cyclic $*$-representation of A. Let $\Gamma = \{\gamma_1, \ldots, \gamma_n\}$ be a cyclic set for π. Then the following conditions are equivalent:*

(i) *π is completely positive with respect to $\boldsymbol{K}$.*

(ii) *π is n-positive with respect to $\boldsymbol{K} \cap \boldsymbol{M}_n(\mathsf{A})$.*

(iii) *$f_{\pi,\Gamma}$ is non-negative on $\boldsymbol{K} \cap \boldsymbol{M}_n(\mathsf{A})$.*

Proof. (i) $\rightarrow$ (ii) $\rightarrow$ (iii) is obvious. In order to prove the implication (iii) $\rightarrow$ (i), we set $\boldsymbol{K}_m := \boldsymbol{K} \cap \boldsymbol{M}_m(\mathsf{A})$ for $m \in \mathbb{N}$ in Proposition 11.2.5. By assumption, $\boldsymbol{K}$ is m-admissible. Hence $\boldsymbol{K}_n(m) \subseteqq \boldsymbol{K} \cap \boldsymbol{M}_m(\mathsf{A})$. Therefore, by (iii), Proposition 11.2.5 shows that π is m-positive w.r.t. $\boldsymbol{K}_m$ for all $m \in \mathbb{N}$. This gives (i). □

It is not difficult to see that each functional $f_{\pi,\Gamma}$ defined by (6) is a positive linear functional on the $*$-algebra $\boldsymbol{M}_n(\mathsf{A})$. The next proposition shows that all positive linear functionals on $\boldsymbol{M}_n(\mathsf{A})$ arise in that way.

Proposition 11.2.7. *Let $n \in \mathbb{N}$ and let f be a positive linear functional on the $*$-algebra $\boldsymbol{M}_n(\mathsf{A})$. There exists a closed n-cyclic $*$-representation π of A and a subset $\Gamma = \{\gamma_1, \ldots, \gamma_n\}$ of $\mathcal{D}(\pi)$ which is cyclic for π such that $f = f_{\pi,\Gamma}$.*

Proof. We define a linear mapping of A into $B(\mathbb{C}^n)$ by $\Phi(a)(\psi_r, \psi_s) := f([a\delta_{rk}\delta_{sl}])$, where $a \in A$ and ψ_r is the basis vector $(\delta_{rj}: j = 1, \ldots, n)$ of $\mathbb{C}^n$, $r, s = 1, \ldots, n$. Then f coincides with the linear functional f defined in Proposition 11.1.7. Since $\mathcal{P}(\boldsymbol{M}(A)) \cap \boldsymbol{M}_n(A) = \mathcal{P}(\boldsymbol{M}_n(A))$ by the characterization of $\mathcal{P}(\boldsymbol{M}(A))$ given at the beginning of this section, f is non-negative on $\mathcal{P}(\boldsymbol{M}(A)) \cap \boldsymbol{M}_n(A)$. Therefore, by Proposition 11.1.7 applied with $\boldsymbol{K} = \mathcal{P}(\boldsymbol{M}(A))$, Φ is a completely positive map of A on $\mathbb{C}^n$ with respect to $\mathcal{P}(\boldsymbol{M}(A))$; so Theorem 11.2.2 applies. Then $\pi := \pi_\Phi$ and $\Gamma := \{V_\Phi\psi_1, \ldots, V_\Phi\psi_n\}$ have the desired properties. □

The following theorem combines Theorem 11.1.5, Theorem 11.2.2 and Proposition 11.2.4. It will be the crucial result for the applications given in the next three sections.

Theorem 11.2.8. *Suppose that* B *is a $*$-algebra with unit element* 1 *and* $\boldsymbol{K}$ *is an m-admissible wedge in* $\boldsymbol{M}(B)$. *Let* A *be a $*$-subalgebra of* B *with* $1 \in A$. *Suppose that* A *is cofinal in* B *with respect to the wedge* $\boldsymbol{K} \cap B$ *(i. e., given* $b \in B_h$ *there is an* $a \in A_h$ *such that* $a \in \boldsymbol{K}$ *and* $a - b \in \boldsymbol{K}$*). Let* π *be a $*$-representation of* A *which is completely positive with respect to* $\boldsymbol{K} \cap \boldsymbol{M}(A)$.

Then there exists a closed $*$-representation π_1 *of* B *which is completely positive with respect to* $\boldsymbol{K}$ *such that* $\mathcal{D}(\pi) \subseteq \mathcal{D}(\pi_1)$, $\pi(a) = \pi_1(a) \upharpoonright \mathcal{D}(\pi)$ *for all* $a \in A$ *and such that* $\pi_1(B)\,\mathcal{D}(\pi)$ *is dense in* $\mathcal{D}(\pi_1)\,[t_{\pi_1}]$.

Proof. First we apply Theorem 11.1.5 with $E = B$, $E_0 = A$, $\boldsymbol{K}(E) = \boldsymbol{K}$ and $\Phi_0 = \pi$. Then there is a linear map Φ of B into $B(\mathcal{D}(\pi))$ which is completely positive with respect to $\boldsymbol{K}$ and satisfies $\Phi \upharpoonright A = \pi$. In particular, the map Φ is completely positive (w.r.t. $\mathcal{P}(\boldsymbol{M}(B))$); so the Stinespring dilation $\{\pi_\Phi, V_\Phi\}$ is well-defined according to Theorem 11.2.2. From the equality

$$\langle\varphi, \psi\rangle = \langle\pi(1)\,\varphi, \psi\rangle = \Phi(1)\,(\varphi, \psi) = \langle\pi_\Phi(1)\,V_\Phi\varphi, V_\Phi\psi\rangle = \langle V_\Phi\varphi, V_\Phi\psi\rangle$$

for $\varphi, \psi \in \mathcal{D}(\pi)$ we see that V_Φ is an injective linear mapping of $\mathcal{D}(\pi)$ into $\mathcal{D}(\pi_\Phi)$ which preserves the scalar product. For notational simplicity, we consider $\mathcal{D}(\pi)$ as a subspace of $\mathcal{D}(\pi_\Phi)$ by identifying φ with $V_\Phi\varphi$, $\varphi \in \mathcal{D}(\pi)$. Then, of course, $\pi_\Phi(B)\,\mathcal{D}(\pi)$ is dense in $\mathcal{D}(\pi_\Phi)\,[t_{\pi_\Phi}]$ by Theorem 11.2.2.

We prove that $\pi(a) = \pi_\Phi(a) \upharpoonright \mathcal{D}(\pi)$ for all $a \in A$. Fix $a \in A$. Let P be the projection of $\mathcal{H}(\pi_\Phi)$ onto $\mathcal{H}(\pi)$. We have $\langle\pi(a)\,\varphi, \psi\rangle = \Phi(a)\,(\varphi, \psi) = \langle\pi_\Phi(a)\,\varphi, \psi\rangle$ for $\varphi, \psi \in \mathcal{D}(\pi)$; so $P\pi_\Phi(a) \upharpoonright \mathcal{D}(\pi) = \pi(a)$. Combining the latter with the fact that π and π_Φ are $*$-representations, we obtain

$$\begin{aligned} \|(I - P)\,\pi_\Phi(a)\,\varphi\|^2 &= \|\pi_\Phi(a)\,\varphi\|^2 - \|P\pi_\Phi(a)\,\varphi\|^2 \\ &= \langle\pi_\Phi(a^+a)\,\varphi, \varphi\rangle - \langle P\pi_\Phi(a)\,\varphi, P\pi_\Phi(a)\,\varphi\rangle \\ &= \langle\pi(a^+a)\,\varphi, \varphi\rangle - \langle\pi(a)\,\varphi, \pi(a)\,\varphi\rangle = 0 \quad \text{for} \quad \varphi \in \mathcal{D}(\pi). \end{aligned}$$

Thus $\pi(a) = \pi_\Phi(a) \upharpoonright \mathcal{D}(\pi)$.

By Proposition 11.2.4, π_Φ is completely positive with respect to $\boldsymbol{K}$, since $\boldsymbol{K}$ is m-admissible. Setting $\pi_1 := \pi_\Phi$, the proof is complete. □

In concrete applications the wedge $\boldsymbol{K}$ is often of the following form: Suppose $\mathfrak{R}$ is a distinguished family of $*$-representations of A. Define

$$\boldsymbol{M}_n(A; \mathfrak{R})_+ := \{[a_{kl}] \in \boldsymbol{M}_n(A)_h : \sum_{k,l=1}^{n} \langle\pi(a_{kl})\,\varphi_l, \varphi_k\rangle \geqq 0 \text{ for all } \pi \in \mathfrak{R} \text{ and all vectors } \varphi_1, \ldots, \varphi_n \in \mathcal{D}(\pi)\} \tag{6}$$

for $n \in \mathbb{N}$, and $M(\mathsf{A}; \mathfrak{R})_+ := \bigcup_{n\in\mathbb{N}} M_n(\mathsf{A}; \mathfrak{R})_+$. Roughly speaking, $M(\mathsf{A}; \mathfrak{R})_+$ is the set of all matrices in $M(\mathsf{A})_h$ which are mapped into positive matrices by the representations in $\mathfrak{R}$. From the calculations before Proposition 11.2.4 we see immediately that $M(\mathsf{A}; \mathfrak{R})_+$ is an m-admissible wedge in $M(\mathsf{A})$.

An important special form of wedges $M(\mathsf{A}; \mathfrak{R})_+$ is used in Section 11.4. In this case A is the enveloping algebra $\mathscr{E}(\mathfrak{g})$ of a finite dimensional Lie algebra $\mathfrak{g}$ and $\mathfrak{R}$ is the family of all G-integrable $*$-representations of $\mathsf{A} \equiv \mathscr{E}(\mathfrak{g})$, where G is a Lie group which has $\mathfrak{g}$ as its Lie algebra.

We now describe another special form of wedges $M(\mathsf{A}; \mathfrak{R})_+$ which is needed later (see e.g. Theorem 11.4.4 and Corollary 11.6.2). Suppose that A is an O*-algebra $\mathcal{A}$ and $\mathfrak{R}$ consists of the identity representation of $\mathsf{A} \equiv \mathcal{A}$ only (i.e., the representation π with $\mathcal{D}(\pi) := \mathcal{D}(\mathcal{A})$ and $\pi(a) := a$, $a \in \mathcal{A}$). In this case we write $M_n(\mathcal{A})_+$ for $M_n(\mathsf{A}; \mathfrak{R})_+$ and $M(\mathcal{A})_+$ for $M(\mathsf{A}; \mathfrak{R})_+$. That is, we have

$$M_n(\mathcal{A})_+ = \{[a_{kl}] \in M_n(\mathcal{A})_h : \sum_{k,l=1}^{n} \langle a_{kl}\varphi_l, \varphi_k\rangle \geqq 0 \text{ for all vectors } \varphi_1, \ldots, \varphi_n \in \mathcal{D}(\mathcal{A})\} \tag{7}$$

for $n \in \mathbb{N}$ and $M(\mathcal{A})_+ = \bigcup_{n\in\mathbb{N}} M_n(\mathcal{A})_+$.

The wedge $M_n(\mathcal{A})_+$ can also be interpreted as follows. Let $\mathcal{D}_n(\mathcal{A})$ be the set of all vectors $(\varphi_1, \ldots, \varphi_n)$ in the Hilbert space $\mathcal{H} \oplus \cdots \oplus \mathcal{H}$ (n times) with $\varphi_1, \ldots, \varphi_n \in \mathcal{D}(\mathcal{A})$. We consider $M_n(\mathcal{A})$ as an O*-algebra on the domain $\mathcal{D}_n(\mathcal{A})$ by identifying the matrix $[a_{kl}] \in M_n(\mathcal{A})$ with the operator on $\mathcal{D}_n(\mathcal{A})$ defined by

$$[a_{kl}]\,(\varphi_1, \ldots, \varphi_n) := \left(\sum_{k=1}^{n} a_{1k}\varphi_k, \ldots, \sum_{k=1}^{n} a_{nk}\varphi_k\right),$$

$\varphi_1, \ldots, \varphi_n \in \mathcal{D}(\mathcal{A})$. Then the wedge $M_n(\mathcal{A})_+$ defined by (7) is nothing but the cone $M_n(\mathcal{A})_+$ (in the sense of Definition 2.6.1) of the O*-algebra $M_n(\mathcal{A})$. From this we see in particular that $M_n(\mathcal{A})_+$ and $M(\mathcal{A})_+$ are cones.

Let Φ be a linear mapping of the O*-algebra $\mathcal{A}$ into $B(\mathfrak{X})$, where $\mathfrak{X}$ is a vector space. We say that Φ is *strongly n-positive* if Φ is strongly positive with respect to $M_n(\mathcal{A})_+$ and that Φ is *completely strongly positive* if Φ is completely positive with respect to $M(\mathcal{A})_+$. By this definition, a $*$-representation of $\mathcal{A}$ or a linear functional on $\mathcal{A}$ is strongly 1-positive if and only if it is strongly positive according to Definition 2.6.1.

Remark 6. The general wedge $M(\mathsf{A}; \mathfrak{R})_+$ defined above can be reduced to the preceding special case, since obviously $M(\mathsf{A}; \mathfrak{R})_+ = \left\{[a_{kl}] \in M(\mathsf{A})_h : [\varrho_{\mathfrak{R}}(a_{kl})] \in M\big(\varrho_{\mathfrak{R}}(\mathsf{A})\big)_+\right\}$, where $\varrho_{\mathfrak{R}}$ denotes the direct sum of all representations in $\mathfrak{R}$.

11.3. A First Application: Integrable Extensions of $*$-Representations of Commutative $*$-Algebras

Throughout this section A will denote a commutative $*$-algebra with unit.

A matrix $[p_{kl}] \in M(\mathbb{C}[\mathsf{x}_1, \ldots, \mathsf{x}_n])$ is said to be *positive definite* if for each $(\lambda_1, \ldots, \lambda_n)$ $\in \mathbb{R}^n$ the matrix $[p_{kl}(\lambda_1, \ldots, \lambda_n)]$ is positive semi-definite, i.e., $\sum_{k,l=1}^{\infty} p_{kl}(\lambda_1, \ldots, \lambda_n)\, \alpha_l \overline{\alpha_k} \geqq 0$

for arbitrary complex numbers α_l, $l \in \mathbb{N}$. (Note that the sum is in fact a finite sum, since $[p_{kl}]$ is a finite matrix.)

Definition 11.3.1. Suppose that $\mathsf{Y} \equiv \{y_j : j \in J\}$ is a subset of A_h such that $\mathsf{Y} \cup \{1\}$ generates the $*$-algebra A. Let $\boldsymbol{M}(\mathsf{A}; \mathrm{int})_+$ be the set of all matrices in $\boldsymbol{M}(\mathsf{A})_h$ of the form $[p_{kl}(y_{j_1}, \dots, y_{j_m})]$, where $m \in \mathbb{N}$, $[p_{kl}]$ is a positive definite matrix of $\boldsymbol{M}(\mathbb{C}[\mathsf{x}_1, \dots, \mathsf{x}_m])$ and $j_1, \dots, j_m \in J$. Let $\boldsymbol{M}_n(\mathsf{A}; \mathrm{int})_+ := \boldsymbol{M}(\mathsf{A}; \mathrm{int})_+ \cap \boldsymbol{M}_n(\mathsf{A})$ for $n \in \mathbb{N}$ and $\mathsf{A}_+^{\mathrm{int}} := \boldsymbol{M}(\mathsf{A}; \mathrm{int})_+ \cap \mathsf{A} \equiv \boldsymbol{M}_1(\mathsf{A}; \mathrm{int})_+$.

Lemma 11.3.2. (i) *$\boldsymbol{M}(\mathsf{A}; \mathrm{int})_+$ is independent of the special set Y occurring in Definition* 11.3.1.

(ii) *$\boldsymbol{M}(\mathsf{A}; \mathrm{int})_+$ is an m-admissible cone in $\boldsymbol{M}(\mathsf{A})$.*

(iii) *$\mathsf{A}_+^{\mathrm{int}}$ is an m-admissible cone in A.*

Proof. (i): Let $\tilde{\mathsf{Y}} = \{y_{\tilde{j}} : \tilde{j} \in \tilde{J}\}$ be another subset of A_h such that $\tilde{\mathsf{Y}} \cup \{1\}$ generates the $*$-algebra A. We denote the corresponding sets from Definition 11.3.1 by $\boldsymbol{M}(\mathsf{A}; \mathrm{int})_{+,\mathsf{Y}}$ and $\boldsymbol{M}(\mathsf{A}; \mathrm{int})_{+,\tilde{\mathsf{Y}}}$. Suppose $[p_{kl}(y_{j_1}, \dots, y_{j_m})] \in \boldsymbol{M}(\mathsf{A}; \mathrm{int})_{+,\mathsf{Y}}$, where $[p_{kl}]$ is a positive definite matrix of $\boldsymbol{M}(\mathbb{C}[\mathsf{x}_1, \dots, \mathsf{x}_m])$ and $j_1, \dots, j_m \in J$. Since $\tilde{\mathsf{Y}} \subseteq \mathsf{A}_h$ and $\tilde{\mathsf{Y}} \cup \{1\}$ generates the $*$-algebra A, there are $r \in \mathbb{N}$, indices $\tilde{j}_1, \dots, \tilde{j}_r \in \tilde{J}$ and polynomials $q_1, \dots, q_m \in \mathbb{C}[\mathsf{x}_1, \dots, \mathsf{x}_r]$ with real coefficients such that $y_{j_s} = q_s(y_{\tilde{j}_1}, \dots, y_{\tilde{j}_r})$, $s = 1, \dots, m$. Define $\tilde{p}_{kl} := p_{kl}(q_1, \dots, q_m) \in \mathbb{C}[\mathsf{x}_1, \dots, \mathsf{x}_r]$. The matrix $[\tilde{p}_{kl}]$ is obviously positive definite and hence $[p_{kl}(y_{j_1}, \dots, y_{j_m})] = [\tilde{p}_{kl}(y_{\tilde{j}_1}, \dots, y_{\tilde{j}_r})] \in \boldsymbol{M}(\mathsf{A}; \mathrm{int})_{+,\tilde{\mathsf{Y}}}$, so $\boldsymbol{M}(\mathsf{A}; \mathrm{int})_{+,\mathsf{Y}} \subseteq \boldsymbol{M}(\mathsf{A}; \mathrm{int})_{+,\tilde{\mathsf{Y}}}$. By symmetry, $\boldsymbol{M}(\mathsf{A}; \mathrm{int})_{+,\mathsf{Y}} = \boldsymbol{M}(\mathsf{A}; \mathrm{int})_{+,\tilde{\mathsf{Y}}}$.

The simple proofs of (ii) and (iii) are omitted. □

Remark 1. By Lemma 11.3.2, (i), we could have taken $\mathsf{Y} = \mathsf{A}_h$ in Definition 11.3.1 and also in the proof of Theorem 11.3.3 below. (This would simplify the notation in this proof.) We prefered not to do this, since for concrete algebras such as $\mathbb{C}[\mathsf{x}_1, \dots, \mathsf{x}_n]$ it seems to be better to work with a fixed (small) set of hermitian generators.

Remark 2. By the above definition, $\mathbb{C}[\mathsf{x}_1, \dots, \mathsf{x}_n]_+^{\mathrm{int}} = \{p \in \mathbb{C}[\mathsf{x}_1, \dots, \mathsf{x}_n] : p(t_1, \dots, t_n) \geqq 0 \text{ for all } (t_1, \dots, t_n) \in \mathbb{R}^n\}$.

Example 11.3.3. Let B be the $*$-algebra $C[0, 1]$ and let A be the $*$-subalgebra of B formed by the functions $p(e^t)$, where $p \in \mathbb{C}[\mathsf{x}]$. Since $e^t = p(e^{t/2})$ with $p(\mathsf{x}) = \mathsf{x}^2$, we have $e^t \in \mathsf{B}_+^{\mathrm{int}}$. Obviously, e^t is not in $\mathsf{A}_+^{\mathrm{int}}$. Hence $\mathsf{A}_+^{\mathrm{int}} \neq \mathsf{B}_+^{\mathrm{int}} \cap \mathsf{A}$. ○

The abbreviation "int" (for "integrable") is suggested by the following theorem which is the main result in this section.

Theorem 11.3.4. *For every $*$-representation π of A, the following two statements are equivalent:*

(i) *π is completely positive with respect to the cone $\boldsymbol{M}(\mathsf{A}; \mathrm{int})_+$.*

(ii) *There exists an integrable $*$-representation π_1 of A in a possibly larger Hilbert space such that $\pi \subseteq \pi_1$.*

If (i) *is satisfied, then the $*$-representation π_1 in* (ii) *can be chosen such that $(\pi_1; \mathcal{M})$ is an induced extension of π (in the sense of Definition* 8.5.3), *where $\mathcal{M}$ is the commutative von Neumann algebra $\pi_1(\mathsf{A})''$ on $\mathcal{H}(\pi_1)$.*

Remark 3. In Theorem 11.3.4, $\mathcal{M} \equiv \pi_1(\mathsf{A})''$ is also the von Neumann algebra which is generated by the spectral projections of the self-adjoint operators $\overline{\pi_1(a)}$, $a \in \mathsf{A}_h$.

Corollary 11.3.5. *A self-adjoint representation of* A *is integrable if and only if it is completely positive with respect to the cone* $\boldsymbol{M}(\mathsf{A}; int)_+$.

Proof. By Proposition 9.1.17, a self-adjoint subrepresentation of an integrable representation of A is itself integrable. Using this fact the assertion follows at once from Theorem 11.3.4. □

Proof of Theorem 11.3.4. We first prove the implication (i) → (ii) which is the main assertion of the theorem.

Suppose that π is completely positive w.r.t. $\boldsymbol{M}(\mathsf{A}; \mathrm{int})_+$, and let $\mathsf{Y} \equiv \{y_j : j \in J\}$ be as in Definition 11.3.1. Let F be the $*$-algebra of all functions from $\mathbb{R}^J$ into $\mathbb{C}$ with the usual pointwise algebraic operations. Let P be the set of all f in F for which there exist an $n \in \mathbb{N}$, a polynomial $p \in \mathbb{C}[\mathsf{x}_1, \ldots, \mathsf{x}_n]$ and indices $j_1, \ldots, j_n \in J$ such that $f(\lambda) = p(\lambda_{j_1}, \ldots, \lambda_{j_n})$ for all $\lambda \equiv (\lambda_j : j \in J) \in \mathbb{R}^J$. We shall simply write $p \equiv p(\lambda_{j_1}, \ldots, \lambda_{j_n})$ for such a function. Let R be the $*$-subalgebra of F generated by P and the elements $(p \pm \mathrm{i})^{-1} \in \mathsf{F}$ for $p = p^+ \in \mathsf{P}$. We define a $*$-representation ϱ of P by

$$\varrho(p(\lambda_{j_1}, \ldots, \lambda_{j_n})) := \pi(p(y_{j_1}, \ldots, y_{j_n})) \tag{1}$$

for $n \in \mathbb{N}$, $p \in \mathbb{C}[\mathsf{x}_1, \ldots, \mathsf{x}_n]$ and $j_1, \ldots, j_n \in J$. Since π is completely positive w.r.t. $\boldsymbol{M}(\mathsf{A}; \mathrm{int})_+$, it is obvious that ϱ is completely positive w.r.t. $\boldsymbol{M}(\mathsf{P}; \mathrm{int})_+$. From the special form of the algebras P and R we see easily that $\boldsymbol{M}(\mathsf{P}; \mathrm{int})_+ = \boldsymbol{M}(\mathsf{R}; \mathrm{int})_+ \cap \boldsymbol{M}(\mathsf{P})$, so ϱ is completely positive w.r.t. $\boldsymbol{M}(\mathsf{R}; \mathrm{int})_+ \cap \boldsymbol{M}(\mathsf{P})$. Further, P is cofinal in R w.r.t the cone $\mathsf{R}_+^{\mathrm{int}}$. Thus ϱ satisfies the assumptions of Theorem 11.2.8 in case $\mathsf{A} := \mathsf{P}$, $\mathsf{B} := \mathsf{R}$ and $K := \boldsymbol{M}(\mathsf{R}; \mathrm{int})_+$. Let ϱ_1 be the corresponding $*$-representation of R which exists by Theorem 11.2.8.

For $n \in \mathbb{N}$, $p \in \mathbb{C}[\mathsf{x}_1, \ldots, \mathsf{x}_n]$ and $j_1, \ldots, j_n \in J$, we define

$$\pi_0(p(y_{j_1}, \ldots, y_{j_n})) := \varrho_1(p(\lambda_{j_1}, \ldots, \lambda_{j_n})). \tag{2}$$

We check that this definition is unambiguously, that is, we show that $p(y_{j_1}, \ldots, y_{j_n}) = 0$ in A implies that $\varrho_1(p(\lambda_{j_1}, \ldots, \lambda_{j_n})) = 0$ on $\mathcal{D}(\varrho_1)$. Indeed, if $f \in \mathsf{R}$ and $\varphi \in \mathcal{D}(\varrho)$, then

$$\varrho_1(p(\lambda_{j_1}, \ldots, \lambda_{j_n}))\, \varrho_1(f)\, \varphi = \varrho_1(f)\, \varrho_1(p(\lambda_{j_1}, \ldots, \lambda_{j_n}))\, \varphi$$
$$= \varrho_1(f)\, \varrho(p(\lambda_{j_1}, \ldots, \lambda_{j_n}))\, \varphi = \varrho_1(f)\, \pi(p(y_{j_1}, \ldots, y_{j_n}))\, \varphi = 0,$$

where the second equality follows from the fact that $\varrho \subseteq \varrho_1 \upharpoonright \mathsf{P}$ (by Theorem 11.2.8) and the third follows from (1). Since $\varrho_1(\mathsf{R})\, \mathcal{D}(\varrho)$ is dense in $\mathcal{D}(\varrho_1)$ $[t_{\varrho_1}]$ by Theorem 11.2.8, this gives $\varrho_1(p(\lambda_{j_1}, \ldots, \lambda_{j_n})) = 0$. Hence π_0 is a well-defined $*$-representation of A on $\mathcal{D}(\pi_0) \equiv \mathcal{D}(\varrho_1)$. Combining (1) and (2) with the relation $\varrho \subseteq \varrho_1 \upharpoonright \mathsf{P}$ (by Theorem 11.2.8) we conclude that $\pi \subseteq \pi_0$. Set $\pi_1 := \widehat{\pi_0}$. Of course, $\pi \subseteq \pi_1$.

We prove that π_1 is integrable. Let $a \in \mathsf{A}_\mathrm{h}$. Then a is of the form $a = p(y_{j_1}, \ldots, y_{j_n})$ with $p = p^+ \in \mathbb{C}[\mathsf{x}_1, \ldots, \mathsf{x}_n]$ and $j_1, \ldots, j_n \in J$. Since $(p \pm \mathrm{i})^{-1} \in \mathsf{R}$, Proposition 8.1.19 shows that the operator

$$\varrho_1(p) = \pi_0(p(y_{j_1}, \ldots, y_{j_n})) \equiv \pi_0(a) \tag{3}$$

is essentially self-adjoint and

$$\overline{\varrho_1((p \pm \mathrm{i})^{-1})} = (\overline{\varrho_1(p)} \pm \mathrm{i})^{-1} = (\overline{\pi_0(a)} \pm \mathrm{i})^{-1} = (\overline{\pi_1(a)} \pm \mathrm{i})^{-1}. \tag{4}$$

Therefore, by Theorem 9.1.2, $\pi_1 \equiv \widehat{\pi_0}$ is integrable, and the implication (i) → (ii) is proved.

From Theorem 9.1.7, $\mathcal{M} = \pi_1(\mathsf{A})''$ is abelian. We show that $(\pi_1, \mathcal{M})$ is an induced extension of π. For this it remains only to prove that $\mathcal{M}\mathcal{D}(\pi)$ is dense in $\mathcal{D}(\pi_1)$ $[\mathsf{t}_{\pi_1}]$. Suppose that $p_1 = p_1^+$, $p_2 = p_2^+$ and p_3 are elements of P. Let a_1, a_2 and a_3 denote the corresponding elements of A which are obtained when we replace λ_j by y_j, $j \in J$. Let $k, l \in \mathbb{N}$. Set $f := (p_1 + \mathrm{i})^{-k} (p_2 - \mathrm{i})^{-l} p_3$. By (3) and (4), we have for $\varphi \in \mathcal{D}(\varrho) \equiv \mathcal{D}(\pi)$

$$\begin{aligned} \varrho_1(f)\,\varphi &= \varrho_1\big((p_1 + \mathrm{i})^{-1}\big)^k \varrho_1\big((p_2 - \mathrm{i})^{-1}\big)^l \varrho_1(p_3)\,\varphi \\ &= \big(\overline{\pi_1(a_1)} + \mathrm{i}\big)^{-k} \big(\overline{\pi_1(a_2)} - \mathrm{i}\big)^{-l} \pi_1(a_3)\,\varphi. \end{aligned} \tag{5}$$

Since π_1 is integrable, for each $a \in \mathsf{A}_\mathrm{h}$ the self-adjoint operator $\overline{\pi_1(a)}$ (by Theorem 9.1.2) is affiliated with the von Neumann algebra $\mathcal{M} \equiv \pi_1(\mathsf{A})'' = \big(\pi_1(\mathsf{A})'_{\mathrm{ss}}\big)'$, cf. Proposition 7.2.10. Hence $\big(\overline{\pi_1(a_1)} + \mathrm{i}\big)^{-k}$ and $\big(\overline{\pi_1(a_2)} - \mathrm{i}\big)^{-l}$ are in $\mathcal{M}$. Further, $\pi_1(a_3)\,\varphi = \pi(a_3)\,\varphi$, since $\pi \subseteq \pi_1$. Because of these facts, (5) shows that $\varrho_1(f)\,\mathcal{D}(\varrho) \subseteq \mathcal{M}\mathcal{D}(\pi)$. The *-algebra R is the linear span of elements f of the above form, so $\varrho_1(\mathsf{R})\,\mathcal{D}(\varrho) \subseteq \mathcal{M}\mathcal{D}(\pi)$. By Theorem 11.2.8, $\varrho_1(\mathsf{R})\,\mathcal{D}(\varrho)$ is dense in $\mathcal{D}(\varrho_1)$ $[\mathsf{t}_{\varrho_1}]$ and hence in $\mathcal{D}(\pi_0)$ $[\mathsf{t}_{\pi_0}]$, since $\pi_0(\mathsf{A}) \subseteq \varrho_1(\mathsf{R})$. Because of $\pi_1 = \widehat{\pi_0}$, this implies that $\mathcal{M}\mathcal{D}(\pi)$ is dense in $\mathcal{D}(\pi_1)$ $[\mathsf{t}_{\pi_1}]$.

Now we prove that (ii) implies (i). We suppose $n \in \mathbb{N}$ and $[a_{kl}] \in \boldsymbol{M}_n(\mathsf{A}; \mathrm{int})_+$. Then there exist $m \in \mathbb{N}$, a positive definite matrix $[p_{kl}] \in \boldsymbol{M}_n(\mathbb{C}[\mathsf{x}_1, \ldots, \mathsf{x}_m])$ and indices $j_1, \ldots, j_m \in J$ such that $a_{kl} = p_{kl}(y_{j_1}, \ldots, y_{j_m})$ for all k, l. Since π_1 is integrable by (ii), we know from Theorem 9.1.2 that the operators $\overline{\pi_1(y_{j_1})}, \ldots, \overline{\pi_1(y_{j_m})}$ are self-adjoint and that their spectral projections mutually commute. Let $\{e_r(\lambda) : \lambda \in \mathbb{R}\}$ be the spectrae resolution of $\overline{\pi_1(y_{j_r})}$, $r = 1, \ldots, m$. From the spectral calculus of strongly commuting self-adjoint operators and from the fact that $\pi \subseteq \pi_1$ we obtain

$$\begin{aligned} &\sum_{k,l=1}^{n} \langle \pi(a_{kl})\,\varphi_l, \varphi_k \rangle \\ &= \int \sum_{k,l=1}^{n} p_{kl}(\lambda_1, \ldots, \lambda_m)\, \mathrm{d}\langle e_1(\lambda_1) \ldots e_m(\lambda_m)\,\varphi_l, \varphi_k \rangle \end{aligned} \tag{6}$$

for $\varphi_1, \ldots, \varphi_n \in \mathcal{D}(\pi)$. The expression in (6) is the limit of sums of terms $\sum_{k,l=1}^{n} \alpha_{kl} \langle e\varphi_l, \varphi_k \rangle$, where e is a certain projection on $\mathcal{H}(\pi_1)$ and $[\alpha_{kl}]$ is a positive semi-definite matrix of $\boldsymbol{M}_n(\mathbb{C})$. From the finite dimensional version of the spectral theorem there are a unitary matrix $[\gamma_{kl}] \in \boldsymbol{M}_n(\mathbb{C})$ and non-negative numbers $\delta_1, \ldots, \delta_n$ such that $\alpha_{kl} = \sum_{r=1}^{n} \overline{\gamma_{rk}} \gamma_{rl} \delta_r$. Then

$$\sum_{k,l=1}^{n} \alpha_{kl} \langle e\varphi_l, \varphi_k \rangle = \sum_{r=1}^{n} \delta_r \langle e\psi_r, \psi_r \rangle \geqq 0, \quad \text{where} \quad \psi_r := \sum_{l=1}^{n} \gamma_{rl} \varphi_l.$$

Therefore, the expression in (6) is non-negative. This shows that π is n-positive w.r.t. $\boldsymbol{M}_n(\mathsf{A}; \mathrm{int})_+$ for each $n \in \mathbb{N}$, so π is completely positive w.r.t. $\boldsymbol{M}(\mathsf{A}; \mathrm{int})_+$. □

Remark 4. If the *-representation π of A is n-cyclic, then it suffices (by Corollary 11.2.6) to assum. in Theorem 11.3.4, (i), that π is n-positive with respect to $\boldsymbol{M}_n(\mathsf{A}; \mathrm{int})_+$.

We now derive some further corollaries.

Corollary 11.3.6. *A positive linear functional ω on A is $\mathsf{A}_+^{\mathrm{int}}$-positive if and only if there exists an integrable *-representation π_1 of A which extends π_ω. If π_ω is self-adjoint, then ω is $\mathsf{A}_+^{\mathrm{int}}$-positive if and only if π_ω is integrable.*

Proof. By Corollary 11.2.6, ω is $\mathsf{A}_+^{\text{int}}$-positive if and only if π_ω is completely positive w.r.t. $\boldsymbol{M}(\mathsf{A}; \text{int})_+$; so the assertions follow immediately from Theorem 11.3.4 and Corollary 11.3.5. □

Remark 5. Let $\mathfrak{X}$ be a vector space, and let Φ be a linear mapping of the polynomial algebra $\mathsf{A} := \mathbb{C}[\mathsf{x}]$ into $B(\mathfrak{X})$. If Φ is completely positive (w.r.t. $\mathcal{P}(\boldsymbol{M}(\mathsf{A}))$, then Φ is also completely positive w.r.t. $\boldsymbol{M}(\mathsf{A}; \text{int})_+$. In particular, this means that each $*$-representation of $\mathsf{A} = \mathbb{C}[\mathsf{x}]$ is completely positive w.r.t. $\boldsymbol{M}(\mathsf{A}; \text{int})_+$. We sketch a proof of this assertion. From the fact that a symmetric operator has always a self-adjoint extension in a larger Hilbert space it follows easily that each $*$-representation of $\mathsf{A} \equiv \mathbb{C}[\mathsf{x}]$ and so π_Φ has an integrable extension. From Theorem 11.3.4, π_Φ is completely positive w.r.t. $\boldsymbol{M}(\mathsf{A}; \text{int})_+$. Hence Φ is completely positive w.r.t. $\boldsymbol{M}(\mathsf{A}; \text{int})_+$ by Proposition 11.2.4. □

Remark 6. Let $\mathsf{A} := \mathbb{C}[\mathsf{x}_1, \ldots, \mathsf{x}_n]$ with $n \geqq 2$, and let $\mathcal{A} := \mathbb{C}[x_1, \ldots, x_n]$ be the O*-algebra from Example 2.6.11. That is, $\mathcal{A}$ is the image of A under the faithful $*$-representation π of A defined by $\pi(\mathsf{x}_l) := x_l$, $l = 1, \ldots, n$. From Remark 2 and 2.6/(1), we have $\pi(\mathsf{A}_+^{\text{int}}) = \mathcal{A}_+$. Therefore, Example 2.6.11 also describes the $\mathsf{A}_+^{\text{int}}$-positive linear functionals on A (by Statement 1) and it gives an explicit example (in Statement 3) of a positive linear functional on A which is not $\mathsf{A}_+^{\text{int}}$-positive.

Corollary 11.3.7. *Suppose $\mathfrak{X}$ is a vector space. If the $*$-algebra A is symmetric, then each completely positive linear mapping Φ of A into $B(\mathfrak{X})$ is completely positive with respect to $\boldsymbol{M}(\mathsf{A}; \text{int})_+$. In particular, each positive linear functional on A is $\mathsf{A}_+^{\text{int}}$-positive when A is symmetric.*

Proof. From Corollary 9.1.4, π_Φ is integrable. By Theorem 11.3.4 and Proposition 11.2.4, Φ is completely positive w.r.t. $\boldsymbol{M}(\mathsf{A}; \text{int})_+$. □

Corollary 11.3.8. *Every hermitian character (cf. p. 21) ω on A is $\mathsf{A}_+^{\text{int}}$-positive and a pure state of A, i.e., $\omega \in \text{ex}\, \mathcal{Z}(\mathsf{A})$.*

Proof. Since ω is a hermitian character, ω is a state of A and $\dim \mathcal{H}(\pi_\omega) = 1$. By the latter, $\pi_\omega(\mathsf{A})'_{\mathrm{w}}$ is trivial, so that ω is pure by Corollary 8.6.7. Since all operators $\pi_\omega(a)$, $a \in \mathsf{A}$, are bounded, π_ω is integrable. By Corollary 11.3.6, ω is $\mathsf{A}_+^{\text{int}}$-positive. □

We give a second "elementary" proof of this corollary.

Second proof of Corollary 11.3.8.

Let $a \in \mathsf{A}_+$. By Definition 11.3.1 applied with $\mathsf{Y} \equiv \mathsf{A}_{\mathrm{h}}$, a is of the form $a = p(a_1, \ldots, a_n)$, where $a_1, \ldots, a_n \in \mathsf{A}_{\mathrm{h}}$ and p is a polynomial from $\mathbb{C}[\mathsf{x}_1, \ldots, \mathsf{x}_n]$ which is non-negative on $\mathbb{R}^n$. Since ω is a hermitian character, $\omega(1) = 1$ and $(\omega(a_1), \ldots, \omega(a_n)) \in \mathbb{R}^n$, so $\omega(a) = p(\omega(a_1), \ldots, \omega(a_n)) \geqq 0$. Hence ω is $\mathsf{A}_+^{\text{int}}$-positive. In order to prove that $\omega \in \text{ex}\, \mathcal{Z}(\mathsf{A})$, suppose $\omega = \lambda\omega_1 + (1 - \lambda)\, \omega_2$ with $\omega_1, \omega_2 \in \mathcal{Z}(\mathsf{A})$ and $0 < \lambda < 1$. Let $a \in \mathsf{A}_{\mathrm{h}}$. Then we have

$$\lambda\omega_1(a)^2 + (1 - \lambda)\, \omega_2(a)^2 \leqq \lambda\omega_1(a^2) + (1 - \lambda)\, \omega_2(a^2) = \omega(a^2) = \omega(a)^2$$
$$= \lambda^2\omega_1(a)^2 + (1 - \lambda)^2\, \omega_2(a)^2 + 2\lambda(1 - \lambda)\, \omega_1(a)\omega_2(a),$$

where we used the Cauchy-Schwarz inequality. Therefore,
$\lambda(1 - \lambda)\, (\omega_1(a) - \omega_2(a))^2 \leqq 0$ which gives $\omega_1(a) = \omega_2(a)$. Hence $\omega_1 = \omega_2$ and $\omega \in \text{ex}\, \mathcal{Z}(\mathsf{A})$. □

The next proposition is needed in Chapter 12.

Proposition 11.3.9. *If ω is an extreme point of the convex set of all $\mathsf{A}_+^{\text{int}}$-positive states of A,* *then ω is a character and an extreme point of the set of all states of* A, *that is,*

$$\text{ex}\left((\mathsf{A}_+^{\text{int}})^* \cap \mathcal{Z}(\mathsf{A})\right) \subseteq \text{ex}\,\mathcal{Z}(\mathsf{A}).$$

Proof. We use some notation from the proof of Theorem 11.3.4, (i) → (ii). Let P and R be as defined there. Define
$v_0(p(\lambda_{j_1}, \ldots, \lambda_{j_n})) := \omega(p(y_{j_1}, \ldots, y_{j_n}))$, where $p \in \mathbb{C}[x_1, \ldots, x_n]$, $n \in \mathbb{N}$ and $j_1, \ldots, j_n \in J$. From the assumption $\omega \in \text{ex}\left((\mathsf{A}_+^{\text{int}})^* \cap \mathcal{Z}(\mathsf{A})\right)$ we conclude that $v_0 \in \text{ex}\left((\mathsf{P}_+^{\text{int}})^* \cap \mathcal{Z}(\mathsf{P})\right)$. Recall that P is cofinal in R with respect to the wedge $\mathsf{R}_+^{\text{int}}$. From Lemma 1.3.2 it follows that there is an $\mathsf{R}_+^{\text{int}}$-positive state v on R such that v extends v_0 and v is an extremal point of $(\mathsf{R}_+^{\text{int}})^*$.

We show that the restriction $\pi_v \upharpoonright \mathsf{P}$ is an irreducible integrable representation of P. Proposition 8.1.19 shows that for any $p \in \mathsf{P}_h$ the operator $\overline{\pi_v(p)}$ is self-adjoint and

$$\overline{\pi_v((p \pm \mathrm{i})^{-1})} = \left(\overline{\pi_v(p)} \pm \mathrm{i}\right)^{-1}. \tag{7}$$

Further recall that the *-algebra R is generated by P_h and by the elements $(p \pm \mathrm{i})^{-1}$, $p \in \mathsf{P}_h$. Since the operators $\pi_v((p \pm \mathrm{i})^{-1})$ are bounded by (7), the graph topologies of $\pi_v(\mathsf{P})$ and $\pi_v(\mathsf{R})$ coincide. Thus $\pi_v \upharpoonright \mathsf{P}$ is closed, since π_v is, too. By Theorem 9.1.2, $\pi_v \upharpoonright \mathsf{P}$ is integrable. Let e be a projection in $\pi_v(\mathsf{P})_s'$. Then e commutes with $\pi_v(p)$ for any $p \in \mathsf{P}_h$ and so with $\pi_v((p \pm \mathrm{i})^{-1})$ by (7). Consequently, $e \in \pi_v(\mathsf{R})_s'$. Hence $v_e(a) = \langle e\pi_v(a)\varphi_v, \varphi_v\rangle = \langle \pi_v(a)\, e\varphi_v, e\varphi_v\rangle \geqq 0$ for all $a \in \mathsf{R}_+^{\text{int}}$, since v is $\mathsf{R}_+^{\text{int}}$-positive. Thus $v_e \in (\mathsf{R}_+^{\text{int}})^*$. Similarly, $v - v_e \equiv v_{I-e} \in (\mathsf{R}_+^{\text{int}})^*$. Since v is an extremal point of $(\mathsf{R}_+^{\text{int}})^*$, it follows that $v_e = \lambda v$ and so $e = \lambda \cdot I$ for some $\lambda \in [0, 1]$. Because e was a projection, $e = 0$ or $e = I$. This shows that $\pi_v \upharpoonright \mathsf{P}$ is irreducible.

Being integrable and irreducible, $\pi_v \upharpoonright \mathsf{P}$ must act on a one-dimensional Hilbert space by Corollary 9.1.11. Therefore, v and so ω is a character. From Corollary 11.3.8, $\omega \in \text{ex}\,\mathcal{Z}(\mathsf{A})$. □

11.4. A Second Application: Integrable Extensions of *-Representations of Enveloping Algebras

Throughout this section, G is a Lie group with Lie algebra $\mathfrak{g}$ and $\mathcal{E}(\mathfrak{g})$ is the enveloping algebra of $\mathfrak{g}$, cf. Section 1.7. We shall use some notation and facts from Sections 10.1 and 10.2.

Definition 11.4.1. For $n \in \mathbb{N}$, let $M_n(\mathcal{E}(\mathfrak{g}); G)_+$ be the set of all matrices $[a_{kl}] \in M_n(\mathcal{E}(\mathfrak{g}))_h$ such that for each unitary representation U of G the following condition is fulfilled:

$$\sum_{k,l=1}^{n} \langle \mathrm{d}U(a_{kl})\, \varphi_l, \varphi_k\rangle \geqq 0 \quad \text{for all vectors} \quad \varphi_1, \ldots, \varphi_n \in \mathcal{D}^\infty(U). \tag{1}$$

Set $M(\mathcal{E}(\mathfrak{g}); G)_+ := \bigcup_{n \in \mathbb{N}} M_n(\mathcal{E}(\mathfrak{g}); G)_+$.

It is clear that $M(\mathcal{E}(\mathfrak{g}); G)_+$ coincides with the wedge $M(\mathsf{A}; \mathfrak{R})_+$ defined by 11.2/(6) when $\mathsf{A} := \mathcal{E}(\mathfrak{g})$ and $\mathfrak{R}$ is the family of all G-integrable representations of $\mathcal{E}(\mathfrak{g})$. Hence $M(\mathcal{E}(\mathfrak{g}); G)_+$ is an m-admissible wedge in $M(\mathcal{E}(\mathfrak{g}))$. Letting $U := U_{lr}$ (cf. Example 10.1.8), we conclude that $M(\mathcal{E}(\mathfrak{g}); G)_+$ is a cone.

Remark 1. In Proposition 12.3.6 we show that the same wedge will be obtained if we require (1) only for irreducible unitary representations of G.

Remark 2. When $G = \mathbb{R}^d$, $\mathcal{E}(\mathfrak{g})$ coincides with the polynomial algebra $\mathbb{C}[\mathsf{x}_1, \ldots, \mathsf{x}_d]$ in the usual way. In this case $\boldsymbol{M}(\mathcal{E}(\mathfrak{g}); G)_+$ is equal to the cone $\boldsymbol{M}(\mathbb{C}[\mathsf{x}_1, \ldots, \mathsf{x}_d]; \mathrm{int})_+$ from Definition 11.3.1. To verify this, we set $\mathsf{Y} = \{\mathsf{x}_1, \ldots, \mathsf{x}_d\}$ in Definition 11.3.1. Then the equality of both wedges follows from Remark 1 and the following well-known fact: The irreducible unitary representations of $G = \mathbb{R}^d$ are precisely those of the form U_λ with $\lambda \equiv (\lambda_1, \ldots, \lambda_d) \in \mathbb{R}^d$, where $\mathrm{d}U_\lambda(p) = p(\lambda_1, \ldots, \lambda_d)$ for $p \in \mathbb{C}[\mathsf{x}_1, \ldots, \mathsf{x}_d]$ and $\mathcal{H}(U_\lambda) = \mathbb{C}$.

We now state the main result of this section.

Theorem 11.4.2. *Suppose that π is a $*$-representation of $\mathcal{E}(\mathfrak{g})$ which is completely positive with respect to the wedge $\boldsymbol{M}(\mathcal{E}(\mathfrak{g}); G)_+$. Then there exists a unitary representation V of G on a possibly larger Hilbert space such that $\pi \subseteq \mathrm{d}V$.*

From this theorem we obtain the following corollary.

Corollary 11.4.3. *Suppose that the Lie group G is connected. Then a self-adjoint $*$-representation of $\mathcal{E}(\mathfrak{g})$ is G-integrable if and only if it is completely positive with respect to $\boldsymbol{M}(\mathcal{E}(\mathfrak{g}); G)_+$.*

Proof. By definition each G-integrable $*$-representation of $\mathcal{E}(\mathfrak{g})$ is trivially completely positive w.r.t. $\boldsymbol{M}(\mathcal{E}(\mathfrak{g}); G)_+$. In order to prove the sufficiency, we recall from Proposition 10.2.19 that a self-adjoint subrepresentation of a G-integrable representation is itself G-integrable, since G is connected; hence the assertion follows at once from Theorem 11.4.2. □

We will derive Theorem 11.4.2 from the following theorem which states the main result in a slightly different form.

Theorem 11.4.4. *Let U be a unitary representation of the Lie group G. Suppose that π is a $*$-representation of the O^*-algebra $\mathrm{d}U(\mathcal{E}(\mathfrak{g}))$ which is completely positive with respect to the wedge $\boldsymbol{M}(\mathrm{d}U(\mathcal{E}(\mathfrak{g})))_+$. Then there exists a unitary representation V of G such that $\pi \circ \mathrm{d}U \subseteq \mathrm{d}V$. If G is connected and π is self-adjoint, then $\pi \circ \mathrm{d}U$ is G-integrable.*

Proof of Theorem 11.4.2 (granted Theorem 11.4.4). Recall from Example 10.1.8 that the differential $\mathrm{d}U_{lr}$ of the left regular representation U_{lr} of G is a faithful $*$-representation of $\mathcal{E}(\mathfrak{g})$. Thus $\pi_0 := \pi \circ (\mathrm{d}U_{lr})^{-1}$ is a $*$-representation of the O*-algebra $\mathrm{d}U_{lr}(\mathcal{E}(\mathfrak{g}))$. From the definition of $\boldsymbol{M}(\mathcal{E}(\mathfrak{g}); G)_+$ it is obvious that a matrix $[\mathrm{d}U_{lr}(a_{kl})]$ is in $\boldsymbol{M}(\mathrm{d}U_{lr}(\mathcal{E}(\mathfrak{g})))_+$ when $[a_{kl}]$ is in $\boldsymbol{M}(\mathcal{E}(\mathfrak{g}); G)_+$; so π_0 is completely positive with respect to $\boldsymbol{M}\mathrm{d}(U_{lr}(\mathcal{E}(\mathfrak{g})))_+$ and Theorem 11.4.4 applies to π_0 and U_{lr}. Letting V be the corresponding unitary representation of G, we have $\pi \equiv \pi_0 \circ \mathrm{d}U_{lr} \subseteq \mathrm{d}V$ which is the assertion of Theorem 11.4.2. □

The rest of this section is devoted to the proof of Theorem 11.4.4. Since some steps of this proof are of interest in itself, they are stated separately as lemmas and proved in a somewhat stronger form than is really needed.

Lemma 11.4.5. *Let $\{x_1, \ldots, x_d\}$ be a basis of $\mathfrak{g}$ and let V be a homomorphism of G into the unitaries of a Hilbert space $\mathcal{H}$ such that $V(e) = I$. Suppose that $\lim_{t\to 0} V(\exp tx_k)\,\varphi = \varphi$ in $\mathcal{H}$ for all $\varphi \in \mathcal{H}$ and $k = 1, \ldots, d$. Then V is continuous relative to the strong-operator topology on $\mathcal{H}$ and thus a unitary representation of G.*

Proof. Put $g(t) := \exp t_1x_1 \ldots \exp t_dx_d$ for $t = (t_1, \ldots, t_d) \in \mathbb{R}^d$ and $V_k(s) := V(\exp sx_k)$

for $k = 1, \dots, d$ and $s \in \mathbb{R}$. The estimate

$$\left\|\left(V(g(t)) - V(e)\right)\varphi\right\|$$
$$= \left\|V_1(t_1) \dots V_{d-1}(t_{d-1})\left(V_d(t_d) - I\right)\varphi + V_1(t_1) \dots V_{d-2}(t_{d-2})\left(V_{d-1}(t_{d-1}) - I\right)\varphi + \cdots\right.$$
$$\left. + \left(V_1(t_1) - I\right)\varphi\right\| \leqq \left\|\left(V_1(t_1) - I\right)\varphi\right\| + \cdots + \left\|\left(V_d(t_d) - I\right)\varphi\right\|$$

shows that for any $\varphi \in \mathcal{H}$ the map $g \to V(g)\varphi$ of G into $\mathcal{H}$ is continuous at e. □

Suppose U is a unitary representation of G. As noted at the beginning of Section 10.1, the operators $U(g)$, $g \in G$, leave $\mathcal{D}^\infty(U)$ invariant. Let $\mathcal{A} := \mathrm{d}U(\mathcal{E}(\mathfrak{g}))$ and let $\mathcal{B}$ denote the O*-algebra on $\mathcal{D}^\infty(U)$ which is generated by the operators $\mathrm{d}U(x)$, $x \in \mathcal{E}(\mathfrak{g})$, and $\tilde{U}(g) := U(g) \upharpoonright \mathcal{D}^\infty(U)$, $g \in G$. We keep this notation in the next two lemmas.

Lemma 11.4.6. *$\mathcal{A}$ is cofinal in $\mathcal{B}$ with respect to the cone $\mathcal{P}(\mathcal{B})$.*

Proof. We have to show that for each $b \in \mathcal{B}_\mathrm{h}$ there is an $a \in \mathcal{A}_\mathrm{h}$ such that $a \in \mathcal{P}(\mathcal{B})$ and $a - b \in \mathcal{P}(\mathcal{B})$. Since $U(g)\,\mathrm{d}U(x)\,\varphi = \mathrm{d}U(\mathrm{Ad}g\,(x))\,U(g)\,\varphi$ for $x \in \mathcal{E}(\mathfrak{g})$, $g \in G$ and $\varphi \in \mathcal{D}^\infty(U)$ by Lemma 10.1.12, $\mathcal{B}$ is the linear span of elements $c = \mathrm{d}U(x)\,\tilde{U}(g)$, where $x \in \mathcal{E}(\mathfrak{g})$ and $g \in G$. It therefore suffices to prove the assertion for elements b of the form $b = c + c^+$, since $\mathcal{B}_\mathrm{h}$ is the real linear span of these elements and $\mathcal{P}(\mathcal{B})$ is a wedge. Set $a := \mathrm{d}U(x)\,\mathrm{d}U(x)^+ + I$. Obviously, $a \in \mathcal{P}(\mathcal{B})$. We have

$$\left(\mathrm{d}U(x^+) - \tilde{U}(g)\right)^+ \left(\mathrm{d}U(x^+) - \tilde{U}(g)\right) = \mathrm{d}U(x)\,\mathrm{d}U(x)^+ - \tilde{U}(g)^+\,\mathrm{d}U(x)^+ - \mathrm{d}U(x)\,\tilde{U}(g)$$
$$+ \tilde{U}(g)^+\tilde{U}(g) = \mathrm{d}U(x)\,\mathrm{d}U(x)^+ - c^+ - c + I$$
$$= a - b \in \mathcal{P}(\mathcal{B}).\ \square$$

The heart of the proof of Theorem 11.4.4 is contained in

Lemma 11.4.7. *Suppose that π_1 is a $*$-representation of $\mathcal{B}$ such that $\pi_1(b) \geqq 0$ when $b \in \mathcal{B}_+$. Define $\varrho(x) := \pi_1(\mathrm{d}U(x))$ for $x \in \mathcal{E}(\mathfrak{g})$. Then there is a unitary representation V of G on the Hilbert space $\mathcal{H}(\pi_1)$ such that $\mathrm{d}V = \hat{\varrho}$.*

Proof. Define $V(g) := \pi_1(\tilde{U}(g))$, $g \in G$. Since π_1 is a $*$-representation of $\mathcal{B}$, we have $V(g)^+\,V(g) = V(g)\,V(g)^+ = I$ for $g \in G$; so $V(g)$ extends by continuity to a unitary operator on $\mathcal{H}(\pi_1)$ which we denote by the same symbol. Using once more that π_1 is a $*$-representation we conclude that $g \to V(g)$ is a homomorphism of G into the unitaries of $\mathcal{H}(\pi_1)$ and $V(e) = I$.

Fix $x \in \mathfrak{g}$. Set $a(t) := \tilde{U}(\exp tx) - I - t\,\mathrm{d}U(x)$, $t \in \mathbb{R}$. By Corollary 10.2.11, $\overline{\mathrm{d}U(ix)}$ is a self-adjoint operator on $\mathcal{H}(U)$. Let $e(\lambda)$, $\lambda \in \mathbb{R}$, denote the spectral projections of this operator. Recall that $U(\exp tx) = \exp t\,\overline{\mathrm{d}U(x)}$, $t \in \mathbb{R}$, by Corollary 10.2.13. Since $|\mathrm{e}^{-it\lambda} - 1 + it\lambda| \leqq \lambda^2t^2$ for all real λ and t, it follows from the functional calculus for self-adjoint operators that

$$\langle a(t)^+\,a(t)\,\varphi, \varphi\rangle = \|a(t)\,\varphi\|^2 = \int |\mathrm{e}^{-it\lambda} - 1 + it\lambda|^2\,\mathrm{d}\,\|e(\lambda)\,\varphi\|^2 \leqq t^4 \int \lambda^4\,\mathrm{d}\,\|e(\lambda)\,\varphi\|^2$$
$$= t^4\left\|\left(\overline{\mathrm{d}U(ix)}\right)^2\varphi\right\|^2 = \langle t^4\,\mathrm{d}U(x^4)\,\varphi, \varphi\rangle$$

for $\varphi \in \mathcal{D}^\infty(U)$ and $t \in \mathbb{R}$. That is, $t^4\,\mathrm{d}U(x^4) - a(t)^+\,a(t) \in \mathcal{B}_+$ for all $t \in \mathbb{R}$. Therefore, by assumption, $\pi_1(t^4\,\mathrm{d}U(x^4) - a(t)^+\,a(t)) \geqq 0$, i.e.,

$$\langle \pi_1(a(t)^+\,a(t))\,\psi, \psi\rangle = \|\pi_1(a(t))\,\psi\|^2 = \left\|\left(V(\exp tx) - I - t\varrho(x)\right)\psi\right\|^2$$
$$\leqq \langle \pi_1(t^4\,\mathrm{d}U(x^4))\,\psi, \psi\rangle = t^4\langle \varrho(x^4)\,\psi, \psi\rangle = \left(t^2\,\|\varrho(x^2)\,\psi\|\right)^2$$

for $\psi \in \mathcal{D}(\pi_1) \equiv \mathcal{D}(\varrho)$. Thus, for $\psi \in \mathcal{D}(\varrho)$ and $t \in \mathbb{R}$, we have

$$\left\|t^{-1}\big(V(\exp tx) - I\big)\psi - \varrho(x)\psi\right\| \leqq t\,\|\varrho(x^2)\psi\|. \tag{1}$$

In particular, (1) implies that $\lim_{t\to 0} V(\exp tx)\psi = \psi$ in $\mathcal{H}(\varrho)$ for all $\psi \in \mathcal{D}(\varrho)$. Since $\mathcal{D}(\varrho)$ is dense in $\mathcal{H}(\varrho)$, the latter is true for all $\psi \in \mathcal{H}(\varrho)$. Then, by Lemma 11.4.5, the map $g \to V(g)$ is a unitary representation of G on $\mathcal{H}(\varrho)$.

We prove that $\mathrm{d}V = \hat{\varrho}$. Again let $x \in \mathfrak{g}$. By definition, $\partial V(x)$ is the infinitesimal generator of the one-parameter unitary group $t \to V(\exp tx)$. Therefore, we conclude from (1) that each vector $\psi \in \mathcal{D}(\varrho)$ is in $\mathcal{D}(\partial V(x))$ and $\varrho(x)\psi = \partial V(x)\psi$. Since $\varrho(x)$ leaves $\mathcal{D}(\varrho)$ invariant, this leads to $\mathcal{D}(\varrho) \subseteq \bigcap_{k=1}^{d} \bigcap_{n\in\mathbb{N}} \mathcal{D}\big(\partial V(x_k)^n\big)$, when $\{x_1, \dots, x_d\}$ is a basis of $\mathfrak{g}$. Theorem 10.1.9 says that the latter set is equal to $\mathcal{D}^\infty(V)$; so $\mathcal{D}(\varrho) \subseteq \mathcal{D}^\infty(V)$. Since $\varrho(x)\psi = \partial V(x)\psi = \mathrm{d}V(x)\psi$ for $x \in \mathfrak{g}$ and $\psi \in \mathcal{D}(\varrho)$, we have $\varrho \subseteq \mathrm{d}V$. By construction the operators $V(g)$, $g \in G$, leave the domain $\mathcal{D}(\varrho)$ invariant. Therefore, by Theorem 10.1.14, $\mathcal{D}(\varrho)$ is dense in $\mathcal{D}^\infty(V)\,[\mathrm{t}_{\mathrm{d}V}]$. This gives $\mathrm{d}V = \hat{\varrho}$. □

Proof of Theorem 11.4.4. Let $\mathcal{A}$ and $\mathcal{B}$ be as defined above. First we note that $\mathsf{A} := \mathcal{A}$, $\mathsf{B} := \mathcal{B}$, $\boldsymbol{K} := \boldsymbol{M}(\mathcal{B})_+$, and π satisfy the assumptions of Theorem 11.2.8. Indeed, by Lemma 11.4.6, $\mathcal{A}$ is cofinal in $\mathcal{B}$ with respect to $\mathcal{P}(\mathcal{B})$ and hence with respect to $\mathcal{B}_+ \equiv \boldsymbol{K} \cap \mathsf{B}$. From the definitions it is clear that $\boldsymbol{M}(\mathcal{B})_+ \cap \boldsymbol{M}(\mathcal{A}) = \boldsymbol{M}(\mathcal{A})_+ \equiv \boldsymbol{M}\big(\mathrm{d}U(\mathcal{E}(\mathfrak{g}))\big)_+$. Hence, by the assumptions of Theorem 11.4.4, π is completely positive with respect to $\boldsymbol{K} \cap \boldsymbol{M}(\mathsf{A})$, and the assumptions of Theorem 11.2.8 are fulfilled. Let π_1 be the representation of $\mathsf{B} \equiv \mathcal{B}$ from Theorem 11.2.8. Since π_1 is completely positive with respect to $\boldsymbol{M}(\mathcal{B})_+$, we have $\pi_1(b) \geqq 0$ when $b \in \mathcal{B}_+$. Thus, by Lemma 11.4.7, there is a unitary representation V of G on $\mathcal{H}(\pi_1)$ such that $\varrho \subseteq \mathrm{d}V$, where $\mathcal{D}(\varrho) := \mathcal{D}(\pi_1)$ and $\varrho(x) := \pi_1\big(\mathrm{d}U(x)\big)$, $x \in \mathcal{E}(\mathfrak{g})$. Since $\pi(a) \subseteq \pi_1(a)$ for $a \in \mathcal{A}$ by Theorem 11.2.8, we obtain $\pi \circ \mathrm{d}U \subseteq \mathrm{d}V$, and the proof of the main assertion of Theorem 11.4.4 is complete.

If in addition π is self-adjoint and G is connected, then, of course, $\pi \circ \mathrm{d}V$ is also self-adjoint and hence G-integrable by Proposition 10.2.19. □

Remark 3. In the notation of the preceding proof, the linear space spanned by the vectors $\pi_1\big(\bar{U}(g)\big)\varphi$, where $g \in G$ and $\varphi \in \mathcal{D}(\pi)$, is dense in $\mathcal{D}(\pi_1)$ relative to the graph topology of $\pi_1\big(\mathrm{d}U(\mathcal{E}(\mathfrak{g}))\big)$.

11.5. A Third Application: Completely Centrally Positive Operators

Throughout this section, A denotes a $*$-algebra with unit and a denotes a fixed element of A.

Let π be a $*$-representation of A. If a is a hermitian element of the center of A, then there is, in general, no $*$-representation π_1 of A such that $\pi \subseteq \pi_1$ and such that $\overline{\pi_1(a)}$ is affiliated with the von Neumann algebra $\pi_1(\mathsf{A})'_{\mathrm{ss}}$, see Example 11.5.8 below. In this section we give a necessary and sufficient condition in terms of complete positivity with respect to a certain wedge in $\boldsymbol{M}(\mathsf{A})$ that such an extension exists.

If $x \in \mathsf{A}$ and $A \equiv [a_{kl}] \in \boldsymbol{M}(\mathsf{A})$, we let xA denote the matrix $[xa_{kl}]$.

Definition 11.5.1. Let $\boldsymbol{K}(\mathsf{A}; a)$ be the set of all matrices $A \in \boldsymbol{M}(\mathsf{A})$ of the form $A = A_0 + aA_1 + \dots + a^m A_m$, where $m \in \mathbb{N}_0$ and $A_0, A_1, \dots, A_m$ are matrices in $\boldsymbol{M}(\mathsf{A})$ such that $A_0 + \lambda A_1 + \dots + \lambda^m A_m \in \mathcal{P}\big(\boldsymbol{M}(\mathsf{A})\big)$ for all real numbers λ. For $n \in \mathbb{N}$, let $\boldsymbol{K}_n(\mathsf{A}; a) := \boldsymbol{K}(\mathsf{A}; a) \cap \boldsymbol{M}_n(\mathsf{A})$.

Remark 1. If $A = A_0 + aA_1 + \cdots + a^m A_m$ is as in Definition 11.5.1, then we conclude easily that $A_0, \ldots, A_m \in \boldsymbol{M}(\mathsf{A})_{\mathrm{h}}$.

Remark 2. Obviously, $\boldsymbol{K}(\mathsf{A}; a)$ is a wedge in the vector space $\boldsymbol{M}(\mathsf{A})$. However, $\boldsymbol{K}(\mathsf{A}; a)$ is not contained in $\boldsymbol{M}(\mathsf{A})_{\mathrm{h}}$ in general. If a is a hermitian element in the center of A, then it follows immediately (using Remark 1) that $\boldsymbol{K}(\mathsf{A}; a)$ is an m-admissible wedge in $\boldsymbol{M}(\mathsf{A})$.

Definition 11.5.2. Let $\mathfrak{X}$ be a vector space and let Φ be a linear map of $\mathcal{A}$ into $B(\mathfrak{X})$. Let $n \in \mathbb{N}$. We say that $\Phi(a)$ is *centrally n-positive* if Φ is n-positive with respect to $\boldsymbol{K}_n(\mathsf{A}; a)$ (i.e., $\Phi_{(n)}$ maps $\boldsymbol{K}_n(\mathsf{A}; a)$ into $\boldsymbol{K}_n(B(\mathfrak{X}))$) and that $\Phi(a)$ is *completely centrally positive* if Φ is completely positive with respect to $\boldsymbol{K}(\mathsf{A}; a)$ (i.e., $\Phi_{(\infty)}$ maps $\boldsymbol{K}(\mathsf{A}; a)$ into $\boldsymbol{K}(B(\mathfrak{X}))$).

In other words, $\Phi(a)$ is centrally n-positive if and only if

$$\sum_{k,l=1}^{n} \Phi\left(\sum_{r=0}^{m} a^r a_{kl}^{(r)}\right) (\varphi_l, \varphi_k) \geqq 0$$

for arbitrary vectors $\varphi_1, \ldots, \varphi_n \in \mathfrak{X}$ and matrices $A_r = [a_{kl}^{(r)}] \in \boldsymbol{M}(\mathsf{A})$, $r = 0, \ldots, m$, $m \in \mathbb{N}_0$, for which $A_0 + \lambda A_1 + \cdots + \lambda^m A_m \in \mathcal{P}(\boldsymbol{M}(\mathsf{A}))$ for any $\lambda \in \mathbb{R}$. If this holds for all $n \in \mathbb{N}$, then $\Phi(a)$ is completely centrally positive.

Remark 3. It should be noted that the central n-positivity depends not only on the element $\Phi(a)$ itself, but also on the map Φ.

Before we state the main theorem, we prove two preliminary lemmas. The first one justifies the word "centrally" in Definition 11.5.2.

Lemma 11.5.3. *Suppose π is a $*$-representation of* A. *If $\pi(a)$ is centrally 1-positive, then $\pi(a)$ is a symmetric operator contained in the center of the O*-algebra $\pi(\mathsf{A})$.*

Proof. Suppose $x \in \mathsf{A}_{\mathrm{h}}$. Since $\lambda^2 x^2 \pm 2\lambda x + 1 = (\lambda x \pm 1)^+ (\lambda x \pm 1) \in \mathcal{P}(\mathsf{A})$ for all $\lambda \in \mathbb{R}$ and $\pi(a)$ is centrally 1-positive, we have $\pi(a^2x^2 + 2ax + 1) \geqq 0$ and $\pi(a^2x^2 - 2ax + 1) \geqq 0$ on $\mathcal{D}(\pi)$. Hence $\pi(ax)$ is a symmetric operator. Putting $x = 1$ we see that $\pi(a)$ is a symmetric operator. For general $x \in \mathsf{A}_{\mathrm{h}}$, we have

$$\pi(a)\,\pi(x) = \pi(ax) = \pi(ax)^+ = \big(\pi(a)\,\pi(x)\big)^+ = \pi(x)^+\,\pi(a)^+ = \pi(x)\,\pi(a).$$

This yields $\pi(a)\,\pi(x) = \pi(x)\,\pi(a)$ for all $x \in \mathsf{A}$; so $\pi(a)$ belongs to the center of $\pi(\mathsf{A})$. □

Lemma 11.5.4. *Let π be a $*$-representation of* A. *If $\pi(a)$ is a symmetric operator such that $\overline{\pi(a)}$ is affiliated with the von Neumann algebra $\pi(\mathsf{A})'_{\mathrm{ss}}$, then $\overline{\pi(a)}$ is a self-adjoint operator.*

Proof. Upon replacing π by $\tilde{\pi}$ if necessary, we can assume without loss of generality that π is closed. Then $\pi(\mathsf{A})'_{\mathrm{ss}} \subseteqq \pi(\mathsf{A})'_{\mathrm{s}}$ by Proposition 7.2.9, (ii). Let Q_+ and Q_- be the projection of $\mathcal{H}(\pi)$ onto the deficiency spaces of the symmetric operator $\overline{\pi(a)}$ for $z = \mathrm{i}$ and $z = -\mathrm{i}$, respectively. Since $\overline{\pi(a)}$ is affiliated with $\pi(\mathsf{A})'_{\mathrm{ss}}$, Q_+ and Q_- are in $\pi(\mathsf{A})'_{\mathrm{ss}}$ and so in $\pi(\mathsf{A})'_{\mathrm{s}}$. In particular, Q_+ and Q_- leave $\mathcal{D}(\pi)$ invariant. Let $\varphi \in \mathcal{D}(\pi)$. Since $\pi(a) \subseteqq \pi(a)^*$, we have $\pi(a)\,Q_\pm\,\varphi = \pi(a)^*\,Q_\pm\,\varphi = \pm\mathrm{i}Q_\pm\varphi$. Because $\pi(a)$ is symmetric, $Q_\pm\varphi = 0$. Thus $Q_+ = Q_- = 0$, and $\overline{\pi(a)}$ is self-adjoint. □

Theorem 11.5.5. *Suppose π is a $*$-representation of* A *and a is a hermitian element of* A. *Then the following two statements are equivalent:*

(i) *$\pi(a)$ is completely centrally positive.*

(ii) *There exists a $*$-representation π_1 of* A *such that $\pi \subseteq \pi_1$ and such that the operator $\overline{\pi_1(a)}$ is affiliated with the von Neumann algebra $\pi_1(\mathsf{A})'_{ss}$.*

If (i) *is valid, then the $*$-representation π_1 in* (ii) *can be chosen such that π_1 is closed and $(\pi_1, \mathcal{M})$ is an induced extension of π (in the sense of Definition* 8.5.3), *where $\mathcal{M}$ is the commutative von Neumann algebra $\left(\overline{\pi_1(a)}\right)''_s$.*

Remark 4. If $\overline{\pi_1(a)}$ is affiliated with $\pi_1(\mathsf{A})'_{ss}$, then $\overline{\pi_1(a)}$ is self-adjoint by Lemma 11.5.4 and so $\left(\overline{\pi_1(a)}\right)''_s$ is equal to the commutative von Neumann algebra which is generated by the spectral projections of $\overline{\pi_1(a)}$.

Proof of Theorem 11.5.5. We first prove that (i) implies (ii).

Let A_1 [resp. B_1] be the $*$-algebra of all mappings $\lambda \to x(\lambda)$ of the real line into A of the form

$$x(\lambda) = \sum_{k=1}^{n} q_k(\lambda)\, x_k \tag{1}$$

with $n \in \mathbb{N}$, $x_k \in \mathsf{A}$ and $q_k(\lambda)$ a complex polynomial in λ [resp. a polynomially bounded continuous function in λ] for $k = 1, \dots, n$. The algebraic operations of A_1 and B_1 are defined to be the pointwise operations. Then A_1 is a $*$-subalgebra of B_1 which contains the unit element of B_1.

We check that A_1 is cofinal in B_1 w.r.t. the wedge $\mathcal{P}(\mathsf{B}_1)$. We let $x(\lambda) = x(\lambda)^+ \in \mathsf{B}_1$. We can express $x(\lambda)$ as in (1) with $x_k = x_k^+ \in \mathsf{A}$ and q_k real polynomially bounded continuous functions. We take a real polynomial $p_k(\lambda)$ such that $|q_k(\lambda)| \leqq p_k(\lambda)$ for all $\lambda \in \mathbb{R}$. Set $y(\lambda) := \sum_{k=1}^{n} \frac{1}{2}\, p_k(\lambda)\, (x_k^2 + 1)$. It is easy to see that $y(\lambda) - x(\lambda) \in \mathcal{P}(\mathsf{B}_1)$. Since obviously $y(\lambda) \in \mathcal{P}(\mathsf{B}_1)$, this shows that A_1 is cofinal in B_1 w.r.t. $\mathcal{P}(\mathsf{B}_1)$.

We define

$$\varrho\big(x(\lambda)\big) = \sum_{k=1}^{n} \pi\big(q_k(a)\, x_k\big) \tag{2}$$

when $x(\lambda) \in \mathsf{A}_1$ is as in (1). Since $\pi(a)$ is in particular centrally 1-positive, Lemma 11.5.4 says that $\pi(a)$ is a symmetric operator in the center of $\pi(\mathsf{A})$. Using this fact it is straightforward to verify that ϱ is a well-defined $*$-representation of A_1 on $\mathcal{D}(\varrho) := \mathcal{D}(\pi)$. Since $\pi(a)$ is completely centrally positive by (i), it follows immediately from Definition 11.5.2 that ϱ is completely positive (with respect to $\mathcal{P}\big(\boldsymbol{M}(\mathsf{B}_1)\big) \cap \boldsymbol{M}(\mathsf{A}_1)$). Therefore, by Theorem 11.2.8, there is a closed $*$-representation ϱ_1 of B_1 such that

$$\mathcal{D}(\varrho) \subseteq \mathcal{D}(\varrho_1) \quad \text{and} \quad \varrho(x) = \varrho_1(x) \restriction \mathcal{D}(\varrho) \quad \text{for} \quad x \in \mathsf{A}_1 \tag{3}$$

and such that $\varrho_1(\mathsf{B}_1)\, \mathcal{D}(\varrho)$ is dense in $\mathcal{D}(\varrho_1)\, [\mathsf{t}_{\varrho_1}]$.

We consider A as a $*$-subalgebra of A_1 by identifying $x \in \mathsf{A}$ with the "constant" mapping $x(\lambda) = x$, $\lambda \in \mathbb{R}$. Let π_1 be the $*$-representation of A defined by $\mathcal{D}(\pi_1) := \mathcal{D}(\varrho_1)$ and $\pi_1(x) := \varrho_1(x)$, $x \in \mathsf{A}$. By the above definition of ϱ, we have $\pi(x) = \varrho(x)$ for $x \in \mathsf{A}$. Combined with (3), this gives $\pi \subseteq \pi_1$.

Next we show that $\overline{\pi_1(a)}$ is a self-adjoint operator which is affiliated with the von Neumann algebra $\pi_1(\mathsf{A})'_{ss}$. First we check that $\pi_1(a) = \varrho_1(\lambda \cdot 1)$. Suppose $q(\lambda)$ is a polynomially bounded continuous function on $\mathbb{R}$, $x \in \mathsf{A}$ and $\varphi \in \mathcal{D}(\varrho)$. From (2) and (3), we have

$$\begin{aligned}\big(\pi_1(a) - \varrho_1(\lambda\cdot 1)\big)\,\varrho_1\big(q(\lambda)\,x\big)\,\varphi &= \varrho_1\big((a-\lambda\cdot 1)\,q(\lambda)\,x\big)\,\varphi\\ &= \varrho_1\big(q(\lambda)\,1\big)\,\varrho_1\big((a-\lambda\cdot 1)\,x\big)\,\varphi\\ &= \varrho_1\big(q(\lambda)\,1\big)\,\varrho\big((a-\lambda\cdot 1)\,x\big)\,\varphi\\ &= \varrho_1\big(q(\lambda)\,1\big)\,\pi(ax-ax)\,\varphi = 0.\end{aligned}$$

Since B_1 is the linear span of such elements $q(\lambda)\,x$, we get $\pi_1(a)\,\psi = \varrho_1(\lambda\cdot 1)\,\psi$ for $\psi \in \varrho_1(\mathsf{B}_1)\,\mathcal{D}(\varrho)$. Because the latter is dense in $\mathcal{D}(\varrho_1)\,[\mathsf{t}_{\varrho_1}]$, this implies that $\pi_1(a) = \varrho_1(\lambda\cdot 1)$. Recall that $(\lambda - z)^{-1}\cdot 1 \in \mathsf{B}_1$ for any $z \in \mathbb{C}\setminus\mathbb{R}$. Therefore, it follows from Proposition 8.1.19 that the operator $\overline{\varrho_1(\lambda\cdot 1)} \equiv \overline{\pi_1(a)}$ is self-adjoint and

$$A_z := \overline{\varrho_1\big((\lambda - z)^{-1}\cdot 1\big)} = \big(\overline{\varrho_1(\lambda\cdot 1)} - z\big)^{-1} \equiv \big(\overline{\pi_1(a)} - z\big)^{-1} \tag{4}$$

for $z \in \mathbb{C}\setminus\mathbb{R}$. Since $(\lambda - z)^{-1}\cdot 1$ belongs to the center of B_1, we have $A_z \in \varrho_1(\mathsf{B}_1)'_{\mathrm{s}} \subseteq \pi_1(\mathsf{A})'_{\mathrm{s}}$. Since $A_z^* = A_{\bar z}$ by (4), we also have $A_z^* \in \pi_1(\mathsf{A})'_{\mathrm{s}}$. Thus, by Proposition 7.2.9, (ii), $A_z \in \pi_1(\mathsf{A})'_{\mathrm{ss}}$. Again by (4), this implies that $\overline{\pi_1(a)}$ is affiliated with the von Neumann algebra $\pi_1(\mathsf{A})'_{\mathrm{ss}}$. Thus (ii) is proved.

It remains to show that π_1 has the other properties stated in the theorem.

Since A_1 is cofinal in B_1 w.r.t. $\mathcal{P}(\mathsf{B}_1)$ as shown above, $\varrho_1(\mathsf{A}_1)$ is cofinal in $\varrho_1(\mathsf{B}_1)$ w.r.t. $\mathcal{P}\big(\varrho_1(\mathsf{B}_1)\big)$ and hence w.r.t. $\varrho_1(\mathsf{B}_1)_+$. Thus $\mathsf{t}_{\varrho_1(\mathsf{A}_1)} = \mathsf{t}_{\varrho_1(\mathsf{B}_1)}$ by Corollary 2.6.6. Since $\varrho_1(\mathsf{A}_1) = \pi_1(\mathsf{A})$ because of $\varrho_1(\lambda\cdot 1) = \pi_1(a)$ and ϱ_1 is a closed representation of B_1, it follows that π_1 is also closed.

Finally we prove that $(\pi_1, \mathcal{M})$ is an induced extension of π. This means we have to show that $\mathcal{M} \subseteq \pi_1(\mathsf{A})'_{\mathrm{s}}$ and that $\mathcal{M}\mathcal{D}(\pi)$ is dense in $\mathcal{D}(\pi_1)\,[\mathsf{t}_{\pi_1}]$.

Let $\mathcal{M}_0$ be the $*$-algebra generated by I and A_z, where $z \in \mathbb{C}\setminus\mathbb{R}$. It is well-known that an operator in $\mathbb{B}\big(\mathcal{H}(\pi_1)\big)$ commutes with the self-adjoint operator $\overline{\pi_1(a)}$ if and only if it commutes with A_z for all $z \in \mathbb{C}\setminus\mathbb{R}$. That is, $\big(\overline{\pi_1(a)}\big)'_{\mathrm{s}} = \mathcal{M}_0'$ which gives $\mathcal{M} = \mathcal{M}_0''$; so $\mathcal{M}$ is the weak-operator closure of $\mathcal{M}_0$ in $\mathbb{B}\big(\mathcal{H}(\pi_1)\big)$. Since $A_z \in \pi_1(\mathsf{A})'_{\mathrm{s}}$ for $z \in \mathbb{C}\setminus\mathbb{R}$ as noted above, we have $\mathcal{M}_0 \subseteq \pi_1(\mathsf{A})'_{\mathrm{s}}$. By Proposition 7.2.9, $\pi_1(\mathsf{A})'_{\mathrm{s}}$ is weak-operator closed in $\mathbb{B}\big(\mathcal{H}(\pi_1)\big)$, because π_1 is a closed $*$-representation. Therefore, we get $\mathcal{M} \subseteq \pi_1(\mathsf{A})'_{\mathrm{s}}$.

Let $C(\widehat{\mathbb{R}})$ be the C^*-algebra of all continuous functions f on $\mathbb{R}$ for which $\lim_{|\lambda|\to+\infty} f(\lambda)$ exists, endowed with the usual supremum norm. Let C_0 be the $*$-subalgebra of $C(\widehat{\mathbb{R}})$ which is generated by the functions $(\lambda - z)^{-1}$, $z \in \mathbb{C}\setminus\mathbb{R}$, and by the constant function 1. From the Stone-Weierstraß theorem (applied to the Alexandroff compactification of $\mathbb{R}$) we conclude that C_0 is dense in $C(\widehat{\mathbb{R}})$. The map $f \to \varrho_2(f) := \overline{\varrho_1\big(f(\lambda)\,1\big)}$ is, of course, a $*$-representation of the C^*-algebra $C(\widehat{\mathbb{R}})$ on $\mathcal{H}(\varrho_1)$. Consequently, $\varrho_2(C_0)$ is norm dense in $\varrho_2\big(C(\widehat{\mathbb{R}})\big)$. Since $\mathcal{M}_0 = \varrho_2(C_0)$ by (4), this implies that $\varrho_2\big(C(\widehat{\mathbb{R}})\big) \subseteq \mathcal{M}$. We use this fact in order to prove that $\mathcal{M}\mathcal{D}(\pi)$ is dense in $\mathcal{D}(\pi_1)\,[\mathsf{t}_{\pi_1}]$. Because $\varrho_2\big(C(\widehat{\mathbb{R}})\big) \subseteq \mathcal{M}$ and $\varrho_1(\mathsf{B}_1)\,\mathcal{D}(\varrho) \equiv \varrho_1(\mathsf{B}_1)\,\mathcal{D}(\pi)$ is dense in $\mathcal{D}(\varrho_1)\,[\mathsf{t}_{\varrho_1}] \equiv \mathcal{D}(\pi_1)\,[\mathsf{t}_{\pi_1}]$, it is sufficient to prove that $\varrho_1(\mathsf{B}_1)\,\mathcal{D}(\pi) \subseteq \varrho_2\big(C(\widehat{\mathbb{R}})\big)\,\mathcal{D}(\pi)$. Let $q(\lambda)$ be a polynomially bounded continuous function on $\mathbb{R}$, $x \in \mathsf{A}$ and $\varphi \in \mathcal{D}(\pi)$. There are a polynomial $p(\lambda)$ and a function $f \in C(\widehat{\mathbb{R}})$

such that $q(\lambda) = f(\lambda)\, p(\lambda)$, $\lambda \in \mathbb{R}$. Then

$$\varrho_1\big(q(\lambda)\, x\big)\, \varphi = \varrho_1\big(f(\lambda)\, 1\big)\, \varrho_1\big(p(\lambda)\, x\big)\, \varphi = \varrho_2(f)\, \varrho\big(p(\lambda)\, x\big)\, \varphi = \varrho_2(f)\, \pi\big(p(a)\, x\big)\, \varphi \in \varrho_2(f)\, \mathcal{D}(\pi),$$

where we used (2) and (3). Since $\varrho_1(\mathsf{B}_1)\, \mathcal{D}(\pi)$ is the linear hull of such vectors $\varrho_1\big(q(\lambda)\, x\big)\, \varphi$, we have shown that $\varrho_1(\mathsf{B}_1)\, \mathcal{D}(\pi) \subseteq \varrho_2\big(C(\widehat{\mathbb{R}})\big)\, \mathcal{D}(\pi)$. Together with the preceding, we have proved that $(\pi_1, \mathcal{M})$ is an induced extension of π.

Now we prove the implication (ii) $\rightarrow$ (i). Let π_1 be as stated in (ii). There is no loss of generality to assume that π_1 is closed. Then we have $\pi_1(\mathsf{A})'_{\mathrm{ss}} \subseteq \pi_1(\mathsf{A})'_{\mathrm{s}}$. By Lemma 11.5.4, $\overline{\pi_1(a)}$ is a self-adjoint operator. Let $e(\lambda)$, $\lambda \in \mathbb{R}$, be the spectral projections of this operator. Since $\overline{\pi_1(a)}$ is affiliated with $\pi_1(\mathsf{A})'_{\mathrm{ss}}$, we have $e(\lambda) \in \pi_1(\mathsf{A})'_{\mathrm{ss}}$ and so $e(\lambda) \in \pi_1(\mathsf{A})'_{\mathrm{s}}$ for all real λ. Suppose $n \in \mathbb{N}$, $A \in \boldsymbol{K}_n(\mathsf{A}; a)$ and $\varphi_1, \ldots, \varphi_n \in \mathcal{D}(\pi)$. We can write A as $A = A_0 + aA_1 + \cdots + a^m A_m$, where $m \in \mathbb{N}_0$ and $A_0, \ldots, A_m \in \boldsymbol{M}(\mathsf{A})$ are such that $A_0 + \lambda A_1 + \cdots + \lambda^m A_m \in \mathcal{P}\big(\boldsymbol{M}(\mathsf{A})\big)$ for any real λ. Let $A_r \equiv [a_{kl}^{(r)}]$, $r = 0, \ldots, m$. From $\pi \subseteq \pi_1$ and from the spectral theorem we obtain

$$\sum_{k,l=1}^{n} \left\langle \pi\left(\sum_{r=0}^{m} a^r a_{kl}^{(r)}\right) \varphi_l, \varphi_k \right\rangle = \sum_{k,l=1}^{n} \sum_{r=0}^{m} \int \lambda^r \, \mathrm{d}\langle e(\lambda)\, \pi_1(a_{kl}^{(r)})\, \varphi_l, \varphi_k\rangle. \tag{5}$$

Approximating the integrals in (5) by Riemann sums, the above expression is the limit of sums of the form

$$\sum_{k,l=1}^{n} \sum_{r=0}^{m} \sum_{j=1}^{s} \lambda_j^r \langle \big(e(\lambda_{j+1}) - e(\lambda_j)\big)\, \pi_1(a_{kl}^{(r)})\, \varphi_l, \varphi_k\rangle$$
$$= \sum_{j=1}^{s} \sum_{k,l=1}^{n} \left\langle \pi_1\left(\sum_{r=0}^{m} \lambda_j^r a_{kl}^{(r)}\right) \varphi_{jl}, \varphi_{jk} \right\rangle, \tag{6}$$

where $\varphi_{jk} := \big(e(\lambda_{j+1}) - e(\lambda_j)\big)\, \varphi_k$ for $j = 1, \ldots, s$ and $k = 1, \ldots, n$. Here we used that $e(\lambda) \in \pi_1(\mathsf{A})'_{\mathrm{s}}$ for $\lambda \in \mathbb{R}$. Recall that $A_0 + \lambda A_1 + \cdots + \lambda^m A_m \equiv \left[\sum_{r=0}^{m} \lambda^r a_{kl}^{(r)}\right]$ is in $\mathcal{P}\big(\boldsymbol{M}(\mathsf{A})\big)$ for real λ. Therefore, since π_1 is a $*$-representation of A and hence completely positive (see Remark 2 in 11.2), the sums in (6) are non-negative. Consequently, the expression in (5) is non-negative. This proves that $\pi(a)$ is completely centrally positive. $\square$

Corollary 11.5.6. *Suppose that π is a self-adjoint representation of A and a is a hermitian element of A. Then $\pi(a)$ is completely centrally positive if and only if the operator $\overline{\pi(a)}$ is affiliated with the von Neumann algebra $\pi(\mathsf{A})'$. If the latter is true, then $\overline{\pi(a)}$ is a self-adjoint operator.*

Proof. Since π is self-adjoint, we have $\pi(\mathsf{A})' = \pi(\mathsf{A})'_{\mathrm{ss}}$ by Proposition 7.2.10. Thus the if part follows at once from Theorem 11.5.5 ,(ii) $\rightarrow$ (i), by letting $\pi_1 := \pi$. Now suppose that $\pi(a)$ is completely centrally positive. Let π_1 be the $*$-representation which exists by Theorem 11.5.5, (ii). We have shown in the proof of Theorem 11.5.5 that the operator $A_z = \big(\overline{\pi_1(a)} - z\big)^{-1}$ is in $\pi_1(\mathsf{A})'_{\mathrm{s}}$ for $z \in \mathbb{C}\setminus\mathbb{R}$. Since $\pi \subseteq \pi_1$, we have $\widetilde{A}_z := \mathrm{pr}_{\mathcal{H}(\pi)} A_z \in \pi(\mathsf{A})'_{\mathrm{w}} = \pi(\mathsf{A})'$ by Proposition 7.2.16. Using this fact and $\pi(a) \subseteq \pi_1(a)$, we obtain

$$\big(\pi(a) - z\big)\, \widetilde{A}_z \varphi = \widetilde{A}_z\big(\pi(a) - z\big)\, \varphi = P_{\mathcal{H}(\pi)}\big(\overline{\pi_1(a)} - z\big)^{-1}\, \big(\pi(a) - z\big)\, \varphi$$
$$= P_{\mathcal{H}(\pi)}\varphi = \varphi$$

for $z \in \mathbb{C}\backslash\mathbb{R}$ and $\varphi \in \mathcal{D}(\pi)$. By Proposition 8.1.19 this implies that $\overline{\pi(a)}$ is a self-adjoint operator and $\widetilde{A}_z = \left(\overline{\pi(a)} - z\right)^{-1}$ when $z \in \mathbb{C}\backslash\mathbb{R}$. Because $\widetilde{A}_z \in \pi(\mathsf{A})'$, $\overline{\pi(a)}$ is affiliated with the von Neumann algebra $\pi(\mathsf{A})'$. □

Remark 5. Suppose that the $*$-algebra A is commutative. Then it is not difficult to see that $\boldsymbol{K}(\mathsf{A}; a) \subseteq \boldsymbol{M}(\mathsf{A}; \mathrm{int})_+$ for every $a \in \mathsf{A}_\mathrm{h}$, cf. Definition 11.3.1. Therefore, if a linear map Φ of A into some $B(\mathfrak{X})$ is n-positive w.r.t. $\boldsymbol{M}_n(\mathsf{A}; \mathrm{int})_+$, then $\Phi(a)$ is centrally n-positive for each $a \in \mathsf{A}_\mathrm{h}$.

Lemma 11.5.7. *For any positive linear functional ω on the $*$-algebra $\mathsf{A} = \mathbb{C}[\mathsf{x}_1, \mathsf{x}_2]$, the following assertions are equivalent:*

(i) $\pi_\omega(\mathsf{x}_1)$ *is completely centrally positive.*
(ii) $\pi_\omega(\mathsf{x}_1)$ *is centrally 1-positive.*
(iii) ω *is $\mathbb{C}[\mathsf{x}_1, \mathsf{x}_2]_+^{\mathrm{int}}$-positive.*

Proof. (i) → (ii) is trivial. We prove that (ii) implies (iii). Let $p \in \mathbb{C}[\mathsf{x}_1, \mathsf{x}_2]_+^{\mathrm{int}}$. Fix $\lambda \in \mathbb{R}$. It is clear that $p(\lambda, \mathsf{x}_2) \in \mathbb{C}[\mathsf{x}_2]_+^{\mathrm{int}}$, cf. Remark 2 in 11.3. Since $\mathbb{C}[\mathsf{x}_2]_+^{\mathrm{int}} = \mathcal{P}(\mathbb{C}[\mathsf{x}_2])$ (cf. Example 2.6.11), it follows that $p(\lambda, \mathsf{x}_2) \in \mathcal{P}(\mathbb{C}[\mathsf{x}_2]) \subseteq \mathcal{P}(\mathbb{C}[\mathsf{x}_1, \mathsf{x}_2])$. Since we can write p as $p(\mathsf{x}_1, \mathsf{x}_2) = \sum_{r=0}^{m} \mathsf{x}_1^r p_r(\mathsf{x}_2)$ with $p_0, \ldots, p_m \in \mathbb{C}[\mathsf{x}_2]$, this shows that $p \in \boldsymbol{K}_1(\mathbb{C}[\mathsf{x}_1, \mathsf{x}_2]; \mathsf{x}_1)$. Because $\pi(\mathsf{x}_1)$ is centrally 1-positive, $\langle \pi_\omega(p)\,\varphi_\omega, \varphi_\omega\rangle = \omega(p) \geqq 0$ which proves that ω is $\mathbb{C}[\mathsf{x}_1, \mathsf{x}_2]_+^{\mathrm{int}}$-positive.

Finally we verify (iii) → (i). From Corollary 11.2.6 (applied with $\boldsymbol{K} = \boldsymbol{M}(\mathbb{C}[\mathsf{x}_1, \mathsf{x}_2]; \mathrm{int})_+$) we conclude that π_ω is completely positive w.r.t. $\boldsymbol{M}(\mathbb{C}[\mathsf{x}_1, \mathsf{x}_2]; \mathrm{int})_+$, so $\pi_\omega(\mathsf{x}_1)$ is completely centrally positive by Remark 5. □

Remark 6. Lemma 11.5.7 allows to construct $*$-representations π of $\mathbb{C}[\mathsf{x}_1, \mathsf{x}_2]$ for which $\pi(\mathsf{x}_1)$ is completely centrally positive as well as those for which $\pi(\mathsf{x}_1)$ is not centrally positive. Indeed, it suffices to set $\pi = \pi_\omega$, where ω is a positive linear functional on $\mathbb{C}[\mathsf{x}_1, \mathsf{x}_2]$ which is $\mathbb{C}[\mathsf{x}_1, \mathsf{x}_2]_+^{\mathrm{int}}$-positive in the former case and which is not $\mathbb{C}[\mathsf{x}_1, \mathsf{x}_2]_+^{\mathrm{int}}$-positive in the latter case; see also Remark 6 in 11.3.

We close this section with another example where $\mathsf{A} = \mathbb{C}[\mathsf{x}_1, \mathsf{x}_2]$.

Example 11.5.8. Suppose that π is a non-integrable self-adjoint representation of $\mathsf{A} \equiv \mathbb{C}[\mathsf{x}_1, \mathsf{x}_2]$ such that the operators $\overline{\pi(\mathsf{x}_1)}$ and $\overline{\pi(\mathsf{x}_2)}$ are self-adjoint. (Such examples have been constructed in Section 9.4.) Then the operator $\pi(\mathsf{x}_1)$ is not affiliated with the von Neumann algebra $\pi(\mathsf{A})'$. (Indeed, otherwise, $\left(\overline{\pi(\mathsf{x}_1)} - \mathrm{i}\right)^{-1} \in \pi(\mathsf{A})'$ and hence $\left(\overline{\pi(\mathsf{x}_1)} - \mathrm{i}\right)^{-1}$ commutes with $\pi(\mathsf{x}_2)$. By Lemma 1.6.2, the self-adjoint operators $\overline{\pi(\mathsf{x}_1)}$ and $\overline{\pi(\mathsf{x}_2)}$ strongly commute, so π would be integrable by Corollary 9.1.14.)

Therefore, by Corollary 11.5.6, $\pi(\mathsf{x}_1)$ is not completely centrally positive. By Theorem 11.5.5, there is no $*$-representation π_1 of A such that $\pi \subseteq \pi_1$ and $\overline{\pi_1(\mathsf{x}_1)}$ is affiliated with the von Neumann algebra $\pi_1(\mathsf{A})'_{\mathrm{ss}}$. (By Corollary 8.3.13 each extension π_1 of the self-adjoint $*$-representation π splits into a direct sum $\pi_1 = \pi \oplus \pi_0$. Using this fact the latter assertion can also be obtained directly without appealing to Theorem 11.5.5.) ○

11.6. Strongly 1-Positive $*$-Representations which are not Strongly 2-Positive

In the first subsection we prove some auxiliary results. They are needed for the construction of some special $*$-representations of the polynomial algebra $\mathbb{C}[x_1, x_2]$.

Closedness of the Wedges $M_2(\mathcal{A}; 1)_+$ and $\mathcal{P}(\mathcal{A})$ for Certain O*-Algebras

Throughout this subsection we assume that $\mathcal{A}$ is an O*-algebra which is the union of an increasing sequence $(\mathcal{A}_k : k \in \mathbb{N})$ of finite dimensional linear subspaces $\mathcal{A}_k$, $k \in \mathbb{N}$. (Of course, this implies that the $*$-algebra $\mathcal{A}$ is countably generated.)
Consider the following two conditions:

(I) If $a \in \mathcal{A}$, $c \in \mathcal{A}_+$ and $a^+ca = 0$, then $a = 0$ or $c = 0$.

(II) If $\sum_{j=1}^{m} a_j^+ c_j a_j \in \mathcal{A}_k$ with $k, m \in \mathbb{N}$ and $a_j \in \mathcal{A}$, $c_j \in \mathcal{A}_+$ for $j = 1, \ldots, m$, then we have $a_j^+ c_j a_j = 0$ or $a_j \in \mathcal{A}_k$ and $c_j \in \mathcal{A}_k$ for all $j = 1, \ldots, m$.

Let $M_2(\mathcal{A}; 1)_+$ be the wedge of all finite sums of terms A^+cA, where $c \in \mathcal{A}_+$ and $A \in M_{1,2}(\mathcal{A})$. Recall that τ_{st} denotes the finest locally convex topology on a vector space.

Theorem 11.6.1. *Suppose that the O*-algebra $\mathcal{A}$ and the sequence $(\mathcal{A}_k : k \in \mathbb{N})$ satisfy the conditions* (I) *and* (II). *Then the set $M_2(\mathcal{A}; 1)_+$ is closed in the locally convex space $M_2(\mathcal{A})[\tau_{st}]$.*

Proof. The proof will be divided into four steps. Suppose $k \in \mathbb{N}$. To avoid trivial cases, we can assume that $\mathcal{A}_k \neq \{0\}$.
Statement 1: There is a finite subset $\mathcal{D}_k$ of $\mathcal{D}(\mathcal{A})$ such that $\|\cdot\|_k := \sup \{|\langle \cdot \varphi, \varphi\rangle| : \varphi \in \mathcal{D}_k\}$ is a norm on $\mathcal{A}_k$.

Proof. Let $\mathcal{W}$ be the unit sphere with respect to any norm on $\mathcal{A}_k$. If $a \in \mathcal{W}$, then there is a vector $\varphi_a \in \mathcal{D}(\mathcal{A})$ such that $\langle a\varphi_a, \varphi_a\rangle \neq 0$. The sets $\mathcal{W}(a) := \{b \in \mathcal{W} : \langle b\varphi_a, \varphi_a\rangle \neq 0\}$, $a \in \mathcal{W}$, form an open cover of $\mathcal{W}$. By the compactness of $\mathcal{W}$ there is a finite subcover, say $\{\mathcal{W}(a_1), \ldots, \mathcal{W}(a_m)\}$. Then $\mathcal{D}_k := \{\varphi_{a_1}, \ldots, \varphi_{a_m}\}$ has the desired property. □

From now on we equip $\mathcal{A}_k$ with the norm $\|\cdot\|_k$. Let $\mathcal{W}_k$ be the unit sphere of the normed space $\mathcal{A}_k$.

Statement 2: There are numbers $s_k \in \mathbb{N}$, $s_k \geq k$, and $\delta_k > 0$ such that $(a, b, c) \to a^+cb$ is a continuous mapping of $\mathcal{A}_k \times \mathcal{A}_k \times \mathcal{A}_k$ into $\mathcal{A}_{s_k}$ and such that $\|a^+ca\|_{s_k} \geq \delta_k \|a\|_k^2 \|c\|_k$ for $a \in \mathcal{A}_k$ and $c \in \mathcal{A}_k \cap \mathcal{A}_+$.

Proof. The first assertion follows immediately from the fact that $(\mathcal{A}_n : n \in \mathbb{N})$ is an increasing sequence of finite dimensional spaces that exhaust $\mathcal{A}$. For the second assertion, we can assume that $\mathcal{W}_k \cap \mathcal{A}_+$ is non-empty, since otherwise the assertion is trivial. The set $Q_k := \{(a, c, a) : a \in \mathcal{W}_k$ and $c \in \mathcal{W}_k \cap \mathcal{A}_+\}$ in $\mathcal{A}_k \times \mathcal{A}_k \times \mathcal{A}_k$ is compact, so is their image under the continuous mapping $(a, b, c) \to a^+cb$. Hence there are $a_0 \in \mathcal{W}_k$ and $c_0 \in \mathcal{W}_k \cap \mathcal{A}_+$ such that $\|a_0^+c_0a_0\|_{s_k} = \inf \{\|a^+ca\|_{s_k} : (a, c, a) \in Q_k\}$. Since $a_0 \neq 0$ and $c_0 \neq 0$, $a_0^+c_0a_0 \neq 0$ by (I). The assertion follows by setting $\delta_k := \|a_0^+c_0a_0\|_{s_k}$. □

Let E_k be the vector space of all matrices in $M_2(\mathcal{A})$ whose entries are in $\mathcal{A}_k$, equipped with the norm defined by $|||[a_{rs}]|||_k := \sum_{r,s=1}^{2} \|a_{rs}\|_k$. The main step in this proof is

Statement 3: $M_2(\mathcal{A}; 1)_+ \cap E_k$ is closed in E_k.

Proof. Let d_k be the dimension of E_k. First we note that each element X of $M_2(\mathcal{A}; 1)_+ \cap E_k$ is a sum of d_k terms of the form A^+cA, where $c \in \mathcal{A}_+$ and $A \in M_{1,2}(\mathcal{A})$. Indeed, let $X = \sum_{j=1}^{m} A_j^+ c_j A_j$. If $m < d_k$, then we add zeros. Suppose $m > d_k$. Then there is a non-zero $(\lambda_1, \ldots, \lambda_m) \in \mathbb{R}^m$ such that $\sum_{j=1}^{m} \lambda_j A_j^+ c_j A_j = 0$. Without loss of generality, $\lambda_m \geqq |\lambda_j|$ for $j = 1, \ldots, m-1$. Then $X = \sum_{j=1}^{m-1} A_j^+ \tilde{c}_j A_j$, where $\tilde{c}_j := (1 - \lambda_j/\lambda_m) c_j$ for $j = 1, \ldots, m-1$. Continuing this reasoning, we arrive at d_k terms.

Now let $X \equiv [x_{rs}] \in E_k$ be in the closure of $M_2(\mathcal{A}; 1)_+ \cap E_k$ in E_k. Then there is a sequence $(X_n : n \in \mathbb{N})$ in $M_2(\mathcal{A}; 1)_+ \cap E_k$ which converges to X. By the preceding, we can write $X_n \equiv [x_{rs}^{(n)}]_{r,s=1,2} = \sum_{j=1}^{d_k} A_{nj}^+ c_{nj} A_{nj}$ with $c_{nj} \in A_j$ and $A_{nj} \equiv (a_{nj}, b_{nj}) \in M_{1,2}(\mathcal{A})$ for $j = 1, \ldots, d_k$. Then all elements a_{nj}, b_{nj} and c_{nj} are in $\mathcal{A}_k$. Indeed, since $X_n \in E_k$, we have for $n \in \mathbb{N}$

$$x_{11}^{(n)} = \sum_{j=1}^{d_k} a_{nj}^+ c_{nj} a_{nj} \in \mathcal{A}_k. \tag{1}$$

If $a_{nj}^+ c_{nj} a_{nj} = 0$, then $a_{nj} = 0$ and $c_{nj} = 0$ by (I). If $a_{nj}^+ c_{nj} a_{nj} \neq 0$, then $a_{nj} \in \mathcal{A}_k$ and $c_{nj} \in \mathcal{A}_k$ by (1) and (II). The same argument with $x_{22}^{(n)}$ in place of $x_{11}^{(n)}$ shows that $b_{nj} \in \mathcal{A}_k$.

Without loss of generality we can assume that either $c_{nj} = 0$ and $A_{nj} = (0, 0)$ or $\|c_{nj}\|_k = 1$ for all n and k. (Otherwise we replace c_{nj} by 0 and A_{nj} by $(0, 0)$ when $A_{nj}^+ c_{nj} A_{nj} = 0$ and c_{nj} by $c_{nj} \|c_{nj}\|_k^{-1}$ and A_{nj} by $A_{nj} \|c_{nj}\|_k^{1/2}$ when $A_{nj}^+ c_{nj} A_{nj} \neq 0$.) We have $x_{11} = \lim_n x_{11}^{(n)}$ in $\mathcal{A}_k$ and hence in $\mathcal{A}_{s_k}$. Let $j \in \{1, \ldots, d_k\}$. From (1) and from the definition of the norm $\|\cdot\|_{s_k}$ it follows that $\|a_{nj}^+ c_{nj} a_{nj}\|_{s_k} \leqq \|x_{11}^{(n)}\|_{s_k}$ for $n \in \mathbb{N}$, so $\{a_{nj}^+ c_{nj} a_{nj} : n \in \mathbb{N}\}$ is a bounded set in the space $\mathcal{A}_{s_k}$. By Statement 2 and the assumption stated at the beginning of this paragraph, this implies that the set $\{a_{nj} : n \in \mathbb{N}\}$ is bounded in $\mathcal{A}_k$. Similarly, $\{b_{nj} : n \in \mathbb{N}\}$ is bounded in $\mathcal{A}_k$. By construction the set $\{c_{nj} : n \in \mathbb{N}\}$ is bounded in $\mathcal{A}_k$. Thus there exists a subsequence $(m_n : n \in \mathbb{N})$ of the sequence of natural numbers such that the sequences $(a_{m_n j} : n \in \mathbb{N})$, $(b_{m_n j} : n \in \mathbb{N})$ and $(c_{m_n j} : n \in \mathbb{N})$ converge in $\mathcal{A}_k$. Let a_j, b_j and c_j denote their limits. Using (1), we get $x_{11} = \lim_n x_{11}^{(m_n)} = \sum_{j=1}^{d_k} a_j^+ c_j a_j$ in A_{s_k}. For x_{12}, x_{21} and x_{22} we obtain the corresponding expressions which show that $X = \sum_{j=1}^{d_k} A_j^+ c_j A_j$, where $A_j := (a_j, b_j)$ for $j = 1, \ldots, d_k$. Since $c_j \in \mathcal{A}_+$ for $j = 1, \ldots, d_k$, this shows that $X \in M_2(\mathcal{A}; 1)_+$. □

Statement 4: $M_2(\mathcal{A}; 1)_+$ is closed in $M_2(\mathcal{A})$ $[\tau_{st}]$.

Proof. It is clear that $E := M_2(\mathcal{A})$ $[\tau_{st}]$ is the strict inductive limit of the increasing sequence $(E_k : k \in \mathbb{N})$ of finite dimensional normed spaces E_k, $k \in \mathbb{N}$. Hence the strong dual E' of E is a reflexive Frechet space. We apply the Krein-Šmulian theorem to this space. Let U be a 0-neighbourhood in E'. Then the polar U^0 of U in E is bounded and hence contained in some E_k, $k \in \mathbb{N}$, by a property of the strict inductive limit. From Statement 3 we conclude that $M_2(\mathcal{A}; 1)_+ \cap U^0$ is closed in E_k and so is $\sigma(E, E')$-closed in E. Therefore, the Krein-Šmulian theorem (SCHÄFER [1], IV, 6.4) shows that $M_2(\mathcal{A}; 1)_+$ is $\sigma(E, E')$-closed in E which gives the assertion of the theorem. □

Corollary 11.6.2. *Let $\mathcal{A}$ be as in Theorem 11.6.1. If $M_2(\mathcal{A}; 1)_+ \neq M_2(\mathcal{A})_+$, then there exists a closed 2-cyclic $*$-representation of $\mathcal{A}$ which is strongly 1-positive, but not strongly 2-positive.*

Proof. Since $M_2(\mathcal{A}; 1)_+ \neq M_2(\mathcal{A})_+$, there is a matrix $B \in M_2(\mathcal{A})_+$ which is not in $M_2(\mathcal{A}; 1)_+$. From Theorem 11.6.1, $M_2(\mathcal{A}; 1)_+$ is closed in $M_2(\mathcal{A})\,[\tau_{st}]$ and so is in the real locally convex space $M_2(\mathcal{A})_h\,[\tau'_{st}]$, where τ'_{st} is the induced topology on $M_2(\mathcal{A})_h$ of the topology τ_{st} from $M_2(\mathcal{A})$. Obviously, $M_2(\mathcal{A}; 1)_+$ is a convex set in $M_2(\mathcal{A})_h$. By the separation theorem for convex sets (see e.g. SCHÄFER [1], II, 9.2) there is a real linear functional g on $M_2(\mathcal{A})_h$ such that $g(B) < \inf\{g(A)\colon A \in M_2(\mathcal{A}; 1)_+\} = 0$, where the latter equality follows from the fact that $M_2(\mathcal{A}; 1)_+$ is a wedge. By Lemma 1.3.1, $f(X_1 + iX_2) := g(X_1) + ig(X_2)$, $X_1, X_2 \in M_2(\mathcal{A})_h$, defines a linear functional on the complex vector space $M_2(\mathcal{A})$. Since $\mathcal{P}(M_2(\mathcal{A})) \subseteq M_2(\mathcal{A}; 1)_+$, f is a positive linear functional on the $*$-algebra $M_2(\mathcal{A})$. Let π be the closed 2-cyclic $*$-representation of $\mathcal{A}$ which exists by Proposition 11.2.7. If $K_1 := \mathcal{A}_+$, then $M_2(\mathcal{A}; 1)_+$ is the wedge $K_2(1)$ defined before Proposition 11.2.5. Since f is non-negative on $M_2(\mathcal{A}; 1)_+$, it follows therefore from Proposition 11.2.5 that π is strongly 1-positive. Since $B \in M_2(\mathcal{A})_+$ and $f(B) < 0$, Proposition 11.2.5 applied with $K_2 := M_2(\mathcal{A})_+$ shows that π is not strongly 2-positive. □

Some arguments of the two preceding proofs can be used to obtain (under some weaker assumptions) similar results for the cone $\mathcal{P}(\mathcal{A})$. Recall we assumed that $(\mathcal{A}_k\colon k \in \mathbb{N})$ is an increasing sequence of finite dimensional linear subspaces of the O*-algebra $\mathcal{A}$ whose union is $\mathcal{A}$. Now we need the following condition:

(III) If $\sum_{j=1}^{m} a_j^+ a_j \in \mathcal{A}_k$ with $k, m \in \mathbb{N}$ and $a_j \in \mathcal{A}$, then $a_j \in \mathcal{A}_k$ for all $j = 1, \ldots, m$.

Theorem 11.6.3. *Suppose that the O*-algebra $\mathcal{A}$ and the sequence $(\mathcal{A}_k\colon k \in \mathbb{N})$ satisfy condition* (III). *Then the cone $\mathcal{P}(\mathcal{A})$ is closed in the locally convex space $\mathcal{A}[\tau_{st}]$. If, in addition, $\mathcal{P}(\mathcal{A}) \neq \mathcal{A}_+$, then there exists a (closed cyclic) $*$-representation of $\mathcal{A}$ which is not strongly positive.*

A proof of this theorem can be given by appropriate modifications in the proofs of Theorem 11.6.1 and of Corollary 11.6.2; we omit the details. Of course, for the second assertion we can use directly the GNS construction instead of Proposition 11.2.7.

Corollary 11.6.4. *Suppose that A is one of the following $*$-algebras:*

(i) *the polynomial algebra $\mathbb{C}[x_1, \ldots, x_n]$, $n \in \mathbb{N}$,*

(ii) *the enveloping algebra $\mathcal{E}(\mathfrak{g})$ of a finite dimensional Lie algebra $\mathfrak{g}$,*

(iii) *the Weyl algebra $\mathbb{A}(p_1, q_1, \ldots, p_n, q_n)$, $n \in \mathbb{N}$.*

Then $\mathcal{P}(\mathsf{A})$ is a closed cone in $\mathsf{A}[\tau_{st}]$.

Proof. First suppose $\mathsf{A} = \mathbb{A}(p_1, q_1, \ldots, p_n, q_n)$. Let A_k, $k \in \mathbb{N}$, be the vector space of all elements in A whose degree with respect to the basis in 2.5/(3) is at most k. Recall that the Schrödinger representation π provides a $*$-isomorphism of A on an O*-algebra $\mathcal{A} := \pi(\mathsf{A})$. Using the commutation relations 2.5/(2) it is easy to check that $\mathcal{A}$ and $(\mathcal{A}_k := \pi(\mathsf{A}_k)\colon k \in \mathbb{N})$ satisfy (III). Thus $\mathcal{P}(\mathcal{A})$ is closed in $\mathcal{A}[\tau_{st}]$ by Theorem 11.6.3. Since

π is a $*$-isomorphism and hence $\pi(\mathcal{P}(\mathsf{A})) = \mathcal{P}(\pi(\mathsf{A}))$, $\mathcal{P}(\mathsf{A})$ is a cone and closed in $\mathsf{A}[\tau_{st}]$.

The proof for $\mathscr{E}(\mathfrak{g})$ is similar when we use the $*$-isomorphism dU_{l_r} of Example 10.1.8 and the basis $\{x^n : n \in \mathbb{N}_0^d\}$ obtained from the Poincaré-Birkhoff-Witt theorem, cf. 1.7. (i) is the special case $\mathfrak{g} = \mathbb{R}^n$ of (ii). □

A Strange $*$-Representation of the Polynomial Algebra $\mathbb{C}[x_1, x_2]$

We begin with two algebraic auxiliary lemmas.

Lemma 11.6.5. *Let $p, q \in \mathbb{C}[x_1, x_2]$, $p \not\equiv 0$, $q \not\equiv 0$. Suppose that*

$$0 \leqq p(t_1, t_2)\,|q(t_1, t_2)|^2 \leqq 1 + t_1^4 t_2^2 \quad \textit{for all} \quad (t_1, t_2) \in \mathbb{R}^2. \tag{2}$$

Then we have

(i) *$p(x_1, x_2)\, q(x_1, x_2) = \alpha x_1^k x_2^l$ with $\alpha \in \mathbb{C}$ and $k, l \in \mathbb{N}$, $k \geqq 2$,*
or
(ii) *there are polynomials $p_1, q_1 \in \mathbb{C}[x]$ such that $p(x_1, x_2) = p_1(x_1^2 x_2)$ and $q(x_1, x_2) = q_1(x_1^2 x_2)$.*

Proof. By (2), there are polynomials $r_0, r_1, r_2, s_0, s_1 \in \mathbb{C}[x]$ such that $p(x_1, x_2) = r_0(x_1) + r_1(x_1)\, x_2 + r_2(x_1)\, x_2^2$ and $q(x_1, x_2) = s_0(x_1) + s_1(x_1)\, x_2$. Setting $t_2 = 0$ in (2), we conclude that $r_0 \equiv 0$ (case 1), $s_0 \equiv 0$ (case 2) or r_0 and s_0 are constant (case 3). Case 3 is divided into case 3.1: $r_2 \not\equiv 0$ and case 3.2: $r_2 \equiv 0$.

Case 1: Since $p(t_1, t_2) \geqq 0$ for all (t_1, t_2) belonging to the dense subset $\{(t_1, t_2) : q(t_1, t_2) \neq 0\}$ of $\mathbb{R}^2$ and so for all $(t_1, t_2) \in \mathbb{R}^2$, we conclude that $r_1 \equiv 0$. Since $p \not\equiv 0$, $r_2 \not\equiv 0$. Thus $s_1 \equiv 0$ because of the degree of $1 + x_1^4 x_2^2$ with respect to x_2. For large t_2 it follows from (2) that $0 \leqq r_2(t_1)\, |s_0(t_1)|^2 \leqq t_1^4$ for $t_1 \in \mathbb{R}$ which leads to case (i) of our assertion.

Case 2: Since $q \not\equiv 0$, $s_1 \not\equiv 0$. Thus $r_1 \equiv r_2 \equiv 0$ because of the degree of $1 + x_1^4 x_2^2$ with respect to x_2. A similar reasoning as in case 1 leads to case (i) of the assertion.

Case 3.1: Similarly as in case 1 we obtain $s_1 \equiv 0$ and $0 \leqq r_2(t_1)|s_0|^2 \leqq t_1^4$ for $t_1 \in \mathbb{R}$. Since $q \not\equiv 0$, the constant s_0 is non-zero. Therefore, $r_2(x_1) = \alpha_2 x_1^4$ for some $\alpha_2 \geqq 0$. From (2) and $s_0 \neq 0$ we have $p(t_1, t_2) \geqq 0$ on $\mathbb{R}^2$. This yields $4\,|r_1(t_1)|^2 \leqq r_0 r_2(t_1) = r_0 \alpha_2 t_1^4$ for $t_1 \in \mathbb{R}$. Hence $r_1(x_1) = \alpha_1 x_1^2$ for some $\alpha_1 \geqq 0$, and we are in case (ii) of the assertion.

Case 3.2: Combined with (2), $r_2 \equiv 0$ implies that $r_1 \equiv 0$. Hence $p = r_0$ is a non-zero constant. Taking large $t_2 \in \mathbb{R}$ in (2), we get $r_0\, |s_1(t_1)|^2 \leqq t_1^4$ for all $t_1 \in \mathbb{R}$. Thus $s_1(x_1) = \alpha_1 x_1^2$ with some $\alpha_1 \in \mathbb{C}$, and we are again in case (ii) of the assertion. □

Let $M_2(\mathbb{C}[x_1, x_2]; \mathrm{int}; 1)_+$ denote the set of all finite sums of terms $A^+ p A$ with $p \in \mathbb{C}[x_1, x_2]_+^{\mathrm{int}}$ and $A \in M_{1,2}(\mathbb{C}[x_1, x_2])$.

Lemma 11.6.6. *The matrix*

$$B := \begin{bmatrix} 1 + x_1^4 x_2^2 & x_1 x_2 \\ x_1 x_2 & 1 + x_1^2 x_2^4 \end{bmatrix}$$

is in $M_2(\mathbb{C}[x_1, x_2]; \mathrm{int})_+$, but not in $M_2(\mathbb{C}[x_1, x_2]; \mathrm{int}; 1)_+$.

Proof. To prove that $B \in M_2(\mathbb{C}[x_1, x_2]; \mathrm{int})_+$, it suffices to check that for all $(t_1, t_2) \in \mathbb{R}^2$

the principal minors $D_1 = 1 + t_1^4t_2^2$ and $D_2 = (1 + t_1^4t_2^2)(1 + t_1^2t_2^4) - t_1^2t_2^2$ are non-negative. For D_1 this is clear. For D_2 we have $D_2 \geqq t_1^2t_2^2(t_1^2 + t_2^2 - 1) + 1$ and the latter polynomial is non-negative on $\mathbb{R}^2$ by Statement 2 in Example 2.6.11.

We show that B is not in $M_2(\mathbb{C}[x_1, x_2]; \text{int}; 1)_+$. Assume the contrary. Then there are matrices $A_j = (a_j, b_j) \in M_{1,2}(\mathbb{C}[x_1, x_2])$ and polynomials $p_j \in \mathbb{C}[x_1, x_2]_+^{\text{int}}$ such that $B = \sum_{j=1}^m A_j^+ p_j A_j$. Comparing the entries in this identity, we obtain

$$1 + x_1^4x_2^2 = \sum_{j=1}^m p_j a_j^+ a_j, \quad 1 + x_1^2x_2^4 = \sum_{j=1}^m p_j b_j^+ b_j \tag{3}$$

and

$$x_1x_2 = \sum_{j=1}^m p_j a_j^+ b_j. \tag{4}$$

Let j be such that $p_j a_j^+ b_j \not\equiv 0$. Then, by (3), the assumptions of Lemma 11.6.5 are fulfilled in case $p = p_j$, $q = a_j^+$ and in case $p = p_j$, $q = b_j$ when we change the roles of x_1 and x_2. From this lemma we conclude that the term x_1x_2 occurs in $p_j a_j^+ b_j$ only with vanishing coefficient. This contradicts (4). □

Now we can state and prove the main result in this subsection.

Theorem 11.6.7. *There exists a $*$-representation π of the polynomial algebra $\mathbb{C}[x_1, x_2]$ which is 1-positive with respect to $M_1(\mathbb{C}[x_1, x_2]; \text{int})_+ \equiv \mathbb{C}[x_1, x_2]_+^{\text{int}}$, but not 2-positive with respect to $M_2(\mathbb{C}[x_1, x_2]; \text{int})_+$. π can be chosen to be 2-cyclic and closed.*

Proof. Let π_0 be the $*$-representation of $\mathsf{A} := \mathbb{C}[x_1, x_2]$ on the domain $\mathcal{D}(\pi_0) := \{\varphi \in L^2(\mathbb{R}^2) : t^k\varphi(t) \in L^2(\mathbb{R}^2) \text{ for all } k \in \mathbb{N}_0^2\}$ in the Hilbert space $L^2(\mathbb{R}^2)$ which is defined by $(\pi_0(p)\,\varphi)(t) := p(t)\,\varphi(t)$, where $p \in \mathbb{C}[x_1, x_2]$, $\varphi \in \mathcal{D}(\pi_0)$ and $t \in \mathbb{R}^2$. Set $\mathcal{A} := \pi_0(\mathsf{A})$. Let $\mathcal{A}_k$ be the set of all $\pi_0(p)$, where $p \in \mathbb{C}[x_1, x_2]$ has degree at most k. It is not difficult to check that the O*-algebra $\mathcal{A}$ and the sequence $(\mathcal{A}_k : k \in \mathbb{N})$ satisfy conditions (I) and (II). From the corresponding definitions we conclude easily that a matrix $[a_{kl}]$ is in $M_1(\mathsf{A}; \text{int})_+ \equiv \mathsf{A}_+^{\text{int}}$ [resp. $M_2(\mathsf{A}; \text{int}; 1)_+$, $M_2(\mathsf{A}; \text{int})_+$] if and only if the matrix $[\pi_0(a_{kl})]$ is in $M_1(\mathcal{A})_+ \equiv \mathcal{A}_+$ [resp. $M_2(\mathcal{A}; 1)_+$, $M_2(\mathcal{A})_+$]. Hence $M_2(\mathcal{A}; 1)_+ \neq M_2(\mathcal{A})_+$ by Lemma 11.6.6 and so Corollary 11.6.2 applies. If π_1 denotes the $*$-representation of $\mathcal{A}$ which exists by Corollary 11.6.2, then the $*$-representation $\pi := \pi_1 \circ \pi_0$ of $\mathsf{A} \equiv \mathbb{C}[x_1, x_2]$ has the required properties. □

A by-product of the preceding theorem is

Corollary 11.6.8. *The $*$-representation π from Theorem 11.6.7 cannot be decomposed as a direct sum of cyclic $*$-representations.*

Proof. Assume to the contrary that π is the direct sum of cyclic $*$-representations π_i, $i \in I$. Since π and so each π_i is 1-positive w.r.t. $\mathbb{C}[x_1, x_2]_+^{\text{int}}$, Corollary 11.2.6 implies that each π_i is 2-positive w.r.t. $M_2(\mathbb{C}[x_1, x_2]; \text{int})_+$. But then π would be 2-positive w.r.t. $M_2(\mathbb{C}[x_1, x_2]; \text{int})_+$ which contradicts Theorem 11.6.7. □

Remark 1. There is a similar result as Theorem 11.6.7 for the O*-algebra $\mathcal{A} := \mathbb{A}(p_1, q_1)$ of Example 2.5.2. In this case it can be shown (see FRIEDRICH/SCHMÜDGEN [1]) that the matrix

$$\begin{bmatrix} (N-1)(N-2) & 3^{-1/2}(a^+)^3 \\ 3^{-1/2}a^3 & N+1 \end{bmatrix}$$

belongs to $M_2(\mathcal{A})_+$, but not to $M_2(\mathcal{A}; 1)_+$. Here we set $a := 2^{-1/2}(q_1 + ip_1)$ and $N := a^+a$. From this fact and Corollary 11.6.2 (note that $\mathcal{A}$ also satisfies the assumptions of this corollary) it follows that there exists a strongly 1-positive *-representation of $\mathcal{A}$ which is not strongly 2-positive.

Notes

The pioneering work for this chapter is POWERS [2]. The concept of complete positivity with respect to a general wedge, the extension Theorem 11.1.5, the dilation Theorem 11.2.2 and the three applications developed in Section 11.3—11.5 are due to POWERS [2]. However, our presentation differs from the one of Powers, some results have been generalized, and additional material has been included. For instance, all three applications are formulated as results on the existence of certain extensions, and they are derived from Theorem 11.2.8. (POWERS [2] gives only the versions for self-adjoint representations which are stated as Corollaries 11.3.5, 11.4.3 and 11.5.6).

Integrable extensions of representations of commutative *-algebras were also studied by BORCHERS/YNGVASON [2]. The results on n-positive representations such as Proposition 11.2.5 and the whole of Section 11.6 are taken from FRIEDRICH/SCHMÜDGEN [1]. In case of C^*-algebras the extension theorem for completely positive mappings is due to ARVESON [1] and the (Stinespring) dilation theorem appeared in STINESPRING [1].

12. Integral Decompositions of $*$-Representations and States

The principal goal of this final chapter is to contribute to the following two problems: to decompose a $*$-representation as a direct integral of irreducible $*$-representations and to decompose a positive linear functional as an integral over pure states. Loosely speaking, for most of our results concerning these problems some nuclearity assumptions play a crucial role. We briefly explain our approach to the first problem. Let π be a $*$-representation of a $*$-algebra A on a separable Hilbert space. We decompose the Hilbert space $\mathcal{H}(\pi)$ into a direct integral $\int_\Lambda^\oplus \mathcal{H}_\lambda \, d\mu(\lambda)$ of Hilbert spaces relative to a maximal abelian subalgebra of the von Neumann algebra $\pi(\mathsf{A})'_{ss}$. Then all operators $\overline{\pi(a)}$, $a \in \mathsf{A}$, are decomposable, and the families of components in $\mathcal{H}_\lambda$, $\lambda \in \Lambda$, will be irreducible a.e. The main difficulty that arises now lies in the definition of the corresponding $*$-representation π_λ of A on $\mathcal{H}_\lambda$. For this we apply a technique which is usually known under the name "nuclear spectral theorem". To be more precise, we assume that there exists another scalar product on the domain $\mathcal{D}(\pi)$ such that the canonical embedding map $\mathfrak{j}$ of the associated Hilbert space $\mathcal{K}$ into $\mathcal{H}(\pi)$ is a Hilbert-Schmidt mapping. Then there are Hilbert-Schmidt mappings $\mathfrak{j}_\lambda$ of $\mathcal{K}$ into $\mathcal{H}_\lambda$ for each $\lambda \in \Lambda$ such that $\mathfrak{j}\varphi$ coincides with the field $\lambda \to \mathfrak{j}_\lambda\varphi$ for any vector φ in $\mathcal{D}(\pi)$. We then define π_λ by $\pi_\lambda(a)\, \mathfrak{j}_\lambda\varphi = \mathfrak{j}_\lambda(a\varphi)$ for $a \in \mathsf{A}$ and $\varphi \in \mathcal{D}(\pi)$.

The first three sections of this chapter are concerned with direct integrals of measurable fields of closed operators and $*$-representations. In Section 12.1 we give a rather detailed study of decomposable closed linear operators relative to a direct integral of Hilbert spaces. The localization technique indicated above is developed in Section 12.2. Direct integral decompositions of $*$-representations are defined and investigated in Section 12.3.

The second problem is studied in Section 12.4. Our approach is based on Choquet theory of boundary integrals on compact convex sets. Since the state space of the $*$-algebra A is not weakly compact in general, we apply this theory to a *cap* of the cone of (all or some) positive linear functionals on A. Thus the essence of the proofs is to show that the positive linear functional is contained in some cap. In the last subsection of Section 12.4 integrals over states are considered and the orthogonality of the measure is characterized. In Section 12.5 the moment problem over a real nuclear locally convex Hausdorff space is treated. We present two proofs for the existence of a solution, the first one uses the main result of Section 12.4 and the second is based on the Bochner-Minlos theorem.

12.1. Decomposable Closed Operators

In the next three sections we frequently use the direct integral $\int_\Lambda^\oplus \mathcal{H}_\lambda \, d\mu(\lambda)$ of a measurable field $\lambda \to \mathcal{H}_\lambda$ of Hilbert spaces over a measure space (Λ, μ). We refer to part II of DIXMIER [1] (or to KADISON/RINGROSE [2], ch. 14) for the definition and basic properties of this notion and of other related concepts such as decomposable or diagonalizable (bounded) operators. In order to avoid all possible difficulties which can occur when dealing with general measure spaces, we always assume in these three sections that Λ is a locally compact σ-compact Hausdorff space and μ is the completion of a positive regular Borel measure on Λ. At a certain stage (see Section 12.2) we assume in addition that Λ is a metric space.

Let $\mathcal{H} = \int_\Lambda^\oplus \mathcal{H}_\lambda \, d\mu(\lambda)$ be a direct integral of Hilbert spaces which will be fixed in what follows.

Recall that by definition all spaces $\mathcal{H}_\lambda$, $\lambda \in \Lambda$, are separable and $\mathcal{H}$ is also separable (DIXMIER [1], p. 164 and 172). We mention some general notation and terminology we use. The scalar product of $\mathcal{H}_\lambda$ is denoted by $\langle .,. \rangle_\lambda$ and I_λ is the identity map of $\mathcal{H}_\lambda$. If no confusion can arise, we omit the lower subscript under the integral sign. By a statement like $K = \int^\oplus K_\lambda \, d\mu(\lambda)$ we always mean that the field $\lambda \to K_\lambda$ of (bounded or closed) operators or of closed subspaces is measurable (relative to the field $\lambda \to \mathcal{H}_\lambda$) and that the equality $K = \int^\oplus K_\lambda \, d\mu(\lambda)$ it true. The elements of $\mathcal{H}$ are considered as vector fields (although, they are, of course, equivalence classes of those) and we write $\varphi(\lambda)$ for the value of $\varphi \in \mathcal{H}$ at $\lambda \in \Lambda$. If $\varphi(\lambda)$ is defined on Λ up to a μ-null set N, then by saying that φ is in $\mathcal{H}$ we mean that the field $\lambda \to \varphi(\lambda)$ obtained by setting $\varphi(\lambda) = 0$ on N is in $\mathcal{H}$. For $\varphi \in \mathcal{H}$ and $f \in L^\infty(\Lambda; \mu)$, let $f\varphi$ be the element of $\mathcal{H}$ defined by $(f\varphi)(\lambda) := f(\lambda), \varphi(\lambda)$, $\lambda \in \Lambda$. A subset of a Hilbert space is called *total* if its linear span is dense in the space.

Next we restate some results from DIXMIER [1] as a reference.

Lemma 12.1.1. *If $\mathcal{N}$ is an abelian von Neumann algebra acting on a separable Hilbert space, then there is a direct integral $\int_\Lambda^\oplus \mathcal{H}_\lambda \, d\mu(\lambda)$ of non-zero Hilbert spaces $\mathcal{H}_\lambda$, $\lambda \in \Lambda$, such that $\mathcal{N}$ is unitarily equivalent to the algebra of all diagonalizable operators relative to this direct integral. Here Λ can be chosen to be a compact metric space and μ to be the completion of a regular Borel measure with support Λ.*

Proof. DIXMIER [1], part II, ch. 6, Theorem 2. □

Lemma 12.1.2. *Let $(\varphi_n : n \in \mathbb{N})$ be a sequence in $\mathcal{H} = \int_\Lambda^\oplus \mathcal{H}_\lambda \, d\mu(\lambda)$.*

(i) *If $\{\varphi_n(\lambda) : n \in \mathbb{N}\}$ is total in $\mathcal{H}_\lambda$ a.e., then the set $\{f\varphi_n : f \in L^\infty(\Lambda; \mu)$ and $n \in \mathbb{N}\}$ is total in $\mathcal{H}$.*

(ii) *If $\{\varphi_n : n \in \mathbb{N}\}$ is total in $\mathcal{H}$, then $\{\varphi_n(\lambda) : n \in \mathbb{N}\}$ is total in $\mathcal{H}_\lambda$ a.e.*

Proof. DIXMIER [1], part II, ch. 1, Propositions 7 and 8. □

Lemma 12.1.3. *For $\lambda \in \Lambda$, let $\mathcal{G}_\lambda$ be a closed linear subspace of $\mathcal{H}_\lambda$. The following statements are equivalent:*

(i) *There is a sequence $(\varphi_n : n \in \mathbb{N})$ of measurable vector fields (relative to the field $\lambda \to \mathcal{H}_\lambda$) such that $\{\varphi_n(\lambda) : n \in \mathbb{N}\}$ is total in $\mathcal{G}_\lambda$* a.e.

(ii) *$\lambda \to \mathcal{G}_\lambda$ is a measurable field of closed linear subspaces.*

(iii) *$\lambda \to P_{\mathcal{G}_\lambda}$ is a measurable field of operators.*

If one of these conditions is valid and $\mathcal{G} := \int^\oplus \mathcal{G}_\lambda \, d\mu(\lambda)$, then $P_\mathcal{G} = \int^\oplus P_{\mathcal{G}_\lambda} \, d\mu(\lambda)$.

Proof. Dixmier [1], part II, ch. 1, Proposition 9; in fact, (i) is taken as a definition for (ii) there. The proof of the last assertion is straightforward and therefore omitted. □
The following simple lemma is often needed in the sequel.

Lemma 12.1.4. *Let $\lambda \to \mathcal{G}_\lambda$ and $\lambda \to \mathcal{K}_\lambda$ be measurable fields of closed linear subspaces of $\mathcal{H}$. Set $\mathcal{G} = \int^\oplus \mathcal{G}_\lambda \, d\mu(\lambda)$ and $\mathcal{K} = \int^\oplus \mathcal{K}_\lambda \, d\mu(\lambda)$. Then:*

(i) $\mathcal{G}^\perp = \int^\oplus \mathcal{G}_\lambda^\perp \, d\mu(\lambda)$.

(ii) $\mathcal{G} \vee \mathcal{K} = \int^\oplus \mathcal{G}_\lambda \vee \mathcal{K}_\lambda \, d\mu(\lambda)$, *where, as usual, "$\vee$" denotes the closed linear span of the subspaces.*

(iii) $\mathcal{G} \cap \mathcal{K} = \int^\oplus \mathcal{G}_\lambda \cap \mathcal{K}_\lambda \, d\mu(\lambda)$.

(iv) $\mathcal{G} \subseteq \mathcal{K}$ *if and only if* $\mathcal{G}_\lambda \subseteq \mathcal{K}_\lambda$ *a.e.*

(v) $\mathcal{G} = \{0\}$ *if and only if* $\mathcal{G}_\lambda = \{0\}$ *a.e.*

Proof. (i) follows immediately from Lemma 12.1.3 and the relation $I - P_\mathcal{G} = \int^\oplus (I_\lambda - P_{\mathcal{G}_\lambda}) \, d\mu(\lambda)$.

(ii): Suppose $\{\varphi_n : n \in \mathbb{N}\}$ and $\{\psi_n : n \in \mathbb{N}\}$ are total subsets of $\mathcal{G}$ and $\mathcal{K}$, respectively. From Lemma 12.1.2, (ii), it follows that $\{\varphi_n(\lambda), \psi_n(\lambda) : n \in \mathbb{N}\}$ is total in $\mathcal{G}_\lambda \vee \mathcal{K}_\lambda$ a.e.; so the field $\lambda \to \mathcal{G}_\lambda \vee \mathcal{K}_\lambda$ is measurable by Lemma 12.1.3. Set $\mathcal{X} := \int^\oplus \mathcal{G}_\lambda \vee \mathcal{K}_\lambda \, d\mu(\lambda)$. Applying Lemma 12.1.2, (i), we see that $\{f\varphi_n, f\psi_n : f \in L^\infty(\Lambda; \mu)$ and $n \in \mathbb{N}\}$ is total in $\mathcal{X}$. But this set is also total in $\mathcal{G} \vee \mathcal{K}$; hence $\mathcal{X} = \mathcal{G} \vee \mathcal{K}$.

(iii) follows at once from (i), (ii), and the identities $\mathcal{G} \cap \mathcal{K} = (\mathcal{G}^\perp \vee \mathcal{K}^\perp)^\perp$ and $\mathcal{G}_\lambda \cap \mathcal{K}_\lambda = (\mathcal{G}_\lambda^\perp \vee \mathcal{K}_\lambda^\perp)^\perp$.

(iv): Suppose $\mathcal{G} \subseteq \mathcal{K}$. Take a total set $\{\varphi_n : n \in \mathbb{N}\}$ in $\mathcal{G}$. Then $\{\varphi_n(\lambda) : n \in \mathbb{N}\}$ is total in $\mathcal{G}_\lambda$ a.e. which yields $\mathcal{G}_\lambda \subseteq \mathcal{K}_\lambda$ a.e. The opposite direction is trivial.

(v): Set $\mathcal{K}_\lambda := \{0\}$ for all $\lambda \in \Lambda$ and apply (iv). □

Corollary 12.1.5. *Let $\lambda \to \mathcal{G}_\lambda$ be a measurable field of closed linear subspaces of $\mathcal{H} \oplus \mathcal{H} = \int^\oplus \mathcal{H}_\lambda \oplus \mathcal{H}_\lambda \, d\mu(\lambda)$ and let $\mathcal{G} = \int^\oplus \mathcal{G}_\lambda \, d\mu(\lambda)$. Then $\mathcal{G}$ is the graph of a closed linear operator a in $\mathcal{H}$ if and only if a.e. $\mathcal{G}_\lambda$ is the graph of a closed linear operator a_λ in $\mathcal{H}_\lambda$.*

Proof. From Lemma 12.1.4, (iii), $\mathcal{G} \cap (\{0\} \oplus \mathcal{H}) = \int^\oplus \mathcal{G}_\lambda \cap (\{0\} \oplus \mathcal{H}_\lambda) \, d\mu(\lambda)$. Therefore, by Lemma 12.1.4, (v), $\mathcal{G} \cap (\{0\} \oplus \mathcal{H}) = \{(0, 0)\}$ if and only if $\mathcal{G}_\lambda \cap (\{0\} \oplus \mathcal{H}_\lambda) = \{(0, 0)\}$ a.e. This gives the assertion. □

For the next definition we recall that the graph gr a of a closed operator a on a Hilbert space $\mathcal{G}$ is a closed linear subspace of $\mathcal{G} \oplus \mathcal{G}$.

Definition 12.1.6. For every $\lambda \in \Lambda$ let a_λ be a closed linear operator in the Hilbert space $\mathcal{H}_\lambda$. The field $\lambda \to a_\lambda$ is said to be *measurable* if the field $\lambda \to \operatorname{gr} a_\lambda$ of closed linear subspaces of $\mathcal{H} \oplus \mathcal{H} = \int^{\oplus} \mathcal{H}_\lambda \oplus \mathcal{H}_\lambda \, d\mu(\lambda)$ is measurable. If the field $\lambda \to a_\lambda$ is measurable, then, by Corollary 12.1.5, there is a (unique) closed operator a in the Hilbert space $\mathcal{H}$ such that $\operatorname{gr} a = \int^{\oplus} \operatorname{gr} a_\lambda \, d\mu(\lambda)$. The operator a is said to be *decomposable* and is denoted by $a = \int^{\oplus} a_\lambda \, d\mu(\lambda)$. If the field $\lambda \to a_\lambda$ is measurable and all operators a_λ, $\lambda \in \Lambda$, are scalars (i.e., $a_\lambda = f(\lambda) I_\lambda$ with $f(\lambda) \in \mathbb{C}$), then the operator $a = \int^{\oplus} a_\lambda \, d\mu(\lambda) \equiv \int^{\oplus} f(\lambda) I_\lambda \, d\mu(\lambda)$ is called *diagonalizable*.

Suppose $a \equiv \int^{\oplus} a_\lambda \, d\mu(\lambda)$ is a decomposable closed operator. By the preceding definition, $\mathcal{D}(a)$ is the set of all $\varphi \in \mathcal{H}$ such that $\varphi(\lambda) \in \mathcal{D}(a_\lambda)$ a.e. and the field $\lambda \to \psi(\lambda) := a_\lambda \varphi(\lambda)$ is square integrable (or equivalently, belongs to $\mathcal{H}$), and the operator a acts by $a\varphi := \psi$.

Remark 1. From Dixmier [1], p. 179, it is clear that the above definition of measurability of the field $\lambda \to a_\lambda$ is equivalent to the usual one when all operators a_λ are bounded and everywhere defined on $\mathcal{H}_\lambda$. Further, a bounded operator a is decomposable or diagonalizable in the usual sense (i.e., as defined in Dixmier [1]) if and only if it has this property according to Definition 12.1.6. One way to see this is to compare Proposition 12.1.7 with the corresponding results for bounded operators.

Remark 2. Let b be a closed operator in a Hilbert space $\mathcal{G}$. The projection $Q(b)$ of $\mathcal{G} \oplus \mathcal{G}$ onto gr b can be written as a 2×2 matrix $[q_{kl}(b)]_{k,l=1,2}$ with entries in $\mathbb{B}(\mathcal{G})$. This matrix is called the *characteristic matrix* of the operator b.

From Lemma 12.1.3 and Definition 12.1.5, a field $\lambda \to a_\lambda$ of closed operators is measurable if and only if the field $\lambda \to Q(a_\lambda)$ of projections is measurable (relative to the direct integral $\mathcal{H} \oplus \mathcal{H} = \int^{\oplus} \mathcal{H}_\lambda \oplus \mathcal{H}_\lambda \, d\mu(\lambda)$). Obviously, this is equivalent to the measurability of the four fields $\lambda \to q_{kl}(a_\lambda)$, $k, l = 1, 2$ (relative to $\mathcal{H} = \int^{\oplus} \mathcal{H}_\lambda \, d\mu(\lambda)$). The latter has been frequently taken as definition in place of the one given in Definition 12.1.6 (for instance, in Nussbaum [1]).

Proposition 12.1.7. *Let $\mathcal{N}$ be the abelian von Neumann algebra of bounded diagonalizable operators and $\mathcal{R}$ the von Neumann algebra of bounded decomposable operators on $\mathcal{H} = \int^{\oplus} \mathcal{H}_\lambda \, d\mu(\lambda)$. Suppose a is a closed operator in $\mathcal{H}$.*

(i) *a is decomposable if and only if $\mathcal{N} \subseteq (a)'_s$ (or equivalently, if a is affiliated with $\mathcal{R}$).*

(ii) *a is diagonalizable if and only if $\mathcal{R} \subseteq (a)'_s$ (or equivalently, if a is affiliated with $\mathcal{N}$).*

Proof. (i): It is obvious from the above definition that $\mathcal{N} \subseteq (a)'_s$ when $\mathcal{N}$ is decomposable. Suppose now that $\mathcal{N} \subseteq (a)'_s$. We take a subset $\{\varphi_n : n \in \mathbb{N}\}$ of $\mathcal{D}(a)$ such that $\{(\varphi_n, a\varphi_n) : n \in \mathbb{N}\}$ is dense in gr a. Let $\mathcal{G}_\lambda$ denote the closed linear span of the set $\{(\varphi_n(\lambda), (a\varphi_n)(\lambda)) : n \in \mathbb{N}\}$ in $\mathcal{H}_\lambda \oplus \mathcal{H}_\lambda$. From Lemma 12.1.3, the field $\lambda \to \mathcal{G}_\lambda$ of closed linear subspaces of $\mathcal{H}_\lambda \oplus \mathcal{H}_\lambda$ is measurable. Since, of course, $(\varphi_n, a\varphi_n) \in \mathcal{G}$

$:= \int^{\oplus} \mathcal{G}_\lambda \, d\mu(\lambda)$ for $n \in \mathbb{N}$, we have $\operatorname{gr} a \subseteq \mathcal{G}$. Suppose $f \in L^\infty(\Lambda; \mu)$. Upon changing f on a μ-null set we can assume that $f(\cdot)$ is finite on Λ. Then $x_f := \int^{\oplus} f(\lambda) I_\lambda \, d\mu(\lambda) \in \mathcal{N} \subseteq (a)'_s$ and hence $fa\varphi_n = x_f a\varphi_n = ax_f\varphi_n = af\varphi_n$ for $n \in \mathbb{N}$. From this it follows that $\Gamma := \{(f\varphi_n, fa\varphi_n) : f \in L^\infty(\Lambda; \mu)$ and $n \in \mathbb{N}\}$ is contained in $\operatorname{gr} a$. Since Γ is also total in $\mathcal{G}$ by Lemma 12.1.2, (i), we get $\mathcal{G} \subseteq \operatorname{gr} a$. Thus $\operatorname{gr} a = \mathcal{G}$. By Corollary 12.1.5, there are a μ-null set N and closed operators a_λ, $\lambda \in \Lambda \mid N$, such that $\mathcal{G}_\lambda = \operatorname{gr} a_\lambda$. Upon replacing $\mathcal{G}_\lambda$ by $\{0\}$ and setting $a_\lambda = 0$ if $\lambda \in N$, we can assume the latter for all $\lambda \in \Lambda$. Then, by Definition 12.1.6, a is decomposable.

(ii): It is clear that $\mathcal{R} \subseteq (a)'_s$ if the operator a is diagonalizable. To prove the converse, assume that $\mathcal{R} \subseteq (a)'_s$. Since $\mathcal{N} \subseteq \mathcal{R}$, part (i) implies that a is decomposable, i.e., a is of the form $a = \int^{\oplus} a_\lambda \, d\mu(\lambda)$. We have to show that $a_\lambda = f(\lambda) I_\lambda$ for some $f(\lambda) \in \mathbb{C}$ a.e. By a well-known technique in direct integral theory (see e.g. Dixmier [1], part II, ch. 2) it suffices to prove the latter in case where $\lambda \to \mathcal{H}_\lambda$ is the constant field corresponding to a (fixed) Hilbert space $\mathcal{K}$. That is, we can assume without loss of generality that $\mathcal{H}$ is the Hilbert space $L^2_{\mathcal{K}}(\Lambda; \mu)$ of all $\mathcal{K}$-valued square integrable mappings of Λ into $\mathcal{K}$. Since $\mathcal{K}$ is separable (because $\mathcal{H}$ is), $\mathbb{B}(\mathcal{K})$ admits a countable dense subset $\mathcal{X}$ in the weak-operator topology. For $x \in \mathcal{X}$, let $\tilde{x}$ denote the operator in $\mathcal{R}$ defined by $(\tilde{x}\varphi)(\lambda) := x\varphi(\lambda)$, $\varphi \in \mathcal{H}$ and $\lambda \in \Lambda$. Let $\Gamma \equiv \{(\varphi_n, a\varphi_n) : n \in \mathbb{N}\}$ be a total set in $\operatorname{gr} a$. Since $\mathcal{X}$ and Γ are countable, there is a μ-null set N such that

$$xa_\lambda\varphi_n(\lambda) = (xa\varphi_n)(\lambda) = (ax\varphi_n)(\lambda) = a_\lambda x\varphi_n(\lambda)$$

for all $\lambda \in \Lambda \setminus N$ and $n \in \mathbb{N}$. Since $\{(\varphi_n(\lambda), a_\lambda\varphi_n(\lambda)) : n \in N\}$ is total in $\operatorname{gr} a_\lambda$ a.e. by Lemma 12.1.2, (ii), the preceding implies that $xa_\lambda \subseteq a_\lambda x$ for all $x \in \mathcal{X}$ a.e. By Lemma 7.2.8, $(a_\lambda)'_s$ is weak-operator closed in $\mathbb{B}(\mathcal{K})$. Hence we get $xa_\lambda \subseteq a_\lambda x$ for all $x \in \mathbb{B}(\mathcal{K})$, i.e., the operator a_λ is affiliated with $\mathbb{B}(\mathcal{K})'$ a.e. Therefore, $a_\lambda = f(\lambda) I_\lambda$ with $f(\lambda) \in \mathbb{C}$ a.e., and a is diagonalizable.

The statements in the parentheses are equivalent to $\mathcal{N} \subseteq (a)'_s$ and $\mathcal{R} \subseteq (a)'_s$, respectively, since $\mathcal{R}' = \mathcal{N}$ and $\mathcal{N}' = \mathcal{R}$ (Kadison/Ringrose [1], 14.1.10). □

Remark 3. Suppose that $\lambda \to a_\lambda$ is a measurable field of closed operators and a is a closed operator in $\mathcal{H}$ such that $(a\varphi)(\lambda) = a_\lambda\varphi(\lambda)$ a.e. for all $\varphi \in \mathcal{D}(a)$. From this we cannot conclude that a is decomposable. (For instance, let a be a restriction of $\int^{\oplus} a_\lambda \, d\mu(\lambda)$ such that $\mathcal{D}(a)$ is not invariant under $\mathcal{N}$.) However, if in addition $\mathcal{D}(a)$ is invariant under the operators in $\mathcal{N}$, then we have $\mathcal{N} \subseteq (a)'_s$ and hence a is decomposable.

Proposition 12.1.8. *Suppose* $a = \int^{\oplus} a_\lambda \, d\mu(\lambda)$ *and* $b = \int^{\oplus} b_\lambda \, d\mu(\lambda)$. *Then:*

(i) $a \subseteq b$ *if and only if* $a_\lambda \subseteq b_\lambda$ *a.e.*

(ii) $a = b$ *if and only if* $a_\lambda = b_\lambda$ *a.e.*

(iii) $\ker a = \int^{\oplus} \ker a_\lambda \, d\mu(\lambda)$, $\overline{a\mathcal{H}} = \int^{\oplus} \overline{a_\lambda\mathcal{H}_\lambda} \, d\mu(\lambda)$, $\overline{\mathcal{D}(a)} = \int^{\oplus} \overline{\mathcal{D}(a_\lambda)} \, d\mu(\lambda)$. $\mathcal{D}(a)$ *is dense in* $\mathcal{H}$ *if and only if* $\mathcal{D}(a_\lambda)$ *is dense in* $\mathcal{H}_\lambda$ *a.e.*

(iv) a^{-1} *exists if and only if* a_λ^{-1} *exists a.e., and then* $a^{-1} = \int^{\oplus} a_\lambda^{-1} \, d\mu(\lambda)$.

(v) a^* *exists if and only if* a_λ^* *exists a.e., and then* $a^* = \int^{\oplus} a_\lambda^* \, d\mu(\lambda)$.

(In (iv) *and* (v) *we set* $a_\lambda^{-1} = 0$ *and* $a_\lambda^* = 0$ *on the null set where* a_λ^{-1} *and* a_λ^* *are not defined.)*

Proof. (i): Clearly, $a \subseteq b$ if and only if $\operatorname{gr} a \equiv \int^{\oplus} \operatorname{gr} a_\lambda \, d\mu(\lambda) \subseteq \operatorname{gr} b \equiv \int^{\oplus} \operatorname{gr} b_\lambda \, d\mu(\lambda)$. By Lemma 12.1.4, (iv), the latter is equivalent to $a_\lambda \subseteq b_\lambda$ a.e.
(ii) follows at once from (i).
(iii): From Lemma 12.1.4, (iii), $\ker a \oplus \{0\} = \operatorname{gr} a \cap (\mathcal{H} \oplus \{0\}) = \int^{\oplus} \operatorname{gr} a_\lambda \cap (\mathcal{H}_\lambda \oplus \{0\}) \, d\mu(\lambda) = \int^{\oplus} (\ker a_\lambda \oplus \{0\}) \, d\mu(\lambda)$ which gives the first equality. Let $\{(\varphi_n, a\varphi_n) : n \in \mathbb{N}\}$ be a total subset of $\operatorname{gr} a$. Since $\operatorname{gr} a = \int^{\oplus} \operatorname{gr} a_\lambda \, d\mu(\lambda)$ by definition, Lemma 12.1.2,(ii), ensures that $\{(\varphi_n(\lambda), a_\lambda \varphi_n(\lambda)) : n \in \mathbb{N}\}$ is total in $\operatorname{gr} a_\lambda$ a.e. Hence $\{\varphi_n(\lambda) : n \in \mathbb{N}\}$ is total in $\overline{\mathcal{D}(a_\lambda)}$ a.e. By Lemma 12.1.3, the field $\lambda \to \overline{\mathcal{D}(a_\lambda)}$ is measurable. Set $\mathcal{X} := \int^{\oplus} \overline{\mathcal{D}(a_\lambda)} \, d\mu(\lambda)$. Since the set $\{f\varphi_n : f \in L^\infty(\Lambda; \mu)$ and $n \in \mathbb{N}\}$ is total in both $\overline{\mathcal{D}(a)}$ and $\mathcal{X}$ by construction or by Lemma 12.1.2,(i), we have $\overline{\mathcal{D}(a)} = \mathcal{X}$. A similar reasoning yields the assertion for the range. The final statement in (iii) follows from Lemma 12.1.4,(i), and the equality $\overline{\mathcal{D}(a)} = \int^{\oplus} \overline{\mathcal{D}(a_\lambda)} \, d\mu(\lambda)$.
(iv): By (iii) and Lemma 12.1.4,(v), $\ker a = \{0\}$ if and only if $\ker a_\lambda = \{0\}$ a.e., i.e., a^{-1} exists if and only if a_λ^{-1} exists a.e. For a Hilbert space $\mathcal{K}$, define $T(\varphi, \psi) := (\psi, \varphi)$, $\varphi, \psi \in \mathcal{K}$. Then we have $\operatorname{gr} a^{-1} = T(\operatorname{gr} a) = T\left(\int^{\oplus} \operatorname{gr} a_\lambda \, d\mu(\lambda)\right) = \int^{\oplus} T(\operatorname{gr} a_\lambda) \, d\mu(\lambda) = \int^{\oplus} \operatorname{gr} a_\lambda^{-1} \, d\mu(\lambda)$, so that $a^{-1} = \int^{\oplus} a_\lambda^{-1} \, d\mu(\lambda)$.
(v): By (iii), a^* exists (i.e., $\mathcal{D}(a)$ is dense in $\mathcal{H}$) if and only if a_λ^* exists (i.e., $\mathcal{D}(a_\lambda)$ is dense in $\mathcal{H}_\lambda$) a.e. The desired equality follows in the same way as for the inverses if we replace the mapping T by S, where $S(\varphi, \psi) := (\psi, -\varphi)$, $\varphi, \psi \in \mathcal{K}$. □

The last proposition in this section will show that each set of closed linear operators in a separable Hilbert space can be decomposed into irreducible components.

A set $\mathcal{A}$ of closed linear operators in a Hilbert space $\mathcal{G}$ is said to be *irreducible* if there exists no closed subspace $\mathcal{K}$ of $\mathcal{G}$ other than $\mathcal{G}$ and $\{0\}$ such that every operator $a \in \mathcal{A}$ can be written as a direct sum $a = a_1 \oplus a_2$, where a_1 and a_2 are linear operators on $\mathcal{K}$ and $\mathcal{G} \ominus \mathcal{K}$, respectively. It is easy to see that $\mathcal{A}$ is irreducible if and only if there are no projections other than I and 0 in $\mathcal{A}'_{\text{ss}}$ or equivalently if the von Neumann algebra $\mathcal{A}'_{\text{ss}}$ consists of scalars only.

Now suppose that $\mathcal{B}$ is a (non-empty) set of closed linear operators in a separable Hilbert space $\mathcal{H}$ and $\mathcal{N}$ is an abelian von Neumann algebra on $\mathcal{H}$ contained in $\mathcal{B}'_{\text{ss}}$. By Lemma 12.1.1, there is a unitary isomorphism U of $\mathcal{H}$ onto a direct integral $\int^{\oplus} \mathcal{H}_\lambda \, d\mu(\lambda)$ of Hilbert spaces $\mathcal{H}_\lambda, \lambda \in \Lambda$, such that $U\mathcal{N}U^{-1}$ coincides with the algebra of all bounded diagonalizable operators on $\int^{\oplus} \mathcal{H}_\lambda \, d\mu(\lambda)$. Suppose $b \in \mathcal{B}$. Since $U\mathcal{N}U^{-1} \subseteq (UbU^{-1})'_{\text{s}}$ by the assumption $\mathcal{N} \subseteq \mathcal{B}'_{\text{ss}}$, we conclude from Proposition 12.1.7,(i), that the operator UbU^{-1} is decomposable and hence of the form $\int^{\oplus} b_\lambda \, d\mu(\lambda)$. Clearly, the operators b_λ are determined by $b \in \mathcal{B}$ up to a null set only. We fix one choice of b_λ, $\lambda \in \Lambda$, for each $b \in \mathcal{B}$ and let $\mathcal{B}_\lambda$ denote the set of all b_λ when b ranges over $\mathcal{B}$. The following proposition shows that the components $\mathcal{B}_\lambda$ will be irreducible a.e. if we choose $\mathcal{N}$ to be maximal abelian in the von Neumann algebra $\mathcal{B}'_{ss}$.

Proposition 12.1.9. *Retain the above assumptions and notation:*

(i) *If $\mathcal{N}$ is maximal abelian in $\mathcal{B}'_{ss}$, then $\mathcal{B}_\lambda$ is irreducible a.e.*

(ii) *If the set $\mathcal{B}$ is countable and $\mathcal{B}_\lambda$ is irreducible a.e., then $\mathcal{N}$ is maximal abelian in $\mathcal{B}'_{ss}$.*

Proof. In order to simplify the notation we identify $\mathcal{H}$ and $\int^\oplus \mathcal{H}_\lambda \, d\mu(\lambda)$ via the unitary mapping U. Further, we use the notation $Q(b) \equiv [q_{kl}(b)]$ introduced in Remark 2 above. Let $b \in \mathcal{B}$. It is not difficult to check that an operator $x = x^* \in \mathbb{B}(\mathcal{H})$ is in $(b)'_s$ if and only if $x \oplus x$ commutes with the projection $Q(b)$ on $\mathcal{H} \oplus \mathcal{H}$. Carrying out the matrix multiplication we see that $x \in (b)'_s$ is equivalent to $x \in \{q_{kl}(b) : k, l = 1, 2\}'$. From this we obtain that $\mathcal{B}'_{ss} = \{q_{kl}(b) : b \in \mathcal{B} \text{ and } k, l = 1, 2\}'$. Similarly, we get $(\mathcal{B}_\lambda)'_{ss} = \{q_{kl}(b_\lambda) : b \in \mathcal{B} \text{ and } k, l = 1, 2\}'$. Since $Q(b) = \int^\oplus Q(b_\lambda) \, d\mu(\lambda)$ by Lemma 12.1.3 and Definition 12.1.6, we have $q_{kl}(b) = \int^\oplus q_{kl}(b_\lambda) \, d\mu(\lambda)$ for $b \in \mathcal{B}$ and $k, l = 1, 2$. Thus the assumptions of Corollary 1 in DIXMIER [1], p. 196, are satisfied. Part (i) of this result states that $(v(\mathcal{B}_\lambda)'_{ss}) = \mathbb{B}(\mathcal{H}_\lambda)$ a.e. provided that $\mathcal{N}$ is maximal abelian in $\mathcal{B}'_{ss}$. Part (ii) asserts that the converse is true if $\mathcal{B}$ is countable. From this the assertions follow. □

12.2. Localization of Decomposable Operators

In this section we suppose that $\mathcal{H} = \int_\Lambda^\oplus \mathcal{H}_\lambda \, d\mu(\lambda)$ is a fixed direct integral of non-zero Hilbert spaces $\mathcal{H}_\lambda$, $\lambda \in \Lambda$, $\mathcal{K}$ is a Hilbert space with scalar product $(\cdot, \cdot)$ and norm $|||\cdot|||$ and $\mathfrak{j}$ is a Hilbert-Schmidt mapping of $\mathcal{K}$ into $\mathcal{H}$. For Propositions 12.2.2 and 12.2.3 we also assume that Λ is a metric space.

We refer to GELFAND/WILENKIN [1], I, § 2, or to WEIDMANN [1], ch. 6, for the facts about Hilbert-Schmidt mappings we use in this section.

Proposition 12.2.1. *For each $\lambda \in \Lambda$ there exists a Hilbert-Schmidt operator $\mathfrak{j}_\lambda$ of $\mathcal{K}$ into $\mathcal{H}_\lambda$ such that for every $\varphi \in \mathcal{K}$ the vector field $\lambda \to \mathfrak{j}_\lambda \varphi$ belongs to $\mathcal{H}$ and $(\mathfrak{j}\varphi)(\lambda) = \mathfrak{j}_\lambda \varphi$ a.e. on Λ. Furthermore, we have*

$$\|\mathfrak{j}\|_2^2 = \int_\Lambda \|\mathfrak{j}_\lambda\|_2^2 \, d\mu(\lambda), \tag{1}$$

where $\|\cdot\|_2$ denotes the Hilbert-Schmidt norm.

Proof. For notational simplicity we assume that $\mathcal{K}$ is infinite dimensional. Since $\mathfrak{j}$ is a Hilbert-Schmidt operator of $\mathcal{K}$ into $\mathcal{H}$, there exists an orthonormal sequence $(\varphi_n : n \in \mathbb{N})$ of $\mathcal{K}$ and a sequence $(\psi_n : n \in \mathbb{N})$ of vectors in $\mathcal{H}$ such that $\sum_{n=1}^\infty \|\psi_n\|^2 < \infty$ and $\mathfrak{j}\varphi = \sum_{n=1}^\infty (\varphi, \varphi_n) \psi_n$ for $\varphi \in \mathcal{K}$. Let φ_i, $i \in I$, be vectors in $\mathcal{K}$ such that the set $\{\varphi_n, \varphi_i\}$ is an orthonormal basis of $\mathcal{K}$. Then,

$$\|\mathfrak{j}\|_2^2 = \sum_n \|\mathfrak{j}\varphi_n\|^2 + \sum_i \|\mathfrak{j}\varphi_i\|^2 = \sum_n \|\psi_n\|^2 = \sum_{n=1}^{\infty} \int_\Lambda \|\psi_n(\lambda)\|_\lambda^2 \, d\mu(\lambda)$$
$$= \int_\Lambda \left(\sum_{n=1}^{\infty} \|\psi_n(\lambda)\|_\lambda^2 \right) d\mu(\lambda) < \infty. \quad (2)$$

Hence there is a μ-null set N such that $C_\lambda := \sum_{n=1}^{\infty} \|\psi_n(\lambda)\|_\lambda^2 < \infty$ for all $\lambda \in \Lambda \setminus N$. We set $\mathfrak{j}_\lambda = 0$ if $\lambda \in N$. Now let $\lambda \in \Lambda \setminus N$. From

$$\sum_n |(\varphi, \varphi_n)| \, \|\psi_n(\lambda)\|_\lambda \leqq \Big(\sum_n |(\varphi, \varphi_n)|^2\Big)^{1/2} \Big(\sum_n \|\psi_n(\lambda)\|_\lambda^2\Big)^{1/2} \leqq \|\varphi\| \, C_\lambda^{1/2}, \quad \varphi \in \mathcal{K},$$

we conclude that $\mathfrak{j}_\lambda := \sum_{n=1}^{\infty} (\cdot, \varphi_n) \, \psi_n(\lambda)$ is a well-defined bounded linear operator from $\mathcal{K}$ into $\mathcal{H}_\lambda$. We have

$$\sum_n \|\mathfrak{j}_\lambda \varphi_n\|_\lambda^2 + \sum_i \|\mathfrak{j}_\lambda \varphi_i\|_\lambda^2 = \sum_n \|\psi_n(\lambda)\|_\lambda^2 = C_\lambda < \infty.$$

From this it follows that $\mathfrak{j}_\lambda$ is a Hilbert-Schmidt operator and $\|\mathfrak{j}_\lambda\|_2^2 = C_\lambda$ for $\lambda \in \Lambda \setminus N$. Putting the latter into (2) and using the fact that $\mu(N) = 0$ we obtain (1).

Let $\varphi \in \mathcal{K}$. We show that $(\mathfrak{j}\varphi)(\lambda) = \mathfrak{j}_\lambda \varphi$ a.e. Set $S_k\varphi := \sum_{n=1}^{k} (\varphi, \varphi_n) \, \psi_n$, $k \in \mathbb{N}$. Since the sequence $(S_k\varphi: k \in \mathbb{N})$ converges to $\mathfrak{j}\varphi$ in $\mathcal{H}$, there exists a subsequence $(S_{k_m}\varphi: m \in \mathbb{N})$ such that $\big((S_{k_m}\varphi)(\lambda): m \in \mathbb{N}\big)$ converges to $(\mathfrak{j}\varphi)(\lambda)$ in $\mathcal{H}_\lambda$ a.e. (Dixmier [1], part II. ch. 1, Proposition 5). But, by the definition of $\mathfrak{j}_\lambda$, the sequence $\big((S_{k_m}\varphi)(\lambda): m \in \mathbb{N}\big)$ converges to $\mathfrak{j}_\lambda\varphi$ a.e. Hence $(\mathfrak{j}\varphi)(\lambda) = \mathfrak{j}_\lambda\varphi$ a.e. □

We define a positive Borel measure as follows. For a Borel subset M of Λ we set $\nu(M) := \int_M \|\mathfrak{j}_\lambda\|_2^2 \, d\mu(\lambda)$, where $\mathfrak{j}_\lambda$ are the Hilbert-Schmidt operators from Proposition 12.2.1.

For $\lambda \in \Lambda$ and $\varepsilon > 0$, let $W_\varepsilon(\lambda)$ denote the closed ball in the metric space Λ with radius ε centered at λ. If M is a Borel set in Λ, we let $E(M) := \int_M^{\oplus} I_\lambda \, d\mu(\lambda)$.

Proposition 12.2.2. *Suppose that the Hilbert space $\mathcal{K}$ is separable and $\mathfrak{j}\mathcal{K}$ is dense in $\mathcal{H}$. Then we have:*

(i) *The measure ν on Λ is finite and equivalent to μ.*

(ii) *There is a μ-null set N such that $\nu\big(W_\varepsilon(\lambda)\big) > 0$ for $\varepsilon > 0$ and*

$$\lim_{\varepsilon \to +0} \|\mathfrak{j}_\lambda\|_2^2 \, \nu\big(W_\varepsilon(\lambda)\big)^{-1} \big\langle E\big(W_\varepsilon(\lambda)\big) \mathfrak{j}\varphi, \mathfrak{j}\psi \big\rangle = \langle \mathfrak{j}_\lambda\varphi, \mathfrak{j}_\lambda\psi \rangle_\lambda \quad (3)$$

for all $\lambda \in \Lambda \setminus N$ and $\varphi, \psi \in \mathcal{K}$.

Proof. (i): By (1), $\nu(\Lambda) = \int_\Lambda \|\mathfrak{j}_\lambda\|_2^2 \, d\mu(\lambda) = \|\mathfrak{j}\|_2^2 < \infty$. Let M be a ν-null set. Since $\|\cdot\|_2$ is a norm, we get $\mathfrak{j}_\lambda = 0$ a.e. on M and hence $E(M) \, \mathfrak{j}\mathcal{K} = \{0\}$. Because $\mathfrak{j}\mathcal{K}$ is dense in $\mathcal{H}$, $E(M) = 0$. Since all Hilbert spaces $\mathcal{H}$ are non-zero by assumption, this leads to $\mu(M) = 0$; so μ is absolutely continuous with respect to ν. By the above definition, ν is absolutely continuous with respect to μ. Thus ν and μ are equivalent.

(ii): Since $\mathcal{K}$ is separable, there is a countable dense subset $\{\xi_n : n \in \mathbb{N}\}$ of $\mathcal{K}$. Let Λ_s denote the support of ν and let $\lambda \in \Lambda_s$. Then we have $\nu(W_\varepsilon(\lambda)) > 0$ for any $\varepsilon > 0$. The equality (3) for all $\varphi, \psi \in \mathcal{K}$ is equivalent to the fact that $T_{\varepsilon,\lambda} := \|\mathfrak{j}_\lambda\|_2^2 \, \nu(W_\varepsilon(\lambda))^{-1} \, \mathfrak{j}^* E(W_\varepsilon(\lambda)) \, \mathfrak{j}$ converges weakly to $T_\lambda := \mathfrak{j}_\lambda^* \mathfrak{j}_\lambda$ in the Hilbert space $\mathcal{K}$ as $\varepsilon \to +0$. For any φ any ψ in $\mathcal{K}$, we have

$$\nu(W_\varepsilon(\lambda))^{-1} \left|\left(\mathfrak{j}^* E(W_\varepsilon(\lambda)) \, \mathfrak{j}\varphi, \psi\right)\right| = \nu(W_\varepsilon(\lambda))^{-1} \left| \int\limits_{W_\varepsilon(\lambda)} \langle \mathfrak{j}_\gamma \varphi, \mathfrak{j}_\gamma \psi \rangle_\gamma \, \mathrm{d}\mu(\gamma) \right| \leq$$

$$\nu(W_\varepsilon(\lambda))^{-1} \int\limits_{W_\varepsilon(\lambda)} \|\mathfrak{j}_\gamma\|_2^2 \, |||\varphi||| \, |||\psi||| \, \mathrm{d}\mu(\gamma) = |||\varphi||| \, |||\psi||| \, .$$

This shows that the set $\{T_{\varepsilon,\lambda} : \varepsilon > 0\}$ is uniformly bounded in $\mathbb{B}(\mathcal{K})$. To prove (3), it therefore suffices to show that $\lim\limits_{\varepsilon \to +0} (T_{\varepsilon,\lambda} \xi_k, \xi_n) = (T_\lambda \xi_k, \xi_n)$ for all $k, n \in \mathbb{N}$. Fix $k, n \in \mathbb{N}$. Define $f_{kn}(\gamma) := 0$ if $\gamma \in N_0 := \{\alpha \in \Lambda : \mathfrak{j}_\alpha = 0\}$ and $f_{kn}(\gamma) := \langle \mathfrak{j}_\gamma \xi_k, \mathfrak{j}_\gamma \xi_n \rangle_\gamma \, \|\mathfrak{j}_\gamma\|_2^{-2}$ if $\gamma \in \Lambda \setminus N_0$. The function f_{kn} is in $L^1(\Lambda; \nu)$ because

$$\int\limits_\Lambda |f_{kn}(\gamma)| \, \mathrm{d}\nu(\gamma) \leq \int\limits_\Lambda \|\mathfrak{j}_\gamma \xi_k\|_\gamma \, \|\mathfrak{j}_\gamma \xi_n\|_\gamma \, \mathrm{d}\mu(\gamma) \leq \int\limits_\Lambda \|\mathfrak{j}_\gamma\|_2^2 \, |||\xi_k||| \, |||\xi_n||| \, \mathrm{d}\mu(\gamma)$$

$$\leq \|\mathfrak{j}\|_2^2 \, |||\xi_k||| \, |||\xi_n|||$$

by (1). Therefore, by a general measure-theoretic result (cf. Federer [1], Theorem 2.9.8), there exists a ν-null set N_{kn} such that for all $\lambda \in \Lambda_s \setminus (N_{kn} \cup N_0)$

$$\|\mathfrak{j}_\lambda\|_2^{-2} \, (T_{\varepsilon,\lambda} \xi_k, \xi_n) \equiv \nu(W_\varepsilon(\lambda))^{-1} \left(\mathfrak{j}^* E(W_\varepsilon(\lambda)) \, \mathfrak{j}\xi_k, \xi_n\right)$$

$$\equiv \nu(W_\varepsilon(\lambda))^{-1} \int\limits_{W_\varepsilon(\lambda)} \langle \mathfrak{j}_\gamma \xi_k, \mathfrak{j}_\gamma \xi_n \rangle_\gamma \, \mathrm{d}\mu(\gamma)$$

$$\equiv \nu(W_\varepsilon(\lambda))^{-1} \int\limits_{W_\varepsilon(\lambda)} f_{kn}(\gamma) \, \mathrm{d}\nu(\gamma)$$

converges to $f_{kn}(\lambda) \equiv \|\mathfrak{j}_\lambda\|_2^{-2} \, (T_\lambda \xi_k, \xi_n)$ as $\varepsilon \to +0$. Set $N := (\Lambda \setminus \Lambda_s) \cup N_0 \cup \bigcup\limits_{k,n=1}^{\infty} N_{kn}$. By construction we have $\nu(\Lambda \setminus \Lambda_s) = \nu(N_0) = \nu(N_{kn}) = 0$ for $k, n \in \mathbb{N}$. Thus N is a μ-null set, since ν and μ are equivalent. By the preceding proof we have shown that $\nu(W_\varepsilon(\lambda)) > 0$ for $\varepsilon > 0$ and (3) is valid for all $\lambda \in \Lambda \setminus N$. □

Now we shall apply Proposition 12.2.2,(ii), in order to "localize" decomposable operators in a direct integral of Hilbert spaces. For this we need the following condition on a linear subspace $\mathcal{D}$ of a Hilbert space $\mathcal{H}$.

(HS) There exists a Hilbert space $\mathcal{K}$ that contains $\mathcal{D}$ as a linear subspace and is itself a linear subspace of the vector space $\mathcal{H}$ such that the canonical embedding $\mathfrak{j}$ of $\mathcal{K}$ into $\mathcal{H}$ is a Hilbert-Schmidt mapping of the Hilbert space $\mathcal{K}$ into the Hilbert space $\mathcal{H}$.

Remark 1. Let $\mathcal{D}$ be a dense linear subspace of a Hilbert space $\mathcal{H}$. (For this remark we do not assume that $\mathcal{H}$ is of the form set out at the beginning of this section.) Suppose that (HS) is satisfied. Then $\mathcal{K}$ is separable, since $\mathfrak{j}^*$ is a Hilbert-Schmidt operator of $\mathcal{H}$ into $\mathcal{K}$ and the range of $\mathfrak{j}^*$ is dense in $\mathcal{K}$ because of $(\mathfrak{j}^* \mathcal{H})^\perp = \ker \mathfrak{j} = \{0\}$. Since $\mathfrak{j}$ is a continuous map of $\mathcal{K}$ into $\mathcal{H}$, $\mathcal{K}$ is also separable relative to the norm of $\mathcal{H}$. Since $\mathcal{D} \subseteq \mathcal{K}$ and $\mathcal{D}$ is dense in $\mathcal{H}$, it follows that $\mathcal{H}$ is separable and $\mathfrak{j}\mathcal{K}$ is dense in $\mathcal{H}$.

Proposition 12.2.3. *Let $\mathcal{D}$ be a dense linear subspace of the Hilbert space $\mathcal{H} = \int_\Lambda^\oplus \mathcal{H}_\lambda \, \mathrm{d}\mu(\lambda)$, and let $\mathcal{A}$ be an O*-family on $\mathcal{D}$. Suppose* (HS) *is satisfied and $a\mathcal{D} \subseteq \mathcal{H}$ for all $a \in \mathcal{A}$. Suppose that the von Neumann algebra $\mathcal{N}$ of bounded diagonalizable operators is contained in the commutant $\mathcal{A}'_{\mathrm{ss}}$.*

There exists a μ-null set N such that the following statements are true when we define $J_\lambda(a)\, \mathrm{j}_\lambda\varphi := \mathrm{j}_\lambda a\varphi$, $\mathcal{D}_\lambda := \mathrm{j}_\lambda(\mathcal{D})$ if $\lambda \in \Lambda \setminus N$, $a \in \mathcal{A}$ and $\varphi \in \mathcal{D}$, $J_\lambda(a) := 0$, $\mathcal{D}_\lambda := \{0\}$ if $\lambda \in N$ and $a \in \mathcal{A}$ and $\mathcal{A}_\lambda := J_\lambda(\mathcal{A})$ if $\lambda \in \Lambda$. Here j_λ, $\lambda \in \Lambda$, are the operators from Proposition 12.2.1.

(i) *Suppose $\lambda \in \Lambda$. For each $a \in \mathcal{A}$, $J_\lambda(a)$ is a well-defined linear operator on $\mathcal{D}_\lambda$. Further, $\mathcal{A}_\lambda$ is an O*-family on $\mathcal{D}_\lambda$, and J_λ is a *-preserving map of $\mathcal{A}$ onto $\mathcal{A}_\lambda$. If $\mathcal{A}$ is an O*-algebra, then J_λ is a *-representation of $\mathcal{A}$ on $\mathcal{D}_\lambda$. $\mathcal{D}_\lambda$ is dense in $\mathcal{H}_\lambda$ if $\lambda \in \Lambda \setminus N$.*
(ii) *If $a \in \mathcal{A}$ and $a \geqq 0$ on $\mathcal{D}$, then $J_\lambda(a) \geqq 0$ on $\mathcal{D}_\lambda$ for $\lambda \in \Lambda$.*
(iii) *For each $a \in \mathcal{A}$, $\lambda \to \overline{J_\lambda(a)}$ is a measurable field of closed operators and $\bar{a} = \int^\oplus \overline{J_\lambda(a)}\, \mathrm{d}\mu(\lambda)$.*

Proof. Recall that $\varphi(\lambda) \equiv (\mathrm{j}\varphi)(\lambda) = \mathrm{j}_\lambda\varphi$ a.e. for any $\varphi \in \mathcal{D}$ by Proposition 12.2.1. Since $\mathcal{D}$ is dense in $\mathcal{H}$ and $\mathcal{H}$ is separable, there exists a countable subset of $\mathcal{D}$ that is dense in $\mathcal{H}$. From these facts and Lemma 12.1.2,(ii), it follows that $\mathrm{j}_\lambda(\mathcal{D})$ is dense in $\mathcal{H}_\lambda$ a.e. As noted in Remark 1, condition (HS) implies that the assumptions of Proposition 12.2.2 are fulfilled. Thus there exists a μ-null set N for which the statement of Proposition 12.2.2, (ii), holds and such that $\mathcal{D}_\lambda := \mathrm{j}_\lambda(\mathcal{D})$ is dense in $\mathcal{H}_\lambda$ if $\lambda \in \Lambda \setminus N$.

(i): We can assume that $\lambda \in \Lambda \setminus N$, since otherwise the assertions are trivial. Let $a \in \mathcal{A}$, and let $\varphi, \psi \in \mathcal{D}$. Since $E(W_\varepsilon(\lambda)) \in \mathcal{N} \subseteq \mathcal{A}'_{\mathrm{ss}}$ and $\mathcal{A}'_{\mathrm{ss}} \subseteq \mathcal{A}'_{\mathrm{w}}$ by Proposition 7.2.10, we have $E(W_\varepsilon(\lambda)) \in \mathcal{A}'_{\mathrm{w}}$ for any $\varepsilon > 0$. From this and formula (3) we obtain

$$\begin{aligned} \langle J_\lambda(a)\, \mathrm{j}_\lambda\varphi, \mathrm{j}_\lambda\psi\rangle_\lambda &= \langle \mathrm{j}_\lambda(a\varphi), \mathrm{j}_\lambda\psi\rangle_\lambda = \lim_{\varepsilon \to +0} \nu(W_\varepsilon(\lambda))^{-1} \|\mathrm{j}_\lambda\|_2^2 \langle E(W_\varepsilon(\lambda))\, a\varphi, \psi\rangle \\ &= \lim_{\varepsilon \to +0} \nu(W_\varepsilon(\lambda))^{-1} \|\mathrm{j}_\lambda\|_2^2 \langle E(W_\varepsilon(\lambda))\, \varphi, a^+\psi\rangle \\ &= \langle \mathrm{j}_\lambda\varphi, \mathrm{j}_\lambda(a^+\psi)\rangle_\lambda = \langle \mathrm{j}_\lambda\varphi, J_\lambda(a^+)\, \mathrm{j}_\lambda\psi\rangle_\lambda. \end{aligned} \tag{4}$$

Since $\mathcal{D}_\lambda$ is dense in $\mathcal{H}_\lambda$ because of $\lambda \in \Lambda \setminus N$, we conclude from (4) that $J_\lambda(a)\, \mathrm{j}_\lambda\varphi = 0$ provided that $\mathrm{j}_\lambda\varphi = 0$; so $J_\lambda(a)$ is a well-defined linear operator on $\mathcal{D}_\lambda$. Further, we see from (4) that $\mathcal{A}_\lambda = J_\lambda(\mathcal{A})$ is an O*-family on $\mathcal{D}_\lambda$ and that $J_\lambda(a)^+ = J_\lambda(a^+)$ for each $a \in \mathcal{A}$. It is clear that J_λ is a *-representation when $\mathcal{A}$ is an O*-algebra.

(ii): Again we can assume that $\lambda \in \Lambda \setminus N$. Let $\varphi \in \mathcal{D}$. From $E(W_\varepsilon(\lambda)) \in \mathcal{N} \subseteq \mathcal{A}'_{\mathrm{ss}}$, we have $E(W_\varepsilon(\lambda))\, \bar{a} \subseteq \bar{a}E(W_\varepsilon(\lambda))$ and hence

$$\begin{aligned} \langle E(W_\varepsilon(\lambda))\, a\varphi, \varphi\rangle &= \langle E(W_\varepsilon(\lambda))\, a\varphi, E(W_\varepsilon(\lambda))\, \varphi\rangle \\ &= \langle \bar{a}E(W_\varepsilon(\lambda))\, \varphi, E(W_\varepsilon(\lambda))\, \varphi\rangle \geqq 0 \quad \text{for} \quad \varepsilon > 0. \end{aligned}$$

Combined with (3), this gives

$$\langle J_\lambda(a)\, \mathrm{j}_\lambda\varphi, \mathrm{j}_\lambda\varphi\rangle_\lambda = \lim_{\varepsilon \to +0} \nu(W_\varepsilon(\lambda))^{-1} \|\mathrm{j}_\lambda\|_2^2 \langle E(W_\varepsilon(\lambda))\, a\varphi, \varphi\rangle \geqq 0.$$

Thus $J_\lambda(a) \geqq 0$.

(iii): Fix $a \in \mathcal{A}$. From the assumptions, $\mathcal{N} \subseteq (\bar{a})'_s$. Therefore, by Proposition 12.1.7, the operator $\bar{a}$ is decomposable, i.e., we have $\bar{a} = \int^{\oplus} a_\lambda \, d\mu(\lambda)$ for some measurable field $\lambda \to a_\lambda$ of closed operators. The proof will be complete once we have shown that $a_\lambda = \overline{J_\lambda(a)}$ a.e.

Let $\mathrm{gr}_{\mathcal{K}}\, a$ and $\mathrm{gr}_{\mathcal{H}}\, \bar{a}$ denote the graphs of a and $\bar{a}$ equipped with the norms of $\mathcal{K} \oplus \mathcal{K}$ and $\mathcal{H} \oplus \mathcal{H}$, respectively. Since $\mathcal{K}$ is separable (see Remark 1) and so is $\mathrm{gr}_{\mathcal{K}}\, a$, there is a countable subset $\{\zeta_n : n \in \mathbb{N}\}$ of $\mathcal{D}$ such that $\Gamma := \{(\zeta_n, a\zeta_n) : n \in \mathbb{N}\}$ is dense in $\mathrm{gr}_{\mathcal{K}}\, a$ and hence in $\mathrm{gr}_{\mathcal{H}}\, \bar{a}$. For each $n \in \mathbb{N}$, we have

$$a_\lambda \varphi_n = (a\varphi_n)(\lambda) = \mathfrak{j}_\lambda a\varphi_n = J_\lambda(1)\, \mathfrak{j}_\lambda \varphi_n \quad \text{a.e.} \tag{5}$$

Let $\lambda \in \Lambda \setminus N$. Since $\mathfrak{j}_\lambda : \mathcal{K} \to \mathcal{H}_\lambda$ is a Hilbert-Schmidt mapping, $\mathfrak{j}_\lambda$ maps $(\mathcal{D}, |||\cdot|||)$ continuously into $(\mathcal{D}_\lambda, \|\cdot\|_\lambda)$. From the density of Γ in $\mathrm{gr}_{\mathcal{K}}\, a$ it follows that the set $\{(\mathfrak{j}_\lambda\zeta_n, \mathfrak{j}_\lambda a\zeta_n) : n \in \mathbb{N}\} = \{(\mathfrak{j}_\lambda\zeta_n, J_\lambda(a)\, \mathfrak{j}_\lambda\zeta_n) : n \in \mathbb{N}\}$ is dense in $\mathrm{gr}\, J_\lambda(a)$ (in the norm of $\mathcal{H} \oplus \mathcal{H}$). Therefore, by (5), we get $J_\lambda(a) \subseteq a_\lambda$ and hence $\overline{J_\lambda(a)} \subseteq a_\lambda$ a.e. Since Γ is dense in $\mathrm{gr}_{\mathcal{H}}\, \bar{a}$, we conclude from Lemma 12.1.2,(ii), that $\{(\mathfrak{j}_\lambda\zeta_n, a_\lambda \mathfrak{j}_\lambda\zeta_n) : n \in \mathbb{N}\}$ is dense in $\mathrm{gr}\, \bar{a}_\lambda$ (again in the norm of $\mathcal{H}_\lambda \oplus \mathcal{H}_\lambda$) a.e. Applying (5) once again we obtain $a_\lambda \subseteq \overline{J_\lambda(a)}$ a.e. Thus $a_\lambda = \overline{J_\lambda(a)}$ a.e. □

Remark 2. The preceding proof shows that part (i) of Proposition 12.2.3 is valid if we only assume that $\mathcal{N} \subseteq \mathcal{A}'_w$ instead of $\mathcal{N} \subseteq \mathcal{A}'_{ss}$.

Remark 3. Retain the assumptions and the notation of Proposition 12.2.3. The following simple continuity result might be useful sometimes. Suppose that $\mathfrak{x} \to a_{\mathfrak{x}}$ is a mapping of a topological space $\mathfrak{X}$ into the O*-family $\mathcal{A}$ such that for arbitrary $\varphi \in \mathcal{D}$ and $\psi \in \mathcal{K}$ the function $\mathfrak{x} \to (a_{\mathfrak{x}}\varphi, \psi)$ is continuous on $\mathfrak{X}$. Then the function $\mathfrak{x} \to \langle J_\lambda(a_{\mathfrak{x}})\, \mathfrak{j}_\lambda\varphi, \mathfrak{j}_\lambda\psi\rangle_\lambda$ is continuous on $\mathfrak{X}$ for any $\lambda \in \Lambda$, $\varphi \in \mathcal{D}$ and $\psi \in \mathcal{K}$. The proof of this statement follows at once from the identity $\langle J_\lambda(a_{\mathfrak{x}})\, \mathfrak{j}_\lambda\varphi, \mathfrak{j}_\lambda\psi\rangle_\lambda = \langle \mathfrak{j}_\lambda a_{\mathfrak{x}}\varphi, \mathfrak{j}_\lambda\psi\rangle_\lambda = (a_{\mathfrak{x}}\varphi, \mathfrak{j}_\lambda^*\mathfrak{j}_\lambda\psi)$ which holds for each $\lambda \in \Lambda \setminus N$.

12.3. Decomposition of $*$-Representations

In this section A will denote a $*$-algebra with unit. First we define the direct integral of $*$-representations. Let $\mathcal{H} = \int^{\oplus} \mathcal{H}_\lambda \, d\mu(\lambda)$ be a (fixed) direct integral of Hilbert spaces. For each $\lambda \in \Lambda$, let π_λ be a $*$-representation of A on a linear subspace $\mathcal{D}(\pi_\lambda)$ of $\mathcal{H}_\lambda$. We say that the mapping $\lambda \to \pi_\lambda$ is a *measurable field of $*$-representations* if $\mathcal{D}(\pi_\lambda)$ is dense in $\mathcal{H}_\lambda$ a.e. and if $\lambda \to \overline{\pi_\lambda(a)}$ is a measurable field of closed operators for each $a \in \mathsf{A}$.

Suppose $\lambda \to \pi_\lambda$ is a measurable field of $*$-representations of A. Let $\mathcal{D}(\pi)$ be the set of all vectors φ in $\mathcal{H}$ such that $\varphi(\lambda) \in \mathcal{D}(\pi_\lambda)$ a.e. and the field $\lambda \to \pi_\lambda(a)\, \varphi(\lambda)$ belongs to $\mathcal{H}$ (i.e., the field is square integrable with respect to μ) for all $a \in \mathsf{A}$, and let $\mathcal{H}(\pi)$ be the closure of $\mathcal{D}(\pi)$ in $\mathcal{H}$. We define $\big(\pi(a)\, \varphi\big)(\lambda) := \pi_\lambda(a)\, \varphi(\lambda)$ for $a \in \mathsf{A}$ and $\varphi \in \mathcal{D}(\pi)$. Using the assumption that each π_λ is a $*$-representation of A it follows easily that π is a $*$-representation of A in the Hilbert space $\mathcal{H}(\pi)$. We verify (for instance) that π preserves the multiplication and the involution. Let $a, b \in \mathsf{A}$ and let $\varphi, \psi \in \mathcal{D}(\pi)$. From the above definition, we have

$$\begin{aligned}\big(\pi(a)\, \pi(b)\, \varphi\big)(\lambda) &= \pi_\lambda(a)\, \big(\pi(b)\, \varphi\big)(\lambda) = \pi_\lambda(a)\, \pi_\lambda(b)\, \varphi(\lambda) = \pi_\lambda(ab)\, \varphi(\lambda) \\ &= \big(\pi(ab)\, \varphi\big)(\lambda) \quad \text{a.e.,}\end{aligned}$$

i.e., $\pi(a)\,\pi(b) = \pi(ab)$, and

$$\langle \pi(a)\,\varphi, \psi\rangle = \int_\Lambda \langle \pi_\lambda(a)\,\varphi(\lambda), \psi(\lambda)\rangle_\lambda \,\mathrm{d}\mu(\lambda) = \int_\Lambda \langle \varphi(\lambda), \pi_\lambda(a^+)\,\psi(\lambda)\rangle_\lambda \,\mathrm{d}\mu(\lambda)$$
$$= \langle \varphi, \pi(a^+)\,\psi\rangle.$$

Definition 12.3.1. The $*$-representation π defined above is called the *direct integral of the field* $\lambda \to \pi_\lambda$. We write $\pi = \int^\oplus \pi_\lambda \,\mathrm{d}\mu(\lambda)$.

From Definition 12.1.6 we obtain the following slight reformulation of the above definition. The space $\mathcal{D}(\pi)$ consists precisely of all $\varphi \in \mathcal{H}$ for which $\varphi(\lambda) \in \mathcal{D}(\pi_\lambda)$ a.e. and φ is in the domain of the operator $\int^\oplus \overline{\pi_\lambda(a)}\,\mathrm{d}\mu(\lambda)$ for all $a \in \mathsf{A}$. For each $a \in \mathsf{A}$, $\pi(a)$ is the restriction to $\mathcal{D}(\pi)$ of the operator $\int^\oplus \overline{\pi_\lambda(a)}\,\mathrm{d}\mu(\lambda)$.

The following simple example shows that the linear space $\mathcal{D}(\pi)$ is not dense in $\mathcal{H}$ in general even not if all operators $\pi_\lambda(a)$, $a \in \mathsf{A}$, are bounded and $\mathcal{D}(\pi_\lambda) = \mathcal{H}_\lambda$ for $\lambda \in \Lambda$.

Example 12.3.2. Suppose A is the $*$-algebra of all measurable functions on the interval $[0, 1]$ (under equality everywhere) with the usual pointwise algebraic operations. We consider the Hilbert space $\mathcal{H} := L^2(0, 1)$ as a direct integral of one-dimensional Hilbert spaces $\mathcal{H}_\lambda := \mathbb{C}$ with respect to the Lebesgue measure μ on $\Lambda := [0, 1]$. Let π_λ, $\lambda \in \Lambda$, be the $*$-representation of A on $\mathcal{D}(\pi_\lambda) := \mathcal{H}_\lambda \equiv \mathbb{C}$ defined by $\pi_\lambda(f) := f(\lambda)$, $f \in \mathsf{A}$. It is clear that $\lambda \to \pi_\lambda$ is a measurable field of $*$-representations. The operator $\int^\oplus \overline{\pi_\lambda(f)}\,\mathrm{d}\mu(\lambda)$ is obviously the multiplication operator by the function $f \in \mathsf{A}$. It is not difficult to see that the intersection of the domains of these operators is $\{0\}$. But $\mathcal{D}(\pi)$ is contained in this intersection by definition; so we get $\mathcal{D}(\pi) = \{0\}$. ○

Remark 1. If $\pi = \int^\oplus \pi_\lambda \,\mathrm{d}\mu(\lambda)$ and $\mathcal{D}(\pi)$ is dense in $\mathcal{H}$, then it follows immediately from the above definition that the algebra $\mathcal{N}$ of bounded diagonalizable operators is contained in the strong commutant $\pi(\mathsf{A})'_\mathrm{s}$.

Theorem 12.3.3. *Let π be a $*$-representation of A. Suppose there is a subrepresentation ϱ of π with $\pi \subseteq \hat{\varrho}$ such that $\mathcal{D} := \mathcal{D}(\varrho)$ and $\mathcal{H} := \mathcal{H}(\pi) = \mathcal{H}(\varrho)$ satisfy condition* (HS).

Then there exist a compact metric space Λ, a positive measure μ (which is the completion of a regular Borel measure) on Λ with support Λ, a measurable field $\lambda \to \mathcal{H}_\lambda$ on non-zero Hilbert spaces $\mathcal{H}_\lambda$, a measurable field $\lambda \to \pi_\lambda$ of closed $*$-representations of A and an isometry U of $\mathcal{H}(\pi)$ onto the Hilbert space $\int^\oplus \mathcal{H}_\lambda \,\mathrm{d}\mu(\lambda)$ such that:

(i) π_λ *is irreducible* a.e.

(ii) $U\overline{\pi(a)}\,U^{-1} = \int^\oplus \overline{\pi_\lambda(a)}\,\mathrm{d}\mu(\lambda)$ *for all* $a \in \mathsf{A}$.

(iii) $U\varrho U^{-1} \subseteq \int^\oplus \pi_\lambda \,\mathrm{d}\mu(\lambda) \subseteq U\hat{\pi}U^{-1}$.

If π is closed and the graph topology of $\pi(\mathsf{A})$ is metrizable, then we have in addition that

(iv) $U\pi U^{-1} = \int^\oplus \pi_\lambda \,\mathrm{d}\mu(\lambda)$.

Proof. We choose a maximal abelian von Neumann subalgebra $\mathcal{N}$ of the von Neumann algebra $\varrho(\mathsf{A})'_{\mathrm{ss}}$. As noted in Remark 1 in 12.2, condition (HS) implies that $\mathcal{H} = \mathcal{H}(\pi)$ is separable. Thus, by Lemma 12.1.1, there exist Λ, μ, $\lambda \to \mathcal{H}_\lambda$ and U as stated in the above theorem such that $U\mathcal{N}U^{-1}$ is the algebra of bounded diagonalizable operators in the direct integral $\int^\oplus \mathcal{H}_\lambda \, \mathrm{d}\mu(\lambda)$. For notational simplicity we shall identify $\mathcal{H}(\pi)$ and $\int^\oplus \mathcal{H}_\lambda \, \mathrm{d}\mu(\lambda)$ via the unitary mapping U. Since $\mathcal{D} = \mathcal{D}(\varrho)$ satisfies (HS) and $\mathcal{N} \subseteq \varrho(\mathsf{A})'_{\mathrm{ss}}$, Proposition 12.2.3 applies to the O*-algebra $\mathcal{A} := \varrho(\mathsf{A})$. Define $\varrho_\lambda(a) := J_\lambda(\varrho(a))$ and $\mathcal{D}(\varrho_\lambda) := \mathcal{D}_\lambda$ for $a \in \mathsf{A}$ and $\lambda \in \Lambda$, where J_λ and $\mathcal{D}_\lambda$ are as in Proposition 12.2.3. Since $\varrho \subseteq \pi \subseteq \hat{\varrho}$, we have $\overline{\pi(a)} = \overline{\varrho(a)}$ for $a \in \mathsf{A}$. From this fact and the properties stated in Proposition 12.2.3 we see immediately that $\lambda \to \varrho_\lambda$ and so $\lambda \to \pi_\lambda := \hat{\varrho}_\lambda$ is a measurable field of $*$-representations and (ii) is satisfied. From the preceding definitions it is clear that $\varrho \equiv \pi \restriction \mathcal{D}(\varrho) \subseteq \int^\oplus \pi_\lambda \, \mathrm{d}\mu(\lambda)$. Suppose φ is in the domain of $\int^\oplus \pi_\lambda \, \mathrm{d}\mu(\lambda)$. Then, for all $a \in \mathsf{A}$, φ belongs to the domain of the operator $\int^\oplus \overline{\pi_\lambda(a)} \, \mathrm{d}\mu(\lambda) = \overline{\pi(a)}$ by (ii). From this we conclude that $\varphi \in \mathcal{D}(\hat{\pi})$ and $\int^\oplus \pi_\lambda \, \mathrm{d}\mu(\lambda) \subseteq \hat{\pi}$. This gives (iii). To prove (i), we apply Proposition 12.1.9 with $\mathcal{B} := \{\overline{\pi(a)} : a \in \mathsf{A}\}$. Since $\mathcal{N}$ is maximal abelian in $\mathcal{B}'_{\mathrm{ss}} = \pi(\mathsf{A})'_{\mathrm{ss}} = \varrho(\mathsf{A})'_{\mathrm{ss}}$, it follows then that $\mathcal{B}_\lambda \equiv \{\overline{\pi_\lambda(a)} : a \in \mathsf{A}\}$ (by (ii)) is irreducible and hence $(\mathcal{B}_\lambda)'_{\mathrm{ss}} = \pi_\lambda(\mathsf{A})'_{\mathrm{ss}}$ consists only of scalar multiples of the identity a.e. This implies (i) (see Lemma 8.3.5,(i) $\leftrightarrow$ (iv)).

Suppose now in addition that π is closed and that the graph topology t_π is metrizable. Then t_ϱ is metrizable and there is a sequence $(a_n : n \in \mathbb{N})$ in A with $a_1 = 1$ such that $\{\|\cdot\|_{\varrho(a_n)} : n \in \mathbb{N}\}$ is a directed family of seminorms which generates the topology t_ϱ. Let $\lambda \in \Lambda$. By Proposition 12.2.3,(i) and (ii), J_λ is a strongly positive $*$-representation of the O*-algebra $\mathcal{A} \equiv \varrho(\mathsf{A})$. Therefore, since $\pi_\lambda = \hat{\varrho}_\lambda = \widehat{J_\lambda \circ \varrho}$, $\{\|\cdot\|_{\pi_\lambda(a_n)} : n \in \mathbb{N}\}$ is also a directed family of seminorms which generates the graph topology of $\pi_\lambda(\mathsf{A})$. Thus $\mathcal{D}(\pi_\lambda) = \mathcal{D}(\hat{\pi}_\lambda) = \bigcap_{n \in \mathbb{N}} \mathcal{D}(\overline{\pi_\lambda(a_n)})$ by Proposition 2.2.12. Suppose $\varphi \in \mathcal{D}(\pi)$. Then $\varphi \in \mathcal{D}(\overline{\pi(a_n)})$ for all $n \in \mathbb{N}$. By (ii), there is a μ-null set N_n such that $\varphi(\lambda) \in \mathcal{D}(\overline{\pi_\lambda(a_n)})$ if $\lambda \in \Lambda \setminus N_n$. Setting $N := \bigcup_{n \in \mathbb{N}} N_n$, we have $\mu(N) = 0$ and $\varphi(\lambda) \in \bigcap_{n \in \mathbb{N}} \mathcal{D}(\overline{\pi_\lambda(a_n)}) = \mathcal{D}(\pi_\lambda)$ if $\lambda \in \Lambda \setminus N$. Since φ is, of course, in the domain of $\overline{\pi(a)} = \int^\oplus \overline{\pi_\lambda(a)} \, \mathrm{d}\mu(\lambda)$ for any $a \in \mathsf{A}$, this shows (by the second definition above) that φ is in the domain of $\int^\oplus \pi_\lambda \, \mathrm{d}\mu(\lambda)$. Combined with $\int^\oplus \pi_\lambda \, \mathrm{d}\mu(\lambda) \subseteq \pi$ (by (iii) and by the assumption that π is closed), we get $\pi = \int^\oplus \pi_\lambda \, \mathrm{d}\mu(\lambda)$. $\square$

The next proposition describes a class of $*$-representations for which Theorem 12.3.3 applies.

Proposition 12.3.4. *Let π be a $*$-representation of A. Suppose that there are a countable subset Γ of $\mathcal{D}(\pi)$ that is cyclic for π and a nuclear locally convex topology τ on A such that for each $\varphi \in \Gamma$ the map $T_\varphi : a \to \pi(a)\varphi$ of $\mathsf{A}[\tau]$ into $\mathcal{H}(\pi)$ is continuous. Then the $*$-representation π satisfies the assumption of Theorem 12.3.3 with $\varrho := \pi \restriction \pi(\mathsf{A})\Gamma$.*

We refer to Schäfer [1], III, 7, or to Pietsch [1] for the concept of nuclear locally convex spaces. Recall that $\pi(\mathsf{A})\Gamma$ means l.h. $\{\pi(a)\varphi : a \in \mathsf{A} \text{ and } \varphi \in \Gamma\}$.

Proof. Let $\Gamma = \{\varphi_n : n \in \mathbb{N}\}$. We write $t_{\|\cdot\|}$ for the topology of the Hilbert space norm of $\mathcal{H}(\pi)$. Put $\mathcal{D}_n := \pi(\mathsf{A})\,\varphi_n$ for $n \in \mathbb{N}$. Since the map T_{φ_n} is continuous, the quotient topology τ_n of $\mathsf{A}[\tau]$ on $\mathcal{D}_n \cong \mathsf{A}/\ker T_{\varphi_n}$ is finer than the topology $t_{\|\cdot\|}$ on $\mathcal{D}_n$, $n \in \mathbb{N}$. Hence $T : (\psi_n) \to \sum_n \psi_n$ is a continuous map of $\mathcal{D}_\Sigma$ into $(\mathcal{D}(\varrho), \|\cdot\|)$, where $\mathcal{D}_\Sigma$ is the direct sum of the locally convex spaces $\mathcal{D}_n[\tau_n]$, $n \in \mathbb{N}$. (Note that $\sum_n \psi_n$ is in fact a finite sum, because for any $(\psi_n) \in \mathcal{D}_\Sigma$ we have $\psi_n = 0$ if n is sufficiently large.) By definition, $\mathcal{D}(\varrho)$ is the range of T. The continuity of T implies that the quotient topology τ_0 on $\mathcal{D}(\varrho) \cong \mathcal{D}_\Sigma/\ker T$ is finer than the topology $t_{\|\cdot\|}$. The class of nuclear locally convex spaces is stabil under quotients by closed subspaces and under countable direct sums (SCHÄFER [1], III, 7.4). Therefore, $\mathcal{D}(\varrho)\,[\tau_0]$ is nuclear, because $\mathsf{A}[\tau]$ was assumed to be nuclear. Since the norm $\|\cdot\|$ (from $\mathcal{H}(\pi)$) is continuous on the nuclear locally convex space $\mathcal{D}(\varrho)\,[\tau_0]$, there is a Hilbertian norm $|||\cdot|||$ on $\mathcal{D}(\varrho)$ satisfying $\|\cdot\| \leqq |||\cdot|||$ such that the canonical embedding $\mathfrak{j}$ of the Hilbert space $\mathcal{K}$ which is the completion of $(\mathcal{D}(\varrho), |||\cdot|||)$ into $\mathcal{H}(\pi)$, the completion of $(\mathcal{D}(\varrho), \|\cdot\|)$, is nuclear. This follows directly from the definition of nuclearity applied to the space $\mathcal{D}(\varrho)\,[\tau_0]$. Since $\mathfrak{j}$ is in particular a Hilbert-Schmidt mapping, this shows that $\mathcal{D} := \mathcal{D}(\varrho)$ and $\mathcal{H} := \mathcal{H}(\pi)$ satisfy condition (HS). Since Γ is cyclic for π, we have $\pi \subseteq \hat{\varrho}$. $\square$

Some of the results obtained so far are summarized in the following theorem.

Theorem 12.3.5. *Suppose π is a closed $*$-representation of A on a separable Hilbert space $\mathcal{H}(\pi)$ such that the graph topology of $\pi(\mathsf{A})$ is metrizable. Suppose that there is a nuclear locally convex Hausdorff topology τ on A such that for each vector $\varphi \in \mathcal{D}(\pi)$ the map $a \to \pi(a)\varphi$ of $\mathsf{A}[\tau]$ into $\mathcal{H}(\pi)$ is continuous. Then π is unitarily equivalent to the direct integral $\int^{\oplus} \pi_\lambda\, \mathrm{d}\mu(\lambda)$ of a measurable field $\lambda \to \pi_\lambda$ of closed $*$-representations of A such that π_λ is irreducible a.e.*

Proof. Since $\mathcal{A} := \pi(\mathsf{A})$ is an O*-algebra on a separable Hilbert space with metrizable graph topology, Proposition 2.3.3 says that the locally convex space $\mathcal{D}_\mathcal{A} \equiv \mathcal{D}(\pi)\,[t_\pi]$ is separable. Hence there exists a countable subset of $\mathcal{D}(\pi)$ that is cyclic for π; so Proposition 12.3.4 applies. The assertion now follows from Theorem 12.3.3 (see statements (i) and (iv) there). $\square$

Remark 2. The preceding theorem applies (for instance) to each closed $*$-representation π in separable Hilbert space $\mathcal{H}(\pi)$ of a countably generated $*$-algebra A with unit. In this case we let τ be the finest locally convex topology on A. Then the continuity of the maps $a \to \pi(a)\,\varphi$ is obvious. Further, $\mathsf{A}[\tau]$ is nuclear and the graph topology of $\pi(\mathsf{A})$ is metrizable, since A is countably generated.

Remark 3. We briefly consider the assumption of Theorem 12.3.5 that the mapping $a \to \pi(a)\,\varphi$ of $\mathsf{A}[\tau]$ into $\mathcal{H}(\pi)$ is continuous for all $\varphi \in \mathcal{D}(\pi)$. If $\mathsf{A}[\tau]$ is barrelled and the $*$-representation π is weakly continuous (i.e., all functionals $\omega_\varphi(\cdot) := \langle \pi(\cdot)\,\varphi, \varphi\rangle$, $\varphi \in \mathcal{D}(\pi)$, are continuous on $\mathsf{A}[\tau]$), then it follows from the second statement in Proposition 3.6.5 that this assumption is satisfied. Theorem 3.6.8 shows that this assumption holds for every $*$-representation π when $\mathsf{A}[\tau]$ is a Frechet topological $*$-algebra.

We conclude this section by proving a result which was already noted in Remark 1 of 11.4.

Proposition 12.3.6. *Let G be a Lie group with Lie algebra $\mathfrak{g}$ and let $\mathscr{E}(\mathfrak{g})$ be the enveloping algebra of $\mathfrak{g}$ (cf. Section 1.7). Let $n \in \mathbb{N}$. If a matrix $[a_{kl}] \in M_n(\mathscr{E}(\mathfrak{g}))$ satisfies the condition stated in 11.4/(1) for all irreducible unitary representations U of G, then it satisfies the same condition for every unitary representation U of G.*

Proof. Suppose U is an arbitrary unitary representation of G. By writing U as a direct sum of cyclic representations U_i, $i \in I$, and using the equality $\mathrm{d}U = \sum_{i \in I} \oplus \mathrm{d}U_i$, it follows that it suffices to assume that U is cyclic. Since the infinitesimal representation $\mathrm{d}U$ depends only on the restriction of U to the connected component of the unit of G, there is no loss of generality in supposing that G is connected. Then the Lie group G is separable. Therefore, since U is cyclic, the Hilbert space $\mathscr{H}(U)$ is separable, and the decomposition theory of unitary representations of separable locally compact groups (see e.g. Dixmier [2], 18.7.6, or Kirillov [1], 8.4) applies to U. By this theory, $\mathscr{H}(U)$ can be written as a direct integral of Hilbert spaces $\int_\Lambda^\oplus \mathscr{H}_\lambda \, \mathrm{d}\mu(\lambda)$, and there are irreducible unitary representations U_λ, $\lambda \in \Lambda$, of G in $\mathscr{H}(U_\lambda) := \mathscr{H}_\lambda$ such that $U(g) = \int^\oplus U_\lambda(g) \, \mathrm{d}\mu(\lambda)$ for each $g \in G$.

Next we note that if V is a unitary representation of G, then we have for each $x \in \mathfrak{g}$ and $\varphi \in \mathscr{H}(V)$

$$(I - \partial V(x))^{-1} \varphi = \int_0^\infty V(\exp tx) \, \mathrm{e}^{-t} \varphi \, \mathrm{d}t. \tag{1}$$

Indeed, since $V(\exp tx) = \exp t \, \partial V(x)$ by definition, (1) is a well-known formula which relates a unitary group to the resolvent of its generator (Kato [1], IX, § 1,3.).

Fix $x \in \mathfrak{g}$. Since each operator $U(g)$, $g \in G$, is decomposable and $(I - \partial U(x))^{-1} \in U(G)''$ by (1), $(I - \partial U(x))^{-1}$ is decomposable. Combining the equality $U(\exp tx) = \int^\oplus U_\lambda(\exp tx) \, \mathrm{d}\mu(\lambda)$ with (1), we obtain

$$(I - \partial U(x))^{-1} = \int^\oplus (I_\lambda - \partial U_\lambda(x))^{-1} \, \mathrm{d}\mu(\lambda) \quad \text{for} \quad x \in \mathfrak{g}. \tag{2}$$

Let $\varphi \in \mathscr{D}^\infty(U)$. We prove that $\varphi(\lambda) \in \mathscr{D}^\infty(U_\lambda)$ a.e. Let $\{x_1, \ldots, x_d\}$ be a basis for $\mathfrak{g}$, $n \in \mathbb{N}$ and $k \in \{1, \ldots, d\}$. Then $\varphi = (I_\lambda - \partial U(x_k))^{-n} \psi$ for some $\psi \in \mathscr{H}(U)$. By (2) there is a μ-null set N_{kn} such that $\varphi(\lambda) = (I - \partial U_\lambda(x_k))^{-n} \psi(\lambda) \in \mathscr{D}(\partial U_\lambda(x_k)^n)$ if $\lambda \notin N_{kn}$. Hence $\varphi(\lambda) \in \bigcap_{k=1}^d \bigcap_{n \in \mathbb{N}} \mathscr{D}(\partial U_\lambda(x_k)^n)$ and so $\varphi(\lambda) \in \mathscr{D}^\infty(U_\lambda)$ by Theorem 10.1.9 for $\lambda \in \Lambda \setminus N$, where N is the μ-null set $\bigcup_{k,n} N_{kn}$.

From (2) and Proposition 12.1.8,(iv), we have

$$\partial U(x) = \int^\oplus \partial U_\lambda(x) \, \mathrm{d}\mu(\lambda) \quad \text{for} \quad x \in \mathfrak{g}. \tag{3}$$

Recall that by definition $\mathrm{d}U(x) = \partial U(x) \upharpoonright \mathscr{D}^\infty(U)$ and $\mathrm{d}U_\lambda(x) = \partial U_\lambda(x) \upharpoonright \mathscr{D}^\infty(U_\lambda)$ for $x \in \mathfrak{g}$. Therefore, it follows from (3) and the preceding paragraph that for each $\varphi \in \mathscr{D}^\infty(U)$ and $x \in \mathscr{E}(\mathfrak{g})$ there is a μ-null set N (depending on φ and x) such that $\varphi(\lambda) \in \mathscr{D}^\infty(U_\lambda)$ and $(\mathrm{d}U(x) \varphi)(\lambda) = \mathrm{d}U_\lambda(x) \varphi(\lambda)$ for all $\lambda \in \Lambda \setminus N$. Thus we have

$$\sum_{k,l=1}^n \langle \mathrm{d}U(a_{kl}) \varphi_l, \varphi_k \rangle = \int \sum_{k,l=1}^n \langle \mathrm{d}U_\lambda(a_{kl}) \varphi_l(\lambda), \varphi_k(\lambda) \rangle_\lambda \, \mathrm{d}\mu(\lambda)$$

for $\varphi_1, \ldots, \varphi_n \in \mathcal{D}^\infty(U)$. Since, by assumption, the matrix $[a_{kl}]$ satisfies 11.4/(1) for irreducible representations, the right-hand side of this equality is non-negative, so is the left-hand side, and 11.4/(1) is proved for U. □

12.4. Integral Representation of Positive Linear Functionals

In the first subsection we collect the results from Choquet theory which are needed for the extremal decomposition of positive linear functionals in the second subsection. In a third subsection some general properties of integrals of states are studied.

Preliminaries on Choquet Theory

In this subsection E denotes a real locally convex Hausdorff space.

We recall some standard terminology. Suppose X is a compact Hausdorff space. A *Baire subset* of X is a set in the σ-algebra generated by the compact G_δ-subsets of X. G_δ-sets are defined as the countable intersections of open sets. Note that the compact G_δ-subsets of X are precisely the zero sets $\{x \in X : f(x) = 0\}$ of the continuous functions f on X. Each Baire set is obviously a Borel set. Recall that ex K is the set of extreme points of a convex set K.

Now we state the two fundamental results from Choquet theory which we shall apply in the next subsection.

Lemma 12.4.1. (CHOQUET) *Suppose that K is a metrizable compact convex subset of E. Then* ex K *is a G_δ-set (hence a Borel set) of K and for every point $x \in K$ there exists a positive regular Borel measure ν_x on K such that ν_x is concentrated on* ex K *(i.e., $\nu_x(K \setminus \mathrm{ex}\, K) = 0$) and $f(x) = \int_K f(y)\, \mathrm{d}\nu_x(y)$ for all $f \in E'$.*

Proof. ALFSEN [1], p. 36, Corollary I.4.9, or CHOQUET [1], p. 140, Theorem 27.6, and p. 138, Corollary 27.3, or the original paper of CHOQUET [2]; cf. also PHELPS [1], § 3. □

Lemma 12.4.2. (BISHOP-DE LEEUW) *Suppose K is a compact convex subset of E. For every point $x \in K$ there is a positive measure ν_x on the σ-algebra generated by* ex K *and by the Baire subsets of K such that ν_x is concentrated on* ex K *and $f(x) = \int_K f(y)\, \mathrm{d}\nu_x(y)$ for all $f \in E'$.*

Proof. PHELPS [1], p. § 4, Theorem, or the original paper of BISHOP/DE LEEUW [1]; cf. also ALFSEN [1], p. 39, Theorem I.4.14. □

Remark 1. Lemma 12.4.1 is no longer true if the metrizibility assumption is omitted. There exist a compact convex set K and a point $x_0 \in K$ such that ex K is a Borel set and $\nu(\mathrm{ex}\, K) = 0$ for any positive regular Borel measure that represents x_0 (i.e., for which $f(x_0) = \int_K f(y)\, \mathrm{d}\nu(y)$ for all $f \in E'$). Such an example is given in BISHOP/DE LEEUW [1]; see also PHELPS [1], p. § 4 or ALFSEN [1], I.4.

The preceding results do not apply directly to the state space of topological $*$-algebras because this space is not weakly compact in general. To overcome this difficulty, it is common to use the concept of a cap (see e.g. CHOQUET [1], § 30). A *cap* of a wedge C in E is a non-empty compact convex subset K of C such that $C \setminus K$ is also convex.

The following two simple lemmas are needed later on.

Lemma 12.4.3. *Let K be a non-empty compact subset of a wedge C in E. The set K is a cap of C if and only if there is a positively homogeneous additive map $h: C \to [0, +\infty]$ such that $K = \{x \in C: h(x) \leqq 1\}$. Moreover, if K is a cap of C, then $0 \in K$.*

Proof. First suppose there exists an h as stated above. Let $x, y \in C \setminus K$. Then $h(x) > 1$ and $h(y) > 1$, so that $h\left(\frac{1}{2}(x+y)\right) = \frac{1}{2} h(x) + \frac{1}{2} h(y) > 1$, i.e., $\frac{1}{2}(x+y) \in C \setminus K$. This shows that $C \setminus K$ is convex. Similarly K is convex. Thus K is a cap of C.

Suppose now that K is a cap of C. We first check that $0 \in K$. Assume the contrary, that is, $0 \in C \setminus K$. Then there is a non-zero $x \in K$. Since K is compact, $\lambda x \in C \setminus K$ for some $\lambda > 1$. This implies $x \in C \setminus K$, because $C \setminus K$ is convex. This contradiction proves that $0 \in K$. Define h by $h(x) := \inf \{\lambda \in (0, +\infty]: \lambda^{-1} x \in K\}$ for $x \in C$, where we set, of course, $(+\infty)^{-1} := 0$. Since K is a closed convex set containing 0, it is well-known (see e.g. SCHÄFER [1], II, 1.4) that $K = \{x \in C: h(x) \leqq 1\}$ and h is positively homogeneous and subadditive. It remains to prove that $h(x) + h(y) \leqq h(x+y)$ for $x, y \in C$. Since $h(z) = 0$ implies $z = 0$, we can assume that $h(x) > 0$ and $h(y) > 0$. Suppose $0 < \alpha < h(x)$ and $0 < \beta < h(y)$. Then $\alpha^{-1} x \in C \setminus K$ and $\beta^{-1} y \in C \setminus K$ and hence $x + y \in (\alpha + \beta) C \setminus K$, because $C \setminus K$ is convex. Therefore, $(\alpha+\beta)^{-1}(x+y) \notin K$ which gives $\alpha + \beta \leqq h(x+y)$. Letting $\alpha \uparrow h(x)$ and $\beta \uparrow h(y)$, we get $h(x) + h(y) \leqq h(x+y)$. □

Lemma 12.4.4. *Let K be a cap of the wedge C in E. Suppose f is a linear functional on E such that $\ker f \cap K = \{0\}$ and $f(x) \geqq 0$ for $x \in C$. If y is a non-zero extreme point of K, then $f(y)^{-1} y$ is an extreme point of the convex set $B := \{x \in C: f(x) = 1\}$.*

Proof. Let h be as in Lemma 12.4.3. First we note that $h(x) \neq 0$ for all non-zero $x \in K$, since otherwise $K \supseteqq \{\lambda x: \lambda \geqq 0\}$ which contradicts the compactness of K. In particular, $h(y) \neq 0$. We have $h(y)^{-1} y \in K$, $0 \in K$ (by Lemma 12.4.3) and $y = h(y)\left(h(y)^{-1} y\right) + \left(1 - h(y)\right) 0$ with $0 < h(y) \leqq 1$ (by $y \in K$). Since $y \in \operatorname{ex} K$ and $y \neq 0$, the latter is only possible if $h(y) = 1$.

We show that $f(y)^{-1} y \in \operatorname{ex} B$. Let $f(y)^{-1} y = \lambda y_1 + (1 - \lambda) y_2$ with $y_1, y_2 \in B$ and $0 < \lambda < 1$. Since h is additive and positively homogeneous, $1 = h(y) = \lambda f(y) h(y_1) + (1 - \lambda) f(y) h(y_2)$. Hence $h(y_1) < +\infty$ and $h(y_2) < +\infty$, since $\lambda \in (0, 1)$. From $y_k \in B$, we have $y_k \neq 0$ and so $h(y_k) \neq 0$ for $k = 1, 2$ as noted above. Put $z_k := h(y_k)^{-1} y_k$ for $k = 1, 2$. Then $z_1, z_2 \in K$ and $y = \lambda f(y) h(y_1) z_1 + (1 - \lambda) f(y) h(y_2) z_2$. The latter is a convex combination with $\lambda f(y) h(y_1) \in (0, 1)$. Hence we conclude from $y \in \operatorname{ex} K$ that $z_1 = z_2 = y$. Therefore, $f(y) = f(z_k) = h(y_k)^{-1} f(y_k) = h(y_k)^{-1}$ (by $y_k \in B$) for $k = 1, 2$ which in turn yields that $y_1 = y_2$. Thus we have proved that $f(y)^{-1} y \in \operatorname{ex} B$. □

Extremal Decomposition of Positive Linear Functionals

In this and the following subsection A will denote a $*$-algebra with unit.

In what follows we briefly write σ for the weak topology $\sigma(\mathsf{A}^*, \mathsf{A})$ on the algebraic dual A^* of A or on a subset of A^*, and we equip the set $\mathcal{Z}(\mathsf{A})$ of all states of A with the topology σ. For a topology τ on A, $\mathcal{Z}_\tau(\mathsf{A})$ is the subspace of $\mathcal{Z}(\mathsf{A})$ formed by the τ-continuous states. Recall that A_h^* is the real vector space of the hermitian linear functionals on A.

Theorem 12.4.5. *Let τ be a nuclear locally convex Hausdorff topology on the $*$-algebra A and let $\mathcal{Q}$ be a wedge in A_h which contains $\mathcal{P}(\mathsf{A})$. For every linear functional $\omega_0 \not\equiv 0$ on A, the following two statements are equivalent:*

(i) *ω_0 is $\mathcal{Q}$-positive and the seminorm r defined by $r(a) := \omega_0(a^+a)^{1/2}$, $a \in \mathsf{A}$, is continuous on $\mathsf{A}[\tau]$.*

(ii) *There exist a metrizable compact subspace Ω of $\mathcal{Z}_\tau(\mathsf{A})$, a positive regular Borel measure μ on Ω and a Borel subset Ω_0 of Ω such that:*

(ii.1) $\omega_0(a) = \int_{\Omega_0} \omega(a)\, d\mu(\omega)$ *for all* $a \in \mathsf{A}$.

(ii.2) $\Omega_0 \subseteq \operatorname{ex}(\mathcal{Q}^* \cap \mathcal{Z}(\mathsf{A}))$.

(ii.3) *There are a function $\xi \in L^2(\Omega_0; \mu)$ and a continuous seminorm q on $\mathsf{A}[\tau]$ such that $\omega(a^+a)^{1/2} \leq |\xi(\omega)|\, q(a)$ for $a \in \mathsf{A}$ and $\omega \in \Omega_0$.*

The crucial step in the proof of this theorem is contained in

Lemma 12.4.6. *Suppose statement* (i) *of Theorem* 12.4.5 *is fulfilled. Then there exists a metrizable cap K of the wedge $\mathcal{Q}^*$ in the real locally convex space $\mathsf{A}_h^*[\sigma]$ such that $\omega_0 \in K$ and $K \subseteq \mathsf{A}[\tau]'$.*

Proof. For a seminorm p on A, we let $\widehat{\mathsf{A}}_p$ denote the completion of the normed space $\mathsf{A}_p := \mathsf{A}/\ker p$ endowed with the factor norm of p. Because r is a continuous seminorm on the nuclear space $\mathsf{A}[\tau]$, there exists a continuous Hilbertian seminorm q on $\mathsf{A}[\tau]$ such that $r \leq q$ on A and the canonical embedding $\mathfrak{j}$ of $\widehat{\mathsf{A}}_q$ into $\widehat{\mathsf{A}}_r$ is nuclear. Since r and q are Hilbertian seminorms, $\widehat{\mathsf{A}}_r$ and $\widehat{\mathsf{A}}_q$ are Hilbert spaces. Let $(\cdot,\cdot)_r$ and $(\cdot,\cdot)_q$ denote the corresponding scalar products. Set $V := \{\omega \in \mathcal{Q}^*\colon \omega(a^+a) \leq r(a)^2 \text{ for all } a \in \mathsf{A}\}$. By the assumption $\mathcal{P}(\mathsf{A}) \subseteq \mathcal{Q}$, each $\omega \in \mathcal{Q}^*$ is a positive linear functional on A. Hence, by the Cauchy-Schwarz inequality, we have $|\omega(a)|^2 \leq \omega(a^+a)\,\omega(1) \leq r(a)^2 r(1)^2$ for $a \in \mathsf{A}$ and $\omega \in V$. Therefore, V is a σ-closed convex subset of the equicontinuous set $\{\omega \in \mathsf{A}[\tau]'\colon |\omega(a)| \leq r(1)\, r(a) \text{ for } a \in \mathsf{A}\}$ and hence σ-compact.

Let W be the real linear span of V in A_h^* and let $W_+ := \bigcup_{\lambda \geq 0} \lambda V$. Let $a \to \bar{a}$ denote the quotient map of A into $\mathsf{A}_q \equiv \mathsf{A}/\ker q$. Suppose $\omega \in W$. Since V is convex, there are $\omega_1, \omega_2 \in V$ and $\lambda_1, \lambda_2 \in \mathbb{C}$ such that $\omega = \lambda_1\omega_1 + \lambda_2\omega_2$. For $a, b \in \mathsf{A}$, we have

$$\begin{aligned} |\omega(b^+a)| &\leq |\lambda_1|\, \omega_1(b^+b)^{1/2}\, \omega_1(a^+a)^{1/2} + |\lambda_2|\, \omega_2(b^+b)^{1/2}\, \omega_2(a^+a)^{1/2} \\ &\leq (|\lambda_1| + |\lambda_2|)\, r(b)\, r(a) \leq (|\lambda_1| + |\lambda_2|)\, q(b)\, q(a). \end{aligned}$$

From this we deduce that the map $\mathsf{A}_q \times \mathsf{A}_q \ni (\bar{a}, \bar{b}) \to \omega(b^+a)$ defines, unambiguously, a continuous sesquilinear form on the normed space A_q. Hence there exists a bounded linear operator T_ω on the Hilbert space $\widehat{\mathsf{A}}_q$ such that $\omega(b^+a) = (T_\omega\bar{a}, \bar{b})_q$, $a, b \in \mathsf{A}$. In particular, ω_0 belongs to W. From $(T_{\omega_0}\bar{a}, \bar{b})_q = \omega_0(b^+a) = (\bar{a}, \bar{b})_r = (\mathfrak{j}\bar{a}, \mathfrak{j}\bar{b})_q$ for $a, b \in \mathsf{A}$ we conclude that $\mathfrak{j}^*\mathfrak{j} = T_{\omega_0}$. Since $\mathfrak{j}\colon \widehat{\mathsf{A}}_q \to \widehat{\mathsf{A}}_r$ is nuclear, T_{ω_0} is a nuclear operator of $\mathbb{B}(\widehat{\mathsf{A}}_q)$. By the definition of V, we have $0 \leq T_\omega \leq T_{\omega_0}$ for $\omega \in V$. From this and

$T_{\omega_0} \in \mathbb{B}_1(\widehat{\mathsf{A}}_q)$ it follows that $T_\omega \in \mathbb{B}_1(\widehat{\mathsf{A}}_q)$ for $\omega \in W$. From $\omega_0 \not\equiv 0$, $(T_{\omega_0}\tilde{1}, \tilde{1})_q = \omega_0(1) \neq 0$. Since $T_{\omega_0} \geqq 0$, this gives $\operatorname{Tr} T_{\omega_0} \neq 0$. We define a map $h\colon \mathcal{Q}^* \to [0, +\infty]$ by $h(\omega) := (\operatorname{Tr} T_{\omega_0})^{-1} \operatorname{Tr} T_\omega$ if $\omega \in W_+$ and $h(\omega) := +\infty$ if $\omega \in \mathcal{Q}^* \setminus W_+$. From the definition of V we see immediately that h is additive and positively homogeneous. Define $K := \{\omega \in \mathcal{Q}^* : h(\omega) \leqq 1\}$. It is clear that $\omega_0 \in K$. In order to prove that K is a cap for $\mathcal{Q}^*$ in $\mathsf{A}_h^*[\sigma]$, it is sufficient to show by Lemma 12.4.3 that K is σ-compact. By definition, $K = \{\omega \in V : \operatorname{Tr} T_\omega \leqq \operatorname{Tr} T_{\omega_0}\}$; so $\{T_\omega : \omega \in V\}$ is bounded in $\mathbb{B}(\widehat{\mathsf{A}}_q)$. Hence the weak-operator topology of $\mathbb{B}(\widehat{\mathsf{A}}_q)$ coincides on the set $\{T_\omega : \omega \in V\}$ with the locally convex topology which is generated by the family of seminorms $|(\cdot \tilde{a}, \tilde{b})_q|$, where a and b range over the dense subset A_q of $\widehat{\mathsf{A}}_q$. Since $\omega(b^+a) = (T_\omega \tilde{a}, \tilde{b})_q$, $a, b \in \mathsf{A}$, it follows that the map $\omega \to T_\omega$ of $V[\sigma]$ into $\mathbb{B}(\widehat{\mathsf{A}}_q)$ is continuous if $\mathbb{B}(\widehat{\mathsf{A}}_q)$ carries the weak-operator topology. Therefore, $K \equiv \{\omega \in V : \operatorname{Tr} T_\omega \leqq \operatorname{Tr} T_{\omega_0}\}$ is σ-closed in V. Since V is σ-compact as stated above, K is also σ-compact; so K is a cap of $\mathcal{Q}^*$ in A_h^*.

It remains to verify that $K[\sigma]$ is metrizable. Since $\mathsf{A}[\tau]$ is nuclear, the Hilbert space $\widehat{\mathsf{A}}_q$ is separable (Pietsch [1], 4.4.9). Hence, by Lemma 5.2.8, there is a countable subset $\{a_n : n \in \mathbb{N}\}$ of A such that $\{\tilde{a}_n : n \in \mathbb{N}\}$ is dense in $\widehat{\mathsf{A}}_q$. Let σ_0 denote the locally convex topology on W defined by the seminorms $\omega \to |\omega(a_n^+ a_m)|$, $n, m \in \mathbb{N}$. If $\omega \in W$ satisfies $\omega(a_n^+ a_m) \equiv (T_\omega \tilde{a}_m, \tilde{a}_n)_q = 0$ for all $n, m \in \mathbb{N}$, then $T_\omega \equiv 0$ and so $\omega = 0$. That is, σ_0 is a Hausdorff topology on W. Since $\sigma_0 \subseteqq \sigma$ on W, σ_0 and σ coincide on the σ-compact set K. Thus $K[\sigma] \equiv K[\sigma_0]$ is metrizable. □

Proof of Theorem 12.4.5.

(i) → (ii): We apply Choquet's theorem (Lemma 12.4.1) to the (compact convex) set K from Lemma 12.4.6 with $x := \omega_0$ and $E := \mathsf{A}_h^*[\sigma]$. By this result there exists a positive regular Borel measure ν on K such that $\nu(K \setminus \operatorname{ex} K) = 0$ and

$$\omega_0(a) = \int_K \omega(a) \, d\nu(\omega) \tag{1}$$

for $a \in \mathsf{A}_h$. Here we used the fact that for each $a \in \mathsf{A}_h$ the map $\omega \to \omega(a)$ is a continuous (real) linear functional on $\mathsf{A}_h^*[\sigma]$. By linearity, (1) extends to all $a \in \mathsf{A}$. The preceding proof that the cap K is σ-compact also shows that the set $K_1 := \{\omega \in \mathcal{Q}^* := h(\omega) = 1\}$ is σ-compact. Define a map T of $K \setminus \{0\}$ onto a subspace Ω of $\mathcal{Z}_\tau(\mathsf{A})$ by $T(\omega) = \omega^{\mathsf{I}} := \omega(1)^{-1}\,\omega$. Then T provides a homeomorphism of K_1 onto $\Omega \equiv T(K \setminus \{0\})$. Hence Ω is compact and metrizable, since K_1 is also. We define a Borel measure μ on Ω by $\mu(\cdot) := \nu(T^{-1}(\cdot))$. From (1) and $\nu(K \setminus \operatorname{ex} K) = 0$, we obtain $\omega_0(a) = \int_{\operatorname{ex} K \setminus \{0\}} \tilde{\omega}(a)\, \omega(1)\, d\nu(\omega)$ $= \int_{\Omega_0} \tilde{\omega}(a)\, d\mu(\tilde{\omega})$ for all $a \in \mathsf{A}$, where $\Omega_0 := \{\tilde{\omega} : \omega \in \operatorname{ex} K \text{ and } \omega \neq 0\}$. This proves (ii.1). Note that Ω_0 is a Borel subset of Ω, since ex K is a Borel set of K by Lemma 12.4.1.

To verify (ii.2), we essentially use that K is a cap of $\mathcal{Q}^*$. Let f be the linear functional on $E = \mathsf{A}_h^*[\sigma]$ defined by $f(\omega) := \omega(1)$. From Lemma 12.4.4, if ω is a non-zero extreme point of K, then $f(\omega)^{-1}\,\omega \equiv \tilde{\omega}$ is an extreme point of $\mathcal{Q}^* \cap \mathcal{Z}(\mathsf{A})$; so $\Omega_0 \subseteqq \operatorname{ex}\left(\mathcal{Q}^* \cap \mathcal{Z}(\mathsf{A})\right)$.

Define $\xi(\tilde{\omega}) := \omega(1)^{-1/2}$ if $\omega \neq 0$ and $\omega \in \operatorname{ex} K$. From

$$\int_{\Omega_0} |\xi(\tilde{\omega})|^2 \, d\mu(\tilde{\omega}) = \nu(\operatorname{ex} K \setminus \{0\}) \leqq \nu(K) < \infty$$

we see that $\xi \in L^2(\Omega_0; \mu)$. By the proof of Lemma 12.4.6, $\omega(a^+a) \leqq r(a)^2$ for $a \in \mathsf{A}$ and $\omega \in K$. Hence $\tilde{\omega}(a^+a) = \omega(1)^{-1}\, \omega(a^+a) \leqq |\xi(\tilde{\omega})|^2\, r(a)^2$ for $a \in \mathsf{A}$ and $\tilde{\omega} \in \Omega_0$. Thus (ii.3) is proved.

(ii) $\to$ (i): (ii.1) and (ii.2) imply that ω_0 is $\mathcal{Q}^*$-positive. The continuity of the seminorm r follows easily from (ii.1) and (ii.3). □

Theorem 12.4.7. *Let $\mathcal{Q}$ be a wedge in A_h such that $\mathcal{Q}^* \subseteqq \mathsf{A}_\mathrm{h}^*$. Suppose there exists a countable subset $\{a_n : n \in \mathbb{N}\}$ of $\mathcal{Q}$ such that for every $a \in \mathsf{A}_\mathrm{h}$ there are numbers $n \in \mathbb{N}$ and $\alpha > 0$ satisfying $\alpha a_n - a \in \mathcal{Q}$.*

Then for each $\mathcal{Q}$-positive linear functional $\omega_0 \not\equiv 0$ on A there exist a topological subspace Ω of $\mathcal{Z}(\mathsf{A})$, a subset Ω_0 of Ω and a positive measure μ on the σ-algebra generated by Ω_0 and by the Baire subsets of Ω such that $\Omega_0 \subseteqq \mathrm{ex}\,(\mathcal{Q}^* \cap \mathcal{Z}(\mathsf{A}))$ and $\omega_0(a) = \int\limits_{\Omega_0} \omega(a)\, \mathrm{d}\mu(\omega)$ for all $a \in \mathsf{A}$.

The proof of this theorem is similar to the above proof of Theorem 12.4.5,(i) $\to$ (ii), when we use Lemma 12.4.2 and Lemma 12.4.8 below instead of Lemmas 12.4.1 and 12.4.6, respectively. We do not carry out these details.

Lemma 12.4.8. *Retaining the assumptions of Theorem 12.4.7, ω_0 is contained in some cap K of the wedge $\mathcal{Q}^*$ in the space $\mathsf{A}_\mathrm{h}^*[\sigma]$.*

Proof. We choose positive numbers δ_n, $n \in \mathbb{N}$, such that $\sum\limits_{n=1}^{\infty} \delta_n \omega_0(a_n) \leqq 1$. We then define an additive and positively homogeneous map $h\colon \mathcal{Q}^* \to [0, +\infty]$ by $h(\omega) := \sum\limits_{n=1}^{\infty} \delta_n \omega(a_n)$, $\omega \in \mathcal{Q}^*$, and set $K := \{\omega \in \mathcal{Q}^* : h(\omega) \leqq 1\}$. Suppose $a \in \mathsf{A}_\mathrm{h}$. By assumption, there are numbers $n, m \in \mathbb{N}$, $\alpha > 0$ and $\beta > 0$ such that $\alpha a_n - a \in \mathcal{Q}$ and $\beta a_m + a \in \mathcal{Q}$. Then $|\omega(a)| = \max\,\{\omega(a), \omega(-a)\} \leqq \max\,\{\alpha\omega(a_n), \beta\omega(a_m)\} \leqq \max\{\alpha\delta_n^{-1}, \beta\delta_m^{-1}\}$ for any $\omega \in K$. Since $\mathsf{A} = \mathsf{A}_\mathrm{h} + \mathrm{i}\mathsf{A}_\mathrm{h}$, this shows that K is bounded in $\mathsf{A}^*[\sigma]$. Hence the polar K° of K in the dual pairing of A and A^* is a 0-neighbourhood for the finest locally convex topology τ_{st} on the vector space A. Therefore, by the Alaoglu-Bourbaki theorem (Schäfer [1], III, 4.3), the bipolar $K^{\circ\circ}$ is σ-compact in $\mathsf{A}^* \equiv \mathsf{A}[\tau_{\mathrm{st}}]'$. Since $K \subseteqq K^{\circ\circ}$ and K is obviously σ-closed in A^*, it follows that K is σ-compact. Thus K is a cap of $\mathcal{Q}^*$ in $\mathsf{A}_\mathrm{h}^*[\sigma]$ by Lemma 12.4.3. By construction, $h(\omega_0) \leqq 1$ and so $\omega_0 \in K$. □

Remark 2. Let ω_0, μ, Ω and Ω_0 be as in Theorem 12.4.5,(ii), or in Theorem 12.4.7. For the statement of these results we can assume without loss of generality that $\mu(\Omega \setminus \Omega_0) = 0$. Since $\mu(\Omega) = \int\limits_{\Omega_0} \Omega_0\omega(1)\, \mathrm{d}\mu(\omega) = \omega_0(1) < \infty$, then the measure μ is finite on Ω.

Remark 3. In case $\mathcal{Q} = \mathcal{P}(\mathsf{A})$ the statement (ii) in Theorem 12.4.5 provides a decomposition of the positive linear functional ω_0 as an integral over pure states, since then $\Omega_0 \subseteqq \mathrm{ex}\, \mathcal{Z}(\mathsf{A})$.

Remark 4. There is an important and rather general situation in which Theorem 12.4.7 applies: if A is an O*-algebra $\mathcal{A}$ with metrizable graph topology $\mathrm{t}_{\mathcal{A}}$ and $\mathcal{Q}$ is the cone $\mathcal{A}_+$. (The second assumption follows then from Corollary 2.6.7.)

Remark 5. It should be noted that the states in the set Ω_0 $\left(\subseteqq \mathrm{ex}\,(\mathcal{Q}^* \cap \mathcal{Z}(\mathsf{A}))\right)$ in Theorems 12.4.5 and 12.4.7 are only extreme points of the set of *$\mathcal{Q}$-positive* states in general. It is therefore natural to ask for conditions which ensure that they are also extreme points of the set $\mathcal{Z}(\mathsf{A})$ of *all* states. Proposition 11.3.9 is a result of this kind.

Remark 6. Let $\mathsf{A} := \mathbb{C}[x_1, \ldots, x_n]$ with $n \geqq 2$. Theorem 12.4.5 can be used to conclude that there are pure states on A which are not characters. Indeed, let ω_0 be a positive linear functional on A which is not $\mathsf{A}_+^{\text{int}}$-positive. (Such functionals exist by Example 2.6.11; see e.g. Remark 6 in 11.3.) We apply Theorem 12.4.5 with $\mathcal{Q} = \mathcal{P}(\mathsf{A})$ and τ the finest locally convex topology on A. Since characters are strongly positive (by Corollary 11.3.8), not all ω in Ω_0 $\big(\subseteqq \operatorname{ex} \mathcal{Z}(\mathsf{A})\big)$ can be characters.

Remark 7. Example 12.4.9 below shows that the conclusions of Theorems 12.4.5 or 12.4.7 are not true in general, if A is an O*-algebra $\mathcal{A}$, $\mathcal{Q} = \mathcal{A}_+$ and ω_0 is a strongly positive linear functional on $\mathcal{A}$. This means that additional assumptions as in these theorems or as the metrizibility of the graph topology $t_{\mathcal{A}}$ (cf. Remark 4) are indeed needed.

Example 12.4.9. Let A be the Arens algebra $L^\omega(0, 1)$; cf. Example 2.5.5. We identify each $f \in L^\omega(0, 1)$ with the corresponding multiplication operator on the domain $\{\varphi \in L^2(0, 1) : f \cdot \varphi \in L^2(0, 1) \text{ for all } f \in L^\omega(0, 1)\}$ in the Hilbert space $L^2(0, 1)$. Thus A becomes an O*-algebra. Let $\mathcal{Q} := \mathsf{A}_+$ and let ω_0 be an arbitrary vector state on A. Since obviously $\mathsf{A}_+ = \mathsf{A}_+^{\text{int}}$, Proposition 11.3.9 says that each element of $\operatorname{ex} \big(\mathcal{Q}^* \cap \mathcal{Z}(\mathsf{A})\big)$ must be a character on A. But A has no characters as shown in Example 2.5.5. Hence $\operatorname{ex} \big(\mathcal{Q}^* \cap \mathcal{Z}(\mathsf{A})\big)$ is empty, and ω_0 cannot be an integral over a subset of $\operatorname{ex} \big(\mathcal{Q}^* \cap \mathcal{Z}(\mathsf{A})\big)$. ○

Integrals of States and Orthogonal Measures

Recall that A denotes a $*$-algebra with unit.

In this subsection we suppose that Ω is a subset of $\mathcal{Z}(\mathsf{A})$, $\mathfrak{S}$ is a σ-algebra in Ω and μ is a positive finite measure on $\mathfrak{S}$ such that the function $\omega \to \omega(a)$ on Ω is in $L^1(\Omega; \mu)$ for each $a \in \mathsf{A}$. We define a positive linear functional ϑ on A by $\vartheta(a) := \int_\Omega \omega(a) \, \mathrm{d}\mu(\omega)$, $a \in \mathsf{A}$. (Note that this covers the situations described by Theorem 12.4.5,(ii), or by the assertion of Theorem 12.4.7, since we can assume therein that $\mu(\Omega \setminus \Omega_0) = 0$; see Remark 2. But we do not assume that Ω is contained in $\operatorname{ex} \mathcal{Z}(\mathsf{A})$.)

Proposition 12.4.10. *There exists a unique $*$-preserving contractive positive linear map $f \to T(f)$ of the W*-algebra $L^\infty(\Omega; \mu)$ into $\pi_\vartheta(\mathsf{A})'_{\mathrm{w}}$ satisfying*

$$\langle T(f) \, \pi_\vartheta(a) \, \varphi_\vartheta, \pi_\vartheta(b) \, \varphi_\vartheta \rangle = \int_\Omega f(\omega) \, \omega(b^+ a) \, \mathrm{d}\mu(\omega) \tag{2}$$

for all $a, b \in \mathsf{A}$ and $f \in L^\infty(\Omega; \mu)$. Moreover, the map $f \to T(f)$ is continuous if $L^\infty(\Omega; \mu)$ carries the weak topology $\sigma\big(L^\infty(\Omega; \mu), L^1(\Omega; \mu)\big)$ and $\pi_\vartheta(\mathsf{A})'_{\mathrm{w}}$ the weak-operator topology.

Proof. Suppose $f \in L^\infty(\Omega; \mu)$. Define a sesquilinear form $\mathfrak{c}_f$ on $\mathcal{D}_\vartheta \times \mathcal{D}_\vartheta$ by

$$\mathfrak{c}_f\big(\pi_\vartheta(a) \, \varphi_\vartheta, \pi_\vartheta(b) \, \varphi_\vartheta\big) := \int_\Omega f(\omega) \, \omega(b^+ a) \, \mathrm{d}\mu(\omega), \quad a, b \in \mathsf{A}.$$

For $a, b \in \mathsf{A}$, we have

$$\begin{aligned} \left| \int f(\omega) \, \omega(b^+ a) \, \mathrm{d}\mu(\omega) \right| &\leqq \|f\|_\infty \int |\omega(b^+ a)| \, \mathrm{d}\mu(\omega) \leqq \|f\|_\infty \int \omega(b^+ b)^{1/2} \, \omega(a^+ a)^{1/2} \mathrm{d}\mu(\omega) \\ &\leqq \|f\|_\infty \Big(\int \omega(b^+ b) \, \mathrm{d}\mu(\omega)\Big)^{1/2} \Big(\int \omega(a^+ a) \, \mathrm{d}\mu(\omega)\Big)^{1/2} \\ &= \|f\|_\infty \vartheta(b^+ b)^{1/2} \, \vartheta(a^+ a)^{1/2} = \|f\|_\infty \, \|\pi_\vartheta(b) \, \varphi_\vartheta\| \, \|\pi_\vartheta(a) \, \varphi_\vartheta\|, \end{aligned}$$

where $\|\cdot\|_\infty$ is the norm of $L^\infty(\Omega; \mu)$. This shows that $\mathfrak{c}_f$ is well-defined and bounded in the Hilbert space norm of $\mathcal{D}_\vartheta$. Hence $\mathfrak{c}_f$ is represented by a unique bounded operator

$T(f)$ on $\mathcal{H}_\vartheta$; so (2) is fulfilled by definition. From the previous inequality we see that the map $f \to T(f)$ is contractive. From (2) it is clear that T is linear, $*$-preserving and positive. When $a, b, c \in \mathsf{A}$ and $f \in L^\infty(\Omega; \mu)$, we have

$$\langle T(f)\, \pi_\vartheta(c)\, \pi_\vartheta(a)\, \varphi_\vartheta, \pi_\vartheta(b)\, \varphi_\vartheta\rangle = \int f(\omega)\, \omega(b^+ca)\, d\mu(\omega) = \int f(\omega)\, \omega\big((c^+b)^+a\big)\, d\mu(\omega)$$
$$= \langle T(f)\, \pi_\vartheta(a)\, \varphi_\vartheta, \pi_\vartheta(c^+)\, \pi_\vartheta(b)\, \varphi_\vartheta\rangle.$$

Therefore, $T(f) \in \tilde{\pi}_\vartheta(\mathsf{A})'_{\mathrm{w}}$ and hence $T(f) \in \pi_\vartheta(\mathsf{A})'_{\mathrm{w}}$, since π_ϑ is the closure of $\tilde{\pi}_\vartheta$; cf. 8.6. It remains to prove the continuity assertion. Since the function $\omega \to \omega(b^+a)$ is in $L^1(\Omega; \mu)$ for any $a, b \in \mathsf{A}$, we conclude from (2) that for arbitrary vectors $\varphi, \psi \in \mathcal{D}_\vartheta$ the linear functional $f \to \langle T(f)\, \varphi, \psi\rangle$ on $L^\infty(\Omega; \mu)$ is continuous in the topology $\sigma(L^\infty, L^1)$. Now let $\varphi, \psi \in \mathcal{H}_\vartheta$. Since $\mathcal{D}_\vartheta$ is dense in $\mathcal{H}_\vartheta$ and T is contractive, the preceding implies that $f \to \langle T(f)\, \varphi, \psi\rangle$ is $\sigma(L^\infty, L^1)$-continuous on the unit ball of $L^\infty(\Omega; \mu)$. Because the measure μ is finite, $L^\infty(\Omega; \mu)$ is the dual of $L^1(\Omega; \mu)$. Hence the continuity of $f \to \langle T(f)\, \varphi, \psi\rangle$ on the whole of $L^\infty(\Omega; \mu)$ follows from the Krein-Šmulian theorem (Schäfer [1], IV, 6.4). □

Definition 12.4.11. We say that μ is an *orthogonal measure* on $(\Omega, \mathfrak{S})$ if for every set $M \in \mathfrak{S}$ the positive linear functionals $\omega_M(\cdot) := \int_M \omega(\cdot)\, d\mu(\omega)$ and $\omega_{\Omega \setminus M}(\cdot) := \int_{\Omega \setminus M} \omega(\cdot)\, d\mu(\omega)$ on A are orthogonal in the sense of Definition 8.6.13.

Proposition 12.4.12. *Suppose that* $\pi_\vartheta(\mathsf{A})'_{\mathrm{s}} = \pi_\vartheta(\mathsf{A})'_{\mathrm{w}}$. *Then the following conditions are equivalent:*

(i) μ *is an orthogonal measure on* $(\Omega, \mathfrak{S})$.

(ii) T *is a* $*$*-isomorphism of* $L^\infty(\Omega; \mu)$ *onto a* $*$*-subalgebra of* $\pi_\vartheta(\mathsf{A})'_{\mathrm{w}}$.

(iii) $T(fg) = T(f)\, T(g)$ *for all* $f, g \in L^\infty(\Omega; \mu)$.

Proof. Let χ_M denote the characteristic function of a set M.

(i) → (ii): Suppose $M \in \mathfrak{S}$. We have

$$\omega_M(\cdot) = \int_\Omega \chi_M(\omega)\, \omega(\cdot)\, d\mu(\omega) = \langle T(\chi_M)\, \pi_\vartheta(\cdot)\, \varphi_\vartheta, \varphi_\vartheta\rangle = \vartheta_{T(\chi_M)}$$

and similarly $\omega_{\Omega\setminus M} = \vartheta - \vartheta_{T(\chi_M)}$. Since μ is orthogonal by (i), $\omega_M \perp \omega_{\Omega\setminus M}$. Therefore, by Remark 6 in 8.6, $T(\chi_M)$ is a projection on $\mathcal{H}_\vartheta$.

If f and g are the characteristic functions of disjoint sets in $\mathfrak{S}$, then $f(\cdot) \leqq (1 - g)(\cdot)$ on Ω and so $T(f) \leqq T(1 - g) = I - T(g)$. Hence $T(f)\, T(g) = 0$, since $T(f)$ and $T(g)$ are projections. Now let f and g be characteristic functions of arbitrary sets in $\mathfrak{S}$. From $f = fg + f(1 - g)$, $g = fg + (1 - f)\, g$ and the preceding, we obtain

$$T(f)\, T(g) = T(fg)^2 + T(fg)\, T\big((1 - f)\, g\big) + T\big(f(1 - g)\big)\, T(fg)$$
$$+ T\big(f(1 - g)\big)\, T\big((1 - f)\, g\big) = T(fg)^2 + 0 + 0 + 0 = T(fg),$$

where the latter is true because $T(fg)$ is a projection. The relation $T(f)\, T(g) = T(fg)$ clearly extends to functions f and g in the linear span of characteristic functions of sets in $\mathfrak{S}$. From the continuity assertion in Proposition 12.4.10 it follows that $T(f)\, T(g) = T(fg)$ for all $f, g \in L^\infty(\Omega; \mu)$. Using this fact we have

$$\|T(f)\, \varphi_\vartheta\|^2 = \langle T(f)^*\, T(f)\, \varphi_\vartheta, \varphi_\vartheta\rangle = \langle T(|f|^2)\, \varphi_\vartheta, \varphi_\vartheta\rangle = \int_\Omega |f(\omega)|^2\, d\mu(\omega)$$

for $f \in L^\infty(\Omega; \mu)$. Thus $T(f) \neq 0$ if $f \neq 0$ which shows that T is injective. We stated already in Proposition 12.4.10 that T is linear and $*$-preserving, so T is a $*$-isomorphism.

(ii) $\to$ (iii) is trivial.

(iii) $\to$ (i): Let $M \in \mathfrak{S}$. By (iii), we have $T(\chi_M)\big(I - T(\chi_M)\big) = T(\chi_M)\, T(\chi_{\Omega\setminus M}) = T(\chi_M \chi_{\Omega\setminus M}) = 0$. Since $T(\chi_M)$ is self-adjoint, $T(\chi_M)$ is a projection. From Remark 6 in 8.6, the functionals $\vartheta_{T(\chi_M)} \equiv \omega_M$ and $\vartheta - \vartheta_{T(\chi_M)} \equiv \omega_{\Omega\setminus M}$ are orthogonal. Hence μ is an orthogonal measure on $(\Omega, \mathfrak{S})$. □

Corollary 12.4.13. *If $\pi_\vartheta(\mathsf{A})'_s = \pi_\vartheta(\mathsf{A})'_w$ and μ is an orthogonal measure on $(\Omega, \mathfrak{S})$, then $\{T(f) : f \in L^\infty(\Omega; \mu)\}$ is an abelian von Neumann subalgebra of $\pi_\vartheta(\mathsf{A})'_w$.*

Proof. Being $*$-isomorphic to $L^\infty(\Omega; \mu)$ by Proposition 12.4.12, (i) $\to$ (ii), $\mathcal{N} := \{T(f) : f \in L^\infty(\Omega; \mu)\}$ is an abelian C^*-algebra. By the continuity of T, the unit ball of $\mathcal{N}$ is compact and hence closed in $\mathbb{B}(\mathcal{H}_\vartheta)$ in the weak-operator topology. The Kaplansky density theorem implies that $\mathcal{N}$ is weak-operator closed in $\mathbb{B}(\mathcal{H}_\vartheta)$ and hence a von Neumann algebra. □

Remark 8. If we keep the positive linear functional ϑ fixed, then the map T depends, of course, essentially on μ and on Ω. We have avoided this dependence in the notation.

The next result gives (under additional assumptions) a criterion for the orthogonality of the measure μ in terms of the representations π_ϑ and π_ω, $\omega \in \Omega$. For this we let the measure space be of the form set out in the first paragraph of Section 12.1. Besides from this technical assumption, we suppose that there exists a locally convex topology τ on A such that the following three conditions are valid:

(α) *$\mathsf{A}[\tau]$ is separable.*

(β) *For all $a \in \mathsf{A}$, the map $x \to ax$ is continuous on $\mathsf{A}[\tau]$.*

(γ) *For all $\omega \in \Omega$, the seminorm $r_\omega(a) := \omega(a^+a)^{1/2}$, $a \in \mathsf{A}$, is continuous on $\mathsf{A}[\tau]$.*

We need some preliminary constructions. From (α), there exists a countable subset $\{b_n : n \in \mathbb{N}\}$ of A that is dense in $\mathsf{A}[\tau]$. For $a, b \in \mathsf{A}$, $\omega \in \Omega$ and $n \in \mathbb{N}$, we have $\|\pi_\omega(a)\big(\pi_\omega(b)\varphi_\omega - \pi_\omega(b_n)\varphi_\omega\big)\| = r_\omega\big(a(b - b_n)\big)$. Hence by ($\beta$) and ($\gamma$), $\Gamma_\omega := \{\pi_\omega(b_n)\varphi_\omega : n \in \mathbb{N}\}$ is a dense subset of $\mathcal{D}_\omega$ and so of $\mathcal{D}(\pi_\omega)$ in the graph topology of $\pi_\omega(\mathsf{A})$. Let ψ_n, $n \in \mathbb{N}$, denote the vector field on Ω defined by $\psi_n(\omega) := \pi_\omega(b_n)\,\varphi_\omega$, $\omega \in \Omega$. It is clear that the functions $\omega \to \langle \psi_n(\omega), \psi_m(\omega)\rangle = \omega(b_m^+ b_n)$ are μ-measurable on Ω for all $n, m \in \mathbb{N}$ and that the set $\Gamma_\omega = \{\psi_n(\omega) : n \in \mathbb{N}\}$ is dense in $\mathcal{H}_\omega$ for all $\omega \in \Omega$. From this we conclude that $\omega \to \mathcal{H}_\omega \equiv \mathcal{H}(\pi_\omega)$ is a measurable field of Hilbert spaces over Ω with respect to the fundamental sequence $(\psi_n : n \in \mathbb{N})$ of μ-measurable vector fields. Let

$$\mathcal{H}_\mu := \int_\Omega^\oplus \mathcal{H}_\omega \,\mathrm{d}\mu(\omega).$$

Suppose that $a \in \mathsf{A}$. Since Γ_ω is dense in $\mathcal{D}_\omega[\mathrm{t}_{\pi_\omega}]$ as noted above, the set $\mathcal{S}_\omega := \big\{\big(\psi_n(\omega), \overline{\pi_\omega(a)}\,\psi_n(\omega)\big) : n \in \mathbb{N}\big\}$ is dense in $\mathrm{gr}\,\overline{\pi_\omega(a)}$ for $\omega \in \Omega$. Therefore, by Lemma 12.1.3 and Definition 12.1.6, $\omega \to \overline{\pi_\omega(a)}$ is a measurable field of closed operators. Hence $\omega \to \pi_\omega$ is a measurable field of $*$-representations. Let ϱ_μ denote the direct integral of this field.

It is obvious that $\Gamma := \{f\psi_n : n \in \mathbb{N}$ and $f \in L^\infty(\Omega; \mu)\}$ is a subset of $\mathcal{D}(\varrho_\mu)$. Since

$\mathcal{G}_\omega$ is dense in gr $\overline{\pi_\omega(a)}$ for $\omega \in \Omega$ and $\overline{\varrho_\mu(a)} \subseteq \int^\oplus \overline{\pi_\omega(a)}\, d\mu(\omega)$ by definition, Lemma 12.1.2, (i), shows that $\{(f\psi_n, \overline{\varrho_\mu(a)}\, f\psi_n) : n \in \mathbb{N}$ and $f \in L^\infty(\Omega; \mu)\}$ is total in gr $\overline{\varrho_\mu(a)}$. This implies that the linear span of Γ is dense in $\mathcal{D}(\varrho_\mu)$ relative to the graph topology of $\varrho_\mu(\mathsf{A})$. (Moreover, it follows that $\overline{\varrho_\mu(a)} = \int^\oplus \overline{\pi_\omega(a)}\, d\mu(\omega)$ for $a \in \mathsf{A}$.) In particular, we see that $\mathcal{D}(\varrho_\mu)$ is dense in $\mathcal{H}_\mu$, since Γ is also total in $\mathcal{H}_\mu$ by Lemma 12.1.2,(i).

Let ψ_μ denote the vector in $\mathcal{D}(\varrho_\mu)$ with $\psi_\mu(\omega) := \varphi_\omega$, $\omega \in \Omega$.

Proposition 12.4.14. *Suppose that $\pi_\vartheta(\mathsf{A})'_s = \pi_\vartheta(\mathsf{A})'_w$. Retaining also the above assumptions (i.e., conditions (α), (β) and (γ)) and notations, the following three statements are equivalent:*

(i) *The measure μ is orthogonal on $(\Omega, \mathfrak{S})$.*

(ii) *The vector ψ_μ is cyclic for φ_μ.*

(iii) *The vector ψ_μ is weakly cyclic for φ_μ.*

Proof. Let $f \in L^\infty(\Omega; \mu)$ and $a, b \in \mathsf{A}$. We abbreviate the bounded diagonalizable operator $\int^\oplus f(\omega)\, I_\omega\, d\mu(\omega)$ by x_f. By (2), we have that

$$\left.\begin{aligned} \langle T(f)\, \pi_\vartheta(a)\, \varphi_\vartheta, \pi_\vartheta(b)\, \varphi_\vartheta\rangle &= \int_\Omega f(\omega)\, \omega(b^+a)\, d\mu(\omega) \\ &= \int_\Omega f(\omega)\, \langle \pi_\omega(a)\, \varphi_\omega, \pi_\omega(b)\, \varphi_\omega\rangle\, d\mu(\omega) \\ &= \langle x_f \varrho_\mu(a)\, \psi_\mu, \varrho_\mu(b)\, \psi_\mu\rangle. \end{aligned}\right\} \quad (3)$$

Setting $a = b$ and $f(\cdot) \equiv 1$ in (3), we get $\|\pi_\vartheta(a)\, \varphi_\vartheta\|^2 = \|\varrho_\mu(a)\, \psi_\mu\|^2$, so the map U defined by $U(\pi_\vartheta(a)\, \varphi_\vartheta) := \varrho_\mu(a)\, \psi_\mu$, $a \in \mathsf{A}$, extends by continuity to an isometry, again denoted by U, of $\mathcal{H}_\vartheta$ into $\mathcal{H}_\mu$.

(i) $\to$ (ii): As shown in the discussion preceding Proposition 12.4.14, the linear span of the set Γ is dense in $\mathcal{D}(\varrho_\mu)$ $[t_{\varrho_\mu}]$. Therefore, in order to prove that ψ_μ is cyclic for ϱ_μ, it is sufficient to show that for any $g \in L^\infty(\Omega; \mu)$, $n \in \mathbb{N}$, $x \in \mathsf{A}$ and $\varepsilon > 0$ there exists a $y \in \mathsf{A}$ such that $\|\varrho_\mu(x)\, (g\psi_n - \varrho_\mu(y)\, \psi_\mu)\| > \varepsilon$. We fix g, n, ε, x and y. Clearly,

$$(\varrho_\mu(x)\, g\psi_n)\, (\omega) = \pi_\omega(x)\, g(\omega)\, \pi_\omega(b_n)\, \varphi_\omega = (x_g \varrho_\mu(xb_n)\, \psi_\mu)\, (\omega) \text{ a.e.}$$

By (3), we have

$$\begin{aligned} \|\varrho_\mu(x)\, (g\psi_n - \varrho_\mu(y)\, \psi_\mu)\|^2 &= \|x_g \varrho_\mu(xb_n)\, \psi_\mu - \varrho_\mu(xy)\, \psi_\mu\|^2 \\ &= \langle x_{|g|^2}\, \varrho_\mu(xb_n)\, \psi_\mu, \varrho_\mu(xb_n)\, \psi_\mu\rangle - \langle x_g \varrho_\mu(xb_n)\, \psi_\mu, \varrho_\mu(xy)\, \psi_\mu\rangle \\ &\quad - \langle x_{\bar g} \varrho_\mu(xy)\, \psi_\mu, \varrho_\mu(xb_n)\, \psi_\mu\rangle + \langle \varrho_\mu(xy)\, \psi_\mu, \varrho_\mu(xy)\, \psi_\mu\rangle \\ &= \langle T(|g|^2)\, \pi_\vartheta(xb_n)\, \varphi_\vartheta, \pi_\vartheta(xb_n)\, \varphi_\vartheta\rangle \\ &\quad - \langle T(g)\, \pi_\vartheta(xb_n)\, \varphi_\vartheta, \pi_\vartheta(xy)\, \varphi_\vartheta\rangle \\ &\quad - \langle T(\bar g)\, \pi_\vartheta(xy)\, \varphi_\vartheta, \pi_\vartheta(xb_n)\, \varphi_\vartheta\rangle + \langle \pi_\vartheta(xy)\, \varphi_\vartheta, \pi_\vartheta(xy)\, \varphi_\vartheta\rangle. \end{aligned}$$

Since μ is orthogonal by (i) and $\pi_\vartheta(\mathsf{A})'_s = \pi_\vartheta(\mathsf{A})'_w$ by assumption, Proposition 12.4.12 gives $T(|g|^2) = T(g)^*\, T(g)$. Moreover, $T(\bar g) = T(g)^*$. Putting these two facts into the last calculation, we get

$$\|\varrho_\mu(x)\,(g\psi_n - \varrho_\mu(y)\,\psi_\mu)\|^2 = \|T(g)\,\pi_\vartheta(xb_n)\,\varphi_\vartheta - \pi_\vartheta(xy)\,\varphi_\vartheta\|^2$$
$$= \left\|\pi_\vartheta(x)\,\big(T(g)\,\pi_\vartheta(b_n)\,\varphi_\vartheta - \pi_\vartheta(y)\,\varphi_\vartheta\big)\right\|^2, \tag{4}$$

where the last equality is true because $T(g) \in \pi_\vartheta(\mathsf{A})'_{\mathrm{w}} = \pi_\vartheta(\mathsf{A})'_{\mathrm{s}}$. Since $T(g)\,\pi_\vartheta(b_n)\,\varphi_\vartheta \in \mathcal{D}(\pi_\vartheta)$ and φ_ϑ is cyclic for π_ϑ (by definition), there is a $y \in \mathsf{A}$ such that the expression in (4) is less than ε^2.

(ii) $\to$ (iii) is trivial.

(iii) $\to$ (i): Since the range of U contains $\varrho_\mu(\mathsf{A})\,\psi_\mu$ and $\varrho_\mu(\mathsf{A})\,\psi_\mu$ is dense in $\mathcal{H}_\mu$ by (iii), U is an isometry of $\mathcal{H}_\vartheta$ onto $\mathcal{H}_\mu$. Let $f \in L^\infty(\Omega; \mu)$. From (3), we have

$$\langle x_f \varrho_\mu(a)\,\psi_\mu, \varrho_\mu(b)\,\psi_\mu\rangle = \langle T(f)\,\pi_\vartheta(a)\,\varphi_\vartheta, \pi_\vartheta(b)\,\varphi_\vartheta\rangle$$
$$= \langle UT(f)\,U^{-1}\,\varrho_\mu(a)\,\psi_\mu, \varrho_\mu(b)\,\psi_\mu\rangle$$

for $a, b \in \mathsf{A}$. Hence $x_f = UT(f)\,U^{-1}$, since $\varrho_\mu(\mathsf{A})\,\psi_\mu$ is dense in $\mathcal{H}_\mu$. As $x_{fg} = x_f x_g$ we therefore have $T(fg) = T(f)\,T(g)$ for $f, g \in L^\infty(\Omega; \mu)$, so μ is orthogonal by Proposition 12.4.12. □

12.5. The Moment Problem over Nuclear Spaces

Throughout this section V will denote a real nuclear locally convex Hausdorff space.

First we briefly describe the construction of the completed symmetric tensor algebra $\underline{S}(V)$ over V.

Let $V_{\mathbb{C}}$ be the complexification of the real locally convex space V, equipped with the continuous involution defined by $(v + iw)^+ := v - iw$ for $v, w \in V$. Thus $V_{\mathbb{C}}$ is a $*$-vector space. For $k \in \mathbb{N}$, let $\underline{V}_k$ be the completion of the k-fold projective tensor product $V_k := V_{\mathbb{C}} \otimes_\pi \cdots \otimes_\pi V_{\mathbb{C}}$, where $V_1 := V_{\mathbb{C}}$. We denote by $S_k(V)$ the subset of symmetric tensors in V_k and by $\underline{S}_k(V)$ the closure of $S_k(V)$ in $\underline{V}_k$, endowed with the topology induced from $\underline{V}_k$. Set $S_0(V) \equiv \underline{S}_0(V) := \mathbb{C}$. Let $S(V)$ and $\underline{S}(V)$ be the direct sums of the locally convex spaces $S_n(V)$, $n \in \mathbb{N}_0$, and $\underline{S}_n(V)$, $n \in \mathbb{N}_0$, respectively. Since nuclearity is preserved under countable direct sums and projective tensor products (Schäfer [1] III, 7.4 and 7.5), $S(V)$, $\underline{S}(V)$ and each $\underline{S}_n(V)$, $n \in \mathbb{N}$, are nuclear locally convex spaces. We shall identify $v_k \in \underline{S}_k(V)$ with the vector $(\delta_{nk} v_k)$ in $\underline{S}(V)$; so $S_k(V)$ and $\underline{S}_k(V)$ are linear subspaces of $S(V)$ and $\underline{S}(V)$, respectively. Set $S^n(V) := S_0(V) + \cdots + S_n(V)$ and $\underline{S}^n(V) := \underline{S}_0(V) + \cdots + \underline{S}_n(V)$, $n \in \mathbb{N}$. For $v = \sum_l v_{l1} \otimes \cdots \otimes v_{lk} \in V_k$, let $s(v)$ denote the element $\frac{1}{k!} \sum_l \sum_{\vartheta \in P_k} v_{l\vartheta(1)} \otimes \cdots \otimes v_{l\vartheta(k)}$ of $S_k(V)$, where P_k is the set of all permutations of $\{1, \ldots, k\}$. We define the product of two elements $v = (v_n)$ and $w = (w_n)$ of $S(V)$ by $v \cdot w := \sum_{n,m} s(v_n \otimes w_m)$ with the obvious interpretations $v_0 \otimes w_m = v_0 w_m$, $v_n \otimes w_0 = w_0 v_n$ and $s(v_0 \otimes w_0) = v_0 w_0$. With this product, $S(V)$ becomes a commutative algebra with unit element $\mathbb{1} := (1, 0, \ldots)$, The involution of $V_{\mathbb{C}}$ extends in a unique way to an algebra involution of $S(V)$; so $S(V)$ is a $*$-algebra. (Of course, for the latter no topology on V is needed.)

Let q be a continuous seminorm on $V_{\mathbb{C}}$. For $k \in \mathbb{N}$, let q^k denote the continuous extension of the seminorm $q \otimes_\pi \cdots \otimes_\pi q$ (k times) on V_k to $\underline{V}_k$. We set $q^0 := |\cdot|$. It is well-known that q^k, $k \geqq 2$, has the cross-property, i.e., $q^k(w_1 \otimes \cdots \otimes w_k) = q(w_1) \ldots q(w_k)$ for all $w_1, \ldots, w_k \in V_{\mathbb{C}}$. From the definitions of the seminorms q^k and of the multiplication in $S(V)$ it follows easily that for all $v_n \in S_n(V)$, $w_m \in S_m(V)$, $n, m \in \mathbb{N}$,

$$q^{n+m}(v_n w_m) \leqq q^n(v_n)\, q^m(w_m). \tag{1}$$

This implies that the multiplication of $S(V)$ is continuous as a map of $S^n(V) \times S(V)$ into $S(V)$ for any $n \in \mathbb{N}$. From this we conclude immediately that the multiplication of $S(V)$ extends by continuity to $\underline{S}(V)$ such that $\underline{S}(V)$ becomes a topological algebra. Moreover, (1) remains valid for $v_n \in \underline{S}_n(V)$ and $w_m \in \underline{S}_m(V)$, and the map $(v, w) \to vw$ of $\underline{S}^n(V) \times \underline{S}(V)$ into $\underline{S}(V)$ is continuous for $n \in \mathbb{N}$. Note that the multiplication of $\underline{S}(V)$ is not jointly continuous in general. The continuity of the involution of $V_{\mathbb{C}}$ gives the continuity of the involution of $S(V)$. Thus the involution of $S(V)$ also extends by continuity to $\underline{S}(V)$, and $\underline{S}(V)$ will be a topological $*$-algebra. Summing up, $S(V)$ and $\underline{S}(V)$ are both *nuclear topological $*$-algebras with unit.*

The following simple lemma is needed later.

Lemma 12.5.1. (i) *If ω_0 is a continuous positive linear functional on $S(V)$, then the seminorm $r(a) := \omega_0(a^+a)^{1/2}$, $a \in S(V)$, is continuous on $S(V)$.*

(ii) *The map $\omega \to \omega^{\text{!}} := \omega \restriction V$ is a bijection of the continuous hermitian characters on $S(V)$ onto $V^{\text{!}}$, the dual of the (real!) locally convex space V.*

(iii) *$S(V)_+^{\text{int}}$ is dense in $\underline{S}(V)_+^{\text{int}}$ relative to the topology of $\underline{S}(V)$.*

Proof. (i): Since ω_0 is continuous on $S(V)$, there are continuous seminorms q_n, $n \in \mathbb{N}_0$, on $V_{\mathbb{C}}$ such that $|\omega_0(v_{2n})| \leqq q_n^{2n}(v_{2n})$ for all $v_{2n} \in S_{2n}(V)$. By the continuity of the involution in $V_{\mathbb{C}}$, there is no loss of generality if we assume that the seminorms q_n are invariant under the involution. Now the assertion follows from

$$r(v) \leqq \sum_n r(v_n) = \sum_n \omega_0(v_n^+ v_n)^{1/2} \leqq \sum_n q_n^{2n}(v_n^+ v_n)^{1/2} \leqq \sum_n q_n^n(v_n)$$

for all $v = (v_n) \in S(V)$, where the last inequality is true by (1) and by the invariance of q_n under the involution.

(ii): We verify (for instance) that the map is surjective. Let $\omega^{\text{!}} \in V^{\text{!}}$. Then $\omega^{\text{!}}$ extends uniquely to a homomorphism ω of $S(V)$ into $\mathbb{C}$ satisfying $\omega(1) = 1$. Clearly, ω is continuous on $S(V)$, since $\omega^{\text{!}} \in V^{\text{!}}$. Because $\omega^{\text{!}}$ is real on V by assumption, the character ω is hermitian.

(iii): We use Definition 11.3.1 with $\mathsf{Y} = \underline{S}(V)_{\mathrm{h}}$. Suppose $a \in \underline{S}(V)_+^{\text{int}}$. Then there are elements $y_1, \ldots, y_n \in \underline{S}(V)_{\mathrm{h}}$ and a polynomial $p \in \mathbb{C}[\mathsf{x}_1, \ldots, \mathsf{x}_n]$ with non-negative values on $\mathbb{R}^n$ such that $a = p(y_1, \ldots, y_n)$. Let d be the degree of p, and let $k \in \mathbb{N}$ be such that $y_1, \ldots, y_n \in \underline{S}^k(V)$. It is clear that $p(z_1, \ldots, z_n) \in \underline{S}^{kd}(V)$ for all $z_1, \ldots, z_n \in \underline{S}^k(V)$. From this and the fact that the multiplication of $\underline{S}(V)$ is jointly continuous when restricted to a fixed space $\underline{S}^m(V)$, we conclude that $a = p(y_1, \ldots, y_n)$ can be approximated arbitrarily close by elements $p(z_1, \ldots, z_n)$ with $z_1, \ldots, z_n \in S^k(V)$. Since $p(z_1, \ldots, z_n) \in S(V)_+^{\text{int}}$, the assertion follows. □

By a *measure* on $V^\dagger$ we mean in the following a measure on the σ-algebra generated by the cylinder sets of $V^\dagger$. Recall that a *cylinder set* of $V^\dagger$ is a set of the form

$$\{\omega^\dagger \in V^\dagger : (\omega^\dagger(v_1), \ldots, \omega^\dagger(v_n)) \in M\},$$

where $v_1, \ldots, v_n \in V$ and M is a Borel set in $\mathbb{R}^n$.

The main implication in the next theorem (or in the next corollary) is what is usually called the *solution of the moment problem over the nuclear space* V.

Theorem 12.5.2. *Suppose V is a real nuclear locally convex Hausdorff space. For every linear functional ω_0 on $S(V)$, the following two assertions are equivalent:*

(i) *ω_0 is $S(V)_+^{\text{int}}$-positive, and ω_0 is continuous on $S(V)$.*

(ii) *There exists a positive measure ν on $V^\dagger$ such that the following is true:*

(ii.1) *For arbitrary $n \in \mathbb{N}$, $v_1, \ldots, v_n \in V$ and $p \in \mathbb{C}[x_1, \ldots, x_n]$, the function $\omega^\dagger \to p(\omega^\dagger(v_1), \ldots, \omega^\dagger(v_n))$ on $V^\dagger$ is in $L^1(V^\dagger; \nu)$ and*

$$\omega_0(p(v_1, \ldots, v_n)) = \int_{V^\dagger} p(\omega^\dagger(v_1), \ldots, \omega^\dagger(v_n))\, d\nu(\omega^\dagger). \tag{2}$$

(ii.2) *There are a function $\zeta \in L^2(V^\dagger; \nu)$, a ν-null set N, and continuous seminorms q_n, $n \in \mathbb{N}$, such that $\zeta(\cdot)$ is finite on $V^\dagger \setminus N$ and $|\omega^\dagger(v)| \leqq |\zeta(\omega^\dagger)|^{1/n} q_n(v)$ for all $\omega^\dagger \in V^\dagger \setminus N$, $v \in V$ and $n \in \mathbb{N}$.*

Proof. (i) $\to$ (ii): We apply Theorem 12.4.5,(i) $\to$ (ii), to the nuclear topological $*$-algebra $S(V)$ with $Q := S(V)_+^{\text{int}}$. Since $\mathcal{P}(S(V)) \subseteqq S(V)_+^{\text{int}}$, ω_0 is a positive linear functional. Lemma 12.5.1,(i), combined with the continuity of ω_0 ensures that assumption (i) of Theorem 12.4.5 is satisfied. It is clear that the map $T: \omega \to \omega^\dagger := \omega \upharpoonright V$ of Ω into $V^\dagger$ is continuous in the corresponding weak topologies $\sigma(S(V)^\dagger, S(V))$ and $\sigma(V^\dagger, V)$. Let M be a set in the σ-algebra generated by the cylinder set of $V^\dagger$. From the definition of T it follows that $T^{-1}(M)$ is a Borel set of Ω; so we get a positive measure ν on $V^\dagger$ when we define $\nu(M) := \mu(T^{-1}(M))$.

We prove (ii.1). Let $v_1, \ldots, v_n \in V$ and $p \in \mathbb{C}[x_1, \ldots, x_n]$. By Proposition 11.3.9 and Theorem 12.4.5,(ii), each $\omega \in \Omega_0$ is a character on $S(V)$ and hence $\omega(p(v_1, \ldots, v_n)) = p(\omega^\dagger(v_1), \ldots, \omega^\dagger(v_n))$. Therefore, the function $\omega^\dagger \to p(\omega^\dagger(v_1), \ldots, \omega^\dagger(v_n))$ is in $L^1(V^\dagger; \nu)$, since $\omega \to \omega(p(v_1, \ldots, v_n))$ is in $L^1(\Omega_0; \mu)$. By Theorem 12.4.5,(ii.1),

$$\omega_0(p(v_1, \ldots, v_n)) = \int_{\Omega_0} \omega(p(v_1, \ldots, v_n))\, d\mu(\omega) = \int_{V^\dagger} p(\omega^\dagger(v_1), \ldots, \omega^\dagger(v_n))\, d\nu(\omega^\dagger).$$

We verify (ii.2). Let ξ and q be as in Theorem 12.4.5,(ii.3). There is no loss of generality to assume in Theorem 12.4.5,(ii), that $\mu(\Omega \setminus \Omega_0) = 0$ and that ξ is everywhere finite on Ω_0. We shall do this. Define $N := V^\dagger \setminus T(\Omega_0)$, $\zeta(\omega^\dagger) := \xi(\omega)$ for $\omega \in \Omega_0$ and $\zeta(\omega^\dagger) := 0$ otherwise. Then $\nu(N) = 0$ by $\mu(\Omega \setminus \Omega_0) = 0$, and $\zeta \in L^2(V^\dagger; \nu)$ by $\xi \in L^2(\Omega_0; \nu)$. Since q is continuous on $S(V)$, there are continuous seminorms q_n, $n \in \mathbb{N}$, on $V_{\mathbb{C}}$ such that $q(v_n) \leqq q_n^n(v_n)$ for all $v_n \in S_n(V)$. Using Theorem 12.4.5, (ii.3), and (1), we get

$$|\omega^\dagger(v)|^{2n} = \left|\omega((v^n)^+ v^n)\right| \leqq |\xi(\omega)|^2\, q(v^n)^2 \leqq |\xi(\omega)|^2\, q_n^n(v^n)^2$$
$$\leqq |\zeta(\omega^\dagger)|^2\, q_n(v)^{2n} \quad \text{for} \quad \omega^\dagger \in V^\dagger \setminus N, v \in V \quad \text{and } n \in \mathbb{N}.$$

(ii) $\to$ (i): From (2) we see at once that ω_0 is $S(V)_+^{\text{int}}$-positive. We show the continuity of ω_0 on $S(V)$. First we recall from Lemma 12.5.1,(ii), that for any $\omega^\dagger \in V^\dagger$ there is a unique character ω on $S(V)$ such that $\omega^{\mathsf{I}} = \omega \restriction V$. Suppose for a moment we have proved that

$$|\omega(v_k)| \leqq |\zeta(\omega^{\mathsf{I}})|\, 2^k q_k^k(v_k) \quad \text{for all} \quad \omega^{\mathsf{I}} \in V^{\mathsf{I}} \setminus N,\, v_k \in S_k(V) \quad \text{and} \quad k \in \mathbb{N}. \tag{3}$$

Since ω is a character and $\nu(N) = 0$, it follows from (2) and (3) that

$$|\omega_0(v_k)| = \left|\int\limits_{V^{\mathsf{I}}} \omega(v_k)\, \mathrm{d}\nu(\omega^{\mathsf{I}})\right| \leqq \left(\int\limits_{V^{\mathsf{I}}} |\zeta(\omega^{\mathsf{I}})|\, \mathrm{d}\nu(\omega^{\mathsf{I}})\right) 2^k q_k^k(v_k)$$

for $v_k \in S_k(V)$ and $k \in \mathbb{N}$. Since $\zeta \in L^2(V^{\mathsf{I}}; \nu)$ by (ii.2), and since $\omega_0(1) = \nu(V^{\mathsf{I}}) < \infty$ by (2), it follows that $\zeta \in L^1(V^{\mathsf{I}}; \nu)$. Therefore, the preceding inequality yields the continuity of ω_0 on $S(V)$.

We prove (3). Since $V_{\mathbb{C}}$ is the complexification of V, we can assume without restriction of generality that $q_k(u) \leqq q_k(u + \mathrm{i}w)$ and $q_k(w) \leqq q_k(u + \mathrm{i}w)$ for $u, w \in V$. Let $\varepsilon > 0$ be given. From the definition of $q_k^k(v_k) = (q_k \otimes_\pi \cdots \otimes_\pi q_k)(v_k)$ we can find a representation $v_k = \sum_l x_{l1} \otimes \cdots \otimes x_{lk}$ with $x_{ln} \in V_{\mathbb{C}}$ such that

$$\sum_l q(x_{l1}) \dots q(x_{lk}) \leqq q_k^k(v_k) + \varepsilon.$$

From $v_k \in S_k(V)$, we have

$$v_k = s(v_k) = \sum_l s(x_{l1} \otimes \cdots \otimes x_{lk}) = \sum_l x_{l1} \dots x_{lk}.$$

We write x_{ln} as $x_{ln} = u_{ln} + \mathrm{i}w_{ln}$ with $u_{ln}, w_{ln} \in V$. From the preceding and (ii.2), we get

$$\begin{aligned}
|\omega_0(v_k)| &= \left|\sum_l \left(\omega^{\mathsf{I}}(u_{l1}) + \mathrm{i}\omega^{\mathsf{I}}(w_{l1})\right) \dots \left(\omega^{\mathsf{I}}(u_{lk}) + \mathrm{i}\omega^{\mathsf{I}}(w_{lk})\right)\right| \\
&\leqq \sum_l |\zeta(\omega^{\mathsf{I}})| \left(q_k(u_{l1}) + q_k(w_{l1})\right) \dots \left(q_k(u_{lk}) + q_k(w_{lk})\right) \\
&\leqq |\zeta(\omega^{\mathsf{I}})|\, 2^k \sum_l q_k(x_{l1}) \dots q_k(x_{lk}) \leqq |\zeta(\omega^{\mathsf{I}})|\, 2^k\left(q_k^k(v_k) + \varepsilon\right).
\end{aligned}$$

Letting $\varepsilon \to +0$, this gives (3). $\square$

Corollary 12.5.3. *Let V be as above. A continuous linear functional ω_0 on $\underline{S}(V)$ is $\underline{S}(V)_+^{\text{int}}$-positive if and only if there is a positive measure ν on V^{I} such that condition* (ii.1) *in Theorem* 12.5.2 *is satisfied.*

Proof. The necessity follows at once from Theorem 12.5.2,(i) $\to$ (ii). We verify the sufficiency. By (2) the restriction $\omega_0 \restriction S(V)$ is $S(V)_+^{\text{int}}$-positive on $S(V)$. From the continuity of ω_0 and Lemma 12.5.1,(iii), we conclude that ω_0 is $\underline{S}(V)_+^{\text{int}}$-positive itself. $\square$

Remark 1. The n-dimensional classical Hamburger moment problem can be considered as the special case $V = \mathbb{R}^n$ of the nuclear moment problem. In this case $S(V)$ is $*$-isomorphic to the polynomial algebra $\mathbb{C}[x_1, \dots, x_n]$, and the topology of $\underline{S}(V)$ corresponds to the finest locally convex topology on $\mathbb{C}[x_1, \dots, x_n]$.

The preceding approach to the nuclear moment problem was essentially based on Theorem 12.4.5 and so on Choquet theory (i.e., on Lemma 12.4.1). We conclude this section by presenting another (and simpler) approach which uses the Bochner-Minlos theorem in place of Theorem 12.4.5.

A function F on V is called *positive definite* if for arbitrary $n \in \mathbb{N}$ and elements $v_1, \ldots, v_n \in V$ the matrix $[F(v_k - v_l)]_{k,l=1,\ldots,n}$ of $M_n(\mathbb{C})$ is positive semi-definite, i.e., $\sum_{k,l=1}^{n} F(v_k - v_l)\, \overline{\alpha_l}\alpha_k \geqq 0$ for all $\alpha_1, \ldots, \alpha_n \in \mathbb{C}$.

The following result is the *Bochner-Minlos theorem*.

Lemma 12.5.4. *Let V be a real nuclear locally convex Hausdorff space. For every complex-valued continuous positive definite function F on V there exists a finite positive measure ν on $V^\dagger$ such that $F(v) = \int_{V^\dagger} e^{i\omega^\dagger(v)}\, d\nu(\omega^\dagger)$ for all $v \in V$.*

Proof. Gelfand/Wilenkin [1], p. 322, Proposition 2, or Maurin [1], p. 302, Theorem 13. □

Lemma 12.5.5. *For $n \in \mathbb{N}$, $v_1, \ldots, v_n \in V$ and $p \in \mathbb{C}[x_1, \ldots, x_n]$, let $J(p(v_1, \ldots, v_n))$ denote the function on $V^\dagger$ defined by $J(p(v_1, \ldots, v_n))\,(\omega^\dagger) := p(\omega^\dagger(v_1), \ldots, \omega^\dagger(v_n))$, $\omega^\dagger \in V^\dagger$. The set $P(V^\dagger)$ of such functions is a $*$-algebra with the usual pointwise algebraic operations and J is a $*$-isomorphism of $S(V)$ onto $P(V^\dagger)$. Further, if $P(V^\dagger)_+$ is the cone of all non-negative functions in $P(V^\dagger)$, then we have $J(S(V)_+^{\rm int}) = P(V^\dagger)_+$.*

Proof. That $P(V^\dagger)$ is a $*$-algebra and J is a $*$-homomorphism of $S(V)$ onto $P(V^\dagger)$, is clear. Suppose $a \in S(V)$, $a \neq 0$. There are elements $v_1, \ldots, v_n \in V$ and a polynomial $p \in \mathbb{C}[x_1, \ldots, x_n]$ such that $a = p(v_1, \ldots, v_n)$. There is no loss of generality to assume the elements $v_1, \ldots, v_n$ to be linearly independent in V. Then each $(\lambda_1, \ldots, \lambda_n) \in \mathbb{R}^n$ is of the form $(\lambda_1, \ldots, \lambda_n) = (\omega^\dagger(v_1), \ldots, \omega^\dagger(v_n))$ for some $\omega^\dagger \in V^\dagger$ and hence $p(\lambda_1, \ldots, \lambda_n) = p(\omega^\dagger(v_1), \ldots, \omega^\dagger(v_n)) = J(a)\,(\omega^\dagger)$. Since $a \neq 0$ and so $p \not\equiv 0$, this shows that $J(a) \neq 0$. Thus J is injective. Further, if $J(a)\,(\omega^\dagger) \equiv p(\omega^\dagger(v_1), \ldots, \omega^\dagger(v_n)) \geqq 0$ for all $\omega^\dagger \in V^\dagger$, then, by the preceding, p is non-negative on $\mathbb{R}^n$ and hence $a = p(v_1, \ldots, v_n)$ is in $S(V)_+^{\rm int}$; so $P(V^\dagger) \subseteq J(S(V)_+^{\rm int})$. The opposite inclusion is trivially true by the definition of $S(V)_+^{\rm int}$. □

Now we can give a second proof for the existence of a solution of the moment problem over a real nuclear space. More precisely, we will prove the following statement which is the main assertion of Theorem 12.5.2:

As above, let V be a real nuclear locally convex Hausdorff space. If ω_0 is a $S(V)_+^{\rm int}$-positive continuous linear functional on $S(V)$, then there exists a positive measure ν on $V^\dagger$ such that condition (ii.1) *of Theorem* 12.5.2 *is valid.*

Proof. Let $F(V^\dagger)$ be the $*$-algebra of all $\sigma(V^\dagger, V)$-continuous functions f on $V^\dagger$ for which there is a function $g \in P(V^\dagger)$ such that $|f(\omega^\dagger)| \leqq g(\omega^\dagger)$ for all $\omega^\dagger \in V^\dagger$. Let $F(V^\dagger)_+$ be the functions in $F(V^\dagger)$ that are non-negative on $V^\dagger$. From Lemma 12.5.5 it follows that $\vartheta_0(J(a)) := \omega_0(a)$, $a \in S(V)$, defines unambiguously a linear functional ϑ_0 on $P(V^\dagger)$ which is non-negative on $P(V^\dagger)_+ = P(V^\dagger) \cap F(V^\dagger)_+$. Since $P(V^\dagger)$ is cofinal in $F(V^\dagger)$ with respect to the cone $F(V^\dagger)_+$, we can extend ϑ_0 (by Lemma 1.3.2) to an $F(V^\dagger)_+$-positive linear functional on $F(V^\dagger)$ which we denote again by ϑ_0.

Define $F(v) := \vartheta_0(e^{i\omega^\dagger(v)})$, $v \in V$. If $v_1, \ldots, v_n \in V$ and $\alpha_1, \ldots, \alpha_n \in \mathbb{C}$, then we have

$$\sum_{k,l=1}^{n} F(v_k - v_l)\, \overline{\alpha}_l \alpha_k = \vartheta_0 \left(\left| \sum_{k=1}^{n} \alpha_k\, e^{i\omega^\dagger(v_k)} \right|^2 \right) \geqq 0.$$

This shows that F is a positive definite function on V. We prove that F is continuous. Let $r(a) := \omega_0(a^+a)^{1/2}$, $a \in S(V)$. Since ω_0 is continuous on $S(V)$, the seminorm r is continuous on $S(V)$ by Lemma 12.5.1,(i). If $u, v \in V$ and $\omega^\text{l} \in V^\text{l}$, we have

$$|e^{i\omega^\text{l}(u)} - e^{i\omega^\text{l}(v)}| \leqq |\omega^\text{l}(u - v)|$$

and so

$$|F(u) - F(v)|^2 = |\vartheta_0(e^{i\omega^\text{l}(u)} - e^{i\omega^\text{l}(v)})|^2 \leqq \vartheta_0(1)\, \vartheta_0(|e^{i\omega^\text{l}(u)} - e^{i\omega^\text{l}(v)}|^2)$$

$$\leqq \vartheta_0(1)\, \vartheta_0\big(\omega^\text{l}(u - v)^2\big) = \vartheta_0(1)\, \omega_0\big((u - v)^2\big) = \vartheta_0(1)\, r(u - v)^2,$$

where we used the Cauchy-Schwarz inequality and the $F(V^\text{l})_+$-positivity of ϑ_0. Combined with the continuity of r, this proves that F is continuous on V. By Lemma 12.5.4, there is a positive measure ν on V^l such that

$$\vartheta_0(e^{i\omega^\text{l}(v)}) \equiv F(v) = \int_{V^\text{l}} e^{i\omega^\text{l}(v)}\, d\nu(\omega^\text{l}) \quad \text{for} \quad v \in V. \tag{4}$$

Roughly speaking, the assertion will be obtained by differentiation from (4). To be precise, we shall prove that for arbitrary $n \in \mathbb{N}$ and $v, v_1, \ldots, v_n \in V$ the function $J(v_1, \ldots, v_n)\,(\omega^\text{l}) \equiv \omega^\text{l}(v_1) \ldots \omega^\text{l}(v_n)$ is in $L^1(V^\text{l}; \nu)$ and

$$\vartheta_0\big(\omega^\text{l}(v_1) \ldots \omega^\text{l}(v_n)\, e^{i\omega^\text{l}(v)}\big) = \int_{V^\text{l}} \omega^\text{l}(v_1) \ldots \omega^\text{l}(v_n)\, e^{i\omega^\text{l}(v)}\, d\nu(\omega^\text{l}). \tag{5}$$

Let $(\varepsilon_k : k \in \mathbb{N})$ be a fixed positive sequence which converges to zero. Suppose $v_1, \ldots, v_n \in V$. We first show that the function $J(v_1, \ldots, v_n)$ is in $L^1(V^\text{l}; \nu)$. We abbreviate $h_{rk}(\omega^\text{l}) := \varepsilon_k^{-1}(e^{i\omega^\text{l}(\varepsilon_k v_r)} - 1)$ for $\omega^\text{l} \in V^\text{l}$, $r = 1, \ldots, n$ and $k \in \mathbb{N}$. Then we have $|h_{rk}(\omega^\text{l})| \leqq |\omega^\text{l}(v_r)|$ on V^l, and the function $|h_{1k}(\omega^\text{l}) \ldots h_{nk}(\omega^\text{l})|^2$ is a linear combination of terms of the form $e^{i\omega^\text{l}(u)}$ with $u \in V$. Therefore, it follows from (4) and the $F(V^\text{l})_+$-positivity of ϑ_0 that for $k \in \mathbb{N}$

$$\int_{V^\text{l}} |h_{1k}(\omega^\text{l}) \ldots h_{nk}(\omega^\text{l})|^2\, d\nu(\omega^\text{l}) = \vartheta_0(|h_{1k} \ldots h_{nk}|^2)$$

$$\leqq \vartheta_0\big(\omega^\text{l}(v_1)^2 \ldots \omega^\text{l}(v_n)^2\big). \tag{6}$$

Obviously, the sequence $(|h_{1k} \ldots h_{nk}|^2 : k \in \mathbb{N})$ converges pointwise on V^l to the function $|J(v_1, \ldots, v_n)|^2$. Hence we conclude from (6) and Fatou's lemma that

$$\int_{V^\text{l}} |J(v_1, \ldots, v_n)\,(\omega^\text{l})|^2\, d\nu(\omega^\text{l}) \leqq \vartheta_0\big(\omega^\text{l}(v_1)^2 \ldots \omega^\text{l}(v_n)^2\big) < \infty.$$

Thus $J(v_1, \ldots, v_n) \in L^2(V^\text{l}; \nu)$. Since $\nu(V^\text{l}) = F(0) < \infty$, this gives $J(v_1, \ldots, v_n) \in L^1(V^\text{l}; \nu)$. (5) will be proved by induction on n. Assume that (5) is true for some $n \in \mathbb{N}$ and arbitrary $v, v_1, \ldots, v_n \in V$. Now let $v, v_1, \ldots, v_{n+1} \in V$. We set $h_k(\omega^\text{l}) := \varepsilon_k^{-1}(e^{i\omega^\text{l}(\varepsilon_k v_{n+1} + v)} - e^{i\omega^\text{l}(v)})$ for $\omega^\text{l} \in V^\text{l}$ and $k \in \mathbb{N}$. It is clear that

$$|\omega^\text{l}(v_1) \ldots \omega^\text{l}(v_n)\, h_k(\omega^\text{l}) - \omega^\text{l}(v_1) \ldots \omega^\text{l}(v_{n+1})\, ie^{i\omega^\text{l}(v)}|$$

$$\leqq \varepsilon_k\, |\omega^\text{l}(v_1) \ldots \omega^\text{l}(v_n)\, \omega^\text{l}(v_{n+1})^2| \quad \text{for} \quad \omega^\text{l} \in V^\text{l} \quad \text{and} \quad k \in \mathbb{N}.$$

Employing once more the Cauchy-Schwarz inequality and the $F(V^\dagger)_+$-positivity of ϑ_0, this implies that

$$\left|\vartheta_0\big(\omega^\dagger(v_1)\dots\omega^\dagger(v_n)\,h_k(\omega^\dagger)\big)-\vartheta_0\big(\omega^\dagger(v_1)\dots\omega^\dagger(v_{n+1})\,\mathrm{i}e^{\mathrm{i}\omega^\dagger(v)}\big)\right|^2$$
$$\leqq \vartheta_0(1)\,\vartheta_0\big(|\omega^\dagger(v_1)\dots\omega^\dagger(v_n)\,h_k(\omega^\dagger)-\omega^\dagger(v_1)\dots\omega^\dagger(v_{n+1})\,\mathrm{i}e^{\mathrm{i}\omega^\dagger(v)}|^2\big)$$
$$\leqq \vartheta_0(1)\,\varepsilon_k^2\vartheta_0\big(\omega^\dagger(v_1)^2\dots\omega^\dagger(v_n)^2\,\omega^\dagger(v_{n+1})^4\big),$$

so

$$\lim_{k\to\infty}\vartheta_0\big(\omega^\dagger(v_1)\dots\omega^\dagger(v_n)\,h_k(\omega^\dagger)\big)=\vartheta_0\big(\omega^\dagger(v_1)\dots\omega^\dagger(v_{n+1})\,\mathrm{i}e^{\mathrm{i}\omega^\dagger(v)}\big)$$

On the other hand, we have

$$|\omega^\dagger(v_1)\dots\omega^\dagger(v_n)\,h_k(\omega^\dagger)|\leqq|\omega^\dagger(v_1)\dots\omega^\dagger(v_n)\,\omega^\dagger(v_{n+1})|\equiv|J(v_1,\dots,v_{n+1})\,(\omega^\dagger)|$$

for $\omega^\dagger\in V^\dagger$ and $k\in\mathbb{N}$. Since the latter function is in $L^1(V^\dagger;\nu)$, Lebesgue's dominated convergence theorem applies and yields

$$\lim_{k\to\infty}\int_{V^\dagger}\omega^\dagger(v_1)\dots\omega^\dagger(v_n)\,h_k(\omega^\dagger)\,\mathrm{d}\nu(\omega^\dagger)=\int_{V^\dagger}\omega^\dagger(v_1)\dots\omega^\dagger(v_n)\,\omega^\dagger(v_{n+1})\,\mathrm{i}e^{\mathrm{i}\omega^\dagger(v)}\,\mathrm{d}\nu(\omega^\dagger).$$

Since

$$\vartheta_0\big(\omega^\dagger(v_1)\dots\omega^\dagger(v_n)\,h_k(\omega^\dagger)\big)=\int_{V^\dagger}\omega^\dagger(v_1)\dots\omega^\dagger(v_n)\,h_k(\omega^\dagger)\,\mathrm{d}\nu(\omega^\dagger)$$

by the induction hypothesis, the equality of the two previous limits gives (5) in case $n+1$. Using (4) instead of the induction hypothesis, the same reasoning proves (5) in case $n=1$. Thus the induction proof is complete.

From (5) applied with $v=0$ and from the definition of ϑ_0, we obtain $\omega_0(v_1,\dots,v_n)=\vartheta_0\big(\omega^\dagger(v_1)\dots\omega^\dagger(v_n)\big)=\int_{V^\dagger}\omega^\dagger(v_1)\dots\omega^\dagger(v_n)\,\mathrm{d}\nu(\omega^\dagger)$ for all $v_1,\dots,v_n\in V$. Setting $v=0$ in (4), we obtain $\omega_0(1)=\vartheta_0(1)=\int_{V^\dagger}\mathrm{d}\nu(\omega^\dagger)$. Condition (ii.1) in Theorem 12.5.2 follows now by linearity. □

Notes

12.1. Decomposition theory for unbounded closed operators was treated by NUSSBAUM [1] who obtained the main results of this section. Nussbaum defined the measurability of a field of closed operators by the measurability of the field of characteristic matrices. As in RICHTER [1], our definition is based on the measurability of the field of graphs.

12.2. The so-called "nuclear spectral theorem" is an important technical tool in order to construct expansions in eigenfunctions of self-adjoint operators, see MAURIN [1], ch. II, or GELFAND/WILENKIN [1], ch. 1, § 4. The version of this theorem we need is stated as Proposition 12.2.1. Propositions 12.2.2 and 12.2.3 are from RICHTER [3].

12.3. A decomposition theory for (strongly continuous) *-representations of nuclear separable topological *-algebras was developed by BORCHERS/YNGVASON [1]. They also used the nuclear spectral theorem combined with an extension theory which is of interest in itself. The decomposition of *-representations of countably generated *-algebras was previously studied by BORISOV [1]. Our approach follows largely the paper RICHTER [3]. It uses the localization technique of Section 12.2. Note that there is no unique terminology in the literature what a decomposition of a *-representation into irreducible components means. Our Definition 12.3.1 which differs from the ones used by the above mentioned authors requires a closer connection between the $*$-representation and its components.

12.4. Borchers and Yngvason also applied their decomposition theory of $*$-representations to the extremal decomposition of states. HEGERFELDT [1] was the first who used the Choquet theory. However, he applied this theory to a proper, metrizable and weakly complete cone. A result closely to the main assertion of our Theorem 12.4.5 can also be derived from Theorem 20 in THOMAS [1]. Our approach presented in the second subsection is taken from RICHTER [2]. It is based on the concept of a cap. The material in the third subsection appears to be new in the unbounded case. For C^*-algebras these results are known and can be found in SKAU [1]; cf. also TAKESAKI [1], ch. IV, § 6.

12.5. The solution of the nuclear moment problem was given simultaneously and by different methods in BORCHERS/YNGVASON [2] and in CHALLIFOUR/SLINKER [1], see also HEGERFELDT [1].

Additional References:

12.1. DIXON [2].
12.3. DEBACKER-MATHOT [1].
12.4. NUSSBAUM [2].
12.5. DUBIN/HENNINGS [1].

Bibliography

Books, Monographs and Lecture Notes

AKHIEZER, N. I.

[1] The Classical Moment Problem. Oliver and Boyd, Edinburgh, 1965.

ALFSEN, E.

[1] Compact Convex Sets and Boundary Integrals. Springer-Verlag, Berlin, 1971.

BARUT, A. O. and RACZKA, R.

[1] Theory of Group Representations and Applications. PWN, Warsaw, 1977.

BERESANSKIĬ, J. M. (Березанский, Ю. М.)

[1] Самосопряженные операторы в пространствах функций бесконечного числа переменных. Наукова думка, Киев, 1978.

BERS, J., JOHN, F., and SCHECHTER, M.

[1] Partial Differential Equations. Interscience, New York, 1964.

BIRMAN, M. S. and SOLOMJAK, M. Z. (Бирман, М. Ш. и Соломяк, М. З.)

[1] Спектральная теория самосопряженных операторов в Гильбертовом пространстве. Ленингр. университет, Ленинград, 1980.

BOURBAKI, N.

[1] Topologie generale. Hermann, Paris, 1951.

CHOQUET, G.

[1] Lectures on Analysis. Vol. II, Benjamin Inc., New York, 1969.

DIXMIER, J.

[1] Von Neumann Algebras. North-Holland Publ. Comp., Amsterdam, 1981.

[2] C^*-Algebras. North-Holland Publ. Comp., Amsterdam, 1977.

[3] Enveloping Algebras. Akademie-Verlag, Berlin, 1977.

DOUGLAS, R. G.

[1] Banach Algebra Techniques in Operator Theory. Academic Press, New York, 1972.

DUNFORD, N. and SCHWARTZ, J. T.

[1] Linear Operators. Vol. II, Interscience, New York, 1963.

FEDERER, H.

[1] Geometric Measure Theory. Springer-Verlag, Berlin, 1969.

GELFAND, I. M. and WILENKIN, N. J.

[1] Verallgemeinerte Funktionen. Vol. IV, DVW, Berlin, 1964.

GILLMAN, L. and JERISON, M.

[1] Rings of Continuous Functions. Van Nostrand, Princeton, 1960.

GROTHENDIECK, A.
[1] Produits tensoriels topologiques et espaces nucléaires. Mem. Amer. Math. Soc. 16, Providence, 1955.
[2] Topological Vector Spaces. Gordon and Breach, New York, 1973.

HALMOS, P. R.
[1] Measure Theory. Van Nostrand, Princeton, 1950.
[2] A Hilbert Space Problem Book. Van Nostrand, Princeton, 1967.

JARCHOW, H.
[1] Locally Convex Spaces. B. G. Teubner, Stuttgart, 1981.

JORGENSEN, P. E. T. and MOORE, R. T.
[1] Operator Commutation Relations. D. Reidel Publ. Comp., Dordrecht, 1984.

JUNEK, H.
[1] Locally Convex Spaces and Operator Ideals. Teubner-Texte. Vol. 56, B. G. Teubner, Leipzig, 1983.

JURZAK, J. P.
[1] Unbounded Non-commutative Integration. D. Reidel Publ. Comp., Dordrecht, 1986.

KADISON, R. V. and RINGROSE, J. R.
[1] Fundamentals of the Theory of Operator Algebras. Vol. I, Academic Press, New York, 1983.
[2] Fundamentals of the Theory of Operator Algebras. Vol. II, Academic Press, New York, 1986.

KATO, T.
[1] Perturbation Theory for Linear Operators. Springer-Verlag, Berlin, 1966.

KIRILLOV, A. A.
[1] Elements of the Theory of Representations. Springer-Verlag, Berlin, 1976.

KÖTHE, G.
[1] Topologische lineare Räume. Vol. I, Springer-Verlag, Berlin, 1960.
[2] Topological vector spaces. Vol. II, Springer-Verlag, Berlin, 1979.

MAURIN, K.
[1] General Eigenfunction Expansions and Unitary Representations of Topological Groups. PWN, Warsaw, 1968.

MICHAEL, E. A.
[1] Locally Multiplicatively-convex Topological Algebras. Mem. Amer. Math. Soc. 11, Providence, 1953.

NEUMARK, M. A.
[1] Normierte Algebren. DVW, Berlin, 1959.

PERESSINI, A. L.
[1] Ordered Topological Vector Spaces. Harper and Row, New York, 1967.

PHELPS, R. R.
[1] Lectures on Choquet's Theorem. Van Nostrand, Princeton, 1966.

PIETSCH, A.
[1] Nuclear Locally Convex Spaces. Springer-Verlag, Berlin, 1966.

REED, M. and SIMON, B.
[1] Methods of Modern Mathematical Physics. Vol. I, Academic Press, New York, 1972.
[2] Methods of Modern Mathematical Physics. Vol. II, Academic Press, New York, 1975.

RIESZ, F. and SZ.-NAGY, B.
[1] Vorlesungen über Funktionalanalysis. DVW, Berlin, 1956.

RUDIN, W.
[1] Functional Analysis. McGraw-Hill, New York, 1973.

SAMOILENKO, J. S. (Самойленко, Ю. С.)
[1] Спектральная теория наборов самосопряженных операторов. Наукова думка, Киев, 1984.

SCHÄFER, H.
[1] Topological Vector Spaces. Springer-Verlag, Berlin, 1972.

SHOHAT, J. A. and TAMARKIN, J. D.
[1] The Problem of Moments. Amer. Math. Soc., Providence, 1943.

STRATILA, S. and ZSIDO, L.
[1] Lectures on von Neumann Algebras. Abacus Press, Tunbridge Wells, 1979.

SZ.-NAGY, B. and FOIAS, C.
[1] Analyse harmonique des operateurs de l'espace de Hilbert. Académiai Kiadó, Budapest, 1970.

TAKESAKI, M.
[1] Theory of Operator Algebras I. Springer-Verlag, New York, 1979.

TOPPING, D. M.
[1] Lectures on von Neumann Algebras. Van Nostrand, New York, 1971.

VARADARAJAN, V. S.
[1] Lie Groups, Lie Algebras, and their Representations. Springer-Verlag, New York, 1964.

WARNER, G.
[1] Harmonic Analysis on Semi-simple Lie Groups I. Springer-Verlag, Berlin, 1972.

WEIDMANN, J.
[1] Lineare Operatoren in Hilberträumen. B. G. Teubner, Stuttgart, 1976.

WLOKA, J.
[1] Partielle Differentialgleichungen. B. G. Teubner, Stuttgart, 1982.

ZELAZKO, W.
[1] Selected Topics in Topological Algebras. Lecture Notes, Aarhus University, 1971.

Articles

ALCANTARA, J. and DUBIN, D. A.
[1] I^*-algebras and their applications. Publ. RIMS Kyoto Univ. **17** (1981), 179–199.
[2] States on the current algebra. Rep. Math. Phys. **19** (1984), 13–26.

ALLAN, G. R.
[1] On a class of locally convex algebras. Proc. London Math. Soc. (3) **17** (1967), 91–114.

ANTOINE, J.-P. and KARWOWSKI, W.
[1] Partial $*$-algebras of closed linear operators in Hilbert space. Publ. RIMS Kyoto Univ. **21** (1985), 205–236. Addendum, ibid., **22** (1986), 507–509.

ARAKI, H. and JURZAK, J.-P.
[1] On a certain class of $*$-algebras of unbounded operators. Publ. RIMS Kyoto Univ. **18** (1982), 1013–1044.

ARENS, R.
[1] The space L^ω and convex topological rings. Bull. Amer. Math. Soc. **52** (1946), 931–935.

ARNAL, D. and JURZAK, J.-P.
[1] Topological aspects of algebras of unbounded operators. J. Funct. Analysis **24** (1977), 397 to 425.

ARVESON, W. B.
[1] Subalgebras of C^*-algebras. Acta Math. **123** (1969), 141–224.

ASCOLI, R., EPIFANIO, G., and RESTIVO, A.
[1] On the mathematical description of quantized fields. Commun. Math. Phys. 18 (1970), 291 to 300.

BEHNCKE, H.
[1] Topics in C^*- and von Neumann algebras. In: Lecture Notes in Math. No. 247, 1–54, Springer-Verlag, Berlin, 1972.

BERESANSKIĬ, I. A. (Березанский, И. А.)
[1] Индуктивно рефлексивные локально выпуклые пространства. Докл. АН СССР 182 (1968), 20–22.

BERG, C., CHRISTENSEN, J. P. R., and JENSEN, C. U.
[1] A remark on the multidimensional moment problem. Math. Ann. 243 (1979), 163–169.

BHATT, S. J.
[1] Structure of normal homomorphisms on a class of unbounded operator algebras. Technical Report No. 13, 1–12, Sardar Patel Univ., 1985.

BISHOP, E. and DE LEEUW, K.
[1] The representation of linear functionals by measures on sets of extreme points. Ann. Inst. Fourier (Grenoble) 9 (1959), 305–331.

BOAS, R. P.
[1] The Stieltjes moment problem for functions of bounded variation. Bull. Amer. Math. Soc. 45 (1939), 399–404.

BORCHERS, H. J.
[1] On the structure of the algebra of field operators. Nuovo Cimento 24 (1962), 214–236.
[2] Algebraic aspects of Wightman field theory. In: Statistical Mechanics and Field Theory, R. N. SEN and C. WEIL (Editors), Halsted Press, New York, 1972.

BORCHERS, H. J. and YNGVASON, J.
[1] On the algebra of field operators. The weak commutant and integral decomposition of states. Commun. Math. Phys. 42 (1975), 231–252.
[2] Integral representations for Schwinger functionals and the moment problem over nuclear spaces. Commun. Math. Phys. 43 (1975), 255–271.
[3] Necessary and sufficient conditions for integral representations of Wightman functionals at Schwinger points. Commun. Math. Phys. 47 (1976), 197–214.

BORISOV, N. V. (Борисов, Н. В.)
[1] Структура канонических переменных в теории квантовых систем с конечным и бесконечным числом степеней свободы. ТМФ 19 (1974), 27–36.

BORISOV, N. V. and REICHERT, A. M. (Борисов, Н. В. и Рсихерт, А. М.)
[1] Самосопряженные представления универсальной обертывающей алгебры и алгебры Ли. Вестник ЛГУ 22 (1974), 43–50.

BROOKS, R. M.
[1] Some algebras of unbounded operators. Math. Nachr. 56 (1973), 47–62.

BRUHAT, F.
[1] Sur le représentations induites des groupes de Lie. Bull. Soc. Math. France 84 (1956), 97–205.

CARTIER, P. and DIXMIER, J.
[1] Vecteurs analytiques dans les représentations des groupes de Lie. Amer. J. Math. 80 (1958), 131–145.

CHALLIFOUR, J. L. and SLINKER, S. P.
[1] Euclidean field theory I. The moment problem. Commun. Math. Phys. 43 (1975), 41–58.

CHERNOFF, P. R.
[1] Quasi-analytic vectors and quasi-analytic functions. Bull. Amer. Math. Soc. **81** (1975), 637–646.

COLLINS, H. S. and RUESS, W.
[1] Duals of spaces of compact operators. Studia Math. **74** (1982), 213–245.

DEBACKER-MATHOT, F.
[1] Integral decomposition of unbounded operator families. Commun. Math. Phys. **71** (1980), 47–58.

DIXMIER, J. and MALLIAVIN, P.
[1] Factorisations de functions et de vecteurs indefiniment différentiables. Bull. Sci. Math. (2) **102** (1978), 305–330.

DIXON, P. G.
[1] Generalized B^*-algebras. Proc. London Math. Soc. (3) **21** (1970), 693–715.
[2] Unbounded operator algebras. Proc. London Math. Soc. (3) **23** (1971), 53–69.

DIXON, P. G. and FREMLIN, D. H.
[1] A remark concerning multiplicative functionals on LMC algebras. J. London Math. Soc. (2) **5** (1972), 231–232.

DUBIN, D. A. and HENNINGS, M. A.
[1] Regular tensor algebras. Preprint, Open University, Milton Keynes, 1986.

FLATO, M. and SIMON, J.
[1] Separate and joint analyticity in Lie group representations. J. Funct. Analysis **13** (1973), 268–276.

FLATO, M., SIMON, J., SNELLMAN, H., and STERNHEIMER, D.
[1] Simple facts about analytic vectors and integrability. Ann. Scient. de l'École Norm. Sup. **5** (1972), 423–434.

FRIEDRICH, J.
[1] A note on the two-dimensional moment problem. Math. Nachr. **121** (1985), 285–286.
[2] Über Eigenschaften von Paaren kommutierender selbstadjungierter Operatoren. Dissertation A, Leipzig, 1984.
[3] On first order partial differential operators on bounded regions of the plane. Math. Nachr. **131** (1987), 33–47.

FRIEDRICH, J. and SCHMÜDGEN, K.
[1] n-Positivity of unbounded $*$-representations. Math. Nachr. **141** (1989), 233–250.

FRIEDRICH, M. and LASSNER, G.
[1] Angereicherte Hilberträume, die zu Operatorenalgebren assoziiert sind. Wiss. Z. KMU Leipzig, Math.-Naturw. R. **27** (1978), 245–251.

FRÖHLICH, J.
[1] Application of commutator theorems to the integration of representations of Lie algebras and commutation relations. Commun. Math. Phys. **51** (1977), 135–150.

FUGLEDE, B.
[1] On the relation $PQ - QP = -iI$. Math. Scand. **20** (1967), 79–88.
[2] Conditions for two self-adjoint operators to commute or to satisfy the Weyl relation. Math. Scand. **51** (1982), 163–178.
[3] Commuting self-adjoint partial differential operators and a group theoretic problem. J. Funct. Analysis **16** (1974), 101–121.

GÅRDING, L.
[1] Note on continuous representations of Lie groups. Proc. Nat. Acad. Sci. U.S.A. **33** (1947), 331–332.

[2] Vecteurs analytiques dans les representations des groupes de Lie. Bull. Soc. Math. France **88** (1960), 73–93.

GOODMAN, R.

[1] Analytic and entire vectors for representations of Lie groups. Trans. Amer. Math. Soc. **143** (1969), 55–76.

[2] Analytic domination by fractional powers of a positive operator. J. Funct. Analysis **3** (1969), 246–264.

[3] One parameter groups generated by operators in an enveloping algebra. J. Funct. Analysis **6** (1970), 218–236.

GROTHENDIECK, A.

[3] Sur les espaces (F) et (DF). Summa Brasil. Math. **3** (1954), 57–123.

GUDDER, S. P.

[1] A Radon-Nikodym theorem for *-algebras. Pacific J. Math. **70** (1979), 141–149.

GUDDER, S. P. and HUDSON, R. L.

[1] A noncommutative probability theory. Trans. Amer. Math. Soc. **245** (1978), 1–41.

GUDDER, S. P. and SCRUGGS, W.

[1] Unbounded representations of *-algebras. Pacific J. Math. **80** (1977), 369–382.

HARISH-CHANDRA

[1] Representations of semi-simple Lie groups I. Trans. Amer. Math. Soc. **75** (1953), 185–243.

HEGERFELDT, G. C.

[1] Extremal decomposition of Wightman functions and of states on nuclear *-algebras by Choquet theory. Commun. Math. Phys. **45** (1975), 133–135.

HILBERT, D.

[1] Über die Darstellung definiter Formen als Summe von Formenquadraten. Math. Ann. **32** (1888), 342–350.

INOUE, A.

[1] On a class of unbounded operator algebras. Pacific J. Math. **65** (1976), 77–95.

[2] On a class of unbounded operator algebras II. Pacific J. Math. **66** (1976), 411–431.

[3] A commutant of an unbounded operator algebra. Proc. Amer. Math. Soc. **69** (1978), 97–102.

[4] Unbounded generalizations of left Hilbert algebras. J. Funct. Analysis **34** (1979), 339–362.

[5] Unbounded generalizations of left Hilbert algebras. II. J. Funct. Analysis **35** (1980), 230–250.

[6] Standard representations of commutative *-algebras. Fukuoka Univ. Sci. Reports **14** (1984), 61–66.

[7] Self-adjointness of unbounded *-representations. Preprint, Fukuoka, 1984.

[8] Self-adjointness of *-representations generated by positive linear functionals. Proc. Amer. Math. Soc. **93** (1985), 643–647.

INOUE, A. and OTA, S.

[1] Derivations on algebras of unbounded operators. Trans. Amer. Math. Soc. **261** (1980), 567 to 577.

INOUE, A. and TAKESUE, K.

[1] Self-adjoint representations of polynomial algebras. Trans. Amer. Math. Soc. **280** (1983), 393–400.

INOUE, A., UEDA, H. and YAMAUCHI, T.

[1] Commutants and bicommutants of algebras of unbounded operators. J. Math. Phys. **28** (1987), 1–7.

JORGENSEN, P. E. T.

[1] Representations of differential operators on a Lie group, and conditions for a Lie algebra to generate a representation of the group. J. d'Analyse Math. **43** (1983/84), 251–288.

[2] The integrability problem for infinite-dimensional representations of finite-dimensional Lie algebras. Expo. Math. **4** (1983), 289–306.

JURZAK, J.-P.
[2] Simple facts about algebras of unbounded operators. J. Funct. Analysis **21** (1976), 469–482.

KADISON, R. V.
[1] Algebras of unbounded functions and operators. Expo. Math. **4** (1986), 3–33.

KRÖGER, P.
[1] Derivationen in $L^+(D)$. Wiss. Z. KMU Leipzig, Math.-Naturw. R. **24** (1975), 525–528.
[2] On EC^*-algebras. Preprint, Leipzig, 1977.
[3] Über die Ordnungstopologie auf Op*-Algebren. Preprint, Leipzig, 1978.

KUNZE, W.
[1] Zur algebraischen und topologischen Struktur der GC^*-Algebren. Dissertation A, Leipzig, 1975.
[2] Halbordnung und Topologie in GC^*-Algebren. Wiss. Z. KMU Leipzig, Math.-Naturw. R. **31** (1982), 55–62.

KÜRSTEN, K.-D.
[1] Ein Gegenbeispiel zum Reflexivitätsproblem für gemeinsame Definitionsbereiche von Operatorenalgebren im separablen Hilbert-Raum. Wiss. Z. KMU Leipzig, Math.-Naturw. R. **31** (1982), 49–54.
[2] Lokalkonvexe *-Algebren und andere lokalkonvexe Räume von auf einem unitären Raum definierten linearen Operatoren. Dissertation B, Leipzig, 1985.
[3] The completion of the maximal Op*-algebra on a Frechet domain. Publ. RIMS Kyoto Univ. **22** (1986), 151–175.
[4] Two-sided closed ideals of certain algebras of unbounded operators. Math. Nachr. **129** (1986), 157–166.
[5] Duality for maximal Op*-algebras on Frechet domains. Publ. RIMS Kyoto Univ. **24** (1988), 585–620.

LASSNER, G.
[1] Topological algebras of operators. Rep. Math. Phys. **3** (1972), 279–293.
[2] Mathematische Beschreibung von Observablen-Zustandssystemen. Wiss. Z. KMU Leipzig, Math.-Naturw. R. **22** (1973), 103–138.
[3] Über die Realisierbarkeit topologischer Tensoralgebren. Math. Nachr. **62** (1974), 89–101.
[4] Topologien auf Op*-Algebren. Wiss. Z. KMU Leipzig, Math.-Naturw. R. **24** (1975), 465–471.
[5] The β-topology on operator algebras. Coll. Intern. CNRS No. 274, Marseille, 1979.
[6] Quasi-uniform topologies on local observables. Math. Aspects of Quantum Field Theory I. Acta Univ. Wratislaviensis No. 519, Wroclav, 1979.
[7] Topological algebras and their applications in quantum statistics. Wiss. Z. KMU Leipzig, Math.-Naturw. R. **30** (1981), 572–595.
[8] Algebras of unbounded operators and quantum dynamics. Physica A **124** (1984), 471–479.

LASSNER, G. and LASSNER, G. A.
[1] On the continuity of entropy. Rep. Math. Phys. **15** (1980), 41–46.

LASSNER, G. and TIMMERMANN, W.
[1] Normal states on algebras of unbounded operators. Rep. Math. Phys. **3** (1972), 295–305.
[2] Classification of domains of operator algebras. Rep. Math. Phys. **9** (1976), 205–217.

LASSNER, G. and UHLMANN, A.
[1] On positive functionals on algebras of test functions for quantum fields. Commun. Math. Phys. **7** (1968), 152–159.

LÖFFLER, F. and TIMMERMANN, W.
[1] The Calkin representation for a certain class of algebras of unbounded operators. Rev. Roum. Pur. et Appl. 31 (1986), 891–903.
[2] Singular states on maximal Op*-algebras. Publ. RIMS Kyoto Univ. 22 (1986), 671–687.

MATHOT, F.
[1] Topological properties of unbounded bicommutants. J. Math. Phys. 26 (1985), 1118–1124.

MITJAGIN, B. S. (Митягин, Б. С.)
[1] Геометрия линейных пространств и линейных операторов. В кн.: Теория операторов в функциональных пространствах, с. 213–239, Наука, Новосибирск, 1977.

MITJAGIN, B. S. and ZOBIN, N. M.
[1] Contre-exemple a l'existence d'une base dans un espace de Fréchet nucléare. C. R. Acad. Sci. Paris 279 (1974), 255–256 and 325–327.

NELSON, E.
[1] Analytic vectors. Ann. Math. 70 (1959), 572–615.

NELSON, E. and STINESPRING, W. F.
[1] Representation of elliptic operators in an enveloping algebra. Amer. J. Math. 81 (1959), 547–560.

NG, SHU-BUN and WARNER, S.
[1] Continuity of positive and multiplicative functionals. Duke Math. J. 39 (1972), 281–284.

NUSSBAUM, A. E.
[1] Reduction theory for unbounded closed operators in Hilbert space. Duke Math. J. 31 (1964), 33–44.
[2] On the integral representation of positive linear functionals. Trans. Amer. Math. Soc. 128 (1967), 460–473.
[3] A commutativity theorem for unbounded operators. Trans. Amer. Math. Soc. 140 (1969), 485–493.
[4] Quasi-analytic vectors. Ark. Math. 6 (1967), 179–191.

OTA, S.
[1] Unbounded representations of a *-algebra on indefinite metric space. Preprint, Fukuoka, 1985.

POULSEN, N. S.
[1] On C^∞-vectors and intertwining bilinear forms for representations of Lie groups. J. Funct. Analysis 9 (1972), 87–120.
[2] On the canonical commutation relations. Math. Scand. 32 (1973) 112–122.

POWERS, R. T.
[1] Self-adjoint algebras of unbounded operators. Commun. Math. Phys. 21 (1971), 85–124.
[2] Self-adjoint algebras of unbounded operators. II. Trans. Amer. Math. Soc. 187 (1974), 261 to 293.

RICHTER, P.
[1] Zur Reduktionstheorie von Operatorenalgebren. Dissertation A, Leipzig, 1979.
[2] Zerlegung positiver definiter Kerne und Entwicklung nach gemeinsamen verallgemeinerten Eigenfunktionen für Familien streng kommutierender symmetrischer Operatoren. Wiss. Z. KMU Leipzig, Math.-Naturw. R. 31 (1982), 63–68.
[3] Zur Zerlegung von Darstellungen nuklearer Algebren. Wiss. Z. KMU Leipzig, Math.-Naturw. R. 33 (1984), 63–65.

RUELLE, D.
[1] On the asymptotic condition in quantum field theory. Helv. Phys. Acta 35 (1962), 147–163.

Samoilenko, J. S. (Самойленко, Ю. С.)
[2] Локально зависимые представления семейств нормальных операторов. В кн.: Операторы математической физики и бесконечномерный анализ, с. 110—114, Киев, 1979.

Schmüdgen, K.
[1] Beiträge zur Theorie topologischer *-Algebren und ihrer Realisierungen als Operatorenalgebren. Dissertation A, KMU, Leipzig, 1973.
[2] The order structure of topological *-algebras of unbounded operators. Rep. Math. Phys. 7 (1975), 215—227.
[3] Der beschränkte Teil in Operatorenalgebren. Wiss. Z. KMU Leipzig, Math.-Naturw. R. 24 (1975), 473—490.
[4] Lokal multiplikativ konvexe Op*-Algebren. Math. Nachr. 85 (1978), 161—170.
[5] On trace representation of linear functionals on unbounded operator algebras. Commun. Math. Phys. 63 (1978), 113—130.
[6] An example of a positive polynomial which is not a sum of squares of polynomials. A positive, but not strongly positive functional. Math. Nachr. 88 (1979), 385—390.
[7] A proof of a theorem on trace representation of strongly positive linear functionals on Op*-algebras. J. Operator Theory 2 (1979), 39—47.
[8] Uniform topologies on enveloping algebras. J. Funct. Analysis 39 (1980), 57—66.
[9] On topologization of unbounded operator algebras. Rep. Math. Phys. 17 (1980), 359—371.
[10] Two theorems about topologies on countably generated Op*-algebras. Acta Math. Acad. Sci. Hungar. 35 (1980), 139—150.
[11] Graded and filtrated *-algebras I. Graded normal topologies. Rep. Math. Phys. 18 (1980), 211—229.
[12] Unbounded *-representations. Unpublished manuscript, Leipzig, 1979/80.
[13] On the Heisenberg commutation relation II. Publ. RIMS Kyoto Univ. 19 (1983), 601—671.
[14] On domains of powers of closed symmetric operators. J. Operator Theory 9 (1983), 53—75. Correction: ibid, 12 (1984), 199.
[15] On restrictions of unbounded symmetric operators. J. Operator Theory 11 (1984), 379—393.
[16] On commuting unbounded self-adjoint operators. Acta Sci. Math. Szeged 47 (1984), 131—146.
[17] On commuting unbounded self-adjoint operators. III. Manuscripta Math. 54 (1985), 221—247.
[18] On commuting unbounded self-adjoint operators. IV. Math. Nachr. 125 (1986), 83—102.
[19] A note on commuting unbounded self-adjoint operators affiliated to properly infinite von Neumann algebras. Bull. London Math. Soc. 16 (1986), 287—292.
[20] Topological realizations of Calkin algebras on Frechet domains of unbounded operator algebras. Z. Anal. Anw. 5 (1986), 481—490.
[21] Unbounded commutants and intertwining spaces of unbounded symmetric operators and *-representations. J. Funct. Analysis 71 (1987), 47—68.
[22] Strongly commuting self-adjoint operators and commutants of unbounded operator algebras. Proc. Amer. Math. Soc. 102 (1988), 365—372.
[23] Spaces of continuous sesquilinear forms associated with unbounded operator algebras. Z. Anal. Anw. 7 (1988), 309—319.
[24] A note on the strong operator topology of countably generated 0*-vector spaces. Z. Anal. Anw. 8 (1989), 425—430.

Schmüdgen, K. and Friedrich, J.
[1] On commuting unbounded self-adjoint operators. II. J. Integral Equ. and Operator Theory 7 (1984), 815—867.

Segal, I. E.
[1] A class of operator algebras which are determined by groups. Duke Math. J. 18 (1951), 221—265.

[2] Hypermaximality of certain operators on Lie groups. Proc. Amer Math. Soc. **3** (1952), 13–15.
[3] A theorem on the measurability of group-invariant operators. Duke Math. J. **26** (1959), 549–552.
[4] An extension of a theorem of L. O'Raifeartaigh. J. Funct. Analysis **1** (1967), 1–21.

Sherman, T.
[1] Positive linear functionals on $*$-algebras of unbounded operators. J. Math. Anal. Appl. **22** (1968), 285–318.
[2] A moment problem on R^n. Rend. Cir. Mat. Palermo **13** (1964), 273–278.

Simon, B.
[1] The theory of semi-analytic vectors: a new proof of a theorem of Masson and McClary. Indiana Univ. Math. J. **20** (1971), 1145–1151.

Simon, J.
[1] On the integrability of finite-dimensional Lie algebras. Commun. Math. Phys. **28** (1972), 39–42.

Skau, C.
[1] Orthogonal measures on the state space of C^*-algebra. In: Algebras in Analysis, pp. 272–303, J. H. Williamson (Editor), Academic Press, New York, 1975.

Slinker, S. P.
[1] On commuting self-adjoint extensions of unbounded operators. Indiana Univ. Math. J. **27** (1978), 629–636.

Stinespring, W. F.
[1] Positive functions on C^*-algebras. Proc. Amer. Math. Soc. **6** (1955), 211–216.

Takesue, K.
[1] Spatial theory for algebras of unbounded operators. Rep. Math. Phys. **21** (1985), 347–355

Taskinen, J.
[1] Counterexamples to "probleme des topologies" of Grothendieck. Ann. Acad. Sci. Fenn. Ser. A I Math. Diss. **63** (1986), 1–25.
[2] (FBa)- and (FBB)-spaces. Preprint, Helsinki, 1987.

Thomas, E.
[1] Integral representations in convex cones. Preprint, Groningen, 1977.

Timmermann, W.
[1] On an ideal in algebras of unbounded operators. Math. Nachr. **91** (1979), 347–355.
[2] Ideals of algebras of unbounded operators. Math. Nachr. **92** (1979), 99–100.

Todorov, T. S.
[1] On an explicit connection between representations whose generating functionals satisfy a dominance relation. In: Proceedings of the Conference on Complex Analysis and Applications, Varna, 1985.

Uhlmann, A.
[1] Über die Definition der Quantenfelder nach Wightman und Haag. Wiss. Z. KMU Leipzig, Math.-Naturw. R. **11** (1962), 213–217.
[2] Some general properties of $*$-algebra representations. Preprint, Leipzig, 1971.
[3] Properties of the algebra $L^+(D)$. Preprint E2-8149, Dubna, 1974.

Vasiliev, A. N. (Васильев, А. Н.)
[1] К теории представлений топологической (не-банаховой) алгебры с инволюцией. ТМФ **2** (1970), 153–168.
[2] Алгебраические аспекты аксиоматики Вайтмана. ТМФ **3** (1970), 24–56.

VORONIN, A. V., SUSHKO, V. N., and HORUZHY, S. S. (Воронин, А. В., Сушко, В. Н. и Хоружий, С. С.)

[1] Алгебра неограниченных операторов и вакуумный суперотбор в квантовой теории поля. ТМФ **59** (1984), 28–48.

WAELBROECK, L.

[1] Le calcul symbolique dans les algèbras commutatives. J. Math. Pures Appl. (9) **33** (1954), 147–186.

WOGEN, W.

[1] On generators for von Neumann algebras. Bull. Amer. Math. Soc. **75** (1969), 95–99.

WORONOWICZ, S. L.

[1] The quantum problem of moments I. Rep. Math. Phys. **1** (1970), 135–145.

[2] The quantum problem of moments II. Rep. Math. Phys. **1** (1971), 175–183.

WYSS, W.

[1] On Wightman's theory of quantized fields. In: Lectures in theoretical physics. Boulder, 1968, Gordon and Breach, New York, 1969.

XIA, DAO-XING (Ся, До-Шин)

[1] О полунормированных кольцах с инволюцией. Изв. АН СССР **23** (1959), 509–523.

YNGVASON, J.

[1] On the algebra of test functions for field operators. Commun. Math. Phys. **34** (1973), 315–333.

Added in Proof:

ANTOINE, J.-P. and MATHOT, F.

[1] Partial *-algebras of closed operators and their commutants. — I. General structure. Ann. Inst. H. Poincaré **46** (1987), 299–324.

ANTOINE, J.-P., MATHOT, F. and TRAPANI, C.

[1] Partial *-algebras of closed operators and their commutants. — II. Commutants and bicommutants. Ann. Inst. H. Poincaré **46** (1987), 325–351.

BHATT, S. J.

[2] An irreducible representation of a symmetric star algebra is bounded. Trans. Amer. Math. Soc. **292** (1985), 645–652.

DUBIN, D. A. and HENNINGS, M. A.

[2] Symmetric tensor algebras and integral decompositions. Preprint, Open University, Milton Keynes, 1988.

DUBIN, D. A. and SOTELO-CAMPOS, J.

[1] A theory of quantum measurement based on the CCR algebra $L^+(W)$. Z. Anal. Anw. **5** (1986), 1–26.

HORUZHY, S. S. (Хоружий, С. С.)

[1] Введение в алгебраическую квантовую теорию поля. Наука, Москва, 1986.

INOUE, A.

[9] An unbounded generalization of the Tomita-Takesaki theory II. Publ. RIMS Kyoto Univ. **23** (1987), 673–726.

[10] Self-adjointness of the *-representation generated by the sum of two positive linear functionals. Proc. Amer. Math. Soc., *to appear*.

JORGENSEN, P. E. T.

[3] Operators and Representations Theory. North-Holland, Amsterdam, 1988.

JUNEK, H.
[2] Maximal Op*-algebras on DF-domains. Z. Anal. Anw. **9** (1990), *to appear.*

KATAVOLOS, K. and KOCH, I.
[1] Extension of Tomita-Takesaki theory to the unbounded algebra of the canonical commutation relations. Rep. Math. Phys. **16** (1979), 335–352.

KNAPP, A.
[1] Representation Theory of Semisimple Groups – An Overview Based on Examples. Princeton University Press, Princeton, 1986.

KÜRSTEN, K.-D.
[6] On Hilbertizable quasi-Frechét spaces I. Preprint, Leipzig, 1988.
[7] On commutatively dominated Op*-algebras with Frechét domains. Preprint, Leipzig, 1988.

KÜRSTEN, K.-D. and MILDE, M.
[1] Calkin representations of unbounded operator algebras acting on non-separable domains. Preprint, Leipzig, 1989.

LÖFFLER, F. and TIMMERMANN, W.
[3] On the structure of the state space of maximal Op*-algebras. Publ. RIMS Kyoto Univ. **22** (1986), 1063–1078.

MALLIOS, A.
[1] Topological Algebras: Selected Topics. North-Holland, Amsterdam, 1986.

NGUYEN, N.
[1] Commutants of self-adjoint *-representations of the polynomial algebra in two variables. Math. Nachr., *to appear.*

TAKESUE, K.
[2] Standard representations induced by positive linear functionals. Mem. Fac. Sci. Kyushu Univ. **37** (1983), 211–225.

TIMMERMANN, W.
[3] On commutators in algebras of unbounded operators. Z. Anal. Anw. **7** (1988), 1–14.

VAN DAELE, A. and KASPAREK, A.
[1] On the strong unbounded commutant of an O*-algebra. Proc. Amer. Math. Soc. **105** (1989), 111–116.

Symbol Index

Locally Convex Spaces and Related Constructs

Ordered $*$-Vector Spaces

Operators on Hilbert Space

$*$-Algebras and Positive Linear Functionals

O-Families and Graph Topologies

Spaces of Operators Associated with O-Families

Topologies on Spaces of Operators and Related Matters

Commutants

Representations

Representations of Enveloping Algebras

Matrix Spaces and Wedges

Decomposition Theory

Further Notations

Subject Index

Printed in the United States
By Bookmasters